The Future of the World's Climate

The Future of the World's Climate

Ann Henderson-Sellers

Kendal McGuffie

ELSEVIER

AMSTERDAM • BOSTON • HEIDELBERG • LONDON • NEW YORK • OXFORD • PARIS
SAN DIEGO • SAN FRANCISCO • SINGAPORE • SYDNEY • TOKYO

Elsevier
225 Wyman Street, Waltham, MA 02451, USA
The Boulevard, Langford Lane, Kidlington, Oxford OX5 1GB, UK
Radarweg 29, PO Box 211, 1000 AE Amsterdam, The Netherlands

Notice
No responsibility is assumed by the publisher for any injury and/or damage to persons or property
as a matter of products liability, negligence or otherwise, or from any use or operation of any methods,
products, instructions or ideas contained in the material herein.

British Library Cataloguing in Publication Data
A catalogue record for this book is available from the British Library

Library of Congress Cataloging-in-Publication Data
A catalog record for this book is available from the Library of Congress

ISBN: 978-0-12-386917-3

For information on all Elsevier publications
visit our website at books.elsevier.com

Working together to grow
libraries in developing countries

www.elsevier.com | www.bookaid.org | www.sabre.org

ELSEVIER   BOOK AID
            International   Sabre Foundation

Dedicated to Stephen H. Schneider
(February 11, 1945 - July 19, 2010)

Contents

Introduction
Future Climate

Section I
The Anthropocene

Section IV
Learning Lessons

At a time when the world is becoming aware of the risks of climate change and, especially, when we are beginning to realize the magnitude of the overwhelming upheaval we will have to achieve to bequeath to our children a well-preserved environment, the evidence patiently accumulated by the young science of climatology provides indispensable reference points for those who want to take action.

It happens that more and more people refuse to give up and wish to react. The truth is now widely accepted and the need to implement something on a very large scale is understood throughout the world, from China to the United States, from Brazil to the Sahel, and from Europe to Bangladesh; everyone well knows that we cannot escape from a 'high-carbon' economy simply through a few multilateral negotiations or a few regulations voted here and there.

Profound changes can only be obtained by associating in our movement as many individual players as possible, and mobilizing the public as broadly as possible.

Together, as political leaders or anonymous champions of the environment cause, we all need more than ever a solid foundation of knowledge that can be shared, not only because it shows us the path forward, but especially because it creates the common verifiable understandable base that alone can drive the global planet-wide movement we all wish to call.

As this major work proves, climatology today is able to provide these data on the basis of solid results. It finds them through examination of past climates and scientific modelling of future phenomena. It assembles them through observation of the skies and exploration of the ground. It owes them mainly to the expertise and determination of all those who build this science and make it progress day after day despite pitfalls.

It is therefore legitimate for me to pay tribute here to their essential contribution.

Like many of those who preceded us, we look to the sky. But, unlike earlier civilizations that observed the clouds with a view that combined magic with hope and fear, our today's anxious, disillusioned spirits seek elements of understanding for effective action. I have no doubt they will find them in the impressive work we have here.

H. S. H. Prince Albert II of Monaco
August 2011

This book started out as an update to *Future Climates of The World: A Modelling Perspective*, volume 16 of the prestigious World Survey of Climatology series. However, the book you are holding (either physically or virtually via your e-reader) is not a revision, but an altogether new work, which brings together in one place a comprehensive view of the future of the Earth's climate. As one of the authors writes, detailed modelling presupposes detailed understanding and, as a consequence, understanding the future climate must rely heavily on gaining a greater understanding of the forcings, processes, and responses that have created the climate of the past.

This book comprises a state-of-the-art compendium of our current understanding of climatic prediction and of the way in which understanding depends crucially on improvements in, and improved understanding of, climatic models. The present volume is aimed at a similar audience to its predecessor: the generally interested reader, climatology undergraduates, and graduates in any discipline. Today, as climate change becomes so urgent and so prevalent in the media, this book is also a means of wider public education and dissemination.

All the chapters were subjected to extensive peer review, with authors further developing their contribution in response to three or more reviewers of international standing. We need to acknowledge these reviewers, who provided helpful and constructive input to the authors. The reviewers were, in alphabetical order, Ray Arritt, Roger Barry, Michael Benton, Nathan Bindoff, Tony Brazel, Simon Brewer, Pep Canadell, Kate Crowley, Tom Delworth, Andy Dessler, John Dodson, Sarah Doherty, Annica Ekman, Matt England, Graham Farquhar, Kirsten Findell, Martin Heimann, Regine Hock, Irene Hudson, Andreas Jonsson, Johan Jungclaus, Masa Kageyama, Wolfgang Kiessling, Anders Levermann, Pak Sum Low, Ian Lowe, Dan Lunt, Richard Matear, Supriya Mathew, John McGregor, Gerald Mills, Igor Polyakov, Julia Pongratz, Mattias Prange, Colin Prentice, Mike Raupach, Anni Reissell, Martin Rice, John Schellnhuber, Marko Scholtze, Robert Spicer, Michel Verstraete, Xuebin Zhang, and Francis Zwiers.

The wonderful imagery on the cover and in Chapter 1 was created by Leon Cmielewski and Josephine Starrs (Starrs and Cmielewski, 2010). On the cover the lyrics from John Lennon's 'Nobody Told Me' (a song he wrote in 1980) wind through the Ganges Delta's mangrove swamp. This image portrays the words 'days like these' embedded into imagery of the Sundarbans, the river system at the mouth of the Ganges Delta, where flooding is displacing local people and destroying vulnerable ecologies. Lennon's upbeat song cites contemporary dichotomies such as "Everybody's talking and no one says a word. Everybody's making love and no one really cares… Strange days indeed".

'A Living Body' (Figure 1.4 in Chapter 1) incorporates a quotation from Tom Trevorrow of the Ngarrindjeri people from the Coorong in South Australia who, on the steps of the South Australian Parliament in 2009, described the Coorong in South Australia as "a living body". The words have been digitally manipulated into the sandhill landscape of Younghusband Peninsula in South Australia's Coorong region. Both art creations are images adapted from the Incompatible Elements video installation by Josephine Starrs and Leon Cmielewski, exhibited at Performance Space Sydney, October 2010 (see *http://lx.sysx.org*).

In the growing trans-discipline of 'climate' there are many sets of terminologies. These overlap and intertwine in a way that can be confusing. There are a multitude of acronyms; a diverse array of commonly used words and phrases with meaning particular (and sometimes even peculiar) to climate cliques. As climate prediction (the primary focus of this book) is scientific, there is a desire to use SI units. However, this, and much of our scientific terminology, is not language normally heard in everyday life. As climate has become 'popular' and political, there has been a need and a desire to formulate findings in terms (and units) that regular conversations might employ. As editors we found ourselves caught between, for example, temperatures in K (correct SI), °C (widely understood), and °F (still used in the United States).

Climate modellers have stopped caring whether GCM stands for General Circulation Model or Global Climate Model, but is GCCM (the name of a Global Climate–Chemistry Model) generally recognized? We need to discuss BC (black carbon) as part of the complex story of aerosols and atmospheric chemistry but worry that this abbreviation might be read as 'Before Christ'. This prompts a still trickier issue: that of describing time. This challenge is in two parts: the large number of 'convenient' units of time employed by different types of climate scientists — days, months, years, decades, centuries, thousands of years (ka), millions of years (Ma), and billions of years (Ga). We should all use (but we don't) only seconds to precisely specify a time interval using strict SI terminology (e.g., do we mean a sidereal day, a solar day, or do we mean 86,400 seconds). The second issue with time is from where to count it backwards (for palaeoclimate) and forwards (for prediction)? The common assumption is that the 'present' referred to in Before Present (BP) relates to 'now' (say, 2011) but, in fact, the 'present' is conventionally defined as 1950 AD to aid comparisons of records. Another hiccup is AD, which some prefer to denote as CE (the common era, current era) with BC having the alternative of BCE.

Seasons are a pest, as those of us from the Southern Hemisphere have ours 'inverted' and anyone living in the tropics will not recognize 'summer' and 'winter' as being as important as, say, wet and dry seasons. Here, we have tried to be specific and say 'boreal' or 'austral' where required and employ the conventional definitions of the boreal summer (June, July, and August) [and its (seasonal) complement, the austral summer — December, January, and February] as these months are commonly used for averaging climate model output and for data—model comparisons.

There are shorthand expressions that prevail in climate circles and we attempt to define them carefully and to spell them out on first occurrence. Perhaps the most insidious of these is 'doubled CO_2'. This term arose in the early history of GCMs when the only greenhouse gas represented in models was carbon dioxide (CO_2) and the only way to try to determine the sensitivity of a model to a prescribed perturbation was to make a discontinuous change — in the very early simulations this was doubling of the CO_2 concentration in the atmosphere — at a stroke. While such simple models and simple tests are long gone, their ghosts haunt us still in definitions of climate sensitivity, in the form of the phrase 'doubled CO_2'. Today, this generally means doubling 'effective CO_2' — another term with another abbreviation, CO_2e. We list all the abbreviations and acronyms used in the book and present it before the text begins. We have also collected all the cited literature in a full bibliography at the end of the book.

Is 'greenhouse warming' the same as 'global warming'? In this book, they are mostly synonyms, but there is a further trap in the form of 'enhanced global (or greenhouse) warming'. Climate practitioners all know the use of the word 'enhanced' in this context is the absolute opposite of the meaning of enhanced in the vernacular. This is undoubtedly a very serious problem in what has been termed the 'climate change code' (e.g., Hassol, 2008). Climate change scientists are surely to blame: the phrase 'enhanced greenhouse warming' has been in the science literature for at least a decade and probably more than 20 years. There is *no excuse for this terminology, which only confuses* 'enhancement' meaning 'getting better' with the actual reality of this enhancement meaning a rapidly deteriorating climate (at least from the human perspective).

There are other examples of poor word usage such as 'bias', 'debate', and 'positive' (as in positive feedback and trend) and, perhaps most dangerous of all, 'theory'. All of these, and many other, code terms have a day-to-day meaning that is very different from their climate use (Henderson-Sellers, 2010a). In this book, we strive to avoid these pitfalls of terminology.

This book is dedicated to the memory, and to the continuation of the work, of Stephen H. Schneider, and is one of many remembrances of Steve. We (the editors) last saw Stephen a couple of weeks before he died. During our dinner together, we shared our holiday experiences from the Lamington National Park and were totally outdone by Steve's tales of he and his wife Terry stalking lyrebirds for hours and hours during their honeymoon there. The two of them outdid us on all counts, and hearing Steve's recollections illuminated the great happiness he shared with Terry. He was so very thrilled to be a grandfather and described his joy over "Becca's delivery" and discussed Adam's archaeological studies at length, exhibiting a father's pride in his son's working vacation to Turkey.

Stephen was, undoubtedly, a world-leading scientist — one of, if not *the* best in our field. Steve amazed us all again and again by his forward thinking and his unbending clarity. Personally, we miss him very much and we know a great many around the world share this sense of real loss. All of us also share a forceful intention to, as Phil Fearnside said, "redouble our efforts in the fight he led so well." This book will, we hope, be a contribution to this goal.

The editors are very grateful for the continued support of Macquarie University and the University of Technology, Sydney, and we also wish to thank the Department of Applied Environmental Sciences and the International Meteorological Institute at the University of Stockholm, where we finalized this book. Hiking through the snow and hail to our visitor Stockholm offices, we frequently thought of Nobel Laureate Svante Arrhenius, the "lonely Swedish scientist [who] discovered global warming — as a theoretical concept, which most other experts declared implausible." (Weart, 2004). That was in 1896. Time races past!

Finally, we both wish to thank Brian for assisting, proofreading, hyphenating (lots), and generally supporting us as we collected and edited the chapters herein. Thank you!

Ann Henderson-Sellers and Kendal McGuffie
Stockholm University, Sweden
May 2011

ABBREVIATION	DEFINITION
$^{206}Pb/^{238}U$	Ratio of the daughter isotope (Lead-206) to the parent isotope (Uranium-238) used for geochronological dating.
$^{40}Ar/^{39}Ar$	Ratio of the isotope of Argon-40 to Argon-39, used in geochronological dating.
A1	Emissions scenario family from the IPCC Special Reports on Emissions Scenarios (SRES). A1FI — An emphasis on fossil-fuels (Fossil Intensive); A1B — A balanced emphasis on all energy sources; A1T — Emphasis on non-fossil energy sources.
A2	Emissions scenario family from the IPCC Special Reports on Emissions Scenarios (SRES). A2 describes a very heterogeneous world with continuously increasing global population and regionally oriented economic growth that is more fragmented and slower than in other storylines.
AAR	Accumulation-Area Ratio
ABC	Atmospheric Brown Cloud
AC&C	Atmospheric Chemistry and Climate
ACARE	Advisory Council for Aeronautics Research in Europe
ACC	Antarctic Circumpolar Current
ACCENT	Atmospheric Composition Change The European Network of Excellence
ACCENT-MIP	Model Intercomparison Project in ACCENT
ACF	Autocorrelation Function
ACW	Antarctic Circumpolar Wave
AD	Anno Domini
AeroCom	Aerosol-Comparison project to improve understanding of the global aerosol and its impact on climate
AGAGE	Advanced Global Atmospheric Gases Experiment
AGCM	Atmospheric General Circulation Model
AGCM−CM	Atmospheric General Circulation Model−Column Model

ABBREVIATION	DEFINITION
AGU	American Geophysical Union
AIE	Aerosol Indirect Effect
AIM	Asian-Pacific Integrated Model
AIMES (IGBP)	Analysis, Integration and Modelling of the Earth System
ALE	Atmospheric Lifetime Experiment
AML	Atmosphere−Mixed-Layer model
AMO	Atlantic Multi-decadal Oscillation
AMOC	Atlantic Meridional Overturning Circulation
AMV	Atlantic Multi-decadal Variability
AOD	Aerosol Optical Depth
AOGCM	Atmosphere−Ocean General Circulation Model
AP	After Present-day
APN	Asia-Pacific Network (for Global Change Research)
AR	Assessment Report of the IPCC − FAR (First AR), SAR (Second AR), TAR
AR4	4th assessment report of the Intergovernmental Panel on Climate Change
AR5	5th assessment report of the Intergovernmental Panel on Climate Change
ASTER	Advanced Spaceborne Thermal Emission and reflection Radiometer
ATLANTA	ATlanta Land-use ANalysis: Temperature and Air-quality
ATTICA	European Assessment of Transport Impacts on Climate Change and Ozone Depletion
AURA	NASA EOS chemistry mission
B1	Emissions scenario family from the IPCC Special Reports on Emission Scenarios (SRES). B1 storyline and scenario family describe a convergent world with the same global population as in the A1 storyline but with rapid changes in economic structures toward a service and information economy, with reductions in material intensity, and the introduction of clean and resource-efficient technologies.

ABBREVIATION	DEFINITION
B2	Emissions scenario family from the IPCC Special Reports on Emission Scenarios (SRES). B2 storyline and scenario family describe a world in which the emphasis is on local solutions to economic, social, and environmental sustainability, with continuously increasing population (lower than A2) and intermediate economic development.
BA	Bølling-Allerød interstadial
BC	Black Carbon
BCAS	Bangladesh Centre for Advanced Studies
BCM	Bergen Climate Model
BERN	University of Bern Climate Model
BF	Bergeron-Findeisen
BIDS	Bangladesh Institute of Development Studies
BP	Years Before Present (conventionally defined as 1950 AD)
BSRN	Baseline Surface Radiation Network
C4MIP	Coupled Climate Carbon Cycle Model Intercomparison Project
CAC	Climate Action Coogee
CAIR	Clean Air Interstate Rule
CAMP	Central Atlantic Magmatic Province
CCAFS	Climate Change, Agriculture and Food Security
CCAM	Conformal-Cubic Atmospheric Model
CCl	Commission for Climatology
CCM	Climate−Chemistry Model
CCM1	Community Climate Model 1
CCMVal	SPARC Chemistry-Climate Model Validation Activity
CCN	Cloud Condensation Nuclei
CCSM	(NCAR) Community Climate System Model (CCSM3 − version 3)
CDKN	Climate and Development Knowledge Network (DfID, UK)
CDNC	Cloud Droplet Number Concentration
CET	Central England Temperature (record)
CFC	Chlorofluorocarbon
CGCM	Coupled General Circulation Model
Chron 29R	A division of the geomagnetic polarity timescale
CIF	Climate Investment Funds (administered by the Multilateral Development Banks)
CLE	Current Legislation (of Emissions)
CLIO	Primitive-equation ocean model (part of the ECBilt-CLIO combination)
CLIVAR	WCRP project on Climate Variability and Predictability

ABBREVIATION	DEFINITION
CLM4	Community Land Model Version 4
CMIP(3 or 5)	Coupled Model Intercomparison Project (Phases 3 & 5) [linked to AR4 & AR5]
CO_2e	Carbon dioxide equivalent (of all GHGs)
COAG	Council of Australian Governments
COP	Conference of the Parties to the UNFCCC (the United Nations Framework Convention on Climate Change) referred to by number (e.g. COP13 or COP15)
CORDEX	Co-ordinated Regional Climate Downscaling Experiment
CPR	Continuous Plankton Recorder
CPRS	Carbon Pollution Reduction Scheme (Australia)
CRU UEA	Climatic Research Unit of the University of East Anglia, UK
CSM	(NCAR) Climate System Model
CTM	Chemical Transport Models
D&A	Detection and Attribution
DFA	De-trended Fluctuation Analysis
DfID	Department for International Development, UK
DGVM	Dynamic Global Vegetation Model
DIC	Dissolved Inorganic Carbon
DIVERSITAS	An international programme on biodiversity science
DJF	December−January−February
DMS	Dimethyl Sulfide
DOC	Dissolved Organic Carbon
DOCG	Denominazione di Origine Controllata e Garantita
DSP	Dominant Social Paradigm
DU	Dobson Unit (measurement of atmospheric ozone)
ECA	European Climate Assessment
ECBilt	Netherlands Centre for Climate Research intermediate complexity atmospheric model (part of the ECBilt-CLIO combination)
ECHAM	European Centre Hamburg Model
ECHAM5-HAM:	ECHAM model extended by the microphysical aerosol model HAM
ECMWF	European Centre for Medium-Range Weather Forecasts
EDC	EPICA Dome C ice core
EDGAR	Emissions Database for Global Atmospheric Research
ELA	Equilibrium-Line Altitude

ABBREVIATION	DEFINITION
EMEP	Co-operative Programme for Monitoring and Evaluation of the Long-range Transmission of Air Pollutants in Europe
EMIC	Earth System Model of Intermediate Complexity
ENSEMBLES	EU Project on ensemble modelling and forecasts
ENSO	El Niño—Southern Oscillation
EPA	Environment Protection Authority
EPICA	European Project for Ice Coring in Antarctica
ERA	ECMWF Reanalysis
ES	Earth System
ESA	Earth System Analysis
ESS	Earth System Science
ESSP	Earth System Science Partnership
ETCCDI	Expert Team on Climate Change Detection and Indices
EU	European Union
EU ETS	EU Emission Trading Scheme
EUCAARI	European Integrated project on Aerosol Cloud Climate and Air Quality Interactions
EVT	Extreme Value Theory
FACE	Free Air CO_2 Enrichment
FAO	UN Food and Agriculture Organisation
Final SLR	Total Sea-Level Rise between 2001 and 2100
FLUXNET	Global network of micro-meteorological towers using eddy covariance to measure exchanges of CO_2, water vapour & energy between biosphere & atmosphere
FoE EWNI	Friends of the Earth — England, Wales and Northern Ireland
FTIR	Fourier Transform Infrared Spectrometry
GAGE	Global Atmospheric Gases Experiment
GAINS	Greenhouse Gas and Air Pollution Interactions and Synergies
GARP	Global Atmospheric Research Programme
GCCM	Global Climate—Chemistry Model
GCM	Global Climate Model or General Circulation Model
GCP	Global Carbon Project
GDP	Gross Domestic Product
GEC	Global Environmental Change
GECAFS	Global Environmental Change and Food Systems
GECHH	Global Environmental Change and Human Health

ABBREVIATION	DEFINITION
GEV	Generalized Extreme Value (distribution)
GFDL	Geophysical Fluid Dynamics Laboratory
GHG(s)	Greenhouse Gas(es)
GI	Greenland Interstadial
GISS	Goddard Institute for Space Studies
GL	Gigalitre (10^9 litres); a cubic hectometre
GLASS	Global Land Atmosphere System Study
GOME	Global Ozone Monitoring Experiment
GPD	Generalized Pareto Distribution
GPP	Gross Primary Production
GRACE	Gravity Recovery and Climate Experiment
GS	Greenland Stadial
GSSP	Global Stratotype Section and Point
GTP	Global Temperature change Potential
GWP	Global Warming Potential
GWSP	Global Water System Project
HadISST1	Sea-ice and sea-surface temperature dataset from the Hadley Centre
HFCs	Hydrofluorocarbons
HIRHAM-MOM	HIgh Resolution HAMburg—Modular Ocean Model
HKHT	Hindu—Kush—Himalaya—Tibetan
hPa	Hectopascal, measure of (usually atmospheric) pressure
HYMN	HYdrogen, Methane and Nitrous oxide: Trend variability, budgets and interactions with the biosphere [project]
IAC	Inter-Academy Council, USA
IAM	Integrated Assessment Model
IAPSAG	International Aerosol Precipitation Science Assessment Group
IASI	Infrared Atmospheric Sounding Interferometer
ICO	International Carbon Office
IDS	Institute of Development Studies, University of Sussex
IEA	International Energy Agency
IGBP	International Geosphere—Biosphere Programme
IHDP	International Human Dimensions Programme on Global Environmental Change
IIASA	International Institute for Applied Systems. Analysis, Austria
iLEAPS	Integrated Land Ecosystem—Atmosphere Processes Study

ABBREVIATION	DEFINITION
IMAGE	Integrated Model to Assess the Global Environment
IN	Ice Nuclei
IOD	Indian Ocean Dipole
IPCC	Intergovernmental Panel on Climate Change
IPSL	Institut Pierre Simon Laplace
IS92a	IPCC Scenario 92a [one of six alternative emissions scenarios (a-f) published in 1992].
ISM	Indian Summer Monsoon
ISO	International Organization for Standardization
ITCZ	Inter-Tropical Convergence Zone
JCOMM	Joint (World Meteorological Organization/Intergovernmental Oceanographic Commission) Technical Commission for Oceanography and Marine Meteorology
JJA	June—July—August
JSC	Joint Scientific Committee
K-Pg event	Cretaceous—Paleogene geological event
ka	Thousand years
KCM	Kiel Climate Model
LAI	Leaf Area Index
LBA	Large-Scale Biosphere Atmosphere Experiment in Amazonia
LDCs	Least Developed Countries
LGA	Local Government Authority
LGM	Last Glacial Maximum
LIA	Little Ice Age
LIPS	Large Igneous Provinces
LMDzINCA	Laboratoire de Météorologie Dynamique — INteraction Chimie-Aérosols [model]
LOVECLIM	LOch-Vecode-Ecbilt-CLio-agIsm Model
LPJ	Lund-Potsdam-Jena (a DGVM)
LSM	Land Surface Model
LUCID	Land-Use and Climate, IDentification of robust impacts
LULCC	Land-Use induced Land-Cover Change
LWC	Liquid Water Content
Ma	Million years
MBE	Mid-Brunhes Event
MCA	Medieval Climate Anomaly
MCF	Methyl Chloroform
MCTEX	Maritime Continent Thunderstorm Experiment
MCW	Mid-Century Warming
MDBA	Murray—Darling Basin Authority

ABBREVIATION	DEFINITION
MDBC	Murray—Darling Basin Commission
MDG-F	Millennium Development Goal Achievement Fund
MEGAN	Model of Emissions of Gases and Aerosols from Nature
METROMEX	Metropolitan Meteorological Experiment
MFR	Maximum Feasible Reduction (of emissions)
MiniCAM	Mini Climate Change Assessment Model
MIS	Marine Isotopic Stage
MIT	Massachusetts Institute of Technology
MJO	Madden—Julian Oscillation
MM5	The Pennsylvania State University/National Center for Atmospheric Research Mesoscale Model (Version 5)
MOC	Meridional Overturning Circulation (also known as the thermohaline circulation or THC)
MODIS	Moderate Resolution Imaging Spectroradiometer
MOPITT	Measurements Of Pollution In The Troposphere
MOS	Model Output Statistics
MOZART	Model for Ozone and Related chemical Tracers — a chemistry transport model
MP	Member of Parliament
MPI	Maximum Potential Intensity
MPIM	Max Planck Institute for Meteorology
MSA	Methane Sulfonate
NADW	North Atlantic deep water
NAO	North Atlantic Oscillation
NARCCAP	North American Regional Climate Change Assessment Program
NASA	National Aeronautics and Space Administration
NCAR	National Center for Atmospheric Research
NCCARF	National Climate Change Adaptation Research Facility
NCOS	National Carbon Offset Scheme (Australia)
NDACC	Network for the Detection of Atmospheric Composition Change
NECP	Net Ecosystem Production
NEP	New Environmental Paradigm
NGERS	National Greenhouse and Energy Reporting Act (Australia)
NGO	Non-Governmental Organization
NGRIP	North Greenland Ice Core Project

ABBREVIATION	DEFINITION
NH	Northern Hemisphere
NMVOC	Non-Methane Volatile Organic Compounds
NOAA	National Oceanic and Atmospheric Administration
NOAA-GMD	National Oceanic and Atmospheric Administration — Global Monitoring Division
NPI	North Pacific Index
NPP	Net Primary Productivity
NRC	National Research Council (USA)
NWP	Numerical Weather Prediction
OA	Organic Aerosols
OC	Organic Carbon
OCMIP	Ocean Carbon Model Intercomparison Project
ODA	Official Development Assistance
OECD	Organisation for Economic Coordination and Development
OGCM	Ocean General Circulation Model
OMI	Ozone Monitoring Instrument
ORCHID	Opportunities and Risks from Climate Change and Disasters (IDS/TERI/BCAS/IIASA/BIDS/UEA-DEV screening tool)
OsloCTM2	Oslo Chemical Transport Model (version 2)
p-TOMCAT	Toulouse Off-line Model of Chemistry And Transport (parallel version)
P&Ps	Policies and Plans
PAHs	Polycyclic aromatic hydrocarbons
PAN	PeroxyAcetyl Nitrate
PAR	Photosynthetically Active Radiation
PBAP	Primary Biogenic Aerosol Particles
PCMDI	Program for Climate Model Diagnosis and Intercomparison
pCO_2	Partial pressure of CO_2 (in the atmosphere or ocean)
pdf	Probability Distribution Function or Probability Density Function
PDO	Pacific Decadal Oscillation
PDSI	Palmer Drought Severity Index
PDV	Pacific Decadal Variability
PETM	Paleocene—Eocene Thermal Maximum
PFC	Perfluorocarbons
pH	Chemical measure of acidity of a solution
PIRCS	Project to Intercompare Regional Climate Simulations
PMF/PMT	Penalized Maximal F or t test
PMIP1(2)	Palaeo-climate Modelling Intercomparison Project (comparison 1 or 2)

ABBREVIATION	DEFINITION
POA	Primary Organic Aerosols
POC	Particulate Organic Carbon
POT	Peaks Over Threshold
PP	Perfect Prognosis
ppbv	Parts per billion (10^9) by volume
ppmv	Parts per million (10^6) by volume
PRUDENCE	Prediction of Regional scenarios and Uncertainties for Defining EuropeaN Climate change risks and Effects
PV	Potential Vorticity
QUANTIFY	Quantifying the Climate Impact of Global and European Transport Systems Project
RACMO2	Regional Atmospheric Climate Model
RCM	Regional Climate Model
RCP	Representative Concentration Pathways
REA	Reliability Ensemble Averaging
REAS	Regional Emission inventory in ASia
RECCAP	REgional Carbon Cycle Assessment and Processes
RETRO	REanalysis of the TROpospheric chemical composition over the past 40 years
RF	Radiative Forcing
RH	Relative Humidity
RSL	Roughness Sub-Layer
Rubisco	RuBisCO or enzyme E.C. 4.1.1.39
SAM	Southern Annular Mode
SANZ	Sustainable Aotearoa New Zealand
SAR	Second Assessment Report (see IPCC)
SAT	Surface Air Temperature
SCIAMACHY	SCanning Imaging Absorption spectroMeter for Atmospheric CartograpHY
SDL	Sustainable Diversion Limit
SEA	Strategic Environmental Assessment
SES	Socio-Ecological System
SH	Southern Hemisphere
SiB2	Simple Biosphere Model Version 2
SL-G	Short-Lived Gases
SLE	Sea-Level Equivalent (in the units m SLE and mm SLE)
SLP	Sea-Level Pressure
SLR	Sea-Level Rise
SOA	Secondary Organic Aerosols
SOM	Self-Organizing Map
SPARC	Stratospheric Processes And their Role in Climate
SPECMAP	Spectral Mapping (the standard chronology for oxygen isotope records)
SPEEDY	Simplified Parameterizations primitivE Equation DYnamics model

ABBREVIATION	DEFINITION
SPI	Standard Precipitation Index
SRES	(IPCC) Special Report on Emissions Scenarios
SSA	Singular Spectrum Analysis
SSH	Sea-Surface Height
SSR	Surface Solar Radiation
SSS	Sea-Surface Salinity
SST	Sea Surface Temperature
SSTA	Sea Surface Temperature Anomaly
START	Global change SysTem for Analysis, Research, and Training
SVF	Sky View Factor
SWNA	Southwest North America
T21	Spectral truncation at wavenumber 21 (roughly $5.6° \times 5.6°$)
TAR	Third Assessment Report (of the IPCC)
TAV	Tropical Atlantic Variability
TEAP	Technology and Economic Assessment Panel of the Montreal Protocol
TEB	Town Energy Balance Model
TEM	Terrestrial Ecosystem Model
TERI	The Energy Research Institute, India
TEX86	A palaeo-temperature proxy based on the distribution of crenarchaeotal membrane lipids.
TF HTAP	Task Force on Hemispheric Transport of Atmospheric Pollutants
TgS	Tera (10^{12}) grams of sulfur
THC	Thermohaline Circulation
TM3, TM4	Koninklijk Nederlands Meteorologisch Instituut (KNMI) Chemistry Transport Models
TOA	Top-Of-the-Atmosphere
TOGA	Tropical Ocean − Global Atmosphere
TOMS	Total Ozone Mapping Spectrometer
TPU	Triose−Phosphate Utilization
Tr−J event	Triassic−Jurassic geological event
UCL	Urban Canopy Layer
UEA-DEV	School of International Development, University of East Anglia
UHI	Urban Heat Island
UK	United Kingdom
ULSM	Urban Land-Surface Models

ABBREVIATION	DEFINITION
UNDP	United Nations Development Programme
UNECE	United Nations Economic Commission for Europe
UNEP	United Nations Environment Programme
UNFCCC	United Nations Framework Convention on Climate Change
USA	United States of America
UTLS	Upper Troposphere − Lower Stratosphere
UV-A, -B	Ultra-Violet light types (wavelength range 400 nm−315 nm for UV-A); (wavelength range 315 nm−280 nm for UV-B)
VECODE	Vegetation Continuous Description model
VOC	Volatile Organic Compounds
WAIS	West Antarctic Ice Sheet
WAM	West African Monsoon
WBCSD	World Business Council for Sustainable Development
WCED	World Commission for Environment and Development
WCRP	World Climate Research Programme
WG	Working Group (of the IPCC) either WG1, WG2 or WG3
WGCM (WCRP)	Working Group on Coupled Modelling
WMGG	Well-Mixed Greenhouse Gases
WMO	World Meteorological Organization
WOCE	World Ocean Circulation Experiment
WRF	Weather Research and Forecasting
WRI	World Resources Institute
WSSD	World Summit on Sustainable Development
WSUD	Water Sensitive Urban Design
XBT	Expendable Bathythermograph
YDC	Younger Dryas Chronozone
$\delta^{13}C$	Ratio of stable isotopes of carbon ^{13}C to ^{12}C
$\delta^{18}O$	Ratio of stable isotopes of oxygen ^{18}O to ^{16}O
Ω_{arag}	Calcium carbonate saturation state (aragonite saturation state)

"Men seem to be born with a debt to humanity they can never pay," John Steinbeck wrote. The quality of the person's gift to humanity in payment of that debt, he concluded, was the measure of the soul. Dr. Stephen H. Schneider repaid more of that debt than any person I have ever known.

In his 40-year career, Steve Schneider was at the forefront of the scientific and public battle lines in climate science. He pioneered our understanding of aerosol (soot and dust) particle and cloud-feedback effects on the climate. He continually advocated an interdisciplinary approach to many aspects of climate change and founded and edited the first interdisciplinary climate journal *Climatic Change*. And he blazed a bold new path in engaging the public and media about climate change, urging his fellow scientists to do likewise. He wrote many popular books on climate change, authored hundreds of scientific papers, and gave thousands of public lectures.

Following his passing, I read the amazing outpouring of tributes, memories, and stories about Steve from colleagues, students, and laypeople across the globe. I remember thinking, where do we go from here without such a great man? And then, I realized. The stories were the answer. Every story represented the spark of a human connection, a life Steve had touched with his teaching.

Steve cared deeply about teaching the next generation. I first remember sitting across the desk from Steve, trying desperately to absorb the many layers of brilliance that came at me like a machine gunner that never ran out of ammunition. He was a patient and rigorous teacher. When we wrote problem sets for his climate science class at Stanford University, he delighted in pressing the students, often subtly, to see if they truly and thoroughly understood the material. Steve and his wife Terry Root always had students over to their house for class discussions, dinners, and guitar nights.

Dozens of Stanford students separately approached Steve saying that they wanted to attend the United Nations Framework Convention on Climate Change international negotiations in Copenhagen in December 2009. Many professors of his caliber and schedule would have said, "Wonderful, do it, but you're on your own." Not Steve. He and Terry arranged to teach a class entitled 'Approaching Copenhagen'. Nearly 50 Stanford students, undergraduates and graduates, went to the conference with Steve and Terry as leaders. Once again, Steve went out of his way to ensure that every student had a place, an internship, and contacts so that they could get the fullest experience possible. He arranged world-renowned speakers to brief the students in the mornings. He ran himself ragged, staying up often until three in the morning guiding students and doing his other work, only to wake up at six to get to the conference center to meet the arranged speakers.

Around Copenhagen, the climate discourse took a turn for the worse. Following the manufactured controversy of the "Climategate" emails and the subsequent campaign to discredit climate scientists, I remember Steve's weariness with how ugly things had become. He wrote scores of emails to journalists in his typical style of all capital letters, where you could almost feel the force of his personality behind the words. He had a strange pride but disturbed sadness in his eyes as he showed me the hate email he got. His voice would fall when he wondered privately if it would take an environmental catastrophe of epic proportions to get the world to deal with climate change.

Above all else he was amazingly, incomprehensibly selfless. He gave his time freely to students, colleagues, reporters, and donated years of his life to defending science and contributing to the IPCC. He made sure that he provided his students with the best opportunities that he could, writing articles, giving talks, connecting them with colleagues. He exemplified what it meant to be a mentor.

People orbited around Steve. We were pulled in by his gravity, his brilliance that illuminated our lives. He was one of the most well-known professors on campus, because no one could leave his presence and forget what it felt like.

His star has gone out, but his light lives on. He gave us all something extraordinary. To his students, he gave a sextant, a compass, and a mental map to navigate, to chart, and hopefully someday to illuminate the universe. He gave us his love and infectious joy and passion for this planet. To his colleagues and peers, he gave a lasting legacy of interdisciplinary studies of climate change, the highest of academic standards, and the steadfast belief that scientists can and must communicate their science. To the world and billions who will never know him but whose lives he touched, he gave his life, his guidance, and the fire of inspiration to make the world a better place.

He left us to carry on with his fight. It is a fight that could not be nobler. A fight to push back darkness, ignorance, and greed. A fight where we must light the fusion of intellectual passion and human compassion within ourselves. A fight for the economic well-being, the happiness, and the sustainability of future humans on this planet.

William R. L. Anderegg

From *Throught & Action*, Fall 2010. Copyright by the National Education Association and used by permission. All rights reserved.

Future Climate

The world's climate is changing: it has always changed and always will. Intense public discussions on this topic and especially on the policy implications in terms of human activities are testimony to the importance of this issue. The complexity of the systems we are studying is so immense that only great leaders manage to research, describe results, and communicate their implications: one such was Stephen H. Schneider. This book is dedicated to Steve's triumph as a climate futurologist and to his wish that we continue to explain our research findings urgently and also as clearly as possible.

Now it's up to us

We can feel
the tears
in our words
we can feel
the fears
in our world.

We can feel
the urgency
of our time
but we harbor
hope
in our hearts
and minds.

We hold on
to his memory
we are awed
by his legacy.

Let's teach
our children well
let's allow them
to live to tell.

A story of
our reckoning
our transformation
a world saved
by profound
awakening
and cooperation.

In his dream
we trust
in ourselves?
we must!

For
now
it's up
to us.

— Marilyn Cornelius, inspired by the community at the Steve Schneider 2011 Symposium, Boulder, Colorado, 25 August 2011 (reprinted with permission)

The only chapter in this section, Chapter 1, ***Seeing Further: The Futurology of Climate*** by Ann Henderson-Sellers, introduces 'The Future of the World's Climate', the topic of this book, and relates its contents to the author's personal reminiscences of Stephen's life and work as a vigorous, passionate, and globally admired climate scientist and world-class communicator.

Dear Steve:

Many people have been offering chronicles of your amazing life, but it will take an historian to put all of your accomplishments and contributions into their proper context. When that is done, we are all "virtually certain" that you will be correctly heralded as one of the giants of science and the paragon of unflagging civic engagement on a global level. Rather than looking back to see what we have lost, we, the members of the Editorial Board of Climatic Change—the journal that will be the lasting and living embodiment of all you that stood for—have chosen to hear your call that we must always look forward. You set the standard of excellence by example for an entire community of scholars from a multitude of disciplines. We understand completely that it is now up to us to make certain that your legacy survives so that the planet has a fighting chance to endure humans' persistent abuse. We pledge to try, as a community, to fill the enormous gaps that your passing has left. To that end, we will conduct honest science. We will communicate our findings with passion and integrity. We will avoid hyperbole, but we will not shrink from exploring the full range of possible futures, including the "dark tails" where the real danger lies. We will do whatever we can whenever we can to reach out to students from around the world so that the next generations of climate change scholars can move the frontiers of understanding farther than we could ever have imagined. We will insist that people listen and act accordingly. And we will never forget that the voice that we all hear in the back of our minds is yours— "get it right, get it out there, and never let "mediarology," political chicanery, or intellectual vandalism get you down". We will also always be there for Terry and the rest of your family. When your grandson and subsequent grandchildren for that matter come of age, they will learn that they are surrounded by the "Friends of Steve"—thousands strong who will offer support and comfort from every continent on earth. We will share your legacy with them, and they will be very proud.

Editorial Board of Climatic Change
Climatic Change (2010) 102:1-7

Chapter 1

Seeing Further: The Futurology of Climate

Ann Henderson-Sellers

Department of Environment and Geography, Macquarie University, Sydney, Australia

1.1. THE FUTURE OF OUR CLIMATE: INTRODUCTION AND OUTLINE

"We honor Steve Schneider by caring about the strange and beautiful planet on which we live, by protecting its climate, and by ensuring that our policymakers do not fall asleep at the wheel."

Ben Santer, July 2010 (Santer, 2010)

There is no doubt that the world's climate is changing: it has changed in the past and will in the future. What makes today different is people: our impact on climate, the public discussion about climate, and the human consequences of all the aspects of climate change. Observed increases in greenhouse gases are responsible for almost all of the recent measured warming of the surface and the atmosphere and for disturbing the global hydrological cycle. Longer timescale astronomical and geological forcing adds a variety of contrasting and competing disturbances that probably have little effect on the timescales of human-caused climate change, but are very important to understand since they are frequently invoked in the mass media and in public policy fora. Overlapping these climate changes are the internal dynamics of biology, oceanography, geochemistry, and very many additional 'ologies' and 'istries'. The complexity, immensity, and inter-connectedness of these systems are such that many of us fear the rapid approach to thresholds and tipping points whose consequences we cannot, or do not fully, foresee. In making

or seeking predictions about the future of the Earth's climate we are all becoming futurologists[1].

The study of climate, once a minor province of those augmenting the description of an apparently constant and benign 'background' climatology, has metamorphosed into a politically charged arena replete with contestants and commentators. Among the most coherent of those of us who champion clarity about the future of the world's climate was Stephen H. Schneider (1945–2010). Steve offered leadership in many aspects of climate, including preparedness to interact with journalists and fluency in describing the findings of science in the mass media discussions with climate contrarians[2]. This book is one modest avenue to pursuing and sustaining Steve's goal of explaining current climate research findings clearly and fearlessly. This chapter introduces and outlines the book and tries to relate its topics to my recollections of Stephen's interests and achievements as an exemplary climate scientist (Table 1.1).

Describing the world's future climates is among the most significant and most difficult scientific and communication

---

1. Those studying the future live dangerously, e.g., Hoyle, 2011; *http://www.bbc.co.uk/news/magazine-12058575*.
2. Climate 'contrarians' challenge what they see as a false consensus of mainstream climate science. The term is perceived to be less value-laden than other commonly used terms such as 'denier' or 'sceptic' (e.g., O'Neil and Boycoff, 2010).

The Future of the World's Climate. DOI: 10.1016/B978-0-12-386917-3.00001-4

TABLE 1.1 A Friend's View of Stephen Schneider's Leadership in Climate and Links to This Book

Book Section	Schneider Theme	Steve's Illustration	Chapter Topics	Area: Example Schneider Reference	Watch Steve Schneider
I. The Anthropocene	Precautionary principle	Why most people have fire insurance for their homes even though the real risk is very small	2. Policy and politics 3. Urban climates 4. Land use and people modifying it	Climate economics: Are the economic costs of stabilizing the atmosphere prohibitive? Azar and Schneider, 2002	Argues for climate precaution http://www.youtube.com/watch?v=7rj1QcdEqU0 Urgent need for action http://solveclimate.com/news/20100720/final-word-stephen-schneider-climate-scientist-and-warrior-1945-2010-video
II. Time and Tide	Multiple benefit action	Always seek win-win (or better) solutions despite prediction strengths and shortcomings	5. Climatic feedbacks 6. Ocean climates 7. Change and variability 8. Glaciers and climate 9. Regional climates	Climate feedbacks: Cloudiness as a global climatic feedback mechanism, Schneider, 1972	Science as a Contact Sport – discussing his book http://www.youtube.com/watch?v=8rIyOSrvb1Y
III. Looking Forward	Mediarology*	Everything is connected to everything else and the timing of climate events	10. Climate extremes 11. Terrestrial carbon and nitrogen cycles 12. Atmospheric chemistry 13. Aerosols and future climates	Dust and aerosol cooling; Atmospheric carbon dioxide and aerosols — effects of large increases on global climate, Rasool and Schneider, 1971	Misrepresentation of climate science http://www.5min.com/Video/Stephen-Schneider-on-Misleading-Sound-Bites-Foster-Climate-Denial-516907669
IV. Learning Lessons	Limits on the concept of falsifiability	How some contrarians make apparently persuasive arguments for inaction	14. Palaeoclimate records 15. Astronomical forcing 16. Cometary impacts and large-scale volcanism	Taking on the sceptics: Misleading math about the Earth: Schneider, 2002	Global warming – what we know http://academicearth.org/lectures/global-warming-stephen-schneider
V. Understanding the Unknowns	Tipping points	Not all the stuff we currently don't understand will turn out to be beneficial	17. Climate surprises 18. Earth system science	The patient from hell, Schneider and Lane, 2005; Scientists on Gaia, Schneider and Boston, 1991	Risk assessment in climate change http://www.eoearth.org/article/Stephen_Schneider_Video_lecture?topic=54271

"The Anthropocene" includes: "People, Policy, and Politics in Future Climates" , (Taplin, 2012, this volume); "Urban Climates" (Cleugh and Grimmond, 2012, this volume); and "Human Effects On Climate Through Land-Use-Induced Land-Cover Change" (Pitman and de Noblet-Ducoudré, 2012, this volume). "Time and Tide" contains five chapters: "Fast and Slow Feedbacks in Future Climates" (Harvey, 2012, this volume); "Variability and Change in the Ocean" (Sen Gupta and McNeil, 2012, this volume); "Climatic Variability on Decadal to Century Timescales" (Latif and Park, 2012, this volume); "The Future of the World's Glaciers" (Cogley, 2012, this volume); and "Future Regional Climates" (Evans et al., 2012, this volume). "Looking Forward" comprises: "Climate and Weather Extremes: Observations, Modelling, and Projections" (Alexander and Tebaldi, 2012, this volume); and "Interaction between Future Climate and Terrestrial Carbon and Nitrogen" (Dickinson, 2012, this volume); "Atmospheric Composition Change: Climate–Chemistry Interactions" (Isaksen et al., 2012, this volume); and "Climate–Chemistry Interaction: Future Tropospheric Ozone and Aerosols" (Wang et al., 2012, this volume); and "Learning Lessons" contains: "Records from the Past, Lessons for the Future: What the Palaeorecord Implies about Mechanisms of Global Change" (Harrison and Bartlein, 2012, this volume); "Modelling the Past and Future Interglacials in Response to Astronomical and Greenhouse Gas Forcing" (Berger and Yin, 2012, this volume); "Catastrophe: Extraterrestrial Impacts, Massive Volcanism, and the Biosphere" (Belcher and Mander, 2012, this volume). Finally, "Understanding the Unknowns" comprises: "Future Climate Surprises" (Lenton, 2012, this volume); and "Future Climate: One Vital Component of Trans-disciplinary Earth System Science" (Rice and Henderson-Sellers, 2012, this volume).

*Stephen invented the term 'mediarology' to describe science's interactions with the mass media and wrote about this in his book, Are We Entering the Greenhouse Century? (Schneider, 1989): http://stephenschneider.stanford.edu/Publications/PDF_Papers/GlblWarming4rev.pdf.

challenges of our time. New developments and research on climate change are reported seemingly on a weekly basis (e.g., one example week: Lacis et al., 2010; Dessler and Davis, 2010; Sherwood and Huber, 2010). This book, *The Future of the World's Climate*, is still timelier than its earlier incarnation, published in 1995 (Henderson-Sellers, 1995a): offering a state-of-the-art overview of our understanding of future climates. The book comprises 17 up-to-the-minute reviews of current literature that follow this introductory chapter. This book differs very significantly from two other types of collections on climate: it avoids the consensus constraint demanded in Intergovernmental Panel on Climate Change (IPCC) assessments being the peer-reviewed views of individual authors. Perhaps more importantly, through its dedication to Stephen Schneider and links to his life's work, we hope this book offers direction for these challenging times; that is, it is not only policy-relevant (as is IPCC) but may also be read as indicative of preferred policy directions.

The study of climate seems today to be dominated by anthropogenic global warming[3] but these predictions must be placed in their geological, palaeoclimatic, and astronomical context to create a complete picture of the Earth's future climate. All these topics are discussed in this book so that the integrated implications of individual effects and their interactions on the Earth's future climate can be fully understood. The authors are drawn from all over the world and from the highest regarded peer-reviewed groups. The tools we employ in our attempts to predict future climates are the same as those used by everyone attempting to understand the complexities of our world: observation, reconstruction of past events, interpretation (often in process studies), and then extrapolation (i.e., model use). In the case of climate, observations comprise surface and satellite data retrievals; past events can be reconstructed from the geological record and also from the historical record (both of observations and of proxy information); and extrapolation can take many forms but is increasingly being synthesized with simulations from computer-based models, themselves derived from interpretation of past events and studies of current climatic and weather processes.

This book is organised into five sections: The Anthropocene; Time and Tide; Looking Forward; Learning Lessons; and Understanding the Unknowns (Table 1.1). Because every chapter draws on both observational data and on model simulations, there are no separate chapters on climate observations or on climate modelling. Overall, this

is a rich book containing more than 400 tables, diagrams, illustrations, and photographs. More than 95% of the text is fully new, rewriting the first edition, and all has been thoroughly peer-reviewed.

This chapter first discusses the fraught nature of anthropogenic global warming (Section 1.2). The section following this is on the complexity of future climate (Section 1.3) and then the chapter moves to examining an area of southern Australia, the Coorong, where Steve once worked, as an example of a climate change 'ground zero' (Section 1.4). Section 1.5 points the way into the 'future of the world's climate': the topic and text of this book.

1.2. GLOBAL WARMING: CLIMATE'S 'ELEPHANT IN THE ROOM'

"This extensive analysis of the mainstream versus sceptical/ contrarian researchers suggests a strong role for considering expert credibility in the relative weight of and attention to these groups of researchers in future discussions in media, policy, and public forums regarding anthropogenic climate change."
Stephen Schneider (with colleagues) in Anderegg et al. (2010)

1.2.1. Informing the Public on the Greenhouse 'Debate'

Just before his death, Stephen Schneider wrote a paper with colleagues (Anderegg et al., 2010) that analysed the relative contributions to the scientific literature of those of us who work in climate research as compared to people who have taken positions on climate (often against policy changes). Steve, along with many others, fought in today's pejorative climate change battles. He also rose above the contrarians (sceptics or deniers) (Anderegg et al., 2010 cf. O'Neil and Boycoff, 2010). He understood and explained the complexity and the inadequacy of all aspects of climate models (e.g., cloud feedbacks, nuclear winter, economics of abatement of emissions). He wrote and spoke on topics as diverse as politics, cancer treatments, and wildlife conservation, and he engaged with detractors and followers and, importantly, with the way his scientific findings were portrayed in the media and in politics (cf. Table 1.1 — last column).

When the first edition of this book was developed, human-induced global warming was important but not all pervasive. These days, climate science has almost become a synonym for anthropogenic warming, and the latter can appear to be predominantly, perhaps almost exclusively, political. The public policy disputes concerning global warming have recently become heated and frequently very divisive (Mastrandrea and Schneider, 2010 cf. Pearce, 2010). Those of us who research the prediction of the future climate of the Earth are frequently challenged to be clear

---

3. While 'global warming' usually refers to anthropogenic disturbance — see also footnote 4 — 'climate change' can also encompass 'natural' disturbances as well as those due to human activities. This confusion is compounded by the fact that two leading international agencies (UNFCCC and IPCC) have different definitions for 'climate change'— UNFCCC considering only human effects on climate while IPCC incorporates 'natural' as well as anthropogenic impacts.

about our different roles as scientists and as citizens and, these days, can find ourselves accused of being policy advocates (Schneider, 2009).

Speaking in July 2010 (Figure 1.1) at a national climate adaptation conference in Australia (NCCARF, 2010), Stephen argued that expertise is a reason for participation in policy advocacy, not a reason to be denied a voice. He likened the role of real climate researchers to that of airline pilots or cancer doctors: professionals whom we expect to undertake their role expertly. In a time of crisis, the judgement of such experts is especially important: a pilot to deal with an emergency landing and the doctor to recommend and administer life-prolonging treatment. In all these situations, we do not call for a vote or demand that non-experts' opinions be given weight equal to that of the expert. Policy (on chemotherapy, a blazing wing engine, and the price and cost of carbon pollution) is a choice that experts can and should advise and act upon. However, the validity and usefulness of such analogies warrant discussion. In the non-greenhouse examples, the fate of a small number of individuals (from one patient to a few hundred passengers) is entirely in the hands of a single specialist, who is the only person with accepted responsibility. Few question the authority and competence of this designated person because no-one else would dare to manipulate the scalpel or the plane controls. However, in climate, there are many thousands of individuals designated or self-promoted (many with profound conflicts of interest) with the responsibility of making decisions and taking actions on behalf of millions of people. Even if the scientists may legitimately claim climate competence, we certainly do not have the authority or a mandate to decide and act. This is not to say that scientists should remain silent. Steve

FIGURE 1.1 Stephen Schneider and the author sharing a good bottle of Aussie red on the Gold Coast, Australia (June 2010).

Schneider wrote and spoke on this topic with fervour, carefully articulating that to suggest that scientists call the shots is paramount to denying democracy but that, on the other hand, democratic decision-making demands much greater and more effective public education and outreach by all researchers (cf. Cribb, 2011).

1.2.2. Global Warming 'Just a Theory'

Lewis Carroll asked, 'Why is a raven like a writing desk?' The question is posed in his book *Alice's Adventures in Wonderland* at the famous Mad Hatter's Tea Party (Carroll, 1865). Although Alice and the other guests attempt to puzzle the riddle out, Carroll's Mad Hatter never gives the answer. In fact, he later admits, "I haven't the slightest idea." Nowadays, I often ponder on a similar enigma, 'why is gravitation like the greenhouse?' The elements of a short story help to illustrate the point of the comparison in Table 1.2.

A few hundred years ago a 'scientific theory' was born…
It was welcomed because it explained something that had puzzled people…
After a while its value increased because it was used to make predictions…
Of course some people raised objections about how it worked…
And some predictions made using it were wrong…
Scientists corrected it and made it still more useful…
Some of their 'improvements' were hard to understand…
Even so it was repeatedly validated…
Calling it 'flawed' failed to diminish its application and value…
Today this 'theory' is very useful, still incomplete and almost universally accepted.

The question is not whether this 'story' outline is about gravitation or global warming: it is applicable to both (Table 1.2). Surprisingly perhaps, these two theories have very similar histories with greenhouse lagging gravitation by between 50 and 100 years. Gravitation was 'discovered' by Sir Isaac Newton in 1687; it explained a trying scientific puzzle of those times — Kepler's observation of elliptical orbits of the planets — and it successfully predicted (in 1824) the existence of a 'new' planet, later named Neptune, from the observed perturbations of the orbit of Uranus. However, gravitation was not a very satisfactory 'theory' — the inventor himself strongly disliked the 'action at a distance' he had to invoke to make it work. Despite its shortcomings, it has been validated innumerable times and is used to guide our everyday lives. We remember learning at school Newton's three laws of motion: Every body remains in a state of rest or uniform motion (constant velocity) unless acted upon by an external unbalanced force; a body of mass, m, subject to a force, F, undergoes an acceleration directly proportional to the force and inversely proportional to the mass, $F = ma$; and the gravitational force applied by one body on another is equal in strength

TABLE 1.2 Comparing the Histories of the Scientific Theories of Gravitation and Global Warming (Greenhouse)

Characteristics	Gravitation	Greenhouse
Theory 'born'	300+ years old (Newton, 1687)	~180 years old (Pouillet, 1838) or older (Mariotte, 1681)
Explains what	Kepler's elliptical orbits	Earth's surface temperature
Predicted	Neptune (1824) from Uranus	Venus 'hot house' (Goody, 1952) (see Goody, 2002)
Objections (to the way it works)	Newton disliked 'action at a distance'	Clouds, especially feedbacks, are very poor cf. Lindzen et al. (2001)
Observation vs. theory prediction conflicts	Precession of Mercury's perihelion, 1859 (Le Verrier, 1859)	Wide variety of values deduced for so-called climate sensitivity*: 3K (IPCC), 1K (Lindzen), 3–4K (Hansen)
Improved	Special Relativity (1905): $E = mc^2$	Aerosols included in IPCC TAR (2001a); carbon feedbacks included in IPCC AR4 (2007a)
Hard to understand	Einstein's General Relativity (1915): curvature of space–time	Tyndall's laboratory proof that gases in the air absorb heat (1859) (Tyndall, 1861)
Accepted	*Layperson:* everyday mechanics is Newtonian *Scientific:* Eddington proves General Relativity 1919 and space agencies use in space programmes	*Layperson:* Earth is habitable because of greenhouse gases *Scientific:* Human disturbance demonstrated by Keeling's CO_2 observations from 1958; and by temperature increases at least since 1990
Still shaky about	Action at distance persists in quantum entanglement	On very long timescales, negative climate feedbacks dominate
Public entertainment	Stephen Hawking and *Star Trek*	Jim Hansen (350.org) and *Star Trek*

*A better example of conflict than, for example, the media-pounded error by IPCC (AR4 WG2) glacier melt-date responses, 2009.

and opposite in direction to the force exerted by the latter on the former or, more colloquially, "for every action there is an equal and opposite reaction".

Still this 'theory' of gravitation made incorrect predictions — the most famous of which was the rate of precession of the perihelion of the orbit of Mercury determined by Le Verrier in 1859. Other scientists — most notably Albert Einstein — 'improved' the science of gravitation. We all know and can quote the famous result from his Special Theory of Relativity (1905) $E = mc^2$, even if we have no idea how matter can become energy. Gravitation was not really 'understood' until Einstein further improved his theory — in the form of General Relativity (1915) — for which he invoked the curvature of space–time. Very few non-physicists have much idea what this is about or even grasp that Sir Arthur Eddington proved this 'theory' of gravitation almost a century ago, in 1919. In ignorance but without qualms, we enjoy fairground rides, buy 'Big Bang' books by the wonderful Stephen Hawking, and watch episodes of *Star Trek* premised on the science of gravitation. Indeed, Stephen Hawking and Stephen Schneider alike deserve acclaim for their many interesting and profound attempts to explain complex concepts to the general public so they can understand enough to, perhaps, express an informed layperson's opinion.

The 'theory' of global (or greenhouse[4]) warming is usually[5] considered to have been 'born' around 150 years later than that of gravitation: in 1838 when French physicist Claude Pouillet[6] described how the Earth's atmosphere increases the surface temperature (Van der Veen, 2000). As for gravitation, this 'theory' of greenhouse was confirmed by observations — for example, John Tyndall conducted a set of laboratory experiments, as long ago as 1859, demonstrating that water vapour and carbon dioxide absorb infrared radiation. Like gravitation, greenhouse was used to make predictions that were later observed as fact — for example, in the 1950s, Richard Goody anticipated that the surface temperature of Venus would be very high, a fact not verified until 1967 when the *Venera 4* spacecraft successfully entered Venus' atmosphere and measured its surface temperature. In common with gravitation, greenhouse has stumbled more

---

4. The terms 'greenhouse warming' and 'global warming' are used interchangeably in this chapter as is common (but not good practice) in this literature.

5. The fact of solar radiation passing unimpeded through transparent materials while heat did not was understood at least as early as 1681. Van der Veen (2000) disputes the terminology claim as being due to Fourier even though the latter does invoke greenhouses in 1824.

6. Many sources incorrectly attribute this to Fourier, who had (in 1824) described heating, but only by conduction (Van der Veen, 2000).

FIGURE 1.2 The numerals '350' spelled out by people holding blue umbrellas on the steps of the Sydney Opera House on Saturday, 24 October, 2009. This demonstration was replicated around the world underlining public enthusiasm for action against human-induced climatic change and arguing for a limit to atmospheric loading of CO_2. James Hansen first suggested the proposed upper limiting level of 350 ppmv. Today (April, 2011) the atmospheric CO_2 concentration is 392 ppmv — *http://co2now.org/ (Source: Photo courtesy of Kendal McGuffie.)*

than once. All practitioners recognize that clouds are poorly parameterized in climate models and that this weakness is a serious handicap to increasing confidence in future climate predictions (e.g., Doherty et al., 2009; Knutti, 2010). This shortcoming (somewhat analogous to 'action at a distance') is both well-known among climate researchers (e.g., Henderson-Sellers, 2008) and a vehicle for science-based challenges to model predictions (cf. Lindzen et al., 2001).

Our understanding of greenhouse (or global warming) has been improved and augmented by painstaking observations (e.g., Charles Keeling's careful observations of CO_2 increase from 1958 — see Sundquist and Keeling, 2009) and almost unsurpassed international co-operation (IPCC, 1990a, 1990b, 1990c, 1995a, 1995b, 1995c, 2001a, 2001b, 2001c, 2001d, 2007a, 2007b, 2007c, 2007d) but it has also been plagued by real challenges such as cloud parameterization and detection and attribution of human effects (see Santer et al., 1996), by mistakes, and by poor responses to clerical errors (e.g., IPCC Working Group 2's statements in 2009 about the likely melt-date of Himalayan glaciers — see also Cogley, 2012, this volume). The lack of aerosols in climate models early in the IPCC process was not rectified until the third assessment (IPCC, 2001a) and awaits Fifth Assessment Report (AR5) for chemical interactivity. Like gravitation, the full science of greenhouse is still incomplete: the inherent nature of science. For example, an important challenge to our understanding of the climate system is how negative feedbacks prevail over very long timescales, as demonstrated by the uncanny stability of the Earth's surface temperature over its full history of

4.5 billion years (e.g., Lovelock, 1979, 1989), while positive feedbacks seem likely to destabilize the current climate as a result of perturbations due to greenhouse warming (Lenton, 2012, this volume). In common with its scientific cousin, greenhouse has 'starred' in *Star Trek* and has made naturally reclusive scientists famous, for example, James Hansen, for community-championed action on the 350 ppmv (parts per million by volume) target for greenhouse gas concentration in our atmosphere (see Figure 1.2).

Greenhouse disputes are magnified in the media and in the blogosphere. A recent quote from Professor Roger Pielke, Jr. of the University of Colorado at Boulder, "If these two guys (Lindzen and Emanuel) can't agree on the basic conclusions of the social significance of [climate change science], how can we expect 6.5 billion people to?" is one of extraordinarily many examples (Pielke, 2010). Problems persist for greenhouse: a most painful one being what Professor Myles Allen of Oxford University has termed our 'holy grail of climate sensitivity'. This is the quest to determine the expected rise in temperature due to a doubling of CO_2 that is discussed in full in Chapter 5 (Harvey, 2012, this volume). This sensitivity (which Harvey terms *fast feedback sensitivity*) has been valued at about 3 K[7] by the IPCC (2007d); as less than 1 K by Professor Richard Lindzen of MIT (from volcanic eruption observations); and as 3–4 K by NASA's Dr James Hansen

---

7. Frequently this 'climate sensitivity' is quoted in °C not K; here we use the SI unit.

based on evidence from ice cores. Harvey (2012, this volume) presents comprehensive support for 'fast feedback' climate sensitivity being between 1.5 K and 4.5 K, and not possibly greater than 6 K. He further adds that slow feedbacks, especially concerning the carbon cycle, are likely to increase the climate sensitivity by 25–50% (Harvey, 2012, this volume).

We take gravitation, 'only a *theory*' about which scientists have and still do (in small ways) disagree, for granted. Despite valid evidence being raised against gravitation, the public seem content that the science evolves, is improved and yet, while still incomplete, is *useful*. Most of us do not really understand why 'action at distance' disturbed Newton and we are also mostly unaware that this phenomenon persists in quantum entanglement today (Table 1.2). We would be astonished to see a polarized debate in our mass media about the reality of gravitation and would probably feel uncomfortable if our political leaders were expected to determine the 'truth' of this 'theory'. We do not expect this but we have come to expect individual views on greenhouse and many vehemently insist that their interpretations are as valid as those of climate scientists. Why do we find it laughable to expect the mass media to solicit opinions about gravitation from the unskilled, but apparently avidly read; why do we watch and listen to such opinions about global warming? Perhaps this is because greenhouse (not gravitation) is about guilt, lifestyle, and money! (e.g., Hamilton, 2010). This book recognizes that our lounge-room elephant is real, is certainly political, and is a serious challenge to careful, valuable, and incremental science.

1.2.3. Schneider and Climate Connectedness

In one of his early books, Stephen Schneider pointed out that in climate "everything is connected to everything else" (Schneider and Mesirow, 1976). This book deals with greenhouse (or global) warming as one, important, component of the whole evolution of the Earth's climate (cf. Table 1.1). Global warming is, as Steve said, connected to all the other aspects of climate change. Today, a crucially important aspect of this inter-connectedness is between people and climate (Taplin, 2012, this volume). The chapters here are written at the close of a decade (the 'noughties'), which has been the warmest ever observed (Arndt et al., 2010); a fact not just accepted by the mass media but described as 'undeniable'. We wrote in a year (2010) when Pakistan floods and a severe Russian heatwave killed thousands, made millions homeless, and forced up food prices globally as wheat exports were stopped by Russia and virtual water trade became more precarious and more expensive. In Australia, declared to be the 'most vulnerable' of the developed nations to the impacts of global warming (Garnaut, 2008), the Murray–Darling

Basin is arguably one of the world's first cases for climate change 'triage' (see Section 1.4). There is overlap and synergy between topics and among this book's chapters and sections. However, two chapters of this book particularly consider the complexity of climate and so echo Steve's connectedness characteristic. These are Chapter 2, which links people and climate policies, and Chapter 5, which reviews timescales of climate change. Here, these are briefly introduced.

Chapter 2, "People, Policy, and Politics in Future Climates" by Roslyn Taplin, addresses climate change and societal responses arising in international fora, especially the United Nations Framework Convention on Climate Change (UNFCCC). The chapter reviews global climate change politics and the scientific collaboration via the IPCC. Although United Nations' agencies and national governments have given significant attention to climate change concerns for two decades, the challenges of implementation of effective governance approaches to address climate change remain immense and are becoming a much greater barrier to action than scientific understanding. Chapter 2 discusses communities and individuals, and how they are addressing this climate change challenge. The author moves on to discuss other international activity such as the European Union's role and support for the Kyoto Protocol regime and its implementation; the evolving position of the USA; and attitudes of the range of developing and rapidly industrializing nations, including the pros and cons of the Asia Pacific Partnership on Clean Development and Climate. Taplin (2012, this volume) argues that implementation of a learning governance approach would give countries better capacity to address climate change and assist the polycentric systems approach to coping with climate change.

Chapter 5, "Fast and Slow Feedbacks in Future Climates" by Danny Harvey, reviews the latest evidence concerning an important and contentious issue: the magnitude of the climate sensitivity. The chapter reviews fast feedback processes (involving water vapour, temperature lapse rate, clouds, and seasonal snow and ice) and the potential slow feedback processes that could amplify the fast-feedback sensitivity, especially the magnitude of climate–carbon cycle feedbacks. Linear feedback analysis is used for both fast feedbacks and for the inclusion of carbon cycle and climate–carbon cycle feedbacks, along with illustrative quantitative examples to demonstrate the dependence of future peak temperatures in response to anthropogenic CO_2 emissions on the interaction between fast and slow feedback processes. Harvey (2012, this volume) concludes that the long-held consensus that the climate sensitivity based on fast feedback processes is very likely to fall between 1.5 K and 4.5 K, with essentially no possibility of a fast-feedback sensitivity greater than 6 K, is correct. He also determines that positive climate–carbon cycle feedbacks are likely to enhance the climate sensitivity by 25–50%, excluding potential feedbacks related to short but intense emissions of CO_2 and/or methane from yedoma soils and methane clathrates. However, much greater amplification of the future warming could occur through release of CO_2 and methane from thawing yedoma (loess) soils and from destabilization of marine methane clathrates. Excluding greenhouse gas releases from yedoma soils and clathrates, the likely range of peak global mean warming associated with an extremely stringent CO_2 emission scenario (global fossil fuel emissions reaching zero by 2080 and cumulative emissions of 800 GtC) is 1.5–4.5 K, while the likely range for a business-as-usual scenario with cumulative emissions of 1760 GtC is 2.5–6.0 K. Inclusion of greenhouse gas releases from yedoma soils has the potential to turn what would be a 3.5 K peak warming into a 5.5 K peak warming and still higher peaks if failure of policy results in a business-as-usual emission scenario.

Climate connectedness carries us further than these chapter topics, however. The modern scientific recognition of climate as one component of a much more holistic geophysiological system that is the Earth is, in large part, due to James Lovelock (e.g., Lovelock, 1979, 1989, 2006) and Lyn Margulis (e.g., Lovelock and Margulis, 1974). Stephen Schneider played a crucial role in fostering discussion of Lovelock's Gaia hypothesis (Schneider and Boston, 1991). The first, and important, means for this was the Gaia meeting (a Chapman Conference) that Steve hosted in 1988. The American Geophysical Union (AGU) Chapman conference on the Gaia Hypothesis held at San Diego in 1988 gave rise to a series of papers in the AGU's flagship journal (e.g., Kirchner, 1989; Lovelock, 1989) and also the full collection of talks (Schneider and Boston, 1991). Building on this work, others have developed Gaia into many areas of climate science (e.g., Lenton, 1998). These concepts are furthered in this book, especially in Chapters 17 and 18, which are now introduced.

Chapter 17, "Future Climate Surprises" by Timothy M. Lenton, addresses the surprising 'punches' that future climate change may soon throw our way. Reviewing the categorization of climate surprises based on concepts from dynamical systems theory, tipping elements and tipping points are described and the case of noise-induced transitions added. The climate subsystems that could undergo surprise changes are introduced in three broad categories: the melting of large masses of ice, changes in atmospheric and ocean circulation, and the loss of biomes. Lenton (2012, this volume) next focuses on how science can help societies cope with climate surprises, starting with how to assess the risk they pose, by combining information on their likelihood (under a given scenario) and their impacts. The prospects for removing the element of surprise — by achieving early warning of approaching thresholds — are considered in detail. Finally, the available response and

recovery strategies are assessed. In future, societies may well be faced with unwelcome climate surprises.

Only a couple of weeks before he died, Steve regaled Australia's (then) Chief Scientist, Dr Penny Sackett, and I with the claim that he was the person who introduced the AGU to the term 'epistemology'. At the same conference, he referred back to my idea of the three Gaian monkeys. Chapter 18 brings these monkeys up-to-date. The Gaia hypothesis postulates that the climate and chemical composition of the Earth's surface environment self-regulates into a state tolerable for the biota (Lovelock and Margulis, 1974; Lovelock, 1979). Since the early, definitional, papers, Gaia has matured and is now seen as the scientific 'parent' of Earth system science (cf. Table 1.1) that views the evolution of the biota and of their environment as a single, tightly coupled process, with the self-regulation of climate and chemistry as an emergent property (Kump and Lovelock, 1995). Gaia is now a theory able to make predictions that are falsifiable[8], for example, that oxygen is and has been regulated during the existence of land plants, within a few percent of its present level (cf. Kirchner, 1989).

Chapter 18, "Future Climate: One Vital Component of Trans-Disciplinary Earth System Science" by Martin Rice and Ann Henderson-Sellers, argues that the future climate is just one component of a larger, complex system, including people and their practices, as well as the physical and biogeochemical aspects of the Earth. The authors place our future climate in this 'bigger picture' that has many names, including Gaia and Earth system science. Chapter 18 recognizes that climate has been used frequently as a way of illustrating emergent Earth system behaviour and explores how future climates are integral to this whole-of-Earth system. As future climate challenges will have to be communicated clearly and effectively to the public (who demand or refuse policy direction changes), Rice and Henderson-Sellers (2012, this volume) also consider the clear communications goal of Stephen Schneider, for whom this book is dedicated, and the aim of James Lovelock, the creator of the Gaia hypothesis around which their chapter is organised. Rice and Henderson-Sellers (2012, this volume) explore today's climate research integrity paradox; evaluate policy in terms of media-mediation or people's ponzi; and consider Gaian governance before closing with a section on Earth system analysis and the value of trans-disciplinarity.

Chatting to Steve in June 2010, he strongly supported this book's broadening of climate into Earth systems research. He emphasized his endorsement by pointing out that the journal *Climatic Change*, which he began, edited, and nurtured for decades, had always adopted inter- and

---

8. Falsifiability was said by Karl Popper to be essential. He asserted that a hypothesis, proposition, or theory is scientific only if it is, among other things, falsifiable (e.g., Popper, 1959).

trans-disciplinary research as its guiding mantra and encouraged probing of the broadest definition of 'climate' (Schneider, 1977; and Schneider and Oppenheimer, 2009).

1.3. THE COMPLEXITY OF THE FUTURE OF THE WORLD'S CLIMATE

"No one, and I mean no one, had a broader and deeper under-standing of the climate issue than Stephen. More than anyone else, he helped shape the way the public and experts thought about this problem — from the basic physics of the problem, to the impact of human beings on nature's ecosystems, to developing policy."

Michael Oppenheimer, July 2010 (Washington Post, 2010)

1.3.1. Changing Climates

I well remember my first meeting with Stephen Schneider: it was in 1975 at a conference on climate history held at the University of East Anglia. This meeting followed my puzzling over the second research paper on climate that I ever tried to understand. It was by Stephen and was entitled 'Cloudiness as a Global Climatic Feedback Mechanism' (Schneider, 1972). In 2010, as the authors of this book wrote their chapters, these two topics — the history of the world's climate and the way feedbacks moderate and exacerbate the climate system — still dominate our thinking, as was the case when the first edition was pub-lished. I wrote then in the introductory chapter, "Human beings are curious: we seek to understand and hence to predict. We are dependent upon the climate and so our desire to predict is not driven solely by curiosity but by a need to predict future weather, its variability and other characteristics: future climates" (Henderson-Sellers, 1995b, p. 1). These words still ring true.

Unlike the first edition of this book, entitled *Future Climates of the World: a Modelling Perspective*, this edition has no subtitle because, these days, it is well-accepted that future climate predictions are accomplished using models. As one of the two editors, I no longer feel the need, as I did in 1995, to apologize for material that is "much more ephemeral than the climatic facts and figures in Volumes 1—15". *The Future of the World's Climate* is a different book from the first edition. The contents have changed dramatically — other than my Introduction, only three chapters remain by the original authors (Wang, Cleugh, and Berger) and all have been joined by new co-authors. Seven chapters are on topics not covered in the original, while the remaining seven chapters are on issues that appeared in 1995 but are now written by different authors with new views.

My 1995 introductory text went on, "So far our predictive skills are rather poor. In addition, technology and the harnessing of natural resources appear to have decreased the need to predict the future climate of indus-trial, housing, or even agricultural developments. However, although oil and gas pipelines can be laid across perma-frost, indoor climates can be controlled independently of the outdoors, airports operate year-round, and irrigated land can grow a wide variety of crops, there are still massive costs, in lives and revenue, associated with weather extremes and with climatic shifts. Marginal changes in climate can be 'managed' although this may be costly: financially in developed nations but in terms of human life in the less-developed nations" (Henderson-Sellers, 1995b, p. 1). These statements are also still correct but the sense with which we read them has evolved over the 15 years since they were written. Today, I believe our predictive skills are good enough to allow us to foresee climate change on roughly a timescale of the next 30—50 years (Meehl et al., 2009; Boer, 2010). Today, we recognize that costs of weather extremes are increasing because of population pressure, poor development, and so on, but also, many argue, in some part due to our pollution of the atmosphere (IPCC, 1990a, 1990b, 1990c, 1995a, 1995b, 1995c, 2001a, 2001b, 2001c, 2001d, 2007a, 2007b, 2007c, 2007d). Today, we are beginning to estimate the costs of climate change and acknowledge the losses it may already be causing and will certainly further exact (Stern, 2006; Garnaut, 2008). The main themes of the chapters in this book (Table 1.1) are summarized in Sections 1.2 and 1.4, and below. The chapters outlined here reflect this growing understanding of the complexity of climate and the speed of some aspects of climate change.

Chapter 3, "Urban Climates and Global Climate Change" by Helen Cleugh and Susan Grimmond, looks at the future climate of where most people live: cities. Cities and global climate change are closely linked: cities are where the bulk of greenhouse gas emissions take place through the consumption of fossil fuels; they are where an increasing proportion of the world's people live; and they also generate their own climate, including the well-known urban heat island. Chapter 3 explains why understanding the way cities affect the cycling of energy, water, and carbon to create an urban climate is a key element of climate change mitigation and adaptation strategies, espe-cially in the context of rising global temperatures and ongoing air quality challenges in many cities. Cleugh and Grimmond (2012, this volume) argue that, as climate models resolve finer spatial scales, they will need to represent those areas in which more than 50% of the world's population already live to provide climate projec-tions that are of greater use to planning and decision-making. They also underline that many of the processes that are instrumental in determining urban climate are the very factors causing global anthropogenic climate change, namely regional-scale land-use changes; increased energy use; and increased emissions of chemically-active and climatically-relevant atmospheric constituents. Cleugh and Grimmond (2012, this volume) make the case that addressing the challenges of managing urban environments in a changing climate requires understanding the energy, water, and carbon balances for an urban landscape and, importantly, their interactions and feedbacks and links to human behaviour and controls. Cities are crucial for better understanding and are themselves active 'agents' in both causing and in mitigating anthropogenic climate change. This topic has close links with Chapter 4 (Pitman and de Noblet-Ducoudré, 2012, this volume), described in Section 1.4.2.

Chapter 6, "Variability and Change in the Ocean" by Alexander Sen Gupta and Ben McNeil, considers the world's ocean as an integral part of the climate system. The ocean has absorbed over 80% of the additional heat and about one third of the additional carbon dioxide that has been added to the climate system over the last few decades. This absorption strongly moderates the impacts that would otherwise have occurred. However, as Sen Gupta and McNeil (2012, this volume) point out, these services come at a cost. The ocean has become warmer to depths of hundreds of metres, increasing surface stratification and causing thermal expansion and sea-level rise. The ocean chemistry has changed with the water becoming less alkaline (ocean acidification) and with lower aragonite saturation. In addition, an intensification of the atmospheric hydrological cycle and changes in surface winds have driven regional changes in ocean salinity and certain circulation patterns. While there are far fewer chemical and biological observations of the global ocean, significant trends in oxygen concentration, primary productivity, and the characteristics of higher species are becoming apparent. Climate models capture these changes when driven by anthropogenic greenhouse gas emissions. They also demonstrate that much more dramatic changes are in store as we increasingly perturb the system. A shortcoming of current models reviewed by Sen Gupta and McNeil (2012, this volume) is that they do not represent key carbon cycle feedbacks that could potentially have a large effect on the ocean's ability to moderate climate change in the future. This chapter is closely linked to Chapter 7.

Chapter 7, "Climatic Variability on Decadal to Century Timescales" by Mojib Latif and Wonsun Park, reviews a central challenge of climate research: the need to understand the dynamics of these fluctuations and predict regional-scale climate variability and change over timescales of decades or even longer. Latif and Park (2012, this volume) point out that, while the twentieth century has featured considerable global warming of the order of 0.7 K, the Earth will continue to warm over the next several decades under any plausible policy scenario. Chapter 7 underlines that natural variability means that the future climate trajectory will not follow a steady path. The

regional-scale natural variability and change that must accompany global warming will have profoundly important impacts. These will be comparable to, or perhaps even worse than, twentieth century fluctuations including the 1930s' US Dust Bowl drought, the Sahelian drought of the 1970s and 1980s, the ongoing droughts in the southwestern USA and in southwestern Australia, and decadal-scale changes in Atlantic hurricane activity. The origin of decadal to centennial variability is the challenge tackled in Chapter 7 that emphasizes the role of the world's oceans in future climates. Another major subsystem of the climate (with the oceans and the land) is the cryosphere: the frozen water of the Earth. This is discussed in Chapter 8.

Chapter 8, "The Future of the World's Glaciers" by Graham Cogley, considers the small glaciers (those other than the ice sheets), which, in the author's words, are "insensitive but reliable indicators of climatic change". Observations show that these so-called 'small' glaciers have been losing mass since the mid-nineteenth century, with a marked acceleration since 1970. The total mass stored in small glaciers was between about 500 and 700 mm of equivalent sea-level rise in the late twentieth century. Measurement of glacier mass balance provides information that is independent of weather stations and bathythermographs. However, because the mass balance and the near-surface air temperature are both driven by the radiation and energy balances, the correlation between them is very good and is the basis for most projections of the evolution of glaciers. Projections show that over the twenty-first century, a cumulative loss of about 90–160 mm of equivalent sea-level rise can be expected with reasonable confidence, but that poorly quantified and poorly understood phenomena make the probability distribution function asymmetrical. Cogley (2012, this volume) concludes that, while substantially lesser losses are very improbable, since the projections do not quantify adequately the delayed response of glacier dynamics to the forcing and the role of losses by frontal ablation (calving and melting at terminuses at sea level), greater losses cannot be ruled out with the same confidence.

1.3.2. Challenges in Climate Science

The challenges in climate prediction today range from anthropic to geological upsets and also include long-term changes and their causes. Arguably the dominant subsystem of today's climate is people. Thus, the definition of climate, and more broadly Earth system, science must be widened to fully embrace humans, their activities, and their societies. Indeed, the Anthropocene is now accepted as a geological era, created and characterized by human disturbance. People are climate culprits and — at the same time — deeply susceptible to extremes of weather and future climate changes. Seven chapters of the book, which are introduced here, consider these human–climate interactions in detail.

Chapter 10, "Climate and Weather Extremes: Observations, Modelling, and Projections" by Lisa Alexander and Claudia Tebaldi, reviews weather and climate extremes in the context of future climate change, discussing some of the difficulties in defining and modelling events using both statistical and dynamical models. Alexander and Tebaldi (2012, this volume) consider as wide a range of types of extreme events although much of their focus is on the types of events that are statistically robust, cover a wide range of climates, and have a high signal-to-noise ratio for use in 'detection and attribution' studies. The importance of data quality and consistency is addressed as underpinning any analysis of climate extremes. In addition, the statistical analysis of extremes is discussed in some detail along with the challenges of scale differences that exist when comparing observations and climate model output, including a discussion of downscaling techniques. Alexander and Tebaldi (2012, this volume) describe the current state of the monitoring, understanding, and physical modelling of the characteristics of climatic extremes, including projected changes, focusing both on the great progress that has taken place in the last decade and in the challenges that still lie ahead.

Chapter 11, "Interaction between Future Climate and Terrestrial Carbon and Nitrogen" by Robert E. Dickinson, begins by noting that the terrestrial components of the climate system store about five times as much carbon as does the atmosphere and annually removes about one-fourth of the carbon dioxide added by human activities. These exchanges with the atmosphere are highly dynamic, depending on the assimilation of carbon by vegetation and on plant and soil respiratory processes. Chapter 11 discusses the uptake rate, which depends on climate, atmospheric carbon dioxide, and availability of nutrients, especially nitrogen, and the respiratory losses in the soil column, which depend on soil temperature, humidity, and oxygen levels. Nitrogen availability, in turn, depends on a balance between sources and sinks that are closely coupled to soil–carbon cycling and climate influences. Dickinson (2012, this volume) describes how carbon and nitrogen cycling are now included in many climate models, some of which are used to predict future climate. However, these models make various assumptions and approximations and provide a wide divergence of conclusions as to relative magnitudes and importance of various terms. This chapter highlights current observational and modelling research on the processes involved in their use. Other questions discussed include an evaluation of how nitrogen affects carbon storage, what determines storage of nitrogen in the soil and what are the mechanisms coupling atmospheric temperature, the moisture of the atmosphere, the soil, and CO_2. The final section reviews current results and

insights from coupled carbon–climate models, including dealing with the role of other nutrients and the fate of Arctic and tropical peat.

Stephen Schneider worked with his wife, Terry Root, on the development of strategies to combat and live with anthropogenerated climate change: strategies that must embrace the natural world and crops and pests, as well as human infrastructure and societies (e.g., Root and Schneider, 1995, 2006). Our need to understand better how climate–phenology relationships impact on the prediction of land-surface exchanges, and hence on land-management decision and policy making (Morisette et al., 2009), is one new challenge of climate prediction (see Pitman and de Noblet-Ducoudré, 2012, this volume). Interactive atmospheric chemistry (gases, solids, and droplets) is another.

Chapter 12, "Atmospheric Composition Change: Climate–Chemistry Interactions" by Ivan Isaksen and colleagues (Isaksen et al., 2012, this volume), is republished here from the journal *Atmospheric Environment* (with permission). It covers chemically active climate compounds that are either primary compounds like methane (CH_4), removed by oxidation in the atmosphere, or secondary compounds like ozone (O_3), sulfate, and organic aerosols, both formed and removed in the atmosphere. Emissions of pollutants change the atmospheric composition contributing to climate change by modifying subsystem components (e.g., through changes in temperature, dynamics, the hydrological cycle, atmospheric stability, and biosphere–atmosphere interactions), and affect the atmospheric composition and oxidation processes in the troposphere. Chapter 12 summarizes progress in our understanding of processes of importance for climate–chemistry interactions, and their contributions to changes in atmospheric composition and climate forcing. Isaksen et al. (2009, 2012, this volume) show that land-based emissions have a different effect on climate than ship and aircraft emissions, and different measures are needed to reduce the climate impact. Several areas where climate change can affect the tropospheric oxidation process and the chemical composition are identified. This can take place through enhanced stratospheric–tropospheric exchange of ozone, more frequent periods with stable conditions favouring pollution build-up over industrial areas, enhanced temperature-induced biogenic emissions, methane releases from permafrost thawing, and enhanced concentration through reduced biospheric uptake. During the last 5–10 years, new observational data have been made available and used for model evaluation and the study of atmospheric processes. Although there are significant uncertainties in the modelling of composition changes, access to new observational data has improved modelling capability. Emission scenarios for the coming decades have a large uncertainty range, in particular with respect to regional trends, leading to a significant uncertainty in estimated regional composition changes and climate impact. The topics covered in Chapter 12 and in Chapter 13 are closely intertwined.

Chapter 13, "Climate−Chemistry Interaction: Future Tropospheric Ozone and Aerosols", is by Wei-Chyung Wang, Jen-Ping Chen, Ivar S. A. Isaksen, I-Chun Tsai, Kevin Noone, and Kendal McGuffie. The authors build on the important fact that the atmosphere contains radiatively−chemically important trace gases (in addition to CO_2), which are changing and expected to change more in the future. The radiative heating and cooling of the atmosphere is affected by perturbations of CO_2, primary aerosols, and chemically-active greenhouse compounds (CH_4, N_2O, CFCs), and by secondary compounds (tropospheric ozone, sulfate, and organic aerosols) that are formed in the atmosphere through a variety of chemical and physical processes. While the concentrations of these secondary compounds are dominated by the surface emissions of their gas phase precursors, they are also closely coupled to meteorological parameters: temperature, wind, and the hydrological cycle (moisture, precipitation, and clouds), and to solar radiation (photo-dissociation). Therefore, climate changes due to global warming from anthropogenic greenhouse additions may perturb atmospheric concentrations of chemically active climate compounds and thus the oxidation capacity of the atmosphere, providing feedback to the climate system. As the climatic states and surface precursor emissions exhibit distinctively different regional characteristics, these climate−chemistry interactions will play an important role in future climate changes that will disturb different areas differently. Chapter 13 also reviews future changes including scenarios of surface emissions, changes in concentrations and the associated effects on climate, concluding with a discussion of uncertainties in atmospheric composition and hence in future climates and climate impacts. The links made in this chapter between vegetation, people, and atmospheric chemistry serve as an excellent introduction to Chapters 14 and 15.

Chapter 14, "Records from the Past, Lessons for the Future: What the Palaeorecord Implies about Mechanisms of Global Change" by Sandy P. Harrison and Patrick J. Bartlein, documents past variations in climate on multiple time and space scales. At one end of these authors' timescale is the gradual cooling over the past 70 million years that resulted in the Earth shifting from a warm, ice-free state to a predominantly cold, glaciated state. At the other extreme, the Dansgaard−Oeschger cycles are rapid shifts between warm and cold states that can occur very quickly, sometimes within decades, and were most marked in regions bordering the North Atlantic. These very different types of climate change are explained in terms of insolation forcing and the differentially lagged response of components of the Earth system. Biophysical and biogeochemical feedbacks associated with changes in the hydrosphere, and the marine and terrestrial biosphere, are responsible for amplification of initial climate changes and can result in extremely rapid climate changes. Although the past does not provide direct analogues for potential future climate changes, Harrison and Bartlein (2012, this volume) argue that it does provide insight into the mechanisms of climate change of similar magnitude to, and at time and space scales congruent with, the changes that might occur in response to anthropogenic forcing. Palaeo-environmental records of the response of physical and biological systems to past climate changes provide targets for the evaluation of the Earth system models used to project potential future climate change. Chapter 14 illustrates the wealth of information available from the palaeorecord, how these records should be interpreted, and how they illuminate the complex mechanisms of climate change and interactions between components of the Earth system. Harrison and Bartlein (2012, this volume) conclude by drawing attention to the lessons that the study of the past provides both for future developments in climate science and for understanding and predicting future climate change. This review links closely with the more detailed look at external and internal forcings of climate in Chapters 15 and 16.

Chapter 15, "Modelling the Past and Future Interglacials in Response to Astronomical and Greenhouse Gas Forcing" by André Berger and Q. Z. Yin, uses an Earth system model of intermediate complexity (an EMIC) to understand climate at the peaks of the interglacials of the last 800,000 years in response to insolation and greenhouse gas forcings. Their simulations show that the greenhouse gas concentrations play a dominant role on the variations of the annual temperature averaged over the globe and over the southern high latitudes, whereas insolation plays a dominant role in the northern high latitudes. The interglacial marine isotope stage 11, assumed to be a good analogue of the present interglacial, is warm only because of its high greenhouse gas concentrations, the insolation contributing to a cooling. Chapter 15 describes how, for the warmest interglacials (known as marine isotope stage 9 and marine isotope stage 5), the greenhouse gas concentrations and insolation changes reinforce one another. Berger and Yin (2012, this volume) point out that precession changes are important for predicting the temperature of the northern high latitudes, while, in response to insolation, the annual mean temperatures averaged over the globe and over southern high latitudes are highly linearly correlated with obliquity. During local winters, the response over the polar oceans, although the available energy is small, is larger than during the local summers due to the summer remnant effect. The careful analysis of detailed climate simulations and long-term palaeo-evidence allows the chapter authors to draw important conclusions: that the sensitivity to doubled CO_2 is the highest for the coolest interglacial and that marine isotope stage 19, which has the same greenhouse gas concentrations and insolation pattern as marine isotope stage 1, appears to be the best analogue for our current interglacial.

Chapter 16, "Catastrophe: Extraterrestrial Impacts, Massive Volcanism, and the Biosphere" by Claire Belcher and Luke Mander, reviews two intervals of major environmental change in the geological past that are associated with mass extinction of life. Although catastrophic events are rare in Earth history, life has been abruptly devastated five times in the past 600 million years. The first climate catastrophe occurred at the transition from the Triassic to the Jurassic Period (Tr–J; ~200 million years ago (Ma)) and the second at the end of the Cretaceous Period (K–Pg; ~65 Ma). Belcher and Mander (2012, this volume) use the K–Pg event to highlight the possible environmental and biological consequences of the impact of an extraterrestrial body with the Earth, and the Tr–J to highlight the possible environmental and biological consequences of massive volcanism. Such cataclysmic events are believed to have greatly influenced Earth's climate such that the resulting climate change led to extinctions on a global scale. Analysing the possible environmental and climatic effects of each of these climate catastrophes will improve our understanding of the potential risk facing life on Earth should such cataclysmic events occur in Earth's future.

The first incarnation of this book was published fairly soon after the First Assessment Report of the Intergovernmental Panel on Climate Change (IPCC, 1990a, 1990b, 1990c). This new book draws on the very early results of simulations undertaken for the Fifth Assessment Report of the IPCC due to be published in 2014. We also take a wider brief than the IPCC, with its concern about human modification of the climate. The IPCC has amassed a wealth of information but, despite all we have learned, we cannot yet simply 'switch on or log in and request the latest climate forecast'. We are closing the gap between weather and climate forecasting but, so far, there remains a leap of understanding yet to be ventured, or dared.

Our world is in danger and we, people, are the main cause. We know we need to make decisions very urgently. Back in 1988, when I was a very new Australian immigrant, I was invited to speak at 'Greenhouse '88' — a conference on climate change held in Sydney. Among the other invited speakers was Stephen Schneider. Ian Lowe, now an Emeritus Professor of science, technology, and society at Griffith University in Queensland and President of the Australian Conservation Foundation, remembers this meeting and Steve's part in it:

Stephen Schneider was one of the world's foremost climate scientists. He was also a wonderful communicator of the science. Australia's Commission for the Future, when I was its Director in 1988, invited Professor Schneider to play the leading role in

Greenhouse '88, a nationally-linked conference to draw public attention to the clear conclusions emerging from climate science. His clear and authoritative presentations contributed greatly to public understanding of the problem, in turn altering the political climate; as a result, Australia adopted a leading role in the late 1980s and early 1990s in developing responses to climate change, culminating in the agreement at Rio in 1992 of the Framework Convention. He [Steve] has continued to be a powerful voice of reason in the debate about climate change. His untimely death is a real loss, not just to the scientific community but to the wider world in which he was so positively engaged.

Lowe (2010)

Then, 1988, Stephen clearly explained that it was "time to act". The wonderfully prophetic poster for this conference (Figure 1.3) has the tag line,

If we live as if it matters, and it doesn't matter, it doesn't matter. If we live as if it doesn't matter, and it matters, then it matters.

This is not only still true today, but is still more true! This book helps us understand how and why we are changing the

FIGURE 1.3 Poster advertising the CSIRO 'Greenhouse '88' Conference. Whether the Sydney Opera House is submerged or this depicts circling sharks remains a personal perception choice. The tag line is even truer today than in 1988. *(Source: Reproduced by permission of the CSIRO.)*

planet's climate, how anthropogenic forcing interlinks with non-human (natural) climate changes, and what steps we can take to mitigate and to try to live with future changes. Steve would, I believe, have supported the breadth of disciplines represented in this book (Table 1.1). He himself sought ever-wider circles of cause, effect, and consequence in his climate research expanding from clouds and models to aerosols, economics through media-reporting, impacts on biodiversity, to uncertainty statements and the standing, credibility, and reputation of scholars (e.g., Santer and Solomon, 2010).

1.4. CLIMATE FUTURE OF THE COORONG: COMMUNICATING FROM GLOBAL 'GROUND ZERO'

"The planet feels hotter now, and certainly more at risk. The world is smaller for the death of Stanford University climatologist Stephen H. Schneider. And certainly a whole lot less intelligent and decent."

Bud Ward, July 2010 (Ward, 2010)

1.4.1. From Global to Local

This book is dedicated to the memory and, I hope, also to the continuance of the work of Professor Stephen Schneider. Steve died on 19 July, 2010, and has been mourned by his family and friends, by climate researchers worldwide, and by all those who depended on his perceptive insight, his intelligence, wise counsel, and his apparently unfailing energy. As well as being saddened by his passing, Steve's demise has evoked what Paul Ehrlich called "the outpouring immediately after his death from fellow scientists and former students testifying to Steve's generosity in mentoring others" (Ehrlich, 2010, p. 776) and Ehrlich goes on "I knew how open he was with students, but I didn't fully realize the scale of his giving." For those of us who wish to keep alive Steve's determination to persuade the public and politicians of the truth of anthropogenic climate change and the urgent need to respond to it, this book offers one modest means of sharing our understanding and communicating more effectively.

I wished to augment the honouring of Stephen's memory in this book by devoting a section to one of the very many areas of interest he championed. South Australia's Coorong (Figure 1.4) could be said to exemplify all Stephen strove for: raising awareness; beating political neglect; recognizing the integrated nature of biodiversity, climate, and human disturbance; and people's need for comprehensive information. Back in 2006, as the then 'Thinker in Residence' for the South Australian govern-ment, Stephen Schneider proposed that research be undertaken to "understand South Australia's vulnerability in terms of exposure to reduced flows, quality of water supplies and risks of algal blooms"... and... "potential physical risks associated with reduced flows to the lower lakes, rising sea levels, storm surges, and greater sand movement" (Schneider, 2006, p. 46).

Compounding the bleak climate future for the Coorong is the fact that there is very little time to act. On 12 August, 2010, I was asked by a journalist (from the UK's *Independent* newspaper) whether I believed that the Coorong was, or would soon become, the "climate change's ground zero"? Kathy Marks was covering the Australian 2010 general election and was interested in the fact that, the previous day, both major parties had announced very large sums of money to be spent on 'fixing' the Murray River (Marks, 2010). It is true that the urgency of the environmental disaster envel-oping the mouth of the Murray River today (Figure 1.5) has brought to Australia what might be the world's first climate change triage challenge. The question is: Do we save the Coorong at all costs, do we leave it to become something different as is already happening, or do we invest heavily in 'palliative care', using up resources without changing the region's ultimate fate? Indeed, is this the world's first climate change 'ground zero'? I wished I knew. I wished Stephen were still alive to ask.

Stephen Schneider was a leader among the world's most lucid science communicators and made a specialty of ensuring that mass media journalists received valuable messages and comprehensible information. He researched at the cutting edge of climate science, one aspect of which is the very great challenge of downscaling from global simulations to local and useful predictions. One chapter (Chapter 9) in this book updates the prediction capabilities we have at regional scales. Chapter 9, "Future Regional Climates" by Jason Evans, John McGregor, and Kendal McGuffie, deals with the very challenging downscaling issue: from global to regional climates. It includes sections on methods for downscaling climate predictions including statistical, dynamical, and stochastic processes. The authors describe the sources of uncertainty such as the appropriate selection of a human greenhouse gas emission scenario (cf. Taplin, 2012, this volume), global climate model (GCM) uncertainties, the choices of downscaling method, bias correction, and building ensembles. Evans, McGregor, and McGuffie complete their chapter by exploring the uncertainties in future regional climates, including, inter alia, the prediction capabilities we have for the future hydrology of large basins such as the Murray–Darling River. Analysing the range of regional prediction techniques, Chapter 9 examines ways to quantify the 'most likely' future regional climate and the related uncertainty, so that informed investigations can be made into the future climate change impact on natural and anthropogenic systems.

FIGURE 1.4 'A Living Body' digitally manipulated image of part of the Coorong created by Leon Cmielewski and Josephine Starrs (Starrs and Cmielewski, 2010). The words 'a living body' are quoting Tom Trevorrow of the Ngarrindjeri people who, on the steps of the South Australian Parliament in 2009, described the Coorong in South Australia as 'a living body'. This speech can be viewed on YouTube: Our Water Our Rights, steps of Parliament House, Adelaide, South Australia, 10 October, 2009, organised by the Water Action Coalition; speaker Tom Trevorrow. The words have been digitally manipulated into the sand hill landscape of Younghusband Peninsula in South Australia's Coorong region. (*Source: Photography/Cadastral Data supplied by Customer Service Centre, Client Services, Department of Environment and Natural Resources.*)

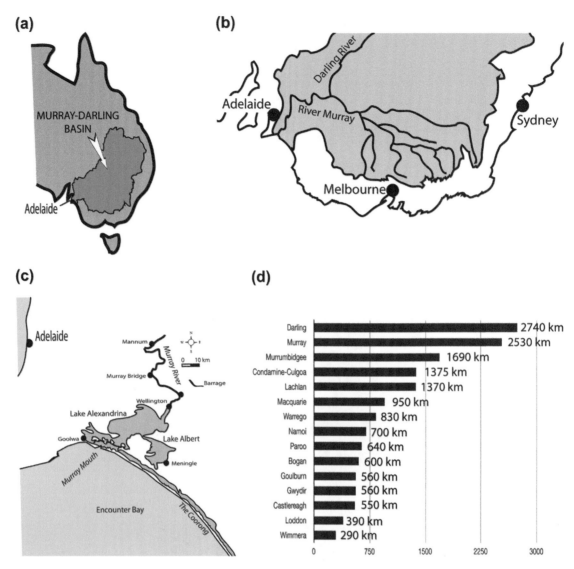

FIGURE 1.5　The Murray—Darling: (a) Basin location; (b) lower Basin showing the confluence of the two major rivers; (c) Mouth of the Murray and the Coorong; and (d) whole Basin's component river systems with lengths. *(Source: All redrawn from: (a) and (d) MDBA website; (c) 'The Living Murray' figure 6.1, MDBC, 2005; (b) MDBC Annual Report 2004—5, fig 3.1.)*

The year ending June 2008 was the sixth lowest water inflow year in the 117-year record with only 2220 gigalitres[9] of inflow, about 25% of the long-term average of 8900 gigalitres flowing into the Murray River system according to the Murray—Darling Basin Commission. Since the 2009/10 El Niño dissipated in early May 2010, the climate shifted into weak La Niña conditions, bringing considerable precipitation to eastern Australia and the Murray—Darling Basin (Figure 1.6a). Despite this late respite, the first decade of this century delivered the lowest ever observed precipitation to this region (Figure 1.6b). The water cycling, already impacted by our additional global burden of greenhouse gases, is seriously disrupted in the

Murray—Darling Basin by land-use change (see Pitman and de Noblet-Ducoudré, 2012, this volume) and by subtler but more insidious anthropogenic climate change and other human impacts.

1.4.2. Witnessing the World's First Climate 'Ground Zero'

Frighteningly, the dire situation of 2009/10 will worsen with global warming: hydrologists predict that we must subtract a further 15% streamflow for every future 1 K greenhouse increase in maximum temperature in southeastern Australia. The IPCC, in its Fourth Assessment (2007b), states with 'high confidence' that "as a result of reduced precipitation and increased evaporation, water security problems are

9. One gigalitre (GL) is 10^9 litres or a cubic hectometre.

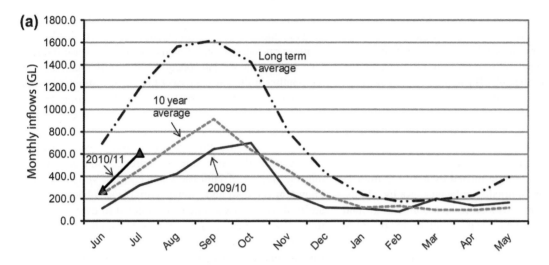

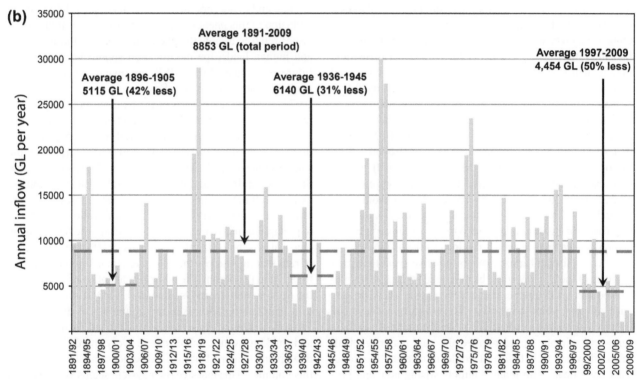

FIGURE 1.6 (a) Murray River system monthly inflows in gigalitres (GL) (excluding the Snowy and Darling) as reported on 4 August, 2010, by the Murray–Darling Basin Authority for the preceding water year. The 10-year average is the immediately past decade. i.e., 2000–2010. The massive drought culminating in 2009–10 is clearly seen, as is the beginning of recovery in mid-2010. *(Source: Murray–Darling Basin Authority report for week ending 4 August, 2010.)* (b) Inflows to the Murray River system (gigalitres (GL) per year) (excluding Darling inflows and Snowy releases and shown in 'water years') for the period July 1891 to June 2009. The inflow over the most recent drought and particularly the last couple of years (cf. part (a)) is lower than during earlier droughts. *(Source: Murray–Darling Basin Authority and CSIRO, 2010: Climate variability and change in south-eastern Australia: A synthesis of findings from Phase 1 of the South Eastern Australian Climate Initiative (SEACI)).*

projected to intensify by 2030 in southern and eastern Australia" (IPCC, 2007b). Professor Garnaut's (2008) report takes a longer and still more depressing view of the Murray–Darling Basin's future under unmitigated global warming. He says that, by the end of the century, climate change will lead to "increased frequency of drought, combined with decreased median rainfall and a nearly complete absence of runoff in the Murray–Darling Basin"...and... "ended irrigated agriculture for this region, and depopulation" (Garnaut, 2008, p. 125).

The 2010 Australian general election prompted a bidding war between the Coalition opposition and the Gillard

Labor government. Tony Abbott and the Nationals' Senate leader, Barnaby Joyce, unveiled their $750 million plan at the mouth of the Murray in South Australia on Wednesday, 11 August, 2010. They also promised to reinstate the Howard government's 10-point Basin plan, valued at a cool $10 billion (including $3 billion for water buy-back), in early 2007. In response, the then Federal Water Minister, Penny Wong, announced a $13.7 million plan to send 56 billion litres of water down the Snowy River during the next two years if Labor won a second term. This, Wong said, was "on top of $50 million promised by Labor at the last election". Perhaps my journalist colleague, Kathy Marks, was correct, the Coorong seemed to be a political 'ground zero'[10] prompting a defiant, 'win at all costs' response (cf. Taplin, 2012, this volume)

The general public, and particularly those who pay taxes, deserve to be involved in determining, or at the least understanding, how their money is being used to respond to climate change. Reclaim the past, change the present, or face the future? The grim reality for southeastern Australia may be that past profligate sales of water rights combined with present agricultural practices, exacerbated by future climate change, may have already doomed the Murray mouth (Figure 1.5). While arguments rage about farmers higher up the Murray River, a study in late 2010 showed that the Coorong has economic value of around A$3.2 billion. Morrison and Hatton MacDonald (2010) project that restoring the Coorong from its present poor environmental health to close to pristine conditions would add around A$4.3 billion to its value. The Coorong is our first test case: an internationally recognized (Ramsar[11]-listed) ecological wetland of which the lower 60% is already too salty for the fish and aquatic plants to revive even if the total reserves of the Murray could be delivered to the river mouth immediately. Of course, we cannot supply the Coorong with all this water now even if we had it because the merciless evaporative demand of southeastern Australia's semi-arid climate would remove the vast majority of all released water well before it could relieve the Coorong. This book deals with many conflicts arising now and predicted for the near future from the tensions created between land-use needs and climate change. Specifically, the last chapter to be introduced (Pitman and de Noblet-Ducoudré, 2012, this volume) tackles the difficult topic of climate land-use change interactions.

Chapter 4, "Human Effects on Climate through Land-Use-Induced Land-Cover Change" by Andy Pitman and Nathalie de Noblet-Ducoudré, provides the background to understanding the role of land-use-induced land-cover changes on regional and global climate. Land-use-induced land-cover change has occurred for millennia, and on a globally significant scale over the last few hundred years. It changes the nature of the landscape, its functioning, and causes the emissions of carbon dioxide and other gases. It has a clear and unambiguous impact on the local climate. By focusing on the physical and biogeophysical changes that occur as the landscape is modified, Chapter 4 argues that the impact of land-use change is not yet sufficiently known at regional and global spatial scales because there are limits to the experiments conducted to date to explore this issue. Pitman and de Noblet-Ducoudré (2012, this volume) review the key contributions to understanding the role of land-use-induced land-cover changes on large-scale climate and make specific recommendations relating to the inclusion of this phenomenon in future climate simulations. Issues relating to how the phenomena of vegetation change and human landscape modification should be included in future assessments by the IPCC are reviewed and proposals made for use by IPCC in AR5.

The lakes and wetlands of the Murray mouth are dying because of two centuries of human-induced land-use change (i.e., agriculture better suited to Europe); uncoordinated and often inconsistent management (among the states of Queensland, New South Wales, ACT, Victoria, and South Australia, and multiple federal systems); and an extreme climate. Although large (over a million square kilometres), the Murray−Darling Basin's ratio of discharge to precipitation is extremely low − perhaps the lowest in the world at less than 0.05, that is, for every 100 mm (or inches) of rain that falls across the Basin less than 5 mm (or inches) makes it to the ocean. The other climate plague of Australia, the El Niño−Southern Oscillation (ENSO), further subjects this national bread-basket to unique flood and drought variability driving up evaporative demand in dry periods and making water management very hard in massive downpours (Figure 1.6b) (cf. Alexander and Tebaldi, 2012, this volume). This massive river system seems very likely to also be suffering drought because of climate change. CSIRO (2010) say of the Murray−Darling Basin, "the 13 percent reduction in rainfall during 1997−2006 led to a streamflow decrease of 44 percent." (CSIRO, 2010, p. 2). They go on to evaluate the factors believed to be responsible for this climate shift in which they include: the Pacific Ocean El Niño−Southern Oscillation (ENSO), the Indian Ocean impacts through the Indian Ocean Dipole (IOD), and the effects of high latitude atmospheric circulation via the Southern Annular Mode (SAM) (for details see Latif and Park, 2012, this volume). Most importantly, CSIRO (2010, p. 2) say that, while "none of them appears to be able to explain the decline in autumn rainfall... changes in the Hadley Cell... are leading to increasing surface pressure

---

10. "The basin is ailing, and scientists fear that as climate change grips the driest inhabited continent, its main foodbowl could become a global warming ground zero" (Marks, 2010).

11. The Ramsar Convention (adopted in the Iranian city of Ramsar in 1971) is the only global environmental treaty that deals with a specific ecosystem − see *http://www.ramsar.org*.

across the region… which is found to be well correlated with decreasing rainfall in all seasons except summer, including the decrease in autumn rainfall." These changes (in the Hadley Circulation and in the subtropical ridge pressure) are linked to global warming.

The rapidly deteriorating state of the Murray's Lower Lakes (Figure 1.4) is a product of years of river regulation and the extreme drought conditions of the past decade (Figure 1.6). There are engineering solutions, including the Wellington Weir, being considered by the South Australian government. This temporary weir across the almost dry Murray, 30 or 40 km back from the ocean, would allow flooding of some Coorong lakes with seawater (Figure 1.5). Australia's 'Storm Boy' ecological wonderland would shift from Ramstar internationally recognized to an ignominious reclassification as a Montreux[12]-listed wetland, and consign Australia's environment to the international shame file. While the Murray−Darling Basin shares quasi-continental status with the Amazon, Mississippi, and Yangtze Basins, it alone exhibits a uniquely extreme hydrological variability as a result of climate. To have managed over 200 years of successful agriculture in the face of such climate demands says much for Australia's farmers and growers. Now, the political challenge is whether to support the current industry through what may be increasingly frequent prolonged droughts or to try other options or different practices (see MDBA, 2010a, 2010b).

In Bali in December 2007, the world accepted the reality of global warming and began the search for climate change action (Taplin, 2012, this volume). Our politicians and business leaders are now attempting to balance mitigation (reducing atmospheric greenhouse gases) against adaptation (living in warmer and hydrologically more extreme conditions). Responses, at least in democracies, should follow from community-based prioritization comparing loss reduction investment against funding to improve wins. As the reality of climate change bites, a third response strategy may be required: triage. Triage is applied in an emergency to allow the most productive use of inadequate support. It is unpleasant even to consider for most of us. However, as the world's population comes to grips with climate change disasters, it will be necessary to stratify responses three ways. There will be severe climate disruptions, which will be left untreated because they will be recognized as still able to recover autonomously. People living in these places and businesses affected by their local climate deterioration will not gain support. There will be a small number of climatically induced emergencies where taxpayers' money can genuinely reduce suffering and

where investment is deemed worthwhile. These communities and enterprises will be 'saved' at the expense of others. Finally, and with frightening speed perhaps, we shall be faced with situations like the Coorong today in South Australia where the judgement might be to let it change: creating a climate 'ground zero' memorial.

While I was writing this introductory chapter, the Murray−Darling Basin Authority published the first and second volumes of its 'Guide to the Proposed Basin Plan−Volumes 1 and 2' (MDBA, 2010a, 2010b). The sustainable diversion limits (or SDLs) proposal for surface water is an additional 3000−4000 GL per year to the environment. (The current diversion limit is about 13,700 GL per year.) This SDL range will produce an estimated long-term average flow of 7100−7700 GL per year through the Murray mouth. This means that the amount of water available to the environment would be 22,100−23,100 GL per year or 67−70% of all inflows, compared with 58% or 19,100 GL per year as is the case today. Volume 1 of the report goes on to point out that this additional 3000−4000 GL per year represents "a Basin-scale average 22−29% reduction in current diversions for consumptive purposes (from all diversions; i.e., watercourse diversions, floodplain harvesting, and interceptions such as farm dams and forestry), or an average 27−37% reduction if the reduction is sourced only from watercourse diversions" (MDBA, 2010a, p. xxiii).

Australia's mass media reported on the new Murray plan with gusto in early October 2010[13] (cf. Rice and Henderson-Sellers, 2012, this volume). Commentators perceived the story to be news of violence and vehemence at consultation meetings and discussed the competing 'rights' of land-owners, especially farmers dependent on irrigation, and the environment. After I read the first volume of the Murray−Darling Basin Authority Guide, I was struck by the degree to which the media missed the real question which, in my view, is "can the massive water diversion being proposed by the Murray−Darling Basin Authority against the competing claims for land-owner compensation, continued livelihoods, support of regional towns and communities and all the Australian tax-payer investment, sustain the Murray River and hence the Coorong?" I tried to look forward − say to 2030 − and to foresee how the climate, biodiversity and social and economic needs might co-evolve (cf. Schneider, 2006; Drought Policy Review, 2008).

I searched the Murray−Darling Basin future water plan for details of the assumptions made about future climate change and was confused by what I found (MDBA, 2010a). The plan says, correctly, that regional climate change is hard to predict and indicates our lack of skill at the scales

---

12. Montreux Record is the list of Ramsar sites where change in ecological character has occurred, is occurring, or is likely to occur as a result of technological development, pollution, or other human interference (established by Resolution 5.4).

13. For example 'Murray plan could cost 10,000 jobs' (Nine News, 15 October, 2010) and 'The Murray−Darling Basin Authority has bowed to public pressure and ordered a new study' (ABC, 18 October, 2010).

vital for even this very large river basin (Evans et al., 2012, this volume). However, it says, "While there is uncertainty associated with different predictions of the magnitude of climate change effects by 2030, there is general agreement that surface-water availability across the entire Basin is much more likely to decline than to increase. Recent updates suggest the Basin-wide change in surface-water availability for the period from 1990 to 2030 will be about 10%. This means the latest climate change modelling suggests that, *under a median 2030 prediction, conditions are likely to be around 10% drier than past experience.*" (Italics added for emphasis; MDBA, 2010a, p. 33.) As far as I can discern, no change was assumed in the rainfall distribution (only in the annual total); that is, they shifted the pdf (probability density function) but did not change its shape (Smith and Chandler, 2010 cf. Alexander and Tebaldi, 2012, this volume). In the section about the approach taken on climate change, it says, "As it is not yet possible to separate the effects of climate change from overall variability in water availability in the Basin, the Authority has adopted the full historical record (1895 to 2009) as the assessment baseline" (MDBA, 2010a, pp. 33−34.) Despite proposing a review to test the degree of change in case it is 'unwarranted', the same section also notes that, "surface-water water resource plans will also be required to show *how they would manage conditions that include a repeat of extremely dry periods such as the 2000−10 drought,*" (italics added for emphasis; MDBA, 2010a, pp. 33−34). (See Figure 1.6a.)

Now comes the catch that no journalist saw, or at least none I have found chose to cover, or even mention, namely that the Murray−Darling Basin Authority water reclamation plan does not even deal with climate change as far out as 2030. It adopts its preferred prediction (the median − a 10% decrease of available water by 2030) but then chooses to make recommendations based only on less than one third of this, that is, a 3% decrease. "Therefore the Authority has determined that the percentage of the remaining change due to climate change not already in the modelling is 3% of the entire water resource, and so *a 3% reduction in the current diversion limit is an appropriate allowance for the effect of climate change,*" (italics added for emphasis; MDBA, 2010a, p. xxvi).

This massive under-resourcing is said to be justified as follows, "Given the Basin Plan will apply to water resource planning in the Basin for successive 10-year periods commencing between 2012 and 2019, and the plan must be reviewed by around 2021 if not before, *the Authority considers that incorporation in the first Basin Plan of the full effect of the 10% predicted decline in average annual water availability under median 2030 conditions is unwarranted*" (italics added for emphasis; MDBA, 2010a, pp. 33−34). Thus, even though any more climate change will render the estimates of the water required for adequate environmental

flows flawed, the plan claims that it hopes its recommendations will make the Basin more resilient, "Reductions in water use under the SDLs will drive improvements in water use efficiency, which will make agricultural production more resilient to shocks and prepared for climate change impacts in the future," (MDBA, 2010a, p. 181). The Basin Plan further states, "*Declining flows associated with climate change will make it increasingly difficult to achieve the salt export target in the future. The scenarios with greater environmental water will provide increased capacity to achieve the target on an ongoing basis.*" (Italics added for emphasis; MDBA, 2010a, p. 118.)

The abysmal failure of the Australian media to discover and pursue the real Coorong story in 2010 is a salutary but by no means an isolated example of the dismal state of public understanding of climate change (cf. Rice and Henderson-Sellers, 2012, this volume). Surely, community comprehension would be greater if journalists were well-briefed in climate science and had adequate time to contemplate the intricacies of the challenges of global warming combining with more local human disturbances of land-use and water availability. Maybe, as Steve himself might have argued, society could be still better served if more scientists were willing and able to explain the rapidly unfolding climate crises: in the Murray−Darling Basin and around the world. After all, if we live as if it doesn't matter, and it matters, then it matters.

1.5. FUTUROLOGY OF CLIMATE

"A remarkable combination of ambition, commitment, communication skills and unbelievable energy... Schneider had a knack for explaining difficult concepts in a language that most members of Congress would understand... The world needs more Steve Schneiders."

John Holdren, December 2010 (Holdren, 2010)

I have termed the diabolical challenge of the Coorong a *trilemma* because Australia faces three possible options. We might try to save the Murray mouth whatever it takes, even knowing the climate will continue to remove ever more water; we can allow agricultural practices to continue as long as the changing climate, as fully as we know it, permits, thus condemning the Coorong to Montreux rather than Ramstar status; or we can invest in shifting affected rural communities towards our future climate while also attempting to manage some minimal flow to the mouth of the Murray. By October 2010, I knew my journalist friend was correct: the Coorong is climate change's ground zero (Henderson-Sellers, personal communication to Kathy Marks, 22 October, 2010).

The previous section (Section 1.4) honours Stephen's memory in two ways: by focusing on one of the very many areas he championed and by highlighting a positive and informative interaction between a journalist and a climate scientist. It also, sadly, underlines the failure of the broader mass media to identify, understand, and pursue the important aspects of a climate change story. South Australia's Coorong was overcoming previous political neglect but, in recognizing the integrated nature of biodiversity, climate, human disturbance, and economic need, the new Murray–Darling Basin Authority Plan has stirred up a highly contentious but inevitable debate. Compounding the bleak climate future for the Coorong is the fact that there is very little time to act (Productivity Commission, 2009). The Council of Australian Governments (COAG) recently agreed that the federal government would approve a new cap for the Murray–Darling Basin in 2011. As part of that agreement, existing state water plans will continue until they expire: in New South Wales this is in 2014 while Victorian plans do not expire until 2019. So, at best we cannot change until 2020 when we know that the combination of El Niño droughts and further global warming will very likely have shifted the southeastern climate of Australia into still worsened lower river flows. Moving into this future, will environmentalists, economists, farmers, journalists, and politicians agree on a future water management plan that is economically wise or socially ethical? Of course, the Murray–Darling region is just one of the very many highly stressed areas of the Earth today. We know we are changing the climate but we cannot know all the consequences of this rash activity (e.g., Lenton, 2012, this volume).

Climate futurology has matured over the past few decades. Modelling groups now create ranges of possible climatic futures and the IPCC assesses these and generates consensus reviews from time to time. However, as Wigley and Raper (2001) explain, any climate prediction has the potential to be misleading unless it is accompanied by some guidance as to what the range of climate change foreseen means in terms of probability. They suggest a four-step procedure for providing such guidance: identify the main sources of uncertainty; represent each of these input uncertainties as a probability distribution function; draw samples from this probability distribution function to drive the model that is to generate the projection; and, finally, construct an output probability distribution function from these results. Within climatic subsystems, this process is just becoming possible although, as Cogley (2012, this volume) points out for glaciology at present, this approach fails at the second step because "research cannot yet even sketch distribution functions for some of the least well known uncertainties." Other Earth subsystems, such as plant phenology, are newer and still more problematic and here we need effective and focused research just to allow us to understand processes well enough to write down the controlling concepts in a useful and practicable way.

With his wife, Terry Root, Steve researched phenological responses to greenhouse-induced climate shifts; the sensitivity of plant seasonal lifecycles to climate impacts; and disturbance to the land-derived carbon cycle. These effects compound the carbon feedbacks on climatic change (Dickinson, 2012, this volume) and, if combined with analysis of inter-annual ecosystem carbon exchange, together they may provide a new route to assessing climate sensitivity (Desai, 2010; Harvey, 2012, this volume). Phenology features in a number of the chapters of this book: Cleugh and Grimmond (2012, this volume) note the changes already occurring to urban ecosystems and Sen Gupta and McNeil (2012, this volume) identify similarly disturbing changes to oceanic ecosystems. These are both insidious indicators of anthropogenic changes already occurring and they flag potential feedback features that we have barely begun to identify, much less include, in our predictive models (cf. Rice and Henderson-Sellers, 2012, this volume; Lenton, 2012, this volume).

This book is not a consensus; rather, it is a coherent compilation of a set of current reviews of climate science. The authors writing here identify many hazards in our climate system, from future astronomical and geological disturbances of great power to the challenge of climatically accommodating the more than 50% of people who are urban dwellers. What adaptations do we need to manage the impact of global climate change in cities? Future urban climates will affect our health and wellbeing — for example, modifying the thermal stress due to elevated nighttime minimum temperatures and increased frequency and duration of heatwaves. Feedbacks will amplify global climate change by increasing energy consumption in warm climates, and the effects of reduced rainfall in subtropical latitudes will limit water availability and reduce the uptake of CO_2 by urban vegetation. Reduced urban air quality due to photochemistry and increase in particulates will have negative impacts on visibility and on the health of urban dwellers, even impacting indoor air quality.

Earth system science links these disturbing glimpses of city life to the rising threat of coastal erosion (Rice and Henderson-Sellers, 2012, this volume). All coasts are becoming more vulnerable as sea level rises. The latest identified region of this danger is the Arctic (Lantuit et al., 2011). Around a third of the world's coasts are located in the Arctic permafrost — not rock, but frozen soft substrate vulnerable to rapid erosion once the permafrost begins to melt (cf. Lenton, 2012, this volume). The continuous decline in Arctic sea-ice is now jeopardizing previously protected high-latitude coasts opening these areas that have remained stable for millennia to rapid, perhaps catastrophic, changes (cf. Harvey, 2012, this volume).

This book is essential for those at the forefront of the public policy debate on the urgency of the challenge of current climate change because, while taking a balanced view, it contains some very frightening outlooks including:

A global mean warming of 10–12 K beyond AD 2100 is a distinct possibility… global mean warming of this magnitude would render portions of the world currently occupied by over half of the human population to be uninhabitable by humans (in the absence of access to 100% reliable air-conditioning by the entire populations in the affected regions) due to the periodic occurrence of 6-hour mean wet-bulb temperatures in excess of the practical physiological limit of 33°C.

(Harvey, 2012, this volume)

…almost all models suggest that the waiting time for a late twentieth century, 20-year extreme, 24-hour precipitation event will be reduced to substantially less than 20 years by mid-twenty-first and by much more by the late twenty-first century, indicating an increase in frequency of the extreme precipitation at continental and subcontinental-scales.

(Alexander and Tebaldi, 2012, this volume)

…human activities might fundamentally alter the dynamics of the climate system, switching it out of its recent mode of roughly 100,000-year glacial cycles and back into an earlier (Pliocene-like) state in which Greenland lives up to its name and remains unglaciated, West Antarctica only sporadically has an ice sheet, East Antarctica holds less ice, and sea levels are in the long term around 25 metres higher.

(Lenton, 2012, this volume)

The question of to what extent and in what voice climate researchers should speak about these climate futures remains a personal and a public challenge. Stephen Schneider was an iconic example of a courageous person, prepared always to state his views and defend them by explaining his reasoning.

Stephen (Figure 1.7) said of himself, "Being stereotyped as the 'pro' advocate versus the 'con' advocate as far as action on climate change is concerned is not a quick ticket to a healthy scientific reputation as an objective interpreter of the science. This is all part of the problem I have, somewhat whimsically, called 'mediarology': the problematic world of communications," (Schneider web page, 2010b). He conquered this challenge in a way that few others have managed. One journalist's view honours this achievement and, indeed, the whole of Stephen Schneider's life and work:

Schneider (was) unique in the level of respect he earned not only from his science colleagues, but also from those in the news media trying most conscientiously to cover the issues in ways consistent with sound science and quality journalism. He was an unrelenting critic of lazy climate journalism, and in so being he endeared himself to those reporters most serious about their work.

Cristine Russell (Russell, 2010)

FIGURE 1.7 Stephen Schneider and the author at Steve's 60th birthday party, Stanford, USA (February 2005).

Gavin Schmidt, writing on the website Real Climate in 2011, describes Steve's first ever letter to *The New York Times* (in 1971). In this, Schneider clearly identifies an op-ed piece (by Guccione) as, "often inaccurate and certainly misleading". Schmidt (2011) says of Steve's 1971 letter, it "is a good rebuttal that makes some points that are still apropos today", pointing out the highly selective character of the cited metrics and highlighting the bogus nature of the arguments. Schneider, in 1971, ends on two points that were exactly right in 1971 (Schmidt, 2011):

[S]erious scientific studies have indicated that CO_2 and dust [aerosols] can affect climate, albeit in opposite directions. We do not yet know the magnitude of these influences well enough to be certain which, if either, of those effects might predominate.

What we do need is an accelerated program of scientific research along with improved international cooperation.

Schmidt (2011) closes, "In 1971, Schneider was correct to say that more research was needed to see which of the effects would predominate, but *in 2011, it is very clear that greenhouse gases have, and will continue to do so.*" (Italics added for emphasis.)

In an interview with *The New York Times* just a year later, in 1972, Schneider paraphrased Mark Twain: "*Nowadays,*" he said, "*everybody is doing something about the weather but nobody is talking about it.*" These sentiments echo those of John Lennon, incorporated into the cover image of this book created by Leon Cmielewski and Josephine Starrs (Starrs and Cmielewski, 2010). The Ganges River delta photograph has his phrase, "days like these", embedded into imagery of the Sundarbans, the river system at the mouth of the Ganges, where flooding is displacing local people and destroying vulnerable ecologies. The lyrics from John Lennon's 'Nobody Told Me' underlines, then (January 1984), contemporary dichotomies including, "Everybody's talking and no one says a word. ..." and, "They're starving back in China, so finish what you got". We were then witnessing, and still are, "strange days indeed"; but now none of us can any longer claim that "nobody told me".

From his first interaction with the mass media on climate in 1971 until his death in 2010, Stephen Schneider worked to reverse the paralysing feeling that, although we are all talking, far too few of us are acting. Steve both spoke fluently about climate and climate change and acted on the views he held. We all hope that this book, coming as it does midway between the IPCC's Fourth and Fifth Assessments, will redefine and illuminate the many challenges that Stephen championed throughout his life and assist those who wish to follow where he led in acting quickly and effectively to avoid nasty, even catastrophic, future climate surprises.

ACKNOWLEDGEMENTS

I am grateful to the Australian Research Council for financial support and to Michel Verstraete, Irene Hudson, and one anonymous person for positive and beneficial reviews. Most of all, I am truly grateful for the inspiration Steve Schneider was for me personally: a very long-term colleague and someone I thought of as a 'best friend'.

The Anthropocene

People are now a major cause of climate change. This section discusses how humans modify climate on all scales. It also explores how inadvertent human actions, including some response strategies, can be as important as clearly evident planetary pollution.

A global mean warming of 10–12 K beyond AD 2100 is a distinct possibility. … global mean warming of this magnitude would render portions of the world currently occupied by over half of the human population to be uninhabitable by humans due to the periodic occurrence of 6-hour mean wet-bulb temperatures in excess of the practical physiological limit of 33°C.

Harvey, 2012, this volume

Chapter 2, ***People, Policy, and Politics in Future Climates*** by Roslyn Taplin, addresses climate change and the current responses stemming from the international climate regime. Although UN agencies and national governments have given significant attention to climate change concerns for more than two decades, the challenges of implementing effective governance approaches to address climate change are immense. Coverage includes global climate change politics and the scientific collaboration via the Intergovernmental Panel on Climate Change as well as communities and individuals, and how they are addressing climate change. It is argued that implementation of Milbrath's learning governance approach would give countries better capacity to address climate change and assist the polycentric systems approach to coping with climate change that Ostrom has proposed.

Chapter 3, ***Urban Climates and Global Climate Change*** by Helen Cleugh and Susan Grimmond, explores the very close linkage between cities and global climate change. Cities are where the bulk of greenhouse gas emissions take place through the consumption of fossil fuels; they are where an increasing proportion of the world's people live; and they also generate their own climate — commonly characterized by the urban heat island. In this way, understanding the way cities affect the cycling of energy, water, and carbon to create an urban climate is a key element of climate mitigation and adaptation strategies, especially in the context of rising global temperatures and deteriorating air quality in many cities. As climate models resolve finer spatial scales, they will need to represent those areas in which more than 50% of the world's population already live to provide climate projections that are of greater use to planning and decision-making. Finally, many of the processes that are instrumental in determining urban climate are the same factors leading to global anthropogenic climate change, namely regional-scale land-use changes; increased energy use; and increased emissions of climatically relevant atmospheric constituents. Cities are therefore both a case study for understanding and an agent in mitigating anthropogenic climate change. Chapter 3 reviews and summarizes the current state of understanding of the physical basis of urban climates, as well as our ability to represent these in models. Addressing the challenges of managing urban environments in a changing climate requires understanding the energy, water, and carbon balances for an urban landscape and, importantly, their interactions and feedbacks, together with their links to human behaviour and controls.

Chapter 4, *Human Effects on Climate through Land-Use-Induced Land-Cover Change* by Andy Pitman and Natalie de Noblet-Ducoudré, provides the background to understanding the role of land-use-induced land-cover changes on regional and global climate, focusing on the physical and biogeophysical changes that occur as the landscape is modified. Land-use-induced land-cover change has occurred for millennia and on a globally significant scale over the last few hundred years. It changes the nature of the landscape and its functioning, and causes the emissions of carbon dioxide and other gases. It has a clear and unambiguous impact on the local climate. It is argued that its impact is not yet sufficiently known at regional and global spatial scales because there are limits to the experiments conducted to date to explore this issue. The key contributions to understanding of the role of land-use-induced land-cover changes on large-scale climate are reviewed and assessed, leading to recommendations relating to the inclusion of this phenomenon in future climate simulations and in future assessments by the Intergovernmental Panel on Climate Change.

Several people have said: 'Well, isn't it a good thing that our industrial progress has produced not just carbon dioxide but sulfur aerosols, which cool us back down?' And I've always said I didn't like the idea of using acid rain to solve global warming, because those aerosols are not only bad for ecosystems when they rain acids into the lakes and streams and soils, but they're also part of the air pollutants which, when we breathe, we know from statistical tests, leads to increased lung and respiratory disease and what we call excess deaths, which sounds very clinical unless somebody in your family happens to be susceptible to that kind of air pollution. Some people want to shove it in the stratosphere–what we call geo-engineering. That at least wouldn't have health effects. But the aerosol offset is only partial. And even if it would offset the global warming almost completely, it's not going to leave the world's climate unchanged, because there'll be pockets in the world that'll actually be cooler, then other pockets much warmer, so you'll have blobs of warming and blobs of cooling. And that's a change, because our water supplies, our agriculture, and our ecosystems, they live locally, not globally. They don't care about 2 degree global mean change. They care about what happens in their region. And having regional aerosols offsetting some of the global effects is not going to prevent regions from still being disturbed. And we're still going to have climate disturbance if we try to solve global warming by regional air pollution, to say nothing of the health effects and the environmental effects of that air pollution.

Stephen H. Schneider, PBS interview

Chapter 2

People, Policy and Politics in Future Climates

Roslyn Taplin

Institute of Sustainable Development and Architecture, Bond University, Gold Coast, Australia

Chapter Outline

"We are trying to undo some of the harm we have done, and as climate change worsens we will try harder, even desperately, but until we see that the Earth is more than a mere ball of rock we are unlikely to succeed."

(Lovelock, 2010, p. 3)

(Copyright © 2010 James Lovelock. Reprinted by permission of Basic Books, a member of the Perseus Books Group.)

2.1. INTRODUCTION: HUMAN AND ECOLOGICAL SYSTEMS AND PARADIGM CHANGE

The influence that global human society is perpetrating on future climates of the Earth is related to population, politics, policy, and ethics. We are at a fork in the road, or a turning point where the environmental decisions of citizens, public officials, political leaders, business and labour leaders, diplomats, and scientists are critical. Climate change has been on the agenda of international agencies, governments around the globe, and local authorities since the late 1980s. Now, around 20 years later, many local and national initiatives have been employed to mitigate greenhouse gas emissions. However, international co-operation via the United Nations Framework Convention on Climate Change (UNFCCC, 1992) and its Kyoto Protocol is greatly lacking as evidenced by the outcome from Copenhagen in 2009. The many local initiatives that have been taken up are praiseworthy, but in the face of national inaction in some countries such as the United States (USA) and my home country, Australia, they are not enough.

Political scientists Elinor Ostrom and Oran Young have made contrasting assessments of human capacity to curb emissions on a global scale. Ostrom (2009a, 2010) argues that climate change does not have to be addressed as a global collective action problem (Olsen, 1965) alone but believes that a polycentric approach of "...providing diverse public goods and services...[via] many centers of decision making that are formally independent of each other" (Ostrom, 2010) is a feasible way of 'coping with' climate change. She observes that:

Given that the recognition of the danger of climate change among citizens and public officials is still relatively recent, and given the

The Future of the World's Climate. DOI: 10.1016/B978-0-12-386917-3.00002-6

debates about who is responsible for causing the problem and for finding solutions, one cannot expect that an effective polycentric system will be constructed in the near future. But given the slowness and conflict involved in achieving a global solution, recognizing the potential of building even more effective ways of reducing energy use at multiple levels is an important step forward.

(Ostrom, 2009a, p. 38)

However, Young (2009, pp. 27—28) cautions that:

...the growing prominence of civil society networks...is not a basis for concluding that such networks can go a long way toward meeting the demand for governance, especially in addressing large-scale issues such as climate change...

Ostrom's (2010) theory certainly is very dependent on societies and individuals across the globe transforming from what has been referred to as dominant social paradigm (DSP) views to new environmental paradigm (NEP) thinking.

DSP and NEP beliefs and values were first discussed and delineated in the 1980s by Catton and Dunlap (1978), Dunlap and van Leire (1978), Milbrath (1984, 1989), Cotgrove and Duff (1980), and, more recently, have been considered by Dunlap et al. (2000), Sterling (2001), Birkeland (2002), and Dunlap (2008). Milbrath (1984) carried out an extensive survey of the environmental beliefs and values of the general public, public officials, business and labour leaders, and environmentalists in the UK, Germany, and the USA to ascertain DSP and NEP values (see Table 2.1, showing the contrasts between competing paradigms). He found that the DSP legitimizes competition, the market economy, maximization of wealth, and unchecked exploitation of resources whereas the NEP focuses on valuation of nature, compassion towards other species and other people, acceptable risk, limits to growth, and participatory decision making (Milbrath, 1984). The Bruntland Report, *Our Common Future,* outlined this situation as involving "dangerously rapid consumption of finite resources" (WCED, 1987, Part I,

TABLE 2.1 Contrasts between Competing Paradigms

New Environmental Paradigm	Dominant Social Paradigm
1. High valuation on nature a. nature for its own sake — worshipful love of nature b. wholistic relationship between humans and nature c. environmental protection over economic growth	1. Lower valuation on nature a. use of nature to produce goods b. human domination of nature c. economic growth over environmental protection
2. Generalized compassion towards a. other species b. other peoples c. other generations	2. Compassion for only those near and dear a. exploitation of other species for human needs b. lack of concern for other people c. concern for this generation only
3. Careful plans and actions to avoid risk a. science and technology not always good b. halt to further development of nuclear power c. development and use of soft technology d. government regulation to protect nature and humans	3. Risk acceptable in order to maximize wealth a. science and technology a great boon to humans b. swift development of nuclear power c. emphasis on hard technology d. de-emphasis on regulation — use of the market — individual responsibility for risk
4. Limits to growth a. resource shortages b. increased needs of an exploding population c. conservation	4. No limits to growth a. no resource shortages b. no problem with population c. production and consumption
5. Completely new society a. serious damage by humans to nature and themselves b. openness and participation c. emphasis on public goods d. co-operation e. simple lifestyles f. emphasis on worker satisfaction	5. Present society okay a. no serious damage to nature by humans b. hierarchy and efficiency c. emphasis on market d. competition e. complex and fast lifestyles f. emphasis on jobs for economic needs
6. New politics a. consultation and participation b. emphasis on foresight and planning c. willingness to use direct action d. new party structure along a new axis	6. Old politics a. determination by experts b. emphasis on market control c. opposition to direct action — use of normal channels d. left—right party axis — argument over means of production

(Source: Reprinted by permission from Envisioning a Sustainable Society: Learning Our Way Out edited by Lester W. Milbrath, the State University of New York Press ©1989, p. 119, State University of New York. All rights reserved.)

para. 14) by the developed world while a "large number of people…live in absolute poverty — that is who are unable to satisfy even the most basic of their needs…in developing countries." (WCED, 1987, Part III, para. 29).

When leaders are totally focused on DSP values that result in exploitation of Earth system resources and enhancement of the market economy, then communities and individuals with NEP concerns can take some sustainability and climate change mitigation and adaptation actions (e.g., local initiatives in the USA and similarly in Australia). However, top-down political and organisational leadership often has a more significant influence on regions and countries than bottom-up initiatives, for example, as evidenced by climate policy implementation in the European Union (EU) in comparison to the USA and blocking of grassroots action on a Climate Protection Bill in Australia (Hall et al., 2010). In some cases top-down leadership can be influenced by public support for green issues, as is evidenced in the EU, while in others that is not the situation.

With the predominance of the DSP over the last half century and going further back in history since the Industrial Revolution, the environment has been a 'second string' or 'second cousin' to economic and financial planning and decision making at all levels in polycentric systems (Walker, 1994). This indicates the degree of systemic change that is needed for good Earth system governance to prevail (see also Rice and Henderson-Sellers, 2012, this volume).

Ostrom's (2009a, 2009b) polycentric theory of sustainability and coping with climate change stemming from the bottom-up in socio-ecological systems (SESs) is explored in this chapter to assess if there is reassuring evidence of Earth system environmental and climate change decisions arising from action from the grassroots. SESs are complex and involve humans and their interactions with each other and the natural and built environment systems in which they live on the Earth. These systems thus involve people, policy, and politics. Taking into account Ostrom's theory together with Milbrath's (1984, 1989) DSP and NEP theoretical understanding, current and potential changes in decision making on climate change with regard to resource systems, resource units, governance systems, and users are reviewed in this chapter.

This chapter looks at climate change and the responses stemming from the international climate regime. It traverses from global climate change politics and the scientific collaboration via the Intergovernmental Panel on Climate Change (IPCC) to communities and individuals, and how they are addressing climate change. Various examples of the many mitigation and adaptation initiatives taken by countries and local communities on climate change to date are discussed, as well as the decision-making implications of progress with understanding climate science, impacts, and adaptation implications. From a planetary perspective, the impact of humanity on the Earth system associated with climate change is a critical issue (Lovelock, 1979, 2010; Harding, 2006). The key question is how may people living in differing political systems and places on Earth all contribute to mitigating climate change impacts via their multiple efforts.

2.2. THE CHALLENGES OF GOVERNANCE FOR MITIGATION OF CLIMATE CHANGE

Law and policy directed towards mitigation of, and adaptation to, climate change is the primary focus of the United Nations Framework Convention on Climate Change (UNFCCC), which has near universal membership of 193 countries, and the EU, as a regional economic integration organisation. UN agencies and national governments have given significant attention to climate change concerns for two decades (see Figure 2.1). Both developed and developing nations have formulated approaches to climate change on a spectrum of enacting domestic law and formulating policy to providing decision-making guidance. Countries have reported on their efforts in lengthy National Communications to the UNFCCC and as Low (2010) reflects, preparation of National Communications is "…not only a reporting process, but also a national capacity-building process" for developing countries. Sustainability, or sustainable development, is the axis around which these approaches have been developed and such efforts have involved aspirations to: (a) implement the precautionary principle; (b) stay within source and sink constraints; (c) maintain natural capital at or near current levels; (d) implement the polluter pays principle; and (e) maintain inter-generational and intra-generational equity.

These ambitions are embedded in the UNFCCC and at the national, state/province, and local decision-making level in many countries. Nevertheless, many nations are not seriously attempting to keep their resource use within source and sink constraints and to maintain their natural capital. This reflects the conflicts between economic development goals and the need to take action on climate change, that is, the priority, or lack of priority, that climate change is given.

O'Riordan and Voisey (1998) observed more than a decade ago with regard to sustainability that implementation ranges from 'very weak' where there is tokenistic policy integration to 'very strong' where comprehensive shifts in economic, environmental, cultural governance approaches, and thinking have been implemented. A similar situation exists for climate change; cases of policy implementation range from very weak symbolic action to very strong action (e.g., national mandatory renewable energy targets).

2.3. A GOVERNANCE APPROACH TO ADDRESS CLIMATE CHANGE

Good environmental governance is fundamental to addressing climate change and the possibility of a transition

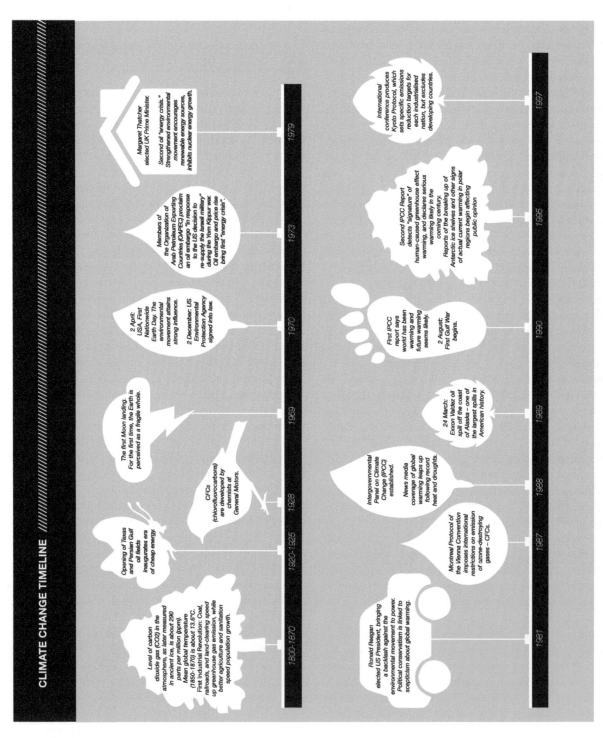

FIGURE 2.1 Climate change timeline. (*Source: Researched by Eleanor Rhode and designed by Vikki Peter for the National Theatre's Programme for Greenland, 2011 © National Theatre.*)

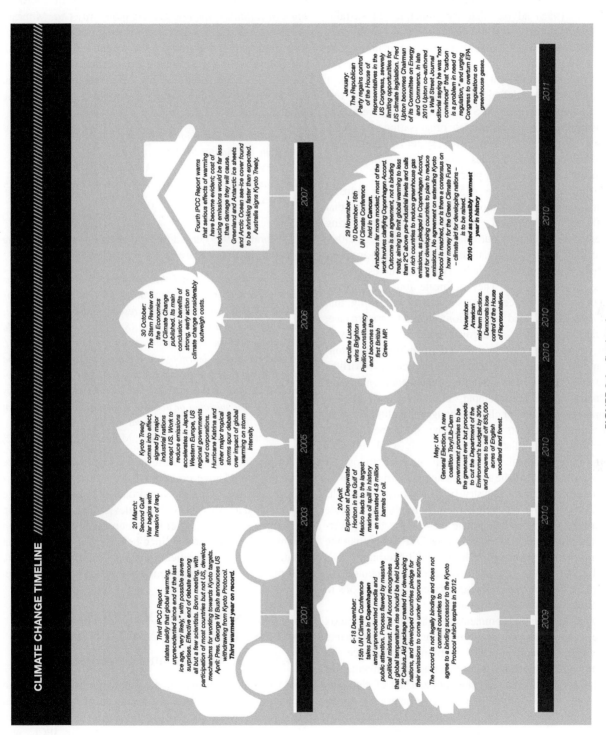

FIGURE 2.1 Continued

to sustainability. This applies equally for local communities, countries, regions, and, ultimately, global society. The elements required for good environmental governance have been considered by several scholars.

Bossel (1998) advises that the principles of sustainability, subsidiarity, self-organisation, and sufficiency ('four Ss') are needed for governance for sustainability. The subsidiarity principle of power devolving to local government is promoted in Local Agenda 21 (UNCED, 1992).

Ostrom (2009a) defines the components of a governance system within an SES as comprising government organisations, non-governmental organisations (NGOs), network structure, property-rights systems, operational rules, collective-choice rules, constitutional rules, and monitoring and sanctioning processes.

Griffiths et al. (2007, p. 415), in researching industry governance systems in relation to climate change, found that "…different institutional governance systems…bring different climate change outcomes…" and that "…neither firm nor state-centric explanations can account for variations in responses to transforming industry systems in response to climate change issues." They delineate four overlapping institutional governance systems currently operating that affect mitigation and adaptation: market governance, corporate governance, state governance, and joint governance where "the state is involved in industry decision-making and industry value chains are integrated." (Griffiths et al., 2007, p. 419). In critiquing dependence on market governance (via, for example, emissions trading) and corporate governance for climate change mitigation, they say "…the short term costs of adaptation…are borne by those groups that are pluralistically or politically weak − either labour, environmental groups…community groups…or occupy a weak place in the supply chain (the tourism industry as opposed to the coal industry [in Australia])." (Griffiths et al., 2007, p. 424). Thus their findings implicitly indicate that barriers to addressing climate change via good governance are predominantly centred in the DSP rather than the NEP in developed countries (Milbrath, 1989; Sterling, 2001). Arguably, this is the case in many developing countries as well where lack of financial resources constrains actions in both mitigation as well as adaptation.

Young (2009, p. 14) comments that there is a "…paradox of the rising demand for governance for sustainable development coupled with declining confidence in government as a mechanism for meeting this demand." He says that really "there is no standardized list of elements that belong to the domain of good governance but many lists include stakeholder involvement, accountability, democracy, respect for civil liberties and legitimacy" (Young, 2009, p. 32). Young is concerned that the "speed of changes in biophysical systems can easily exceed the rate at which it is possible to (re)construct and implement governance systems" (2009, p. 19).

Arguably, the best practical proposal for governance to address sustainability and hence climate change has been made by Milbrath (1989). In response to his concerns about the overriding DSP influence on governance, in *Envisioning a Sustainable Society: Learning our Way Out*, Milbrath (1989) argued for overlaying a learning governance structure on existing traditional governmental structures with the potential of strengthening governments in the move towards a sustainable society. He recommended that a learning governance structure should have four basic pillars:

An education and information system
A systemic and futures thinking capacity
An intervention capability for stakeholders
A long range sustainability impact assessment capacity for policy initiatives

(Milbrath, 1989, p. 282)

Milbrath (1989) urged that the advantages of a learning governance structure are:

*First, **if government can better anticipate the future, it has a better chance of creatively dealing with such problems as climate change**, overpopulation, resource shortages, species extinction and ecosystem damage…Second, government could better undertake the new functions that are being assigned to it: encouraging social learning, facilitating quality of life, and helping society to become sustainable. Third, the learning structure could improve the processes and outcomes of government by helping to avoid mistakes or to avoid governmental aid to special interests at the expense of the general welfare.*

(bold added for emphasis; Milbrath, 1989, pp. 301−2)

Implementation of Milbrath's learning governance structure would give countries better capacity to address climate change. Aspects of approaches to climate change action to date that have been devised to work within existing governance systems, resource systems, and with scientists and users are outlined below and are discussed in relation to the theoretical perspectives on polycentric systems, social paradigms, and governance outlined above.

2.4. SCIENCE AND POLITICS IN THE INTERNATIONAL CLIMATE REGIME

2.4.1. IPCC Science and Governance

As Milbrath (1989) indicated, systems for information and education, as well as futures thinking capacity are essential for good governance for sustainability (cf. Rice and Henderson-Sellers, 2012, this volume). The IPCC was formed to fulfil the role of a new form of institution for information provision; it has been in existence now for over 20 years. Established in 1988, the IPCC was conceived by the World Meteorological Organisation (WMO) and the

United Nations Environment Programme (UNEP) as an institution to address the serious concerns of scientists around the globe that a 'consensus' on the scientific work being done on climate change needed to be developed and co-ordinated for communication with politicians and decision makers in government, and business and industry at the local, country, regional, and international levels (Bolin 2007). The IPCC's First Assessment Report (FAR) was released in 1990 (IPCC, 1990a, 1990b, 1990c) and this was used as a basis for the drafting and development of the international treaty – the UNFCCC. The ultimate objective of this Convention is based on scientific understanding of the need to mitigate greenhouse gases, as well as the need for adaptation; Article 2 states that the UNFCCC's aim is:

...to achieve, in accordance with the relevant provisions of the Convention, stabilization of greenhouse gas concentrations in the atmosphere at a level that would prevent dangerous anthropogenic interference with the climate system. Such a level should be achieved within a time-frame sufficient to allow ecosystems to adapt naturally to climate change, to ensure that food production is not threatened, and to enable economic development to proceed in a sustainable manner.

(UNFCCC, 1992, Art. 2)

Subsequent to the 1990 report, three further major reports (IPCC, 1995a, 1995b, 1995c (SAR); IPCC 2001a, 2001b, 2001c (TAR); IPCC 2007a, 2007b, 2007c (AR4)) have been prepared by the IPCC on climate change science, impacts, and responses over the last two decades. Sub-reports have been prepared by:

Working Group One (WG I) [which] assesses the physical scientific aspects of the climate system and climate change.

Working Group Two (WG II) [which] assesses the vulnerability of socio-economic and natural systems to climate change, negative and positive consequences of climate change, and options for adapting to it.

Working Group Three (WG III) [which] assesses options for mitigating climate change through limiting or preventing greenhouse gas emissions and enhancing activities that remove them from the atmosphere.

(IPCC, 2010)

Also, many supplementary reports on issues including emissions scenarios, greenhouse inventory methods, and regional climate change impacts have been compiled to contribute to the information base and education for decision-makers and others. The next major IPCC report (AR5) is currently in preparation and planned to be released in 2013–2014. The report, like this book, will be dedicated to the memory of Steve Schneider (see Figure 2.2). According to Dr. Pachauri, the AR5 Working Group I report will be completed in September 2013 "...followed rapidly thereafter by the reports of Working Groups II and III respectively" and the "Synthesis Report of the AR5 will be completed in November 2014." (Pachauri, 2010a, p. 4).

In the IPCC's AR4, the peer reviewed published papers of thousands of scientific experts has been summarized, synthesized, and written in a manner palatable for the educated layperson needing information about climate change. Accordingly, it is a report incorporating the range of scientific understanding and opinion on climate research. The IPCC reports have been referred to somewhat confusingly as 'consensus reports', but this does not mean that every item of scientific information in

FIGURE 2.2 AR5 Synthesis Report to be dedicated to Steve Schneider's memory: announcement at IPCC Synthesis Report Scoping Meeting held in Liège, Belgium, August 2010. *(Source: Image courtesy of Jean-Pascal van Ypersele.)*

the reports is agreed by every scientist involved in the IPCC. Rather, chapters are overseen by lead authors and are reviewed by expert reviewers appointed by the IPCC, as well as government-appointed reviewers. Contributions were made over six years from greater than 2500 scientific expert reviewers, 800 contributing authors, and 450 lead authors from 130 countries to compile the 2007 report. The IPCC AR4 'Summary for Policymakers' prepared by WG I included a statement expressing marked concern about the anthropogenic greenhouse effect, albeit couched in carefully worded scientific language:

*Most of the observed increase in global average temperatures since the mid-20th century is **very likely due to** the observed increase in anthropogenic GHG concentrations. This is an advance since the...[TAR's] conclusion that "most of the observed warming over the last 50 years is likely to have been due to the increase in greenhouse gas concentrations." **Discernible human influences** now extend to other aspects of climate, including ocean-warming, continental-average temperatures, temperature extremes, and wind patterns.*

(bold added for emphasis; IPCC, 2007a, p. 10)

The 2007 report, using the strongest language to date in emphasizing the consequences of inaction on climate change, though still cautiously worded as can be seen from the above quotation, became subject to critique by individuals and organisations unconvinced by the work of climate science and impact researchers. Part of this related to the so-called 'Climategate', where hacked emails to, and from, colleagues at the Climate Research Unit at the University of East Anglia were illegally released to the internet, resulting in much media coverage, a criminal investigation into the hacking, and a UK Parliamentary Inquiry by the House of Commons Science and Technology Committee in March (UK House of Commons Science and Technology Committee, 2010). The University of East Anglia also commissioned two independent inquiries in 2010: the Lord Oxburgh Scientific Assessment Panel in April (2010) and the Muir Russell Review in July (Russell et al., 2010). All three of the inquiries supported the honesty and integrity of scientists in the Climate Research Unit.

Part of the debate also stemmed from a small range of 'errors' identified in the 2007 AR4 report. IPCC Chair Pachauri (2010b) has commented that the "Climate science attacks were vast and co-ordinated" and that "six months of emails were scrutinized to refute three errors". He has also commented with regard to the issue of uncertainty about climate change that scientists are ignorant of how their information is received by policymakers and communities saying, "there is a public perception of uncertainty meaning 'scientists don't know' versus the researchers thinking they are communicating that 'there is a range of possibilities' " (Pachauri, 2010b).

2.4.2. IPCC 2010 Review

Due to the negative attention on the IPCC, an IPCC review process was commissioned by UN Secretary General Ban Ki-Moon and IPCC Chair Pachauri in March and completed at the end of August 2010. It was carried out under the auspices of the US InterAcademy Council (IAC) and chaired by Harold Shapiro, a former President of Princeton University.

The report states that it was prepared "In response to some sustained criticism and a heightened level of public scrutiny of the Fourth Assessment Report" (IAC, 2010). The mandate of the IAC review committee was "to review the *processes and procedures of the IPCC* and make recommendations for change that would enhance the authoritative nature of the IPCC reports" (italics added for emphasis, IAC, 2010, p. vii).

Together with recommendations on the IPCC organisational processes, the review process for future reports, and communication with stakeholders, a central recommendation on uncertainty was "All Working Groups should use the qualitative level-of-understanding scale in their Summary for Policy Makers and Technical Summary, as suggested in IPCC's uncertainty guidance for the Fourth Assessment Report. This scale may be supplemented by a quantitative probability scale, if appropriate." (IAC, 2010, p. xv). *The Economist* (30 August, 2010) observed, "...the Shapiro committee offers sharp criticisms of the way the panel organises itself and calls for reforms." This was somewhat unwarranted as the IAC committee's remarks were more moderate and constructive.

Hulme and Mahoney (2010, p. 1) commented that "the IPCC's influence on the construction, mobilisation and consumption of climate change knowledge is considerable" but cautions with regard to consensus that "without a careful explanation about what it means, this drive for consensus can leave the IPCC vulnerable to outside criticism." There has certainly been increased pressure on climate scientists and scrutiny of their science since the 2009 Climategate incident and the AR4 errors were revealed (Hulme, 2009; Liverman, 2010). The challenges for IPCC according to van Ypersele (2010) are to "restore confidence in climate science and to improve policy relevance without becoming policy prescriptive". Insisting on this is not a realistic demand for the organisation.

2.4.3. IPCC as an SES Contributor

However, the IPCC can be seen as part of the Earth system or a global SES that sources knowledge from local SESs and contributes knowledge back to local SESs (Rice and Henderson-Sellers, 2012, this volume). Contributors to, and reviewers of, the IPCC reports, which assess scientific literature, essentially contribute to the development of information to drive action on climate change — which really is directed towards the transformation of societies to being more environmentally sustainable.

Given the information that the IPCC provides, it is not surprising that the organisation has become a target. The IPCC forms a vital part of the education and information system and the systemic and future thinking capacity that Milbrath (1989) advocated as necessary for a learning governance structure.

Hulme (2009, pp. 106–7) cautioned that there are three limits to the performance of science in provision of climate change knowledge for decision making: firstly, "…scientific knowledge about climate change will always be incomplete, and it will always be uncertain. Science always speaks with a conditional voice…"; secondly, "…without trust and/or participation, scientific knowledge about climate change is unlikely to prove robust enough to be put to good use [in policymaking]"; and thirdly, "we must be honest and transparent about what the science can tell us and what it can't…climate change will always require judgments beyond the reach of science." Liverman (2010, slide 10) commented science is coming into debates about who should pay and how much and that increased pressure on science will start to change the science/policy interface.

Despite the heroic efforts of people such as Stephen Schneider who, throughout his life, strove to assist and inform journalists to improve media coverage of climate issues (e.g., Henderson-Sellers, 2012, this volume), Yearley (2009, p. 389) concludes "…there is a need for (i) greater understanding of the social dimensions of the scientific community that studies climate change and (ii) more social science reflection on the roles of social science in climate change models and projections." On reflection, the situation of IPCC scientists being drawn into polarized discussions was inevitable due to their findings clashing with DSP business-as-usual perspectives.

Oreskes and Conway in *Merchants of Doubt* explain "…how a cadre of influential scientists have clouded public understanding of the scientific facts [on climate change] to advance a political and economic agenda" and demonstrate that the same figures who were:

…skeptical of acid rain, the ozone hole, and global warming strove to 'maintain the controversy' and 'keep the debate alive' by fostering claims that were contrary to the mainstream of scientific evidence and expert judgment.

(Oreskes and Conway, 2010, p. 241)
(© 2010 by Oreskes and Conway. Reprinted by permission of Bloomsbury Press)

Oreskes and Conway highlight the role of the media in this false debate as:

…the media became complicit as they reported the controversy as if it was a legitimate debate. Often the media did so without informing readers, viewers and listeners that the 'experts' being quoted…were affiliated with partisan think tanks funded by

industries, or were simply habitual contrarians who perhaps enjoyed the attention garnered by outlier views.

(Oreskes and Conway, 2010, p. 241)
(© 2010 by Oreskes and Conway. Reprinted by permission of Bloomsbury Press)

2.4.4. IPCC Projections, Tipping Points, and Policy-Making

Tipping points have become much discussed in relation to concerns about climate change, both by scientists and the popular media (see, for example, Lenton et al., 2008; Hansen, 2009; Lenton, 2012, this volume). The term, adopted from the medical literature in relation to its use in regard to epidemics and popularized by Gladwell (2000), refers to "…non-linear transitions where 'a small change can make a big difference'" (Lenton, 2009, para. 2). Lenton (2009) distinguishes between tipping points and tipping elements saying that tipping elements are "the components of the Earth System that can be switched – under particular conditions – into a qualitatively different state by small perturbations" (Lenton et al., 2008, p. 1786). Such tipping elements may be related to Greenland ice sheet melting, Atlantic thermohaline circulation, and coverage of summer and winter Arctic sea-ice. The IPCC has reported on the first two in AR4 but has not directed its reporting on tipping points to date having focused on projections of global means of temperature, precipitation, and sea level – in relation to scenarios for future human behaviour (Kaya and Yokobori, 1997) and their impacts on greenhouse gas and aerosol emissions (IPCC, 2000). Lenton (2009, para. 1) comments that although the IPCC AR4 projections "…are rarely straight lines, the underlying system and its responses appear 'linear' to the report's audiences". Lovelock (2010) goes further saying:

I have little confidence in the smooth rising curve of temperature that modellers predict for the next ninety years. The Earth's history and simple climate models based on the notion of a live and responsive Earth suggest that sudden change and surprise are more likely…business and government appear to be accepting uncritically a belief that climate change is easily and profitably reversible.

(Lovelock, 2010, pp. 4–5)
(Copyright © 2010 James Lovelock. Reprinted by permission of Basic Books, a member of the Perseus Books Group.)

Lenton (2009, para. 16) advises that "Climate policy, especially mitigation policy, should be more concerned with what tipping elements might be triggered by human activities in the future, and whether their tipping points can be avoided." This ties in with the ultimate objective of the UNFCCC "…to prevent dangerous anthropogenic interference with the climate system." (UNFCCC, 1992, Art. 2). Hansen (2009, p. 276) comments:

We already know that we should reduce atmospheric carbon dioxide to, at most, 350 ppm.

Key quantities we should watch to assess the status of potential climate tipping points are (1) the mass balance of the West Antarctic and Greenland ice sheets, including ice shelves and the principal outlet glaciers of the ice sheets, (2) the percentage of fossil fuel carbon dioxide emissions that remains in the air, and (3) changes of atmospheric methane.

Reasons given by Hansen for concentrating on these drivers are:

(1) If the ice sheets become more mobile, discharging more ice to the ocean, it will bode ill for future sea level and storms. (2) The percentage of fossil fuel carbon dioxide remaining in the air has averaged about 56 percent for decades, the other 44 percent being taken up by the land and ocean. If the ability of the land or ocean to soak up carbon dioxide decreases, that could cause global warming to accelerate, which could amplify other feedbacks. (3) Methane is important because of the possibility of an increasing discharge from frozen methane.

(Hansen, 2009, p. 276)
(© 2009 by Hansen. Reprinted by permission of Bloomsbury USA)

Schneider (2010a) also warned about tipping points "We know they're there but we don't know what they are…there may be warming of 1.8, 2.2, 4, or 6°C by 2100." He cautioned that decision-makers should take into account scientific data 'outliers' in policy development rather than focusing on IPCC data in the 'fat end of the tail'.

2.5. THE ROLE OF THE UNFCCC AND KYOTO PROTOCOL

The international climate regime comprising the UNFCCC, the Kyoto Protocol, and all the associated activities with these agreements has been unsuccessful largely because of the lack of political will of the governments of the developed countries as Parties to the Convention being dependent on the DSP. The UNFCCC was originally drafted as a framework convention and accordingly had no targets and timetables. Because of pressure from the USA and other Parties unwilling to take concerted action, it took a whole decade, from 1992 when the Convention was launched at Rio, for the Kyoto Protocol to come into force. Currently, the USA still is not a Party. Commentators including Tompkins and Amundsen (2008) conclude the UNFCCC is limited in its ability to drive national action on climate change and has not inspired its member countries. However, this is not completely the case as significant implementation of market-based mitigation via the Kyoto Protocol mechanisms — Emissions Trading, Joint Implementation, and the Clean Development Mechanism — has occurred. However emissions trading, while embraced by the EU, has not been overwhelmingly popular. Implementation of the New Zealand Emissions Trading Scheme

began in mid-2010 but many countries that foreshadowed they would implement emissions trading including Japan, Korea, and Australia have not followed through.

The outcomes of the UNFCCC meetings, COP15 in Copenhagen in 2009 (see Figure 2.3), and COP16 in Cancun in 2010, have been weak and reflected DSP priorities. The Copenhagen Accord developed as a compromise at COP15 has no legal relationship with the UNFCCC and the Kyoto Protocol and differs markedly in some aspects. As Siegele (2010) points out, Paragraph 8 of the Copenhagen Accord defines a new category of countries as "the most vulnerable developing countries", which is a new group not defined under the UNFCCC. Mandatory reporting and verification were the sticking points between USA and China at Copenhagen because of baselines, additionality, and sustainable development. Noble (2010) commented that the assumption in international negotiations that all countries are speaking the same 'language' is erroneous and that co-learning needs to come about between North and South, South and South, and South and North. He suggested that adaptation financing for developing countries from the developed world is likely to occur via the development of new mechanisms (Noble, 2010). Stern (2010) was optimistic about Copenhagen Accord prospects whereas the analysis of Levin and Bradley (2010) indicates:

Existing pledges by developed countries, when added together, could represent a substantial effort for reducing Annex I emissions by 2020 — a 12 to 19% reduction of emissions below 1990 levels depending on the assumptions made about the details of the pledges. But they still fall far short of the range of emission reductions — 25 to 40% — that the IPCC notes would be necessary for stabilizing concentrations of CO_2e at 450 ppm, a level associated with a 26 to 78% risk of overshooting a 2°C goal. If the pledges are not ratcheted up even beyond the highest pledges, this analysis shows that the additional reductions required between 2020 and 2050 would be significant…

(Levin and Bradley, 2010, p. 2)

Unfortunately, the Copenhagen Accord appears very unlikely to result in the co-operation needed to reduce greenhouse gas emissions by the amounts necessary to stop warming at the 2°C threshold that stems from EU policy determined politically rather than scientifically. Small island states argued at Copenhagen that 1.5°C is a more realistic threshold based on their vulnerability. This is because it is likely that there are regional variations in terms of temperature threshold. In addition, temperature threshold in combination with other climatic factors, such as changes in precipitation and large-scale atmospheric circulation patterns, as well as non-climatic factors, such as socio-economic development paths, will all contribute to determining the final magnitude and timing of adverse impacts in certain areas. The risk can be very low or high depending on how the other factors

FIGURE 2.3 Roslyn Taplin *Copenhagen 2009 #2* (2010), Indian ink and crayon on paper, 81 × 99 cm. *(Source: Image courtesy of David Toyer.)*

develop and interact (see Harvey, 2012, this volume). Thus, other than a certain temperature threshold, it is more important to consider, assess, and capture the multidimensional threshold for both climatic and non-climatic factors that will cause the impacts (IPCC WGII, 2007b, p. 625).

Somewhat more positive outcomes manifested in the 'Cancun Agreements' did occur at COP16. These included:

1. An agreement to peak emissions at an overall 2°C target and recognition by Parties of the need to consider strengthening this long-term global goal in relation to a global average temperature rise of 1.5°C
2. Formalizing, to some degree, details of what countries promised to do to mitigate climate change under the Copenhagen Accord within the UN architecture
3. Agreeing more systematized measurement, reporting, and verification of emissions
4. Establishing a Green Climate Fund to support projects, programs, policies, and other activities in developing countries
5. Agreeing to a framework to slow, halt, and reverse destruction of forests (REDD+) and agreeing the rules for delivering REDD+ and for monitoring progress
6. Setting up the mechanisms to help developing countries access low-carbon technology; a registry will be set up to record developing country Nationally Appropriate Mitigation Actions (NAMAs) seeking international support and to facilitate matching of finance, technology, and capacity-building support from developed country Parties for these NAMAs
7. Establishing the Cancun Adaptation Framework to assist developing countries to adapt to climate change via planning, prioritizing and implementing adaptation actions, vulnerability and adaptation assessments, strengthening institutional capacities, building resilience of socio-economic and ecological systems, including through economic diversification, and sustainable management of natural resources
8. Inclusion of carbon capture and storage (CCS) in the CDM (Baines, 2010; Pew Center, 2010)

The UNFCCC has hailed this progress stating the Cancun Agreements:

...form the basis for the largest collective effort the world has ever seen to reduce emissions, in a mutually accountable way, with national plans captured formally at international level under the banner of the United Nations Framework Convention on Climate Change, ...include the most comprehensive package ever agreed by Governments to help developing nations deal with climate change. This encompasses finance, technology and capacity-building support [and]...include a timely schedule for nations under the Climate Change Convention to review the progress they make towards their expressed objective of keeping the average global temperature rise below two degrees Celsius. This includes an agreement to review whether the objective needs to be strengthened in future, on the basis of the best scientific knowledge available.

(UNFCCC, 2011a, paras. 3—5)

Nevertheless, this action will still not slow climate change sufficiently. Also there is still no agreement on the second commitment period of the Kyoto Protocol. This reflects the domination of the complex global SES by leadership swayed by the prevailing influence of the DSP. Pockets of NEP action to implement various forms of climate mitigation action exist and are important (as discussed in Section 2.7: Bottom-Up Approaches) but their growth may not be sufficient to provide an effective polycentric approach to address climate mitigation (Ostrom, 2009a).

2.6. TOP-DOWN ACTIONS STEMMING FROM INSIDE AND OUTSIDE UNFCCC/KYOTO

It must be acknowledged that many initiatives undertaken at the international level, at the country level, and within countries have arisen from the climate regime even though the international negotiations on the second phase of Kyoto appear to have reached an impasse. There have been many initiatives of a very constructive nature implemented by international organisations, national governments, and corporations over the last decade. Some of the more recent programmes are mentioned as examples here.

2.6.1. Greenhouse Gas Accounting

The International Organisation for Standardization (ISO), the World Resources Institute (WRI), and the World Business Council for Sustainable Development (WBCSD), all international NGOs that promote links between the public and private sectors, have worked to establish sound international carbon accounting frameworks for quantification and reporting and for validation and verification of greenhouse emissions of business and industry and governments related to the organisational level, projects, products, and supply chain. The WRI's and WBCSD's Greenhouse Gas Protocol (GHG Protocol, 2011) has been the result of the voluntary efforts of hundreds of companies and their professional staff in the design and testing of the various aspects of the framework. The ISO 14064 Standard (ISO, 2006) that gives internationally-agreed broad guideline requirements, established in 2006, is compatible with the GHG Protocol, published by WRI and WBCSD in 2004, which provides detailed guidance with complementary accounting tools and processes. Industries in developing as well as developed countries have begun to adopt the procedures, with groups of companies in Brazil and India adopting the GHG Protocol in 2008 and cement companies in China adopting the standard in 2009. Ironically, the GHG Protocol has much more widespread global adoption than ISO 14064.

The most recent GHG Protocol initiative has been development of two new accounting standards in 2009—2010 for assessment of emissions associated with product lifecycle and supply chains. They were drafted via a global, collaborative multistakeholder process in 2009, with participation from over 1000 volunteer representatives from industry, government, academia, and NGOs. The draft standards were tested in 2010 by 62 companies across 17 countries including China for expected finalization and release in September 2011 (GHG Protocol, 2011).

2.6.2. Development Initiatives on Climate Change

In 2009, the OECD (Organisation for Economic Co-operation and Development) released a Policy Guidance on Integrating Climate Change into Development Co-operation due to concerns that "many development policies, plans, and projects currently fail to take into account climate variability, let alone climate change" (OECD, 2009, p. 11). The objectives of this Policy Guidance are to:

...promote understanding of the implications of climate change on development practices and the associated need to mainstream climate adaptation in development co-operation agencies and partner countries; identify appropriate approaches for integrating climate adaptation into development policies at national, sectoral, and project levels and in urban and rural contexts; identify practical ways for donors to support developing country partners in their efforts to reduce their vulnerability to climate variability and climate change.

(OECD, 2009, p. 12).

Prior to this in 2008, two Climate Investment Funds, a Clean Technology Fund, and a Strategic Climate Fund, were established in a development bank partnership of the World Bank, the African Development Bank, the Asian Development Bank, the European Bank for Reconstruction and Development, and the Inter-American Development

Bank to provide loans to "…assist developing countries fill financing gaps in supporting efforts aimed at climate mitigation or strengthened resilience to climate change impacts." (CIF, 2010, para. 2). Under the Strategic Climate Fund, a Pilot Program on Climate Resilience was started in late 2008. The G8 have pledged US$6 billion as an ODA (official development assistance) contribution to these funds. Thus, these are not 'new and additional financial resources' under Article 4.3 of the UNFCCC. Private and public sector adaptation projects in LDCs (least developed countries) based on the UNFCCC required National Adaptation Programs of Action (NAPAs) are eligible to apply for funding. However, many of these nations have preferred not to borrow.

Spain and the UK have also been proactive with supporting climate change initiatives in developing countries. In 2006, the UNDP/Spain Millennium Development Goal Achievement Fund (MDG-F) was established with the contribution of €528 million to the UN. A current project is assessing climate change vulnerabilities in Colombia's indigenous communities (MDG-F 2010). Additionally, the UK Department for International Development (DfID, 2010) launched a £50 million Climate and Development Knowledge Network (CDKN) in March 2010 to support developing countries in tackling the challenges posed by climate change (CDKN, 2010). The five-year initiative is managed by a consortium led by PricewaterhouseCoopers LLP. The initiative comprises part of the UK's 'fast start funding' commitment post-Copenhagen.

2.6.3. EU Member Countries Policies and Programs

The EU has demonstrated continuing leadership in, and support for, the climate regime and Kyoto implementation since implementation of the EU Emission Trading Scheme (ETS) in 2005. Grubb et al. (2009) comment in their report on the EU ETS that "GDP impacts are small" and that "Industry can benefit" (p. 3). Also all EU member countries including Germany, the United Kingdom (UK), and the Netherlands (discussed below as examples) have undertaken national level climate change measures as well.

Germany has adopted an ambitious national climate protection goal of a 40% reduction of greenhouse gas emissions by 2020 compared to 1990 levels. Its Climate Initiative entails programmes on emissions reduction, energy efficiency, and the expansion of renewable energies, and provision of a consumer information service for citizens, enterprises, municipalities, and schools have been undertaken using funds raised by the auctioning of emissions allowances. With its federal system of multi-level governance, one of Germany's major climate change challenges with regard to adaptation seems to be awareness-raising among the general population, local government, and stakeholders in business and industry (Daschkeit and Mahrenholz, 2010).

As well as implementing climate change mitigation measures in a proactive manner, the Netherlands is seriously tackling climate change adaptation in areas below sea level to achieve practical outcomes for its citizens and their property including building an impressive flood defence system. de Bruin et al. (2009, p. 63) comment that the research behind this work very usefully "…adds to the IAM [Integrated Assessment Model] and climate change literature by explicitly including adaptation in an IAM, thereby making the tradeoffs between adaptation and mitigation visible".

The UK Government also has been active with development of various adaptation policies and measures including flood defence as well as mitigation policy implementation. Its advisory body, Natural England, recently released a report for the UK on Climate Change Adaptation Indicators for the Natural Environment that utilized a local community-led approach and involved workshops with stakeholders to determine indicators for monitoring and surveillance of the natural environment's resilience in adapting to climate change (Natural England, 2010).

2.6.4. Climate Change Vacillation by the USA and Australia

The USA and Australia have been much less proactive than EU countries on mitigation due to political opposition and concerns about the potential negative economic impacts; nevertheless, local and state-based climate change action is taking place in both countries. The USA remains a Kyoto non-ratifier while Australia ratified the Protocol in late 2007.

The current US administration has shown some signs of renewed interest in the UN climate regime with the American Clean Energy and Security Act (Waxman–Markey Bill) being passed in the House of Representatives in mid-2009. It is a step towards establishing an economy-wide cap and trade programme and creating other incentives and standards for increasing energy efficiency and low-carbon energy consumption. Also, attendance by President Obama at the Copenhagen COP15 later in 2009 indicated the consideration given to climate change by the USA. However, passage of a Senate counterpart to the Waxman–Markey Bill, the American Power Act (Kerry–Lieberman Bill) has not occurred to date. Nevertheless, 300 cities across the USA have adopted climate policies (Schneider, 2010a) and regional emissions trading schemes have been established by nine northeastern states in relation to power generation (the Regional Greenhouse Gas Initiative) and by seven states in the western US and four provinces in western Canada (the Western Climate Initiative). California has also legislated the California Global Warming Solutions Act 2006.

A decision-making study by the USA National Research Council (NRC, 2009) involved 90 scientists across four panels providing "policy advice, based on science, to guide the nation's response to climate change." The US Government's request to the NRC was to "…investigate and study the serious and sweeping issues relating to global climate change and make recommendations regarding what steps must be taken and what strategies must be adopted in response to global climate change, including the science and technology challenges thereof." Hansen (2009, p. 276) has lamented, with regard to this type of activity:

The problem with governments is not scientific ability — the Obama administration, for example, appointed some of the best scientists in the country to top positions in science and energy. Instead the government's problem is politics, politics as usual.

Successive Australian governments have had a similar pattern of putting politics before climate change and environmental concerns. Australia legislated a National Greenhouse and Energy Reporting (NGERS) Act in 2007 but has had difficulties in introducing emissions trading legislation or a proposed Carbon Pollution Reduction Scheme (CPRS). Two different versions of the proposed CPRS legislation were rejected in the Australian Senate in August and then December 2009. In April 2010, the Government announced plans to delay its introduction of a Carbon Pollution Reduction Scheme until at least 2013 although a voluntary National Carbon Offset Scheme (NCOS) is proceeding together with the implementation of the NGERS legislation. A new Government formed under Prime Minister Julia Gillard after a federal election in August 2010 announced, not long after assuming office, the formation of a high level Multi-Party Climate Change Committee chaired by the Prime Minister to examine the best way forward to price carbon; in July 2011 a decision-making outcome was announced by the Gillard Government that Australia will introduce a carbon tax on 1 July 2012 and that in three to five years a full emissions trading scheme will be introduced.

On the adaptation front, Australia has made more headway with the establishment of a National Climate Change Adaptation Research Facility to lead "the research community in a national interdisciplinary effort to generate the information needed by decision-makers in government and in vulnerable sectors and communities to manage the risks of climate change impacts" (NCCARF, 2010). Also an Australian Government sponsored Local Adaptation Pathways program has had positive outcomes including raising awareness with local government authorities (LGAs); engaging people across the community; and establishing processes that can be replicated in other LGAs. The programme has also experienced limitations in that most actions that have been identified are too high level to be costed, there are no links to implementation, the capacity of councils to

progress actions identified is very low, and there is little help with decision support for councils.

2.6.5. The Asia Pacific Partnership on Clean Development and Climate

The Asia Pacific Partnership (APP) for Clean Development and Climate was formed in July 2005. Aspects of Taplin and McGee's (2010) discussion of the development and implications of the APP are recounted here. The APP is a multilateral agreement between Australia, Canada, China, India, Japan, the Republic of Korea (Korea), and the USA. In international legal terms, the Partnership is soft law, that is, a non-binding memorandum of understanding agreement directed at international co-operation on development, energy, environment, and climate change issues (Taplin and McGee, 2010). Co-operation includes transfer of cleaner technologies and facilitation of practices that might enable reduction of GHG emissions. Unlike the Kyoto Protocol, this agreement allows member states to voluntarily set their own goals for reducing GHG emissions without mandatory enforcement mechanisms or binding commitment to achievement of those goals. The Partnership also encourages national goals to be based around greenhouse intensity targets, that is, reducing greenhouse emissions per unit of economic output. Its inception was driven by the USA and Australia with the Partnership subsequently being joined by Canada in October 2007. It has been hailed as a new model for an international climate agreement and as an alternative to the Kyoto Protocol. Questions about the effectiveness of the APP in comparison the Kyoto Protocol and whether the Partnership is a model for multilateral action on climate change that could replace the Protocol were raised in the past few years especially in the months preceding COP15. However, as an opposing model to that of Kyoto, it is in contravention of the UN Framework Convention on Climate Change's (UNFCCC's) principle of common but differentiated responsibilities and hence a symbolic contributor to the crumbling of climate governance (Taplin and McGee, 2010). Co-operation rather than competition ideally should be the future for the relationship between the APP and the UNFCCC/Kyoto. Industry already has involvement with clean development, emissions trading, and technology transfer under the UN climate regime so some synchronicity between the APP and Kyoto is preferable. For example, APP nations could be involved in multilateral emissions trading outside the Kyoto Protocol architecture. Ultimately, the hurdles of the USA not being a Kyoto ratifier, equity issues, additionality questions, capacity building, trade barriers, intellectual property, funding, and assessment of emissions reductions need to be overcome for the APP to be a complement to Kyoto and not a barrier to effectiveness.

2.6.6. New Zealand's Policy Development

The New Zealand Government has relied almost exclusively on voluntary and informational instruments for climate change mitigation and shied away from adopting regulations and economic measures (Bührs, 2008) until a number of amendments to emission trading legislation over seven years culminated in the Climate Change Response (Moderated Emissions Trading) Amendment Act 2009. Implementation began in mid-2010 with plans for full implementation for all sectors by 2015. While New Zealand's national climate change adaptation strategy is guided by the Resource Management Act 1991, the responsibility for reducing vulnerability and adapting to climate change resilience is devolved to local government. Iati (2008), for example, has emphasized the potential for the involvement of civil society in New Zealand in the form of environmental NGOs in adaptation implementation. However, little NGO response or intervention beyond mitigation is apparent to date, for example, from examination of New Zealand's Environmental Defence Society and Greenpeace New Zealand websites. Notwithstanding that, the work of SANZ (2009) indicates that there are New Zealand NGOs that are advocating that climate change can be addressed by bottom-up as well as top-down approaches in New Zealand.

2.6.7. Developing Nations

There is a huge range of developing country plans and proposals for both climate change mitigation and adaptation. In particular, development and implementation of adaptation measures at the local level are essential to protect vulnerable communities. A much-needed centre for research and training in Bangladesh that will focus on community adaptation for vulnerable groups in least developed countries, the International Centre for Climate Change and Development, was opened at the Independent University in September 2009. The UK Department for International Development (DfID) has contributed £500,000 for its establishment. It is led by Saleemul Huq of the International Institute of Sustainable Development (IIED), London. The IIED is "...responsible for overall management, including both finances and faculty members" (Nelson 2009, para, 5) of the centre.

Another initiative has been the ORCHID (Opportunities and Risks from Climate Change and Disasters) screening process for climate change project investment in Bangladesh and India by DfID. Development of screening methods has been via a partnership between:

1. The Energy and Resources Institute (TERI)
2. Bangladesh Centre for Advanced Studies (BCAS)

3. International Institute for Applied Systems Analysis (IIASA)
4. Bangladesh Institute of Development Studies (BIDS)
5. School of International Development, University of East Anglia (IDS, 2011).

Tanner et al. (2007), in discussing the ORCHID approach, say:

Climate risk screening tackles the impact of climate-related events on poverty and poverty reduction programmes. It addresses the need for adaptation to reduce the risks posed by climate change to people's lives and livelihoods. ORCHID is a risk management-based methodological approach to portfolio screening, stressing both risks and opportunities in tackling climate change.

(Tanner et al., 2007, p. 4)

Overall, there has been considerable recent effort by developing nations, including involvement in CDM projects as well as recent climate policy measures taken by individual countries, making it impossible to survey them all here. Two national policy examples are the January 2010 commitments to the UNFCCC by China and India:

China will endeavor to lower its carbon dioxide emissions per unit of GDP by 40−45% by 2020 compared to the 2005 level, increase the share of non-fossil fuels in primary energy consumption to around 15% by 2020 and increase forest coverage by 40 million hectares and forest stock volume by 1.3 billion cubic meters by 2020 from the 2005 levels.

(UNFCCC, 2011b, para. 2)

and:

India will endeavour to reduce the emissions intensity of its GDP by 20−25% by 2020 in comparison to the 2005 level.

(UNFCCC, 2011c, para. 2)

2.7. BOTTOM-UP APPROACHES: CIVIL SOCIETY PARTICIPATION AND INFLUENCE

Bottom-up approaches have also been growing in numbers and strength. Legislative development in the UK is a key example. With the passing of its Climate Change Act in 2008, the UK became the first country to enact national climate change legislation. Even more remarkable is that this Act developed from grassroots action initialized by the NGO, Friends of the Earth — England, Wales, and Northern Ireland (FoE EWNI). Aspects of Hall and Taplin's (2007) discussion of the development of the Bill are recounted here.

In 2005, FoE EWNI initiated a campaign called 'Big Ask' to pressure the UK Government to introduce a Climate Change Bill. The draft Bill was devised by a group of ten NGOs and written by FoE EWNI and WWF UK. It stated that the UK's GHG emissions should reduce to 20% below the UK's 1990 baseline by 2010 and decrease a further 3% annually until 2050. The draft Bill required the UK Prime Minister to develop a strategy to reduce emissions and report annually to Parliament on the progress of these cuts. If emissions exceeded the target by more than 10%, the NGOs proposed that the Prime Minister and relevant Department Secretary have their salary reduced by 10% per annum (which not surprisingly did not eventuate in the final Act). FoE EWNI orchestrated a nationwide grassroots campaign and was confident throughout the process that it would result in legislation because "FoE has had a track record in getting new Bills through Parliament" (Hall and Taplin, 2007, p. 13). In its initial form as an Early Day Motion, it was promoted by over 200 local groups around the UK. FoE EWNI estimated that 130,000 citizens contacted their MPs in support of the draft Bill, resulting in signed support from 200 MPs prior to the 2005 national election. Further support and development over a three-year period of the Bill ensued and eventually resulted in the Climate Change Act being passed in November 2008 with 400 MPs voting in favour of it. The Act's legally binding target is at least an 80% cut in greenhouse gas emissions by 2050, to be achieved through action in the UK and abroad and a reduction in emissions of at least 34% by 2020 against a 1990 baseline. Although these targets are not as stringent as those first drafted by the NGOs, they are certainly impressive goals arising from civil society action.

A similar grassroots campaign for the Australian climate legislation based on the UK model was mounted in 2007 by 65 locally based citizens' climate groups including Climate Action Coogee (CAC) in Sydney (Hall et al., 2010). CAC wrote its own Climate Protection Bill that required Australian greenhouse gas emissions to be reduced to 30% below the 1990 emission levels by 2020, and 80% below 1990 levels by 2050. Widespread grassroots endorsement and political awareness of the Bill resulted. It was submitted to the House of Representatives as a Private Members Bill in November 2008 without success.

2.8. PROSPECTS FOR THE FUTURE

2.8.1. Institutional Change

Do we need 'new institutional processes' for adaptation to future climates as advocated by Adger et al. (2005)? Will polycentric systems solve climate policy problems as Ostrom (2010) has suggested? The current situation is that we have a sea of climate change related laws, regulations, policies, and plans at all scales — local being more effective than global. As Schneider (2010a) observed "we've had a lot of good local stuff now — we need to align both local and global." What is desirable is that climate change considerations are built into all levels of government, business, and organisational decision making.

The process of positive resilience building, via institutional change more broadly with regard to sustainability, is discussed by Dovers (2005, pp. 174—85). He proposes four strategies to change institutional systems over time: legal change through legislative review and statutory expression of principles; a strategic environmental assessment (SEA) regime (assessment of all policy and programme development in non-environmental sectors); inclusive national councils for sustainability; and an in-government commissioner for environment or sustainability (Dovers, 2005, pp. 176—77). However, he says:

The key factor [in reshaping the institutional system over time]…is obvious: environment and sustainability, although strongly supported in policy rhetoric, is but one social and policy goal amidst many, and moreover one that does not have the same priority of longer-standing social and economic concerns…Deeper reforms of this sort are unlikely to be achieved quickly as the slow evolution and still uncertain status of SEA demonstrates.

(Dovers, 2005, p. 179)

Dovers' institutional strategy proposals have some resonance with Milbrath's (1989) recommendations for a learning governance structure, that is, systemic and futures thinking capability via national councils for sustainability and in-government environmental/sustainability commissioners, and long-range sustainability impact assessment capacity for policy initiatives via SEA. There are a positive models in some settings for these governance attributes, for example, with SEA implementation with the EU's Directive on Strategic Environmental Assessment (Directive 2001/42/EC). Interestingly, an EU review of the SEA Directive in 2009 found:

Climate change issues are considered in SEA on a case-by-case basis, and mainly in relation to P&P [policies and plans] with a potential significant impact on climate, such as energy or transport P&P. However, a trend to pay more attention to climate change considerations in the other P&P is emerging.

(EUR-Lex, 2009, para. 4.4).

The EU is addressing these climate change shortcomings as a response to this review of the SEA Directive.

2.8.2. Social Learning

Learning about climate change is, for individual citizens, communities, political office holders, government officers, a necessary and integral part of a transition towards sustainability. As Tilbury and Wortman (2004) emphasize, different stakeholders with different values and beliefs need to come together — education is a crucial part of the process of moving towards sustainability and addressing climate change. Sterling (2001) also advocates that there must be "shifts from mechanistic to ecological thinking" (Sterling 2001, p. 53). Education for sustainability focuses on critical thinking challenging false consciousness and assisting values clarification, power relationships, and participatory learning activities, and underlines quality of life and futures thinking, using action research as a means for change (Tilbury and Wortman, 2004).

Gifford (2008) talks about the "dragons of inaction" and the "dragon of denial" with regard to the psychological aspects of climate change for the general public. He proposes five ways to assist citizens slay their dragons: assist them to understand the impact of their transportation, home heating, and so on, on carbon emissions; help them to find ways to make good choices more attractive (leadership examples); make climate change links to what is happening now (bring home climate change to local environment, i.e., by local recording of changes by bird watchers, gardeners, etc.); give them opportunities to be involved in local neighbourhood groups, statewide groups, federal groups, and church groups addressing climate change and sustainability. He says: "…what I tell people is if you're not at the table your contribution to policy will never happen" (ABC Radio National, 4 September, 2010).

Those in business and government also need to learn about climate change. A Canadian survey of 503 senior respondents in municipal, provincial government agencies and businesses in November 2009 revealed that confusion still exists in understanding between mitigation and adaptation action on climate change:

About half of the respondents report that their organisation is currently doing something to adapt to climate change. When describing their actions, the majority of respondents cited climate change mitigation activities (e.g., focus on reducing greenhouse gases), while a minority described adaptation activities (e.g., focus on reducing climate change impacts).

(Environics Research Group, 2010, p. iii)

Canadian adaptation researcher Stewart Cohen (2010) advocates a "shared learning model" where researchers and local practitioners work together to produce relevant information for local decision making with the result that the local people involved become climate change "extension agents". Cohen (2010, p. 131) concludes, "We need to find new business and management models with academia, government, and NGOs that can sustain shared learning on climate change beyond individual projects and short-term opportunities."

Milbrath (1989) in *Envisioning a Sustainable Society: Learning Our Way Out* stressed that learning is the path to a sustainable society. Influenced by the work of Boulding (1978; 1985) who argued that evolutionary learning is the major way that societies advance, Milbrath (1989) said:

We must learn how to become conscious of our ways of knowing…As we do so, we will come to recognize the key role that

society plays in knowledge development, in development of beliefs about how the world works, and in value clarification (or obfuscation). Recognizing that our beliefs and values are culturally derived frees us to reexamine them to see if they can be revised to serve us better.

(Milbrath, 1989, p. 85)

Aspects of Milbrath's learning governance approach for climate change are apparent at the international level via the UNFCCC and the IPCC, which provide systemic information and resources on climate change and contribute to systemic and futures thinking capacity. In some countries, this is replicated at the national level. However, long-range climate change impact assessment for all policy initiatives and stakeholder intervention capability (e.g., via the EU's SEA Directive and the UNECE Aarhus Convention on Access to Information, Public Participation in Decision Making and Access to Justice in Environmental Matters) are rare.

Ostrom's (2010) polycentric approach for dealing with climate change can arguably only happen effectively with transformation of beliefs and values of individuals and with a learning governance framework in place.

2.9. FUTURE UNKNOWNS: LIVING ON A WARMER EARTH?

What are the prospects for people living on a warmer Earth? Lovelock's (2010) and Hansen's (2009) perspectives on this question are not optimistic (see also Lenton, 2012, this volume). Similarly, Ann Henderson-Sellers argues in Chapter 1 of this book that triage may be required (Henderson-Sellers, 2012, this volume). The political and policy challenges are immense. These include dealing with the debates about trust in the reliability of climate science, scientific uncertainty, together with arguments about lifestyle, economic affluence, and the need to maintain business-as-usual. The role of the media is also critical as is education of communities and decision-makers (cf. Rice and Henderson-Sellers, 2012, this volume).

Arguably, the fundamental obstacle to strong action to slow climate change is the predominance of DSP (dominant social paradigm) thinking tied in with the primacy of economic goals. This indicates the degree of change that is needed for NEP (new environmental paradigm) attitudes to be reflected in political decision making and good Earth system governance to prevail. Climate change deniers also have added confusion about the science. As Naomi Oreskes and Erik Conway have said:

For the past 150 years, industrial civilization has been dining on the energy stored in fossil fuels and the bill has now come due. Yet we have sat around the dinner table denying that it is our bill, and doubting the credibility of the man [sic] who delivered it.

...No wonder the merchants of doubt have been successful. They've permitted us to think we could ignore the waiter, while we haggled about the bill. The failure of the United States to act on global warming as well as the long delays between when the science was settled and when we acted on tobacco, acid rain and the ozone hole are prima facie empirical evidence that doubt-mongering works.

(Oreskes and Conway, 2010, p. 266)

Ultimately government is the final risk manager. Any effective climate policy must include mitigation and adaptation. Schneider (2010a) counselled, "Adaptation should be the number one focus and a moral obligation, then mitigation is number two...adaptation and mitigation are complementary". He advised we "need to move to bottom-up analysis then couple bottom-up with top-down" Schneider (2010a).

People living in differing political systems and places on Earth via their multiple efforts can all contribute to adapting to and mitigating climate change impacts. Good learning governance approaches for sustainability and climate change to support this are essential.

Chapter 3

Urban Climates and Global Climate Change

Helen Cleugh[a,b] and Sue Grimmond[c]

[a]Corresponding author, [b]CSIRO Marine and Atmospheric Research, Aspendale, Australia, [c]Department of Geography, King's College London, London, UK

Chapter Outline

3.1. INTRODUCTION: LIVING IN CITIES

3.1.1. Overview

As humanity prepares to meet the challenge of global climate change, understanding the role of cities will be critical because they are where the majority of the world's population live, and are where the bulk of human emissions of heat, gases, and particulates that affect climate originate. Cities are highly modified systems that are both formally and informally managed and so provide an institutional governance and innovative capacity to effect change (Taplin, 2012, this volume). In creating their own distinctive and measureable climate, cities provide a tangible demonstration that human activities can modify climate and are therefore an analogue for studying global anthropogenic climate change caused by increasing levels of greenhouse gases (GHG) in the atmosphere.

Some of the challenges for mitigating and adapting to global climate change involve managing the impact of urban climate on water supply, health, and energy demand, namely:

- Energy — carbon: Reducing consumption of fossil fuels, and emissions of CO_2 and heat, for space heating/cooling, transport, and industry will involve redesign of

The Future of the World's Climate. DOI: 10.1016/B978-0-12-386917-3.00003-8

urban areas at all scales (building to city-scale), and managing the effect of cities on the local and regional climate via the urban energy balance (i.e., altering the receipt of solar radiation and/or sensible heat flux through the use of different surface materials including vegetation).

- Climate – health – air quality: Heat-related mortality and air quality are risks to the health and wellbeing of urban dwellers. Mitigating excessive urban heating for many cities will be a key adaptive measure to reduce the impact of global climate change on the health of urban dwellers and urban ecosystems. Adaptive measures need to reduce, rather than increase, energy demand to achieve both energy security, especially during peak demand periods, and reduced GHG emissions. Maintaining urban air quality requires lowering emissions of pollutants and GHG, and managing the effects of cities on the local and regional weather and climate.
- Water – energy – carbon: Against the backdrop of rising global temperatures, rising sea levels, and intensification of the hydrological cycle (where some regions will become drier, others will experience increased storm frequency and rainfall intensity; e.g., Alexander and Tebaldi, 2012, this volume), management of urban water needs to be coupled to strategies that manage the effects of cities on energy use, net carbon emissions, and health.

These challenges are all linked by the interacting cycles of energy, water, and carbon. The urban system is characterized by strong feedbacks and a mix of human and biophysical drivers and processes. Managing this system requires an integrated approach and a simulation capability that can quantify these coupled cycles, their links, and feedbacks, together with socio-economic and biophysical processes.

Tremendous growth in urban climate process studies over the last two decades, and especially since the earlier publication of Cleugh (1995a), has led to a step change in the quantified understanding of the biophysical basis of urban climates and an increase in the development and evaluation of urban climate models. These advances mean that the urban climate science community is now well-placed to contribute to issues that lie at the nexus between urban climates and global climate change. The goal of this chapter is to facilitate this by: (i) synthesizing the current state of knowledge of urban biophysical processes and urban climates; (ii) reviewing current generation weather and climate models and their ability to represent these processes; (iii) exploring how cities contribute to global warming and thus how urban design can be used to *mitigate*, and what adaptive measures are needed to *manage*, the impact of global climate change; and (iv) identifying the knowledge and capability gaps that currently impede our ability to comprehensively address

the points raised in (iii). This chapter is complementary to both Taplin (2012, this volume), which addresses the more social and political dimensions of urbanisation, and Pitman and de Noblet-Ducoudré (2012, this volume), which explains the climate effects of land-cover modification.

3.1.2. Why Are Urban Climates Important to Future Climates?

Urbanisation is an agent of change at local, regional, and global-scales. The conversion of a landscape to urban land cover causes large changes to the local and regional energy, carbon, and water balances; as well as air quality and climate. At regional to global-scales, the flows of materials (water, carbon, pollutants, etc.) and energy are modified by the new ecosystem: they are imported and consumed within the city and the resulting waste emitted and exported to the surroundings. Urban areas therefore have a much greater environmental impact than the 2%[1] of the total global land area they occupy. By way of illustration, Decker et al. (2000) estimate that the area required to assimilate and process the material flows through cities is 400–1000 times larger than the area they directly occupy. The clearest demonstration of their global impact, however, is the significant contribution that cities make to anthropogenic climate change through large net emissions of GHG. This is why understanding the interactions between the urban environment and global climate change is so important and yet, although there has been significant progress in quantifying local and regional effects of cities over the last three decades, much less is known about the feedbacks between urbanisation and global climates, despite some relevant and important contributions in recent years (e.g., Mills, 2007; McCarthy et al., 2010). Aspects of particular relevance to the urban climate–global climate change nexus are outlined below.

3.1.2.1. Cities As an Analogue for Global Climate Change

Associated changes of the surface and atmospheric properties result in measureable changes such that a distinctly different microclimate develops. Many of these attributes also contribute to global, anthropogenic climate change, namely: land-cover changes with consequent changes to the surface energy, water, and carbon balances; increased energy use and emissions of waste heat; and increased emissions of radiatively-active atmospheric constituents, especially CO_2. The warmer temperatures and enhanced

---

1. The uncertainty about the exact fraction of the global land mass that can be defined as urban is acknowledged. Many authors suggest a figure between 1% and 2.5%; we use 2% here as an indicative figure only.

CO_2 concentrations demonstrate, at the scale of a city, the impact of future global processes. Changes, such as air temperature, are often larger in cities than the global temperature increases predicted to result from a doubling of atmospheric CO_2 through this century. Many have suggested that cities are a useful 'laboratory' in which to study anthropogenic climate change.

3.1.2.2. Climate Impacts in Cities Result from Urban and Global Climate Changes

The climate for a specific city results from the superposition of the local urbanisation effects and the global climate change, and this has two very important implications. Firstly, the detection of global trends in near-surface air temperature requires that the record is checked for biases that could result from urban development near to the measurement station (Jones, 2010). Secondly, the climate change impacts experienced will result from both global and local influences. This may amplify the impact of some climate elements, for example, the additional thermal stress and more frequent heatwaves because of the urban heat island (UHI) effect superimposed upon a global trend of increasing nighttime minimum temperatures, which increases the vulnerability of urban areas (people and infrastructure) to global climate change. The corollary of this is that designing cities to reduce the strength of the urban warming effect will provide greater 'headroom' for cities to adapt to the impacts of global climate change.

The future world will not only be a warmer world due to anthropogenic climate change, but it will also be an urban world, with over two thirds of the population expected to reside in cities by 2050 (Box 3.1). Using these projections, more than 6 billion people will experience global climate change through an urban filter — this is the same as the world's entire population in 2004[2]. This growth is mostly expected in less developed regions, most notably in Asia and Africa, which experience tropical and subtropical climates (Pearlmutter et al., 2007; Roth, 2007). This is critical as much of our current urban climate knowledge is from temperate, mid-latitude cities. Furthermore, many of the world's most populous cities are close to sea level, and so their people and infrastructure are vulnerable to the combined effects of global and urban climate change, sea-level rise, storm surges, and coastal inundation (e.g., Rice and Henderson-Sellers, 2012, this volume). Cities in Mediterranean-type or in arid climate regimes are likely to experience increased extremes such as heatwaves and deterioration in air quality due to increased levels of dust, smoke, and photochemical smog.

Adaptation to manage the impacts of climate change on health and wellbeing for urban dwellers, infrastructure, and ecosystems must take this superposition of urban climate and global climate change into account.

3.1.2.3. Mitigating Climate Change by Reducing GHG Emissions

While not all cities have disproportionately large per capita GHG emissions (Dodman, 2009), they nonetheless are a dominant source in the global carbon budget by virtue of the fact that over half of the world's population is concentrated in cities. This explains the often quoted statistic that cities are responsible for 75% of the world's energy consumption and over 80% of global GHG (e.g., Churkina, 2008) emissions. Cities are hot spots of atmospheric pollutants that affect the climate and air quality locally and regionally, and contribute to the global burden of atmospheric constituents that are climate drivers.

Quantifying and attributing urban GHG emissions is complex and urbanisation is not necessarily the direct cause of growing GHG emissions; for example, Satterthwaite (2009) argues that GHG emissions are driven more by consumption than population growth and size. Nonetheless, reducing urban GHG emissions remains an effective and

---

BOX 3.1 An Urban Future

In 2009, the fraction of the world's population living in urban areas exceeded 50% for the first time in history. The number of urban inhabitants (3.5 billion) is projected to almost double to 6.3 billion by 2050, which would constitute 69% of the anticipated global population. In these four decades, urban areas are expected to absorb all of the projected global population growth as well as continuing to receive migrants from rural areas[2].

The growth will predominantly take place in the cities and towns of less developed regions; with the urban populations in Asia, Africa, and Latin America/Caribbean projected to grow by 1.7, 0.8, and 0.2 billion people, respectively. Developed regions will continue to urbanise, with the fraction of the population living in cities projected to rise from the current 75% to 86% by 2050. By comparison, 66% of the population in less developed regions will inhabit urban areas[2].

The urban population is dispersed across cities of a wide range of sizes. The current 21 megacities (urban agglomerations with at least 10 million inhabitants) account for ~10% of the urban population and over half of the world's urban population reside in communities with less than 500,000 inhabitants[2].

2. These data are from the 2009 Revision of World Urbanisation Prospects, published by the United Nations. The population projections are based on assumptions regarding fertility rates, as described in that report.

essential strategy to mitigate global climate change. Cities provide a great opportunity through their institutional planning and governance capacity to enable mitigation measures and policies to be effectively implemented. The concentration of people and resources means there is opportunity for innovation and adoption of new technologies (Dodman, 2009).

Changes in urban design, form, and function that reduce energy use and the emission of heat, GHG, particulates, and reactive gases will be an important way to mitigate the drivers of global climate change. Mitigating global climate change is very likely to require transformational changes in cities and these are a potential key agent in efforts to mitigate climate change. Urban planning needs to be informed by a quantified understanding of the role for 'climate sensitive urban design' to mitigate climate change via reduced energy consumption.

3.2. LOCAL AND REGIONAL URBAN CLIMATES: THE BIOPHYSICAL BASIS

3.2.1. Urban Morphology

Urbanisation involves changes in both the form and function of the landscape. As a city grows, the need to house, feed, transport, and support the livelihood of a growing number of people concentrated in a relatively small area

brings along changes in land use: residential, transport, and manufacturing activities replace the previous landscape (Mills et al., 2010).

Urban development replaces natural or agricultural vegetation with buildings that are typically impermeable and inflexible to the movement of water and air. The sharp edges inject highly turbulent wakes and vortices into the airflow; and the buildings are often arranged in relatively regular rows/grids that channel and accelerate airflow. Water and carbon stores, as well as exchanges across the soil—plant—atmosphere continuum, are modified by the inclusion of impervious surfaces and buildings that divert the flows of rain, air, and sunlight from pervious pathways. External environmental drivers on the urban energy, water, and carbon exchanges are less influenced by the solar cycle and seasonal changes in temperature and rainfall. This is because of the regulation and consumption of energy required to maintain the metabolism of the urban system.

These changes to the radiative, thermal, aerodynamic, and moisture-holding properties of the landscape modify the heat, momentum, and mass balances such that urban areas develop their own characteristic climate (Figure 3.1). The urban landscape differs from the microscale, with individual buildings and trees in the urban canopy, to the entire city at the meso-scale. A city-scale urban ecosystem is created through the imposition of the built morphology on the existing landscape: a network of

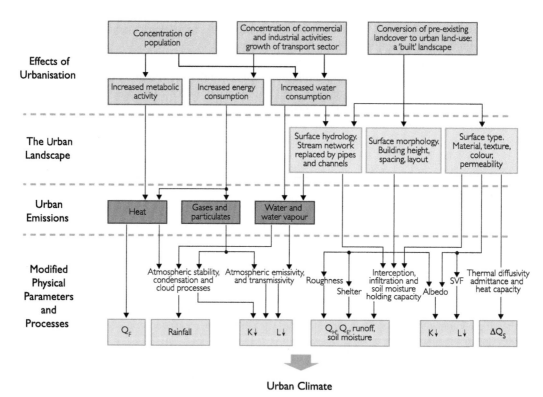

FIGURE 3.1 Effects of urbanisation on processes affecting climate at local- and regional-scales. *(Source: Modified from Cleugh, 1995a.)*

TABLE 3.1 Spatial Scales in the Urban Environment

| Units | Urban Features | Spatial Dimensions | | Scale | Atmospheric Layer |
		Width (m)	Length (m)		
Building	Single building, tree or garden	10	10	Micro	Urban canopy layer *including* roughness sublayer
Canyon	Urban street and bordering buildings	30	300		
Block	City block, park, factory complex	500	500	Local	Inertial sublayer, or constant flux, layer. Also referred to as the surface layer
Land-use zone or neighbourhood	Residential, industrial, commercial, etc.	5000	5000		
City	Urban area	> 25,000	> 25,000	Meso	Urban boundary layer *including* surface layer

(Source: Cleugh, 1995a, adapted from Oke, 1984.)

pipes, channels, and canals replace or supplement the natural drainage systems such as creeks and rivers; the fluxes of materials and energy are driven by non-biophysical drivers; and an urban climate is created. Cities also include large areas of 'greenspace', such as lawns, parks, gardens, shrubs, and trees, which in some cases mean that there is more vegetation than previously; consequently, many western cities suburbs meet the definition of a forest (Oke, 1989).

Cities indirectly modify atmospheric variables, such as temperature, humidity, airflow, and rainfall, whose ensemble average defines the climate. This happens at micro-scale ($< 10^2$ m), local or neighbourhood-scale ($10^2 – 10^4$ m), and meso ($10^3 – 10^5$ m) spatial-scales (Table 3.1, Figure 3.2a, 3.2b, and 3.2c):

Micro-scale: is influenced by individual buildings, trees, and gardens, which are the roughness elements forming the urban canopy layer (UCL).

Local or **neighbourhood-scale**: is influenced by land-use zones with similar land-cover characteristics, for example, a residential area dominated by single storey buildings or an industrial zone dominated by warehouses.

Meso-scale: is influenced by the whole city in contrast to the surrounding non-urban area.

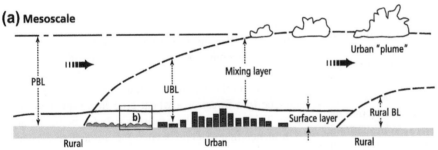

FIGURE 3.2 Scales in the urban environment. *(Source: Piringer et al., 2002. With kind permission from Springer Science+Business Media)*

As implied by Table 3.1, each of these scales is associated with an atmospheric layer where changes at the relevant scale are felt. For clarity, these are briefly defined in Box 3.2 and are illustrated in Figure 3.2. If measurements are to represent the averaged influence of a neighbourhood, then they need to be made above the UCL and, ideally, above the RSL.

3.2.1.1. Individual Buildings

These are the fundamental units to create the urban climate through their influence on airflow and energy exchanges. They are characterized by the following properties:

- Sharp-edged, rigid, and impermeable bluff bodies, causing the airflow to be highly turbulent.
- Larger thermal capacity than vegetation.
- Transmission of solar radiation to the ground is restricted, which changes the receipt of direct and diffuse solar radiation. Material characteristics control where in the immediate building 'envelope' radiation is absorbed, reflected, and transmitted.
- A regulated internal climate typically with an additional energy source.
- Heat, moisture, and gases are exchanged by conduction through the solid exterior shell and convection through gaps and vents. This exterior shell limits the exchange of moisture and gases, while heat is transported into the urban atmosphere as a point source from building infrastructure (e.g., chimneys above roof height) or an areal source within the urban canopy. Humans regulate the timing and magnitude of these emissions.

3.2.1.2. Trees and Gardens

Trees and gardens also contribute to the roughness elements that comprise the UCL, the atmospheric layer that extends from ground level to the mean building height (Oke, 1976; Cleugh, 1995a). The microclimate in this layer is very much influenced by the structure and properties of the buildings, vegetation, and urban blocks. The resulting microclimate is highly variable in space and time, but it is

important to be able to quantify this using measurements and models, as this is the climate experienced by people, plants, and animals.

3.2.1.3. Urban Canyons

The urban canyon, formed by two rows of buildings separated by a street or alley, is typically regarded as the fundamental repeated unit at the micro-scale. Created by the presence of transport corridors, it is characterized by the building height (H) and width (W) of space between. In the same way that leaf area index (LAI) is a fundamental descriptor of plant canopy structure, the canyon aspect ratio (H/W) and related sky view factor (SVF) affect radiative and convective heat exchanges, as well as airflow in the urban canopy and above. The *city block*, the clustering of several urban canyons and manufacturing or retail complexes that are of similar dimensions, is the second distinctive element in the urban canopy (Mills et al., 2010).

Urban roughness elements are more sparsely distributed than typical (closed) plant canopies, as illustrated by the characteristic spacing of 20–50 m found in suburban Vancouver (Schmid and Oke, 1992). This makes urban canopies aerodynamically rough, with momentum roughness lengths varying from 0.3–1.5 m in low-to-medium height/density suburban land use to over 2 m in high-rise, high density urban land use (Grimmond and Oke, 1999a). This relatively open canopy structure and mix of horizontal surfaces with contrasting properties (e.g., dry roads and wet lawns) lead to active horizontal exchanges of heat and mass via micro-scale advection. The natural landscape most similar to an urban canopy would be a savanna, where the spaces between the roughness elements are as important as the roughness elements themselves (Oke, 1989).

City blocks and urban canyons typically cluster into land-use zones that extend for several kilometres (Figure 3.2). By simplifying the complex urban landscape at the neighbourhood-scale, these provide a useful construct to model and observe urban climates. A land-use zone, with similar land-cover characteristics, is the urban analogue to a forest or a wheat crop in a vegetated landscape. This land use, like a crop or a forest, begins to have an influence that extends beyond the immediate cluster of buildings into the overlying boundary layer.

At the largest spatial-scale, the nature of this boundary layer reflects the influence of the entire city, such that it is referred to as the UBL.

3.2.2. Momentum Fluxes and Turbulence

These building and urban canopy characteristics strongly affect the turbulent and mean airflow, as well as the exchange of momentum between the urban canopy and the atmosphere above (Roth, 2000; Britter and Hanna, 2003; Collier, 2006). From field measurement campaigns and

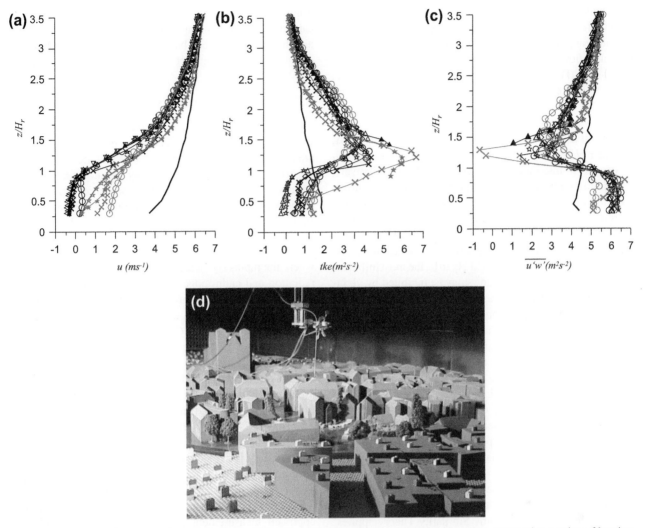

FIGURE 3.3 Profiles of: (a) mean along-wind velocity, (b) turbulent kinetic energy, and (c) turbulent shear stress measured at a variety of locations (symbols) in and above a wind tunnel model of Nantes. (d) illustrates the layout and variation in the height of the buildings in this UCL. The solid black line is the profile measured upwind of the urban canopy, that is, it represents a non-UBL. The y-axis in all parts is height (z), which is scaled by the mean building height in the model canopy (H_r). Some profiles were conducted near buildings that exceed this mean height, and this accounts for some of the variation near $z/H_r \sim 1$. *(Source: Kastner-Klein and Rotach, 2004. With kind permission from Springer Science+Business Media)*

wind tunnel experiments, the important characteristics of UCL turbulence are:

- The flow regime within the urban canopy, and extending 2−3 times the mean building height above, depends on the building dimensions and spacing. It is therefore highly variable and complex. This layer is known as the RSL (see Roth (2000) for further description) and features large variability in the mean and turbulent airflow properties due to both form drag and wake diffusion effects. This is shown in Figure 3.3, in particular for all heights (z) less than twice the height of the roughness elements (H_R), (i.e., $z/H_R \sim 2$). This means that any measurement within the RSL is unlikely to represent the averaged properties of the underlying urban neighbourhood, whereas above the RSL, 'blending' of these properties occurs.

- The RSL is characterized by an increase in the vertical momentum flux ($\overline{u'w'}$ in Figure 3.3) and turbulent kinetic energy from within the urban canopy, reaching its maximum value just above roof height (Figure 3.3), although other studies have found this maximum to occur at up to twice the roof height.

- This maximum indicates a highly turbulent shear layer, which is created near the top of the urban canopy and has markedly different turbulent properties than the overlying boundary layer (Roth, 2000) and large turbulent intensities. This is very clearly seen by the large turbulent kinetic energy and turbulent shear stress terms near $z = H_R$ (Figure 3.3). There are strong similarities between the urban turbulent shear layer and those seen to form immediately above a plant canopy (Finnigan, 2000) and near porous windbreaks (Cleugh and Hughes, 2002).

- Three flow regimes characterize the building–canyon spacing (Hussain and Lee, 1980; Grimmond and Oke, 1999a). For an idealized canopy where the building spacing is isotropic and there is little variation in mean roof height, these regimes are: (i) *skimming*, where the canyon aspect ratio (*H/W*) < 1; (ii) *isolated roughness*, when *H/W* >> 1; and (iii) *wake interference*, which occurs at intermediate values for *H/W*. Qualitatively, skimming flow means that the air layers above the urban canopy are decoupled from the flow within, while isolated roughness flow means that the flow is readjusting to an equilibrium. The greatest coupling between flow above and within an urban canopy will occur with wake interference flow and this is also likely to be the most turbulent and variable. Skimming flow may be inhibited by variations in roof height (Britter and Hanna, 2003).
- The mean wind speed profiles in Figure 3.3a illustrate some of these characteristic features in the way that they are clustered into two groups: those where the wind speed is near or less than zero (black curves in Figure 3.3a) demonstrate a skimming flow regime, and a second set where the wind speeds are greater than zero and more variable (grey curves in Figure 3.3a), representing a more wake interference flow regime at intersections and more open parts of the urban canopy.
- Theoretically, the inertial sublayer (or constant flux layer) is where fluxes are constant with height, and the height dependency of the mean horizontal wind speed can be described by the well-known log law. This is also where the Monin–Obukhov Similarity Theory applies. In urban canopies, this layer may become 'squeezed' out as a result of the complex flow characterizing the RSL and the outer boundary layer. Figure 3.3 shows that above $z/H_R \sim 3$, the spatial variation in mean wind speed, turbulent kinetic energy, and turbulent shear stress are less than 2%, 10%, and 15%, respectively, which means that the airflow is fully adjusted to the underlying surface above this height.

Despite the relatively thin inertial sublayer, Kastner-Klein and Rotach (2004) found the log law was a good fit to the measured wind speeds above the urban canopy.

UCL turbulence in most real cities is influenced by a mix of buildings (as illustrated by Figure 3.3d) and urban vegetation – especially trees. Compared to buildings, trees are porous and flexible, which means that they can absorb momentum and reduce the turbulent intensity of the airflow. Adding trees into a dense urban canopy with tall buildings (i.e., where the aspect ratio is large) will tend to reduce the turbulent shear stress and roughness length as a result. Seasonal variations in leaf area can bring about similar seasonal variations in

canopy layer turbulence. Adding tall trees into a suburban canopy, where the building heights may be lower than the trees and the spacing between buildings much larger, can have the opposite effect (Finnigan et al., 1994).

3.2.3. Urban Energy Exchanges

Much of the physical basis for urban climates is the energy, water, and carbon exchanges between the urban canopy and atmosphere (Figure 3.1), and the way that these are altered by urbanisation. The urban energy balance for a control volume containing an urban canyon or city block (approximately equal to the box in Figure 3.2b) is:

$$Q^* + Q_F = Q_H + Q_E + \Delta Q_S \qquad (3.1)$$

where the inputs (Q^* and Q_F) are the fluxes of net all-wave radiation and anthropogenic heat, and the outputs include energy used to evaporate and transpire water through evapotranspiration (Q_E) and energy used to heat the air and urban volume (Q_H and ΔQ_S). Compared to plant canopies, the urban energy balance is regulated by both biophysical and human factors.

Equation (3.1) is applicable at any spatial-scale in the urban environment, but it assumes that the lateral transport of energy due to advection is negligible. This latter assumption is particularly difficult to justify in urban energy balance measurement programmes, and there is a real role for future integrated measurement and modelling programmes to assess the magnitude and variability of this term (see Pigeon et al. (2007) for further discussion).

The last decade has seen a surge in large-scale, multi-approach measurement programmes that have provided a new level of information on energy, and to a lesser extent water and carbon, budget dynamics in urban environments (see Table 3.2 for an extensive list of urban energy balance studies). These have integrated multiple measurement and modelling approaches for a broad range of cities that span climate, geography, population, density, and urban design. This step change in data and analysis has progressed our understanding such that a generalized picture can now be presented of the urban energy budget at the canopy or local (i.e., neighbourhood-scale). [See reviews by Arnfield (2003) and Grimmond et al. (2007) for more details; and Arnfield (2003) for a similar review of the energy balance within the urban canopy volume including within-UCL energy exchanges.]

The typical changes to the energy balance in suburban land use compared to a control, non-urban site (in this case, a well-watered grass field) are captured in Figure 3.4, namely: (i) more energy is partitioned into the sensible heat fluxes; (ii) large reductions in the latent heat flux; and (iii) a lag in the timing of the sensible heat flux in urban areas such that the difference is larger in the late afternoon and not in phase

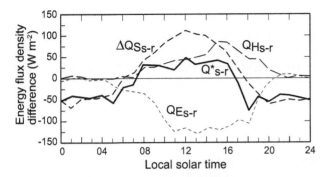

FIGURE 3.4 Effect of suburban land use on the energy balance, plotted as suburban–rural differences. *(Source: Cleugh and Oke, 1986. With kind permission from Springer Science+Business Media)*

with the radiative forcing. The last one arises because the suburban turbulent heat flux remains positive, and an energy source into the urban atmosphere, after the solar radiation input ceases and net all-wave radiation falls below zero.

3.2.3.1. Urban Radiation Balance

The downwelling shortwave radiation (diffuse plus direct beam radiation) is reduced in urban areas, varying from minor reductions (~5%) in those cities with low aerosol concentrations to much larger reductions (up to ~30%) in cities that have high levels of particulate pollution, for example, Hong Kong (Stanhill and Kalma, 1995) and Mexico City (Jauregui and Luyando, 1998); see also references in Stanhill and Cohen (2009). Air pollutants in the urban atmosphere alter the transmission of solar radiation, causing increased diffuse radiation and attenuation of both UV-B and photosynthetically active radiation (Arnfield, 2003).

With increased emissivity due to the presence of GHG (especially CO_2), aerosols, elevated temperature, and often humidity levels (see discussion below on the UBL), the urban atmosphere has the potential to create an enhanced greenhouse effect, very similar to that occurring at the global-scale, which increases the flux of downwelling longwave. As for solar radiation, these changes in longwave radiation vary from city to city.

The net result of lower downwelling shortwave, and increased longwave, radiation fluxes is little change in the total incoming all-wave radiation (Oke, 1988; Cleugh, 1995a; Arnfield, 2003). Any reduction in solar radiation receipt is very likely to be offset by the lower albedo of the urban canopy, which has been well-established through many measurement programmes to be between 0.10 and 0.15 for suburban and urban landscapes (Cleugh, 1995a; Jin et al., 2005). These effects, and the larger active area for radiation absorption, means that urban areas absorb more shortwave radiation than a non-urban landscape subjected to the same global radiation. Similarly, the reduced SVF in urban canyons impedes longwave radiation losses and so the receipt of net longwave radiation is typically increased.

The net effect of these changes in the component radiation fluxes is that differences in the net all-wave radiation flux between urban and plant canopies are, in general, surprisingly small (Cleugh, 1995a; Arnfield, 2003; Lietzke and Vogt, 2009; and examples for particular cities are included in Christen and Vogt, 2004; Rotach et al., 2005; Loridan and Grimmond, 2011). This has implications for the sensitivity of the local-scale urban climate to changes in building materials and urban designs that modify the radiation balance; for example, recent suggestions on reducing urban heating through modification of the albedo of buildings (see discussion in Section 3.3).

This local-scale behaviour is quite different from the radiation regime for an individual element (such as a person or a tree) or facet (a wall or roof or pavement) of the three-dimensional urban canopy that will be highly variable in space or time — for example, due to intense direct beam radiation reflected by, or deep shading caused by, tall buildings.

3.2.3.2. Urban Heat Storage and Anthropogenic Fluxes

Urban heat storage and anthropogenic fluxes differ greatly between plant and urban canopies. The change in heat storage, ΔQ_S, in all elements (air, biomass, soil, and built-component) of the UCL is a dominant term in the urban energy balance that, unfortunately, is very difficult to measure (Offerle et al., 2005a). It can be approximated by the sum of the heat fluxes conducted into/out of the solid/air interfaces (i.e., walls and rooftops; pavement and roads; trees, lawns and gardens) (Arnfield, 2003), or is estimated as a residual in the energy balance when all other terms are measured (Roberts et al., 2006, for a more detailed discussion of measuring ΔQ_S).

From estimates using the residual approach, ΔQ_S by day consumes between 20% and 30% of the net all-wave radiation in suburban land use and up to half of the net radiation in heavily urbanised or industrial sites such as those found in Mexico City (Oke et al., 1999) and industrial areas of Vancouver (Voogt and Grimmond, 2000). By night, the net loss of radiation from the urban canopy is typically balanced by the heat storage term. These studies also found the daily variation ΔQ_S to be out of phase with the net all-wave radiation, leading to a non-linear $\Delta Q_S/Q^*$ relationship (see Figure 3.5). Analyses using numerical models (Arnfield and Grimmond, 1998) confirmed this 'hysteresis' behaviour for $\Delta Q_S/Q^*$ in urban canyons and showed that thermal admittance and canyon geometry were important factors controlling this behaviour.

Recent studies (Masson et al., 2008; Allen et al., 2010; Sailor, 2011) provide insight into the magnitude and variability of the anthropogenic heat flux (Q_F), which results from the emission of waste heat from stationary and mobile

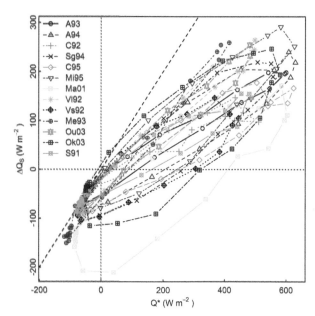

FIGURE 3.5 Variation in the relation between ΔQ_S and Q^* showing hysteresis behaviour. *(Source: Modified from Grimmond and Oke, 2002 and Loridan and Grimmond, 2011).* Sites and methods for removal of anthropogenic heat flux from the observations are described in Loridan and Grimmond (2011). See Table 3.2 for site identity.

sources, and demonstrates the importance of this additional heating term in the surface energy balance and climate of urban areas. It is difficult to measure this term directly and so it is estimated from energy use statistics or from surrogates for energy use such as traffic numbers. Other studies estimate Q_F as a residual in the urban balance (Offerle et al., 2005a; Pigeon et al., 2007). CO_2 fluxes can also be used as an indicative measure of energy consumption (Offerle et al., 2005a).

While Q_F might be of order 10 W m^{-2} on average, this can increase by one or even two (e.g., Ichinose et al., 1999) orders of magnitude in winter periods for the densest parts of urban areas (for a comprehensive summary of data see Allen et al. (2010) and Sailor (2011)). For example, Masson et al. (2008) found that, although the net all-wave radiation is close to zero in the winter months in Toulouse (France), the anthropogenic heat flux provides an ongoing input of energy, about two thirds the size of the summertime net all-wave radiation, to sustain positive sensible (i.e., exchanging heat from within the urban canopy to the atmosphere) and latent heat fluxes throughout the winter. Table 3.2b provides values as a fraction of the all-wave incoming radiation for a variety of cities and through the seasons (based on Allen et al., 2010 and Loridan and Grimmond, 2011).

Given the potentially important role of the anthropogenic heat flux in urban climates, the difficulties in quantifying its magnitude and, importantly, that it is driven by energy consumption and urban form — all variables that can be managed and modified to achieve different outcomes — indicate that further research will be important.

3.2.3.3. Turbulent Latent and Sensible Fluxes

Turbulent heat fluxes are potentially large because of the aerodynamically rough nature of the urban canopy. Their relative magnitudes (quantified by the Bowen ratio, $\beta = Q_H/Q_E$) and temporal variability are driven primarily by water availability, which, in turn, depends on the amount of vegetation in the urban canopy and the input of water through precipitation and irrigation (Figure 3.6). Measurements reveal that, while urban areas reduce this term *in general*, there are large variations across and within cities (Figure 3.7). Indeed, in well-watered urban gardens and parks, the evapotranspiration rate can be larger than in surrounding rural landscapes (Oke, 1979; Oke and McCaughey, 1983; Grimmond et al., 1993; Spronken-Smith et al., 2000).

As is typical of aerodynamically rough canopies, the day-to-day variability in β will be influenced by the aerodynamic drivers of evapotranspiration, that is, vapour pressure deficit and wind speed. The effect of this is twofold: firstly, evapotranspiration in urban areas is not as well-characterized by the equilibrium evaporation model, where energy is the only driver of variability, and, secondly, there can be large temporal variability in β (Cleugh and Oke, 1986; Cleugh, 1990).

Despite large variations in the sensible and latent heat fluxes from day to day, across different cities, and between different land-use zones within cities (Figures 3.6, 3.7, and 3.8), urban landscapes can be characterized by $\beta > 1$ and enhanced stored and convective heat fluxes (Table 3.2). These enhanced sensible heat fluxes, along with the anthropogenic heat flux, contribute to enhanced heating in the urban canopy and boundary layer, and to the UHI.

The role of evapotranspiration is particularly important in determining the energy, water, and carbon balance in, and hence managing the climate and hydrology of, urban areas. The ability to manipulate this term has been exploited as a potentially powerful way to passively mitigate urban heating and hence energy consumption (Mitchell et al., 2008).

Roth's (2007) survey of urban climate studies in tropical and subtropical cities confirms these basic features of the urban energy balance. This review also reveals a strong influence of urban vegetation and water availability on energy partitioning, regardless of climate zone, and suggests that vegetation may therefore be an appropriate design option to reduce heat storage by day and hence mitigate the UHI.

3.2.4. Urban Water Balance

The urban water balance, expressed in unit mass (mm or kg) per unit time is:

$$P + I + F_W = E + R + \Delta S_W \qquad (3.2)$$

where the input terms are precipitation P (including dew, fog, hail, rain, and snow), the piped water supply (I), which

TABLE 3.2 Energy Partitioning for Urban and Rural Land Use for: (a) Daytime Periods, and (b) Midday Periods (Cleugh, 1995a; Loridan and Grimmond, 2011)

(a) Daytime periods

Location	Land Use Type	Year or Season[1]	β	ΔQ_S/Q*	Q_E/Q*	Q_H/Q*	Author
Vancouver, Canada	Industrial	1992 [VI92] summer	4.4	0.48	0.10	0.42	Voogt and Grimmond (2000)
Vancouver, Canada	Residential	1983 summer	1.28[A]	0.22[B]	0.34[A]	0.44	Cleugh and Oke (1986)
Vancouver, Canada	Residential	1992 late summer [Vs92]	2.9	0.17	0.22	0.62	Grimmond and Oke (1999c)
Tucson, USA	Residential	1990 summer	2.1	0.23	0.25	0.52	Grimmond and Oke (1995)
Mexico City, Mexico	Dense urban, mixed	1985 late dry season	1.12[A]	0.36[B]	0.30[A]	0.34	Oke et al. (1992)
Mexico City, Mexico	Old central city, dense urban	1993 [Me93] mid dry season	9.9	0.58	0.04	0.38	Oke et al. (1999)
Sacramento, USA	Residential	1991 [S91] summer	1.3	0.26	0.33	0.41	Grimmond et al. (1993)
Los Angeles, USA	Residential	1993 summer	1.2	0.30	0.31	0.39	Grimmond and Oke (1995)
		1994 [A94] summer	1.4	0.31	0.26	0.43	Grimmond et al. (1996)
		1994 [Sg94] summer	2.2	0.29	0.22	0.49	
Miami, USA	Residential	1995 [Mi95] summer	1.6	0.30	0.27	0.42	Newton et al. (2007)
Chicago, USA	Residential	1995 [C95] summer	1.2	0.17	0.37	0.46	Grimmond and Oke (1999c)
Basel, Switzerland	Residential	2002	2.55	0.34	0.22	0.54	Christen and Vogt (2004)
	Residential and commercial	2002	2.47	0.37	0.24	0.55	
Łódź, Poland	CBD	2002	1.83	0.32[D] (0.29)	0.23	0.44	Offerle et al. (2006a)
	Industrial		1.61	0.41[D] (0.31)	0.21	0.36	
	Residential		0.80	0.32[D] (0.23)	0.37	0.30	

(Continued)

TABLE 3.2 Energy Partitioning for Urban and Rural Land Use for: (a) Daytime Periods, and (b) Midday Periods (Cleugh, 1995a; Loridan and Grimmond, 2011)—cont'd

(a) Daytime periods

Location	Land Use Type	Year or Season[1]	β	$\Delta Q_S/Q^*$	Q_E/Q^*	Q_H/Q^*	Author
Montreal, Canada	Dense residential a) With snow	2005	7.82		0.08	0.32	Lemonsu et al. (2008)
	b) No snow		10.39		0.04	0.44	
Rural							
Vancouver, Canada	Short grass, rural	1983	0.46	0.04[C]	0.66	0.30	Cleugh and Oke (1983)
Łódź, Poland	Rural	2002	0.41	0.24[D] (0.16)	0.54	0.22	Offerle et al. (2006a)
Sacramento, USA	Irrigated rural	1991	0.18	0.07[C]	0.63	0.11	Grimmond et al. (1993)
	Dry rural		22.23	0.12[C]	0.04	0.80	

(b) Mean midday ratios (±3 h of solar noon) (from Loridan and Grimmond, 2011). Long-term sites are separated in two-month subsets (JF: January–February; MA: March–April; MJ: May–June; JA: July–August; SO: September–October; ND: November–December). Note that these ratios use incoming all-wave radiation ($Q\downarrow$), not net all-wave radiation as in a).

Location	Land Use Type	Year or Season[1]	$Q\uparrow/Q\downarrow$	$\Delta Q_S/Q\downarrow$	$Q_E/Q\downarrow$	$Q_H/Q\downarrow$	$Q_F/Q\downarrow$	Original Author
Marseille, France	CBD	2001 [Ma01] summer	0.520	0.133	0.066	0.281	0.027	Grimmond et al. (2004a)
Oklahoma City, USA	Residential	2003 [Ok03] summer	0.609	0.099	0.139	0.154	0	Grimmond et al. (2004b)
Chicago, USA	Residential	1992 [C92] summer	0.526	0.152	0.176	0.146	0.055	Grimmond and Oke (1995)
Ouagadougou, Burkina Faso	Residential	2003 [0u03] winter	0.645	0.171	0.039	0.145	0	Offerle et al. (2005b)
Melbourne, Australia	Residential	2006: All seasons						Coutts et al. (2007a, b)
		Mb06_JA	0.661	0.174	0.075	0.089	0.043	
		Mb06_SO	0.585	0.161	0.111	0.143	0.031	
		Mb06_ND	0.564	0.151	0.12	0.164	0.03	
		Mb06_JF	0.535	0.189	0.082	0.193	0.03	
		Mb06_MA	0.6	0.195	0.064	0.141	0.037	
		Mb06_MJ	0.721	0.152	0.071	0.056	0.047	

(b) Mean midday ratios (±3 h of solar noon) (from Loridan and Grimmond, 2011). Long-term sites are separated in two-month subsets (JF: January–February; MA: March–April; MJ: May–June; JA: July–August; SO: September–October; ND: November–December). Note that these ratios use incoming all-wave radiation ($Q\downarrow$), not net all-wave radiation as in a).

Location	Land Use Type	Year or Season[1]	$Q\uparrow/Q\downarrow$	$\Delta Q_S/Q\downarrow$	$Q_E/Q\downarrow$	$Q_H/Q\downarrow$	$Q_F/Q\downarrow$	Original Author
Łódź, Poland	CBD	2002: All seasons						
		Łó02_JF	0.791	0.063	0.066	0.08	0.04	Offerle et al. (2006b)
		Łó02_MA	0.63	0.151	0.059	0.161	0.024	
		Łó02_MJ	0.551	0.157	0.118	0.174	0.012	
		Łó02_JA	0.577	0.14	0.108	0.175	0.012	
		Łó02_SO	0.696	0.087	0.084	0.133	0.018	
		Łó02_ND	0.871	−0.023	0.049	0.104	0.042	
Helsinki, Finland	Institutional	2009: All seasons						
		He09_JF	0.907	−0.038	0.032	0.099	0.067	Vesala et al. (2008)
		He09_MA	0.654	0.111	0.053	0.182	0.058	
		He09_MJ	0.555	0.128	0.129	0.188	0.024	
		He09_JA	0.577	0.093	0.184	0.147	0.023	
		He09_SO	0.69	0.073	0.122	0.114	0.038	
		He09_ND	0.945	−0.009	0.034	0.03	0.103	
Tokyo, Japan	Residential and commercial	2001: All seasons						
		TK01_JF	0.578	0.278	0.041	0.102	0.118	Moriwaki and Kanda (2004)
		TK01_MA	0.523	0.267	0.044	0.166	0.075	
		TK01_MJ	0.573	0.197	0.094	0.136	0.072	
		TK01_JA	0.557	0.187	0.114	0.141	0.078	
		TK01_AO	0.605	0.239	0.063	0.093	0.082	
		TK01_ND	0.612	0.254	0.046	0.088	0.097	

Key to superscripts: 1: [SSYY] refers to datasets in Figures 3.6 and 3.7; A: Latent heat fluxes and Bowen ratio values are estimated as a residual; B: ΔQ_S is modelled, otherwise it is computed as a residual in the surface energy balance; C: At rural sites this is Q_G/Q^* and is measured; D: Lodz data are based on detailed surface temperature measurements to directly determine ΔQ_S rather than as a residual to the energy balance; figures in brackets show results after daytime turbulent fluxes are adjusted so that storage is zero over the period (see Offerle et al., 2006a).

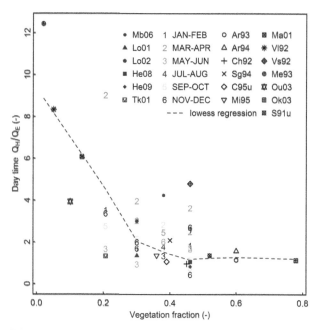

FIGURE 3.6 Midday (± 3 h of solar noon) ensemble mean Bowen ratio ($\beta = Q_H/Q_E$), plotted against fraction of greenspace. See Table 3.2 for identification of each city. *(Source: Modified from Grimmond and Oke, 2002 and Loridan and Grimmond, 2011).* Lowess regression (dashed line) performed using summer months and short-term datasets only (July–August for Ł601–Ł602 and He08–He09, and January–February for Mb06) (Southern Hemisphere).

is consumed both indoors and externally, and water released from combustion processes (F_W). The output terms are evapotranspiration (E), runoff (R), which includes both wastewater and stormwater components, and changes in

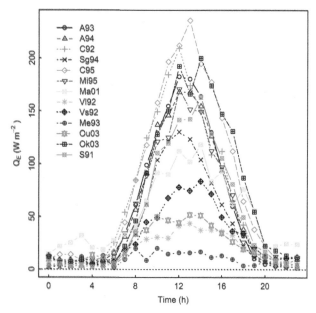

FIGURE 3.7 Diurnal variation of the ensemble mean latent heat flux (Q_E). See Table 3.2 for identification of each city. *(Source: Modified from Grimmond and Oke, 1999b.)*

stored water (ΔS_W) in the urban canopy (air, biomass, soil, and built components).

Writing the water balance in this way (Equation (3.2)) assumes a control volume that extends vertically from the soil up into the urban canopy airspace. Following Mitchell et al. (2001) and Cleugh et al. (2005), Equation (3.2) applies equally to a single household, neighbourhood (equivalent to the neighbourhood, or local, scale defined above), or a catchment that may be defined by topography and/or the pipe supply network.

All terms in Equation (3.2) can be affected by urbanisation, but arguably the dominant changes are the enhanced input of water through I, and increases in R due to the increased fraction of impervious cover in urban areas and increased volumes of water used internally and externally (Grimmond and Oke, 1986). While urban evapotranspiration is reduced in general, corresponding to the reduced amount of vegetation, a more secure and enhanced water supply (through I) can sustain large rates of evapotranspiration as suggested by the energy balance results in Section 3.2.3. This means that E is often the largest output term in the urban water balance, exceeding the runoff (e.g., Grimmond and Oke, 1991) and, given that cities typically have a more secure water supply, E will be higher in suburban compared to rural landscapes (in the absence of water restrictions).

This link between urban climates and urban hydrology results from evapotranspiration being common to the energy and water balance, as a mass flux (E) in the latter and as the energy required to change water from liquid to vapour (Q_E) in the former (i.e., $Q_E = L_v E$ where L_v is the latent heat of vaporization). Thus the urban microclimate can be modulated by manipulating the water balance — for example, sustaining evapotranspiration through importing water (I) and/or harnessing and reusing runoff (R). So, managing the urban water balance is a key potential strategy to mitigate excessive urban heating. Furthermore, this link provides energy and mass conservation constraints on estimates of urban evapotranspiration and means that the recent progress in quantifying urban evapotranspiration will improve hydrological models and assist in their evaluation.

A continuous water balance model for urban hydroclimates (Grimmond and Oke, 1991, Järvi et al., 2011), with a modified Penman–Monteith model of evapotranspiration and the Rutter–Shuttleworth model for interception, when combined with a fully calibrated urban water cycle model that simulates both internal and external components of the urban water balance (this model code is called *Aquacycle*, and is described in Mitchell et al., 2001), allowed Mitchell et al. (2008) to test the proposition that water can be used to modulate the urban microclimate through urban evapotranspiration.

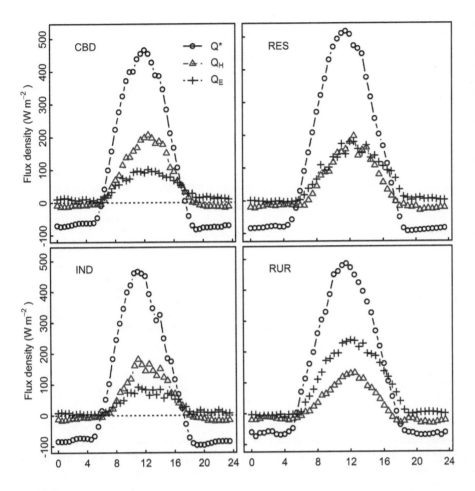

FIGURE 3.8 Ensemble averages of the surface energy balance for three different land-use zones in a European city (RES: residential, IND: industrial, LTM: light manufacturing, RUR: rural), from measurements during the dry period to show intra-urban differences. The x-axis refers to solar time, in hours. *(Source: Modified from Offerle et al., 2006a. © American Meteorological Society. Reprinted with permission).*

3.2.5. Urban Carbon Balance

The role of urbanisation in perturbing the carbon budget, both regionally and globally, has received considerable attention in the last decade as a result of an increased focus on tracking the drivers of global climate change and parallel improvements in the ability to measure directly the net CO_2 exchanges in urban landscapes (e.g., Pataki et al., 2006). Our knowledge of the urban carbon balance is not as advanced as for energy and water. As for the energy and water balance, a simple representation of the CO_2 balance in an urban area is:

$$F_{CP} + F_{CF} = F_{CR} + F_{CM} + \Delta S_C \qquad (3.3)$$

where the inputs (to the urban canopy or surface) are F_{CP}, the CO_2 assimilated through photosynthesis by the urban vegetation, and F_{CF}, the imported CO_2, which is primarily fossil fuel. The outputs are the CO_2 emissions resulting from heterotrophic and autotrophic respiration (F_{CR}) and from human activities (F_{CM}), which includes emissions from transport, space heating/cooling, industries, and manufacturing. The change in storage (ΔS_C) represents either a net gain or loss of CO_2 in the urban system. The control volume for this balance, although relatively straightforward in theory, in practice requires careful definition of the boundaries.

Again, this simple balance has both anthropogenic and biogenic components (Table 3.3). The different processes, drivers, and source/sink distributions (Table 3.3) all affect the magnitude and variability of the terms in the carbon balance and the urban contribution. In most urban environments, the anthropogenic CO_2 sources are much larger than the biogenic ($F_{CM} >> F_{CR}$). When plants are actively growing (and thus sequestering CO_2) there may be a balance with respiration (i.e., $F_{CP} \cong F_{CR}$) or even a net uptake. A disturbance to the vegetated areas, for example, through harvesting or fire, will contribute to the net emission of CO_2 because photosynthetic uptake is reduced.

Changes in the urban climate, especially the urban heat island, the quality of the solar radiation, irrigation, enhanced levels of CO_2, and atmospheric deposition of nitrogen, are all likely to modulate the biological processes of photosynthesis and respiration in urban vegetation.

The UHI, in particular, has long been suspected of changing the phenology of urban vegetation (i.e., time of leaf growth and leaf-fall) and therefore the net CO_2

TABLE 3.3 Key Features of the Biogenic and Anthropogenic Components of the Urban CO$_2$ Balance

	Component of Urban CO$_2$ Balance:	
	Biogenic	Anthropogenic

Key processes	• Photosynthesis (F_{CP})
	 • Respiration (F_{CR})
	   − Autotrophic
	   − Heterotrophic
	
	 • Disturbance (F_{CD})
		• Fossil fuel consumption and emission of CO$_2$ due to urban functions (F_{CM}):
		   − Transport
		   − Industrial processing, manufacturing
		   − Residential heating and cooling
		 • Disturbance (F_{CD}) − a net sink or source depending on stage of urban development

Drivers (not independent)	• Phenology
	
	 • Radiation (PAR)
	
	 • Water, nutrient availability
	
	 • Temperature
	 • Atmospheric CO$_2$
		• Temperature
		   − Diurnal, seasonal, annual cycles
		 • Human activities
		   − Day/night; weekday/weekend; holidays
		 • Economic basis
		   − Manufacturing industries
		 • Wealth (income) and consumption
		 • Urban transport system
		 • Urban design and age
		 • Disturbances (land-use change, fire)

Source/sink distribution	• Fractional or active area of urban greenspace
	 • Heterogeneous ('patchy') spatial distribution
	   − vertical and horizontal
	   − parks, lawns, trees
		• 'Footprint' may extend beyond city's geographical boundary, difficult to define
		 • Mobile and stationary sources
		 • Within urban canopy and boundary layer

sequestration in an urban ecosystem. Using data sources as diverse as satellite remote sensing (Zhang et al., 2004) and herbarium records (Neil et al., 2010), studies have found significant differences for individual cities and cities in individual regions. Gazal et al. (2008) explored this link for seven cities across Asia, Europe, Africa, and North America, using Landsat satellite imagery to characterize the urban radiative surface temperatures and careful protocols to determine budburst dates in native deciduous trees. They found that there were significant urban−rural differences in the timing of budburst, but there was no conclusive evidence of a distinct urban response. Each city responded differently and, in Alaska, budburst was delayed rather than advanced. This suggests that urban warming can cause changes in phenology, but there is no generalized response across cities and climate regimes. There is evidence that the enhanced CO$_2$ concentrations found in urban areas do result in increased above-ground biomass and plant height (Ziska et al., 2004).

Urbanising a vegetated landscape will typically increase the anthropogenic inputs and outputs (i.e., F_{CF} and F_{CM}) and reduce the photosynthesis and respiration (F_{CP} and F_{CR}). The decline in net uptake for the conterminous USA was estimated to be 1.6% (current day−pre-urban) by Imhoff et al. (2004).

Studies that explore the many issues associated with CO$_2$ exchanges in urban environments fall roughly into four groups: (i) estimation of CO$_2$ emissions based on

inventories of energy consumption; (ii) direct measurements of the net exchange of CO$_2$ between the urban canopy and the atmosphere; (iii) quantifying space/time patterns of CO$_2$ concentrations across urban areas; and (iv) assessing CO$_2$ emissions as part of air quality assessments for a range of air pollutants, with a focus on chemical composition. The direct measurement of CO$_2$ (i.e., ii) is most relevant here. To date, there are about 30 observational studies of net CO$_2$ exchanges in urban areas (Velasco and Roth, 2010), largely from mid-latitude, North American and European cities, with five from tropical and subtropical cities.

These studies measure the net exchange that results from all the inputs and outputs (emissions) in Equation (3.3), thereby providing a 'top-down' view of urban effects on the carbon balance for an urban neighbourhood. The typical daily net CO$_2$ exchange for most of these urban studies is of the size order, but with the opposite sign, of a tropical rainforest (which Velasco and Roth (2010) showing a peak uptake of around 3 t CO$_2$ km^{-2} h^{-1})[3].

The diurnal and seasonal variation in net CO$_2$ fluxes (emissions) is dominated by the cycle of human activities and emissions from vehicles and space heating (Figure 3.9 and Table 3.3). The net emissions are larger during

---

3. This is from figure 5g in Velasco and Roth where the units are given as tons per km^2 per hour. Whether this is a US ton (907.18 kg) or a metric tonne (1000 kg) is not explained.

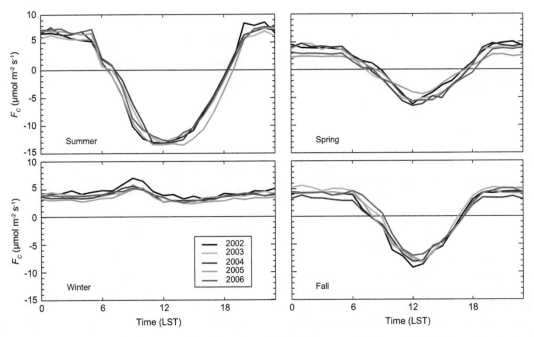

FIGURE 3.9 Diurnal and seasonal variation in measured net CO_2 exchange in the suburban area of Baltimore, USA. The data collection, processing, and gap-filling methods are described in Crawford et al. (2011).

week-days and during cold periods (such as the winter months when additional heating is required). For example, Moriwaki and Kanda (2004) found that domestic heating contributed 62% to the total CO_2 in a residential area of Tokyo. Therefore, net CO_2 losses from an urban canopy are lower at night than during the day, which is in stark contrast to plant canopies.

Biogenic processes become more important in the leaf-on period when the area is more highly vegetated (Figure 3.9). The relative importance of anthropogenic influences to the magnitude of the net CO_2 fluxes varies with fractional vegetation cover. In heavily vegetated suburbs, such as in Baltimore (Crawford et al., 2011) and Salt Lake City (Pataki et al., 2009), the urban canopy can be a sink not a source for CO_2 during some periods. However, long-term measurements in two residential suburbs in Melbourne, Australia (Coutts et al., 2007a) and at the Baltimore site (Crawford et al., 2011) found quite highly vegetated urban areas were a net source of CO_2 at annual timescales. Therefore, not unexpectedly, less vegetated residential areas (Tokyo: Moriwaki and Kanda, 2004), residential–commercial areas (Mexico City: Velasco et al., 2009), and city centres (Marseilles: Grimmond et al., 2004a; Firenze: Matese et al., 2009; Basel: Vogt et al., 2006; London: Helfter et al., 2011) are all net sources of CO_2.

The urban greenspace does sequester some of the CO_2 emissions from vehicles and other urban sources but current measurements (e.g., Coutts et al., 2007a; Crawford et al., 2011) suggest this is insufficient to offset these emissions. The annual emissions of CO_2 (cited by Velasco and Roth, 2010) for Tokyo, Mexico City, Melbourne, and Copenhagen, varied between 8 and 15 kg CO_2 m^{-2} y^{-1}, (i.e., 80 and 150 t CO_2 ha^{-1} y^{-1}). These emissions are about 10 times the CO_2 sequestered by a mature and productive forest. It is important to note that the scale for these various studies does vary — only the figures for Copenhagen were for the entire city, all others were for individual suburbs within the city. This result confirms the conclusion of Pataki et al. (2009), who suggest that the largest effects of urban tree planting on mitigating climate change will not lie in carbon sequestration, but rather in the way that urban greenspace affects the urban energy balance.

Across the studies, no correlation was found between population density and daily CO_2 fluxes (Velasco and Roth, 2010). While the daily CO_2 fluxes (emissions) were larger in the city core areas compared to the suburbs, this was not the case in terms of per capita emissions. This is independent evidence corroborating the view that it is not population *per se* that drives emissions from urban areas, but consumption.

3.2.6. Summary: Coupling Energy, Water, and Carbon in Urban Areas

Understanding, quantifying, and then managing the direct and indirect effects of cities on the climate is informed by considering the water, energy, and carbon budgets and their coupling:

- Evapotranspiration, common to water and energy budgets, is a mechanism by which urban microclimate changes can be passively controlled.
- Transpiration is tightly coupled to the biological process of photosynthesis (F_{CP}) in the carbon balance. CO_2 assimilation by stomata occurs along the vapour pressure gradient that exists across the stomatal aperture. This link between the water, energy, and carbon budgets (via E, Q_E, and F_{CP}) provides an important constraint on estimates of the stomatal conductance parameters needed for urban land-surface models (ULSM).
- There is a very strong influence of temperature (i.e., the urban energy balance) on the urban carbon balance (i.e., respiration, emissions from space heating).
- There is an equally strong link between CO_2 emissions, due to human activities, and the anthropogenic heat flux. Calculations for each term are typically derived from the same database. Alternatively, if directly-measured CO_2 fluxes such as those presented in Velasco and Roth (2010) can be separated into the biogenic and anthropogenic components, then the latter can be used as an independent constraint on estimates of Q_F.

These links demonstrate clearly the potential to mutually constrain the budgets, and the individual fluxes, if the carbon, water, and energy balances are simulated as a coupled system.

3.2.7. Direct Urban Climate Effects

These changes (Sections 3.2.2 to 3.2.5) lead to a chain of processes whose emergent property is the urban climate. Urban effects on the climate, especially air temperature, have been observed for settlements with as few as 4500 inhabitants. Typical changes in a series of climate elements, for a hypothetical mid-latitude city of a million inhabitants, are shown in Table 3.4 (from Oke, 1994). In the following, we briefly explain these main effects, noting that some of this is a summary of the earlier energy balance discussion. It also should be noted that these effects are manifested at each of the spatial-scales explained earlier, that is, within the UCL, at the urban neighbourhood-scale (at a height in the surface layer, similar to the measurement height described in Section 3.2.1), and in the UBL.

3.2.7.1. Radiation

The reduction in the quantity and quality of solar radiation received varies widely with geographical location and the level of reactive gases and particulates in the urban atmosphere. These reductions are offset by an enhanced

TABLE 3.4 Urban Climate of a Hypothetical Mid-Latitude City

Variable	Change	Magnitude
UV radiation	Much less	25%–90%
Solar radiation	Less	1%–25%
Longwave radiation	More	5%–40%
Visibility	Reduced	
Evapotranspiration	Less	~50%
Heat fluxes	More	~50%
Turbulence	Greater	10%–50%
Wind speed	Decreased	57%–30% at 10 m
Wind direction	Altered	1–10°
Temperature	Warmer	1–3°C annual mean Up to 12°C on individual nights
Humidity	Day − less; night − more	
Cloud	More urban haze, more cloud in lee of city	
Precipitation	Less snow (converted to rain) Probably more precipitation in lee of city	

(Source: Oke, 1994.)

downward flux of diffuse shortwave and longwave radiation, from a warm and often polluted urban sky and relatively warm buildings, as well as reductions in the albedo of the urban surface. Observations show that urban net all-wave radiation (for urban surfaces, at the local or neighbourhood-scale) is therefore often very similar to non-urban landscapes. Shading and specular effects dominate the shortwave radiation climate *within* urban canyons, causing large space/time variability in the receipt of solar radiation. Table 3.2b provides examples of differences in partitioning of the incoming and outgoing all-wave radiation for a few cities and the variations by season.

3.2.7.2. Surface Energy Balance

Urban landscapes typically partition more net radiation into the stored and convective heat fluxes, which warm the urban fabric and air, and less into latent heat. Anthropogenic heat sources enhance this additional heating and are a contributing factor to the observed UCL heat island. While urban surfaces generally partition a greater fraction of available energy into sensible heat, observations show that the evaporation fluxes can nonetheless be large − in direct correlation with the proportion of urban greenspace (Figure 3.6) and degree of garden irrigation. This

(a)

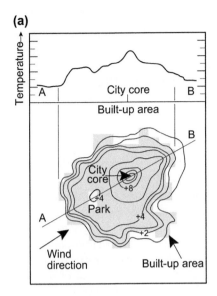

(b)

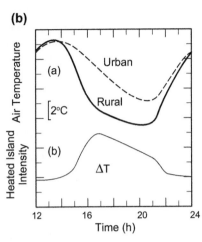

FIGURE 3.10 The UHI: (a) shown as an idealized plan view and transect of near surface air temperature across a city *(Source: Reprinted from Cleugh, 1995a; with permission from Elsevier.)*; and (b) shown as the difference in rural–urban cooling of air temperature, which contributes to the strength of the heat island, defined as $\Delta T_{U\text{-}R}$ *(Source: Oke, 1987.)*

demonstrates the potential for vegetation to passively modify microclimates and to sequester CO_2.

3.2.7.3. Airflow

The mean and turbulent airflow within urban canyons is highly complex. Winds flowing perpendicular to the long axis of an urban canyon can generate a recirculating vortex in the canyon airspace, especially if the height-to-width ratio of the canyon is small, which effectively decouples the air within the canyon from that above. As the mean flow moves from normal to more oblique approach angles, spiralling vortices will develop until, with the flow parallel to the urban canyon axis, along-canyon jetting occurs. Turbulent gusts, with length-scales equivalent to, and greater than, building height also intermittently penetrate and ventilate the floor of the urban canyons.

In the UBL there are two airflow regimes. Firstly, under light wind conditions a localized thermal circulation develops over the warm city core (Hidalgo et al., 2008), which can strengthen the winds as air accelerates into the city centre where it converges, rises, and descends over the surrounding rural areas. Secondly, stronger regional winds (above ~ 3 m s^{-1}) are usually slowed by the extra friction created by the additional roughness of urban areas. This extra friction and heating generated by the UCL can modify the occurrence and amount of rainfall, thunder and lightning, tornadoes, and frontal movement (see below).

3.2.7.4. Urban Heat Island

This most widely studied and recognized urban climate phenomenon, the UHI, refers to the observation that urban areas are generally warmer than the surrounding

region (Figure 3.10a). Characteristically, this is a nocturnal phenomenon, although daytime heat islands are observed. The UHI strength (defined as $\Delta T_{U\text{-}R}$, where U and R refer to urban and rural surface air temperatures, respectively) can be as large as $\sim 10°C$ on individual nights and is typically at its peak in the city core, declining with distance towards the urban/rural boundary.

On a daily timescale, the UHI results from a divergence in the relative cooling rates between urban and surrounding non-urban air temperatures (Figure 3.10b) and can develop in any settlement. It is best expressed under clear skies and calm winds, such that urban areas are characterized by higher daily minimum and higher mean temperatures.

Two key results to emerge from the enormous number of studies over the last two decades (see reviews by Arnfield, 2003; Souch and Grimmond, 2006; Yow, 2007) are that, firstly, great care must be taken in the method used to observe and analyse the UHI; and, secondly, there are, in fact, many UHIs[4]. Subterranean urban heating effects have been observed (e.g., Allen et al., 2003) along with UCL heat islands, which are most commonly described (e.g., Figure 3.10) and can be distinguished from the UBL heat island (Oke, 1976). A UHI is often observed in the radiative surface temperature (Jin et al., 2005). That the UCL air temperature and radiative surface temperature heat islands differ in their strength and timing is important in terms of interpreting reported results of UHI processes and characteristics.

---

4. Voogt, J. *http://www.epa.gov/heatisld/resources/pdf/EPA_How_to_measure_a_UHI.pdf*. (Accessed 25 August, 2011.)

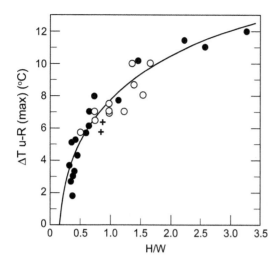

FIGURE 3.11 UHI strength ($\Delta T_{U\text{-}R}$) is influenced strongly by the ratio of building height to width (H/W). Symbols refer to observations from 31 cities in North America (solid circles), Europe (open circles) and Australasia (crosses). *(Source: figure 5 in Mills, 2004 at http://www.urban-climate.com/UHI_Canopy.pdf/, accessed 15/10/2010.)*

The UCL heat island strength varies with population, building density (i.e., the aspect ratio of the urban canyon), the fraction of the urban canopy that is built, the season (stronger in the dry season in subtropical and tropical regions), and latitude (weaker and less pronounced at lower latitudes), with the SVF being a primary factor (Figure 3.11) that acts as a surrogate measure of many of these characteristics.

The main factors leading to the UHI observed in the urban canopy and boundary layer are summarized in Table 3.5.

3.2.7.5. Urban Effects on the Boundary Layer and Rainfall

In most large cities, the atmosphere above the roof level is influenced by all of the changes described previously, such that an UBL extends above the city's roofline to occupy the lower kilometre or so of the atmosphere. This UBL is deeper (e.g., Collier, 2006), warmer, sometimes drier, and more turbulent than that upwind of the city, with the strength of these heat and moisture effects varying with wind speed and season, as well as the size and location of the city. This often neutrally-stratified (Lemonsu and Masson, 2002) UBL is advected downwind and, in so doing, transports this urban influence to the areas downwind of the city.

Whether cities can modify rainfall is a question that has been actively researched for over four decades. It is

---

TABLE 3.5 Factors that Contribute to the Urban Canopy and Boundary Layer Heat Island

Energy Balance Term	Urban Feature	Effect on Urban Canopy Layer
Increased net shortwave radiation absorption in urban canopy	Canyon geometry increases surface area and multiple reflection	Greater absorption of shortwave radiation in urban canopy
Increased downwelling longwave radiation	Polluted and warmer boundary layer	Greater absorption and re-emission
Reduced emitted longwave radiation	Canyon geometry	Reduced SVF
Anthropogenic heat flux (Q_F)	Heat emissions from domestic, industrial, and transport sources	Additional heat source
Increased heat storage (ΔQ_S)	Building materials and urban geometry	Increased thermal admittance
Reduced convective heat loss from canyon (Q_H)	Canyon geometry	Increased sheltering for flow normal to long canyon axis
Reduced latent heat losses (Q_E)	Building materials	Less vegetation and more impervious surfaces reduce water availability
Energy Balance Term	**Urban Feature**	**Effect on Urban Boundary Layer**
Net shortwave radiation	Increased aerosols	Increased absorption and scattering
Anthropogenic heat flux (Q_F)	Heat and moisture releases from chimney stacks	Elevated heat and moisture source
Increased input of sensible heat (Q_H)	Canopy layer heat island	Heat source
Increased sensible heat and momentum fluxes	Aerodynamically rough and dry urban canopy	Increased turbulence and entrainment of warm, dry air above UBL

(Source: Adapted from Oke, 1982; Cleugh, 1995a and Mills, 2004.)

well-recognized that urban areas have the potential to affect rainfall because of changes in the atmospheric dynamics in the UBL, as well as changes in the microphysical mechanisms associated with cloud and rainfall development (Lowry 1998; Shepherd, 2005; Souch and Grimmond, 2006; Hidalgo et al., 2008) described below.

3.2.7.6. Atmospheric Thermodynamics

The thermodynamics of the urban atmosphere can be modified by enhanced convective and mechanical turbulence due to the rough urban surface and the UCL heat island that decreases the stability of the UBL and amplifies the uplift created by convergent airflow over the city under light synoptic wind regimes. Studies combining modelling and measurements to attribute the dominant factors suggest that it is the UHI that has the greatest influence in initiating moist deep convection (Baik et al., 2001; Rozoff et al., 2003). Under calm synoptic conditions, these effects can initiate convective thunderstorms (Bornstein and Lin, 2000). The roughness of the urban area can perturb storm tracks and synoptic fronts as they move across urban areas (Cleugh, 1995a; Bornstein and Lin, 2000) and urban industrial processes can result in the input of additional moisture to the UBL.

3.2.7.7. Microphysical Processes

Emissions of aerosols from human activities can grow into hygroscopic (due to high levels of black carbon: Ramanathan and Carmichael, 2008) cloud condensation nuclei and modify microphysical processes associated with the growth of cloud and rain droplets. These processes are complex; indeed increased presence of cloud condensation nuclei can actually suppress rainfall but modify cloud amounts.

Experimental studies such as METROMEX (Changnon, 1981) and ATLANTA (Bornstein and Lin, 2000) were established to test the hypothesis that urbanisation affects both the weather systems that deliver rainfall and the quantities of rainfall, within and downwind of urban areas. These large experimental campaigns, and many other studies, have documented cloud and rainfall anomalies that lend support to this hypothesis. For example, METROMEX found that urbanisation increased precipitation from 5% to 25% over background values during the summer months, within and 50–75 km downwind of an urban area (Shepherd, 2005).

New measurement technologies, such as satellite-based observations of rainfall (Shepherd et al., 2002) and lightning sensor networks (Orville et al., 2001), have improved the capability to quantify the effect of urbanisation on precipitation. Hidalgo et al. (2008) conclude that these data are consistent in their evidence of increased frequency of low cloud amounts, cloud-to-ground lightning, and rainfall amounts based on lightning sensor networks and satellite observations of rainfall.

Despite this, attributing these observed anomalies to urbanisation remains a challenge, in part due to the methodological issues identified by Lowry (1998). This will require numerical models that can adequately simulate both the urban land cover and processes together with cloud and precipitation processes. To date, such modelling efforts have been limited by a failure to capture all of these processes equally well; hence, further progress will require improvement in the models (Shepherd et al., 2010).

3.3. CITIES AND GLOBAL CLIMATE CHANGE

Many of the processes that are instrumental in determining urban climate are the same factors leading to global anthropogenic climate change, namely: regional-scale land-use changes; increased energy use; and increased emissions of climatically-relevant atmospheric constituents. Cities are therefore both a case study for understanding, and an agent in mitigating, anthropogenic climate change. Given the knowledge base described in Sections 3.1 and 3.2, this section explores how the urban environment can be managed to mitigate and adapt to global climate change by addressing two questions:

1. Can global climate change be *mitigated* by manipulating those processes that contribute to urban climates, such as climate-sensitive urban design to reduce energy consumption?
2. What *adaptation* measures are needed to manage the impact of global climate change in cities, with a particular focus on:
   - Thermal stress and air quality, which both have serious effects on the health and wellbeing of urban dwellers and are likely to be exacerbated under climate change.
   - Feedbacks that will amplify or dampen the drivers of global climate change, such as increased energy consumption in warm climates that may be offset by reduced energy required for space heating in cold climates; and the implications for water availability and uptake of CO_2 by urban vegetation of reduced rainfall in subtropical latitudes under global climate change scenarios.
   - The rising importance of indoor air quality, as buildings are designed to become more energy efficient and, as a consequence, more airtight.
   - Vulnerability of the urban system, for example urban infrastructure and transport systems, electricity demand, morbidity and mortality due to extreme temperatures.

3.3.1. Using Urban Design to Mitigate Global Climate Change

As noted above, although they occupy less than 2% of the global land area, cities contribute to global climate change because they are large consumers of energy and significant emitters of GHG and aerosols, which perturb the global radiation balance. The Intergovernmental Panel on Climate Change (IPCC) Fourth Assessment claimed that redesign of residential and commercial buildings has the potential to cut 29% of projected GHG emissions by 2020 (IPCC, 2007d). As already described, cities also contribute to local and regional climate changes by changing the land cover and therefore the budgets of energy, water, and carbon.

Design options can be used to manage the negative impacts of inadvertent urban climate change; this is referred to as *climate sensitive urban design*. A generalized classification of these options, using the urban energy balance as a framework, is provided in Table 3.6. Any mechanism or design option that reduces energy demand for space heating and cooling will also reduce CO_2 emissions from fossil fuel consumption. Many of these design options are already well-understood and in use by architects, builders, and planning authorities, for example, many government bodies have instituted energy ratings for new and existing residences where the calculations take into consideration design features at the building-scale such as double-glazed windows, curtains, aspect, window area, insulation (roof, walls), building materials, and so on.

Our focus here is more at the urban neighbourhood-scale and on two approaches that have the potential to significantly affect the energy balance of the UCL at a local/neighbourhood-scale: i) reducing urban heating by modifying the albedo of rooftops and pavements in urban areas; and ii) the use of urban greenspace to offset urban heating, reduce energy consumption via shade and shelter, and sequester carbon.

Manipulating the albedo of urban areas to reduce urban heating and energy use is under active discussion in the urban climate literature. Akbari et al. (2009) argued that, because roofs occupy about 20% ± 5% of the area of cities and pavements a further 40% ± 5%, there is considerable potential to offset daytime urban heating by increasing the albedo of these surfaces. The shortwave albedo of most roof surfaces is 20% (this is for an individual roof and will be reduced for an urban neighbourhood by the relative fraction of roof to total area), and of most pavements is 10%−15%. Utilizing reflective materials could increase the albedo for these surfaces to ~50% (commercial buildings) and ~30% (sloped residential roofs) in order to reduce the net receipt of shortwave radiation in the urban canopy. This

TABLE 3.6 Summary of Potential Climate Sensitive Urban Design Options, and Those Being Actively Explored for Implementation

Energy Balance Element	Urban Design Mechanism and Spatial-Scale	Design Options Under Consideration
1. Solar and longwave radiation	Building (within UCL) scale: − Spacing, orientation, and height − Roof, wall, and pavement albedo Urban-scale (see 4. below)	• Increasing the albedo, via building layout or urban fabric • Shading to reduce radiation load on pedestrians and buildings (external and internal) • Aspect to control solar receipt
2. Sensible heat and storage fluxes; anthropogenic heat losses	Building (within UCL) scale: − Building materials − Shelter Urban-scale (see 4. below)	• Use of insulation and building materials to reduce heat losses • Provide via buildings or vegetation
3. Latent and sensible heating	Use urban greenspace throughout UCL to modify the Bowen ratio	• Trees, green roofs, parks, lawns: passive cooling via evapotranspiration
4. Layout at local scale (i.e., neighbourhoods) to optimize one or all energy balance components	• Canyon aspect ratio: − Shortwave receipt (shading) and absorption (albedo) − Longwave budget through SVF (recall Figure 3.11) − Modify heat losses through airflow and ventilation • Urban trees for shading in summer and shelter in winter; and CO_2 sequestration • Urban parks to maximize passive cooling via urban evapotranspiration − Optimal size and mix of park and its composition (grass, trees) − Species to maximize CO_2 sequestration − Location to make most use of WSUD	

would reduce the strength of a key radiative driver of the UHI (Table 3.5), potentially offsetting excessive urban heating and mitigating risks to urban health while also reducing energy consumed for air-conditioning.

At the urban-scale, Krayenhoff and Voogt (2010) simulated the sensitivity of daytime near-surface air temperatures to changes in urban albedo using a one-dimensional (1-D) boundary layer model (thus neglecting large-scale advection) coupled to a land-surface scheme (Town Energy Balance model, TEB: Masson, 2000) that represents urban canopy processes. This model has been evaluated using urban surface energy balance observations from Basel (Rotach et al., 2005). The authors caution that the simulations, which were done for Chicago — a mid-latitude city with marked warm/cold seasons and a history of severe heatwave events — are likely to be maximum sensitivities as the periods considered have minimal synoptic flow or meso-scale circulations (such as sea or lake breezes). Simulations showed the largest effects of albedo changes on near-surface air temperatures in summer, as expected. Cooling the air temperature by up to 4.8°C by day, for a high-density neighbourhood with a large roof area, required an increase in albedo from 6% to 65%. This extremely large sensitivity is unrealistic because the change in roof albedo is unrealistically large while advection, which could reduce this sensitivity to only 1°C, was neglected. For Chicago, the annual energy use is also unlikely to change because the reduction in the number of 'cooling degree days' was offset by an increase in 'heating degree days' in winter.

Synthesizing the modelling studies to date, Krayenhoff and Voogt (2010) concluded that the sensitivity of maximum near-surface air temperature (T_{max}) to a 10% increase in albedo is unlikely to exceed 1°C except under very stagnant synoptic flow. A typical seasonal sensitivity for T_{max} in summer is closer to 0.5°C for a 10% increase in urban albedo (this value is for the urban canopy, not individual roof surfaces). In other words, for a moderate density urban neighbourhood, where roofs occupy 25% of the area, the albedo would need to be increased by 40% to reduce the summer maximum air temperature by 0.6°C. In moderate density suburbs, where the canyon aspect ratio allows solar radiation to penetrate to the canyon floor, the canopy-layer air temperature is more sensitive to changes in ground-surface albedo than roof albedo. These results are consistent with the results from Oleson et al. (2010), who coupled an urban canyon and global climate model to simulate the effect of changing urban albedo on the energy balance and air temperatures globally, and found that the annual mean heat island decreased by 33%; and the daily maximum and minimum temperatures decreased by 0.6°C and 0.3°C, respectively.

Manipulating the albedo of urban areas at a neighbourhood-scale, based on this evidence, is therefore a reasonable strategy to manage excessive daytime urban heating. The approach of combining simulation models with observations should be extended to explore the sensitivity of nocturnal temperatures, as this is more likely to be associated with increased mortality in heatwave events (Section 3.3.2).

Akbari et al. (2009) explored the global-scale impact of changing urban albedo on emissions of GHG, as a result of reduced energy consumption for space heating and cooling. Assuming that a 10% increase in albedo across all urban areas is feasible, and that urban areas cover 1.2% of the global land area, they simulated an increase of 0.03% in the Earth's albedo and a radiative forcing (cooling) of 4.4×10^{-2} W m^{-2}, equivalent to offsetting 44 Gt of emitted CO_2. While these numbers are very preliminary, and are based on many assumptions, the temperature sensitivity found in the analysis of Menon et al. (2010) using a global climate model is consistent with the sensitivity found (at an urban neighbourhood/seasonal space/timescale) by Krayenhoff and Voogt (2010). Menon et al. (2010) revised the estimated CO_2 offset to 57 Gt of emitted CO_2. These values (i.e., 44 and 57 Gt) are the same order of magnitude as the mean annual emissions of CO_2 from fossil fuel consumption in the period 2000–2008 (7.7 Gt C per year: Le Quéré et al., 2009). However, these estimates need to be interpreted in light of the simulations of Oleson et al (2010), which revealed that the increase in space heating, globally, was greater than the decrease in air-conditioning as a result of installing white roofs globally.

Huang et al. (1987) simulated the effects of shade, shelter, and transpiration on air temperatures and energy consumption in urban neighbourhoods in three cities in California (USA). Although the approaches to modelling urban evapotranspiration (potential evaporation was used rather than actual), urban mixing layer heights, and the neighbourhood-scale microclimate could be improved with current knowledge, their results are nonetheless important because they simulate the individual effects of shade, shelter, and evapotranspiration at building and neighbourhood-scales. Evapotranspiration had the largest effects on air temperature and energy use; increasing the general tree cover in a neighbourhood by 25% led to a 2°C–3°C reduction in T_{max} (near-surface) for each of the cities studied (as potential evaporation is simulated, these are likely to be maximum changes). These temperature changes were estimated to yield a 25%–40% reduction in the annual energy consumed for cooling, and most of the change (over 80% in their analysis) resulted from the effects of evapotranspiration, not shading. Furthermore, there were significant reductions in the peak power consumption required for cooling (12%–30%) with the 25% increase in tree cover.

Taha (1996), in his simulation of the effects of urban vegetation on photochemical smog in California, reported similar reductions in air temperatures. Simple 1-D boundary layer analysis by Oke (1989) also demonstrated that manipulating the percentage of greenspace in urban neighbourhoods can achieve changes in the UBL T_{max} $\sim 2°C$. A review of studies measuring the effect of urban parks, trees, and vegetated roofs on near-surface air temperature concluded that the average cooling effect by day was $0.94°C$, and $1.15°C$ by night, although there is uncertainty about the general applicability of this figure, especially across different-sized parks and at larger spatial-scales (Bowler et al., 2010).

There are relatively few studies quantifying the micro-climate effects of urban-scale changes in evapotranspiration, yet the urban energy balance measurements across 10 cities shown in Figure 3.7 demonstrated a fivefold increase in the maximum daytime Q_E for an eightfold increase in green-space, suggesting that mixing greenspace into urban neigh-bourhoods could achieve significant offsets in urban heating.

However, it is the net effect of these changes in urban greenspace and form on the energy balance and microcli-mate that is important for climate sensitive urban design. The results from simultaneous measurements of the summertime energy and water balance terms for two suburbs (one with 30% and the other with 10% tree/shrub cover) in Los Angeles showed that while urban evapo-transpiration was larger in the more treed suburb, the net all-wave radiation and all energy flux terms were increased (including the sensible heat terms) as a result of the lower albedo and surface temperatures (Grimmond et al., 1996).

Given these results, the potential role for urban evapo-transpiration simulated by Huang et al. (1987), and the advances that have occurred in our knowledge and simu-lation capability in the two decades since, the time is right and the requirement even greater for new studies that rigorously assess the impact of using urban greenspace to manipulate the energy partitioning and modulate the urban microclimate.

Of course, the strategy of using urban greenspace to modulate urban climates requires adequate moisture to sustain evapotranspiration. This is clear from Figure 3.6, where the variation in Bowen ratio depends on the fraction of *irrigated* greenspace. This is potentially a real challenge for cities located in semi-arid or Mediterranean climate regimes where the summer periods — when cooling is most needed — are often characterized by drought and limited water avail-ability (e.g., Coutts et al., 2007a). This scenario is projected to worsen with global climate change as most global climate model projections indicate a declining trend in rainfall in subtropical regions (IPCC, Fourth Assessment, 2007d).

This challenge could be addressed by deliberate management of the urban water cycle to provide an adequate supply of water to sustain evapotranspiration during low rainfall periods. This is where climate sensitive urban design meets Water Sensitive Urban Design (WSUD). WSUD (Wong and Brown, 2009) refers to design options that reduce and reuse urban runoff, that is, storm-water and wastewater.

The primary goal of WSUD is to manage the risks to the security of future urban water supplies that result from: (i) climate change, which *reduces supply* through reduced rainfall and *increases demand* with rising global tempera-tures and increased frequency of extreme temperature events (e.g., Warner, 2009); (ii) threats to water quality due to pollution, bushfires, and so on; and (iii) competition from other users, especially agricultural and industrial. Climate change is a common factor to all risks.

The possibility of climate change mitigation as a goal of WSUD is often mentioned but not well-quantified in the WSUD literature, despite the obvious benefits to both urban hydrology and urban climate simulations of these linkages. An exception is the study by Mitchell et al. (2008), who simulated a range of WSUD features to assess their relative effect on the imported water, stormwater, and evapotrans-piration for a small suburb in Canberra (Australia). Using vegetated roofs to augment the fraction of greenspace increased annual evapotranspiration by 11%. While the WSUD features (grass swales and a wetland) achieved the goal of reducing stormwater, the presence of vegetated roofs maintained the urban evaporation rate despite a reduction in water being used to irrigate gardens. A simple 1-D boundary layer analysis found that this enhanced evapotranspiration yielded a cooling effect of the order of $0.5°C$ in maximum air temperature.

Cleugh et al. (2005) also used *Aquacycle* to quantify the effects of suburban design (layout, population, style, and size of housing, etc.) on all water balance components (recall Equation (3.2)) including urban evapotranspiration. They showed that urbanisation of a catchment in Canberra (Australia) had brought about the following changes in the urban water balance:

- Imported water (I) augmented the annual rainfall by 26%, on average, to sustain urban evapotranspiration (E) at a rate very similar to the pre-urban land use despite there being 21% of the catchment covered by impervious materials. The magnitude of the annual average urban evapotranspiration was almost twice the runoff (both stormwater and wastewater).
- The use of imported water contributed to a 250% increase in wastewater and stormwater runoff.
- Just 30% of the annual runoff, or 100% of the waste-water stream, is sufficient to meet the external water required to maintain the urban greenspace if it could be harvested and was of adequate quality.

Using *Aquacycle* to explore the effect of differing urban layouts, they found that moving from a suburban design characterized by a relatively small house and large block, to a design where the individual house size increased and block size decreased, had the effect of slightly reducing the greenspace area, changing the volume of imported water and increasing (reducing) the output of water via runoff (urban evapotranspiration).

While these results are very preliminary, they indicate the need to more rigorously quantify the multiple benefits of using water-sensitive and climate-sensitive urban design strategies in an integrated way, at both building and suburb-scales.

Urban areas in regions such as southern Australia and the southwest of the USA are facing ongoing water shortages as a result of the combined effects of climate variability and, increasingly, climate change. The response in Australian cities has been to impose water restrictions in the face of dwindling urban supplies. That such a response could exacerbate the negative impacts of urban heating highlights the importance of implementing WSUD to sustain urban evapotranspiration and provide some level of carbon sequestration.

3.3.2. Adapting to Global Climate Change in Cities

The impact of thermal stress, exacerbated by the UHI, on the health of urban dwellers has become a major concern after a surge in heatwave events accompanied by a large number of deaths occurred in Europe, USA, and Australia over the last two decades. By way of example, August 2003 was Europe's hottest summer in over 500 years, with between 22,000 and 45,000 heat-related deaths occurring in a two-week period across Europe. There is a real risk that the likelihood of more frequent and/or more severe heatwaves in a future warmer world (Patz et al., 2005) will be exacerbated by effects of urbanisation on the local and regional climate, which, as described earlier, can elevate nighttime minimum temperatures by as much as 10°C on individual nights.

Nicholls et al. (2008) found that heat-related mortality in people aged over 65 increased by 15%−17% when the mean temperature (= the mean of the afternoon maximum, and subsequent night's minimum, temperatures) exceeded 30°C in Melbourne (Australia), or when the nighttime minimum exceeded 24°C. This indicates the importance of nighttime minima to heat-related mortality and the risk of it being worsened by urban heating. Although the majority of heat-related mortality studies have been conducted in temperate climates, similar relationships are beginning to emerge from subtropical cities (Patz et al., 2005). Furthermore, Nicholls et al. (2008) found, in agreement with other studies, that this rise in mortality is not simply advancement, that is, people who might have died anyway.

As an example, the city of Phoenix (USA) has experienced three decades of rapid urbanisation that has converted about 2500 km² of desert and irrigated agricultural lands to urban land use. From 1961 to 2008, ten 'extreme heat events' were recorded in metropolitan Phoenix, with 40% of these events occurring in the last five years of the record. In a simulation study designed to quantify the extent to which urbanisation has contributed to the frequency and severity of these extreme heat events, Grossman-Clarke et al. (2010) found that the change from predominantly irrigated agricultural land to mostly commercial and residential land use caused an average increase in the nighttime minima of 5°C−8°C (range: 4°C to 10°C). The increase in daytime maximum temperatures depended on whether the land-use change was from irrigated agriculture (2°C−4°C) or desert (~1°C). They concluded that urbanisation had contributed to the observed increased frequency and severity of summertime extreme heat events and that these changes were independent of a trend due to global warming or to changes in the urban density.

The factors causing heat stress in cities reflect the three fundamental spatial-scales in the urban environment: (i) building-scale − heat load and ventilation as a result of building design, orientation, and building materials, and the positioning of individual trees or gardens; (ii) local or neighbourhood-scale heating of the urban canopy airspace will be influenced by the amount and distribution of urban greenspace, water availability, and the characteristics of the urban canyons; and (iii) city-scale − weather and climate driven by the synoptic and climate regime, and meso-scale circulations (such as valley or sea breezes) and the nature of the city. This demonstrates that the impact of heatwaves in urban areas can be managed through modifications to urban form, to provide optimal ventilation, shading, and nighttime radiative cooling; and to urban greenspace, to enhance shade and evapotranspiration.

Excessive urban heating, due to an UHI superimposed on the effects of global warming on mean and extreme temperatures, also poses significant risks to urban infrastructure such as energy supply and transport. Some of these effects will amplify other impacts; for example, a surge in energy demand during extreme temperatures can result in interruptions to energy supply from the grid that, in turn, exacerbates the health risks already associated with the extreme temperatures. The occurrence of positive feedbacks, such as that just described, makes urban areas and their inhabitants particularly vulnerable to global climate change − especially for people living in cities in developing nations and/or coastal cities. This means that the principles of climate-sensitive urban design, and the options discussed in Section 3.3.1, are also important adaptive measures to manage health and infrastructure risks in cities that are associated with global climate change. This is because design options that mitigate local urban heating − both in the mean and in the extremes − will serve to reduce the additional risk imposed by global warming. Climate-sensitive urban design is therefore a potentially important climate adaptation, as well as mitigation, strategy.

3.3.3. Managing Air Quality Risks in a Warmer and Urbanised World

Managing air quality to reduce adverse impacts on health remains an important challenge in cities across the globe. Air quality in cities will continue to be influenced by urban processes that affect the turbulent transport and dispersion of airborne pollutants; and in the future these effects will take place in a background of rising global air temperatures and ozone levels. This means that there is a growing need to ensure that actions taken to mitigate and adapt to climate change are developed in the knowledge that poor air quality still remains a large threat to ecosystem and human health in many urban areas of the world. Five issues that illustrate the importance of explicitly linking urban climate, air quality, and global climate change in the future are:

1. Air quality management strategies will increasingly sit alongside climate change mitigation strategies. These strategies are well-aligned in many cases, for example, reducing fuel consumption through better vehicle, road, and public transit systems will be beneficial for climate change mitigation and urban air quality, but the rising use of diesel as a fuel, with lower GHG emissions but higher emissions of particulates, or the fact that some air pollution control technologies may increase CO_2 emissions (e.g., electric vehicles using electricity from coal-fired power stations), are examples of approaches that can lead to contradictory outcomes. Developing optimum solutions for both global climate change and air quality will require lifecycle analyses and management of the tradeoffs between local/regional goals and national/global goals.

   It is also worth noting that regional air quality does have a global climate impact too, as evidenced by the 'masking' of global warming in the middle of the twentieth century due to the effects of aerosols. Aerosols still remain the radiative forcing term with the greatest uncertainty (IPCC Fourth Assessment, 2007d).

2. The increased risk of photochemical smog that results from global warming and growing background levels of ozone will be exacerbated by increased levels of smoke and dust aerosols because of the indirect effects of climate change on the frequency and intensity of bushfires and dust storms in some regions. These particulate pollution sources are not amenable to regulatory control and, furthermore, are often located outside city or state jurisdictional boundaries. This means that air quality management and approaches will need to move from a local/regional focus to perhaps a regional/national-scale.

3. Most operational air quality models do not adequately include urban-scale processes and effects, despite the first-order forcing that the urban energy balance has on atmospheric mixing and that atmospheric transport is very much affected by urban canopy and boundary layer turbulence. In terms of personal exposure, information is needed about processes deep within the UCL — yet this is a region that will be most poorly represented in most meso-scale atmospheric models for air quality applications. In light of this, and given the preceding point, multiscale models or model systems will be needed to ensure that the right simulation tools are available. Moreover, strategies are needed to guide the development of these model systems to ensure they are fit-for-purpose and include an appropriate level of urban processes, similar to that discussed in Baklanov et al. (2009).

4. For urban dwellers, heatwave conditions that cause heat-related stress and mortality may coincide with environmental conditions that are conducive to poor air quality, through photochemical smog and particulate pollution from dust and smoke, causing respiratory problems. The net effect is an amplification of the environmental stressors affecting their health and wellbeing. Adaptation strategies need to account for the coincidence of these events that can quickly exceed thresholds. In particular, strategies involving urban greenspace, which are key for climate-sensitive urban design options at the local-scale, will need to account for the possible risks to air quality of increased emissions of volatile organic compounds (VOCs).

5. An adaptive response to global climate change may be that individuals spend more time indoors and that houses will increasingly be designed to be more thermally efficient to reduce energy consumption and GHG emissions. This will mean that the traditional domain of air quality assessments may have to move indoors. Some early work has begun to provide baseline assessments of indoor air quality in residential (Melbourne) and commercial (London) districts.

3.4. CURRENT STATE-OF-THE-ART IN SIMULATING URBAN CLIMATES

There is a vast array of models available to study urban climates. They can be divided into three major types: hardware, statistical, and physically-based models.

3.4.1. Hardware Models

Hardware models are routinely used within wind tunnels in the process of design of large buildings. However, such models have also been used to investigate a wide range of processes at a range of scales (e.g., the influence of green infrastructure, influence of geometry on albedo, influence of morphology on wind flow). These studies have provided important information to understand processes that can be used to develop the algorithms for numerical models. For example, the development of the resistance network in a numerical land-surface model (Harman et al., 2004) is based on the hardware modelling of Barlow et al. (2004). Kanda (2006) reviews a wide range of indoor and outdoor hardware model studies that have explored turbulent fows, scalar dispersion, local transfer coefficients, radiative processes, and surface energy exchanges.

3.4.2. Statistical Models

Statistical models have been used as the initial basis for many urban climate models. A classic example, to estimate the size of the urban heat island (UHI) (canopy layer air temperature), related the size of the maximum UHI to population (Oke, 1973). As more data have become available, it has become apparent that one set of fitted parameters does not provide a good predictor of all UHIs. Using a hardware model, and reanalysing the then available empirical data, another statistical model was proposed to estimate UHI_{max}, this time using a physical characteristic of the urban area: the SVF (Oke, 1981). This example demonstrates a limitation of statistical models to the dependence on the available data for the development. Use of a model in conditions beyond the development dataset (i.e., extrapolation) is unwise as the model behaviour is not tested and the statistical function may not be at all related to the processes being modelled.

Other issues related to the non-independence of meteorological data (observations at one time and place may not be independent of observations at a proximal time and place) and climatic variables commonly have non-normal distributions (e.g., Gumbel for precipitation; Weibull for temperature). This has resulted in extensive research and techniques of statistical climatology. These problems aside, statistical models are useful tools for exploring relations between process variables and environmental parameters. Given the complexity of the urban environment they are a useful tool to tease out understanding. Also, statistical models provide a tool for gap-filling when datasets have missing observations. Statistical modelling techniques in general are reviewed in a variety of books (e.g., Wilks, 1995) and papers (e.g., Robeson, 2005).

3.4.3. Physically-Based Models

A wide range of physically-based numerical models for urban atmospheric processes exist. The models range from those concerned only with individual processes to those that try to account for a wider range of interacting processes. Similarly, there are models that resolve details of an individual building to those for an entire urban area. This type of model plays the most important role in climate and weather forecasting. Recently, there have been a number of review papers (e.g., Best, 2006; Masson, 2006; Martilli, 2007, Grimmond et al., 2009).

The physically-based numerical models can be differentiated by the degree of parameterization of the processes that are included. Computational fluid dynamics (CFD) models have a more complete description of the Navier–Stokes equations. Typically CFD models are constrained by computer resources, which impacts the mesh or grid spacing, which needs to be finer than the smallest scales of motion to be resolved. They range in terms of decreasing computational needs from direct numerical simulation (DNS), through large eddy simulation (LES) to Reynolds-averaged Navier–Stokes (RANS). DNS solves all the spatial and temporal-scales without any empirical information, so computational resources are greatest and are currently generally restricted to simple geometries and low Reynolds numbers (Coceal et al., 2007). Current computer resources allow the reduced space and time grids of LES, relative to DNS, to be feasible (Xie and Castro, 2009). CFD codes can simulate flow and dispersion over and inside street canyons with additional modifications to examine the transport of reactive air pollutants. A meteorological profile at the upwind edge of their geographical domain, usually 1 to 5 km on a side, is required. This may be obtained from a larger-scale meteorological model through nesting or as unlinked input. Urban-adjusted CFD models have reasonable agreement with field observations and allow the flow patterns around specific buildings to be seen, such as large eddies in the lee of tall buildings.

More commonly for climate purposes, a greater level of parameterization of the subgrid processes is used (Best, 2006; Masson, 2006; Martilli 2007). These ULSM, like CFD models, range in complexity and can be run as standalone models or embedded within meso-scale or global-scale models. Recently there has been a large increase in the number of models that simulate urban areas within such models because of increased computing capacity resulting in decrease in grid size (Grimmond et al., 2010b, 2011). This combined with the growth in urban areas make the representation of the urban land surface more frequently required. For a ULSM, the objective is to model the bulk, or area-average, response for the grid cell rather than the response from individual roughness elements (buildings, trees, etc.) within the UCL. The lowest layer of the meso-scale or global climate model typically is approximately the 'blending height', which may be taken to be above the RSL (Figure 3.2); that is, a local-scale or integrated response of the surface (Masson et al., 2002).

Given the wide range of urban features that could be included, not unsurprisingly a wide degree of complexity of urban characteristics has been incorporated (Figure 3.12). The simplifications include what fluxes are resolved. Some models do not include the anthropogenic heat flux and some do not include the latent heat flux. There is clear evidence that not accounting for the latent heat flux, even in the most built-up areas, is inappropriate (Grimmond et al., 2010b, 2011). A wide range of approaches is taken to addressing the urban morphology from the simplest of assuming effectively a 'concrete slab' to, at the most complex, allowing for changing material characteristics in three dimensions while also allowing for complex geometry that will influence radiative and aerodynamic exchanges (Figure 3.12).

A recent international model comparison ('Urban-PILPS') has demonstrated that there is no individual model that is best at modelling all of the individual fluxes (Grimmond et al., 2010b, 2011). So, although there has been rapid improvement in capability in the last decade, there is a clear need for further improvements. These will be essential for the wide range of model applications that are needed. As no model is likely to predict all conditions perfectly, another approach is to run an ensemble of models. This allows for a 'cloud' or 'range of results' to be produced and outlier and central tendencies to be identified. This is found to often produce the best results (Grimmond et al., 2011).

A summary of the inter-comparison by Grimmond et al. (2011) provides useful insight into the ability of the current generation ULSM. Model performance was generally best for radiative fluxes and least good for latent heat fluxes (Grimmond et al., 2010b, 2011). The mean root mean square error (RMSE) for 31 models was 17 W m^{-2} for outgoing shortwave radiation for the daytime flux over a year when no information was provided about the site beyond it being urban (Grimmond et al., 2011). Again, without any knowledge of the site, the 31 models had a mean RMSE for emitted longwave radiation of 18 W m^{-2} (all hours, annual). Under the same conditions, the mean RMSE for Q^*, Q_H, and Q_E was 28, 66, and 51 W m^{-2}, respectively. This has implications for the use of the models. Firstly, it suggests there may be instances when the models are predicting the correct fluxes for the wrong reason. Secondly, care needs to be employed when using the models to ensure that they are capable of modelling all of the exchanges of interest. For example, models used to evaluate the impact of changing a large number of urban surface properties (e.g., reflectivity, vegetation amount) as mechanisms to mitigate the size of the urban heating effect, need to model realistically and account appropriately for shortwave and longwave exchanges.

It is anticipated with the improved capabilities of these models that there will be applications to address a wide

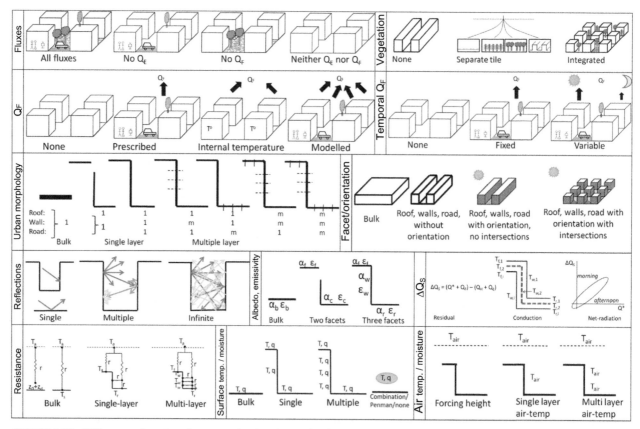

FIGURE 3.12 Different configurations for representing the urban surface in atmospheric models. (*Source: Modified from Grimmond et al., 2011.*)

variety of inter-linked issues (e.g., energy and water use, energy exchanges, and air quality, etc.). The feedback needs to be explored between human behaviour and biophysical processes (cf. Rice and Henderson-Sellers, 2012, this volume). For example, many communities are advocating the addition of green infrastructure. However, to maintain the greenspace, it is often necessary to provide water. The application of water is a function of who is responsible for the management of the vegetation, individual behaviour, and local government policy (Taplin, 2012, this volume). Introducing more greenspace will alter the energy exchanges by changing the radiative and latent heat fluxes, and in turn modify water balance exchanges. The presence of vegetation will also alter CO_2 and other atmospheric chemistry exchanges (e.g., VOCs). It becomes apparent very quickly that, when using models to explore the wide range of possible mitigation and adaptation strategies that are being considered for cities, it is very important to account for the wide range of processes. However, model-use requires care in order to ensure that the strengths/weaknesses of the individual models are accounted for, while obtaining insight to these interactions.

With time, it is expected that more complex models will be developed (e.g., Chen et al., 2011) that allow for more complete description of the processes and can be linked to more complete air quality or chemistry transport models. Ideally there will be two-way nesting not only in terms of the spatial-scale, but also in terms of processes that are responsive to each other. If, for example, we are to achieve low/zero carbon emission cities, changes in transportation and building heating will be required. This will impact air quality and anthropogenic heat exchanges. In high latitudes, a reduction in anthropogenic heat fluxes during wintertime periods may result in higher frequency of stable or neutrally stable atmospheric conditions and therefore greater frequency of lower boundary layer heights. This, in turn, may result in greater local air pollution episodes and increase air-quality-related health impacts. Thus the air-quality-related health impacts may change with time.

3.5. CITIES AND THE FUTURE CLIMATE

The links between urbanisation and global climate change are complex (Grimmond, 2007), but important. Cities are direct and indirect drivers of climate change through the effect of urbanisation and land-use change, and through GHG emissions from the fossil fuels consumed for transport, space heating and cooling, industrial processing (which provides employment and sustains economic growth), and so on. In the future, as is the case today, cities will be where the majority of the world's population live and the source of the bulk of human emissions of heat, gases, and particulates that affect climate and air quality.

Cities modulate the local climate by changing the radiation balance of the atmosphere, through emission of heat and GHG, and by modifying the energy, water, and carbon balance of the urban surface via land-cover and land-use change. The urban heat island, which results from these changes, is one of the most well-documented examples of inadvertent climate modification by humans. Excessive urban heating in many cities may amplify the impact of global warming on human and ecosystem health and lead to even greater emission of GHG through increased energy demand. Urban ecosystems, including the human inhabitants, are likely to be more vulnerable to the effects of global climate change.

The last two decades have seen major scientific advances in understanding the biophysical basis of urban climates, including conceptual models, field observations, process studies, and the development and evaluation of urban climate models (Grimmond et al., 2010b). This chapter has described the state of this knowledge and has assessed the modelling and simulation approaches that are needed for the urban climate science community to contribute to the 'grand challenges' that lie at the nexus between urban climates and global climate change, namely: (i) reducing energy demand, consumption, and GHG emissions; (ii) managing the combined effects of global climate change, urban heating, and air quality on ecosystems and human health in urban areas; and (iii) managing the urban water supply, in the context of a warming and drying climate in many regions of the globe, in a way that reinforces those strategies for designing sustainable cities.

There is still a great deal to be done in urban climate science. While process-based understanding has improved dramatically, it is not as advanced for urban as for other ecosystems such as forests and crops or for oceans. The two-way interaction between cities and global climate change has not really been assessed in a systematic and rigorous way. There is growing demand for urban climate information in the design and management of more sustainable cities.

We therefore conclude by describing the priority observations, research, and enhanced knowledge and models that are needed to more effectively manage urban areas in a changing environment. These recommendations (Grimmond et al., 2010a) have been endorsed by the WMO World Climate Conference 3 (2009) and NRC (National Research Council, 2010):

- Observations and data:
  - *Improving urban climate observation networks*: Urban climate stations and networks should be greatly improved, including vertical information where possible. Quality observations are needed in and near rapidly growing cities in developing regions.

- *Acquire and maintain standardized information on city form*: International archives of urban climate, morphological, and land-cover data should be established, along with protocols for standardizing the way that information is reported and archived.
- Research programmes:
  - *Climate research for hot cities*: Strengthen observational networks and establish urban climate research programmes for tropical cities where population growth is greatest and vulnerability to excess heat and inundation is highest.
  - *Meet the needs of urban decision-makers*: Engage urban decision-makers in the design of urban research programmes, for example, data that shows the correspondence between the urban landscape and climate effects.
  - *Urban climate modelling*: Improved numerical models should be developed to forecast weather and air quality, and provide climate projections that include cities. A focus should be to incorporate urban land-surface schemes into global climate models, to downscale regional climate predictions and projections to the urban-scale, and to assess their impact on urban health, safety, and management.
- Understanding local, regional, and global climate linkages:
  - *Develop integrated hierarchal models* that can provide useful predictions at urban planning-scales.
  - *Integrate urban climate knowledge* into the practice of sustainable urban planning. Link urban climate effects with broader environmental contexts and set within broader social and economic contexts.
  - *Encourage cross-disciplinary research* on urban climates and their effects together with a dialogue between researchers, practitioners, and decision-makers.

Chapter 4

Human Effects on Climate Through Land-Use-Induced Land-Cover Change

Andy Pitman[a] and Nathalie de Noblet-Ducoudré[b]

[a] *Centre of Excellence for Climate System Science, The University of New South Wales, Sydney, Australia,* [b] *Laboratoire des Sciences du Climat et de l'Environnement, Unité Mixte CEA-CNRS-UVSQ, France*

Chapter Outline

4.1. INTRODUCTION: LAND CHANGE AND CLIMATE

The modification of the landscape by humans and their ancestors is very ancient. The history of deforestation from prehistoric times through to the present was explored by Williams (2003). In a remarkable account, he highlighted the fundamental connectivity between human development and large-scale land modification. Large-scale landscape modification affects climate. There is a direct effect on atmospheric carbon dioxide (CO_2) concentrations, on the energy availability at the surface, on the partitioning of energy and water at the surface, and on the emissions of aerosols and various volatile organic components. However, while it is indisputable that large-scale landscape modification affects climate (according to theory and local-scale observations), there is a necessity to identify the footprints of land-cover changes on climate, to define 'large-scale', what is meant by 'modification', and what scale the 'effect on climate' is and how it is realized. In terms of scale, 'large-scale' is used here to mean of the order of a hundred kilometres — the scale of a climate model grid element, a large catchment, or a major city (cf. Cleugh and Grimmond, 2012, this volume). In terms of climate modification, 'climate' means a longer-term statistic (e.g., decades), but does not necessarily imply

averages. 'Modification' means a measurable impact on climate statistics that can be formally attributed to the land-cover change and is detectable above the background noise induced by natural variability.

There are two ways that large-scale landscape modification can occur. Humans can clear forests with fire, a chain-staw, tractor or buldozer. They also grow crops or build cities. This is a direct modification that is very obvious to all of us. Emissions of CO_2 from the burning of fossil fuels can also indirectly affect the large-scale landscape via the CO_2 fertilization effect (Collatz et al., 1991; Field et al., 1995; Sellers et al., 1996) that might lead to a change in leaf size, vegetation height, and stomatal function, via competition species types and via evolution of the nature of the vegetation itself. This indirect effect, among others, is subtler than deforestation but globally integrated may be more important (Dickinson, 2012, this volume).

This chapter provides an overview of how humans affect climate through land surface modification. We focus only on the physical and biogeophysical impacts because Dickinson (2012, this volume) highlights the biogeochemical effects of land-cover change while Harvey (2012, this volume) reviews fast and slow feedbacks on climate. To simplify the text, the term LULCC (land-use-induced land-cover change) will be used throughout; also simply termed anthropogenic land-cover change. Firstly, the scale of human modification will be highlighted to place this perturbation in context. Secondly, the theory that explains why landscape modification should affect climate will be explored. Thirdly, evidence of how landscape modification affects climate will be presented with the implications for the future climate discussed.

4.2. THE SCALE OF HUMAN MODIFICATION

Humans have undertaken intensive LULCC at a scale commonly underestimated by scientists working in other fields of climate science. Several groups have undertaken reconstructions of the scale of LULCC — work that underpins current efforts to determine the impact of this change on climate (e.g., Defries et al., 1995; Ramankutty and Foley, 1999; Klein Goldewijk, 2001; Hurtt et al., 2006; Pongratz et al., 2008). Figure 4.1a shows that by 1500 large areas of Western Europe had been partially cleared for agriculture and for timber. LULCC intensified, particularly in Western Europe, through to 1800. Indeed, by 1750, 7.9–9.2 million km^2 (6%–7% of the global land surface; note that all these percentages are calculated over land excluding Greenland and Antarctica) were in cultivation (Forster et al., 2007) although only in Western Europe and perhaps parts of northern China had the intensity of LULCC led to more than ~60% agricultural cover for a given region (Figures 4.1b, 4.1c, 4.1d). By 1990, 45.7–51.3 million km^2 (35%–39% of global land surface) was in cultivation, forest cover had decreased by about 11 million km^2, and intensive LULCC had impacted parts of the USA, much of Western Europe, India, northern China, and elsewhere. Large areas of the Southern Hemisphere underwent LULCC throughout the nineteenth century. By 2000 (Figure 4.1f) the fingerprint of human activity through LULCC has only *not* affected a few desert regions, the central Amazon and Congo Basins, and the Arctic and Antarctic (not shown). Figure 4.1 needs to be interpreted qualitatively. A careful examination points to some changes, such as over Australia between 1800 and 1900, which seem unlikely, but the overall pattern of changes are probably reliable at large scales. Williams (2003) provides a detailed account of these global and regional changes as well as their underlying causation.

The easiest and so far most traditional way to quantify the impact of these changes is by means of the surface albedo (the fraction of incoming solar radiation reflected by a surface), which can be translated into a change in radiative forcing. Forests are dark to the wavelengths of visible light and absorb very efficiently (albedos range from around 0.05–0.2). Croplands are commonly much more reflective (around 0.2–0.25 when they are snow-free, but much higher

in the presence of snow: Bonan (2002)). Thus, the global net seasonal and annual impact of LULCC is an increase in albedo. While there is considerable uncertainty in the precise impact of LULCC on radiative forcing, Forster et al. (2007) suggest a reduction of 0.2 W m^{-2} ± 0.2 W m^{-2} on the global average. While this appears to be a large range (perhaps as large as 0.4 W m^{-2} or as small as 0.0 W m^{-2}), available evidence points to it being very small compared to well-mixed greenhouse gases (GHGs) (about +2.5 W m^{-2}).

Reasoning of this kind has led to the role of LULCC being ignored in most climate projections. The reports by the Intergovernmental Panel on Climate Change (IPCC, e.g., the Fourth Assessment Report, Solomon et al. (2007)) highlight LULCC as an area of uncertainty but the model simulations assessed in the Fourth Assessment Report do not include LULCC in terms of how it modifies the surface (only direct emissions of CO_2 are accounted for). Pitman et al. (2009) discuss the implications of omitting LULCC from the IPCC assessments and conclude that at the global-scale there is no evidence that this matters to global metrics, although a more recent study by Davin and de Noblet-Ducoudré (2010) demonstrated that oceanic feedbacks have the potential to enhance the LULCC impacts. Overall, IPCC AR4 statements of the likely amount of global warming associated with a given emission scenario are not flawed globally due to the lack of inclusion of LULCC. However, while GHGs are globally well-mixed, LULCC is highly concentrated at present in Western Europe, the eastern US, China, and India (Figure 4.1f). Thus, while the global impact may be negligible, and the impact on global means may be small, the global impact is contributed from a fraction of the land surface strongly coincident with human population (cf. Cleugh and Grimmond, 2012, this volume). Furthermore, it has been pointed out (Pielke et al., 2002; Davin et al., 2007) that radiative forcing is not a good way to measure the impact of LULCC on surface climate. A regionally-significant impact on climate can be achieved by LULCC without any change in albedo if the partitioning of energy or rainfall is modified. LULCC changes the seasonality of heat and moisture fluxes, changes the probability of extremes, and may provide a local perturbation that triggers a change in the atmosphere sufficient to cause changes remote from the perturbation. Thus, while the global impact of LULCC on mean radiative forcing and mean global climate sensitivity may well be negligible, this finding is really only relevant in theoretical studies since humans and the Earth's terrestrial flora and fauna live, source their food, and source their water at local and regional-scales.

The question is therefore not whether LULCC affects the global climate (it does via release of, for example, CO_2; see Dickinson, 2012, this volume). Rather, it is whether it affects climate at any space or timescale of significance for living bodies; and clearly 'significant' is a value judgement. A major change in climate over one region may be

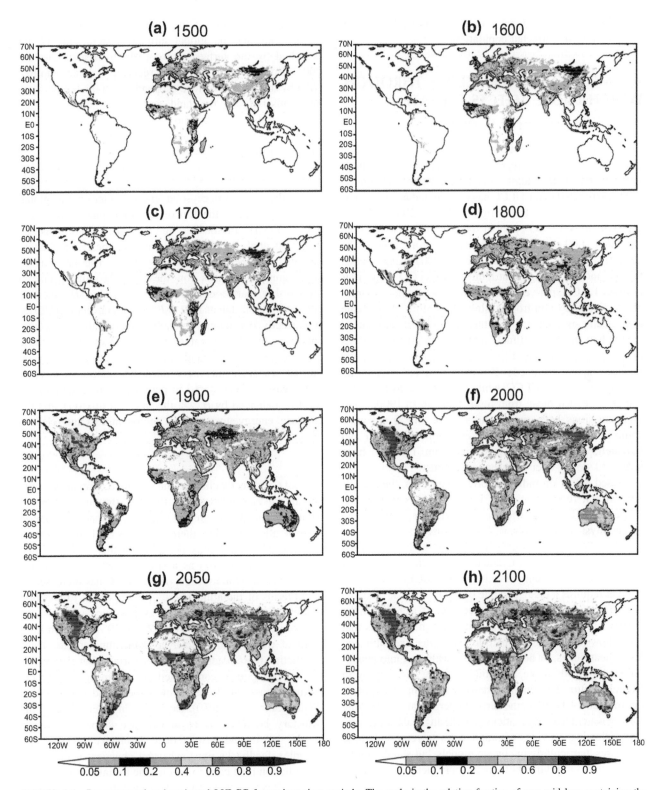

FIGURE 4.1 Reconstructed and projected LULCC for various time periods. The scale is the relative fraction of any grid box containing the sum of pasture or crops. *(Source: These data were downloaded from the Land Use Harmonization web site at* http://luh.unh.edu*).*

insignificant to the global climate scientist, but highly significant to the affected population.

4.3. MECHANISMS/PROCESSES THROUGH WHICH LULCC AFFECTS CLIMATE

4.3.1. The Terrestrial Carbon Balance

The Earth's land surface stores very large amounts of carbon and, where LULCC is represented by deforestation, large amounts of CO_2 are released into the atmosphere. Houghton (2003) estimated that LULCC has contributed net emissions of 156 GtC during the period 1850 to 2000. This is approximately 12−35 ppmv of the total CO_2 rise between 1850 and 2000 of ~90 ppmv (Matthews et al., 2004). The latest IPCC assessment (Denman et al., 2007) suggests that the net land-use change contribution to total CO_2 emissions is about 1.6 GtC per year comprising large tropical sources offset slightly by non-tropical sinks. Emissions of this scale are globally significant, add to the radiative forcing induced via the burning of fossil fuels, and the sum greatly exceeds the global significance of LULCC on radiative forcing via surface changes in albedo. However, to avoid overstating the importance of LULCC, while CO_2 emissions from this source exceeded those from fossil fuels up to 1900, fossil fuel emissions have risen rapidly since ~1950 and now comprise ~85% of total annual emissions (Raupach and Canadell, 2010), while emissions from LULCC have increased slowly over the twentieth century.

Based solely on radiative forcing concerns, the IPCC's non-explicit representation of LULCC seems reasonable. However, LULCC also affects climate via other physical and biogeophysical mechanisms, but the specific mechanisms depend on the nature of the perturbation, the climate of the region undergoing LULCC, and how the region's land cover may recover after being abandoned. For LULCC to affect climate requires a change to one of the fluxes that links the land to the atmosphere. This can occur due to a change in the albedo and thereby the net radiation, via mechanisms that affect how net radiation is partitioned between sensible and latent heat fluxes, or how rainfall is partitioned into evaporation, runoff, drainage, and soil moisture, and thereby in latent heat flux. To formalize these relationships and potential impacts of LULCC the surface energy balance and the surface water balance equations provide an ideal framework. The following provides common ways to represent these processes, particularly in climate models, but should not be interpreted as implying these are the only methods available.

4.3.2. The Surface Energy Balance

The shortwave radiation emitted by the Sun is reflected, absorbed, or transmitted by the atmosphere. An amount of energy ($S\downarrow$) reaches the Earth's surface and some is reflected (depending on the albedo, α). Of 100 units of energy entering the global climate system, 46−47 are absorbed by the surface (Rosen, 1999; Trenberth et al., 2009). Infrared radiation is also received ($L\downarrow$) and emitted ($L\uparrow$) by the Earth's surface (depending on the temperature and emissivity of the land and atmosphere). The net balance of the incoming and reflected shortwave radiation, and the incoming and emitted longwave radiation at the Earth's surface is the net radiation (R_n):

$$R_n = S\downarrow(1 - \alpha) + L\downarrow - L\uparrow \qquad (4.1)$$

Of the 100 units of solar energy entering the global climate system, at the top of the atmosphere over land, about 30−31 (according to the latest estimates by Trenberth et al., 2009) are exchanged as sensible and latent heat fluxes − the turbulent energy fluxes. The land surface significantly influences how these 30−31 units of energy are further partitioned between sensible (H: thermal convection) and latent heat (λE: moist convection) fluxes. A perfectly dry desert typically releases all 31 units of energy as H, while a water surface releases most of those 31 units as λE. The land surface also stores energy on diurnal, seasonal, and longer timescales (thousands of years in the case of heat stored in permafrost for example). R_n must be balanced by H, λE, the soil heat flux (G), and the chemical energy stored during photosynthesis and released by respiration (F):

$$R_n = H + \lambda E + G + F \qquad (4.2)$$

LULCC directly changes the albedo, α, that perturbs the amount of energy available to drive the surface energy balance. From Equation (4.1), it is clear that a change in α must affect R_n and, therefore, affect all the terms in Equation (4.2). It is this perturbation that is highlighted in Forster et al. (2007) as a small reduction in radiative forcing because, on average, past LULCC has increased α.

In terms of the atmosphere, it is important to partition R_n between H and λE as well as possible. λE cools the surface very efficiently, moving 2.5×10^6 J for each kilogram of water evaporated away from the surface. Thus, a reduction in λE leads to a warmer surface, but also means less water vapour is contributed to the atmosphere, which, in turn, tends to decrease cloudiness and precipitation (Seneviratne et al., 2010). Decreases in H tend to cool the planetary boundary layer and reduce convection (Betts et al., 1996) and also reduce boundary layer depth. Complex feedbacks exist whereby changes in clouds or precipitation feed back to modify the initial perturbation to albedo. Precisely how LULCC affects the partitioning of R_n is, therefore, critical to understanding the role of LULCC on the dynamics of the atmosphere and on the coupled atmosphere−ocean system.

The two key equations that describe H and λE can be written in many ways. The link between surface characteristics and the turbulence that drives the exchange of H and λE and CO_2 is the aerodynamic resistance for heat (r_{ah}) and water (r_{aw}). Following Bonan (2002):

$$r_{ah} = \frac{1}{k^2 u_a} \left[\ln\left(\frac{z-d}{z_{0m}}\right) - \varphi_{m(\zeta)} \right] \left[\ln\left(\frac{z-d}{z_{0h}}\right) - \varphi_{h(\zeta)} \right]$$

(4.3)

$$r_{aw} = \frac{1}{k^2 u_a} \left[\ln\left(\frac{z-d}{z_{0m}}\right) - \varphi_{m(\zeta)} \right] \left[\ln\left(\frac{z-d}{z_{0w}}\right) - \varphi_{w(\zeta)} \right]$$

(4.4)

Quantities such as k (von Kármán constant) are independent of the nature of the surface. Other quantities include u_a (wind speed at height z) and the roughness length for momentum (z_{0m}). The functions $\varphi_m(\zeta)$, $\varphi_h(\zeta)$, and $\varphi_w(\zeta)$ account for the influence of atmospheric stability on the turbulent fluxes. This leaves three terms: z (height in the surface layer), d (the displacement height), and z_{0h}, z_{0w} (the roughness lengths for heat and moisture). The variables $z_{0h} + d$ and $z_{0w} + d$ are the effective heights at which heat and moisture are exchanged with the atmosphere. The key fact linking Equations (4.3) and (4.4) to LULCC is that changing a landscape affects $z_{0h} + d$ and $z_{0w} + d$. A typical value of $z_{0h} + d$ and $z_{0w} + d$ for a grassland or crop is a few tens of centimetres depending on the type of grass or crop, although this is highly dependent on the type of crop. For a forest, $z_{0h} + d$ and $z_{0w} + d$ is around 8–10 metres. Thus, changing the nature of the vegetation changes (commonly by two orders of magnitude) $z_{0h} + d$ and $z_{0w} + d$, which, through Equations (4.3) and (4.4), directly affect r_{ah} and r_{aw}, which, in turn, affect H and λE. The H can be represented as a quasi-diffusive process:

$$H = \frac{T_s - T_r}{r_{ah}} \rho c_p$$

(4.5)

where T_s is the surface temperature (itself a function of the surface energy balance), T_r is a reference temperature above the surface, ρ is the air density, and c_p is the specific heat of air. Thus, a change in r_{ah} directly affects H — something that must also affect λE through Equation (4.2). However, λE fluxes are underpinned by a more complex process than H as it involves most of the challenges of simulating H, plus all those processes that provide the water to evaporate (canopy interception, root water uptake, soil moisture diffusion, etc.). While there are several ways to represent λE, the aerodynamic approach is commonly used (Sellers, 1992):

$$\lambda E = \left(\frac{e_s - e_r}{r_{surf} + r_{aw}} \right) \frac{\rho c_p}{\gamma}$$

(4.6)

where e_s is the saturated vapour pressure at the surface from which evaporation is occurring, e_r is the vapour pressure at a reference height, and γ is the psychrometric constant.

Thus, as with H, the change in $z_{0e} + d$ and thereby the change in r_{aw} must directly impact λE. However, Equation (4.6) contains an additional term, r_{surf}, the surface resistance to the transfer of water from the surface to the air, which is highly variable depending on the land-cover type. In general, from terrestrial surfaces, λE can be thought of as the sum of three fluxes: a flux from the ground (λE_g), a flux from the leaves (λE_v), and a flux from the air within the canopy to the air above the canopy (λE_a):

$$\lambda E_g = \left(\frac{e_g - e_{ac}}{r_{ac}} \right) \frac{\rho c_p}{\gamma}$$

(4.7)

$$\lambda E_v = \left(\frac{e_v - e_{ac}}{r_c} \right) \frac{\rho c_p}{\gamma}$$

(4.8)

$$\lambda E_a = \left(\frac{e_{ac} - e_a}{r_a} \right) \frac{\rho c_p}{\gamma}$$

(4.9)

where the subscripts for λE and e refers to the ground (g), canopy surface (v), canopy airspace (ac), and the air above the canopy (a). Different vegetation types have different levels of ground cover. A dense forest may cover nearly 100% of the surface, making λE_g negligible. Removing this forest completely makes λE_v irrelevant and λE_g dominant. These fluxes differ because the resistance terms used in the formulations of Equations (4.7), (4.8), and (4.9) represent different pathways for the water molecules. Both r_{ac} and r_a are physically based resistances: r_c, the canopy resistance, includes biological processes that are active in managing losses of water via the stomates to balance gains of CO_2. Since each type of vegetation has a specific optimization to balance water loss with carbon gain, each type of vegetation has evolved strategies that lead to different behaviour of r_c. Most simply (Dickinson et al. 1991):

$$r_c = \frac{<r_s>}{L}$$

(4.10)

where the angled brackets denote an inverse average over the range of the canopy leaf area index, L such that:

$$< () > = \frac{L}{\int^L dL/()}$$

(4.11)

and the stomatal resistance, r_s, is:

$$r_s = r_{s\,min} f_1(T) f_2(D) f_3(PAR) f_4(\theta) f_5(\Psi)\ldots$$

(4.12)

where each of the functions f range from zero to one and represents a dependence of r_s on temperature (T), vapour pressure deficit (D), photosynthetically active radiation (PAR), soil moisture availability (θ), and nutrient stress (Ψ). This approach, based on Jarvis (1976), is commonly used in land-surface schemes coupled to climate models, but it is limiting because it does not couple the carbon, water, and energy balances at the level of the leaf. More recently (e.g., Collatz et al., 1991; Bonan, 1995; Arora, 2002), models have included parameterizations based on Ball et al. (1987) that link the response of stomatal conductance (g_s) to the rate of net CO_2 uptake (A_n), the relative humidity (h_s), and the CO_2 partial pressure at the leaf surface:

$$\frac{1}{r_s} = g_s = m\frac{A_n h_s P}{c_s} + b \qquad (4.13)$$

where m is a constant and b is a coefficient obtained from observations ($b = 0.01$ and 0.04 for C3 and C4 plants respectively), P is atmospheric pressure, and c_s is the CO_2 partial pressure at the leaf surface. A_n, in turn, is defined as the lesser of w_c (the Rubisco-limited rate of photosynthesis and w_j (the light-limited rate allowed by the rate of regeneration of the Ribulose-1,5-Bisphosphate (RuBP) molecule):

$$A_n = \min(w_c, w_j) - R_d \qquad (4.14)$$

and R_d is the dark respiration rate. Both w_c and w_j vary as a function of the vegetation type. Therefore, LULCC affects A_n, which, in turn, affects g_s and r_s, and therefore r_c and therefore λE and therefore H. Furthermore, LULCC affects L, which affects the scaling of r_s to r_c (Equation (4.10)) and λE and therefore H.

There are two remaining major vegetation-related parameters that affect the exchange of energy and water between the land and the atmosphere. The leaf area index is important in intercepting rainfall and the subsequent redistribution of intercepted rainfall to throughfall, stemflow, or re-evaporation. The amount of rainfall interception is based on attributes including canopy characteristics, rainfall intensity and duration (Dickinson et al., 1986), and wind. The importance of rainfall interception is linked to the timescales over which rainfall is returned to the atmosphere. Intercepted water that fails to fall through to the soil re-evaporates rapidly, commonly on timescales of minutes to hours, due to the large aerodynamic roughness of canopies, high ventilation, and large surface area in contact with the atmosphere. In contrast, precipitation that infiltrates into the soil tends to remain there for much longer periods (days to months) and may reach rivers or groundwater to be effectively lost to the atmosphere, potentially for years to centuries. A key rationale for incorporating canopy interception into land-surface models (LSMs) is, therefore, to capture the higher time frequency response driven by canopy interception (Scott et al., 1995), which could trigger changes in the diurnal cycle of clouds and affect radiation, in comparison to the slower timescale response of soil-based processes.

Precipitation interception (I) by canopies varies depending on the vegetation type. Rutter et al. (1972) suggest it is commonly 20%−40% of annual rainfall in conifers and 10%−20% in hardwoods. Many models of canopy interception have been proposed, but many parallel

Sellers et al. (1986) where interception of rainfall (P) is assumed to be similar to the treatment of the transmission of the solar beam for spherically distributed leaves, which takes into account the factor of leaf angle distribution:

$$I = P(0.25(1 - e^{L(p_1 + p_2)})) \qquad (4.15)$$

where p_1 and p_2 are vegetation-type dependent parameters that depend on leaf angle parameters (Sellers et al., 1986). Note the use of L in the exponent; thus, I is dependent on L, which is dependent on vegetation type. Typically, LULCC, in the form of deforestation, reduces L but the indirect impact of human activity via CO_2 fertilization has the potential to increase L.

The final piece of the jigsaw is the response of vegetation to the risk of moisture stress. Some plants respond to moisture stress by dieback. Grasses in particular are shallow-rooted and tend to simply die under severe moisture stress and regrow from seed when moisture is again available. Trees tend to allocate more carbon to grow roots that penetrate more deeply into the soil, potentially to tap groundwater (Jackson et al., 1996; Canadell et al., 1996). As a result, trees can withstand dry periods and maintain transpiration and carbon uptake and thereby maintain a cooler and moister local climate (Bonan, 2002). Following LULCC in the form of deforestation, therefore, an environment of relatively low seasonality in the moisture flux (because the trees can tap deep soil water) is replaced by crops or grasses that may evaporate actively in spring and early summer, but as the soil dries the large evaporative flux is replaced with a large sensible flux with implications for the overlying atmosphere. Kleidon and Heimann (2000) used a relatively coarse resolution climate model to explore the role of deep roots in tropical forests. They suggest that when deep roots are correctly represented, deforestation can have a significant impact on the simulation of the atmosphere and these can lead to teleconnections via changes in moisture transport (cf. McGuffie and Henderson-Sellers, 2004).

Thus, LULCC affects *both* the energy available to drive the system via a change in the albedo and changes in the way net radiation is partitioned between sensible and latent heat fluxes because it affects the capacity of the soil−vegetation system to evaporate. Changes in LAI, roughness, stomatal conductance, photosynthesis, and root−soil water interactions all combine to affect the balance of H and λE. *This is not merely on the annual or decadal timescale.* LULCC also directly impacts the seasonality of these fluxes. Deforestation tends to lead to a drier surface that is warmer, with a high seasonality in λE and a higher probability of severe moisture stress late in the season. Drier surfaces tend to have a larger diurnal temperature cycle with higher maximum temperatures while, if sustained for longer periods, can lead to a higher probability of heatwaves that are suppressed in a moister and cooler forested environment.

Figure 4.2 illustrates the impact of LULCC in the tropics using observational data measured over two sites in Amazonia: a near-undisturbed forest (the Tapajos National Forest: Miller et al. (2009)) and a cleared forest now covered with pasture (Para, Brazil: Fitzjarrald and Sakai (2010)). These sites are ~80 km apart and therefore not precisely comparable, but are exposed to similar large-scale meteorology. Figure 4.2a shows the impact of LULCC on temperature measured at 2 m above the surface for the average of three days in January. The pasture site is warmer throughout the day, by ~1°C at night and by more than 2°C

during the day. The two sites are contrasting in terms of the surface energy balance. The pasture site experiences relatively low latent heat (Figure 4.2b) and high sensible heat fluxes (Figure 4.2c) while the forested site experiences high latent heat and low sensible heat. This is not an isolated occurrence; Figure 4.3a shows that the forest site is cooler throughout the year (2002 is shown here) by 1.5°C–2°C. The capacity of the forests to maintain a high evaporative flux through the year, and through any period of lower rainfall, is shown in Figure 4.3b. The pasture experiences high evaporation early in the year (as high or higher than the forest) but, from July, while the forest maintains a flux of ~100 W m^{-2}, the evaporation from grasses drops to ~20 W m^{-2} and the sensible heat increases, highlighting a moisture-limited surface (Figure 4.3c). Sensible heat remains low through the year in the forest (Figure 4.3c).

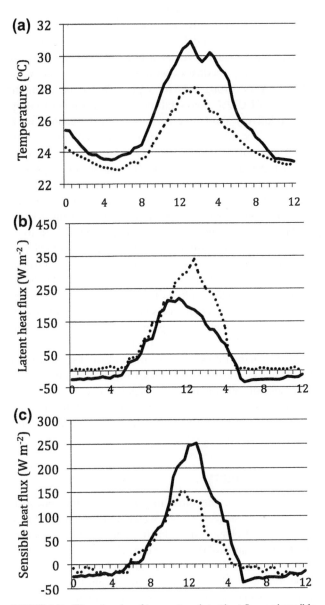

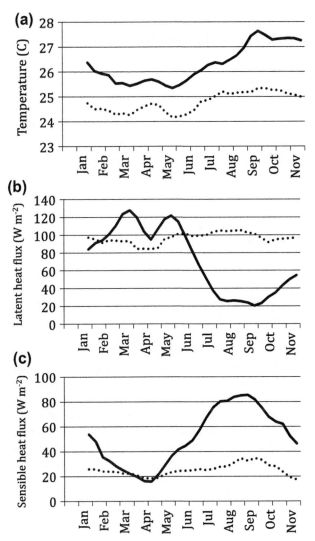

FIGURE 4.2 Diurnal cycles of temperature, latent heat flux, and sensible heat flux for a near-undisturbed forest (the Tapajos National Forest: Miller et al. (2009) − dotted line) and a cleared forest now covered with pasture (Para, Brazil: Fitzjarrald and Sakai (2010) − solid line). Data are averaged over three January days.

FIGURE 4.3 As Figure 4.2, but averaged over a full annual cycle. Lines are smoothed to highlight the differences between the near-undisturbed forest (dotted line) and the cleared forest (solid line).

The scale of this difference, particularly in the second half of the year, is considerable and is sustained to an extent that could likely affect cloud cover, boundary layer depth, and atmospheric moisture content, if spatially significant.

Figures 4.2 and 4.3 represent one measured example of the impact of LULCC over a relatively short period of time. It is important not to extrapolate this impact over the scale of the LULCC (Figure 4.1), since the way fluxes respond to deforestation (for example) depend very much on the background climate (Davin and de Noblet-Ducoudré, 2010) and the characteristics of the native vegetation being removed. Nor should one assume that Figures 4.2 and 4.3 are representative of all days or seasons in the tropics. It is also important not to prejudge the case that these changes would perturb the atmosphere in a climatological sense or assume that, if clouds and atmospheric moisture are affected, this would necessarily have a wider scale impact.

4.3.3. The Surface Water Balance

Rainfall (P) falling on the surface is partitioned between evaporation (E) and runoff (R) with changes in soil moisture (S) and snow mass (N):

$$P = E + R + \Delta S + \Delta N \qquad (4.16)$$

However, this can be expanded to account for interception loss I, the difference between evaporation pathways (through the soil, E_s, or the vegetation E_t) and the partitioning of total runoff between fast (R_{surf}) and slow (R_{drain}) components:

$$P - I = E_s + E_t + R_{surf} + R_{drain} + \Delta S + \Delta N \quad (4.17)$$

Land-cover change affects I, the partitioning of E between E_s and E_t via resistances to the flow of water, canopy cover, and the distribution of roots. Land-cover change affects the intensity of rainfall hitting the surface because (in the case of deforestation) of reduced I and likely increased R_{surf}. If more rainfall generates fast runoff then this water tends to flow into rivers and is rapidly removed from a catchment in comparison with water that infiltrates and increases soil moisture. Changes in the rate of infiltration and changes in the production of surface runoff and drainage have been shown to have clear impacts on the amount of soil water available for plants, and therefore on the values of all evaporative fluxes. These potentially affect P via atmospheric feedbacks (changes in atmospheric moisture and convergence), as illustrated by Ducharne et al. (1998) through sensitivity studies changing the formulation of infiltration/runoff/drainage processes. Moreover, different land-cover types also have distinctly different root distributions (Canadell et al., 1996; Jackson et al., 1996) that can lead to very different strategies to extract water from the soil, and therefore to very different amounts of water transferred back to the atmosphere through evapotranspiration (Kleidon and Heimann 2000). These processes, though, have not yet received enough attention in most LSMs (Feddes et al., 2001) and such omission may very well lead to an underestimate of the impacts of LULCC on climate.

4.3.4. The Snow–Climate Feedback

Over 50% of Eurasia and North America can be seasonally covered by snow (Robinson et al., 1993), leading to significant spatial and temporal fluctuations in surface conditions. The properties of snow (e.g., high albedo, low roughness length, and low thermal conductivity) lead to global-scale impacts (Vernekar et al., 1995). Snow is also one of the key feedbacks within the climate system and plays a very strong role as a positive feedback. In the context of LULCC, snow falling on a forest tends to drop, over relatively short time periods, from the canopy on to the soil. The forest, therefore, remains radiatively dark, absorbing solar radiation to stay relatively warm. If the forest is removed and replaced with grasses or crops, then these tend to be senescent in winter and small amounts of snow can transform a region from being relatively dark due to the low reflectivity of most organically rich soils to being highly reflective once snow accumulates enough to mask the vegetation. The snow–albedo feedback, coupled with changes in vegetation, is therefore dramatically different depending on the nature of the vegetation. In a landmark paper, Betts (2000) explored this issue to show that forestation in regions subject to seasonal snow cover led to significant regional warming, marginally due to a lower albedo through summer and, most importantly, due to the reduction in the impact of snow in winter and spring caused by the masking effect of the trees. Conversely, deforestation that typically warms in the tropics can cool in the higher latitudes where snow becomes a stronger force through winter and spring, as discussed in Davin and de Noblet-Ducoudré (2010).

4.3.5. Summary

The coupling of the three major roles of the terrestrial surface — the surface energy, water, and carbon budgets — is intimate. A change in λE due to a change in net radiation, or the mechanisms that influence λE, affects the surface energy balance. Since λE (W m^{-2}) and E (mm, mm per unit time, or kg m^{-2}) are linked via the latent heat of vaporization, changes in λE affects the surface water balance and, therefore, the partitioning of P between E and R. Since λE can occur as transpiration, λE is coupled into the carbon balance (see Dickinson, 2012, this volume). Changes in transpiration can affect carbon uptake, but changes in atmospheric CO_2 can also reduce stomatal conductance, reduce transpiration, reduce total evaporation, and therefore have the potential to increase runoff via the surface water balance (Gedney et al., 2006; Betts et al., 2007).

4.4. LINKS BETWEEN LULCC AND CLIMATE

A change in the surface energy balance *must* affect the overlying atmosphere. A reduction in λE reduces the flow of moisture into the atmosphere, reducing the likelihood of rainfall and reducing cloud cover. An increase in H tends to warm the atmosphere, deepening the boundary layer and likely increasing advection of heat and moisture. A large-scale reduction in vegetation cover also reduces the aerodynamic drag on the atmosphere, increasing wind speeds and increasing the distances from, or to which, moisture and heat can be advected. However, while LULCC must affect the overlying atmosphere, the observational evidence that this is climatologically significant is not extensive. The absence of observational evidence does not mean there is no climatological influence (as reported below), but disentangling the LULCC impact from the influence of, for example, increased GHGs, changes in aerosols concentration, and natural variability remains very challenging.

One region that has been extensively studied is Western Australia. In a body of work covering almost two decades, Tom Lyons has established a regionally significant impact of LULCC on the micrometeorology of specific Australian areas (Lyons et al., 1993; 1996; Lyons, 2002), more specifically on the characteristics of the boundary layer and on convection. Changes in the regional meteorology have been linked to LULCC using a combination of observations and modelling, and the perturbation appears to have affected cloudiness resulting from changes in sensible heat fluxes (Ray et al., 2003), at least on short timescales. There is further evidence of LULCC affecting observed temperatures over the USA (Bonan, 2001) and at local-scales there is strong evidence in many regions. Irrigation, a form of land use that does not necessarily imply changes in land cover, has also been recognized as a significant way to impact the diurnal amplitude of temperature, its daily and even seasonal values, and evapotranspiration (de Ridder and Gallée, 1998; Boucher et al., 2004; Sacks et al., 2009). However, almost every other observationally-based example of LULCC affecting climate is a short-term case study or is focused on perhaps the most intensive form of LULCC — urbanisation (see Arnfield, 2003). Urbanisation does affect climate (Karl et al., 1988; Gallo et al., 1999; Hale et al., 2008; Cleugh and Grimmond, 2012, this volume).

In a recent review (Seneviratne et al., 2010), the potential role of soil moisture is separated into three classes. Soil moisture and LULCC are not directly comparable of course, but both can represent a major and coherent land-based anomaly, and therefore some insight sourced from analyses of soil moisture might be applicable to LULCC. Firstly, Seneviratne et al. (2010) note that higher rainfall usually leads to higher soil moisture. Secondly, they ask whether higher soil moisture leads to higher evapotranspiration. They point out that this relationship would only hold in regions defined as transitional — that is, where a combination of energy limitation and water limitation defines the evapotranspiration. Clearly, over a water surface, the availability of water does not limit evaporation; rather, the availability of energy to drive evaporation provides the constraint. Similarly, over a hot desert, energy is unlikely to be the limit to evaporation; rather, the lack of water represents the limitation. At some points in-between, 'transitional' situations arise that are more complex than these examples and both water and energy can provide limits to evapotranspiration. We can, therefore, ask a similar question to Seneviratne et al. (2010): does LULCC (represented by deforestation) lead to lower evapotranspiration? As for soil moisture, the answer depends on the detail of the controls on evapotranspiration. Depending on the specific environment, LULCC (deforestation) could decrease evapotranspiration, but it could also increase the release of latent heat (i.e., evapotranspiration) if albedo is decreased, for example, via the planting of a particularly dark crop (Ridgwell et al., 2009). LULCC could also increase evapotranspiration if selective logging increases the effective roughness by introducing small cleared areas within a forest (Mei and Wang, 2010). Thirdly, Seneviratne et al. (2010) ask whether higher evapotranspiration leads to higher rainfall. They note that this is not straightforward and, while a large number of modelling studies have identified a possible feedback between higher evapotranspiration and higher rainfall (Beljaars et al., 1996; Koster et al., 2000; Betts, 2004; Pal and Eltahir, 2008), others have pointed to the convective instability and/or cloud formation being stronger over dry surfaces (Ek and Mahrt, 1994; Findell and Eltahir, 2003; Hohenegger et al., 2009). The point here is that while LULCC (deforestation) tends to decrease evapotranspiration that might decrease rainfall, it is not necessarily always the case that it does. Indeed, regional warming, generally associated with decreased evapotranspiration, may lead to increased advection of air masses and potentially of water vapour, thereby favouring rainfall or changes in deep convection that might also trigger increased precipitation (Polcher, 1995).

An additional challenge relates to the 'hotspots' identified originally by Koster et al. (2004) (Figure 4.4). These are regions where soil moisture anomalies were shown to have a direct impact on the overlying atmosphere. Persistence patterns in soil moisture and evapotranspiration (Seneviratne et al., 2006; Teuling et al., 2006) or the correlation of evapotranspiration with temperature

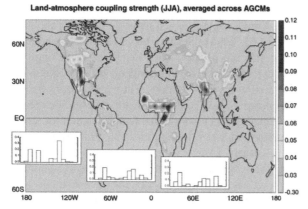

Land-atmosphere coupling strength (JJA), averaged across AGCMs

FIGURE 4.4 A diagnostic of the strength of the land-atmosphere coupling for boreal summer averaged across 12 atmosphere-only climate models. Regions coloured orange and red represent 'hot-spots' where the coupling is strong. (*Source: Koster et al., 2004. Reprinted with permission from AAAS.*)

(Seneviratne et al., 2006) can be used as indirect estimates of soil moisture—evapotranspiration coupling. Betts (2004) identified several tight coupling relationships between the soil moisture and boundary layer structure, cloud cover, and radiation using re-analysis in selected river basins. These relate to soil moisture, but similar 'hotspots' probably relate to LULCC. This implies that in some regions, LULCC at a given scale may not perturb climate because it happens to occur in a region of weak land—atmosphere coupling. Conversely, a smaller perturbation may have a significant impact on the atmosphere because it occurs in a tightly coupled region. LULCC might be expected to have the main impact where:

(a) land—atmosphere coupling is strong;
(b) rainfall recycling is strong; and
(c) most rainfall originates via local evaporation.

We neither know where these regions coincide, nor do we know if LULCC changes land—atmosphere coupling strength.

In the absence of strong observationally-based evidence, modelling provides useful insight to guide us on the impacts of LULCC on climate. There is a vast literature using regional and, to a lesser extent, global-scale models to explore the impact of LULCC. Some of the best examples have focused on Amazonian LULCC. They include Henderson-Sellers and Gornitz (1984), Dickinson and Henderson-Sellers (1988), Polcher and Laval (1994), Polcher (1995), McGuffie et al. (1995), Sud et al. (1996), Chase et al. (2000), and Avissar and Werth (2005). The impact of temperate deforestation on climate has been explored by Bonan (1997, 1999) and Oleson et al. (2004), while the impact of global LULCC has been considered by (among others) Chase et al. (2000), Betts (2000), Betts (2001), Govindasamy et al. (2001), Zhao et al. (2001), Bounoua et al. (2002), Zhao and Pitman (2002), Findell

et al. (2007, 2009), Lawrence and Chase (2010), Pielke et al. (1998, 2002), Davin and de Noblet-Ducoudré (2010), and Pongratz et al. (2010). Mahmood et al. (2010) provide detailed reviews of much of this work, but focus mainly on the evidence in support of the impact of LULCC rather than any confounding evidence or weaknesses that apply to the reviewed studies. In general, we see a suite of warning signs that are worth watching for when reading the literature on LULCC. These do not mean the results identified by these warning signs are wrong, but it means they need to be considered with appropriate caution.

1. *Simulations conducted over a few days.* These commonly highlight a response by the atmosphere but impacts over a few days may be compensated a few days later to generate a minimal net impact. One certainly cannot extrapolate from a simulation of a few days to conclude a climatologically significant response.

2. *Simulations focused on a specific meteorological event* (e.g., thunderstorms; Shepherd, 2005). While LULCC might trigger an event (Gero and Pitman, 2006), one cannot infer a climatologically significant result from single events. Even well-conducted studies (e.g., Beljaars et al., 1996) may be examples of a special case too rare to be climatologically significant.

3. *Simulations that convert large regions from tropical forest to short grassland, or convert all global forests to grassland* (Davin and de Noblet-Ducoudré, 2010). These are important in examining the potential or theoretical impact of LULCC. Some cases are among the most famous climate model experiments conducted (Henderson-Sellers and Gornitz, 1984) but, as Figure 4.1 suggests, it is not a plausible outcome in the twenty-first century. The focus herein is on perturbations that are plausible in the same way as emission scenarios used in AR4 are plausible, while a climate simulation to 2100 using a CO_2 concentration of 3500 ppmv would be seen as a sensitivity study.

4. *With exceptions, simulations conducted with a single climate model.* Climate models vary greatly (Randall et al., 2007) and project very different responses to a perturbation. Koster et al. (2004) showed that the land is coupled to the atmosphere with very different strengths and this may lead to major differences between models in the impact of LULCC. In the same way as the IPCC would not base conclusions on one climate model, we should not base our conclusions of the significance of LULCC on any single model.

5. *Simulations that have assessed statistical significance using tests that do not take autocorrelation into account* (e.g., a Student's t-test). These tend to overestimate the significance of a change in the climate (see Findell et al. (2006) for an explanation) and generate excessive false positives ('false' in a statistical sense).

6. *Simulations using first-generation LSMs* (Pitman, 2003). These are likely to overestimate the impact of a perturbation (Chen et al., 1997) because they do not represent the coupled energy, water, and carbon cycles (cf. Dickinson, 2012, this volume).

7. *Simulations that have been updated using newer versions of a given climate model, where simulations with fixed sea surface temperatures (SST) were redone using mixed layer models, or where an experiment has been repeated more robustly than earlier examples.* Old experiments with older models are not necessarily wrong but, if newer and more physically realistic versions are available, it is likely wise to consider them.

8. *Simulations conducted at a coarse spatial resolution.* The relationship between model resolution and the scale of impact on the atmosphere is not clear, but early evidence (Hohenegger et al., 2009) points to a strong sensitivity between soil moisture anomalies and impact associated with the intensity of convection. While this does not necessarily imply a strong sensitivity between LULCC, convection, and model resolution, it does suggest that results from coarse resolution models should be treated cautiously. Defining the scale at which a model is 'coarse' in this context is not currently possible.

In the following we have tried to identify some selected publications that are less affected by these weaknesses. There are no individual studies that avoid all these weaknesses, so if these conditions were strictly enforced there would be virtually no results from models to discuss. However, there are some publications that have avoided key weaknesses that make them particularly useful to look at more closely. The following presents our judgement of where the science of LULCC and its impact on the atmosphere and climate currently stands — in the sense that these represent the most recent efforts to rigorously examine the impact of LULCC and provide guidance on how to resolve major methodological weaknesses.

4.4.1. Hasler et al. (2009)

In a well-constructed experiment, Hasler et al. (2009) undertook multimodel simulations to address the single-model weakness of earlier work by Avissar and Werth (2005) and Werth and Avissar (2002, 2005). They used three climate models, each at $4° \times 5°$ resolution and undertook 52-year simulations using fixed SSTs, omitting the first 4 years to avoid spin-up issues. The use of three models makes this an important paper that made a significant step forward in addressing results generated with a single climate model, which is why it is discussed here. The models used were either relatively old (two versions of the GISS climate model) or were used at a spatial resolution

significantly degraded from the IPCC version of the model (CCSM was used at $1.4° \times 1.4°$ in the AR4: Randall et al. (2007)). It is not easy to judge how independent these three model estimates are, or how sensitive these three models are to a land perturbation. A version of the CCSM was used by Koster et al. (2003, 2006) that appears to be relatively sensitive to terrestrial processes (noting that this does not make it wrong). The imposed LULCC change was massive (but typical of earlier work by many authors, see figure 1 of their paper) including a conversion of regions larger than the Amazon Basin, the Congo, and all of South East Asia (including India, southern China, northern Australia, and Indonesia) from forest to grass and shrubs.

Hasler et al. (2009) assess the skill of each model in simulating seasonal rainfall: a very necessary step that is not always undertaken. While the rainfall climatology is not unreasonable, some large biases are clear in all three models. The authors do not show the actual changes in rainfall induced by LULCC as maps of mm d^{-1} in the three models and therefore it is not possible to determine whether the changes are very small or very large relative to the biases in the model. The authors also do not attempt to diagnose whether there is a link between the intensity of the rainfall response to LULCC and the magnitude of the present-day bias. Rather, they focus on the mean ensemble to track the strongest shared response between the models.

Hasler et al. (2009) show where rainfall has either significantly increased or decreased for at least 3 months of the year across an ensemble of the three models, as well as for individual models (figure 5 of their paper). They use a statistical test (Student's t-test) at a 95% significance level, *ignoring autocorrelation*. They used fixed SSTs and this was their rationale for ignoring autocorrelation. They find a strong response to LULCC of two kinds. Firstly, over regions of the tropics where LULCC has been perturbed, they capture a clear impact of LULCC (reduced rainfall rates) over Amazonia, the Congo, and parts of South East Asia. This result is consistent with a mass of literature and is probably robust at these large regional-scales. The result is common to all three models (see figure 5 of their paper) at a magnitude that is large relative to the background variability in each model. They also highlight a fair amount of positive responses remote from the actual perturbations. However, as at a 95% significance level one would expect 5% of all model grid points to give a false positive result (in a statistical sense), it is not convincing that the remote changes found by Hasler et al. (2009) exceed this level. This finding may, therefore, be modified if another statistical method were to be used. However, the authors also provide an analysis of geopotential changes. This builds on earlier analyses by Zhang et al. (1996) and Chase et al. (2000), who linked perturbations in the tropics to remote impacts using changes in atmospheric dynamics as the conduit for these teleconnections. They found seasonally-dependent changes in the northward transport and changes in diabatic heating and potential energy production associated with the changes in rainfall, although these mechanisms differed between the three models. They concluded that they found evidence of a wave-train forced by the changes in the tropics, but that this effect is buried in the natural variability of the climate system and can be more clearly found via multimodel ensembles.

Hasler et al. (2009) concluded that their results suggest that remote impacts of tropical deforestation on rainfall are real, but also note earlier conflicting results (e.g., Findell et al., 2006) and the fact that their results were inconsistent. They correctly note that the use of fixed SSTs tend to dampen internal model variability. While this would hint that using a coupled ocean model would increase the variability and, therefore, reduce the scale of LULCC impact, this is not necessarily the case if LULCC leads to a systematic change in the ocean temperatures (cf. Latif and Park, 2012, this volume).

Despite the fact that interpretation of this work still needs to be undertaken with care, Hasler et al. (2009) remains a rare publication — one that uses several models, long simulations, and an analysis of both local and remote impacts. It strongly reinforces the conclusion that LULCC has a strong regional impact, while noting that the scale of the perturbation imposed was not intended to be realistic. Despite the very sizable perturbation, the conclusion that their results "suggest the existence of a remote effect" is not definitive and may well be the result of their experimental and statistical design. Therefore, it needs to be addressed by many other models, as well as carried out with an additional accounting for oceanic feedbacks.

4.4.2. Findell et al. (2006, 2007, 2009)

In a series of publications, Findell et al. (2006, 2007, 2009) used the GFDL climate model at a $2° \times 2.5°$ resolution. A key development in their analysis was the accounting for autocorrelation and field significance in the statistical testing of the remote responses to LULCC. Simulation lengths were relatively long (around 50 years) and a mix of fixed SSTs and a mixed-layer ocean model was used. The GFDL climate model used in AR4 is among those with a higher sensitivity, but includes a coupled ocean model and is, therefore, only indicative of how the atmosphere-only model would respond. The GFDL model is also relatively strongly coupled to the surface (Koster et al., 2006) and might be expected to show a relatively large effect from LULCC.

Findell et al. (2006) perturbed the LULCC similarly to Hasler et al. (2009) with the broadleaf evergreen forests of South America, central Africa, and Oceania replaced with grasslands. The impact of these changes was a strong

reduction in latent heat flux in the tropics, coincident with LULCC. This led to warming in these regions, a result in common with many earlier experiments. Findell et al. (2006) then explored the impact of LULCC on seasonal 200 hPa temperature and 200 hPa geopotential height, showing that there was a significant effect through the tropics in Northern Hemisphere spring and summer. In terms of precipitation, impacts were discernible in the tropics, but not in the extra-tropics. They conclude that their use of long-term averaging and sound statistical methods allow them to conclude that, while LULCC (limited to the tropics, but greatly over-intense in terms of the actual perturbation) has a significant impact on the tropics, it does not lead to significant extra-tropical impacts *in the GFDL climate model.*

Findell et al. (2007) extended the focus of Findell et al. (2006) from the tropics to a global implementation of LULCC using an observed estimate of historical LULCC. In these experiments, therefore, most of the changes in LULCC are in Europe, the eastern US, India, and China, and the intensity of LULCC is smaller than with full-scale clearance, as represented in many earlier studies. The impact on annual air temperature was statistically significant (accounting for autocorrelation and field significance) over India and Europe only and the change in annual precipitation was only significant over parts of Europe and India (Figure 4.5). Overall, the global impact of LULCC was, therefore, negligible, but regionally a significant impact on the climate was simulated. A common criticism of studies of LULCC is that, while there may be regional impacts, these are insignificant in comparison to major modes of variability in the ocean (e.g., El Niño—Southern Oscillation, North Atlantic Oscillation, Southern Annual Mode — see Latif and Park (2012, this volume)). Findell et al. (2009) built on Findell et al. (2007) to explore this issue using an identical model to compare the impacts of ocean modes of variability with the LULCC perturbation.

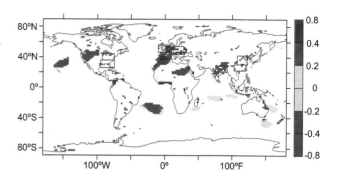

FIGURE 4.5 Annual differences in surface air temperature due to land-cover change. Differences are shaded only where they are statistically significant at the 95% significance level according to a modified t-test. A value of 6.0% passes the 95% significance level. *(Source: Findell et al., 2007. © American Meteorological Society. Reprinted with permission.)*

Their results demonstrated that in terms of global impacts, LULCC was negligible in comparison to the ocean modes tested in terms of air temperature but did have an impact on global precipitation on a par with the ocean modes. Globally, LULCC could be ignored in terms of the non-CO_2 component of climate change. However, over regions of LULCC, the impact of LULCC was notably larger on hydrometeorological quantities than the ocean modes.

In retrospect, this is hardly surprising since the forcing from LULCC is regionally focused, whereas the ocean forcing is by definition remote. However, Findell et al. (2009) demonstrated that, if a climate model is being used to explore surface quantities and the region has undergone significant LULCC, it is necessary to consider LULCC as a significant forcing. Since regions of intense LULCC are coincident with dense human populations and are therefore a key vulnerability to climate, the need to include LULCC in simulations of the regional impacts of (for example) increasing CO_2 is clear. It does not matter whether LULCC has a global teleconnection in this context.

4.4.3. Urban LULCC

Despite the increasing urbanisation in many countries, and the decades of research that point to large urban areas affecting regional climate (Arnfield, 2003), the representation of LULCC in the form of urban landscapes has received little attention in climate modelling or climate projections research. Oleson et al. (2008) are among a small group of authors who have coupled urban landscapes into climate models for innovative research into geo-engineering (Oleson et al., 2010) while several groups have explored the impact of urban landscapes on the scale of individual cities (see Cleugh and Grimmond, 2012, this volume).

McCarthy et al. (2010) used the Hadley Centre climate model and undertook 25-year simulations, including an urban land-surface scheme (Best et al., 2006) linked to the extent of urban coverage (Loveland et al., 2000). McCarthy et al. (2010) also added energy representing urban energy use into the atmosphere above the urban landscapes (either 20 W m^{-2} or 60 W m^{-2}). Their results show strongly regionally-varying impacts of adding in urban landscapes with the effect mainly seen on minimum temperatures and largely caused by the urban land, not the additional energy flux used to represent energy use. The impact on maximum temperatures was small relative to the impact of increasing CO_2 (perhaps an additional 15% of warming) but was significant in minimum temperatures. Expressed as a change in the number of hot nights for major cities, McCarthy et al. (2010) show dramatic increases for all cities included. Figure 4.6 shows a simplified version of McCarthy et al.'s (2010) results omitting estimates of uncertainty and the 60 W m^{-2} simulations. Figure 4.6 shows that the frequency of hot nights is always negligible under $1 \times CO_2$ compared

to $2 \times CO_2$. Furthermore, representing urban landscapes increases the likelihood of very hot nights under $1 \times CO_2$ (e.g., from negligible using rural land cover to 10–20 events on average per year over Los Angeles and Tehran). More dramatically, doubling CO_2 greatly increases the frequency of very hot nights in all cities (Alexander and Tebaldi, 2012, this volume). However, representing the cities in the climate model adds an additional and large increase in this frequency, doubling it in cities including Delhi, Beijing, Los Angeles, and Tehran. That is, omitting the representation of urban landscapes in climate models leads to a systematic underestimate of the changes over major cities. Note, Figure 4.6 suggests that it is the addition of the urban landscape, not the 20 W m^{-2}, that dominates the results in agreement with many other similar evaluations (Cleugh and Grimmond, 2012, this volume).

McCarthy et al. (2010) were the first to explore this issue explicitly. It is important to note that single model simulations were utilized, with a single estimate of urban extent. The best way to parameterize urban landscapes in climate models is at best uncertain and is some two decades less well-developed in comparison to natural landscapes (although rapidly resolving some of these uncertainties: Grimmond et al. (2010)). The conclusion of McCarthy et al. (2010) that urban landscapes increase the vulnerability of populations to global warming seems robust, but the scale of this increase remains necessarily uncertain until multiple-model comparisons can be undertaken.

4.4.4. Land-Use and Climate, Identification of Robust Impacts (LUCID): Pitman et al. (2009)

Given the criticism that most LULCC experiments were conducted with a single climate model, it was natural that a project was set up to attempt to resolve this problem. Under the auspices of the International Geosphere Biosphere Programme's Integrated Land Ecosystem–Atmosphere Processes Study (iLEAPS) and the World Climate Research Programme's Global Land Atmosphere System Study (GLASS), the Land-Use and Climate, Identification of Robust Impacts (LUCID) project was launched. The important word in 'LUCID' is 'robust' — the project was designed to identify those impacts that constitute a genuine signal, as opposed to noise generated by model variability. A second level of 'robustness' was to explore which of the apparent signals were consistent across multiple climate models.

Seven climate models were used to perform the LUCID experiments. Two experiments were conducted using prescribed inter-annually and seasonally varying SST and sea-ice extent using data from the C20C project (HadISST1.1, *ftp://www.iges.org/pub/kinter/c20c/HadISST/*). Firstly, present-day simulations, with all

FIGURE 4.6 Annual frequency of hot nights for $1 \times CO_2$ (left three bars) and $2 \times CO_2$ (right three bars). The *x*-axis shows rural (Rur), urban (Ur), and urban combined with a 20 W m^{-2} additional heat flux (Ur + 20). *(Source: Modified from McCarthy et al., 2010.)*

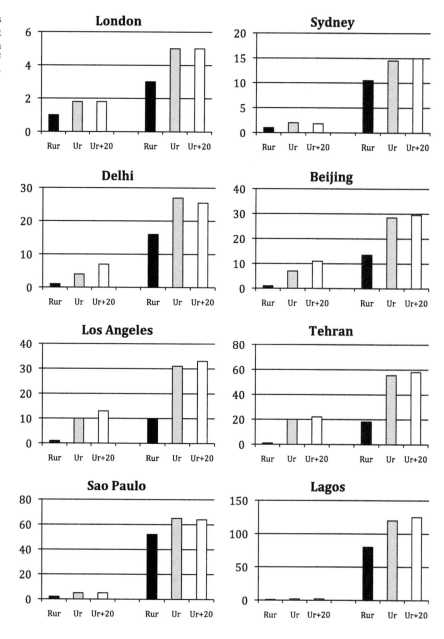

GHGs, land cover, and SSTs prescribed at their present-day values were conducted. The land cover was prescribed using a map reflecting 1992 and the period 1972–2002 is simulated. This was then followed by an identical experiment, except that land cover was changed to reflect 1870 conditions. Critically, to address the signal-to-noise ratio, each modelling group was required to conduct at least five independent realizations. The analysis used the Findell et al. (2007) method of a modified Student's t-test (Zwiers and von Storch, 1995) that accounts for autocorrelation within the time series, reducing the rate of false positives.

The preliminary results from LUCID were presented by Pitman et al. (2009). The key result is shown in Figure 4.7.

Here, for every climate model, there is a strongly, statistically-significant impact of LULCC on the simulated summer latent heat flux over the regions where LULCC was changed. In Figure 4.7, to be significant, values have to be well in excess of the horizontal line drawn at 5% since, with a 95% significance level, 5% of grid points should appear significant by chance. Clearly, over the regions of LULCC for both the latent heat flux and temperature (2 m air temperature) of every model greatly exceeds this 5% threshold. Given that seven models performed this experiment, the conclusion that LULCC affects the latent heat flux and temperature over regions affected by large-scale LULCC is *robust*. However, the direction of summer temperature change was inconsistent among the models,

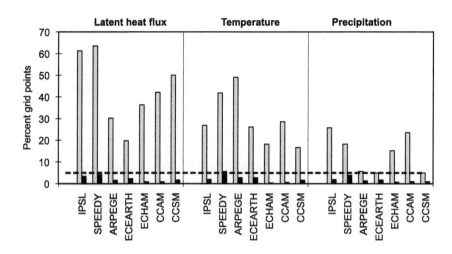

FIGURE 4.7 Percentage of land area that exhibits statistically significant changes in June-July-August (JJA) for latent heat flux, temperature, and precipitation. Light grey bar is the percentage of grid points with statistically significant changes where land cover changes (change in LAI > 0.5) within each climate model. The solid bar is the percentage of grid points with statistically significant changes where land cover is not changed. The horizontal line is the 5% significance level, expected by chance. *(Source: Modified from Pitman et al., 2009.)*

undermining our confidence in the robustness of this result. In the case of rainfall, as noted by Seneviratne et al. (2010) it is less clear whether LULCC affects rainfall. Four of the climate models used in LUCID (IPSL, SPEEDY, ECHAM, and CCAM) show a strongly significant impact on rainfall over regions of LULCC. The other three models do not show impacts above the level expected by chance. In terms of the results from LUCID, therefore, the hypothesis that LULCC affects regional rainfall in statistically significant ways remains unproven.

In Figure 4.7, the percent of land area significantly affected by LULCC *but remote from the actual LULCC* is shown by the solid black bars. For none of latent heat flux, temperature, or rainfall does a single model suggest a scale of impact larger than expected by chance. In terms of LUCID experiments, therefore, there was no evidence to support a hypothesis of a significant impact on these quantities remote from the regions of LULCC.

Is this conclusive? – unfortunately not. Firstly, LUCID used seven climate models and, even if this is a very large sample size for LULCC experiments, it is probably not large enough to form a reliable sample. Secondly, LUCID used fixed SSTs that may have suppressed teleconnections, particularly over the oceans. Thirdly, LUCID used historical LULCC that is mainly focused away from the tropics. Many of the experiments that have identified global teleconnections (e.g., Henderson-Sellers et al., 1993; Zhang et al., 1996; Gedney and Valdes 2000; Werth and Avissar, 2002, 2005) emphasize the global implications of future extensive tropical deforestation. Finally, while LUCID prescribed a common geographical distribution of LULCC, it could not require each model to represent LULCC in identical ways.

However, while LUCID is not conclusive in determining whether LULCC leads to large-scale teleconnections, it is conclusive in several other ways. Firstly, it highlights that any experiments exploring the impact of LULCC on a phenomenon is unreliable if only a few models are used. Thus, while exploration of the impacts of LULCC on the monsoon (Takata et al., 2009) or on extremes (Zhao et al., 2002; Deo et al., 2009; Douglas et al., 2009) are important first steps, the results are not conclusive. Secondly, it demonstrated that precisely how LULCC was implemented in a climate model mattered and that seven modelling groups could choose different, but equally legitimate, ways to do it. This leads to an unresolved research challenge of how best to impose LULCC on a climate model, which is discussed later.

4.4.5. Implications of LULCC for Future Simulations; Feddema et al. (2005)

Several authors have explored the impact of future LULCC under enhanced GHG levels (Costa and Foley 2000; Zhao et al., 2001; Defries et al., 2002; Narisma and Pitman, 2004; Bala et al., 2007; Paeth et al., 2009). Several groups have tried to predict the scale of future LULCC. Figure 4.1g and 4.1h show one estimate derived using the IMAGE model (see *http://www.pbl.nl/en/themasites/ image/index.html*). Relative to 2000, the changes estimated to 2050 are mainly focused on the edges of regions yet to be cleared (the Amazon and Congo Basins) and a general intensification of areas already partially cleared. This continues through to 2100, reducing the size of the Amazon and affecting most of the Congo and other tropical forests in the tropics.

Feddema et al. (2005) linked future changes in GHGs consistent with AR4-style emission scenarios with future changes in LULCC. Using a single climate model (the Parallel Climate Model at a resolution not stated) they added the effects of LULCC into simulations conducted using the A2 (relatively high) and B1 (relatively low) emissions scenarios (Nakicenovic et al., 2000). These simulations were then compared with identical future simulations that lacked estimates of future LULCC (i.e., in common with AR4 projections). They conducted transient simulations.

Feddema et al. (2005) found that the impact of LULCC varied depending on the emissions scenario used. They also showed that the impacts of LULCC were negligible on the global average (due to offsetting regional signals) and that most significant climate effects were associated directly with LULCC in tropical and temperate regions, but were indirectly linked to local LULCC at higher latitudes. Feddema et al. (2005) is a key step forward in the study of LULCC and the interactions with climate because it was the first attempt to integrate LULCC with IPCC emission scenarios. However, the statistical methods used, coupled with the single realizations performed and the use of a single climate model, means their results are preliminary. For example, comparing figure 1a in Feddema et al. (2005), which shows the change in land cover for 2050 under a low emissions future, with the first panel in their figure 2 (Northern Hemisphere summer change in temperature) points to a problem. The changes in LULCC included in the model are very small in the Southern Hemisphere with no changes south of about 40°S. Despite this, among the largest regions displaying warming is Antarctica, which warms by 0.5−1.0°C almost everywhere and by up to 2°C (and statistically significantly) at the continental margin. This is quite a common signal in climate model experiments, as are the large changes over the Arctic. These commonly disappear if multiple realizations are performed and statistical tests accounting for autocorrelation are used. Given that these signals are therefore probably noise, it casts doubt on the other apparently statistically significant signals identified by Feddema et al. (2005). To be clear, this does not mean that Feddema et al. (2005) are wrong; it merely means that one set of simulations cannot be conclusive and their leadership in conducting these simulations needs to invigorate other groups to follow. What is clear from their experiments is that over regions of intense LULCC, in common with many earlier studies, a strong and coherent signal is obtained for temperature. Since the impacts of climate change will be felt most strongly at regional-scales and that in many regions LULCC leads to warming that is additive to the CO_2 signal, the conclusion in Feddema et al. (2005) that their results demonstrate the importance of including land-cover change in forcing scenarios for future climate change studies seems entirely defensible.

4.5. LAND USE AND UNDERSTANDING OUR FUTURE CLIMATE

The context of this chapter is the biogeophysical (not biogeochemical) impacts of LULCC (Dickinson (2012, this volume) covers aspects of biochemistry). In the context of

the biogeophysical processes related to LULCC, we can address the question, "Is LULCC important?" This can be answered definitively, but the answer depends on the context of the question. At one extreme is the question, "Is LULCC important to the global climate sensitivity to a doubling of CO_2?" Answered relative to CO_2, the answer from available numerical experiments seems to be, "No". It cannot be entirely ignored due to the release of CO_2 via LULCC, but suggests that global climate sensitivity would be substantially different if LULCC was accounted for more explicitly is a hypothesis that lacks significant supporting evidence. At the other extreme is the question, "Is LULCC important to the regional climate (or regional climate extremes, or regional climate sensitivity to increasing CO_2)?". The answer to this question is definitively, "Yes". LULCC, in this context can include deforestation (e.g., Pitman et al., 2004; Narisma and Pitman, 2003), agriculture (Lyons et al., 1996); irrigation (de Rosnay et al., 2003; Boucher et al., 2004; Douglas et al., 2009; Sacks et al., 2009), urbanisation (Shepherd, 2005; Trusilova et al., 2009), and fire (Görgen et al., 2006). Even where the impact is currently small, it is likely to grow in the future (e.g., Sahel: Taylor et al., 2002). Evidence from almost every region where intensive LULCC has been explored points to a significant impact (China: Gao et al., 2002, 2003; Europe: Heck et al., 2001; United States: Bonan, 1997, 1999, 2001; Australia: Lyons et al., 1996; Narisma and Pitman, 2003; McAlpine et al., 2009).

Between these two extremes are more difficult areas of the science to navigate, where there are no definitive answers. Some of these have become contentious as groups have worked to win an argument over the realism of global teleconnections. Some groups find clear teleconnections (e.g., Henderson-Sellers et al., 1993; Zhang et al., 1996; Gedney and Valdes, 2000; Werth and Avissar, 2002, 2005), while others do not (Findell et al., 2007, 2009; Pitman et al., 2009). There is an imperfect resolution to this disagreement. Most groups that find teleconnections convert the whole of the Amazon (and perhaps other tropical forests) from dense, dark, and lush forest to short bright grass — a sensitivity experiment of value in understanding the potential response of the climate, but not an attempt at a reasonable perturbation based on observations or projections of future land-cover change. Groups that do not find clear teleconnections tend to explore observed changes in LULCC and the scale of that change is small in comparison with the aforementioned tropical deforestation experiments. Thus, Findell et al. (2007) and Pitman et al. (2009) explore the impact of observed LULCC (i.e., little LULCC in the tropics) *and* use statistics that account for autocorrelation to minimize false positives and do not find teleconnections. This does not mean teleconnections will not occur if LULCC becomes intensive in the future, though Feddema et al. (2005) do not really provide strong evidence that they will. A reasonable set of hypotheses is therefore that:

- LULCC directly affects regions under LULCC in terms of temperature, turbulent energy fluxes, soil moisture, runoff, carbon budget, and perhaps rainfall. Importantly, extremes might be quite strongly affected (e.g., heatwaves, floods, and drought; cf. Alexander and Tebaldi, 2012, this volume).
- LULCC does not affect the global climate sensitivity to future (end of the twenty-first century) increases in atmospheric CO_2 significantly (e.g., Harvey, 2012, this volume).
- Historical LULCC does not affect the climate of regions geographically remote from regions of LULCC if the perturbation imposed is realistic; future changes may generate remote changes depending on the scale of LULCC.

The second and third hypotheses are contentious in that they are expressed negatively and many groups would very strongly disagree (although note that they are expressed as hypotheses). Several recent papers have pointed out methods to explore these hypotheses in rigorous ways that would generate a clearer understanding of the science. Building from statements in Findell et al. (2009) and Pitman et al. (2009), one way of addressing the remaining uncertainties would be a coordinated set of experiments designed along the following lines. We assume that these would be conducted using coupled climate models but this would be computationally expensive and a well-designed regional climate modelling initiative could address many of these questions.

1. A rigorous evaluation of LSMs uncoupled from the host climate models (i.e., off-line) needs to be conducted. Do the LSMs capture the contrast between natural and anthropogenic land cover? Can they simulate crops, recovery post-deforestation, and other significant processes? Do they properly simulate the impact of LULCC on fluxes? Off-line skill in capturing the impact of LULCC is clearly a necessity if the model is to capture the impact in a coupled model; it is, however, an insufficient criterion and can only be a first step in any systematic evaluation of the impact of LULCC on climate.

2. A climate model's sensitivity to LULCC needs to be placed in context. If, for example, a model's climate changes by 10°C due to a tweak in a specific parameterization, then it would suggest that it is an overly sensitive model. There are several simple sensitivity experiments that could be performed to determine the models' background response to a perturbation, but the simplest would be to know the impact of a doubling of atmospheric CO_2 in the model. It is important to obtain

this signal with *precisely* the same model as is used for an LULCC experiment, and this should be conducted with a fully coupled model to obtain a value of climate sensitivity consistent with IPCC estimates. An alternative would be to determine the climate model's response to a known ENSO-like SST anomaly. Teleconnection patterns from ENSO are quite well-known and a model should be able to broadly capture these. If a climate model captures teleconnections induced by an ENSO anomaly well, given it is a tropical anomaly, then it is hard to assert that teleconnections induced in the same model by a tropical anomaly caused by LULCC are unrealistic.

3. The impact of LULCC is likely to be significantly affected by the strength of the land–atmosphere coupling (Koster et al., 2004; Seneviratne et al., 2006). It is, therefore, helpful to know the coherence between each individual model's pattern of coupling strength and LULCC to understand how a model responds to LULCC.

4. Impose an LULCC anomaly *either* based on observed changes to date, or projected into the future. Any such anomaly should be realistic (e.g., Feddema et al., 2005) not 'scorched earth'.

5. While it is computationally expensive, multiple realizations are required to allow the signal from LULCC to be identified against a given climate model's background (natural) variability. LUCID used five realizations, but also used fixed SSTs. If transient experiments are run then, the number of realizations most likely has to exceed ten.

6. While it is organisationally difficult, multiple models should be used. LUCID used seven climate models but this is not enough to form a reliable sample. The number of models used is defined by community interest and resources and needs to be as inclusive as possible. The models used also have to be current versions of a given climate model to allow the relative role of LULCC to be compared with CO_2.

The effort involved in implementing this set of experiments is very considerable and we recognize this. However, unless we resolve the role of LULCC at regional-scales, the ability to project regional-scale changes resulting from other forcings, including CO_2, will remain limited, at least in regions affected by LULCC. If it can be demonstrated that LULCC does not trigger large-scale teleconnections, regional climate models (see Evans et al., 2012, this volume) may provide an alternative methodology, but this would still require an experimental design that would be computationally demanding.

This brings us to the scientific challenge of imposing a realistic LULCC on a model. When imposing an LULCC, two (or more) maps are used to describe the state of the

FIGURE 4.8 Illustration of how 50% of a natural landscape (50% forest, 50% grass) can be converted into very different combinations of vegetation via different, but equally valid, decisions: (a) original natural distribution; (b) each original vegetation type has been proportionally reduced; (c) all forests have been cleared and grasslands have remained untouched (for e.g., grazing); and (d) all grasslands have been converted into cropland while forests have remained untouched.

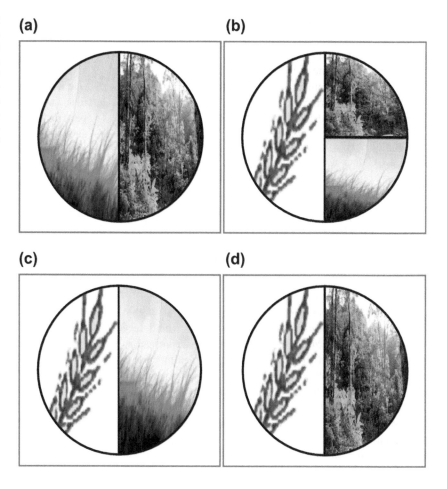

vegetation at various times. It is not straightforward for all modelling groups to use the same map. It is also not straightforward to implement these maps in a homogeneous way. There are multiple issues:

- It is common to reduce the natural vegetation proportionally, hence, if the original landscape was 50% trees and 50% grass (Figure 4.8a) and 50% of this area becomes agricultural then the new fractions would be 25% trees, 25% grass, and 50% crops (Figure 4.8b). But why? It may be as legitimate to remove all the trees and replace just these with crops (Figure 4.8c) on the grounds that a farmer might already use the grasses for grazing and would add productive land by removing the trees. Alternatively, it may be as legitimate to remove all the grasses and replace these with crops. Indeed this would be much easier and quicker for the farmer (Figure 4.8d). A modeller can therefore end up with a 50% tree cover (as in the natural state, Figure 4.8a or the perturbed state, Figure 4.8d), or 25% tree cover (Figure 4.8b) or no tree cover (Figure 4.8c) via equally reasonable judgements. This means that, while common maps might be provided, how the agricultural land is to

be created has to be agreed. This is critical in regions of snow (Betts, 2000), but is also likely to be important elsewhere.

- A similar issue is that it is difficult for most modelling groups to change their standard vegetation maps. Climate models are built with a particular vegetation map in mind and with decisions on how many vegetation types will be represented. A climate model may, in some ways, be calibrated to produce a good climate of the twentieth century with a specific vegetation map and a particular number of vegetation types. If one LSM allows for 5 vegetation types and another allows for 20, and maps are provided with 20 classes of vegetation, the implementation of these maps will mean different things in the 2 different models.

- The representation of crop phenology in LSMs is important. Some LSMs prescribe phenology using present-day satellite-derived datasets; others simulate seasonal phenology with implicit or calculated dates for cropping and harvest; and others simply describe crops as natural grassland, but with a different set of parameter values, without representing harvesting. Some

LSMs only represent natural vegetation and describe pasture and crops as a single type of grass. Some simulate bare soil between harvest and sowing, others simulate grass. These all affect the detail of how a given LULCC is implemented in a climate model. These difficulties are further confounded by disturbance to the phenology caused by global warming (e.g., Schneider and Root, 2002).

Pitman et al. (2009) concluded that the expression of LULCC in a climate model depends on how vegetation types are parameterized, how the LSM tiles the surface (there are several approaches), how land covers are actually implemented, which parameters are fixed, which are time-varying, how these differ between LSMs, and how strongly the surface is coupled to the atmosphere (Seneviratne et al., 2006). An identical land-cover perturbation is, therefore, impossible to impose on all models in this context, but a common set of procedures by which a similar LULCC perturbation can be imposed on multiple climate models is possible. This is what has been suggested by a group of scientists for the upcoming IPCC simulations (CMIP5) and would enable a multimodel approach to assess the impact of LULCC on climate.

This leaves us with a problem. Multiple climate models must be used to explore the scale of the impact of LULCC. The imposition of LULCC on these models needs to be as similar as possible, otherwise differences in the impact would be the result of the way LULCC was implemented rather than the impact of a realistic LULCC signal on climate. At present, we cannot do this — there are too many uncertainties in how to model terrestrial processes, how to implement LULCC, and how the land is coupled to the atmosphere to resolve the real signal from LULCC on climate. However, we *know* that LULCC has a large impact on regional climate in regions of intensive LULCC. So, as noted by Feddema et al. (2005) and by Pitman et al. (2009), it is necessary for regional predictions that LULCC is included in AR5 climate projections, and that these changes should include urbanisation (McCarthy et al., 2010; Cleugh and Grimmond, 2012, this volume).

The logical conclusion from this is that including LULCC in AR5 climate models may increase the differences between models *over regions of intensive LULCC*. Clearly, there is a need to design a common numerical experiment, such as the classic 'doubling CO_2' experiment to diagnose the actual sensitivity of individual climate models to LULCC. Such an experiment would encompass, simultaneously, the way the land-atmosphere interactions are parameterized and the intensity of the coupling. Such an experiment would need to include a standardized land forcing that can be applied to all models and avoid potential differences arising from variations in the implementation and/or parameterizations.

In summary, LULCC is a key component of the Earth's climate in terms of the net carbon balance, emissions of CO_2, regional impacts on radiative forcing, and impacts on the partitioning of available energy between sensible and latent heat. Collapsing the role of LULCC into radiative forcing is an invalid simplification (Davin et al., 2007; Davin and de Noblet-Ducoudré, 2010). Radiative forcing is a poor measure because it is only one part of the impact of LULCC. The substantial effects LULCC has on the partitioning of available energy between sensible and latent heat and, the way precipitation is partitioned between evaporation and runoff, is not captured by changes in radiative forcing. Furthermore, global measures of the impact of LULCC are not appropriate since different regions commonly experience impacts of different sign such that a global average is negligible (Feddema et al., 2005). A global measure, such as radiative forcing, is useful for the change in energy due to elevated CO_2 but, as has been convincingly argued by Pielke et al. (2002), it is not for LULCC.

LULCC, while important, is probably only important at specific scales. It is also particularly difficult to resolve in global climate models. There is no doubt that LULCC has a large impact on some regions (but not all) that have been intensively affected by LULCC. It is also very unlikely that it has a strong effect on the global climate sensitivity. In-between, the question of the scale of remote impacts of LULCC remains unresolved because almost all experiments have used modelling approaches that leave difficult problems of statistical significance and the reasonableness of the imposed changes unresolved. Resolving these problems requires considerable effort and most climate modelling groups invest remarkably small amounts of resources in terrestrial modelling. This is surprising since virtually every important reason for developing climate models relates to terrestrial quantities (such as soil moisture, crop yields, biodiversity, drought, streamflow). The ways forward for the climate modelling communities to resolve the scale of the role of LULCC on climate is clear, but it requires a base-level of effort that is not commensurate with current investment in this area and, therefore, represents a real and continuing challenge for the regional prediction of the impacts of various types of forcing on quantities of direct relevance to humans.

Time and Tide

Feedbacks and timescales of climate change are strongly dependent on the two-thirds of the Earth's surface that is ocean-covered and on the frozen water masses in glaciers, sea-ice, and ice sheets. The hydrosphere and cryosphere are examined as vital components of the integrated climate system.

Almost all models suggest that the waiting time for a late twentieth century, 20-year extreme, 24-hour precipitation event will be reduced to substantially less than 20 years by mid-twenty-first and by much more by the late twenty-first century, indicating an increase in frequency of the extreme precipitation at continental and subcontinental-scales.

Alexander and Tebaldi, 2012, this volume

Chapter 5, *Fast and Slow Feedbacks in Future Climates* by Danny Harvey, reviews the latest evidence concerning the magnitude of the climate sensitivity based on fast feedback processes (involving water vapour, temperature lapse rate, clouds, and seasonal snow and ice) and presents a linear feedback analysis framework for the analysis of the effect of feedbacks on climate sensitivity. Chapter 5 then reviews potential slow feedback processes that could amplify the fast-feedback sensitivity, with an emphasis on theoretical, observational, and modelling studies concerning the magnitude of climate—carbon cycle feedbacks. The linear-feedback analysis framework is extended to include the carbon cycle and climate—carbon cycle feedbacks, along with illustrative quantitative examples to illustrate the dependence of peak temperature warming in response to anthropogenic CO_2 emissions on the interaction between fast and slow feedback processes. This analysis leads to firm conclusions. There is comprehensive and overwhelming evidence supporting the long-held consensus viewpoint that the climate sensitivity based on fast feedback processes is very likely to fall between 1.5 K and 4.5 K, with essentially no possibility of a fast-feedback sensitivity greater than 6 K. Positive climate—carbon cycle feedbacks are likely to enhance the climate sensitivity by 25—50%, excluding potential feedbacks related to short but intense emissions of CO_2 and/or methane from yedoma (loess) soils and methane clathrates. Much greater amplification of the future warming could occur through release of CO_2 and methane from thawing yedoma soils and from destabilization of marine methane clathrates. Excluding greenhouse gas releases from yedoma soils and clathrates, the likely range of peak global mean warming associated with an extremely stringent CO_2 emission scenario (global fossil fuel emissions reaching zero by 2080 and cumulative emissions of 800 GtC) is 1.5—4.5 K, while the likely range for a business-as-usual scenario with cumulative emissions of 1760 GtC is 2.5—6.0 K. Inclusion of greenhouse gas releases from yedoma soils has the potential to turn what would be a 3.5 K peak warming into a 5.5 K peak warming.

Chapter 6, *Variability and Change in the Ocean* by Alex Sen Gupta and Ben McNeil, examines the ocean as an integral part of the climate system. The world's oceans have absorbed over 80% of the additional heat and about one third of the additional carbon dioxide that has been added to the climate system over the last few decades. This strongly moderates the impacts that would otherwise have occurred. However, these services come at a cost. The ocean has become warmer to depths of hundreds of metres, increasing surface stratification and causing thermal expansion and sea-level rise. The ocean chemistry has changed, with the water becoming less alkaline (ocean acidification) and with lower aragonite saturation.

In addition, an intensification of the atmospheric hydrological cycle and changes in surface winds have driven regional changes in ocean salinity and certain circulation patterns. While there are far fewer chemical and biological observations of the global ocean, significant trends in oxygen concentration, primary productivity, and the characteristics of higher species are becoming apparent. Climate models capture many of the observed changes when driven by anthropogenic greenhouse gas emissions. They also demonstrate that much more dramatic changes are in store as we increasingly perturb the system. A shortcoming of these models, however, is that they do not represent key carbon cycle feedbacks that could potentially have a large effect on the ocean's ability to moderate climate change in the future.

Chapter 7, *Climatic Variability on Decadal to Century Timescales* by Mojib Latif and Wonsun Park, takes as its starting point the fact that the twentieth century featured considerable global warming of the order of 0.7°C. The Earth will continue to warm over the next several decades under any plausible scenario, but the climate trajectory will not follow a steady path, due to the presence of natural variability. The origin and future of decadal to centennial variability is the topic of Chapter 7. Global warming will inevitably be accompanied by regional-scale natural variability and change, which can have profoundly important impacts. Examples during the twentieth century include the Dust Bowl drought in the USA of the 1930s, the Sahel drought of the 1970s and 1980s, the ongoing drought in the southwestern US, and the decadal-scale changes in Atlantic hurricane activity. It is a central challenge of climate research to understand the dynamics of these fluctuations and predict such regional-scale climate variability and change over timescales of decades or even longer.

Chapter 8, *The Future of the World's Glaciers* by Graham Cogley, examines glaciers as insensitive but reliable indicators of climatic change. Measurements of glacier mass balance provide information that is independent of weather stations and bathythermographs. However, because the mass balance and the near-surface air temperature are both driven by radiation and energy balances, the correlation between them is very good and is the basis for most projections of the evolution of glaciers. Glaciers other than the ice sheets are the focus here. Observations show that they have been losing mass since the mid-nineteenth century, with a marked acceleration since about 1970. The total mass stored in these glaciers, which is less than 1% of that stored in the ice sheets, was between about 500 and 700 mm of equivalent sea-level rise in the late twentieth century. Projections show that over the twenty-first century a cumulative loss of 90−160 mm of equivalent sea-level rise can be expected with reasonable confidence. However, poorly quantified and poorly understood phenomena, including the delayed response of glacier dynamics to the forcing and losses by frontal ablation (calving and melting at terminuses at sea level), make the probability distribution function asymmetrical. Substantially lesser losses are very improbable, but greater losses cannot be ruled out with the same confidence.

Chapter 9, *Future Regional Climates* by Jason Evans, John McGregor, and Kendal McGuffie, recognizes that the climate of any region depends on phenomena acting at scales ranging from global to local. Predicting future regional climates requires capturing the influence of processes across this full range of scales. While global climate models can robustly capture the large-scale effects, they are not able to resolve the regional- to local-scale effects that can significantly influence a location's climate. In order to account for these smaller-scale processes, future global climates are 'downscaled' using various techniques to produce more realistic regional climates. Chapter 9 describes a number of techniques that have been used to downscale global climate simulations to the regional scale. It explores the uncertainties in the future regional climate predictions associated with the various techniques, and examines ways to quantify the 'most likely' future regional climate and the related uncertainty, so that informed investigations can be made into the impact on natural and anthropogenic systems of future climate change.

The bulk of scientists are pretty straight about saying this is a probability distribution. And right now our best guess is that we're expecting warming on the order of a few degrees in the next century. It's our best guess. We do not rule out the catastrophic 5 degrees or the mild half or one degree. And the special interests, ... from deep ecology groups grabbing the 5 degrees as if it's the truth, or the coal industry grabbing the half degree and saying, "Oh, we're going to end up with negligible change and CO2's a fertilizer," and then spinning that as if that's the whole story–that's the difference between what goes on in the scientific community and what goes on in the public debate.

Stephen H. Schneider, PBS interview

Chapter 5

Fast and Slow Feedbacks in Future Climates

Danny Harvey

Department of Geography, University of Toronto, Ontario, Canada

Chapter Outline

The Future of the World's Climate. DOI: 10.1016/B978-0-12-386917-3.00005-1

5.1. INTRODUCTION: THE SENSITIVE CLIMATE

Changes in climate over the coming centuries will be dominated by the effects of human emissions of greenhouse gases (GHGs) and the resulting increase in GHG concentrations, as these effects will overwhelm the long-term effects of natural causes of climatic change, such as changes in solar luminosity, changes in volcanic activity, internally-driven fluctuations in the exchange of heat between the ocean surface layer and the deep ocean, and slow changes in the Earth's orbit (which operate over periods of thousands of years, as discussed by Berger and Yin, 2012, this volume). Human emissions of aerosol precursors as a by-product of fossil fuel use and biomass burning also have a strong effect on climate (largely a cooling effect), but these are particles that last in the atmosphere only for days, requiring a continuous emission source for a sustained effect. The carbon dioxide (CO_2) added to the atmosphere, on the other hand, requires on the order of 100,000 years to be fully removed, while most other GHGs have atmospheric lifespans on the order of 100 years or less. Thus, when the fossil fuel era comes to an end (either due to exhaustion of the fossil fuel resource or through deliberate actions to make the transition to renewable energy), aerosol-cooling effects will drop out and we will be largely left with the accumulated increase in atmospheric CO_2.

The temperature response to increases in the concentration of CO_2 and other GHGs is thus of crucial importance to future climate. The eventual response to a *given* increase in GHG concentrations can be broken into the product of *radiative forcing* and *climate sensitivity*, as outlined below. However, as climate changes, the natural fluxes of CO_2 and methane (CH_4) into and out of the atmosphere will also change, amplifying or diminishing the initial changes in CO_2 and CH_4 concentration. These subsequent changes constitute a *climate–carbon cycle feedback*. In this chapter, that nature of the radiative forcing due to increasing CO_2 is discussed, followed by a comprehensive and critical review of the evidence concerning the magnitude of the climate sensitivity and of climate–carbon cycle and other slow feedbacks. An analytical framework is presented for the quantification of individual climate and climate–carbon cycle feedback processes and on how to combine them in assessing their effect on the eventual climate response. The chapter closes with some illustrative scenarios of global mean temperature change for low and high future GHG emissions, climate sensitivity, and strength of the positive climate–carbon cycle feedback.

5.1.1. Radiative Forcing

Temperatures within the atmosphere and of the surface tend to adjust themselves such that the absorption of solar radiation is balanced by the emission of infrared radiation to space in the global and annual mean. If we now *impose* a change in the absorption of solar radiation (through, for example, a change in the solar luminosity or the concentration of reflective aerosols in the atmosphere) or in the emission of infrared radiation (by, for example, changing the concentration of GHGs), the temperatures will re-adjust. Because an imposed change in the radiative fluxes drives a subsequent change in temperature, the imposed change is called the *radiative forcing*. Changes in surface temperature respond to the change in net radiation at the tropopause (the boundary between the troposphere and stratosphere), rather than at the surface itself or at the top of the atmosphere. This is because the surface and troposphere are tightly coupled through radiative and non-radiative heat exchanges and so respond together to the net energy input into the combined surface–troposphere system, whereas the stratosphere is largely decoupled from the troposphere (due to the strong inversion at the base of the stratosphere, which inhibits vertical motions across the tropopause) and so can respond independently of the troposphere. Thus, the additional downward emission of infrared radiation from the stratosphere to the troposphere when CO_2 increases adds to the troposphere–surface warming while causing the stratosphere to cool. The cooling of the stratosphere occurs quickly (within months) and subtracts slightly from the initial increase in downward emission at the tropopause. Thus, the relevant forcing for the surface temperature response is the change in net radiation at the tropopause after allowing for adjustment of stratospheric temperatures, but holding all other temperatures constant[1]. This is

---

1. After adjustment of stratospheric temperature, the change in net radiation at the tropopause is the same as the change in net radiation at the top of the atmosphere.

referred to as the *adjusted* radiative forcing, but will usually be referred to as just "the forcing" here. A CO_2 doubling produces a radiative forcing of about 3.5—4.0 W m^{-2} in most models (Williams et al., 2008, their table 2). In the absence of stratospheric adjustment, the forcing would be about 4.0—4.5 W m^{-2}.

5.1.2. Climate Sensitivity and Feedback Processes

The term 'climate sensitivity' refers to the ratio of the steady-state (or 'equilibrium') increase in the global and annual mean surface air temperature to the global and annual mean radiative forcing. It is standard practice to include only the fast feedback processes, including changes in water vapour, in the calculation of climate sensitivity, but to exclude possible induced changes in the concentrations of other GHGs.

Changes in climate in response to changes in GHG concentrations and other driving factors can be computed using relatively simple climate models in which the climate sensitivity is prescribed and the radiative forcing is computed from the concentrations of individual GHGs using simple formulae based on the results of detailed calculations (as reviewed in Harvey et al., 1997), or they can be computed using 3D atmospheric general circulation models (AGCMs) coupled to a slab that represents the surface layer (mixed layer) of the ocean only (giving atmosphere—mixed layer or AML models). Alternatively, they can be computed using coupled 3D atmospheric and oceanic general circulation models (AOGCMs). In simple models, one can simply add up the individual radiative forcings to get the total global mean radiative forcing, and then apply the climate sensitivity obtained for a CO_2 doubling in simulating the global mean temperature response. AML models and AOGCMs, on the other hand, have the flexibility to respond in separate ways to different forcing mechanisms.

The climate sensitivity is essentially the same for different well-mixed GHGs (and also for ozone if an appropriately adjusted forcing is used), so it is appropriate to compute an *equivalent CO_2 concentration*, which is the concentration of CO_2 alone that would give the same radiative forcing as that produced by the sum of the forcings due to the increases in individual GHG concentrations. For example, the estimated radiative forcings in 2000, due to all the increases of GHGs since pre-industrial times, sum to 2.5—3.7 W m^{-2} (Forster et al., 2007, their table 2.12) whereas the forcing for a CO_2 doubling is about 3.75 W m^{-2}. Thus, the GHG radiative forcing in 2000 was equivalent to a 67%—99% increase in CO_2 concentration, whereas CO_2 alone had increased by only about 32% by 2000.

Although the climate sensitivity is a ratio (temperature change over forcing), it is common to refer to the eventual global mean warming for a fixed CO_2 doubling (or equivalent CO_2 doubling) as the 'climate sensitivity'. Henceforth, this is how the term 'climate sensitivity' will be used here.

5.1.2.1. Fast and Slow Feedback Process

The fast feedback processes, on which the climate sensitivity depends, include:

- Changes in the amount and vertical distribution of water vapour in the atmosphere
- Changes in the temperature lapse rate
- Changes in the occurrence and properties of clouds
- Changes in the extent of seasonal snow and sea-ice (altering the surface reflectivity or albedo)

In addition to these fast feedbacks, there are a number of slow feedbacks that would further amplify the response to a CO_2 doubling or would lead to further increases in the concentration of CO_2 and other GHGs. These slow processes include:

- Shifts in the distribution and extent of different terrestrial biomes, thereby altering the surface albedo
- Melting and retreat of the Greenland and West Antarctic ice caps (and also the East Antarctic ice sheet if enough warming occurs), thereby adding to the positive surface albedo feedback
- Changes in the natural fluxes of CO_2, CH_4, and other GHGs, and in the production of natural aerosols, leading to changes in their concentrations

This chapter reviews multiple lines of evidence concerning the magnitude of the climate sensitivity as determined by fast-feedback processes, followed by a review of evidence concerning the magnitude of the feedback arising from slow climate—carbon cycle feedbacks and slow shifts in the distribution of vegetation. An analytical framework will be presented that allows the intercomparison and combination of slow and fast feedback processes so as to permit estimation of their interactive effect on long-term changes in global mean temperature.

5.2. FAST-FEEDBACK CLIMATE SENSITIVITY

This section reviews climate sensitivity by first presenting an analytical framework for the analysis of climate feedbacks in computer climate models, as this provides a number of useful insights.

5.2.1. Linear Feedback Analysis

As noted earlier, the temperatures of the atmosphere and surface tend to adjust themselves such that there is a balance between the absorption of energy from the Sun and the emission of infrared radiation to space. That is:

$$(1 - \alpha_p(T))\frac{Q_s}{4} - F(T) = N(T) = 0 \quad (5.1)$$

where $\alpha_p(T)$ is the planetary albedo, Q_s is the average solar flux density (W m^{-2}) on a plane perpendicular to the Sun's rays, $F(T)$ is the emission of infrared radiation to space, $N(T)$ is the net radiation, and T is the global mean surface air temperature. The first term on the left-hand side is the global mean rate of absorption of solar radiation and the second term is the rate of emission of infrared radiation.

If a radiative perturbation ΔR is imposed that is independent of the radiating temperature, the temperature will tend to adjust such that:

$$(1 - \alpha_p(T))\frac{Q_s}{4} + \Delta R - F(T) = N^*(T) + \Delta R = 0 \quad (5.2)$$

where N^* is the net radiation excluding the radiative forcing. Let T_1 be the temperature that satisfies Equation (5.1) and T_2 be the temperature that satisfies Equation (5.2). Furthermore, write T_2 as $T_2 = T_1 + \Delta T$. Then $N^*(T_2) \sim N(T_1) + (dN/dT)\Delta T$ and:

$$N^*(T_2) + \Delta R = N(T_1) + (dN/dT)\Delta T + \Delta R = 0 \quad (5.3)$$

But $N(T_1) = 0$ by definition, so we get:

$$\Delta T_{eq} = -\frac{\Delta R}{dN/dT} = \frac{\Delta R}{\lambda} \quad (5.4)$$

where $\lambda = -dN/dT = dF/dT - dQ/dT$. It is common to call λ the *radiative damping* or *feedback parameter*. dF/dT and dQ/dT can be expanded as:

$$\frac{dF}{dT} = \frac{\partial F}{\partial T} + \sum_{i=1}^{n} \frac{\partial F}{\partial I_i}\frac{dI_i}{dT} \quad (5.5)$$

and

$$\frac{dQ}{dT} = \sum_{i=1}^{n} \frac{\partial Q}{\partial I_i}\frac{dI_i}{dT} \quad (5.6)$$

respectively. The total derivative dF/dT involves the dependence of F on temperature in two ways: a direct dependence on temperature ($\delta F/\delta T$) and an indirect dependence on temperature through the direct dependence of F on various internal variables I_i ($\delta F/\delta I_i$) which, in turn, directly depend on temperature (dI_i/dT). The direct dependence of F on temperature arises through the *Planck function*, which gives the maximum possible rate of emission at each wavelength solely as a function of temperature. The maximum possible emission over all wavelengths, F_{max}, is given by $F_{max} = \sigma T^4$, so $\delta F/\delta T = 4\sigma(T_{eff})^3$ where T_{eff} (the effective radiating temperature of the planet) is the

temperature such that $\sigma(T_{eff})^4 =$ the observed emission of infrared radiation to space (for Earth this is about 240 W m^{-2} in the global and annual mean). Q depends on many individual variables, such as cloud amounts and cloud optical properties, and on atmospheric and surface reflectivity, all of which change with temperature, but does not depend directly on temperature (so there is no $\delta Q/\delta T$ term in Equation (5.6)).

As $N(T) = Q(T) - F(T)$, the total derivative dN/dT can be written as:

$$\frac{dN}{dT} = \frac{\partial N}{\partial T} + \sum_{i=i}^{n} \frac{\partial N}{\partial I_i}\frac{dI_i}{dT} = -\left(\lambda_o + \sum_{i=1}^{n} \lambda_i\right) \quad (5.7)$$

where $\lambda_o = -\delta N/\delta T = \delta F/\delta T$, $\delta N/\delta I_i = \delta Q/\delta I_i - \delta F/\delta I_i$, and $\lambda_i = -(\delta N/\delta I_i)(dI_i/dT)$. The fast feedbacks involve the direct dependence of F on T and the dependence of F and/or Q on changes in water vapour, temperature lapse rate, clouds, and surface albedo. The total radiative feedback parameter λ can thus be written as:

$$\lambda = \lambda_0 + \lambda_{wv} + \lambda_{lr} + \lambda_c + \lambda_a \quad (5.8)$$

where the λ_i represent the inherent strengths of a given feedback process i, with $\lambda_i > 0$ representing a negative feedback and $\lambda_i < 0$ representing a positive feedback. Our physical understanding of the various radiative feedback processes is reviewed in Bony et al. (2006). The overall feedback is negative ($\lambda > 0$, $dN/dT < 0$), as otherwise there could not be a stable climate. The single largest factor contributing to the overall negative feedback is the increase in emission through the Planck function, represented by λ_o, where $\lambda_o = dF/dT|_{Teff} = 4\sigma T_{eff}^3 = 3.76$ W m^{-2} K^{-1} for the idealized case of globally uniform warming[2]. λ_o is referred to as the Planck feedback.

The key assumption underlying the above analysis is that the feedbacks due to individual processes do not interfere with each other and so can be simply added together to get the total feedback. That is, that the feedbacks combine linearly. This is not exactly true when combining cloud feedbacks with other feedbacks, but does not introduce a large error. More importantly, considerable qualitative insight can be gained into how climate sensitivity changes as additional positive feedbacks are added when this simplifying assumption is made, as is shown next.

---

2. At any given location, λ_o is equal to the rate of increase of infrared emission to space with surface temperature when all atmospheric temperatures increase at the same rate as the surface temperature (that is, with fixed lapse rate). Thus, λ_o will be large at low latitudes and small at high latitudes. The global mean value of λ_o is equal to $(\int_A (\delta F \delta T \, dT/d\overline{T}) \, dA)/(\text{global area})$ and, because surface warming is larger at high latitudes compared to low latitudes, the global mean λ_o will be smaller than the value of 3.76 W m^{-2} obtained by evaluating $\delta F \delta T$ once at T_{eff}. The greater the polar amplification of the warming, the smaller the global mean value of λ_o will be.

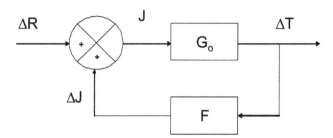

FIGURE 5.1 Block diagram showing a climate feedback loop.

5.2.1.1. Relationship between Total Feedback Strength and Climate Sensitivity

The following analysis is based on Schlesinger (1985). Consider the case where there are no feedbacks except through the direct dependence of infrared emission on temperature (so $dF/dT = \partial F/\partial T$ and $dQ/dT = 0$). The temperature response ΔT can be written as:

$$\Delta T_{eq} = G_o \Delta R \tag{5.9}$$

where $G_o = (\partial F/\partial T)^{-1}$ is the *system gain*. When indirect feedbacks are allowed, the change in temperature leads to a further change, ΔJ, in the net radiation, which then feeds back into the system, as shown in Figure 5.1.

The temperature response is now given by:

$$\Delta T_{eq} = G_o(\Delta R + \Delta J) \tag{5.10}$$

On the assumption that the individual feedbacks are independent of one another, then $\Delta J = \Sigma \Delta J_i$, where each ΔJ_i is the change in net radiation due to the change ΔT_{eq} provoking feedback process i. It can be written as:

$$\Delta J_i = \left(\frac{\partial Q}{\partial I_i} - \frac{\partial F}{\partial I_i}\right)\frac{dI_i}{dT}\Delta T_{eq} = \frac{\partial N_i}{\partial I_i}\frac{dI_i}{dT}\Delta T_{eq} = -\lambda_i \Delta T_{eq} \tag{5.11}$$

If ΔR and ΔJ_i are both positive or both negative, then the feedback involving variable i is a positive feedback since it reinforces the initial heating perturbation. Letting $G = -\Sigma\lambda_i$ (excluding λ_o), we can rewrite Equation (5.10) as:

$$\Delta T_{eq} = G_o(\Delta R + G\Delta T_{eq}) \tag{5.12}$$

Solving for ΔT_{eq}, we obtain:

$$\Delta T_{eq} = \left(\frac{G_o}{1 - f_\lambda}\right)\Delta R \tag{5.13}$$

where $f_\lambda = GG_o = \Sigma f_{\lambda i}$, and:

$$f_{\lambda i} = G_o\frac{\partial N}{\partial I_i}\frac{dI_i}{dT} = G_o\frac{\Delta J_i}{\Delta T_{eq}} = -G_o\lambda_i \tag{5.14}$$

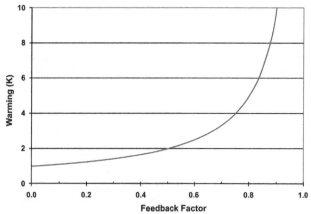

FIGURE 5.2 Equilibrium global mean warming for a CO_2 doubling as a function of the total feedback factor, f_λ.

The variables f_λ and $f_{\lambda i}$ are referred to as the *feedback factor*. The magnitude of $f_{\lambda i}$ is a dimensionless measure of the strength of the feedback involving variable i. The ratio of ΔT_{eq} with all feedbacks to ΔT_{eq} without indirect feedbacks, R_f, is equal to $1/(1 - f_\lambda)$.

Figure 5.2 shows the warming for a CO_2 doubling as a function of f_λ. As $f_\lambda \to 1$, the temperature change approaches infinity; that is, there is a runaway positive feedback. In deriving the results leading up to this figure, we have assumed that: (i) the strength of each feedback is independent of all the other feedbacks; and (ii) the strength of the feedback does not change as the magnitude or sign of ΔR and ΔT_{eq} changes. In other words, we have assumed that the system is *linear*. Nevertheless, the increase in ΔT_{eq} with increasing f_λ (as more positive feedbacks are added) is decidedly non-linear: a given increase in f_λ has a greater absolute effect on ΔT_{eq} the greater the initial value of f_λ. Thus, the effect of a given feedback on ΔT_{eq} depends on what other feedbacks are already present.

Another important implication of this analysis is that the effect of uncertainty in the sign of a given feedback is not symmetric. Rather, adding a positive feedback has a greater effect on climate sensitivity than adding a negative feedback of the same magnitude.

5.2.2. Climate Sensitivities of AML Models and AOGCMs

It is prohibitively expensive to run an AOGCM until it has reached a new equilibrium after applying a radiative forcing (thousands of years of simulated time would be required). Instead, the sensitivity of an AOGCM has been assumed to be approximately equal to the sensitivity of the corresponding AML model. An AML model does not permit simulation of ocean currents, so the heat transport associated with ocean currents is prescribed and held fixed as the climate changes. Because of differences in the radiative feedbacks, the AML climate sensitivity will only approximate the climate sensitivity of the AOGCM. Feedbacks can differ because of differences in the initial amount of ice and snow (if there is too much ice and snow, the potential ice–snow feedback is larger), differences in the initial cloud amounts or distributions, and differences in the cloud feedback due to differences in the spatial pattern of temperature change arising from the fact that ocean currents can change in strength in the AOGCM but are implicitly held fixed in the AML.

However, the equilibrium climate sensitivity expected from an AOGCM itself can be estimated without running the AOCGM to a new equilibrium (Gregory et al., 2004; Williams et al., 2008). The temperature change at any given time during the transient satisfies:

$$\Delta T(t) = \frac{\Delta R - N(T)}{\lambda} \qquad (5.15)$$

(at $t = 0$, just after ΔR has been applied, $N = \Delta R$, while in the final equilibrium, $N = 0$). Thus, the best-fit slope through a plot of N against ΔT provides an estimate of λ, and extrapolation of the curve to the N axis (where $\Delta T = 0$) gives an estimate of ΔR. However, the ΔR obtained in this way does not agree with the expected ΔR based on the change in net radiation at the tropopause when only stratospheric temperatures are allowed to adjust. Call the forcing obtained from the plot of N against T the effective radiative forcing, ΔR_{eff} (the reason ΔR_{eff} differs from ΔR will be explained later). ΔR_{eff} and λ from the slope of N against ΔT can be used in Equation (5.4) to estimate ΔT_{eq}. Conversely, the equilibrium response that is consistent with the partial transient response of the AOGCM can be estimated as the value of ΔT where the best-fit line intercepts the ΔT axis (where $N = 0$).

This procedure is illustrated in Figure 5.3 for an experiment with an AOGCM in which atmospheric CO_2 concentration increased by 1% per year for 70 years (at which the point the concentration had doubled), then was held fixed while the model was integrated a further 500 years or more. The radiative forcing in this model when CO_2 doubles ($\Delta R'$) is 3.8 W m^{-2}, but the data points in Figure 5.3 lie close to a straight line that extrapolates to an effective forcing ΔR_{eff} of only 2.3 W m^{-2}. The slope of the line going through the point (0, ΔR_{eff}) is much smaller (and the corresponding climate sensitivity much larger) than for a line constrained to go through the point (0, $\Delta R'$). Furthermore, the slope of a line from (0, $\Delta R'$) to successive (ΔT, N) data points becomes progressively smaller, which corresponds to the effective transient climate sensitivity increasing during the simulation. However, if the slope of the best-fit line is allowed to intersect the net radiation axis at the effective forcing

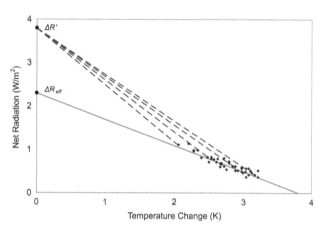

FIGURE 5.3 Example of the determination of the effective radiative forcing and the radiative feedback parameter, from the transient response of an AOGCM. Purple diamonds give model simulated net radiation after various amount of global mean warming. These extrapolate (green line) to a net radiation of 2.3 Wm^{-2} at 0 K warming, which is the effective forcing. Lines (red) drawn from the calculated adjusted forcing (3.8 Wm^{-2}) to successively later net radiation-warming combinations have a progressively less negative slope, which corresponds to an increasing climate sensitivity. (*Source: Williams et al., 2008. © American Meteorological Society. Reprinted with permission.*)

(ΔR_{eff}), the effective climate sensitivity during the transient is roughly constant at the higher, final sensitivity.

The reason for this behaviour is that a CO_2 increase induces a rapid adjustment in tropospheric temperatures even with fixed surface temperatures, and this adjustment reduces the net forcing at the tropopause. An increase in CO_2 alters the vertical variation in radiative net heating or cooling within the atmosphere through emission and absorption of infrared radiation, and this in turn causes rapid local changes to the vertical temperature profile that, in turn, alter the stability, vertical mixing, and moisture profile (Gregory and Webb, 2008). Like stratospheric adjustment, these changes occur in much less than one year. It is the adjusted forcing that the ocean sees, so the transient response of the surface is governed by this adjusted forcing and the radiative feedbacks that are triggered as the coupled ocean–troposphere–land surface system warms in response to the adjusted forcing.

Thus, to properly simulate the transient response, we need to break the response into two parts:

- One part that is rapid (within much less than one year) and that produces a change in net radiation that is subtracted from the initial forcing to give the adjusted forcing
- One part that is gradual and driven by the adjusted forcing

If one were interested in only the equilibrium response, the difference between initial and adjusted forcing would not matter, because there would be a compensating change in the deduced equilibrium climate sensitivity. However,

TABLE 5.1 Comparison of Climate Sensitivities Estimated for Various AOGCMs with the Climate Sensitivity of the Corresponding AML Model

Model	ΔR (W m^{-2})	ΔR_{eff} (W m^{-2})	AOGCM ΔT_{eq} (K)	AML ΔT_{eq} (K)
CCSM3	4.0	2.9 ± 0.4	2.4 ± 0.5	2.7
CGCM3.1	3.3	4.0 ± 0.4	2.8 ± 0.4	3.4
ECHAM5	4.0	3.2 ± 0.3	3.7 ± 0.3	3.4
GFDL	3.5	1.7 ± 0.2	3.2 ± 0.2	2.9
GISS	4.1	3.8 ± 0.3	2.4 ± 0.3	2.7
MIROC3.2	3.1	3.2 ± 0.2	4.3 ± 0.3	4.0
HadCM3	3.8	2.3 ± 0.1	3.8 ± 0.2	3.3
HadGEM1	3.8	2.0 ± 0.1	3.4 ± 0.2	4.4

Also shown is the traditional radiative forcing ΔR (based on adjustment of stratospheric temperatures only) for each model and ΔR_{eff}. AOGCM ΔT_{eq} and ΔR_{eff} were computed from transient runs whereby CO_2 concentration increases by 1% per year for 70 years, then is held fixed at a doubling while the simulation continues for at least another 150 years.
(Source: Williams et al., 2008.)

the distinction is relevant for the transient response, because the adjustment in the forcing is not proportional to the evolving temperature change.

Williams et al. (2008) used this approach to estimate the equilibrium climate sensitivity for eight different AOGCMs and compared these with the climate sensitivities of the corresponding AMLs. Results are summarized in Table 5.1, along with ΔR and ΔR_{eff}. AOGCM sensitivities differ by up to 0.6 K from the corresponding AML sensitivities. However, the range in climate sensitivity is similar for the two groups: 2.7–4.4 K for the AMLs and 2.4–4.3 K for the AOGCMs.

5.2.2.1. Feedback Parameters for Individual Processes in AOGCMs

The feedback factors associated with individual feedback processes in an AML model or an AOGCM can be directly computed from the equilibrium or transient changes produced by the model. To do so, one would first compute the global mean net radiation over the course of one year associated with the equilibrium (unchanging) climate prior to applying a radiative forcing. To compute the feedback factor associated with water vapour changes, for example, one would then repeat the calculation of annual net radiation using the new water vapour amounts produced after some period of simulated time following an increase in CO_2, but using the original values for all other variables including temperature. The difference in global mean net radiation gives ΔJ_{wv}, from which the product λ_{wv} and f_{wv} can be computed using the difference in global mean temperature. This is referred to as the partial radiative

perturbation method, and agrees to within a few percent with a more accurate method developed by Soden et al. (2008).

Figure 5.4a compares the radiative feedback parameters for different feedback processes for 14 different AOGCMs computed in this way. There is a large spread in the calculated feedbacks for water vapour and for lapse rate changes. However, the water vapour and lapse rate feedbacks are negatively correlated, as can be seen in Figure 5.4b, which plots the water vapour feedback parameter against the lapse rate feedback parameter. The negative correlation arises because relative humidity (RH) in the upper troposphere tends to be constant as temperatures change. Thus, a decrease in lapse rate as the climate warms (which occurs in the tropics) serves as a negative feedback (because temperatures in the upper troposphere then increase faster than at the surface, leading to stronger radiative damping to space) but, subject to constant RH, the amount of water vapour in the upper troposphere will increase more than if the lapse rate were constant, which produces a stronger positive water vapour feedback. Thus, the uncertainty in the combined water vapour + lapse rate feedback is much less than the uncertainty in the individual feedbacks (as seen from Figure 5.4a). The albedo feedback is positive, but modest. There is a large uncertainty in the cloud feedback, which ranges from near zero to strongly positive. The final column of Figure 5.4a shows the net feedback for each model. Combined with the λ_o for each model, the overall feedback parameter ranges from 0.88 W m^{-2} K^{-1} to 1.63 W m^{-2} K^{-1}, which, assuming a radiative forcing of 3.8 W m^{-2}, corresponds to a climate sensitivity ranging from 4.3 K to 2.3 K.

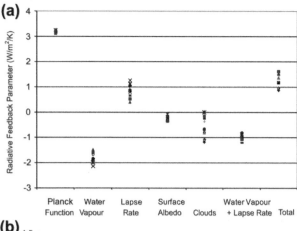

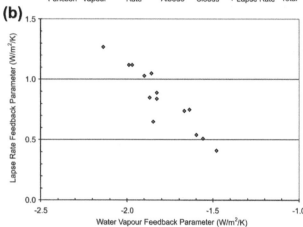

FIGURE 5.4 (a) Radiative feedback parameters for changes in water vapour, lapse rate, surface albedo, clouds, and water vapour + lapse rate, as estimated from 14 AOGCMs. (b) Relationship between water vapour and lapse rate feedback parameters in 14 AOGCMs. *(Source: Soden and Held, 2006.)*

5.2.3. Observational Validation of the Water Vapour Feedback in AOGCMs

The water vapour feedback is a consistently strong positive feedback in AGCMs, with absolute humidity increasing with warming by an amount that roughly holds RH constant. This matches observations. Observed trends in both surface (Dai, 2006) and lower and middle tropospheric (McCarthy et al., 2009) humidity are very close to what is required to maintain constant RH. In the upper troposphere, where water vapour is particularly effective as a GHG, Gettelman and Fu (2008) find that RH, averaged over the region from $30°S$ to $30°N$, increased by $2 \pm 2\%$ K^{-1} at 250 hPa over the period September 2002 to February 2007. Similar trends are obtained in simulations with an AGCM driven with observed changes in sea surface temperature (SST). Minschwaner and Dessler (2004), in contrast, found RH to be decreasing by $4.8 \pm 1.7\%$ K^{-1} at 215 hPa between $\pm 20°$ latitude over the period 1992–1997. Seventeen AOCGMs driven by GHG and other forcings simulated

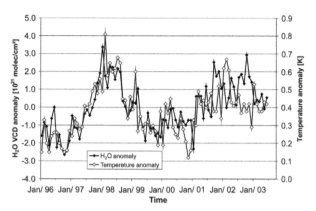

FIGURE 5.5 Variation in globally averaged precipitable water and surface temperature under clear sky conditions for the period 1996–2003. *(Source: Wagner et al., 2006 © Copyright 2006 American Geophysical Union. Reproduced/modified by permission of American Geophysical Union.)*

a decrease over the period 1980–2000, ranging from 1% K^{-1} to 4% K^{-1} (Minschwaner et al., 2006), so there is a tendency in the models for absolute humidity in the upper tropical troposphere to increase slightly faster than observed over the period 1992–1997.

Wagner et al. (2006) assessed global trends in total column precipitable water under mainly clear sky conditions over the period 1996–2003, as measured by the Global Ozone Monitoring Experiment. They find that variations in globally averaged precipitable water and surface temperature are highly correlated at monthly and longer timescales (Figure 5.5). However, over Northern Hemisphere (NH) continents, there is much scatter in the correlations, and trends in precipitable water can be opposite to trends in surface temperature in some regions.

It is possible to compare the water vapour, lapse rate, and cloud feedbacks predicted by AOGCMs to operate during short-term oscillations, such as the seasonal cycle or inter-annual variability, with those observed in the real atmosphere. However, such a comparison is only a partial test of the correctness of the processes in AOGCMs that determine the feedbacks in response to long-term forcing, because the feedbacks that operate in the long-term response to an increase in GHGs are likely to be quite different from the short-term feedbacks (due to different changes in spatial temperature gradients and in the atmospheric circulation in response to the seasonal cycle, El Niño oscillations, and long-term climatic change). Dessler and Wong (2009) compared the water vapour feedback over the period 1979–2002, as simulated by 10 AGCMs with that computed from re-analysis data[3] for the period 1979–2002; the AGCMs were driven by observed variations in SST and sea-ice which, apart from the long-term

---

3. These are the fields simulated by meteorological forecast models that are continuously nudged towards intermittent observations.

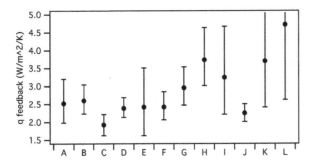

FIGURE 5.6 Comparison of the global mean water vapour radiative feedback parameter as estimated by 10 AGCMs when forced with observed variations in SST and sea-ice extent for the period 1979–2002 (cases A to J), and as computed from observations using two different re-analysis datasets (cases K and L). *(Source: Dessler and Wong, 2009. © American Meteorological Society. Reprinted with permission.)*

trend, are largely due to El Niño oscillations. The water vapour feedback (λ_{wv}) in the models (1.5–2.8 W m^{-2} K^{-1} for full uncertainty range, 1.7–2.4 W m^{-2} K^{-1} based on best estimates only) is similar to, but probably modestly weaker than, the re-analysis dataset (1.7–4.4 W m^{-2} K^{-1}, with best estimates of 2.5 and 2.8 W m^{-2} K^{-1} based on two different re-analysis datasets) (Figure 5.6).

Another indication of the reliability of water vapour processes in AGCMs is the excellent agreement found between model-simulated and observed variations in emitted radiation at wavelengths sensitive to the amount and distribution of water vapour in the troposphere (the 6.3 μm brightness temperature) following the eruption of Mount Pinatubo in 1991 (Soden et al., 2002).

In summary, computer climate models simulate an increase in absolute humidity at all levels in the atmosphere with increasing temperature that is sufficient to hold RH approximately constant. This is true both for inter-annual variations driven by observed changes in SST and for longer term changes, and makes the water vapour feedback in models a significant positive feedback. Comparison of AOGCM or AGCM simulations with a variety of observations (or with re-analysis data) indicates that models either slightly overestimate or slightly underestimate the strength of the water vapour feedback compared to the real climate system, depending on the observational dataset used and time period considered.

5.2.4. Climate Sensitivity Deduced from Historical Temperature Trends

The variation in global mean surface temperature over the past 150 years, for which there are good observations (Brohan et al., 2006), depends on the total radiative forcing, the rate of absorption of heat by the oceans, and the climate sensitivity. The total radiative forcing is the sum of forcings due to increasing GHG concentrations; changes in the

atmospheric loading of various aerosols (sulfate, nitrate, reflective organic compounds, and soot); changes in land use (which alter the surface albedo); changes in the solar luminosity; and variations in the activity of volcanoes. With the exception of soot, aerosols have a cooling effect on climate by increasing the reflection of sunlight and so offset the heating effect of increasing GHGs to some extent. The most important uncertainty in the radiative forcing is the uncertainty in the forcing due to aerosols. Based on calculations from first principles, the aerosol-cooling tendency could exceed the total forcing from GHGs. However, the true aerosol forcing must have been less than the GHG forcing, as otherwise there would be no net heating and thus no means of explaining the highly unusual warming of the past century.

For a given variation in net radiative forcing over the past century, the surface temperature response and the oceanic heat uptake will be larger the greater the climate sensitivity, but the surface temperature response will be smaller the larger the effective vertical diffusion coefficient in the upper ocean while the oceanic heat uptake will be larger. Thus, if too large a product of radiative forcing times climate sensitivity is offset by too large an ocean diffusivity, the oceanic heat uptake will be too large. Additional constraints arise from the fact that the effective vertical diffusion coefficient for heat (K_v) must be consistent with the effective vertical diffusion coefficient needed for the correct simulation (in non-diffusive models) of the total inventory and vertical distribution of natural and atomic-bomb ^{14}C in the oceans[4].

With regard to the uncertain aerosol forcing, the larger the aerosol offset, the smaller the net forcing, and so the greater the permitted climate sensitivity for a given oceanic heat uptake while still producing the same observed surface warming. However, the magnitude of the permitted aerosol cooling is constrained by the fact that most of the aerosol cooling is in the NH. If too large a climate sensitivity is offset by too large an aerosol cooling, this will suppress the warming in the NH more than observed (even if the correct global average warming is simulated). On this basis, Harvey and Kaufmann (2002) concluded that aerosols offset, at most, about half of the GHG heating so far, and that the climate sensitivity is likely to be 2–3 K, with a possible extreme range of 1–5 K. They did not systematically vary K_v in their analysis but, rather, chose the K_v deduced by Harvey (2001) and confirmed that their simulated uptake of heat by the oceans fell within observational uncertainties.

Thus, the key unknowns are the climate sensitivity, the aerosol forcing, and the effective diffusion coefficient for

---

4. For reasons explained in Harvey and Huang (2001), these coefficients differ for heat, CO_2, and ^{14}C, but the differences are constrained by physical considerations.

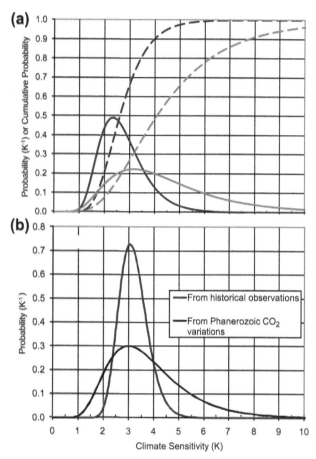

FIGURE 5.7 (a) Illustration of a pdf for climate sensitivity with a 5%−95% probability range of 1.5−4.5 K and a mode of 2.5 K (red curves) or a 5%−95% range of 1.8−9.2 K and a mode of 3.2 K (blue curves) (solid lines are probabilities and dashed lines are cumulative probabilities). (b) pdfs derived by Annan and Hargreaves (2006) with 5 constraints from recent observed changes (red curve), and as derived by Royer et al. (2007) based on the simulation of Phanerozoic CO_2 variations (purple curve).

In figure (b) legend: From historical observations / From Phanerozoic CO_2 variations

the oceanic uptake of heat. The observational constraints are the variation in global mean temperature, the variation in NH−SH temperature difference, the change in oceanic heat content, and the amount and distribution of natural and bomb-produced ^{14}C in the oceans. Uncertainty in the observations can be translated into a probability distribution function (pdf) for climate sensitivity, which gives the probability of the climate sensitivity falling within different climate sensitivity intervals. Two climate sensitivity pdfs are illustrated in Figure 5.7a, one with a 5%−95% probability range of 1.5−4.5 K and a mode (most likely climate sensitivity) of 2.4 K, and the other with 5%−95% range of 1.8−9.2 K and a mode of 3.2 K. The case with a long tail extending to the high climate sensitivities represents a substantially greater societal risk and is representative of many of the pdfs derived in the mid-2000s, but has only a slightly higher mode than the first case. As noted by

Annan and Hargreaves (2006), the long tails are a result of considering, in each study, only a subset of the total body of evidence that could be used to constrain the climate sensitivity. Annan and Hargreaves (2006) show that, when all the evidence is combined together, the long tail is greatly reduced. They estimate that there is less than a 5% probability that the climate sensitivity exceeds 4.5 K and negligible probability that it exceeds 6 K. Their pdf (and a further pdf developed from the simulation of inferred CO_2 variations during the past 420 Ma, which will be discussed later) is shown in Figure 5.7b.

Forest et al. (2008) calculated a 3D pdf associated with joint variation in climate sensitivity, K_v, and aerosol radiative forcing. The pdf was constructed by selecting permissible ranges for all three parameters, choosing some sampling interval for each of the three parameters, and performing simulations of twentieth century climate change for all possible combinations of the three parameters (chosen from the sampled values within the permitted ranges). A given set of parameters is rejected if the corresponding simulation of upper-air temperature, surface temperature, and oceanic heat content is inconsistent with the observed variation in these parameters, given observational uncertainties and the magnitude of internal variability (unrelated to external forcing) as estimated by AOGCMs. Figure 5.8 shows the marginal pdfs for each of these unknown, but adjustable, parameters. The marginal pdf for climate sensitivity (for example) is obtained as follows: the probability of each climate sensitivity interval is given as the proportion of the simulations using a climate sensitivity in that interval and all other combinations of the other two variables that were not rejected, normalized to give a total probability of unity. The 5%−95% probability ranges deduced in this way are 2.0−5.0 K for climate sensitivity, 0.04−4.1 $cm^2 s^{-1}$ for the effective K_v, and −0.27 to −0.7 $W m^{-2}$ for the aerosol forcing in 1995. Figure 5.9 shows a 2D pdf for climate sensitivity and K_v. The most probable combination is a climate sensitivity of 2.5 K and an effective K_v of 0.7 $cm^2 s^{-1}$.

5.2.5. Climate Sensitivity Deduced from Observed Short-Term Temperature Changes

Climate sensitivity has also been estimated based on a variety of short-term observations: the temperature response to the seasonal cycle of solar irradiance, to volcanic eruptions, and to the recent decline in aerosol radiative forcing. In particular:

- Knutti et al. (2006) estimated that there is a 90% probability of the climate sensitivity falling between 1.5−2.0 and 5−6.5 K, with a most likely climate sensitivity of 3.0−3.5 K, based on the temperature response to the seasonal cycle in solar irradiance.

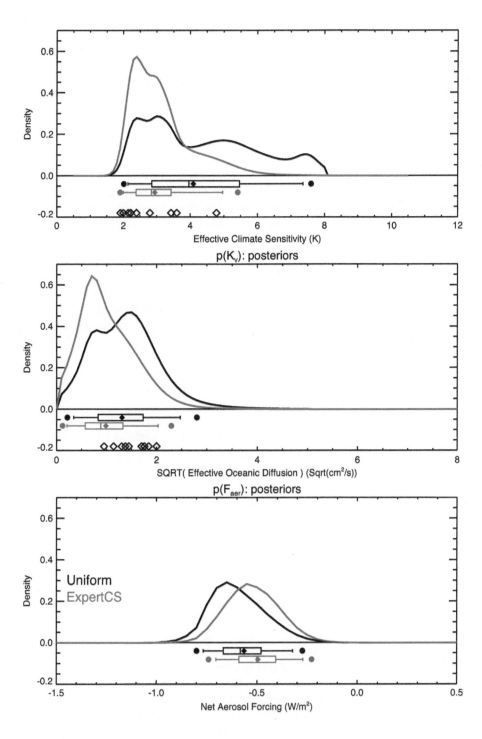

FIGURE 5.8 One-dimensional pdfs for climate sensitivity, effective K_v, and aerosol radiative forcing in 2000 or so as deduced by Forest et al. (2008). The deduced pdfs depend on an initial assumption concerning the shape of the pdf, and results are shown initially assuming the same probability for all values of a parameter within the specified range of probabilities (black curves, uniform priors) or assuming an initial pdf based on expert judgement for climate sensitivity (expert priors, blue curves). The whisker plots indicate boundaries for percentiles 2.9–97.5 (dots), 5–95 (vertical bars at ends), 25–75 (box ends), and 50 (vertical bar in box). Means are indicated by solid diamonds.

- Wigley et al. (2005) concluded that the observed peak temperature response and subsequent rate of recovery to a composite of various volcanic eruptions during the past century appear to rule out climate sensitivities of less than 1.5 K, but cannot rule out sensitivities greater than 4.5 K.
- Bender et al. (2010) estimated a climate sensitivity of 1.7–4.1 K based on the ratio of the integrated

temperature response to the integrated radiative perturbation following the eruption of Mount Pinatubo.
- Chylek et al. (2007) computed a climate sensitivity of 2.1–3.4 K based on an estimate in the change in aerosol optical depth over the period 2000–2005, combined with data on the associated change in global mean temperature and a parameterization of the oceanic heat uptake.

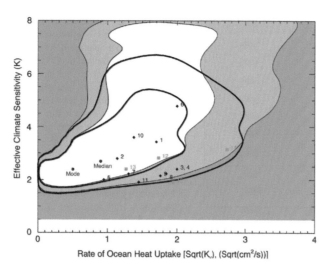

FIGURE 5.9 The joint climate sensitivity, K_v, pdf as deduced by Forest et al. (2008). Diamonds and squares represent parameter combinations required in the MIT 2D ocean model to replicate the behaviour of specific AOGCMs. The inner and outer boundaries of the lightly shaded area represent 10% and 1% significance levels, respectively, for the rejection of specific parameter combinations using uniform priors (defined in the caption to Figure 5.8). The thick black lines show the same thing using expert priors.

Other analyses of the observed response to volcanic eruptions lead to the conclusion that the climate sensitivity is very low, but these analyses have been found to have various flaws.

5.2.6. Climate Sensitivity Deduced from Past Climates and Forcings

If global mean temperature changes and global mean radiative forcing (including that due to slow feedback processes) can be estimated for various times in the geological past when temperatures were substantially different than today, then the climate sensitivity can be estimated. This has been done for the last glacial maximum (LGM) and for warm conditions during the middle Pliocene and at the Palaeocene—Eocene boundary (see also Harrison and Bartlein, 2012, this volume).

5.2.6.1. Late Glacial Maximum

Hoffert and Covey (1992) deduced a climate sensitivity of 1.4—3.2 K based on conditions during the LGM, about 20,000 years ago, while Lea (2004) deduced a sensitivity of tropical SSTs of 4.4—5.6 K.

One of the largest sources of uncertainty in the radiative forcing during the LGM is the negative forcing due to an increase in the content of dust aerosols in the glacial atmosphere. Dust concentrations in Antarctic ice cores indicate large (up to a factor of 50) increases in the rate of dust deposition over Antarctica during the LGM, compared

to the present (Lambert et al., 2008). The uncertainty in LGM aerosol forcing affects the lower limit to the possible climate sensitivity, but not the upper limit. This is because the aerosol forcing adds to the other forcings, so if the aerosol forcing is zero, the total forcing is small and the required climate sensitivity is large, but the required climate sensitivity is an upper limit because the aerosol forcing could not have been positive. On this basis, Schneider von Deimling et al. (2006) estimated an upper limit for climate sensitivity during the LGM of about 5.3 K. Conversely, the larger the aerosol forcing, the smaller the climate sensitivity, but there is no clear upper limit to the possible aerosol forcing and thus no clear lower limit to the climate sensitivity based on glacial conditions.

5.2.6.2. Pliocene Climate Sensitivity

Pagani et al. (2010) estimated a rather high climate sensitivity of 6.1—10.0 K for the middle Pliocene (4 Ma) based on rather convincing indications that the atmospheric CO_2 concentration was only 365—415 ppmv and global mean temperatures were about 2.5—3.0 K warmer than pre-industrial values[5]. This sensitivity includes the effect of slow changes in the distribution of vegetation and the extent of ice sheets, and possible increases in non-CO_2 GHGs as a feedback from climatic change, but does not include positive climate—carbon cycle feedbacks involving CO_2 because the CO_2 concentrations are taken as a given when inferring climate sensitivity.

5.2.6.3. Palaeocene—Eocene Thermal Maximum (PETM)

Just prior to the broad early Eocene temperature peak was a sharp climatic aberration referred to as the Palaeocene—Eocene Thermal Maximum (PETM), occurring on the Palaeocene—Eocene boundary at 55 Ma. This involved a 4—8 K increase in surface and deep-sea temperature in less than 10,000 years (Sluijs et al., 2006). Temperatures subsequently recovered over a period of 200,000 years. This was accompanied by a significant negative deviation in the carbon isotope ratio of marine carbonate sediments (Figure 5.10). The leading theory to explain both the temperature and C-isotope deviations is the release of ~2000—2600 Gt of ^{13}C-depleted CH_4 from CH_4 clathrate deposits (which exist today and are a cause of concern over possible future releases, as explained later). Another brief (one to several million years) warming of 3—6 K (in the southwest Pacific Ocean) developed over a period of 400,000 years during the middle Eocene in clear association with a CO_2 increase by a factor of 2—3 (Bijl et al., 2010).

---

5. A recent workshop converged on a consensus that Pliocene CO_2 concentration was likely 400 ± 50 ppmv (Schneider and Schneider, 2010).

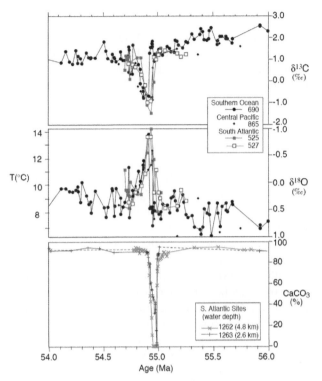

FIGURE 5.10 Co-variation of $\delta^{13}C$ in marine carbon sediments (an indicator of a CO_2 pulse into the atmosphere), the $\delta^{18}O$ in marine sediments (an indicator of seawater temperature), and the percentage of $CaCO_3$ in marine sediments (an indicator of ocean acidity) at the Palaeocene–Eocene transition. *(Source: Zachos et al., 2008. Copyright 2008 National Academy of Sciences, U.S.A.)*

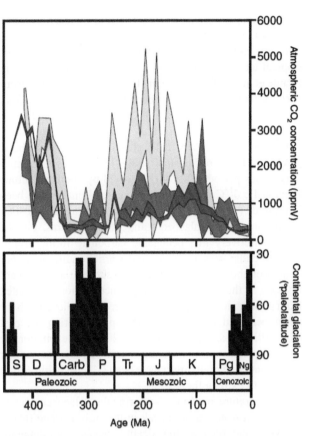

FIGURE 5.11 Inferred variation in CO_2 from carbonate sediments (upper panel) and occurrence of glacial epochs (lower panel) over the past 420 Ma. Blue: model-simulated CO_2 variation; yellow: recent proxy-based estimates of CO_2 variation; red: revised proxy-based estimates of CO_2 variation that take into account the seasonality of carbonate formation. *(Source: Breecker et al., 2010. Reprinted by permission from Macmillan Publishers Ltd.)*

The negative $\delta^{13}C$ perturbation seen during the PETM (Figure 5.10) and associated warming can be used to constrain both the magnitude of carbon release and the climate sensitivity at this time. Pagani et al. (2006) worked out combinations of carbon release and climate sensitivity that satisfy the observed excursion in the $\delta^{13}C$ of $-3\permil$ to $-5\permil$ in marine carbonate sediment, while producing an assumed global mean warming of 5 K during the PETM. If the source of the carbon was CH_4 from CH_4 clathrate (with a $\delta^{13}C$ of $-60\permil$), then the permitted release is 1800–3500 GtC. To explain a 5 K warming, the required climate sensitivity is 5.5–7.8 K. On the other hand, if the source of the carbon was terrestrial and marine organic carbon (with a $\delta^{13}C$ of $-20\permil$), the carbon release required to produce a $\delta^{13}C$ excursion of $-3\permil$ to $-5\permil$ is 5500 to 35,000 Gt and the required climate sensitivity is only 2.3–4.5 K. However, release of 5500–35,000 Gt of organic carbon in such a short time currently defies explanation. Furthermore, Zeebe et al. (2009) estimated the maximum possible release of carbon released during the PETM to be about 3000 Gt, based on two lines of evidence: the shallowing of the carbonate compensation depth (the transition zone between deep water that is unsaturated with respect to calcium carbonate and shallower water that is supersaturated) and the reconstructed deep-sea

$[CO_3^{2-}]$ gradient between different ocean basins. This supports Pagani et al.'s (2006) inference of a high climate sensitivity unless other, unknown, radiative forcings were also at work during the PETM.

5.2.7. Evidence from the Co-variation of Temperature and CO_2 Over Geological Time

There is widespread evidence that variation in atmospheric partial pressure of CO_2 (pCO_2) played an important role in climatic changes during the past 450 million years (see Belcher and Mander, 2012, this volume), a period for which there are various proxy indicators for both atmospheric CO_2 concentration and temperature. CO_2 concentrations have been high during periods of particularly warm climate (such as during most of the Mesozoic era and the Eocene epoch) and low during times of cold climate, including the major epochs with periodic glaciations (namely, the Permo-carboniferous and late Cenozoic glaciations), as illustrated in Figure 5.11 and Figure 14.1 in Harrison and Bartlein, (2012, this volume). In particular, atmospheric pCO_2 was considerably higher than present during the early Silurian period (423–443 Ma), according to various proxy indicators and carbon cycle modelling, and at the same time, tropical temperatures were 6–8 K warmer than at present (Came et al., 2007). During the Permian and Carboniferous periods, there seems to have been two glacial phases (from 326–312 Ma and 302–290 Ma), each associated with low CO_2 (300–350 ppmv, according to Park and Royer, 2011; but cf. Berger and Yin, 2012, this volume). Atmospheric pCO_2 rose to a broad peak of 1000–1500 ppmv in the mid-Cretaceous period (100 Ma), followed by another decline. Previously, $\delta^{13}C$ ratios in pedogenic carbonates had been interpreted as implying CO_2 concentrations as high as 4000 ppmv during the Palaeozoic and Mesozoic eras, but the recent discovery that pedogenic carbonate forms preferentially under seasonally dry and warm conditions, rather than throughout the year, has led to a substantial downward revision in the atmospheric CO_2 concentrations estimated from pedogenic carbonates, which are now believed to have never persisted above 1500–2000 ppmv during the past 420 million years (Breecker et al., 2010). The smaller the past CO_2 concentration during warm periods, the greater the climate sensitivity must have been (all else being equal) in order to explain the warmth.

Temperature and CO_2 rose during the Palaeocene epoch (of the Cenozoic era) to a peak in the early Eocene (about 52 Ma), and then a slow decline began. Simultaneous boron, Mg/Ca, and oxygen isotope measurements from the same samples document a decline in temperatures across the Eocene/Oligocene boundary (at 33.7 Ma), along with a decline in CO_2 concentration from 900–1300 ppmv 34.4 Ma to about 700–900 ppmv about 33.5 Ma. At this point, a threshold was reached allowing glaciation of Antarctica to begin (Pearson et al., 2009), as indicated by a step-like increase in the $^{18}O/^{16}O$ ratio that can be explained by the geologically rapid accumulation of isotopically-light ice on land[6]. Continued decrease in atmospheric CO_2 permitted small-scale ice sheets to form first in the NH during the late Miocene (6–10 Ma) (Zachos et al., 2001), while extensive glaciation covering most of Greenland occurred during the late Pliocene, around 3 Ma, at a CO_2 concentration of about 400 ppmv. Tripati et al. (2009) show that atmospheric CO_2 transitions, as estimated from boron/calcium ratios in foraminifera, closely match inferred climate transitions during the mid-Miocene (5–10 Ma), late Pliocene (5.3–2.4 Ma), and early Pleistocene (1.4–0.9 Ma)[7].

5.2.7.1. Ordovician Glaciations

An apparent exception to the CO_2 temperature correlation presented above is the evidence for short-lived glaciations during the late Ordovician and early Silurian periods (at 460 and 440 Ma), which was also a time of generally high atmospheric CO_2. Early analyses of palaeosol proxies (reviewed by Royer, 2006) suggested CO_2 concentrations of almost 6000 ppmv during the Ordovician period, but more recent analyses and modelling indicate concentrations of only 700–1200 ppmv at 450 Ma, rising to a peak of 1500–1800 ppmv (5–6 times pre-industrial) at 400 Ma (Park and Royer, 2011). Climate model simulations by Herrmann et al. (2004) indicate that a combination of the unusual configuration of the supercontinent Gondwanaland, which was tangent to the South Pole at that time, a solar luminosity that was about 5% lower than at present and either a lowering of sea level from mid-Ordovician levels or ocean heat transport that is reduced by 50% can account for Ordovician glaciations with CO_2 concentrations as high as 15 pre-industrial times. The confounding influences of orbitally induced variations in solar radiation and GHG concentrations are reviewed by Berger and Yin (2012, this volume).

5.2.8. Climate Sensitivity Deduced from Slow Variations in Atmospheric CO_2 Concentration

The atmospheric CO_2 concentration varied in the geological past over periods of million years in response to imbalances between volcanic outgassing and changing rates of weathering due to variations in plate tectonic activity and associated mountain uplift. The magnitude of the CO_2 variation depends in part on a negative feedback between CO_2 concentration and removal rates by chemical weathering, whereby higher atmospheric CO_2 leads to a warmer climate that, in turn, accelerates chemical weathering, leading to more rapid removal of CO_2 from the atmosphere. This feedback, in turn, depends on how large the climate sensitivity is: a small climate sensitivity results in a weaker negative feedback on CO_2 concentration through enhanced weathering and a larger variation in CO_2 concentration on million-year and longer timescales.

Simulations of CO_2 variations during the past 420 million years by Royer et al. (2007) indicate that if climate sensitivity is less than about 1.5 K, impossibly high CO_2 peaks are obtained at certain times while, if it is greater than about 3 K, unreasonably low values are obtained at other times. The main sources of uncertainty are: (i) uncertainty in the real variation in atmospheric CO_2 concentration during the past 420 Ma, which is based on various proxies[8];

---

6. The separation of India from Antarctica, permitting the replacement of north–south ocean currents that transported heat poleward with east-west currents, was likely another factor in the development of ice on Antarctica.

7. The Ba/Ca-based pCO_2 variation continues through to the present and has been validated over the past 800,000 years with the pCO_2 variation that has been directly measured in polar ice cores.

---

8. Examples include $^{13}C/^{12}C$ isotope ratios in phytoplankton and carbonate concretions in palaeosols, stomatal density of fossil leaves, boron/calcium ratios, and boron isotope ratios.

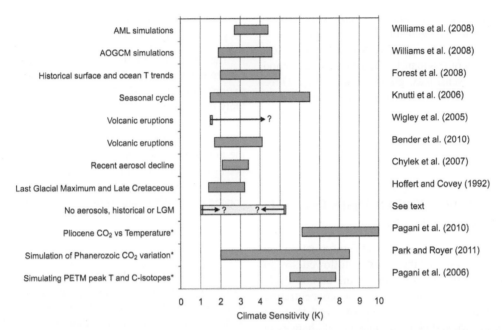

FIGURE 5.12 The range of climate sensitivities calculated using the various estimation methods discussed in this chapter. Methods marked with an asterisk (*) include the effect of changes in vegetation distribution and the full response of ice sheets, but do not include the effect of positive climate–carbon cycle feedbacks as the GHG concentrations are taken as a given in computing the climate sensitivity.

and (ii) uncertainties in four factors other than climate sensitivity that also affect the link between atmospheric CO_2 concentration and the rate of removal by weathering[9]. Assuming that all values of the various parameters in the carbon cycle model are equally likely within their specified uncertainty range, a pdf for climate sensitivity can be generated[10]. The resulting pdf is shown in Figure 5.7b. Interestingly, the most probable climate sensitivity from this analysis (3 K) is the same as that deduced by Annan and Hargreaves (2006) from historical data, although — not surprisingly — the uncertainty is much larger, resulting in a very long tail extending out to high climate sensitivities.

Park and Royer (2011) updated the earlier analysis by: (i) using revised weathering rates; (ii) using updated and expanded proxy CO_2 concentration indicators (which reduced the highest CO_2 concentrations previously inferred

to have occurred); and (iii) allowing for a greater climate sensitivity during glacial epochs (260–340 Ma and 0–40 Ma) than during non-glacial epochs during the past 420 Ma. The most probable climate sensitivity deduced for non-glacial epochs remains around 3 K, but the most probable climate sensitivity for glacial epochs (including the present) is about 6 K, with a 99% probability that the glacial climate sensitivity is >2 K[11]. This greater sensitivity is due in part to the slow albedo feedback arising from changes in the extent of polar ice sheets. The 5%–95% probability range for the glacial climate sensitivity is 3.0–15 K, although little credence can be placed on the upper limit. Like the non-glacial climate sensitivity and 6.1–10.0 K sensitivity range inferred by Pagani et al. (2010) from Pliocene climates, the glacial climate sensitivity would include the effect of changes in the distribution of vegetation zones and any feedbacks between climate and CH_4, N_2O, ozone, and aerosols, but would not include feedback between climate and the CO_2 concentration, since the CO_2 concentrations are given.

5.2.9. Conclusion Concerning the Fast-Feedback Climate Sensitivity

The ranges of climate sensitivities as estimated by the various methods reviewed here are summarized in Figure 5.12. A remarkably wide range of evidence supports

---

9. These factors are the temperature coefficient for the weathering of calcium and magnesium silicates, the CO_2 fertilization of plant-assisted weathering, the weathering rate ratio of early biota to that of modern trees, and the weathering rate ratio of gymnosperms to angiosperms.

10. The pdf is generated as follows: for each climate sensitivity considered, perform 10,000 simulations (arising from 10 different values for each of four uncertain parameters, giving $10^4 = 10,000$ possible combinations) of the variation in atmospheric CO_2 concentration over the past 420 Ma and count the number of simulations where the statistical misfit between the simulated and proxy CO_2 variation is no greater than the statistical uncertainty in the proxy CO_2 concentrations themselves. The number of simulations where this condition is satisfied for different climate sensitivities indicates the relative probabilities of the various climate sensitivities.

11. The climate sensitivity pdf in this case is essentially the same as that shown in Figure 5.7b, but with the temperature scale multiplied by two.

the long-standing viewpoint that the traditional climate sensitivity (based on fast feedback processes only) is very likely to fall between 1.5 K and 4.5 K. No method indicates a climate sensitivity of less than 1.5 K, but only one method requires a climate sensitivity greater than 4.5 K. Three methods that include slow feedback processes related to changes in the distribution of vegetation and glacier ice caps, as well as other as yet unidentified processes, indicate a substantially larger climate sensitivity, with a consensus climate sensitivity range of 6–8 K.

5.3. SLOW FEEDBACK PROCESSES RELATED TO THE CARBON CYCLE

This section begins with background information on carbon cycle processes in the oceans, followed by an assessment of how the oceanic part of the carbon cycle responds to increases in atmospheric CO_2 concentration in the absence of any change in climate, and then an assessment of known and potential impacts of changes in climate on the oceanic part of the carbon cycle. The same three-step review is presented for the terrestrial part of the carbon cycle (see also Dickinson, 2012, this volume).

In this and subsequent sections, selected results from the simulations of the uptake of anthropogenic CO_2 by the oceanic and terrestrial parts of AOGCMs or other Earth system models are presented. These studies typically used one of the scenarios for emissions of GHGs that was developed by the Intergovernmental Panel on Climate Change (IPCC) in its *Special Report on Emission Scenarios* (Nakicenovic et al., 2000). These are referred to in the remainder of this chapter by the standard designators A1F1, A2, B1, and B2, having cumulative fossil fuel CO_2 emissions to 2100 of about 2300 GtC, 2000 GtC, 1200 GtC, and 1400 GtC, respectively[12].

5.3.1. Oceanic Carbon Cycle Processes

Carbon can occur in the oceans in the following forms:

- As dissolved inorganic carbon (CO_2, HCO_3^-, and CO_3^{2-}), referred to as DIC
- As organic carbon (fleshy material), both in particulate and dissolved form (POC and DOC)
- As $CaCO_3$ (calcium carbonate), which occurs in two mineral forms: aragonite, used as the structural material of corals, pteropods (high-latitude zooplankton) and some molluscs, and calcite, used as the structural material of the foraminifera (animals ranging in size

from less than 1 mm to several cm), coccolithophores (a group of phytoplankton), and many other organisms

Less than 1% of the DIC in the ocean surface layer (referred to as the *mixed layer*) is in the form of CO_2, the vast majority occurring as HCO_3^- (~90%) and CO_3^{2-} (~10%). Only the DIC in the form of CO_2 can be exchanged with the atmosphere. There will be a net flow of CO_2 from the atmosphere to the ocean if the partial pressure of CO_2 in the atmosphere exceeds the partial pressure of CO_2 in the ocean mixed layer, and vice versa. The mixed-layer CO_2 partial pressure pCO_2 depends on the mixed-layer CO_2 concentration $[CO_2]$ and *solubility* α according to:

$$pCO_2 = \frac{[CO_2]}{\alpha} \qquad (5.16)$$

Colder water has a larger solubility than warmer water, so for a given $[CO_2]$, the pCO_2 will be smaller for colder water, which allows cold water to hold more CO_2 and thus more total DIC without exceeding the atmospheric pCO_2.

The production of organic tissue through photosynthesis removes CO_2 from the mixed layer through the reaction:

$$CO_2 + H_2O \rightarrow CH_2O + O_2 \qquad (R5.1)$$

while the net reaction involving construction of construction of $CaCO_3$ releases CO_2 to the mixed layer:

$$Ca^{2+} + 2HCO_3^- \rightarrow CaCO_3 + CO_2 + H_2O \qquad (R5.2)$$

The rate of photosynthesis minus respiration by photosynthetic organisms themselves is referred to as *net primary production* (NPP). Most marine organisms construct skeletal material out of silica (SiO_2) rather than $CaCO_3$, and the construction of SiO_2 has a negligible effect on the concentration of CO_2 in seawater. Much of the organic tissue and calcium carbonate that are produced in the mixed layer is recycled within the mixed layer when organisms die, but some of it sinks into the deeper ocean layers. This is referred to as *export production*. The organic matter largely decomposes within the top 1000 m of the ocean, adding CO_2 to the subsurface water that then slowly diffuses back to the surface. Calcium carbonate dissolves and removes CO_2 from seawater over a much greater depth range. This downward transfer of biogenic carbon (organic matter and $CaCO_3$) is referred to as the *biological pump* (see Sen Gupta and McNeil, 2012, this volume).

Thus, the main biological parameters affecting the concentration of CO_2 in surface water are: (i) the overall magnitude of the biological pump; (ii) the ratio of organic matter to $CaCO_3$ production in the mixed layer which, in turn, depends on the proportions of total primary production by calcareous and non-calcareous organisms and the ratio of

---

12. These sums include the historical cumulative emission of 214 GtC to 1989 plus the 1990–2100 cumulative emissions given in Nakicenovic et al. (2000).

organic matter to $CaCO_3$ production by calcareous micro-organisms; (iii) the proportion of organic matter and $CaCO_3$ production that sinks from the mixed layer into deeper water; and (iv) the depth intervals over which most of the sinking organic matter and $CaCO_3$ decomposes or dissolves.

Carbon is also transferred vertically through convective mixing at high latitudes, through the large-scale over-turning circulation (which includes sinking of cold water at high latitudes), and through diffusive mixing. Convection and overturning cause a net downward transfer of carbon due to the fact that the DIC concentration is greater in the sinking water than in the rising water which, in turn, is related to the fact that sinking waters are cold and thus have a higher solubility for CO_2. This downward transfer is therefore referred to as the *solubility pump*. Combined with the biological pump, this causes the DIC concentration to be much higher in the deep ocean than in the mixed layer. In steady state, the downward transfers are balanced by an upward transfer through diffusion.

Over most of the ocean, photosynthesis is limited by the availability of nutrients, particularly nitrogen (N). Nutrient concentrations are greater in the deep ocean than in the mixed layer (due to their continuous release by the decay of falling organic matter and continuous consumption during photosynthesis in the mixed layer). However, convective mixing and the upwelling branches of vertical overturning cells add nutrients to the mixed layer, thereby facilitating greater NPP as long as light is not a limiting factor.

5.3.2. Ocean Carbon Cycle Feedback Processes

In response to anthropogenic emissions of CO_2 into the atmosphere, the atmospheric partial pressure rises above that of the ocean mixed layer, causing a net transfer of CO_2 into the surface water. The absorption of CO_2 by the mixed layer is rapidly followed by the chemical reactions:

$$CO_{2(gas)} + H_2O_{(liquid)} \rightarrow H_2CO_3(aq) \qquad (R5.3)$$

$$H_2CO_{3(aq)} \rightarrow H^+ + HCO_3^- \qquad (R5.4)$$

$$CO_3^{2-} + H^+ \rightarrow HCO_3^- \qquad (R5.5)$$

To the extent that these reactions proceed to completion, the net reaction is:

$$H_2O + CO_2 + CO_3^{2-} \rightarrow 2HCO_3^- \qquad (R5.6)$$

Reaction (R5.4) would tend to increase the acidity of seawater as CO_2 is added, except that CO_3^{2-} consumes the H^+ that is released by Reaction (R5.4), so that there is no change of pH as long as the occurrence of Reaction (R5.4) is balanced by Reaction (R5.5). The carbonate ion thus acts as a buffer, inhibiting changes in pH to the extent that it is

available. However, the supply of CO_3^{2-} in the surface layer of the ocean is limited, so as more CO_2 is absorbed by the ocean, the H^+ concentration — and hence acidity — of ocean water increases (some of the added H^+ reacts with OH^-, so the OH^- concentration decreases). At the same time as ocean acidity increases, the concentration of CO_3^{2-} decreases, thereby reducing the degree of supersaturation of surface water with respect to aragonite and calcite and reducing the ability of the mixed layer to absorb addi-tional CO_2.

In sum, the direct response of the ocean to the addition of anthropogenic CO_2 to the atmosphere is to absorb some of the added CO_2, thereby reducing the increase in CO_2 concentration. The oceanic part of the carbon cycle (like the terrestrial biosphere) thus acts as a negative feedback on the increase in atmospheric pCO_2[13]. This negative feedback will be altered in a variety of ways related to the increase in the dissolved CO_2 and H^+ concentrations and decrease in the OH^- and CO_3^{2-} concentrations, as dis-cussed by Riebesell et al. (2009) and Sen Gupta and McNeil (2012, this volume).

5.3.3. Ocean Climate—Carbon Cycle Feedback Processes

Warming of the climate will affect the oceanic uptake of CO_2 through direct effects on the solubility of seawater, through direct effects on the biological pump, and through indirect effects on the biological pump related to changes in vertical mixing. These are briefly discussed here.

Riebesell et al. (2009) estimate that the reduction in mixed layer solubility, as climate warms over the next century under a business-as-usual emission scenario, will reduce the cumulative uptake of anthropogenic CO_2 by 9%–15% by 2100. Cao and Jain (2005) estimate that warming reduced the oceanic uptake of CO_2 by 7%–9% during the 1980s and by 6%–8% over the entire period 1765–1990.

Warmer temperatures will directly affect the biological pump through:

- Induced changes in rates of biological processes, including respiration in the mixed layer (López-Urrutia et al., 2006; Vázquez-Domínguez et al., 2007) and of falling organic carbon (Kwon et al., 2009)
- Induced changes in species composition (ecosystem structure), thereby affecting overall NPP (independently of changes in nutrient supply), the fraction of NPP that is exported from the mixed layer to the deep ocean, the depth over which falling organic particles decompose, the C:N ratio of falling organic carbon, and the ratio of

---

13. This is a carbon cycle feedback, rather than a climate—carbon cycle feedback, as it does not involve changes in temperature.

organic carbon to carbonate export (Bopp et al., 2005; Omta et al., 2006)

Warming is initially greater at the ocean surface than below the surface, thereby making the surface layer lighter relative to the underlying water and reducing the intensity of convective and diffusive mixing. This, in turn, has the following effects:

- Reduced upward transfer of nutrients, thereby reducing NPP and the strength of the biological pump in regions (such as tropical and middle latitudes) where the biological pump is limited by nutrients (this will affect the large background biopump flux as well as any change in the biological pump due to the direct effect of higher dissolved CO_2 and warmer temperatures)
- Reduced downward mixing of phytoplankton, resulting in a shallower average depth of phytoplankton and therefore a greater availability of light and, in regions (such as high latitudes) where light is a limiting factor, an increase in the strength of the biological pump
- Reduced upward transfer of DIC, thereby reducing the outgassing of natural CO_2
- Reduced downward mixing of anthropogenic CO_2

As well as reducing convective and diffusive mixing, global warming will alter the thermohaline overturning circulation by altering the large-scale density gradient (which depends on the spatial variation in temperature and salinity). In 18 out of 19 AOGCMs examined by Meehl et al. (2007), there is a transient weakening in the thermohaline overturning circulation in the Atlantic Ocean as the climate warms (see Latif and Park, 2012, this volume). Circulation changes are projected to reduce the uptake of anthropogenic CO_2 by 3%–20% through their effect on the solubility pump (Riebesell et al., 2009).

5.3.4. Observed Climate-Related Changes in Oceanic CO_2 Uptake and Related Variables

Various observations indicate that the warming climate has already started to alter the uptake of anthropogenic CO_2 by the oceans or to alter key processes related to the uptake of anthropogenic CO_2, as reviewed by Sen Gupta and McNeil (2012, this volume) and highlighted below.

5.3.4.1. Estimates of Recent Changes in CO_2 Uptake by the Oceans

As reviewed by Le Quéré et al. (2010), numerous studies over the past decade show that mixed layer pCO_2 has been increasing faster than atmospheric pCO_2 in many parts of the ocean, implying a decreasing oceanic CO_2 sink (in these regions) in spite of increasing atmospheric pCO_2. Among these studies is the analysis of Schuster and Watson (2007), who estimate that uptake of CO_2 by the North Atlantic

between 20°N and 65°N decreased by 0.25 GtC per year between 1994 and 1995 and 2002 and 2005. This is attributed to a decline in rates of wintertime mixing between surface and subsurface waters, which, in turn, is linked to a shift in the North Atlantic Oscillation (NAO). Inverse calculations with atmospheric models indicate that the net uptake of CO_2 in the Southern Ocean (the ocean south of 30°S) weakened by about 0.2 GtC per year over the period 1981–2004 compared to the increase in uptake that would be expected from the increase in atmospheric CO_2 concentration during this time period (Le Quéré et al., 2007). Le Quéré et al. (2010) estimate that climatic change and variability reduced the cumulative CO_2 oceanic uptake over the 1981–2007 time period by 12% compared to what it would have been otherwise (from 66.7 GtC to 59.4 GtC). The rise in atmospheric pCO_2 during this time period would have caused the oceanic sink to increase by 0.32 GtC per year per decade, but climatic change and variability are estimated to have decreased the sink by 0.20 GtC per year per decade — an offset of ~63%. Most of this decrease is estimated to be due to wind-induced changes in ocean circulation. Changes in wind caused increased upwelling of CO_2-rich deep water in the equatorial Pacific and Southern Ocean and, as a consequence, increased outgassing of CO_2. The increased wind and upwelling and reduced CO_2 uptake in the Southern Ocean, in turn, have been successfully simulated by a climate–carbon cycle AOGCM that accounts for the observed depletion of stratospheric ozone (Lenton et al., 2009).

Although the reduced North Atlantic uptake of CO_2 may be a temporary feature due to natural variability in the NAO, the observed changes indicate that changes in winds and ocean mixing do alter the net air–sea CO_2 flux. Similarly, although much of the observed increase in wind speeds over the Southern Ocean and the associated decrease in net CO_2 uptake is likely to be due to the temporary loss of polar stratospheric ozone, global warming is projected to intensify Southern Ocean winds throughout the twenty-first century (Shindell and Schmidt, 2004; Fyfe and Saenko, 2006).

5.3.4.2. Inter-annual and Longer Variations in NPP and Chlorophyll

On an annual to decadal timescale, a strong correlation between global oceanic NPP (as inferred from satellite-estimated chlorophyll mass) and climate variability has been observed. Behrenfeld et al. (2006) and Martinez et al. (2009) found that chlorophyll abundance and SST variations at annual to decadal timescales are inversely related over much of the oceans. This can be explained by enhanced stratification when surface waters are warming, thereby suppressing the upward flux of nutrients through mixing. Consistent with this, Polovina et al. (2008) report that the extent of low-chlorophyll regions in the Atlantic and Pacific oceans increased by 15% between 1998 and 2006, but the increase is much larger than would be expected based on model projections of increasing ocean stratification.

Based on thousands of records of ocean surface water transparency made with Secchi disks since 1899, Boyce et al. (2010) estimate that the global abundance of phytoplankton has declined by an average of 0.9% per year over the period 1950−2003. A decrease in phytoplankton abundance over the 1980−2000 period in the North Atlantic is consistent with the analysis of seawater pCO_2 data by Lefèvre et al. (2004); they find that the seawater pCO_2 increased faster than atmospheric pCO_2 between 1982−1998, which would have led to a weakening CO_2 sink, and they suggest that decreasing biological activity is the explanation.

5.3.4.3. Observed Relationship between Temperature and the Size Distribution of Phytoplankton

It is a well-known principle in aquatic ecology that warmer temperatures favour smaller-sized species and, within a given species, result in smaller mean body size (Daufresne et al., 2009).

In agreement with this principle, Morán et al. (2010) find, based on data collected from research cruises from 1994 to 2005 in the northwest Atlantic, that the proportion of total phytoplankton biomass as picophytoplankton (unicellular organisms in the 0.2−2.0 μm size range) in different locations increases with increasing water temperature throughout the −0.6−22°C temperature range, while the mean picophytoplankton cell size decreases. Assuming that the adjustment of phytoplankton to changing temperature over time is the same as the adjustment over space, global warming should lead to a gradual decrease in phytoplankton size. This, in turn, would tend to reduce the export of carbon into the deep ocean, thereby slowing the oceanic absorption of anthropogenic CO_2.

There is evidence that recent warming has already altered species body size in marine ecosystems. Based on data from the Continuous Plankton Recorder (CPR) survey, which has operated monthly in the North Atlantic Ocean and North Sea since 1946, Beaugrand et al. (2008, 2010) find that the zone of higher copepod diversity south of 45−55°N in the North Atlantic Ocean has shifted northward over time as SST has warmed, but that mean copepod body size has decreased.

5.3.5. Climate−Ocean-Sink Feedbacks as Projected by Models

The oceanic part of the carbon cycle can be simulated using a variety of different models, the most complex being 3D

ocean general circulation models (OGCMs) in which the distribution of temperature, DIC, alkalinity, and other chemical properties is computed and the mixed layer pCO_2 and the atmosphere−ocean CO_2 flux is computed from these. The OGCMs contain subroutines that compute biological processes (NPP, respiration, and the export of carbon to the deep ocean) based on temperature, availability of light and nutrients, and on other factors.

Crueger et al. (2008) assessed the impact of climate feedbacks on the oceanic uptake of CO_2 as simulated by the Max Planck Institute Earth system model. The feedback processes in the model include:

- Temperature effects on solubility
- Reduction in the vertical mixing of CO_2 and nutrients
- Changes in the large-scale overturning circulation
- Reductions in the extent of sea-ice, which allow for air−sea gaseous CO_2 exchange in regions formerly covered with sea-ice, and allow greater absorption of solar radiation, thereby stimulating the biological pump
- Changes in wind speed, which influence the air−sea gaseous exchange in ice-free regions

The net result of these feedbacks is negligible until about 2050, after which the oceanic uptake for simulations with and without climatic effects on the oceanic uptake begins to diverge. By 2100, the oceanic uptake is 5.0 GtC per year without climatic feedbacks and 4.5 GtC per year with feedbacks for the A2 emission scenario − a reduction of about 10%. Averaged over 100 years, the difference would be about 25 GtC, which corresponds to a difference in atmospheric CO_2 concentration of about 12 ppmv, neglecting any response of the terrestrial biosphere to the reduced oceanic uptake.

Chuck et al. (2005), using the HadOCC model with an emission scenario that leads to 701 ppmv atmospheric CO_2 by 2100, find the following impacts on atmospheric CO_2:

- Increase surface temperature by 4 K: +22 ppmv
- Double the maximum growth rate of phytoplankton: −3.9 ppmv
- Change calcite production rate by ±50%: ±4.3 ppmv
- Decrease particle flux to deep ocean by 25%: +16 ppmv
- Increase particle flux to deep ocean by 25%: −8.4 ppmv

Steinacher et al. (2010) simulated changes in primary productivity and the export of particulate carbon to the deep ocean using four different ocean biogeochemistry models (IPSL, MPIM, CSM1.4, and CCSM3). Simulated NPP for present conditions ranges from 24 GtC per year to 49 GtC per year, with only one model falling within the observational estimate of 35−70 GtC per year. All four models show a decrease in NPP over the course of the next century under a high-emissions scenario, with the decrease ranging from 2% to 13% of current NPP. The decrease is driven by reduced overall delivery of nutrients to the surface due to increased stratification and a slower thermohaline over-turning circulation. However, as Chuck et al. (2005) find that decreasing the particle flux to the deep ocean by 25% increases atmospheric CO_2 by only 16 ppmv, a large effect is not expected from the changes in NPP.

In summary, feedbacks between climate and the oceanic part of the carbon cycle are not expected to be large, even in response to large (50%) changes in the rate of calcite production or large (25%) changes in the export of partic-ulate organic carbon.

5.3.6. Terrestrial Carbon Cycle Processes

The key processes in the terrestrial part of the carbon cycle are gross photosynthesis (referred to as *gross primary production* or GPP), respiration by plants themselves (autotrophic respiration), and respiration by decomposers and higher animals (heterotrophic respiration). GPP minus autotrophic respiration gives NPP, while NPP minus heterotrophic respiration gives *net ecosystem production* (NEcP). Positive NEcP over a period of many decades is required in order to offset losses due to periodic forest fires or insect outbreaks, and so is not necessarily indicative of a long-term carbon sink (Dickinson, 2012, this volume). The net balance over a period of time spanning many fire and insect-outbreak cycles is referred to as *net biome productivity*.

Decomposition of soil organic matter is the process of: (i) ingestion of organic matter by soil micro-organisms

and their use as an energy source, leading to the release of CO_2 as the organic matter is respired by micro-organisms and secretion of unrespired organic matter in altered form (exudates); and (ii) breakdown of organic materials by enzymes that are secreted by micro-organisms. Soil organic matter is a mixture of thousands of different carbon compounds, each with its own inherent rates of reaction with enzymes. Organic compounds can be phys-ically protected from enzymes if the compounds are inside soil aggregates, and they can be chemically protected if they are adsorbed onto mineral surfaces. The response of soil respiration rate to an increase in temperature can be much less or much greater than expected based on the inherent sensitivity of reaction rates to changes in temperature, depending on how other environmental constraints on decomposition change. These constraints include:

- Physical and chemical protection (discussed above)
- Availability of moisture (both drought and flooding reduce rates of decomposition, the latter by limiting the availability of oxygen, such that only the slower anaerobic decomposition can proceed)
- Freezing

Table 5.2 summarizes estimates of the size of various terrestrial carbon pools. The amount of carbon in soils and litter (>3500 Gt) is estimated to be many times that in living plants above the soil (500−600 GtC) and many times the current atmospheric CO_2 content (780 GtC). Very

TABLE 5.2 Estimated Distribution of Carbon in the Terrestrial Biosphere

Carbon Pool	Global Size (Gt)	Source
Living plants	450−650	Prentice et al. (2001)
Litter, fine (5-yr mean turnover time)	160	Matthews (1997)
Litter, coarse (13-yr mean turnover time)	150	Matthews (1997)
Upland mineral soils	2300	Davidson and Janssens (2006)
Tropical peatlands	70	Page et al. (2004)
NH Permafrost region		
Peatlands, 0−3 m depth	280	Tarnocai et al. (2009)
Other soils, 0−3 m depth	750	Tarnocai et al. (2009)
Yedoma soils, >3 m depth	400	Tarnocai et al. (2009)
Deltas, >3 m depth	240	Tarnocai et al. (2009)
Total	1670	
SH non-tropical peatlands	13−18	Yu et al. (2010)
Total	~4000−5000	

The reduced total assumes that 'other soils' in NH permafrost regions overlaps with 'upland mineral soils'. All estimates have large uncertainties, but serve to illustrate the likely relative importance of different carbon stores.

large pools are found in NH high-latitude wetlands and frozen in permafrost, with the latter susceptible to release when the permafrost thaws as the climate warms.

5.3.7. Terrestrial Carbon Cycle Feedback Processes

In this section the feedback between atmospheric CO_2 concentration and the fluxes between the atmosphere and terrestrial biosphere are briefly outlined. These constitute a carbon cycle feedback rather than a climate–carbon cycle feedback, but serve as a useful starting point for the discussion of climate–carbon cycle feedbacks.

Higher atmospheric CO_2 tends to stimulate rates of photosynthesis. Early experiments involved growing seedlings in glasshouses that were maintained at different atmospheric CO_2 concentrations. Beginning in the 1990s, outdoor experiments were conducted in which elevated CO_2 concentrations were maintained over an experimental plot by continuously releasing CO_2 from storage canisters. These are referred to as *Free Air Concentration Enhancement* or *FACE* experiments, some of which (at various sites in the USA, Europe, and New Zealand) have now been carried out for more than a decade (e.g., Seiler et al., 2009; Norby et al., 2010; van Kessel et al., 2006; Rütting et al., 2010). These studies confirm that higher CO_2 stimulates increased photosynthesis and storage of carbon, but saturation of the photosynthesis response occurs at a CO_2 concentration of around 500–600 ppmv, which is much lower than expected based on leaf-level physiology (Canadell et al., 2007). In addition, carbon–nitrogen (C–N) interactions can substantially reduce the stimulation of photosynthesis due to higher atmospheric CO_2 concentration after a few years (Norby et al., 2010). More detailed information can be found in Dickinson (2012, this volume).

Higher atmospheric CO_2 has been observed to increase rates of soil respiration in various grassland ecosystems (Wan et al., 2007). This could be due to an increase in plant photosynthesis and the availability of soil carbon, or due to an increase in soil moisture due to reduced stomatal conductance, both a direct result of higher atmospheric CO_2 concentration and leading to enhanced microbial activity. Evidence that the increase in respiration is due to reduced soil moisture loss is provided by the fact that Pendall et al. (2003) report an increase in soil respiration by 25% under elevated CO_2 during the moist season (when water would be less limiting) and by 85% during the dry season in a short-grass prairie in North America. Wan et al. (2007) found that CO_2 enhancement of soil respiration is greater under elevated temperature than ambient temperature. They speculate that this may be due to warmer conditions having a greater drying effect, so that the impact of higher CO_2 in reducing water loss and hence in stimulating respiration is greater. In experiments with tree seedlings, Tingey et al. (2006) report no effect of higher CO_2 on soil respiration, but they do find that elevated CO_2 increases the sensitivity of soil respiration to increases in temperature[14].

Higher CO_2 is expected to initially increase rates of N fixation, followed by down-regulation after a few years (Hungate et al., 2004). The initial increase in N fixation is related to the fact that N fixation is energetically expensive and limited by carbon, as discussed by Dickinson (2012, this volume).

In summary, a higher atmospheric CO_2 concentration usually stimulates higher rates of photosynthesis in the short term, and may stimulate greater rates of N fixation, although both effects generally decrease over time. Higher CO_2 also sometimes increases the rate of soil respiration by increasing the soil moisture content. Warmer temperatures sometimes increase the effect of higher atmospheric pCO_2 on soil respiration and, conversely, higher atmospheric pCO_2 increases the effect of temperature on soil respiration.

5.3.8. Terrestrial Climate–Carbon Cycle Feedback Processes

Climate–carbon cycle feedbacks through the terrestrial biosphere involve changes in the rate of photosynthesis and respiration due to changes in temperature and soil moisture, in the rate of N fixation (which ultimately affects the rate of photosynthesis), in the production of CH_4 from wetlands, and in the release of CO_2 and CH_4 from thawing permafrost soils. Other potential feedbacks involve an increase in the frequency and severity of fires and insect outbreaks.

5.3.8.1. Feedbacks Involving GPP, NPP, and Foliar Respiration

Where temperatures are currently below the optimum for GPP and water and nutrients are not limiting factors, warming will increase GPP but, where temperatures are already above the optimum temperature, further warming will reduce GPP. Warmer temperatures tend to increase both autotrophic and heterotrophic respiration but, in both cases, there is a tendency for the respiration rates to decrease back towards the initial respiration rates over time — a process called *acclimation*. The degree of acclimation of foliar respiration to temperature increases varies widely between species, but can approach 100% (such that there is no increase in respiration with temperature) (Ow et al., 2010).

---

14. That is, Q_{10} (defined later) is larger at higher CO_2.

5.3.8.2. Increased Respiration of Soil Carbon

Warmer temperatures will tend to directly increase the rate of soil respiration by accelerating the rates of chemical reactions, and will indirectly affect respiration through changes in soil moisture and through thawing of permafrost in permafrost regions. The effect of temperature on biological processes is commonly expressed as a Q_{10} factor — the factorial increase in the rate of a process for each 10 K increase in temperature. That is, the rate of some process as a function of the warming ΔT is given by:

$$R(\Delta T) = R_o(Q_{10})^{\Delta T/10} \qquad (5.17)$$

where R_o is the initial rate. Observed Q_{10} values are typically 2.0–3.0. As discussed by Davidson and Janssens (2006), the Arrhenius equation[15] implies that the sensitivity of decomposition to increasing temperature will be larger the more resistant the material in question is to decomposition, and will decrease as temperature increases (a saturation effect). Many studies, reviewed by and including Craine et al. (2010), confirm that respiration rates are generally more sensitive to temperature changes the more resistant the material (and hence the lower the initial respiration rate). Accurate projection of future soil respiration rates at a given location requires correct partitioning of the soil carbon into fractions that are and are not resistant to decomposition, choosing the correct initial respiration rates, and choosing the correct temperature dependencies.

Addition of fresh carbon (from increased NPP) can provoke the decomposition of old, resistant carbon by providing readily available energy for soil microbes. This is referred to as a *priming* effect of fresh carbon on respiration of old carbon (Kuzyakov, 2002).

In many experiments in which soils are heated over a period of months or years so as to mimic the effect of global warming, soil respiration rates have been observed to initially increase, but then to decrease over time. That is, there is an *acclimation of soil respiration* to higher temperature, similar to that observed for foliar respiration. With 100% acclimation, there would be no long-term change in respiration rates in response to warmer temperatures. On the other hand, in some experiments (e.g., Malcolm et al., 2008; Dorrepaal et al., 2009; Reth et al., 2009), no acclimation has been observed after periods of 8–10 years.

Acclimation of decomposers, where observed, could be the result of: (i) a decrease in the availability of substrates; or (ii) genuine acclimation of decomposers, fungi, or roots.

In experiments by Reth et al. (2009) over a period of 8–10 years, an apparent acclimation is due to reduced availability of easily decomposed substrates in the warmed soils, rather than a genuine acclimation.

5.3.8.3. Dieback of Middle- and High-Latitude Forests

Dieback of forests could occur as a result of climatic changes (particularly decreases in precipitation) that render any forest ecosystem unviable, or as a result of climate changes that require the transition from one forest type to another, with dieback of the existing forest occurring before the new forest has established itself. As an example of the former, parts of the existing boreal forest in western Canada may revert to grassland ecosystems, whereas the existing boreal forest in eastern Canada will probably be replaced with temperate broadleaf forests (Notaro et al., 2007). The ability for the boreal forest to re-establish itself north of the existing boreal forests will in many places be limited by poor soils. Insect damage will most likely interact with the direct effects of climatic change in provoking forest dieback.

Black spruce is the dominant species in the vast boreal forest, and is exceptionally sensitive to changes in temperature. Way and Sage (2008) grew black spruce seedlings from seeds that were taken from southern Ontario (at the southern limit of the boreal forest) and placed in 3.8 litre pots that were filled with peat moss, watered as needed to maintain a moist growing medium, and fertilized weekly. Twelve days after germination, half of the seedlings were moved to a greenhouse with day/night temperatures of 30°C/24°C (HT, high temperature, corresponding to summer temperatures expected by 2100) and the other half were maintained at 22°C/16°C (LT, low temperature, representative of current summer conditions). The HT seedlings, were 20% shorter, 58% lighter, and had a 58% lower root:shoot ratio than the LT seedlings. Mortality at the end of the growing season was negligible for LT seedlings but reached 14% for HT seedlings. Net photosynthesis at growth temperature was 19%–35% lower in HT than in LT trees. Note that these impacts do not include possible effects of drought associated with warmer temperatures, as the growing medium was maintained in a moist state for both temperature treatments.

The incidence of forest fires in boreal forests is projected to increase strongly as the region warms. For example, Balshi et al. (2009) project an increase in CO_2 emissions from fires in the North American boreal forest by a factor of about 2.5–4.5 during the twenty-first century, using the TEM ecosystem model driven by climate changes from the Canadian AOCGM (CGCM2). The net effect of increasing incidence of fires (taking into account regrowth after fires) is a decrease in C storage by 2100 by about

---

15. The Arrhenius equation for reaction rate constant, k, as a function of activation energy, E_a, and temperature, T, is $k = Ae^{-E_a/RT}$ Here, R is the gas constant and A is the number of collisions that may or may not result in a reaction.

19 Gt. Harrison and Bartlein, (2012, this volume) review palaeo-evidence for the relationship between large-scale fires and climate change.

5.3.8.4. Amazon Ecosystem Collapse

Modelling studies indicate that the Amazon rainforest might be close to a tipping point (threshold) whereby increased occurrence and severity of drought could lead to widespread collapse of the forest, which would revert to savanna or grassland (Phillips et al., 2009; Malhi et al., 2009). Reduced growth and increased tree mortality were associated with a single severe drought in 2005, which caused the emission of 1.2–1.6 GtC (Phillips et al., 2009). Satellite data indicate a 33% increase in the extent of forest fires in 2005 compared to the 1999–2005 mean (Aragão et al., 2007). Even though we might be close to a threshold temperature change, with a strong decrease in the equilibrium forest cover with further warming, the lag in the rainforest ecosystem response is such that very little change might occur until long after the tipping point has been passed (cf. Lenton, 2012, this volume). This is illustrated in Figure 5.13, which shows the simulated variation in

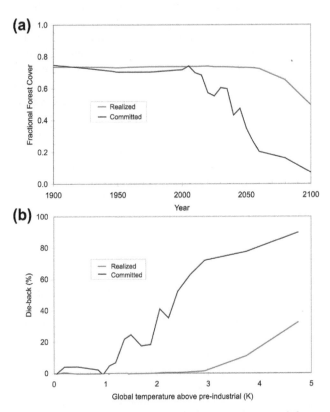

(a)

(b)

Amazon forest cover over time and the dieback as a function of the global mean temperature change, along with the committed reduction in forest cover for each year, for the HadCM3LC model driven by the A2 emission scenario. According to this model, if the climate were stabilized at 2 K global mean warming by 2050, there would be very little change in Amazon forest cover at that point in time, but significant dieback would occur over the next 100–200 years.

The major source of uncertainty concerning the fate of the Amazon lies in the projected change in climate, particularly rainfall, rather than in the ecosystem models (Poulter et al., 2010).

Throughfall exclusion experiments, in which up to 60% of the rainfall is prevented from entering the soil using plastic panels that are installed beneath the understorey, indicate negligible impacts from 3–4 years of rainfall deprivation (due to the ability of deep tap roots to access deep soil moisture), but increasingly severe impacts thereafter (Nepstad et al., 2007; da Costa et al., 2010). Salazar et al. (2007) used monthly temperature and precipitation changes over the next century as projected by 15 different AOGCMs to project changes in South American biomes using the LPJ (Lund–Potsdam–Jena) dynamic global vegetation model (DGVM) (Figure 5.14)[16]. For the A2 emission scenario, 75% of the models project that at least 50% of the Amazon remains by the end of the century, 75% of the models show that 20% is converted to savanna, while there is no consensus on the fate of the remaining 32%. These simulations all neglect the continuing impact of human-induced land use and land-cover changes (Pitman and de Noblet-Ducoudré, 2012, this volume).

Not included in current projections of the impact of climatic change in the Amazon are likely increases in the incidence of fire, the impacts of which could be worse than those of drought stress (Barlow and Peres, 2008). Fragmentation of intact forests through the construction of roads and logging operations increases the susceptibility of forests to fire by creating dry, fire-prone forest edges, and introduces humans that can start fires (Laurance, 2000). Conversely, with limits on human encroachment and vigorous efforts to contain fires when they are started, it is possible — given the inertia of the ecosystem — that large-scale collapse could be deferred until after 2200 (Malhi et al., 2009), by which time it might have been possible to reduce GHG heating sufficiently to avoid collapse (by creating negative emissions in various ways).

FIGURE 5.13 (a) Fractional tree cover in the current Amazon rainforest that remains as climate changes under the A2 emissions scenario according to the HadCM3LC model, and the committed state corresponding to each year. (b) The same as in (a) but plotted as the dieback as a function of the global mean warming. *(Source: Jones et al., 2009. Reprinted by permission from Macmillan Publishers Ltd.)*

16. A DGVM is a model that simulates the spatial distribution of different plant functional types, as well as the cycling of carbon within each plant functional type together with exchanges of energy and moisture with the atmosphere.

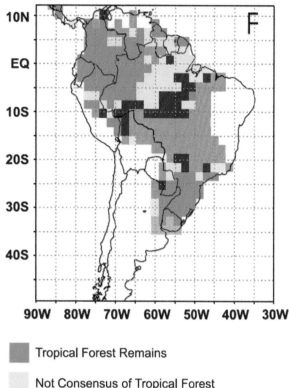

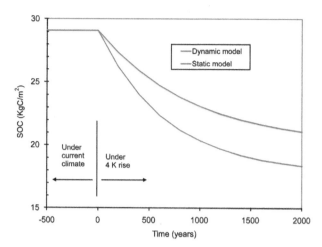

FIGURE 5.15 Simulated decrease in peatland carbon (SOC, soil organic carbon) for a site in northern Manitoba following a sudden 4 K surface warming. *(Source: Ise et al., 2008 © Reprinted by permission from Macmillan Publishers Ltd.)*

FIGURE 5.14 Fraction of 15 AOGCM climate change simulations that provoke various transitions in the type of biome at various grid cells in South America by 2100 for the A2 emissions scenario. *(Source: Salazar et al., 2007 © Copyright 2007 American Geophysical Union. Reproduced/modified by permission of American Geophysical Union.)*

5.3.8.5. Methane Emissions from Wetlands

CH_4 is produced in wetlands through anaerobic decomposition of organic matter. The rate of production of CH_4 depends on the availability of decomposable organic matter (and so may increase as NPP increases), the water level (which determines the extent of anaerobic conditions), and temperature. Shindell et al. (2004) ran a detailed process-based model on a $1° \times 1°$ (latitude–longitude) grid to calculate present-day wetland emissions of CH_4 as a function of water-table level, soil temperature, and NPP. Local correlations between monthly mean emissions and monthly mean soil temperature and moisture were then used in combination with soil temperature and moisture output from the GISS (Goddard Institute for Space Studies) AGCM to calculate CH_4 emissions for present conditions and for the climate in equilibrium with a doubling of atmospheric CO_2

(given a global mean warming of 3.4 K). Global wetland emissions increase by 78%, from 156 Tg per year to 277 Tg per year. Most of the increase is due to emissions from tropical wetlands, although emissions from high northern latitude wetlands nearly triple during the summer.

The feedback factor associated with the calculated increase in wetland emissions can be computed as follows: using formulae for the lifespan and radiative forcing due to CH_4 given in Harvey (2011), the steady-state CH_4 concentration assuming fixed natural and anthropogenic emissions of 185 TgC per year and 196 TgC per year, respectively, is 1860 ppbv[17]. The steady-state concentration with an additional CH_4 emission of 120 TgC per year is 2600 ppbv, which produces an additional radiative forcing of 0.375 W m^{-2} (this accounts for the overlap with N_2O and includes an extra forcing of 0.130 W m^{-2} due to extra tropospheric ozone and stratospheric water vapour). This extra forcing gives a feedback parameter $\lambda_{wetland}$ of -0.110 W m^{-2} K^{-1}. This, in turn, would increase a climate sensitivity of 1.50 K to 1.57 K (a 5% enhancement) or a climate sensitivity of 4.5 K to 5.18 K (a 15% enhancement).

Ise et al. (2008) used a detailed process-based model of the biogeochemistry, hydrology, and temperature of a peatland in northern Manitoba to study the impact of warming. Figure 5.15 shows the change in the carbon content over time when a 4 K warming is suddenly imposed at year zero. About 63% (1/e) of the final response occurs over a period of about 350 years, with the full response to the warming perturbation requiring about 1000 years. Two cases are shown in Figure 5.15: one (static) where the soil depth is held fixed even as organic matter decays, and one

---

17. In transient calculations performed in the Excel spreadsheets that accompany Harvey (2011), the transient CH_4 concentration reaches 1730 pbbv by 2000, in line with observations.

(dynamic) in which it is allowed to change. In the latter case, decreasing the soil depth reduces the water-holding capacity, which provokes further drying and loss of carbon. The inherent temperature sensitivity of respiration in the model is a Q_{10} of 2.0, but the effective Q_{10} (taking into account losses due to drier conditions) is 2.5 for the static model and 3.6 for the dynamic model.

5.3.8.6. Feedbacks Involving Permafrost Soils

Simulations of climate change using an AOGCM indicate that the majority of the world's permafrost could be lost this century under business-as-usual emission scenarios: the global extent of permafrost (excluding Greenland and Antarctica) decreases from around 10 million km^2 in 2000 to about 4 million km^2 in 2100 under the B1 emission scenario and to about 1 million km^2 under the A2 emission scenario (Lawrence and Slater, 2005) (see also Lenton, 2012, this volume). Permafrost regions are estimated to contain in excess of 1600 GtC (Table 5.2), much of which could potentially be released to the atmosphere once the permafrost thaws.

Of particular concern are the organic-rich soils in Siberia — known as *yedoma* in the Russian literature and as loess in the English-language literature — which contain carbon that accumulated during the glacial periods of the Pleistocene and which decomposes quickly when the soils are thawed. Khvorostyanov et al. (2008a) used a detailed permafrost model to simulate the impact of local warming on yedoma soils to a depth of 12 m. They estimate that intense mobilization of carbon would begin when the regional warming reaches about 9 K and continue for several hundred years. Based on an estimated initial total carbon stock for the yedoma region of 375 Gt and using the regionally-averaged soil temperature, they find:

- Assuming normal hydrological conditions, the soil carbon stock decreases to about 340 Gt by the time the warming reaches 9 K, then drops to 130 GtC over the next 100 years (an average loss of 2.1 GtC per year) and to 60 GtC over the following 100 years (an average loss of 0.7 GtC per year), with 12% of the total lost carbon (about 40 Gt) emitted as CH_4.
- Assuming the upper metre of the soil to be always saturated, the soil carbon stock decreases to about 360 Gt by the time the warming reaches 9 K, then drops to 180 GtC over the next 100 years (an average loss of 1.8 GtC per year) and to 140 GtC over the following 100 years (an average loss of 0.4 GtC per year), with all of the lost carbon emitted as CH_4[18].

Inasmuch as the mean annual warming over land at the latitudes (60°N to 70°N) where yedoma soils occur is about

twice the global mean warming (see figure 10.6 of Meehl et al., 2007), this implies that significant carbon release would occur in association with a global mean warming of about 4.5 K. Mobilization and release of soil carbon begins with much less warming, but is initially quite slow. Intense carbon release is triggered by internal heat generation by decomposers, and is irreversible once it starts (see also Lenton, 2012, this volume).

Thaw lakes are another source of CH_4 from permafrost regions. They are produced when the water along the edges of the lakes causes massive subsurface ice wedges to melt, in turn causing the ground surface to subside in a self-perpetuating cycle (this is referred to as thermokarst erosion). Walter et al. (2006) conclude that the CH_4 flux from the lakes in their study region in Siberia is 5 times greater than previously estimated, and increased by about 60% between 1974 and 2000.

Schaefer et al. (2011) used a detailed 1D permafrost model, placed at each grid point in a global grid covering permafrost regions, to estimate the potential CO_2 fluxes associated with the A1B emission scenario. They simulate cumulative emissions of 70–140 GtC by 2100 (0.7–1.4 GtC per year) and 150–250 GtC by 2200. These are global emissions, but are based on only the top 3 m of permafrost soils and do not account for the positive feedback on C release through internal heat generation by the decomposition process nor for releases from regions with discontinuous permafrost or from thaw lakes. Khvorostyanov et al. (2008a,b) simulate a larger flux (1.8–2.1 GtC per year averaged over the first 100 years of active release) from only one (admittedly carbon intense) region, but the vast majority of the C released in their simulations is from below the 3 m depth (see their Figure 5.4c). Given the estimated size of the yedoma soil C pool and widespread observations (summarized by Schaefer et al., 2011) of warming by 1–3 K at depths of 10–20 m over that past few decades, the large fluxes simulated by Khvorostyanov et al. (2008a,b) are plausible.

5.3.8.7. Summary

There is the potential for large positive feedbacks between the climate and the terrestrial part of the carbon cycle. Warmer conditions will tend to increase GPP and soil respiration, although the former might be limited by lack of moisture or nutrients, or reversed when temperatures exceed optimal conditions, while decreasing availability of easily respirable soil carbon will limit the long-term increase in soil respiration. However, significant dieback of mid-latitude and tropical forests could occur, with large transient emissions of CO_2. Increased CH_4 emissions from wetlands could increase climate sensitivity by 5%–15%. The largest potential fluxes involve release of carbon from thawing permafrost, much of which could be in the form of CH_4.

---

18. By comparison, the fossil fuel CO_2 emission in 2007 was 8.0 GtC (Boden and Marland, 2010).

5.3.9. Terrestrial Climate–Carbon Cycle Feedback: Local and Large-Scale Observations

The previous section summarized laboratory and field experiments and modelling studies. In this section, observations concerning recent changes in photosynthesis, NPP, drought, fires, and insect outbreaks in natural ecosystems and their links to recent changes in climate or climate variability are summarized. The feedback between climate and the terrestrial component of the carbon cycle is likely to have been negative up to the present; that is, changes in climate so far have probably stimulated net absorption of CO_2 from the atmosphere by the terrestrial biosphere. However, there is evidence of an emerging positive feedback consistent with the expectations that were reviewed in the previous section.

5.3.9.1. Trends in Photosynthesis and Net Primary Productivity

The global rate of photosynthesis and global net biome productivity have clearly increased over the past half century or more. This is evident from satellite and ground-based data, as reviewed by Boisvenue and Running (2006), from the need for a terrestrial biosphere sink in order to account for all of the CO_2 emitted by humans (Boppet al., 2002), from vertical CO_2 profiles (Stephenset al., 2007), and from inversion modelling (Gurney et al., 2008).

However, in spite of the overall stimulation of photosynthesis, there is observational evidence that the trend may slow down or even reverse:

- Model analysis of the satellite-based observed greening north of 25°N over the period 1980–2000 indicates that increasing atmospheric CO_2, temperature, and precipitation accounted for 49%, 31%, and 13% of the greening, but that the greening trend will weaken or disappear with continued warming (Piao et al., 2006)[19].
- More recent satellite observations indicate a reversal or at least a stabilization of the upward trend between 2000–2009, due largely to decreasing NPP in the SH that is linked to a widespread SH drying trend (Zhao and Running, 2010).
- Carnicer et al. (2011) report a doubling in the percentage of tree crown defoliation (from about 12% to 25%) in southern Europe over the period 1990–2007 in parallel with increasing mean annual temperatures and decreasing rainfall.
- Based on detailed repeated censuses between 1981 and 2005 on 50-ha forest plots in Panama and Malaysia (where pronounced warming has occurred), Feeley et al.

(2007) report that stem growth rates within distinct size categories decreased over time (with community level rates of decrease in stem growth of 1.2% per year in the Panamanian plots and 6.2% per year in the Malaysian plots).

- The warm and dry climate of the early 1980s in Alaska caused the Arctic tundra there to switch from a carbon sink to a carbon source. During the 1990s, summer sink activity resumed, indicating a previously undemonstrated capacity of this ecosystem to acclimatize to decadal (and longer) warming (Oechel et al., 2000). However, the tundra ecosystem studied is still a net source of CO_2 to the atmosphere on an annual basis due to winter release of CO_2.
- In a large (>1500) sample of white spruce trees in Alaska, over 40% displayed decreased growth with warmer temperatures over the last few decades and less than 40% displayed increased growth, the differences being due to differences in water stress at different sites (Juday et al., 2005).
- A divergence of populations, which used to vary in unison prior to the twentieth century, similar to that noted above for Alaska, has also been observed in the Mackenzie delta and other subarctic locations (Pisaric et al., 2007).

5.3.9.2. Response of NPP to Inter-annual Variability

The extreme heatwave in Europe in the summer of 2003 led to a strong reduction in both photosynthesis and respiration due to drought stress, with a net source of 0.5 GtC that year (Ciais et al., 2005; Reichstein et al., 2007). Conversely, there was an exceptional uptake of CO_2 at six monitored forest sites in Europe (spanning 44°N to 62°N) in the spring of 2007 in response to an exceptionally warm spring. However, as discussed by Delpierre et al. (2009), when a warm spring is followed by a summer drought, the two effects cancel each other out, but if the warm spring hastens the drought (through the earlier resumption of transpiration after the winter), there would be a tendency towards net carbon release. Warmer and drier conditions in the tropics during recent El Niño events are calculated to have suppressed NPP by 1.21 GtC per year and enhanced heterotrophic respiration by 0.56 GtC per year (Qian et al., 2008).

5.3.9.3. Tree Mortality and Insect Outbreaks

Allen et al. (2010) present tables summarizing 88 instances of drought and heat-induced tree mortality that have been reported in the scientific literature during the last two decades or so. The portion of the literature with the keyword 'forest' that also has the keywords 'forests

---

19. Land-use change, N deposition, and ozone were not included in the analysis.

AND mortality AND drought' increased from 0.1% in the mid-1980s to 0.3% by 2008—2009, although this may not be indicative of a real trend in tree mortality. Rates of tree mortality in old forests across a broad region of the western USA have doubled to tripled during the past four decades, while the rate of establishment of new trees has increased by only about 50% (van Mantgem et al., 2009). This has been accompanied by warming at rates of 0.3—0.5 K per decade at the elevational range of the forests, which caused an increase in summer drought.

Widespread losses in forests have occurred in recent years due to insect outbreaks that are related at least in part to warmer temperatures. Examples include:

- Mountain pine beetle outbreaks in western Canada, killing millions of trees over an area of 130,000 km^2 in 6 years, including areas at higher elevations and more northern latitudes than in the past (Kurz et al., 2008)
- Drought and insect-driven mortality affecting 12,000 km^2 of piñon pine, killing 40%—97% of trees at some sites in less than 3 years (Breshears et al., 2005; McDowell et al., 2008)
- Other insect outbreaks from Mexico to Alaska (Raffa et al., 2008)

More frequent outbreaks of insects are expected in tropical forests under warmer and drier conditions (Coley, 1998). Boreal forests are observed to be more susceptible to insect pests than other forests, perhaps because of their low genetic diversity.

5.3.9.4. Increased Incidence and Severity of Forest Fires

Temperature variations have been a driving factor behind variations in the area of forest burning per decade over the past century. In Canada, both temperature and area burned decreased modestly from 1920 to 1970, then increased sharply from 1970 to 2000, while, in Alaska, decadal burned area decreased from the 1940s to the 1980s, then rose to over 150% that of the 1940s by the 2000s. Gillett et al. (2004) examined the statistical relationship between forest area burned (in 5-year blocks) and temperature in Canada since 1920 and found that temperature variations can explain 59% of the variation in five-year totals. Kasischke et al. (2010) find temperature and precipitation variations to be the main factor behind Alaskan trends, although a decrease in fire suppression activities in the 1990s and 2000s was a contributing factor to the recent increase in decadal burn area.

Figure 5.16 shows data on the variation in forest fire extent or frequency during the past few decades in selected regions. The area of forest burned per decade increased by a factor of 4 in Canada and Alaska combined from the

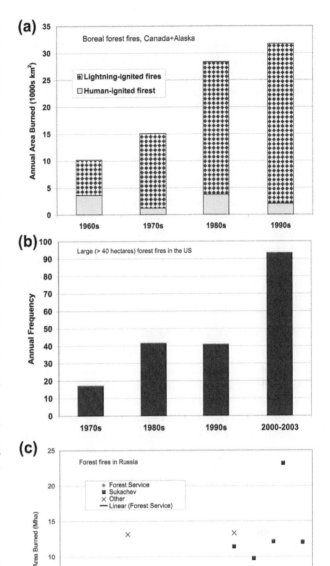

FIGURE 5.16 Trends in: (a) the area of boreal forest burned annually in Canada + Alaska; (b) in the annual frequency of large (>40 ha) forest fires in the lower 48 US states; and (c) the annual area of forest burned in Russia. (*Sources: (a) Data from Kasischke and Turetsky, 2006; (b) data from Westerling et al., 2006; and (c) Soja et al., 2007. Reprinted from Soja et al., 2007; with permission from Elsevier*)

1960s to the 1990s (Stocks et al., 2003; Kasischke and Turetsky, 2006), while the number of large (>400 ha) fires per year in the lower 48 states of the USA increased by more than a factor of 6 after the 1980s (Running, 2006; Westerling et al., 2006). There has also been a dramatic increase in recent years in the frequency and severity of fires in Russia (Soja et al., 2007).

5.3.9.5. Fires in Peatlands

Turetsky et al. (2002) crudely estimate that there has been a 2.5-fold increase in the annual area of peatlands burning in Canada in recent decades compared to the long-term average since the last glaciation. In Indonesia, peat fires triggered by dry conditions associated with a strong El Niño released 0.4–2.2 GtC in 1997 (Page et al., 2002). An increase in lighting strikes could also increase the incidence of fire. The frequency of lightning strikes on the tundra of the North Slope in Alaska has apparently increased by a factor of 10 in the last 10 years and, of 26 recorded fires since 1950, close to one third took place over the 3 years preceding 2009 (Qui, 2009). Based on the record of varying charcoal abundance in lake sediments, Hu et al. (2010) conclude that the large fires in 2007 in central Alaska were unprecedented within the last 5000 years (and see also Harrison and Bartlein, 2012, this volume).

5.3.9.6. Empirical Estimate of the Global-Scale Climate–Soil Carbon Feedback

Given estimates of anthropogenic CO_2 emissions during the past century (the only significant uncertainty pertaining to land-use emissions; see Pitman and de Noblet-Ducoudré, 2012, this volume) and observed yearly CO_2 concentrations, the CO_2 airborne fraction (defined as the annual mass increase in atmospheric CO_2 divided by the annual emission) can be computed and compared with what would be expected from a carbon cycle model with no climate–carbon cycle feedback. Rafelski et al. (2009) did just this, using a model with separate fast and slow soil carbon pools. The rate of growth of fossil fuel emissions fell from 4.3% per year before 1980 to 1.5% per year after 1980, which should have caused a decrease in the airborne fraction. Instead, the airborne fraction has been nearly constant since 1958. For a wide variety of assumptions concerning land-use emissions and N fertilization, Rafelski et al. (2009) find that this can be explained by a warming-related increase in natural CO_2 emissions that can be obtained assuming a Q_{10} for respiration of fast soil carbon of 1.5–5.0 (depending on assumptions concerning land use emissions, N fertilization, and oceanic CO_2 uptake)[20].

5.3.9.7. Summary

There is widespread evidence that the terrestrial biosphere sink is beginning to weaken in regions where warmer temperatures and/or reduced rainfall have led to drought

stress. Growth rates have slowed in water-stressed northern boreal sites and in some tropical forest sites, and there is an increasing incidence of crown defoliation, insect outbreaks related to heat and drought, forest dieback, and forest and peatland fires. The combination of a slowing in the rate of growth of anthropogenic emissions after 1980, combined with a near-constant airborne fraction indicate that the overall effect of warmer temperatures has been to cause a net emission of CO_2 from the terrestrial biosphere.

5.3.10. Destabilization of Methane Clathrate

CH_4 clathrate (or CH_4 hydrate) is an ice-like compound formed when water freezes in the presence of sufficient CH_4 and other gases, such that these gases become trapped within the water molecule lattice. The temperature and pressure combinations found at the sea floor are such that CH_4 clathrate is stable there, although the sediment temperature increases with depth, such that below a certain depth any clathrate would melt. The amount of CH_4 in clathrate at a given location depends on the thickness of the stability zone, the sediment porosity and the fraction of the pore space occupied by clathrate. The possible origins, geology, properties, past destabilization events, and other characteristics of CH_4 clathrate are reviewed in Archer (2007).

Estimates of the global inventory of CH_4 in clathrates range from 700 to over 10,000 GtC. As the ocean seafloor warms in response to the warming climate, heat will diffuse into the sea floor, raising temperatures and destabilizing clathrates from the bottom of the stability zone upward (Harvey and Huang, 1995). Upwardly migrating CH_4 could reform clathrates within the overlying stability zone and, once this zone becomes impermeable or if it is already impermeable, gas would build up below the stability zone. Thawing of clathrates at the base of the stability zone would initially be slow and would continue for centuries. However, if pressure builds up between the thinning stability zone, a point may be reached where large amounts of CH_4 are released abruptly due to fracturing of the overlying clathrate layer, or due to sliding of sloping sediments. Nevertheless, due to the fact that the warming would need to penetrate to the base of the stable zone before thawing of CH_4 clathrates would begin, and that the released CH_4 would then re-enter the stable zone, Harvey and Huang (1995) concluded that destabilization of CH_4 clathrates would not significantly enhance the global mean warming until several centuries into the future, and that significant effects on climate would occur only if the global mean warming had already reached 3–5 K.

More recently, Archer et al. (2009) estimated that a warming of 3 K would be sufficient to eventually

---

20. A $Q_{10} > 1.0$ for the slow soil carbon reservoir is likely but cannot be determined by this analysis because of the much smaller respiration flux (and correspondingly, the much longer turnover time) for the slow carbon reservoir.

release half of the CH$_4$ in marine clathrates. The major unknown is the fraction of CH$_4$ that would escape to the atmosphere; much of it could be oxidized to CO$_2$ in the water column before reaching the atmosphere if it is released slowly enough. Archer et al. (2009) estimate the releasable CH$_4$ from a 3 K global mean warming to be 35–940 GtC, released over a period of several thousand years. The effect of CH$_4$ clathrates in this case would be to slow the decline of atmospheric CO$_2$ from its peak, rather than causing a larger peak concentration. However, Reagan and Moridis (2007) conclude that, although deep hydrate deposits are resistant to rapid destabilization, shallow deposits, such as those found in Arctic regions and in the Gulf of Mexico, can rapidly dissociate, producing significant fluxes within the next 100 years in response to a seafloor warming of 3 K over the next 100 years.

Indeed, destabilization of shallow marine clathrates may have already started on the continental slope west of Spitsbergen, where sediments apparently warmed by 3 K over the past century (Reagan and Moridis, 2009). Westbrook et al. (2009) report observations of more then 250 plumes of gas bubbles emanating from the seabed of the West Spitsbergen continental margin. While at least some of this is from natural seepage of pre-existing CH$_4$, not from dissociation of CH$_4$ clathrate, some is due to dissociation. Low-level (~8 TgC per year) release of CH$_4$ from sediments to the atmosphere is occurring from the East Siberian Arctic shelf[21], but it is unknown whether this is entirely a natural background flux or partly the beginning of temperature-induced CH$_4$ release (Shakhova et al., 2010a,b). The total of all natural emissions from various geological sources (including clathrates) is estimated to be about 35 TgC per year (Kvenvolden and Rogers, 2005), which would have contributed to the background (prehuman) atmospheric CH$_4$ concentration.

Release of CH$_4$ from clathrates is suspected of being a major factor in several dramatic warmings in the geological past, including at the transition between the Palaeocene and Eocene time periods — the so-called Palaeocene–Eocene Thermal Maximum (PETM; see Belcher and Mander, 2012, this volume) — at 55 Ma (Sluijs et al., 2007), during the mid-Jurassic at 183 Ma, (Hesselbo et al., 2000) and a massive release at 635 Ma that may have ended the last 'snowball Earth' period (Kennedy et al., 2008). Periodic minor and occasional large CH$_4$ releases are thought to have occurred during warm periods that interspersed glaciations during the last two million years (Kennett et al., 2000). Evidence from the PETM indicates that the suspected release of CH$_4$ was preceded by a slow warming that reached 5 K in what is now Wyoming (Secord et al., 2010).

5.4. COUPLED CLIMATE–CARBON CYCLE MODEL RESULTS AND LINEAR FEEDBACK ANALYSIS

In Section 5.2.1, a linear feedback analysis was presented of climate sensitivity, showing how to quantify the inherent strength of individual feedback processes. A key conclusion was that the impact on climate sensitivity of adding a given feedback process depends on the strength of the pre-existing feedback processes. We now extend this analysis to incorporate carbon cycle processes, following closely the presentation in Gregory et al. (2009). The incorporation of climate–carbon cycle feedback in the analysis presented below, however, pertains only to feedbacks that vary in proportion to the change in global mean temperature. Thus, non-linear responses such as an abrupt increase in the emissions of CH$_4$ from thawing permafrost or clathrates are not included.

5.4.1. Effect of the Oceans in Limiting the Transient Temperature Response

As the climate warms in response to a positive radiative forcing, the emission of infrared radiation to space increases, serving as a break on subsequent warming. During the initial part of the transient response, a net heat flux to the deep ocean arises that also serves as a break on subsequent surface warming. In the equilibrium climate state that is reached many centuries after the radiative forcing has been stabilized, the net heat flux to the ocean will be zero because the deep ocean will have finished warming. However, during the early period of rapidly and continuously increasing radiative forcing, minimal deep ocean warming will have occurred, so the heat flux F_D to the deep ocean will be approximately proportional to the surface warming. Due to the very small heat capacity of the atmosphere and minimal heat flow into the land surface (because of the small thermal conductivity of the subsurface), the net radiation N is almost exactly equal to the net heat flux into the deep ocean. Thus:

$$N \approx F_D \approx \kappa \Delta T \qquad (5.18)$$

where κ is a proportionality constant (equivalent to a damping coefficient) that applies only during a period of steadily increasing ΔT. From Equations (5.15) and (5.18) it follows that the transient temperature change is:

$$\Delta T(t) \approx \frac{\Delta R(t)}{\lambda + \kappa} = \frac{\Delta R(t)}{\rho} \qquad (5.19)$$

---

21. By comparison, the total (natural + anthropogenic) CH$_4$ emission over the period 2000–2004 is estimated to have been 582 TgC per year (Denman et al., 2007).

TABLE 5.3 Radiative Feedback Parameter (λ) and Ocean Heat Flux Parameter (κ) for 19 AOGCMs as Deduced by Gregory and Forster (2008), the Sum of the Two (ρ), and the Ratio $\lambda/(\lambda + \kappa)$

Model	λ	κ	ρ	$\lambda/(\lambda + \kappa)$
CCSM3	1.84	0.67	2.51	0.73
CGCM3.1(T47)	1.28	0.55	1.83	0.70
CNRM-CM3	1.6	0.58	2.18	0.73
CSIRO-Mk3.0	1.6	0.83	2.43	0.66
ECHAM/MPI-OM	1.01	0.66	1.67	0.60
GFDL-CM2.0	1.96	0.64	2.6	0.75
GFDL-CM2.1	1.74	0.73	2.47	0.70
GISS-EH	1.46	0.77	2.23	0.65
INM-CM3.0	1.77	0.48	2.25	0.79
IPSL-CM4	1.03	0.7	1.73	0.60
MIROC3.2(hires)	0.87	0.56	1.43	0.61
MIROC3.2(medres)	0.97	0.81	1.78	0.54
MRI-CGCM2.3.2	1.23	0.41	1.64	0.75
PCM	2.08	0.45	2.53	0.82
UKMO-HadCM3	1.09	0.53	1.62	0.67
UKMO-HadGEM1	1.27	0.56	1.83	0.69
Ensemble	1.4 ± 0.6	0.6 ± 0.2	2.0 ± 0.7	0.70

The ensemble uncertainties are the 5%–95% probability range assuming that the model results are normally distributed. Units for λ, κ, and ρ: W m^{-2} K^{-1}

where $\rho = \lambda + \kappa$. Thus, the effect of mixing of heat into the oceans in delaying the temperature response to radiative forcing can be represented in two ways: either through incorporation of $N(T)$ into Equation (5.15) or through the incorporation of κ in Equation (5.19).

Table 5.3 gives values of λ and κ as deduced by Gregory and Forster (2008) for 19 AOGCMs that were forced with a scenario of CO_2 increasing by 1% per year for 70 years, where λ was estimated from a regression of $\Delta R - N(T)$ against ΔT and κ was estimated from a regression of $N(T)$ against ΔT. Their results are plotted in Figure 5.17. The ratio of the realized change in global mean temperature to the equilibrium change (for which $\kappa = 0$) is equal to $\lambda/(\lambda + \kappa)$, and this ratio is given for each model as the final column of Table 5.3. On average, the global mean temperature at the end of 70 years of a 1% per year CO_2 increase is equal to 70% of the response that would occur with the same radiative forcing in the absence of the deep ocean, with most models giving a response of 60%–75% of the equilibrium response. However, by reducing the transient temperature response, the oceans also reduce the effect of the climate—carbon cycle feedback, so that (as illustrated later) the impact of the oceans in limiting

transient warming is much greater than implied by the $\lambda/(\lambda + \kappa)$ ratio when the net climate—carbon cycle feedback is strongly positive.

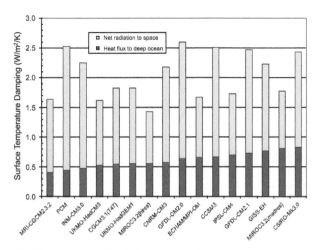

FIGURE 5.17 Overall radiative and ocean heat flux damping parameters, λ and κ, as deduced by Gregory and Forster (2008) for 19 AOGCMs. *(Source: Copyright 2008 American Geophysical Union. Reproduced/modified by permission of American Geophysical Union.)*

5.4.2. Climatic Change As a Feedback on the Carbon Cycle

The cumulative CO_2 emission, C_E, can be written as the sum of the CO_2 that resides in the atmosphere (C_A) and is taken up by the land (C_L) and oceans (C_O). That is:

$$C_E = C_A + C_L + C_O \qquad (5.20)$$

The uptake of carbon by the land and ocean can be broken down into components that depend on the increase in atmospheric CO_2 and those that depend on the change in temperature. Thus, $C_L = C_{L\beta} + C_{L\gamma}$ and $C_O = C_{O\beta} + C_{O\gamma}$, where the subscripts β and γ denote these two components. Combining the terrestrial and marine portions, we can write:

$$C_E = C_A + C_\beta + C_\gamma \qquad (5.21)$$

where $C_\beta = C_{L\beta} + C_{O\beta}$ and $C_\gamma = C_{L\gamma} + C_{O\gamma}$. Higher atmospheric CO_2 stimulates uptake by the land and oceans. Thus, $C_\beta > 0$, while warming stimulates release of CO_2, so that $C_\gamma < 0$, but $C_\beta + C_\gamma > 0$, so that $C_A < C_E$. Following Gregory et al. (2009) and Friedlingstein et al. (2006), we assume that C_β is proportional to the atmospheric CO_2 increase and that C_γ is proportional to the change in global mean surface air temperature, and that the adjustments in $C_\beta + C_\gamma$ to changing C_A and ΔT are instantaneous. That is, $C\beta = \beta C_A$ and $C\gamma = \gamma \Delta T$, where β is dimensionless (GtC uptake per GtC atmospheric CO_2 increase) and γ has units of GtC K^{-1} ($\beta > 0$ and $\gamma < 0$). Then:

$$C_E = C_A + \beta C_A + \gamma \Delta T \qquad (5.22)$$

$\gamma < 0$ means that carbon is released from the land-plus-oceans as the climate warms, which (intuitively and according to Equation 5.22), requires a larger C_A for a given C_E. The radiative forcing can be written as the sum of components due to CO_2 (R_C) and due to all other forcings together (R_N), so from Equation (5.19), we obtain:

$$R_C(C_A) + R_N = \rho \Delta T \qquad (5.23)$$

From Equations (5.22) and (5.23):

$$R_C(C_A) + R_N = \rho \frac{C_E - (1 + \beta)C_A}{\gamma} \qquad (5.24)$$

R_C varies with the logarithm of CO_2 concentration according to:

$$R_C(C_A) = F_{2x} \frac{\ln[(C_1 + C_A)/C_1]}{\ln 2} \approx \phi C_A \qquad (5.25)$$

where F_{2x} is the forcing for a CO_2 doubling, C_1 is the pre-industrial CO_2 concentration, and ϕ is the average forcing (W m^{-2} (GtC)$^{-1}$). From Equations (5.24) and (5.25), and neglecting R_N, it follows that:

$$C_E = C_A\left(1 + \beta + \frac{\phi\gamma}{\rho}\right) = C_A(1 + \beta + u_\gamma) = u C_A \qquad (5.26)$$

where $u_\gamma = \phi\gamma/\rho$. As discussed by Gregory et al. (2009), Equation (5.26) is analogous to Equation (5.4) in the form:

$$\Delta R = \Delta T(\lambda_o + \lambda_{wv} + \lambda_{lr} + \lambda_c + \lambda_a) = \lambda\Delta T \qquad (5.27)$$

In Equation (5.26), C_E is the forcing, C_A is the response, and u is a carbon cycle response parameter that is analogous to the climate feedback parameter λ. It is the sum of contributions related to increasing storage in the atmosphere ($u_o = 1$), a concentration–carbon cycle feedback parameter β, and a climate–carbon cycle feedback parameter u_γ. Negative contributions to λ constitute a positive feedback since they result in a larger ΔT being required to balance ΔR, and, similarly, a negative contribution to u (namely, u_γ) constitutes a positive feedback. Equations (5.4), (5.26), and (5.27) are resistance representations of the temperature or carbon cycle response, in that the components of λ and u add linearly but the larger the total λ or u, the smaller the response.

An expression analogous to the gain representation of the temperature response (Equation 5.13) can be derived for the carbon cycle response. A commonly considered quantity is the cumulative airborne fraction, A, which is the fraction of cumulative emission that resides in the atmosphere at any given time, that is, C_A/C_E. From Equation (5.26):

$$A = \frac{C_A}{C_E} = \frac{1}{1 + \beta + u_\gamma} \qquad (5.28)$$

If there is no concentration–carbon cycle or climate–carbon cycle feedback, then $A = A_o = 1$ and $C_A = C_E$. By analogy to Equation (5.13), we can write:

$$C_A = A C_E = \left(\frac{A_o}{1 - f_c}\right)C_E = \left(\frac{1}{1 - f_c}\right)C_E \qquad (5.29)$$

where $f_c = -\beta - u_\gamma$. When $f_c = 0$, $C_A = A_o C_E$, in the same way that when $f_\lambda = 0$, $\Delta T = G_o \Delta R$.

The upper portion of Table 5.4 gives separate land and ocean β and γ values as determined by Friedlingstein et al. (2006) from transient simulations with 11 coupled climate–carbon cycle models that were involved in the initial *Coupled Climate Carbon Cycle Model Intercomparison Project* (C4MIP). The β and γ values are applicable to the A2 emissions scenario in the year 2100. For this scenario and timeframe, β_L is on average about 20% larger than β_O (that is, the terrestrial biosphere takes up 20% more CO_2 than the oceans). On the other hand, γ_L is over twice γ_O, meaning that over two-thirds of the positive climate–carbon cycle feedback involves the terrestrial biosphere. For both carbon-cycle and climate–carbon cycle feedbacks, the uncertainty (as represented by the standard deviations of the model β and γ values) is over twice as large for the terrestrial as for the oceanic component of the carbon cycle. The impact of increased emissions of CH_4 is

TABLE 5.4 Land, Ocean, and Overall Carbon Cycle Feedback Parameters (β, GtC/GtC) and Climate–Carbon Cycle Feedback Parameters (γ, GtC K^{-1}) Based on the Response by 2100 of 11 Coupled Climate–Carbon Cycle Models to the A2 Emissions Scenario without C–N Coupling (First 11 Data Rows) and of 3 Models with C–N Coupling (Lower)

Model	Land-Surface Biosphere Scheme	β_L	β_O	β	γ_L	γ_O	γ
Terrestrial biosphere modules without C–N coupling							
HadCM3LC	TRIFFID	0.61	0.37	0.99	−177	−24	−201
IPSL-CM2C	SLAVE	0.75	0.75	1.50	−98	−30	−128
IPSL-CM4-LOOP	ORCHIDEE	0.61	0.51	1.13	−20	−16	−36
CSM-1	CASA	0.51	0.42	0.94	−23	−17	−40
MPI	JSBACH	0.66	0.51	1.17	−65	−22	−87
LLNL	IBIS	1.31	0.42	1.74	−70	−14	−84
FRCGC	Sim-CYCLE	0.56	0.56	1.13	−112	−46	−158
UMD	VEGAS	0.09	0.70	0.80	−40	−67	−107
UVic-2.7	TRIFFID	0.56	0.51	1.08	−98	−43	−141
CLIMBER	LPJ	0.51	0.42	0.94	−57	−22	−79
BERN-CC	LPJ	0.75	0.61	1.36	−105	−39	−144
Mean		**0.63**	**0.53**	**1.16**	**−79**	**−31**	**−110**
Standard deviation		0.29	0.12	0.27	46	16	51
Terrestrial biosphere modules with C–N coupling							
MIT-IGSM2.3	TEM	0.26	0.63	0.89	13	−14	−1
CCSM3.0	CLM3-CN	0.19	0.38	0.56	20	−10	10
IPSL-CM4-LOOP	ORCHIDEE	0.30	0.51	0.81	−51	−16	−67
Mean		**0.25**	**0.50**	**0.75**	**−6**	**−13**	**−19**
Standard deviation		0.05	0.13	0.17	39	3	42

(Source: Friedlingstein et al., 2006 except for MIT-IGSM2.3 (from Plattner et al., 2008), CCSM3.0 (from Thornton et al., 2009) and second IPSL-CM4-LOOP line from Zaehle et al., 2010). Definitions of model and land-surface scheme acronyms and original references are given in Friedlingstein et al., 2006.)

not included in this analysis, but would have the effect of increasing γ_L. At the end of a period of increasing emissions, the value of β should be smaller the greater the cumulative emission due to partial saturation of the terrestrial and oceanic CO_2 sinks but, once CO_2 emissions cease, β_O and the overall β will increase over time as more CO_2 is taken up by the deep ocean[22]. The first expectation has been confirmed by Plattner et al. (2008) for the BERN model, for which both β_L and β_O in 2100 are about 40% smaller for the A2 emissions scenario (2000 GtC cumulative CO_2 emission to 2100) than for the B1 emissions scenario (1200 GtC cumulative emissions).

---

22. Recall that $\beta =$ (uptake by the biosphere and oceans)/(atmospheric increase, C_A), so as $C_A \to 0$, $\beta \to \infty$.

5.4.3. The Carbon Cycle As a Climate Feedback

The preceding analysis presented climatic change as a feedback on the carbon cycle by eliminating ΔT from Equations (5.22) and (5.23), with C_E the forcing and C_A the response. Alternatively, we can eliminate C_A. From Equations (5.22), (5.23), and (5.25), we get:

$$R_C(C_E) = \phi C_E = \phi(C_A + \beta C_A + \gamma \Delta T)$$
$$= \rho \Delta T(1 + \beta) + \phi \gamma \Delta T \qquad (5.30)$$

or

$$\phi C_E = \Delta T(\rho + r_\beta + r_\gamma) \qquad (5.31)$$

TABLE 5.5 Climate, Carbon Cycle, and Climate–Carbon Cycle Feedback Parameters and Cumulative Airborne Fraction Based on the Response of Various Coupled Climate–Carbon Cycle Models by 2100 to the A2 Emissions Scenario

Model	DGVM?	ρ W m^{-2} K^{-1}	β GtC/GtC	γ GtC K	u_γ -	r_β W m^{-2} K^{-1}	r_γ W m^{-2} K^{-1}	A -	f_c -
Terrestrial biosphere modules without C–N coupling									
HadCM3LC	Yes	1.56	0.99	−201	−0.62	1.54	−0.98	0.73	−0.37
IPSL-CM2C	No	1.59	1.50	−128	−0.39	2.39	−0.62	0.47	−1.11
IPSL-CM4-LOOP	Yes	1.43	1.13	−36	−0.12	1.62	−0.17	0.50	−1.01
CMS-1	?	2.72	0.94	−40	−0.07	2.55	−0.19	0.54	−0.87
MPI	?	1.26	1.17	−87	−0.34	1.48	−0.42	0.54	−0.83
LLNL	Yes	1.52	1.74	−84	−0.27	2.64	−0.41	0.40	−1.47
FRCGC	No	1.75	1.13	−158	−0.44	1.97	−0.77	0.59	−0.69
UMD	Yes	1.84	0.80	−107	−0.28	1.47	−0.52	0.66	−0.52
UVic-2.7	Yes	1.64	1.08	−141	−0.42	1.77	−0.68	0.60	−0.66
CLIMBER	Yes	1.95	0.94	−79	−0.20	1.83	−0.38	0.57	−0.74
BERN-CC	Yes	2.24	1.36	−144	−0.31	3.06	−0.70	0.49	−1.05
Mean		**1.77**	**1.16**	**−110**	**−0.31**	**2.03**	**−0.53**	**0.55**	
Standard Deviation		0.41	0.27	51	0.16	0.54	0.25	0.09	-
Terrestrial biosphere modules with C–N coupling									
MIT-IGSM2.3			0.89	13	0.04	1.57	0.06	0.52	−0.92
CCSM3.0			0.56	20	0.06	1.00	0.10	0.62	−0.62
IPSL-CM4-LOOP	Yes	1.43	0.81	−51	−0.17	1.15	−0.25	0.61	−0.63

A 'Yes' in the column 'DGVM' indicates that the distribution of vegetation types can shift in response to changing CO_2 concentration and climate, so that the related changes in surface albedo are included in the climate radiative feedback factor λ (which is contained in ρ). *(Source: Gregory et al., 2009 for models where the terrestrial biosphere component has C–N coupling, and sources as given in Table 5.4 for the other models.)* A ø value of 0.0049 W m^{-2}GtC^{-1} was used in computing u_γ and r_γ. The mean value of ø from the first 11 models was used here in computing u_γ and r_β for the MIT-IGSM2.3 and CCSM3.0 models.

where $r_\beta = \rho\beta$ and $r_\gamma = \phi\gamma$. Now, C_E is the ultimate forcing and ΔT is the response. ϕC_E is the radiative forcing that would occur with no carbon cycle or climate–carbon cycle feedbacks; $r_\beta\Delta T = \phi\beta C_A$ is the reduction in radiative forcing due to carbon cycle feedback and $-r_\gamma\Delta T = -\phi\gamma\Delta T$ is the increase in radiative forcing due to climate–carbon cycle feedback. The net forcing is offset by an increased heat flux to space or the deep ocean, $\rho\Delta T$. The beauty of this formulation is that the heat flux damping, carbon cycle feedback, and climate–carbon cycle feedback are now represented with metrics (ρ, r_β, and r_γ) having the same units (W m^{-2} K^{-1}), so that they can be directly compared with one another.

Table 5.5 gives the overall values for β and γ (from Table 5.4), along with ρ values and the corresponding values of u_β, u_γ, and A as computed by Gregory et al. (2009), and f_c as computed here. The models that allow the vegetation distribution to shift in response to changing atmospheric CO_2 concentration and climate are indicated in Table 5.5 as, in these models, λ and hence ρ includes the contribution from surface albedo changes due to changes in the distribution of different biomes. The negative carbon cycle feedback (represented by r_β) is on average slightly stronger than the combined radiative and ocean heat flux damping (represented by ρ) and, on average, is about four times stronger than the positive climate–carbon cycle feedback

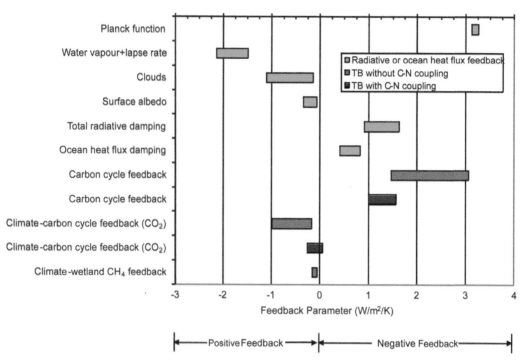

FIGURE 5.18   Range of radiative feedback or damping parameters (λ_i, from Soden and Held (2006) and shown in Figure 5.4a), heat flux damping to the deep ocean (κ, from Gregory and Forster (2008)), and carbon cycle and climate–carbon cycle feedback parameters (r_β and r_γ, as computed by Gregory et al. (2009) for models without C–N coupling in the terrestrial biosphere (TB) component, and from their table 10 for models with C–N coupling).

(represented by r_γ), but the uncertainty in the carbon cycle feedback (as represented by the standard deviation of the model r_β values) is almost twice as large as the uncertainty in the climate–carbon cycle feedback (as represented by the standard deviation of the model r_γ values) in spite of the very large variation in γ among the models.

Figure 5.18 compares the climate feedbacks (λ_i, from Figure 5.4a) and heat-flux damping to the deep ocean (κ, from Gregory and Forster, 2008) with r_β and r_γ as computed by Gregory et al. (2009). As seen from Figure 5.18 and noted above, carbon-cycle feedback is a fourfold stronger feedback than the climate–carbon cycle feedback, although the latter feedback could be as strong as the combined surface-albedo, water-vapour, and lapse-rate feedbacks. The midrange climate–carbon cycle feedback is comparable to the midrange cloud feedback (both being positive). The negative carbon cycle feedback ranges from 1/3 to 100% as strong as the negative Planck function feedback (λ_o). The coefficient for heat flux damping to the deep ocean (κ) ranges from 0.4–0.8 W m^{-2} K^{-1}.

5.4.4. Role of Carbon–Nitrogen (C–N) Coupling

The quantitative feedback factors discussed above are based on terrestrial biosphere models that neglect C–N

interactions. However, the stimulation of photosynthesis by higher CO_2 is likely to be restrained at some point by limitations in the supply of N. Conversely, accelerated decomposition of soil carbon due to warming will lead to a greater release of N and other nutrients back to the soil, thereby stimulating photosynthesis and, in so doing, offsetting some or all of the reduction in soil carbon. The last three data rows of Tables 5.4 and 5.5 give the feedback parameters for three models that use a terrestrial biosphere submodule that allows C–N interactions. These models are TEM (Terrestrial Ecosystem Model), the Community Land Model version 3.0 (CLM3.0), which is embedded in the NCAR (National Center for Atmospheric Research) AGCM, and the CM4 land model that is coupled to the IPSL AGCM. The average β_L value for the three models with C–N coupling is less than half that of the models without C–N coupling, while the average γ_L value is less than one tenth that of the models without C–N coupling.

In TEM, the total N content in the simulated ecosystem is constant as CO_2 and temperature increase. CLM3.0 has prognostic N inputs, including biological N fixation and losses due to wildfires, while IPSL-CM4 has variable N losses by leaching and through trace gas emissions. In the steady state, N losses (from forest fires, for example) will be balanced by N inputs (largely N fixation, with some atmospheric deposition). An increased incidence of fires would lead to net N loss, but an increase in CO_2 tends to increase the rate of N fixation (which in turn stimulates NPP) by providing more energy to the roots (where N fixers reside). However, N fixation also requires various trace elements, particularly iron, vanadium, and molybdenum (see Dickinson, 2012, this volume). Thus, the initial increase in N fixation in CO_2 enhancement experiments is followed by down-regulation after a few years (Hungate et al., 2004).

Temperature also affects the rate of N fixation. Current surface air temperatures in the tropics are close to the optimum for N fixation but substantially below the optimum in the extra-tropics. Thus, warming will reduce N availability in the tropics and increase it in the extra-tropics (Wang and Houlton, 2009). When: (i) changes in N availability due to changes in the rate of N fixation caused by the warming climate; and (ii) increases in N deposition due to the increase fossil fuel use that ultimately drives the change in climate in the C4MIP models are taken into account, Wang and Houlton (2009) found that the increase in carbon storage on land simulated by 9 out of 11 C4MIP models (which, as already noted, do not account for N limitations) is greater than what is feasible based on an upper-bound estimate of the amount of available N. The carbon that is absorbed in the models in excess of the available N would in fact be distributed between the atmosphere and oceans. The additional warming due to non-absorption of the excess carbon is estimated by Wang and Houlton (2009) to be 0.18–0.52 K by 2050 and 0.15–0.65 K by 2100 for these models.

In summary, accounting for C–N interactions dramatically alters the response of carbon storage to both increasing atmospheric pCO_2 and temperature, but a proper accounting must include the effects of additional N losses due to fires and changes in the rate of N fixation induced by changes in atmospheric pCO_2 and temperature, as well as the effect of extra N that is released to the soil solution from increased decomposition of organic matter. Thus, it is too early to conclude that γ_L is close to zero, although it is likely to be in the lower half of the range for the models in Table 5.4 without C–N interactions.

5.4.5. Combination of Climate Sensitivity and Carbon Feedback Gain Formulation

This subsection is an extension of the analysis presented in Gregory et al. (2009). Re-introducing R_N, Equations (5.24) and (5.25) give:

$$C_E = C_A \left(1 + \beta + \frac{\phi\gamma}{\rho} \right) + \frac{\gamma}{\rho} R_N \quad (5.32)$$

instead of Equation (5.26), from which we obtain:

$$C_A = \frac{C_E - \frac{\gamma}{\rho} R_N}{1 + \beta + \frac{\phi\gamma}{\rho}} \quad (5.33)$$

Because $\gamma < 0$, R_N increases C_A for a given C_E, while γ in both the numerator and denominator of Equation (5.33) increases C_A (i.e., climate–carbon cycle feedback increases C_A through the warming that is induced by the initial emission of CO_2 and through the warming induced by the non-CO_2 forcing). Equation (5.13) becomes:

$$\Delta T = \left(\frac{G_o}{1 - f_\lambda} \right) (R_C + R_N) \quad (5.34)$$

where, now, $f_\lambda = -G_o(\lambda_{wv} + \lambda_{lr} + \lambda_c + \lambda_a + \kappa) = -G_o(\rho - \lambda_o)$[23]. From Equations (5.25), (5.33), and (5.34) we obtain:

$$\Delta T = \left(\frac{G_o}{1 - f_\lambda} \right) \left(\phi C_E \left(\frac{1 - \left(\frac{\gamma}{\rho} \right) \left(\frac{R_N}{C_E} \right)}{1 + \beta + \frac{\phi\gamma}{\rho}} \right) + R_N \right) \quad (5.35)$$

which can be rewritten as:

$$\Delta T = \left(\frac{G_o}{1 - f_\lambda} \right) \left(\phi C_E \left(\frac{1 - (\gamma/\rho)(R_N/C_E)}{1 - f_c} \right) + R_N \right) \quad (5.36)$$

where $f_c = -\beta - u_\lambda = -\beta - \phi\gamma/(\lambda + \kappa)$[24]. For the case where $R_N = 0$, Equation (5.36) simplifies to:

$$\Delta T = \phi \left(\frac{G_o}{1 - f_\lambda} \right) \left(\frac{1}{1 - f_c} \right) C_E \quad (5.37)$$

Over a period of time, κ decreases as the deep ocean warms (increasing f_λ and tending to increase ΔT because there is a smaller heat flux into the deep ocean), while β increases as an increasing amount of the emitted CO_2 is absorbed by the oceans (decreasing f_c and tending to decrease ΔT). Over a period on the order of 100,000 years, $\beta \to \infty$ (the airborne fraction goes to zero because all of the emitted CO_2 is eventually absorbed by the oceans or consumed through enhanced rock weathering), which drives ΔT to zero.

The ratio of the temperature response with climate–carbon cycle feedback to the response without this feedback is equal to:

$$R = \frac{1 + \beta}{1 + \beta + \phi\gamma/\rho} \quad (5.38)$$

---

23. As f_λ now includes the transient heat flux to the ocean, Equation (5.34) gives the instantaneous, rather than the equilibrium, ΔT.
24. Recall that $\lambda_i < 0$ and $\gamma < 0$ are positive feedbacks.

This ratio is >1 because $\gamma < 0$. From Equation (5.38), it can be seen that the impact of a given climate−carbon cycle feedback (represented by γ) is greater the larger the climate sensitivity (i.e., the smaller ρ).

5.4.6. Applying Climate Sensitivity to Future Climate Policy Strategies

Taplin (2012, this volume) reviews the current situation regarding international negotiations on fossil fuel emissions reduction. All policy decisions have climate consequences, which can be quantitatively assessed using climate sensitivity analysis. We illustrate the application of the preceding analysis to scenarios with cumulative fossil fuel emissions by 2100 of 680 GtC, 1100 GtC, and 1640 GtC (corresponding to the most stringent climate policy scenario, the least stringent climate policy scenario, and the high business-as-usual scenario of Harvey (2010b), respectively), plus 120 GtC cumulative emission from land-use changes in each case[25]. The feedback parameters depend, to some extent, on the specific emission scenario, but we will use lower and upper bounds from the estimates reviewed here (Table 5.5). It is assumed that the carbon cycle and climate−carbon cycle feedbacks combine linearly (that is, they do not interfere with each other), which is only approximately true. In spite of these limitations, a number of useful insights can be gained through the application of the linear feedback analysis.

Figure 5.19 shows the (peak) temperature response for the three emissions scenarios as computed from Equation (5.35) for λ chosen to give fast-feedback climate sensitivities of 2 or 4 K, for $\beta = 0.9$ or 1.7, and for $\gamma = -40$ GtC K or -200 GtC K. These temperatures are computed iteratively, since ΔT depends on ϕ and C_E through Equation (5.35), ϕ depends on C_A through Equation (5.25), and C_A depends on ϕ through Equation (5.33). A non-CO_2 radiative forcing R_N of 1.0 W m^{-2} is assumed in all cases. Also given in Figure 5.19 for each (γ, β) combination are the ratios of the temperature change with climate−carbon cycle feedback to the corresponding case without climate−carbon cycle feedback.

The lowest part of each bar in Figure 5.19 gives the temperature response when $\kappa = 0.8$ W m^{-2} K^{-1} (the upper limit of the uncertainty range given in Table 5.3), the next segment of each bar gives the additional temperature response if $\kappa = 0.4$ W m^{-2} K^{-1} (the lower limit of the

uncertainty range in Table 5.3) instead of $\kappa = 0.8$ W m^{-2} K^{-1}, and the top segment gives the additional temperature response if there is no heat flux to the deep ocean. The last case is unrealistic but serves to illustrate the overall importance of heat absorption by the oceans in limiting the warming. Figure 5.20 gives the corresponding peak atmospheric CO_2 concentrations. The ratio of temperature response with and without the ocean heat flux is smaller than $\lambda/(\lambda + \kappa)$, especially for cases with a strong positive climate−carbon cycle feedback (large negative γ). This is because the CO_2 concentration itself is smaller the larger the heat flux to the ocean (as can be seen from Figure 5.20), due to weaker climate−carbon cycle feedback the more that the transient warming is limited.

As seen from Figure 5.19, the relative impact of climate−carbon cycle feedback is largest for the scenario with the lowest cumulative emission (because the CO_2 radiative forcing varies with the logarithm of CO_2 concentration), for smaller negative carbon cycle feedback, and for larger climate sensitivity (as expected from Equation 5.38).

The parameter values and the peak CO_2 concentration for the business-as-usual emissions scenario with high climate sensitivity, weak carbon cycle feedback, and strong climate−carbon cycle feedback correspond closely to those of the Hadley Centre model reported by Cox et al. (2000, 2004), whereby atmospheric CO_2 reaches a concentration of about 750 ppmv by 2100 without climate−carbon cycle feedback and about 1000 ppmv with climate−carbon cycle feedback.

The climate−carbon cycle feedback parameter used in the above calculations does not include the effect of CH_4 or CO_2 release from thawing permafrost or melting of clathrates. The impact of potential GHG releases from thawing yedoma soils was assessed by Harvey (2010b) using the coupled climate−carbon cycle model of Harvey (2001) and is presented here in Figure 5.21. In line with the simulations by Khvorostyanov et al. (2008a) and observational evidence cited earlier, it is assumed that the rate of carbon release increases linearly from 0 GtC per year at 0.8 K global mean warming (i.e., starting from the present) to 0.3 GtC per year at 3.3 K global mean warming, then increases to 3 GtC per year at 4.3 K global mean warming and continues until the cumulative emission reaches 300 GtC, then abruptly stops. It is further assumed that 25% of the released C is emitted to the atmospheric in the form of CH_4, and the balance as CO_2. Figure 5.21a shows the variation in global mean temperature for an emissions scenario that gives a peak global mean warming of 3.38 K with a climate sensitivity of 3 K and in the absence of yedoma feedback. However, assuming that carbon release from yedoma soils begins slowly, it unleashes a positive feedback sufficient to push the peak warming to 5.6 K. This, in turn, is large enough to unleash a CH_4−clathrate feedback sufficient to cause an additional

---

25. The cumulative fossil fuel emissions to 2100 in the business-as-usual scenarios of Harvey (2010b) range from 820−1640 GtC. These are much less than typical business-as-usual scenarios, which see cumulative emissions (beyond 2100) of about 5000 GtC, but are based on recent evidence that the useable coal resource is far less than is traditionally assumed and that the global coal supply will probably peak within the next few decades (see Harvey 2010a, section 2.5.3; Patzek and Croft, 2010). The cumulative fossil fuel CO_2 emission to 2010 is about 375 GtC.

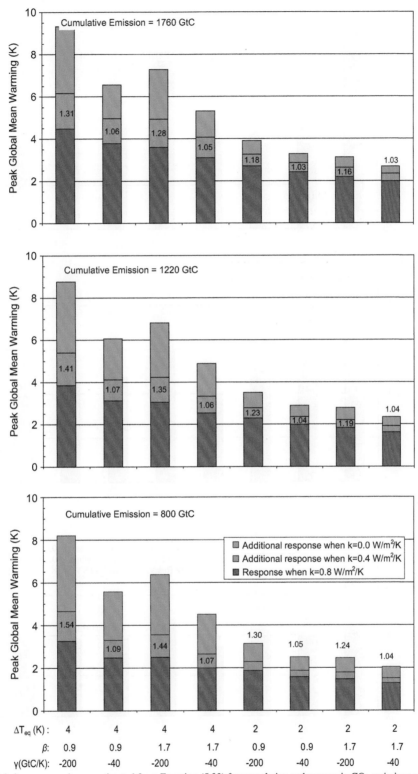

FIGURE 5.19 Peak global mean warming as estimated from Equation (5.33) for cumulative anthropogenic CO_2 emissions of 800 GtC, 1220 GtC, and 1760 GtC for various climate sensitivities (ΔT_{2x}), carbon cycle feedback parameters (β), and climate–carbon cycle feedback parameters (γ). Shown is the warming when $\kappa = 0.8$ W m^{-2} K^{-1}, the additional warming when $\kappa = 0.4$ W m^{-2} K^{-1} instead of $\kappa = 0.8$ W m^{-2} K^{-1}, and the additional hypothetical warming with no oceanic uptake of heat.

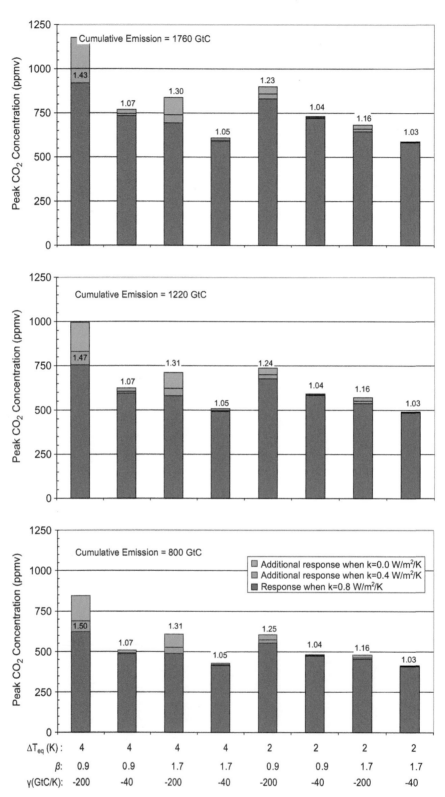

FIGURE 5.20 As for Figure 5.19 except showing peak atmospheric CO_2 concentrations based on C_A as estimated from Equation (5.31).

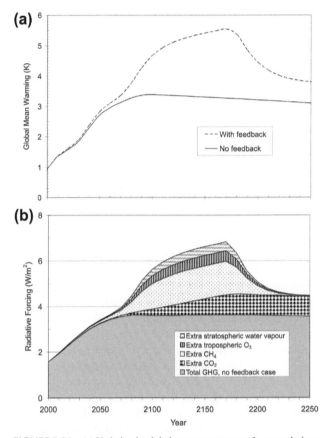

FIGURE 5.21 (a) Variation in global mean temperature for an emissions scenario—climate sensitivity combination that produces a peak warming of 3.3 K in the absence of carbon release from yedoma soils ('No feedback') and the result when a gradual release begins at 0.8 K warming, as explained in the text ('With feedback'). (b) Contributions to the total radiative forcing for the case with yedoma feedback.

1−2 K warming over the course of 300−500 years (Harvey and Huang, 1995). Figure 5.21b shows the different contributions to the extra radiative forcing caused by emissions of CO_2 and CH_4 from thawing yedoma soils. The total radiative forcing is more than doubled by the end of the carbon emissions, but the temperature response does not quite double due to the delay in warming caused by oceans combined with the rapid decline in those forcings related to CH_4 release once emissions cease.

5.5. OTHER SLOW AND LESS-CONSIDERED FEEDBACKS

Other feedbacks that may alter the long-term sensitivity to GHG increases are less frequently considered in the literature. These include biological effects (physiological and plant-type distribution changes), oceanic and marine biospheric changes, and Greenland ice-cap melting (Lenton, 2012, this volume).

5.5.1. Enhanced Land Surface Warming Due to the Physiological Effect of Higher CO_2

The increase in atmospheric CO_2 leads to a restriction in stomatal opening, reducing evaporative cooling and thereby increasing the local warming of the land surface compared to the case without physiological effects of higher CO_2 (Sellers et al., 1996; Betts et al., 1997). The reduction in evapotranspiration also triggers changes in the atmospheric water vapour content and cloudiness, which alters the radiative energy fluxes and thereby alters the final global-scale temperature change.

According to a simulation with the HadCM3LC model, the resulting decrease in evapotranspiration following an instantaneous doubling or quadrupling of atmospheric CO_2 rapidly leads to a decrease in low cloud cover that adds about 15% to the original CO_2 forcing (that is, a 15% increase in ΔR_{eff}) with no statistically significant change in the radiative feedback parameter λ (Doutriaux-Boucher et al., 2009). Thus, ΔT also increases by about 15%, irrespective of the overall feedback strength − unlike the case where there is a perturbation in λ itself, in which case the impact on temperature depends on the pre-existing feedback strength (as seen in Figure 5.2). In simulations using the NCAR Community Atmospheric Model and Community Land Model, Cao et al. (2010) find that partial stomatal closure increases the average surface air warming over land from 2.86 K to 3.33 K (a 16% increase). This is due to reduced evaporative cooling and reduced low-level cloud cover.

5.5.2. Shifts in the Distribution of Plant Functional Types

Several different models have been developed that simulate both the distribution of vegetation types and the amount of carbon stored as biomass and in the soil, as well as the physical coupling with the atmosphere through exchanges of heat, moisture, and momentum. These models are referred to as dynamic global vegetation models (DGVMs). Experiments with DGVMs indicate that eventual shifts in vegetation in response to the initial warming enhance the global mean temperature response to increasing CO_2 by about 10%−13% (O'ishi and Abe-Ouchi, 2009; O'ishi et al., 2009; Gregory et al., 2009).

Matthews et al. (2007) carried out experiments with a coupled AOGCM-DGVM in which an increase in atmospheric CO_2 to 770−926 ppmv but with no CO_2 radiative effects induced a global mean surface warming of about 0.2 K, solely as a result of changes in the distribution of different vegetation types. Matthews and Keith (2007) state that the warming caused by vegetation shifts induced by higher CO_2 largely offsets the cooling effect from the uptake of CO_2 induced by higher atmospheric CO_2 in the fully-coupled version of their model. This effect is equivalent to changing f_c from $-(\beta_L + \beta_O + u_\lambda)$ to $-(\beta_O + u_\lambda)$; that is, increasing the magnitude of the positive f_c.

5.5.3. Decrease in the Extent of the Greenland Ice Cap

On a timescale of many centuries to a few thousand years, complete melting of the Greenland ice sheet could occur in association with global mean warming of as little as $1-2$ K[26]. This would replace a high albedo (0.7) surface with a low-albedo (0.25) surface over an area of 1.7 million km^2. The average solar irradiance on the Greenland ice sheet is about 100 W m^{-2}, so the change in the global mean absorption of solar radiation resulting from complete disappearance of the Greenland ice sheet would be about 0.15 W m^{-2}[27]. However, this feedback would not increase the peak warming in response to anthropogenic emissions of CO$_2$ because significant retreat of the Greenland ice cap would not have occurred until long after atmospheric CO$_2$ concentration had peaked and then begun to decline (Lenton, 2012, this volume).

5.5.4. Delayed Ocean Circulation Changes and Cloud Feedback

As noted earlier (Section 5.2.6.2), a rather large climate sensitivity ($6-10$ K) has been deduced for the early Pliocene ($3-5$ Ma). Prior to 3 Ma, there was a permanent El Niño condition, with greater warming in the eastern equatorial Pacific than in the western equatorial Pacific relative to today (Fedorov et al., 2006). This would have resulted in less low-level cloud prior to 3 Ma and hence a lower planetary albedo, and might explain the relatively warm conditions of the early Pliocene with almost the same external forcing factors as today (Barreiro et al., 2006). The critical question is, would initial warming due to anthropogenic emissions of GHGs provoke a flip to a permanent El Niño condition that would trigger a jump in the climate sensitivity? The answer to this question depends on the reason for the permanent El Niño condition during the early Pliocene. Fedorov et al. (2006) rule out closure of the Panamanian seaway as a trigger for the

transition away from the permanent El Niño condition, since it occurred too early (between 4.0 and 4.5 Ma). Rather, the loss of the permanent El Niño condition seems to be related to high latitude cooling, which would have produced a shallower thermocline at low latitudes, thereby permitting tropical winds to bring cold waters to the surface in the eastern equatorial Pacific and other upwelling regions (Fedorov et al., 2004).

The reverse process could occur if extra-tropical waters become less saline as the climate warms, since sufficiently large freshening in the extra-tropics can induce a permanent El Niño. In simulations with an idealized OCGM, Fedorov et al. (2004) find that a gradual increase in the surplus of precipitation over evaporation in the northern Pacific Ocean induces an abrupt transition to a permanent El Niño state once a critical threshold is passed. The transition occurs on a timescale of decades. Alternatively, a reduction in the rate of heat loss from the oceans to the atmosphere could deepen the tropical thermocline, thereby reducing the near-surface upwelling of cold water (Fedorov et al., 2006). After GHG concentrations are stabilized, it is possible that the climate could at first appear to be adjusting to a relatively low ($1.5-4.5$ K) climate sensitivity, only to make a rapid transition to a higher sensitivity once a critical threshold involving the subsurface ocean temperature structure is passed. However, to the extent that the higher sensitivity is the result of a one-time threshold crossing, it would not be applicable to the radiative forcing in excess of that needed to cross the threshold.

5.5.5. Collapse of Marine Bioproductivity and Cloud Feedback

One of the unsolved mysteries concerning past climates is the cause of exceptionally warm conditions (35°C tropical SST, 10°C polar temperatures) during the middle Cretaceous (about 100 Ma), particularly if CO$_2$ concentration never exceeded about $5 \times$ pre-industrial (1500 ppmv). At 35°C, marine bioproductivity would have been greatly reduced compared to at present, leading to greatly reduced emission of dimethylsulfide, a key precursor to cloud condensation nuclei (CCN). This in turn would have lowered cloud reflectivity, constituting a strong positive feedback. Kump and Pollard (2008) tested the impact of a tenfold to 100-fold decrease in CCN amounts in combination with $4 \times$ CO$_2$ in an AGCM that explicitly computes cloud optical properties. They find that there are fewer and optically thinner clouds, reducing global mean planetary albedo from 0.30 to 0.24. With today's solar luminosity, this would produce an increase in global mean absorption of solar radiation by about 20 W m^{-2}, which is three times the radiative forcing due to a quadrupling of atmospheric CO$_2$.

---

26. Evidence supporting this possibility is twofold: (i) sea level during the last interglacial, when global mean temperature was $1 - 2$ K warmer, is estimated to have been $6-9$ m above present sea level (Kopp et al., 2009), an amount that requires a substantial or total loss of Greenland ice,- depending on the concurrent loss of West Antarctic ice; and (ii) the trend line for the area of the Greenland ice sheet subject to at least one day of melting per year is a 45% increase from 1979–2005 (Fettweis et al., 2007) in association with only 0.8 K global mean warming since the late nineteenth century, while the ice sheet made the transition in the early 2000s from annually increasing to annually decreasing ice mass (Velicogna, 2009).

27. This rough estimate neglects changes in the atmospheric masking of surface albedo changes due, for example, to a change in cloudiness.

5.6. CLIMATE FEEDBACKS AND THE FUTURE CLIMATE

The climate response to anthropogenic emissions of CO_2 involves negative feedbacks between atmospheric CO_2 concentration and the absorption of CO_2 by the oceans and terrestrial biosphere (carbon cycle feedbacks), a variety of fast (days to months) radiative feedbacks (climate feedbacks), coupled climate–carbon cycle feedbacks that unfold over a period of decades to centuries, and other slow feedbacks involving the response of the Greenland and Antarctic ice sheets and slow changes in the ocean circulation and subsurface temperature field (which can alter surface temperature patterns and cloud feedbacks).

There is comprehensive and widespread evidence that the traditional climate sensitivity, based on fast feedback processes only, is very likely (90% probability) to fall between the long-held consensus limits of 1.5 K and 4.5 K. In scenarios where the effective CO_2 concentration increases exponentially by 1% per year until it doubles, the realized global mean temperature change at the time of a CO_2 doubling is typically 60%–75% of the equilibrium warming, with an average over 16 AOGCMs of 70%. Extremely stringent scenarios for reductions in emissions of all GHGs (with CO_2 emission reaching zero by 2085) still result in peak GHG concentrations close to the equivalent of a CO_2 doubling; anything less will result in an even greater equivalent CO_2 concentration. As these scenarios require reducing CO_2 emissions from fossil fuel use and biomass burning to near zero, they would also lead to the elimination of the current cooling effect of aerosols. Thus, the minimum expected peak warming — excluding the effects of positive climate–carbon cycle feedbacks — can be estimated based on the minimum peak equivalent CO_2 concentration, the fast feedback climate sensitivity, and the ratio of realized to equilibrium warming. This gives a minimum peak warming of 1.0 K to 3.5 K.

Climate–carbon cycle feedbacks involving the ocean sink are likely to be small. Feedbacks involving photosynthesis, respiration, and the dieback of maladapted terrestrial ecosystems, as estimated from various models, have ranged from small (sufficient to amplify the temperature response by 4%–9%) to very large (sufficient to amplify the temperature response by about 15%–50%), with greater amplification if the negative carbon cycle feedback is small and the climate sensitivity large. Inclusion of soil C–N interactions greatly reduces the strength of both the stimulation of carbon uptake by higher CO_2 and the positive climate–carbon cycle feedback, but there are still many uncertainties concerning most processes involved in both carbon cycle and climate–carbon cycle feedbacks.

Much greater amplification of the future warming could occur through release of CO_2 and CH_4 from thawing yedoma soils (where significant fluxes could arise within 100 years and then be sustained for another 100 years) and from destabilization of marine CH_4 clathrates (where non-negligible fluxes could begin within 100 years and potentially catastrophic fluxes could arise within a few centuries). If global mean warming reaches 3–4 K, yedoma feedback could push the warming a further 2 K, at which point CH_4 releases from marine clathrates could be large enough to add another 2 K. Some time during this process, a threshold involving the temperature of upwelling water in the eastern equatorial Pacific could be crossed, leading to an abrupt (within decades) transition to a permanent El Niño state and a potentially large positive cloud feedback that might add another 1–2 K global mean warming[28]. Yet further warming could occur through collapse of the marine biosphere (which would be likely at this point) and concomitant reductions in cloud albedo leading to a reduction in global mean planetary albedo from 0.30 to 0.24.

A global mean warming of 10–12 K beyond AD 2100 is a distinct possibility if the fast-feedback climate sensitivity is near the upper end of the 1.5–4.5 K uncertainty range, if the slow positive feedbacks are near the high end of the possible range that has been identified here, and if the recoverable fossil fuel resource is as large as 4000–5000 GtC (as is commonly assumed) and most of it is used. As shown by Sherwood and Huber (2010), global mean warming of this magnitude would render portions of the world currently occupied by over half of the human population to be uninhabitable by humans (in the absence of access to 100% reliable air-conditioning by the entire populations in the affected regions) due to the periodic occurrence of 6-hour mean wet-bulb temperatures in excess of the practical physiological limit of 33°C. Very serious social and political upheaval that could provoke military conflict (Dyer, 2008) would occur long before this point is reached, and could be an outcome of more moderate fossil fuel emissions (1000–2000 GtC cumulative emission) and less than worst-case fast and slow feedback climate sensitivities. Further research could perhaps provide a better assessment of the likelihood of catastrophic climate–carbon cycle and other slow feedbacks as a function of the cumulative fossil fuel CO_2 emission and of the underlying fast-feedback climate sensitivity.

ACKNOWLEDGEMENTS

This paper benefited from comments by Josep Canadell and three anonymous reviewers.

---

28. This is the temperature response associated with 350–450 ppmv CO_2 during the early Pliocene beyond that expected based on the fast-feedback climate sensitivity, and is assumed to be a one-time increment in temperature response as GHG concentrations continue to increase.

Chapter 6

Variability and Change in the Ocean

Alex Sen Gupta and Ben McNeil

Climate Change Research Centre, University of New South Wales, Sydney, Australia

Chapter Outline

6.1. INTRODUCTION: CLIMATE VARIABILITY

The ocean is host to a vast diversity of life; it is of immense social and economic value to us and it plays a crucial role in our climate system over synoptic to multimillennium timescales. Consequently, it is of great importance to understand changes that oceans have gone through in the past and how they might change in the future as we increasingly perturb the climate system. In the context of anthropogenic climate change, the ocean acts as an important buffer. The majority of extra heat entering the climate system through enhanced greenhouse forcing (~80% since 1960; Levitus et al., 2005) and almost half of the anthropogenic carbon dioxide (CO_2) since pre-industrial times (Sabine, 2004; or ~30% since 1960, Canadell et al., 2007) have been sequestered and stored in the ocean. While mitigating against more rapid temperature rise in the atmosphere, the oceanic absorption of heat and

CO_2 has impacts on the ocean environment. For example, the absorption of heat has caused thermal expansion of the ocean, which is a major contributor to global sea-level rise (e.g., Domingues et al., 2008); temperature and stratification increase can directly (via physiological changes) and indirectly (via changes in ocean mixing and nutrient supply) affect marine species. The sequestration of anthropogenic CO_2 has measurably altered ocean chemistry, reducing water pH and aragonite saturation (Raven et al., 2005). An increasing number of studies are showing that such chemical changes can have major impacts on both marine calcifiers and higher marine animals (Fabry et al., 2008; Doney et al., 2009). As a result of the physical and biochemical changes that have occurred in the ocean, some evidence is emerging that the ability of the ocean to mitigate atmospheric temperature rise, via the sequestration and removal to the deep ocean of CO_2, is declining (Canadell et al., 2007; Le Quéré et al., 2009).

The Future of the World's Climate. DOI: 10.1016/B978-0-12-386917-3.00006-3

Compared to pre-industrial times, atmospheric CO_2 concentrations have increased by over 30%, which, together with increases in other greenhouse gases (GHGs) have driven a rise in globally-averaged ocean surface temperatures of >0.6°C (IPCC, 2007a), lagging slightly behind increases in atmospheric temperatures. Over the same time, average surface pH has reduced from ~8.2 to 8.1, corresponding to a ~30% increase in hydrogen ion content (Raven et al., 2005). The oceanic changes already observed are likely to be compounded in the future, given that GHG emissions are still growing and show little sign of an imminent slowdown (e.g., Le Quéré et al., 2009). It is unlikely, however, that future change will scale in the same way as historical change, as the radiative forcing perturbation increasingly diverges from pre-industrial levels. Feedbacks within the system mean that future responses will probably be non-linear and may result in tipping points or abrupt changes to components of the system (e.g., Schneider, 2004; Lenton, 2012, this volume). While some of this non-linearity is inherent in the climate models (e.g., water vapour and ice—albedo feedbacks), we know that certain feedbacks are poorly represented (in particular those related to clouds) or entirely missing. The climate models used to inform AR4 projections do not explicitly include carbon cycle interactions or feedbacks related to terrestrial or marine biological systems. Some of the models taking part in AR5 will include ocean and terrestrial carbon cycle components (Taylor et al., 2009).

The goals in this chapter are threefold. Our first goal is to examine the large-scale changes to the oceans that have been observed over recent decades and examine any link to anthropogenic climate change (Section 6.2). Two fundamental challenges exist here. Firstly, compared to terrestrial or atmospheric systems the ocean is relatively poorly sampled, particularly in the Southern Hemisphere and as we go further back in time prior to the advent of remote measurement. In Section 6.2.1, we describe some of the important new observing systems and the challenges they face. Secondly, the oceans are highly dynamic and, like the atmosphere, exhibit natural variability on a wide range of timescales. Given the brevity of the observational records, it therefore becomes difficult to disentangle long-term natural variability from change that is driven by increased greenhouse forcing, particularly at regional and smaller scales. In Section 6.2.2, we briefly discuss the primary modes of variability that are important for the ocean.

The second goal is to examine projections of future physical and chemical changes in the ocean (Section 6.3). This is primarily based on the intercomparison of multiple coupled climate models. Such a review is necessarily biased towards changes in the physical environment, given that most climate models are yet to include sophisticated biochemical components. There is, however, a more limited set of models that provide insights into chemical and biological change, and attempts have been made to infer how ecosystems may respond to given projections of the physical system. We focus our discussion of projections on three key areas: (i) the tropical Pacific Ocean, where changes in the El Niño—Southern Oscillation (ENSO) can have major regional and global repercussions (see Latif and Park, 2012, this volume); (ii) the Southern Ocean, which plays an important role in both the global overturning circulation and the two-way transfer of carbon between the ocean and atmosphere; and (iii) sea-level rise, where some post AR4 research points towards considerably larger changes than those derived from climate models.

The third goal is to discuss how the observed and projected changes to the ocean will affect the regulatory function of the ocean within the climate system. We review a number of feedback mechanisms, including those associated with the carbon pumps (physical and biological processes that can transport carbon from the surface ocean to the deep ocean or sediment) that can modulate the ability of the ocean to sequester atmospheric CO_2 (Section 6.4). In Section 6.5, we draw conclusions and briefly examine what the next generation of climate models have in store.

6.2. OBSERVED OCEAN VARIABILITY AND CHANGE

6.2.1. Observing the Global Ocean

Our understanding of the ocean and its role in the climate system is underpinned by observations. However, large parts of the ocean remain poorly sampled, partly because the ocean can be a hostile and inaccessible place, making it a costly challenge to take observations. In addition, unlike in meteorology where there is obvious day-to-day justification for the collection of large-scale, high-resolution atmospheric data (e.g., weather forecasting, aviation), this is not the case for the ocean.

Over recent decades, there have been considerable improvements in our ability to observe the ocean as new technologies have been deployed, especially under the auspices of various international collaborations. A number of particular projects have led to major advances in our understanding of the ocean's role in climate. These include:

The Tropical Ocean − Global Atmosphere (TOGA) study (Hayes et al., 1991), which has implemented an array of upper ocean moorings along the equatorial Pacific Ocean. This has provided valuable insights into ENSO mechanisms and has led to the ability to produce long-term forecasts of ENSO-related climate variability.

The World Ocean Circulation Experiment (WOCE), an unprecedented collaboration of 30 countries during the 1990s to produce a three-dimensional view of the structure of the ocean (Siedler et al., 2001), using a variety of technologies, including an extensive network of hydrographic sections, current moorings, Expendable Bathythermograph (XBT) deployments, floats, and so on.

Rapid Climate Change (RAPID) mooring array, an array of moorings spanning the North Atlantic at 26.5°N that provides an accurate measure of the upper and lower branches of the Atlantic Meridional Overturning Circulation (AMOC or 'thermohaline circulation'). This circulation carries heat, CO_2, and other properties throughout the global ocean. It includes both upper and deep ocean circulation pathways, connected by regions of sinking or ascent at high latitudes. As such, it plays a major role in modulating the climate system. In fact, it has been implicated as a major driver of rapid climate change at certain times in the palae-orecord (e.g., McManus et al., 2004).

Over recent years, two new technologies in particular have produced step-changes in our ability to observe the global ocean:

Satellite technology now gives us near-global coverage of the surface ocean. Variables that are now routinely measured from space include sea surface temperature (SST), sea-ice cover, wind speed (that can be used to derive the surface flow in the upper few tens of metres of the ocean), sea-level anomaly (which provides information about the depth-integrated ocean circulation), and ocean colour (which can be used to infer primary productivity and total biomass). Global coverage of ocean colour has only been available since the mid-1990s as part of projects like the Sea-viewing Wide Field-of-view Sensor (SeaWiFS); although limited measurements from other instruments were available from 1978 to 1986 (McClain, 2009). In 2009, a satellite mission, *Soil Moisture Ocean Salinity*, was launched that utilizes the sensitivity of microwave emissions to the dielectric properties of water (which is sensitive to salinity) to provide the first exploratory global sea-surface salinity (SSS) measurements from space. Further satellite missions are planned for the near future (Lagerloef et al., 2009).

ARGO profiling floats provide near-global coverage of the subsurface ocean. Over the last decade, a working array of ~3000 floats has been built up (*http://wo. jcommops.org/cgi-bin/WebObjects/Argo*). Floats typically descend to ~1000 m—2000 m measuring temperature, salinity, and pressure, remain submerged for ~8—10 days, and then ascend to the surface where the stored data are relayed via satellite to a data centre. In its lifetime of a few years, each float can provide

hundreds of profiles. Recent improvements to some of the floats include an ice-detection algorithm and interim storage that means that the ARGO network is being extended to the poorly sampled sub-sea-ice regions (Klatt et al., 2007). In addition, the advent of new stable oxygen sensors (Tengberg et al., 2006) has led to pilot deployments of ARGO floats with oxygen measuring capability, with limited results now becoming available (e.g., Riser and Johnson, 2008).

While there has been an order of magnitude jump in our observing capability over recent years, we are still limited to relatively short data records over most of the ocean. Satellite-derived SST extends back as far as the 1970s (Guan and Kawamura, 2003) and altimetry with accuracies of a few cm since the early 1990s (Fu et al., 1994), but large-scale global coverage of the subsurface ocean, provided by ARGO, is available for less than a decade. Furthermore, there are large discrepancies in the type of data available. Ocean temperature is the most well-observed ocean property, due a long history of hydrographic and XBT measurements and, more recently, the inclusion of satellite data. Even so, data gaps mean that identifying reliable long-term trends is still not possible in some regions (Deser et al., 2010; see also Figure 6.1). Salinity has also been routinely measured. However, as a result of XBT temperature measurements in the salinity database is only a fraction of the size of the global temperature record (de Boyer Montegut et al., 2007).

Ocean chemical (e.g., oxygen, pH, nutrients) and biological (e.g., chlorophyll/productivity) properties are much more poorly sampled. A striking consequence of this is that over 28,000 significant biological changes were noted for the terrestrial system in the IPCC AR4, whereas only 85 were noted for marine and freshwater systems (Richardson and Poloczanska, 2008). There is a vital need to maintain our existing observational network and to improve it in some important key areas. In particular, the Southern Ocean is very poorly sampled due to its remoteness and its inhospitable nature.

6.2.2. Natural Modes of Variability

In addition to data scarcity, examining long-term ocean change and its possible link to anthropogenic climate change is made much more challenging because of the signal-to-noise problem. The oceans are dynamic and, like the atmosphere, exhibit natural variability on a wide range of timescales: ocean 'weather systems' or eddies are pervasive throughout the ocean, there is strong seasonality in ocean properties and circulation and, on longer timescales, the ocean exhibits intrinsic inter-annual variability, spanning a few years to multiple decades. In addition to this intrinsic variability, the ocean is also affected by non-anthropogenic

FIGURE 6.1 (a) Linear trend in SST for HadSST2 dataset, 1900—2009 (blank areas indicate $5° \times 5°$ regions where there are less than 5 observations per decade at some point in the time series) and (b) NOAA OI SST dataset (1982—2009, which includes satellite data). There are no missing values for the latter dataset as observations have been interpolated to provide full spatial coverage. Units are °C per decade. *(Source: HadSST2,* http://hadobs.metoffice.com/hadsst2/; *NOAA OI SST,* http://www.emc.ncep. noaa.gov/research/cmb/sst_analysis/.)

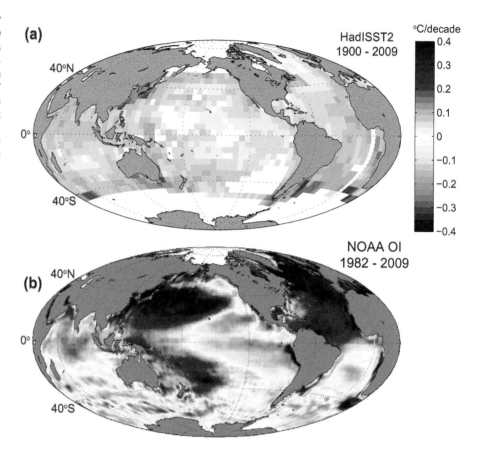

external factors, including solar variability and volcanoes. Given that natural variability may possess significant low-frequency components, long observational records and sophisticated statistical techniques are often required to extract any long-term signals from anthropogenic interference (see Latif and Park, 2012, this volume).

Climate variability is largely understood in terms of recurring regional patterns (or 'climate modes') related to natural internal dynamics of the ocean and atmosphere, and much effort goes into understanding the mechanisms behind, and the impacts of, these modes. A particular mode is associated with a distinct set of responses in a variety of climate properties both locally (where the dynamic interactions are taking place) and remotely (via changes propagated to other regions via the atmosphere and ocean). A further complication in separating anthropogenic signal from climate noise exists since some evidence suggests that the spatial pattern of anthropogenic climate change may project on to existing modes of variability (Corti et al., 1999; Stone et al., 2001; Latif and Park, 2012, this volume). A clear example of this is the long-term historical trends in highlatitude atmospheric circulation that exhibit strong similarities to the positive phase of the naturally occurring Southern and Northern Hemisphere annular modes (Stone

et al., 2001; Brandefelt and Källén, 2004). Before turning to the observed long-term changes, we introduce the dominant 'modes', how they impact ocean properties, and whether there is evidence for any long-term changes in these modes.

6.2.2.1. El Niño—Southern Oscillation (ENSO)

ENSO is a coupled oscillation in the tropical Pacific Ocean and atmosphere, characterized in the ocean by a shift in the position of the very warm water that makes up the tropical Western Pacific Warm Pool. During the El Niño phase of ENSO, as the tropical easterly winds relax, the equatorial thermocline flattens and the Warm Pool waters flood eastwards, shifting the centre of maximum atmospheric convection. This perturbation in the tropical Pacific propagates via the atmosphere and ocean to remote regions around the world (Bjerknes, 1969), causing significant changes to regional weather conditions (including tropical cyclones and the monsoon systems) and to ocean ecosystems and fisheries (e.g., Holland 2009 and references therein). The tropics represent the major source of natural CO_2 outgassing from the ocean to the atmosphere and ENSO-related changes strongly modulate this flux (Feely et al., 1999). As such, changes in the frequency or strength of this mode would have major impacts on global and regional climate.

A multidecadal increase in the strength of ENSO events has been reported over the last 30 years with the two largest recorded events having occurred in 1982 and 1998 (Fedorov and Philander, 2000). The shortness of the instrumental record, however, precludes attributing these changes to anthropogenic warming, especially given that ENSO exhibits substantial low frequency modulation (see PDO in Section 6.2.2.2). Furthermore, in a multimillennial climate model simulation, Wittenberg (2009) finds substantial low frequency natural variability, with multi-decadal to inter-centennial periods of both, almost no ENSO activity, and much enhanced ENSO activity. If this is representative of the real climate, it suggests that untan-gling human-induced trends from natural variability would be extremely hard without very long climate records.

Since the 1970s, the 'typical' ENSO pattern, with maximum El Niño (La Niña) warming (cooling) centred in the eastern Pacific has often been replaced by a distinctly different pattern where maximum warming (cooling) is shifted to the central Pacific (Latif and Park, 2012, this volume). This modified variability, termed El Niño Modoki (Ashok et al., 2007), produces very distinct changes to the Walker Circulation (Ashok et al., 2007; Wang and Hendon, 2007) and remote responses compared to canonical ENSO events (e.g., Ashok et al., 2007; Taschetto, et al., 2009, 2010).

Air–sea fluxes of CO_2 in the tropical Pacific (particu-larly in the eastern and central basin) dominate inter-annual variations in oceanic CO_2 and may constitute up to 30% of atmospheric variability (Feely et al., 2002). Upwelling of deep CO_2-rich water in the eastern Pacific drives a net CO_2 flux to the atmosphere. During El Niño events, the weaker winds, a reduction in upwelling, and the eastward expan-sion of the Warm Pool waters can considerably reduce normal outgassing (from ~0.8–1 PgC per year to 0.2 PgC–0.4 PgC per year: Feely et al., 2002). Decadal vari-ability in ENSO activity can cause significant low frequency variations in the build-up of atmospheric CO_2 levels. Thus, long-term climate changes in the tropical Pacific may be associated with significant modulation of the global carbon cycle.

6.2.2.2. Pacific Decadal Oscillation (PDO)

Closely linked to low frequency modulations of ENSO is the Pacific Decadal Oscillation (PDO: Mantua et al., 1997) or closely related Inter-Decadal Pacific Oscillation (Power et al., 1998). (These phenomena are defined differently but express essentially the same thing.) The associated pattern of surface ocean warming and cooling is similar to ENSO (although the temperature anomaly tends to be broader in a north–south sense). However, changes in the PDO phase

occur on multidecadal, rather than inter-annual, timescales. Untangling the influence of this low-frequency variability and anthropogenic long-term change requires care (see discussion of SST in Section 6.2.3). The widespread impact of this mode on marine and terrestrial systems has led to transitions between its phases being described as 'regime shifts'. A number of Pacific marine species are thought to be sensitive to changes in the PDO (see review by Mantua and Hare, 2002). Increases in tropical wind strength and equatorial upwelling associated with the most recent PDO phase change in the late 1990s are consistent with increased CO_2 outgassing from the tropical Pacific in recent years (Feely et al., 2006). The brevity of the data record precludes the identification of any long-term trends in the PDO (Henson et al., 2010). Pacific decadal variability is further explored in Latif and Park (2012, this volume).

6.2.2.3. Southern Annular Mode (SAM)

The Southern Annular Mode (SAM) or Antarctic Oscilla-tion is the dominant pattern of natural variability in the Southern Hemisphere outside the tropics. The SAM is characterized by a poleward intensification (equatorward weakening) of the mid-latitude westerly winds that extends from the surface to the upper jet, in its positive (negative) phase. It has an important influence on rainfall over high-latitude countries (e.g., Gillet et al., 2006; Hendon et al., 2007), Southern Ocean circulation, SST, and sea-ice concentration (e.g., Sen Gupta and England, 2006) as well as biological productivity (Lovenduski and Gruber, 2005). Modelling studies suggest that inter-annual changes in the SAM can account for ~40% of the inter-annual variability in CO_2 flux in the Southern Ocean. During a positive SAM, increased winds drive greater upwelling of natural carbon from the deep ocean to the surface, diminishing the ability for the ocean to absorb CO_2 (Lenton and Matear, 2007). The SAM has shown a robust long-term trend towards a more positive state over recent decades, with the strongest surface trend occurring in austral summer (Thompson and Solomon, 2002; Gillett and Thompson, 2003). Observa-tional and modelling studies have attributed the trend to anthropogenic factors, in particular a combination of stratospheric ozone depletion and enhanced GHG forcing (e.g., Fyfe et al., 1999; Arblaster and Meehl, 2006; Cai and Cowen, 2007). Observational (Le Quéré et al., 2007) and modelling studies (Lenton et al., 2009) suggest the ocean circulation changes associated with the SAM trend have partially offset the increase in the Southern Ocean sink of CO_2 that would otherwise have occurred due to rising atmospheric CO_2 concentrations.

6.2.2.4. North Atlantic Oscillation (NAO)

The North Atlantic Oscillation (NAO) is a prominent mode of variability with important impacts from the polar to subtropical Atlantic and surrounding landmasses (Latif and Park, 2012, this volume). It is most pronounced during boreal winter. The NAO is associated with changes in wind strength and direction, heat and moisture transport, and the frequency and strength of storms (see Hurrell et al., 2003). It also plays an important role in modulating ocean properties, such as SST, mixed layer depth, and, on long time-scales, basin-wide changes in circulation (Visbeck et al., 2003) including modulation of North Atlantic overturning (e.g., Biastoch et al., 2008). Low frequency biological activity has been related to the NAO variability. In the subpolar regions for example, the spring phytoplankton bloom is delayed by 2–3 weeks during positive NAO phases, related to changes in wind-driven mixing (Henson et al., 2009). A positive trend in the winter NAO index over the last 40 years of the twentieth century has been attributed in part to increases in GHGs (see the review by Gillet et al., 2003). However, the NAO is subject to strong multidecadal variability (particularly in winter) and the NAO index has subsequently decreased to relatively quiescent conditions over the last decade (e.g., *http://www.cpc.ncep.noaa.gov/ products/precip/CWlink/pna/new.nao.shtml*). Modelling studies suggest that the phase of the NAO can drive sub-basin-scale changes in surface CO_2 concentrations, driven largely by modulations in the advection of dissolved inorganic carbon by the North Atlantic Current. These NAO-driven changes exceed anything driven by secular trends in atmospheric CO_2 (Thomas et al., 2008).

6.2.2.5. Atlantic Multidecadal Oscillation (AMO)

Multidecadal variability is also present in the Atlantic in the form of the Atlantic Multidecadal Oscillation (AMO) (Kerr, 2000; see also Latif and Park, 2012, in this volume). The AMO appears to be intimately tied to variations in the Atlantic overturning circulation, over timescales of ~60 years. Changes in the AMO are manifested as slow changes to Atlantic SST (Knight et al., 2005). The timescales involved in this natural oceanic oscillation mean that great care must be taken in distinguishing the low-frequency changes associated with this mode and any human-induced signal. This can be a non-trivial task, as can be seen in the recent controversy regarding the relative importance of the AMO compared to global warming on the melting of Swiss glaciers (Huss et al., 2010a, 2010b; Leclercq et al., 2010; Cogley, 2012, this volume)

6.2.3. Surface Temperature and Salinity

SST is of fundamental importance to the climate system. SST plays a major role in determining the rate of heat, moisture, and gas flux between the ocean and atmosphere. It represents a proxy for upper ocean mixed-layer temperature and, as a result of the large thermal inertia of this upper layer, makes it an important quantity for monitoring

climate change. SST is also a major factor affecting marine ecosystems.

Recent estimates suggest that globally-averaged SST has increased by ~0.68°C per century (linear trend, 1901–2004, HasSST2), with more rapid warming occurring in the later part of the time series (Rayner et al., 2006). It should be noted that such estimates incorporate sparse data in certain extended regions, including much of the Southern Ocean (Figure 6.1). The spatial pattern of warming is highly non-uniform. It is well-known that ocean temperature increases lag behind the land surface (as a consequence of different thermal capacities and the ocean's greater ability to lose energy via latent heat fluxes: Sutton et al., 2007; Dommenget, 2009). Furthermore, the high-latitude Arctic regions have amplified warming (as a result of sea-ice interactions: IPCC, 2007a). In addition to this, there is a high degree of spatial variability in trends across the oceans. Deser et al. (2010) analysed a variety of global SST (and air temperature) datasets to identify robust spatial trends over a period of time sufficient to average over much of the inherent natural variability (1900–2008). Warming has been occurring in most datasets over all areas where a trend could be computed, except in the North Atlantic where there is a significant cooling, possibly a consequence of long-term changes in the NAO or the overturning circulation (Deser et al., 2010). Insufficient data were available for high-latitude Arctic and Southern Ocean regions.

The various SST products present conflicting trends in the central and eastern Pacific (Vecchi et al., 2008; Deser et al., 2010), an area important for ENSO processes. However, examination of independent air temperature measurements are consistent with the uninterpolated SST datasets that more closely reflect the in-situ data, suggesting that warming is occurring across the entire tropical Pacific.

It is interesting to examine equivalent trends for a shorter time period. Figure 6.1b shows a 30-year trend from the NOAA Optimum Interpolation SST V2 dataset (Reynolds et al., 2002), which includes satellite data. While the globally-averaged rate of warming is larger than for the longer period, significant cooling is evident over large parts of the central and eastern Pacific Decadal Oscillation. The pattern in the Pacific is consistent with the observed downward swing in the phase of the PDO (Section 6.2.2.2). The cooling in the Southern Ocean is consistent with the observed intensification of the mid-latitude westerlies (Section 6.2.2.3), which force cold high-latitude surface water northwards (see Sen Gupta and England 2006, their figure 12).

Salinity, like temperature, plays a key role in ocean processes, affecting water density, water column stability and the overturning circulation. It also provides valuable information for understanding the marine hydrological cycle (e.g., Durack and Wijffels, 2010; Helm et al., 2010). The large-scale salinity distribution is largely determined by local precipitation minus evaporation rates and modified by ocean circulation and mixing, and at coastal locations by freshwater runoff or sea-ice formation or melt.

As tropospheric temperatures increase, theoretical arguments suggest a resultant intensification of the hydrological cycle (e.g., Held and Soden, 2006). Over the ocean, such changes are difficult to measure directly due to the scarcity of sampling stations, which are primarily located on land areas, and the highly heterogeneous nature of the rainfall distribution. Examination of historical salinity data, however, is able to resolve changes that are uncertain in the rainfall datasets. Even prior to the expansion of salinity observations afforded by ARGO, studies were able to identify a freshening (reduced salinity) at high latitudes and increased subtropical water salinity (Antonov et al., 2002; Boyer et al., 2005). More recent studies incorporate the independent ARGO dataset. Hosoda et al. (2009), Helm et al. (2010), and Durack and Wijffels (2010) demonstrate these results to be robust. Using subsurface salinity data, Helm et al. (2010) find a salinity increase within the near surface (~100 m) salinity maximum and a freshening of the intermediate depth salinity minimum (for ~1970−2005). The upper waters are primarily sourced from the subtropics (regions of relatively weak net precipitation) while intermediate waters are sourced at higher latitudes of the North Atlantic and Southern Ocean (regions of relatively high net precipitation). These changes imply an intensification of the hydrological cycle. Durack and Wijffels (2010) have constructed a new salinity database that attempts to remove the effect of inter-annual ENSO variability. The 1950−2008 trend (Figure 6.2) bears a strong resemblance to the mean salinity field and, in turn, to the mean freshwater flux field (see Durack and Wijffels, 2010, their figure 5). This is again indicative of a spin-up of the hydrological cycle. In precipitation-dominated low-salinity regions, including the tropical west Pacific Warm Pool, under the

Atlantic ITCZ, and at high latitudes (particularly in the Southern Ocean and the North Pacific), significant freshening has occurred. Conversely, in the subtropical gyres, located in regions of mean atmospheric subsidence, salinity has been increasing. While globally-averaged surface salinity remains nearly unchanged, the normal gradient between the relatively fresh Pacific basin and salty Atlantic basin has increased (Durack and Wijffels, 2010).

Considerable interest has been focused on the Pacific Warm Pool. In the region of the warmest waters (>28.5°C), temperatures have increased on average ~0.3°C and salinity has dropped by ~0.34 psu over the last 50 years (1955−2003: Cravatte et al., 2009). The magnitude and distribution of the observed salinity changes are consistent with the long-term warming signal (as theory suggests that net precipitation will change in proportion to the SST change; Held and Soden, 2006; Cravatte et al., 2009). An important consequence of these changes is a considerable expansion of Warm Pool area, mean depth and volume, and an eastward shift of the Warm Pool edge (as defined by either the 34.6‰ or 34.8‰ isohalines or the 29°C isotherm) of between 15° and 20° of longitude (Cravatte et al., 2009). This region plays an important role in ENSO dynamics (e.g., Picaut et al., 1997). In addition, the distribution of important marine species (e.g., skipjack tuna) are highly sensitive to the location of the Warm Pool boundary (Lehodey et al., 1997, 2010).

6.2.4. Heat Content and Sea Level

Given that ~80% of the total additional heat related to anthropogenic warming now resides in the ocean (Levitus et al., 2005; Bindoff et al., 2007), ocean heat content is an important measure of the change in the Earth's radiative balance (Harvey, 2012, this volume). It is also a major contributor (via thermal expansion) to sea-level rise (e.g., Domingues et al., 2008). In addition to problems of data scarcity, estimating global heat content has been subject to data bias issues, most importantly in relation to

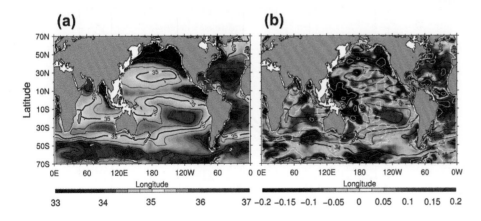

FIGURE 6.2 The 1950−2000 mean salinity (a, ppt) and linear trend in surface salinity (b, ppt per 50 years). Salinity change contours every 0.2 are plotted in white. Regions where the resolved linear trend is not significant at the 99% confidence level are stippled in grey. (*Source: Durack, P., and S. Wijffels, 2010, their figure 7.5. © American Meteorological Society. Reprinted with permission.*)

the XBT devices, which have historically been the most common instruments for measuring subsurface temperature (>70% of the global temperature archive: Wijffels et al., 2008). XBTs are dropped from ships while attached to a thin wire and fall from the surface to the deep ocean, measuring temperature. These devices estimate depth using a fall-rate equation and the time elapsed from release. It has long been recognized that errors in the fall-rate calculation

produce significant biases in heat content estimates. A number of recent studies have used different techniques to correct for the XBT bias (see review by Lyman et al., 2010). Results from three such studies are shown in Figure 6.3 (Levitus et al., 2009). Linear trends for 1969−2003 range from 0.24 to 0.41 times 10^{22} J per year of additional heat (Ishii and Kimoto, 2009; Levitus et al., 2009; Domingues et al., 2008). Based on the higher estimate of Domingues

FIGURE 6.3 (a) Global 0 m−700 m ocean heat content from three recent studies *(Source: Levitus et al., 2009 © Copyright 2009 American Geophysical Union. Reproduced/modified by permission of American Geophysical Union.).* (b) Breakdown of total sea-level rise into upper ocean (0 m−700 m, red) and deep (orange) thermal expansion, ice sheets (blue), and terrestrial storage (green). (c) Observed sea-level rise from two independent studies (black, yellow dashed) and satellite altimetry (red-dashed) compared to the sum of the contributions from the middle panel (blue). *(Source: Panels (b) and (c) from Domingues, et al., 2008 (their figure 3). Reprinted by permission from Macmillan Publishers Ltd.)*

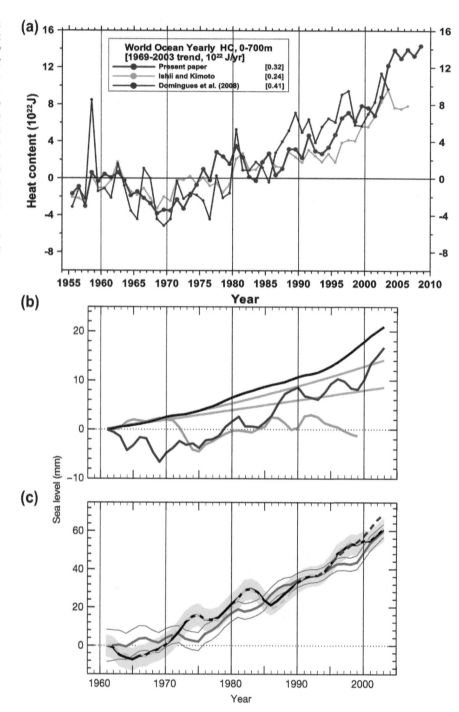

et al. (2008), this corresponds to an additional globally-averaged air—sea heat flux of just under 0.4 Wm^{-2} with about 90% of the additional heat stored in the upper 300 m. This additional heat corresponds to a sea-level rise of ~0.5 mm per year due to upper-ocean thermal expansion. Domingues et al. (2008) examine literature-based estimates for the various factors contributing to sea-level rise, including: (i) ~0.2 mm per year from Antarctica and Greenland since 1993 (with little data to constrain contributions prior to this time); (ii) ~0.5 mm per year from glaciers and ice caps for 1961—2003 (Dyurgerov et al., 2005), the importance of glaciers being dealt with in greater detail in Cogley (2012, this volume); (iii) terrestrial water storage, which lacks any long-term trend but exhibits large inter-annual variability; and (iv) penetration of heat into the deep ocean, particularly at high southern latitudes in regions of deep convective mixing (Rintoul, 2007; Johnson and Doney, 2006; Johnson et al., 2007, 2008). Figure 6.3b shows time series for these contributing components (Domingues et al., 2008). Despite recognized uncertainty in these terms, the observed estimate of total sea-level rise and the sum of independently estimated components (Figure 6.3c) show very similar trends ~1.6 mm per year (1961—2003), providing some confidence that the sea-level budget is approaching closure.

In addition to changes in global inventories of heat content, regional changes in temperature within different water masses are also apparent. Levitus et al. (2005), updated in Bindoff et al. (2007), reported on the zonally-integrated heat uptake into the top 1500 m of the global ocean. A number of interesting features are evident, and are discussed below.

In the Northern Hemisphere, there is a deep warming signal (centred at 40°N, Figure 6.4) associated with a warming of the North Atlantic subtropical gyre, the Atlantic current, and the Gulf Stream. In the subpolar region, there is an associated cooling and a strong freshening that affects the properties of Labrador Sea Water, an important component of the North Atlantic Deep Water that forms the Atlantic branch of the global overturning circulation. However, on inter-annual timescales, changes in water mass properties in the far North Atlantic are sensitive to the state of the NAO, which modulates the intensity of deep mixing and the circulation of the subpolar gyre (Sarafanov, 2009). Since the mid-1990s, the observed trends in salinity and temperature that took place over the preceding ~30 years and reported in Levitus et al. (2005), have largely reversed in tandem with the change in the phase of the NAO (Holliday et al., 2008). This highlights the difficulty in separating natural variability from anthropogenic factors.

In the Southern Hemisphere, a deep warming signal is evident, particularly in the vicinity of the Antarctic Circumpolar Current (Levitus et al., 2005; and Figure 6.4).

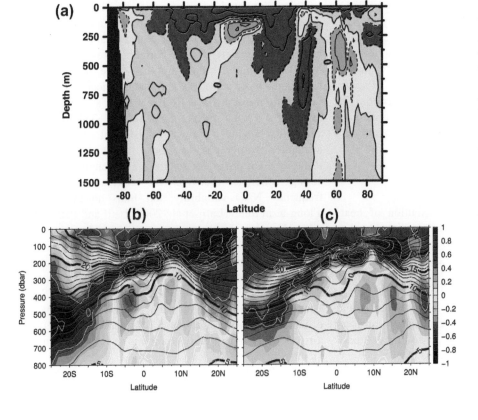

FIGURE 6.4 (a) Linear trend in the zonally-averaged temperature for the world ocean (1955—2003, contour interval is 0.05°C per decade; dark red or blue shading indicates regions where the magnitude of warming or cooling exceeds 0.025°C, respectively *(Source: modified from IPCC, 2007a and Levitus et al., 2005 © Copyright 2005 American Geophysical Union. Reproduced/modified by permission of American Geophysical Union.).* Linear trend (1950—2008) in subsurface temperature at (b) 160°E and (c) 160°W using the new dataset of Durack and Wijffels (2010). *(Source: Figure courtesy of Paul Durack. © American Meteorological Society. Reprinted with permission.)*

This pattern of warming has been confirmed in more recent studies that include large amounts of new ARGO data (Gille, 2008; Böning et al., 2008). The mechanism driving these and other changes are still a matter of debate. As noted by Gille (2008), the pattern of change is consistent with a southward displacement of the ACC and the observed southward shift in the overlying westerly winds (Thompson and Solomon, 2002). The observed mid-latitude warming is also well-reproduced in the current coupled climate models. While the simulated response in the models is indeed driven by a southward shift of the ACC, the southward displacement is not purely adiabatic (i.e., an additional input of energy is required), and a large amount of heat is absorbed to the south of the ACC and advected northwards in the surface Ekman layer (Cai et al., 2010). However, as Cai et al. (2010) and others have pointed out, the lack of meso-scale eddy activity in climate models may undermine any conclusions with regards to the processes driving change in the ACC (see Section 6.2.5).

In the tropics, there is a distinctive cooling in the thermocline straddling the equator, sandwiched between strong surface and weaker deep warming (Figure 6.4a). This pattern is evident in both the Pacific and the Indian Oceans (Levitus et al., 2005) and appears to be related to a shoaling of the tropical thermocline associated with a weakening of the Walker circulation and the equatorial trade winds (Mcphaden and Zhang, 2004; Vecchi et al., 2006; Han et al., 2006). Over the last decade, there appears to have been a partial recovery associated with changes in the PDO (Mcphaden and Zhang, 2004). However, analysis that includes the most recent ocean observations and explicitly attempts to extract natural variability from the long-term trend (Durack and Wijffels, 2010) still shows a strong subsurface cooling along the equatorial Pacific (Figure 6.4b, 6.4c). Furthermore, the subsurface cooling is a robust feature of the coupled climate models that include anthropogenic forcing, suggesting that this pattern is indeed a fingerprint of human-induced change.

6.2.5. Ocean Circulation

A key role of the ocean is the redistribution of heat between the low and high latitudes. This is mediated by the large-scale circulation and the meso-scale activity that permeates the ocean. The large-scale circulation is often compartmentalized into an upper ocean circulation, including the expansive subtropical and subpolar gyres and the overturning (or thermohaline) circulation that includes some of the upper ocean pathways, but also comprises deep circulation fed by the sinking of dense high-latitude water. The circulation is controlled by a number of factors including: (i) fluxes of heat and freshwater to and from the ocean; (ii) wind forces, which are the major driver of upper

ocean circulation and play an important role in ocean mixing; and (iii) tidal forces, which drive the daily tidal currents and also play an important role in ocean mixing. Significant long-term trends have been observed in both heat fluxes, freshwater fluxes, and wind fields in certain regions (Trenberth and Josey, 2007; Bindoff et al., 2007), leading to the expectation of changes to the ocean circulation.

A key metric that has been the focus of much scientific and public interest is the strength of the North Atlantic overturning circulation. The upper branch of this circulation brings large amounts of heat, via the Gulf Stream and Atlantic Current, from low latitudes to the North Atlantic. This plays an important role in warming the climate of northern Europe via heat lost to the atmosphere from the current as it travels northwards. Evidence also suggests that dramatic changes in the overturning circulation strength and location have played an important role in abrupt climate change (e.g., Clark et al., 2002; McManus et al., 2004). Model simulations also suggest that the warming and freshening in the North Atlantic, which would accompany anthropogenic climate change, could cause a significant slowdown of this circulation in the future (Meehl et al., 2007, their figure 10.15).

Considerable excitement accompanied the release of findings, based on five repeat hydrographic sections in the North Atlantic at 25°N (Bryden et al., 2005), which suggested a 30% slowdown in the overturning circulation between 1957 and 2004, a rate considerably larger than anything simulated by climate models. However, a slowdown of this magnitude should have left a strong cooling signature in SST (~1°C) in the North Atlantic, which was not observed (Kerr, 2005). Based on results from the RAPID continuous monitoring array (see Section 6.2.1), Cunningham et al. (2007) found a mean (±standard deviation) transport of ~$19 \pm 6 \times 10^6 m^3 s^{-1}$ with a full range of 4.0×10^6 to $34.9 \times 10^6 m^3 s^{-1}$. This degree of variability precludes drawing robust results from the few available hydrographic sections. Independent analysis, using an SST proxy for overturning strength, also finds strong coherence between the circulation and low-frequency NAO variability, but finds no evidence of a long-term trend (Latif et al., 2006; Latif and Park, 2012, this volume).

While the lack of pertinent observations precludes direct estimation, model experiments and indirect observational evidence suggest an increase in the transport of warm salty water from the Indian Ocean to the South Atlantic via the 'Agulhas leakage', another important part of the thermohaline circulation. Biastoch et al. (2009) use a realistically-forced eddy-resolving ocean model (in the vicinity of South Africa) nested in a global model (for 1958–2004) to simulate the circulation changes. The observed southward shift of the westerlies (Section 6.2.2.3) drives a simulated southward shift of ~2° in the position of the Southern Hemisphere subtropical gyre. This, in turn, leads to a significant increase in the leakage of water into the Atlantic. The hindcast simulation also shows an increase in salinity of the Brazil current that is consistent with observations. A significant portion of this leaked water is not entrained back into the subtropical gyre and instead moves northward, leading to a 25% increase in the simulated transport of salt into the Northern Hemisphere. Biastoch et al. (2009) postulate that such changes could moderate future changes in the North Atlantic overturning circulation. The advection of the greater amounts of salty (Indian Ocean) water into the North Atlantic could act to partially offset any future freshening in the North Atlantic, potentially helping to stabilize the sinking branch of the overturning circulation (Weijer et al., 2002).

The mechanism behind this southward shift of the circulation relates to the southward intensification of the mid-latitude westerlies. While the latitude of maximum wind stress change sits well south of the subtropical gyres, it is the wind stress curl (in particular north—south gradients in the east—west wind) that, via Sverdrup dynamics, control the gyre circulations (Saenko et al., 2005; Cai et al., 2005). The maximum change in the wind stress curl occurs considerably further north over the subtropical gyres. Using historical wind observations, Cai (2006) computes the theoretical change in the depth-integrated ocean circulation. He finds a southward shift of the 'supergyre' that encompasses all the Southern Hemisphere subtropical gyres. Using a number of independent observational datasets including satellite altimetry, hydrography, and drifting floats, Roemmich et al. (2007) find a robust intensification of the South Pacific subtropical gyre (1993—2004) consistent with the wind stress changes. Altimetry also suggests an intensification of the gyres in the other basins. Using a unique 60-year time series (1944—2002) off southeastern Tasmania, Hill et al. (2008) show dramatic trends in temperature (2.23°C per century, well above global average increases) and salinity (0.34 psu per century) with both the decadal variability and the long-term trend consistent with changes in the wind field driving a southward intensification of the East Australian Current. The circulation and temperature change have also been implicated in important changes to local ecosystems. For example, it has permitted the invasion of the spiny sea urchin (*C. rodgersii*) from mainland Australia to Tasmania, where overgrazing by the invading urchins has led to large localized species losses (Ling et al., 2009).

Theoretical arguments suggest that the southward shift in the maximum winds should also drive a southward shift in the location of the Antarctic Circumpolar Current (ACC) (Saenko et al., 2005, and references therein). Based on hydrographic and ARGO data, Sallée (2008) finds coherence between the SAM and the position of the fronts

defining the ACC at certain longitudes, depending on the relative position of the mean westerlies, the ACC, and topographic constraints. Recently, Sokolov and Rintoul (2009) documented a zonally-averaged southward shift in the position of the ACC fronts (as determined by examining sea-surface height) of ~0.6° of latitude since the early 1990s. The southward shift is consistent with the observed southward displacement of density surfaces and large positive trends in temperature centred at ACC latitudes (Gille, 2008; Böning et al., 2008). While there is some observational evidence that supports a coherence between the Southern Annular Mode (see Section 6.2.2.3) and ACC transport through the Drake Passage on inter-annual timescales based on moorings (Meredith et al., 2004) and satellite gravity measurements (Böning et al., 2010), observations are insufficient to determine any long-term trends. The issue of ACC intensification and the associated spin-up of the Southern Ocean overturning is a matter of considerable contention as CMIP3 climate models and eddy-resolving ocean models produce different responses to intensified winds (Meredith and Hogg, 2006; Screen et al., 2009). In the climate models, increased westerly winds drive increased northward Ekman transport and an acceleration of the ACC, while in eddy-resolving models much of the additional energy input goes into an intensification of the eddy field, rather than into the mean circulation.

6.2.6. Oxygen

Dissolved oxygen plays a vital role for biological processes both directly through its effect on organism physiology and indirectly through its effect on the nitrogen and carbon cycles. Low oxygen conditions (hypoxia) can lead to significant die-offs in animal species (e.g., Chan et al., 2008) and result in extensive 'dead zones' in the ocean (Keeling et al., 2010). Patterns of oxygen concentrations also hold important clues for understanding changes in ocean biogeochemistry and circulation (e.g., Joos et al., 2003).

Dissolved oxygen concentrations are controlled by air—sea fluxes (which are strongly related to temperature-dependent solubility), ocean mixing (which helps to ventilate subsurface waters), internal advection, and biochemical processes, including primary production in the upper ocean and the bacterial breakdown of organic matter in the deep ocean. Recent attention has focused on the oxygen minimum zones, regions of poor ventilation, and/ or high oxygen utilization that exist in the eastern tropical Pacific and Atlantic and the northern Indian and Pacific Oceans at intermediate depths. These regions can be hypoxic for certain species (i.e., low oxygen levels cause stress and ultimately death if maintained for extended periods). The threshold for hypoxia varies considerably depending on the species, although a relatively consistent lower bound exists ($pO_2 < 0.8$ kPa) beyond which only species with specific low oxygen adaptation can survive (Seibel, 2011). While observations are much sparser than for physical variables, time series have been constructed that show a robust vertical expansion of the Atlantic and Pacific oxygen minimum zones (Stramma et al., 2008). These changes are consistent with model simulations with interactive ocean carbon cycles (e.g., Matear et al., 2000; Matear and Hirst, 2003). An extreme expansion in the tropical North Atlantic of 85% is observed since the 1960s, together with a decrease in minimum concentrations. Brandt et al. (2010) use extensive cruise data to both confirm the reduction in oxygen concentration (15 μmol kg^{-1} between 1972–1985 and 1999–2008), and to propose a mechanism for the change. They show that ventilation of the tropical North Atlantic is from eastward-flowing jets of relatively high-oxygen concentration. Although no direct current measurements are available, hydrographic calculations suggest a weakening of the jets. While lack of observations precludes the determination of any trend in the northern Indian Ocean oxygen minimum zones, a time series has been constructed in the eastern tropical region, where no significant trend was found (Stramma et al., 2008).

Another region that has exhibited robust oxygen changes is the North Pacific, a vast region of relatively low concentration. Station data from off the Japan coast and in the Alaskan gyre, together with a number of repeat sections across the Northern Hemisphere basin, all show consistent declines in oxygen concentration across subpolar and eastern subtropical regions and in many marginal seas around Japan and western North America (Keeling et al., 2010, and references therein). Deutsch et al. (2006) use a hindcast, physical–biogeochemical model to show that the large changes in the lower subpolar thermocline are primarily due to weakened ventilation of dense water in the North Pacific. While significant fluctuations in biological export production are simulated, the influence of this on oxygen levels is confined to the upper thermocline. Earlier biogeochemical modelling suggests that, in general, thermocline concentrations are influenced primarily by solubility at the surface source region, while deeper oxygen decreases are more influenced by reductions in interior ventilation (Matear and Hirst, 2003).

Given the biological and oceanographic importance of oxygen, a concerted effort is under way to expand the observational network, including pilot deployments that include oxygen sensors on some of the ARGO floats. Initial data will be carefully compared with shipboard measurements, with large global deployments being subsequently planned (Gruber et al., 2007).

6.2.7. Carbon and Biogeochemistry

The ocean holds about 50 times the amount of carbon than the atmosphere and, therefore, has a profound influence on

atmospheric CO_2 levels on both short and long timescales (cf. Harvey, 2012, this volume). The ocean has two different mechanisms by which it can affect atmospheric CO_2 via modulations in the surface-ocean CO_2 concentration and resulting changes in air–sea CO_2 fluxes. These will be discussed below.

6.2.7.1. Anthropogenic Ocean CO_2 Budget

Detailed records since the start of the Industrial Revolution give us the ability to determine the amount of CO_2 emitted globally from anthropogenic fossil fuel use and land-use changes. In this time, about 244 ± 20 PgC has been emitted into the atmosphere (Sabine et al., 2004). In 2008, annual anthropogenic CO_2 emissions were 8.7 ± 0.5 PgC, which corresponds to a rate increase of 40% since 1990 (Le Quéré et al., 2009). Atmospheric CO_2 measuring stations are located across the world and have accurately determined the in situ concentrations of CO_2 in the atmosphere over the last 60 years. Comparing the estimated concentration of CO_2 solely from fossil fuel emissions and the actual concentration in the atmosphere shows a large difference (Figure 6.5), with a shortfall in the atmospheric concentration of at least 55 ppmv. On average, only 43% of the total anthropogenic CO_2 emissions each year between 1959 and 2008 has remained in the atmosphere (Le Quéré et al., 2009). A large part of the 'missing' atmospheric CO_2 has been absorbed by the ocean (Sabine, 2004).

The ocean has a very large capacity to absorb atmospheric CO_2 due to both its volume and its ability to

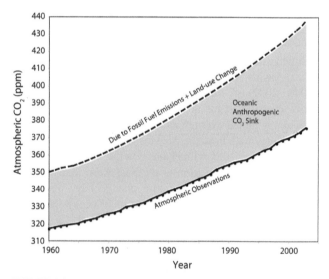

FIGURE 6.5 Average annual atmospheric CO_2 concentrations (dotted line) and the atmospheric CO_2 concentrations that should be in the atmosphere solely due to fossil fuel emissions and land-use change (dashed line). Blue shading is the proportion of the anthropogenic CO_2 absorbed since the Industrial Revolution. Most of this CO_2 is absorbed by the ocean, with the terrestrial biosphere playing a minor role with regard to net changes over this timeframe (*Source: Sabine et al., 2004.*)

redistribute CO_2 into other inorganic forms of carbon. A wide variety of measuring programmes and methods have accurately determined the anthropogenic CO_2 sink (Sabine, 2004), which indicates that, since the Industrial Revolution, the ocean has absorbed about half of the total anthropogenic CO_2 emissions (120 ± 15 PgC up until 1995). More recent observational constraints suggest the oceanic sink was 2.2 ± 0.4 PgC per year between 1990 and 2000 (Keeling and Garcia, 2002; McNeil et al., 2003; Dickinson, 2012, this volume).

6.2.7.2. The Solubility CO_2 Pump

Oceanic CO_2 enters the ocean via air–sea gas exchange. The dissolved CO_2 then dissociates into bicarbonate (HCO^{3-}) and carbonate ions (CO_3^{2-}) as shown in Equation (6.1). The vast majority of oceanic inorganic carbon exists as either bicarbonate (~90%) or carbonate (~9%), with CO_2 only 1% of the total inorganic carbon pool. The presence of inorganic carbon in three different forms means that the ocean has an immense capacity to alter oceanic CO_2 equilibration and to influence atmospheric CO_2 on long timescales. In fact, even if humans release all of the estimated 5000 PgC of current reserves of fossil fuel carbon (both conventional and unconventional), the ocean will eventually (2000–20,000 years) absorb between 65%–80% of all anthropogenic CO_2 after equilibration (Archer et al., 2009).

$$CO_2 + H_2O \Leftrightarrow H_2CO_3 \Leftrightarrow HCO_3^- + H^+$$
$$\Leftrightarrow CO_3^{2-} + 2H^+ \quad (6.1)$$

The solubility of CO_2 in seawater is dependent on ocean temperature, with higher dissolution in cooler seawater (Weiss, 1974). If the oceanic CO_2 distribution were based on solubility alone, a quasi-linear relationship between the partial pressure of carbon dioxide (pCO_2) and SST would exist. Oceanic CO_2, however, is only weakly correlated with SST (Takahashi et al., 2002), which demonstrates the importance of other mechanisms in driving CO_2 variations in the ocean. The ocean overturning circulation is one of those important mechanisms for sequestering atmospheric CO_2. Surface waters exchange CO_2 with the atmosphere. As water is transferred from the surface ocean to the deep ocean via density transformation, there is also a transfer of large amounts of CO_2 that become sequestered (or isolated) from the atmosphere for long timescales (>1000 years). The rate of sequestration is therefore strongly tied to the rate of overturning. The transfer of CO_2 into the interior of the ocean via physical mixing of ocean waters is termed the ocean's solubility carbon pump and has been an important contributor to atmospheric CO_2 variations in the past (Sigman and Boyle, 2000; Figure 6.6).

Recent observational evidence suggests that the efficacy of the physical pump may have changed over recent decades in certain regions. In the Southern Ocean, using a network of observational stations combined with a carbon model, Le Quéré et al. (2007, 2008) found a net reduction in CO_2 uptake since 1981. These authors suggest that the stronger overturning circulation, which might accompany the intensified winds (see Section 6.2.5), would allow more

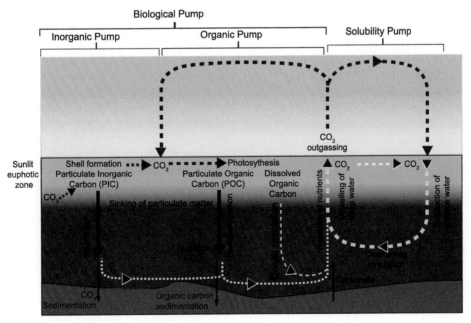

FIGURE 6.6 Schematic of the Oceanic Biological and Physical Carbon Pumps. (*Source: Modified from* http://en.wikipedia.org/wiki/File: CO2_pump_hg.svg.)

anthropogenic CO_2 to be sequestered. At the same time, however, it would also bring to the surface deep water that has high concentrations of old natural CO_2. Le Quéré et al. (2007, 2008) find this latter process to be dominating, resulting in a net decrease in Southern Ocean CO_2 sequestration. However, this is at odds with some modelling studies (Law et al., 2008) and shows some sensitivity to the choice of stations (Zickfeld et al., 2008).

6.2.7.3. The Biological CO_2 Pump

The other ocean mechanism that can significantly alter the atmospheric CO_2 concentrations is the biological CO_2 pump (Figure 6.6). Via photosynthesis in the sunlit 'euphotic' zone, phytoplankton utilize CO_2 to form biomass. This primary productivity has the effect of reducing the concentration of CO_2 in the surface ocean and creates a concentration gradient, which causes an additional air to sea flux of CO_2 into the ocean. Primary production in the ocean fixes about 45 PgC per year of biomass, which is about eight times the annual anthropogenic CO_2 emissions (~8 PgC per year). However, 75% of the fixed organic carbon in the surface ocean is respired (~34 PgC per year) back into the atmosphere while the remaining ~11 PgC per year is biologically exported into the interior of the ocean (Laws et al., 2000). This biological carbon pump has been in quasi-steady state throughout the twentieth century and the first decade of the twenty-first century. Although biological productivity drives very large gross CO_2 fluxes across the air–sea interface, the steady-state fluxes have resulted in little influence on the net changes in atmospheric CO_2 during the industrial period (Matear and Hirst, 1999). The process that could potentially drive significant changes to atmospheric CO_2 is related to biological carbon export. When phytoplankton utilize CO_2 to create organic material, a portion of this organic material is grazed on by zooplankton and higher marine organisms. Following death or aggregate formation, this organic material sinks from the surface ocean into the deeper ocean. This process, known as carbon export, is an efficient way of sequestering CO_2 from the atmosphere. If the rate of carbon export is equal to the physical supply of carbon by upwelling or mixing, then there is no net change on atmospheric CO_2. Any change in the balance between the physical supply of carbon and the biological carbon pump could in principle drive large changes in atmospheric CO_2 concentrations. Variation in the strength of this pump is a critically important factor in controlling CO_2 variations between glacial and interglacial periods (Martin, 1990). We will discuss later how non-linear changes to the biological pump are starting to emerge as an important factor for future change.

Calcium carbonate ($CaCO_3$) is a mineral produced by many marine organisms (coccolithophores, foraminifera,

pteropods, etc.). The production of particulate inorganic carbon (PIC) in the form of $CaCO_3$ via calcification has a smaller, but not insignificant, role to play in modulating atmospheric CO_2 on centennial timescales. Unlike photosynthesis, the process of $CaCO_3$ production actually acts to raise the surface-dissolved CO_2, and thus retards the sequestration of atmospheric CO_2. Of the 11 PgC per year of biological carbon export in the form of particulate organic carbon (POC), about 1 PgC per year is in the form of $CaCO_3$ or PIC. Although the biological carbonate pump does not play a significant role in the short-term fluctuations of atmospheric CO_2, carbonate processes and changes start to play an important role over multimillennial timescales, the timescale over which ocean/terrestrial weathering becomes important (Zondervan et al., 2001).

6.2.7.4. Ocean Acidification

The absorption of nearly half of the anthropogenic CO_2 load since the Industrial Revolution has resulted in chemical changes within the ocean, that pose potential threats to certain marine organisms (Fabry et al., 2008; Gattuso et al., 1998; Langdon et al., 2000). As anthropogenic CO_2 enters the ocean, the pH of the seawater decreases (i.e., acidity increases) and the surface ocean carbonate ion (CO_3^{2-}) concentration decreases (see Equation (6.1)). These changes, together known as 'ocean acidification', have direct consequences for the marine $CaCO_3$ cycle and those species that interact, exploit, and secrete the mineral. Ocean observations already show surface pH to be lowered by 0.1 (Raven et al., 2005; Figure 6.10). Palaeoproxy data suggests that current pH levels are more extreme than anything experienced over the glacial–interglacial cycles of the last ~2 million years, with levels projected for the end of the century likely to be unprecedented for ~40 million years. Moreover, the rates of change that are likely to occur during this century are probably two orders of magnitude greater than those experienced over the glacial–interglacial cycles (Pelejero et al., 2010). At these rates of change, the natural buffering mechanisms that can act to moderate pH changes on millennium timescales are unable to keep up. The role of CO_2 in the palaeorecord is discussed in more detail in Berger and Yin (2012, this volume).

Marine biological calcification requires adequate availability of seawater carbonate ions (Raven et al., 2005), which is often represented using the $CaCO_3$ saturation state of seawater: $\Omega_x = \frac{[Ca^{2+}][CO_3^{2-}]}{\lambda_x}$, where λ_x is the solubility coefficient appropriate for the particular form of $CaCO_3$ under consideration (i.e., x is either calcite or aragonite). Aragonite is the more soluble form of $CaCO_3$ and is secreted by many marine organisms. The decrease in Ω from ocean acidification reduces the ability of some species (e.g., corals, calcifying algae, echinoderms, molluscs) to form their skeletal material and, therefore, future ocean acidification is expected to adversely affect these species and their associated communities and ecosystems (Raven et al., 2005). Experiments have shown that calcifying marine organisms are highly susceptible to changes in the aragonite saturation state (Ω_{arag}) (Fabry et al., 2008; Langdon et al., 2000). Furthermore, direct dissolution of aragonite occurs when $\Omega_{arag} < 1$. The Southern and Arctic Oceans are most vulnerable to reaching the corrosive point for aragnonite. Recent work suggests these regions will start to become corrosive for aragonite once atmospheric CO_2 reaches 450–470 ppmv (McNeil and Matear, 2008; Steinacher et al., 2009). Despite the changes occurring within the ocean carbonate system, much uncertainty remains over the likely impacts to the marine ecosystem. Although some aragonitic species are likely to be detrimentally impacted by ocean acidification, along with those that graze on these species, other species that were predated on will likely 'win'. The ecosystem-wide effects of ocean acidification are still uncertain.

6.2.8. Ocean Biology

Climate variability and change affect ocean ecosystems through a large variety of mechanisms (see review by Drinkwater et al., 2010). Temperature changes, for example, may affect physiology and growth rates, activity rates (e.g., swimming speed), reproductive fecundity, phenology, and species distribution. Surface temperature changes also drive changes in vertical stratification and mixing, which, in turn, affect the upwards entrainment of nutrients from the deeper ocean and light availability for phytoplankton at high latitudes. Changes in ocean circulation can affect dispersal of larvae and fish eggs and their subsequent recruitment (Drinkwater et al., 2010). Compounding these changes related to the physical environment are changes in ocean chemistry, which affect both calcification and physiology in a variety of marine species (Fbry et al., 2008). Below, we examine some examples of biological changes that have been associated with long-term anthropogenic warming.

Phytoplankton form the basis for the vast majority of marine ecosystems and make up approximately half of the planet's primary productivity (Field et al., 1998). They are sensitive to their physical and chemical environments and so are vulnerable to long-term climate change. They also represent an important feedback onto the climate system, modulating the absorption of energy in the upper ocean and affecting CO_2 sequestration (e.g., Murtugudde et al., 2002; Sabine et al., 2004). Large-scale measurement of both the rate of phytoplankton production and the standing stock are generally achieved through the measurement of chlorophyll concentration. While measurements of chlorophyll have been taken since the start of the twentieth century (Jeffrey et al., 1997), synoptic global-scale measurements have only been available since the mid-1990s (Section 6.2.1).

Based on 9 years of satellite data (1998–2007), Polovina et al. (2008) find a dramatic increase in the area of the productivity minimum zones. These are regions of oceanic downwelling and poor nutrient supply, collocated with the subtropical gyres. They find area increases of between 0.8% and 4.3% per year along with corresponding increases in SST. These changes are consistent with increased stratification and a weakening of the nutrient supply from the deep ocean. Vantrepotte and Melin (2009) also note a significant reduction in productivity within the subtropical gyres, but find both increasing and decreasing trends in other locations. Behrenfeld et al. (2006) also find a strong negative relationship between SST and chlorophyll, but show that the relatively short satellite record is dominated by inter-annual variability, particularly associated with ENSO. Henson et al. (2010) demonstrate that the brevity of the data record precludes unequivocal attribution of satellite-derived change, which would require ~40 years of sustained observations.

The transparency of the water column can also be related to surface chlorophyll concentrations and can be derived from satellite, in situ optical measurements and Secchi disk measurements (Boyce et al., 2010). Secchi disks are simple black-and-white patterned disks that are lowered down through the water column until their pattern becomes indistinct to the surface observer. The associated depth can be converted to a chlorophyll estimate in low-turbidity open-ocean regions. Such measurements have been routinely made since the late nineteenth century. Boyce et al. (2010) have collated available transparency measurements to obtain centennial-scale estimates of primary productivity trends. Based on a set of climatically similar regions, the authors show that in 80% of these regions, productivity shows a long-term decline, with a globally-averaged rate of decline of ~1% per year. They also show a correspondence between productivity and SST, suggesting that stratification associated with global warming is a major driver in these changes (at mid and low latitudes). Considerable controversy surrounds this result however, as biases may be introduced when combining Secchi disk and in situ chlorophyll measurements (Mackas, 2011; Rykaczewski and Dunne, 2011; McQuatters-Gollop et al., 2011; Boyce et al., 2011).

A growing number of studies have found widespread poleward or downward shifts in the distribution of various marine species (Drinkwater et al., 2010, and references therein). Perry et al. (2005), for example, examined a large number of fish species in the North Atlantic and found that half of the species had moved significantly poleward over a 25-year period (independently of changes in fishing pressure). Hiddink and Hofstede (2008) find that this range expansion has led to increased species richness in the North Sea, a consequence of lower latitudes having greater species richness combined with the northward shift in the species distribution.

Another sensitive indicator of climate-related change is phenology — the seasonal timing of important events within a species lifecycle (e.g., migration, reproduction, feeding). However, the numbers of species for which sufficiently long time series are available are limited, with few data outside the Northern Hemisphere (and the North Atlantic in particular: Ji et al., 2010). Investigating a large number of planktonic species in the North Atlantic over the second half of the twentieth century, Edwards and Richardson (2004) find that larval release and development, which is related to ocean temperature, occurred significantly earlier, while the lifecycle of many diatom species, which form the nutritional basis for many of the higher animals and whose lifecycle timing is more closely tied to light availability, has remained relatively stable. This timing mismatch produces a reduction in the efficiency of energy transfer through the food chain and may have significant impact on regional fisheries (Edwards and Richardson, 2004). Focusing on zooplankton in the North Pacific, and species of pelagic fish and seabirds feeding on them, Mackas et al. (2007) also find shifts of up to a few weeks in zooplankton phenology between warm and cold years. They also postulate the strong likelihood of ecosystem disruption, as the phenology changes are species-specific. Temperature-related phenological changes have been observed in a number of other species, although these changes are usually related to inter-annual or regime shift changes (Drinkwater et al., 2010), with insufficient data to attribute changes to long-term anthropogenic warming.

Most of the evidence linking lowered pH with changes to biological systems is based on laboratory-based experiments (e.g., Fabry et al., 2008). However, a more limited number of studies provide direct in situ evidence. For example, Moy et al. (2009) find that planktonic foraminifera in the Southern Ocean have shell weights that are 30%–35% lighter than background Holocene levels, consistent with increased acidification. De'ath et al. (2009) show that a particular group of corals that span the Great Barrier Reef exhibit a 14% reduction in calcification since 1990 that is unprecedented for at least 400 years (based on the available coral proxy record). Again, they find this to be consistent with increased stress from both temperature and declining saturation state.

6.3. PROJECTIONS FOR THE FUTURE

As detailed in the preceding sections, the ocean has changed profoundly over the last few decades. While low-frequency natural variability plays an important role in these changes, human-induced changes are also apparent for many properties.

To obtain quantitative estimates of how the ocean might respond to further greenhouse emissions requires the use of coupled climate models. Most of these models, those that comprise the CMIP3 repository used for informing the IPCC AR4, lean strongly towards the physical climate system. At the core of a climate model lies the numerics to describe the dynamics and thermodynamics of the ocean and atmosphere. Forcing by GHGs in the atmosphere is prescribed and, as such, explicit feedbacks associated with changes in the carbon or other cycles and with ecosystem changes are precluded (Harvey, 2012, this volume). A smaller number of coupled climate–carbon cycle models are also available (comprising the Coupled Climate–Carbon Cycle Model Intercomparison Project C4MIP; Friedlingstein et al., 2006). While incorporating components of the ocean and terrestrial carbon cycles, the physical components of these models tend to be less sophisticated than the CMIP3 models and simulations are generally carried out at coarser resolutions. While carboncycle feedbacks may only represent second-order corrections to future change, this may not be the case. In the C4MIP models, for instance, feedback effects associated with the land and the ocean will increase atmospheric CO_2 concentrations by between 20 ppmv to 200 ppmv (full inter-model spread) by 2100, equivalent to an additional $0.1°C$ to $1.5°C$ warming (Friedlingstein et al., 2006). Such feedbacks increase the likelihood of surprises—of tipping points or abrupt change (Smith et al., 2009; Lenton, 2012, this volume).

The IPCC AR4 (IPCC, 2007a) presents a thorough review of climate model-derived projections. While we cannot update all aspects of oceanic change dealt with in the IPCC report, we will review some of the latest work on a selection of key ocean regions and processes. This will be followed by an examination of some of the potential oceanic feedback mechanisms absent from the CMIP3 models that will be incorporated some of the models employed for simulations assessed in the future by the IPCC.

6.3.1. Tropical Pacific

Given that ENSO has such widespread importance, any changes to the tropical Pacific are of great interest. As such, the assessment of the fidelity of ENSO in coupled climate models and projected changes have received significant attention in the scientific literature (Collins et al., 2010, and references therein). These studies have generally found little agreement across the models with respect to how the strength or frequency of ENSO will evolve in the future (e.g., van Oldenborgh et al., 2005; Guilyardi, 2006). This is because characteristics of ENSO events are associated with a number of processes that compete to either dampen, or enhance, ENSO variability. Yang and Zhang (2008), for example, used model experiments to demonstrate the competing effects of a reduced Walker circulation, which tends to weaken variability, and an increase in the equatorial stratification, which acts to enhance variability. While their model exhibited enhanced ENSO variability as a result of the dominance of the latter process, this does not hold true for all climate models. A number of additional feedback processes that affect ENSO strength are discussed by Collins et al. (2010). Using a long model simulation, Wittenberg et al. (2009) also suggest that a large natural inter-centennial modulation of ENSO could also make the isolation of any anthropogenic signal very hard.

Despite lack of clarity on the future evolution of ENSO, the models do show considerable agreement in their projections of changes to the mean state. Many of the projected oceanic changes stem from the robust simulated slowdown of the equatorial trade winds, which are, in turn, a consequence of the tropospheric warming (Held and Sodden, 2006). The warmer atmosphere is able to hold additional moisture; this moisture is precipitated out as the air rises over the Warm Pool in the western Pacific and in the convergence zones. A result of this, a robust projected western Pacific freshening, occurs across the models

(Ganachaud et al., 2011). The rate of precipitation increase is, however, constrained by the rate at which the additional latent heat released by the condensing water can be radiated away and the rate of moisture increase exceeds the rate of precipitation increase. As a result, the upward transport of humid air must weaken. This drives a general slowdown of the atmospheric overturning circulation, including the surface trade winds (Held and Sodden, 2006; Vecchi et al., 2006). As the upper ocean equatorial circulation is primarily driven by local wind changes, this also causes a slowdown of the complex array of equatorial currents and counter-currents and the tropical overturning circulation (Ganachaud et al., 2011, and references therein). One exception to this is the Equatorial Undercurrent, a fast eastward flowing jet that sits between depths of 200 m and 400 m and plays an important role in transporting dissolved iron to the iron-limited eastern Pacific (Mackey et al., 2002). The undercurrent shows a robust projected increase in strength, driven primarily by an increased western boundary current input in the Southern Hemisphere (Luo et al., 2009; Ganachaud et al., 2011).

Most of the models show a tropical warming pattern that is largest near the equator (Figure 6.7), particularly in the central Pacific. Dinezio et al. (2009) show that the pattern of warming stems from a complex interaction of processes. This includes: (i) reduced poleward divergence

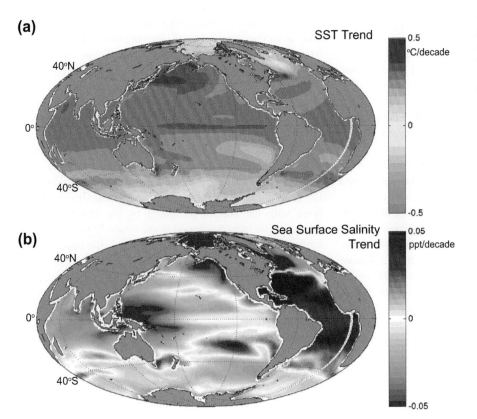

FIGURE 6.7 CMIP3 multimodel mean linear trend (2000–2099) in SST ((a) °C per 100 years) and surface salinity ((b) psu per 100 years) for the 'business as usual' SRES A2 emissions scenario.

of surface waters, which slows the removal of heat; (ii) cloud and evaporative processes in the west having a cooling effect; (iii) reduced westward flow that alters the lateral advection of heat; and (iv) altered vertical heat transport in the east, which causes a relative cooling as a result of increased thermal stratification, despite weakened upwelling. The causes for projected SST pattern formation have been extended to the wider extra-tropics (and other basins) by Xie et al. (2010). They show, for example, that the very weak simulated warming projected in the southeastern extra-tropical Pacific (Figure 6.7) is driven by an increase in the wind strength that allows extra latent heat to escape from the ocean.

Despite the weakening of the trade winds and the associated thermocline changes, it is a misnomer to describe the changes as El Niño-like, as is often done. In particular, the projected changes are not accompanied by a projected decrease in the zonal SST gradient, an important part of the Bjerknes feedback that characterizes ENSO variability. Consequently, the mechanisms driving the projected tropical Pacific changes, which are ultimately tied to long-term warming, are distinct from those that drive inter-annual ENSO variability (Vecchi and Soden, 2007; Collins et al., 2010).

Bell et al. (2011) provide a comprehensive review of the projected physical changes in the tropical Pacific and the effects these changes are likely to imply for biological systems and fisheries.

6.3.2. Southern Ocean

The Southern Ocean is a region where profound changes are occurring rapidly and major additional changes are projected. Many of these changes are intimately tied to possible climate feedbacks associated with the ocean carbon cycle. However, it is also a region where major uncertainties in the science and, therefore, in model projections are known to exist.

Assessment of the CMIP3 models shows qualitatively consistent projections for the ocean, despite sometimes large quantitative inter-model differences (Sen Gupta et al., 2009, and references therein). The changes are driven by a projected intensification of the hydrological cycle, a southward intensification of the westerlies (and the associated wind stress curl), and an increase in the radiative heat flux into the ocean, primarily southward of ~50°S. The shift in the wind drives a southward migration and intensification of the Antarctic Circumpolar Current (ACC), with an associated increase in northward Ekman transport and increased upwelling of circumpolar deep water. The shift in the maximum wind stress curl occurs at lower latitudes over the subtropical gyres, causing them to shift southwards and, in some cases, intensify. The southward shift in the mid- and high-latitude lateral circulation is

manifested as a large increase in the depth-integrated temperature between the southern limbs of the subtropical gyre and the core of the ACC (Sen Gupta et al., 2009). The projected shift in the circulation and enhanced deep mid-latitude warming is consistent with changes already observed over recent decades (see Section 6.2.4). Despite most of the extra heat entering the ocean at high latitudes, this heat is advected northwards and subducted (Cai et al., 2010). Hence, the high-latitude ocean is projected to warm at a substantially lower rate than that further to the north (Figure 6.7). Increases in high-latitude precipitation, together with increased Antarctic runoff, causes a significant freshening of the Southern Ocean (Figure 6.7). This high-latitude warming and freshening and the intensification of the westerlies is consistent with a simulated slow-down of Antarctic Bottom Water formation and subsequent northward export at abyssal depths (Sen Gupta et al., 2009; Saenko et al., 2011). Like Bottom Water, Mode and Intermediate Water formed at mid-latitudes in the Southern Ocean also play an important role in the removal of surface water into the ocean interior (subduction) and, therefore, the sequestration of anthropogenic CO_2. Projections suggest that, as the climate warms, the rate of subduction of intermediate water and denser mode waters will be reduced, while there is greater uncertainty with regard to the lighter Mode Waters (Downes et al., 2010; Saenko et al., 2011). As noted by Saenko et al. (2011), the role of ocean eddies (not included in the climate models) may affect these results.

The simulated wind-driven intensification of the ACC and associated spin-up of the overturning circulation (making up the Deacon circulation) in climate models has been the subject of considerable contention recently. High-resolution ocean models that can simulate small-scale ocean eddies (e.g., Meredith and Hogg, 2006; Screen et al., 2009; Spence et al., 2010) and some empirical evidence (Böning et al., 2008) suggests that the additional wind-derived energy from the strengthened westerlies would energize eddy activity. This would act to oppose the enhanced Ekman transport and so dampen any change in the strength of the circulation on long timescales. A second issue with the picture presented by the climate models relates to the projected trend in the westerly winds. Recent evidence, based on models with more realistic depictions of the stratosphere, suggests that the poleward intensification of the wind field may halt, or even reverse, at least during austral summer, as a result of the recovery of stratospheric ozone over Antarctica (Son et al., 2008). Only a few of the AR4 models incorporate any form of ozone recovery and those that do tend to underestimate the resulting modification in the atmospheric circulation compared to models with more realistic stratospheres (Son et al., 2008).

Each of the Southern Ocean water masses transports significant quantities of heat and dissolved gases into the

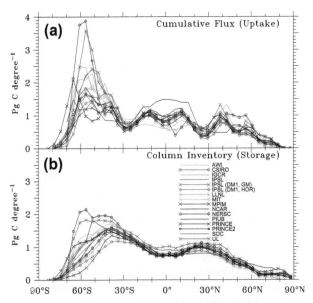

FIGURE 6.8 Latitudinal variations in anthropogenic CO_2 uptake (a), and inventory (b) from 1765 to 1996 for models taking part in the Ocean Carbon Model Intercomparison Project. *(Source: Orr, 2002, their figure 1.18.)*

ocean interior. Independent modelling studies suggest that the Southern Ocean is the most important oceanic region for anthropogenic CO_2 uptake and the region likely to undergo most change due to global warming (Sarmiento and Le Quéré, 1996; Sarmiento et al., 1998; Matear and Hirst, 1999; Caldeira and Duffy, 2000). This conclusion was further supported by results from 13 different OGCMs in the Ocean Carbon Model Intercomparison Project (OCMIP; Orr, 2002). Figure 6.8 shows OCMIP-2 results. The Southern Ocean (south of 40°S) is important in all models, absorbing from 30%–50% of the total global

anthropogenic CO_2 taken up by the ocean. Any changes in Southern Ocean uptake could therefore have a large effect on the global CO_2 budget.

6.3.3. Sea-Level

Based on climate model projections, the IPCC provides a range of values for globally-averaged sea-level rise of 18 cm to 59 cm over the next 100 years (depending on the choice of emissions scenario) with an additional 10 cm–20 cm for rapid dynamical changes in ice flow of the ice sheets (Meehl et al., 2007). There is an important proviso attached to these estimates, however. The CMIP3 models do not represent processes related to ice-sheet dynamics, while recent observational evidence suggests a growing importance for these processes (Rignot, 2006; Allison et al., 2009b). Furthermore, model estimates of sea-level rise tend to underestimate recent observations (Rahmstorf et al., 2007; Allison et al., 2009a).

A number of recent studies have attempted to provide an independent estimate of future sea level based on various semi-empirical techniques that use historical records to calibrate simple statistical models (Rahmstorf, 2007; Horton et al., 2008; Vermeer and Rahmstorf, 2009; Jevrejeva et al., 2010; Grinsted et al., 2009). These studies generally suggest a future sea-level rise with upper bounds that are larger than the IPCC values (for the same amount of globally-averaged warming; see Figure 6.9). One shortcoming of these statistical models is that they only take into account processes that are operating within the calibration period. For example, non-linear changes in ice flow that may occur at higher temperatures will not be accounted for (Rahmstorf, 2010). Using mass budget estimates, Pfeffer et al. (2008) consider the case of a rapid, but physically

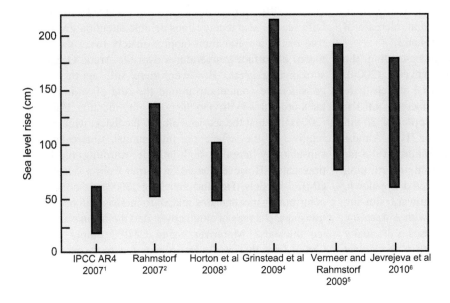

FIGURE 6.9 Estimates of projected sea-level rise from the IPCC (AR4), based on output from the CMIP3 climate models, semi-empirical sea-level models, and mass budget estimates (see text). *(Source: Rahmstorf, 2010, his figure 1; all projections span the B1 to A1FI range of emissions scenarios. Reprinted by permission from Macmillan Publishers Ltd.)*

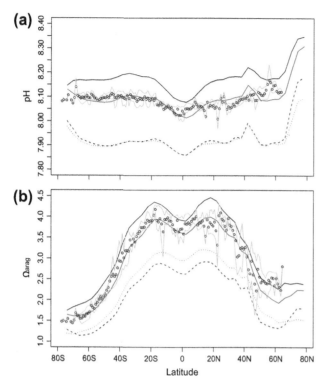

(a)

pH

(b)

Ω_{arag}

80S   60S   40S   20S   0   20N   40N   60N   80N

Latitude

FIGURE 6.10 Zonally-averaged change in surface-ocean CO_2 chemistry from 1880 to 2100: (a) pH and (b) Ω_{arag}. Solid line is for the year 1880, dashed line is for the year 2100 under the control experiment, and the dotted line is for the year 2100 under the climate change experiment. The blue circles show the average calculated values using the global carbon observations (1991–1998) *(Source: Key et al., 2004, using dissociation constants from Dickson and Millero, 1987)*. The grey solid lines around the observations are the 25th and 75th quantiles to illustrate the range of natural variability. The red line is the model predictions for the year 1995 and serves as a direct comparison to the observations. *(Source: Modified from McNeil and Matear, 2007, their figure 1.)*

plausible acceleration of ice-sheet melt to put an upper bound on possible sea-level rise of 2 m. Using more conservative rates of increase they still find an increase of 0.8 m, which is in excess of the IPCC estimates.

Recent research has focused on understanding the regional variations of future sea-level rise. Yin et al. (2009), for example, find that in a selection of CMIP3 models there is a particularly large projected sea-level increase off the northeast coast of North America (of the order of 20 cm above the global mean in the A2 scenario for 2100). Almost all the models project warming and freshening in the far North Atlantic that drives a slowdown in the overturning circulation (Meehl et al., 2007). The projected slowdown, characterized by the weakening of southward flowing North Atlantic Deep Water, is associated with a warming along the flow pathway. This, in turn, causes a thermally driven sea-level rise, which, together with a dynamical adjustment, produces the rapid projected sea-level rise response (Yin et al., 2009).

Timmerman et al. (2010) examine regional changes in the Tropical Pacific. Using a idealised ocean model forced by the wind projections from a selection of CMIP3 models, they were able to attribute most of the spatial variability in the projected sea-level rise pattern to purely wind-driven convergences and divergences in the surface circulation (i.e., Ekman pumping anomalies). These wind-driven adjustments cause regional deviations of up to 30% of the projected global mean sea-level rise (Timmerman et al., 2010).

Gravitational readjustment of the ocean due to changes in the mass of land-locked ice can also significantly affect regional sea level. This is particularly pertinent for the large masses associated with the Antarctic and Greenland ice sheets. The gravitational attraction of these ice sheets tends to draw water towards them. As the ice sheet melts, the attraction is reduced and there is a redistribution in sea level, together with a small adjustment in the Earth's axis of rotation (Mitrovica et al., 2009). For Antarctica, mass-losses drive an absolute sea-level reduction within ~2000 km of the ice sheet, despite an increase in the global average. In addition, regions to the east and west of North America and sites bordering the Indian Ocean experience sea-level rise as much as 30% greater than would be expected from a uniform redistribution of water from the ice-sheet melt (Mitrovica et al., 2009).

6.4. OCEAN BIOGEOCHEMICAL FEEDBACKS

The current generation of climate models include many of the processes required to investigate physical feedbacks within the climate system (see Harvey, 2012, this volume). Primary among these are water vapour feedbacks, which play a dominant role in determining climate sensitivity (see Harvey, 2012, this volume). Ocean feedbacks associated with sea-ice also receive considerable attention given their role in the amplification (approximately twice the global mean) of surface temperatures over the Arctic Ocean and surrounding areas. However, there still appears to be considerable contention around the role of various feedback processes in driving the polar amplification. Hall et al. (2004) find that the sea-ice–albedo feedback, while having relatively little effect on inter-annual timescales, can substantially amplify high-latitude warming on longer timescales. However, based on output from a selection of CMIP3 models, Holland and Bitz (2003) describe other contributing mechanisms related to increases in ocean heat transport, changes in cloud cover, and the mean state of the ice thickness. Moreover, using CMIP3 output, Winton (2006) finds that, while ice–albedo feedbacks play a role, feedbacks associated with changes in cloud cover, long-wave and shortwave radiation are equally important. Based on an analysis of the observed seasonal variability of the polar amplification, Screen and Simmonds (2010) find that ocean heat losses associated with reduced ice cover have historically played a major part in polar temperature amplification in autumn and winter.

Another oceanic feedback process that receives considerable attention relates to the Atlantic overturning circulation and the associated poleward transport of heat. Changes to this circulation driven by density changes in the high-latitude North Atlantic (usually associated with increased freshwater input) have been implicated in rapid climate change in the palaeorecord (e.g., McManus et al., 2004). In the context of climate change over the next century, while most climate models project some slowdown of the overturning, none show a collapse of the circulation (Meehl et al., 2007) and any associated Northern Hemisphere cooling is more than offset by increased radiative forcing.

While representing many of the physical processes operating in the climate system, the CMIP3 models do not explicitly include climate feedbacks that involve changes in the carbon cycle: CO_2 (and other GHG) levels in the atmosphere are prescribed in the models, based on a set of future emissions scenarios (Nakicenovic et al., 2000). These feedbacks may substantially alter the efforts required to reach CO_2 stabilization. For example, based on the results of the C4MIP models, emissions may need to be cut by an additional 30% when trying to reach a 450 ppmv CO_2 stabilization, when the effects of changes in the land and ocean sinks are factored in (Friedlingstein et al., 2006; Friedlingstein, 2008). While the feedbacks associated with terrestrial processes are more important in the C4MIP models (Friedlingstein et al., 2006), large uncertainties in ocean carbon cycle processes still abound.

6.4.1. Solubility Carbon Pump

Given the ability of the ocean to modulate anthropogenic CO_2 in the atmosphere, it is fundamentally important to understand the ocean's stability and the extent to which it can continue acting as a significant sink for anthropogenic CO_2. Climate change itself will impact physical oceanic processes in two main ways that will alter the capacity of the ocean to absorb anthropogenic CO_2. Firstly, business-as-usual emissions scenarios are projected to warm the upper ocean by up to 4°C by the end of the century, which reduces the solubility of CO_2. The warming effect on CO_2 solubility reduces the cumulative absorption by ~10%–14% over the next century (Matear and Hirst, 1999; Plattner et al., 2001; Sarmiento et al., 1998), equivalent to a ~20 ppmv increase in the atmosphere. Secondly, surface warming combined with regional freshening (at high latitudes and in tropical regions) is projected to cause enhanced stratification of the upper ocean, with impacts varying according to region. Stronger density gradients through the water column inhibit vertical mixing, impeding the transfer of CO_2 into the interior of the ocean where it can be sequestered on long timescales. This circulation feedback is projected to also slow the uptake of anthropogenic CO_2 by between 5% and 17% (Matear and Hirst, 1999; Plattner et al., 2001; Sarmiento et al., 1998; Friedlingstein et al., 2006). Both direct effects of ocean warming on solubility and the reduction in the rate of overturning are projected to slow the cumulative uptake of anthropogenic CO_2 by at least 15% by the end of this century, resulting in higher atmospheric CO_2 accumulation (Matear and Hirst, 1999; Plattner et al., 2001; Sarmiento et al., 1998). Despite our understanding of the stability of the oceanic carbon sink, many uncertainties remain, particularly with regard to model uncertainties and regional differences, particularly in the Southern Ocean.

6.4.2. The Biological Pump

The oceanic biological pump, as described earlier (the export of organic carbon from the euphotic zone into the ocean interior: Figure 6.6), is a fundamental process regulating atmospheric CO_2 (Martin, 1990; Volk and Hoffert, 1985). The biological carbon pump has remained close to a steady state in the ocean for at least the last 10,000 years prior to industrialization, since atmospheric CO_2 levels changed little over this period (Petit et al., 1999). However, future non-linear changes to the biological pump (via climate change or ocean acidification) may be associated with a significant future ocean carbon/climate feedback (Riebesell et al., 2009). Higher biological drawdown at the surface would lower CO_2 and allow the ocean to take up additional carbon from the atmosphere (and vice versa for lower biological drawdown). However, the magnitude and extent to which the biological carbon pump will respond to both higher CO_2 and climate change (nutrient supply) is still uncertain, although recent research has brought new insight into this evolving field of research.

6.4.2.1. Traditional View of Biological Carbon Feedback

Nutrients and light are fundamental when considering the biological carbon pump, since photosynthetic growth at the surface is dependent on them. In most of the ocean, the amount of light available to phytoplankton will not be the limiting factor in the future, so most of the traditional views on the biological carbon pump have a nutrient-centred viewpoint. This comes about because, in vast areas of the surface ocean (e.g., the subtropical gyres), nutrient levels are very low, thereby limiting photosynthetic growth. In some regions, however, (such as the Southern Ocean) there are ample macronutrients. Here, light availability and micronutrients (such as iron) also limit growth. Inorganic carbon has never been limiting in the ocean, so the traditional view of biological feedbacks ignores carbon as being important in the alteration of the ocean's biological pump. In this view, the biological carbon pump is a balance between the upward supply of nutrients from the interior of the ocean and the downward export of carbon via POC. This balance can be altered by physical climate change since the upward supply of nutrients to the euphotic zone is likely to be significantly reduced under enhanced upper ocean stratification. Hence, stratification in some regions will reduce the biological carbon export. However, in some regions, where nutrient supply is not limiting, stratification may actually induce higher carbon export (for a finite period of time) (Matear and Hirst, 1999; Plattner et al., 2001; Sarmiento et al., 1998). This increase in carbon export is because warming may directly increase the rate of photosynthetic activity by 1%–8% (Sarmiento et al., 2004) and stratification creates more stable conditions for biological growth and export. Taken together, the traditional view of changes to biological carbon dynamics under global warming is nutrient-centred and will be regionally specific.

6.4.2.2. A Potential New Paradigm for the Biological Pump

A recent mesocosm experiment identified a potentially profound new paradigm for understanding ocean biological carbon feedbacks in a high CO_2 world. It was found that rising CO_2 levels may stimulate the biological carbon pump, thereby providing a negative feedback on future climate change (Riebesell et al., 2007). This study suggested that enhanced biological carbon consumption in a high CO_2 ocean would increase the ratio of carbon to nutrients (nitrogen and phosphorus) for the exported particulate organic matter (POM). This means that carbon would be acting as a limiting nutrient. This has significant ramifications for understanding future carbon feedbacks. Based on a simple box model, Riebesell et al. (2007) estimated that enhanced carbon export of POM with rising CO_2 would increase oceanic uptake of CO_2 from the atmosphere by between 74 PgC and 154 PgC by 2100 under the IS92a scenario from a present-day range of biological carbon export of between 8 PgC and 16 PgC per year, respectively. If correct, this would represent a large negative feedback on future atmospheric CO_2. For example, the positive feedback related to changes in the solubility pump due to ocean warming is ~50 PgC per year over the same period of time (Plattner et al., 2001). The biological carbon feedback could potentially offset these solubility-related changes threefold, thus enhancing the combined CO_2 drawdown by the carbon pumps over time.

Riebesell et al.'s (2007) estimate of the atmospheric drawdown of CO_2 uses a highly simplified box model approach. Many details of their ocean model are also missing, which makes it difficult to confirm their results. Furthermore, the physical resupply of carbon and nutrients from the deep and intermediate ocean to the surface ocean may not be accurately simulated using simple box models, as demonstrated by Oschlies et al. (2008). As with biological carbon pump perturbations using iron fertilization or macronutrient fertilization, the atmospheric uptake efficiency of the biological carbon pump is quite inefficient over large spatial-scales. This inefficiency dampens the impact of a stimulated biological pump associated with high atmospheric CO_2. It is important to emphasize, however, that the Riebesell et al. (2007) mesocosm is still an area of contention, since other mesocosm work on coccolithophore species *Emiliania huxleyi* did not find equivalent results (Iglesias-Rodriguez et al., 2008). Despite this, non-linear biological changes have the potential to dramatically alter the ocean's role in modulating atmospheric CO_2 and will be the subject of many mesocosm and modelling experiments to better constrain the biological carbon pump.

There are other interactions between the marine carbon and nutrient cycles that have the potential to affect CO_2 feedbacks. These are reviewed in Matear et al. (2010). They find that feedbacks associated with the biological pump are an order of magnitude smaller in the ocean than on land on century-scale timescales, although they note that considerable uncertainty remains.

6.4.3. Ocean Acidification Feedbacks

Projections of future decreases in both pH and Ω (the $CaCO_3$ saturation state) due to higher atmospheric CO_2 levels have traditionally been obtained from ocean-only models (Caldeira and Wickett, 2003; Kleypas et al., 1999), although these modelling studies have not considered the effect of climate-change feedbacks on the carbon chemistry of the ocean. Recently, however, Orr et al. (2005) explored the significance that physical climate change (i.e., changes over and above those directly associated with CO_2 change) plays on the extent of ocean acidification. Using three separate climate models, they found that the physical components of climate change have little impact on the projected future decreases of pH, while having a significant impact on Ω. That is, climate change has the potential to alter the surface-ocean carbon characteristics. As discussed earlier, the three main processes that will influence carbon feedbacks are circulation-driven changes; biologically-induced changes (both organic and inorganic); and ocean warming (direct warming effect and CO_2 solubility-driven effect). Recent work has found that pH and Ω_{arag} respond differently to these climate change feedbacks due to non-linearities in ocean CO_2 chemistry (McNeil and Matear, 2007). In simulations with physical climate change suppressed (i.e., with CO_2 increases only), pH is projected to decline by ~0.3 (Figure 6.10). With physical climate change feedbacks also included, there is little additional impact on the projected changes because of non-linearities in seawater CO_2 chemistry (McNeil and Matear, 2007). For Ω_{arag}, with physical climate change suppressed, surface values are projected to decrease substantially across the ocean, although inclusion of climate-change feedbacks buffers the simulated decline by as much as 15% (Figure 6.10). This buffering is due to a temperature-driven chemical change in the surface ocean that causes increased CO_2 outgassing. The buffering of Ω_{arag} is most pronounced in regions that have a higher ability to absorb anthropogenic CO_2, that is, subtropical regions.

6.4.3.1. Calcium Carbonate Ballast Feedback

Chemical changes associated with ocean acidification (lowering pH and Ω) are expected to have the largest impact on the $CaCO_3$ cycle in the ocean. $CaCO_3$ is produced by a range of phytoplankton (e.g., coccolithophores and pteropods), marine organisms (e.g., molluscs, oysters), and coral reefs. Acidification causes a broad-scale reduction in calcification and an increase in the dissolution of the $CaCO_3$ mineral (see Section 6.2.7.4). Klaas and Archer (2002) found that $CaCO_3$ may act as a ballast for sinking POC export. The POC is protected within the sinking $CaCO_3$ shells of microorganisms, thereby transporting organic matter to much greater depths. Reduced calcification or enhanced dissolution of $CaCO_3$ in the surface ocean, via ocean acidification, could therefore inhibit the organic matter flux into the deeper ocean interior, acting as a large positive feedback on atmospheric CO_2 (Heinze, 2004; Riebesell et al., 2009). The importance of this $CaCO_3$ feedback is most likely to be regionally and species-specific. For example, corrosive conditions for aragonite are expected to occur seasonally when atmospheric CO_2 reaches 450 ppmv in the Southern Ocean (McNeil and Matear, 2008). Although this ballasting effect would probably begin in the Southern Ocean, the biological proportion of $CaCO_3$ to POC is very small in comparison to the subtropical oceans, so the effect is likely to be small. The magnitude and spatial effects of this ballasting feedback is therefore poorly understood, as is our understanding of the global feedback related to $CaCO_3$ ballasting changes (Riebesell et al., 2009).

6.4.4. Other Climate Feedbacks

In addition to carbon-cycle–related feedbacks, other chemical cycles may be influenced by, and in turn influence, future climate change. For example, nitrous oxide, another important GHG, is produced in the bacterial breakdown of organic matter with air–sea fluxes constituting an important source to the atmosphere (Suntharalingam and Sarmiento, 2000). Interestingly, model studies suggest that any benefits related to the increased sequestration of CO_2 resulting from iron fertilization initiatives may be offset by increased remineralization and release of nitrous oxide to the atmosphere (Jin and Gruber, 2003). This result could presumably apply to any increase in ocean productivity.

The ocean also releases aerosols, the rate of which is sensitive to the physical and biological environment. Sea-salt aerosol, for example, affects cloud formation and rainfall and is sensitive to factors including wind speed and surface temperature. Recent model experiments indicate that, in the Arctic, as sea-ice extent reduces, there is an increase in sea-salt aerosol flux. This in turn causes an increase in cloud albedo and an associated reduction in incoming radiation (Struthers et al., 2010). As such, the air–salt–sea-ice interaction may constitute a negative feedback to Arctic warming.

Dimethylsulfide (DMS) produced primarily by phytoplankton emissions is a major contributor to atmospheric aerosols (~43% of global sulfate aerosols: Chin and Jacob, 1996). Changes in DMS production are related to multiple physical and biological factors and, as such, both negative and positive climate feedbacks have been suggested, as reviewed in Rice and Henderson-Sellers (2012, this volume). Despite considerable research, the effect of DMS on climate remains uncertain (Ayers and Cainey, 2007; Carslaw et al., 2009).

6.5. OCEANIC VARIABILITY AND CHANGE

6.5.1. Oceans and the Future Climate

The ocean changes on a wide range of timescales, with certain natural climate cycles varying over periods of multiple decades or longer. Nevertheless, we are now able to identify changes to the ocean that are not consistent with natural variability. Figure 6.11 presents a number of robust changes that have been attributed to anthropogenic influences via increases in atmospheric CO_2 and other GHGs (including ozone). Robust physical changes include: (i) a surface- intensified warming across most of the global ocean with an associated increase in stratification and ocean heat content; (ii) broad-scale coherent trends in surface salinity, associated with an intensification of the hydrological cycle; (iii) a dramatic reduction in Arctic sea-ice extent (and probably thickness); and (iv) an acceleration in global sea-level rise, driven by both thermal expansion and land-ice melt. In addition, there is some evidence to suggest modified water mass properties and pathways, an acceleration of ice melt from both Greenland and the Antarctic, together with modified circulation, associated with large-scale wind changes.

FIGURE 6.11 Schematic showing observed changes in physical, chemical, and biological components of the climate system, pertinent to the ocean, and the links between them. Changes that are supported by overwhelming observational and theoretical evidence are shown as darker bubbles. Changes based on less conclusive evidence are shown as lighter bubbles.

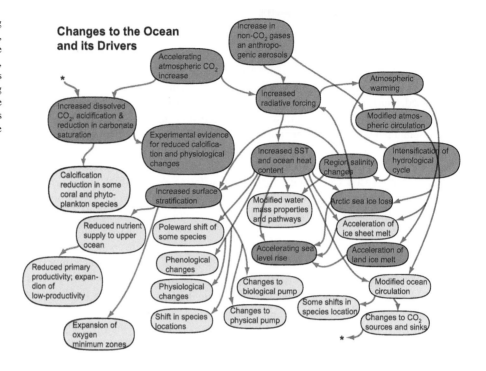

Combined with these physical changes has been a robust increase in dissolved CO_2 concentration in the upper ocean and an associated lowering of pH and $CaCO_3$ saturation. The observational record of chemical and biological change is considerably more limited than for many of the physical variables. For example, worldwide, only a handful of sites have continuous biogeochemical monitoring that starts before 1990 (Chavez et al., 2011). Maintaining and expanding our current observational network, together with biogeochemical process studies at key locations, must be a priority if we are to adequately understand the biological and chemical changes to the ocean and more accurately project the future evolution of our climate system. Nevertheless, evidence is growing for a number of other long-term changes to the ocean. These include: (i) long-term decreases in primary productivity and an expansion of the low-productivity subtropical gyre zones; (ii) reduced calcification in certain phytoplankton and corals (in certain regions); (iii) regional expansion of the oxygen minimum zones; (iv) poleward shifts in the location of some species; and (v) physiological and phenological changes in some species (Figure 6.11).

6.5.2. Future Unknowns

Carbon cycle feedbacks have played an important role in mitigating against an increased greenhouse effect in the past. However, evidence suggests that the efficacy of these feedbacks is changing and may change more dramatically in the future as the climate system is increasingly perturbed (Le Quéré et al., 2007; Friedlingstein, 2008). Marine carbon cycle feedbacks can be examined in terms of various carbon 'pumps'. Physical ocean changes affect the solubility pump and the transfer via ocean overturning, of CO_2 saturated surface water, into the deep ocean. Warming-related responses to the solubility pump stem from a number of factors including the reduced ability for warmer water to absorb CO_2; and increased stratification, which acts to isolate the surface and deep ocean. At the same time, regional wind-driven changes can generate enhanced overturning, increasing the transport of old, natural CO_2 from the deep to the surface ocean. Some evidence suggests that the wind-driven changes appear to be dominating in the Southern Ocean (Le Quéré et al., 2007), although considerable uncertainty exists with regards to future projections. While there is little evidence that the biological pump has changed significantly over recent decades, carbon cycle models suggest that changes to this pump may be more important in the future. Physical and chemical changes to ocean productivity would alter the export of carbon from the surface into the isolated ocean interior. Complexities related to how biological processes respond, especially in light of increased ocean acidification, and changes in nutrient cycles and to the intriguing possibility that CO_2 may actually have a fertilizing effect add additional uncertainty.

Based on the CMIP3 climate models, all of the physical changes (discussed above) are projected to intensify substantially over the coming century. However, we know

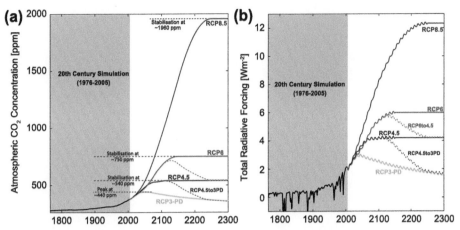

FIGURE 6.12 Global CO_2 concentration (a) and total radiative forcing (b) for the twentieth century, RCP3-PD (peak in radiative forcing at ~3 W m^{-2} before 2100 and decline) (van Vuuren et al., 2007), RCP4.5 (stabilization without overshoot pathway to 4.5 W m^{-2} at stabilization after 2100; Clarke et al., 2007), RCP6 (stabilization without overshoot pathway to 6 W m^{-2} at stabilization after 2100; Hijioka et al., 2008), and RCP8.5 (rising radiative forcing pathway leading to 8.5 W m^{-2} in 2100; Riahi et al., 2007) [RCP = Representative Concentration Pathways.] Two additional mitigation pathways RCP4.5 to RCP3-PD and RCP6 to RCP4.5 are shown, which are not part of the CMIP5 recommendations, but may be run by some modelling groups. *(Source:* http://www.pik-potsdam.de/~mmalte/rcps/.*)*

that many important processes within the models are poorly represented or entirely absent, adding uncertainty to any quantitative assessment of change. In particular, carbon cycle and other biogeochemical feedbacks are excluded from the CMIP3 models. Some of these uncertainties will hopefully be resolved in the next generation of models (taking part in CMIP5) that will provide input to the IPCC AR5.

Although at the time of writing there are few new results available for CMIP5, the new simulations are likely to be a major boost for understanding the future response of the climate system, including carbon cycle feedbacks. While peer-reviewed assessments are not yet available, there are anecdotal reports of significant improvements in the fidelity of many of the models in reproducing the current climate, for example in the representation of ENSO (a major failing in the CMIP3 models; e.g., Collins et al., 2010). The number of climate models will expand from ~24 to ~40 (from CMIP3 to CMIP5[1]) and the experiments being run will diverge considerably from those undertaken as part of CMIP2 and CMIP3. In particular the SRES pathways (Nakicenovic et al., 2000) that form the basis for previous projections have been replaced by a set of Representative Concentration Pathways. These cover both unmitigated and alternative mitigation scenarios (Figure 6.12) and are based on a better understanding of recent environmental changes, economics, and emerging technologies (Moss et al., 2010). In addition to all forcing experiments, some modelling groups are conducting single forcing runs to identify the

impacts of given climate forcings (e.g., CO_2 only, anthropogenic aerosol only). A new core set of experiments will also examine decadal climate forecasts where climate simulations are initialized from observed ocean and ice conditions. The long-term persistence of anomalies within the ocean and ice systems has the potential to provide additional skill when producing decadal projections.

Some of the new models will now also include chemical and biological components, for both the ocean and the land, which will provide new insights into feedback processes absent in the CMIP3 models (CLIVAR, 2009; Taylor et al., 2009). Historical simulations and future projections (based on the high emissions RCP8.5: Figure 6.12) will be supplemented by experiments where: (i) the physical effects of climate change are suppressed (i.e., changes due to CO_2 increase alone are isolated); and (ii) warming occurs in response to increasing CO_2, but the CO_2 change is 'hidden' from the carbon cycle model components (Taylor et al., 2009). These experiments should provide more quantitative information regarding the strength of certain carbon cycle feedbacks and, as a result, provide more credible estimates of future change.

ACKNOWLEDGEMENTS

We would like to thank the four scientific reviewers, Richard Matear, Nathan Bindoff, Martin Heimann, and one anonymous reviewer for their valuable comments.

---

1. WCRP has opted to skip CMIP4 to reduce confusion with C4MIP and to mesh the numbering of the IPCC assessments and its model inter-comparisons.

Chapter 7

Climatic Variability on Decadal to Century Timescales

Mojib Latif and Wonsun Park

Leibniz Institute of Marine Sciences, Kiel University, Kiel, Germany

7.1. INTRODUCTION: OCEANS AND FUTURE CLIMATE

Natural climate fluctuations and anthropogenic climate change go side by side. Surface air temperature (SAT) during the twentieth century (Figure 7.1), for instance, displays a gradual warming of about 0.7°C with superimposed short-term fluctuations on inter-annual to decadal timescales. The long-term upward trend contains the climate response to enhanced atmospheric greenhouse gas levels, in particular those of carbon dioxide (Figure 7.1) that have gradually increased from the pre-industrial value of about 280 ppmv to about 390 ppmv in 2010. The temperature trend presumably also contains a natural component, a warming or cooling, reflecting centennial-scale natural variability. It is commonly believed, however, that the natural contribution to twentieth century global warming is smaller than the anthropogenic contribution (IPCC, 2007a).

Although greenhouse gas concentrations rose monotonically over the twentieth century (Figure 7.1), global SAT obviously did not, exhibiting clear fluctuations from year to year and from decade to decade. The temperature ups and downs around the long-term warming trend largely reflect shorter-term natural variability. The El Niño−Southern Oscillation (ENSO) phenomenon, originating in the equatorial Pacific and affecting sea surface temperature (SST)

there, is one of the major factors affecting inter-annual variability in the tropics and globally. The positive phase of the quasi-periodic cycle is termed El Niño. Such events are typified by strong warming of the order of a few degrees in the central and eastern tropical Pacific Ocean, with cooling over portions of the subtropics and mid-latitudes and the tropical western Pacific. These differences in regional SST,

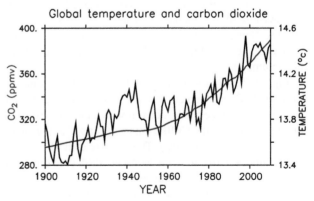

FIGURE 7.1 Globally-averaged annual mean surface air temperature (SAT) (HadCRUT3, blue curve) and the smoothed atmospheric CO_2 concentration (red curve), 1900−2010. *(Source: HadCRUT3 data were obtained from http://www.cru.uea.ac.uk/cru/data/temperature. CO_2 data is a merged product of Law Dome reconstruction and Mauna Loa observation obtained from http://chartsgraphs.wordpress.com/climate-data-links/.)*

The Future of the World's Climate. DOI: 10.1016/B978-0-12-386917-3.00007-5

which last for several months, drive anomalous diabatic heating in the upper atmosphere, thereby perturbing the global atmospheric circulation and generating global tele-connections. As a consequence, some regions become warmer and drier, while other regions cool and become wetter (cf. Alexander and Tebaldi, 2012, this volume). Hurricane activity over the Atlantic is also strongly affected by ENSO through changes in upper-level vertical wind shear. The major El Niño warming of the boreal winter of 1997/1998 helped to make the year 1998 one of the two warmest years on record globally since instrumental measurements began in 1850 (Figure 7.1). The other record year, 2010, was also strongly influenced by an El Niño peaking at the end of 2009.

Natural decadal-scale variability is presumed to be responsible for the observation that the warming trend during the last decade (2000–2009) was considerably less than that in the preceding decades. Thus, natural climate variability introduces an element of irregularity as global warming evolves. Natural climate variations are of two types: internal and external. Internal variability is produced by the climate system itself, due to its chaotic nature, whereas external fluctuations need a forcing, in mathematical terms a change in the boundary conditions (see Harvey, 2012, this volume). Volcanic eruptions and fluctuations in solar output are examples of external drivers of climate (e.g., Belcher and Mander, 2012, this volume; Berger and Yin, 2012, this volume). The eruption of Mount Pinatubo in the Philippines in 1991 caused a relatively short-lived drop in global SAT of about 0.15°C in 1992 (Figure 7.1) while an increase of the solar radiation reaching the Earth may have contributed, together with internal processes, to the mid-century warming (MCW) during 1920–1940 (Figure 7.1), as described by Hegerl et al. (1997). Zhou and Tung (2010) described a consistent response of global SST to solar variations during the last 150 years. The anthropogenic influence on climate through the emissions of greenhouse gases and aerosols and through land-use changes is also considered as external within this framework (e.g., Pitman and de Noblet-Ducoudré, 2012, this volume).

Figure 7.2 in its upper panel repeats the globally-averaged SAT and depicts also the SAT simulated by an ensemble of climate models driven with estimated external forcing taken from the CMIP3 database (*http://www-pcmdi. llnl.gov/ipcc/about_ipcc.php*). The average over all models is sometimes referred to as the 'consensus' (dashed line) and a measure of the externally driven climate change, provided that both the models and the forcing are realistic. The long-term upward trend in the multimodel average is largely consistent with the observed trend and has been attributed primarily to the increase in the concentrations of long-lived atmospheric greenhouse gases (e.g., Hegerl et al., 1997; IPCC, 2007a). The deviation of the observations from the

multimodel mean would be a measure of the internal variability in the data. The MCW and the subsequent cooling, for example, would be mostly due to internal processes within this framework. In contrast, almost all of the warming during the recent decades would be externally driven, since the multimodel mean reproduces the observed SAT well, a picture that has been challenged in some studies speculating about a larger share of internal variability (e.g., Semenov et al., 2010).

Also shown in Figure 7.2 are observed North Atlantic (0–60°N) and eastern tropical Pacific (150–90°W and 20°S–20°N) averaged SST, again together with the corresponding model simulations. Differences between observed and multimodel mean SST are clearly greater at regional-scales, in comparison to globally averaged SAT, possibly indicating a more prominent role of internal variability. North Atlantic (0–60°N) SST exhibited superimposed on a long-term warming trend warming and cooling periods coherent with global SAT. In particular, the northern North Atlantic (40–60°N) SST does not even show any statistically significant warming trend during the twentieth century and is dominated by multidecadal variability (not shown). The eastern Pacific SST also shows decadal-scale changes and some correspondence to global changes, but these changes have a shorter timescale and are less prominent compared to inter-annual variability.

Of direct relevance to society, decadal to inter-decadal fluctuations are also found in atmospheric circulation patterns, precipitation, and climate extremes (lower three panels of Figure 7.2). For example, the North Atlantic Oscillation (NAO: Hurrell, 1995; Hurrell et al., 2003), an oscillation in sea-level pressure (SLP) between Iceland and the subtropical North Atlantic and the most important mode of atmospheric variability in the North Atlantic sector, underwent pronounced decadal-scale variations. These were associated with strong changes in wintertime storminess, as well as European and North American surface temperature and precipitation, and thus had major economic impacts. A long-term trend, however, cannot be identified in the NAO. Large inter-decadal fluctuations were also seen in summertime Sahel rainfall, with profound consequences for people living in the region. For example, the drought of the 1970s to 1980s caused the death of at least 100,000 people and displaced many more. Again, a clear trend during the twentieth century is hard to detect.

The detection of anthropogenic climate change against the natural background variability is obviously a complicated issue (e.g., Rosenzweig et al., 2008). To some extent, we need to 'ignore' the natural fluctuations, if we want to 'see' the human influence on climate during the twentieth century. Of particular importance in this context is the decadal-scale variability. Had forecasters extrapolated the MCW into the future, they would have predicted far more warming than actually occurred. Likewise, the subsequent

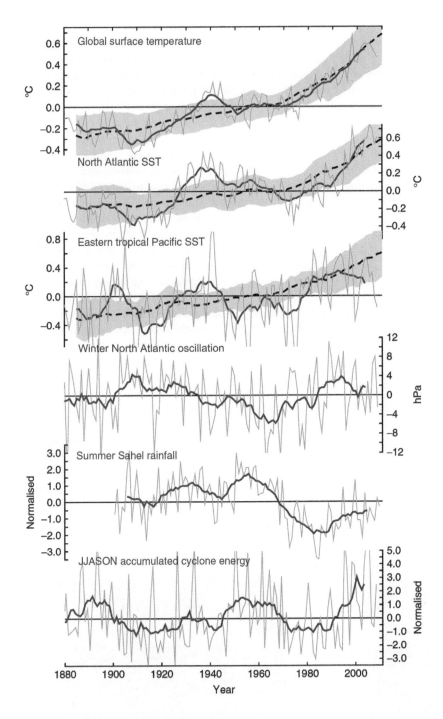

FIGURE 7.2 Observed (blue) indices of climate variability and the ensemble mean of 21 climate model simulations (black dashed) that account only for external factors. Shown are global, North Atlantic (0–60°N), and eastern tropical Pacific (150–90°W and 20°S–20°N) average surface temperature; December to February NAO index; June to October precipitation averaged over the Sahel; and June to November (JJASON) accumulated cyclone energy (ACE) index of Atlantic hurricane activity. Thick (thin) lines show 11-year running (annual/seasonal) means, and grey shading shows the 90% confidence interval computed from model spread. *(Source: Model data are from CMIP3 database used for the Intergovernmental Panel on Climate Change (IPCC) Fourth Assessment Report. From Keenlyside and Ba, 2010.)*

cooling trend during 1950–1970, if used as the basis for a long-range forecast, could have erroneously supported the idea of a rapidly approaching ice age. The simplest approach to filter out the short-term climate fluctuations is that of fitting a smooth function to the data, with or without taking into account any physical reasoning. A physically-based method is to make use of climate models (e.g., Ting et al., 2009) to estimate the external contribution to climate change, as shown in Figure 7.2 (upper three panels).

However, models suffer from biases and there remains some uncertainty when following this approach.

Climate change detection and attribution is about identifying variability and trends in the climate system and ascribing these changes to specific causative factors, whether natural or human-induced. The detection of the anthropogenic climate signal, especially on regional-scales, thus requires at least the analysis of sufficiently long records, because we can be easily fooled by the short-term natural fluctuations. Such long records are generally not available from data, so that the natural variability is often estimated from climate models (but cf. Harrison and Bartlein, 2012, this volume). However, as climate models can suffer from large biases, we need to improve them to better describe the natural climate fluctuations in order to tackle the detection/attribution problem in the best possible manner. This concerns especially the decadal to centennial timescale fluctuations, because these variations evolve on similar timescales to global warming, thereby exacerbating early detection of anthropogenic signals. The ongoing discussion about a potential anthropogenic signal in the Atlantic hurricane activity (Figure 7.2) is an example.

Consequently, natural variability is one important source of uncertainty in climate change projections. Uncertainty in projections for the twenty-first century arises from three distinct sources: natural variability, model uncertainty, and scenario uncertainty. Using projections from a suite of climate models, these sources of uncertainty were separated and quantified by Hawkins and Sutton (2009). Natural variability was defined by the authors only in terms of its internal component, ignoring the unpredictable externally-forced variability over the course of the twenty-first century, which cannot be taken into account in climate change projections. The relative importance of the three sources of uncertainty varies with prediction lead-time and with spatial and temporal averaging scale. Figure 7.3 shows that, for time horizons of many decades or longer and the global-scale, the dominant sources of uncertainty are model uncertainty (blue colour) and scenario uncertainty (green colour). For time horizons of

a few decades and for regions, however, the dominant sources of uncertainty are model uncertainty and internal variability (orange colour). In general, the importance of internal variability increases on shorter time and space-scales (Figures 7.3a and 7.3b), but the next several decades will continue to be strongly influenced by decadal variability. Some part of this may be predictable (e.g., Collins et al., 2006; Latif et al., 2006a), and near-term climate projections could be potentially improved by initializing the models with an estimate of the current climate state.

The physics behind the decadal to centennial variability is not well understood. This is due to the lack of a sufficient database and to model biases. Here, we describe some basic concepts that rely heavily on the findings obtained either from theoretical concepts or simulations with a hierarchy of climate models, but which may shed some light on the dynamics of decadal to centennial variability. Instrumental data cover only about the last 150 years and resolve just a few realizations of decadal variability and hardly any centennial variability, which is much too short to come up with any statistically meaningful findings about this type of long-term natural variability. Nevertheless, the instrumental data have been proven useful in describing some salient features of decadal variability, such as spatial structure. Palaeo data (e.g., from tree rings, corals, or oxygen isotopes from ice cores) are scarce and uncertainty in the reconstructions is large, as discussed below, but the proxies are helpful in describing local variations and, to a lesser extent, Northern Hemisphere averaged SAT changes during the last millennium, and to identify some dominant timescales of natural variability (Harrison and

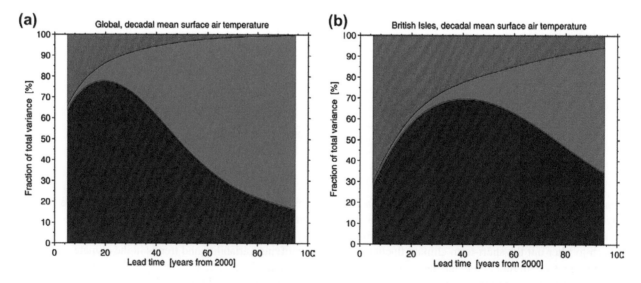

FIGURE 7.3 The relative importance of each source of uncertainty in decadal mean SAT projections is shown as a function of lead-time by the fractional uncertainty for the global (left) and regional (British Isles) mean relative to the warming from the 1971–2000 mean. Green regions represent scenario uncertainty, blue regions represent model uncertainty, and orange regions represent the internal variability component. *(Source: Hawkins and Sutton, 2009. © American Meteorological Society. Reprinted with permission.)*

Bartlein, 2012, this volume). The joint assessment of climate model simulations, instrumental, and palaeo data is one means of making progress in our understanding of decadal to centennial variability, and a few examples of such a strategy are presented below.

This chapter primarily reviews analysis and modelling of oceans. It builds on the context created regarding the importance of oceans in future climates in Sen Gupta and McNeil (2012, this volume). Section 7.2 provides some aspects of tropical Pacific Decadal Variability (PDV), while Section 7.3 describes basin-scale decadal variability in the Pacific and Atlantic sectors. Section 7.4 deals with centennial variability during the last millennium. The null hypothesis for the generation of decadal to centennial variability, the stochastic climate model scenario, is introduced in Section 7.5 by describing some theoretical concepts and extended-range integrations of a variety of climate models that were either explicitly driven with stochastic forcing or contain stochastic forcing as an integral part. This chapter concludes with a summary of the current knowledge on decadal to centennial variability and a discussion of the future needs to obtain a better understanding of decadal to centennial variability (Section 7.6).

7.2. TROPICAL DECADAL VARIABILITY

The tropical climate system features a number of internal climate modes from intra-seasonal to decadal timescales. In the tropics, the 30–60 day oscillation (Madden–Julian Oscillation, MJO: Madden and Julian, 1971) is a prominent intra-seasonal mode of atmospheric variability. ENSO is the most important example on inter-annual timescales

(Figure 7.4a). Obviously, the two modes have rather different timescales and they are based on different physical mechanisms. However, MJO and ENSO do interact with each other. Mode interactions is generally an important mechanism for decadal to centennial variability, not only in the tropics. Furthermore, the modes can be entrained and synchronized by external forcing, such as decadal to centennial variability of the solar radiation reaching the Earth. All this makes it extremely difficult to detect the modes in data and to understand the dynamics and interactions among each other and response to external forcing: the reason for model simulations being heavily used to understand some of these issues.

The decadal variability in the tropical Pacific is seen in the evolution of eastern and western central equatorial Pacific sea surface temperature anomalies (SSTAs) from 1856 onward in terms of the Niño3 and Niño4 index, respectively (Figure 7.4). Historically, El Niño events occur about every 3 to 7 years and alternate with the opposite phase of below-average temperatures in the tropical Pacific which, in analogy to El Niño, are called La Niña events. What is important here is that the nature of ENSO and tropical climate in general (see also Figure 7.2) has considerably varied over time, as exemplified, for example, by the slow increase in eastern and central equatorial Pacific SSTA variance towards the end of the twentieth century (Figure 7.4a).

Many studies have documented decadal and longer-term variability of tropical Pacific climate (e.g., Gu and Philander, 1997; Zhang et al., 1997; Rodgers et al., 2004; McPhaden and Zhang 2002; Lohmann and Latif, 2005; Merryfield and Boer, 2005; Zhang and McPhaden, 2006; Lee and McPhaden, 2008; Meehl et al., 2009). For instance,

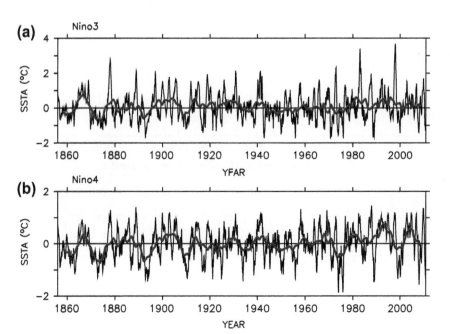

FIGURE 7.4 (a) Niño3 SST anomaly (SSTA) index Jan 1856–Dec 2010 [°C] (150–90°W, 5°S–5°N); (b) Niño4 SST anomaly index Jan 1856–Dec 2010 [°C] (160°E–150°W, 5°S–5°N). The thick red lines denote the 5-year running means and highlight the decadal variability. *(Source: The data are from* http://gcmd.nasa.gov/records/GCMD_indices_nino.html.)

two recent studies (Yeh et al., 2009; Lee and McPhaden, 2010) show that the canonical El Niño has become less frequent and that a different kind of El Niño has become more common during the late twentieth century, in which warm SST anomalies in the central Pacific are flanked on the east and west by cooler SSTAs. This type of El Niño, termed the central Pacific El Niño (CP-El Niño), differs from the canonical eastern Pacific El Niño (EP-El Niño) in both the location of maximum SST anomalies and tropics—mid-latitude teleconnections. It has been speculated that the change in ENSO pattern may be a consequence of global warming. Yeh et al. (2011), by analysing multimillennial control integration with a global climate model (Kiel Climate Model, KCM), have, however, recently shown that such a change in the ENSO pattern can also be internally created, that is, without external forcing.

While ENSO statistics such as spatial expression, SSTA variance, or skewness can vary on decadal timescales due to changes in the background mean state, stochastic forcing, or non-linear processes, or in response to external forcing, ENSO variability itself is also superimposed on lower frequency decadal variability. Lohmann and Latif (2005), for example, have analysed quasi-decadal variability in the equatorial Pacific by means of observations and numerical model simulations. The two leading modes of the observed SST variability in the western equatorial Pacific as defined by the Niño4 index (Figure 7.4b) are a quasi-decadal mode with a period of about 10 years (Figure 7.5a) and the ENSO mode with a dominant period of about 4 years (Figure 7.5b). The SST anomaly pattern of

the quasi-decadal mode is 'ENSO-like' (not shown). The quasi-decadal mode, however, explains most of the variance in the western equatorial Pacific and off the Equator, in contrast to ENSO, which explains most of the variance in the east and at the Equator. The two most recent realizations of the quasi-decadal mode, for instance, can be clearly seen in Figure 7.4b. It was found from forced ocean general circulation model (OGCM) and coupled general circulation model (CGCM) simulations that the variability of the shallow subtropical—tropical overturning cells (STCs) is an important factor in driving the quasi-decadal variability in the tropical Pacific SST (e.g., Merryfield and Boer, 2005), which gives rise to some predictability of equatorial Pacific SST variations, since STC strength anomalies lead those of equatorial SST by one to two years (Lohmann and Latif, 2005).

The ENSO mode itself also exhibits pronounced decadal-scale variability, which is seen as an amplitude modulation (Figure 7.5b). The processes behind this amplitude modulation are still highly controversial. However, no clear trend in the ENSO amplitude is seen in the corresponding reconstruction, so that the interpretation of the observed climate change in the tropical Pacific during recent decades could be as follows (Lohmann and Latif, 2005): the more frequent and stronger El Niño events during the 1980s and 1990s are due to the positive swing in the decadal mode, which simply shifted the 'working point' for ENSO towards the positive side. This means, in particular, that the El Niño of 1997—98 may have become a record event only because there was an upward trend in

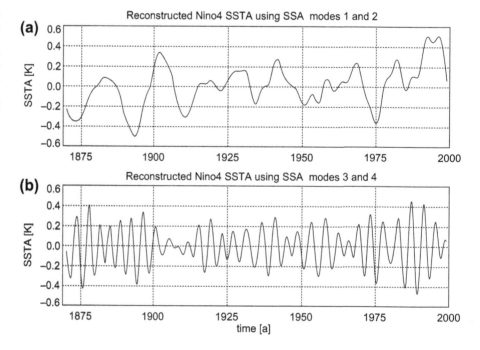

FIGURE 7.5 Reconstruction of the monthly observed Niño4 SST anomalies from an SSA using: (a) modes 1 and 2 (25%) and (b) modes 3 and 4 (20%). Modes 1 and 2 describe the decadal mode, and modes 3 and 4 ENSO. *(Source: Lohmann and Latif, 2005. © American Meteorological Society. Reprinted with permission.)*

the decadal mode and because the decadal mode was additionally in a positive extreme phase. Thus, one can speculate that, while the background conditions on which ENSO operated had changed, ENSO itself did not. The equatorial Pacific SST has the potential to affect climate worldwide through global teleconnections, which makes the decadal variability in the tropical Pacific so important to understand decadal variability on a basin-scale and even globally.

In the tropical Indian Ocean, a subsurface cooling has been reported corresponding to a shoaling[1] of the thermocline and increasing vertical stratification. Most models suggest this trend is likely to be associated with the observed weakening of the Pacific trade winds and transmitted to the Indian Ocean by the Indonesian throughflow. Abram et al. (2008) used a suite of coral oxygen isotope records to reconstruct a basin-wide index of the Indian Ocean Dipole (IOD) behaviour since 1846. The IOD is a pattern of internal variability with anomalously low SST off Sumatra and high SST in the western Indian Ocean, with accompanying wind and precipitation anomalies (Saji et al., 1999). An increase in the frequency and strength of IOD events during the twentieth century was found, associated with enhanced seasonal upwelling in the eastern Indian Ocean. The impacts of this IOD intensification are evidenced by divergent east–west trends in rainfall across the Indian Ocean basin. Through the twentieth century, IOD seasonal rainfall has progressively decreased in western Indonesia and increased in eastern Africa. Cai et al. (2009) show that the increase in the number of positive IOD events during the recent decades, which also potentially accounts for much of the observed austral winter and spring rainfall reduction over southeastern Australia since 1950, is consistent with projected future climate change and, hence, with what is expected from global warming. However, Ihara et al. (2008) found that the period 1950–2004 was characterized by strong and frequent occurrences of positive IOD events in September–November associated with El Niño events.

7.3. DESCRIPTION OF EXTRA-TROPICAL DECADAL VARIABILITY

The extra-tropical decadal-scale variability can be described in different ways. We begin with some aspects of decadal to multidecadal SST variability associated with two phenomena: the Pacific Decadal Oscillation or Variability and the Atlantic Multidecadal Oscillation or Variability. We use the more general terms Pacific Decadal Variability (PDV) and Atlantic Multidecadal Variability (AMV) since instrumental records are not long enough to

determine whether the variability has a well-defined period rather than a simpler character such as red noise. Indices of PDV and AMV are defined by using the Principal Component (PC) of the leading Empirical Orthogonal Function (EOF) of North Pacific SST (PDV) or in terms of area-averaged North Atlantic SST (AMV), respectively (Figures 7.6a, 7.6c: see captions for details). The robustness of the AMV signal has been addressed using palaeoclimate records and similar fluctuations have been documented through the last four centuries (e.g., Delworth and Mann, 2000). Over the northern North Atlantic, the multidecadal SST signal explains about one third of the total variance computed from annual data (e.g., Latif et al., 2004; Knight et al., 2005). Thus, the decadal-scale variability in SST has large amplitude relative to inter-annual variability in this region and is 'potentially predictable' (e.g., Boer and Lambert, 2008). Over the North Pacific, decadal-scale variability is less pronounced relative to inter-annual variability and probably less predictable.

Phase changes of the PDV are associated with pronounced changes in temperature and rainfall patterns across North and South America, Asia and Australia, as well as important ecological consequences, including major shifts in distribution and abundance of zooplankton and important commercial species of fish (Hare and Mantua, 2000). Furthermore, ENSO teleconnections on inter-annual timescales around the Pacific basin are significantly modified by the PDV, as reported by Gershunov and Barnett (1998). They show that the North Pacific inter-decadal variability exerts a modulating effect on ENSO teleconnections over the United States. As a consequence, seasonal climate anomalies over North America, for instance, exhibit rather large variability between years characterized by the same ENSO phase. This lack of consistency reduces potential statistically based ENSO-related climate predictability. These results suggest that confidence in ENSO-based long-range climate forecasts for North America should reflect decadal-scale climatic anomalies in the North Pacific.

We lack a good understanding of whether and how decadal variability in the tropics and mid-latitudes interact in the Pacific sector and how they are related to PDV. Different competing proposals have been made: firstly, that decadal variability in the tropics is internal, that is, a linear stochastic scenario applies (e.g., Eckert and Latif, 1997; Neelin et al., 1998) and/or non-linear processes within the equatorial region (e.g., Jin et al., 1994; Rodgers et al., 2004) produce the decadal variability. Secondly, that tropical decadal variability is driven by processes in the subtropics and mid-latitudes (e.g., Pierce et al., 2000; Merryfield and Boer, 2005; Lohmann and Latif, 2005). Thirdly, that the decadal variability in the North Pacific originates in the North Pacific itself (e.g., Wu et al., 2003; Latif and Barnett, 1994). Fourthly and finally, the decadal-scale variability in

---

1. Becoming less deep.

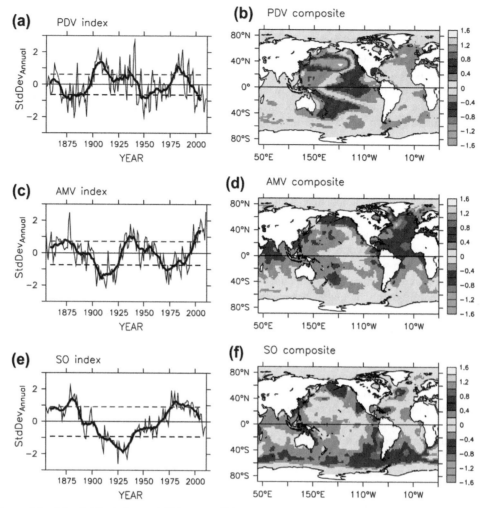

FIGURE 7.6 Observed decadal variability indices and SST patterns: [°C] (a) PDV index as defined by the first EOF of North Pacific (20–60°N) SST, and (b) SST pattern computed as the composite difference between positive and negative phases (the dashed lines denote the thresholds as defined by one standard deviation of the 11-year running mean values); (c) AMV index as defined as linearly detrended North Atlantic (0–60°N) average SST and (d) SST pattern (computed as in b); (e) Southern Ocean SST index (SO index) as defined as the average over the region 50–70°S, and (f) SST pattern (computed as in b). Thick (thin) lines indicate 11-year running (annual) mean. Note that the scale of the panels on the left is given in terms of the standard deviation of the annual mean time series.

the Pacific basin constitutes a coupled air–sea mode, involving the tropics and the extra-tropics, a fast bridge through the atmosphere, and a slow bridge through the ocean, which sets the timescale (e.g., Gu and Philander, 1997). Which of these scenarios applies is a topic of ongoing research. The tropics, however, are likely to play a role in producing decadal variability of basin-scale through global atmospheric teleconnections, which can synchronize variability in the two hemispheres (see also Figure 7.14).

The slow changes in the Atlantic SSTs as expressed by the AMO index (Figure 7.6c) have affected regional climate trends over parts of North America and Europe (e.g., Sutton and Hodson, 2005), Northern Hemisphere temperature anomalies (e.g., Zhang et al., 2007), Arctic

sea-ice (e.g., Venegas and Mysak, 2000; Deser et al., 2000), Arctic Ocean conditions (Bengtsson et al., 2004), Atlantic hurricane activity (e.g., Goldenberg et al., 2001; Zhang and Delworth, 2006), and fisheries production in the northern North Atlantic and North Sea (e.g., Drinkwater, 2006; Beaugrand, 2004). In addition, tropical Atlantic SST anomalies associated with AMV have contributed to shifts in the Intertropical Convergence Zone (ITCZ) and to rainfall anomalies over the Caribbean and the Nordeste region of Brazil, and severe multiyear droughts over parts of Africa, including the Sahel (see Figure 7.2) (e.g., Folland et al., 1986; Bader and Latif, 2003; Giannini et al., 2003; Lu and Delworth, 2005). Tropical Atlantic SST variations are also a factor in producing drought conditions over portions of North America, although tropical Pacific SST variations appear to play a more dominant role (e.g., Schubert et al., 2004; Seager et al., 2005, 2007).

Bjerknes (1964) concluded from his early analysis of the observations in the mid-latitudinal Atlantic region that the atmosphere drives the ocean at inter-annual timescales, while, at the decadal to multidecadal timescales, it is the ocean dynamics that produce long-term variability in the ocean that may feed back onto the atmosphere. Numerous modelling studies, uncoupled and coupled, have examined potential links between the AMV and the Atlantic Meridional Overturning Circulation (AMOC: see also Section 7.5), but the nature of the observed relationship is unclear owing to a lack of long-term, continuous records of the AMOC. There is clear evidence, however, of decadal variability in the heat and freshwater content of the Atlantic Ocean (Lozier et al., 2008; Curry and Mauritzen, 2005), as well as evidence of ocean circulation changes in recent decades, which have been inferred from observations of sea surface height (SSH) and SST (e.g., Häkkinen and Rhines, 2004; Latif et al., 2006b). Some of the decadal ocean circulation changes can be traced back to decadal variations of the NAO using ocean models (Section 7.5.5). All this suggests that ocean dynamical processes play a prominent role in driving decadal SST variability in the Atlantic and we shall revisit this issue later in the chapter when addressing the decadal to centennial variability simulated by climate models.

It is commonly assumed that both PDV and AMV largely reflect internal modes of climate variability, although external forcing may have contributed to the SST changes. For example, anthropogenic aerosol forcing has been suggested to explain part of the cooling in the North Atlantic following the MCW during 1950–1970 (e.g., Stott et al., 2000; Rotstayn and Lohmann, 2002). The internal nature of both PDV and AMV, however, is supported by climate models that simulate similar decadal to inter-decadal variability independently of external forcing. AMV is characterized by basin-wide fluctuations in North Atlantic SST that exhibited a 60-year to 80-year periodicity during the instrumental record (Figures 7.2 and 7.6c). PDV has both a strong bidecadal and multidecadal component (Figure 7.6a) with a period of the latter slightly less than that of AMV. PDV exhibits a V-shaped pattern of SST anomalies in the tropical Pacific and opposite-signed anomalies in the western extra-tropical North and South Pacific. AMV is characterized by monopolar SST anomalies in the North Atlantic and anomalies of opposite sign in the South Atlantic. This characteristic SST anomaly pattern in the Atlantic was originally derived from historical SST data by Folland et al. (1986) by means of EOF analysis. It is commonly referred to as the inter-hemispheric SST contrast. Whether the PDV and the AMV are related cannot yet be answered. On the one hand, it has been shown by d'Orgeville and Peltier (2007) that PDV and AMV are

lag-correlated during the instrumental record. On the other hand, Park and Latif (2010) have recently reported, by analyzing a millennial CGCM control integration, that PDV and AMV can be regarded to first order as independent phenomena in their model.

The SST anomaly patterns (Figures 7.6b, 7.6d) associated to the PDV and AMV index are subject to large uncertainties if derived from data, given the very few realizations of decadal-scale variability during the instrumental record. When computing the deviation of SAT observed during the decade 1999–2008 from the SAT expected from the long-term linear trend during the period 1950–2008 (Figure 7.7), an interesting pattern emerges in the Pacific and Atlantic. The pattern is consistent with the contemporaneous occurrence of PDV and AMV and shows similarities with the PDV and AMV patterns simulated by many climate models. The multidecadal SST fluctuations in the North Atlantic for instance, are anti-correlated with those in the South Atlantic in climate models (see, for example, Park and Latif, 2008; also Section 7.5.8). This suggests that the current climate is still significantly influenced by the natural modes of the climate system, even on decadal timescales.

We performed an analysis of Southern Ocean SST variability consistent with that of PDV and AMV. It should be mentioned, however, that Southern Ocean SST variability is not independent of PDV and AMV. For example, decadal to centennial AMOC variations are associated with opposite SST changes in the North and South Atlantic in climate models. An index was defined as the SST averaged over the region 50°–70°S (Figure 7.6e) and this SO index exhibits a rather strong centennial variability with a maximum around 1880, a minimum around 1930, and another maximum around 1980. The SST has considerably cooled thereafter. The time evolution of the SO index during the twentieth century projects on that of globally averaged SAT (Figure 7.1), exhibiting a clear warming trend. There are, however, some noticeable differences. The cooling tendency during the recent decades is one example. In fact, the SO index during the last years has even been below normal, in clear contrast to globally-averaged SAT.

The associated spatial pattern (Figure 7.6f) projects (as expected from the SO index) on to the pattern of global warming, but also depicts interesting differences. Firstly, there is a strong north–south asymmetry, with more warming in the Southern Hemisphere. Secondly, the strongest signals are seen off the coast of South America and in the mid-latitudinal southern Atlantic. Thirdly, both the Kuroshio and Gulfstream/North Atlantic Current Regions are characterized by negative anomalies. These are preliminary results whose robustness is yet to be tested. Statistical significance cannot be assigned to the results, since the SST data cover only

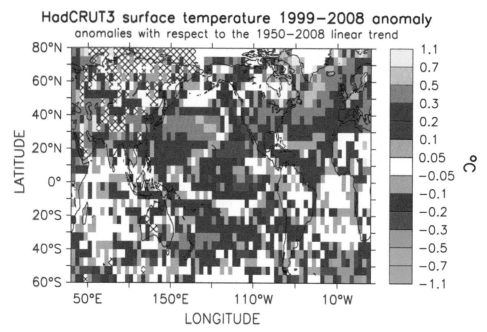

HadCRUT3 surface temperature 1999–2008 anomaly
anomalies with respect to the 1950–2008 linear trend

FIGURE 7.7 Decadal SAT anomalies (°C) 1999–2008 relative to the local long-term linear trend 1950–2008. The figure shows regions that warmed considerably more (red) or less (blue) compared to the 'normal' linear SAT trend. The signatures of PDV and AMV are obvious. *(Source: Courtesy of N. S. Keenlyside.)*

one realization of the centennial variability. Yet these results are interesting enough to warrant further observational analysis and model studies. Furthermore, the SST anomaly pattern associated with the SO index is strikingly similar to that simulated in the KCM at centennial timescales, as described in Section 7.5.8 and shown in Figure 7.18a.

The North Pacific and North Atlantic sector SLP variability is associated with the major pressure systems in these two regions, the Aleutian Low, and the Icelandic Low and Azores High, respectively (see Sen Gupta and McNeil, 2012, this volume). However, it is unclear how the above-described changes in the SST of the two basins relate to SLP. An EOF analysis was performed to derive the leading mode of winter SLP (DJF) variability over the North Pacific (Figure 7.8). It describes the variability associated with the Aleutian low-pressure system. Decadal to multidecadal variability in the atmospheric circulation in the North Pacific sector can be also inferred from area-averaged wintertime SLP as expressed by the North Pacific Index (NPI), which is highly correlated with PC1 (Figure 7.8b). The decadal-scale variability in the North Pacific is more prominent than in the tropical Pacific, and fluctuations in the strength of the wintertime Aleutian low-pressure system co-vary with North and tropical Pacific SST (e.g., Trenberth and Hurrell, 1994; Mantua et al., 1997). For the North Pacific, a clear impact of equatorial Pacific SST at inter-annual timescales during ENSO extremes has been reported such that anomalously warm (cold) SST in the

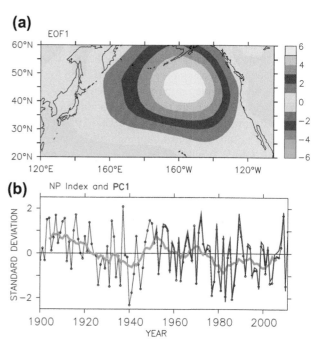

FIGURE 7.8 (a) The first EOF of winter (DJF: December, January, February) SLP [hPa] of NCEP data over the North Pacific sector (20–60°N, 120–260°E) for the period of 1951–2010. The first mode accounts for 48% of DJF SLP variability. (b) Corresponding PC1 (red line) and the NP Index (NPI, black line) together with its 11-year running mean (green line), as defined by Trenberth and Hurrell (1994). The NPI is the area-weighted sea-level pressure (SLP) over the region 30–65°N, 160°E–140W and from December through February. *(Source: Monthly NPI data are from http://www.cgd.ucar.edu/cas/jhurrell/indices.data.html.)*

equatorial Pacific drives an anomalously strong (weak) Aleutian Low, which, in turn, forces anomalously cold (warm) SST in the North Pacific through anomalous surface turbulent heat fluxes. We can expect that this link carries over to decadal or longer timescales. Statistically significant correlations of North Pacific decadal SLP changes are also found with the South Pacific SLP and SST, so that decadal to multidecadal variability is clearly on a basin-scale in the Pacific.

The leading mode of internal wintertime atmospheric variability in the North Atlantic sector is the NAO. Its spatial pattern, as given by the leading EOF of observed wintertime SLP, is characterized by an SLP dipole, with centres near Iceland and the Azores (Figure 7.9a). The NAO index, as defined by the principal component of the leading EOF (Figure 7.9b), is characterized by strong inter-annual variability; but, like the NPI, it also exhibits considerable decadal variability. For instance, there has been a clear decadal upward trend in the NAO from the 1960s to the 1990s, which had many impacts in and around the North Atlantic. Swings in the NAO typically produce changes in wind speed and direction over the Atlantic that significantly alter the transport of heat and moisture. During positive NAO index winters, enhanced westerly flow across the North Atlantic moves relatively warm and

moist maritime air over much of northern Europe and far downstream across Asia, while stronger northerlies carry cold air southward and decrease land and SSTs over the northwest Atlantic. Temperature variations over North Africa and the Middle East (cooling), as well as south-eastern North America (warming), associated with the stronger clockwise flow around the subtropical Atlantic high-pressure centre are also notable. Finally, the long-term changes in atmospheric circulation associated with the NAO index have contributed substantially to the winter warming of the Northern Hemisphere during recent decades.

There is ample evidence that most of the atmospheric circulation variability in the form of the NAO arises from the internal, non-linear dynamics of the extra-tropical atmosphere. Consistent with this picture, the spectrum of the NAO index is almost white, with only a modest redness. Furthermore, decadal NAO trends observed during the twentieth century are largely consistent with those inter-nally simulated by climate models in control integrations without external forcing (e.g., Semenov et al., 2008). However, it is still highly controversial what the role of external or boundary (e.g., SST) forcing in driving low-frequency NAO changes is. Recent research has also sug-gested some role of stratospheric processes in influencing the low-level atmospheric circulation, specifically the NAO (e.g., Scaife et al., 2005). Other studies suggest a feedback by stratospheric ozone as a potential amplifier of the NAO response to solar forcing (e.g., Shindell et al., 2001; Kuroda et al., 2008).

In the Southern Hemisphere decadal variability has not been studied as extensively as over the Northern Hemi-sphere, where the NAO strongly projects on a pattern that is referred to as the Northern Hemisphere annular mode. The annular modes are zonally symmetric hemispheric-scale patterns of internal atmospheric variability (Thompson and Wallace, 2000). There are two annular modes: the northern annular mode (NAM) and the southern annular mode (SAM). Both annular modes explain more of the week-to-week, month-to-month, and year-to-year variance in the extra-tropical atmospheric flow than any other atmospheric phenomenon. For example, both NAM and SAM explain on the order of ~20%–30% of the total variance in the geopotential height and wind fields of their respective hemispheres, depending on the level and timescale considered. The annular modes are characterized by north–south shifts in atmospheric mass between the polar regions and the mid-latitudes. In the wind field, the annular modes describe north–south oscillations in the extra-tropical zonal wind with centres of action located ~55°–60° and ~30°–35° latitude. By convention, the high index polarity of the annular modes is defined as lower than normal pressures over the polar regions and westerly wind anomalies along ~55°–60° latitude.

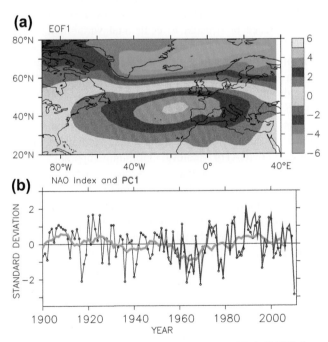

FIGURE 7.9 (a) The first EOF of winter (DJF) SLP [hPa] of NCEP data over the North Atlantic sector (20–80°N, 90°W–40°E) for the period of 1951–2010. [The first mode accounts for 48% of DJF SLP variability.] (b) Corresponding PC1 (red line) and the monthly PC-based NAO index during the twentieth century (black line) together with its 11-year running mean (green line). *(Source: The twentieth century data are from http://www.cgd.ucar.edu/cas/jhurrell/indices.data.html.)*

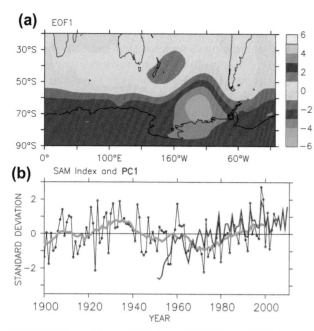

(a) EOF1

(b) SAM Index and **PC1**

FIGURE 7.10 (a) The first EOF of annual SLP [hPa] of NCEP data over the Southern Hemisphere (20−80°S) for the period of 1951−2010. [The first mode accounts for 35% of annual SLP variability.] (b) Corresponding PC1 (red line) and the annual SAM index from Visbeck (2009) (black line) together with its 11-year running mean (green line). *(Source: The SAM index is from* http://www.ifm-geomar.de/~sam.*)*

The variability in the Southern Hemisphere has been studied a great deal in terms of the SAM. It is shown in Figure 7.10 by means of the leading SLP EOF from NCEP/NCAR re-analysis using annual data. Due to the southward shift of the storm track, a high SAM index is associated with anomalously dry conditions over southern South America, New Zealand, and Tasmania and wet conditions over much of mainland Australia and South Africa. The stronger westerlies above the Southern Ocean also increase the insulation of Antarctica. As a result, there is less heat exchange between the tropics and the poles, leading to a cooling of Antarctica and the surrounding seas. However, in this situation the Antarctic Peninsula warms due to a western wind anomaly bringing maritime air on to the Peninsula. Indeed, the ocean surrounding the Antarctic Peninsula is, in general, warmer than the Peninsula itself and stronger westerly winds mean more heat transport on to the Peninsula.

Atmospheric pressure observations have been used (by Visbeck, 2009) to estimate monthly and annually-averaged indexes of the SAM back to 1884. The annual SAM-index exhibits pronounced decadal-scale changes (Figure 7.10b), which are likely to have affected the large-scale Southern Hemisphere ocean circulation. However, it remains unclear how exactly the Southern Ocean circulation responds to atmospheric changes. Several papers have specifically investigated the decadal changes in the Indian Ocean.

Alory et al. (2007), for instance, estimated the linear trends in ocean temperature from 1960 to 1999. Widespread surface warming has been found. The warming is particularly large in the subtropics and extends down to 800 m around 40°−50°S. Models suggest the deep-reaching subtropical warming is related to a southward shift of the subtropical gyre driven by a strengthening of the westerly winds associated with the upward trend in the SAM index (Figure 7.10b).

7.4. EVIDENCE OF CENTENNIAL VARIABILITY

We have described inter-annual and longer-term oceanic and atmospheric modes of variability that are also relevant to the understanding of centennial-scale variability. The climate during the last millennium has seen prominent century-long changes. Figure 7.11, taken from the summary paper of Mann et al. (2009), displays the Northern Hemisphere averaged SAT, together with averages for specific regions during the last millennium. We note that the regions chosen to describe the different modes are not exactly those used here. Multidecadal to centennial variability is obvious in the reconstructions. It should be mentioned, however, that there exists a large uncertainty concerning the phase and amplitude of the variability during the last millennium. Anecdotal evidence confirms the existence of decadal to centennial variability. Norse seafaring and colonization around the North Atlantic at the end of the ninth century clearly indicates that the regional North Atlantic climate was warmer during medieval times than during the cooler Little Ice Age (LIA) of the fifteenth to nineteenth centuries.

As palaeoclimatic records have become more numerous, it has been understood that Medieval Climate Anomaly (MCA) or Medieval Optimum temperatures were warmer over the Northern Hemisphere than during the subsequent LIA and also comparable to temperatures during the early twentieth century (Figure 7.11). However, uncertainty is large, as shown by the shading in the figure and illustrated by the discussion about the so-called 'hockey stick' (e.g., Kerr, 2006), the term often used in the public discourse to describe the first Northern Hemisphere surface temperature reconstruction by Mann et al. (1999). The various studies reconstructing Northern Hemisphere averaged SAT differ in methodology, and in the underlying proxy data utilized, but most of them reconstruct the same basic pattern of cool LIA, warmer medieval period, and still warmer late twentieth and twenty-first century temperatures (IPCC, 2001a, 2007a). Thus, it appears that the late twentieth and early twenty-first centuries are likely to have been the warmest period that the Northern Hemisphere has seen during the last millennium.

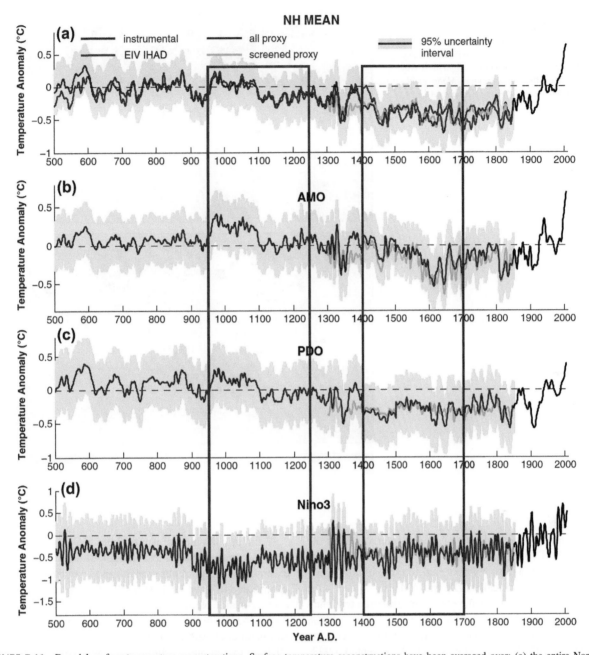

FIGURE 7.11 Decadal surface temperature reconstructions. Surface temperature reconstructions have been averaged over: (a) the entire Northern Hemisphere (NH), (b) AMV region, (c) PDV region, and (d) Niño3 region. Shading indicates 95% confidence intervals. The intervals best defining the MCA and LIA based on the Northern Hemisphere mean (NH MEAN) series are shown by red and blue boxes, respectively. For comparison, results are also shown for parallel ('screened') reconstructions that are based on a subset of the proxy data that pass screening for a local temperature signal. The 'Northern Hemisphere mean Errors in Variables (EIV) reconstruction is also shown for comparison. Note that the definition of the boxes is not identical to that used in this chapter. *(Source: Mann et al., 2009. Reprinted with permission from AAAS.)*

The regional patterns of MCA and LIA (Figures 7.12a, 7.12b) and the magnitude of the changes remain an area of active research because the data become sparse going back in time prior to the last four centuries. The temperature pattern during MCA (Figure 7.12a) supports the notion of a regionally confined strong warming that was mostly restricted to the North Atlantic (Mann et al., 2009). The

South Atlantic has been anomalously cold according to the Mann et al. (2009) reconstruction, which, together with the North Atlantic warming, is reminiscent of a positive AMV phase. Large areas over the Northern Hemisphere (e.g., Eurasia) were anomalously cold and a PDV-type pattern can be seen in the reconstruction of the MCA temperature pattern. Overall, there is some similarity to the

FIGURE 7.12 Reconstructed surface temperature pattern for MCA (950 to 1250) and LIA (1400 to 1700). Shown are the mean surface temperature anomalies. They are defined relative to the 1961–1990 reference period mean. Statistical skill is indicated by hatching. *(Source: Mann et al., 1999. Reprinted with permission from AAAS.)*

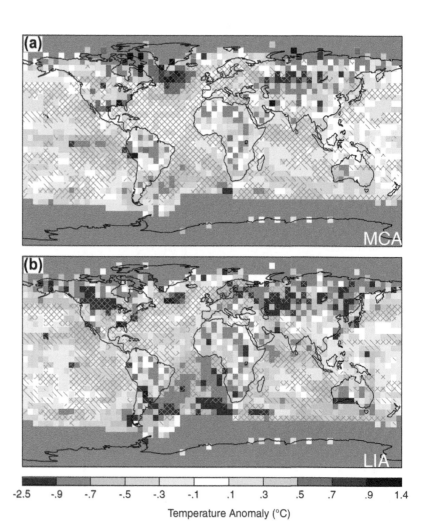

Temperature Anomaly (°C)

SAT pattern shown in Figure 7.7 that was derived from the last decade. This supports a mechanism for MCA that takes into account, in addition to external forcing, internal variability modes: specifically PDV and AMV.

Mann et al. (2009), on the other hand, report a tendency of more frequent La Niña events during MCA (Figure 7.12a). This, however, may simply reflect the involvement of PDV. Trouet et al. (2009) presented a 947-year multidecadal NAO reconstruction and found a persistent positive NAO phase during the MCA. Supplementary reconstructions based on climate model results and proxy data indicate a clear shift to weaker NAO conditions into the LIA. Trouet et al. (2009) conclude that this NAO shift is one aspect of a global MCA–LIA climate transition that probably was linked to prevailing La Niña-like (or PDV) conditions and stronger AMOC during the MCA. They argue that the relaxation from this particular ocean–atmosphere state into the LIA has been globally contemporaneous, suggesting a notable and persistent reorganisation of large-scale oceanic and atmospheric circulation patterns.

Although the mechanism for both MCA and LIA is still highly controversial, most model studies suggest that external forcing, both solar and volcanic, strongly contributed to drive, at least partly, the two most important century-scale Northern Hemisphere climate anomalies during the last millennium (e.g., Shindell et al., 2001, 2003; Ammann et al., 2007; Jungclaus et al., 2010; Ottera et al., 2010; Harrison and Bartlein, 2012, this volume). Shindell et al. (2001) proposed a feedback by altered stratospheric ozone levels in response to changes in solar radiation, specifically the ultraviolet (UV). They were able to reproduce in a model study the unusually cold temperatures over Europe and anomalously warm temperatures over the northern North Atlantic during the Maunder Minimum (Figures 7.11, 7.12b), a period between 1645 and 1715 when sunspots became exceedingly rare, as noted by solar observers of the time. Estimated changes in UV radiation and ozone appear to have shifted the NAO into an extended negative phase, which is consistent with the study of Trouet et al. (2009). The modelled changes in ozone were the key in producing the enhanced response of the NAO that eventually forced the strong changes in the surface climate. Ocean circulation feedbacks, however, were not considered in the model. A prolonged negative NAO phase, for instance, would tend to slow down the AMOC, which would cool the North Atlantic and weaken the positive SAT anomaly simulated in this region.

Zanchettin et al. (2010) recently investigated the long-term relative importance of internally and externally driven variability in a climate model and their possible interferences on global SAT. Multidecadal SAT fluctuations were found to be mostly associated with internal variability. Externally-forced perturbations acted predominantly on centennial timescales tending to overwhelm and distort the internal multidecadal variability. Externally-forced perturbations also tended to correspond to major changes in the coherence among internal climate processes, and between them and global SAT. However, reconstructions of the individual external forcing agents exhibit large uncertainty (see, for example, the recent discussion in Jungclaus et al., 2010), and the results of a climate model driven with different forcing data differ largely between each other (e.g., Ammann et al., 2007).

The above discussion demonstrates that an understanding of century-scale climate changes during the last millennium cannot be obtained without an understanding of the internal modes of the climate system and their response to external forcing. For instance, both MCA and LIA exhibit temperature changes in the Atlantic that project well on to AMV, indirectly implying a role of the AMOC in shaping the regional patterns in the Atlantic sector. Furthermore, MCA shows a signature in the Pacific that is symmetric about the Equator and consistent with PDV. Changes in ENSO and NAO have been also discussed in this context. We return to this point in the Section 7.5.8 where we discuss the internally generated and forced centennial variability in one particular climate model to understand some aspects of the interaction between external forcing and internal climate modes.

7.5. THE STOCHASTIC CLIMATE MODEL: THE NULL HYPOTHESIS FOR CLIMATE VARIABILITY

The climate system displays variability over a broad range of timescales, from monthly to millennial and longer timescales. It is impossible to describe the full range of climate variability deterministically with one model, since the governing mathematical equations are complex and analytical solutions are not known. A numerical solution of the complete set of equations is possible, but not feasible for very long timescales of many millennia, because the necessary computer resources are not available and will not become available in the near future. What is therefore

needed is a hierarchical approach: the application of complex models for short timescales and reduced models for long timescales. However, the hierarchical approach is intellectually challenging and by no means entirely satisfying, as the omission or simplified representation of important physics, such as small-scale/high-frequency processes, is not justified in many cases given the highly non-linear dynamics of the climate system.

The climate system is comprised of components with very different internal timescales. Weather phenomena, for instance, have typical lifetimes of hours or days, while the deep ocean needs many centuries to adjust to changes in surface boundary conditions. Hasselmann (1976) introduced an approach to modelling the effect of the fast variables on the slow by analogy to Brownian motion. He suggested treating the former not as deterministic variables but as stochastic variables, so that the slow variables evolve following dynamical equations with stochastic forcing. The chaotic components of the system often have well-defined statistical properties and these can be built into approximate stochastic representations of the high-frequency variability. The resulting models for the slow variables are referred to collectively as stochastic climate models, although the precise timescale considered to be 'slow' may vary greatly from model to model. We describe in the following some types of stochastic models that may help to explain some aspects of the internally-driven decadal to centennial variability. A summary of the models is given schematically in Figure 7.13.

We consider in this section only the coupled ocean—sea-ice—atmosphere system, thereby neglecting all other potentially important feedbacks (e.g., vegetation and ozone feedbacks). The atmosphere is regarded as the fast component and the ocean—sea-ice system as the slow component. In the simplest case, the atmosphere is treated as a white-noise process, that is, the spectrum of the atmospheric forcing, such as the air—sea heat flux, is white, which means that its amplitude is frequency-independent (see, for example, figure 14.7 in Harrison and Bartlein, 2012, this

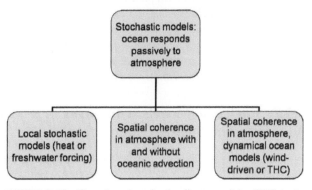

FIGURE 7.13 Hierarchy of stochastic climate models. (THC is the thermohaline circulation.) *(Source: Reprinted from Latif and Keenlyside, 2011; with permission from Elsevier.)*

volume). This is justified to some extent, because typical spectra obtained from atmospheric quantities in the mid-latitudes (e.g., the NAO index) indeed exhibit a 'white' character. Internal atmospheric decadal variability is implicitly included in the white-noise representation.

7.5.1. The Zero-Order Stochastic Climate Model

We now assume the simplest case in which linear dynamics govern the slow system and a local model in which the atmospheric forcing at one location drives only changes in the ocean—sea-ice system at this very point; neither the atmosphere nor the ocean—sea-ice system exhibit spatial coherence. The ocean—sea-ice system defined in this way integrates the weather noise, and the resulting spectrum of a typical variable, say SST, is red, which means that the power increases with a timescale corresponding to the inverse of the square root of frequency. To avoid a singularity at zero frequency, a damping was introduced by Hasselmann (1976).

Frankignoul and Hasselmann (1977) have shown that observed SST variability is consistent with such a local model in parts of the mid-latitudes, away from coasts and fronts, whereas the simple stochastic model fails in regions where meso-scale eddies or advection are important. Hall and Manabe (1997) analysed data from four mid-latitude ocean weather stations and explained differences in SST and sea surface salinity (SSS) spectra by the simple model and report that a complex climate model does reproduce this behaviour. We note again that the zero-order model is linear and no coupling of different timescales is implied. Thus, the ocean—sea-ice system simply amplifies the variance present in the atmosphere at long timescales, which can explain the origin of decadal variability in certain ocean regions. Barsugli and Battisti (1998), by extending the Hasselmann (1976) model, constructed a simple stochastically forced, one-dimensional, linear, coupled energy balance model and obtained important insight into the nature of coupled interactions in the mid-latitudes. They concluded that the experimental design of an atmospheric model coupled to a mixed-layer ocean model would provide a reasonable null hypothesis against which to test for the presence of distinctive decadal variability (see also Section 7.5.4).

7.5.2. Stochastic Models with Mean Advection and Spatial Coherence

Several further refinements have been proposed since Hasselmann (1976) first introduced stochastic climate models. Lemke et al. (1980) applied a dynamical model based on white noise atmospheric forcing, local stabilizing relaxation, and lateral diffusion and advection to explain sea-ice variability. Longitudinally-dependent forcing, feedback, lateral diffusion, and advection parameters were derived by fitting the model to the observed cross-spectral matrix of the sea-ice anomaly fields. Lemke et al. (1980) inferred that diffusion and advection of sea-ice anomalies were important in sea-ice dynamics. In particular, the model advection patterns agreed reasonably well with the observed ocean surface circulation in the Arctic Ocean and around Antarctica. Frankignoul and Reynolds (1983) described the use of a local stochastic model, including the effects of advection by the observed mean current, to predict the statistical characteristics of observed SST anomalies in the North Pacific on timescales of several months. They find that mean advection has only a small effect in general although, in regions of large currents, the advection effects were important at lower frequencies. Finally, Herterich and Hasselmann (1987) have fitted a more general non-local stochastic model, incorporating advection and diffusion, to observed SST anomalies over the same region. Their analysis, however, supported previous models in which the origin of mid-latitudinal SST anomalies on timescales of months to a few years can basically be attributed to local stochastic forcing by the atmosphere.

A major advance in understanding decadal variability was the inclusion of spatial coherence in the atmosphere. Atmospheric variability on timescales of a month or longer is dominated by a small number of large-scale spatial patterns whose time evolution has a significant stochastic component (Davis, 1976). One prominent example is the above-described NAO with its dipolar SLP anomaly pattern (Figure 7.9). One may expect the atmospheric patterns to play an important role in ocean—sea-ice—atmosphere interaction, with advection potentially playing a role in this coupling.

Following the results of Saravanan and McWilliams (1997), obtained with a more complex but idealized coupled ocean—atmosphere model, a one-dimensional stochastic model of the interaction between spatially coherent atmospheric forcing patterns and an 'advective' ocean was developed by Saravanan and McWilliams (1998). Their model equations are simple enough to allow analytical treatment. The model solution can be separated into two different regimes: a slow-shallow regime where local damping effects dominate advection and a fast-deep regime where non-local advection effects dominate thermal damping. An interesting feature of the fast-deep regime is that the ocean shows preferred timescales, although there is no underlying oscillatory mechanism either in the ocean or in the atmosphere. Furthermore, the existence of the preferred timescale in the ocean does not depend on the existence of an atmospheric response to SST anomalies. Rather, it is determined by the advective velocity scale associated with the upper ocean and the length scale associated with low-frequency atmospheric variability. This mechanism is often referred to as 'spatial resonance' or 'optimal forcing'. For the extra-tropical North Atlantic basin, this timescale would be of the order of a decade. Interestingly, Deser and Blackmon (1993), Sutton and Allen (1997), and Alvarez-Garcia et al. (2008) find such a decadal timescale in surface observations of the North Atlantic. However, the findings of the studies differ in several aspects, especially concerning the derived propagation characteristics. The stochastic–advective mechanism may also underlie the Antarctic Circumpolar Wave (ACW: e.g., White and Peterson, 1996) in the Southern Hemisphere, as shown in the model study of Weisse et al. (1999) who drove an ocean–sea-ice general circulation model by spatially coherent, but temporally white surface, forcing.

7.5.3. Stochastic Wind Stress Forcing of a Dynamical Ocean

So far we have considered only thermohaline forcing, that is, heat and freshwater forcing, not varying ocean dynamics. Frankignoul et al. (1997) used a simple linear model to estimate the dynamical response of the extra-tropical ocean to spatially-coherent stochastic wind stress forcing with a white frequency spectrum. The barotropic fields are governed by a time-dependent Sverdrup balance, the baroclinic ones by the long Rossby wave equation. At each frequency, the baroclinic response consists of a forced response plus a Rossby wave generated at the eastern boundary. For forcing without zonal variation, the response propagates westward at twice the Rossby wave phase speed. The baroclinic response is spread over a continuum of frequencies, with a dominant timescale determined by the time it takes a long Rossby wave to propagate across the basin, thus increasing with the basin width. The baroclinic predictions for a white wind-stress-curl spectrum are broadly consistent with the frequency spectrum of sea-level changes and temperature fluctuations in the thermocline observed near Bermuda.

The 'complete' wind-forced problem was studied by Taguchi et al. (2007). They used a much more complex (non-linear) model and realistic wind forcing to investigate decadal variability associated with the wind-driven gyre circulation in the North Pacific. They studied low-frequency variability of the Kuroshio Extension (KE) using a multidecadal (1950–2003) hindcast by a high-resolution (0.1°), eddy-resolving, global ocean general circulation model. SSH variability in the western North Pacific east of Japan is explained by two modes. The first mode represents a southward shift and, to a lesser degree, an acceleration of the KE jet associated with the 1976/77 shift in basin-scale winds. The second mode reflects quasi-decadal variations in the intensity of the KE jet. Both the spatial structure and time series of these modes derived from the hindcast are in close agreement with satellite altimeter observations. KE variability could be decomposed into broad- and frontal-scale components in the meridional direction – the former following the linear Rossby wave solution and the latter closely resembling ocean intrinsic modes. Taguchi et al. (2007) concluded that basin-scale wind variability excites broad-scale Rossby waves, which propagate westward, triggering intrinsic modes of the KE jet and reorganising SSH variability in space. Stochastic wind stress forcing may thus explain a substantial part of the decadal variability of the wind-driven oceanic gyres.

Schneider et al. (2002) previously showed that there is evidence in data that first baroclinic-mode Rossby waves carry thermocline depth perturbations to the west in the North Pacific, which then results in a shift of the Kuroshio Extension and subsequently in SST anomalies. Schneider and Cornuelle (2005) showed that the PDV can be recovered from a reconstruction of North Pacific SST anomalies, based on a first-order autoregressive model and forcing by variability of the Aleutian Low and ENSO, together with oceanic zonal-advection anomalies in the KE region. The latter is a result of oceanic Rossby waves that are forced by North Pacific Ekman pumping. Latif and Barnett (1994) explained PDV by a coupled air–sea mode that originates in the mid-latitudes and also involves baroclinic Rossby wave propagation as part of the gyre adjustment to changing winds.

7.5.4. Hyper-climate Mode

We describe now a special case: a coupled ocean–atmosphere model in which the atmosphere is represented by a state-of-the-art atmospheric general circulation model (AGCM). The ocean is represented by a vertical column model (CM) in which the individual levels communicate only by vertical diffusion (see also Alexander and Penland, 1996). Such a coupled model (Dommenget and Latif, 2008) displays a number of features of observed decadal variability in SST, especially in the Pacific. Since horizontal ocean dynamics are not considered – oceanic Rossby waves, for instance, are excluded from the model – the nature of air–sea interactions is greatly simplified in the model. However, some important aspects of the space–time structure of the observed decadal SST variability can be explained, introducing a level of realism to our discussion that permits the treatment of basin-scale decadal variability, connecting processes in the North Pacific to changes in the South Pacific.

The concept of a global hyper-climate mode was introduced by Dommenget and Latif (2008), in which surface heat flux variability associated with regional atmospheric variability patterns (such as variability associated with the Aleutian Low) is integrated by the large heat capacity of the extra-tropical oceans, leading to a continuous increase of SST variance towards longer timescales. Atmospheric teleconnections and coupled feedbacks (associated with anomalous heat flux or wind mixing, such as the wind−evaporation−sea surface temperature (WES) feedback) spread the extra-tropical signal to the tropical regions. Once SST anomalies have developed in the tropics, global atmospheric tele-connections spread the signal around the world creating a global hyper-climate mode. Calculations with a further reduced stochastic model suggest that a hyper-climate mode can vary on timescales longer than 1000 years depending on the atmospheric damping in relation to the oceanic heat capacity, which controls the timescale at which the spectrum will flatten (Dommenget and Latif, 2008).

The pattern of the leading SSTA EOF derived from the AGCM-CM that was integrated for 800 years is timescale-dependent (Figure 7.14). On inter-annual timescales (Figure 7.14a), the leading EOF mode obtained from high-pass filtering the data has loadings basically restricted to the North Pacific. However, at longer timescales, the loadings extend to the equatorial (Figure 7.14b) and South Pacific (Figure 7.14c) and even beyond the Pacific region, even-tually becoming nearly symmetric with respect to the Equator when only variability on timescales of 40 years and longer is retained. The SST anomaly pattern in the Pacific simulated at multidecadal timescales in the AGCM-CM is remarkably similar to those derived from observations and from long control integrations with sophisticated coupled ocean−atmosphere general circulation models. Thus, the hyper mode mechanism could underlie the PDV, whose basin-wide structure is reasonably well reproduced by the simplified model. The comparison of Figure 7.14c with Figures 7.6b and 7.7 yields indeed some striking similari-ties, for example, a strong symmetry about the equator. Active ocean dynamics, together with consideration of full large-scale ocean−atmosphere coupling may modify the hyper mode, especially in the tropics, and influence the regional expression of the associated variability. Equatorial ocean dynamics such as those operating in ENSO, would enhance the variability in the eastern and central equatorial Pacific. Such feedbacks would make the model certainly more realistic but are not at the heart of the mechanism that produces the hyper mode. Park and Latif (2010), for instance, argue that PDV is largely consistent with the hyper mode picture in their model, a distinct period (i.e., 45 years) being introduced through gyre adjustment processes. If, however, the pure hyper mode scenario applies to the real world, the decadal predictability potential associated with this type of variability would be only modest and not exceed that expected from an autoregressive process of the first order. Indeed, decadal potential predictability derived

from climate models in the North Pacific is found to be considerably less than in the North Atlantic (Boer and Lambert, 2008).

7.5.5. Stochastically-Driven AMOC Variability

The Atlantic decadal to centennial variability, in conjunction with the AMOC, plays a central role in decadal predict-ability in the Atlantic sector. Competing mechanisms have been proposed for the AMOC variability. One idea is that low-frequency AMOC variability, consistent with a simple stochastic scenario, is driven by the low-frequency portion of the spectrum of atmospheric forcing. This hypothesis was first tested in the uncoupled mode by Mikolajewicz and Maier-Reimer (1990) describing results from a multimillen-nial integration with the Hamburg Large-Scale Geostrophic (LSG) ocean general circulation model forced by spatially correlated white-noise freshwater flux anomalies. In addition to the expected red-noise character of the oceanic response, the model simulated enhanced variability predominantly in the Atlantic basin in a frequency band around 320 years. This has been attributed by Pierce et al. (1995) to the movement between two model states: one characterized by strong convection and an active overturning circulation and the other with a halocline around Antarctica closing (capping off) the water column, thus preventing convection. The physical mechanism that forces the model from the quiescent state to an actively convecting one is subsurface (300 m) heating around Antarctica, which destabilizes the water column; the ultimate source of this heat being advected North Atlantic Deep Water (NADW). This leads to a teleconnection between forcing conditions in the North Atlantic and the thermohaline structure of the Southern Ocean. The mechanism that shuts off convection is surface freshening, primarily by precipitation, in the region pole-ward of the Antarctic Circumpolar Current (ACC).

Weisse et al. (1994) describe decadal variability with a timescale of the order of 10 to 40 years in the North Atlantic in the same experiment. It involves the generation of salinity anomalies in the Labrador Sea and the subse-quent discharge into the North Atlantic. The generation of the salinity anomalies is mainly due to an almost undis-turbed local integration of the white-noise freshwater fluxes. The timescale and damping term of the integration process are determined by the flushing time of the well-mixed upper layer. The decadal mode affects the AMOC and represents a discharge process that depends non-linearly on the modulated circulation structure rather than a regular linear oscillator. It should be mentioned, however, that the (uncoupled) stochastically forced LSG model integrations were performed with mixed boundary conditions, which may considerably distort the physics of the coupled ocean−atmosphere system and lead to an

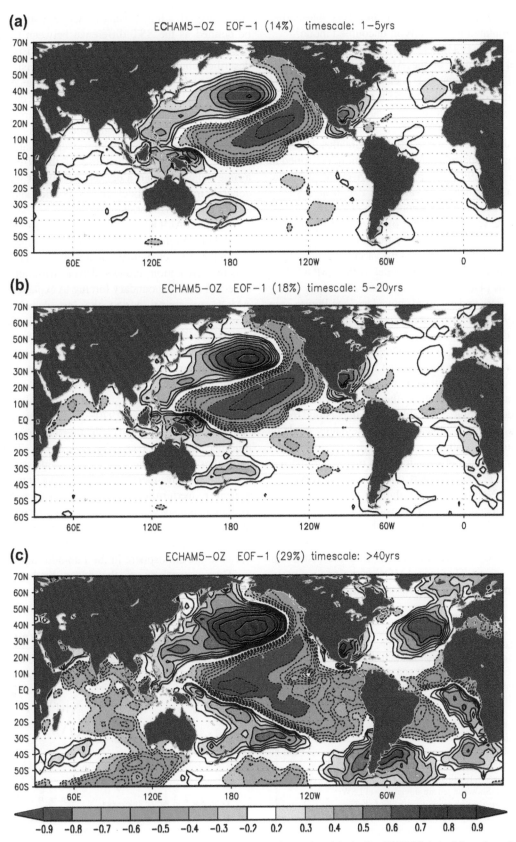

FIGURE 7.14 Correlation maps of annual SST with the principal component time series of the leading SST EOF derived from the variability in three timescale ranges from the simple climate model (AGCM-CM) described in Section 7.5.4. Filtering was performed by applying and/or subtracting running means. Shown is (a) the correlation with the leading EOF mode of the high-pass filtered (1–5 years), (b) band-pass filtered (5–20 years), and (c) low-pass filtered (larger than 40 years) SSTs. The hyper mode is fully developed at multidecadal timescales (bottom panel). *(Source: Dommenget and Latif, 2008 © Copyright 2008 American Geophysical Union. Reproduced/modified by permission of American Geophysical Union.)*

overly strong variability. Furthermore, SST variability cannot be simulated realistically in uncoupled mode.

Delworth et al. (1993) were the first to report pronounced multidecadal variability of the AMOC in a multicentury control run, with a coupled ocean–atmosphere general circulation model that allowed SST to respond fully to AMOC changes. The simulated SST variability in the North Atlantic is consistent with observed AMV in both spatial structure and timescale. Delworth and Greatbatch (2000) show that the multidecadal AMOC fluctuations described in Delworth et al. (1993) are primarily driven by a spatial pattern of surface heat flux variations that bear a strong resemblance to the NAO. No conclusive evidence is found that the AMOC variability is part of a dynamically coupled atmosphere–ocean mode in this particular model. Griffies and Tziperman (1995) interpreted the variability in terms of a stochastically forced four-box model of the AMOC. This box model was placed in a linearly stable, thermally dominant mean state under mixed boundary conditions (see Stommel, 1961). A linear stability analysis of this state reveals one damped oscillatory overturning mode, in addition to purely decaying modes. Direct comparison of the variability in the box model and that in a coupled ocean–atmosphere general circulation model reveals common qualitative aspects, supporting the hypothesis that the coupled model's AMOC variability can be understood by the stochastic excitation of a linear damped oscillatory mode of the overturning circulation.

Analyses of ocean observations and model simulations by Latif et al. (2006b) support this picture (see also Sen Gupta and McNeil, 2012, this volume). They suggest that there have indeed occurred considerable multidecadal changes in AMOC strength during the last century that may have driven AMV. AMOC variations, however, were indirectly determined and reconstructed from the history of observed Atlantic SST. Since AMOC variations are associated in climate models with variations in the meridional

heat transport, a fingerprint of relative AMOC strength can be defined as the SST difference between the North and South Atlantic making use of the inter-hemispheric contrast described above. Latif et al. (2004) previously showed that this approach worked well for multidecadal AMOC variations in a climate model. The observed changes in the dipole SST index are driven by the low-frequency variations of the NAO through changes in Labrador Sea convection (Figure 7.15). Multidecadal variations in the dipole SST index follow those of the NAO index with a time delay of about a decade, consistent with the ocean general circulation model studies by Eden and Jung (2001) and Eden and Willebrand (2001).

As direct AMOC observations exist only for the last few years and the reliability of AMOC-strength reconstructions by SST can be challenged, many studies have used ocean general circulation models driven with an estimate of observed surface boundary forcing to explain decadal-scale AMOC variability during the last decades. Here, we describe results from Alvarez-Garcia et al. (2008) for the period 1958–2000. Multichannel Singular Spectrum Analysis (MSSA) was used to extract the dominant space–time modes of the ocean model data in the North Atlantic poleward of the Equator. The leading MSSA mode is multidecadal. It displays, for instance, prolonged negative SST anomalies during 1970–1980 covering the whole North Atlantic (not shown) reflecting a negative phase of the multidecadal cycle (see also Figure 7.2). The cold SST anomalies are preceded by a basin-wide cell of negative anomalies in the meridional streamfunction, and thus by a weaker overturning about 5 years before (Figure 7.16). The anomalously weak overturning is a result of an anomalously weak NAO with its low extreme during the 1960s (Figure 7.15) and the associated reduced heat loss of the ocean to the atmosphere in the Labrador Sea at this time.

These snapshots of the ocean model's overturning streamfunction five years apart from each other, as

FIGURE 7.15 Time series of the boreal winter [December–March (DJFM)] NAO index (hPa, shaded curve), a measure of the strength of the westerlies and heat fluxes over the North Atlantic and the Atlantic dipole SST anomaly index (°C, black curve), a measure of the strength of the AMOC. The NAO index is smoothed with an 11-year running mean; the dipole index is also smoothed with an 11-year running mean filter. Multidecadal changes of the AMOC as indicated by the dipolar SST index lag those of the NAO by about a decade, supporting the notion that a significant fraction of the low-frequency variability of the AMOC is driven by that of the NAO. Shown in red are annual data of LSW thickness (m), a measure of convection in the Labrador Sea, at ocean weather ship Bravo, defined between isopycnals σ 1.5 = 34.72 − 34.62, following Curry et al. (1998). (*Source: Reprinted from Latif and Keenlyside, 2011; with permission from Elsevier.*)

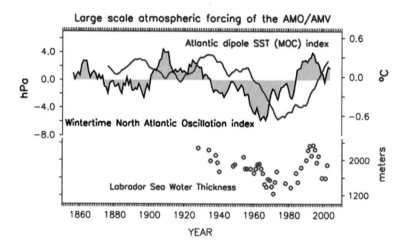

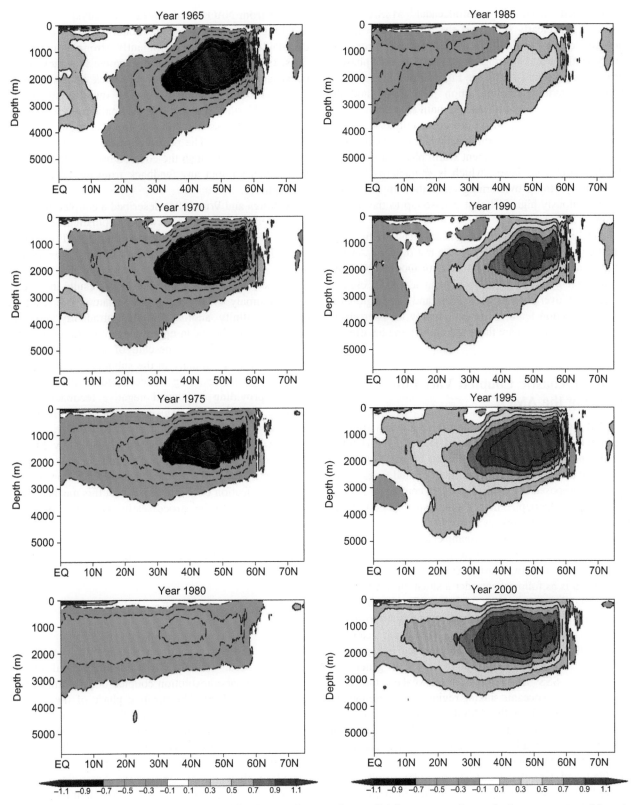

FIGURE 7.16 Snapshots (annual means) of the meridional overturning streamfuncton (Sv) five years apart from each other, as reconstructed from the multidecadal SSA mode obtained from the results of a forced simulation with an OGCM using observed boundary forcing. The transition from an anomalously weak overturning during the 1970s to an anomalously strong overturning during the 1990s is clearly seen. *(Source: Alvarez-Garcia et al., 2008. © American Meteorological Society. Reprinted with permission.)*

reconstructed from the multidecadal MSSA mode (Figure 7.16), clearly show how the negative overturning anomalies develop in the 1960s and subsequently propagate slowly southward. Following the height of the cold phase in surface temperature (~1975, see Figure 7.6c), the tendency in the overturning streamfunction is reversed and the negative anomalies start to weaken, until they are replaced by positive overturning anomalies in the mid-1980s in the north. The positive overturning anomalies expand southward and initiate the subsequent warm phase in the North Atlantic during the 1990s, which is characterized by an anomalously strong overturning circulation in the model and anomalously high SST that persists up to the present (Figures 7.2 and 7.6c).

Modelling studies have suggested that variability in Atlantic Ocean circulation, including not only the AMOC but also variability in the tropical Atlantic Ocean (TAV) and wind-driven gyres, may all be playing a role in climate impacts both over the USA and Europe and indeed globally. These circulation features are presumably all interrelated and variations in each are thought to affect and be affected by the others.

7.5.6. Stochastic Coupled Variability Involving the AMOC

Coupled air—sea modes, in which the feedback from the ocean to the atmosphere is necessary to produce variability at a distinctive timescale above the background red spectrum (as it is the case for ENSO at inter-annual timescales), were also proposed to explain decadal to centennial AMOC variability. Such coupled modes have also to be considered in a stochastic framework, as we expect them to be damped and not self-sustained. Timmermann et al. (1998) described coupled variability with a 35-year period in a multicentury integration of the ECHAM3/LSG climate model. The mechanism is as follows: consider a situation in which the North Atlantic is covered by positive SST anomalies, that is, a positive AMV phase. The corresponding atmospheric response involves a strengthened NAO that leads to anomalously weak evaporation and Ekman transport off Newfoundland and in the Greenland Sea and the generation of negative SSS anomalies. These weaken the deep convection in the oceanic sinking regions and subsequently reduce the strength of the AMOC, leading to a reduced poleward heat transport and eventually the formation of negative SST anomalies, which completes the reversal to the negative AMV phase.

Eden and Greatbatch (2003) describe results from a simple stochastic atmospheric feedback model coupled to a realistic model of the North Atlantic. A north—south SST dipole, with its zero line centred along the subpolar front, drives the atmosphere model, which, in turn, forces the ocean model by patterns of surface fluxes

derived from NAO-based regression analysis as in Eden and Jung (2001). The coupled model simulates a damped decadal oscillation for sufficiently strong coupling. It consists of a fast, wind-driven, positive feedback of the ocean and a delayed negative feedback orchestrated by the onset of an anomaly in the overturning in the subpolar North Atlantic. This anomaly transports more or less heat across the subpolar front, changing the sign of the SST dipole. The positive feedback turns out to be necessary to distinguish the coupled oscillation from that in a model without any feedback from the ocean to the atmosphere.

Vellinga and Wu (2004) described a coupled feedback on centennial timescales from a sophisticated climate model (HadCM3). They report that the ITCZ both strengthens and moves northwards if AMOC is anomalously strong. The increased freshwater flux into the ocean associated with a stronger ITCZ results in a freshwater anomaly in the equatorial Atlantic. The resulting negative salinity anomaly is then gradually advected northwards by the mean upper ocean circulation into the subpolar region on a timescale of a few decades. A negative salinity anomaly in the subpolar region reduces the density here resulting in decreased deep convection, thereby providing a delayed negative feedback on the AMOC. A similar mechanism on centennial timescales seems also to operate in another climate model, the Kiel Climate Model (KCM), as recently shown by Menary et al. (2011). However, in a third climate model (ECHAM5/MPI-OM) this feedback is less prominent. The quantification of the importance of this mechanism in comparison with other processes in explaining the origin of centennial variability is a topic of ongoing research. The results shown in Figures 7.6e and 7.6f, for example, suggest a Southern Ocean control of the centennial variability.

7.5.7. Stochastically Forced Southern Ocean Variability

Hall and Visbeck (2002) investigated the decadal to centennial ocean variability in a multimillennial integration with a coarse-resolution coupled ocean—atmosphere model (GFDL-R15). The positive phase of the SAM is associated with an intensification of the surface westerlies over the circumpolar ocean around 60°S, and a weakening farther north. The changes in the Ekman drift through mass continuity generate anomalous upwelling along the margins of the Antarctic continent and downwelling around 45°S. The anomalous flow diverging from the Antarctic continent increases the vertical tilt of the isopycnals in the Southern Ocean, so that a more intense Antarctic Circumpolar Current (ACC) is also closely associated with positive SAM. In addition, the anomalous divergent flow advects sea-ice farther north, resulting in an increase in sea-ice coverage. Finally, positive SAM drives increases in poleward heat transport at about 30°S, while decreases occur in the circumpolar region. Ocean and sea-ice anomalies of opposite sign occur when the SAM is negative. The ocean and sea-ice fluctuations associated with the SAM constitute a significant fraction of simulated ocean variability poleward of 30°S year-round in the coupled model. By and large the variability is consistent with the stochastic climate model scenario. The simulated SAM index is basically white, while ocean spectra are red. In particular, the spectrum of the Drake Passage transport increases up to centennial timescales.

Böning et al. (2008) challenge the applicability of coarse-resolution models to study the impact of intensifying westerlies on the Southern Ocean circulation. They analysed the Argo network of profiling floats and historical oceanographic data to detect coherent hemispheric-scale warming and freshening trends that extend to depths of more than 1000 m. The warming and freshening is partly related to changes in the properties of the water masses that make up the Antarctic Circumpolar Current, which are consistent with the anthropogenic changes in heat and freshwater fluxes suggested by climate models. However, they did not find an increase in the tilt of isopycnals across the ACC, implying that the transport in the ACC and meridional overturning in the Southern Ocean are insensitive to decadal changes in wind stress.

7.5.8. Forced AMOC Variability

Finally, we turn to the AMOC response to external forcing as simulated by one particular climate model, the KCM (Park and Latif, 2011, submitted). The internal decadal to centennial variability is briefly described first to aid the interpretation of the forced response. Park and Latif (2008) have previously shown that KCM simulates considerable AMOC variability in this timescale range. Northern Hemisphere and even globally averaged SAT exhibit strong internal variability in KCM, especially on centennial timescales. The standard deviation of 10-year and 100-year means of Northern Hemisphere SAT obtained from KCM's control integration correspond well to those calculated from the Jones and Mann (2004) reconstruction (see details in Latif et al., 2009). Although the realism of the model can be questioned for several reasons (e.g., coarse resolution), the model results demonstrate that internal variability cannot be ruled out *a priori* to explain the decadal to centennial variability seen in palaeoclimatic reconstructions of Northern Hemisphere SAT during the last millennium.

The spectral characteristics of globally and Northern Hemisphere averaged SAT variability, as well as AMOC variability is depicted by means of wavelet analysis of the 4200-year-long control integration (Figure 7.17). Both global (Figure 7.17a) and Northern Hemisphere (Figure 7.17b) SAT exhibits distinct centennial-scale variability well above the red background throughout the analysed period. The AMOC index, defined as the maximum of the overturning streamfunction at 30°N, shows enhanced variability on multidecadal and multi-centennial variability (Figure 7.17c), as previously shown by Park and Latif (2008). Interestingly, KCM simulates multicentennial overturning variability, which shares some aspects with that simulated in uncoupled mode by Mikolajewicz and Maier-Reimer (1990) described above.

The spatial SAT pattern associated to the AMOC variability in two specific frequency bands (multicentennial and multidecadal) were shown by Park and Latif (2008) and are repeated here (Figure 7.18a, 7.18b). The associated SAT pattern for the multicentennial band clearly projects on global and Northern Hemisphere SAT (Figure 7.18a). It has a strong Southern Ocean bias, suggesting that processes in this region drive the multicentennial AMOC variability. The globally and Northern Hemisphere averaged SAT, however, is not strongly related to AMOC on the multidecadal timescales, as indicated by the two wavelet spectra, which do not show enhanced variability on these timescales. This is consistent with the dipolar SST anomaly pattern associated with the multidecadal AMOC variability (Figure 7.18b) which does not strongly project on Northern Hemisphere and global SAT.

Further analysis of KCM's internal AMOC variability indicates that three distinct timescales can be identified: one multidecadal with a period of about 60 years, one quasi-centennial with a period of about 100 years, and one multicentennial with a period of about 300–400 years. Most variance is explained by the multicentennial mode and the least by the quasi-centennial. We performed SSA on the AMOC index to obtain the AMOC variability at the three frequencies and computed regression patterns for the zonally-integrated global overturning streamfunction for each of the three timescales using the corresponding SSA reconstructions (Figure 7.19). The longer the timescale, the more important the Southern Ocean becomes, confirming the above results (Figure 7.18). What is important here is the quite complicated frequency structure of the decadal to centennial AMOC variability in KCM, with the variability distributed over several modes and a relatively large frequency range.

Two further 1000-year simulations were also performed, in which the solar constant varies sinusoidally with a period of 100 years and amplitude of 1 and 2 W m^{-2}, respectively. Latif et al. (2009) describe similar experiments with a forcing period of 1000 years. We note that the effective forcing is only one quarter of the change in the solar constant, as the energy intercepted by Earth from

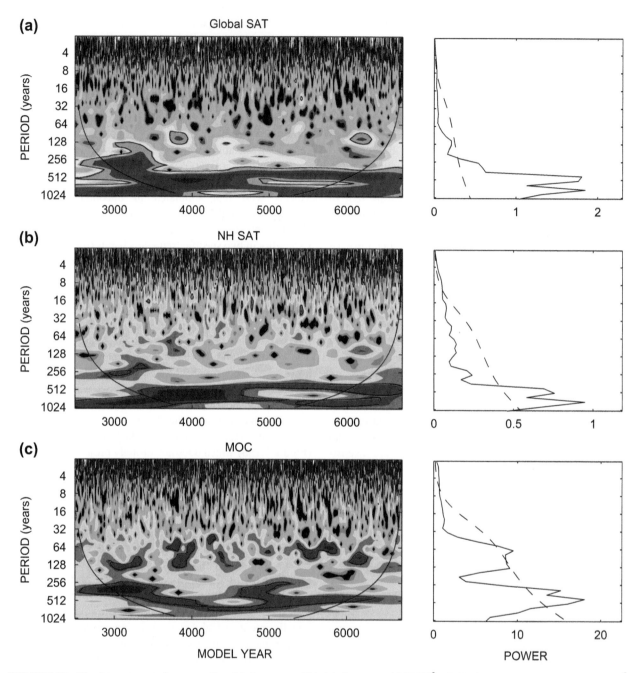

FIGURE 7.17 Wavelet spectrum and corresponding global spectrum of (a) globally-averaged SAT [°C^2], (b) Northern Hemisphere averaged SAT [°C^2], and (c) AMOC index [Sv^2] simulated by the KCM in a 4200-year control run. The variance increases from blue to red colour.

the solar beam is spread over the full Earth surface. Variations in the solar constant over the last centuries have recently been argued to be much smaller than previously thought. For example, Jungclaus et al. (2010) used, as standard solar forcing in their millennial integrations, a total increase of only 0.1% (i.e., about 1.3 W m^{-2}) from the Maunder Minimum (1645–1715) to today, which is at the lower limit of the available reconstructions, so that our forcing amplitude does not seem to be overly strong. It

should be mentioned, however, that we do not attempt to simulate climate variations observed during the last millennium. Rather, we are interested in the response characteristics of the coupled system under idealized forcing.

The response of the AMOC to the periodically varying solar forcing is highly non-linear. We performed wavelet analyses of the AMOC index from all three (one control and two forced) experiments (Figure 7.20) using now only

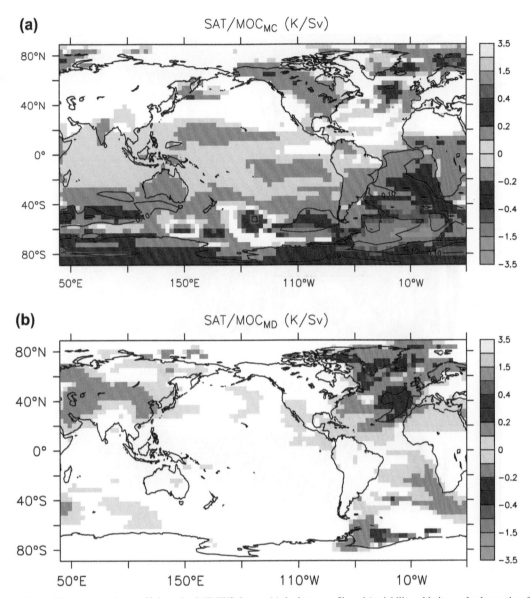

FIGURE 7.18 Maps of linear regression coefficients for SAT [K/Sv] upon (a) the low-pass filtered (variability with timescales larger than 90 years was retained) AMOC strength highlighting the multicentennial variability and (b) the band-pass filtered (variability with timescales of 30–90 years was retained) AMOC strength highlighting the multidecadal variability. Shading denotes the regression values; contours denote explained variances (contour interval amounts to 0.1). *(Source: Redrawn after Park and Latif, 2008 © Copyright 2008 American Geophysical Union. Reproduced/modified by permission of American Geophysical Union.)*

a 1000-year period of the control run to ease comparison. The most striking result is that AMOC variability becomes strongly modified by the external forcing in the forced run with an amplitude of 2 W m^{-2}. While the wavelet analysis of the control run (Figure 7.20a) depicts multitimescale behaviour, the variability is channelled into a relatively narrow band close to the frequency of one internal AMOC mode (Figure 7.20c); it is the quasi-centennial AMOC mode with a period of about 100 years (Figure 7.19c) that explains only a relatively small fraction of the variance in the control run that is excited by the forcing. This result is

also reflected in the spatial SAT response pattern (not shown) that strongly projects on the pattern of the quasi-centennial AMOC mode exhibiting a structure similar to that of the multidecadal mode (Figure 7.18b) but with stronger amplitude in the South Atlantic. This resonance behaviour indicates that it is not necessarily the most unstable (or least damped) mode that is excited by external forcing, implying that we need to understand the full modal structure of the unforced variability to understand the forced variability in a non-linear system. This is a challenge for climate models. If the internal modes of the system are

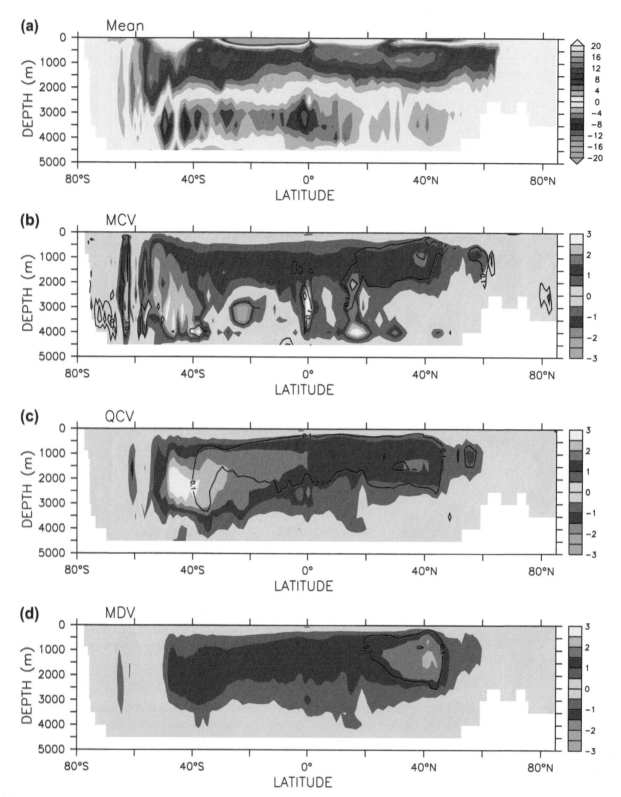

FIGURE 7.19 (a) Mean global streamfunction [Sv] simulated in a multimillennial control integral with KCM. Associated regression patterns of the global zonally-averaged overturning streamfunction [Sv] for the three timescales of AMOC variability identified by SSA applied to the control integration. Shown is the overturning for (b) multicentennial variability (MCV), (c) quasi-centennial variability (QCV), and (d) multidecadal variability (MDV). Contour lines represent explained variance.

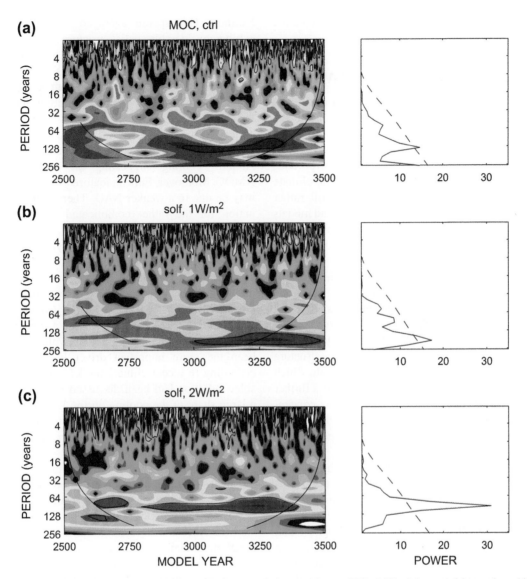

FIGURE 7.20 (a) Wavelet spectrum of the AMOC strength simulated during model years 2500–3499 of the control integration with KCM (see Figure 7.17c) and the corresponding global spectrum, (b) wavelet spectrum of the AMOC strength simulated during a 1000-year-long run with KCM forced by periodically changing solar constant with an amplitude of 1 W m^{-2} and the corresponding global spectrum, and (c) wavelet analysis of the AMOC strength simulated during a 1000-year-long run with KCM forced by periodically changing solar constant with an amplitude of 2 W m^{-2} and the corresponding global spectrum. The period of the forcing amounts to 100 years. The variance increases from blue to red colour.

not well-represented, we cannot expect the response to external forcing to be realistically captured (cf. Berger and Yin, 2012, this volume).

The model run with the weaker forcing amplitude of 1 W m^{-2} exhibits most pronounced variability at a period considerably longer than the forcing period (Figure 7.20b), but the statistical significance can be questioned given a time interval of only 1000 years. This is supported by the wavelet spectrum of the chosen 1000-year segment of the control run (Figure 7.20a), which does not display the peaks at those frequencies that are seen in the analysis of the full 4200 years of the control run (Figure 7.17c). These

results are preliminary, but clearly show that moderate external forcing can definitely exert a strong influence on the AMOC. This could be an important element in understanding regional climate change and is entirely consistent with the recent findings of Ottera et al. (2010), who studied the climate variability of the past 600 years with the Bergen Climate Model (BCM). They show that the phasing of the fluctuations in the North Atlantic during the past 600 years is, to a large degree, governed by changes in the solar and volcanic forcing. Volcanoes play a particularly important part in the phasing of the variability through their direct influence on tropical SST, on the leading mode of Northern

Hemisphere atmospheric circulation (the NAO), and on the AMOC.

7.6. SUMMARY: FUTURE UNKNOWNS

We have reviewed observations and models of decadal to centennial variability. Emphasis has been put on the dynamics of such fluctuations and the mechanisms that could potentially underlie decadal to centennial variability. Both internal and external mechanisms can produce significant decadal to centennial variability in climate models. However, our current knowledge is still rather limited because observational records are short and models often disagree about the mechanisms involved. We need to understand the dynamics of the internal modes of the climate system as much as possible because it is plausible to assume that any external forcing will excite these modes to some degree and that these will eventually shape the climate response, especially on regional-scales. It is not necessarily the dominant modes that were observed during one particular time period that will be excited during another, but the modes that best fit the space—time structure of the forcing. This makes it difficult to anticipate which modes will be excited by external forcing such as a further gradual increase of atmospheric greenhouse gas concentrations during this century, because some of the excited modes may be strongly damped under present-day conditions.

Two decadal-scale phenomena have been the focus of research during the recent years: PDV and the AMV. Results from climate models suggest the two are internal modes of the climate system and are governed by rather different dynamics. Both models and data suggest that PDV is associated with the adjustment of the ocean gyres to fluctuating wind stress. Off-equatorial baroclinic Rossby wave propagation appears to be an important element in this. However, it cannot be ruled out that a much simpler stochastic mechanism applies in which ocean dynamics do not play a prominent role and the ocean can be simply represented by a mixed-layer model. Such a model is able to simulate basin-scale variability consistent with observations. In the Atlantic, however, most evidence from climate models points to an important role of ocean dynamics, specifically the AMOC in producing AMV, consistent with rather early assertions made on the basis of observations. The NAO may play a crucial role in driving the AMOC on these long time-scales. However, many open questions remain concerning the dynamics of PDV and AMV. We currently cannot answer the question of whether PDV and AMV are oscillatory in nature. Furthermore, it is not clear whether or not they are linked in some way.

On centennial timescales, the MCA and LIA were the strongest events during the last millennium in the Northern Hemisphere averaged SAT. Both anomalies exhibit considerable spatial structure and research starts to interpret these changes as the forced response to external drivers in terms of the internal modes of the system. Data coverage is, however, poor during both periods, so caution should be exercised when interpreting these patterns. MCA exhibits a relatively complex surface temperature pattern with a PDV or La Niña-like pattern in the Pacific, a strong North Atlantic warming and moderate South Atlantic cooling suggesting an anomalously strong AMOC. Moreover, Eurasia featured strong cooling, which may be due to a weaker NAO. There is no simple explanation for the rather complicated SAT pattern reconstructed for MCA.

LIA also features an AMV-type pattern in the Atlantic. This, however, is superimposed on a generally anomalously cold planet relative to the present, which may argue for a stronger role of external forcing than during MCA, since both volcanic (by definition) and solar forcing were associated with a negative radiative forcing during this time. Strong volcanic forcing has been the more important driver in creating the LIA, according to recent climate model simulations. Reduced solar radiation had been discussed previously to explain a large share of the cooling during LIA. However, recent reconstructions yield only modest changes of solar radiation during the last millennium, specifically during the Maunder Minimum.

One should also keep in mind that the solar and volcanic influences have a competing effect on the AMOC. While increased solar radiation, which is often discussed in the context of MCA, would force global warming and tend to slow the AMOC with reduced warming or even cooling in the North Atlantic, global cooling in response to enhanced volcanic activity would tend to spin up the Atlantic overturning circulation and lead to a relative warming in the North Atlantic. On the other hand, changes in atmospheric circulation patterns like the NAO, forced by external drivers, could also exert an influence on the AMOC. The net effect on the AMOC is, thus, hard to estimate if one does not consider the complete coupled system, including potential feedbacks from vegetation (Dickinson, 2012, this volume) and atmospheric chemistry (cf. Harvey, 2012, this volume).

One important question in this context is whether features such as MCA or LIA can be internally generated without taking into account external factors. Some climate models support this. If the Northern Hemisphere temperature variability simulated in control runs is compared with the SAT variability that ranges at the lower limit of the available reconstructions, internal processes can indeed explain the fluctuations during the last millennium. Thus we need to reduce the uncertainty in the reconstructions of climate variability during the last few millennia in order to better assess the role of internal variability.

However, we need also to improve our models, since they play a central role in understanding decadal to centennial variability. Biases are still large in state-of-the-art climate models. Errors in SAT, for instance, can amount up to almost 10°C in certain regions in individual models. Hot-spots in this respect are, for example, the eastern tropical and subtropical oceans exhibiting a large warm bias, and the North Atlantic and North Pacific generally suffering from a large cold bias, with all regions exhibiting significant decadal to centennial variability during the instrumental record.

Likewise, significant discrepancies to observations exist in the variability. Many models, for instance, fail to simulate a realistic ENSO. Other models do not simulate PDV and AMV with the periodicities observed during the instrumental record. Thus, it cannot be assumed that current climate models simulate the decadal to centennial variability realistically. Much higher resolution is certainly one important step to improve models, as has been shown in several recent studies. The more realistic inclusion of vegetation and chemical feedbacks is another step in this direction. The biases in the mean state and variability also limit our ability to predict the potential impacts of global warming.

This survey of observed and model-simulated decadal to centennial climate variability makes it clear that, on the regional-scales on which most planning decisions are made, anthropogenic climate change signals will be strongly modulated by natural (internal and external) climate variations, and especially those driven by the slowly varying oceans. Figure 7.21 shows the evolution of the AMOC in an ensemble of idealized greenhouse warming simulations with the KCM. A clear trend toward a weakening AMOC is simulated in the ensemble mean. However, the ensemble spread is rather large, and the mean decline exceeds the two standard deviation limit only after about 40 years. Thus, the expected

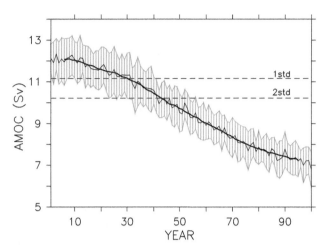

FIGURE 7.21 Response of the AMOC [Sv] in an ensemble of 22 idealized (1% per year CO_2 increase) greenhouse warming experiments with KCM. Each of the 22 integrations was initialized with different initial conditions but uses identical external forcing. The thin black line shows the ensemble mean evolution of the AMOC with annual resolution, the thick black line is the 11-year running mean, and the green vertical bars denote the ensemble spread at each year. The dashed blue lines represent the one and two standard deviation range of the ensemble within the first 10 years.

anthropogenic weakening of the AMOC and related regional climate change patterns may not be detectable during the next few decades.

This non-uniformity of change highlights the challenges of regional climate change that has considerable spatial structure and temporal variability. Moreover, it illustrates the need to predict not just the change in global mean temperature, but also the patterns of important variables around the globe as accurately as possible. Predictability studies with climate models show that some of the decadal-scale variability is potentially predictable at a lead-time of a decade or so. A robust global ocean observing system will be crucial to understand such variability and provide society with useful decadal predictions.

Chapter 8

The Future of the World's Glaciers

Graham Cogley

Department of Geography, Trent University, Ontario, Canada

Chapter Outline

8.1. INTRODUCTION: CLIMATE AND THE CRYOSPHERE

8.1.1. Glaciers in the Context of Climatic Change

Glaciers are perennial masses of ice, and possibly firn and snow, originating on land by the recrystallization of snow and showing evidence of past or present flow (Cogley et al., 2011). They range in size from tiny but numerous glacierets, which differ from perennial snowpatches only in showing evidence of flow, to the Greenland Ice Sheet (area 1.7 Mm^2; 1 $Mm^2 = 10^6$ $km^2 = 10^{12}$ m^2) and the Antarctic Ice Sheet (area 12.3 Mm^2). Collectively, glaciers account for the bulk of the cryosphere, which also contains the world's ice shelves (another 1.5 Mm^2 of thick but floating ice), sea ice, lake and river ice, ground ice (in permafrost), and seasonal snowpacks. Glaciers have radiation balances, energy balances, and mass balances, like any study volume, and they exchange radiation, energy, and mass with other compartments of the near-surface environment, principally the atmosphere and the ocean, but also the aquifers in which much of the Earth's stock of water is stored. In addition, they interact with the Earth's mantle because of its viscoelastic response to loading at the surface, and with the rest of the Earth as a whole through the effect on the gravitational field of their exchanges of mass with the ocean (Peltier, 2004).

Changes in the extent and mass of glacier ice are, in a naive view, 'caused' by changes in the climate. In truth, the direction of causality is not necessarily well-defined. Frozen water is the brightest, and liquid water among the darkest, of the materials at the Earth's surface. Sufficiently large changes in the relative proportions of the two phases can feed back on the climate that appears to have caused the

The Future of the World's Climate. DOI: 10.1016/B978-0-12-386917-3.00008-7

changes. The phenomenon of ice—albedo feedback has been a fundamental feature of our understanding of the Earth's radiation and energy balances since it was first elucidated by Sellers (1969) and Budyko (1968).

A measurement of the mass balance of a glacier involves no assumptions about temperature. Glaciers are therefore environmental and climatic monitoring devices that are physically independent of the weather stations and of bathythermographs. The greater number of temperature measurements, the availability of abundant information about temperature in the form of atmospheric reanalyses and climate projections, and the observed correlation of mass balance with temperature, which is invariably found to be strong, combine to make temperature-index modelling (Hock, 2003) a natural way to explore present and future glacier changes. However, glaciers would yield reliable information about environmental change, and in particular about changes in the radiation balance, even if the thermometer had never been invented.

The information is of course different in kind from that in temperature records. The glacier's annual mass balance, apart from possible losses by iceberg calving, is a response to climatic forcing integrated over the year, over the area covered by the glacier and over its vertical extent. Thus, the glacier is an insensitive, but not therefore a less valuable, indicator of climatic change. Glaciers also sample different parts of the Earth's surface, being found at higher altitudes and latitudes than most weather stations.

8.1.2. Glaciers in the Context of Socio-Economic Change

The mass of fresh water stored in glaciers was about 25.5 million Gt in the late twentieth century, with all but about 0.2 million Gt in the two ice sheets. Of the ice in glaciers other than the ice sheets, most is in uninhabited or thinly peopled drainage basins at high polar latitudes. Nevertheless, there are many catchments in which glacier ice is a significant human resource. For example, glaciers fuel hydroelectric power stations in Norway and other countries; yield fresh water for irrigation and domestic consumption in Canada, Peru, Pakistan, and China; attract tourists to Switzerland, Kenya, Nepal, and the Antarctic; and are icons of the wilderness for billions of city dwellers — you have to traverse one glacier or another if you aspire to climb Mt Everest or K2 and, although they are likely to disappear in the next few decades, glaciers persist today at the summit of Kilimanjaro.

In populated drainage basins, the glaciers may pose a significant hazard to those living downstream. *Jökulhlaup* is an Icelandic word that translates as 'glacier burst' and refers to sudden, very large increases in the discharge of glacier meltwater streams. Some glaciers dam meltwater lakes in marginal valleys. If the meltwater becomes so deep

that the ice dam floats, there is an abrupt, potentially catastrophic increase in discharge, followed by a recession and ultimately a return to low rates of flow. Subglacial volcanic eruptions are especially hazardous because of the large volumes of meltwater that result, and sometimes because of lahars, rapid and damaging mudflows. The lahars of the 1985 eruption of Nevado del Ruiz in Colombia were responsible for most of the 23,000 resulting deaths. Lliboutry et al. (1977) and Arnao (1998) document the known history of glaciogenic lahars and avalanches in Peru that, in sum, have killed tens of thousands of people. Many of these catastrophes are due to the failure of moraines damming proglacial lakes, a hazard that impends also over many glacierized valleys in the Himalaya (Fujita et al., 2009). The hazard can be understood in terms of glacier retreat from advanced positions attained during the Little Ice Age and marked today by the terminal moraines. Avalanches from valley walls and calving events at the glacier terminus generate lake-water surges that trigger failure. Richardson and Reynolds (2000) noted that the frequency of moraine-dam failures in central Asia appears to be increasing. It is unlikely that the increase in frequency could be shown to be statistically significant, but the hazard remains serious and the explanation for it remains obvious.

The most significant way in which glaciers impact society indirectly is through their contributions to the water balance of the ocean or in other words, in the modern context, to sea-level rise. A sea-level rise of as little as a metre may prove fatal to the aspirations of countries whose highest point is only a few metres above the mean sea level of the date at which they became independent. The same sea-level rise would be colossally expensive for older and for more highly-developed countries. And if a substantial portion of the nearly 26 million Gt of water now in glaciers were transferred to the ocean, we would have to choose between the end of civilization as we know it and the physical relocation of civilization to a position at least 80 metres higher than at the present day.

8.1.3. Scope

In what follows, we will first set out the elements of an understanding of what glaciers are and of how they exchange energy with their surroundings (Section 8.2). The focus will then narrow in Section 8.3 to exchanges of mass, because there is far more information on mass balances than on energy balances and because our physical understanding assures us that there is an equivalency, to a good approximation, between mass exchange and energy exchange. Section 8.4 is about the tools with which glaciologists model those attributes of glaciers that are difficult or impossible to measure, or that require to be projected as responses to anticipated environmental change. The initial conditions for these models, including present-day states and rates of change, are of basic importance, and they are the subject of Section 8.5. Section 8.6 is an assessment of recent assessments of future mass losses from glaciers, while Section 8.7 offers concluding reflections.

To keep this survey within manageable limits, the two ice sheets are considered only briefly, and mainly for comparative purposes. For brevity and clarity, the glaciers other than the ice sheets are sometimes referred to as 'small glaciers', but with no implication that they are small by any measure other than that provided by the ice sheets.

8.2. ELEMENTS

8.2.1. Glacier Geography and Physiography

Glaciers are bounded by continuous outlines, parts of which, called divides because the ice flow diverges there, may be shared with adjacent glaciers. Thus, an ice cap, which may be a unit for climatological purposes, is sometimes subdivided for glaciological purposes into several glaciers. The glacier margin is that part of the outline that is in contact not with an adjacent glacier, but with unglacierized terrain or possibly with sea or lake water. Where the ice flow produces an elongate tongue, as in valley glaciers and often in outlet glaciers of ice caps and larger bodies, the margin is called the terminus. Some glaciers have floating tongues, and some nourish ice shelves; here, the grounding line, where the condition for flotation is satisfied, is a significant part of the margin. The condition for flotation of ice of thickness h in water of depth

d is $d = h\,\rho_i/\rho_w$, where ρ_i is the density of the ice and ρ_w the density of the water.

In any vertical column through a glacier, the uppermost parts are likely to consist not of glacier ice but of firn and snow. Snow is solid precipitation that has accumulated during the current balance year. By convention, snow becomes firn instantaneously at the end of the balance year. Firn, then, is snow that has survived at least one season of ablation (mass loss), but has not been compacted to a density near that of pure ice.

Typical glaciers have vertical extents of hundreds and sometimes thousands of metres. Differences of mean temperature of several kelvins between the lowest and highest points on the glacier surface are therefore usual. The components of the local or point mass balance (the change of mass of a vertical column extending through the glacier, over a given span of time) vary greatly between the minimum and maximum elevations (Figure 8.1), so it is instructive and useful to subdivide the glacier into zones (Figure 8.2). Most glaciers have an accumulation zone at higher elevation, where annual mass gain, mainly by snowfall, exceeds annual mass loss, mainly by melting, and an ablation zone at lower elevation, where annual mass loss, possibly including losses by calving, exceeds annual mass gain. Processes of accumulation (mass gain) and ablation (mass loss), other than those just mentioned, include sublimation and deposition of vapour, redistribution by the wind, and avalanching, but these are seldom dominant and often unimportant.

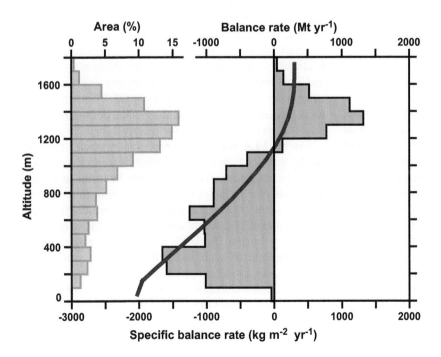

FIGURE 8.1 Hypsometry (distribution of area with altitude; yellow histogram), specific balance rate (red line), and balance rate (blue histogram) of White Glacier in arctic Canada (area 39.4 km²). The specific balance, $b(z)$, is the local sum of accumulation, $c(z)$, and ablation, $a(z)$. Balance data are averages for 1959/60 to 2008/09. The glacier-wide specific balance (the area-weighted average of the red line) was -175 kg m^{-2} yr^{-1} on average over the measurement period. The blue histogram is the product of the yellow histogram and the red line and shows that balance magnitudes are largest where relative area is least.

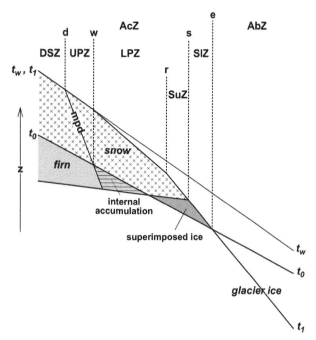

FIGURE 8.2 Glacier zonation and its balance-related aspects on a representative cold or polythermal glacier. At the start of each mass-balance year, the glacier surface is at the line $t_0 - t_0$. It evolves (schematically, the effects of ice flow being neglected) to $t_w - t_w$ at the end of the accumulation season, when the mass of the glacier reaches its annual maximum, and then to $t_1 - t_1$ at the end of the ablation season, when it becomes the summer surface of the next balance year. 'mpd' is the maximum depth to which meltwater percolates before refreezing. The zones are AbZ: ablation zone (the zone below e); AcZ: accumulation zone (all the zones above e); SIZ: superimposed ice zone; SuZ: slush zone, a part of LPZ; UPZ: upper percolation zone; LPZ: lower percolation zone or wet-snow zone; DSZ: dry-snow zone. The zones are separated by the lines e: equilibrium line; s: snowline; r: runoff limit or slush limit (position variable, depending especially on the surface slope); w: wet-snow line (intercept of mpd on summer surface, separating UPZ and LPZ); d: dry-snow line (surface outcrop of mpd). *(Source: "Glossary of Glacier Mass Balance and related Terms", IHP-VII Technical Documents in Hydrology, No2, Figure no 16, p.101, Cogley et al., © UNESCO/IHP 2011. Used by permission of UNESCO.)*

On some glaciers, particularly those in cold climates, it makes sense to subdivide the accumulation zone further, as in Figure 8.2. The dry-snow zone, for example, is the part of the glacier where there is no surface melting. Most glaciers have no dry-snow zone, but the dry-snow zone extends over almost all of the Antarctic Ice Sheet and (until recently) about half of the Greenland Ice Sheet. The superimposed ice zone is the part of the glacier where, at the end of summer, the surface consists of superimposed ice, which is meltwater that has percolated to the base of the winter snowpack and refrozen. The superimposed ice zone, if present, separates the ablation zone from the percolation zone, its upper boundary being the snowline and its lower boundary the equilibrium line. This distinction can be important, for example in attempts to assess mass balance

remotely. Superimposed ice is mass gained during the current mass-balance year, but is extremely difficult to tell apart from the old glacier ice that is exposed below the equilibrium line.

The mass-balance year is only roughly as long as a calendar year. It extends from one annual minimum of mass at the start of winter (September or October, March or April, depending on latitude and hemisphere) to the next annual minimum at the end of the following summer.

Glaciers are distributed widely across the Earth's surface, with an overwhelming concentration near the poles (Figure 8.3). The annual equilibrium line, as observed at the end of each balance year, separates the accumulation zone from the ablation zone. Its mean altitude, abbreviated ELA, is both a fundamental physiographic attribute of the glacier and a valuable descriptor of glacier climate and geography. As shown in Figure 8.4, the ELA reaches sea level at and poleward of about $60° - 65°$S, and approaches, but does not reach, sea level at the North Pole. It rises above 6 km in the desert belts and is depressed in the region through which the Inter-Tropical Convergence Zone migrates seasonally, to below 5 km in places. At equivalent extra-tropical latitudes, the ELA is up to several hundred metres lower in the Southern Hemisphere than the Northern. In the mid-latitudes, there are moderate but visible depressions of the ELA that can be understood as responses to enhanced accumulation beneath the polar front of each hemisphere. The distribution of warmest-month temperature at the ELA reinforces this interpretation. The greater the annual accumulation, the greater the annual ablation must be in order to remove the annual accumulation exactly. Greater ablation requires warmer temperatures and, therefore, a lower ELA. (By definition, surface accumulation and ablation are equal at the ELA.) In a warming climate, the ELA is expected to rise. However, the annual accumulation is also expected to increase in places where it is already substantial, and the annual ablation by melting and sublimation is expected to become more negative where evaporation is already substantial. This expectation translates into a more curvaceous version of Figure 8.4, but it has not yet been detected (or searched for).

As the ELA rises during a period of greater radiative energy gain, or falls during a period of increased accumulation, so the glacier responds, not only by losing or gaining mass, but also by adjusting its size dynamically. Glacier dynamics is not the central concern of the present study, but it is a profoundly important subplot in the narrative of glacier change (cf. Lenton, 2012, this volume). The glacier always seeks equilibrium with its climatic forcing, but always requires years to millennia, depending on its size, to attain that equilibrium (Jóhannesson et al., 1989). So the climatic information in records of glacier change is integrated not just over the glacier's horizontal and vertical extent, and not just over seasonal time, but also over secular time.

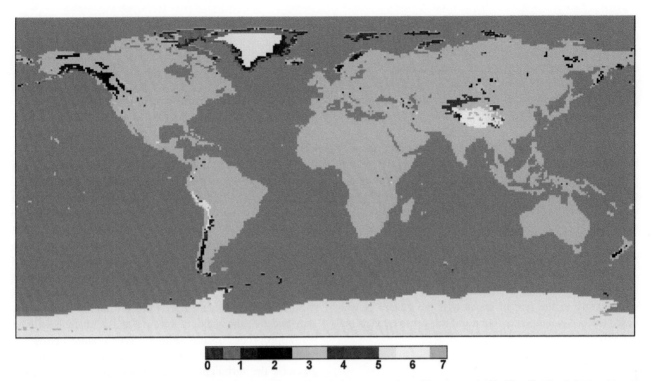

FIGURE 8.3 Geographical variation of the equilibrium-line altitude (ELA) in km, as approximated by averages within $1° \times 1°$ cells of glacier midrange altitudes from the World Glacier Inventory and other sources. The midrange altitude is the average of the glacier's minimum and maximum altitudes, and approximates the ELA well at this scale (Cogley and McIntyre, 2003). Ice sheets, for which the ELA is not estimated, are light pink; ice-free land is light grey; and glacierized cells lacking information on their ELA are black.

The accumulation-area ratio, abbreviated AAR and symbolized as α, is closely related to the ELA and to the mass balance. It is equal to the area, S_c, of the accumulation zone divided by the area, S, of the glacier,

$\alpha = S_c /S$. The AAR is bounded between 0 and 1. An α of 1 requires that the glacier lose mass entirely by frontal ablation, as is nearly true of the Antarctic Ice Sheet. An α of 0 implies that the ELA is above the maximum

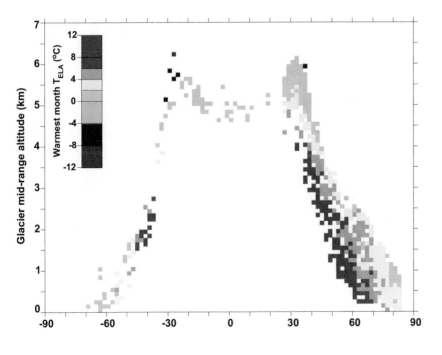

FIGURE 8.4 The ELA as a function of latitude. The information of Figure 8.3 is assembled into cells spanning $2°$ in latitude and 100 m in altitude, then coloured by averaging climatological averages from a reanalysis for 1968–1996 (Kalnay et al., 1996) of the free-air temperature of the warmest month at ELAs falling within each cell.

elevation of the glacier, which will dwindle and eventually vanish if the state $\alpha = 0$ persists. On glaciers with no frontal ablation, annual AAR is always well-correlated linearly with annual mass balance. Smaller AAR implies more negative mass balance. The AAR plays an essential part in one of the projections of glacier evolution during the twenty-first century, to be discussed in Section 8.6.1.

Glacier ice that is at its pressure-melting point is said to be temperate, while glacier ice below the pressure-melting point is said to be cold. Neglecting seasonal warming and cooling near the surface, a column of ice that is partly temperate and partly cold is said to be polythermal. The three adjectives are often applied to entire glaciers. The adjectival distinction is significant. For example, temperate ice is likely to transmit surface meltwater. However, should it percolate to a cold part of the column, the meltwater will refreeze. Furthermore, basal sliding is likely to contribute significantly to the rate of ice flow if the glacier is wet-based (its bed is temperate) but not if it is dry-based (its bed is cold).

8.2.2. The Radiation Balance

Glaciers have radiation balances exactly like those of other natural and constructed surfaces:

$$R_n = (1 - \alpha_s)K_\downarrow + I_a - \varepsilon \sigma T_s^4 \qquad (8.1)$$

where the net radiation, R_n, is the sum of the absorbed solar irradiance (the incident irradiance, $K_\downarrow$, multiplied by 1 minus the surface albedo α_s) and the incident downwelling radiation from the atmosphere, I_a, minus the surface emittance. The surface emittance is proportional to the fourth power of surface temperature T_s. The Stefan—Boltzmann constant, σ, is equal to 5.67×10^{-8} W m^{-2} K^{-4}. Snow and ice are usually treated as black bodies, that is, their emissivity, ε, is assumed to be 1.

Glaciers are distinctive among natural surfaces in that their surface albedo is typically high, although debris cover can eliminate this difference. Exposed glacier ice is usually darker than snow. Snow, if fresh, may absorb three times less solar radiation than the ice that it covers, and its seasonal disappearance is followed by a marked shift in the energy balance to a more absorbent regime in which, other things being equal, melting is accelerated. For present purposes, however, the main radiative peculiarity of glaciers is that their temperature is no greater than the pressure-melting temperature, T_m, which is equal to 273.15 K at the surface, so that the surface emittance is no greater than about 316 W m^{-2}. It can be much lower in winter and at night, and net gain of infrared radiative energy is not unusual on glaciers.

8.2.3. The Energy Balance

The energy balance at the surface of a glacier is:

$$R_n + H + L_v E + G_s + L_f A_s = 0 \qquad (8.2)$$

where all the quantities are flux densities (that is, in W m^{-2}), positive when directed towards the surface. R_n is the net radiation from Equation (8.1). H and $L_v E$ are turbulent fluxes of sensible and latent heat respectively from the air; $L_v = 2.834$ MJ kg^{-1} is the latent heat of sublimation. G_s is the conductive flux of heat from beneath the surface. Finally, $L_f A_s$ represents the energy used for surface melting, with $L_f = 0.334$ MJ kg^{-1} the latent heat of fusion and A_s the surface meltwater flux (kg m^{-2} s^{-1}).

Because the air above glaciers is often warmer than the freezing point in summer, and is often a heat source fuelling radiative cooling of the surface in winter and at night, the sensible heat flux is generally directed downwards. The latent heat flux is often directed downwards also because the vapour pressure at the surface will be that appropriate to saturation at a temperature below T_m, or at T_m if meltwater is present. On the lower parts of glaciers, the turbulent fluxes are enhanced by katabatic drainage of cooled air from high elevations. The katabatic wind, as well as being persistent and directionally constant, can be extremely strong and an important component of the local climate (cf. Evans et al., 2012, this volume).

The heat exchanged with the interior of the glacier drives an annual variation of temperature that is confined to the upper 10 m—15 m. However, in summertime, if and when an isothermal surface layer at the freezing point has been established, the heat flux, G_s, must dwindle to zero and any surplus from the atmospheric terms in Equation (8.2) will be used for melting. This surplus is responsible for most of the ablation on most glaciers, exceeding $-10,000$ kg m^{-2} yr^{-1} (about -100 W m^{-2}) near glacier terminuses in lower latitudes. It may be responsible for advective heat transfer, $\rho_w L_f u_p$, where u_p is the vertical percolation rate, to the interior of the glacier if the meltwater refreezes at depth instead of running off.

Glaciers also have basal energy balances. The energy sources at grounded glacier beds are frictional heat and geothermal heat. Geothermal heat fluxes are of the order of 0.05 to 0.10 W m^{-2}. Frictional heating derives from the loss of potential energy in the ice column as it moves downslope. Its rate can be expressed as the product of basal velocity and basal shear stress, yielding typical fluxes of 0.01 to 1 W m^{-2}. These fluxes are small, 0.1 W m^{-2} being equivalent to -10 kg m^{-2} yr^{-1} of basal melting, but they represent heat sources without compensating sinks, so they tend to make cold glaciers steadily warmer. Once the basal ice has reached its pressure-melting temperature, upward conduction into the body of the glacier ceases and the available energy is used to melt ice. This accounts for one of glaciology's bigger surprises: that the beds of many glaciers, including large parts of the beds of the ice sheets in Greenland and Antarctica, are wet. The pressure-melting point of water saturated with air decreases at -0.86 K km^{-1}, such that, beneath 4000 m of ice, T_m is 3.4 K below its value at the surface (Cogley et al., 2011).

The basal energy balance can be profoundly different when the ice is afloat (Holland and Jenkins, 1999). The frontal energy balance may become significant when a grounded terminus is in contact with sea water or lake water. Most of the concern centres on tidewater glaciers, that is, those in contact with seawater. The principal controls on the basal energy balance are the properties 'imported' by the meso-scale water flow. Warm water, more or less salty, can be advected to the neighbourhood of the glacier by currents. The resulting convective heat transfer to the ice can be substantial. The ice-water contact is at the temperature T_m, which depends on the pressure of the overlying ice (or the water depth at a grounded terminus) and the salinity of the water. There are two complications. Firstly, salt is coupled to heat because freezing increases and melting decreases the salinity of the seawater, altering both T_m and the buoyancy of the water. Secondly, the water flow itself is driven substantially by variations of temperature and salinity. The meltwater is buoyant because it is fresh, and flows upwards along the base of the rapidly-thinning ice to where a lesser pressure implies higher T_m and lesser sensible heat transfer. Sometimes, the meltwater flows into an environment that is actually colder than T_m, and ice begins to form, accreting as 'marine ice'.

Direct measurements are very difficult, but Rignot and Jacobs (2002) have measured West Antarctic basal melting rates indirectly at 23 grounding lines. These rates pertain only to areas of limited extent, but they are extremely high. The greatest recorded magnitude, 425 W m^{-2} at Pine Island Glacier, entails the melting of 44 m of ice per year because conduction of heat upwards into the glacier is unlikely to be significant (Cogley, 2005). More recently, Rignot et al. (2010) have reported rates of frontal melting at outlet glaciers in Greenland that are highly variable, but are comparable with rates of calving. The significance of this rapid basal and frontal melting for the force balance of the ice extends far upglacier, because it 'pulls' grounded ice across the grounding line.

A concern with energy fluxes at the bottom of the glacier, hundreds or thousands of metres below the surface, may seem out of place in a book about the climate. But, whether the basal ice is grounded or afloat, lack of understanding of these basal energy exchanges is the fundamental reason for our inability to describe, let alone to predict, how the ice sheets and other tidewater glaciers respond to climatic forcing.

8.3. GLACIER MASS BALANCE

8.3.1. Terms in the Mass-Balance Equation

The mass balance, ΔM, is the change in the mass of the glacier, or part of the glacier, over a stated span of time. It is the sum of accumulation, C (all gains of mass), and ablation, A (all losses of mass, treated as negative quantities):

$$\Delta M = C + A \qquad (8.3)$$

However, the relevant processes can be distinguished more clearly if it is written as:

$$\Delta M = B + A_f \qquad (8.4)$$

where, in the terminology of Cogley et al. (2011), B is the climatic-basal mass balance and A_f is frontal ablation. The climatic-basal balance is the sum over the extent of the glacier of the surface mass balance, the internal mass balance, and the basal mass balance. Frontal ablation is the sum of losses at the glacier margin by calving, subaerial melting, and sublimation and subaqueous melting.

The surface balance is more often measured than the other balance components, some of which it is not practical to measure while some are zero on many glaciers. The so-called glaciological method of in situ measurement of the surface mass balance relies on stakes and snow pits, and yields no information about the internal and basal balances. In geodetic measurements, the balance is estimated by repeated mapping. The change of glacier volume is obtained as the difference of glacier surface elevation, which is multiplied by an assumed average density to obtain the change of mass. Again, the measurement is ambiguous as to the internal and basal balances, and in particular as to internal accumulation, which can produce a decrease of volume due solely to the increase of density following from refreezing. Gravimetric observations with the GRACE satellites (e.g., Arendt et al., 2008) measure mass change directly, but have spatial resolution of the order of 200 to 300 km and do not resolve the components of the climatic-basal balance, or indeed of the total balance.

Accumulation by snowfall may be augmented by avalanching and vapour deposition, and may be diminished by wind scouring if the scouring is followed by either sublimation or transfer of the remobilized snow across the glacier outline (e.g., Bintanja, 1998). Avalanching can become a dominant term in the balance of very small glaciers in sheltered hollows. Such glaciers may, thus, be less reliable guides to the climate than more exposed glaciers, or at least different guides. An important anticipated consequence of the anthropogenically enhanced greenhouse effect is a more intense hydrological cycle, in which evaporation and precipitation both increase, but in geographically disparate ways. There is some evidence for this in observations (X.B. Zhang et al., 2007; Wentz et al.,

2007). Wentz et al. (2007), for example, show that both precipitable water and precipitation have increased recently, consistent with an expectation that, although the temperature has risen, the relative humidity should not change much. They point out, however, that, for reasons that remain unclear, the climate models tend to simulate more moderate increases in precipitation. Increased snowfall on glaciers is, nevertheless, an expected consequence of climatic change and has been detected on the ice sheets (e.g., Helsen et al., 2008). Trends in accumulation on smaller glaciers have not yet been studied.

Surface ablation occurs mainly by melting, although it should be noted that standard measurement methods would not distinguish between melting and sublimation if the latter were significant. The leading fact about surface ablation is that it is found to be proportional to elevation. Barring possible net losses due to sublimation or removal by the wind, ablation is zero at the dry snow line and decreases downwards along the surface to a minimum at the terminus. (Recall that ablation is negative.) The variation with elevation is often linear, as is that of surface temperature, helping to explain the success of temperature-index models at simulating ablation.

Internal accumulation is dominated by the refreezing of percolating meltwater in firn. It is negligible in temperate glaciers but is a difficult bias to control for in cold and polythermal glaciers, on which it is not detected by glaciological measurements. The main energy source for internal ablation is the potential energy released by downward motion of the flowing ice and of surface meltwater. This energy, and possibly some kinetic energy, is used to melt the walls of englacial conduits, and can sometimes contribute moderately to the climatic-basal mass balance of temperate glaciers. The magnitude of internal ablation has to be calculated rather than measured (Trabant and March, 1999), although it is often neglected.

There is no published estimate of the relative global extents of cold, polythermal, and temperate ice, but the temperature at 10 m to 15 m below the surface can be approximated by the mean annual surface air temperature, T_s, if there is no refreezing. Some internal accumulation should be expected wherever T_s is below the freezing point, which suggests that it is likely to be non-negligible in the upper parts of many glaciers that would be labelled as temperate. Its importance is generally acknowledged in modelling of the mass balance, but the parameterizations of refreezing are typically very simple and the values chosen for the parameters have very limited observational support. The most commonly made assumption is that the firn and the surviving snow have a capacity for refrozen meltwater that is proportional to the temperature. All surface meltwater goes to satisfy that capacity and, when it is reached, all subsequently produced meltwater runs off. An ambiguity between observations and models should be noted

with respect to the terms 'internal accumulation' and 'refreezing'. The latter includes refreezing in the snow, which is generally detected in glaciological observations, while the former does not; the models generally do not distinguish between the two.

The basal mass balance is not measured but, at least for grounded ice, it can be calculated with a fair level of confidence. It is usually estimated to be a small contributor to the climatic-basal balance of grounded ice, although basal ablation and accumulation are major terms for floating ice.

Measurements of frontal ablation are difficult and are not made routinely. Calving, the mechanical detachment of blocks of ice, is an irregular and, at present, an unpredictable process. Except for dry calving from cliff-like termini on land, it is zero for land-terminating glaciers. It may be a significant balance component for lake-terminating glaciers but, on the global scale, calving is essentially a feature of tidewater glaciers and ice shelves. Where it has been evaluated reliably, for example by monitoring ice discharge through gates just upglacier from the calving margin, as well as advance or retreat of the margin, it has been found to be equal to a substantial fraction of the magnitude of total ablation.

There are two ambiguities to be aware of in respect of calving. Firstly, it is not necessarily the same as discharge across the grounding line, which may well be of more interest, especially if the concern is with changes of sea level. Discharge at the grounding line is more easily measured, for example by radar interferometry together with thickness measurements by radio echo sounding (Rignot et al., 2008). Secondly, it is sometimes said that estimates of average glacier mass balance are based only on measurements of the climatic-basal balance (often, just the surface balance) and neglect calving. Here, the verb 'neglect' is inaccurate and should be replaced by 'under-represent'. For example, the three independent assessments that were combined by Kaser et al. (2006) are not based on any measurements that neglect calving. In terms of Equation (8.4), all of the measurements employed are of ΔM, not of B, and the problem is that there are many fewer measurements on glaciers with non-zero A_f. Cogley (2009a) produced circumstantial evidence that calving glaciers have more negative mass balance than others at the present day, but it remains challenging to produce routine large-scale estimates that are not biased by the shortage of measurements of frontal ablation.

Subaerial frontal melting and sublimation have been little studied. They are likely to be insignificant by comparison with the subaqueous frontal melting documented by Rignot et al. (2010). Heat transfer from water to ice can be far more efficient than transfer from the air. Evidence for the incursion of warmer seawater into sub-ice cavities in West Antarctica was first produced by Rignot and Jacobs (2002), but more dramatic incursions into Greenland fiords have been documented (and dated) recently by Holland et al. (2008) — see also Straneo et al. (2010). These publications have exposed a major gap in understanding of the role of the ocean as a trigger, via frontal ablation, for rapid change in the mass balance and dynamics of tidewater glaciers. Most land-terminating glaciers do not have the cliff-like terminus that is a prerequisite for separating the frontal balance from the surface balance. Of those that do, the plateau glaciers on Kilimanjaro are notable for the protracted retreat of their vertical walls (Cullen et al., 2006).

8.3.2. Definitions and Units

The time span of a mass balance is never separable from the mass change and must always be stated, but whether to express the mass change as a magnitude or a rate (by dividing the change by the time span) is a matter of expediency. In this study, most balances are expressed as rates. The year is a natural choice for the unit of time, although the day is sometimes more convenient.

Cogley et al. (2011) explain the units that are used in glaciology. In particular, 'specific' means 'per unit area'. A specific mass balance is a mass balance per unit of horizontally-projected area. Specific units (such as kg m^{-2}) are valuable in comparing point mass balances in different parts of the glacier, in comparing glaciers one with another, and in assessing the relations between mass-balance components and the mass and energy flux densities that constitute the forcing for the mass balance. The often-seen mm w.e. (millimetre of water equivalent) is numerically identical to the kilogram per square metre.

In climatology, and in studies of the ice sheets, dates usually contain calendar years. As a reminder that mass-balance calculations for glaciers other than the ice sheets nearly always refer to mass-balance years, beginning in September or October in the Northern Hemisphere, dates that refer to mass-balance years are given here in the form *yyss/tt*, where *tt* is greater by 1 than *ss* and denotes the calendar year in which the mass-balance year ends.

For the analysis of mass exchange between the glaciers and other stores of water in the hydrosphere, such as the ocean, specific quantities must be converted to totals in units such as kg or Gt. As a convenience, however, many of the mass balances and masses herein are expressed as sea-level equivalents, m SLE or mm SLE. A mass of water or ice can be converted to m SLE, itself a specific unit, by dividing by the product of the density of fresh water, 1000 kg m^{-3}, and the area of the ocean, 362.5 Mm^2 (Cogley et al., 2011), with a change of sign, when necessary. More explicitly, 1 Gt = 10^{12} kg = 1.0/362.5 mm SLE. This conversion fails to allow for the viscoelastic response of the solid earth to differential loading, but that response

has a timescale of several thousand years and is of no consequence for present purposes. The conversion also assumes that the ice is not grounded below sea level and therefore already partly displacing ocean water; that the area of the ocean is not altered by shoreline or grounding-line migration; and that mass lost by glaciers is not gained by aquifers or lakes.

8.4. MODELLING TOOLS

8.4.1. Volume–Area Scaling

Sustained mass loss implies shrinkage of the glacier, reduction of its thickness, and eventual disappearance. Given measured or modelled rates of mass loss, and initial thickness (with a reasonable assumption about bulk density), the mass loss, shrinkage, and thinning can be tracked by simple book-keeping, for which the most frequently used tool is volume–area scaling.

Volume–area scaling is glaciological jargon for the observed tendency of glacier volume to be proportional to glacier area. It is a misnomer, because the independent quantity that is found to be proportional to area is mean thickness (Figure 8.5). Objections to the statistical propriety of correlating area with volume, because volume is the product of mean thickness and area, are therefore without merit. In volume–area scaling, a two- or sometimes three-parameter relationship is fitted to measurements of mean thickness and area. There are, however, at most a few hundred such measurement pairs, although the

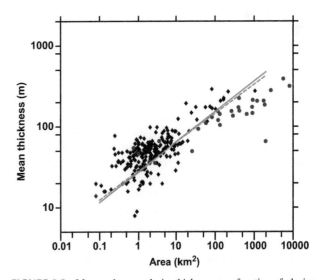

FIGURE 8.5 Measured mean glacier thickness as a function of glacier area for a compilation from the literature of 226 valley and mountain glaciers (diamonds) and 30 ice caps (circles). The most commonly employed scaling relationships are shown as lines (dashed: Chen and Ohmura, 1990; solid: Bahr et al., 1997); they were derived from smaller samples ($n = 61$ for Chen and Ohmura; $n = 144$ for Bahr et al.) excluding ice caps, for which a different scaling is appropriate.

area is known and tabulated for more than 10^5 glaciers. Scaling relationships such as those seen in Figure 8.5 have, thus, become the basis for estimating the volume and mass of glaciers whose thickness has not been measured.

Bahr et al. (1997) offer an elegant argument, finding that mean thickness ought to be proportional to the 3/8ths power of area. Measured mean thicknesses, H, are well-described by a relation with area, S, of the form:

$$H = cS^\beta \qquad (8.5)$$

which is the basis for the volume–area relation:

$$V = SH = cS^\gamma \qquad (8.6)$$

in which $\gamma = 1 + \beta$, and c and β are the fitted parameters. There is good evidence that estimates of the exponent β from observations, near to 0.36 (see also Chen and Ohmura, 1990), are consistent with the theoretical expectation, although the factor c appears to be more variable.

Equation (8.5) can be inverted. Suppose that a known mean thickness H changes by a known amount ΔH:

$$H_* = H + \Delta H \qquad (8.7)$$

We can write:

$$S_* = (H_*/c)^{1/\beta} \qquad (8.8)$$

for the new area S_* implied by the thickness change and the parameter set (c, β). If $\Delta H < -H$, the glacier disappears. ΔH itself is obtained from the measured or modelled mass balance ΔM as:

$$\Delta H = \Delta M/\rho \qquad (8.9)$$

where $\rho = 900$ kg m^{-3} is the density usually assumed for the glacier.

One limitation of scaling relations such as Equation (8.5) is their simplicity. Glaciers of the same area can vary greatly in thickness (Figure 8.5) and other ways of parameterizing mean thickness have been proposed. For example, Haeberli and Hoelzle (1995) related thickness to the surface slope and the basal shear stress, itself parameterized in terms of the elevation range of the glacier. Although there are good dynamical reasons for expecting such a relationship, the observational support, traceable to Maisch and Haeberli (1982), is not strong. A fruitful practical approach would be to estimate thickness with a multivariate relationship, but this has not yet been attempted.

8.4.2. Temperature-Index Models

Temperature-index models are models of mass balance in which surface ablation is estimated as a function of temperature, often near-surface air temperature measured either on the glacier or at the nearest weather station. Temperatures may also be taken from upper-air soundings, meteorological datasets (as in the promising recent work of Hirabayashi et al., 2010), or climate models. Most often, the chosen index of temperature is the positive degree-day sum.

The positive degree-day sum, φ, over any timespan is the integral of the excess of temperature, T, above the melting point, T_m:

$$\varphi = \int_{t_1}^{t_2} \max[0, T(t) - T_m] \, dt \qquad (8.10)$$

In practice, $T(t)$ is available as a series of averages over some time step, or instantaneous values at some interval, of near-surface air temperature, T_i ($i = 1, \ldots, n$), and the expression, with T_i in degrees Celsius, becomes:

$$\varphi = \Delta t \sum_{i=1}^{n} \max[0, T_i] \qquad (8.11)$$

where Δt is expressed in days. The ablation, a, by melting and sublimation over the span, at a point on the glacier surface, is assumed to be linearly proportional to φ:

$$a = -f\varphi \qquad (8.12)$$

(By glaciological convention, lower-case letters are used for mass-balance components of parts of a glacier, while upper-case letters are used for the whole glacier or for collections of glaciers.) The degree-day factor, f, is a parameterization, and therefore a simplification, of the energy balance. It is usually treated as one or more constants. Hock (2003) has reviewed published determinations of the degree-day factor (Figure 8.6). The median for snow is 4.5 kg m^{-2} (K d)$^{-1}$ and for ice it is 7.6 kg m^{-2} (K d)$^{-1}$, greater by a factor

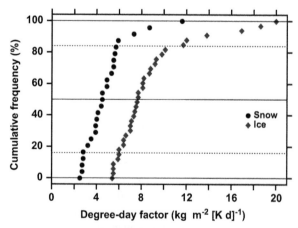

FIGURE 8.6 Degree-day factors for snow and ice estimated from field measurements and summarized in table 1 of Hock (2003). The median (50th percentile) of the cumulative frequency distribution is shown as a solid horizontal line and the 16th and 84th percentiles are shown dotted.

of 1.7. The strongly skewed upper tail of the distributions in Figure 8.6 represents measurements in Greenland, Svalbard, Switzerland, and the Himalaya, and the reasons for it appear to be diverse and not well understood. The body of the distribution, below the 84th percentile, is almost linear for both snow and ice. Its range on the f axis, about 3 and about 5 kg m^{-2} (K d)$^{-1}$ for snow and ice respectively, can be conjectured to reflect, for 'normal' circumstances, a combination of measurement uncertainty and the extent to which the temperature-index approach simplifies the energy balance.

Ohmura (2001) has explained the success of temperature-index models. During the ablation season, the glacier surface is at the melting temperature, the near-surface air is warmer than the surface and the sensible heat flux (Section 8.2.3) is directed downwards. However, Ohmura stressed the role, firstly, of the atmospheric infrared radiation as an energy source for ablation and, secondly, of the near-surface air temperature as a predictor of the atmospheric radiation, much of which comes from the lowest 1 km of the atmosphere. The near-surface air temperature can thus be a very good predictor of the largest energy source for ablation and can also be strongly correlated with the sensible heat flux. There is also usually a correlation with the solar irradiance. Sicart et al. (2008), however, present counter-examples to all of these generalizations, noting that, although atmospheric radiation is indeed the main energy source in most cases, it tends to vary little and its variations are due mainly to variations in cloud cover.

In the context of glaciological modelling, temperature-index models are valuable because temperatures are among the most widely available of climate model outputs, and their attributes, including their defects, have been analysed intensively. The temperature-index models go far towards solving the problem of coupling climatic forcing, at large spatial scales, to the response of the glaciers, all of which are smaller than a GCM grid cell. (The two ice sheets are special cases.) There are residual problems, the greatest of which is that climate model topography is smooth but the glaciers have large vertical ranges. The solution adopted for this problem (e.g., Radić and Hock, 2011) is to apply a suitable lapse rate, and usually also a bias correction, to climate model surface temperatures. Sometimes, the climate model's free-air temperatures are interpolated to glacier surface elevations.

Glacier models that rely on climate models for forcing normally derive accumulation from the climate model precipitation. The fraction of precipitation that is snow is estimated as a function of interpolated temperature. Sometimes, a lapse rate of precipitation is also implemented, although the tendency of precipitation to increase with elevation is less well-constrained either by physical understanding or by observations.

8.4.3. Energy-Balance Models

Hock (1999) showed that the performance of a temperature-index model improved significantly when a radiative term and a correction for shading by topography were introduced. This model is representative of several that are intermediate in complexity between the simple, but sufficiently successful, temperature-index models and fully-fledged energy-balance models that solve Equation (8.2) explicitly. Energy-balance models, however, are more demanding than temperature-index models and are not often used in long or large-scale simulations. The minimum of additional required information beyond near-surface air temperature includes net radiation, wind speed, humidity, and vertically-resolved subsurface temperature.

All modern GCMs, of course, have land-surface schemes that compute the components of Equation (8.2). The difficulty is that the glaciers are not resolvable at the scale of a GCM grid cell. The GCMs cannot resolve the strong dependence, $\partial a / \partial z$, in specific units, of the ablation on elevation (Figure 8.1), nor can they resolve the widespread tendency for $\partial c / \partial z$ to be significant, that is, for accumulation to be greater at higher elevations; hence, the need for off-line calculations of the glacier response to climatic forcing.

Bougamont et al. (2007) provide a cautionary example of the possible limitations of the temperature-index approach. Comparing a positive degree-day model and an energy-balance model of the surface mass balance of the Greenland Ice Sheet, they find that the positive degree-day model produces almost twice as much meltwater runoff as the energy-balance model. Much of the difference is due to the more detailed simulation of subsurface temperature, and consequently the greater refreezing, in the energy-balance model. In the near absence of observational support, however, it is difficult to judge between the different parameterizations of refreezing.

8.4.4. Mass-Balance Sensitivity

Changes in the climatic mass balance, B, can be understood in terms of sensitivity to both temperature and precipitation. It is always found (e.g., Oerlemans and Reichert, 2000) that at a given location the sensitivity to precipitation, $\partial B / \partial P_r$, where $P_r = P/P_{ref}$ is the actual precipitation normalized by the precipitation of some reference period, contributes less to change in B than does the sensitivity to temperature, $\partial B / \partial T$. Also, it is generally found that there is a non-zero sensitivity of precipitation to temperature, the climate growing wetter as it grows warmer.

The sensitivity to temperature is a close relative of the degree-day factor, f, discussed above (Equation (8.12)), but it is important to be aware that $\partial B / \partial T$ is estimated sometimes from the output of temperature-index or

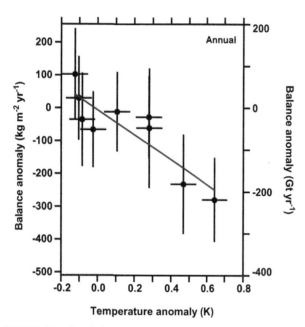

FIGURE 8.7 Correlation of the anomaly (relative to the 1961–1990 average) in pentadal (5-yearly) mean annual climatic mass balance, *B*, (Kaser et al., 2006) with the corresponding anomaly in *T*, surface air temperature over land (CRUTEM3; Trenberth et al., 2007). With 2-standard-error uncertainties, the fitted line suggests a proportionality $\partial B/\partial T$ of -379 ± 170 kg m^{-2} yr^{-1} K^{-1} or (with the Kaser et al. area of 0.785 Mm^2) -297 ± 133 Gt yr^{-1} K^{-1} for the era of glaciological balance measurements (1960/61–2003/04). (*Source: Steffen et al., 2008, with permission.*)

energy-balance models and sometimes from a comparison of measurements of *B* and *T*. In the latter case, changes in variables other than temperature are often neglected as possible causal agents. Furthermore, at large (for example, global) spatial scales the sensitivity may be estimated over a reference period, as in Figure 8.7, but is unlikely to remain unchanged over a longer simulation period. One reason is that the simulated *B* will be an integral over a varying, usually a diminishing, total glacierized area. In Figure 8.7, even the shrinkage of the glaciers over the 45-year reference period has been neglected, for lack of adequate information. Nevertheless, sensitivities estimated for the recent past do tend to agree well. Raper and Braithwaite (2006) and Meehl et al. (2007) quote a total of six sensitivities to global temperature that were obtained in diverse ways. They range from -320 to -410 kg m^{-2} yr^{-1} K^{-1}, with an average of -373 kg m^{-2} yr^{-1} K^{-1} that is almost identical to the sensitivity of Figure 8.7.

8.4.5. Models of Glacier Dynamics

Over decades or longer spans of time, detailed simulations of glacier evolution require accurate simulation not just of the energy and mass balances, but also of the force balance. There are subtle links between the force balance and the

mass balance, because the ice flows from where there is net accumulation to where there is net ablation, although the interplay is governed by glacier size and by the timescale of glacier motion. Typical resulting timescales (Greve and Blatter, 2009) are 10^2 years for a medium-sized valley glacier, 5×10^2 years for an ice shelf and 10^4 years for an ice sheet.

The force balance is not treated here; see, for example, Cuffey and Paterson (2010) for a standard development. At present, the much longer glaciological timescales rule out any direct computational coupling between glacier-dynamics models and climate models that take time steps of the order of 30 minutes. One critical dynamical problem, the mismatch between the ice-shelf and ice-sheet time-scales, or more generally the floating-ice and grounded-ice timescales (Section 8.7.2), is responsible in large part for the present lack of trustworthy projections of ice-sheet responses to twenty-first century climatic forcing.

Volume–area scaling and temperature-index modelling of the mass balance lead, when taken together, to coherent responses of glacier size to glacio-climatic forcing. On the whole, they appear to be reasonably successful substitutes for detailed simulation of the force balance. Nevertheless, relatively simple models of glacier dynamics (Oerlemans, 2001) are also available and offer valuable insight, not limited to the validation of yet simpler models. For example, Reichert et al. (2002) coupled a mass-balance model and an ice-flow model to study two European glaciers that have long records of length fluctuations. The mass-balance model was coupled via seasonal mass-balance sensitivities to GCM output from a control simulation in which external forcing (solar, volcanic, and anthropogenic) was excluded, leaving only internal climatic variability. They demonstrated that the observed acceleration of retreat during the twentieth century is a highly improbable response to internal variability.

8.5. RECENT AND PRESENT STATES OF THE WORLD'S GLACIERS

8.5.1. Kinds of Change

Broadly speaking there are four ways in which we can measure glacier changes. In the climate of recent centuries, nearly all glaciers have been getting smaller, and the sign of the changes is not expected to change in the foreseeable future.

- *Retreat*: Retreat is reduction of the glacier's length. It is easy to measure, so there are many measurements, and difficult to relate to other measures of change (but see Section 8.5.2).
- *Shrinkage*: Shrinkage is reduction of the glacier's area. It requires a measurement effort greater than that for

retreat, but the measurements can be made readily by airborne and particularly satellite remote sensing. Discriminating glaciers from seasonal snow, and debris-covered ice from ice-free terrain, presents difficulties, but the quantity and information content of shrinkage measurements are growing steadily.

- *Volume change*: To measure volume change, measurements are required of the change not just of area, but also of mean thickness. Accurate geodetic measurement of thinning is challenging, but is contributing in increasing quantity and value to our understanding of the recent evolution of glaciers.

- *Mass balance*: The preferred measure of change in most contexts is the mass balance. The change of mass can be estimated from the change of volume if a suitable density can be assumed for the mass gained or lost, but until recent years most measurements of mass balance were made by the glaciological method (Section 8.3.1), which is time-consuming and expensive. Although in situ glaciological measurements cannot match the spatial coverage of geodetic measurements, or the temporal resolution (monthly or better) of gravimetric measurements by the GRACE satellites, they continue to provide the bulk of annually-resolved observational estimates of change.

8.5.2. Evolution of Glacier Mass Balance Since the Little Ice Age

Worldwide observations of glacier terminuses dating from as long ago as the sixteenth century have been compiled by Leclercq et al. (2011). The density of observations is sufficient for detailed analysis only after 1800. Leclercq et al. normalized the glacier length changes and calibrated them against normalized mass changes. The sample of single-glacier length changes is rather coherent when regionalized. It shows stability or terminus advance until the culmination of the Little Ice Age in the mid-nineteenth century, with retreat thereafter in all regions. A well-defined global signal emerges (Figure 8.8) when mass change is scaled to length change with a single glacier-shape parameter. Terminus advance and mass gain become retreat and loss respectively in about 1850. Leclercq et al. estimated the glacier mass loss between 1850 and 2001 (the start date of the regularized projections discussed in Section 8.6.1) as 88 mm ± 22 mm SLE (Section 8.3.2), at an average rate of about 0.6 mm SLE yr^{-1}. Depending on the total mass estimated for 2001 (Section 8.6.1), the implied fractional loss since the culmination of the Little Ice Age is of the order of one eighth or somewhat greater.

Detecting recent change in glacier mass balance is thus the problem of detecting, against a background of natural variability, a shift from a moderate rate of loss, as the cryosphere emerges from the Little Ice Age, to a faster rate

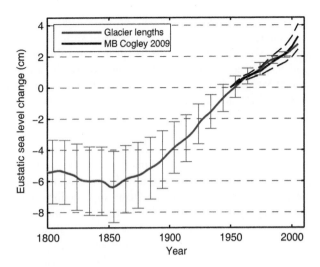

FIGURE 8.8 Reconstruction (red), from a scaling of mass change to length change, of the cumulative small-glacier contribution to sea-level change relative to an arbitrary zero in 1950. The cumulative mass-balance record (black, with upper and lower estimates shown as dashed lines) is from Cogley (2009). *(Source: Reprinted with permission from Leclercq et al., 2011. With kind permission from Springer Science+Business Media.)*

of loss, at least if anthropogenic radiative forcing has played its expected role. The problem is difficult at the century timescale of Figure 8.8, in part because, as will be seen in Section 8.5.5, there was a shorter-term relative maximum of mass balance in the 1960s and 1970s.

8.5.3. Measurements of Shrinkage

Reasonably accurate glacier outlines have been reconstructed for dates as old as the seventeenth century from terminal moraines and similar relict landforms. The oldest maps of adequate accuracy for the measurement of areas were made in Switzerland in the 1860s. Numerous more recent measurements have been published, but they remain uncompiled, so a quantitative global picture of the magnitude and variability of shrinkage rates is not readily available.

Cogley (2008), however, presented a preliminary compilation of 685 measurements of the shrinkage of single glaciers and 271 measurements of regional shrinkage. In the sample as a whole, typical shrinkage rates in the late twentieth century are $-0.10\%\ yr^{-1}$ to $-0.40\%\ yr^{-1}$ and, for glaciers or regions with two or more measurements (that is, three or more area measurements), it is found more often than not that the shrinkage rate has accelerated. But these generalizations conceal some important sources of variability that are difficult to handle. The most difficult is the strong but not universal tendency for regional rates to vary strongly with initial glacier size. Bigger glaciers are found to shrink at much slower rates. The smallest, and on average fastest-shrinking, glaciers have highly variable rates. However, some studies document an absence of this size dependence of shrinkage rate, and the size distribution (the frequency distribution of glacier area, as in Figure 8.9a) is not known at all for several extensively glacierized regions, so that any estimate of the globally-averaged rate would be very uncertain.

Such an estimate of the globally-averaged rate of glacier shrinkage would nevertheless be valuable. Apart from their intrinsic value as indicators of environmental change, shrinkage rates are important as the basis for providing more sharply-focused initial conditions for mass-balance projections such as those of Section 8.6.1. Depending on which region is chosen as a pattern, the global shrinkage rate might be $-2500 \, km^2 \, yr^{-1}$ (strong size dependence, as exhibited by the European Alps or the Queen Elizabeth Islands of arctic Canada) or it might be only $-500 \, km^2 \, yr^{-1}$ (no size dependence, as indicated by some studies in western North America). The initial areas in Table 8.1 derive from inventories spread over several decades in the late twentieth century, and the information cannot be corrected with confidence to the year 2001, or any other year.

The rates just quoted translate to a fractional reduction of extent over 50 years of from less than 5% to as much as 20%. Recently, Dyurgerov (2010), drawing on regional shrinkage rates where available, estimated the global total shrinkage for 1961–2006 as -5.6%, a rate of $-0.13\% \, yr^{-1}$.

8.5.4. Present-Day Extent and Thickness

Incomplete as it is, the World Glacier Inventory is the most reliable global source of physiographic information about glaciers. At present, the version containing the most glaciers, currently about 134,000, is that of Cogley (2009b). It accounts for nearly one half of the global extent of small-glacier ice, recording a varying number of attributes for each glacier. The inventory does not include glacier outlines or hypsometric curves explicitly, although glacier outlines are the central feature of the Global Land Ice Measurements from Space initiative (Raup et al., 2007), a resource of growing importance. Hypsometry can be derived readily by using the outlines to mask a digital elevation model, such as the ASTER Global Digital Elevation Model or that of the Shuttle Radar Topography Mission (Hayakawa et al., 2008). The stringent accuracy requirements for measurements of elevation change (Section 8.3.1) can be relaxed for the modelling of glacier–climate relations.

Figure 8.9a shows the frequency distribution of glacier area from the incomplete inventory. This distribution, although it has no value for the estimation of size distributions in regions with missing information, is useful for the purpose of putting glacier changes in perspective. For example, the evolution of glaciers in the present climate and that projected for the twenty-first century implies a shift of the histogram to the left, by the shrinkage and eventual vanishing, and occasionally by the fragmentation, of individual glaciers. That glaciers are vanishing is a journalistic commonplace and a correct observation, but those that have already vanished are those that were never very big in the first place. They probably formed during the Little Ice Age, and at present occupy the smallest-size classes of the histogram. While those classes are not negligible, most of the extent is accounted for by much larger glaciers that will be much longer-lived. The main reason for their expected longevity is seen in Figure 8.9b, which shows that the concentration of volume in larger glaciers is even more marked, by virtue of Equation (8.6), than that of extent.

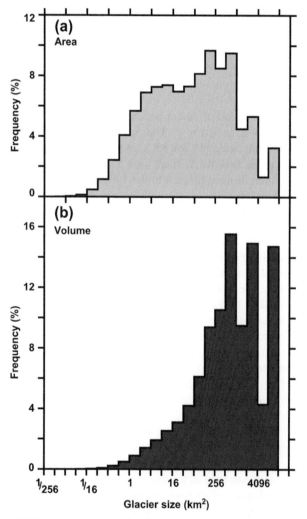

FIGURE 8.9 (a) Frequency distribution of the areas of 129,108 glaciers in the World Glacier Inventory (Cogley, 2009b). Total area covered is 0.343 Mm², representing slightly less than half of the total small-glacier area. (b) Areal frequency distribution of glacier volumes, obtained from panel a by volume–area scaling (Equation (8.6)) with the parameters of Bahr et al. (1997).

The most objective estimate of the total area of glaciers outside the ice sheets to have appeared so far, 0.741 mm $\pm$ 0.068 Mm^2, is that of Radić and Hock (2010). They generated the global total from the inventory of Cogley (2009b), supplemented by the older work of Cogley (2003) and by a much older estimate for peripheral glaciers in Antarctica. This Antarctic estimate includes the ice rises, bodies of grounded ice within the ice shelves. Nothing is known of their mass balance, but they are mostly at such high latitudes that they experience little or no surface ablation today. In addition to this source of uncertainty, the uncertainty assumed by Radić and Hock, $\pm 10\%$ for most of the glaciers, should be taken seriously. It is an antidote to the temptation to believe that areas are so well-known that errors can be neglected. In fact, the measurements themselves, or more accurately the mapmaking and image interpretation on which they are based, are quite uncertain. Inability to correct for shrinkage over the timespan covered by the measurements (Section 8.5.4) introduces further uncertainty.

Radić and Hock (2010) constructed glacier size distributions for regions without complete inventories by an upscaling procedure. By volume—area scaling they then estimated the total small-glacier mass as 600 mm $\pm$ 70 mm SLE (Table 8.1). This is dwarfed by the ice in the two ice sheets. Lythe and Vaughan (2001) estimated the mass of the Antarctic Ice Sheet as 22.6 million Gt, or about 62.5 m SLE. The latter figure becomes about 57 m SLE when ice below sea level is replaced by seawater. The mass of the Greenland Ice Sheet, based on the volume calculated by Bamber et al. (2001), is 2.69 million Gt or, without allowance for a small volume that lies below sea level, 7.4 m SLE.

The corresponding mean thicknesses are 2000 m for the Antarctic Ice Sheet and 1715 m for the Greenland Ice Sheet. The total volume and area of Radić and Hock (2010) yield a mean thickness of 294 m for the other glaciers.

8.5.5. Recent Evolution of Glacier Mass Balance

Measurements of glacier-wide mass balance by geodetic methods go back to the mid-nineteenth century and by glaciological methods to the mid-twentieth century. However, the number and distribution of the measurements is adequate for objective global estimation only since about 1960. Three recent global estimates (Ohmura, 2004; Dyurgerov and Meier, 2005; Cogley, 2005) were synthesized by Kaser et al. (2006) (see also Lemke et al., 2007) into a 'consensus estimate', the term reflecting the broad agreement between its components. The agreement is unsurprising because the component datasets are nearly the same, and the variation between them is due mainly to different treatments of the problem of spatial bias. The

measurements come disproportionately from some mountain ranges, while other regions of extensive glacierization contribute few or no measurements.

The most prominent feature of the consensus estimate is a relative maximum of mass balance centred near 1965—1970. The few measurements before 1960 indicate a more negative global average and, although the indications are weak, they are consistent with other lines of evidence (Section 8.5.2). It is clearer that in recent decades the global average has been growing more negative, although not steadily. The relative maximum of the 1960s and 1970s coincides with an episode of cooling, or slower warming, that is well-documented in temperature records (e.g., Jones and Moberg, 2003), and this coincidence underpins and quantifies (Figure 8.7) the concept of mass-balance sensitivity to temperature.

The most recent global estimates are those of Dyurgerov (2010) and Cogley (2009a). The Dyurgerov estimate is the first to allow explicitly for glacier shrinkage, which makes the global-average total balance less negative, although only moderately so, as time passes. The Cogley (2009a) estimate, summarized in Figure 8.10, resembles the Kaser et al. (2006) estimate, but there is at least one important new detail. In Figure 8.10a, the pentadal estimates are simple unweighted arithmetic averages of the available measurements. The blue circles are averages of geodetic measurements, newly compiled and used for the first time in a global assessment. They provide greater historical depth than the glaciological measurements (red squares) and suggest that the relative maximum at about 1965—1970 was a modest affair. (In the figure, the maximum appears as a minimum because mass balance and sea-level rise are of opposite sign.) More significantly, the balance averages are consistently more negative than the glaciological estimates. Considering that calving glaciers are better represented in the geodetic dataset, the most likely explanation, which has not yet been tested in detail, is that calving glaciers have more negative mass balances. In any case, including the geodetic measurements makes a substantial difference.

Figure 8.10b is the result of a detailed exercise in polynomial spatial interpolation, the aim of which is to correct the uneven representation of glacierized regions in the datasets. It extends back only to the 1950s because the measurements are too few for interpolation at earlier dates. The shapes of the two spatially-corrected histories are similar, but the offset seen in Figure 8.10a survives. (The dark blue circles are actually based on all measurements, combining the glaciological and geodetic datasets, as described by Cogley (2009a).) However, the spatial correction makes the average balances significantly less negative because regions suffering relatively moderate losses are under-represented in the data.

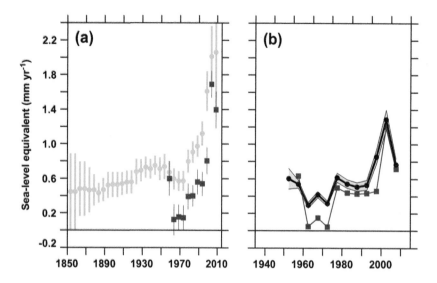

FIGURE 8.10 (a) Pentadal (5-yearly) arithmetic averages of all geodetic (blue circles) and glaciological (red squares) small-glacier mass-balance measurements, 1850–2010. Data updated from Cogley (2009a). Vertical bars represent ±2 standard errors about the mean. (b) Pentadal global averages of mass balance for 1950–2010 estimated with a correction for spatial bias of the datasets as in Cogley (2009a). Red squares rely only on glaciological measurements; dark blue circles with grey confidence envelope (2 standard errors) rely on both glaciological and geodetic measurements. The most recent pentad is incomplete; most reports for the 2009/10 balance year are not yet available.

The pentadal resolution of Figure 8.10 is chosen so as to increase statistical confidence, detrended annual mass balances being serially uncorrelated. However, it also maintains the independence of successive terms in the series while suppressing much of the inter-annual variability and, thus, showing that there is considerable variability in glacier mass balance at decadal scales. In particular, the very negative balances of the early 2000s have not been sustained in the most recent pentad, 2005/06–2009/10, in which glaciers have experienced more moderate although still large losses. This decadal variability means that projections of the simple if-present-trends-continue kind ought to be assessed with caution, at least to the extent of noting carefully the definition of 'present'.

In terms of the energy balance, the mass balance for the pentad of most rapid loss, 2000/01–2004/05, consumed about 6.7 W m^{-2} over the glacierized area, while the 36.4 mm SLE of loss over 1950/51–2008/09 consumed about 3.2 W m^{-2}. As components of the energy balance of the entire Earth, these fluxes are equivalent to 0.010 and 0.005 W m^{-2} respectively, and are small with respect to, for example, the ~0.2 W m^{-2} by which the ocean has been heated during 1950–2009 (Murphy et al., 2009; Sen Gupta and McNeil, 2012, this volume).

Technological advances in radar interferometry; radar and especially laser altimetry and, most recently, gravimetry have made measurements of the mass balances of the ice sheets a reality in the past 20 years. Syntheses with good temporal resolution have been made possible by advances in meso-scale meteorological modelling, which can now provide accurate simulations of accumulation, as well as in the parameterization (Section 8.4.2) or modelling (Section 8.4.3) of the surface energy balance to yield reliable estimates of melting and sublimation. The most recent

assessment of ice sheet mass balance (Rignot et al., 2011) is summarized in Figure 8.11. The climatic mass balance is simulated by a meso-scale climate model and the calving flux at the grounding line is obtained from interferometric measurements of velocity, measurements of ice thickness by radio echo-sounding, and an allowance for migration of the grounding lines of some of the largest outlet glaciers.

There is substantial inter-annual variability in the ice sheet balance series, but Rignot et al. (2011) estimate accelerations during 1992–2009 of −21.9 Gt ± 1.0 Gt yr^{-2} for the Greenland Ice Sheet and −14.5 Gt ± 2.0 Gt yr^{-2} for

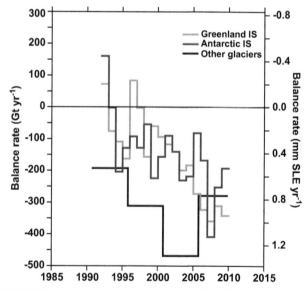

FIGURE 8.11 Annual mass-balance rate, 1992–2009, of the two ice sheets (calendar-year averages of monthly balances from figure 2 of Rignot et al. (2011)), and pentadal mass-balance rate (balance years 1990/91–2008/09) of all other glaciers (from Figure 8.10).

the Antarctic Ice Sheet. For comparison, a trend fitted to the small-glacier balance rates for 1970/71−2008/09 in Figure 8.10 yields an acceleration of -8.0 Gt $\pm$ 0.4 Gt yr^{-2}. The recent moderation of losses from the other glaciers means that the three reservoirs of ice have contributed to sea-level rise at comparable rates over 2006−2009, with the Greenland Ice Sheet contributing the most. The sum of the three contributions for 2006−2009 is -854 Gt yr^{-1} or 2.4 mm SLE yr^{-1}.

8.6. THE OUTLOOK FOR GLACIERS

8.6.1. Future Contributions to Sea-Level Rise

Table 8.1 and Figure 8.12 form the centrepiece of this study, summarizing the principal recent estimates of the global evolution of glaciers over the twenty-first century. The estimates are diverse in methodology. They range from detailed modelling of glacier response to modelled climatic forcing to estimates based on glacier physiography and on expert judgement.

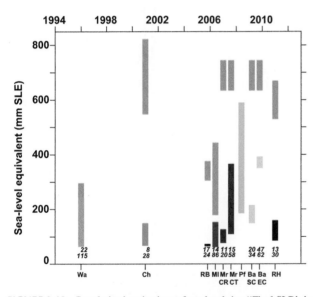

FIGURE 8.12 Regularized projections of sea-level rise ('Final SLR', in colour) due to small-glacier mass balance from 2001 to 2100. Bar height represents range of estimates; bar width is arbitrary. See also Table 8.1. Above each projection (except for Pf) is a grey bar representing Initial mass. Below each projection is given the range of projected ratios of 'Final SLR' to 'Initial mass', as percentages. (That is, the upper and lower numbers are respectively the ratios of the bottom of the coloured bar to the top of the grey bar and of the top of the coloured bar to the bottom of the grey bar.) Tick labels on vertical axis, if divided by 100, give the twenty-first century average rate of rise in mm SLE yr^{-1}. *(Sources: Wa (orange) − Warrick et al., 1996; Ch (green) − Church et al., 2001; RB (black) − Raper and Braithwaite, 2006; Ml (red) − Meehl et al., 2007; MrCR (purple) − Meier et al., 2007, constant rate; MrCT (purple) − Meier et al., 2007, constant trend; Pf (light blue) − Pfeffer et al., 2008; BaSC (yellow) − Bahr et al., 2009, steady climate; BaEC (yellow) − Bahr et al., 2009, evolving climate; RH (dark blue) − Radić and Hock, 2011.)*

In an attempt to eliminate variations that arise from discrepancies of technical detail, the projections of 'Final SLR' (cumulative sea-level rise, 2001−2100) have been 'regularized'. That is, where the authors' original projections were for different durations, they were multiplied by suitable factors near to 1.0 to yield estimates for a span of 100 years. In two cases (Church et al., 2001; Raper and Braithwaite, 2006), the projections were multiplied by equally crude ratios of glacierized areas so as to account for omission of the poorly-known peripheral glaciers in Greenland and Antarctica. Where confidence envelopes for Final SLR were presented as standard errors, and, in some cases, as standard deviations, they have been converted to ranges. Although some of the confidence envelopes were derived from end-to-end error analyses of the corresponding calculations, none of the calculations address all of the known sources of uncertainty, and so none of the confidence envelopes can lay claim to statistical completeness.

It is important, therefore, to realize that the results displayed here may depart from the authors' original intentions, although the regularization does enhance the comparability of the items of information.

The initial areas given in the sources have not been regularized in Table 8.1, and their uncertainty, which is substantial, has not been tabulated. The surveys and maps on which they are based are spread over several decades of the late twentieth century, with some regions having no adequate maps. There is, as yet, no clear picture of how regional shrinkage rates have evolved and it is not practicable to correct the areas objectively so that they represent a single year, such as 2001. The largest source of uncertainty, lack of knowledge of the glaciers peripheral to the two ice sheets, has been addressed in the regularization described above but, although estimates of their area are not very accurate, their mass-balance rates are even more poorly known. A particular ambiguity arises from the inclusion of peripheral glaciers in Antarctica, where the initial area may or may not include the ice rises (Section 8.5.4).

The specific balance rate in Table 8.1 is described as 'notional' because it represents ablation over 100 years divided by the glacierized area at the start of the interval. The area is certain to continue to decrease over the interval and actual balance rates per unit of surviving area will be more negative at each time during the interval. The notional rate, however, is useful as a rough guide to the intensity of mass loss.

8.6.1.1. IPCC Assessment Report 2

Warrick et al. (1996) estimated sea-level rise due to glacier mass balance with a model (Wigley and Raper, 1995) in which glaciers within a number of regions evolve, with a common regional response time, under forcing that is

TABLE 8.1 Regularized Projections of Mass Loss from Small Glaciers Between 2001 and 2100

Source	Initial Area (Mm²)	Initial Mass (mm SLE)	Final SLR (mm SLE)	Notional Balance Rate[a] (kg m^{-2} yr^{-1})	Remarks
Warrick et al., 1996[b] (IPCC SAR[c])	0.680	~270	65 – 280	−346 to −1493	Sensitivity of ice mass to temperature change; range of 3 projections under the IS92a scenario
Church et al., 2001[d] (IPCC TAR[c])	0.540	686 ± 137	69 – 151	−336 to −738	Sensitivity of mass balance to temperature, and volume–area scaling; range of 9 GCMs, IS92a scenario
Raper and Braithwaite, 2006[e]	0.522	342 ± 35	65 – 73	−430 to −480	Degree-day model, triangular hypsometric curve; A1B scenario, 2 GCMs
Meehl et al., 2007[f] (IPCC AR4[c])	0.512–0.546	180 – 444	64 – 155	−440 to −1066	Simple climate model, 6 SRES scenarios
Meier et al., 2007	0.763	690 ± 55	79 – 129	−375 to −645	Constant mass-balance rate
Meier et al., 2007	0.763	690 ± 55	112 – 368	−560 to −1840	Constant mass-balance trend
Pfeffer et al., 2008	n/a[g]	n/a	187 – 592	n/a	Deliberate search for an upper bound, involving expert judgement
Bahr et al., 2009	0.763	690 ± 55	151 – 217	−755 to −1085	Glaciers seek equilibrium with projected steady climate of 2006
Bahr et al., 2009	0.763	690 ± 55	352 – 394	−1760 to −1970	Glaciers seek equilibrium with evolving climate
Radić and Hock, 2011[h]	0.741	600 ± 70	87 – 161	−426 to −788	Degree-day model, triangular hypsometric curve; detailed sensitivity analysis; A1B scenario, 10 GCMs

[a] Final SLR (before regularization) converted to (negative) kg and divided by the product of initial area and the duration of the projection.
[b] Initial area from their table 7.1. Initial mass was assumed to have been 300 mm SLE in 1880; SLR was modelled as ~ 30 mm SLE from 1880 to 2001. Range of Final SLR obtained by multiplying the 2001–2100 SLR of their figure 7.8 by ratios of projections from their figure 7.7.
[c] AR5: SAR, TAR & AR4 are the Second, Third & Fourth Assessment Report.
[d] Initial mass and Final SLR multiplied by 1.37 = 0.741/0.540 to account roughly for omission of glaciers in Greenland and Antarctica.
[e] Initial mass is the estimate of Raper and Braithwaite (2005a) multiplied by 1.42 = 0.741/0.522 to account roughly for omission of glaciers in Greenland and Antarctica. The source's estimate of Final SLR has also been multiplied by 1.42.
[f] Initial area and mass from Lemke et al. 2007 (table 4.3, rows 1–3), but their initial mass multiplied by 1.20 for consistency with their treatment of glaciers in Greenland and Antarctica. Final SLR from their table 10.7. Average of 3 initial areas used in calculating notional balance rate.
[g] n/a: not available.
[h] Initial area and Initial mass from Radić and Hock (2010).

parameterized in terms of a temperature change required to raise the equilibrium line to the top of the glacier. An initial mass is prescribed, although the model's initial mass is inconsistent with the estimate of 500 mm $\pm$ 100 mm SLE tabulated earlier in the same assessment (in its table 7.1). The high end of the range in Figure 8.12 would require removal of more than the model's initial mass, but it was obtained in the present work, indirectly. Warrick et al. (1996) give a Final SLR for glaciers only for the middle of three IS92a projections, and the range as presented here was obtained crudely by scaling the Final SLR by the ratio of low-to-middle and high-to-middle projections of the sea-level rise from all sources. The Final SLR due to the glaciers in the middle projection is 160 mm SLE.

8.6.1.2. IPCC Assessment Report 3

Church et al. (2001) modelled sea-level change from 1990 to 2090 by the methods of van de Wal and Wild (2001). Temperature projections were taken from nine GCMs running under the IS92a emissions scenario. The sensitivity of mass balance to temperature and precipitation can be traced, apparently, to the results of degree-day and energy-balance modelling of the Greenland Ice Sheet. Changes of precipitation were neglected. Glacierized areas were allowed to evolve by volume–area scaling, with separate estimates for each of 15 glacier size classes in 100 regions. The shrinking extent yielded a Final SLR about 25% less than that obtained with the area held constant.

The Church et al. (2001) Final SLR in Table 8.1, 69 mm–151 mm SLE, is the range of results derived from experiments from nine GCMs but, in a different table, Church et al. present a range of 14 mm–315 mm SLE. (These two ranges are regularized as in footnote d of Table 8.1.) The broader range is the envelope of the 2-standard-error confidence regions surrounding each of eight of the experimental results (J.A. Church and J.M. Gregory, pers. comm.; a standard error of 40% was assumed). The minimum rises on which the low end of the broader range is based are all 30 mm SLE or less and, with one exception, all have been overtaken by events: the actual contribution to sea-level rise from 1990/91 to 2008/09 was nearly 20 mm SLE. Thus, the narrower of the two ranges appears in Table 8.1 because it seems more plausible.

8.6.1.3. Raper and Braithwaite (2006)

Raper and Braithwaite (2006) relied, where possible, on the World Glacier Inventory. Working with temperatures from two GCMs and a $1° \times 1°$ grid of glacierized areas, they applied a degree-day model to ideal glaciers of known vertical extent. The glacier ELAs were adjusted, year by year, in accordance with the GCM temperature change. The degree-day model was used to calculate the mass balance, which, in turn, was used to calculate volume change, area

change and change in the area–altitude distribution. The area–altitude distribution of each glacier was assumed to be triangular between its minimum and maximum altitude. Ice caps were considered separately, being assumed to have parabolic profiles. Where the glacier inventory was lacking or incomplete, Raper and Braithwaite (2006) extrapolated from better-known grid cells using a relationship between annual precipitation, summer temperature, and the vertical gradient of modelled mass balance. Thus, the model as a whole stepped through time in response to climatic forcing, with allowance for the dynamical response of the glaciers.

The Raper and Braithwaite (2006) projections are the lowest in Table 8.1. Comparison of their modelling procedure with those of others does not offer an obvious explanation, but it is striking that their estimate of 'Initial mass' (Raper and Braithwaite, 2005a) is well below that of most others. The starting point for their calculation is an old gridded dataset (Cogley, 2003) that was compiled by sparse sampling of small-scale maps, but the probable omission of smaller glaciers implied by the sampling method is unlikely to translate into a large discrepancy with other estimates of glacier volume. The smaller glaciers typically account for only small proportions of regional size distributions and yet smaller proportions of volume distributions, as in Figure 8.9. The lack of reliable glacier size distributions was addressed by modelling well-known regional size distributions in terms of topographic roughness, the latter obtained from TOPO30, a digital elevation model with 30 arc second horizontal resolution (~ 1 km at the Equator). There is no obvious flaw in their treatment of this relationship and its uncertainty, but it is not possible to validate it empirically outside the seven regions with reliable distributions, so a substantial under-representation on average of large valley glaciers remains a possible explanation (Meier et al., 2005), as was acknowledged by Raper and Braithwaite (2005b). Given the regional size distributions, the scaling by which Raper and Braithwaite (2006) proceed to estimate volumes is a widely-used method (Chen and Ohmura, 1990; Bahr et al., 1997). It seems improbable as an explanation of their low Initial mass, and indeed it is the basis of some of the other estimates in Table 8.1.

8.6.1.4. IPCC Assessment Report 4

Meehl et al. (2007) presented sea-level projections as ranges, which they characterized as "5% to 95% intervals", but acknowledged that they were in fact unable to assess likelihoods because of poorly known uncertainties. The projections were based on a simple climate model (updated from Wigley and Raper, 1992). The global sensitivity of mass balance to temperature was taken to be 0.8 mm $\pm$ 0.2 mm SLE yr^{-1} K^{-1}. Meehl et al. (2007) excluded Greenland

and Antarctic glaciers, but multiplied their results by 1.20 to allow for the exclusion. This factor may not be implausible, but it is below the ratio of total glacierized area to the area excluding the Greenland and Antarctic glaciers, which is variously estimated as between 1.30 and 1.45. The factor 1.20 therefore implies that the Greenland and Antarctic glaciers lose mass at less than the global average rate.

The Meehl et al. (2007) ranges may be better regarded as conservative estimates, more 1-sigma-like than 2-sigma-like and therefore more directly comparable with the ranges of the other estimates in Table 8.1. The ranges vary little between the six emissions scenarios from which the climatic forcing was taken. The intense fossil-fuel consumption of scenario A1FI leads to a range of 80 mm−170 mm SLE in Final SLR, while for the middle-of-the-road A1B scenario the range is 80 mm−150 mm SLE and for scenario B1 it is 70 mm−150 mm SLE. Cumulative sea-level rise varies little across the scenarios until the second half of the twenty-first century.

Like that of Raper and Braithwaite (2005a), the Initial mass of Meehl et al. (2007) is relatively low. There were in fact three Initial masses, each deemed equally likely, although it is not clear how the equal likelihood was implemented. One was that of Raper and Braithwaite (2005a), discussed above. The others were from Ohmura (2004) and Dyurgerov and Meier (2005). The Dyurgerov−Meier estimate, 370 mm SLE for glaciers outside Greenland and Antarctica, is closer to other recent determinations. However, the Ohmura estimate, 150 mm SLE for glaciers outside Greenland and Antarctica, is very low. Two thirds of the regional volumes are stated to come from a source that does not, in fact, tabulate volumes, and the largest of the regional estimates, for the Canadian Arctic, is from an early source and cannot be based on full inventory information. In the absence of documentation of the methods, this low estimate must be regarded as doubtful. Noting that a more recent assessment by volume−area scaling (Radić and Hock, 2010) found 193 mm SLE in the incomplete World Glacier Inventory, the Ohmura (2004) estimate can in fact be set aside.

8.6.1.5. Meier et al. (2007)

Meier et al. (2007) argued that, although they account for only about 1% of the total glacier reservoir (about 70 m SLE; Section 8.5.4), glaciers other than the ice sheets can be expected to dominate the cryospheric contribution to sea-level rise between now and 2100. Their estimates of Final SLR are very simple, resting on observations of the recent mass-balance rate and extrapolation to 2100 on two assumptions. Under the constant-rate assumption, the 1995/96−2004/05 balance rate, which they estimate at -527 kg $\pm$ 125 kg m^{-2} yr^{-1} or 1.1 mm $\pm$ 0.24 mm SLE yr^{-1}, is held

constant to 2100. Under the constant-trend assumption, the estimated 1995/96−2004/05 trend of -15.6 m ± 7.4 kg m^{-2} yr^{-1} is projected to be maintained until 2100.

The constant-rate assumption is likely to be very conservative if warming continues, because there is no reason to expect substantial weakening of the correlation of mass balance with temperature. The constant-trend assumption could perhaps be criticized for its simplicity, but a more serious objection is that under both assumptions the shrinkage of the glaciers is neglected. That is, the mass loss in 2100 is estimated to come from the same area of ice as at the beginning of the century. Meier et al. (2007) counter this objection by noting that most of the present-day glacierized area is accounted for by large glaciers, especially high-latitude ice caps, that will not shrink appreciably during the twenty-first century. They also note that under more intense radiative forcing the disappearance of many smaller glaciers in regions that are already warm will tend to be offset by progressively more negative surface mass balances in the coldest polar regions. In particular, meltwater losses from glaciers in southern Antarctica are currently negligible. This latter argument is of doubtful validity, however, because such glaciers are assigned equal weight with other glaciers throughout the span of the projection.

A significant limitation of the Meier et al. (2007) estimates, as of some others in this section, is that they are based on available mass-balance measurements, which under-represent calving glaciers (Section 8.3.1). Adjacent calving and land-terminating glaciers ought not to have very different mass balances, but the timescale of the mass-balance forcing can be much longer when calving is significant (Section 8.7.2), and one cannot be confident that measurements on land-terminating glaciers, on which the dominant term is the surface mass balance, are representative of calving glaciers. There is strong circumstantial evidence, both globally (Cogley, 2009a) and from studies of single glaciers, that calving glaciers have more negative mass balance at present than do land-terminating glaciers.

8.6.1.6. Pfeffer et al. (2008)

The question of differences of mass balance between calving and land-terminating glaciers was addressed by Pfeffer et al. (2008), who were concerned to estimate the plausibility of very large increases of sea level during the twenty-first century. Their main focus was the Greenland Ice Sheet, which is believed to be vulnerable to rapid mass loss by dynamic thinning, that is, accelerated ice discharge from the terminuses of calving outlet glaciers. However, this accelerated frontal ablation is also characteristic of many of the smaller glaciers. The lower Pfeffer et al. (2008) estimate of Final SLR was obtained by allowing the surface mass-balance rate to accelerate, as observed in recent years, and by assuming that frontal ablation would be in the same ratio to surface mass balance, as estimated for the Greenland Ice Sheet. The higher Pfeffer et al. (2008) estimate results from an assumption that frontal ablation increases by an order of magnitude during the first decade and remains high thereafter. It exceeds some of the lower estimates of initial mass in Table 8.1 and requires ice velocities that have only been observed rarely and for short periods (hours) on surging glaciers. Whether such fast rates of flow could become prevalent is not known. Very large losses from glaciers other than the ice sheets would require unprecedented ice discharges, accounting for 506 of the estimated total 592 mm SLE during 2001−2100. Such losses are believed to be extremely unlikely but are not known to be impossible.

8.6.1.7. Bahr et al. (2009)

Bahr et al. (2009) did not attempt to model the evolution of the glaciers in detail. Instead, they tried to establish a lower bound on mass loss by interpreting observations of the AAR (Section 8.2.1), which is measured or calculated routinely as a part of most measurements of annual mass balance.

It was learned long ago that on most land-terminating glaciers, zero balance corresponds to an AAR, α, in the neighbourhood of 0.5 to 0.7, and it has long been accepted that this represents a regularity of glacier shape that arises from characteristic patterns of variation of accumulation and ablation with altitude. Bahr et al. (2009) re-examined published records and found a best estimate of 0.57 for α_0, the AAR corresponding to zero mass balance. They also found that multi-annual averages of observed α are near to 0.44. This difference is consistent with evidence of predominantly negative mass balance but it also offers a way to estimate how much additional mass the present-day glaciers must lose because they are too big for the present-day climate.

The Bahr et al. (2009) analysis relies on volume−area scaling (Section 8.4.1). If ΔV_0 is the volume change required to bring a glacier of volume, V, and AAR, α, to equilibrium, that is, to an AAR of α_0, then:

$$\Delta V_0 = V \left[(\alpha/\alpha_0)^\gamma - 1\right] \qquad (8.13)$$

The result is that glaciers need to shed 26%−27% of their present-day volume if they are to reach a size in equilibrium with the present-day climate. Bahr et al. (2009) address the likelihood that the climate will keep changing by assuming that the observed α will continue to decrease at the rate seen between the 1960s and 1996/97−2005/06. This assumption, rough as it is, leads to an estimate of $\alpha \sim 0.2$ in 2100, so that the glaciers will by then be even further behind the climatic forcing. Projected mass loss

over the century is more than twice as great as for the steady-climate assumption.

The Bahr et al. (2009) estimates in Table 8.1 assume that this equilibrium will be reached precisely at the epoch 2100. The steady-climate estimate also assumes that the climate ceased to change in 2006. Both of these assumptions are improbable, but there is valuable insight in the numbers nevertheless. Firstly, a few to several decades is not unreasonable as a typical timescale for the dynamic adjustment of glaciers to mass-balance forcing. Secondly, we can plausibly interpret the excess of the Bahr et al. (2009) losses over, say, that of Radić and Hock (2011) as a residual response to forcing imposed before 2006 and still to be accomplished after 2100. Thirdly, and most importantly, whatever the date at which the Bahr et al. (2009) Final SLR is realized, it quantifies the notion of committed change in the glaciological context. The glaciers will continue to shrink during the twenty-first century simply because they are not now at equilibrium sizes.

8.6.1.8. Radić and Hock (2011)

The documentation of observational and modelling details in the recent study by Radić and Hock (2011) is more complete than in earlier studies. Necessarily, the documentation is often of shortcomings in the observations and of modelling approximations that are required as responses to those shortcomings. The core of their analysis, which is hierarchical, is an elevation-dependent degree-day model of single-glacier mass balance, with simplified hypsometry interpolated between inventoried minimum, midrange, and maximum glacier elevations. The model is driven by monthly temperatures from the ERA-40 reanalysis and precipitation from a climatological dataset. Model parameters, and model performance in general, are first calibrated against elevation-dependent mass-balance data. The tuned parameters are then extrapolated on a multivariate climatic basis to each glacier in an enlarged version of the World Glacier Inventory (Cogley, 2009b), for which the degree-day model is driven by temperature and precipitation for the period 2001−2100 from 10 GCM experiments run under the A1B scenario. As time advances, the glaciers' sizes are changed using volume−area−length scaling, thus allowing both their areas to shrink and their minimum elevations to rise (Radić et al., 2008). The model thus captures the important feedback of terminus retreat, although it neglects the feedback due to surface elevation change. The inventoried glaciers account for about 40% of the worldwide extent of glaciers. Radić and Hock (2011) define 19 large glacierized regions, for 10 of which the inventory is complete and the regional mass balance is the sum of the single-glacier balances. In the other nine regions, the regional mass balance is obtained by assuming that the ratio of total mass balance between the inventoried glaciers and all glaciers is equal to the ratio of inventoried to total glacierized area. That is, the specific mass balance of the inventoried glaciers is multiplied by the regional glacierized area, the latter obtained from other sources. The global mass balance is the sum of the regional balances.

Although Radić and Hock (2011) is the most detailed projection of the future of glaciers to have appeared to date, with the most detailed treatment of uncertainties, there are some potentially significant omissions. For example, although it affects only the interpretation of the term 'sea-level equivalent', floating ice and ice grounded below sea level, which already displace seawater, are not distinguished from ice grounded above sea level. All meltwater is added immediately to the ocean, thus ignoring possible transfers to aquifers, the soil, and lakes. There is some evidence that transfers to lakes may be significant in enclosed basins in Tibet (Yao et al., 2007). By far the most important omission is the neglect of frontal ablation, which is shared by most other projections. Another feature that the mass-balance model shares with all glacier projections to have appeared so far is that it parameterizes ablation in terms of temperature rather than solving the glacier's energy and radiation balances explicitly. The complexity and attendant uncertainty of energy-balance modelling of glaciers on the global scale are, however, subjects for future research.

Some earlier projections were based on regional-scale analyses, but Radić and Hock (2011) are the first to describe regional variations in detail. They find that regional losses vary greatly. Expressed as percentages of initial mass, the final masses in 2100 range from $92\% \pm 4\%$ in Greenland and $90\% \pm 16\%$ in high-mountain Asia to $28\% \pm 7\%$ in New Zealand and only $25\% \pm 15\%$ in the European Alps. Complete disappearance of ice is projected in one region (the subantarctic islands), where the other nine models project 25%–80% losses. One model yields a projection of slight growth of glaciers in Iceland, one yields growth of 9% in Scandinavia, and four yield growth of up to 8% in high-mountain Asia (where the other six yield losses of up to 34%).

The largest contributions to the global Final SLR come from three regions with relatively moderate percentages of loss but large initial masses: arctic Canada ($27 \text{ mm} \pm 12$ mm SLE from an initial $199 \text{ mm} \pm 30$ mm SLE), Alaska ($26 \text{ mm} \pm 7$ from $68 \text{ mm} \pm 8$ mm) and Antarctica ($21 \text{ mm} \pm 12$ mm from $147 \text{ mm} \pm 64$ mm). These losses are averages across the 10 GCMs used to drive the glacier models. Because the regional projections do not seem to exhibit strong central tendency, the ranges may be more informative about uncertainty in the climatic input projections: 6 mm–43 mm SLE for arctic Canada, 14 mm–36 mm for Alaska, 5 mm–41 mm for Antarctica.

8.6.1.9. Synthesis

The regularized projections agree in suggesting that mass loss over the twenty-first century is likely to contribute several tens to a few hundreds of millimetres of sea-level rise. The most recent and most detailed study (Radić and Hock, 2011) agrees well with the study that is likely to be cited most often in the near future (Meehl et al., 2007).

A notable observation about Figure 8.12 is that the more detailed studies tend to yield lower estimates of glacier mass loss while the higher estimates come from more generalized analyses. It would be a mistake, however, to conclude that the higher estimates are 'wilder' and that a fuller understanding of the problem brings more modest estimates. The modelling of physical processes in detail is possible only when initial and boundary conditions can be prescribed with some confidence. The climatic boundary conditions of this problem are supplied from GCMs, carrying all the uncertainty that is attached to climate projections, and the glaciological initial conditions, in particular the initial (2001) mass, are unsatisfactory in several respects. Moreover, detailed modelling presupposes detailed understanding, and some parts of the physical system are quite poorly understood. The interaction of glaciers with the ocean is the most obvious of these aspects.

An accelerating rate of loss per unit area from a store of diminishing areal extent entails a maximum in the total rate of loss. At first, the shrinkage does not counterbalance the increasing specific loss rate. Later, the continued increase of the loss rate fails to counterbalance the fact that there is less and less area remaining. The date of this maximum of total loss rate is of practical concern. Glacier meltwater is a growing resource for human activities before the maximum but a diminishing resource thereafter. Radić and Hock (2011) project a broad worldwide maximum of meltwater discharge over some decades centred on 2060 to 2080. In regional studies, Xie et al. (2006) assumed warming rates of 0.02 K and 0.03 K yr^{-1} for China and projected a broad maximum of discharge in 2020 to 2040, with rates remaining greater than in 2000 until 2050. Rees and Collins (2006) simulated mass balance in hypothetical glacierized drainage basins in the wetter eastern and drier western Himalaya. They imposed a warming rate of 0.06 K yr^{-1} and found that maxima of annual runoff would be reached around 2050 in the drier and around 2070 in the wetter climate.

It seems likely that glacier meltwater resources will become increasingly scarce beginning sometime in the middle of the twenty-first century, although it is also likely that the date of the inflection will vary by several decades from region to region. The meltwater resource is already scarce in regions such as the Cordillera Blanca of Peru and the basin of the Indus River, which no longer reaches the sea. Here, however, the scarcity is related not just to climatic change but to the growing demand from a growing population, and possibly to imprudent management (cf. Taplin, 2012, this volume).

8.6.2. The Future of Himalayan Glaciers

A number of detailed single-glacier (e.g., Brown et al., 2010; J. Zhang et al., 2007) and regional (e.g., Zemp et al., 2006) projections of glacier evolution are available. The Himalayan region is of particular interest because of the importance of its glaciers as water resources for large populations and because of recent controversy.

Cruz et al. (2007) offered a case study of Himalayan glaciers to illustrate the impact of climatic change on Asia. Unfortunately, the study contained a number of errors. These errors were outlined by Cogley et al. (2010), but they were also the subject of intense scrutiny in the popular media during early 2010 (see Rice and Henderson-Sellers, 2012, this volume). Much of the journalistic and public attention focused on the derivation of the errors, which turned out to be traceable not to the peer-reviewed literature but to inaccurate reporting in an Indian popular science magazine of remarks by an Indian glaciologist that were later characterized as 'speculative'. For this aspect of the story, see Cogley (2011a).

The most significant of the Cruz et al. (2007) errors are in the statement in their section 10.6.2 (p. 493) that "Glaciers in the Himalaya are receding faster than in any other part of the world... and, if the present rate continues, the likelihood of them disappearing by the year 2035 or perhaps sooner is very high if the Earth keeps warming at the current rate." Setting aside some ambiguous repetition, and the fact that Himalayan glaciers are not receding more rapidly than glaciers elsewhere, the main problem here is that simple calculations suggest that the likelihood of disappearance by 2035 is extremely low, not "very high".

While it is easy to show that the speculative remarks that led to the Himalayan controversy cannot be right, it is considerably more difficult to produce reliable estimates of mass loss from Himalayan glaciers. A leading source of difficulty is lack of complete and reliable basic information. In early 2010, there was no complete inventory of the Himalayan glaciers and, where inventories of parts of the region were in existence and were accompanied by glacier outlines, the outlines were often poorly geo-referenced. The lack of accurate digital outlines makes it impossible to construct accurate area—altitude distributions from digital elevation models, and rules out more detailed glacio-climatic models that require hypsometric information as input.

The hypsometric barrier to progress has yet to be removed, but Cogley (2011a) reported on a reconnaissance inventory of the Indian part of Kashmir that fills the last remaining gap in the inventory of the entire mountain range. He went on to project the near future of Himalayan

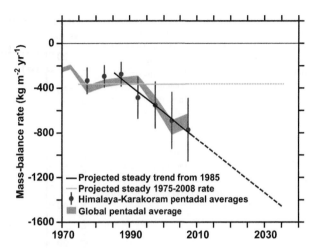

FIGURE 8.13 Constant-rate and constant-trend projections of the specific mass balance of glaciers in the Himalaya. The red circles are from a regional subset of $1° \times 1°$ cells extracted from the global pentadal averages (shown here in grey) of Figure 8.10b. *(Source: Cogley, 2011a, with permission. With kind permission from Springer Science + Business Media B.V.)*

glaciers by relying on the two simple assumptions adopted earlier by Meier et al. (2007) for their global projections: constant rate and constant trend.

The present mass-balance rate estimated by Cogley (2011b) (updating Cogley, 2009a) for the Himalaya—Karakoram region was -364 kg m^{-2} yr^{-1} over 1975/76—2007/08 (Figure 8.13), based on interpolation of a spatial polynomial from sparse local measurements and more distant measurements, the latter discounted by a distance—decay function. The trend (that is, the acceleration of the rate of mass loss) over 1985/86—2007/08 was -24.0 kg m^{-2} yr^{-2}; the beginning of the trend in the mid-1980s was chosen by eye. If the trend persists until 2035, the rate in that year will be about -1440 kg m^{-2} yr^{-1}.

The newly-complete glacier inventory made it possible to model glacier shrinkage as part of the projections, although at present it remains impracticable to incorporate the feedback of changing hypsometry on mass balance. Using the inverted thickness—area scaling equation (Equation (8.8)) and taking the mass-balance projections as measures of thickness change, it was possible to eliminate glaciers when their mean thicknesses diminished to zero. In turn, this made it possible to track the evolution of total glacierized area and total ice mass. The initial mass of each glacier is obtained readily from forward calculations of mean thickness, using Equation (8.5) and taking glacier area from the inventory. The dates of the inventory records vary from 1968 to 2003. The thickness—area scaling was used to synchronize them to 1985, and inventory records without dates were assigned to 1985.

Figure 8.14 shows the evolution of mass relative to mass in 1985, with projections for 2010 and 2035. A source of

FIGURE 8.14 Constant-rate (a) and constant-trend (b) projections of the evolution of total glacier mass in the Himalaya, with allowance for the disappearance of individual glaciers as their mass balance is projected by thickness–area scaling (Section 8.4.1). The quantity plotted is the ratio of total mass to total mass in 1985, near the middle of the range of glacier inventory dates. Five plausible sets of thickness–area scaling parameters are used to illustrate the possible uncertainty of the projections: Bahr et al. (1997); Chen and Ohmura (1990); DeBeer and Sharp (2007); Arendt et al. (2006); Su et al. (1984). *(Source: Cogley, 2011b, with permission.)*

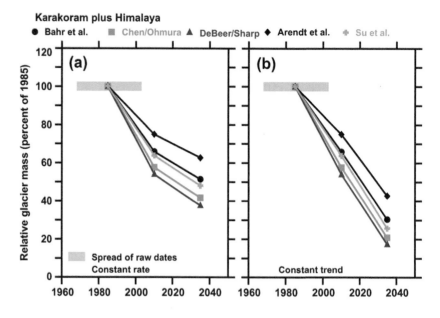

serious uncertainty is concealed by this method of presentation. None of the five sets of scaling parameters can be ruled out as implausible, but they yield total masses in 1985 that vary from 4000 Gt to 8000 Gt. That said, the pattern of evolution is the same for each parameter set. If mass loss proceeds at a constant rate (Figure 8.14a), the proportion of 1985 mass that may have been lost already is between a quarter and a half, while in 2035 the mass remaining will be between one and two thirds. The rate of loss is less from 2010 to 2035 because the glacier size distribution is evolving towards greater dominance by larger glaciers, as a consequence of the disappearance of smaller glaciers. If a constant acceleration of specific mass balance is assumed (Figure 8.14b), the faster specific rate of loss cancels out the impact of the changing size distribution, and the total loss rate remains nearly the same to 2035 as between 1985 and 2010. By 2035, between one fifth and two fifths of the 1985 mass remains.

This analysis is near to the crude end of the spectrum of sophistication, and urgently requires to be followed up with projections that describe more of the physical behaviour of glaciers and subject them to more realistic forcing. However, the analysis does have the merit of offering generalized early corrective guidance to resource managers and policymakers.

8.7. REFLECTIONS: GLACIERS AND THE FUTURE CLIMATE

8.7.1. Basic Information

A recurrent theme of the sections above has been the lack of basic information in readily usable form. For example, the

work reported by Raper and Braithwaite (2005a) was necessary mainly because the World Glacier Inventory is incomplete. Many of the projections described above required similar time-consuming workarounds. Moreover, a problem not yet addressed in any projection is that the World Glacier Inventory is diachronous. Its component regional inventories span several decades, and rates of shrinkage are not well enough known for synchronizing information from different dates in an objective way. The initialization of glaciological projections is thus necessarily fuzzy.

Nevertheless, incremental advances in the completeness and usability of information proceed steadily. For example, digital elevation models of global or near-global extent and with spatial resolution adequate for glaciological modelling have become available in the past decade. They are ready and waiting to be masked by accurate glacier outlines, which are accumulating steadily also, although much labour remains. As the number of glaciers with mass-balance measurements increases unspectacularly, and the technology for mass-balance measurements improves more noticeably, surprising new results appear less often.

On the other hand, there are occasional surprises. Hock et al. (2009) modelled global average mass balance for 1960/61–2003/04 by estimating the sensitivity of mass balance to changes in temperature and precipitation. Glaciers in Antarctica stood out as leading contributors to a more negative global average than that of Lemke et al. (2007), due to their high sensitivity to temperature and the large warming in the vicinity of the Antarctic Peninsula. Local measurements being inadequate, the Antarctic estimates of Lemke et al. (2007) were based on an analogy with the high Arctic of Canada. This example illustrates the growing importance of modelling, not just for investigating glacier futures but for understanding present and recent states of the glaciers. Although it compromises the mutual independence of glaciological and meteorological information about climatic change, modelling is becoming indispensable for the understanding of glacier mass balance. Further incremental advances in modelling methodology, for example for including glaciers in GCM simulations by subgrid-scale parameterization, should be expected and promoted.

8.7.2. Gaps in Understanding

Accuracy in projections of the future of glaciers is, of course, not limited solely by incomplete information. Glaciers share with the rest of the climate system all the dimensions of uncertainty about the future, and glaciology shares with climatology the attribute of incomplete understanding of the system under study.

Two examples of gaps in understanding of how glaciers evolve are the role of internal accumulation in polythermal glaciers (Sections 8.3.1 and 8.4.3) and the dependence of shrinkage rate on initial size (Section 8.5.3). In the first example, the gap is a roadblock because internal accumulation is extremely difficult to measure, at least economically. The result is that there are hardly any measurements, and no routine measurements, and consequently that the only practical approach is to model the phenomenon. However, the models have almost no observational support for calibration, and the problem is thus circular — and persistent. In the second example, the gap takes the form of not knowing why the shrinkage rate depends exponentially on the initial area (that is, why smaller glaciers shrink faster). An exponential dependence on area suggests that the answer might lie in linear behaviour of one horizontal dimension, presumably the glacier length. This phenomenon is made more puzzling by the failure of some glacierized regions to exhibit it and, in more typical regions, by the extreme variability of the shrinkage rates of the smallest glaciers in any representative sample. Their short response times may play a role; but a safe generalization is that the gap in understanding is there because the problem has been little studied hitherto.

The largest single impediment to intellectual progress on the behaviour of glaciers, however, is tidewater instability (Meier and Post, 1987). This term describes unsteady behaviour of a calving glacier that alternates between episodes of slow advance and rapid retreat. Outside the ice sheets, timescales appear to be centuries for advance and decades for retreat. Timescales are not known for the ice streams that discharge much of the ice from the ice sheets, except that some very large ice streams are suffering very rapid retreat at the present day. The conditions permitting advance in the first place, and the triggers for subsequent

unstable retreat, are both poorly understood. Once retreat has begun, however, observation and numerical simulation (Schoof, 2007) agree that, if the bed is grounded below sea level but has a slope opposed to that of the surface, the retreat will continue until the calving front or grounding line reaches a part of the bed that slopes in the same direction as the surface. During this unstable retreat, enhanced calving leads to a positive feedback in which accelerated flow and dynamic thinning extend far upglacier from the part that is grounded below sea level. Mass loss is far greater than, and essentially independent of, the climatic mass balance.

It is difficult, however, to refrain from suspecting a role for climatic change, direct or indirect, in the dramatic changes being exhibited at present by many tidewater glaciers. Steffen et al. (2008, p. 30) assessed the interaction of warm seawater with the peripheries of the ice sheets as "a strong potential cause of abrupt change", an assessment reinforced, for example, by the observations of Rignot et al. (2010). Tidewater instability was described long ago by Weertman (1976, p. 284) as "glaciology's grand unsolved problem". It remains unsolved today, but it is the subject of intense current effort because we cannot assess the probability that large parts of one or both of the ice sheets might collapse into the sea.

8.7.3. The Probability Distribution Function of Glacier Futures: Glimpses of the Known and Unknown

Wigley and Raper (2001) argue that a climatic projection presented as a range may be misleading unless it is accompanied by some guidance as to what the range means in terms of probability. They suggest a four-step procedure for providing such guidance: 1) identify the main sources of uncertainty; 2) represent each of these input uncertainties as a probability distribution function; 3) draw samples from the probability distribution function to drive the model that is to generate the projection; 4) construct an output probability distribution function from the results of step 3. In glaciology at present, this quasi-Bayesian approach breaks down at step 2, because we are unable even to sketch distribution functions for some of the least well-known uncertainties. Step 3 is also problematic in that some of the processes that are at work are understood well enough that models for them are stable and reliable but, for others, especially frontal ablation, it is not yet possible to write down the governing equations in a practicable way.

Nevertheless, the synthesis attempted above does hint at what an output or posterior probability distribution function for mass loss from glaciers might look like. None of the projections are less than 60 mm SLE for the twenty-first century, and the lowest projections appear to have low or very low probability because they underestimate initial glacier mass. The most recent and most detailed projection (Radić and Hock, 2011) is that the twenty-first century contribution to sea-level rise from glaciers other than the ice sheets is more likely than not (in IPCC terminology) to lie between about 90 mm and about 160 mm SLE. It is in general agreement with two earlier IPCC assessments that were more generalized, but were diverse in approach. The highest recent estimates are from calculations that are even more generalized, although they can be seen as tentative descriptions of the upper tail of the probability distribution function.

When the total length of calving glacier margins becomes known more accurately, it is likely that the higher estimate of Pfeffer et al. (2008) will be ruled out. It will probably be found that, even if the present-day calving margins could be maintained in spite of retreat, there is not enough ice tributary to such margins to supply the projected cumulative ice discharge. However, the lower end of the Pfeffer et al. (2008) range will not be so easy to dismiss in this way, and indeed its estimate of frontal ablation addresses a known and significant unquantified omission from all of the detailed projections. For a different reason, the projections of Bahr et al. (2009) are also not dismissible. They address the undoubted fact that some of the response of glaciers to climatic change is 'buried' in their geometric behaviour (delayed shrinkage towards equilibrium sizes), a fact that is addressed only indirectly or, in some cases, not at all in other projections. The higher Bahr et al. (2009) projection was obtained quite crudely and on that ground should be assessed as less probable than the lower projection, but the lower projection suffers from the basic flaw that it assumes a highly improbable unchanging climate.

Table 8.1 and Figure 8.12 present ranges, rather than central values with confidence envelopes. The latter style would mislead by suggesting a known probability distribution function, especially if the confidence region were represented with the '±' symbol. For the future of glaciers, the output probability distribution function envisaged by Wigley and Raper (2001) is not known, except that it seems to be asymmetrical. Small losses are less probable than large losses over the twenty-first century. Asymmetrical probability distributions in the climate system were discussed by Urban and Keller (2009), who suggested that they might arise from incomplete understanding, rather than from the system itself. Whether the apparent glaciological asymmetry will turn out to be intrinsic or just an artefact of the incomplete glaciological understanding described here is an open question.

ACKNOWLEDGEMENTS

I thank Regine Hock and an anonymous reviewer for detailed and constructive comments.

Chapter 9

Future Regional Climates

Jason Evans,[a] John McGregor[b], and Kendal McGuffie[c]

[a] *Climate Change Research Center, University of New South Wales, Sydney, Australia*, [b] *CSIRO Marine and Atmospheric Research, Aspendale, Australia*,
[c] *School of Physics and Advanced Materials, University of Technology, Sydney, Australia*

Chapter Outline

9.1. INTRODUCTION: CLOSE-UP OF CLIMATE CHANGE

The climate in any particular region is produced by an interaction of various phenomena that operate at a range of scales, from the global scales of the general circulation of the atmosphere and oceans, through the regional-scale of tropical cyclones, to the more local scale of the effects of coasts, mountains, and land use. It is the combination of the large-scale and regional/local forcings that produce a region's climate. Large-scale phenomena include features of the global circulation such as the Inter-Tropical Convergence Zone (ITCZ) and subtropical deserts, as well as ocean modes of variability including El Niño–Southern Oscillation (ENSO: Deser and Wallace, 1987), the North Atlantic Oscillation (NAO: Hurrell et al., 2001, 2003), and the Indian Ocean Dipole (IOD: Saji et al., 1999). These

ocean modes have been found to vary on multiyear through decadal to centennial timescales, as is discussed by Latif and Park (2012, this volume). Regional climate phenomena perturb local climates away from what they would be if forced only by the large-scale effects. Examples of some regional-scale phenomena that can have substantial effects on the regional climate are presented in Section 9.2, including tropical cyclones (Frank, 1977), sea and land breezes (Estoque, 1961), monsoons (Webster et al., 1998), orographic precipitation (Roe, 2005), and low-level jets (Bonner, 1968; Banta et al., 2002). All these interacting influences are now being affected by climate change.

Simulating future climates begins at the global scale with Global Climate Models (GCMs), also known as general circulation models. While these models often capture the global-scale phenomena well (Covey et al.,

2003), they are running with a spatial resolution (typically ~100 km) that is too coarse to capture the regional and local phenomena that are important for regional climate. In order to account for these missing phenomena, GCM output may be adjusted to a higher spatial resolution or 'downscaled' in some fashion. This downscaling can be accomplished using a number of techniques that are broadly classified into two groups referred to as dynamical and statistical downscaling. Dynamical downscaling makes use of mathematical representations of the fluid motion of the atmosphere and the various physical processes, in a fashion similar to GCMs, to simulate the climate at scales of a few to tens of kilometres (Giorgi and Mearns, 1999). Statistical downscaling uses long time series of observations to establish statistical relationships with the large-scale features of the climate thought to be represented robustly in GCMs (Wilby et al., 1998). Each of these methods is explored in detail in this chapter.

This chapter first introduces various regional climate phenomena that are important for producing the climate in different regions (Section 9.2). It then goes on to describe various downscaling techniques, the benefits and limitations of each technique, as well as examples of their use (Section 9.3). Producing projections of regional climates is a multistep process with different levels of uncertainty associated with each step. Section 9.4 explores this uncertainty and provides some methods that have been used to identify, quantify, and reduce it. Examples of studies of regional climate change impacts on water resources, ice sheets, and tropical weather systems are presented in Section 9.5, followed by a discussion of future regional climates in Section 9.6.

9.2. REGIONAL-SCALE CLIMATE PHENOMENA

Regional climate phenomena are differentiated from large-scale phenomena by having a limited spatial extent (subcontinental). Many such phenomena exist. In this section, several common important regional phenomena are described, along with examples of places where they play important roles. These phenomena occur at scales that make it difficult, if not impossible, for GCMs to capture them and, hence, GCMs generally fail to simulate accurately climates that depend on them.

9.2.1. Tropical Cyclones

The term tropical cyclone is a generic descriptor for an intense non-frontal synoptic-scale tropical low pressure system. They are regional-scale phenomena, a few hundred kilometres across, with a global-scale influence (Figure 9.1). They are characterized by a deep low pressure centre surrounded by an area of intense destructive winds and intense precipitation. Tropical cyclones are characterized by a warm core associated

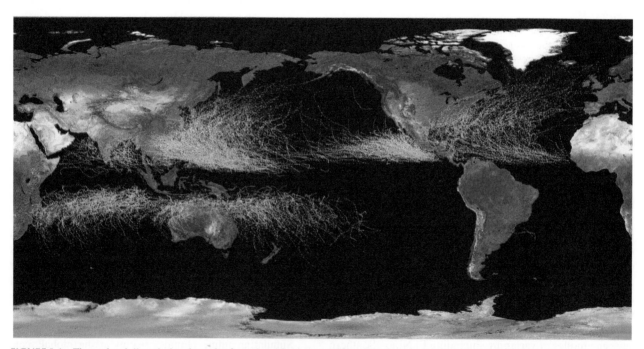

FIGURE 9.1 The tracks of all tropical cyclones that formed worldwide from 1985 to 2005 based on US National Hurricane Center 'best track' data. The points show the locations of the storms at six-hourly intervals and use a colour scheme to show the intensity from 'tropical depression' (blue) to Saffir–Simpson category 5 (red). Storms are classified here according to this uniform scale but, in practice, categories used for forecast purposes vary between tropical cyclone basins. (*Source:* http://commons.wikimedia.org/wiki/File:Global_tropical_cyclone_tracks-edit2.jpg.)

with subsidence in the centre of the storm (Frank, 1977; Glickman, 2000). They are also referred to regionally as hurricanes (in the North Atlantic) and typhoons (in the North Pacific). Tropical cyclones form over some areas of the tropical oceans when convective disturbances become organised in a suitable thermodynamic environment. Tropical cyclones are most frequent in the western North Pacific and the North Atlantic and around 80–90 tropical low pressure areas are classified as tropical cyclones each year around the world (Henderson-Sellers et al., 1998). Figure 9.1 shows the location and intensity of tropical cyclones over the period 1985–2005. The intensities of the storms are colour-coded using the Saffir–Simpson scale from most (red) to least intense (blue). Storms are most frequent in the North Pacific and at their most destructive intensities in the subtropics, although they can continue as intense low pressure areas into mid-latitudes, causing intense rainfall events. For reasons related to the dynamics of the storms, they do not develop or move within about 5° of the Equator and are rare in some ocean basins, notably the South Atlantic and southeast Pacific.

The key features of an individual tropical cyclone can be seen in Figure 9.2, which shows Tropical Cyclone Yasi in the Coral Sea, off northeastern Australia in February 2011. A ring of intense convection, the eyewall, surrounds a calm eye, and outflow at higher levels creates a spiral cirrus 'shield', which conceals a high level of organisation of the convection into spiral bands. Tropical cyclones are characterized by sharp gradients of pressure and there is close coupling between the thermodynamics of precipitation and the larger scale structure of the storm. As such, they are difficult to resolve in all but the highest resolution global models, making them a prime candidate for regional-scale studies (see Section 9.5.3).

Tropical cyclones have their greatest impact when they make landfall, after which they decrease in intensity because of the lack of energy supply through evaporation from the ocean surface (Figure 9.1). For susceptible locations, the presence of tropical cyclones is a defining feature of their climates, causing loss of life and damage to infra-structure and crops (Figure 9.3). These cyclones can provide a large proportion of a region's total precipitation, as well as being responsible for extremes in wind speeds and precipitation intensity.

The modelling of tropical cyclones has a dependence on horizontal resolution. While current GCMs can produce disturbances that bear a reasonable likeness to tropical cyclones (Fudeyasu et al., 2008), simulations using regional models suggest that resolutions of the order of 1 km are required to properly simulate the inner core (Chen et al., 2007), a resolution beyond that of current GCMs. Consequently, a range of downscaling techniques has been used to overcome the spatial resolution limitation (Walsh

FIGURE 9.2 Key features of a tropical cyclone are illustrated in this MODIS image of Tropical Cyclone Yasi over the Coral Sea as it heads towards the coast of Queensland, Australia (00:00 UTC, 1 February 2011). The central eye is surrounded by a ring of intense winds and intense precipitation. Away from the eyewall, rainfall is concentrated in narrow rainbands, which spiral out from the centre. A day later, just before it made landfall, the storm had a well-defined eye and sustained maximum winds of 220 km hr^{-1}. Queensland is in the lower left of the image and Papua, New Guinea in the upper left. Although the storm is large, the physical processes that affect the intensity of the storm occur at scales too small to be resolved explicity by most models. *(Source: NASA Image.)*

et al., 2004; Knutson et al., 2007). These are discussed further in Section 9.5.3.

Over recent years, some Regional Climate Models (RCMs) have been developed specifically to improve the simulation of tropical cyclones, such as the Advanced Research Hurricane Weather Research and Forecasting (WRF) model (Davis et al., 2008). These developments include automatic vortex-following techniques, along with multiple-level, moving, high-resolution nested grids, which allow high resolution in the area of interest without necessitating excessive computations in areas away from the storm. The inclusion of mixed-layer ocean models and the detailed parameterizations of turbulent energy exchanges in high-wind surface environments has increased the ability of these models to capture the characteristics of tropical cyclones.

FIGURE 9.3 Flood-level marks on the walls in this 2007 photograph show the history of tropical cyclone associated water levels at this graphite mill in Madagascar. Such flooding has a major impact on local infrastructure and is a severe constraint to the economic prosperity of the region. The mark to the left of the doorway is for Cyclone Geralda, in February 1994, which killed 70 people in Madagascar and left a reported half a million people homeless. *(Source: Photo - K. McGuffie.)*

9.2.2. Sea Breezes and Monsoons

A sea breeze is a coastal wind generated by the differential heating of the land and the water (Estoque, 1961; Rotunno, 1983). When the land is at a higher temperature than the neighbouring water, the air above it is heated and rises. The air at lower levels is replaced by cooler air advected from the adjacent ocean areas. Coastal temperatures are regularly impacted by sea breezes. If an offshore flow exists when a sea breeze forms, then there will be a region where the two opposing surface winds meet and are forced to ascend; this region is called a sea breeze front (Simpson et al., 1977). If there is enough moisture in the atmosphere, clouds and precipitation can form.

Sea breezes typically impact the land within some tens of kilometres of the coastline, a scale much too small to be resolved in today's global models. They are consistently occurring phenomena that can result in the coastal climate

being several degrees cooler in summer than nearby inland areas, as well as having higher levels of precipitation. While sea breezes affect all coastlines to some degree, there are locations where the effect is larger, often because of the geometry of the coastline. Long, relatively thin peninsulas and islands can form sea breezes from opposing sides that will converge, greatly enhancing the potential for precipitation (Figure 9.4). The inter-relationships between sea breezes, the boundary layer, and larger-scale convection were studied during MCTEX in 1995 (Keenan et al., 2000; Schafer et al., 2001).

The spatial scale of sea breezes is generally too small to be captured by GCMs and hence high-resolution models have been used to study the role that the phenomenon plays in regional climate, at least as far back as Anthes (1978). One of the first studies to quantify the ability of a three-dimensional regional model to simulate a sea breeze was by Steyn and McKendry (1988). Since

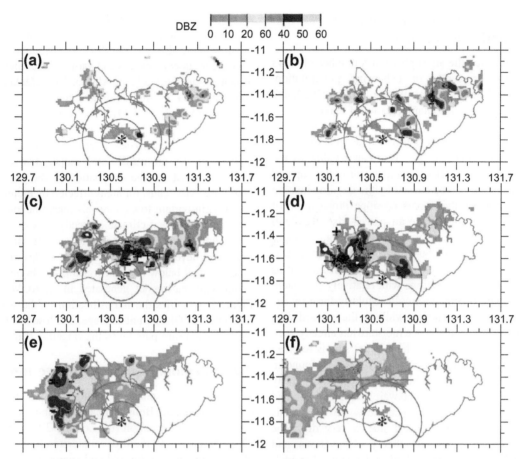

FIGURE 9.4 The local-scale sea breeze can have consequences that feed into larger-scale phenomena. In this sequence of radar images, two sea breeze fronts converge from the north and south coasts of Bathurst Island, Australia during the Maritime Continent Thunderstorm Experiment in 1995. The figure shows horizontal reflectivity at 2 km altitude at (a) 0216, (b) 0315, (c) 0416, (d) 0515, (e) 0614, and (f) 0715. The sea breeze fronts are evident in (a) along the north and south coasts of the island, and the convergence of the fronts prompts the development of more intense convection in the east, which develops, drifts in the environmental flow, and intensifies, forming an intense storm by 0515, which continues west and then dissipates. Carey and Rutledge, 2000. Refer to Carey and Rutledge, 2000, for discussion of other symbols on the figures and to Schafer et al., 2001, for further analysis of boundary layer development associated with these storms. *(Source: Carey, L. D., and S. A. Rutledge, 2000. © American Meteorological Society. Reprinted with permission.)*

then, many studies have used regional-scale dynamical models to investigate aspects of the sea breeze and its interaction with other atmospheric phenomena (Crosman and Horel, 2010).

Monsoon is a name for seasonally reversing winds (Glickman, 2000). As for sea breezes, this phenomenon is caused, at least in part, by the differential heating of land and sea, though on a much larger scale. While sea breezes often occur as part of the diurnal cycle, monsoons are caused by the annual variation in temperature over large land areas relative to the neighbouring ocean surfaces. This temperature difference results in lower pressure over land during summer (driving an onshore wind) and over the water during winter (driving an offshore wind). Factors such as the topography of the land also have a considerable effect on monsoonal winds. Monsoon processes have been studied extensively (Webster et al., 1998; Wang et al., 2003), as has monsoon variability

(Prell and Kutzbach, 1987; Wang et al., 2001). The onshore phase of the monsoon brings large amounts of precipitation and often occurs when the ITCZ is co-located. The major monsoon systems are the India, Asia–Australia, western Africa, and the North and South American monsoons.

Given the dependence of monsoons on the interaction between coastlines and nearby land orography, GCMs struggle to reproduce monsoon systems successfully (Cook and Vizy, 2006). Regional-scale dynamical models have been used extensively to study monsoon systems, beginning with two-dimensional studies (Dudhia, 1989). Recent studies driven by re-analyses have shown that regional models can simulate the monsoons better than coarser global models (Kumar et al., 2010; Zou et al., 2010; Kim and Hong, 2010), as well as providing insights into their physical processes. An example of this is a study over the maritime continent by Qian et al. (2010).

9.2.3. Orographic Precipitation, Rain Shadows, and Foehn Winds

Mountains influence the atmosphere in a number of ways (Smith, 1979). Triggering or enhancing precipitation, hosting perennial snow and glaciers (e.g., Cogley, 2012, this volume), and changing the characteristics of winds all have important implications for climate. Orographic precipitation is caused when moist air is forced to ascend due to the presence of a mountain range. This can be due to the mountains acting as a barrier, forcing the air to flow up and over the range; or it can be caused by the daytime heating of mountain surfaces that forces up-slope propagation of moist air. In either case, as the air parcels rise, they will expand and cool. If parcels cool below the water vapour condensation point, then cloud droplets will form and, if these droplets become large enough, they will fall out as precipitation. Thus, for locations with relatively consistent winds, orographic precipitation will regularly form on the windward side of the mountains, while on the leeward side, the atmosphere will be drier since moisture has already been removed by the precipitation. The leeward side is referred to as being in a 'rain shadow'. This effect can be found in many places around the world, though it is particularly apparent on islands such as Hawaii and New Zealand (e.g., Renwick et al., 1998), where the windward side is densely forested and the leeward side is very dry. Roe (2005) provide a review of the theory, observations, and modelling of orographic precipitation. Dynamical modelling studies have suggested that resolutions on the order of a few kilometres are required to accurately simulate orographic precipitation (Colle et al., 2005).

In rain shadows, dry descending air warms more rapidly than the moist ascending air cools, causing the leeward slopes to be warmer than the equivalent elevations on the windward slopes. These are known as foehn winds. Foehn winds can increase temperatures substantially in a short period of time, leading to the rapid melting of snow or, in warmer climates, to the rapid spread of wildfires (Sharples et al., 2010). If foehn winds occur often, they can produce a lasting climate influence. Central Europe has a warmer climate due to the moist winds blowing off the Mediterranean Sea, over the Alps, and into central Europe as foehn winds (Seibert, 1990; Mayr et al., 2007). Valley effects such as foehn winds are not resolved at GCM scales, but can be successfully simulated with regional models (e.g., Gohm et al., 2004).

9.2.4. Mountain Barrier Jets

A mountain barrier jet is a region of fast-flowing air on the windward side of a mountain range, blowing parallel to the range. The jet is formed when the incoming low-level air is stable and not able to flow over the barrier. In this case, the winds are turned parallel to the range, with a combination of the pressure gradient and Coriolis force acting to accelerate the flow in the along-barrier direction. Mountain barrier jets have been documented windward of the Sierra Nevada mountains in the USA, in Alaska, New Zealand, Taiwan, and Antarctica, as well as windward of the Zagros mountains in the Middle East (Overland and Bond, 1995; Katzfey, 1995; Li and Chen, 1998; Evans and Smith, 2006). These jets can move air masses over long distances relatively quickly. If the air masses are moist, this will produce a wetter climate at the destination of the jet than would otherwise exist. Thus, for areas affected by mountain barrier jets, it is important to capture this phenomenon in order to simulate the present climate correctly, as well as for simulating any future changes in the climate (Evans, 2008). These low-level jets usually occur on scales of tens of kilometres, well below the GCM grid scale. Figure 9.5 shows a mountain barrier jet generated in an RCM simulation. Figure 9.5a shows the intensification of the water vapour flux upwind of the mountain slope. The low-level winds near the mountain turn parallel to the ridgeline (Figure 9.5b).

9.2.5. Regional Climate Change Impacts

Increases in atmospheric greenhouse gases are forcing the climate to change. Being prepared for these changes is the prudent course of action (cf. Lenton, 2012, this volume). This preparation requires understanding the climate change impacts at temporal and spatial scales that can be addressed through policy and management decisions (Taplin, 2012, this volume). In almost all cases, this spatial scale is regional to local. Exactly which climate-related variables are important depends on the impact of interest. This section explores the various types of climate variables that have been investigated in terms of climate change impacts and assesses their dependence on regional-scale phenomena. As such, it complements the analysis of future extremes by Alexander and Tebaldi (2012, this volume).

9.2.5.1. Changes in Means

Changes in the global mean temperature have been extensively studied and are the basis for the term 'global warming'. While, on average, the globe is warming, there are regional differences in the rate of warming. At large-scales, the high northern latitudes are warming faster than the low latitudes (e.g., IPCC, 2007a). Compared to most fields, temperature is a smoothly varying field, where changes at large scales often relate closely to the changes at regional-scales.

Temperature can be strongly affected by some regional-scale phenomena, however. Mean temperature is sensitive to the elevation, so the impact of orography that exists at a spatial scale too small to be resolved by a GCM will be

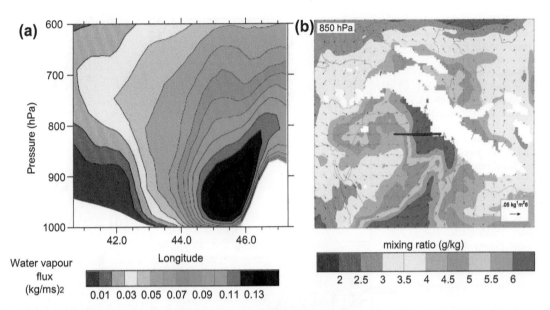

FIGURE 9.5 Mountain barrier jet generated by Zagros Mountains in an MM5 simulation (*Source: adapted from Evans, 2008*). (a) Vertical cross-section of the water vapour flux, (b) horizontal cross-section at 850 hPa of the mixing ratio (coloured contours) and the water vapour flux (vectors). Water bodies are outlined in black, topography above 850 hPa is shown in white, and the vertical cross-section in (a) is taken along the thick red line.

missed entirely by these global models. The temperature can also be heavily impacted by regional phenomena, such as the development of local winds. Changes in the frequency and intensity of these winds can have significant impacts on the change in mean temperature.

Changes in the mean annual or seasonal precipitation have also been extensively studied. Unlike the change in mean temperature, there is no globally-coherent precipitation-change signal. Precipitation is much more sensitive than temperature to regional conditions. Due in part to this reliance on smaller scales, future changes in precipitation as projected by various GCMs can often disagree, even in the direction of the change, as was seen in the GCM ensemble presented in the Fourth Intergovernmental Panel on Climate Change (IPCC) Assessment. There are large land areas where there is notable disagreement between the models as to the sign of the change in precipitation (see figure 10.17 of Alexander and Tebaldi, 2012, this volume).

A number of studies have tried to address this by modelling the climate change in a way that accounts for regional phenomena. Section 9.5.1.1 discusses regionalizing Murray—Darling climate futures, while Evans (2008, 2009, 2010) presents future climate change projections over the Middle East. These projections come from an ensemble of GCMs from the Coupled Model Intercomparison Project 3 (CMIP3) dataset together with those from the climate change projected using a regional climate model, MM5 (Dudhia, 1993; Grell et al., 1994). The changes show clear inter-model differences; the regional model's greater ability to capture the influence of the mountains east of the Mediterranean Sea and the mountain

barrier jet that forms near the Zagros Mountains being a major cause of these differences.

An ensemble of RCM experiments in Europe (the Prediction of Regional Scenarios and Uncertainties for Defining European Climate Change Risks and Effects (PRUDENCE) project; Christensen and Christensen, 2007) showed differences between the climate simulated by the driving GCM and the RCM-simulated climate. Christensen and Christensen (2007) found that, while the different RCMs could have quite different biases compared to current observations, the models projected similar future climate change. For example, the boreal summer precipitation biases for France vary between −47% and +51%, while the future projected changes only vary between −55% and −34%. Wang et al. (2009) present some results from the North American Regional Climate Change Assessment Program (NARCCAP) for the inter-mountain region in western USA (Figure 9.6). This region is of particular importance for water supply to the western United States and GCMs do not simulate the seasonal cycle of precipitation well (Boyle, 1998). Wang et al. (2009) find that, compared to GCMs, the RCMs produce a much improved seasonal cycle and inter-annual variability in precipitation. Due to their better-resolved topography, RCMs typically capture the change in seasonality from one side of the Rocky Mountains to the other (Figure 9.6).

9.2.5.2. Changes in Extreme Events

For many climate change impacts, the changes in extremes are of the most interest (see Alexander and

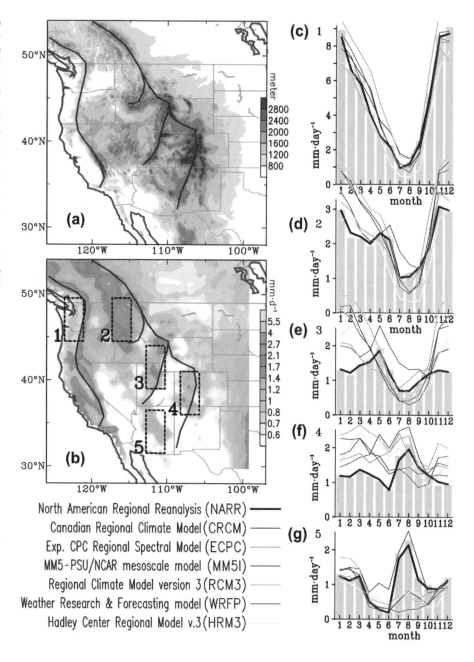

FIGURE 9.6 (a) Orography and (b) cold-season observed rainfall (November–May) of the Intermountain Region. The major mountain ranges are outlined by red lines. (c)–(g) Monthly rainfall histograms of the observations, averaged from the five regions indicated in (b), superimposed with the corresponding precipitation of the NARR (thick black line) and all RCMs (colour lines). Note the precipitation scale in part (c) is twice that in (d)–(g). The abbreviations of the RCMs and their designated colours are given under (b). Observational data from the University of Delaware dataset available from *http://www.esrl. noaa.gov/psd/data/gridded/data.UDel_ AirT_Precip.html. (Source: Wang et al., 2009 © Copyright 2009 American Geophysical Union. Reproduced/modified by permission of American Geophysical Union.)*

Tebaldi, 2012, this volume). Events such as floods, droughts, and heatwaves can have large impacts on both natural and human environments, with the potential to cause extensive damage. Many GCMs project an increase in extremes with the chance of intense precipitation and flooding projected to increase, along with the length of the dry periods between episodic rainfall events, thus also increasing the risk of drought in many areas (e.g., Section 9.5.1). Figure 9.7 shows the projected changes in precipitation from the IPCC Fourth Assessment Report (AR4). The stippling in Figure 9.7 indicates the regions where the majority of models agree on the change; hence, there are large areas of the land surface where the

projected change in the relevant variable remains quite uncertain.

Climate extremes are often the result of large-scale and regional-scale phenomena acting to reinforce each other. Hence, GCMs can have difficulty reproducing the observed climatology of extremes, and their projected future changes in extremes can have significant uncertainty. The spatial-scale issue facing GCMs, when trying to project extremes, is clear when looking at extreme precipitation events. Precipitation intensity varies on scales much smaller than the GCM grid scale (hundreds of kilometres); thus, it is impossible for GCMs to capture the most intense events that would lead to flash flooding and associated impacts

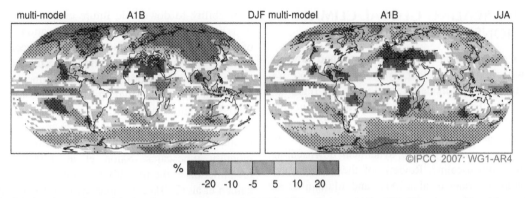

FIGURE 9.7 Relative changes in precipitation (in percent) for the period 2090−2099, relative to 1980−1999. Values are multimodel averages based on the SRES A1B scenario for December to February (left) and June to August (right). White areas are where less than 66% of the models agree in the sign of the change and stippled areas are where more than 90% of the models agree in the sign of the change. *(Source: IPCC, 2007a.)*

such as landslides. For this, it is necessary to capture the small-scale phenomena that contribute to intensifying the precipitation. Even for spatially-widespread extreme events such as drought and heatwaves, regional climate phenomena can play an important role in the intensification of the event. Climate extremes are explored in more detail by Alexander and Tebaldi (2012, this volume).

The STAtistical and Regional dynamic Downscaling of Extremes for European regions (STARDEX) project focused on identifying robust downscaling techniques for extremes (CRU, 2005). It found that, in the majority of cases, no 'best' downscaling method could be identified. It recommended that a range of the most appropriate methods, encompassing different statistical and/or dynamical methods, be used. An example of a spatially-widespread extreme event is the 2003 European heatwave that killed tens of thousands of people. The evolution of this event was studied by Zaitchik et al. (2006), who found that local land-surface feedbacks enhanced the heatwave. Similarly, Fischer et al. (2007) found that the local soil-moisture anomalies were responsible for increasing the heat anomalies by about 40% (but cf. Pitman and de Noblet-Ducoudré, 2012, this volume). When looking at changes in precipitation extremes over Europe, Fowler et al. (2007a, p. 12) found that "RCMs are capable of representing the spatial patterns in extremes that are not resolved by the GCMs". More recently, statistical downscaling methods that use extreme value theory have been developed (Friederichs, 2010) and have been found to perform well when downscaling is applied to a re-analysis.

9.2.5.3. Changes in Ecologically Important Aspects of Climate

When investigating the impact of climate change on vegetation, whether natural or part of an agricultural system, recognizing the non-linear nature and sensitivities

of vegetation growth are important. These dependencies are also species-specific. Thus, to really understand the impact of climate change on a particular species, its climate sensitivities need to be known and explicitly examined in the climate change projections (see also Dickinson, 2012, this volume). These can include climate events, such as the presence of frost days in the early part of the growth cycle, the exceedance of a temperature that causes leaves to die, or whether or not the species receives enough winter chilling to break dormancy in the spring (Schwartz and Hanes, 2010). For agriculture, aspects such as the length of the dry season and the speed at which growing degree days are accumulated become important (Evans, 2009). Many of these aspects are affected by both changes in means and changes in the extremes of various climate variables and, as such, can be influenced by many different regional climate phenomena.

Another aspect of the ecological impact of climate change has been the potential to cause species extinction. A simple example is that of mountain species that will migrate up the mountain as the climate warms, until there is nowhere left to go (Frei et al., 2010). This simple approach does not account for the presence of refugia in the landscape, that is, areas that have a more suitable climate, unlike the larger region, due to regional and local climate phenomena. Ashcroft et al. (2008) studied the environmental factors that influence the distribution of species and found that wind exposure and distance from the coast were both important factors in their region. Wind exposure is a function of both the dominant wind conditions and the location in the landscape of the vegetation, including whether it is in a valley or on a hillside, and what aspect it faces. Using regional/local factors such as these, it may be possible to identify refugia in a future climate and hence plan suitable conservation efforts. There is also much to be learned from palaeoclimate data and species sensitivity to climate change (e.g., Harrison and Bartlein, 2012, this volume; Belcher and Mander, 2012, this volume).

9.3. DOWNSCALING GLOBAL CLIMATE PROJECTIONS

In order to project future regional climate, and how its changes will impact on other systems, a method for downscaling the global model projections to the regional-scale is required. The method needs to take into account the fact that the global projections are unable to capture regional-scale phenomena. Two main approaches to downscaling are currently used: dynamical downscaling and statistical downscaling. Reviews of these techniques can be found in Wang et al. (2004) and Maraun et al. (2010). In this section, these techniques are described, examples of their use given, and their advantages and limitations explored.

9.3.1. Dynamical Downscaling

Dynamical downscaling involves driving an RCM at high resolution using the output of a GCM for the region of interest. Several reviews of RCMs have been undertaken in the last two decades (Giorgi and Mearns, 1991; McGregor, 1997; Mearns et al., 1999; Wang et al., 2004; Rummukainen, 2010). Like a GCM, the RCM simulates explicitly the atmospheric dynamics, thermodynamics, and related physical processes that, together, create what we describe as weather. Running the RCM over an extended period, and calculating the means and probability distributions of the atmospheric states provides the regional climate. RCMs are often driven by winds, temperature, and humidity fields at the lateral boundaries and by sea surface temperatures (SSTs) at the lower boundary (limited area models). These fields are supplied by a driving GCM or by re-analysis data (e.g., NCEP: Kalnay et al., 1996), usually with the expectation that the large-scale fields of the RCM will be consistent with those from the larger-scale model. Extending the domain of the model to be global and achieving a regional focus using variable resolution (Fox-Rabinovitz et al., 2001) avoids the reliance on forcing from another model and the limitations associated with that approach.

RCMs are usually run at resolutions of 50 km or finer, with recent multidecadal simulations at 10 km resolution (Corney et al., 2010; Evans and McCabe, 2010). Because of this higher resolution they are able to better capture the orography, coastlines, and surface characteristics. It also means they are able to capture many of the regional-scale phenomena that the GCMs are unable to resolve. Being physically based, RCMs, in principle, are able to produce spatially and temporally consistent fields of all climate variables.

RCMs have been used to successfully simulate climates for many locations, covering a broad range of climate types; this broad applicability provides a degree of confidence in their ability to downscale future climates. These simulations have been evaluated against observations for a number of variables, usually temperature and precipitation, on various timescales. For example, Evans et al. (2005) investigated the performance through time of many variables, but only for a single grid point. Kostopoulou et al. (2009) examined maximum and minimum temperatures on a seasonal basis. Evans (2009) and Evans et al. (2004) used temperature and precipitation on climatological and monthly timescales. Salon et al. (2008) focused on precipitation at the monthly to annual timescale. Rummukainen et al. (2001) evaluated seasonal to annual temperature and precipitation, as well as SSTs of the Baltic Sea. Solman et al. (2008) studied seasonal means and cycles, inter-annual variability, and extreme events in precipitation and surface air temperatures, while Evans and McCabe (2010) evaluated precipitation and temperature on timescales spanning daily to multi-annual, including ENSO cycles and multiyear droughts. In almost all cases, the RCMs improved on the climates simulated by the driving GCMs, particularly in areas with complex topography. Higher-resolution RCMs tend to perform better than lower-resolution RCMs. There is, however, a fairly wide range of dynamical downscaling methodologies. The more common methodologies will now be described.

9.3.1.1. Limited-Area Models

The majority of RCMs are limited-area models, used from the early 1990s (e.g., Giorgi, 1990; McGregor and Walsh, 1993). This methodology assumes that large-scale climate information can be used to drive the RCM through its lateral boundaries. Whilst there is no requirement that the large-scale RCM climatology must be similar to that of the host GCM, this is often assumed to be the case. For large domains, or in tropical regions where there is little flow through the domain, the RCM may develop its own internal circulations, so that the nested RCM may have significant differences in its climatology from its host model. Hence, it is desirable that an RCM domain is large enough for phenomena to develop that are related to local features of topography, land use, and small-scale atmospheric processes, but small enough that flow patterns do not deviate too much from the driving model (Leduc and Laprise, 2009).

9.3.1.2. Boundary Forcing versus Internal Forcing

The earliest RCMs utilized boundary forcing, whereby at the lateral boundaries large-scale climate information is interpolated in space and time to the RCM resolution. This scale adjustment occurs within a relaxation zone where the boundary conditions and the RCM simulation are merged in a way that dampens numerical noise (Marbaix et al., 2003). While it is known that the exact location and orientation of lateral boundaries can impact RCM simulations, they are usually specified in a pragmatic fashion. Considerations include orienting the domain so that large-scale flow enters the domain as uniformly as possible, and avoiding cutting across mountain ranges that generate sharp precipitation gradients and dynamic atmospheric phenomena.

Starting with Kida et al. (1991), a number of RCMs have applied some form of internal forcing within their domain, to enforce a degree of broad-scale consistency with the host GCM. This is sometimes referred to as 'nudging'. As expected, this technique greatly reduces the possibility of the evolution of flow regimes within the RCM that are disparate from the host GCM. Internal forcing is especially effective for large domains, or for domains where there is little flow through the domain, as is typically the case for tropical domains. The desirability of internal forcing of RCMs has been strongly argued by von Storch et al. (2000). The Regional Spectral Model (RSM) applies internal forcing by means of a scale-selective filter (Kanamaru and Kanamitsu, 2007). Using internal forcing has been shown to eliminate the effects of domain position and geometry on RCM simulations (Miguez-Macho et al., 2004). Some RCMs provide a choice of boundary forcing (one-way nesting) and internal forcing; for example, the Weather Research and Forecasting Model (WRF) provides options for broad-scale forcing of temperature, moisture, winds, and geo-potential height at specified atmospheric levels; this approach for height is possible in non-hydrostatic models, by virtue of height being a prognostic variable.

9.3.1.3. Variable-Resolution Global Models

A growing development is the use of variable-resolution global atmospheric models for dynamical downscaling,

starting with Déqué and Piedlievre (1995). The majority of these applications have employed a 'time-slice' approach, whereby an initial condition is taken from a host GCM, together with the evolving host GCM SSTs and sea-ice. The approach is very effective, and produces realistic regional enhancement of the climatology of the host GCMs, as demonstrated by intercomparison studies over North America (Fox-Rabinovitz et al., 2006).

Variable-resolution models allow for interactions between the regional-scales and the larger (less well-resolved) global scales, analogously to two-way nesting. A variety of methods have been employed to incorporate host GCM information into the variable-resolution conformal-cubic atmospheric model (CCAM; McGregor and Dix, 2008). The methods include far-field forcing by winds from a host GCM (Watterson et al., 2008; Engelbrecht et al., 2009) or weak upper-level forcing of winds and temperatures (Nguyen and McGregor, 2009).

Convolution-based scale-selective filters (Thatcher and McGregor, 2009; Thatcher and McGregor, 2011) may alternatively be used for downscaling to finer resolutions. It is thereby possible to use only internal forcing in the regional simulations so that downscaling is represented solely in terms of a length-scale problem, rather than as a boundary condition problem. The downscaling approach may first utilize the time-slice technique, with subsequent internal forcing for dynamical downscaling to finer scales. The methodology has been used to provide an ensemble of 14 climate simulations over Australia at 0.5° resolution for 1970−2099, driven by the bias-corrected SSTs of six AR4 GCMs, which are further downscaled to 0.1° resolution over Tasmania (Corney et al., 2010). Figure 9.8 shows the present-day rainfall from the host GFDL2.1 GCM, the downscaled 0.5° simulation, the further downscaled 0.1° simulation and observations from the Australian Water Availability Project; the benefits of higher resolution are

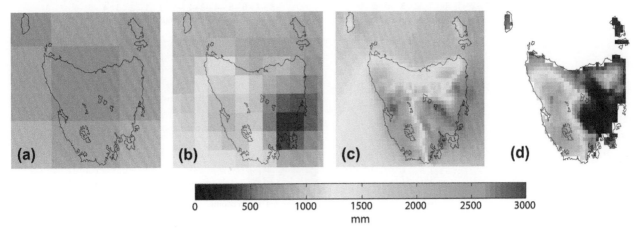

FIGURE 9.8 Mean annual rainfall for Tasmania for the period 1961−1990 from (a) the GFDL2.1 GCM, (b) the 0.5° dynamically downscaled simulation of the same GCM, (c) the 0.1° dynamically downscaled simulation of the same GCM, and (d) AWAP observations. Each resolution increase provides a better simulation of the observed sharp precipitation gradient from west to east across the island. *(Source: Adapted from Corney et al., 2010.)*

seen in the shift from homogeneous precipitation (Figure 9.8a) to a fairly realistic east–west precipitation gradient in Figure 9.8c.

9.3.1.4. Physical Processes and Model Biases

A major assumption of RCMs is that the boundary conditions, provided by a GCM, are accurate and unbiased. Unfortunately, like all models, GCMs, are unrealistic (biased) to some extent (McGuffie and Henderson-Sellers, 2005). While RCMs will add small-scale variability into the GCM fields, they will inherit large-scale biases from the driving GCM. Although the RCM will generally improve on the GCM simulation with its better-resolved topographic forcing, it is unlikely to overcome every deficiency in the GCM fields. This suggests that the driving GCM should be chosen carefully in order to provide the best possible climate simulation.

Even at the higher resolution of RCMs, many physical processes occur on scales smaller than the grid scale. These processes must be represented using a subgrid parameterization. These processes include radiation, convection, cloud microphysics, the planetary boundary layer, and land-surface processes. Each parameterization represents a simplification of the real world and, hence, introduces some uncertainty into the model.

A concern for modelling climate change is that the current generation of coupled GCMs typically exhibit some major biases for present-day climate; for example, the very common problem of an excessive westward-extending cold tongue over the equatorial Pacific. In time-slice ensemble atmosphere-only GCM experiments, Ashfaq et al. (2010) applied a monthly bias-correction to SSTs from coupled GCMs and obtained improved simulation of present-day seasonal precipitation. The regional atmospheric fields have also been reconstructed using a variable-resolution grid for an ensemble of GCMs after SST biases were removed (Corney et al., 2010), which also results in improved simulation of 2 m air temperature and precipitation.

Some RCM groups, for example Kawase et al. (2009), employ a pseudo global warming downscaling method, whereby the boundary condition of the RCM is a composite of 6-hourly re-analysis data and the differences in monthly mean variables between the present and the future climates simulated by GCMs. Knutson et al. (2008) apply a similar method to simulate Atlantic hurricanes (Figure 9.9). Their regional model reproduces the hurricane activity in the North Atlantic basin, including the inter-annual variability. By applying their technique to an ensemble of models from CMIP3 for the climate of the twenty-first century, they found that Atlantic hurricane and

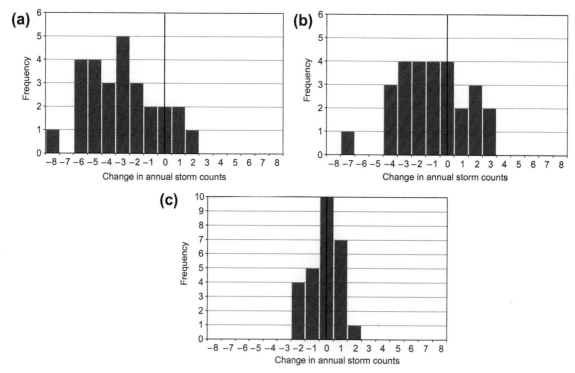

FIGURE 9.9   Frequency distribution of change (warm climate minus control) in storm counts for each year simulated. (a) Distributions for Atlantic basin tropical storms, (b) hurricanes, and (c) major hurricanes (central pressures less than 965 hPa). In the multimodel ensemble climate change experiments, tropical storm counts decrease by 27%, hurricanes by 18%, and major hurricanes by 8% relative to the control. *(Source: Reprinted by permission from Macmillan Publishers Ltd.)*

tropical storm numbers are decreased over this period. It is, however, difficult to evaluate changes in the inter-annual variability, as this technique (Knutson et al., 2008) assumes unchanged variability even in the future climate. Recent downscaling experiments at the US National Center for Atmospheric Research (NCAR), using WRF, apply SST corrections, and also attempt to introduce corrections for the corresponding flow biases (J. Done, personal communication, 21 March 2011).

9.3.1.5. Limitations of Dynamical Downscaling

As mentioned above, RCMs employ subgrid parameterizations for a number of physical processes. For some processes, such as deep convection, several alternative parameterizations have been developed, which indicates the uncertainty that surrounds such parameterizations. More-over, a number of parameters in the model physics are not well-constrained and varying them within reasonable bounds may also produce differences in the simulated climate. It has become common for some RCM groups to select parameterizations and parameter settings for a particular domain. There is, however, a move towards development and adoption of parameterizations that are applicable for a wide range of length scales and for a variety of climatic conditions (Takle et al., 2007). To help achieve this aim, there have been several RCM intercomparison projects, including the Project to Intercompare Regional Climate Simulations (PIRCS; Takle et al., 1999) and NARCCAP (Mearns et al., 2009) over North America, the Regional Climate Model Inter-comparison Project for Asia over Asia (RMIP; Fu et al., 2005), PRUDENCE (Chris-tensen et al., 2007), and the Ensembles-Based Predictions of Climate Changes and Their Impacts Project (ENSEMBLES; van der Linden and Mitchell, 2009) over Europe. Currently under way is the wide-ranging Coordinated Regional Climate Downscaling Experiment (CORDEX) intercom-parison (Giorgi et al., 2009), which is being conducted over 12 regional domains, covering much of the globe.

In terms of future climate simulations, an assumption made in dynamical downscaling is that the same physical parameterizations will continue to remain appropriate. Because RCMs are based on the equations of motion and thermodynamics of the climate system, in principle they are capable of capturing changes irrespective of whether they are small, gradual changes, or abrupt changes in response to non-linearities in the system.

Most climate observations are point-based, measured at a particular meteorological station. However, RCMs can be generally regarded as producing grid-average values, which means that, with a 50 km grid, the RCM values represent an average over an area of 2500 km^2. Thus, care is needed when comparing RCM fields directly to in-situ observa-tions. It is also important, when comparing model temperatures with station observations, to perform a correction for the different elevations, typically making an adjustment of ~6.5 K per km of elevation difference. It is also important to consider, when using RCM data to drive other impacts models, that the impacts models may have been developed to use in situ observations. On the other hand, some impacts models naturally expect spatially-averaged data rather than point data. For example, hydrology models typically require the precipitation aver-aged over the catchment. It should also be noted that satellite data represent a spatial average over some foot-print, rather than a point measurement, and hence may be more comparable to RCM-produced fields.

A difficulty with dynamical downscaling is the computation time required. As the spatial resolution increases, the time step decreases and the computation time may increase dramatically. Performing multidecadal simulations at high resolution may require the use of a supercomputer and access to significant data storage since many terabytes of data can be generated. However, as the speed of modern computers keeps increasing and the cost of data storage systems keeps decreasing, the use of RCMs is becoming more accessible and widespread.

9.3.1.6. Future Development in Dynamical Downscaling

The downscaling skill of any particular RCM generally depends on the location, season, and type of event; hence, some evaluation of this skill should always be performed before it is used in impacts research. Overall, the skill of RCM simulations continues to improve, with some recent evaluations showing good skill (Evans and McCabe, 2010), though predicting extremes remains a significant challenge (cf. Alexander and Tebaldi, 2012, this volume).

The sophistication of RCMs is steadily increasing, as additional parameterizations and modules are incorporated. Several groups now include schemes to model urban interactions within the lower atmospheric layers (cf. Cleugh and Grimmond, 2012, this volume) and researchers are introducing aerosol schemes and chemistry modules (e.g., Wang et al., 2012, this volume). A number of RCMs have incorporated mixed-layer ocean schemes to provide aspects of oceanic feedback, which is important for prop-erly modelling tropical cyclones and for capturing larger-scale phenomena such as the Madden−Julian Oscillation (cf. Latif and Park, 2012, this volume). A few coupled atmosphere−ice−ocean RCMs exist, for example the high resolution coupled model, HIRHAM−MOM, developed by Rinke et al. (2003) for Arctic climate research. The model has been used by Rinke et al. (2006) to study the influence of sea-ice on the atmosphere. They found the sea-ice played a significant role, with changes in the marginal ice zone having the greatest impact on the atmospheric circulation.

9.3.2. Statistical Downscaling

Statistical downscaling involves establishing a mathematical relationship between a large-scale climate field or variable and a local-scale observation. The outcome of statistical downscaling is almost invariably a point estimate of the climate in terms of observed quantities, usually temperature and precipitation. There are many different statistical techniques that have been applied in order to downscale climate data. Prior to discussing the various techniques, we note that they were first classified by Maraun et al. (2010) into Perfect Prognosis (PP), Model Output Statistics (MOS), and Weather Generator approaches.

PP is the most common statistical downscaling approach. It establishes relationships between observed large-scale predictors and observed local-scale predictands. These relationships are then applied to the large-scale predictors simulated by GCMs. This is justified so long as the GCM large-scale fields are simulated realistically — hence the term PP downscaling. Statistical downscaling techniques that form relationships between the climate model-simulated output and local-scale observations, in order to correct either the GCM or RCM errors, are known as MOS. Weather generators are models that generate local-scale weather time series with the same statistical properties as the observed time series. The parameters of these models can be adjusted based on large-scale fields from GCMs or on output from RCMs.

9.3.2.1. Perfect Prognosis (PP)

The basic premise behind PP approaches to statistical downscaling is that GCMs are able to simulate the large-scale atmospheric climate fields realistically, even if fields with high spatial variability, like precipitation, are poorly simulated. Thus, statistical relationships are sought with variables in which there is high confidence, while ignoring those in which there is low confidence. Forming these relationships often uses regression methods, such that the predicting variable, $Y(t)$, based on the large-scale predictors, $X_i(t)$, is given by:

$$Y(t) = f(X_i(t)) + \eta \qquad (9.1)$$

where η is the residual noise. Most PP approaches disregard any residual noise term although some newer PP approaches explicitly provide a noise model to help capture the variability and extremes. Approaches that include a noise model are often referred to as stochastic, while those that do not are termed deterministic.

Building a PP downscaling scheme requires two steps that are often performed together. These are the identification of suitable, observed large-scale predictors, and the development of the statistical relationship between them and the local-scale observations. It is important that predictors that capture the effects of climate change are included in the scheme if it is to be used to downscale future projections. This consideration needs to be kept in mind when identifying predictors based on historical time series that may contain only a small climate change signal. In general, the predictor choice will vary depending on the region and season. Various large-scale predictors for downscaling precipitation have been explored in Wilby and Wigley (2000), along with a comparison of the observed and simulated fields.

Often these predictors are high-dimensional fields of grid-based values. Since these fields frequently have high levels of spatial correlation, the grid-point values are not independent. Thus, it is relatively common to reduce the dimensionality of the predictor field in some way. Common techniques for this include principal component analysis (Hannachi et al., 2007), canonical correlation analysis (Hertig and Jacobeit, 2008; Palatella et al., 2010), maximum covariance analysis (Tippett et al., 2008), support vector machine (Ghosh, 2010; Kim, 2010), and physically-motivated transformations such as using an ENSO index or weather types (Yang et al., 2010). Weather types are circulation patterns or regimes that occur frequently in a location. They can be defined subjectively, by visually inspecting synoptic maps, or objectively using clustering and classification algorithms.

There are many ways to establish the statistical relationship between the predictors and predictands in PP, though in each case the relationship is calibrated using observed variables before being applied to climate model output. Each statistical model has its own set of assumptions and level of complexity. Among the simplest models are linear regression models. These assume that the variables involved are Gaussian-distributed, which is not true of precipitation fields on short timescales, including daily. This assumption has been relaxed in the framework of the generalized linear model. Also, the linearity dependence has been replaced by non-parametric smooth functions in the generalized additive model (e.g., Vrac et al., 2007). All of these methods focus on predicting the mean conditional on a set of predictors. In order to quantify the variance (or higher-order moments) dependence on a set of predictors, vector generalized linear models can be used.

Several non-linear regression techniques have also been applied to the statistical downscaling problem. Such techniques include the application of artificial neural networks to downscale precipitation (e.g., Haylock et al., 2006). Another method that has been applied to downscaling is the analogue method (e.g., Zorita and von Storch, 1999). In this method, a selected metric is used to identify the most similar situation in the historical record and the corresponding local observations are used as the prediction. One major limitation of this approach is that it cannot produce precipitation amounts that have not been observed in the past.

The many different statistical models that can be used in PP downscaling make various assumptions and have various limitations. All PP approaches do, however, share two major assumptions. Firstly, that suitable predictors are well-simulated by the GCM; that is, only fields that have been evaluated and found to perform well should be used. Secondly, that the relationship identified between the predictor and predictands is stationary. That is, it is assumed that, although the climate changes, the relationship between the identified variables does not change. This second assumption is particularly difficult to test and, while there is evidence to the contrary (e.g., Lenton 2012, this volume), its applicability is usually unknown.

9.3.2.2. Model Output Statistics (MOS)

Unlike PP techniques, MOS methods develop statistical relationships between simulated predictors and observed predictands. They are most often applied to climate model-simulated fields of the same variable being predicted. That is, a MOS method can be used to correct the RCM-simulated precipitation field, in order to account for the difference between areal-gridded means and local point observations of precipitation. As such, MOS methods can often be thought of as statistical corrections to RCM-simulated outputs; indeed, they have been used in numerical weather prediction (NWP) for some time (Glahn and Lowry, 1972; Kalnay, 2003).

If the RCM simulation is driven by an atmospheric re-analysis, then there is a direct correspondence between the simulation and observations. In this case, the MOS method will relate the simulated and observed time series through regression techniques. If, on the other hand, the RCM simulation is not driven by a re-analysis, then this direct relationship does not exist between the simulation and observations. In this case, the MOS methods can only be used to link the distributions of the variables.

At their simplest, MOS methods provide a bias correction of the present-day simulated field to match the observations. For variables such as temperature, this is usually a simple arithmetic (e.g., additive) correction while, for precipitation, this is applied as a scaling factor, often calculated and applied separately for each month or season. A more complex approach is quantile matching. In this approach, different intensities are considered individually such that the simulated cumulative distribution function is adjusted to match the observed cumulative distribution function. Similar bias-correction approaches have been further developed to account for persistence in the precipitation fields (Johnson and Sharma, 2009).

9.3.2.3. Weather Generators

Weather generators are statistical models that produce random sequences of climate variables with statistical properties that match those of the observed variables (Wilks and Wilby, 1999). Weather generators were not originally developed with regional downscaling in mind, but rather to generate very long time series to assist in the study of floods, planning for large water engineering projects, and so on. As downscaling tools, they are often used with the statistical relationships being developed from observed data. The same statistical properties are calculated on both a present-day and future climate simulation. The simulated changes in these properties are then applied to the observed parameters (Semenov and Barrow, 1997).

Early weather generators simulated precipitation as a two-step process, modelling the occurrence and intensity of precipitation separately. Precipitation occurrence was often modelled as a first-order Markov chain, while to model the rainfall intensity, the two-parameter Gamma distribution was often used, though many alternatives have been explored in the literature. More recent weather generator approaches have attempted to capture both the temporal statistical properties, as well as the spatial correlation structure. One example of this is the use of non-homogenous and hidden Markov models to probabilistically link weather states and predictors (e.g., Charles et al., 2004).

9.4. SOURCES OF UNCERTAINTY

When studying future regional climate changes, there are many different sources of uncertainty. Here, three main sources of uncertainty are identified. A different, but complementary, way to categorize these sources of uncertainty can be found in Foley (2010). The first source, and one of the largest unknowns, is the future emissions of greenhouse gases. Since this is dependent on human activities and policy actions, the future evolution of greenhouse gas emissions is generally presented as a series of possible emission scenarios or projections (e.g., Taplin, 2012, this volume). These scenarios are then used in GCM simulations to study the impact on climate. The GCM model physics and numerical structure provide the second source of uncertainty. The last main source of uncertainty is the downscaling method itself. In combination, these sources of uncertainty provide a limit to the confidence that can be placed in any particular projection of future regional climate.

9.4.1. Emission Scenarios

Global warming, as it is occurring today, is due largely to the increase in greenhouse gases in the atmosphere. Hence, projecting changes in the emissions of these gases is vital for projecting future climate change. The future emissions of these gases are controlled by a number of factors that are difficult to predict, including future population, economic development, technological development, energy use, and energy sources (cf. Harvey, 2012, this volume). Given all the uncertainties involved, a series of scenarios has been created that aims to cover a range from plausible low-emission scenarios to plausible high-emission scenarios (see also Rice and Henderson-Sellers, 2012, this volume). Since the year 2000, when the emission scenarios used in the IPCC AR4 were published, emissions have tracked close to the highest emission scenarios considered (Canadell et al., 2010).

9.4.1.1. How Are Emissions Scenarios Constructed?

In the IPCC AR4, the emission scenarios are known as the SRES scenarios, since they were published in the IPCC Special Report on Emissions Scenarios (IPCC, 2000). These scenarios were created by first establishing four storylines of the future population growth, economic and technological development, and international co-operation of the world. Each storyline produces a number of emission scenarios, with one designated as the marker scenario. The four (IPCC, 2000) storylines are

- The A1 storyline: rapid economic growth, global population peaks in mid-century, rapid introduction of new and more efficient technologies. Three A1 groups are distinguished by their technological emphasis: fossil-intensive (A1FI), non-fossil energy sources (A1T), or a balance across all sources (A1B).
- The A2 storyline: economic growth varies by region, continuously increasing population, technological change varies by region.
- The B1 storyline: transition towards service and information economy, global population peaks in mid-century, introduction of clean and resource-efficient technologies.
- The B2 storyline: intermediate levels of economic development, continuously increasing population (slower than A2), less rapid and more diverse technological change.

Storylines B1, and A1T represent low-emission scenarios while A2 and A1FI represent high-emission scenarios. By 2100, the range between them is quite large. Currently, the high-emission scenarios are close to 'business-as-usual' scenarios, while the low-emission scenarios could only be reached if there is a global initiative to reduce emissions.

More recently, these emission scenarios were reviewed and new scenarios established for the upcoming IPCC AR5 report (Moss et al., 2008). In these new scenarios, the previous 'marker scenarios' are now referred to as 'Representative Concentration Pathways' (RCPs). These new emission scenarios are defined first in terms of the radiative forcing they have on the climate system, while

the potential future population growth, economic development, and technological change that may lead to this forcing is calculated later (see Rice and Henderson-Sellers, 2012, this volume). Four RCPs will be the main focus of AR5 GCM simulations. These include a high pathway where radiative forcing reaches 8.5 W m^{-2} by 2100 and continues to rise after that; two intermediate 'stabilization pathways' in which radiative forcing stabilizes at 6 W m^{-2} and 4.5 W m^{-2} after 2100; and a low pathway where the radiative forcing peaks at 3 W m^{-2} before 2100 and then declines. The evolution of these RCPs is shown in Figure 9.10. In broad terms the low-emission scenario RCP3.0 (IMAGE 2.9 or 2.6 in Figure 9.10) is similar to the SRES B1 scenario, RCP4.5 (MiniCAM 4.5 in Figure 9.10) is similar to A1B, RCP6.0 (AIM 6.0 in Figure 9.10) is similar to A1FI, and RCP8.5 (MES-A2R 8.5 in Figure 9.10) is a new high-emission scenario.

9.4.2. GCM Uncertainties

GCMs are complex mathematical/computational models that attempt to account for all the physical (and, increasingly, biological and chemical) processes important for climate, including those in the atmosphere, oceans, land, and ice (Trenberth, 1992; Neelin, 2011). The atmospheric component of a GCM is closely related to NWP models. GCMs are the result of many years of collaboration between large numbers of scientists, working in conjunction with computer specialists in a huge collaborative effort (e.g., McGuffie and Henderson-Sellers, 2005). As such, they tend to be developed at major atmospheric or climate science research centres.

GCMs have been shown to be able to simulate the observed recent climate and past climate changes well. At continental and larger scales, they produce quantitatively credible estimates of future climate change. GCMs are the key modelling component that takes various emission scenarios as an input and establishes the climate system response to the resulting radiative forcing (IPCC, 2007a). While GCMs have consistently improved over the last several decades, there still remains uncertainty associated with their simulations.

The first source of uncertainty comes from the climate system itself. There is a significant level of internal variability in the climate system; a common example is the El Niño phenomenon (Cobb et al., 2003; Latif and Park, 2012, this volume). In NWP models, this uncertainty manifests itself largely as a limit to the forecast skill due to the imperfectly known initial conditions (Lorenz, 1964). In climate modelling, a simulation only provides one possible realization among many that may be equally likely, and it requires only small perturbations in the initial conditions to produce different realizations although, averaged over several decades, the different realizations are expected to

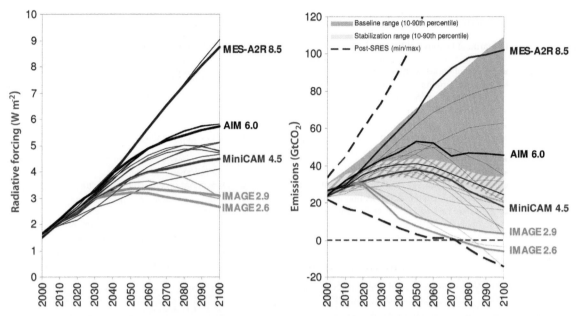

FIGURE 9.10 Radiative forcing compared to pre-industrial (left panel) and energy and industry CO_2 emissions (right panel) for the RCP candidates (coloured lines), and for the maximum and minimum (dashed lines) and 10th to 90th percentile (shaded area) in the post-SRES literature. These percentiles reflect the frequency distribution of existing scenarios and should not be considered probabilities. Blue-shaded area indicates mitigation scenarios; grey-shaded area indicates baseline scenarios. MES-A2R is the MESSAGE model, SRES A2-like scenario. AIM is the Asia-Pacific Integrated Model. MiniCAM is the Mini-Climate Assessment Model. IMAGE is the Integrated Model to Assess the Global Environment. *(Source: Moss et al., 2008: Towards New Scenarios for Analysis of Emissions, Climate Change, Impacts, and Response Strategies, IPCC Expert Meeting Report, Figure 5. Intergovernmental Panel on Climate Change, Geneva, Switzerland.)*

produce very similar mean climate statistics. However, different extremes will be encountered in each realization and this needs to be considered if extremes are of interest (see Alexander and Tebaldi, 2012, this volume).

The second source of uncertainty comes from the parameterization of processes that occur on scales smaller than the grid scale of the model. These processes include cumulus convection, cloud microphysics, radiative transfer in the atmosphere, mixing in the planetary boundary layer, and land-surface processes, among others (Trenberth, 1992). For each of these processes, many different parameterizations have been developed, once again demonstrating an associated uncertainty (e.g., Dickinson, 2012, this volume).

The third source of uncertainty comes from the numerical method chosen to solve the non-linear differential equations of the fluid motion of the atmosphere and the ocean (see e.g., Sen Gupta and McNeil, 2012, this volume). Many different techniques exist for calculating a numerical approximation to the solution of these differential equations. They have different accuracies in various situations and different computational costs. In general, climate models use methods that provide good accuracy for a given computational cost. The accuracy of the technique can also be affected, to a lesser extent, by the precision of the hardware on which the model is run. Thus, even if all the physical processes and initial conditions are known perfectly, there will be uncertainty associated with the

numerical solution of the fundamental differential equations and, in particular, the inherent climate sensitivity of the specific model (e.g., Harvey, 2012, this volume).

While it is difficult to determine precisely how much uncertainty is associated with each of these sources, their overall impact can be estimated by the spread in the climate simulations from many GCMs. Alexander and Tebaldi (2012, this volume) discuss the spread in the global temperature anomaly obtained from 58 simulations produced by 14 different GCMs. While the overall observed trend, including the impact of large volcanic eruptions, is captured well, a spread of about 0.5°C exists throughout the period. This is an indication of the level of uncertainty associated with the simulation of global temperature from GCMs.

9.4.3. Uncertainty from Downscaling Techniques

Each method of downscaling inherits uncertainties associated with the emission scenarios and the host GCMs, as well as having uncertainties of its own (Table 9.1). Each downscaling method contains different uncertainties depending on the assumptions made and the implementation used, as discussed in Section 9.3. For dynamical downscaling, RCM uncertainties have similar causes to those for GCMs: internal climate variability; imperfect

TABLE 9.1 Relative Merits of Dynamical and Statistical Downscaling Techniques

	Dynamical Downscaling	Statistical Downscaling
Advantages	• Physically consistent processes • Resolves smaller-scale atmospheric processes and climate system feedbacks • Provides information over the entire landscape • Provides output for many climate-relevant variables • Can output variables with high temporal resolution	• Computationally inexpensive • Provides information at observation points • Based on accepted statistical procedures • Optimized for observation locations already used by impacts models • Can be applied to existing GCM ensembles
Disadvantages	• Computationally expensive • Limited ensembles are available • Produces very large datasets that the impacts community is not used to dealing with • Does not provide output at the point locations of current observations sites	• Requires long and reliable historical datasets for calibration • Dependent on choice of predictor • Assumes stationarity in predictor–predictand relationship • Climate system feedbacks are not included • Cannot be applied to locations without observations • Only observed variables can be downscaled • Need to account for a tendency to underpredict variance

Note that applying MOS to dynamically downscaled output addresses some of the disadvantages of both.

physical parameterizations; and the use of numerical approximation techniques. In terms of future climate, the only assumptions made with dynamical downscaling are that the physics on which the RCM is based does not change and that the parameterizations of physical processes remain valid. Naturally, this latter assumption becomes easier to challenge as space-scales and timescales become shorter/smaller.

For statistical downscaling, there are many different techniques and the uncertainties inherent in each technique differ. Thus, it is difficult to generalize about these uncertainties and each technique should have its uncertainties explicitly explored and listed when used. Some common causes of uncertainties in these techniques include lack of physical consistency between climate variables; assuming a statistical distribution for individual climate variables; and assuming stationarity of the statistical relationship used. Unlike RCMs, statistical downscaling techniques do not, in general, explicitly consider the physical coherence between climate variables. Thus, they can produce climate variables that evolve in ways that are physically unrealistic. Many of these techniques also assume a particular statistical distribution shape for individual climate variables, for example, a Gaussian distribution for temperature or a Gamma distribution for precipitation. While these distributions are usually fitted to the available data in order to obtain the best fit possible, there are always errors associated with this. Perhaps the largest source of uncertainty in statistical downscaling comes from the assumption of stationarity. That is, these techniques assume that, as the climate changes, the statistical relationships do not. This includes assuming that the statistical distribution associated with each climate variable will not change, that the same

large-scale predictors would be identified as important, and that the same statistical relationship between predictors and predictands exists. It is very difficult to quantify these uncertainties, particularly those associated with stationarity, as future climate conditions can be well outside those in the historical record that were used to calibrate the model. For relatively small and gradual changes, the stationarity assumption is likely to be reasonable. However, if the climate changes become large, or non-linearities (and tipping points) become important, this may not be the case (Schmith, 2008; Lenton, 2012, this volume).

A number of studies have applied both dynamical and statistical downscaling techniques to the same region. In general, they find that no downscaling technique consistently performs better than the others. Most of these studies use dynamical downscaling to about 50 km resolution and evaluate the outcomes at the observation stations used to calibrate the statistical models (Murphy, 1999; Haylock et al., 2006; Landman et al., 2009). A few studies, such as those of Haylock et al. (2006) and Schmidli et al. (2007), have used multiple statistical and multiple dynamical models. Schmidli et al. (2007) applied the downscaling techniques to a domain that included the European Alps. In this study, the statistical methods were applied to a 50 km gridded dataset of climate variables matching the resolution of the dynamical models. They found that both types of models tend to have similar biases, but that the statistical downscaling techniques tended to underestimate the inter-annual variability. Over complex terrain, the dynamical approaches tended to perform better. When examining future projections, they found that all the approaches were in good agreement for winter precipitation, but not for summer precipitation. For summer, all the dynamical models qualitatively agreed with each other, while the statistical models produced a wide variety of predictions. Studies have also found that applying MOS techniques, particularly bias correction, to the dynamically downscaled results produced the best results (Murphy, 1999; Landman et al., 2009; Schoof et al., 2009).

9.4.4. Building Ensembles

Given the need to plan for, and adapt to, the impacts of climate change, techniques for dealing with and/or quantifying the uncertainties discussed in Section 9.4.3 are required. For management and policy decisions, it would be beneficial to have future climate projections associated with probabilities or likelihood of occurrence (e.g., Mathew et al., 2011). Since, even for a given emission scenario, the future climate is unknown, the question is how to estimate both the future climate and the uncertainty around that estimate. This can be done by producing many estimates of the future climate that sample aspects of the uncertainty (cf. Alexander and Tebaldi, 2012, this volume). This collection of future projections is known as an ensemble and it can be used to produce a 'most likely' future climate, along with the uncertainty around that. In order to build a useful ensemble, two main questions need to be answered: how do we adequately sample the uncertainties associated with estimating future regional climates, and how do we identify the 'most likely' future climate and the likelihood of future climates around that?

9.4.4.1. Sampling the Uncertainty

In many ways, the uncertainties in the emission scenarios are due to unknown future policy decisions; that is, policy decisions could be made that would move the world from one emission scenario to another. From a policy viewpoint, the question is whether or not to make those decisions that would move the world from the current high-emission future to a lower-emission future. That is, one needs to know the difference in impacts (benefits/costs) of a high-emission scenario compared to a low-emission scenario. In practice, this translates into producing separate ensembles for each emission scenario (e.g., Hawkins and Sutton, 2009).

Hence, the question becomes: for a given emission scenario, how can we build ensembles that adequately sample the uncertainty in future regional climate projections? The first step is to establish a collection of GCM projections suitable for downscaling. This collection should sample the internal climate variability, the uncertainty in physical process parameterization, and a variety of numerical solution techniques. The second step is to establish a collection of downscaling techniques that adequately

samples the uncertainty associated with those techniques. For dynamical downscaling, this requires covering a similar set of uncertainties to that for GCMs, while for statistical downscaling this requires sampling the various statistical methods available (cf. Giorgi et al., 2009).

There are a number of ways to produce GCM (or RCM) ensembles to address specific forms of the uncertainties. Sampling the internal climate variability with a collection of GCM simulations is usually accomplished by running each simulation with a perturbed initial condition. This is the dominant way of producing ensemble forecasts in NWP. While this method will produce weather systems that diverge in their evolution over days or weeks, when averaged over decades it will usually produce a smaller spread than some other methods of producing ensembles.

In order to sample some of the uncertainty in the parameterizations of physical processes, perturbed-physics ensembles have been created. To create these ensembles, parameters that are not well-constrained in the parameterizations are varied within reasonable ranges. GCMs contain a large number of relatively unconstrained parameters. These parameters can be inter-dependent in non-linear ways, meaning that properly sampling the parameter space would require hundreds of simulations to be performed — a computationally prohibitive exercise. Carefully selecting parameters to which the climate is most sensitive can substantially reduce the number of simulations required, while acknowledging that some important parameter interactions may be missed. The largest perturbed-physics ensemble created to date was for the distributed computing project *Climateprediction.net*, which created a multi-thousand member ensemble in an attempt to span the entire parameter space (Piani et al., 2005). Averaged over decades, these perturbed-physics ensembles tend to produce a larger spread than the perturbed-initial-condition ensembles.

Possibly the most common way to create GCM ensembles today is to create a multimodel ensemble. These ensembles bring together simulations from many different GCMs. In doing so, they sample both the various numerical solution techniques being used and various physical process parameterizations. Multimodel ensembles can also include several realizations from the same GCM, created using perturbed initial conditions or perturbed physics. The use of multimodel ensembles has been greatly facilitated by the CMIP3 archive, organised by the World Climate Research Programme (WCRP) that co-ordinates and oversees the access to and distribution of GCM simulation data created as part of experiments designed to inform the IPCC Fourth Assessment Report. An even larger archive of GCM simulations is currently being prepared as part of CMIP5 that is designed to inform the IPCC Fifth Assessment Report. Thus, both from a viewpoint of sampling the uncertainty range and the availability of GCM simulations, it is desirable to downscale from a multimodel ensemble of GCM simulations.

Since dynamical downscaling (in RCMs) has similar uncertainty issues to GCMs, it follows that a multimodel ensemble of RCMs should also be used to downscale each of the GCMs in order to create regional-scale climate change projections. Producing these RCM ensembles has been managed in large projects, such as NARCCAP (Wang et al., 2009) and ENSEMBLES (van der Linden and Mitchell, 2009), and is currently being pursued as part of the CORDEX (Giorgi et al., 2009). Statistical downscaling covers a wide range of techniques. In order to sample the associated uncertainty, it follows that an ensemble of statistical downscaling techniques should also be used to downscale each GCM (e.g., Zorita and von Storch, 1999; Haylock et al., 2006; Schmidli et al., 2007; Timbal et al., 2008). Hence, a comprehensive attempt to predict a region's future climate, and the uncertainty associated with the prediction (Figure 9.11), would require many GCM projections to be downscaled by many dynamical and statistical methods to produce a very large ensemble of regional climate projections (Hawkins and Sutton, 2009).

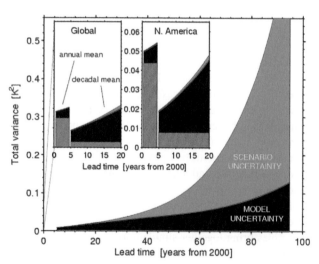

FIGURE 9.11 The relative importance of the three sources of uncertainty changes significantly with region, forecast lead time, and the amount of any temporal meaning applied. Main panel: Total variance for the global mean, decadal mean surface air temperature predictions, split into the three sources of uncertainty. Insets: As in the main panel, but only for lead times less than 20 years for (left) the global mean, and (right) a North American mean. The orange regions represent the internal variability component. For lead times shorter than 5 years, the results are plotted using annual mean data to highlight how the internal variability component is vastly reduced when considering decadal mean data. The uncertainty in the regional prediction is larger than for a global mean. *(Source: Hawkins, Ed, Rowan Sutton, 2009; The potential to Narrow Uncertainty in Regional Climate Predictions. Bull. Amer. Meteor. Soc., 90, 1095−1107. doi: 10.1175/2009BAMS2607.1. © American Meteorological Society. Reprinted with permission.)*

9.4.4.2. The 'Most Likely' Future Climate and Its Probability

Given limitations in time, researcher, and computational resources, it is rarely possible to downscale from a GCM ensemble containing ten or more simulations with multiple downscaling techniques. So, a method must be chosen to determine which GCM simulations to downscale. This choice will often be affected by practical issues like the availability of the required results, but will also be influenced by factors, such as the ability of the GCM to simulate the recent past in the region of interest and the spread in future climates projected by the GCMs. While the ability to successfully simulate the recent past does not translate directly into an ability to simulate future climate, the reverse is very likely true: if a GCM is unable to simulate the recent past, then future climate projections from that model should be given low (or no) credence. Thus, evaluation against recent observations is more an activity performed to identify GCMs that should not be trusted in a particular region, rather than identifying GCMs that perform the best.

There are many different variables and metrics that can be used to evaluate climate model performance against recent observations (e.g., Taylor, 2001). There is no general consensus about which variables and metrics are the best or most appropriate; indeed, preferences may change depending on the application being considered. As a result, evaluation studies over the same region can come to different conclusions as to the relative performance of different GCMs, as can be seen in metadata studies like that of Smith and Chandler (2010). They compared the results from eight different GCM-evaluation studies relevant to southeast Australia, showing the differences in relative performance due to the different variables tested and metrics used. They did, however, identify five GCMs that tended to perform poorly across all assessments. Using such assessments, the poorly performing GCMs can be removed from the ensemble before it is used for future climate prediction.

Given a climate model ensemble, it has been consistently found that the ensemble mean outperforms any individual model in evaluations against observations (Hagedorn et al., 2005). The 'best' or preferred method of creating this ensemble mean remains open. Should all ensemble members be weighted equally? Or should weights be assigned based on performance criteria? Intuitively, models with greater skill deserve higher weighting. In fact, discounting the poorly performing models, as suggested above, is equivalent to giving them little or zero weight. One of the first quantitative methods for deriving such weights was the Reliability Ensemble Averaging (REA) method of Giorgi and Mearns (2002). REA used two criteria: one based on the model performance compared to observations (bias); and the other based on the distance between a model's projected change and the mean projected change, which they refer to as the model convergence criteria. Like all systems that assign weights to an ensemble average, there are a number of counter-arguments. As stated earlier, model performance criteria depend intimately on the variable and metric chosen, while the convergence criteria reward models that are like most other models even though it could be that most of the models perform poorly. More recently, the ENSEMBLES project applied a series of evaluation metrics to create a performance-based weighting system, in order to produce the ensemble mean and probability of an RCM ensemble (Kjellstrom and Giorgi, 2010). The weighted mean was found to perform no better than using equal weights or random weights (Christensen et al., 2010; Déqué and Somot, 2010). Regardless of how it is created, the ensemble mean is often considered the most likely simulated climate. While it does produce a quantitative estimate of the climate, it does not produce any probabilistic information that embodies the uncertainty.

One approach to producing probabilistic predictions of future regional climate change predictions is to use weighted frequency distributions (Déqué and Somot, 2010). In this technique, probability density functions (pdfs) of daily data are produced for each RCM and then combined using performance-based weights to produce an ensemble pdf.

Another approach is to adopt Bayesian methods for producing pdfs. Bayes' theorem states:

$$p(s|data) \propto p(s)p(data|s) \qquad (9.2)$$

where *data* is a collection of observed climate variables, *s* is the simulated variable of interest, and *p* denotes the probability. Using Bayesian terms, the formula says that the posterior probability of *s* is proportional to the prior probability of *s* multiplied by the probability of obtaining the *data* for a given value of *s*, that is, its likelihood. In practice, the application of Equation (9.2) can be complicated and issues surrounding its application to climate projection ensembles can be found in Collins (2007) and the accompanying journal special issue. One of the largest issues in this approach is the need to define a prior distribution based on inadequate data, noting that the choice of prior distribution has a significant impact on the posterior pdf created (see also Harvey, 2012, this volume).

Different techniques have been proposed in this relatively new area of probabilistic climate projections. One example is the work of Tebaldi et al. (2004, 2005), who applied an REA-like approach in the Bayesian framework. They assumed prior distributions that are Gaussian for current and future model projections, centred around the observed climate means, but with model-specific variances.

Implementing the REA criteria of model performance (bias) and convergence in this framework, they were able to solve analytically for the posterior distribution, finding that the model-specific variances were similar to the original REA weights.

More recently, Buser et al. (2010) applied a Bayesian approach to regional climate change simulations in the ENSEMBLES project. They applied the method to seasonal temperature, accounting both for the uncertainty in the RCM output and in the extrapolation of the biases into the future. These biases are included in the formation of the prior distributions. The approach led to several differences in projected regional climate change compared to the unweighted ensemble mean, suggesting the importance of explicitly including probabilities in future projections.

Producing probabilistic projections within a Bayesian framework is an area of active research and a number of alternative approaches to the above can be found. Other approaches use Bayesian hierarchical linear models (Greene et al., 2006) or consider joint probabilities between climate variables, such as temperature and precipitation (Tebaldi and Sanso, 2009). These probabilistic approaches are likely to become more important in impacts assessments as they allow the adoption of risk-based assessment frameworks (New et al., 2007).

9.5. ACHIEVING REGIONAL CLIMATE PREDICTIONS

9.5.1. Water Resources

Studying the way future climate change will impact various natural and anthropogenic systems requires downscaled climate change predictions (Fowler et al., 2007a). These studies have varied from using a single statistically downscaled projection, to using an ensemble of both dynamically and statistically downscaled projections. The climatic stress likely to be imposed on future water resources is politically and socially important (e.g., Henderson-Sellers, 2012, this volume). Of particular interest is the downscaling method used in the future water resource prediction and how uncertainty is dealt with, keeping in mind that the addition of a hydrology model will introduce another source of uncertainty not discussed previously.

Evans and Schreider (2002) used a weather generator approach to estimate the regional climate change impact on river flows in the region surrounding Perth, Australia. In this drought-stressed regional study, a single GCM was downscaled using one method and applied to a single hydrological model. Through the use of the weather generator, 1000-year synthetic time series were generated for precipitation and temperature for the present day, and for two future emission scenarios. Clearly, the use of a single GCM and downscaling technique means that this study did not account for the related uncertainty. However, the weather generator downscaling technique allows the generation of long time series required to perform the flood analysis and, hence, is arguably very appropriate to address the issue.

A more comprehensive sampling of uncertainty using a weather generator technique can be found in Wilby and Harris (2006). They used two statistical downscaling techniques to downscale an ensemble of four GCMs and used the data to drive two hydrological models for the River Thames in England. The two downscaling methods were a simple change-factor method where observed time series are scaled based on changes seen in the GCM output, and a weather generator approach. A Monte Carlo approach was used to quantify the various components of uncertainty affecting the projected hydrological changes. Both GCMs and hydrological models were weighted based on performance criteria. They found that the largest sources of uncertainty were the choice of GCM and downscaling technique, rather than the choice of hydrological model.

Wood et al. (2004) used dynamically downscaled output directly, as well as after a bias correction and spatial disaggregation (MOS) step, to drive a macroscale hydrological model. In this study, a single GCM was downscaled using a single RCM over the Columbia River, USA. They found that the bias-correction step was vital for the production of hydrologically useful simulations. They also found that using the downscaled climate produced larger future hydrological changes than using the GCM simulations alone.

Fowler et al. (2007a) give an example of using probabilistic climate change scenarios to provide probabilistic hydrologic-impact information. They use six dynamically-downscaled scenarios produced by downscaling two GCMs with four RCMS. A Bayesian method is used to derive pdfs of the change in temperature and precipitation. The model ensemble is weighted based on the model performance and convergence criteria using an REA-type approach. A weather generator was then used to take these regional change pdfs down to the catchment scale, producing 100 30-year synthetic climate series that were then run through the hydrologic model applied to the Eden River catchment in the UK. From these pdfs of the change in temperature and precipitation, changes in various aspects of the river flow were calculated (Figure 9.12). These plots show the mean flow and its standard deviation, together with the 5th and 95th percentiles of the changes in the flow predicted for the 2070—2100 period compared to the 1960—1990 period. Each plot shows traces for the four seasons. It can be seen that the largest change occurs in the JJA (boreal summer) season — a significant decrease in the

mean and across the whole distribution compounded by a significant decrease in the standard deviation. Only in the boreal winter (DJF) is there a predicted flow increase — for the mean and the 95th percentile. This suggests significantly lower flows in boreal summer that are inadequately compensated by slight increases in boreal winter flows, while transition seasons suggest marginally decreased flows.

9.5.1.1. Murray—Darling Basin

Regionalization of GCM predictions can be undertaken with more confidence if the 'region' is large, that is, it contains an adequately large number of GCM grid elements. Thus, hydrological downscaling has been undertaken for most of the world's large river basins including the Amazon, Mississippi, Ganges, and the Murray—Darling. The last of these is particularly interesting for two main reasons: the region has suffered a very protracted drought (from 1997—2009), which has created agricultural and urban water supply problems (e.g., Timbal and Jones, 2008). Furthermore, significant political problems have arisen as a result of management conflicts across the basin (cf. Henderson-Sellers, 2012, this volume). Droughts in this important basin are expected to become more frequent and more intense in the future (Garnaut, 2008; Hennessy et al. 2008). Thus, the social and economic impacts of drought, already significant, are likely to increase in the future (Productivity Commission 2009; Drought Policy Review Expert Social Panel, 2008). An overview of recent management and climate challenges for this region is given in CSIRO (2010) and references therein.

Most analyses begin with current GCMs (to date, those from CMIP3 being the most up-to-date). Research has 'graded' GCMs based on evaluation of their current climate simulation in the Australian region (e.g., Smith and Chandler, 2010; Perkins et al., 2009; Perkins and Pitman, 2009) and the impact on projections of excluding models with a poor simulation of current climate has also been examined. As described in this chapter, it is *vital* first to assess how well the GCMs to be employed perform in the region under scrutiny (e.g., Figure 9.13a). However, it must also be acknowledged that the 'best' means of assessing climate model performance is not well-established and, in general, such approaches, although very widely emphasized, are yet to provide major reductions in projection uncertainty for Australia (see e.g., Kirono and Kent, 2010, who go as far as to advise against the approach for future drought assessment). One group's work can be used illustratively. Chiew et al. (2009) use a straightforward scaling technique to downscale output from the 15 GCMs in Figure 9.13a. From this, they investigate the impact on river flow in the Murray—Darling basin and find a significant range of streamflow impacts derived from the various

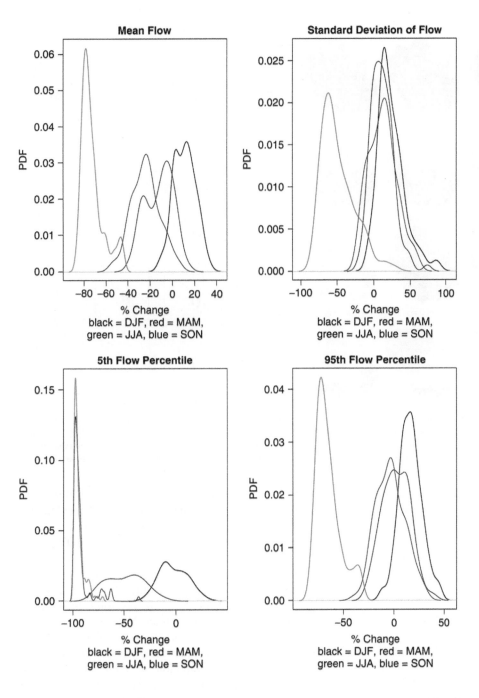

FIGURE 9.12 RCM-projected percentage changes in flow statistics for the 2070–2100 period (change from 1961–1990) for the Eden River Catchment, UK: (a) mean flow, (b) standard deviation of flow, (c) 5th percentile of flow, (d) 95th percentile of flow. Seasonal (boreal) change probabilities are given for winter (DJF, black), spring (MAM, red), summer (JJA, green), and autumn (SON, blue). *(Source: Fowler et al., 2007a.)*

GCMs. Subsequently, Chiew et al. (2010) downscaled output from just three GCMs using one dynamical and four statistical downscaling techniques of varying complexity, finding that the range in runoff produced using the different downscaling techniques was significant (e.g., Figure 9.13b). While these authors, in common with most researchers working on regionalizing climate change predictions, stress the need to place the results within the context of the large range of GCMs' confidence and sensitivity to different hydrological (downscaling) models, their assessments are

in use (e.g., CSIRO, 2010). CSIRO (2010) reports that farmers interviewed as part of their project claimed that using outputs from GCMs assisted them in understanding the potential value of seasonal forecasts. As a result of this perceived value, a statistical model for seasonal forecasting of streamflow has been developed and tested on catchments in the Murray–Darling basin. The use of antecedent streamflow as a predictor was found to be essential, although CSIRO (2010) also report that there is a small, but non-negligible, benefit gained from the inclusion of

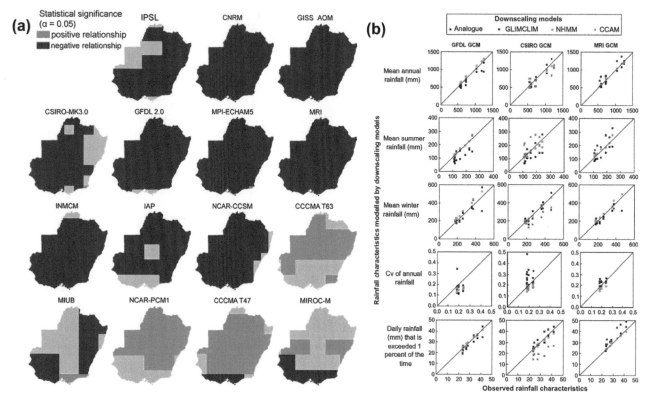

FIGURE 9.13 (a) Statistical significance of the linear relationship between winter rainfall and globally-averaged surface air temperature simulated by 15 GCMs. (b) Comparison of rainfall characteristics downscaled from three GCMs by the four downscaling models with the observed rainfall characteristics over the 1986–2000 period. *(Source: (a) is figure 6 from Chiew et al., 2009 © Copyright 2009 American Geophysical Union. Reproduced/modified by permission of American Geophysical Union. (b) is figure 5 from Chiew et al., 2010. Reprinted from Chiew et al., 2010; with permission from Elsevier.)*

indicators of large-scale climate factors (such as ENSO, IOD, and SAM — see Latif and Park, 2012, this volume).

These examples show the evolution of hydrological impact studies towards the use of multiple downscaling methods, assessed for skill and combined into ensembles, and the probability of the predicted changes being explicitly addressed.

9.5.2. Greenland Mass Balance

The mass balance of the world's glaciers (e.g., Cogley, 2012, this volume) and of the major ice sheets presents a challenge to modellers. Ice sheets have a large spatial extent, but the processes that affect the mass balance occur at much smaller scales and are affected by regional-scale features such as the orographic enhancement of precipitation discussed in Section 9.2 and the effects of katabatic flows (Cogley, 2012, this volume).

In an effort to contribute to a more effective picture of the mass balance of the ice sheets, various authors have used regional models to provide a detailed picture of the processes occurring over the ice sheet and to provide a more detailed picture of accumulation and ablation (e.g., Box and Rinke, 2003; Box et al., 2004; Rignot et al., 2008). Ettema et al. (2009) used the limited-area model

approach discussed in Section 9.3.1 to model the accumulation and ablation at the surface of the Greenland ice sheet between 1958 and 2007. They ran a version of the Regional Atmospheric Climate Model (RACMO2) of van Meijgaard et al. (2008) using an 11 km resolution and lateral boundary conditions imposed from the European Centre for Medium-Range Weather Forecasts ERA-40 re-analysis. Figure 9.14 shows the resulting surface mass balance computed in their study. Note that this surface mass balance is different from the mass balance of the ice sheet, which would include mass loss due to calving of icebergs. The high resolution of their study allowed previous estimates of the mass balance of the ice sheet to be re-assessed. Ettema et al. (2009) suggest that high accumulation zones are systematically undersampled in available observational datasets and that more mass accumulates on the ice sheet than previously believed. Their model also shows the increases in runoff that are consistent with observational studies of glaciers in the region (e.g., Stearns and Hamilton, 2007).

9.5.3. Understanding Tropical Cyclones

Tropical cyclones, as described in Section 9.2.1, are typically too small to be resolved explicitly by even the highest

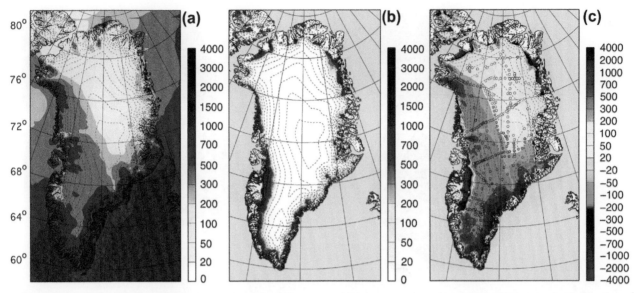

FIGURE 9.14 Modelled annual (1958–2007 average) (a) precipitation, with 20 observations from coastal meteorological stations, (b) runoff, and (c) surface mass balance in kg m^{-2} yr^{-1}, including 500 in situ observations from various published and unpublished sources. Thin, dashed lines are 250 m elevation contours from Bamber et al. (2001). *(Source: Ettema et al., 2009 © Copyright 2009 American Geophysical Union. Reproduced/modified by permission of American Geophysical Union.)*

resolution global models. This, and their devastating impacts, has led to a range of modelling techniques being applied to try to predict their frequency and intensity in future climates (e.g., Henderson-Sellers et al., 1996; Knutson et al., 2010). Some of these modelling approaches have been discussed in Section 9.3.1: most approaches to downscaling for tropical cyclone studies are based on the driving of a regional model with boundary conditions from a global model or re-analysis (e.g., Walsh and Ryan, 2000; Knutson et al., 2008). Other approaches have made use of schemes to relate the large-scale environment to the associated tropical cyclone hazard (e.g., Camargo et al., 2007) by considering an empirical genesis potential or by utilizing a specialized high resolution model driven by a larger scale global model (e.g., Emanuel et al., 2008, 2010). In this section, we briefly consider two approaches, which, in line with Wigley and Raper (2002), are selected to bracket the achievable regional prediction skill (or uncertainty).

9.5.3.1. High-Resolution Process Studies

Detailed modelling of tropical cyclones presupposes detailed understanding of the physical processes involved. Tropical cyclones are intensely dynamic systems, evolving rapidly in response to changes in SST and the presence of islands – the destructive power of the storms is often tied to these rapid changes. At one extreme of the effort to improve the modelling of tropical cyclones are very-high-resolution models aimed at developing greater understanding of the mechanisms that drive, for example, the intensification process and the extreme precipitation associated with the spiral rainbands (Franklin et al., 2005). From models such

as these will come the understanding of physical processes that must subsequently be incorporated in the dynamical downscaling approach.

In another study, Judt and Chen (2010) used a triply-nested model with a 1.67 km resolution in the inner, highest resolution, domain to examine eyewall development and replacement in tropical cyclones, a key factor in rapid intensification of these storms. In this type of model configuration, a variable resolution is employed in a suite of grids that follows the storm. In this way, the detail of the simulation remains focused on the high-resolution part of the storm, around the eye and eyewall, with lower resolutions employing fewer computational resources in the outer regions of the storm (e.g., Wang, 2001). The significant improvement in simulating tropical cyclone processes obtained using these specifically-developed models has made them the preferred choice for many tropical cyclone studies as, illustrated in Figure 9.15 taken from Gentry and Lackmann (2010). Models at resolutions such as these create quite realistic simulations of tropical cyclones (compare the details in Figure 9.15d with the fine structure visible in the satellite image in Figure 9.2). We are still some way off the regular use of models with these high resolutions in studies of future climate.

9.5.3.2. Simple/Empirical Downscaling

The development of tropical cyclones or hurricanes is dependent on a number of factors and the analysis of the large-scale environment can offer some information on the likely intensity of any tropical cyclones that form. In addition to a suitably warm ocean surface and dynamic

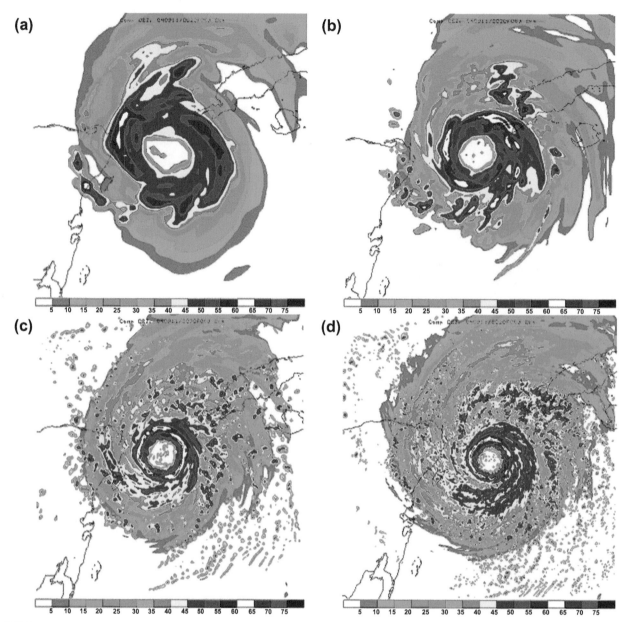

FIGURE 9.15 Results at the same point in time from a series of simulations of Hurricane Ivan off the Yucatan peninsula in the Gulf of Mexico. The figure shows simulated radar reflectivity for (a) 8-km resolution, (b) 4-km resolution, (c) 2-km resolution, and (d) 1-km resolution model runs. *(Source: Gentry and Lackman, 2010. © American Meteorological Society. Reprinted with permission.)*

factors such as low-level cyclonic flow and low vertical wind shear, there must be a suitable thermodynamic environment in which the storm can develop. The nature of this thermodynamic environment determines how intense any storm can become and so imposes a maximum potential intensity (MPI) of the storm. Thermodynamic models of MPI (e.g., Emanuel, 1988; Holland, 1997; Michaud, 2001) have been shown to have skill at a range of timescales for characterizing the intensity of storms, should they occur, in a given thermodynamic environment and, therefore, have the potential to be used to characterize aspects of a future tropical cyclone climate. MPI schemes offer another means for downscaling the results of lower-resolution models to gain information of regional significance. Thermodynamic models consider only the thermodynamic state of the atmosphere—ocean system, and any analysis with these models cannot offer insight into the frequency or likelihood of such storms developing, only the maximum intensity to which any storm could develop. However, Emanuel (2007) hypothesized that, given a large enough sample, the average storm intensity is likely to increase by the same amount as the MPI.

Some authors (e.g., Camargo et al., 2007) have further developed this line of enquiry to formulate a cyclone genesis potential, where other factors such as low-level vorticity and tropospheric wind shear are combined in an empirical way to give a better relationship with cyclone frequency. MPI (Tonkin et al., 2000) and other indices, such as the cyclone genesis index of Camargo et al. (2007), are related to the number of storms observed over a basin, as well as the intensity reached by these storms. Camargo et al. (2007) found that some calibration of their scheme for individual models was needed to achieve the best

relationship. It is clear from these studies that, given a supportive thermodynamic environment, there is no dearth of convective disturbances that can develop into tropical cyclones. T is, the presence of an initial convective disturbance is not a limiting factor. McDonald et al. (2005) concluded that such convective genesis parameters would be likely to provide useful information where models cannot resolve storms.

Figure 9.16 shows results using the thermodynamic MPI scheme of Holland (1997) in a climatological situation, using monthly average results from a selection of

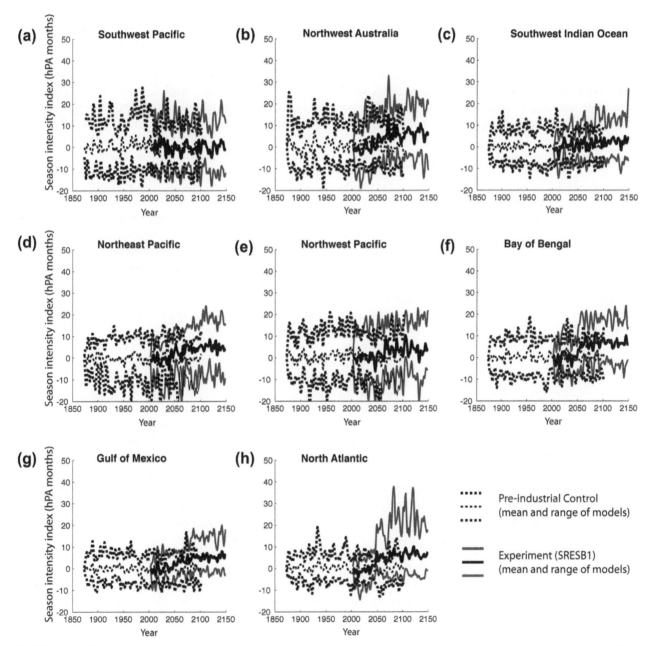

FIGURE 9.16 Time series of changes in cyclone season severity index for the eight tropical cyclone basins. The mean and range of model results is shown for both control and experiment runs. A five year (1 3 4 3 1) weighted filster has been applied to these results.

models participating in CMIP3. Using results for the SRESB1 and pre-industrial scenarios, an index, related to the severity of the cyclone season is computed over the period 1860−2150 and the differences between the experiment and control are shown in Figure 9.16 as a range of model results. Interannual variability in the cyclone season is a well-known feature of the tropical cyclone record (e. g., Vecchi and Knutson, 2008) whether characterized by cyclone intensity or number and this is reflected in the variability of cyclone season intensity derived from the models. Figure 9.16 shows a striking variation between basins for the SRESB1 scenario. The technique predicts a more severe cyclone season, characterized by either longer season length or by the possibility of more intense cyclones, for most basins. Some basins remain unaffected by climate change, at least by this measure. The results suggest that tropical cyclones are a regional phenomenon with a regionally-specific response and care needs to be taken in generalizing response to climate change forcing from one region to another (e.g., Knutti et al. 2010).

9.6. REGIONALIZING FUTURE CLIMATE

An important area of current research is how best to weight climate models in ensembles. If it were possible to know the skill of each model, then using this knowledge to give models with higher skill a greater weight would be a sound approach. The problem is that there is no 'correct' or even 'agreed' method to determine a model's skill. To date, a wide variety of variables and metrics have been used to estimate model skill, producing a wide variety of skill levels for every model and little consensus as to their relative skill (e.g., Taylor, 2001). One possible way to address this is to develop a metric that considers many variables and their properties, and that provides a measure of the overall model performance (Christensen et al., 2010). Another, perhaps more practical, method would be to define metrics based on the intended study such that model skill could be assessed against aspects of the climate system that are important for the application. In this sense, a study of drought would use a metric focused on the persistence of dry periods, while a study of floods would use a metric focused on the recurrence of heavy precipitation events, and in each case different weights would be given to different climate models (e.g., Perkins and Pitman, 2009; Alexander and Tebaldi, 2012, this volume).

Due to practical considerations, downscaling is only done from a subset of the available GCM simulations. How using this subset of GCMs affects the end result, including the uncertainty estimates, remains an open question. Another way to frame the problem is: can a subset of GCMs be chosen that spans the uncertainty range of the full set of GCMs? One way to attempt this

would be to first identify and remove GCMs that have particularly poor skill compared to observations. Once the poorly-performing GCMs have been removed from the analysis, the remaining GCMs can be classified broadly into groups that predict different regional climate changes. This broad classification could be based on mean changes in temperature and precipitation, on model climate sensitivity, or on other climatological features known to be important in the region. From this matrix, representative GCMs from the most common or likely change, as well as representatives from less-likely change groups, can be chosen. This information on how representative the GCM simulations are of the regional climate change projected by the entire GCM ensemble can then be used to place the downscaled results in the context of the larger ensemble.

A major area of future work is in probabilistic predictions of future climate. With the large multimodel GCM datasets developed to support the IPCC process and the additional RCM ensembles through initiatives like COR-DEX (Giorgi et al., 2009), probabilistic predictions at the regional-scale are becoming increasing feasible. An important assumption embedded in all these probabilistic approaches is that the models are structurally independent (Abramowitz and Gupta, 2008; Abramowitz, 2010). This is clearly not the case with many climate models sharing parameterizations, and even core numerical solution algorithms (McGuffie and Henderson-Sellers, 2005). Clearly, successive versions of the same model are also not structurally independent. How to quantify the level of model structural-independence is a wide open question at present but, with the increasingly use of probabilistic methods, this problem needs to be addressed.

Improving the estimates of future regional climates requires improvements in the tools we use to create those estimates. This includes improving the modelling of all processes in both GCMs and RCMs, as well as the basis of statistical models if these are employed in the downscaling. Perhaps the biggest challenge in coming years is the explicit quantification of the chain of uncertainty from the GCMs through the downscaling procedure to the impacts model of interest. Using a Bayesian framework to track pdfs through the system could provide a better indication of uncertainties at the level of the impact and facilitate risk-based assessments to aid decision-makers.

Predicting future regional climates is a policy and research priority. While there continues to be research and development into both dynamical and statistical downscaling techniques, there is a growing focus on how best to use ensembles of downscaled predictions to produce the 'most likely' future regional climate, along with a quantification of the associated uncertainty.

Looking Forward

Climate science has always been about improving predictive skill, especially for extremes that impact humans most severely. Recently, climate models have begun incorporating biological cycles of carbon, nitrogen, and other nutrients and adding aerosols to the chemistry of the atmosphere.

The question is: do we save the Coorong at all costs, do we leave it to become something different as is already happening, or do we invest heavily in 'palliative care', using up resources without changing the region's ultimate fate? Indeed, is this the world's first climate change 'ground zero'?

Henderson-Sellers, 2012, this volume

Chapter 10, *Climate and Weather Extremes: Observations, Modelling, and Projections* by Lisa Alexander and Claudia Tebaldi, emphasizes the importance of synthesizing the study of weather and climate extremes within a climate-change context discussing some of the difficulties in defining and modelling events using both statistical and dynamical models. Although spanning a wide range of types of extreme events, the focus is on the types of events that are statistically robust, cover a wide range of climates, and have a high signal-to-noise ratio for use in 'detection and attribution' studies. The importance of data quality and consistency is addressed as underpinning any analysis of climate extremes. The statistical analysis of extremes is discussed in some detail along with the issues of scale that exist when comparing observations and climate model output, including a discussion of downscaling techniques. The current state of the monitoring, understanding, and physical modelling of the characteristics of climatic extremes is described including projected changes, focusing both on the great progress that has taken place in the last decade and on the challenges that still lie ahead.

Chapter 11, *Interaction between Future Climate and Terrestrial Carbon and Nitrogen* by Robert Dickinson, focuses on the terrestrial components of the climate system, which store about five times as much carbon as does the atmosphere, and annually remove about one-fourth of the carbon dioxide added by human activities. These exchanges with the atmosphere are extremely dynamic, depending on the assimilation of carbon by vegetation and on plant and soil respiratory processes. The rate at which the terrestrial system takes up carbon depends on climate, atmospheric carbon dioxide, and the availability of nutrients, especially nitrogen. Respiratory losses occur in the soil column, depending on soil temperature, humidity, and oxygen levels. Nitrogen availability may be a major factor limiting carbon uptake. It depends on a balance between sources and sinks that are closely coupled to soil carbon cycling and climate influences. Carbon and nitrogen cycling are now included in many climate models, some of which are used to project future climate. The models make various assumptions and approximations and provide a wide divergence of conclusions as to relative magnitudes and importance of various terms. Modelling research that provides a framework for projecting the coupling between future climate and terrestrial carbon storage is combined with observational studies that examine the processes vital for such modelling.

Chapter 12, *Atmospheric Composition Change: Climate–Chemistry Interactions* by Ivar Isaksen, Claire Granier, Gunnar Myhre, Terje Bernsten, Stig Dalsøren, Michael Gauss, Zbigniew Klimont, Rasmus Benestad, Philippe Bousquet, Bill

Collins, Tony Cox, Veronika Eyring, David Fowler, Sandro Fuzzi, Patrick Jöckel, Paolo Laj, Ulrike Lohmann, Michela Maione, Paul Monks, Andre Prevot, Frank Raes, Andreas Richter, Bjørg Rognerud, Michael Schulz, Drew Shindell, David Stevenson, Trude Storelvmo, Wei-Chyung Wang, Michiel van Weele, Martin Wild, and Donald J. Wuebbles (reprinted from *Atmospheric Environment*), recognizes that chemically active climate compounds are either primary compounds such as methane (CH_4), removed by oxidation in the atmosphere, or secondary compounds such as ozone (O_3), sulfate, and organic aerosols, formed and removed in the atmosphere. Anthropogenic climate—chemistry interaction is a two-way process: emissions of pollutants change the atmospheric composition contributing to climate change through the afore-mentioned climate components, and climate change, through changes in temperature, dynamics, the hydrological cycle, atmospheric stability, and biosphere—atmosphere interactions, affects the atmospheric composition and oxidation processes in the troposphere. Chapter 12 summarizes progress in understanding of processes of importance for climate—chemistry interactions, and their contributions to changes in atmospheric composition and climate forcing. Climate changes affecting the tropospheric oxidation processes and the tropospheric chemical composition are identified. A key factor explored is the oxidation potential involving compounds such as O_3 and the hydroxyl radical (OH), including new estimates of radiative forcing. Although there are significant uncertainties in the modelling of composition changes, access to new observational data has improved modelling capability. For example, land-based emissions are found to have a different effect on climate than ship and aircraft emissions, and different measures are needed to reduce the climate impact. Emission scenarios for the coming decades have a large uncertainty range, in particular with respect to regional trends, leading to a significant uncertainty range in estimated regional composition changes and climate impact.

Chapter 13, *Climate—Chemistry Interaction: Future Tropospheric Ozone and Aerosols* by Wei-Chyung Wang, Jen-Ping Chen, Ivar Isaksen, I-Chun Tsai, Kevin Noone, and Kendal McGuffie, explores that fact that climate changes due to global warming from anthropogenic greenhouse additions may perturb atmospheric concentrations of chemically active climate compounds and thus the oxidation capacity of the atmosphere, providing feedback to the climate system. The radiative heating and cooling of the atmosphere is affected by perturbations of CO_2, primary aerosols, and chemically active greenhouse compounds (CH_4, N_2O, CFCs), and by secondary compounds (tropospheric ozone, sulfate, and organic aerosols) that are formed in the atmosphere through a variety of chemical and physical processes. While the concentrations of these secondary compounds are dominated by the surface emissions of their gas-phase precursors, they are also closely coupled to meteorological parameters: temperature, wind, and the hydrological cycle (moisture, precipitation, and clouds), and to solar radiation (photo-dissociation). As the climatic states and surface precursor emissions exhibit distinctively different regional characteristics, these climate—chemistry interactions will play an important role in future climate changes that will disturb different areas differently.

Rapid climate change is one of the widest reaching challenges modern society has faced. Its physical, ecological, economic, political, cultural and ethical dimensions have local, national, and global implications. Significant impacts resulting from rapid climate change are already evident, and pose increasing risks for many vulnerable populations. Society faces immediate choices about how to reduce vulnerabilities, despite inherent uncertainties about the path of future greenhouse gas emissions, the response of the climate system, and the adaptive capacity of human and natural systems.

Stephen H. Schneider (with colleagues, 2010)

Chapter 10

Climate and Weather Extremes: Observations, Modelling, and Projections

Lisa Alexander[a] and Claudia Tebaldi[b]

[a] *Climate Change Research Centre and ARC Centre of Excellence for Climate System Science, University of New South Wales, Sydney, Australia,*
[b] *Climate Central, New Jersey, USA*

Chapter Outline

10.1. INTRODUCTION: EXTREMES OF CLIMATE

10.1.1. Why Study Weather and Climate Extremes?

Weather and climate extremes have significant societal, ecological, and economic impacts across most regions of the world (Easterling et al., 2000). For instance, heatwaves can be devastating for societies that are not used to coping with such extremes (Nicholls and Alexander, 2007). It is estimated that between 25,000 and 70,000 deaths were attributable to the 2003 heatwave in Europe (D'Ippoliti et al., 2010) and of these, about 15,000 deaths in France alone (Poumadere et al., 2005). This event also had effects on water ecosystems and glaciers and led to the destruction of large areas of forests by fire (Gruber et al., 2004; Koppe et al., 2004; Kovats et al., 2004; Schär and Jendritzky, 2004).

At the time of writing, the 'Great Russian Heatwave' of the summer of 2010 and associated wildfires are rivalling this destruction and death toll (Barriopedro et al., 2011).

In addition to the immediate direct impacts of extreme weather events, there may be indirect repercussions, such as insurance losses. During the summer of 2003, European drought conditions caused crop losses of around US$13 billion, while an additional US$1.6 billion in damage was associated with forest fires in Portugal (Schär and Jendritzky, 2004). In the first six months of 2009 alone, insured losses from natural disasters reached US$11 billion worldwide (Munich Re, 2009). Of these losses, US$1.3 billion related to the devastating heatwave and subsequent bushfires that struck southeast Australia at the beginning of 2009, costing 173 lives as a direct result of the fires, and many more lives were lost in the preceding heatwave (VGDHS, 2009).

Other extremes, such as storms and flooding, can also have large impacts. In August 2005, Hurricane Katrina battered the Gulf Coast of the United States, killing an estimated 1300 people during the most active North Atlantic hurricane season on record (Trenberth and Shea, 2006). In the early 2000s, economic damage through flooding in central Europe exceeded €15 billion (RMS, 2003), while, in 2010, the world's largest reinsurer, Munich Re, indicated a drop in its annual profit because of high natural disaster costs, including US$372 million related to Australian flood damage incurred in the final quarter of the year.

In developing countries, while 'insured losses' may be somewhat less than in more developed countries, exposure to extreme events may be much greater. Locations within developing countries will be more exposed, due to rapid urbanisation and other factors changing their vulnerability to extremes. These locations could benefit from research into extreme events, particularly future projections, so that infrastructure and development can be properly planned.

While the importance of the societal and economic impacts of extreme climatic events is undeniable, the study of climate extremes (particularly on a global scale) is still in its infancy. Indeed, the first assessment by the Intergovernmental Panel on Climate Change (IPCC) in the early 1990s did not even address the question of whether extremes of temperature, precipitation, or tropical cyclones had changed (Folland et al., 1992). Due to the significant impacts associated with these events, these gaps in our knowledge served as a catalyst for more urgent analysis of the relationship between extreme weather and climate change. By 1995, the Second Assessment Report (SAR) of the IPCC attempted to specifically address the question of whether the climate had become more variable or extreme (Nicholls et al., 1995). The SAR concluded that, although there was no evidence globally that extreme weather events or climate variability had increased, data and analyses were 'poor and not comprehensive' in spite of changes in extreme weather events observed in some regions where sufficient data were available (Nicholls et al., 1995). There was also little information or conclusive evidence on what effect anthropogenic climate change had had, or would have, on climate extremes. One reason for such ambiguity was that, while there were studies of regional changes in climate extremes, the lack of consistency in the definition of extremes between analyses meant that it was impossible to provide a comprehensive global picture. These ambiguities in the SAR conclusions led to a number of workshops and globally co-ordinated efforts, which have made significant progress in our analysis of extremes (Nicholls and Alexander, 2007). By the time of the IPCC's Third Assessment Report (TAR) and Fourth Assessment Report (AR4) in 2001 and 2007 respectively, some firmer statements could be made about past and future changes in extremes and the attribution of their causes.

However, issues and uncertainties in extremes research remain and international efforts continue to address some of these, particularly where different research groups appear to come to somewhat different conclusions regarding the detection of trends or attribution of cause. One such example is in the field of tropical cyclone research. Regional trends in tropical cyclone frequency have been identified in the North Atlantic, but the fidelity of these trends is widely debated (Holland and Webster, 2007; Landsea, 2007; Mann et al., 2007). This has led to concerted international collaboration to address these uncertainties in order to be able to make firmer statements (e.g., 'likely', 'unlikely', 'low confidence', 'high confidence') about changes. Indeed, in the case of tropical cyclones, it remains uncertain (i.e., 'low confidence') whether past changes in frequency and intensity exceed that expected from natural climate variability after accounting for changes over time in observing capabilities (Knutson et al., 2010).

Even given recent progress in the analysis of weather and climate extremes, we still have very limited ability to monitor these events consistently, to understand what drives them or to have certainty in how they might vary in the future under climate change. Primarily, this is due to observational data limitations and the inability of current state-of-the-art climate models to adequately reproduce the observed variability and trends in extremes, particularly for rainfall (Kiktev et al., 2007; Kharin et al., 2007) and at a regional level (Meehl and Tebaldi, 2004; Meehl et al., 2007a; Sillmann and Roekner, 2008; Alexander and Arblaster, 2009). One reason that the current suite of Global Climate Models (GCMs), and indeed some Regional Climate Models (RCMs) may not adequately simulate observed extremes, is that they are unable to reproduce the environmental conditions leading to extreme events; for example, the spatial scales of GCMs are too coarse to simulate the convective processes that produce short-duration rainstorms. These issues represent a significant limitation for the scientific analysis of climate extremes and, as a consequence, to provide policymakers and stakeholders with advice on how to manage the risks associated with extreme climatic events.

Following the early IPCC reports, it was realized that countries that were reluctant to release raw daily or sub-daily meteorological observations (such as would be required for the study of climatic extremes) would be more willing to exchange climate information if it were in the form of seasonal and/or annual 'climate indices' (e.g., heatwave duration, heavy precipitation events, and number of frost days). These indices of extremes (mostly derived from daily temperature and precipitation) have generally been chosen because they are statistically robust, cover a wide range of climates, and have a high signal-to-noise ratio for use in 'D&A' studies (see Hegerl et al., 2010, and Box 10.1).

BOX 10.1 Detection and Attribution (D&A) of Extremes

'Detection' is the process of demonstrating that climate has changed significantly from what would be expected because of natural variability in some defined statistical sense, while 'attribution' is the process of evaluating the relative contributions of multiple causal factors (anthropogenic and natural external forcings) to this change.

The activity in D&A research in the last few years has indeed shifted towards addressing quantities at regional-scales and other than means. D&A studies of extremes are being pursued and some results are referenced here. Perfect model studies are helping to define the detectability of highly noisy events, like precipitation extremes (Min et al., 2009). Heat extremes carry a stronger signal and Christidis et al. (2005, 2010) found that changes in temperature extremes can be detected and attributed at both global and regional-scales, although there are some discrepancies between the magnitude of model-simulated and observed changes. Zwiers et al. (2011) and Stott et al. (2010) also concluded that anthropogenic forcing has had a detectable influence on extreme temperatures at global and regional-scales. Meehl et al. (2007a) showed that trends in observed extremes indices related to temperature are consistent with model simulations of the same indices when anthropogenic forcings are included, while they are not if the experiments only include natural forcings. More recently, a study of record high and low temperatures in the United States (Meehl et al., 2009) detected behaviour inconsistent with natural variability when considering the larger than expected ratio of record high temperatures to record low temperatures each year. Min et al. (2011) succeeded in identifying a human contribution in the increased intensity of precipitation extremes over the Northern Hemisphere.

The large impact of extremes on social and natural systems has motivated research that goes beyond the attribution of historical records as a whole, aiming at pinpointing the fraction of attributable risk of specific events, the European heatwave of 2003 being a typical example (Stott et al., 2004; Jones et al., 2008). The area of event attribution is very dynamic and new, and is being developed in an attempt to explain, for any given anomalous event, the complex set of causes underlying its occurrence and the role, if any, that anthropogenic forcings may have had in enhancing the chances of this occurrence. A recent paper estimated an increase in the risk of an event like the autumn 2000 flooding in the UK due to enhanced carbon dioxide (CO_2) (Pall et al. 2011).

The contribution of external influences on climate to changes in tropical cyclone activity has been, and continues to be, a focus of interest in the D&A area of research. Lines of research have included, and are still being pursued, model simulations of the sensitivity to greenhouse warming, retrospective simulations of hurricane activity, and examination of past storm data for potential inhomogeneities. Regional model experiments have demonstrated the ability of the numerical simulations to reproduce the observed trend in frequency since 1980 in the Atlantic basin, together with inter-annual variability (Knutson et al., 2007). Some statistical models suggest that Atlantic hurricane activity will increase dramatically with warming Atlantic sea surface temperatures (SSTs), and that a detectable human influence on hurricanes is already present. However, dynamical models contradict this by suggesting only a modest — perhaps negative — sensitivity of Atlantic hurricane activity to greenhouse warming (Knutson et al., 2008) and, thus, no attribution at present, even though some aspects of hurricanes, such as the frequency of the most intense storms and the rainfall from storms, do show increases in the simulations.

The interpretation of changes in extra-tropical storms also poses challenges. Research has been pursued on storminess in northern and central Europe along two directions: that of statistical analysis of past records and that of regional dynamical modelling for both diagnostic studies of the mechanisms driving storminess and future projections under enhanced greenhouse gas forcing. The results are in agreement in highlighting the large magnitude of multidecadal changes, but no systematic trend in the level of activity (Matulla et al., 2008; Zahn and von Storch, 2008). On larger-scales, there is some early evidence that human influence may have determined changes in storminess and wave heights in the northern oceans (Wang et al., 2009).

Groups such as the World Meteorological Organization (WMO), CCl/CLIVAR/JCOMM Expert Team on Climate Change Detection and Indices (ETCCDI)[1], the European Climate Assessment (ECA), and the Asia-Pacific Network (APN) have aimed to provide a framework for defining and analysing these indices, so that the results from different countries could be combined seamlessly (Peterson and Manton, 2008). Table 10.1 lists the indices that are currently recommended by the ETCCDI. Near-global gridded datasets of these indices, derived from daily observed temperature and rainfall and consistently computed, do exist, but they are only available for the latter half of the twentieth century; are generally representative of more 'moderate extremes' or are not extreme at all; do not contain measures of uncertainty, for example, from changing station networks; and are static, that is, they are not updated (Alexander et al., 2006). The current suite of global datasets has allowed some interesting scientific questions to be addressed, but there is now a push to include much more 'extreme extremes' and more rigorous statistical methodologies into climate analyses.

This chapter will cover some of the complexities involved in defining and analysing climate extremes, observed variability, key processes, and driving mechanisms of extremes and future projections. The focus will be primarily on the climate indices that are widely available and have been used within the climate community and, as

---

1. Previously known as the Expert Team on Climate Change Detection, Monitoring and Indices (ETCCDMI)

TABLE 10.1 Extreme Temperature and Precipitation Indices Recommended by the ETCCDI

ID	Indicator Name	Indicator Definitions	Units
TXx	Max Tmax	Monthly maximum value of daily max temperature	°C
TNx	Max Tmin	Monthly maximum value of daily min temperature	°C
TXn	Min Tmax	Monthly minimum value of daily max temperature	°C
TNn	Min Tmin	Monthly minimum value of daily min temperature	°C
TN10p	Cool nights	Percentage of time when daily min temperature < 10th percentile	%
TX10p	Cool days	Percentage of time when daily max temperature < 10th percentile	%
TN90p	Warm nights	Percentage of time when daily min temperature > 90th percentile	%
TX90p	Warm days	Percentage of time when daily max temperature > 90th percentile	%
DTR	Diurnal temperature range	Monthly mean difference between daily max and min temperature	°C
GSL	Growing season length	Annual (1st Jan to 31st Dec in NH, 1st July to 30th June in SH) count between first span of at least 6 days with TG > 5°C and first span after July 1 (January 1 in SH) of 6 days with TG < 5°C	days
FD0	Frost days	Annual count when daily minimum temperature < 0°C	days
SU25	Summer days	Annual count when daily max temperature > 25°C	days
TR20	Tropical nights	Annual count when daily min temperature > 20°C	days
WSDI	Warm spell duration indicator	Annual count when at least 6 consecutive days of max temperature > 90th percentile	days
CSDI	Cold spell duration indicator	Annual count when at least 6 consecutive days of min temperature < 10th percentile	days
RX1day	Max 1-day precipitation amount	Monthly maximum 1-day precipitation	mm
RX5day	Max 5-day precipitation amount	Monthly maximum consecutive 5-day precipitation	mm
SDII	Simple daily intensity index	The ratio of annual total precipitation to the number of wet days (≥ 1 mm)	mm per day
R10	Number of heavy precipitation days	Annual count when precipitation ≥ 10 mm	days
R20	Number of very heavy precipitation days	Annual count when precipitation ≥ 20 mm	days
CDD	Consecutive dry days	Maximum number of consecutive days when precipitation < 1 mm	days
CWD	Consecutive wet days	Maximum number of consecutive days when precipitation ≥ 1 mm	days
R95p	Very wet days	Annual total precipitation from days > 95th percentile	mm
R99p	Extremely wet days	Annual total precipitation from days > 99th percentile	mm
PRCPTOT	Annual total wet-day precipitation	Annual total precipitation from days ≥ 1 mm	mm

The full list of all recommended indices and precise definitions is given at *http://cccma.seos.uvic.ca/ETCCDI/list_27_indices.shtml*.

a result, much of this analysis will pertain to temperature and precipitation. However, our aim is also to include some of the advances in climate extremes research that have been driven by developments in statistical science. Initially, definitional and methodological issues will be addressed. We will then assess the statistical challenges associated with extremes analysis, in addition to summarizing some of the results that have been observed globally for a variety of climatic indicators. We then assess some of the large-scale climatic drivers and modes that impact extremes along with assessing the ability of climate models to simulate these modes and extremes. Finally, we look to the future, providing an overview of some of the projections that have been presented to date in the scientific literature.

10.1.2. Definition of Climate Extremes

IPCC AR4 (IPCC, 2007a) defines an extreme climatic event as one that is rare at a particular place and time of year, but other definitions could include events that cause extraordinary economic and social damage and disruption (e.g., Easterling et al., 2000). Consequently, there is no consistent definition of what constitutes an 'extreme' in the context of climate research (Stephenson, 2008), although the two definitions are not necessarily mutually exclusive. One might consider defining an extreme according to its location within a probability distribution, for example, the highest temperature in a year; or via the impact it has, such as the number of people affected; or via a combination of both criteria.

From a purely mathematical perspective, an extreme might be categorized as the infrequent events at the high and low end of the range of values of a particular variable. The probability of occurrence of values in this range is known as a probability distribution or density function, being for many variables similarly shaped to a 'normal' or 'Gaussian' distribution (i.e., the familiar 'bell' curve). The distribution of temperature can resemble the normal distribution and IPCC TAR illustrated how, in a changing climate, associated changes in the mean and/or variance of a temperature distribution could affect extremes (as shown in Figure 10.1). Figure 10.1a depicts what happens to extremes when just the mean changes (in this case by increasing it). As the IPCC TAR notes, the range between the hottest and coldest temperature in Figure 10.1a does not change, but it may lead to new record high temperatures not previously experienced in this theoretical climate. If the mean remains the same but the variability increases, then this indicates an increase in the probability of both hot and cold extremes, as well as the magnitude of extremes (Figure 10.1b). In the example where both the mean and variability increase (Figure 10.1c), the probability of hot and cold extremes is affected, with more frequent hot events and fewer cold events. This implies that relative changes in extremes are much greater than the relative changes in mean and standard deviation.

An increase in the frequency of one extreme (e.g., the number of hot days) will often be accompanied by a decline in the opposite extreme (in this case the number of cold days, such as frosts). For a 'real world' example, Figure 10.2 shows the daily winter minimum temperature distribution for Melbourne, Australia, for two time periods. Here 1957–1980 and 1981–2005 can be thought of as the 'previous climate' and 'new climate' respectively, given by the theoretical temperature distributions in Figure 10.1. Although the temperature distributions shown here for Melbourne do not exactly match a normal distribution, they do appear to most closely resemble the situation shown in Figure 10.1a, that is, the theoretical distributions where just

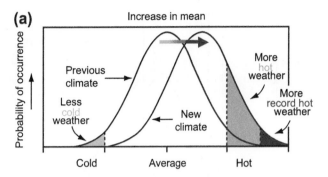

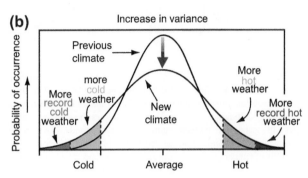

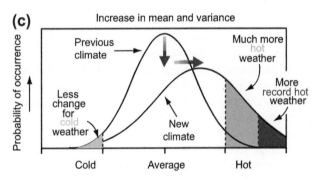

FIGURE 10.1   Schematic showing the effect on temperature extremes when (a) the mean temperature increases, (b) the variance increases, and (c) when both the mean and variance increase for a normal distribution of temperature. *(Source: IPCC, 2001a. Climate Change 2001: The Scientific Basis. Contribution of Working Group I to the Third Assessment Report of the Intergovernmental Panel on Climate Change, Figure 2.32, Cambridge University Press.)*

the mean changes. This case illustrates the effect that a shift in the mean can have on the frequency of extremes (Katz and Brown, 1992). The increase in mean minimum temperature (by 0.7°C in this case) has led to a substantial decrease in the probability of temperatures at the cold end of the distribution. In fact, the threshold at which one might define a 'cold night' (the 5th percentile, say) has increased by 1°C and the figure shows that the number of cold nights by the old definition has approximately halved between the two periods.

Another way to visualize extremes in the context of the distribution of a variable is the boxplot or 'box and whiskers' diagram. This is a convenient way to summarize a distribution or multiple distributions by assessing only

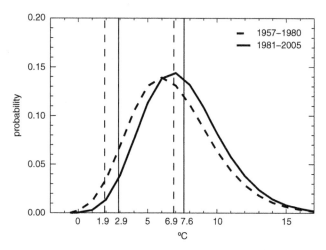

FIGURE 10.2 Example of the effect of an increase in mean temperature on the risk of extremes. Broken (1957–1980) and full (1981–2005) lines show the probability distribution function of daily minimum temperatures in winter for Melbourne, Australia. Vertical thin lines show 5th percentile values and mean minimum temperatures for the two periods. The increase in mean minimum temperature has led to a substantial decrease in the probability of temperatures <1°C (mean number of days <1°C was 1.7 per year in first period, declining to 0.7 per year in second period). (*Source: Adapted from Nicholls and Alexander, 2007.*)

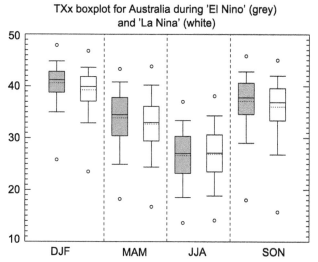

FIGURE 10.3 Boxplots of seasonal daily maximum temperatures (in °C) across Australia for seasons (*from left to right*): DJF, MAM, JJA, and SON between 1957 and 2008. Grey boxes show the inter-quartile range during El Niño events, while the white boxes show the inter-quartile range during La Niña events. Solid horizontal lines within each 'box' represent the median, while the dashed horizontal line is the mean. 'Whiskers' represent the 5th and 95th percentiles and the circles indicate the highest or lowest 'outlier' in each case. TXx (see Table 10.1) seasonal values were calculated on a 1.875 × 1.875 latitude–longitude grid and values for all grid boxes across Australia were included in the analysis.

certain statistical summaries, that is, a set of quantiles. The box represents the 25th to 75th quartile range, lines within the box represent the mean and/or median value, and the whiskers represent the 'extremes' of the distribution. Boxplots are non-parametric: they can display differences between populations without making any assumptions about the underlying statistical distribution. The size of the box and the location of the mean and median lines within the box can help to indicate the spread and skewness of the data, while the whiskers and isolated marks on the diagram help to identify the range and the outliers.

Figure 10.3 shows example boxplots using seasonal extreme daily maximum temperatures across Australia during El Niño and La Niña events. The figure shows that during spring (SON) and summer (DJF) when the El Niño–Southern Oscillation (ENSO; see Section 10.4.1) variability is greatest, extreme maximum temperatures during El Niño events are significantly warmer than during La Niña events. In fact, most marked are the changes in the 'coldest' warmest days seasonally (i.e., bottom 5th percentile). Here, the 5th percentile threshold values are more than 2°C colder during La Niña compared to El Niño in both seasons and this is also the case for the coldest recorded daily maximum temperatures in each season (shown as circles on the plot). The distributions are not significantly different during autumn (MAM) and winter (JJA).

Climate variables very rarely exactly resemble particular distributions; for example, the probability distribution function of daily rainfall looks very different from the classic bell curve. Therefore, combinations of changes in

both mean and variability would lead to different results. In addition to changes in mean ('location') and variance ('scale'), changes in the 'shape' of the distribution can also complicate this simple picture. Daily non-zero precipitation, for example, most commonly follows a Gamma distribution. The probability density function of the Gamma distribution is given by:

$$f(x) = \frac{\left(x\big/\beta\right)^{\alpha-1} e^{-x/\beta}}{\beta\Gamma(\alpha)} \quad \alpha, \beta > 0 \quad (10.1)$$

where x is the precipitation amount, Γ is the Gamma function and α and β are the shape and scale parameters respectively. Equation (10.1) can be integrated to obtain the probability of exceeding a particular threshold. Thus, for example, the 90th percentile of the Gamma distribution could be chosen as a threshold value over which precipitation would be defined as 'intense' or 'extreme'. Figure 10.4 shows an example where Gamma distributions are fitted to daily precipitation totals along with estimates of how 'good' these fits are and the relative shape and scale parameters for each season. In Section 10.2.2, we discuss some of the statistical challenges that exist related to the determination of, and shifts in, the location, scale, and shape parameters.

Thus far, we have only considered the 'top down' (physical) definition of a climate extreme and have not yet considered the 'bottom up' (impact-dictated) definition.

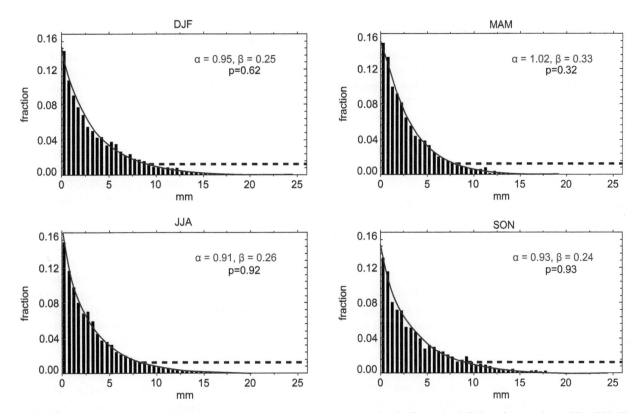

FIGURE 10.4 Seasonal probability distribution functions of 1-day non-zero precipitation for England and Wales using data from 1961–1990. The dashed line represents where the 90th percentile of the Gamma function lies with values 9.9 mm (DJF), 8.2 mm (MAM), 9.4 mm (JJA), and 10.3 mm (SON). Shape (α) and scale (β) parameters of the Gamma distribution are indicated in the top right-hand corner of each graph, top line, while the values on the second line represent the Kolmogrov–Smirnoff 'goodness of fit' test p-values, that is, the probabilities to observe these empirical distributions under the hypothesis that the observations follow a Gamma distribution. Setting a type-1 error of 0.05 (corresponding to a threshold for the probability of erroneously rejecting the hypothesis), the fit is considered 'good' if the p-value lies between 0.025 and 0.975.

While this chapter does not deal specifically with the impact-dictated definition (such as might be considered as a function of vulnerability, exposure, and risk), it is important to understand how the two definitions are related and what different assumptions and techniques are required for analysis. If we take extreme temperatures and health impacts as an example, then one could imagine that extremely high temperatures might directly impact morbidity or mortality rates. Indeed, in the case of heat-waves and particularly warm, humid nights, physiological thresholds have been identified that define critical temperatures from a health perspective for particular regions (Le; Grize et al., 2005; Loughnan et al., 2010). Conversely a high temperature event (say one that occurs in the top 10% of a probability distribution) in a remote location might have very little societal impact.

Some research within epidemiology (e.g., Bambrick et al., 2009) suggests that people living in different regions have different mortality responses to climate, that is, there is not some 'universal threshold' at hot or cold temperatures at which mortality increases. Thus, it is also likely that responses may differ between groups of people living in similar climates, due to their underlying health conditions,

access to services and use of technology (i.e., heating and cooling), and different adaptation measures that may be in place.

An 'extreme impact' can also occur when two or more (even seemingly innocuous) climatic events combine to form another extreme event. This is the so-called joint probability of an extreme event or a 'multivariate' extreme and can include, for example: (i) coastal flooding, which can arise through a combination of a storm surge and heavy rainfall; (ii) strong winds and extreme temperatures, which can combine to drive bushfires; (iii) flash flooding from rains following a drought. The dependence of the events governs the probability of them occurring simultaneously. More precisely, the probability p that two entirely independent events, A and B, occur simultaneously can be expressed as $p(A \cap B) = p(A)\,p(B)$. However when A and B are perfectly correlated, the probability that the two events occur simultaneously is much higher, that is, $p(A \cap B) = p(A) = p(B)$. Multivariate extremes could also take the form of an extreme event occurring at multiple locations. This case is very important for emergency and disaster responses, such as the need to deal with bushfires or floods at many locations at the same time.

The large impacts climate extremes can have, and their tendency to change substantially in frequency with even small changes in average climate may imply that changes in extremes can be the first indication that climate is changing in a way that can affect humans and the environment substantially (Nicholls and Alexander, 2007). On the other hand, problems with data quality and scarcity and uncertainty characterizing estimates of extremes' behaviour can make it very difficult to determine whether or not they are changing (Hegerl et al. 2004; also see Section 10.2.1).

10.2. METHODOLOGICAL ISSUES REGARDING THE ANALYSIS OF EXTREMES

10.2.1. Quality and Homogeneity of Observed Data

The quality of the underlying climatic data becomes particularly important when considering extreme events, as errors are likely to show up as outliers or 'extremes' and may be erroneously included in analyses unless correctly accounted for. Nicholls (1996) observed that a major problem undermining our ability to determine whether extreme weather and climate events were changing was that it is more difficult to maintain the long-term homogeneity of observations required to observe changes in extremes, compared to monitoring changes in means of variables. Ambiguities in defining extreme events, as described in Section 10.1.2, along with difficulties in combining different analyses using different definitions, also complicates analysis on global-scales. This has led to a series of workshops (as introduced in Section 10.1.1) in 'data sparse' regions and has inspired many regional climatic studies using a common framework. The common approach was to select high-quality stations, perform quality control, and investigate trends in extreme events (Peterson and Manton, 2008) — increased efforts being solicited to rescue and digitize data stored on paper in many countries (Page et al., 2004).

The collation and analyses of daily datasets has not been a simple task. One reason is that few countries have the capacity or mandate to freely distribute daily data. Another reason is that data need to undergo rigorous quality control before being used in any extreme analysis, since values are likely to show up erroneously as extreme when incorrectly recorded. Since the IPCC TAR in 2001, the ETCCDI has overseen the development of a standard software package that not only quality controls data, but provides researchers with the opportunity to exchange and compare results. The main purpose of the quality control procedure is to identify errors in data processing, such as negative precipitation or daily minimum temperatures greater than daily maximum temperatures. In addition 'outliers', defined as the departure from the climatological mean value for that day by more than a given number of standard deviations, are identified in daily temperatures. These can then be manually checked and removed or corrected as necessary. The software, RClimDex, developed by the Climate Research Division of Environment Canada (*http://cccma.seos.uvic.ca/ETCCDI/software.shtml*) also calculates a standard set of 27 extremes indices, derived from daily temperature and precipitation and chosen to reflect changes in intensity, frequency, and duration of events (see Table 10.1). While the quality-controlled daily data are rarely exchanged, there have been few obstacles to exchanging the climate extremes indices data, especially given that the software has been made freely available to the international research community.

Errors in other variables, such as pressure and wind speeds, also exist. These can be introduced, for example, by mistakes made when digitizing values from original handwritten records or even when observers incorrectly recorded values. In the case of tropical cyclones, different National Meteorological Services have reclassified cyclone definitions over the years so it has been very difficult to consistently analyse these extreme climatic events through time. In addition, wind speed measurements are very sensitive to instrumentation changes and site moves so quality control is an issue (WASA Group, 1998). As a result, pressure readings are often used as a proxy for wind speeds and in defining cyclones since they are less sensitive to such non-climatic factors. However, even in the case of pressure measurements, data may be keyed 'as read', that is, directly transcribed from original manuscripts without quality control. In general, the majority of errors for pressure readings occur because digits have been transposed (Alexander et al., 2005; Wan et al., 2007), although errors resulting from converting station pressure to mean sea-level pressure and particularly from changes in the surveyed elevation of stations can also be quite large (Alexander et al., 2011). Methods such as analysing the upper and lower percentiles of a probability density function (pdf) of pressure tendencies (i.e., the difference between two subsequent pressure readings) can obviate some of the issues surrounding these errors (Alexander et al., 2010). In order to examine the 'extreme' events in the pdf tails, one can either then hand-check original manuscripts or use objective statistical methods.

However, this type of data quality checking cannot account for sudden non-climatic jumps that may be introduced in the data time series through, for example, the re-survey of a station elevation for pressure measurements, the relocation of an observing site to a more shaded or exposed location in the case of temperature, or the implementation of more accurate recording instrumentation or reporting protocols affecting most variables (e.g., WASA Group, 1998: Peterson et al., 1998). Figure 10.5 shows an example of how some of these sudden jumps can manifest themselves in a time series. Here, a step change in 1995 is

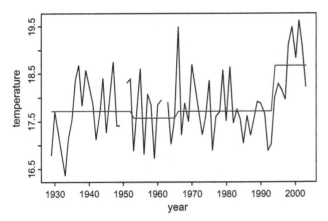

FIGURE 10.5   Time series (black line) and homogeneity test (red line) of annual minimum temperature (in °C) for the station at Rize, Turkey. The red line indicates the location and magnitude of possible step changes in the time series and the linear regression across the homogeneous sections between the possible step-change points. Not all possible step changes are statistically significant or supported by metadata. However, the discontinuity indicated by the large step change in the mid-1990s is not only statistically significant, but also verified by the station history metadata, which indicates that the station relocated in 1995. *(Source: Sensoy et al., 2007. © American Meteorological Society. Reprinted with permission.)*

identified as significant using a statistical technique. Further investigation of station metadata confirms that this station relocated in that year and so the inconsistency is likely to be due to the site move.

An important aspect of the study of extremes is to remove these inconsistencies or 'inhomogeneities' (i.e., artificial changes that cannot be explained by changes in climate) from the underlying data prior to analysis. Such problems can be identified by either a thorough investigation of station metadata, or by using sophisticated statistical techniques that can identify potential inconsistencies in the data. The former method is extremely time-consuming and the type of data required are rarely available (or in a convenient format), so ever more sophisticated statistical techniques are becoming available that can be used with or without access to original station metadata. Zhang et al. (2009) address some of the issues associated with raw meteorological data, highlighting the problems that researchers can come across before they even start to identify extreme events. In this case, they identify biases that are introduced both by rounding observational values and by defining a temperature percentile index. They propose an adjustment for this bias, which restores the precision of the counts of threshold exceedance.

However, the identification, removal, or indeed correction of these types of errors is extremely complex and difficult to do well (Aguilar et al., 2003). Recently, there have been many statistical techniques proposed (e.g., Wijngaard et al., 2003; Wang, 2003, 2008; Menne and Williams, 2005) to identify these step changes in climate series. The ETCCDI have coordinated the development of

standard software, RHTest, using the Wang (2003, 2008) methodology, which can be used in tandem with RClimDex. The RHTestV2 software (Wang, 2008) is well-used and tested and is freely available from *http://cccma.seos.uvic.ca/ETCCDI/software.shtml*. The method is based on the penalized maximal t (PMT) or F test (PMF) and can identify, and adjust for, multiple change points in a time series (see Wang, 2008, and the aforementioned website for more details). PMT requires the use of reference stations for the homogeneity analysis, while PMF can be used when there is no obvious neighbouring or reference sites for comparison. If, for example, there is a non-climatic jump in the data because a barometer is replaced or has changed height, then an inhomogeneity will likely exist in the data as a step change. In the case of the RHTestV2 software, the opportunity then exists to produce an adjusted homogeneous series taking account of these inconsistencies.

In some cases, the step changes could indicate real climatic events. One example is the ENSO, which has a major influence on the climate of many countries, particularly around the Pacific Rim. Another example relates to a major El Niño event in 1941/42, which affected much of the Australian climate. This event produced step changes at multiple sites across the country (Alexander et al., 2010), but an inspection of the metadata for these stations did not suggest that there were any reasons (e.g., change in instrumentation) to infer that these changes were anything other than real. This indicates that care is necessary when applying these techniques automatically.

On regional-scales, there has been some limited success in correcting daily temperatures (e.g., Vincent et al., 2002), precipitation (e.g., Groisman and Rankova, 2001), and pressure (Wan et al., 2007; Alexander et al., 2010) for step-change errors; whereas, globally, given the many different climate regimes, this task has proved too problematical such that generally suspicious data have simply been removed from analyses (Alexander et al., 2006). However, some studies that have corrected for biases in daily climate series have found that previous estimates using unhomogenized data have been conservative in their estimates of the trends over time in temperature extremes (e.g., Della-Marta et al., 2007, with respect to European summer temperatures).

Figure 10.6 indicates how correcting for these inconsistencies can affect the analysis of the extremes. The two stations show that homogenization can change the sign of the trend in either direction. In the case of station X, three significant breakpoints were identified in the daily minimum temperature series (using the PMF methodology from RHTest). Prior to adjusting for these inconsistencies, there was large multidecadal variability in the annual frequency of warm nights with a downward trend. However, after applying the homogenization method to the daily data, the variability in the annual frequency of warm

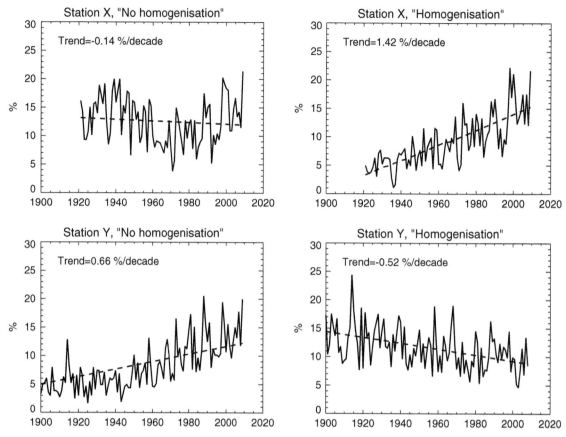

FIGURE 10.6   Example of the annual frequency (in %) of warm nights (using the definition from Table 10.1) using raw data 'No homogenization' and after applying the RHTestV2 penalized maximal F test homogenization algorithm (Wang, 2008) to the daily minimum temperature data. For Station X, dashed lines indicate linear trends that were (per decade) 0.14% in the first case and 1.42% in the second case. In this example, inhomogeneities were detected in 1943, 1951, and 1963. For Station Y, dashed lines indicate linear trends that were (per decade) 0.66% in the first case and −0.52% in the second case. In this example, inhomogeneities were detected in 1949, 1972, 1988, 1991, and 1997.

nights was reduced and the new fitted trend was strongly positive and was statistically significant at the 5% level. For station Y, five breakpoints are detected. The variability of the annual frequency of warm nights pre- and post-homogenization is similar, but the PMF algorithm has produced a strong negative trend post-homogenization, compared to a strong positive trend using the raw data.

In many situations, quality control, consistency checking, and bias correction occur on a case-by-case basis and differ significantly from country to country. For global applications, however, the amounts of data available don't make this type of 'hands-on' analysis viable and automated homogenization procedures are generally required. However, as indicated earlier, care is required when using automated procedures given that they could bias-correct real climate signatures.

10.2.2. Statistical Analysis of Extremes

By their very nature, extreme climate events are rare and this fact challenges their statistical characterization. Indices

such as those shown in Table 10.1 have been very useful in motivating researchers around the world into analysing changes in climate extremes. However, their use might be limited since they provide only little, if any, information about the remaining underlying statistical distribution of the chosen variable. While it is possible to compare trends in the chosen extreme with trends in the mean, this gives little insight into the other attributes of a distribution.

For example, Alexander et al. (2007) showed that across Australia the warming of temperature extremes was greater than the warming in the median in all cases for all seasons. However, this does not necessarily indicate a change in the shape or scale of the frequency distribution; for example, in a normal distribution, an increase in the location parameter with no change in the scale parameter will lead to a greater change in the extreme indices related to the right tail of the distribution than in the mean index (Katz and Brown, 1992). In fact, Caesar et al. (2006) showed that, for Australia as a whole, the warming across percentiles was not consistent with increased variability in either the minimum or maximum daily temperature distribution. Clearly, alternative methods need to be found to analyse shifts in the distributions of extremes, other than those relying on the characterization of a shift in the location of the distribution.

Extreme Value Theory (EVT) is a specific set of theoretical results and methods in statistics pertaining to the characterization of extremes. EVT identifies probability distributions that can be used to describe the statistical properties of quantities like block maxima (e.g., the highest daily temperature recorded over a specified interval, for example, over a year where the block size is 365 days) or threshold exceedances (e.g., the precipitation amounts above a specified large value), under specific assumptions of stationarity of the process generating the data, and in an asymptotic sense; that is, as the block size grows to infinity or as the threshold approaches the (finite or infinite) upper bound. By adopting this framework, the analysis and statistical fits focus on the extreme values in the record, rather than fitting the behaviour of the entire distribution. This permits a better characterization and inference of the so-called tail behaviour (e.g., Coles, 2001), providing, in general, more accurate estimates of rare events. Goodness of fit, however, should always be a concern as, in reality, we always deal with blocks of finite size and finite thresholds. In weather and climate studies, EVT has provided a successful framework within which to characterize extremes far rarer than those described by the suite of global indices described in Table 10.1 (see, for example, Zwiers and Kharin, 1998).

However, these statistical approaches were originally developed for the treatment of isolated extremes in a record of independent and identically distributed observations and are, therefore, less immediately and widely applicable to definitions of extremes that rely on persistence (e.g., definitions of heatwaves or droughts); the joint behaviour of two or more variables; or events that are extremes because of their spatial extent rather than, or in addition to, their intensity at specific locations. Advances in statistical methodologies that try to model dependencies in the time series (Furrer et al., 2010), spatial correlation (Cooley et al., 2007), and multivariate extremes (Durante and Salvadori, 2010; Salvadori and De Michele, 2010, 2011; Cooley et al., 2010) are being proposed in the literature and are extending the use of EVT to the analysis of these more complex types of rare events. The presence of trends in time (a violation of the stationarity assumption), often the case in the analysis of records or simulations in a changing climate, also demands treatments involving either time-varying parameters (e.g., Menéndez and Woodworth, 2010) or the use of covariates (Zhang et al., 2010 and references therein).

In the following, we give a brief explanation of the basic EVT paradigm. Depending on the type of analyses adopted, which can either collect all *peaks-over-threshold* (POT) or simply focus on *block maxima*, different parametric distributions are fitted to the data.

Analysis of block maxima utilizes the Generalized Extreme Value (GEV) distribution, which has a cumulative distribution function:

$$F(x; \mu, \sigma, \xi) = \exp\left\{-\left[1 + \xi\left(\frac{x-\mu}{\sigma}\right)\right]^{-1/\xi}\right\} \quad (10.2)$$

The shape parameter ξ governs the tail behaviour of the distribution. The subfamilies defined by $\xi \to 0$, $\xi > 0$, and $\xi < 0$ correspond, respectively, to the Gumbel, Fréchet, and Weibull families, whose cumulative distribution functions are displayed below:

Gumbel

$$F(x; \mu, \sigma) = e^{-e^{-(x-\mu)/\sigma}} \quad for \ x \in \mathbb{R}. \quad (10.3)$$

Fréchet

$$F(x; \mu, \sigma, \alpha) = \begin{cases} 0 & x \le \mu \\ e^{-((x-\mu)/\sigma)^{-\alpha}} & x > \mu. \end{cases} \quad (10.4)$$

Weibull

$$F(x; \mu, \sigma, \alpha) = \begin{cases} e^{-(-(x-\mu)/\sigma)^{\alpha}} & x < \mu \\ 1 & x \ge \mu \end{cases} \quad (10.5)$$

The fundamental distinction among the three types of GEV consists of the behaviour of the probability distribution for very large values of x: that is, for x tending to infinity. The Weibull distribution (Figure 10.7a; Equation (10.5)) has a finite limit; thus, quantities modelled by it cannot assume infinitely large values, while the other two distributions, Gumbel (Figure 10.7b; Equation (10.3)) and Fréchet (Figure 10.7c; Equation (10.4)), do not have a finite limit beyond which the density is zero, so that, in theory, extremes can be arbitrarily large. Between Fréchet and Gumbel, the difference is in the rate of decay of their tail, with the Gumbel having an exponential rate of decay to zero and the Fréchet being a heavy-tailed distribution with a polynomial rate of decay, that is, a slower one, which translates to a relatively larger probability of extreme values. As already mentioned, these are limiting distributions to which the distribution of the actual block maxima converges as the block size increases, and under assumptions of stationarity and independence of the data within the block.

When analysing data through a POT approach, a different parametric distribution applies: the Poisson–Generalized Pareto Distribution (GPD). For a large enough threshold, u, that in practice has to be determined by data exploration, the values, x, in exceedance of that threshold,

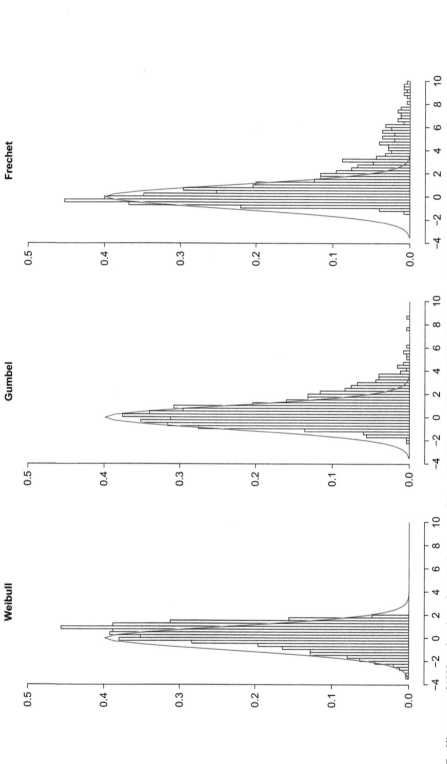

FIGURE 10.7 Histograms of 1000 random draws from the three types of Generalized Extreme Value (GEV) distributions described in Section 10.2.2. The three types have the same functional form, but differ on the basis of the value of the shape parameter, ξ. If the parameter has a negative value, the Weibull distribution results, with a finite upper limit (left plot); if it has a positive value, the Fréchet distribution results (right plot), with a heavy tail decaying at a polynomial rate and thus enhancing the chances of producing very large values; if the parameter tends to zero (in the limit), the Gumbel distribution results, with a lighter tail decaying at an exponential rate, and thus having relatively smaller chances of producing very large values compared to the Fréchet with identical values for parameters μ, σ. In this illustrative case, we fix $\mu = 0$, $\sigma = 1$, and $\xi = -0.5$, 0, and 0.5. For reference, we outline the Normal distribution with same values of the parameters μ, σ.

conditional on such exceedance taking place, are approximately distributed as:

$$H(y) = 1 - \left(1 + \frac{\xi y}{\tilde{\sigma}}\right)^{-1/\xi} \qquad (10.6)$$

where $\tilde{\sigma} = \sigma + \xi(u - \mu)$ and the parameters μ, σ, ξ are the same as the corresponding GEV parameters that would be estimated from the same data when adopting the block maxima approach. The distribution applies under the assumption that the number of exceedances within the block follows a Poisson distribution. The same discussion about the meaning of ξ in Equation (10.6) for the tail behaviour of the distribution (fast decay or exponential, slow decay or polynomial, and finite limit) applies to the Pareto distribution. In the expression of the Pareto distribution, $y = x - u$, and the distribution is intended to be conditional on $x > u$.

Estimating the parameters of GEV or GPD is, of course, challenged by the inherent scarcity of data, relating to extremes, as well as all the other data issues that we have highlighted in the earlier discussion. The uncertainty associated with the estimates (confidence intervals in the classic terminology) is important to always take into account. The uncertainty in the three parameters translates directly into the uncertainty of the commonly used return level and return period concepts. An N-year return level for a quantity modelled by EVT is defined as the size of the extreme that is exceeded on average with probability $1/N$ in any given year. It can also be defined as the extreme that recurs on average once in N years, but this latter definition is hard to interpret in the presence of a trend in the data. Correspondingly, the return period of an extreme event of size q is the number of years N such that, in any given year, the probability of experiencing that extreme is $1/N$. The computation of return periods/return level is a deterministic function of the values of the parameters of the extreme value distribution that models the behaviour of the quantity under study.

Figure 10.8 shows return period estimates (here using 238 years of daily temperature from the Central England Temperature (CET) record (Parker et al., 1992)). Daily CET most closely follows a Weibull distribution, which presupposes a finite limit for the range of the extremes. The Weibull distribution was originally derived to study quantities with a natural lower bound (e.g., zero when modelling the strength of materials), although it often appears to fit maxima as well, when a natural upper bound can be conceived (for example for temperature, as in our example). By fitting a statistical model, we can estimate values outside of those previously experienced in the climate record. For instance, the 500-year return level is estimated as 25.4°C while the maximum daily value ever recorded for daily mean CET was 25.2°C on 29 July 1948. However, one must apply caution to these estimates, as is the case for any kind of extrapolation beyond the length and the range of the data available. From Figure 10.8, conversely, a value of 24.6°C represents a 1 in 100-year event, although analysis of the data shows that a temperature at least as large as this was experienced four times in the 20-year period between 1976 and 1995.

It is obvious that the difficulty in detecting changes or shifts in the behaviour of extremes is compounded by the difficulty of obtaining accurate estimates for the parameters, even under stationary conditions. Also, the concepts of return level and return period acquire a more complex meaning in non-stationary conditions (which we might expect in a climate change scenario), where the expected frequency of a given event is assumed to be changing over time — an expected outcome of the influence of climate change drivers on the statistical characteristics of extremes. Under those conditions, return values can still be interpreted in terms of chances of exceedance every year. For example, a 20-year return value estimated for a decade in a non-stationary period could be expected to be exceeded with a 5% chance every year of that decade or for 5% of the points considered in a global domain if the analysis is performed spatially and the nature of the spatial correlation is properly taken into account. As already mentioned, methods have been proposed to make the parameters of the GEV or GPD a function of time, space, or other covariates to model non-stationarity (Kharin and Zwiers, 2005; Cooley et al., 2007; Nogaj et al., 2007; Kysely et al., 2010).

Perhaps not surprisingly then, to date there have not been formal D&A studies at local-scales that have convincingly associated observed trends in extremes with a human influence (Hegerl et al., 2004). Karoly and Wu (2005) showed that the detection of trends in climate variables is a signal-to noise-problem, and that the noise associated with climate extremes at very small regional-scales is greater than at continental- or global-scales. Min et al. (2009) have developed new perfect model techniques for D&A that can be used to determine the level of the signal-to-noise ratio needed to obtain a statistically significant detection result, specifically for analysis of trends in extremes. The identified challenges notwithstanding, progress is being made in recent D&A studies at global- or large continental-scales, which we describe succinctly in Box 10.1.

10.2.3. Issues of Scale

Extremes can be defined for different space-scales and timescales (from events happening at a point location over sub-daily time periods to events affecting large areas over multiple days and every combination in-between). The need to distinguish extreme behaviour on the basis of the spatial and timescales involved becomes paramount when comparison between observed and modelled

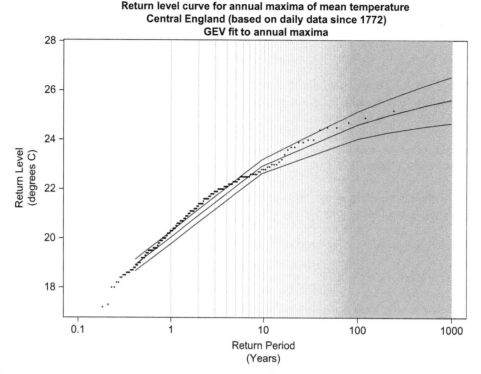

FIGURE 10.8 This plot shows the result of fitting a GEV distribution to the annual maxima obtained from the daily time series of mean temperature values for central England available since 1772 (Parker et al., 1992). The parameters of the GEV fit are $\mu = 20$, $\sigma = 1.4$, $\xi = -0.2$ indicating that the distribution is a Weibull, with a finite upper limit. The return level curve shows, for any value on the x-axis (number of years), the value of temperature that is expected to be recorded, on average, only once in that number of years. For example, according to this analysis, return levels for periods of 10, 50, 100, and f500 years are respectively 22.9°, 24.2°, 24.6°, and 25.4°C. The blue lines indicate the 95% confidence interval around the return level estimates. We note here that the goodness of fit of this curve is less for short return periods (1 through 4 years) than for longer periods (for 5 through over 100-yr return periods). This is an example where perhaps modelling a trend in the parameter of the GEV distribution may be appropriate to improve the fit.

extremes is performed, often in order to assess the processes driving extremes or to investigate future changes. To adequately compare observations and models, observations generally have to be 'gridded', that is, converted to values on a latitude—longitude grid, or climate models have to be 'downscaled', that is, data values relevant for observation sites have to be inferred from gridded values. However, even when comparing gridded to gridded values, care must be taken to distinguish gridded products where the values represent regularly spaced, point locations (for example what is produced by most kriging methods) from gridded products where values represent area averages, as is typically the case for model output or re-analysis products.

At the core of the problem is the fundamental mismatch between the spatial representativeness of in situ observations on the one hand, which are necessarily collected at observation sites (points), and that of climate model output data on the other hand, for which grid-point values are often assumed to represent area mean values (Chen and Knutson, 2008). Scale mismatch (sometimes referred to as the 'problem of a change in support' in the statistical community) more importantly affects phenomena such as

precipitation whose spatial features are discontinuous, or daily or sub-daily phenomena in general because of the unreliability of climate model output at shorter temporal-scales (Stephens et al., 2010). Klein Tank et al. (2009) note that the spatial resolution of climate models is not yet sufficient to easily provide detail on extremes at local and regional levels. They also note that being aware of these scaling issues is important in avoiding the misinterpretation of the results of observed and modelled extremes. While at present it may be difficult to produce observational datasets that are fully comparable with model output, Klein Tank et al. (2009) and others have aimed to provide a set of guidelines and suggestions for the modelling and observational communities that will hopefully make comparison easier in the future.

Gridding techniques have been assessed widely for climatic averages (New et al., 2000; Brohan et al., 2006) but much less attention has been applied to climatic extremes. Methods that have been widely used include angular distance weighting (New et al., 2000), climatologically aided interpolation (Willmott and Robeson, 1995), and kriging (Cressie, 1995). For specific regional domains, attempts have been made to compare gridding

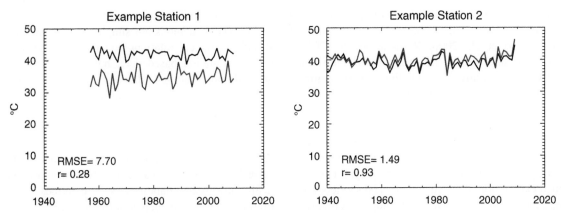

FIGURE 10.9 Examples of annual daily temperature maxima time series where a gridding methodology (in this case Angular Distance Weighting, ADW) has produced a 'bad' reconstruction (Station 1) and a 'good' reconstruction (Station 2). Red lines represent the actual value and black lines represent the simulated value using the ADW technique. The root mean squared error (RMSE) and the correlation coefficient, r, are also shown.

methodologies (e.g., Hofstra et al., 2008) but, to date, multimethod assessments have not been applied on a global-scale for climate extremes. In general, the indices in Table 10.1 have longer 'decorrelation length-scales' than daily or sub-daily data (Alexander et al., 2006), that is, they correlate with each other over longer distances, making them more robust to the choice of gridding technique.

Cross-validation can be used to quantify uncertainty and improve interpolation skill (e.g., Caesar et al., 2006). Uncertainties in observational data can be divided into three groups: (i) instrumental error, the uncertainty of individual station measurements; (ii) sampling error, the uncertainty in a grid box mean caused by estimating the mean from a small number of point values; and (iii) bias error, the uncertainty in large-scale values caused by systematic changes in measurement methods (Brohan et al., 2006). In general, biases introduced by any of these sources of error may not be a critical issue if they apply uniformly across the distribution of the variable recorded, and consistently in time.

Are gridding methods any good at representing extremes? The answer is that it depends on the variable under consideration, the number of stations, and the gridding methodology employed. In Figure 10.9, we show simulated values of extreme annual maximum daily temperature (TXx from Table 10.1) at two stations using the angular distance weighting methodology (Shepherd, 1984). In this case, we use the technique to recreate the annual extremes for each station, pretending that the station in question is at the centre of a grid box to create our 'gridded' value. For one station (station 2), the method is shown to reproduce TXx very well, with high correlation and low error with the real data. However, for station 1, the same technique has given a relatively poor reproduction of TXx, the reconstructed value in this case giving annual values that are several degrees warmer than is the actual case.

Hofstra et al. (2008) compared multiple interpolation methods of daily climate variables over Europe. They found that, irrespective of the method, the main control on spatial patterns of interpolation skill is the density of the station network, with topographic complexity a compounding factor. Hofstra et al. (2010) further found that having too few stations in any interpolation method had a disproportionate effect on the extremes and any derived extremes indices, and also for many regions that the smoothing effect of the interpolation is significant. The effects of these factors may be different depending on how extreme the variable under consideration is whether we are interested in rare extremes at individual points within a gridbox or in a less rare type of event averaged over the gridbox.

Thus far, we have only considered the transformation of point data to grid data, but the converse 'downscaling' is also possible. Downscaling techniques can be divided into four categories: (i) regression methods; (ii) weather pattern (circulation)-based approaches; (iii) stochastic weather generators; and (iv) limited-area climate models (Wilby and Wigley, 1997). The first three of these approaches belong to the category of statistical downscaling, while the physical modelling through RCMs is referred to as dynamical downscaling (see e.g., Evans et al., 2012, this volume).

The advantages of dynamical downscaling through RCMs can be identified in their ability to simulate weather and climate parameters in a physically consistent way, responding to smaller-scale features, such as terrain, land use, proximity to bodies of water, and so on, than GCMs can resolve. The typical resolution of RCM grid boxes is on the order of 20 km to 50 km, but always getting finer with each generation of RCMs. Because of the smaller spatial-scales resolved by RCMs, their output is expected to better represent synoptic conditions and the associated extremes, especially for variables whose spatial fields are characterized by small-scale features, and topography driven variables, such as precipitation and winds (Gao et al., 2006; Rauscher et al., 2010). However, they suffer from the same uncertainties that plague GCMs in terms of modelling approximations (albeit at different scales) and by construction inherit the uncertainty and errors from the GCMs that are used to force them at their boundaries.

Because of their computational expense, it was traditional until a few years ago to only run one pair of GCM+RCM over a domain for a given study, preventing a thorough exploration of the uncertainty and shortcomings affecting the simulation. More recently, international efforts such as PRUDENCE (Deque et al., 2007), NARCCAP (Mearns et al., 2009), ENSEMBLES (Hewitt, 2005), and CORDEX (Giorgi et al., 2009) have spearheaded multimodel approaches to dynamical downscaling by which multiple GCMs and multiple RCMs perform the same set of experiments, allowing for a systematic intercomparison and exploration of the issues related to systematic biases, uncertainty ranges, as well as the relative importance for the simulated differences of initial conditions, boundary drivers, and scenarios. It is fair to say, however, that our understanding of the value added to simulations of extremes by RCMs is still in its infancy. Factors confounding the issue include domain size and location and the driving method (e.g., the use or not of spectral nudging and what variables are nudged), in addition to those already described.

Statistical downscaling can take a host of forms: from simple linear regressions (Wilby and Wigley, 1997); to more complex statistical models accommodating nonlinearities in the regression functions (Chen et al., 2010); to clustering techniques associating large-scale climatological states (explicit, Vrac and Yiou, 2010, or hidden, Khalil et al., 2010) with local variables and to stochastic weather generators (Wilks and Wilby, 1999; Kilsby et al., 2007) that only estimate the parameters of the probability distribution functions of a set of variables and then randomly generate realizations conditional on these parameters. In all cases, the parameters of the statistical models are estimated by using the relationship between gridded and station-based observations and are subsequently applied to downscaled gridded model output, assuming that the relationship remains constant when moving from observed to simulated data and into the future.

Because of the need to estimate relationships accurately and specifically for a given spatial domain, it is rare that statistical downscaling is performed over large regions at once. Usually, even the form of a simple linear regression needs to be adapted to the particular domain of interest by selecting specific sets of relevant predictors. Lately, more basic forms of statistical downscaling have been proposed in order to provide algorithmic solutions to the downscaling of entire sets of GCM experiments at continental- or global-scales. These methods are simpler, in that they do not fit specific statistical relations between large-scale observations and station based data, but usually exploit finely-gridded observational datasets to bias-correct model simulations (assuming constant bias between present and future runs of a model) and spatially disaggregate their coarse output using elevation and climatology information (e.g., Maurer and Hidalgo, 2008). By its very nature, statistical downscaling tends to smooth and dampen the variability of the original data (which would make statistically-downscaled data useless for extremes characterization) and so various techniques of variance inflation have been proposed to overcome this effect. This shortcoming may apply to weather generators too if the distributions used do not have a faithful representation of tail behaviour.

A more direct approach has been taken recently by some studies where the statistical downscaling or weather generator approaches have been applied specifically to extremes, often using the framework of EVT (Wang and Zhang, 2008; Wang et al., 2009; Furrer and Katz, 2008; Kallache et al., 2011).

10.3. OBSERVED CHANGES IN EXTREMES

Section 10.1.2 indicated the difficulty in categorizing extremes consistently, especially on large scales. For this reason, the following sections focus on a small number of categories of extremes that have been reasonably well-monitored on large regional- or global-scales. High-resolution studies of individual locations or catchments will not be considered. For the following categories of extremes, we present current results about observed behaviour, addressing also D&A of observed trends when relevant. Projected changes in these extremes are presented in Section 10.6.

10.3.1. Temperature Extremes

Temperature is generally the best-monitored and most widely available meteorological variable and, in turn, is probably the variable best simulated by current climate models. Temperature percentile indices have now been used widely in many regional studies following the free dissemination of ETCCDI software both at regional workshops and online. This has meant that countries or regions where there has previously been little or no information on climate extremes, such as Asia (Manton et al., 2001; Nicholls and Manton, 2005; Klein Tank et al., 2006, Caesar et al., 2011), the Caribbean region (Peterson et al., 2002), Africa (Easterling et al., 2003; Mokssit, 2003; New et al., 2006; Aguliar et al., 2009); South and Central America (Vincent et al., 2005; Aguilar et al., 2005; Haylock et al., 2006), and the Middle East (Zhang et al., 2005;

Sensoy et al., 2007; Rahimzadeh et al., 2009) have been able to provide information to fill in some of the gaps in our knowledge. This approach was pivotal in 'data mining' in these regions where previously little or no data had been readily available, developing ongoing capacity in these data-sparse regions and enhancing international collaboration (Peterson and Manton, 2008). Modelling groups have also now taken a similar approach through the Joint Scientific Committee (JSC)/CLIVAR Working Group on Coupled Models so that observations and model output can be compared consistently (see Section 10.6). The next sections summarize some of the results that have been obtained from global and regional studies.

10.3.1.1. Hot/Cold Days and Nights

Globally, there have been large-scale warming trends in the extremes of temperature, especially minimum temperature, and the evidence suggests that these trends have occurred since the beginning of the twentieth century (Horton et al., 2001). The first study to attempt a global analysis of temperature extremes under the auspices of the ETCCDI was that of Frich et al. (2002) who showed that there had been significant changes in extreme climate indices, such as increases in warm nights over the last 50 years (see Table 10.1 for definitions of warm and cold days and nights). As an update to the Frich et al. (2002) study, Alexander et al. (2006) found that over 70% of the global land area sampled showed a significant decrease in the annual occurrence of cold nights and a significant increase in the annual occurrence of warm nights (Table 10.2). The spatial distribution of these increases and decreases and where they are significant is shown in Figure 10.10. Some regions have experienced a close to doubling (or halving) of the occurrence of these indices, for example, parts of the Asia-Pacific region (e.g., Choi et al., 2009) or Eurasia (Klein Tank et al., 2006). This implies a positive shift in the distribution of daily minimum temperature throughout the globe, although it says little about the shape or scale of its distribution. Changes in variance and skewness have also been shown to play an important role in these changes (e.g., Ballester et al., 2010; Brown et al., 2008).

Indices based on daily maximum temperature indices have shown similar changes but generally with smaller magnitudes. In the tropics, temperature extremes have warmed at least as fast as other extra-tropical regions. The tendency for warm minimum temperatures to warm faster than warm maximum temperatures is reflected in Figure 10.11d, showing that in most locations across the globe the range between the most extreme maximum and most extreme minimum temperature has decreased over time. In addition, diurnal temperature range (DTR) has decreased over the latter half of the twentieth century in many regions (Vose et al., 2005).

TABLE 10.2 The Percentage of Land Area Sampled for Selected Indicators from table 10.1 with Annual Trends between 1951 and 2003 that Show Field Significance at 5% Level (shown in bold and calculated using a bootstrapping method)

Indicator	Positive Trend (%)	Negative Trend (%)
Maximum Tmin (TNx)	**24.5**	1.1
Minimum Tmax (TXn)	**29.3**	2.6
Minimum Tmin (TNn)	**45.0**	1.9
Cold nights (TN10p)	0.1	**74.0**
Cold days (TX10p)	0.5	**46.0**
Warm nights (TN90p)	**73.1**	0.1
Warm days (TX90p)	**41.0**	0.9
Diurnal temperature range (DTR)	4.2	**39.3**
Frost days (FD)	0.2	**40.6**
Summer days (SU25)	**23.4**	3.7
Ice days (ID)	1.3	**27.4**
Tropical nights (TR20)	**27.6**	1.2
Growing season length (GSL)	**16.8**	0.3
Warm spell duration (WSDI)	**28.8**	0.6
Cold spell duration (CSDI)	10.2	**26.0**
Simple daily intensity (SDII)	**14.6**	2.8

(Source: Results based on Alexander et al., 2006 © Copyright 2006 American Geophysical Union. Reproduced/modified by permission of American Geophysical Union.)

It is obvious from Figures 10.10 and 10.11 that there are still large data-sparse areas on the globe, such as in central Africa. A push to fill in some of this regional detail has shown that trends in temperature extremes are consistent across most regions of the globe. For example, Aguilar et al. (2009) showed that in western central Africa there had been a decrease in cold extremes and an increase in warm extremes over the observational record. Other regional studies, for example, covering Central and Eastern Europe (e.g., Kurbis et al., 2009), North America (e.g., Peterson et al., 2008), and China (You et al., 2011) also seem to confirm these warming signatures in the extremes, with some (e.g., Kurbis et al., 2009) indicating warming trends over century-plus timescales.

Christidis et al. (2005) was perhaps the first study to detect a robust anthropogenic signal in indices of extremely warm nights, although with some indications that the model used overestimated the observed warming. New and ongoing attempts to determine whether, or to what extent,

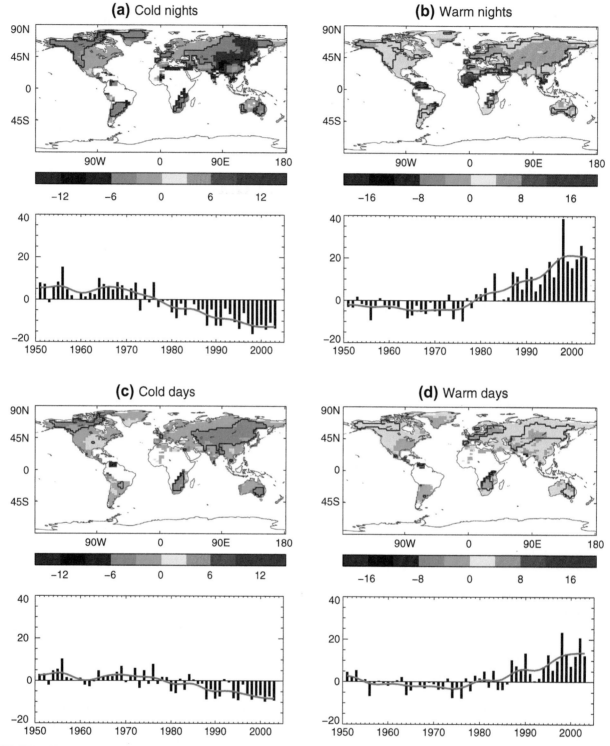

FIGURE 10.10 Trends (in days per decade, shown as maps) and annual time series anomalies relative to 1961–1990 mean values (shown as plots) for 1951–2003 for (a) cold nights, (b) warm nights, (c) cold days, and (d) warm days. Black lines enclose regions where trends are significant at the 5% level. The red curves on the plots highlight decadal variability in the form of a 21-term binomial filter. See Table 10.1 for definitions of these indices. *(Source: Alexander et al., 2006 © Copyright 2006 American Geophysical Union. Reproduced/modified by permission of American Geophysical Union.)*

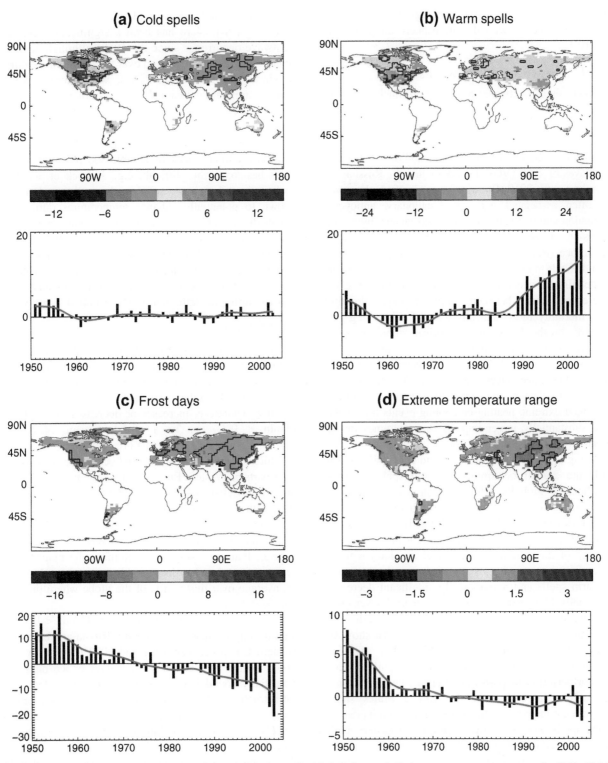

FIGURE 10.11   As Figure 10.10, but for (a) cold spells, (b) warm spells, (c) frost days, and (d) the extreme temperature range. See Table 10.1 for definitions of these indices. *(Source: Alexander et al., 2006 © Copyright 2006 American Geophysical Union. Reproduced/modified by permission of American Geophysical Union.)*

changes in temperature extremes can be attributed to human interference with the atmosphere are reported in Box 10.1.

10.3.1.2. Heatwaves/Coldwaves

The analysis of heatwaves has received much attention in recent decades, primarily following a series of high-profile events with high mortality rates, for example, the Chicago heatwave of 1995 (Karl and Knight 1997), the European summer of 2003 (Poumadere et al., 2005), the Australian heatwaves of 2009 (VGDHS, 2009), and the Russian heatwave in August 2010 (Barriopedro et al., 2011). Heatwaves are often associated with quasi-stationary, anticyclonic circulation anomalies that produce subsidence, light winds, clear skies, warm-air advection, and prolonged conditions at the surface (Black et al., 2004). However, globally, it has been difficult to define a universally valid critical threshold for defining an event as a heatwave (or a 'coldwave') and differences exist between impact-dictated definitions (e.g., mortality rates) and physical definitions (e.g., temperature threshold and duration-driven definitions). Also there is no common definition between countries in how a heatwave is defined. There have, therefore, been fewer studies that have investigated heatwave characteristics compared to changes in warm days or nights, even though prolonged extreme heating or cooling events are likely to have substantially larger impacts than extremes on individual days (Trenberth et al., 2007).

Using the definitions of warm and cold spells in Table 10.1, the annual occurrence of cold spells decreased significantly globally while the annual occurrence of warm spells increased significantly between 1951 and 2003 (Alexander et al., 2006). Table 10.2 indicates that both these results were field significant (i.e., taking account of the spatial correlation of the data) with over 25% of land areas with a statistically significant increase in warm spells and a statistically significant decrease in cold spells. However, the trend in warm spells was greater in magnitude and is related to a dramatic rise in this index since the early 1990s. Figure 10.11a, 10.11b show that, using the definitions from Table 10.1, significant decreases in cold spells occurred predominantly in central and northern Russia, parts of China, and northern Canada, although a large part of the United States showed a significant increase in this index. Significant increases in warm spells were seen over central and eastern United States, Canada, and parts of Europe and Russia. This general increase in warm spells was also highlighted in studies such as Frich et al. (2002) and Klein Tank et al. (2002) although the definition of these indicators varies between each study. Using various definitions, regional studies have also found significant increases in heatwaves and generally significant decreases in cold spells, for example, in China (Ding et al., 2010), Iran (Rahimzadeh

et al., 2009), and Australia (Tryhorn and Riseby, 2006). Over the USA, there has been a significant increase in heatwave occurrence since the 1960s although no significant trend has been detected over the twentieth century as a whole (Kunkel et al., 2008), while across Western Europe summer heatwave duration doubled between 1880 and 2005 (Della-Marta et al., 2007).

More sophisticated definitions for heatwaves have been recently proposed that also include variables such as soil moisture (although lack of reliable soil moisture observations has made this difficult, e.g., Robock et al., 2000). Deficits in soil moisture and drought conditions have been shown to potentially enhance temperatures during heatwaves due to suppressed evaporative cooling (e.g., Fischer et al., 2007; Seneviratne et al., 2006). Recent studies, such as Hirschi et al. (2011) show that the amplification of soil-moisture–temperature feedbacks is likely to have enhanced the duration of extreme summer heatwaves in southeastern Europe during the latter half of the twentieth century.

10.3.1.3. Frosts

While frosts may not be considered extreme in temperate climates since they almost always occur, they can have significant impacts on the emergence or maturation of plants especially if they occur early or late during transition seasons. Therefore, increases or decreases in 'frost days' (Tmin $< 0°C$) can be a useful measure of potential 'extreme' impacts. Figure 10.11c shows that there were significant decreases in the annual occurrence of frost days over parts of Western Europe and a large portion of Russia and China between 1951 and 2003. These results compare well with the results of Kiktev et al. (2003) although the areas of significance vary in some regions. The largest significant negative trend for frost days appeared in the Tibetan Plateau. Using this definition, the annual occurrence of frost days has decreased by approximately 16 days on average over those parts of the globe where this index can be defined.

The annual occurrence of 'ice days' (Tmax $< 0°C$) also significantly decreased over central Russia and eastern and western China between 1951 and 2003 (Alexander et al., 2006). While a reduction in frost days or ice days could potentially sound like a positive outcome, in fact, much infrastructure and many ecosystems rely on cooler temperatures occurring. For example, in the western United States and Canada, destructive colonies of pine beetles have become less vulnerable and have proliferated, aided by, among other factors, the less frequent occurrences of frosts, which kill the insects (Bentz et al., 2010). Melting permafrost can cause extensive infrastructure damage and even the total destruction of certain ecosystems, and could accelerate the rate of carbon loss in other ecosystems, providing a positive feedback for carbon dioxide and methane in the atmosphere (Serreze et al., 2000; Harvey, 2012, this volume).

10.3.2. Precipitation Extremes

Like temperature, precipitation is a well-observed meteorological variable around most places in the globe. It is obviously of importance for supplying water for rivers, reservoirs, and dams but also because extreme rainfall can cause extensive damage. Unlike temperature, however, precipitation has generally received less attention in terms of the representation of extremes in climate model simulations. This is primarily because the processes that produce rainfall are not well-resolved in current climate models (e.g., Stephens et al., 2010) and because the distribution of rainfall is much more 'volatile' (i.e., spatially and temporally variable) than that of temperature. Nonetheless, studies exist that have summarized changes in precipitation extremes on regional and global levels and these are described in the following section.

10.3.2.1. Daily/Sub-daily Extreme Precipitation

Trenberth et al. (2007) concluded that it was likely that annual heavy precipitation events (e.g., 95th percentile) had disproportionately increased compared to mean changes in the second half of the twentieth century over many mid-latitude land regions, even where there had been a reduction in annual total precipitation. Alexander et al. (2006) showed that there have been significant increases of up to 2 days per decade in the number of days in a year with heavy precipitation (using the definition in Table 10.1) in south-central United States and parts of South America (Figure 10.12a). Peterson et al. (2008) also reported that heavy precipitation has been increasing over 1950–2004 in Canada, USA, and Mexico, while Pryor et al. (2009) provided evidence for increases in rainfall intensity above the 95th percentile during the whole of the twentieth century, particularly at the end. In South America, positive trends in extreme rainfall events were detected in the southeastern part of the continent, north central Argentina, northwest Peru, Ecuador, and Sao Paulo, Brazil (Dufek and Ambrizzi, 2008; Marengo et al., 2009; Re and Barros, 2009), although negative trends in extreme winter precipitation were observed in some regions (Penalba and Robledo, 2010). Studies for Europe also indicate that the frequency and intensity of extreme precipitation have increased during the last four decades, although there are some inconsistencies between studies and regions (e.g., Zollina et al., 2008; Bartholy and Pongracz, 2007; Moberg et al., 2006). Mostly, there is much lower confidence in trends in southern Europe and the Mediterranean region. In general, no significant trends in extreme rainfall have been detected over Africa (e.g., Aguilar et al., 2009; New et al., 2006) or the Asia-Pacific region (e.g., Choi et al., 2009), although there are probably trends at subregional-scales.

Alexander et al. (2006) showed that on a global-scale the percentage contribution from very wet days to the annual precipitation total has slightly increased (by 1% since 1951; Figure 10.12b) although the trend is not significant, and there is little significance at the grid box level in this indicator. Trends computed on global averages show little significance; however, there has been a significant increase in most of the precipitation indices given in Table 10.1 except for consecutive dry days and consecutive wet days, reflecting the filtering of noise that makes it easier to detect a significant trend. Note that the wetter regions dominate trends in these indices. Kiktev et al. (2003) and Frich et al. (2002) also found significant increases in the number of days with above 10 mm of rain, but neither of those studies found a significant increase in rainfall intensity. Decreases in the annual consecutive number of dry days are apparent over India, although significant decreases are confined to small regions (Figure 10.12c). Decadal trends in rainfall intensity (that is the annual measure of the average amount of rainfall on a rain day; Table 10.1) indicate a general increase globally (Figure 10.12d), which is in good agreement with other global studies such as Kiktev et al. (2003), although with some variations in where trends are significant.

Most global studies have focused on annual results because of a lack of indices seasonally. However, for indices such as maximum 5-day precipitation totals, there have been increases across most regions of the globe studied, although there are few areas of significant change (Alexander et al., 2006). Significant increases have been observed, however, over south-central United States and northern Russia and Canada in December–January, northern India and southern Brazil in March–May, and southeastern United States and a few other small regions in the United States and Europe in September–November.

In general, statistical tests show changes in precipitation indices that are consistent with a wetter climate. However, when compared with temperature changes, we see a less spatially coherent pattern of change, that is, large areas showing both increasing and decreasing trends and a lower level of statistical significance (Alexander et al., 2006). This is likely to be related to the diverse processes that produce extreme rainfall, but could also be due to the greater sensitivity of gridded rainfall values to changes in the number of observations that are used to calculate grid box values.

Increases have been reported for rarer precipitation events, such as those associated with 50-year return periods, but only a few regions have sufficient data to assess such trends reliably (e.g., Min et al., 2011). Most notably, Groisman et al. (2005) found widespread increases in very intense precipitation (defined as the upper 0.3% of daily precipitation events) across the mid-latitudes. In Europe, there is a clear majority of stations with increasing trends in the number of moderate and very wet days (defined as the exceedance of the 75th and 95th percentiles, respectively) during the second half of the twentieth century (Klein Tank

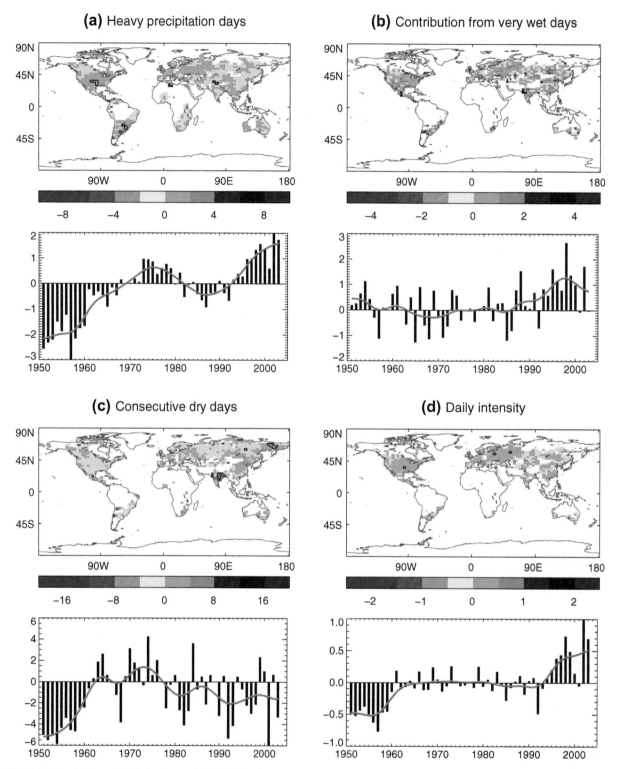

FIGURE 10.12 As Figure 10.10, but for precipitation indices (a) R10 in days, (b) R95pT (i.e., (R95p/PRCPTOT)*100) in %, (c) CDD in days, and (d) SDII in mm per day. See Table 10.1 for definitions of these indices. *(Source: Alexander et al., 2006 © Copyright 2006 American Geophysical Union. Reproduced/modified by permission of American Geophysical Union.)*

and Können, 2003; Haylock and Goodess, 2004). Similarly, Kunkel et al. (2003) and Groisman et al. (2004) found statistically significant increases in heavy (upper 5%) and very heavy (upper 1%) precipitation in the contiguous United States, much of which occurred during the last three decades and is most apparent over the eastern parts of the country. Also, there is evidence for Europe, the United States, and Australia that the relative increase in precipitation extremes is larger than the increase in mean precipitation, this being manifested as an increasing contribution of heavy events to total precipitation (Klein Tank and Können, 2003; Groisman et al., 2004; Alexander et al., 2007).

10.3.3. Complex (Compound) Extremes

Complex or compound extremes are those events related to more than one climate variable, such as storm (wind and rain), and are linked to the multivariate extremes described in Section 10.1.2. They are important because they can have immediate impacts and cause widespread death and destruction. The following subsections outline these events in terms of what is known about their large regional or global trends.

10.3.3.1. Droughts and Floods

Droughts and floods are complex variables that require much more attention than can be adequately given to them in this chapter. Drought indices, for example, have their own well-developed literature and review articles (e.g., Heim, 2002; Dai, 2010). Drought or dryness can be thought of as a period of anomalously dry weather of sufficient duration to cause a serious hydrological imbalance. While drought is a relative term, it can broadly be thought of as falling under three categories: (i) meteorological drought (abnormal precipitation deficit usually relative to some 'normal' amount); (ii) agricultural drought (also soil-moisture drought — a precipitation shortage during the growing season that affects agriculture or ecosystem functions); and (iii) hydrological drought (precipitation shortage affecting surface (e.g., runoff) or subsurface water supply). Another category, 'megadrought', might also be considered for droughts usually lasting a decade or more.

There are still large uncertainties regarding observed trends in droughts worldwide. This is partly due to the differing definitions for drought, but also due to a lack of consistent observations globally, for example, soil moisture. However, if we only consider indices that are based on precipitation deficits, then a measure of the annual maximum consecutive dry days (the CDD index in Table 10.1) may serve as a good proxy for this type of drought. Figure 10.12c shows that there is a mixed picture of drying and wetting across the globe since 1951, although significant changes are confined to small regions. The largest trends are decreases in dry days over India. There

has been a steady decline in consecutive dry days since the 1960s (Frich et al., 2002; Kiktev et al., 2003; Alexander et al., 2006), although, globally-averaged, the trends from 1951 have not been significant. Another widely used precipitation-only drought index is the Standard Precipitation Index (SPI) which, when negative and less than −0.5, corresponds to varying degrees of drought and is generally calculated over varying timescales from 3 months to 12 months (Lloyd-Hughes and Saunders, 2002). In studies over Europe, the SPI has generally indicated no significant trends in observed extreme and moderate drought conditions during the twentieth century (e.g., Lloyd-Hughes and Saunders, 2002; van der Schrier et al., 2006).

Another commonly used index is the Palmer Drought Severity Index (PDSI; Palmer 1965), which uses a simple water balance model to measure moisture balance anomalies (e.g., Dai, 2010). Based on the PDSI model, Dai et al. (2004) and Burke et al. (2006) identified increases in the global land area affected by drought. These studies also found that trends were most affected by temperature and not precipitation change. However, other studies, such as Sheffield and Wood (2008), which was based on soil-moisture simulations from an observation-driven land-surface model, concluded that there was an overall wetting trend during the second half of the twentieth century, which somewhat disagrees with the PDSI-based studies. Thus, there are still large uncertainties regarding the trends in droughts on a global basis.

Regional analyses also report contrasting trends in droughts and there is not enough evidence as yet to suggest 'high confidence' in drying or wetting trends in any continent as a whole. For example, Easterling et al. (2008) find that US droughts are becoming more severe in some regions but there are no clear North America-wide trends. The most severe droughts in the twentieth century occurred in the 1930s and 1950s, where the 1930s Dust Bowl was most intense, and the 1950s' drought, and were most persistent in the United States (Andreadis and Lettenmaier, 2006). Dai et al. (2004) found increases in dryness across most of Europe based on the PDSI but other studies found no statistically significant changes.

SST variability seems to drive droughts in some regions, especially in Africa and western North America, and changes in the atmospheric circulation and precipitation are the predominant drivers in central and southwest Asia. In Australia and Europe, direct relationships to global warming have been inferred through the extreme nature of high temperatures and heatwaves accompanying recent droughts (Nicholls, 2004). More generally, increased temperatures appear to have contributed to increased regions under drought. However, there is also palaeoclimatic evidence that more extreme droughts have occurred over the past millennium (Breda and Badeau, 2008; Kallis, 2008), so recent droughts may not be unusual in a longer-term context (if coincident high temperature extremes are disregarded).

Similarly to droughts, defining floods is complex since there are numerous types including river floods, flash floods, urban floods, pluvial floods, sewer floods, coastal floods, and glacial lake outburst floods. There are also numerous causes of floods, some of the main ones being intense and/or long-lasting precipitation, snow/ice melt, and dam breaks. Observations of floods also suffer from limited spatial and temporal coverage and there are only a limited number of measuring gauges spanning more than 50 years, and even fewer more than 100 years.

IPCC AR4 noted that, while flood damage was increasing (Kundzewicz et al., 2007), there was not a general global trend in the incidence of floods. There is, however, an abundance of evidence that there has been an earlier occurrence of spring-peak river-flows in snow-dominated regions (Rosenzweig et al., 2007). For example, Cunderlik and Ouarda (2009) find that snowmelt spring floods are occurring significantly earlier in southern Canada. While the most evident flood trends appear to be in northern high latitudes, where warming trends have been largest in the observational record there are regions where no evidence of a trend in extreme flooding has been found (e.-g., Shiklomanov et al., 2007, over Russia based on daily river discharge). Other studies for Europe (e.g., Petrow and Merz, 2009; Benito et al., 2005) and Asia (e.g., Jiang et al., 2008; Delgado et al., 2009) show evidence for upward, downward, or no trend in the magnitude and frequency of floods, so we conclude that there is currently no clear and widespread evidence for observed changes in flooding (except for the earlier spring flow in snow-dominated regions).

10.3.3.2. Storms, Cyclones, Winds, and Waves

There has been much debate in the literature (e.g., Holland and Webster, 2007; Landsea, 2007; Mann et al., 2007) about whether there have been trends in the frequency or intensity of tropical cyclones and what role, if any, anthropogenic climate change has played in these trends. Global estimates of the potential destructiveness of hurricanes and the numbers and proportion of tropical cyclones reaching categories 4 and 5 appear to have increased since the 1970s (Webster et al., 2005; Emanuel, 2005), in addition to increased trends in SSTs and other variables that may influence tropical storm development (Hoyos et al., 2006). However, data quality and coverage issues, particularly prior to the satellite era, means that there is low confidence, as yet, and it remains uncertain whether past changes in tropical cyclone activity have exceeded the variability expected from natural causes (Knutson et al., 2010).

While globally and in most ocean basins there appears to have been no significant trend in observed tropical cyclone frequency, regional trends have been identified in the North Atlantic but, as discussed in Section 10.1.1, these conclusions have been somewhat controversial. Detection of trends in tropical cyclone intensity has been even harder to assess due to the sensitivity of these measures to changing technology and improved methodology. Using a homogeneous satellite record, Kossin et al. (2007) have suggested that changing technology has introduced a non-stationary bias that inflates trends in measures of intensity. However, a significant upward trend in the intensity of the strongest tropical cyclones remains even after this bias is accounted for (Elsner et al., 2008). While these analyses have suggested a link between trends in tropical cyclone intensity and climate change, they are constrained to the relatively short period of record of satellite observations and, therefore, cannot provide clear evidence for a longer-term trend.

Outside of the tropics, there has been much more agreement between different analyses performed, and the data used have generally been of higher quality. Nicholls and Alexander (2007, p. 83) state that:

Mid-latitude westerly winds appear to have increased in both hemispheres, and this change appears to be related to changes in the so-called 'annular modes' (defined as zonal mean pressure differences across mid-latitudes, which is related to the zonally averaged mid-latitude westerlies) which have strengthened in most seasons from 1979 to the late 1990s, with poleward displacements of Atlantic and Southern Hemisphere polar front jetstreams. These changes have been accompanied by a tendency toward stronger wintertime polar vortices throughout the troposphere and lower stratosphere. There have been significant decreases in cyclone numbers, and increases in mean cyclone radius and depth, over the southern extra-tropics over the last two or three decades (Simmonds et al., 2003). Geng and Sugi (2001) find that cyclone density, deepening rate, central pressure gradient, and translation speed have all been increasing in the winter North Atlantic.

However, measures of 'storminess' over northern Europe (e.g., Alexandersson et al., 2000; Bärring and von Storch, 2004; Wang et al., 2009) suggest that recent observed changes may not be unusual in terms of long-term variability. Figure 10.13, for example, shows pronounced inter-annual variations in severe autumn storminess across the British Isles since about 1920 with most prominent activity in the 1920s and 1990s. Variability in storm activity in Europe has been linked to variations in the North Atlantic Oscillation (NAO; WASA Group, 1998), which is a measure of the large-scale state of atmospheric circulation in the region. In the case of Figure 10.13, however, the correlation between storms occurring during this period over the British Isles and the NAO is small, although other measures of the NAO have been shown to be significantly correlated (Allan et al., 2009).

Various measures for calculating storminess have been employed. For example, Alexander et al. (2005) and Allan et al. (2009) used an absolute change of 10 hPa over

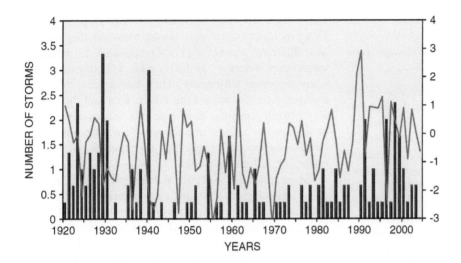

FIGURE 10.13 History of severe storm frequency (left-hand scale) in autumn over the British Isles from 1920–2004. The blue trace is a normalized (relative to 1971–2000 period) North Atlantic Oscillation (NAO) index (right-hand scale). Correlation between this index and severe storm frequency is $r = +0.17$. *(Source: Allan et al., 2009.)*

a 3-hour period as a measure of severe storm events over the British Isles and Iceland, while other studies have focused on analysis of storminess using wind gust characteristics (e.g., Smits et al., 2005), daily pressure tendencies (e.g., Alexandersson et al., 2000; Alexander et al., 2010), and geostrophic winds deduced from station triangles of pressure observations (e.g., Schmidt and von Storch 1993; Matulla et al., 2008; Alexander et al., 2011; Wang et al., 2009). Studies have shown that measures of storminess calculated from pressure observations generally provide a much more homogeneous record for analysis than wind speeds, for example, which are very sensitive to site moves and changing instrumentation (WASA Group, 1998).

In the Southern Hemisphere, studies of storminess have mostly been confined to the last 50 years, when re-analyses data have been available. However, recent studies have been able to utilize recently-available in situ sub-daily pressure data to estimate extreme geostrophic wind speeds or 'geo-winds' over centennial and longer time-scales (e.g., Schmidt and von Storch, 1993). Geo-winds are calculated from inferred geostrophic meridional and zonal wind components using pressure gradients at three station locations to form a triangle (over larger areas, multiple triangles can be combined to form regional averages). These can then be used to calculate upper tail percentiles that can be used as a threshold measure for storminess conditions. In southeast Australia, for example, geo-winds have significantly declined in all seasons between 1885 and 2008 (Alexander et al., 2011). While this study indicates large multidecadal variability, the results are consistent with studies that indicate a southward shift in storm tracks in the Southern Hemisphere (e.g., Hope et al., 2006; Frederiksen and Frederiksen, 2007), which have most likely contributed to the large drying in southwest Western Australia (e.g., Bates et al., 2008) and may also be contributing to persistent drought conditions in southeast Australia.

Rising trends in globally-averaged sea levels have been found (Church and White, 2006) and these could combine with storms to enhance storm surges, posing a major hazard for coastal communities and causing significant coastal damage. Increases in sea-level extremes across the globe have been found to be mostly related to the trend in mean sea level (Woodworth and Blackman, 2004). In addition, significant wave heights have been found to increase and intensify with much of the analysis linking changes in the high-latitude regions of ocean basins to the poleward displacement of the storm tracks discussed earlier (e.g., Wang and Swail, 2001; Sterl and Caires, 2005).

10.3.3.3. Small-Scale Severe Weather Phenomena

Small-scale severe weather phenomena include events that might affect a relatively small area, but can cause a significant amount of damage, such as tornadoes, thunderstorms, hail, and fog. By their nature, they are generally confined to very small spatial-scales and this has made consistent analysis almost impossible. Nicholls and Alexander (2007, p. 83) report some of the problems associated with their analysis:

Evidence for changes in the number or intensity of tornadoes relies on local reports, and Brooks et al. (2003) and Trapp et al. (2005) question the completeness of the tornado record. In many European countries, the number of tornado reports has increased considerably over the last five years (Snow, 2003; Tyrrell, 2003), but it appears likely that the increase in reports is dominated (if not solely caused) by enhanced detection and reporting efficiency. Doswell et al. (2005) highlight the difficulties encountered when trying to find observational evidence for changes in extreme events on local scales connected to severe thunderstorms. In the light of the very strong spatial variability of small-scale severe weather phenomena, the density of surface meteorological observing stations is too coarse to measure all such events. Moreover, homogeneity of existing station series is questionable. While remote sensing techniques allow detection of thunderstorms even in remote areas, they do not always uniquely identify severe weather events from these storms.

Alternatively, measures of severe thunderstorms or hail could be derived by assessing the environmental conditions that are favourable for their formation. Using this technique, Brooks and Dotzek (2008) found no clear trend in severe thunderstorms in a region east of the Rocky Mountains. Using direct measures of hail frequency, some studies have indicated a robust upward trend (e.g., Cao, 2008), while others have found flat or decreasing trends (e.g., Xie et al., 2008). These differing conclusions from the study of small-scale severe weather events show that there is still insufficient evidence to determine whether robust global trends exist in these phenomena (Trenberth et al., 2007).

10.4. CLIMATE PROCESSES AND CLIMATE EXTREMES

10.4.1. Natural Modes of Variability of the Climate System and Their Influence on Extremes' Behaviour

Natural modes of climate variability, such as the ENSO, NAO, and the Southern Annular Mode (SAM), have a significant influence on the variability of atmospheric climate. Of these, ENSO, the most dominant mode of inter-annual variability in the coupled climate system, brings drier/warmer conditions to much of Australia, South East Asia, India, northwest USA, and Canada during positive (El Niño) phases and wetter/cooler conditions in most of these regions during its negative (La Niña) phase (e.g., McBride and Nicholls, 1983; Power et al., 1998; Kiladis and Diaz, 1989; Rasmusson and Carpenter, 1982; Deser and Wallace, 1987, 1990; Ropelewski and Halpert, 1986). In addition to its influence on mean climate conditions, it is also known to be connected with extreme climatic events such as droughts and floods in many parts of the world (e.g., Andrews et al., 2004; Barlow et al., 2001; Meehl et al., 2007b; Zhang et al., 2010).

Previous studies have shown that natural modes of climate variability have a significant influence on the regional response of climate extremes on inter-annual, multidecadal, and longer timescales (e.g., Kunkel, 2003; Nicholls et al., 2005; Scaife et al., 2008; WASA Group, 1998; Wang et al., 2009; Zhang et al., 2010). For instance, Scaife et al. (2008) showed that, using a set of novel experiments where a climate model was forced with observed tropospheric winds, the NAO has a significant impact on extreme temperatures across Europe in winter, particularly frost days.

Perhaps the first study to look at the effect of a number of large-scale climate variability measures on global temperature extremes was Kenyon and Hegerl (2008). Using indices of large-scale climate variation, they showed that different phases of ENSO appeared to influence temperature extremes globally, often affecting cold and warm extremes differently. They found areas of much stronger response around the Pacific Rim and throughout all of North America. They conclude that modes of variability need to be taken into account if we are to provide for reliable attribution of changes in extremes, as well as prediction of future changes.

More recently, Kenyon and Hegerl (2010) applied a similar analysis to precipitation extremes, quantifying the influence of three major modes of variability (ENSO, NAO, and North Pacific Index (NPI)) on indices of extreme precipitation worldwide. They conclude that distinct regional patterns of response to these modes' positive and negative phases are observed, but the most pervasive influence is observed to be from ENSO. Alexander et al. (2009b) investigated the global response of both temperature and precipitation extremes to global 'patterns' of SST variability. Using the method of Self-Organizing Maps (SOM), they categorized observed seasonal SST anomalies between 1870 and 2006 from the HadISST1 dataset (Rayner et al., 2003) into common patterns. Figure 10.14 shows how opposite modes of ENSO (El Niño versus La Niña) can have quite different effects on temperature and precipitation extremes. Globally averaged, there is an increase of about one day per season in the frequency of warm nights during El Niño (compared to 'normal' SST conditions) and, conversely, there is a decrease of about one day per season in the frequency of warm nights during La Niña (compared to 'normal' SST conditions). Figure 10.14e and 10.14f indicate that extreme maximum temperatures are significantly cooler during strong La Niña events than during strong El Niño events over Australia, southern Africa, India, and Canada, while the converse is true for USA and northeastern Siberia.

While the response of precipitation extremes to global SST patterns is less spatially coherent (Figure 10.14g and 10.14h), there are large areas across North America and central Europe that showed statistically significant differences in the response to opposite phases of the ENSO, indicating that the variability of global SST anomaly patterns is important for the modulation of extreme temperature and precipitation globally. Zhang et al. (2010) have also found that ENSO and the Pacific Decadal Oscillation (PDO) have spatially consistent and statistically significant influences on extreme precipitation, while the influence of NAO is regional and is not field significant.

While ENSO is obviously critical in driving extreme events in many regions around the world, there is undoubtedly a tangle of other contributing factors and processes that are also important. If we consider Australia, for example, some of the driving mechanisms

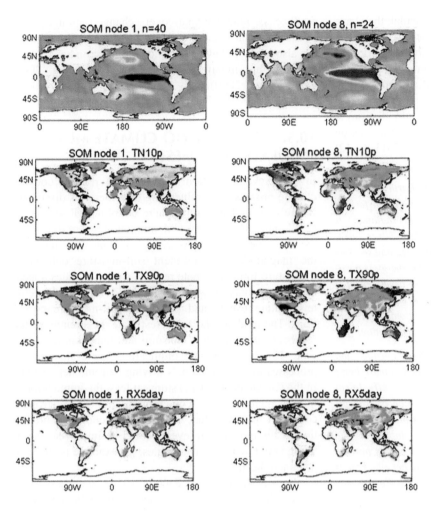

FIGURE 10.14 Influence of SST patterns on temperature and precipitation extremes globally. The left column shows how extremely cold minimum temperatures (TN10p), extremely warm maximum temperatures (TX90p), and annual 5-day maximum consecutive rainfall (RX5day) differ from 'normal' during a strong La Niña event. The right-hand column shows results during a strong El Niño. See Table 10.1 for definitions of these indices. *(Source: Alexander et al., 2009b © Copyright 2009 American Geophysical Union. Reproduced/ modified by permission of American Geophysical Union.)*

of regionally observed trends are still under debate. Alexander and Arblaster (2009) note that, while a number of studies have attributed portions of the drying in southwest Australia to anthropogenic forcing (Cai and Cowan, 2006; Hope, 2006; Timbal et al., 2006), the impact of natural variability (Cai et al., 2005) and land-cover change (Pitman and Narisma, 2005; Timbal and Arblaster, 2006; Pitman and de Noblet-Ducoudré, 2012, this volume) also appear to be reasonably large. An increase in precipitation and associated cooling in northwest Australia (e.g., Nicholls et al., 1997; Power et al., 1998) has been variously ascribed to the poor simulation of aerosols in GCMs (Rotstayn et al., 2007) and continental warming further south, driving an enhancement of the Australian monsoon (Wardle and Smith, 2004), in addition to other large-scale driving mechanisms. Australian precipitation trends appear to be consistent with the decadal variability in tropical Pacific SSTs; thus, if GCMs could capture the zonal gradient of the SST changes, with

a minimum in warming in the central Pacific, they would likely capture the increase in northwest Australian mean precipitation and, by extension, extremes (Alexander and Arblaster 2009). Although Santer et al. (2006) and Gillet et al. (2008) have attributed changes in tropical Atlantic and Pacific SSTs to anthropogenic forcing, the extent to which the pattern of observed trends in tropical SSTs is anthropogenic is unknown.

This type of discussion could be extended to many other regions of the world, and it shows that much further study is required to untangle the contributions to recent changes in climate extremes from natural and anthropogenic variability.

10.4.2. Land–Atmosphere Feedback Processes' Influence on Extremes

Several studies (Seneviratne et al., 2006; Diffenbaugh et al., 2007; Fischer et al., 2007; Durre et al., 2000) have investigated the role of the land surface in suppressing or exacerbating temperature extremes. In particular, they have emphasized the important role of soil-moisture deficits in intensifying or lengthening heatwaves through feedbacks between temperature, evaporation, and precipitation (see also Section 10.3.1.2). The way the land and atmosphere interact affects the 'partitioning' of the net radiation budget (the combination of net solar and net infrared radiation) at the land surface. Soil moisture can have a large impact on this partitioning, as discussed in Alexander (2011, p. 12) who states:

Soil moisture affects how energy absorbed by the Earth's surface is returned to the atmosphere. Some energy is returned in the form of infrared radiation whereas the remainder, net radiation, is predominantly transformed into two turbulent energy fluxes: latent heat flux, manifest as the evaporation of liquid water, and sensible heat flux that directly heats air or land. How the energy is partitioned between those two types of fluxes affects the development of the lowest layer of the atmosphere — the boundary layer — and the hydrological cycle. And understanding this partitioning can help to explain feedbacks between soil moisture and temperature.

The relative contributions of these fluxes are determined by land surface properties, moisture availability and atmospheric conditions. When the land surface has plenty of moisture, latent heat is the dominant flux through soil evaporation or transpiration, commonly referred to as evapotranspiration. High evapotranspiration cools the surface, and increases the concentration of water vapour in the atmosphere. This process enhances cloud formation near the surface. The net effect tends to be cooling, which subsequently reduces the Earth's net radiation budget.

On the other hand, in drought conditions moisture is strongly constrained, and dry landscapes are characterized by a high flux of sensible heat. This produces a deeper, warmer, and drier atmospheric boundary layer that tends to inhibit cloud formation. The sensible heat and associated warming and drying of the atmosphere, combined with less cloud cover, tends to further dry the surface. This increase in the sensible heat flux during dry conditions contributes to the warming of the near-surface atmosphere, and produces a positive land—atmosphere feedback that tends to intensify and lengthen drought conditions.

While there has been limited observational evidence for these relationships and feedbacks, Hirschi et al. (2011) were able to demonstrate a significant influence of soil moisture on the evolution of hot summer temperatures in southeastern Europe, a region where evapotranspiration is limited by the availability of soil moisture. However, they only found a weak relationship in a wetter climate regime between extreme temperature events and antecedent soil-moisture conditions (see also Section 10.5).

In general, processes that affect moisture availability are likely to have the most effect on exacerbating or suppressing extremes. In addition to soil-moisture feedbacks, land-use/land-cover change, ground-water availability, and the physiological response of vegetation are also likely to have a significant impact on how certain extremes are manifest (e.g., Zhao and Pitman, 2002; Collatz et al., 2000).

10.5. HOW WELL DO CLIMATE MODELS SIMULATE EXTREMES?

In the previous section, the complex role of coupled climate processes in influencing extremes was discussed. In the case of land—atmosphere coupling, Hirschi et al. (2011) note that, while RCM simulations seem to capture the link between the severity of hot extremes and the magnitude of antecedent soil-moisture deficits in a moisture-limited climate regime, wetter climates appear to indicate only a weak relationship between extreme temperature events and preceding soil-moisture conditions. The models tend to overestimate the strength of the feedbacks between soil moisture and temperature in this case. This has important implications for evaluating the processes in climate models driving extreme heat events. In the Hirschi et al. (2011) study, at least the implication is that, in regions where temperature extremes are strongly coupled with antecedent soil moisture conditions, prediction of the most severe events could be improved, facilitating better adaptation strategies, for example in the field of human health.

As was discussed in Section 10.2.3, there are many complexities in comparing observations with the output from climate models. Section 10.4.1 indicates that the ability of a GCM to represent coupled climate processes well could significantly affect its ability to simulate extremes. Globally, it has been shown that state-of-the-art global models can reproduce the trends in temperature extremes with some skill, but the models are less skilful at representing trends in precipitation extremes (e.g., Kiktev et al., 2003; Kiktev et al., 2007; Kharin et al., 2007). This may be ascribed, at a fundamental level, to the inability of current GCMs to resolve the convective-scales explicitly, due to the relative coarseness of their grids. Lacking the ability to simulate the environment conducive to these kinds of extremes, it is no surprise that the statistics of extreme precipitation from GCMs do not validate well against observations. However, it should be pointed out that even for extremes that are within GCMs' reach, it is not straightforward, although progress is being made (e.g., Chen and Knutson, 2008) in how to best compare model output (which represents an area mean) with observations of extremes (which are point estimates), as discussed in Section 10.2.3.

A related, but separate issue, is how to combine multiple models' output, in particular when interested in future projections. Much work is being done to develop metrics of model performance (e.g., Gleckler et al., 2008), and to use those to inform how to look at models' future simulations. Over Australia, Perkins et al. (2007) ranked multiple global models by their ability to reproduce the pdfs of observed daily temperature. In doing so, they were able to assign weights associated with the skill of each climate model in reproducing observed temperature. Other work is also attempting to weight models according to their ability to reproduce past climate (Giorgi and Mearns, 2002; Tebaldi et al., 2005, Tebaldi and Sanso, 2008; Smith et al., 2009). The idea is that we may have more confidence in models that are able to reproduce current climate and, therefore, should apply more weight to the output of those models in the future. It is to be noted, however, that very few metrics are found to be predictive of models' future behaviour (but see Qu and Hall, 2006, for a successful example), and it remains unclear how to best combine multiple models in light of this fact, and in light of other statistical characteristics of the model ensemble (see the IPCC guidance paper on multimodel ensembles; Knutti et al., 2010).

Rather than focus on the ability of models to represent the actual extremes, some studies have focused on how well spatial and temporal trends in the models agree with observations. Global studies, comparing observed and modelled trends in climate extremes, have shown reasonably good agreement with temperature trends, but poor agreement (or multimodel disagreement) with observed precipitation patterns or trends (e.g., Kiktev et al., 2007; Kharin et al., 2007). Kiktev et al. (2007) also comment that a 'super ensemble', from multiple climate models, appears to perform better than any individual ensemble member or model, particularly when there is some skill in the contributing ensemble members. However, recent studies show that the regional responses of observed trends in temperature and precipitation extremes can also largely be driven by large-scale processes that might not be adequately simulated in GCMs (Scaife et al., 2008; Meehl et al., 2004; Meehl et al., 2005). On regional-scales (e.g., Sillmann and Roekner, 2008, for Europe; Meehl and Tebaldi, 2004, and Meehl et al., 2007a, for the USA; and Alexander and Arblaster, 2009, for Australia), the modelled trends have been shown to capture observed trends reasonably accurately, although the results are somewhat dependent on the extreme under consideration.

Figure 10.15 shows that the variability and trend, and to some extent spatial pattern of trends, in warm nights over Australia are well-simulated by multiple runs from multiple climate models (these results also generally hold globally; see Kiktev et al., 2003, 2007). However, the multimodel ensemble, on average, underestimates the

trend in heavy precipitation days. Alexander and Arblaster (2009) went on to show that the observed trends in warm nights are inconsistent with natural-only ('unforced') variability. Kiktev et al. (2003) also found that only the inclusion of human-induced forcings in a climate model could account for observed changes in global temperature extremes. Robust anthropogenic changes have been detected globally in indices of extremely warm nights and in the trends of highest and lowest minimum and maximum temperatures, although with some indications that the models underestimate the observed changes in cold nights, overestimate the observed changes in warm days, but also correctly simulate changes in warm nights and cold days (Christidis et al., 2005; Christidis et al., 2010; and Zwiers et al., 2011).

Some of the models' inability to reproduce extremes well may be related to weaknesses in how they represent some fundamental climatic processes. For example, Leloup et al. (2007) showed that none of the GCMs in the IPCC AR4/CMIP3 database were sufficiently good at capturing all aspects of the extent, location, and timing of ENSO events. In addition, Alexander (2009) showed that an atmosphere-only model forced with observed SSTs obtained the opposite response in extreme maximum temperatures over Australia during La Niña and El Niño than observed. This would imply that one or more atmospheric processes and/or interactions within the model are not properly simulated and that this has significant implications for the representation of extremes. For example, the representation of clouds is known to be one of the main sources of uncertainty in GCMs, so improving convection parameterization within models has become a major focus of many modelling groups around the world.

In conclusion, the current inability of GCMs and some RCMs to reproduce trends, variability, or spatial patterns of climate extremes, particularly for precipitation, has serious implications for the use of climate models for impacts assessments.

10.6. THE FUTURE

It is expected that the extremes of climate variables will change at a faster rate than the mean climate (IPCC, 2001a; Katz, 1999; Figure 10.1). In some regions, extremes have already shown amplified responses to changes in means (Folland et al., 2001). Of course, climate extremes have been occurring in a stationary climate, but there is reason for concern when we consider that human-induced climate change may shift extremes towards intensities and frequencies that will challenge the adaptive capacity of the more vulnerable systems. As presented in Section 10.1.2, even a relatively small

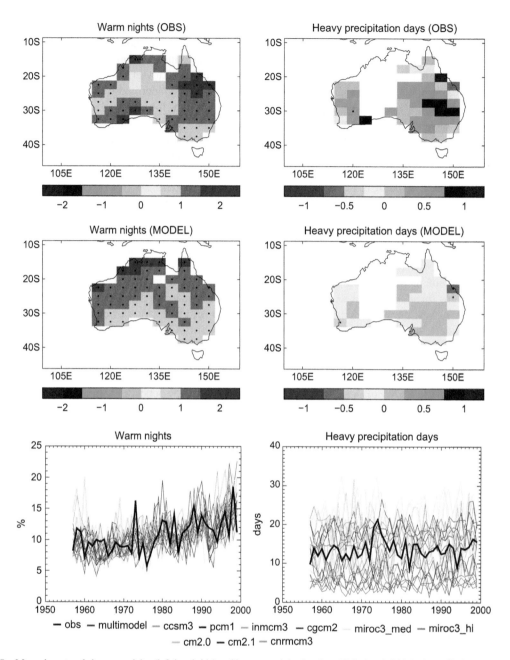

FIGURE 10.15 Maps show trends in warm nights (left-hand side) and heavy precipitation days (right-hand side) in Australia between 1957 to 1999 for observations and CMIP3 multimodel ensemble. Time series show the same indices areally-averaged for observations (black line) and indicated models (coloured lines). Model data are masked, preserving only grid boxes that have observed data. Stippling indicates trend significance at the 5% level. See Table 10.1 for definitions of these indices. *(Source: Alexander and Arblaster, 2009.)*

change in the mean of a variable may lead to substantial changes in the nature (intensity and frequency) of extremes.

The impacts of most extremes are typically felt at a local or regional-scale, so regional studies of climate extremes are of the highest priority for most countries for assessing potential climate impacts. However, when it comes to future projections, we argue that the drivers of change are to be found first from a global perspective, as a function of the forcings that are imposed on the global-scale, and of the uncertain response to this forcing that only GCMs — and ensembles of them — can effectively and coherently represent. Regional studies are then of enormous value to further refine, strengthen and, when possible, add detail to the understanding we gain from global studies.

In addition, given that climate change signals in climate extremes are difficult to detect at a regional-scale (but see Box 10.1 for a summary of those studies that have succeeded in doing so), to understand fully how the climate varies, and the extent to which humans have influenced the climate system, requires a global approach. The study by Tebaldi et al. (2006) was the first to use the multimodel approach to assess potential future changes in climate extremes showing that the twenty-first century would bring global changes in temperature extremes consistent with a warming climate. While that study also showed that global changes in precipitation extremes were consistent with a wetter world with greater precipitation intensity, the consensus and statistical significance of the change (in that study, simply tested grid point by grid point) among the models were weaker when regional patterns were considered. This indicates the importance of combining global results with more regionally-relevant studies to assess the impacts of these changes.

Confidence in predictions of future climate extremes is gained: (i) by assessing the efficacy of model results in the context of observational evidence (are the projected trends already surfacing from observations and/or are they consistent with other changes being experienced in the region?); and (ii) by confirming (or updating) our understanding of the physical system's behaviour (are the projected changes consistent with our understanding of how the climate system should respond when subjected to external forcings?). For some kinds of extremes, like intense heat and precipitation events, it is generally the case that these multiple lines of evidence concur in strengthening our projections of future changes.

Analysis based on the multimodel ensemble CMIP3 and other multinational efforts such as that of Alexander et al. (2006) meant that by the time of the IPCC's TAR and AR4 published in 2001 and 2007, respectively, formal statements focusing specifically on extremes, how they are changing, and how they might change could be included in the reports. Thus, it was both the availability of a large number of simulations that provided output for the analysis of extremes (daily variables/global indices suites), and increased data coverage in space and time, that were crucial in determining the strength of those conclusions.

Most of what follows summarizes the IPCC AR4 statements with regard to expected future changes in extremes. For many of the individual results, detailed references can be found in IPCC AR4, WG 1 [their section 10.3.6 and table 11.2 of that report]. Because of our perspective, which favours global studies as the cornerstone of future projections, the IPCC AR4 arguably remains at present a representative collection of results, as we await the completion of the next round of global climate simulations, CMIP5, under a new set of scenarios that will provide input to the next assessment of IPCC and will be accessible through the Program for Climate Model Diagnosis and Intercomparison

(PCMDI). See *http://cmip-pcmdi.llnl.gov/cmip5/*. However, a couple of more recent studies (Kharin et al., 2007; Orlowsky and Seneviratne, 2011) contribute additional perspectives, as we explain below.

10.6.1. Temperature Extremes

Consistent with the notion of a warming climate, for which we know that shifts in the mean cause corresponding or larger changes in the tails of the distribution (Kharin and Zwiers, 2005, Katz, 1999), and with studies that have already detected anthropogenically-driven changes in global temperature extremes (Christidis et al., 2005, 2010; Stott et al., 2010; Zwiers et al., 2011), models are in strong agreement that all of the land areas will experience, in some form or other, enhanced hot temperature extremes coupled with a decrease in cold temperature extremes. There is consensus on the sign of these changes and significance, regionally and at the global or hemispheric average scale, but, as is the case for other measures of climatic change, the numerical ranges from different models' projections are wide, being sensitive to the different models' parametric and structural formulations.

Longer, more intense, and more frequent heatwaves — variously defined by different studies — are projected for all regions of the world, with some regions projected to see more severe intensification (south and west United States, Mediterranean basin, Western Europe), due to changes in the base-state of the global circulation and because of the interplay of other factors (drying of the climate, decrease in soil moisture).

Decreases in cold spells are also universally predicted, with some regions (western North America, North Atlantic, southern Europe and southern Asia) seeing decreases of lesser magnitude than others, again because of changes in atmospheric circulation.

Changes in indices related to changes in temperature extremes have also been robustly characterized by multimodel results, both by Tebaldi et al. (2006), which was assessed in AR4, and more recently and comprehensively (with a larger number of models and indices) by Orlowsky and Seneviratne (2011). Among these changes, the strongest signals are found in decreasing frost days, lengthening of the growing season, and increasing number of warm nights. Figures 10.16 and 10.17 (left columns) summarize projections at the global average scale and as global geographical patterns of change by the end of this century of five of these indices (see Table 10.1 for definitions). The strength of the change over time and space is clear for all of them, but different scenarios do matter in determining the magnitude of the changes. The lower scenario (SRES B1) appears to extrapolate the current trend of the last decades, while the two higher scenarios (A1B and A2) cause steeper trends in the direction consistent with warming. The spatial patterns

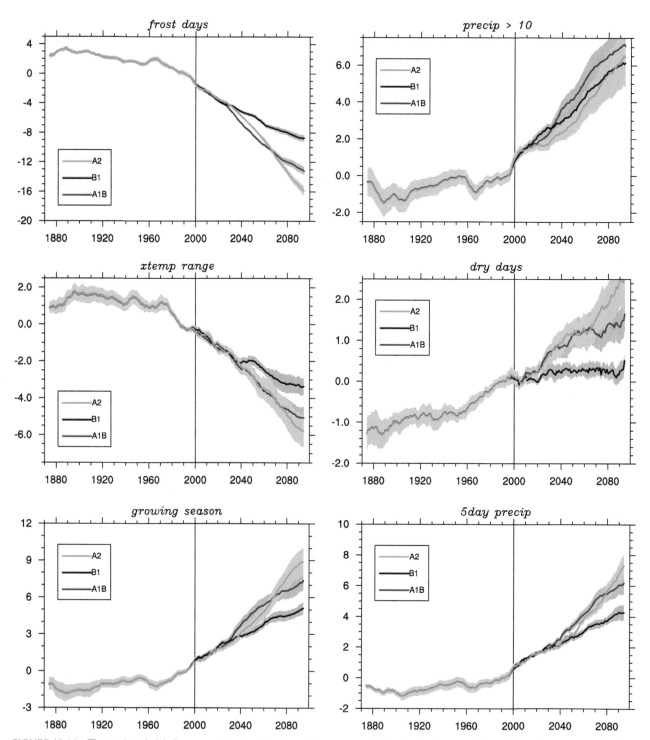

FIGURE 10.16 Time series of globally averaged (land only) values of temperature (left column) and precipitation (right column) extremes indices. Three SRES scenarios are shown in different colours for the length of the twenty-first century: SRES B1 in blue, SRES A1B in green, and SRES A2 in red. The values have been standardized for each model, then averaged and smoothed by a 10-year running mean. The envelope of one standard deviation of the ensemble mean is shown as background shading. See Table 10.1 for definitions of these indices. *(Source: Tebaldi et al., 2006. With kind permission from Springer Science+Business Media.)*

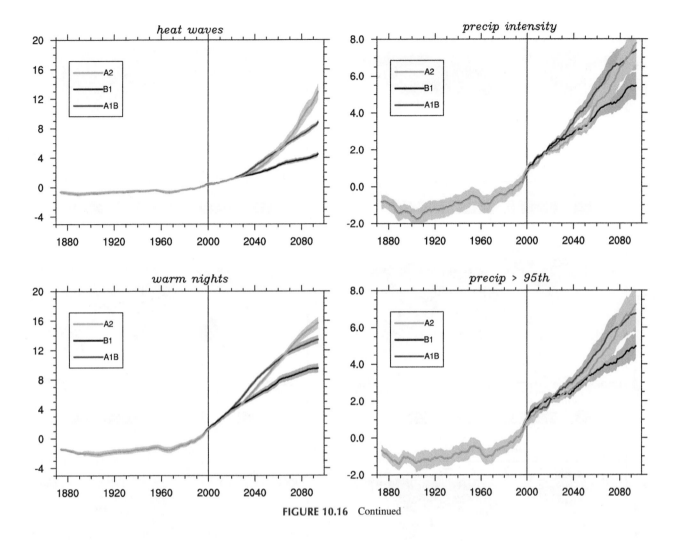

FIGURE 10.16   Continued

of change appear almost everywhere and, for all indices, are strongly significant (as measured by the level of inter-model agreement – denoted by stippled areas – and the size of the change shown here by colour-coding in units of standard deviations). Some of the indices (frost days, growing season length) are not computed for latitudes below 30°N/S because they are not relevant for tropical climates.

Most of these phenomena can be seen as a direct consequence of increasing temperatures (shift in the location of the distribution), but some studies have also documented changes in the future variability of temperature over Europe (e.g., Schär et al., 2004), as well as globally (Hegerl et al., 2004; Kharin et al., 2007). In general, it is found that changes in temperature extremes are not linearly linked to changes in global average temperature and radiative forcings, and may be influenced in their strengths by land–atmosphere feedbacks (Senerivatne et al., 2006; Diffenbaugh et al., 2007).

Kharin et al. (2007) used the EVT framework and estimated changes in the return levels and return periods of annual highest maximum temperatures. They concluded that the 20-year extreme annual daily maximum temperature

will increase by about 2°C by mid-twenty-first century and by about 4°C by late twenty-first century, depending on the region. Correspondingly, waiting time for a late twentieth century 20-year extreme annual daily maximum temperature will be reduced to about 2 years–20 years by mid-twenty-first century and by about 1 years–5 years by late twenty-first century, depending on the region.

10.6.2. Precipitation Extremes

A warmer world enhances the chances of intense precipitation, and trends towards larger precipitation extremes have already been recorded in many regions of the world, as we have already discussed. Consistently, climate models project future episodes of more intense precipitation in the wet seasons for most of the land areas, especially in the Northern Hemisphere and its higher latitude and in the monsoon regions of the world, as well as at a globally-average scale (see Figures 10.16 and 10.17, right columns, and Table 10.1 for definitions). It is again the case that the actual magnitude of the change is dependent on the

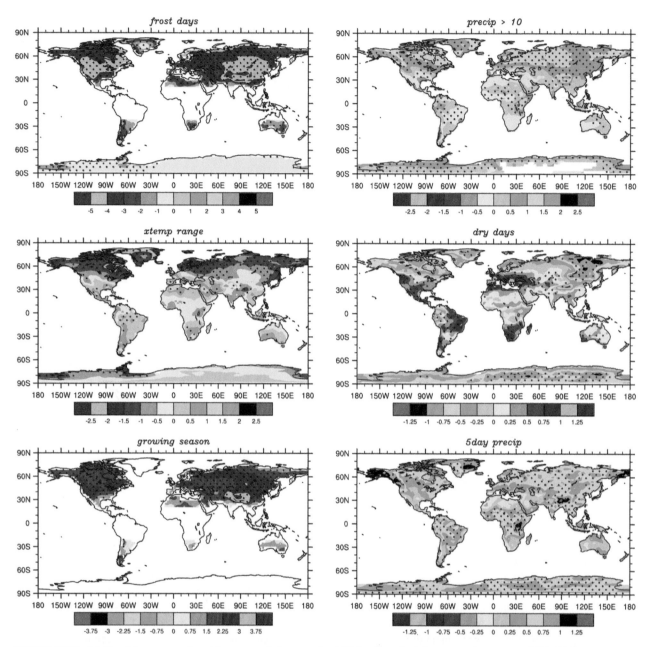

FIGURE 10.17 Multimodel averages of spatial patterns of change under A1B by the end of the twenty-first century (2080–2099) compared to 1980–1999. Each grid-point value for each model has been standardized first, and then a multimodel simple average computed. Stippled regions correspond to areas where at least five of the nine models concur in the statistical significance of the change. See Table 10.1 for definitions of these indices. *(Source: Tebaldi et al., 2006. With kind permission from Springer Science + Business Media)*

model used, but there is strong agreement across the models over the direction of change and our understanding of the causes (warmer air holding more moisture, enhanced water vapour availability fuelling precipitation events, changes in circulation) further help us pin down the large-scale aspects of these changes. Regional details are less robust in terms of the relative magnitude of changes but remain in good accord across models in terms of the sign of the change.

During the dry season, the tendency manifested in the majority of the simulations is for longer dry periods for those regions that are already semi-arid of the mid-latitudes and subtropics like the Mediterranean, the southwest US, southwestern Australia, southern Africa, and a large portion of South America. Agreement in the regional patterns and even in the strength of the future trends at global average scales is less for this index (the latter effect due to mixed results for different regions of

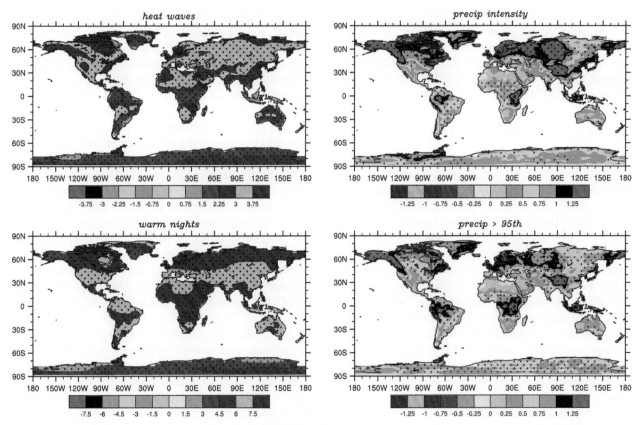

FIGURE 10.17 Continued

the world, which, when averaged, dampen the overall trend).

Kharin et al. (2007) performed a similar analysis to that for temperature extremes on annual maximum 24-hour precipitation rates, finding larger inter-model variability in the results than for temperature extremes, but estimating, as the median model projection, that the extreme 24-hour precipitation rate will increase by about 5%–10% by mid-twenty-first century and by about 10%–20% by late twenty-first century, depending on the region and the emissions scenario. Correspondingly, almost all models suggest that the waiting time for a late twentieth century, 20-year extreme, 24-hour precipitation event will be reduced to substantially less than 20 years by the mid-twenty-first century and much more by the late twenty-first century, indicating an increase in frequency of the extreme precipitation at continental and subcontinental-scales under all three forcing scenarios.

10.6.3. Tropical and Extra-tropical Storms

Difficulties in understanding the factors contributing to the formation and development of tropical storms,

uncertainties in the signal within the observed record, and shortcomings — mostly due to resolution — of the GCM representation of these storms make projection of their future changes particularly challenging. The consensus, gleaned from modelling studies using higher resolutions that are commonly adopted for future projections and have often been run for more idealized scenarios than for the traditional SRES set, is that wind and precipitation associated with tropical cyclones will likely intensify. This suggests an increase in the number of intense storms, but also that the total number of tropical cyclones forming — globally — may in fact decrease, there remaining large uncertainties on the regional distribution of future storms (Knutson et al., 2010).

For extra-tropical storms, similar results and reasoning suggest that there will likely be an intensification of winds and precipitation associated with them, and analysis of future circulation patterns suggests that the storm tracks (i.e., tracks of surface low-pressure systems or individual storms in both hemispheres) will shift poleward (e.g., Bengtsson et al., 2006; Lynch et al., 2006), particularly in winter. This tendency is another example of a phenomenon that has already been detected in observations (e.g., Wang et al., 2006). Wave heights associated with storms are also likely to increase as a direct consequence of stronger winds in the locations where the new storm track will shift to, while they are expected to decrease as a consequence of that displacement in the locations of the old storm track (Wang et al., 2009).

10.7. EXTREMES IN OUR FUTURE CLIMATE

Weather and climate extremes can disrupt both natural and human systems, carrying high costs, in monetary terms and often also in terms of lives lost or other adverse effects. The importance of monitoring and characterizing these events is obvious in a stationary climate, but becomes even more pressing under climate change because of the additional stresses that come when society attempts to adapt to changing, often intensifying, extremes. Recording and analysing extremes poses unique challenges because of the intrinsically rare nature of these events (in space and time) and the fact that, by definition, they happen in conjunction with disruptive conditions. To the challenge of making sense of what has happened and is happening, needs to be added the difficulty in making projections of what lies ahead, given the shortcomings and uncertainties in our current modelling capabilities, especially as our needs require focus on finer and finer spatial and temporal-scales.

This chapter has shown that progress has been made over the past decade or so in the analysis of extreme climatic events. Despite this, progress still needs to be made in understanding the long-term variability of these events, what drives them, or how they might change in the future under climate change (IPCC, 2007a). Klein Tank et al. (2009) note that "The sustainability of economic development and living conditions depends on our ability to manage the risks associated with extreme climate events" and "Knowledge of changes in weather and climate extremes is essential to manage climate-related risks to humans, ecosystems, and infrastructure, and develop resilience through adaptation strategies". Given that "confidence has increased that some extremes will become more frequent, more widespread, and/or more intense during the twenty-first century" (IPCC, 2007a), the demand for information services on weather and climate extremes is growing substantially.

There are still many open questions regarding the definition and analysis of climate extremes (Alexander et al., 2009a). For instance, the role of anthropogenic climate change on the frequency and intensity of tropical cyclones remains uncertain (Knutson et al., 2010). Another important question relates to how to adequately address the scale discrepancy between observations and models (as discussed in Section 10.2.3). Other questions include (but are not limited to) how to improve the simulated representation of extremes, our understanding of the extent of the influence of modes of variability on extremes, and the adequate prediction of the probability of certain extremes. In addition, most of the discussion in this chapter has focused on GCMs, which are generally not useful for local impact studies. A proliferation of high-resolution studies, now ongoing using RCMs, may help to improve the simulation of more high-impact events although, even for regional models, the representation of, for example, precipitation extremes is still often poor (Fowler and Ekstrom, 2009; Fowler et al., 2007). Improvements may be obtained by better observations for comparison, understanding the mechanisms and processes that act on different spatial and temporal-scales, or perhaps a new generation of higher-resolution RCMs (capable of resolving convective processes for summer extremes, for example). Above all, data quality is of the utmost importance in underpinning any research.

A gap also exists between the types of extremes that the 'climate community' focuses on and what we have focused on in this chapter, compared to the types that are of concern for the 'impacts community' (as discussed in Section 10.1.2). As highlighted in Section 10.1.1, climatologists have mainly focused on extremes that are statistically robust, are consistently defined for a wide range of climates, and have a high signal-to-noise ratio for use in D&A studies. This differs generally from the types of extremes of concern to the 'impacts community', for design purposes, for example, which are usually high-impact, regionally-specific events. While the two criteria may not necessarily be mutually exclusive, a gap definitely exists between what is currently possible from one community and what is needed from the other. A comprehensive framework for defining the ensemble of events that can lead to a given impact is also yet to be defined and the clear articulation of these types of events needs to flow between the two communities. A new dialogue has begun between climate modellers and the climate change impacts community (which should also include the latest developments from the statistics community) to help bridge this gap, and one way forward would be to utilize integrated regional climate modelling studies, for example, the EU's PRUDENCE (e.g., Deque et al., 2007) or North America's NARCCAP (Mearns et al., 2009) projects to examine impacts. Ultimately, much more collaboration between the different communities is required if we are to adequately address the scientific problems that still exist together with the needs of stakeholders and policymakers.

Chapter 11

Interaction Between Future Climate and Terrestrial Carbon and Nitrogen

Robert E. Dickinson

University of Texas, Austin, USA

11.1. INTRODUCTION: CYCLING TERRESTRIAL NUTRIENTS

Climate is warming because of increasing radiative forcing. This forcing results from increasing atmospheric concentrations of greenhouse gases, including especially carbon dioxide (CO_2), methane (CH_4), and nitrous oxide (N_2O). Human activities, especially fossil fuel combustion and land clearing, are leading to these increased concentrations (e.g., Pitman and de Noblet-Ducoudré, 2012, this volume). These concentrations will continue to increase, but by how much is uncertain, both because of uncertainties as to future human activities and as to natural biogeochemical processes that involve carbon (C) and nitrogen (N). This chapter addresses, from a modelling viewpoint, how the terrestrial C and N cycles are expected to change because of climate change and what will be the consequences of these changes for the future levels of atmospheric CO_2. Figure 11.1 shows estimates for anthropogenic global CO_2 production and how much of this is stored annually in the atmosphere, oceans, or on land. Although the net storage on land is still estimated as a residual of the other terms, it is also fairly tightly constrained by multiple observational approaches (e.g., Ciais et al., 2010). The warming that the greenhouse gases produce is being mitigated to some extent by processes increasing atmospheric aerosol concentrations. Aerosols can reflect solar radiation, and so cool the climate. These aerosol effects also are uncertain and involve C and N, as well as sulfur and mineral components.

The Future of the World's Climate. DOI: 10.1016/B978-0-12-386917-3.00011-7

FIGURE 11.1 Annual budget of anthropogenic production of carbon (C) and its partitioning into atmosphere, ocean, and land reservoirs. Units are PgC per year, corresponding to Denman and Brasseur (2007) estimates for 2000–2005, except that the land-use source and land sink are reduced by 0.4 to reflect more current literature.

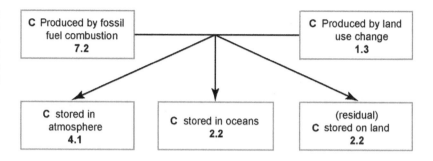

In sum, changes of climate and atmospheric composition from anthropogenic additions of greenhouse gases and aerosols can modify natural sources and sinks for these terms. A recent review (Arneth et al., 2010) suggests that such terrestrial biogeochemical feedbacks could collectively either increase by as much as double or reduce by as much as half the future warming from what it would be in their absence. Such large effects may be unlikely, but are not impossible. Clearly, a better understanding of these feedbacks is needed. The next section addresses the various contributions to future changes in C and N products in the terrestrial system. Subsequent sections review the current understanding of the physical, chemical, and biological processes involved, and the status of their modelling as components of Earth system climate models. Other terrestrial nutrients, such as phosphorus, may also affect future C, but no such role has yet been identified.

11.2. CLIMATE SYSTEM FEEDBACKS

Feedbacks are elements of processes that act to amplify (positive) or reduce (negative) climate change, or contributions to it; in particular in this chapter, C stored in the terrestrial system. Much of the large uncertainty as to details and magnitude of climate change is a result of poorly, or at least inadequately, understood feedbacks (e.g., Harvey, 2012, this volume). As new processes are recognized and added to Earth system models, additional poorly understood feedbacks are identified. "Many active biogeochemical feedback systems exhibit highly non-linear behavior", (Nobre et al., 2010, p. 1390), making lack of their understanding even more problematic because of implied 'tipping points'. Dickinson et al. (2002) discuss aspects of the biophysical feedback of the terrestrial system on climate, as illustrated in Figure 11.2.

11.2.1. Carbon

The terrestrial system extracts C from the atmosphere and stores as much in vegetation as is contained in the atmosphere and about four times as much in soils or peat. Half of

this storage is in the boreal zone (e.g., McGuire et al., 2009, 2010). The stored C is returned to the atmosphere, primarily in the form of CO_2. This return is a consequence of plant and soil (i.e., micro-organism) respiration and also combustion (e.g., forest fires). The processes determining the storage and loss of C are local and heterogeneous but global change of land C, ΔC_L, has been found through modelling to correlate with the global change of atmospheric CO_2, denoted C_A, and global average near-surface land temperature, ΔT_L, such that:

$$\Delta C_L = \beta_C \, \Delta C_A + \gamma_T \, \Delta T_L \qquad (11.1)$$

(Friedlingstein et al., 2006). The subscripts C and T on the coefficients indicate the variable of concern.

The coefficient β_C is generally positive. It is a result of the dependence of plant growth on CO_2, the growth rate of atmospheric CO_2, and the residence time of soil C (Dickinson, 2011). The coefficient γ_T has commonly been found to be negative as controlled by increases of soil and plant respiration that occur with higher temperatures. The factors in Equation (11.1), as estimated by models, change with time and averaging period, and depend on the processes that are included in the model. Positive changes in land C largely translate to negative changes in atmospheric C.

The first term on the right-hand side of Equation (11.1) describes how CO_2 acts as a fertilizer to increase plant growth, and hence to remove C from the atmosphere, and so provides a negative feedback on global warming. The second term describes how much atmospheric warming accelerates C loss processes, hence increasing the atmospheric CO_2, providing a positive feedback on the warming. Bonan and Levis (2010) illustrate issues being examined in the current literature by using the Community Land Model Version 4 (CLM4) to quantify, for that model, various contributions to Equation (11.1). They used historical meteorological data from 1850 to 2004 to prescribe observed increases of CO_2, N deposition, and land-use change. Simulations up to 1973 were used as a control and the changes in C_L from 1973 to 2004 (32 years) were examined. They analysed β_C and γ_T averaged over the

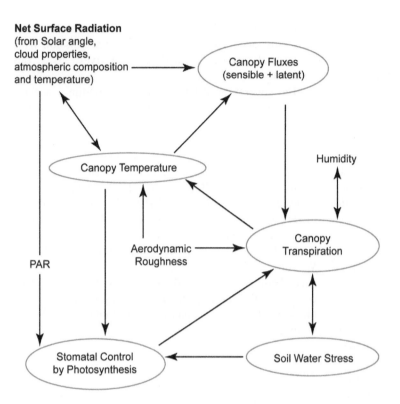

Net Surface Radiation
(from Solar angle,
cloud properties,
atmospheric composition
and temperature)

FIGURE 11.2 Schematic of how canopy photosynthetic capacity feeds back on canopy temperature and water fluxes. PAR is photosynthetically active radiation. *(Source: Dickinson et al., 2002. © American Meteorological Society. Reprinted with permission.)*

32-year period and the contributions from CO_2 change, climate change, N deposition, and land-cover change. The feedback terms were quantified both without and with inclusion of terrestrial N cycling. Inclusion of N in their model greatly reduced the feedback from CO_2 fertilization by requiring plants to assimilate N, as well as CO_2, for their growth. With temperature increase, soil C was lost more rapidly, but also more N was released, promoting plant growth, and so compensating the soil C loss.

Bonan and Levis (2010) found over the 1973–2004 test period in their model that β_C, including N processes, was only 27% of that with only C cycling. Similar results were reported earlier in the papers of Sokolov et al. (2008) and Thornton et al. (2009), that is, they all simulated large reductions in β_C. These last two papers also actually reported a change of the sign of γ_T, that is, the modelled release of N from organic pools to plant-available inorganic pools more than compensated for their temperature-driven C release to the atmosphere, so their inclusion of N had, overall, only a small impact on atmospheric CO_2. Interestingly enough, over their prior 1850–1972 initial period, Bonan and Levis (2010) found that ΔC_L with N was 79% of that without N, suggesting either that the impact of N on β_C increased with time or that it became larger for a smaller averaging time or both. Their reduction of land C from the model simulation with historical temperature increase was only 26% of that gained from C fertilization.

These recent studies can be compared with the C4MIP (Coupled Climate Carbon Cycle Model Intercomparison Project) C only results reported earlier by Friedlingstein et al. (2006), who compared seven coupled models simulating CO_2 and climate change from 1850 to 2100. These earlier models largely agree with the C-only versions of later models as to the magnitude of β_C. However, they reported large impacts of future temperature change, indeed generally larger in magnitude than that of the CO_2 fertilization effect, such that all but one model reported reduced land uptake during the twenty-first century, this reduction increasing with time. Although the Hadley Centre model reported a γ_T of -177 PgC per °C (1850–2100), observational constraints from palaeoclimate suggest a likely range of -2 PgC to -20 PgC per °C (Frank et al., 2010; Friedlingstein and Prentice, 2010; Harrison and Bartlein, 2012, this volume). The latter estimates are comparable to those of Bonan and Levis (2010) who find γ_T to be -12 PgC per °C without N or -0.1 with N (from 1973–2004). The large increases in atmospheric CO_2 reported by 2100 in earlier papers from terrestrial losses engendered the current high priority given to study of the C cycle. The large divergence of model results (C only) may be in part a result of differences in the modelled climate change, rather than in the C cycle processes (e.g., Poulter et al., 2010). However, even for the twentieth century, when modelled climates are similar, the

C4MIP climate models give γ_T varying by a factor of 5. Cox and Jones (2008) report observational evidence in favour of the high end value of the Hadley Centre model, but such high values may now seem less likely than when first published.

Not all models have yet included N in their C cycle. However, those that do all fall at the low end or below the C4MIP models in their inferred impact of temperature change on their C cycles. The C4MIP consensus was that, in the response to warming, the C cycle would provide a substantial positive feedback to further increase atmospheric CO_2, that is, the positive feedback from γ_T would dominate over the negative feedback from β_C. The consensus of the newer models is that with a C cycle, the fertilization effect alone will still act as a negative feedback to reduce atmospheric CO_2, but that addition of N reduces this negative feedback, typically by about half, but from very little to quite a bit more, that is, it acts as a positive feedback to enhance global warming (e.g., Zaehle et al., 2010a), but the climate feedback will be small (cf. Harvey, 2012, this volume). This comparison, between older and newer models, may, however, be biased because only two of the newer models were integrated over the twenty-first century, and the positive feedback from warming in the C4MIP models increased substantially over the twenty-first century. However, the two models that were integrated into the twenty-first century (Thornton et al., 2009; Zaehle et al., 2010a) both showed up to 2100 a dominance of negative feedbacks by CO_2 fertilization, although reduced by the addition of N.

So have all models, by including N, now largely eliminated the temperature feedback term, γ_T? Apparently not, since Zaehle et al. (2010a,b) have reported very little impact of N on γ_T. Why has there been and still is such a divergence in model results? The value of γ_T depends not only on N feedbacks and how soil C is modelled to depend on temperature, but it also depends on the modelled climate change and in particular whether or not forests, especially tropical forests, are impacted by large-scale droughts, which, if occurring, can greatly increase the value of γ_T. To further answer this question, in later sections we turn to examination of the individual processes that are included in current models and assess how well these processes are understood. Firstly, we introduce other important components of biogeochemical feedbacks.

11.2.2. Methane

C discussed in Equation (11.1) refers to oxidized products. However, C can also be released from land in its chemically reduced form, CH_4, as mediated by bacteria and archaea. This release has been increased by human management, for example, through rice paddies and animal husbandry. However, it also can increase as part of natural feedback processes acting on its emission from boreal and tropical wetlands (e.g., Harvey, 2012, this volume).

The generation of CH_4 from soil C within a wetland system depends on both temperature and moisture conditions. Its release also depends on how much is lost by oxidation to CO_2 within the soil, although this term is usually built into the parameterizations for CH_4 release. Its dependence on N cycling is not yet well-understood. Arctic lands (e.g., Baird et al., 2009; McGuire et al., 2009, 2010; Petrescu et al., 2010; Qian et al., 2009) are modelled in future climate scenarios to be warmer and to receive more precipitation, both of which may lead to more CH_4 release, and hence to act as positive feedbacks. Warming may also release CH_4 stored as clathrates in permafrost under land and shallow oceans. A large CH_4 release could abruptly enhance warming but its possibility is difficult to quantify (see Lenton, 2012, this volume).

11.2.3. Aerosols

Aerosols act to absorb and reflect solar radiation, either directly or through their action as cloud condensation nuclei. Absorption is largely determined by black C from fossil fuel and biomass burning, which may have contributed $\sim 70\%$ of the observed warming in the Arctic (Shindell and Faluvegi, 2009).

A large fraction (perhaps half) of global submicron atmospheric aerosols are produced by biogenic emissions of volatile organic compounds (VOC), which are predominantly isoprene. About as much isoprene is emitted as CH_4 and this emission is closely linked to the C metabolism of leaves (Sharkey et al., 2008). Emissions of isoprene from leaves increase with warmer temperatures or higher levels of light. Hence, these emissions may increase with global warming, and so act as a feedback on it (Tsigaridis and Kanakidou, 2007; Heald et al., 2008; Carslaw et al., 2010): negative through this generation of aerosols but positive through their impacts on tropospheric ozone and CH_4 lifetimes. Changing levels of CO_2 and their impacts on leaf stomata also have consequences for isoprene emission. Climate change may indirectly add to aerosols through its enhancement of forest fires.

11.3. BIOGEOCHEMICAL PROCESSES

11.3.1. Leaf Carbon

The assimilation of C by leaves and its loss by soil decomposition are key processes in the determination of terrestrial C storage. Atmospheric CO_2 enters leaves

(a) Overview of Stomatal Functioning

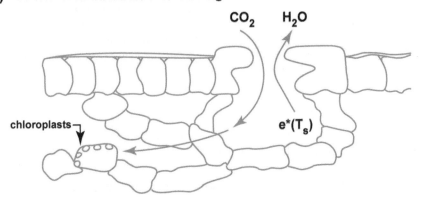

(b) Chloroplast Functioning

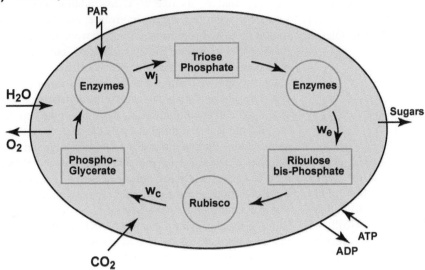

FIGURE 11.3 The currently standard leaf-level model for climate model transpiration and C assimilation: (a) illustrates the transport of water and carbon dioxide (CO_2) through the stomatal opening and the further diffusion of CO_2 through cell walls into photosynthesis sites in the chloroplast; (b) shows, as a schematic blow-up, the modelled (greatly abbreviated) Calvin cycle of photosynthesis. The first step, frequently rate-limiting, joins CO_2 to the Ribulose (5-C sugar) phosphate and is proportional to the concentrations of the Rubisco enzyme. The second series of steps, using photon-derived energy, adds hydrogen from water, and the third series of steps exports sugars and reconstitutes the ribulose phosphate starting point. *(Source: Dickinson et al., 2002, Nitrogen Controls on Climae Model Evapotranspiration, J. Climate, 15, 278–295. doi:10.1175/1520-0442. © American Meteorological Society. Reprinted with permission.)*

through their stomates, accompanied by transpiration, that is, the loss of water from leaf interiors. Most climate models now use a parameterization of these processes based on Farquhar et al. (1980) and reviewed by von Caemmerer (2000) and von Caemmerer et al. (2009). Friend et al. (2009) provide a recent comprehensive review of these parameterizations in global models. These formulations describe light-limited and enzyme-limited rates of C assimilation, which have been expressed in several related forms. Plants are distinguished as C_3 or C_4 depending on their metabolic pathways for photosynthesis, that is, the first step for C_3 plants is to produce a 3-C molecule, while C_4 plants produce a 4-C molecule. The latter step represents an additional metabolic pathway in C_4 plants that acts to concentrate the CO_2 before it enters the photosynthetic pathways. Figure 11.3 illustrates the C_3 plant C assimilation (Dickinson et al., 2002). Following

Oleson et al. (2010), based on Collatz et al. (1991, 1992), the light limited rate, w_j, is written:

$$w_j = (C_i - \Gamma_*)\, 4.6\, \phi\, \alpha/(C_i + 2\Gamma_*) \quad \text{for C}_3 \text{ plants,}$$

(11.2)

and

$$w_j = 4.6\, \phi\, \alpha \quad \text{for C}_4 \text{ plants,} \quad (11.3)$$

where C_i is the CO_2 partial pressure in the leaf's interior airspace, ϕ is the absorbed photosynthetically active radiation (i.e., PAR in W m^{-2}) converted to a photon flux by the 4.6, α is the quantum efficiency, that is, moles CO_2 per mole photons and not necessarily the same for Equation (11.3) as for Equation (11.2), and Γ_* (P_a) is the 'CO_2 photo-compensation point', that is, the C_i value at which C assimilation is balanced by C loss by 'photorespiration'.

The first enzyme limited rate is:

$$w_c = (C_i - \Gamma_*) \, V_{cmax}/[Ci + K_c \, (1 + O_i/K_o)] \tag{11.4}$$

for C_3 plants,

or

$$w_c = V_{cmax} \quad \text{for } C_4 \text{ plants}, \tag{11.5}$$

where V_{cmax} is the limiting rate for large CO_2 and depends primarily on the activity in the leaf of the enzyme Rubisco. Equation (11.5) is the limit of Equation (11.4) as C_i becomes large.

Sharkey (1985) found that Equations (11.2) through (11.5) did not fit data at high light, low temperature, and high CO_2 levels and added a third limiting factor, w_e, as proportional to the maximum rate, TPU, at which triose-phosphate is utilized to regenerate the Rubisco. Although introduced as a separate parameter, Collatz et al. (1991) assumed TPU would scale with V_{cmax}, and so in Oleson et al. (2010):

$$w_e = 0.5 \, V_{cmax} \quad \text{for } C_3 \text{ plants}, \tag{11.6}$$

and

$$w_e = 4000 \, V_{cmax}(C_i/P_{atm}) \quad \text{for } C_4 \text{ plants}, \tag{11.7}$$

where P_{atm} is atmospheric pressure.

The light-limited rates, Equations (11.2) and (11.3), have been simplified from the representation developed by Farquhar et al. (1980) by neglecting limitation by the maximum electron transport rate, J_{max}. This term could be included simply as giving another limiting rate (Kull and Kruijt, 1999). Bonan et al. (2011) include an electron transport rate limitation and parameterize w_e for C_3 plants by approximately 0.18 J_{max}, that is, approximately 0.35 V_{cmax}, using $J_{max} = 1.97 \, V_{cmax}$ at 25°C making w_e occur more frequently as the limiting rate than if Equation (11.6) is used. Another approximation made in Equations (11.2) and (11.4) is the assumption that the CO_2 concentration is the same in the leaf airspace as at the site of C assimilation. However, interpretations of leaf or canopy level assimilation data may require introduction of a mesophyll conductance comparable in effect to that of stomatal conductance (e.g., Harley et al., 1992; Keenan et al., 2010; Gu et al., 2010; von Caemmerer and Evans, 2010), but not yet included in climate model formulations of C assimilation. Gu et al. (2010) have developed a fitting method for deriving the model parameters of leaf C assimilation including TPU and mesophyll conductance from data and review the deficiencies of earlier approaches.

Leaves also lose C by respiration. Especially during daytime, respiration is closely coupled to the physiological processes represented in Equations (11.2) through (11.7). Daytime respiration is sometimes either neglected or set to nighttime respiration, but process modelling (e.g., Buckley and Adams, 2011), in agreement with observations, indicates that daytime leaf respiration can range from 10% to 75% of nighttime, depending, among other things, on whether N is being assimilated in the form of nitrate or ammonium ions. Although daytime C loss by respiration is less than at night, biosynthesis in leaves may still be greater in daytime than at night by use of energy derived from photons absorbed in photosynthesis that are presumably in surplus when the C assimilation is not light-limited (i.e., is light-saturated).

11.3.2. Down-regulation of Leaf Photosynthetic Capacity

The parameter V_{cmax} determines a leaf's photosynthetic capacity, that is, its rate of C assimilation under light-saturated conditions and for a given atmospheric concentration of CO_2. From a modelling viewpoint, V_{cmax} is the most crucial parameter for C assimilation, but it varies not only between species but also between leaves on a single plant, and seasonally for a single leaf. This variability is, in principle, not random but a result of acclimatization to leaf environmental conditions. V_{cmax} depends on the activity of the enzyme Rubisco that catalyses the first step of leaf C assimilation. Because this step is slow, a large amount of Rubisco is needed. Enzymes are relatively unstable and have to be frequently resynthesized with a C cost. Hence, Rubisco is both necessary for the leaf to assimilate C and is a major contributor to the loss of C by leaf respiration. Rubisco, as all enzymes, requires N for its construction, and indeed it accounts for ~25% of the N contained in a leaf. Thus, it is plausible that, as CO_2 increases, leaves adjust downward their Rubisco, hence V_{cmax} and N concentrations, to optimize the C they assimilate. Down-regulation refers to such a decrease of V_{cmax} with increasing atmospheric CO_2.

The optimum level of N also depends on light level and other factors, such as the temperature dependence of V_{cmax} (such enzyme-related processes work faster at higher temperatures) and it is plausible that leaves have mechanisms to adjust their N levels to such an optimum. Such adjustment is most easily recognized in the decrease of leaf N deeper in canopies where light levels are lower (e.g., Field, 1983) and is also seen in seasonal variations of leaf N. Various authors (e.g., Farquhar, 1989; Dewar, 1996; Haxeltine and Prentice, 1996; Franklin, 2007) have developed mathematical models of this optimization. It also can be a consequence of the mechanisms of a dynamic model, for example, that of Dickinson et al. (2002).

Another control on leaf N is its uptake from soil, whose modelling is discussed further in the next subsection. Most biogeochemical models assume an empirically fixed ratio of C to N or a range of this ratio in their various pools and so do not invoke any principles of optimization. Observed increases of the C to N ratio with increased CO_2 have been interpreted in terms of either optimization or in terms of N-limitation. Whether or not plants are N-limited can be established by seeing if, with additional N, they can assimilate more C (e.g., Chapin et al., 1986).

Besides optimization and N-limitation, down-regulation has also been used to refer to the action of the term w_e in Equations (11.6) and (11.7) (i.e., by Maayar et al., 2006), which has more impact on C assimilation at colder temperatures or higher levels of CO_2. One C4MIP model that did not include this term was found to have twice as strong a CO_2 fertilization effect, that is, the value of β_C in Equation (11.1) was twice as large as that of the other models. Thus, this yet relatively poorly characterized triose-phosphate export term (Harley and Sharkey, 1991) appears to be important for the description of future terrestrial removal of C from the atmosphere.

Equations (11.2) through (11.7) show that leaf C assimilation can be limited either by light or by enzymes, that is, by the value of V_{cmax}, and has factors in Equations (11.2) through (11.4) that increase with CO_2 within the leaf. Increases of atmospheric CO_2 increase internal leaf CO_2, and so increase leaf C assimilation when it is Rubisco- or light-limited, commonly even after down-regulation. Transpiration depends on stomatal conductance that is proportional to the product of C assimilation and divided by leaf CO_2. It also depends on humidity, a dependence first estimated observationally by Ball et al. (1987), and whose current theoretical understanding is reviewed in Medlyn et al. (2011). The ratio of CO_2 internal to the leaf to its external value varies with humidity-mediated stomatal closure. With drier air, C_i has to adjust downward to increase the C gradient to compensate for the smaller stomatal conductance. Typically, C_i for C_3 plants is in the range 0.5 to 0.8. The dependence of C assimilation on CO_2 in Equations (11.2) through (11.4) is weaker than linear, so increasing CO_2 reduces transpiration with implications for climate and vegetation growth, that is, less evaporative cooling will occur in moist conditions and plants will be less sensitive to drought. Down-regulation can further decrease transpiration.

To illustrate possible consequences of down-regulation, Bounoua et al. (2010) explored, in a global model with climate forced by a doubling of CO_2, the consequences of several arbitrary, but simple, assumptions: (i) V_{cmax} remained unchanged; (ii) down-regulation occurred such that photosynthesis was consistent with the $1 \times CO_2$ simulation; and (iii) the same as (ii) but with enhanced leaf area such that the $V_{cmax} \times$ leaf area was restored to that of (a). Thus, (c) asks the question as to what would be different if leaves were to use twice as much area to do the same C assimilation. In cases (i) and (iii), C uptake was increased by 35%, but in (ii), not surprisingly, by only

11%. Bounoua et al.'s (2010) most striking result was a large reduction in global warming for their case (iii) compared to (i), that is, over all land by 0.6°C and eastern USA by 1.5°C, the latter more than half the warming for the $2 \times CO_2$ in case (i). They attribute this cooling effect for case (iii) relative to (i) to increased evapotranspiration from an increase in precipitation (from their table, about 15% more over the eastern USA for (iii) than (i)). In sum, the paper of Bounoua et al. (2010), and quite a few earlier studies that adjusted stomatal conductance to mimic the physiological impact of increased CO_2, have highlighted the potential importance of various mechanisms, such as reducing evapotranspiration from stomatal closure by which vegetation can respond to increased CO_2, and so enhance the warming or, in the study of Bounoua et al. (2010) with additional assumptions, reduce the warming. Their quantitative results about the stomatal impacts on global warming may be questionable and require re-examination.

11.3.3. Soil Metabolism

When plant elements are harvested or die, their detritus is used for food by other organisms, mostly by soil micro-organisms, that is, bacteria and fungi. These organisms oxidize the C of plant residues for the energy required for their growth and maintenance, and so release it as CO_2. They also consume other plant minerals such as N but, since these are largely conserved rather than lost to the atmosphere as is C, they eventually become available in the soil for plant nutrition. The decomposition is largely done by release of enzymes into the soil that convert complex organic molecules into simple ones, that is, monomers.

To describe these processes through modelling, the detritus and soil C and N are partitioned into pools that can transfer or exchange C and N or lose it to the atmosphere or to runoff. A key question, then, is the dependence of these exchange rates on climate change, for example, on temperature or humidity conditions.

11.3.4. Nitrogen Cycling and Feedbacks on Carbon

N exists in plants and micro-organisms in reduced form and is first converted to monomers and then 'mineralized' to ammonium ions (NH_4^+) in the soil. 'Nitrifying' bacteria can oxidize these ions to nitrate ions (NO_3^-) and plants can extract either form of N from the soil for their N requirements. Plants are nutrient-limited if N or some other nutrient cannot be supplied from the soil rapidly enough for their optimum growth. At low enough levels of mineralized N, plants and microbes compete, sometimes in symbiosis, for the organic N monomers, primarily amino acids

FIGURE 11.4 Schematic of the model nitrogen (N) reservoirs, internal fluxes between plant and soil, and external exchanges with the atmosphere and to runoff. *(Source: Dickinson et al., 2002, Nitrogen Controls on Climae Model Evapotranspiration, J. Climate, 15, 278–295. doi:10.1175/1520-0442. © American Meteorological Society. Reprinted with permission.)*

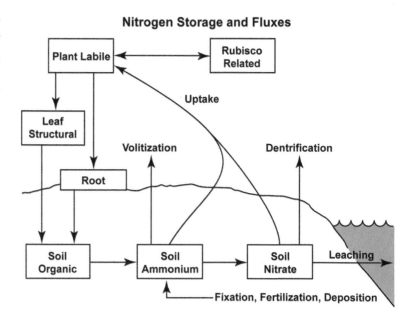

(e.g., Chapin, 1995; Schimel and Bennett, 2004; McFarland et al., 2010; Weedon et al., 2011). The concentrations of N in the soil dissolved in the forms that plants can take up (i.e., monomers or mineral N) depend on processes supplying it or removing it. It is supplied, as already mentioned, by microbiological decomposition of organic matter within, or in contact with, the soil. It is also supplied by N-fixing bacteria and fungi, as discussed in the next section, and by atmospheric deposition or anthropogenic fertilization. It is lost by plant uptake, or in the NO_3^- form by denitrifying bacteria, which uses its oxygen and converts it to N_2 or N_2O. It is also lost in runoff, largely in the more soluble NO_3^- form. Because all of these processes are

complex and local, it is challenging to treat them quantitatively with any degree of certainty in an Earth system model (see Rice and Henderson-Sellers, 2012, this volume). Figure 11.4 illustrates some of the processes involved in terrestrial N cycling (see also Dickinson et al., 2002).

Figure 11.5 illustrates the old and new paradigms for uptake of N by plants. The latter emphasizes the possibility of uptake of organic N, a pathway that can be dominant in high-latitude systems, but is not yet included in most global models.

Much of a plant's N is stored in its leaves, which are commonly renewed on a near annual basis, but whose

FIGURE 11.5 The changing paradigm of the soil N cycle. (a) The dominant paradigm of N cycling up through the middle 1990s. (b) The paradigm as it developed in late 1992. *(Source: Reproduced with permission of Ecological Society of America, from Ecology, "Nitrogen mineralization: Challenges of a changing paradigm", Schimel, J. P. and J. Bennett, 85, 2004; permission conveyed through © Copyright Clearance Center, Inc.)*

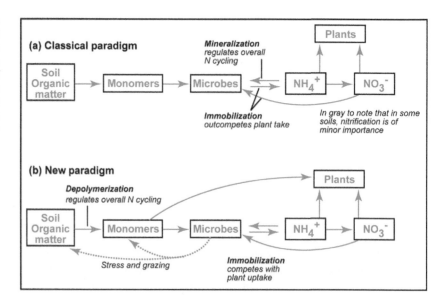

lifetimes can range from 3 months to 10 years. Thus, terrestrial N cycles between soil and plants on approximately an annual timescale. As a closed loop, this cycling should maintain a near constant level of leaf N. With increasing atmospheric CO_2, C_3 plants should increase their C assimilation according to Equations (11.2) through (11.4). However, construction of more plant tissues, especially enzymes, requires additional uptake of N from the soil. Without such an additional supply of N, any additional C assimilated cannot be invested in 'machinery' for more leaf photosynthesis. Thus the enhancement of C storage should be less if such an N limitation occurs, as found in observational studies and recent models to be further discussed later. Another control on N uptake from the soil is the uptake of water in which the soil N is dissolved. Taub and Wang (2008) have suggested that the stomatal closure from increased CO_2, and, hence, the reduced evapotranspiration, and so passive uptake of N from soil water, could be the major cause of reduced N in leaves when CO_2 has increased. Although this feedback can be included in models (e.g., Dickinson et al., 2002), it may only act when the soil is well-supplied with N, and so is weaker than other feedbacks on leaf N, as discussed later.

The limitation of plant growth by lack of N depends on the development of root capacity to extract N from the soil and on the soil-N-dependent rates at which other processes besides plant-decay provide N to soil and remove it. These processes vary in time and between ecosystems. Fires can burn grasslands every few years and, in doing so, volatilize the vegetative N into gaseous N compounds, or leave ashes with nitrate ions that are leached rapidly. Legumes are common in grassland and in a few years restore much of this lost N. Mature boreal forests, on the other hand, have near-closed N cycling, since soil N is almost entirely in organic form or in ammonium ions with only very slow, if any, losses by leaching, nitrification, or volatilization. Tropical forests have many N-fixing trees and may only be N-limited where these are absent. Where fixation or deposition of N from the atmosphere is large enough, or adjustment times are long enough, N limitations and their impacts on the rates of plant growth may be small, but other nutrients may also be limiting.

11.3.5. Nitrogen Fixation

N fixation is the primary biological mechanism for restoring N losses in natural systems. It occurs only in certain C_3 plant species, such as legumes, through a symbiotic relationship to various bacteria, that is, the bacteria fix the N_2 N from air, using photosynthate supplied by the plant as root exudates (root secretions) for energy. It also occurs in some free-living aerobic and anaerobic bacteria. Most of the roughly 20,000 legume species form a relationship with nodule-inducing bacteria, collectively

known as rhizobia (Rogers et al., 2009). As a gardener, I am challenged to grow blue bonnets (a legume and the Texas state flower) in part because of their requirement for rhizobia, which may or may not be in soil in their absence. These rhizobia and other N-fixing bacteria use the enzyme complex nitrogenase to cleave atmospheric N_2 and return it to the plant as NH_4^+. Gutschick (1982) has examined the energy requirements of this process and estimates it takes about 9 g C per g of N to fix the N.

As an enzyme, nitrogenase imposes a respiratory maintenance cost (i.e., cost of turnover) but Gutschick (1982) estimates that this maintenance cost is a relatively small fraction of the total C cost of fixation, which for a legume may be up to 20%−30% of its assimilated C. Consequently, legumes must maintain high levels of C assimilation to fuel this fixation, as can be done by plants adapted to direct Sun. Shade plants apparently cannot afford the C costs of fixation. Thus, N-fixing forbs, that is, wild flower plants such as blue bonnets, grow best in grasslands where they can receive high levels of sunshine. N fixation, depending on enzymes and, in particular, nitrogenase, should quicken with higher temperatures (e.g., Dickinson et al., 2002), thus promoting the growth of legumes in tropical forests relative to boreal forests. Observations show that N fixing shuts down where the N-fixers are not N-limited (Barron et al., 2011).

11.4. OBSERVATIONAL CONSTRAINTS

11.4.1. General Considerations of Rates and Timescales

C and N are exchanged between reservoirs on timescales from minutes to centuries or longer. Different processes determine these exchanges at these different timescales, depending on local environmental variables related to climate, for example, temperature, humidity, soil pH, CO_2 levels in the atmosphere, and oxygen levels in the soil. Figures 11.6 and 11.7 illustrate the process involved in C and N cycling and their coupling.

A common observational target is the correlation between fluctuations in temperature, T, and rates of transfer of C or N between their different reservoirs. The reasoning behind such correlation is often the Arrhenius relationship, which states that such rates should be proportional to $\exp(-E_o/RT)$ where E_o is some chemical energy barrier. Alternatively, it is sometimes assumed that the dependence on T is exponential, as expressed by a Q_{10} parameter, that is, a factor giving the fractional increase of a rate for each $10°C$ increase of temperature. For small changes of T, $\exp(-a/T)$ and $\exp(bT)$ both linearize to $1 + cT$, where $c = b$ or a/T^2, and so both formulations are equivalent.

However, for temperature changes of more than a few degrees, the temperature dependence is likely to be

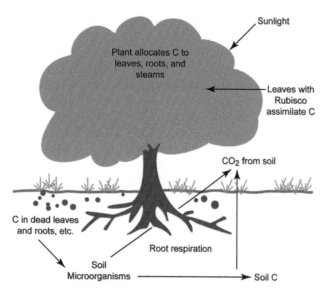

FIGURE 11.6 Simplified version of the terrestrial C cycle. Respiration from leaves and stems is lumped with leaf assimilation and terrestrial losses from fires are not included.

non-linear and perhaps more likely to conform to the Arrhenius expression since it corresponds to a simple physical model. The E_o in this expression is interpreted to be an energy barrier or activation energy that thermal fluctuations must overcome to initiate the transformation,

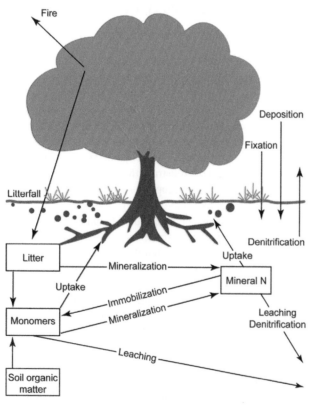

FIGURE 11.7 Terrestrial N cycle, emphasizing soil processes.

that is, it becomes larger for slower reactions implying a stronger dependence on temperature (e.g., Craine et al. (2010) report data on the rate and temperature sensitivity of organic matter decay consistent with this expectation). Both forms of temperature dependence have been incorporated into Earth system climate models, depending on what observational correlation was employed. Another common formulation is an increase to an 'optimum' temperature followed by a decrease at higher temperature (e.g., June et al., 2004).

Observationalists commonly use 2-m air temperature as most available whereas leaf or soil temperatures might be more appropriate. For example, Mahecha et al. (2010b) derive from night-to-night fluctuations in CO_2 emission from FLUXNET stations a Q_{10} for respiration that is about 1.4, not the 2 or larger as usually assumed. What they do not appear to address is the temperatures of the C reservoirs, for example, the temperature a few cm into the soil that will have smaller variations than the air temperature, or how much of the C loss is from processes that may not depend directly on temperature such as plant growth. In any case, theirs and other such studies of short timescale C fluctuations are looking largely at the reservoirs with fast decay rates that are replenished annually or more frequently, and so provide data for model validation primarily at these short timescales. As stated by Luo et al. (2011, pp. 843–844), "the models used for long-term assessments are typically built upon knowledge of ecosystem processes and parameterized by short timescale data. However, ecosystem responses to global change are strongly regulated by long-term slow processes…longer term dynamics are less understood and more difficult to predict".

Measurements by plant physiologists, as used for the parameters of Equations (11.2) through (11.7) (cf. von Caemmerer, 2000), are usually expressed with the Arrhenius temperature dependence. Empirical models of the microbiological decay of plant or soil C, on the other hand, commonly use a Q_{10} dependence. There is no reason to expect such simple exponential decay. However, it is known mathematically that the decay of any substance with time can be approximated by a sum of exponentials, provided the decay is linearly proportional to the amount of decaying substance. Materials undergoing microbiological decay are usually assumed to consist of distinct substrates, for example, labile versus recalcitrant C, with discrete rates of exponential decay, but available data could equally well support a continuum of timescales, for example, as modelled by Feng (2009).

11.4.2. Dependence of Carbon Assimilation on CO_2 and N at Leaf Level

How does plant growth and other plant properties depend on atmospheric CO_2 concentrations and leaf N? This question can be addressed observationally at various levels of ecosystem organisation, from cells of leaves to landscapes or watersheds. Leaf-level C assimilation measurements are commonly conducted using commercially available leaf chambers. These control the light level, CO_2 concentration, and temperature of the air in the chamber but outside the leaf, and measure leaf CO_2 uptake and water loss; from the latter, the leaf stomatal resistance is obtained from which C_i (i.e., leaf internal CO_2) is determined (e.g., von Caemmerer and Farquhar, 1981, appendix 2).

Measurements of these C assimilations, that is, A versus C_i, provide a basis for estimation of the parameters of Equations (11.2) through (11.7), for example, as analysed for 109 species by Wullschleger (1993). He especially noted that annual plants assimilated C at about double the rate of perennials or trees. Table 11.1 shows some values for V_{cmax} and the rate of triose-phosphate utilization (TPU). The latter is multiplied by 6 to examine the accuracy of Equation (11.6) that equates 3 TPU to 0.5 V_{cmax}. Table 11.1 suggests that the factor 0.5 in Equation (11.6) is quite uncertain but may be an upper limit as already mentioned. Bonan et al. (2011) have suggested a new parameterization that somewhat lowers the TPU rate, that is, the 0.5 is reduced to about 0.3. Note that making w_e smaller makes it more frequently limiting; it has the opposite effect to neglecting it.

Evans and Poorter (2001) looked at how leaf area and N change with light levels to maximize C gain, growing the plants at 200 μmol m^{-2} s^{-1} and 1000 μmol m^{-2} s^{-1} of photons. At light saturation, assimilation (i.e., V_{cmax}) was twice as large for the leaves grown at high light levels, but per dry mass stayed the same because the leaves with more light had double the mass per unit area. N per unit mass also stayed the same, but the low-light leaves used a larger fraction of their N for light harvesting. Eichelmann et al. (2005) examined in more detail changes of leaf N between sunlit and shaded leaves. Such studies illustrate the dynamic adjustments made by leaf enzymes to environmental conditions.

11.4.3. Leaf-Level Response to Drought

Leaves can suffer water deficits as imposed by droughts, which affects C assimilation. Chaves (1991) reviewed laboratory studies on this question and conclude that C assimilation could be greatly reduced without much change of a leaf's photosynthetic capacity. In other words, the reduction was mostly or entirely from stomatal closure. More recently, Wilson et al. (2000) examined this question over two years for the light-saturated rates, that is, looked at the V_{cmax}, of five mature tree species and reached a similar conclusion. They also looked at the decrease of assimilation from leaf ageing, and found in this case that reduced C assimilation was mostly a result of reduction of V_{cmax}. Drought limitations vary between species. For example, Taylor et al. (2011) examined how C_3 versus C_4 grasses differed in the limitation of their C assimilation by drought. They found C_4 grasses to be more sensitive to drought, which may seem odd since they grow in drier climates. Presumably, it is their ability to be much more productive than C_3 plants at higher light levels and temperatures that has led to their preference for drier climates.

11.4.4. Temperature Dependence of Carbon Assimilation

As mentioned previously, C_3 and C_4 plants differ in their details of C assimilation. A consequence is that C_4 plants become more efficient than C_3 plants at higher temperatures and lower concentrations of CO_2 (e.g., Ehleringer et al., 1997). Consequently, C_4 plants are largely tropical or semitropical in location.

The individual terms in the leaf C assimilation model, Equations (11.2) through (11.7), have rapid temperature dependences (e.g., the rates at which enzymes catalyse reactions) that are established by laboratory studies. They also may 'acclimatize', that is, adjust to different temperatures on a seasonal or longer timescale, for example, by adjustments of their enzyme concentrations to partially compensate for this fast temperature dependence of individual enzymes (e.g., Bunce, 2008; Gunderson et al., 2010). In colder situations, increased temperatures promote plant growth (e.g., Wu et al., 2011).

With high and increasing temperatures, photosynthetic capacities are reduced, changes that can be reversible, but

TABLE 11.1 Some Values of V_{cmax} and TPU as Analysed by Wallschleger (1993) from a Variety of Plant Species. TPU is Multiplied by 6 to Examine its Resemblance to V_{cmax}

Species	V_{cmax}	6TPU
Arbutus unedo (strawberry tree)	40	38
Chenopodium album (average of 2)	125	68
Soybean	102	48
Tomato (average of 2)	72	90
Bean	57	55
Western Cottonwood	111	48
Quaking Aspen	24	28
Tabebuia rosea (tropical flowering tree)	45	29
Cocklebur weed	144	60

eventually, at high enough temperatures, become irreversible. Lloyd and Farquhar (2008) have reviewed observations and modelled the likely effects of increased temperatures and more CO_2 on the leaves of tropical trees. Irreversible damage begins for $T \geq 45°C$ (Hüve et al., 2011), whereas currently tropical forests may experience leaf temperatures no more than 35°C. Thus, their analysis suggest that global warming could have to be an increase of at least 10°C before it would destroy tropical forests by temperature increase alone. However, leaves, with their low albedos, may, around midday, become substantially warmer than their surrounding air, depending on their rate of evapotranspiration and micrometeorological conditions, such as local wind. Lloyd and Farquhar (2008) indicate that other weaker deleterious effects, such as increased respiration may be balanced by the benefits of increased CO_2 and some acclimatizations with little net effect on the forests. A contrasting conclusion is drawn by Galbraith et al. (2010), who forced three dynamic vegetation models with the climate change determined by the Hadley Centre model; that is, they added the change as a perturbation on present-day climate. They found a large negative effect from temperature change, for example, 50%–80% loss of C for the ICC A1FI scenario, partly compensated by CO_2 fertilization. In considering validation of their model's temperature dependences, they argued the responses were consistent with short timescale leaf observations, but did not take into account the possibility of longer timescale acclimatization.

Temperature can also affect plant uptake of N and P (Lukac et al., 2010) but studies are not advanced enough to draw general conclusions. Jarvis and Linder (2000) concluded that the observed increased growth of boreal forests with temperature was from an increase of N availability, rather than a direct effect of temperature. Studies discussed in the next section likewise indicate that temperature effects on terrestrial storage of C cannot be determined without also considering impacts on nutrient cycling.

11.4.5. Dependence of Plant Growth on CO_2 and N

Studies of how plants respond to CO_2 enrichment were initially conducted in closed plant growth chambers, but the difficulties of sorting out the effects of the chamber are considerable, unless a full system model is used to interpret the data (e.g., as proposed by Manzoni et al., 2011). Since this has yet to be done, experimental studies are now designed to perturb the plant environment as little as possible.

These studies, referred to as 'free-air CO_2 enrichment' or 'FACE', use horizontal or vertical pipes to maintain a level of CO_2 concentrations over the plants higher than ambient. The CO_2 concentrations that have been applied

range between 475 ppmv and 600 ppmv. Their interpretation has lumped together studies at different concentrations without attempting to scale them to a common concentration. Plant functional types have been distinguished as trees, shrubs, grass, legumes, crops, and forbs.

Analyses have been undertaken of the consequent changes in leaf-level C assimilation, whole-plant C, N assimilation relative to C, allocation of C to leaves versus wood and roots, and total-soil respiration. Trees and legumes show the least down-regulation in terms of reduction of V_{cmax}, that is, only about a 10% reduction, whereas grass and shrubs may show 20% (Ainsworth and Long, 2005). Consequently, the C_3 light-saturated rates, see, for example, Equations (11.4) through (11.6), increased most for trees (nearly 50%), while for grass by only 40%. The experimental protocol of FACE appears to preclude distinguishing down-regulation from reductions in V_{cmax} versus from hitting the TPU limit, and only the former has been considered. Forbs down-regulate most, and so show only a small increase in their light-saturated rate. A nine-year study of grassland species at 560 ppmv CO_2 (Crous et al., 2010) was averaged over its last four years. It found forbs to down-regulate by 30%, and so not to increase their photosynthetic performance, and hence to be negatively impacted by the addition of N. Although the three forbs included (goldenrod, thimble weed, and yarrow) are not legumes, the study hypothesized that their negative response was a result of a dependence on soil mycorrhizal symbiosis that is suppressed with additional CO_2 and N. Newton et al. (2010) report on another extended C_3 grassland FACE study (11 years) in which annual harvests were analysed for N. The N harvested progressively declined, a result interpreted in terms of N limitation.

An earlier analysis of chamber data (Wand et al., 1999) suggested that there may be a significant response of C_4 grasses to C enrichment. The limited number of FACE experiments for C_4 grasses, on the other hand, indicate that their response to CO_2 enrichment is small (Leakey et al., 2009), consistent with their mechanisms for increasing CO_2 levels within their leaves. However, their growth may still be enhanced by more CO_2 under dry conditions because of their use of less water with more CO_2. Legumes, because of their N_2-fixing rhizobia, are relatively responsive to CO_2 enrichment, but little affected by addition of N (cf. Rogers et al., 2009).

Experimental studies of the interactions between C and N with increased CO_2 are reviewed in Reich et al. (2006). Some, but not all, studies support the expectation that the response of vegetation to increased CO_2 is reduced with low N. They conclude that leaf-level studies show no effect of N and that the interaction is more likely at the system-scale and "may be manifest in the size, organisation, or turnover rates of canopies and root systems" (Reich et al., 2006, p. 615). Reported impacts of N on above-ground biomass are mixed. Some report no effect but, overall, more studies ($\sim 2/3$) show a greater biomass response to CO_2 with higher N. The leaf area may initially increase, but return to its original level by the end of a growing season. The Reich et al. (2006) review also finds little evidence for a change of the C:N ratio of litter, in contrast to the situation for live leaves. This is possible because leaves recycle (translocate) about half of their leaf N, with some flexibility in so doing for reasons hypothesized by Fisher et al. (2010). The effects of increased CO_2 on the cycling of N in the soil (i.e., the rates of mineralization and immobilization) are also found to be small. N fixation may increase with increased CO_2. Reich et al. (2006) report that this is sometimes found and that fixation generally decreases with higher N levels in the soil. The relative abundance of N_2-fixing plants has been found to increase, decrease, or not change with added CO_2. Additional effects are expected in water-limited systems, but little observational evidence yet exists.

Another review (Dieleman et al., 2010, summarized in Table 11.2) has examined how soil N affects soil C cycling with increased CO_2. Table 11.2 indicates that the increase of ecosystem productivity with more CO_2 enhances root production more than that of leaves, but that the fine root production is much reduced with the addition of N; results consistent with the concept that roots respond dynamically to nutritional deficits. The issue of fine root production was further addressed for a particular site in the study of Iversen and Norby (2008). The site, a sweet gum plantation, had been shown in a FACE system to respond to CO_2 primarily through an increased rate of fine root production. Iversen and Norby (2008) report that when another plot was fertilized, its NPP (net primary productivity) increased by 1/3 but with most of this allocated to wood production. They conclude that the plantation is N-limited and that the increase in fine root production of the FACE site is in

response to this N limitation. Another study (Finzi et al., 2006) found in the Duke forest (loblolly pine) little indication of a N limitation six years into a FACE experiment. At the same location, CO_2 enhancement was found to enhance the production of root exudates that stimulated bacteria to produce enzymes responsible for the decay of soil organic matter (Phillips et al., 2011).

Leakey et al. (2009) make several important points about the response to CO_2 increase, some already mentioned, in reporting six important lessons from FACE. These lessons are (i) the response of leaf uptake is more as enhancement than down-regulation; (ii) N use efficiency increases, that is, more C uptake per leaf N as expected from Equation (11.4); (iii) water use declines as expected from the Ball et al. (1987) model and, more importantly, for modellers, there is no evidence that the empirical coefficient in that model depends on CO_2 levels; (iv) dark respiration (i.e., at night) may increase with increased CO_2 (they report this as a controversial topic because earlier chamber studies reported a decrease, but that conclusion was demonstrated to be an experimental artefact); (v) as mentioned, stimulation of C uptake in C_4 plants primarily occurs with water stress; and (vi) the productivity of crop plants increases less than expected from earlier chamber studies.

11.4.6. A Network for Monitoring the 'Breathing' of the Terrestrial Biosphere

Flux-tower measurement systems began to be installed two decades ago and now there are more than four hundred such research sites over the globe. These sites use the eddy covariance method (i.e., the correlation between fluctuations of a tracer concentration and vertical air velocities) to establish on an hourly basis the exchanges of CO_2 and water vapour between the surface and atmosphere. Baldocchi (2008) provides a comprehensive review of the methodology, assumptions, limitations, and successes of these systems. Net ecosystem C exchange (F_N) consists of the difference between gross C assimilation (F_A) and ecosystem respiration (F_R). The last two terms are estimated from the day–night differences of F_N. Figure 11.8 shows annual measurements of all these terms at Harvard Forest, the earliest established flux-tower site. It shows that, since the measurements began, F_N increased by about 30%, but F_R more than doubled in size, implying F_A grew by even more than F_R.

The flux-towers show seasonal and inter-annual patterns of F_N. In temperate ecosystems, its seasonal pattern follows the seasonal cycle of the Sun. In cold regions, conifer forests lose C in winter and gain it during the growing season. Milder regions can take up C year-round. In semi-arid systems, C uptake occurs during the wet season. Inter-annual variability occurs for different reasons in different

TABLE 11.2 Effects as a Percentage Response to Increased CO_2 by Ecosystem Carbon Pools with and without N Fertilization

	Fertilized	Not Fertilized
Litterfall	20	12
Fine root production	19	52
Above-ground biomass	30	19
Root biomass	50	38
Microbial biomass	2	−5
Soil carbon	14	−5

(Source: Adapted from Dieleman et al., 2010.)

FIGURE 11.8 Multidecadal flux-tower measurements at Harvard Forest of net ecosystem C exchanges (F_N) broken into gross assimilation (F_A) and respiratory loss (F_R). *(Source: Baldocchi, 2008 © Copyright 2008 American Geophysical Union. Reproduced/modified by permission of American Geophysical Union.)*

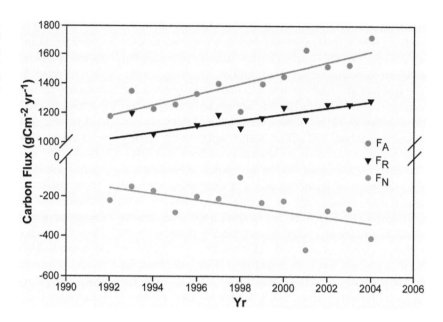

ecosystems, but generally is related to light, temperature, and moisture conditions. Light has been found to be used more efficiently by plants when it is diffuse. Data derived from the flux network has indicated that respiration rates in northern latitudes with cool temperatures are similar to the rates in more equatorward systems with warmer temperatures. Optimal temperatures for C assimilation similarly adjust.

All such findings as discussed above should be used for assessment of or incorporated into the terrestrial component of climate models. Friend et al. (2007) have reviewed the use of the flux-tower data for such development and evaluation of global C models. They find that CO_2 fertilization has been the major driver of recent increases of NPP and that the Amazon Basin largely drives inter-annual variability in global NPP. Mahecha et al. (2010a) also review the use of flux-tower data for evaluating process-based models. They emphasize the importance of comparisons on multiple timescales and how data can be analysed to show variability on different timescales. Schwalm et al. (2010) examined the response to drought using flux-tower data. They found that globally, F_A decreased twice as much as F_R and that croplands had the largest decline in productivity. They also found that, in evergreen forests and wetlands, the productivity could increase.

11.4.7. Atmosphere Concentration as a Global Constraint on Terrestrial Sources and Sinks

Modern high-precision measurements of atmospheric CO_2 by collection in flasks began more than half a century ago and have evolved into a global flask

network. Fung et al. (1983) showed how these data could be used with the winds from a climate model to infer the seasonal exchange of CO_2 with the terrestrial biosphere. This approach, referred to as 'top-down', drew initial attention to the large land-sink for C, whose process details we have addressed in this chapter. This sink is the difference between the terrestrial gain of C from CO_2 fertilization and other imbalances, and the loss from deforestation and other land use changes considered in Section 11.6. Because of the relatively small number of observations, inversion of the flask data provides at best continental-scale resolution (e.g., Le Quéré et al., 2009). CO_2-measuring instruments on orbiting satellites may eventually provide much more detail of atmospheric CO_2 variability, as needed to invert detailed distributions of anthropogenic sources and terrestrial sources and sinks. Other satellite data that provide information on week-to-week variations of terrestrial greenness and leaf areas, could also constrain the natural terrestrial sources and sinks. Such detail could determine shifting patterns of terrestrial C sources and sinks as required to verify efforts to control atmospheric CO_2.

Ideally, such satellite data would be merged with the flask network and flux-tower network data through use of global climate models and data assimilation technology. Randerson et al. (2009) show how currently available global datasets can be used to evaluate the C cycle of climate models. A global climate model should include at some level all the processes described in this chapter, as well as sources from fossil fuel use and land-cover change. The data assimilation would use statistically optimal methods to incorporate observations to improve estimates from such a model (e.g., Knorr et al., 2010; Rayner, 2010).

11.5. MODELLING NITROGEN−CARBON INTERACTIONS

11.5.1. Scaling from Leaf to Canopy

Equations (11.2) through (11.7) apply at leaf level. The intensity of solar radiation at leaf level depends on the details of its attenuation in passing through a canopy and on leaf orientation. The amount of Rubisco, and hence V_{max}, is also less, deeper into the canopy. A numerical accounting of these processes would divide a canopy into an adequate number of layers to account for such heterogeneities if they were known. However, modellers have usually developed analytic expressions for this scaling. The simplest of these is that of Sellers et al. (1996) in their SiB2 land model. They assumed that V_{cmax} would have the same profile as canopy light, in qualitative agreement with observed N profiles. The same argument was independently developed by Dewar (1996) and Haxeltine and Prentice (1996) based on more theoretical reasoning. With this assumption, all leaves have the same ratio of w_j to w_c, and so the leaf-level theory applies to the canopy. Canopy assimilation is then simply given by the product of the assimilation by a top leaf and the depth scale in LAI (leaf area index) of light attenuation. This, and other early models, treated all solar radiation as diffuse, and so its intensity in the canopy was governed by 'Beer's law' of exponential decay and its penetration within the canopy did not depend on the diurnal cycle. However, some (perhaps about half) of the incoming light is directly from the Sun and, where it is incident on a leaf, is unattenuated and only reduced in intensity by its projection on to the leaf normal; the scaling assumption of Sellers et al. (1996) does not apply to direct light.

A standard treatment (Campbell, 1977; Bonan, 1996) is to separate the incident radiation into direct and diffuse beams and apply the theory of Equations (11.2) through (11.7) separately to each. Light is made diffuse by scattering from air molecules, aerosols, and cloud. This diffuse light is commonly modelled as homogeneous, that is, each leaf receives the same amount of diffuse light: the amount averaged over the leaves. This assumption may introduce two difficulties or biases. Firstly, since shaded leaves are typically deeper in the canopy than sunlit leaves, it would be more accurate to apply an average separately over sunlit and shaded leaves, as proposed by de Pury and Farquhar (1997), who analysed in detail the scaling for leaf-level radiation, addressing the intensities at sunlit and shaded leaves. Dai et al. (2004) have applied such improvement. Secondly, at intermediate light levels, higher shaded leaves may be V_{cmax}-limited, but deeper leaves may be light-limited, and at very deep levels, possibly too costly to maintain (provided V_{cmax} declines more slowly in the canopy than does the diffuse light), as addressed by Kull and Kruijt (1998) and Friend (2001). The observed

maximum relative one-sided leaf areas of ∼6 to 10 with corresponding diffuse light levels of 1%−5% of that of full Sun is probably a consequence of this aspect of light limitation. Visible light levels in a canopy are determined by phytochrome, a pigment that detects the ratio of visible to near-infrared light. The latter extends much deeper into the canopy because of leaf scattering. Thus, phytochrome levels deep in a canopy could signal to a plant to quit its development of further leaves. Why N levels in a canopy decline less rapidly than average diffuse light levels is not established. Light levels vary over the diurnal cycle and because of the three-dimensional nature of ecosystems. Optimal use of light may favour use of the higher diffuse values that occur near noon or along gaps or some use of the high intensities from direct Sun.

Sunlit leaves receive full Sun, whatever their depth in the canopy. This illumination only varies because of the distribution of leaf orientation. Leaves oriented in the direction of the Sun are more likely to be light-saturated and those away from the direction of the Sun are more likely to be light-limited. These orientations relative to the Sun change over the diurnal cycle. The average intensity of direct Sun decays in the canopy, depending on Sun angle, but what decays is not the local light intensities, but rather the fraction of leaves that are sunlit. Thus, if deeper leaves have less N, they will be more readily light-saturated where direct sun is incident (Sun flecks), but the area covered by direct Sun will be small so these leaves do a greater proportion of their C assimilation with diffuse light.

What the best formulation is for the scaling from single leaf to canopy has yet to be established. However, some sensitivity studies (e.g., Friend, 2001; Chen et al., 2010) indicate that the greatest sensitivity may be in how the limiting rates, Equations (11.2) through (11.7), are linked together. Sellers et al. (1996) used a smooth interpolation between rates whereas the Community Land Model (CLM) (Dai et al., 2004), following the Land Surface Model (LSM) of Bonan (1996), uses directly the minimum of the three rates. Opinions appear divided as to which formulation is preferable (Bonan et al., 2011, switch CLM to interpolation). However, it is clear that observational constraints at leaf or canopy level need to be interpreted in terms of one or the other, as results will differ, that is, a V_{cmax} obtained by fitting data using one approach is biased if used in a model with the other. In other words, the value of V_{cmax} obtained from fitting data depends on the assumed form for linking together the limiting rates and, in principle, a global model should use values obtained from observations that were interpreted using the same form of the limiting rates as employed in the global model.

The modelling literature indicates less sensitivity to details of the canopy, profile of V_{cmax}. This term has commonly been assumed to scale with N, but in more detail it should scale with Rubisco, which only has some of the leaf N, that is, a leaf can contain varying degrees of N in the absence of Rubisco. That used for other enzymes may scale with Rubisco, but that used for cell walls will not. Thus, it may be necessary to model different leaf and plant pools of N (e.g., as in Kull and Kruijt, 1999; Dickinson et al., 2002), rather than make the more common assumptions that leaf photosynthetic capacity simply varies with N.

11.5.2. Modelling Plant and Soil Carbon and Nitrogen Cycling

C that is assimilated by leaves is allocated to the growth of leaves, roots, and stems. Observational studies show shifts of this allocation in response to increases of CO_2 and N limitations. For example, as already mentioned, doubled CO_2 FACE experiments for some forests show increases primarily in fine roots, but with N added, growth occurs in the wood. Plant C and N are transferred to the soil by death of plant components or whole plants and by other root processes, that is, exudates used to encourage symbiotic behaviour of soil micro-organisms. Many leaves drop annually according to their phenology or are consumed even more rapidly by herbivores (from insects to mammals). On the other hand, the woody components, such as the trunks or large roots of trees, store C for up to many decades.

Thus, a C model consists, in part, of rules that describe this generation of plant debris. After creation, detritus decays at the surface or within the soil as a result of the actions of bacteria and fungi. The standard modelling approach is to divide this plant debris into pools with different lifetimes and assume that some fraction of each pool decays into CO_2 and the rest is transformed into some form of soil C, again described by discrete pools with specified lifetimes. Some models treat the soil bacteria and fungi separately, either combined or as two pools, a distinction of possible significance since the C:N ratio of bacteria is ~ 5 and fungi ~ 10. The soil and micro-organism C pools decay back to CO_2 or convert to other C pools. All these decay rates depend on soil temperature and moisture conditions. With anaerobic conditions, other bacteria take over the decomposition and may produce CH_4 or N_2O. Under dry conditions, the bacteria may become inactive or be transport-limited, that is, enzymes secreted by bacteria must be carried to the C for it to be digested and the products then carried back to the bacteria, both requiring a liquid path from the bacteria to the C.

Soil N is composed of organic and mineral pools. Currently, global models do not distinguish between the complex and monomer N pools, although the latter is characterized from observations as a dominant source of plant N at cold temperatures. The organic N pools are associated with the C pools, for example, by requiring a fixed C:N ratio. These ratios, because they differ for different pools, exert controls on the exchange dynamics of C and N. In particular, the bacteria/fungi pools, and the long-lived soil C from their detritus have smaller C:N ratios, that is, more N than contained in plant materials, so that they can only grow at the expense of soil mineral N, that is, NO_3^- and NH_4^+, in competition with plant uptake. Conversely, decay of these pools to release CO_2 makes N available to plants.

An objective of modelling is to determine how these balances may shift with global climate change or with increases of atmospheric CO_2. Answers obtained depend on changes in a connected chain of processes. How is leaf assimilation changed? How is allocation within a plant changed? How is plant turnover affected? How are soil exchanges affected by soil moisture and temperature conditions?

Soil C models are designed to include constraints from long-term observations of soil C turnover. The decay of ^{14}C from nuclear bomb tests provides such a constraint on long-term C turnover rates. Jenkinson and Coleman (2008) developed a model for C in the first metre of soil from data obtained over 150 years. They point out that prior models (as used in climate modelling) were developed only for the topsoil. In their model, C is transferred downward and becomes more recalcitrant with depth. Yoo et al. (2011) compared soil mixing in a forest versus that of a tilled agriculture field. They found that the bioturbation (e.g., mixing by worms) of the natural system operated on a decadal timescale, leading to a much more vertical stratification in the organic components of this soil compared to that of the agricultural soil.

A key issue is why soil C becomes stabilized, that is, long lived, and what are the implications of this stabilization for the temperature dependence of the decay of long-lived soil C pools. This C stability has been thought to be of chemical origin, as it is on shorter timescales. Consequently, modellers have assumed that Q_{10} rates derived from short-period soil incubation studies also apply to long lifetime soil C pools, but many arguments have been developed against that assumption. For example, Kleber et al. (2011) studied the ^{14}C inferred rates of various soil constituents and found that the most chemically stable had the shortest lifetimes. Reichstein et al. (2005), Davidson and Janssens (2006), Wutzler and Reichstein (2008), and others have all argued that soil C most likely has physical barriers to decomposition that would not have the assumed Q_{10} temperature dependences. A recent, related hypothesis is that the C is locked up in soil micro-aggregates, and so protected from decomposing soil organisms (e.g., Post et al., 2009). Lifetimes of soil C appear to be linked to the soil clay content.

Micro-organisms consume soil organic N in monomer form, after enzymatic decomposition of complex organic C. Its competing plant uptake is not yet included in models. How much N remains available in the long-term depends on the balances between sources and sinks, so it is important that these be carefully described in models. Significant amounts of nitrate and ammonium ions are added by wet and dry deposition from the atmosphere. Other issues are addressed in more detail in the next two subsections.

11.5.3. Modelling Nitrogen Fixation

Can N fixation alleviate to a large extent the N limitations hypothesized to reduce the C fertilization effects for future climate change scenarios? Wang and Houlton (2009) say perhaps not. They find that the C4MIP models would need an additional supply of N of about 1 $gN\,m^{-2}\,yr^{-1}$, which is the upper limit that their N fixation model can supply. However, the only feedback they appear to include in their modelling is the change of fixation with temperature. Because there is no accepted approach to modelling N fixation, most climate models including this term have assigned it largely from observations that suggest a continental average of about 0.6 $gN\,m^{-2}\,yr^{-1}$ to 0.9 $gN\,m^{-2}\,yr^{-1}$. Dickinson et al. (2002) assumed it to be 0.4 $gN\,yr^{-1}$ at 25°C and at low soil ammonium concentrations with an exponential reduction with larger ammonium values, and an exponential increase with temperature (Q_{10} of 2.2). However, such an exponential temperature dependence appears to be invalid. Houlton et al. (2008) have analysed data to suggest a temperature dependence peaking at 25°C. Thornton et al. (2009) use $1.8(1 - \exp(-0.003\,NPP))\,g\,m^{-2}\,yr^{-1}$ where NPP is the net primary production, that is, an expression that increases with NPP consistent with the ecosystem's capacity to generate more root exudate to increase fixation in plants that are more rapidly acquiring C.

Recent studies have emphasized other related feedbacks on the N-fixation modelling. In particular, Gerber et al. (2010) suggest a more elaborate parameterization. They make steady-state fixation proportional to a N demand factor and, outside of the tropics, proportional to ground-level light. Their hypothetical demand term depends on the difference between potential and actual rates of N uptake. Fisher et al. (2010) propose another, more physically based, parameterization. In their parameterization, if a plant can obtain N passively through its uptake of soil water, it does. Otherwise, N is obtained from translocation from dead leaves, or by active uptake from the soil or by fixation. They propose that a plant will choose the pathway with the lowest energy requirement. Their cost functions for leaves and soils are inversely proportional to the N contained in those reservoirs and they assume a temperature dependence for fixation similar to that of Houlton et al. (2008). Fisher et al. (2010) show that their model performs reasonably well at a FACE site, three agro-ecosystem sites, three sites in the Andes, and at a forest in the UK. Although not addressed, perhaps at low light levels plants can largely only afford the passive uptake. However, even passive uptake implies a cost of root construction.

11.5.4. Modelling Nitrification and Leaching Losses

Soil mineral N is either in the form NH_4^+ or NO_3^-. N dissolved in soil water is lost when this water is lost to ground water and runoff, referred to as leaching (Figure 11.7). Since the NO_3^- is much more soluble, this is the form of N most rapidly lost to leaching. Much of the soil NH_4^+ is adsorbed on soil particles (e.g., Mortland, 1955). Besides its loss to nitrification, it can be lost by conversion to NH_3 and released to the atmosphere (volatilization). The latter loss is only significant where soils are alkaline or concentrations of NH_4^+ are very high (usually from anthropogenic activities such as feedlots) so that the effect of soil adsorption is reduced.

Many models lump together all the forms of soil mineral N, thus greatly simplifying the treatment; for example, Thornton et al. (2009) assume that 10% of the mineral N is soluble and lost to leaching. Since soil water is drained in less than a month, such an assumption removes soil mineral N in less than a year, or on about the same timescale as that of exchanges between vegetation and soil. In reality, the soils of boreal ecosystems have very low levels of NO_3^-, and so suffer very little losses except after fires, when the soils receive a flush of nitrate ions from the ashes. The assimilation by plants of N in monomer or ammonium form apparently limits the extent of denitrification.

Plants can use N either as NH_4^+ or NO_3^-, but the latter must be converted back to the former at a high energy cost. Thus, under some conditions, NH_4^+ will be preferentially used. For example, Bloom et al. (2010) report a study showing inhibition of nitrate assimilation for two plant species under CO_2 enrichment. The root uptake mechanisms differ for the two N ions (e.g., Bassirirad, 2000), so it seems that realistic dynamics cannot be achieved without treating them separately in modelling.

11.5.5. What Models Tell Us About How Terrestrial Carbon and Nitrogen Cycles Will Change and Interact with the Atmosphere in Future Climates

The feedbacks of N on the C cycle depend on the extent to which vegetation is N-limited, that is, is not able to extract as much N from the soil as it needs for optimal growth. Such limitation reduces plant growth and, hence, the C it stores. Optimal growth is defined as that given by

Equations (11.2) through (11.7) with V_{cmax} being what it would be without N limitation. The rate of extraction of N from the soil depends on the monomer or mineral N in the soil and the mechanisms of its uptake by plants. Total soil N depends on its fast closed loop with plant N and the slower, previously mentioned, processes that add it to or remove it from the soil. If the timescale of the closed loop is sufficiently fast compared to the timescale on which slow processes equilibriate, then for steady state soil N, will be:

$$N_s = S\tau \qquad (11.8)$$

where N_s is soil N, S represents the net external sources and sinks of N, and τ is the timescale for equilibration. The value of τ depends on the various rates for loss of N_s that depend on the N_s concentration, for example, from runoff (leaching), ammonia volatilization, and denitrification. It also depends on N_s-dependent source terms, in particular fixation. The sink terms shorten τ and the source terms lengthen it. Where this balance between slow sources and sinks can maintain large enough N_s, plants will not be N-limited.

External sources applied to models include atmospheric deposition, largely anthropogenic but with some natural contribution, for example, nitrate ions from lightning and forest fires, or ammonium ions from volatilization from some other location. N fixation has been applied either as a fixed rate depending only on ecosystem productivity (e.g., Sokolov et al., 2008; Thornton et al., 2009; Jain et al., 2009; Wang and Houlton, 2009; Bonan and Levis, 2010) or depending interactively on N (e.g., Dickinson et al., 2002; Zaehle et al., 2010a,b; Gerber et al., 2010; Fisher et al., 2010; Esser et al., 2011), and

independent of T (many formulations) versus depending on T (Dickinson et al., 2002; Houlton et al., 2008, and more recent papers based on it).

Detailed conclusions as to the dependence of future terrestrial storage on N can depend strongly on the assumed magnitude of S and whether or not N fixation is interactive and how its interaction is modelled. Prentice (2001) and Friedlingstein and Prentice (2010) provide an earlier comparison of the different model results as reproduced in Figure 11.9.

The recent literature addressing the role of N in the C cycle component of a climate model can be summarized as suggesting two extreme hypothesis and an intermediate one. The study of Thornton et al. (2009) suggests as one extreme hypothesis that essentially all terrestrial ecosystems are N-limited and that this limitation implies a large reduction in the lowering of the fertilization effect of increasing atmosphere CO_2. The other extreme hypothesis is suggested by the study of Gerber et al. (2010), who proposed that C assimilation is only limited by N after rapid disturbances such as from fires or experienced in FACE experiments and that, for near steady-state systems, inclusion of N in a model would make very little difference. Inclusion of organic monomer N-uptake could further support this hypothesis in boreal systems.

As an intermediate hypothesis, several papers (e.g., Sokolov et al., 2008; Jain et al., 2009; Zaehle et al., 2010a) have found effects of N limitation to be primarily in northern forests and to be absent in the tropics. The tropics may be more phosphate P than N-limited, but attempts to include P in climate models are still at an early stage (e.g., Wang et al., 2010).

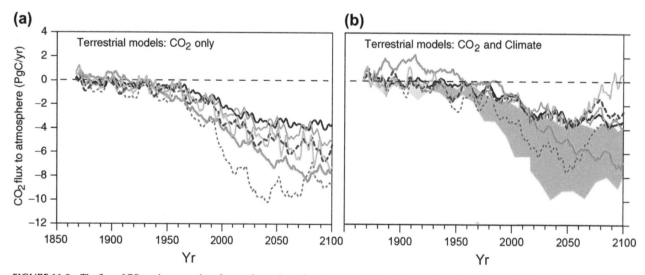

FIGURE 11.9   The flux of CO_2 to the atmosphere from early versions of various terrestrial models, as adapted from Prentice, 2001, p. 218. Panel (a) shows the results of only CO_2 addition, and (b) includes climate change, where the shading indicates the range of results from (a). *(Source: Prentice, 2001. Climate Change 2001: The Scientific Basis. Contribution of Working Group I to the Third Assessment Report of the Intergovernmental Panel on Climate Change, Figure 3.10a & b, Cambridge University Press.)*

11.5.6. Response of Soil Carbon to Future Climate Change

Most models have assumed that the decay rates of soil C pools have temperature dependences independent of time-scale. As already discussed, this assumption is extremely uncertain. Alternative assumptions have generally not been explored and require more evidence from observational studies.

Recent studies have considered another feedback on temperature: the chemical energy released in soil decomposition. Khvorostyanov et al. (2008) developed a detailed, layered model for the melting of permafrost that includes this chemical energy and found that, under some conditions, it could greatly enhance the rate of permafrost melting. Luke and Cox (2011) developed a simple analytic modelling approach to see under what conditions this mechanism could destabilize soil C. To be effective, the mechanism requires a layer near the surface of low thermal conductivity. They estimate that a long-term warming rate of 1°C per decade, not totally impossible, could trigger such a 'compost bomb' (cf. Harvey, 2012, this volume).

11.6. CONSEQUENCES OF LAND-USE AND LAND-COVER CHANGE FOR CARBON AND NITROGEN CYCLES

Historically, human management and use of land cover and its change has affected terrestrial storage of C and N at least as much as has increasing CO_2 and climate change (e.g., Pitman and de Noblet-Ducoudré, 2012, this volume). These changes include timber removal, clearing of forests or shrublands for agriculture or pastures sometimes through use of fire, and conversions of grasslands to agriculture. Management practices also dictate changes in the reverse direction, for example, conversion of agricultural land to forest, encroachment of woody shrubs into grasslands, and promotion of denser forest through fire suppression. The woody encroachment increases above-ground C storage in moist regions and decreases it in dry areas (Knapp et al., 2008) shifting the distribution of above-ground C but with little net effect.

Initial estimates of C flux used 'book keeping' models (Houghton, 2003). Current approaches are to use ecological modelling; for example, Shevliakova et al. (2009, p. 2) attempt to describe "the complex patterns of relocation of permanent agriculture (crops and pastures), shifting cultivation, logging, and the recovery of secondary (cut at least once) lands" over the last 300 years. They describe the global impacts of land-use change over this period on C and find that approximately half the global land surface has been impacted by land-use activities. Shevliakova et al.'s (2009) model tracks the dynamics of natural vegetation, cropland, pastures, and secondary vegetation with external forcing provided by scenarios of land-use transition. Each modelled land grid cell is subdivided into tiles, one tile apiece for primary lands, cropland, and pasture, while secondary vegetation is represented by up to several tiles to describe the distribution of forest stand ages. Forest-age is an easily understood variable that is used to determine other parameters, in particular C storage. The areas of the tiles vary through time. The primary land can be C_3 or C_4 grasses, temperate deciduous trees, tropical trees, or cold region evergreen trees.

Shevliakova et al. (2009) find that croplands increased by transfer from primary and secondary forest and pastures. The net gain of 0.5 PgC per year went into increased storage and C harvests greater than net ecosystem production (approximately equal amounts). Pastures, on the other hand, were able to both store and export C because production gain exceeded harvest losses. This conclusion contrasts with that of an earlier study that suggested pastures to be a large source of C. Much of the exported C, mostly from primary forests to secondary forests, was stored, but some went to clearance for pastures or cropland or to harvests balanced by production.

The Shevliakova et al. (2009) study shows that land-use change has been a significant source of atmospheric CO_2 for at least the last 300 years and, early on, a larger source than the use of fossil fuels. This source grew from about 0.5 PgC per year in the eighteenth century to about 1 PgC per year in the twentieth century, peaking around 1960 at 1.5 PgC per year. Table 11.3 summarizes their results for the twentieth century.

Shevliakova et al. (2009) found that much of the primary forest C was transferred to the secondary forest pool and the rest was lost to harvest or to clearance for pasture and cropland. Except for some longer lived wood, these losses quickly entered the atmosphere. Secondary forests were found to be a net sink for C over the entire simulation. They found that 5/8 of the C lost to the atmosphere was from vegetation and 3/8 from soils. Early clearing was mostly from extra-tropical forests but, for the

TABLE 11.3 Rates in PgC per Year for Terrestrial Carbon Sources and Sinks from Land Use and Land-cover Change According to Shevliakova et al. (2009). Forests Include Shrublands

	Primary Forest	Secondary Forest	Crops	Pastures
Storage change	−3.7	1.6	0.3	0.4
Transfer	−2.5	2.2	0.5	−0.3

last several decades, the C loss from primary forests has been mostly from the tropics. The net loss of 1.2 PgC per year from these forests is consistent with other recent estimates using different approaches, that is, van der Werf et al. (2009), who estimate 1.2 PgC per year from deforestation and 0.3 PgC per year from tropical peatland burning, while Malhi (2010) estimates 1.3 PgC per year from tropical biome conversion.

Arora and Boer (2010) have studied the consequences of land-cover change with a fully coupled model by calculating the differences obtained with and without the use of land-cover change data. They too find pastures to be a substantial sink, such that they infer the net effect in the latter half of the twentieth century of land-use change to be less than half that determined by Shevliakova et al. (2009). They attributed this discrepancy to result from a high bias in the previous study, that is, that it used late twentieth century concentrations of CO_2, which would enhance the biomass of pre-industrial vegetation over that which would be present with the actual CO_2 experienced. Such a large dependence on the assumed CO_2 concentration may be implausible, but determining the actual magnitude of this bias would require studies using both time-varying and fixed CO_2 concentrations. The role of N in determining the impacts of land-cover change on C storage is also yet to be determined in detail, although consequences of some simple scenarios have been examined in previously referenced studies.

11.7. VEGETATION AND THE FUTURE CLIMATE

This chapter has examined the questions involved in determining whether the terrestrial system will continue to absorb a significant fraction of CO_2 released by combustion of fossil fuels. It explains the processes involved and provides observational evidence for how the system works. Future projections can only be made through modelling, but the efficacy of such modelling needs to be evaluated, if possible, in terms of its ability to reproduce past changes and current variability as determined observationally.

Current models are based on plant physiology at leaf-level scale, and through modelling of leaf-level light intensities and canopy structure. These formulations are extended to the global-scale through current maps of global vegetation and dynamic models of how vegetation is altered by the changing climate and by human disturbances.

Some past modelling has suggested that, by 2100, the terrestrial system may become a source rather than a sink of C. However, this conclusion is rather model-dependent. Furthermore, the inclusion of nutrient cycles and, in particular, N, has been shown to have the potential to drastically affect the future terrestrial storage of C, a conclusion that depends on poorly-known details as to how slow sources and sinks of N are controlled by the levels of various forms of soil N and on how soil C responds to temperature change on a decadal timescale. There are, therefore, emerging decadal characteristics of the land component of the Earth system that may contribute variability as does the ocean (see Latif and Park, 2012, this volume).

The future state of the terrestrial C and N cycles cannot be established without taking into account the role of humans and of land-use change. About half the N added globally is from the fertilization of agricultural systems. C added to the atmosphere by land-use change was, a century or more ago, larger than that from burning of fossil fuels and is still a significant fraction of the latter.

In sum, a considerable amount of effort will be required to bring all the pieces of this puzzle together in order to be able to confidently predict future changes. Models need to better incorporate current observational understanding. In particular, the competition between plants and soil bacteria for organic monomers appears to have a dominant role in boreal ecosystem N cycling and needs to be evaluated. Likewise, a more complete and realistic model for the extent of N fixation appears to be needed. The operation of either, or both, of these mechanisms may be the primary determinant of the degree to which N limitation affects future C cycling.

ACKNOWLEDGEMENTS

Graham Farquhar, Colin Prentice, Marko Sholze, and an anonymous reviewer are thanked for reviews that were very insightful and promoted the repair of many deficiencies. The editor, Ann Henderson-Sellers, is thanked for enough encouragement and cajoling to motivate finishing this chapter. Bing Pu and Lisa Helper provided editing improvements. The research was supported by the DOE grant DE-SC0002246.

Chapter 12

Atmospheric Composition Change: Climate–Chemistry Interactions

Ivar S. A. Isaksen,[a,b] Claire Granier,[c,d,e,f] G. Myhre,[a,b] Terje Berntsen,[a,b] Stig B. Dalsøren,[a,b] Michael Gauss,[g] Zbigniew Klimont,[h] Rasmus Benestad,[g] Philippe Bousquet,[i] W. Collins,[j] Tony Cox,[k] Veronika Eyring,[l] David Fowler,[m] Sandro Fuzzi,[n] Patrick Jöckel,[o] Paolo Laj,[p,q] Ulrike Lohmann,[r] Michela Maione,[s] Paul Monks,[t] Andre S. H. Prevot,[u] F. Raes,[v] Andreas Richter,[w] B. Rognerud,[a] Michael Schulz,[x] Drew Shindell,[y] David Stevenson,[z] Trude Storelvmo,[r] Wei-Chyung Wang,[aa] Michiel van Weele,[bb] Martin Wild[r] and Donald J. Wuebbles[cc]

[a]Department of Geosciences, University of Oslo, Oslo, Norway, [b]Center for International Climate and Environmental Research – Oslo (CICERO), Oslo, Norway, [c]Université Pierre et Marie Curie, Paris, France, [d]Service d'Aéronomie CNRS, Paris, France, [e]NOAA Earth System Research Laboratory, Chemical Sciences Division, Boulder, CO, USA, [f]Cooperative Institute for Research in Environmental Sciences, University of Colorado, Boulder, CO, USA, [g]Norwegian Meteorological Institute, Oslo, Norway, [h]International Institute for Applied Systems Analysis, Laxenburg, Austria, [i]Service d'Aéronomie/Institut Pierre-Simon Laplace, Paris, France, [j]Hadley Centre, Met Office, Exeter, UK, [k]Centre for Atmospheric Science, University of Cambridge, Cambridge, UK, [l]Deutsches Zentrum für Luft- und Raumfahrt, Institut für Physik der Atmosphäre, Oberpfaffenhofen, Wessling, Germany, [m]Centre of Ecology and Hydrology, Penicuik, Midlothian, UK, [n]Instituto di Scienze dell'Amtosfera e del Clima, Bologna, Italy, [o]Max Planck Institute for Chemistry, Mainz, Germany, [p]Laboratoire de Météorologie Physique, Observatoire de Physique du Globe de Clermont Ferrand, Université Blaise Pascal, Aubière, France, [q]Université Joseph Fourier – Grenoble, Laboratoire de Glaciologie et Géophysique de l'Environnement, St Martin d'Héres, France, [r]ETH, Institute for Atmospheric and Climate Science, Zurich, Switzerland, [s]Universita' di Urbino, Istituto di Scienze Chimiche "F. Bruner", Urbino, Italy, [t]Department of Chemistry, University of Leicester, Leicester, UK, [u]Laboratory of Atmospheric Chemistry, Paul Scherrer Institute, Villigen PSI, Switzerland, [v]Environment Institute, European Commission Joint Research Centre, Ispra, Italy, [w]Institute of Environmental Physics, University of Bremen, Bremen, Germany, [x]Laboratoire des Sciences du Climat et de l'Environnement, Gif-sur-Yvette, France, [y]NASA Goddard Institute for Space Studies, New York, USA, [z]School of GeoSciences, University of Edinburgh, Edinburgh, UK, [aa]SUNY Albany, Atmospheric Sciences Research Centre, Albany, USA, [bb]Section of Atmospheric Composition, Royal Netherlands Meteorological Institute, De Bilt, Netherlands, [cc]Department of Atmospheric Sciences, University of Illinois at Urbana-Champaign, Illinois, USA

Chapter Outline

The Future of the World's Climate. DOI: 10.1016/B978-0-12-386917-3.00012-9

12.1. INTRODUCTION

The coupling between climate change and atmospheric composition results from the basic structure of the Earth–atmosphere climate system, and the fundamental processes within it. The composition of the atmosphere is determined by natural and human-related emissions, and the energy that flows into, out of, and within the atmosphere. The principal source of this energy is sunlight at ultraviolet (UV), visible, and near-infrared (NIR) wavelengths. This incoming energy is balanced at the top of the atmosphere (TOA) by the outgoing emission of infrared (IR) radiation from the Earth's surface and from the atmosphere. The structure of the troposphere (with temperature generally decreasing with altitude) is largely determined by the energy absorbed at or near the Earth's surface, which leads to the evaporation of water and the presence of reflecting clouds. Through many interactions, the composition and chemistry of the atmosphere are inherently connected to the climate system. The importance of climate–chemistry interactions has been recognized for more than 20 years (Ramanathan et al., 1987).

Atmospheric composition influences climate by regulating the radiation budget. As shown in Figure 12.1, clouds reflect incoming solar radiation (the albedo effect), absorb outgoing surface thermal radiation, and reradiate at the local temperature (the greenhouse effect). It is well-recognized that the thermal structure of the atmosphere is influenced by the presence of small amounts of H_2O, CO_2, CH_4, O_3, and aerosols. The main radiative effect of the gases is through the greenhouse effect, while aerosols may either heat or cool the surface, depending on their optical properties, which affect the solar and thermal radiation. Compounds such as methane, ozone, and different types of secondary particles (sulfate, organic particles, and nitrate) are active chemical compounds in the troposphere; they also have important radiative effects on climate. These compounds are either emitted directly into

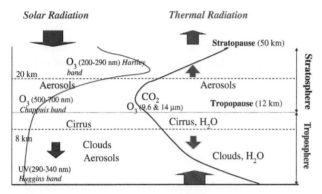

FIGURE 12.1 Atmospheric radiatively-important gases, aerosols, and clouds. For illustrative purpose, the mid-latitude, annual vertical distributions of O_3 (in orange) and temperature (in brown) are taken from IPCC/TEAP (2005).

the atmosphere (methane, primary particles), or formed in the atmosphere (ozone, secondary particles). They are either strongly controlled by chemical oxidation in the atmosphere (e.g., methane and ozone), removed through cloud and precipitation processes (primary and secondary particles), or deposited on the Earth's surface (ozone, particles). Note that the effects on climate from O_3 and secondary particles are non-uniform and will be more difficult to assess than those of the well-mixed greenhouse gases (methane). For example, changes in absolute O_3 densities in the lower stratosphere and upper troposphere have been demonstrated to lead to pronounced impact on surface temperature (Wang and Sze, 1980; Lacis et al., 1990).

Pollutants such as CO, VOC, NO_x and SO_2, which by themselves are negligible greenhouse compounds, have an important indirect effect on climate by altering the abundances of radiatively active gases such as O_3 and CH_4. Furthermore, they act as precursors for sulfate and secondary organic particles. Oxidation in the troposphere, driven by solar UV photo dissociation, result in the formation of the hydroxyl radical (OH), the key oxidant in the troposphere. The OH initiates a large number of chemical reactions affecting the climatically important compounds, such as ozone, methane, and secondary particles.

Although the potential importance of the coupling between climate change and atmospheric chemistry has been recognized for some time, it is only recently that it has been possible to capture the full complexity of the most important interactions. Observations and theory go hand in hand towards understanding this complexity. An important example is the emerging evidence that key interactions can occur through atmosphere–biosphere interactions (Fowler et al., 2009), affecting biogenic emissions, ozone deposition, uptake, and removal of particles in clouds. New observations, in particular from satellite platforms, provide data to verify models and to diagnose the chemical and physical processes, which are important for understanding chemical distributions and processes involved, as well as their long-term trends and sensitivity to climate change.

Tropospheric climate–chemistry interactions deal largely with chemical processes and compounds that show large inhomogeneous global distribution and trends, mainly initiated by inhomogeneous distribution of precursor emissions. There have been extensive studies of climate impact from specific sectors with large spatial variations in emissions and composition changes such as air and ship traffic during the past 3–5 years (Sausen et al., 2005; Eyring et al., 2005a,b, 2007a; Dalsøren et al., 2007). The potential impact from such sectors is discussed.

There are significant differences in the temporal changes of human-related emissions and natural emissions as well as in their geographical distribution. Human-related emissions are characterized by different responses worldwide to regulations and to different growth rates in energy use, in transportation systems, and in industrial behaviour. European and US emissions of pollutants generally show reductions from regulatory actions over the last two to three decades, while emissions in regions such as South East Asia, and in other developing countries, show large increases during recent years. Similarly, certain industrial sectors such as aircraft and shipping show large increases, with potential for further large increases in the future. Natural emissions of key climate precursors (NO_x, CO, biogenic VOCs, and sulfur compounds) are characterized with large year-to-year variations (e.g., biomass burning).

In addition, there is a strong potential for large changes in the emission and deposition of these compounds resulting from future climate change (e.g., due to changes in surface temperature, moisture, atmospheric stability, and biogenic activity). Such climate responses have the potential to affect global chemistry. Although there are a few recent examples indicating possible strong feedback from climate change (Zeng and Pyle, 2003; Collins et al., 2003; Sitch et al., 2007; Solberg et al., 2008), further studies are needed to reveal the full extent of this impact. This synthesis is based on studies following the Working Group I Fourth Assessment Report of the IPCC (IPCC, 2007a). The Atmospheric Composition Change European Network of Excellence (ACCENT) activities and activities obtained in other EU-financed projects have initiated some of the results described. The IPCC report describes progress in understanding of the human and natural drivers of climate change, climate processes and attribution, observed climate changes, and estimates of projected future changes. We will especially focus on developments of significance for climate–chemistry interactions that have appeared after the writing of the IPCC report, with reference to newly published research results. We will, however, also include earlier work where we find it necessary, to give a more comprehensive picture of the scientific development.

12.2. KEY INTERACTIONS IN THE CLIMATE–CHEMISTRY SYSTEM

The physical climate is a major determinant of the atmospheric concentration for all chemically active species, including key climate compounds such as CH_4, O_3, and secondary aerosol compounds such as sulfate and organic aerosols. This applies both locally (more important for short-lived species, e.g., ozone, aerosols) and globally (more important for long-lived species, e.g., methane). The physical state of the atmosphere varies considerably with location, in particular with altitude and latitude. Physical climate exerts its influence in many ways, for example, by affecting natural emissions, photolysis rates, chemical reactions, transport and mixing, and deposition processes. When considering climate–chemistry interactions in the atmosphere, it is not only important to study the emissions and distribution of the climate compounds, but also to consider the chemical processes and the distribution of the compounds determining the tropospheric oxidation.

An overview of important climate–chemistry interactions is given in Figure 12.2 for the chemical active climate compounds O_3, CH_4, and particles (sulfate, secondary organic aerosols (SOA)).

The role of climate for atmospheric chemistry can be demonstrated by considering its influence on the OH. OH is the major component for the overall oxidizing capacity of the troposphere. The primary source of OH is the reaction of excited state oxygen atoms, and it is formed from ozone photolysis, and reaction with water vapour:

$$O_3 + h\nu \rightarrow O(^1D) + O_2 (\lambda < 320 \, nm) \qquad (R12.1)$$

$$O(^1D) + H_2O \rightarrow 2OH \qquad (R12.2)$$

Water vapour concentrations are assumed to increase in a warming climate, thus increasing the rate of Reaction (R12.2), which contributes to ozone loss as it competes with the stabilization of $O(^1D)$ to the ground-state oxygen atom that would return ozone. On the other hand, higher temperatures also cause larger emissions of isoprene, an important ozone precursor, and reduce the stability of peroxyacetyl nitrate (PAN), thus releasing NO_x, which catalyses ozone production. Model studies indicate that the result of these temperature responses would be an increase in summertime ozone over polluted areas and a reduction of background ozone (Jacob and Winner, 2009). The same publication also summarizes possible effects of temperature increase on particle formation, with sulfate aerosols

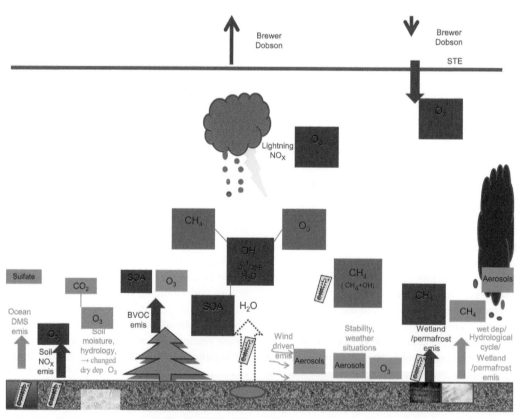

FIGURE 12.2 Important climate–chemistry interactions in the troposphere. The colour denotes the sign of the effects, blue represents a decrease, red an increase, and for green the effect or sign is uncertain or regionally very variable. The sizes of the boxes indicate the importance of the changes for the respective component. The boxes are also placed vertically approximately where the effects are most pronounced.

likely to increase in a warming climate related to faster SO_2 oxidation (Aw and Kleeman, 2003; Dawson et al., 2007; Kleeman, 2007), while nitrate particles and organic semi-volatile components are reduced as they shift from the particle phase to the gas phase (Sheehan and Bowman, 2001; Tsigaridis and Kanakidou, 2007).

12.2.1. Observing Chemistry–Climate Interactions

Direct observation of chemistry–climate interactions is extremely challenging. Composition change is often caused by non-climate factors such as anthropogenic emission trends, while multiple forcing agents have caused climate change in recent decades. This makes it difficult to attribute trends to particular gaseous or aerosol species. The hemispheric asymmetry of observed temperature trends and characteristics of the regional spatial structure of temperature changes have been used, however, to indirectly estimate the effects of Northern Hemisphere (NH) sulfate aerosols (Santer et al., 1996; Kaufmann and Stern, 2002; Stott et al., 2006).

Some effects of climate change on chemistry are observed during short-term climate anomalies periods. For example, as expected from the atmospheric increase in sulfate particles, global temperatures decreased significantly during the years following the 1991 eruption of Mount Pinatubo. Observations show that tropospheric water vapour decreased after the eruption (Soden et al., 2002), in agreement with the response seen in climate models. Additionally, methane emissions from wetlands were reduced during this time (Walter et al., 2001). Observations provide additional evidence for climate-sensitive emissions: Satellite measurements show inter-annual variations in formaldehyde distributions that appear to be tied to changes in biogenic isoprene emissions (Palmer et al., 2006). Ice core records show that emissions of mineral dust can also change dramatically with climate change, at least over glacial–interglacial transitions (Delmonte et al., 2004). Large inter-annual variations in biomass burning have also been observed (van der Werf et al., 2006), some of which have been related to climate (such as anomalies following a large ENSO event).

The influence of composition changes on climate has been estimated using a variety of observations. Changes in the Earth's outgoing energy have been observed at precisely the wavelengths where CO_2, methane, and ozone absorb longwave radiation, demonstrating their impact in the enhancement of the greenhouse effect (Harries et al., 2001). Recently, satellite data have been used to estimate the clear-sky radiative forcing from upper tropospheric ozone over the oceans at 0.48 ± 0.14 W m^{-2} (Worden et al., 2008) (note that this is a total forcing value, not the anthropogenic portion). For aerosols, satellite-based estimates of the direct all-sky forcing at the TOA have yielded values of -0.35 to -0.9 W m^{-2} (Chung et al., 2005; Bellouin et al., 2005; Quaas et al., 2008), with uncertainty ranges that do not overlap, emphasizing the large uncertainties resulting from the use of satellite measurements of optical attenuation to infer aerosol forcing. Such estimates require assumptions about factors such as aerosol composition, shape, scattering properties, and location relative to clouds. Satellite data have also been used to estimate radiative forcing due to indirect effect of aerosols on cloud albedo (the first indirect effect), yielding a value of -0.2 ± 0.1 W m^{-2} (Quaas et al., 2008), substantially lower than that seen in most forward model calculations.

Although there are limitations to forcing estimate from ozone and aerosols using satellite observations, such data usually have better global coverage than traditional data from ground-based and aircraft observations. One limitation with satellite data is that forcing is based on the total abundance of the species rather than on the anthropogenic contribution. Additionally, satellite data are generally limited in spatial coverage, and tend to lack sufficient coverage at high latitudes. Furthermore, radiative forcing, while a useful metric, does not describe fully the climate response to ozone and aerosols, especially at regional-scales. Finally, the information obtained by satellites is not sufficient to characterize the effects of aerosols.

At smaller spatial scales, much more detailed measurements are available, giving more precise estimates of aerosol radiative effects. For example, absorbing aerosols can substantially perturb both total organic aerosols and surface radiative fluxes (Satheesh and Ramanathan, 2000). Aerosols can also have a strong warming effect by enhancing cloudiness at high latitudes during seasons with little sunlight (Garrett and Zhao, 2006; Lubin and Vogelmann, 2006) (see Section 12.6.4.2). Model studies have provided indirect confirmation of the ability of aerosols to modulate local temperatures and precipitation based on comparison of climate observations with simulations with and without observed absorbing aerosols (Menon et al., 2002). These data provide valuable information about chemistry–climate interactions on regional-scales.

12.2.2. Modelling Chemistry–Climate Interactions

A range of numerical models is developed to investigate connections between atmospheric chemistry and the climate system. Past changes in atmospheric composition and climate, and projections of the future state of the atmosphere under various scenarios, use such models to understand observations. The following discussion is restricted to models at the global-scale.

Chemistry-Transport Models (CTMs) are three-dimensional models of atmospheric composition driven by meteorological data. The use of re-analysed data makes it possible to simulate the real evolution of the atmosphere;

output from such models can then be usefully compared to observational data, for example, for model validation. Extensive model validation was performed through the ACCENT project inter-comparisons both for gas phase compounds and for particles (Shindell et al., 2006; Schulz et al., 2006). CTMs that additionally assimilate atmospheric composition data, for example, from surface stations, sondes, or satellites, are used to simulate 'chemical weather'. These models can provide short-term forecasts of the atmospheric composition, for example, air-quality (AQ) forecasts, and can be used in flight route planning during aircraft observation campaigns. Assimilation of atmospheric data is extensively used for estimating emissions of compounds such as CO and CH_4 (Butler et al., 2005; Bousquet et al., 2006; Bergamaschi et al., 2007; Pison et al., 2009).

Alternatively, climate models provide meteorological data — such models are called chemistry—climate models (CCMs). CCMs include different degrees of coupling between the climate and chemistry components.

A CCM simulates the physical state of the atmosphere (dynamics and thermodynamics), largely determined by the flows of energy through the atmosphere and across its lower boundary with the Earth's surface. The chemistry model simulates the chemical state of the atmosphere, integrating the emissions, transport, mixing, physical and chemical transformation, and deposition of key atmospheric components (trace gases and aerosols). All these processes are linked to the physical state of the atmosphere, which to a varying degree is parameterized in the models. Transport and mixing are determined by wind fields and by atmospheric stability. Chemical and photochemical reaction rates are functions of many physical variables, such as levels of UV radiation, pressure, and temperature. Dry deposition to the Earth's surface is controlled by boundary layer turbulence and the surface properties, and the rate of natural emissions of several important species is strongly linked to the physical climate.

While the physical climate strongly influences the chemical climate, as outlined above, there are also influences of the chemical climate on the physical climate in this coupled system. Transmission of radiation through the atmosphere depends on the distribution of radiatively active gases and aerosols. In the stratosphere, where large quantities of UV radiation are absorbed mainly by ozone, significant local heating and reduced penetration into the troposphere occur. Parameterizations of stratospheric interactions in a CCM remain a major research area since the high complexity of CCMs requires a systematic evaluation of the results of the simulations as well as a quantification of uncertainties in the model results. This effort is part of the Chemistry—Climate Model Validation Activity (CCMVal) of the SPARC (Stratospheric Processes and their Role in Climate Programme) (Eyring et al., 2005a,b, 2006, 2007b).

Aerosol—cloud interactions (ACI) represent an area with potential for strong interactions in the climate system. Aerosols determine the cloud radiative properties and thus the radiative heating/cooling, and participate in the precipitation process. Figure 12.3 illustrates the processes linking aerosols, clouds, and climate. Although considerable progress has been made in recent years to include parameterization of aerosol—cloud droplet interaction and explicit microphysics for cloud water/ice content in GCMs, inadequate understanding of the processes contributes to significant uncertainties in model-simulated future climate changes. The individual components shown in Figure 12.3 are currently being developed, but there is still a lack of consistent and accurate parameterization for them, and their implementation into GCMs is being actively pursued (IPCC, 2007a). Further discussions of the interactions are presented in more detail in Sections 12.5.2, 12.5.3, and 12.5.4.

Tropospheric composition also interacts with the physical climate system via radiatively active trace gases and aerosols. In typical 'state-of-the-art' climate modelling (IPCC, 2007a), changes in tropospheric composition are calculated offline and then imposed on the climate, rather than being allowed to develop interactively. Fully coupled CCMs are not yet widely used in studies of tropospheric composition.

12.2.3. Scale Issues

Anthropogenic activity has altered the chemical composition of the atmosphere on local-, regional-, and global-scales. As of today, no single model is capable of reproducing a sufficiently wide range of spatial and temporal scales to address all issues related to air pollution and climate change. Due to computational requirements, there is a limit to the resolution of global models, which currently is 0.5 or 1°. On the other hand, in local and regional models it is difficult to account for processes that occur on the global-scale, that is, outside the modelled domain, but which affect the regional to local scales within the model domain. A well-known example is long-range transport of air pollutants, with intercontinental transport of pollutants and changes in background concentrations.

Most model results reviewed in this report are taken from global models, as a number of global models now have sufficient resolution to cover regional variability as well. However, global-scale models have obvious limitations in representing climate—chemistry interactions occurring on local scales, such as the impact of particle emissions or formation of clouds and precipitation.

Figure 12.4 compares typical spatial and temporal scales of common atmospheric processes with different types of models. For example, deep convection occurs on a spatial scale of a few kilometres and during a few minutes to hours. Global circulation, such as exchange between the

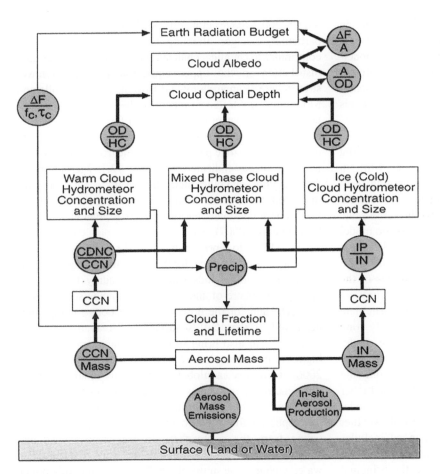

FIGURE 12.3 Flow chart showing the processes linking aerosol emissions/production, cloud condensation nuclei (CCN), cloud droplet number concentration (CDNC), ice nuclei (IN), ice particles (IP), optical depth (OD), hydrometeor (HC), albedo (A), cloud fraction (fc), cloud optical depth (τc), and radiative forcing (ΔF). *(Source: Adapted from IPCC, 2001a. Climate Change 2001: The Scientific Basis. Contribution of Working Group I to the Third Assessment Report of the Intergovernmental Panel on Climate Change, Figure 5.5, Cambridge University Press.)*

NHs and Southern Hemispheres (SHs), occurs on time-scales of several months (troposphere) to years (middle and upper atmosphere). Typical wind speeds of the respective processes are used to relate spatial scales to temporal scales. The figure also shows typical focus areas of today's model systems. For example, global climate models (GCMs) cover the entire globe, that is, tens of thousands of kilometres, and are run for up to several decades to centuries to investigate long-term changes. Urban scale models cover much smaller domains through which

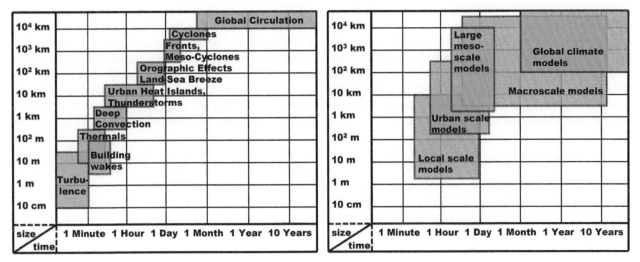

FIGURE 12.4 Characteristic spatial scales versus characteristic timescales. Left panel: common atmospheric phenomena. Right panel: atmospheric models covering different spatial and temporal scales. (See text for more details.)

pollutants are mixed over a much shorter timescale. In order to address all relevant spatial and temporal scales it is necessary to use advanced parameterizations or a modelling system consisting of a hierarchy of models of different scales that are nested into each other. In nesting approaches, particular attention has to be given to the way models communicate information to each other.

One-way nesting accounts only for influences from the larger to the smaller scale, the flow of information going from the coarse model to the finer-scale model, while in two-way nesting, feedback processes from the high resolution to the coarse resolution domain are also considered. The small-scale model uses output from the large-scale model as boundary condition, while the large-scale model uses a combination of output from the small-scale model to define distributions within the domain covered by the small-scale model. For instance at the boundaries of the smaller-scale model, the larger-scale model can provide temporally and spatially varying ozone and particulate matter, which are species that have a sufficiently long lifetime to be transported over the smaller-scale model domain. In two-way nesting, the larger scale model may replace or assimilate species distributions from the smaller-scale model in the grid boxes where a higher resolution is desirable. Depending on the kind of model (e.g., chemical transport model or numerical weather prediction model) the output to be exchanged includes chemical species or meteorological parameters or both. One drawback is that inconsistencies with the principal equations may emerge after the distributions calculated by the large-scale model are updated using output from the small-scale model. This is because after each time step in the large-scale model the species distributions and meteorological parameters are calculated from the principal equations, and any change due to input from another model will induce inconsistencies. Currently available nesting techniques can thus lead to problems such as artificial variability in large-scale models that originates from the calculations of small-scale models, but cannot be resolved in the large-scale models. Although coordination of research among model groups focusing on different scales is well underway (e.g., Moussiopoulos and Isaksen, 2006), there is still a long way to go until scale interactions are fully understood and adequately represented.

12.2.4. Upper Tropospheric Processes

Temperatures in the tropical upper troposphere have increased with about 0.65 K since the 1970s (Allen and Sherwood, 2008), essentially at what is expected from climate models based on observed surface temperature trends. Other analyses suggest that the change in the tropopause height could be a better indication of climate change than the vertical temperature profile (Sausen and

Santer, 2003; Santer et al., 2003a; Santer et al., 2003b). In particular, changes affecting O_3 is important since O_3 contributes to chemical changes and affects the temperature distribution in the upper equatorial troposphere/lower stratosphere (Ramaswamy and Bowen, 1994; Cariolle and Morcrette, 2006). Recent model studies (Hoor et al., 2009) demonstrated that emission from different sectors (Hoor et al., 2009) has a very different effect on the height distribution of ozone perturbation in the troposphere, which is important, as indicated in Figure 12.1. Measurements from satellites and other platforms during recent years have improved the understanding of processes in the upper troposphere and lower stratosphere (e.g., Pan et al., 2007; Strahan et al., 2007).

The ability to represent dynamical processes, in particular convection, is critical for the calculation of the transport of chemical compounds from the boundary layer to the upper troposphere of short-lived compounds such as NO_x and SO_2, and for the downwards transported compounds destroyed at the Earth's surface, such as ozone. Such processes can also bring insoluble organic trace gases to higher altitudes to produce new particles (Kulmala et al., 2006). Currently in global models, this results in large differences in estimates of transport and chemistry. Furthermore, there are limitations in our ability to model stratospheric—tropospheric exchanges (e.g., transport of ozone from the stratosphere to the troposphere), and chemical source gases emitted at the Earth's surface and water vapour to the stratosphere. Because of the importance of NO_x changes to ozone, its distribution and changes in the upper troposphere can be relevant to climate. Models have generally underestimated the amount of NO_x in the upper troposphere compared with observations (Schultz, et al., 1999; Penner et al., 1998). In a more recent analysis (Brunner et al., 2005) based on global model studies in the UTLS region, comparisons were made with observations from two measurement campaigns. A number of processes affect UTLS NO_x, including in situ production from lightning, emissions from aircraft, convection, stratospheric intrusions, and photochemical recycling from long-lived gases, especially peroxyacetyl nitrate (PAN) and nitric acid (HNO_3). Brunner et al. (2005) concluded that it is difficult to reproduce the observed distribution and variations, and that adopted emissions were critical for the modelled results.

Observations have shown that there are specific processes that are significant for the upper tropospheric chemistry. Acetone influences levels of the hydroxyl radical in the upper troposphere (Kotamarthi et al., 1999; Folkins and Chatfield, 2000). Model studies have demonstrated that upper tropospheric OH distribution and changes are sensitive to model assumptions (Shindell et al., 2006; Hoor et al., 2009). Furthermore, new particle formation is active in the upper troposphere (Young et al., 2007; Benson et al., 2008). Emissions from aviation, of sulfur compounds along with small amounts of soot and hydrocarbons, will contribute to the formation of new particles. Once released at cruise altitudes within the upper troposphere and lower stratosphere, these species interact with the background atmosphere and undergo complex processes, resulting in the formation of contrails. Aged contrails may have effects on upper tropospheric cirrus clouds — and these effects may exert spatially inhomogeneous radiative impacts on climate.

The occurrence of wildfires can be affected by climate change. Fires generate updrafts that efficiently transport pollutants into the upper troposphere, a process generally referred to as pyroconvection. Several studies show that this can have a significant impact on the chemical composition in that altitude region (e.g., Damoah et al., 2006; Turquety et al., 2007).

12.3. TRENDS IN EMISSIONS OF CHEMICAL SPECIES AND IN CHEMICALLY ACTIVE GREENHOUSE COMPOUNDS

Understanding of the past developments in emissions of pollutants and greenhouse gases and the driving forces behind the changes in strength of specific sources of radiatively active chemical compounds is crucial for understanding and modelling of the changing atmosphere as well as for the development of future policies. The role and historical trends of specific compounds are discussed in Monks et al. (2009). This article focuses on projections of future emissions.

The growth in the emissions of pollutants associated with the growth of industry, transport, and agriculture started at the end of the nineteenth century, and accelerated strongly after the Second World War. While anthropogenic emissions of SO_2, NO_x, CH_4, and CO_2 merely doubled in the first 50 years of the twentieth century, the next doubling took only about half as long, followed by another doubling in the last quarter of the century (Olivier et al., 2003; Lefohn et al., 1999; Stern, 2005; Smith et al., 2004). During the twentieth century, emissions of black carbon (BC) doubled and primary particulate organic carbon (OC) increased by about 40% (Bond et al., 2007). The largest sources of BC and OC include solid fuel combustion in the residential sector and open biomass burning. Recent year's growth in diesel consumption in transportation caused rapid increase in BC emissions. In the last decades of the twentieth century, a slowdown or even stabilization of global emissions of primary air pollutants has been observed. As discussed in more detail in the companion paper (Monks et al., 2009), this is due to the effect of air pollution legislation introduced in the OECD countries starting in the early 1980s. The current increase, owing to unprecedented growth rates of Asian economies, primarily China and India, is not expected to continue.

12.3.1. Future Emissions

The most important factors determining future emission levels are activity, level of technology development, and penetration of abatement measures.

12.3.1.1. Driving Forces

Activity changes are strongly linked to economic, population, and energy growth, but they are also dependent on the geo-political situation, trade agreements, level of subsidies, labour costs, and so on. While production technology improvements (with respect to emission levels) are also related to economic growth, a far more important factor is environmental legislation. The latter can be a key factor in determining the penetration of abatement measures and consequently the apparent emission factors. Comparison of historical per-capita NO_x emissions in the United States and Europe shows a strong relationship to per-capita income: for example, income above $5000 in the United States led to a strong increase in car ownership. Recently, several developing countries reached such income levels and face a rapid increase in traffic-related emissions and worsening of AQ, especially in megacities. Societal acceptance of measures to improve local AQ has also grown with increasing economic wealth. Therefore, there will be limits to growth in air pollutant emissions in the future (Klimont and Streets, 2007). Transportation is one of the fastest growing sectors, but, owing to ever stricter legislation, its emissions have been growing at a much slower pace or even declined in some industrialized countries. A similar development is expected also in Asia, where many countries either already implement comparable emission limits or prepare respective policies.

Traditionally, national legislation drives the installation of control technology, but in some regions, international (regional or global) agreements have become the key drivers. Examples include the Kyoto Protocol, UNECE CLRTAP (United Nations Economic Commission for Europe Convention on Long-Range Transboundary Air Pollution) Protocols, and EU Directives. At the national level, the economic projections are frequently updated, as are some key activity factors, e.g., population and energy use. Regional or global projections of drivers are updated less frequently, and such work is often driven by policy needs, for example, the global SRES scenarios (Special Report on Emissions Scenarios) (Nakicenovic et al., 2000) developed as part of the IPCC (Intergovernmental Panel on Climate Change) reports, EU energy or agricultural projections, and the work of international agencies such as IEA (International Energy Agency), OECD (Organization for Economic Cooperation and Development), and FAO (United Nations Food and Agriculture Organization).

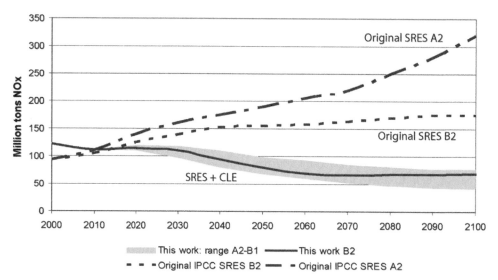

FIGURE 12.5 Change in NO$_x$ future emissions resulting from traffic emissions limits, compared with the original SRES scenarios B2 and A2.

An example of the impact of already committed legislation on NO$_x$ emission estimates for the SRES scenarios compared to the original SRES scenario results (Nakicenovic et al., 2000) is demonstrated in Figure 12.5.

One additional aspect of environmental legislation to be taken into account is the actual level of compliance. For projections, it is assumed that the technical abatement measures will be implemented in a timely manner to comply with the law. As far as performance of the equipment is concerned, approaches vary between studies: some assume that emission factors equal emission standards, while others make explicit assumptions about the probability of failure, for example, the percentage of 'smokers' among the vehicle population. The latter assumptions most often rely on the experience with existing equipment that might not necessarily be representative for the new and future technologies. For a good understanding of the projections, it is of utmost importance to state these assumptions explicitly. There is also a strong interdependence among different air pollutant species, such that many species can be mitigated at the same time by certain kinds of environmental policies, that is, ambitious CO$_2$ reduction targets will result also in significant overall reduction of air pollutants. On the other hand, some specific reduction technologies include a 'pollution penalty', that is, increase in emissions of other compounds, for example, slight increase in fuel consumption when particle traps are installed in vehicles will lead to increase of several pollutants.

12.3.1.2. Global and Regional Future Emission Inventories

There are a number of key studies and papers that provide important information on future emission levels, globally and in certain world regions and countries. The IPCC SRES

scenarios (Nakicenovic et al., 2000) reflect a large, global, long-term effort. Although the SRES scenarios assume improvements in production technology, they do not include some of the expected changes in the future penetration of abatement measures (the impacts of recent legislation); they do not include either some of the aerosols and PM species; and are available only for aggregated regions rather than countries.

There are a number of global projections that have been published in the peer-reviewed literature. For example, Streets et al. (2004) developed a forecast of future BC and OC emissions, drawing on SRES activity data and incorporating the evolution of production and control technology, specifically for non-industrial sectors. Cofala et al. (2007) developed global projections for air pollutants, BC, OC, and methane up to 2030. A longer-term projection (up to 2100) for BC and OC also taking into account CO$_2$ abatement options and policies was prepared by Rao et al. (2005); the activity data are based on the SRES scenarios. The Royal Society Report (2008) assessed possible future changes in global and regional ozone concentrations in 2050 and 2100, given changes in socio-economic factors (Riahi et al., 2006), trends in emissions of precursor gases, as well as climate change projections. As part of its Clean Air Interstate Rule (CAIR), the US EPA has developed near-term emission forecasts of SO$_2$ and NO$_x$ (*http://www.epa.gov/cair*).

Several regional projections are also available. For Europe, the GAINS model (Amann et al., 2004; *http://gains.iiasa.ac.at*) includes projections of air pollutants and greenhouse gases up to 2030, developed in consultation with national experts (Amann et al., 2008; Kupiainen and Klimont, 2007) and the EMEP database (*http://www.emep.int*) contains official projections (up to 2020) for several European countries (Vestreng et al., 2006). For Asia, several studies looked at particular pollutants

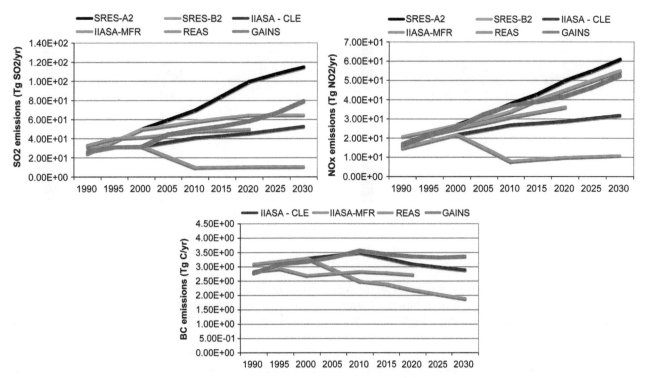

FIGURE 12.6 Evolution of the emissions of SO_2 (left), NO_x (right), and BC (bottom) from different inventories for the 1990—2030 period. *(Source: SRES: Nakicenovic et al., 2000; IIASA: Cofala et al., 2007; REAS: Ohara et al, 2007; GAINS: Klimont et al., 2009. IIASA: Reprinted from Cofala et al., 2007; with permission from Elsevier.)*

(e.g., SO_2, NO_x) but recently more integrated work has been also published, for example, the Japanese Regional Emission Inventory in Asia (REAS) (Ohara et al., 2007) presenting emissions for several pollutants for the period 1980—2020. The GAINS-Asia model has also been updated, superseding previous versions and published projections based on RAINS-Asia (Klimont et al., 2009).

The SRES scenarios are widely used in the evaluation of the future distribution of atmospheric compounds. In the framework of ACCENT and AEROCOM, two other scenarios of the evolution of the emissions of air pollutants (SO_2, NO_x, CO, BC, OC) and methane have been developed (e.g., Dentener et al., 2005; Dentener et al., 2006b; Cofala et al., 2007). The Current Legislation scenario (CLE) is based primarily on national expectations of economic and energy growth and present (end of 2002) emission control legislation. The Maximum Feasible Reduction (MFR) scenario assumes a full implementation of all available current emission reduction technologies. All these scenarios suggest a decrease of most emissions in OECD countries, but large changes in Asia.

In spite of recently introduced environmental legislation specifically targeting transport sector in urban areas as well as power plant sector, the emissions have continued to grow in Asia: the contribution of Asian emissions to the global emissions of SO_2 and NO_x increased from about 30% and 20% in the beginning of the 1990s to 50% and 35% in 2005,

respectively (Cofala et al., 2007). Early inventories and projections suffered from poor data availability, were too optimistic about the pace of introduction and efficiency of environmental legislation, and underestimated the economic growth, and several authors reassessed their previous emission estimates for Asia for several pollutants, for example, Streets et al. (2006a), Ohara et al. (2007), Zhang et al. (2007), Klimont et al. (2009). Figure 12.6 shows the evolution of the emissions of SO_2, NO_x, and BC in Asia. It can be seen from this figure that the SRES A2 scenarios show much higher emissions and seem to underestimate the impact of the current legislation, especially for SO_2, while for NO_x also growth in transport demand was underestimated in the SRES A2 scenario. MFR scenarios suggest a large technical potential for emission reduction after 2010 except BC, where the majority of emissions originate from domestic combustion and end-of-pipe measures are estimated to have only limited applicability within the considered time horizon.

Figure 12.7 illustrates the estimated changes in emissions of SO_2 and NO_x between 2000 and 2030 for a baseline scenario comparable to the IIASA-CLE scenario (Cofala et al., 2007), the basis for the scenarios used in the ACCENT model comparisons (Stevenson et al., 2006). The largest changes occur in the same areas, with China and India showing a strong increase, and Europe, North America, and Japan a decline in emissions (Akimoto, 2003).

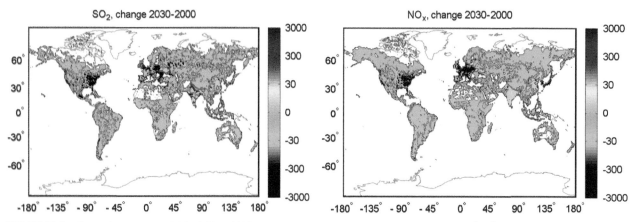

FIGURE 12.7 Change in emissions of SO_2 (left) and NO_x (right) in the 2000–2030 period. Units are ng cm^{-2} s^{-1}. *(Source: GAINS model results, gridding courtesy of Tami Bond.)*

Major uncertainties in emission inventories are associated with inadequate knowledge of open biomass burning (forest fires, agriculture waste burning), biofuel use (heating and cooking), artisanal industry, residential combustion of coal, and agricultural production systems. Due to a lack of comprehensive activity data, there is a tendency to underestimate the emissions of these pollutants. Uncertainties in inventories will vary by region, source, pollutant, and inventory year. Uncertainty estimates for all world regions are not available.

Uncertainties for individual pollutants differ also with the level of experience of compiling an inventory, with reduced uncertainties obtained over time. SO_2 inventories have a long history in Europe and North America and are considered relatively reliable in those regions. Due to the short experience in compiling PM, BC, and OC inventories and the lack of data on the distribution of technology types in key regions, these are even more uncertain. BC and OC inventories have uncertainty ranges of −25% to a factor of two (higher for open burning) (Bond et al., 2004). Typical reported ranges of uncertainty estimates for Europe are: SO_2: ±5%; NO_x: ±14%; NMVOC: 10%−39%; and CO: ±32% (EMEP, 2006). The TRACE-P inventory (Streets et al., 2003) estimated uncertainties in Asian emissions that ranged from ±16% for SO_2 and ±37% for NO_x to more than a factor of four for BC and OC. Within Asia, there was wide variation among countries and regions, with emission uncertainties in Japan being similar to those in Europe, and emissions in south Asia having high uncertainty.

12.3.1.3. Future Inventories from Different Sectors

The relative importance of different sectors for global anthropogenic emissions in 2000 is presented in Figure 12.8 (gaseous species) and Figure 12.9 (BC). The estimates are

based on the GAINS model calculations and for historical years are broadly consistent with the EDGAR FT2000 (van Aardenne et al., 2005) and Bond et al. (2004). The contributions shown in these figures can be markedly different, however, for individual countries and regions and they exclude international shipping and air traffic. The figures also show changes in the source structure in the CLE and MFR scenarios for 2030. The change in size of the pie charts symbolizes1 change (increase or decrease) in emissions compared to the year 2000.

Industrial combustion sources (including power plants) dominate emissions of SO_2, NO_x, and CO_2 in nearly all scenarios, except for 2000 for NO_x where the penetration of reduction measures in transport sector was still relatively low in several world regions leading to high emissions from this sector. It changes in the future when relative contribution of emissions from transport declines from over 50% in 2000 to about 40% in 2030. For CO_2 and SO_2, the industrial combustion share increases by few percentage points from about 63% and 74% to 69% and 79%, respectively.

Emissions of methane (Figure 12.8) are dominated by agriculture, waste, and production and transmission of natural gas (the last two included in 'industrial processes'). Growing demand for gas leads also to an increase in share of the last sector in future emissions; only in the MFR scenario effective control of losses reduces its importance and makes agriculture an even more prominent source.

Primary source of emissions of BC is combustion of solid fuels, mostly biomass, in the domestic sector (Figure 12.9) representing over 60% of BC. Transport also plays an important role, especially for BC due to diesel emissions, representing nearly 20% of total. A relatively high share of industrial combustion in total emissions is associated with assumptions of high emission factors for small industrial boilers (stokers) and furnaces used in the developing countries.

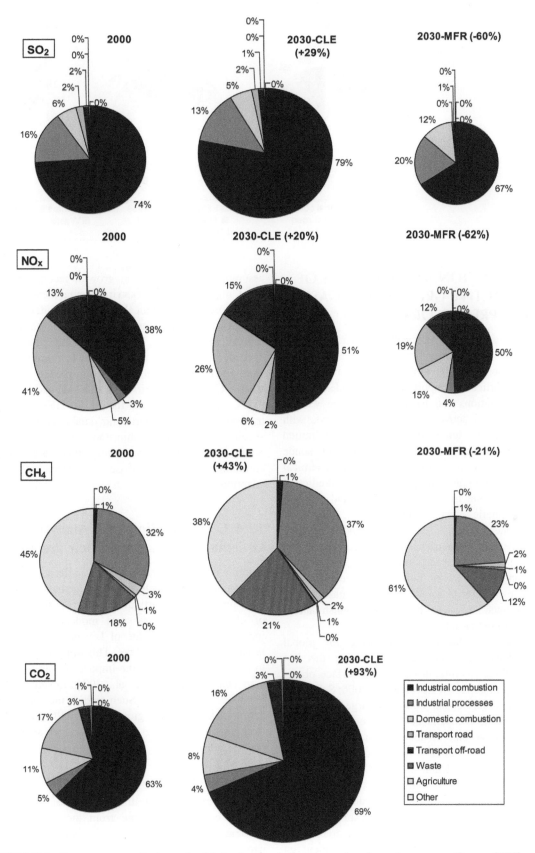

FIGURE 12.8 Source sector contribution to the global emissions of selected aerosols and greenhouse gasses. *(Source: GAINS model.)*

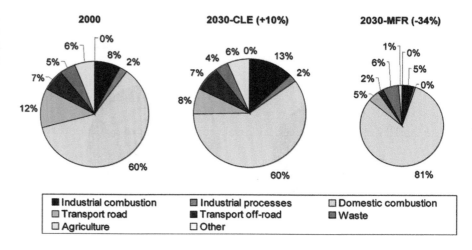

FIGURE 12.9 Source sector contribution to the global anthropogenic (excluding forest fires and savanna burning) emissions of BC. *(Source: GAINS model.)*

12.4. DISTRIBUTION AND CHANGES OF CHEMICAL ACTIVE GREENHOUSE GASES AND THEIR PRECURSORS

Models and observations represent powerful tools to understand past and current behaviour of atmospheric constituents. Observations by satellites, which have a global scale, is of particular importance for understanding large-scale distributions and trends. Also, since we are considering chemical active greenhouse compounds with significant spatial and temporal variations, more limited scale observations represented by surface and sonde observations are necessary for understanding chemical processes and their impact on the distribution. Prediction of future changes due to emission changes and climate-initiated atmospheric changes (temperatures, dynamics, humidity, biospheric response), can only be made if we have reasonable understanding of past and present conditions. We need to know the non-linear responses in composition from emission changes caused by human activities, and the behaviour of the chemical active greenhouse gases (CH_4, O_3, particles). Such knowledge helps us distinguish natural variability from human-induced changes.

The number of observations available for global comparisons have increased manifold during the last decade and include satellite, airborne, and surface observations (Richter et al., 2004; Levelt et al., 2006; Schoeberl et al., 2007; Oltmans et al., 2006). A large number of model/measurement comparisons have been performed, which clearly show that model performance, and our understanding of processes have improved (Isaksen et al., 2005; Textor et al., 2006; Kinne et al., 2006; Schulz et al., 2006; Hoor et al., 2009). Models are also important for improving our understanding of compounds with low spatial and temporal coverage in the observations.

Future changes in the chemically active greenhouse gases and particles are to a large extent dependent on the emission scenarios adopted, which are based on assumptions of population growth, technology development, implementation of environmental measures to reduce emissions, and economical growth. It is interesting to note that growth in emissions since the turn of the century from several sectors (e.g., air traffic, ship transport) have exceeded predictions made only a few years ago significantly. Scenarios commonly used in model predictions are the IPCC SRES scenarios, the EDGAR and RETRO databases, and scenarios from the QUANTIFY project for the transport sector, and particular databases or scenarios for specific sectors such as ship transport and regions such as South East Asia. Unfortunately, there are not yet any common accepted databases or scenarios for future adaption.

12.4.1. Observations and Analysis of Greenhouse Gases and Their Precursors

Satellite observations during the last decade of chemically active compounds such as ozone, methane, CO, and NO_2 have proven to be particularly valuable for the validation of global chemical transport models (CTMs), and for increasing our knowledge of key chemical and physical processes in the troposphere. This improves our capability to reproduce and assess the impact of man induced emission changes (Richter et al., 2004; Levelt et al., 2006; Schoeberl et al., 2007). Such studies include the identification of key source regions, the quantification of sources strengths, and the assessment of changes and trends over past decades.

12.4.1.1. Satellite Observations

The first extended satellite data set of observed tropospheric composition came from the TOMS instrument (Krueger, 1989), which was designed to observe total ozone column but can be used to retrieve tropospheric ozone columns.

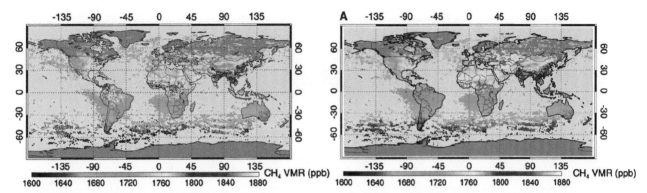

FIGURE 12.10 SCIAMACHY measurements of column-averaged methane mixing ratios in ppb (left) and TM3 model results of column averaged CH₄ mixing ratios (right), averaged for the time period August–November 2003. *(Source: Frankenberg et al., 2005. Reprinted with permission from AAAS.)*

The Global Ozone Monitoring Experiment (GOME) (Burrows et al., 1999) was the first satellite instrument having enough spectral resolution to observe ozone as well as some of its precursors (NO₂, HCHO). These measurements were continued and extended with the SCIAMACHY instrument (Bovensmann et al., 1999) which has a better spatial resolution and NIR observing capability, allowing CO, CH₄, and CO₂ observations. More recently, the OMI instrument (Levelt et al., 2006) on the AURA platform and the GOME-2 (Callies et al., 2000) on MetOp add to the UV/visible data record, with a high spatial resolution and improved coverage. Using measurements in the IR spectral region, the MOPITT instrument (Drummond and Mand, 1996) has been observing CO in the upper and middle troposphere since its launch in December 1999. A number of other tropospheric species including O₃ and several hydro-carbons can also be observed in the upper troposphere by the ACE instrument in solar occultation and in nadir from the TES and IASI instruments. While the IR nadir observations provide some degree of vertical information, they usually have low sensitivity to the lowermost troposphere. However, they greatly extend the list of observable species (Clerbaux et al., 2009 and references therein), and allowing for night-time measurements. All of these datasets are extremely useful in evaluating the modelling capabilities of CTMs. In the next sections, some examples are given for CO, methane, and HCHO. Further discussion, on recent advances in space-borne observations of atmospheric composition and their applications, is given in Laj et al. (2009).

12.4.1.1.1. Methane

Methane (CH₄) concentrations in the atmosphere have increased by at least 2.5-fold during the agro-industrial era (the past three centuries) from about 700 ppb, as derived from ice cores, up to a global average concentration of about 1770 ppb in 2005 based on present-day surface monitoring networks. Satellite observations of CH₄ columns have become possible with the measurements of

the SCIAMACHY instrument (Buchwitz et al., 2005; Frankenberg et al., 2006). The measurements show the hemispheric gradient, the seasonality, and hot-spots over rice paddies and tropical rainforests. Overall, the agreement between satellite measured and modelled CH₄ fields is good but there is indication of a significant underestimation of methane emissions from tropical rainforests (Buchwitz et al., 2005; Frankenberg et al., 2005; Houweling et al., 2006), as seen in Figure 12.10. A recent re-analysis of the satellite data using improved spectroscopic data reduced the difference (Frankenberg et al. 2008).

12.4.1.1.2. Nitrogen Dioxide (NO₂)

Tropospheric NO₂ columns have been derived from measurements of several UV/vis satellite instruments starting with GOME in 1995. As the lifetime of NO₂ in the lower troposphere is short, most of the NO₂ is observed close to the sources and satellite measurements have been used to derive NOₓ emissions, both globally and for specific sources (Beirle et al., 2004; Konovalov et al., 2006; Martin et al., 2004; Muller and Stavrakou, 2005; Richter et al., 2004). Although models represent patterns of observed distributions, there are discrepancies that can be linked to inaccuracies in missions. An example of such discrepancies can be attributed to the denoxification of power plant exhausts during summer in the eastern United States following emissions regulations (Kim et al., 2006) (Figure 12.11). Changes in NOₓ emissions are particularly large in parts of China, as seen from satellite observations (Richter et al., 2005).

Comparisons of satellite NO₂ measurements with model results indicate a possible underestimation of microbial sources of NOₓ (Bertram et al., 2005; Jaegle et al., 2004; Wang et al., 2007). Furthermore, lightning NOₓ emissions derived from satellite measurements are at the low end of other estimates (Beirle et al., 2006; Boersma et al., 2005; Martin et al., 2007) with potential implications for ozone production.

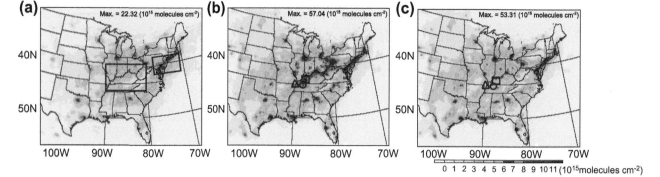

FIGURE 12.11 Spatial distribution of NO_2 columns time averaged over June–August 2004 from (a) SCIAMACHY satellite observations, (b) WRF-Chem reference emission case runs, and (c) WRF-Chem updated emission case runs. *(Source: Kim et al., 2006 © Copyright 2006 American Geophysical Union. Reproduced/modified by permission of American Geophysical Union.)*

12.4.1.1.3. Formaldehyde (HCOH)

Satellite measurements reveal large concentrations of HCHO over ecosystems emitting isoprene and terpenes, in particular over tropical rainforests (Chance et al., 2000; Wittrock et al., 2006). The seasonal and inter-annual cycle has been compared to models and good agreement was found in many regions (De Smedt et al., 2008; Palmer et al., 2003) (see also Figure 12.12). The observations have also been used to estimate emissions of isoprene and other VOCs (Abbot et al., 2003; Fu et al., 2007; Stavrakou et al., 2008). Satellite observations show enhanced formaldehyde distribution over regions affected by biomass burning. Parallel inversion of HCHO and glyoxal (CHOCHO) has confirmed previous HCHO emission estimates but indicated a large missing biogenic source for glyoxal, which must be secondary to explain the satellite observations (Stavrakou et al., 2009a,b).

12.4.1.1.4. Carbon Monoxide (CO)

Comparisons of satellite CO observations from the MOPITT satellite instrument with models reveal substantial differences (Figure 12.13). There is a large underestimate in models of the NH springtime maximum in CO. A similar bias is seen in comparison with surface observations at northern middle and high latitudes (Shindell et al., 2006), suggesting that the discrepancy cannot be attributed purely to biases in the satellite retrievals. Although all models show this bias, there is nonetheless a large range of CO values in the various models. Simulated CO is correlated with OH in the models. Some models produce OH fields inconsistent with observational constraints on OH, but even those models whose OH values are in accord with observation-based estimates are biased in CO, suggesting an underestimate in current emission inventories of CO or its hydrocarbon precursors. Indeed, in a recent study Pison et al. (2009) developed a data assimilation system and applied it to infer CO emissions (for

2004), and came up with significantly higher global emissions than currently used, indicating that underestimated CO is a result of too low emissions. However, the large spread in methane lifetimes in the models (a factor of two; Figure 12.14) indicates that there are also substantial uncertainties in our understanding of the present-day oxidation capacity of the troposphere.

12.4.1.2. Trends at Surface Stations

12.4.1.2.1. Ozone

CTMs indicate that 60%–85% of the present ozone burden in the troposphere results from chemical production in the troposphere (Fusco and Logan, 2003; Lelieveld and Dentener, 2000). Comparisons of present ozone concentrations with values at a few stations in the NH making continuous measurements in the late nineteenth century and early twentieth century suggest that surface ozone approximately doubled over the last century (Vingarzan, 2004). For the last two to three decades the global picture obtained from observations is not uniform. In general, the increases are larger from the 1970s to mid-80s than in recent years, with a more mixed picture. Figure 12.15 shows ozone background trends at Northern Hemispheric mid-latitude stations at the western coasts of the United States and Europe as well as the high-altitude site Jungfraujoch. At the low-altitude station, the background conditions were carefully selected (Parrish et al., 2008). Ozone concentrations have consistently increased at some background stations in Europe and the US since the 1950s. European stations show an increase until the year 2000 from when on the concentrations remained constant. At the west coast of the US the ozone concentrations are still increasing in air coming from the marine boundary layer. For NH mid-latitude stations observations are determined by trends in emission in the source regions. The increase in the 1990s is likely to be affected by large changes in stratospheric ozone concentrations (Isaksen et al., 2005; Ordóñez et al., 2007). Large

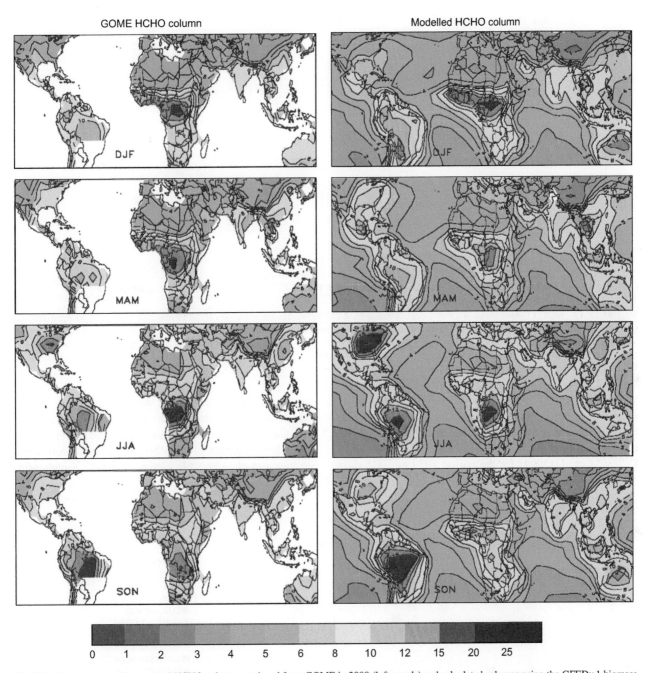

GOME HCHO column

Modelled HCHO column

0   1   2   3   4   5   6   8   10   12   15   20   25

FIGURE 12.12 Seasonally averaged HCHO columns retrieved from GOME in 2000 (left panels) and calculated columns using the GFEDv.1 biomass burning inventory and biogenic emissions from the MEGAN model using ECMWF meteorology (right panels). Units are 1015 molecules cm^{-2}. *(Source: Stavrakou et al., 2008 © Copyright 2008 American Geophysical Union. Reproduced/modified by permission of American Geophysical Union.)*

increases in emissions in Asia and from ocean-going ships during the last 10 years can probably explain some of the increases at the west coasts of the United States and Europe (ship emissions) (Dalsøren et al., 2009a; Dalsøren et al., 2009b).

Stations in different areas of the world might show different trends. Observations in the eastern Mediterranean over a 7-year time period (1997 to 2004) show a significant decline in surface ozone of 3.1% per year, reflecting the

influence of transport from continental Europe with declining emissions of ozone precursors (Gerasopoulos et al., 2005). Observations from northern Canada and the South Pole show a decline both at the surface and in the free troposphere (Logan, 1998; Vingarzan, 2004; Oltmans et al., 2006). This might be linked to decreased ozone flux from the stratosphere (Oltmans et al., 2006), or it could be a result of solar fluxes penetrating to the troposphere where enhanced fluxes seem to reduce ozone in pristine

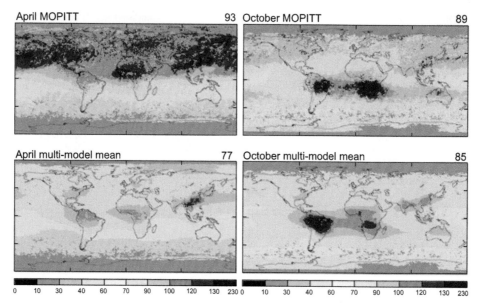

FIGURE 12.13 MOPITT observations and multimodel distributions of CO (ppbv). The top row shows MOPITT observations in 2000 for April (left) and October (right) for the 500 hPa retrieval level. The bottom row shows the equivalent fields from a multimodel ensemble average (26 models participating in the ACCENT modelling project) sampled with the MOPITT averaging kernel and *a priori* CO profiles. Values in the upper right corner are the global mean area-weighted CO (ppbv). Grey areas indicate no data. *(Source: Reprinted from Shindell et al., 2006 © Copyright 2006 American Geophysical Union. Reproduced/modified by permission of American Geophysical Union.)*

background areas (Isaksen et al., 2005). Observational record from other areas in the SH is sparse with no clear trends (Vingarzan, 2004; Oltmans et al., 2006). On the other hand, ship-borne measurements in the Atlantic by Lelieveld et al. (2004) for the period 1979 to 2002 show

significant regional increases, with increasing fossil fuel and biofuel use in Africa as the main source of this trend.

Ozone increases since pre-industrial time are reproduced by models (Berntsen et al., 1997; Wang and Jacob, 1998; Stevenson et al., 1998; Mickley et al., 1999; Hauglustaine

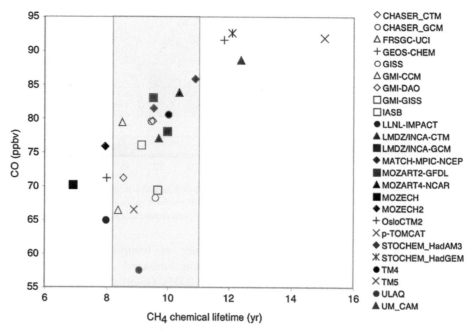

FIGURE 12.14 Annual average methane chemical lifetime versus CO in the broad 500 hPa MOPITT retrieval level. Methane lifetime is inversely proportional to OH; all models used the same prescribed methane values, with only a very small deviation due to temperature differences between models. The shaded area indicates the IPCC-TAR lifetime derived from observations and modelling. *(Source: Reprinted from Shindell et al., 2006 © Copyright 2006 American Geophysical Union. Reproduced/modified by permission of American Geophysical Union.)*

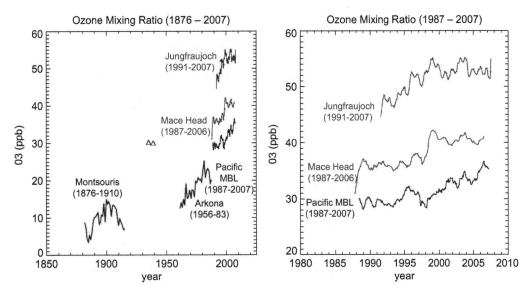

FIGURE 12.15 Observed surface ozone at different Northern Hemispheric surface stations.

and Brasseur, 2001; Shindell et al., 2003; Wong et al., 2004; Lamarque et al., 2005; Gauss et al., 2006), using current and pre-industrial emissions, although models tend to over-estimate ozone for the pre-industrial period.

As part of the EU RETRO project (ECHAM5-MOZ, LMDz-INCA) model studies of the long-term trends for the last 40 years of the twentieth century were performed. In addition, time slice studies for specific years were performed (OsloCTM2 and p-TOMCAT). All models gave significant tropospheric ozone burden increases during the time period studied. The mean year-2000 tropospheric lifetime calculated by these models are in good agreement with the multimodel mean estimate of 22.3 ± 2.0 days of Stevenson et al. (2006). Other model studies investigating shorter time periods, (Karlsdóttir et al., 2000; Lelieveld and Dentener, 2000; Granier et al., 2003; Jonson et al., 2006) were able to reproduce some of the observed ozone trends.

Fusco and Logan (2003) made a thorough estimate of tropospheric ozone distribution and trends. They concluded that surface emissions of NO_x have had the largest effect in the lower troposphere at northern mid-latitudes since 1970. The large emission increase and efficient outflow from Asia makes China responsible for 30% of the increase at mid-latitudes. Methane levels could explain 25% of the global ozone trends. Although reduced stratospheric ozone lead to a 30% decrease in cross-tropopause flux giving significant decreased ozone in the upper and middle troposphere, little impact on surface ozone was found in the summer and fall.

12.4.1.2.2. Methane Monitoring

Pre-anthropogenic methane levels of about 700 ppb constrain the current ratio of global natural (28%—42%) to anthropogenic (58%—72%) emissions to about 1:2. This implies that the present-day anthropogenic emissions exceed 300 Tg yr^{-1} compared to total emission of 500—600 Tg yr^{-1}. Wetlands dominate the natural methane emissions and hydrology changes driven by meteorological variability, climate change (e.g., thawing permafrost), and/or anthropogenic activities (e.g., drainage, deforestation) may have induced significant changes. However, even the sign of such changes is still unknown. While higher arctic temperatures are assumed to increase high-latitude emissions, tropical wetland areas are likely diminishing by rapid land use change. Significant progress is being made in process understanding, for example, in relation to northern peatlands (Wania, 2007) and gas ebullition from northern lakes (Walter et al., 2006a,b). Based on pre-anthropogenic methane levels and fractionation, present-day total natural emissions are likely to exceed 200 Tg yr^{-1}. New estimates on methane soil consumption amount to 28 (9—47) Tg yr^{-1} (Curry, 2007).

Global methane monitoring networks include the in situ networks of NOAA-GMD and ALE-GAGE-AGAGE, and the NDACC FTIR network. The longest records currently span about 25 years. Figure 12.16 shows the NOAA GMD surface measurements from 1984 until 2007, and the yearly variability in changes in global methane. The long-term methane growth rate has decreased mark-edly since the early 1990s, with periods of nearly constant global concentration levels as is seen during the first years of the twenty-first century. The possible causes of this levelling off are subject of intense scientific debate. On top of the declining trend, anomalous events such as in 1991—1993 (ascribed to the Pinatubo eruption) and 1997—1998 (ascribed to an exceptional El Niño) have caused fluctuations in the annual growth rate in

FIGURE 12.16 Global methane time series from the NOAA ESRL network.

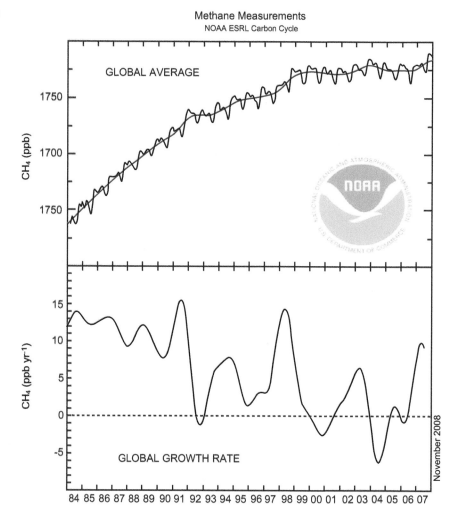

FIGURE 12.16 Global methane time series from the NOAA ESRL network.

response to annual emission anomalies of the order of 10 to 20 Tg.

Atmospheric inversion is a powerful tool to convert atmospheric measurements into estimates of surface emissions. Estimating surface emissions using atmospheric measurements in combination with a CTM and prior estimates of sources and sinks is an efficient approach. This top-down approach is increasingly used also for reactive trace gases (Hein et al., 1997; Petron et al., 2004; Bousquet et al., 2005, 2006; Chen and Prinn, 2006). Using an inversion model, Bousquet et al. (2006) analysed the large atmospheric variations of the growth rate of methane in terms of surface emissions. They found that natural wetlands are largely dominating the inter-annual variability over the 1984–2003 period. Emissions due to biomass burning have an impact on inter-annual variability only during the strong El Niño of 1997–1998. The study of Bousquet et al. (2006) has shown that inverse modelling represents a useful supplement to other estimates of methane emissions that can reduce uncertainties associated with methane trend studies.

12.4.1.2.3. Carbon Monoxide (CO)

CO plays a key role for tropospheric oxidation since the reaction with OH represents the main loss of OH. Studies has shown that since 1980 there has been large regional differences in the emission trends, and some regions like Europe succeeded in implementing efficient abatement measures (Monks et al., 2009; Olivier et al., 2003; EPA, 2003; EMEP, 2004; Dalsøren et al., 2007). NH CO observations indicate an increase of about 1% year^{-1} from the 1950s until the late 1980s, which are consistent with increased emissions (Novelli et al., 2003). Methane oxidation represents an important source of CO, particularly in unpolluted regions, and is likely to have contributed to the observed increase. There were no evident trends for the only available station in the SH. The recent regional reduction in emissions seems to be a likely explanation for measured CO decreases in the US and Europe (EPA, 2003; EMEP, 2004). Anomalous declines in global concentrations were found for the period 1991–1993 and are likely explained by an OH increase after the eruption from Mt Piantubo in 1991

(Novelli et al., 1998). Large enhancements were observed due to unusually large emissions from vegetation fires during the 1997–1998 ENSO period.

12.4.1.2.4. OH Distribution and Trends

Long-term trends in global OH will have a significant impact on methane lifetime and thereby on the global methane trend. Changes in global OH will to a large extent be determined by the ratio of global NO_x to CO emissions (Dalsøren and Isaksen, 2006), with enhanced NO emissions leading to enhanced OH, while enhanced emissions of CO as the main controller of OH, will lead to reduced global OH.

Common to all model-based studies is an overall high uncertainty in estimating OH trends. There are numerous studies of recent global OH changes. Lelieveld et al. (2002) conclude a 60% increase of the overall oxidation capacity during the last century. In contrast to this, Lamarque et al. (2005) found an overall decrease of the OH burden by 8% from 1890 to 1990. Dalsøren and Isaksen (2006) investigated the changes of OH (as well as the changes in the lifetime of methane) from 1990 to 2001 emission changes with a CTM (Oslo CTM2) and found an average OH increase of 0.08% year^{-1} over this time period.

Measurement campaigns with newly developed techniques to directly measure OH provide important insight into the chemical mechanisms more or less representative for a specific region or chemical regime (e.g., Kleffmann et al., 2005; Berresheim et al., 2003; Smith et al., 2006; Emmerson et al., 2007; Bloss et al., 2007; Lelieveld et al., 2008). The most widely used 'tracer of opportunity' to determine global OH is the purely anthropogenic methyl chloroform (MCF, 1,1,1-trichloroethane). It is not quite clear how long MCF can still be measured, since it is currently fading out rapidly after being banned by the Montreal protocol. Note that different oxidation processes in the atmosphere could affect the OH distribution significantly. For instance: It has been shown by Cariolle et al. (2008) that simulated OH is significantly affected, when the results from a recent laboratory study by Butkovskaya et al. (2007) are included, which show that a small fraction of the $NO + HO_2$ yields HNO_3.

12.4.2. Modelling Future Changes

The relatively short atmospheric lifetime of chemically active compounds such as O_3 and secondary aerosols, and pronounced spatial variability of their impact, make it necessary to consider carefully differences in regional development in future emission scenarios. ACCENT PhotoComp inter-comparison (Dentener et al., 2006b; Dentener et al., 2006c; Stevenson et al., 2006; Shindell et al., 2006) is the most comprehensive study, which evaluates the role of emissions and climate change until 2030.

These papers describe results from up to 26 different global models that participated in the ACCENT study.

12.4.2.1. Tropospheric Ozone

In the ACCENT study of ozone changes in 2030 three emissions scenarios were selected to illustrate possible impact of emission changes (climate change not included): An upper estimate was the original IPCC SRESA2 scenario (Nakicenovic et al., 2000), a central case the IPCC B2 + IIASA-CLE (current legislation) scenario (hereinafter B2 + CLE), while a lower case was the IPCC B2 + IIASA MFR (maximum feasible reduction) scenario (Dentener et al., 2005; Cofala et al., 2007; hereinafter B2 + MFR). The scenarios represent the range of possible anthropogenic emissions in the near future. In an additional ACCENT PhotoComp experiment the potential impact of climate change on O_3 by 2030 was studied, using a subset of 10 models.

A summary of the results for tropospheric ozone from the ACCENT PhotoComp study is given in Stevenson et al. (2006). It is shown that the evolution of anthropogenic emissions has a strong impact on tropospheric ozone in 2030. Under the high growth, unregulated A2 scenario, tropospheric ozone increases significantly at northern mid-latitudes with peak changes over South East Asia, where annual tropospheric column (ATC) increases approach 10 DU (1 DU = 2.67×10^{16} molecules cm^{-2}) (>20% increases over year 2000). On the other hand, under the highly regulated B2 + MFR (Maximum Feasible Reduction) scenario, ATC is reduced by 3–4 DU over much of the continental NH. The central scenario, with B2 socio-economics and full implementation of year 2000 legislation (B2 + CLE), shows increases with ATC peaking at 5 DU over south Asia.

Emission changes could have significant impact on surface ozone, with large regional variations. Seasonal mean changes in surface O_3 between 2000 and 2030 for June, July, August (JJA) from the ACCENT study are shown in Figure 12.17. June, July, August is generally the maximum surface O_3 season at polluted northern mid-latitudes, whereas at other locations ozone reaches a maximum in different seasons. In particular, decreases in surface ozone up to 25 ppb can be achieved under the B2 + MFR emission scenario while ozone is predicted to increase by up to 25 ppb in some regions in the A2 emission scenario. The effects of climate change in 2030, while uncertain, are small, compared to potential effects of emission changes. The ACCENT modelling studies show that O_3 in 2030 depends in particular NO_x emissions, but also CH_4, CO, and NMVOC. Therefore, 2030 O_3 levels are largely determined by emission control.

In a recent study organised by the Royal Society (Royal Society, 2008) several new model simulations were performed to evaluate the relative roles of changes in emissions and changes in climate slightly further into the future

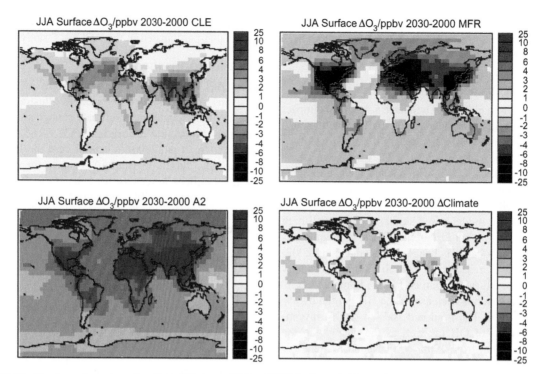

FIGURE 12.17   The mean change in surface O_3 (ppb) between 2000 and 2030 for the period June-July-August (JJA) from the ACCENT PhotoComp study *(Source: Dentener et al., 2006a; Stevenson et al., 2006 © Copyright 2006 American Geophysical Union. Reproduced/modified by permission of American Geophysical Union.).* The top left panel shows the mean change in surface O_3 under the B2 + CLE scenario; the top right panel under the B2 + MFR scenario; and the bottom left under the A2 scenario. The bottom right panel shows the mean change in surface O_3 due to climate change. The results presented are the ensemble mean of 26 models.

(2050) when the effects of climate change on O_3 should be larger and more easily detected. Figure 12.18 shows the change in seasonal mean surface O_3 projected under the new 2050 B2 + CLE emissions scenario, relative to the year 2000. Most regions show reduced or near constant O_3 concentrations by 2050 due to lower future emissions, with improved AQ over much of the developed world. Over the northeast US in the summer, O_3 is reduced by up to 15 ppb. There are also substantial reductions over Europe and Japan. Ozone is generally reduced across most of the NH mid-latitudes by about 5 ppb. One exception is during northern winter, when reduced NO_x emissions result in higher O_3 levels over Europe and North America. This is due to the reduced titration of O_3 by NO. In some parts of Asia and Africa where there is very little current legislation in place and where significant economic growth is projected, increases in O_3 of up to 3 ppb are expected.

Figure 12.19 summarizes global annual mean surface O_3 values for the various scenarios and time horizons considered in the Royal Society study (Royal Society, 2008), the ACCENT PhotoComp study (Dentener et al., 2006b; Stevenson et al., 2006), and from the IPCC Third Assessment Report (Prather et al., 2003). The B + CLE and, in particular, the B2 + MFR scenarios represent relatively successful futures in terms of emissions control

policies. The A2 scenario represents a policy fail situation which demonstrates that background O_3 levels will increase through the next century if NO_x, CH_4, CO, and NMVOC emissions rise (Prather et al., 2003) and control legislation is not implemented. This is illustrated by the potential range of O_3 values in 2100 given in Figure 12.19 and represents the response of surface O_3 to the full range of IPCC SRES scenarios, as estimated by Prather et al. (2003). The high end of this range represents scenarios with very high emissions of O_3 precursors, such as A2 and A1FI. Conversely, the low end represents the more optimistic SRES scenarios, such as B1 and A1 T. Note that the IPCC considered each scenario equally likely.

There have been several recent regional modelling studies which considered future O_3 levels over Europe (Langner et al., 2005; Szopa et al., 2006; Van Loon et al., 2006; Vautard et al., 2006), over North America (Hogrefe et al., 2004; Tagaris et al., 2007; Wu et al., 2008a), and over east Asia (Yamaji et al., 2008). The regional studies together with the global model results presented above which provide regional information albeit on a coarse horizontal resolution (~200 km), demonstrate that future changes in surface O_3 will be spatially heterogeneous, with different drivers in different regions and seasons.

Several studies show that enhanced hemispheric background O_3 in the coming decades will potentially offset

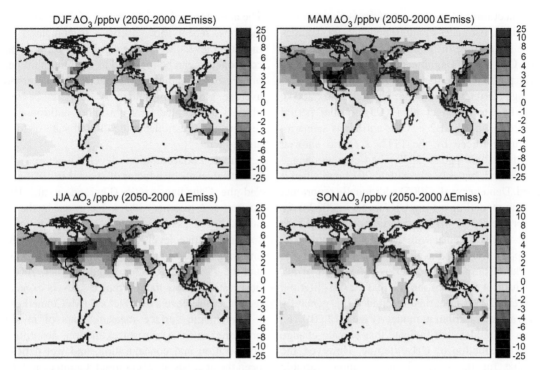

FIGURE 12.18 Projected change in surface O_3 for the four seasons between 2000 and 2050 due to changes in emissions following the IIASA B2 + CLE 2050 scenario. Top left panel: changes in surface O_3 for December-January-February (DJF). Top right panel: March-April-May (MAM). Bottom left: June-July-August (JJA). Bottom right: September-October-November (SON). The results presented are the ensemble mean of five models. *(Source: Royal Society, 2008.)*

efforts to improve regional AQ via reductions in ozone precursor emissions (Collins et al., 2000; Bergin et al., 2005; Dentener et al., 2005; Derwent et al., 2006; Keating et al., 2004; Solberg et al., 2005; Szopa et al., 2006; Yienger et al., 2000). Control of emissions on a global-scale is needed.

12.4.2.2. Projections of OH and CH_4

Over a long time horizon (a period of several decades) methane trends might be strongly affected by changes in the OH distribution. In a coupled climate–chemistry model simulation, Stevenson et al. (2005) found that the expected increase in temperature will increase humidity and result therefore in an increased OH abundance in 2030. The increasing oxidation from enhanced OH will increase aerosol formation and concentration, leading to dominantly negative climate–chemistry feedbacks (less O_3, more OH, a shorter CH_4 lifetime, and more aerosols). This is also in accordance with the CCM study of Lamarque et al. (2005), predicting a warmer and moister climate with a higher OH abundance under reduced aerosol (precursor) emissions. Note that production of SOAs by ozone, which is not included, could alter these results. However, since O_3 is driving the production of secondary biogenic volatile organics, it is unclear what the overall effect will be.

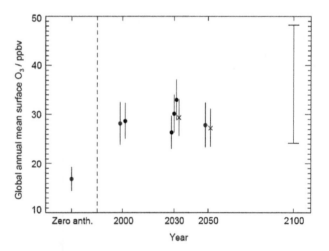

FIGURE 12.19 Multi-model global annual mean surface ozone (ppb) for various scenarios from Royal Society (2008) (red symbols) and the ACCENT PhotoComp study (blue symbols). The bars indicate inter-model standard deviations. The three blue dots for 2030 are the B2 + MFR (lowest value), B2 + CLE, and A2 (highest value) scenarios. The crosses are model results simulated with a future climate, as well as a future emissions scenario (the 2030 emissions scenario is the central one: B2 + CLE). These show a small negative impact of climate change on global annual mean surface O_3 (see Section 12.5.2.2). All other model simulations used a year 2000 climate. The bar shown for 2100 is the estimated range of O_3 responses to the full range of SRES scenarios, as reported by Prather et al. (2003). *(Source: Royal Society, 2008; their figure 5.5.)*

Future projections of CH_4 depend on the evolution of its sources and sinks. Estimates of anthropogenic CH_4 sources are available for a range of future scenarios (Nakicenovic et al., 2000). CH_4 also has significant natural sources. Some of these (in particular, wetlands) are strongly linked to climate change (Gedney et al., 2004; Shindell et al., 2004).

A study by Stevenson et al. (2005) for the period 1990–2030 following the B2 + CLE emissions scenario, with the climate forced by the IS92a scenario showed a CH_4 growth to 2088 ppb in 2030 when climate change was not considered, but only a growth to 2012 ppb when it was included. Dentener et al. (2005) additionally report that using the same modelling set-up, CH_4 levels in 2030 were 1760 ppb for the B2 + MFR scenario (unchanged from 2000). These values, together with the A2 value from IPCC TAR (2163 ppb), were used as boundary conditions in the ACCENT PhotoComp 2030 scenarios (Stevenson et al., 2006). Stevenson et al. (2006) showed that the CH_4 lifetime under different 2030 scenarios responded in generally similar ways across 21 different models (Figure 12.20). The results (from 10 models) all indicate that climate change will lead to a reduction of CH_4 lifetime. However, the absolute value for the CH_4 lifetime is more variable between models, which will have a significant bearing on the future evolution of CH_4.

12.4.3. Aerosol Distribution and Interaction

12.4.3.1. Aerosol Trends

Deep ice core drillings allow the identification of pre-industrial concentration levels of key aerosol components.

The anthropogenic enhancement of these concentrations in the most recent snow deposits as compared to the pre-industrial levels, which are significant for the main aerosol components, can be regarded as the minimum perturbation of aerosol levels in the atmosphere. The perturbation of the average global atmosphere is likely to be larger since glaciers are located far from major anthropogenic source regions and thus reflects the diluted state of the perturbed atmosphere.

Sulfate concentration levels increased from pre-industrial levels by a factor of about 4 in 1980 at Greenland and the Canadian Arctic (Legrand et al., 1997; Goto-Azuma and Koerner, 2001). Ice cores in the Alps show larger increases due to the proximity of the European sources. Nitrate levels in the Arctic from anthropogenic emissions have increased by a factor of 2–3 between 1950 and the 1980s (Goto-Azuma and Koerner, 2001). Ice core observations show that current BC levels exceed the eighteenth century levels by a factor 2 (McConnell et al., 2007).

Systematic surface measurements of in situ aerosol properties such as aerosol extinction, absorption, aerosol composition and concentration, and wet deposition have been the analysis of recent trend. Quinn et al. (2007) report recent trends in aerosol sulfate at Alert, Alaska which show a significant decrease from 1982 to 2004. Peak BC levels at the same location decreased between 1990 and 2000, while absorption at Barrow shows almost no trend between 1990 and 2005.

Major technology developments are expected to lead to a shift in future emissions with a larger fraction emitted in tropical regions with higher levels of BC, primary organic

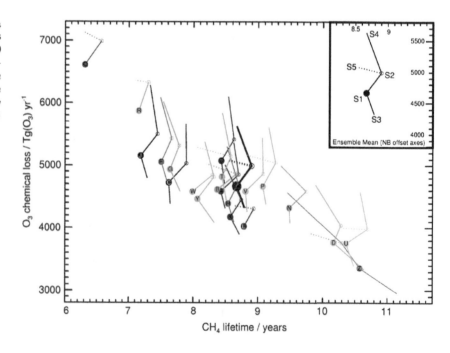

FIGURE 12.20 Response of global CH_4 lifetime to various scenarios across 21 models (S1 = 2000; S2 = 2030 B2 + CLE; S3 = 2030 B2 + MFR; S4 = 2030 A2; S5 = 2030 B2 + CLE + climate change). *(Source: Stevenson et al., 2006 © Copyright 2006 American Geophysical Union. Reproduced/modified by permission of American Geophysical Union.)*

particles and organic aerosol precursors, and aerosol loads (Ramanathan and Carmichael, 2008).

12.4.3.2. Comparisons of Aerosols and Their Precursors

Surface networks from which we can infer information about the global tropospheric distribution of aerosols make use of a diverse set of instruments with subsequent chemical analysis and mass weighting procedures (see Laj et al., 2009). The diversity of the instruments and the spatially unbalanced global distribution of sites with aerosol instrumentation have made satellite retrieved aerosol distributions a major source of information when it comes to describing the global distribution of aerosols. Calibration and precision of the satellite-retrieved aerosol optical depth (AOD) have improved over time with an important increase of available data over land and sea since the late 1990s with the launch of satellite instruments such as POLDER, MODIS, MISR, and CALIOP. Confidence in their products has been established through the synchronous deployment of ground-based Sun photometers (AERONET, GAW, SKYNET, PHOTONS) and lidar systems (EARLINET, MPLNET). Note that the optical instruments can only retrieve the bulk property of column AOD or extinction profiles, which is ambiguous with respect to underlying aerosol composition and size. Intensive in-situ measurements of chemico-physical properties of the aerosol are being frequently used to interpret remote sensing products. The performance of state-of-the-art global aerosol models and other satellites against the bulk AOD derived from the satellite MODIS are shown in the Taylor diagram in Figure 12.21. Considerable scatter of performance can be observed, while the median model outperforms individual models and some other satellite retrievals. Efforts continue in AEROCOM to understand the differences and hence uncertainty in simulated aerosol optical properties. For details about the comparisons, see Kinne et al. (2006) and Textor et al. (2006).

12.4.3.3. Diversity of Simulated Aerosol Loads

Simulation of the global aerosol distributions suggest that the major contributors to global aerosol mass are dust and sea salt. Anthropogenic emissions contribute to the formation of secondary aerosol components such as sulfate, organic matter, soot, and nitrate. Figure 12.22 depicts the global simulated mass distributions for the major aerosol components, given as a median from nine models participating in the AEROCOM project (Schulz et al., 2006). Note that the creation of a median model smoothes the distribution. Due to their short atmospheric lifetime, global particle distribution can be linked to the major source regions. From the distribution patterns of the different aerosol components, one can also infer considerable spatial

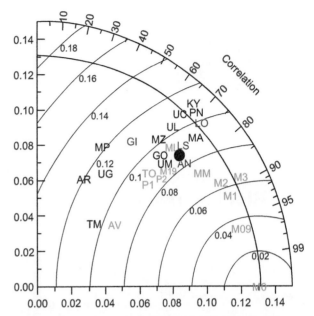

FIGURE 12.21 Taylor plot exhibiting AOD standard deviation (*x*- and *y*-axis) and correlation (angular component), showing year 2000 performance of 16 AEROCOM A models (blue and red symbols: higher and lower resolution CTMs or nudged GCMs; black symbols: not nudged GCMs) against 12 months (April 2000–March 2001) of MODIS satellite derived AOD at 550 nm. Shown is also the performance of other satellite data sets (green symbols: Polder 1 and 2, AVHRR, TOMS MISR, and Modis from other years). The black ball represents the behaviour of the AEROCOM median model constructed from the 16 AEROCOM A model results. *(Source: Data correspond to model results documented in Textor et al., 2006 and Kinne et al., 2006.)*

variability in the aerosol composition. Within the aerosol model inter-comparison initiative AEROCOM, a large range of parameters have been incorporated, including those related to the mass balance, to optical properties, and to the radiative forcing (for further details, see *http:// nansen.ipsl.jussieu.fr/AEROCOM*). The initial documentation showed significant and large diversity for almost all investigated parameters (Textor et al., 2006; Kinne et al., 2006; Schulz et al., 2006). It is hard to explain the differences found in mass burdens and optical properties (Textor et al., 2007). As an example of these findings, Figure 12.23 illustrates the global mass fraction of aerosol found above 5 km altitude in the AEROCOM models in different experiments.

The five major aerosol components are depicted to show differences between the aerosol species. High fractions of sulfate mass above 5 km as compared to total column sulfate are found in all models and point to the slow production of aerosol sulfate in clouds and via OH oxidation. Surface emissions of relative coarse sea salt and dust aerosol particles translate into a relatively small sea-salt and dust-mass fraction in the upper troposphere. The mass fractions for the different model experiments of the same model are shown side by side to

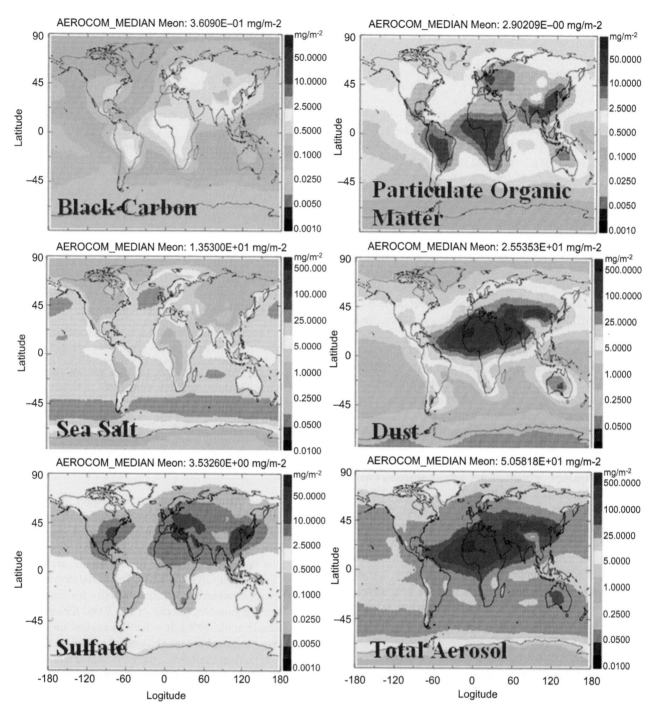

FIGURE 12.22 Aerosol distribution of major components derived from the AEROCOM B simulations (see models in Schulz et al, 2006) of year 2000. Shown is the median model constructed from the modelled distributions of nine aerosol transport models.

illustrate the impact of changing emissions on the vertical distribution.

The diversity among models and the resemblance among model experiments from a given model, as documented by Figure 12.23 with respect to the average vertical transport, have led to several important conclusions: The current global models exhibit significant and important differences in the vertical mixing of aerosol mass. The way different species are affected by vertical mixing processes is varying from model to model. The experiment with harmonized emissions reveals that the internal structure of any aerosol model was far more important for the vertical mixing and aerosol composition than the variations in spatial and temporal emissions imposed through AEROCOM. The

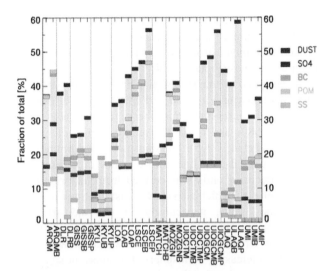

FIGURE 12.23 Aerosol compound mass fraction found above 5 km altitude on global average. AEROCOM A (original model state), B (present-day prescribed emissions), and P (pre-industrial prescribed emissions) experiments are shown if available. Model names for experiments B and P are expanded by experiment abbreviation.

larger diversity in BC export from continental scale regions compared to that of carbon monoxide could be linked to the uncertainty in wet removal and thus long-range transport of aerosols (Shindell et al., 2008a,b). Other processes have been emphasized recently as possible source of diversity, such as humidity growth (Bian et al., 2009), meteorological drivers (Liu et al., 2007), and different aerosol compositions (Myhre et al., 2009). In general the recent works support the idea that differences in radiative forcing originate from different model constructions.

AEROCOM comparisons documented considerable differences with respect to parameterization of several processes, including poleward transport, aerosol water mass, the split between large-scale and convective scavenging, absorption of visible light, secondary sulfate formation, and radiative forcing related parameters. How these are couplings are achieved will have considerable impact on studies of aerosol-climate feedbacks (see e.g., Carslaw et al., 2009). It is clear that a systematic comparison with observations and more thorough analysis of the model structures is needed to reduce the differences in the future.

12.4.4. Observed Brightening and Dimming Trends over the Last 40 Years

The major anthropogenic impact on climate occurs through a modification of the Earth's radiation balance by changing the amount of greenhouse gases and aerosol in the atmosphere. The amount of solar radiation reaching the Earth's surface is thereby particularly important as it provides the primary source of energy for life on the planet and states

a major component of the surface energy balance, which governs the thermal and hydrological conditions at the Earth's surface.

Observational and modelling studies emerging in the past two decades suggest that surface solar radiation (SSR) is not a necessary constant on decadal timescales as often assumed for simplicity, but shows substantial decadal variations. Largely unnoticed over a decade or more, this evidence recently gained a rapid growth of attention under the popular expressions 'global dimming' and 'brightening', which refer to a decadal decrease and increase in SSR, respectively.

Extensive measurements began in the late 1950s during the International Geophysical Year (IGY) in 1957/58. Many of these historic radiation measurements have been collected in the Global Energy Balance Archive (GEBA, Gilgen et al., 1998) at ETH Zurich and in the World Radiation Data Centre (WRDC) of the Main Geophysical Observatory St. Petersburg. In addition, more recently, high-quality surface radiation measurements, such as those from the Baseline Surface Radiation Network (BSRN; Ohmura et al. 1998) and from the Atmospheric Radiation Measurement Program (ARM) have become available. These networks measure surface radiative fluxes at the highest possible accuracy with well-defined and calibrated state-of the-art instrumentation at selected worldwide sites.

12.4.4.1. Surface Solar Dimming from the 1960s to the 1980s

Changes in solar radiation from the beginning of worldwide measurements in the early 1960s until 1990 have been analysed in numerous studies (e.g., Ohmura and Lang, 1989; Gilgen et al., 1998; Stanhill and Cohen, 2001 and references therein; Liepert, 2002; Wild, et al., 2004; Wild, 2009 and references therein). These studies report a general decrease of solar radiation at widespread locations over land surfaces between 1960 and 1990. This phenomenon is now popularly known as 'global dimming'. Figure 12.24 shows the change in global dimming since 1950 for different locations.

Increasing air pollution and associated increase in aerosol concentrations are considered a major cause of the measured decline of SSR (e.g., Stanhill and Cohen, 2001). Changes in cloud amount and optical properties have also been proposed to contribute to the dimming (e.g., Liepert, 2002). Norris and Wild (2007) differentiate between aerosol and cloud impact on radiative changes over Europe. They show that changes in cloud amount cannot explain the changes in surface insolation, pointing to the aerosol direct and indirect effects as major cause of these variations. Alpert et al. (2005) found that the decline in SSR in the 1960 to 1990 period is particularly large in areas with dense population, which also suggests a significant anthropogenic

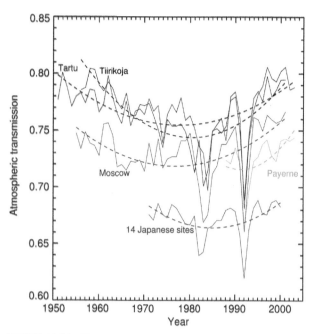

FIGURE 12.24 Time series of annual mean atmospheric transmission under cloud-free conditions determined from pyrheliometer measurements at various sites in Russia (Moscow), Estonia (Tartu–Toravere and Tiirikoja), Switzerland (Payerne), and Japan (average of 14 sites). *(Source: Wild et al., 2005. Reprinted with permission from AAAS.)*

influence through air pollution and aerosols. Several studies (e.g., Dutton et al. 2006) noted a dimming over the 1960 to 1990 period at remote sites, suggesting that the phenomenon is not of purely local nature and air pollution may have far-reaching effects.

12.4.4.2. Surface Solar Brightening from 1980s to Present

The studies on global dimming were all limited to data prior to 1990. ETH Zurich recently undertook a major effort to update the worldwide measured surface radiation data in GEBA for the period from 1990 to present. Wild et al. (2005) evaluated the newly available surface observational records in GEBA and BSRN to investigate the trends in SSR in the more recent years. This analysis showed that the decline in solar radiation at land surfaces seen in earlier data is no longer visible in the 1990s. Instead, the decline levelled off or even turned into a brightening since the mid-1980s at the majority of observation sites. This brightening is not just found under all sky conditions, but also under clear skies, pointing to aerosol as a major cause of this trend reversal (Wild et al. 2005; Norris and Wild 2007). The trend reversal is reconcilable with recently estimated radiation trends derived from independent methods, such as satellite-derived estimates and the independent Earthshine method (Pinker et al. 2005; Hatzianastassiou et al. 2005; Pallé et al., 2005). The

transition from decreasing to increasing solar radiation is in line with a similar shift in atmospheric clear sky transmission determined from pyrheliometer measurements at a number of sites (Figure 12.24).

In addition to the strong signals of major volcanic eruptions (such as Pinatubo and El Chichon), a general tendency of decreasing atmospheric transparency up to the early 1980s and a gradual recovery thereafter were found (Wild et al., 2005). The transition from dimming to brightening is also in line with changes in aerosol and aerosol precursor emissions derived from historic emission inventories, which also show a distinct trend reversal during the 1980s, particularly in the industrialized regions (Streets et al., 2006b). The documented trend reversal in aerosol emission towards a reduction and the associated increasing atmospheric transmission since the mid-1980s may be related to air pollution regulations and the breakdown of the economy in Eastern European countries (Wild et al., 2005; Streets et al., 2006a). The observed reduction of BC and sulfur aerosol measured in the Canadian Arctic during the 1990s is likely related to the decreased emissions in Europe and the Former Soviet Union (Sharma et al. 2004). Since 1990, a reduction of AOD over the world oceans was inferred from satellite data by Mishchenko et al. (2007). This fits well to the general picture of a widespread transition from dimming to brightening seen in the surface radiation observations at the same time.

12.4.4.3. Impact of Dimming and Brightening on the Climate System

A growing number of studies provide evidence that the variations in SSR have a considerable impact on the climate system. Wild et al. (2007) investigated the impact of dimming and brightening on global warming. They present evidence that surface solar dimming was effective in masking and suppressing greenhouse warming, but only up to the mid-1980s, when dimming gradually transformed into brightening. Since then, the uncovered greenhouse effect reveals its full dimension, as manifested in a rapid temperature rise (+0.38°C per decade over land since the mid 1980s). Impacts of the global dimming and brightening transition can be further seen in glaciers and snow cover retreats as well as an intensification of the hydrological cycle, which became evident as soon as the dimming disappeared in the 1990s (Wild, 2009).

Ramanathan and Carmichael (2008) estimate a reduced SSR of 1.7 W m^{-2} due to BC. When including scattering aerosols and the indirect aerosol effect, a reduced surface radiation of 4.4 W m^{-2} has been estimated. Global aerosol models produce weaker changes in the surface radiation (Schulz et al., 2006) than the observational-based method used in Ramanathan and

Carmichael (2008). Kvalevåg and Myhre (2007) show that other components such as gases and contrails (direct and indirect effect) contribute to the reduced SSR. It is further shown that the reduction in SSR calculated by models occurs mostly in industrialized regions, which could explain a large part of the observed dimming.

12.5. CLIMATE IMPACT FROM EMISSION CHANGES

This section describes the climate effects related to the changes in atmospheric constituents, including comparisons between the components of the system and how to define the metrics for mitigation purposes. Although various climate effects of aerosols and greenhouse gases are discussed, focus is on the atmospheric components. Changes in land surface properties are not considered here.

The use of the radiative forcing concept is the most common way to compare the impact leading to climate change. This concept is used because surface temperature changes from GCMs are clearly model dependent (Hansen et al., 1997; Shine et al., 2003). The radiative forcing (RF) is defined as the net change in the irradiance (solar and terrestrial) at the tropopause after allowing the temperature in the stratosphere to adjust to radiative equilibrium (Forster et al., 2007). The surface and tropospheric temperatures are held fixed according to the definition and considered as a response. The impact of the temperature changes on water vapour and clouds is, for example, considered as feedback.

For some of the aerosol effects the RF concept is particularly challenging. This is illustrated in Figure 12.25 showing some of the main atmospheric aerosol effects. The direct aerosol effect and cloud albedo effect can be calculated in a way where only the aerosol abundance and cloud droplet sizes change, according to the radiative forcing concept. In the case of the semidirect aerosol effect and the cloud lifetime effect (see Section 12.5.4), the calculation is much more complex since not only atmospheric concentrations change, but cloud distribution and properties, water vapour abundance, and lapse rate, as well as other factors may change. The uncertainties in the quantification of the semidirect effect and the cloud lifetime effect were the reason why only the direct aerosol and the cloud albedo effect was quantified as radiative forcing mechanisms in IPCC AR4 (Forster et al., 2007). In this paper the semidirect and the cloud lifetime are treated as radiative forcing mechanisms and a quantification of these based on available research is attempted.

The applicability of the radiative forcing concept is that the climate sensitivity relating surface temperature change, as proportional to the radiative forcing, is similar for various climate forcing mechanisms. The climate efficacy

has been defined as the climate sensitivity of a climate forcing mechanism divided by the climate sensitivity for CO_2 (Hansen and Nazarenko, 2004; Hansen et al., 2005; Joshi et al., 2003). For most climate forcing mechanisms the climate efficacy is close to unity (mostly in the range 0.8—1.2) (Forster et al., 2007). The largest deviations from unity of the climate efficacy are found for idealized experiments on BC (Cook and Highwood, 2004; Hansen et al., 1997; Hansen et al., 2005) which is related to semidirect effects. When considering the BC impact on surface albedo of snow and ice, a high climate efficacy is found (Flanner et al., 2007; Hansen and Nazarenko, 2004).

12.5.1. Radiative Forcing from Gases

12.5.1.1. Well-Mixed Greenhouse Gases

The total RF due to the well-mixed greenhouse gases (WMGG) in 2005 relative to 1750 is 2.63 W m^{-2} (Forster et al., 2007). Figure 12.26 shows the percentage contribution of the RF due to CO2 and other WMGG compared to the total RF from WMGG for two past period (pre-industrial to current, 1998—2005) and the future. Estimates for four IPCC SRES scenarios represent the future (2000—2100). The high contribution of CO_2 to the total WMGG RF in the future is evident even in the B1 scenario with a smallest CO_2 increase. Total RF is expected to increase strongly during the twenty-first century. CO_2 is expected to be the dominating contributor to the total WMGG RF also in the future. The average global CO_2 concentration has increased to 384 ppm in 2007 and the RF is increasing at a rate higher than 0.03 W m^{-2} yr^{-1}. It is worth noting that the RF due to CO_2 accounts for 63% of the total RF from WMGG over the industrial period. However, over the 1998 to 2005 time period, the increase of 0.2 W m^{-2} in the RF of CO_2 is the same as the increase in the total WMGG RF (Forster et al., 2007).

The increased methane burden has contributed to about 20% of the anthropogenic climate forcing by greenhouse gases since the pre-industrial era (0.48 $\pm$ 0.05 W m^{-2}; IPCC, 2007a). An emission-based assessment of the climate forcing of methane (Shindell et al., 2005) showed that methane emissions are responsible for most of the increase in global tropospheric ozone, with minor contributions attributed to the increase of CO, VOC, and NO$_x$ emissions. From this estimate the combined methane and tropospheric ozone forcing is about 0.8 W m^{-2}, almost half of the present-day CO_2 forcing of 1.66 W m^{-2}. Although the methane concentration increased between 2006 and 2007 by 0.63%, this had a small impact on the RF.

Currently CFC-12 has the highest RF among the CFCs and has the third highest RF of the WMGG (after CO_2 and CH$_4$). The RF due to N$_2$O has been steadily increasing (Forster et al., 2007) and, as a result of more intensive agricultural practice, may enhance its growth. Since the CFC-12

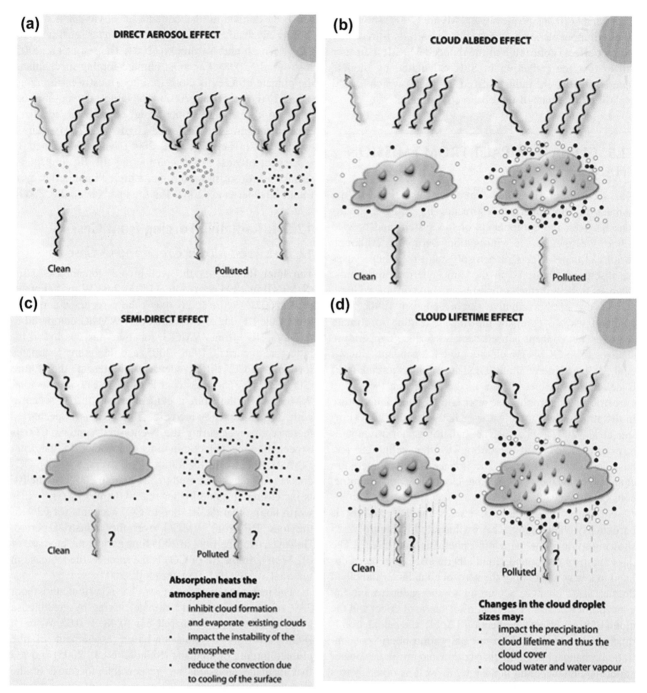

FIGURE 12.25   Illustrations of the (a) direct aerosol effect (with both anthropogenic scattering and absorbing aerosols), (b) cloud albedo effect, (c) semi-direct aerosol effect, and (d) cloud lifetime effect. Black circles represent absorbing aerosols and open circles represent scattering aerosols.

concentration has levelled out over the last few years and is expected to decrease in the future, the RF due N_2O is expected to soon be the third largest among the WMGG.

The future IPCC scenarios show a spread of more than a factor of five for anthropogenic CO_2 emissions; in addition the impact of future warming on the carbon cycle is uncertain. This indicates large uncertainties in concentration changes (Friedlingstein et al., 2006; IPCC, 2007a). A further uncertainty is caused by a recent suggestion that increased ozone may reduce the vegetation uptake of CO_2 (Sitch et al., 2007). Additionally, global dimming and changes in the direct and diffuse solar radiation at the surface may impact the vegetation and thus the CO_2 uptake (Mercado et al., 2009). Human activity has strongly altered the direct and diffuse solar radiation at the surface (Kvalevåg and Myhre, 2007; Mercado et al., 2009; see also Section 12.4.4).

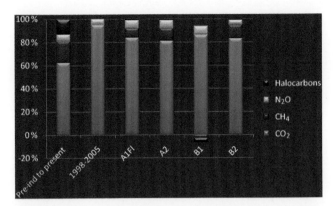

FIGURE 12.26 Percentage contribution of the RF from CO_2, CH_4, N_2O, and halocarbons to the total RF for 2 past time periods and 4 future SRES scenarios. The future scenarios are calculated for the time period 2000 to 2100. All values are taken from Forster et al. 2007 with regard to past changes and IPCC (2001a) for the future.

12.5.1.2. Water Vapour

Water vapour trends are significant in the troposphere (IPCC, 2007a); they represent a feedback to the climate system. IPCC (2007a) suggests an increase of 1.2% per decade for the 1988–2004 period, and an increase of the order of 5% over the twentieth century. As an illustration, Boucher et al. (2004) estimate a forcing of 1.5 W m^{-2} per 1% increase in the current water vapour content. Direct human influence on water vapour is linked also to human practices such as through irrigation (Boucher et al., 2004) as well as to change in vegetation (Forster et al., 2007). However, these water vapour changes are small and difficult to quantify since they involve feedback mechanisms through perturbations of the hydrological cycle (Forster et al., 2007).

The trend in the stratospheric water vapour content is highly uncertain (Randel et al., 2006; Scherer et al., 2008) and its origin is not well-established. A direct human influence on stratospheric water vapour originates from the water vapour emitted by aircraft and by the oxidation of increasing CH4 in the stratosphere. The radiative forcing due to stratospheric water vapour increase from CH4 oxidation has been estimated using two independent approaches by Hansen et al. (2005) and Myhre et al. (2007a). Calculated values of 0.07 and 0.08 W m^{-2}, are obtained respectively. The contribution of the stratospheric water from aircraft exhaust is very small (Sausen et al., 2005), but may be non-negligible in the future (Søvde et al., 2007).

12.5.1.3. Ozone

Within an ACCENT network study Gauss et al. (2006) applied ten global chemistry models to evaluate RF from ozone changes since pre-industrial time. The range of RF due to tropospheric ozone changes obtained by the different models is almost a factor of two. For stratospheric ozone changes, the sign of the RF varied among the models. The

differences in RF due to tropospheric ozone changes were explained by significant differences in the calculated column tropospheric ozone changes (Gauss et al., 2006). Differences in the RF from stratospheric ozone change result mainly from differences in the vertical profile of ozone in the tropics (Forster and Shine, 1997; Hansen et al., 1997). The crossover altitude from an increase of ozone in the troposphere to a reduction in the stratosphere varies among the different models, from a level close to the tropical tropopause to about 25 km (Gauss et al., 2006). This crossover altitude depends on the increase in the ozone amount in the upper troposphere and the transport to the stratosphere. The causes to the differences in the ozone change between the models in the upper troposphere are mainly a result of applying different chemical schemes and parameterizations of boundary layer processes and convection (Gauss et al., 2006).

An indirect RF from ozone precursors emission, previously not accounted for, is the impact on vegetation from ozone enhancement, which has the potential to affect the carbon cycle. This indirect effect is linked to ozone damage on plants and crops resulting in smaller CO_2 uptake and higher atmospheric CO_2 abundance (Sitch et al., 2007). The indirect effect on RF from ozone through this process could be of similar magnitude as the direct RF from ozone changes (see Figure 12.27 taken from Sitch et al., 2007).

12.5.1.4. NO$_2$

NO$_2$ strongly absorbs solar radiation, thereby contributing to RF. Due to its short lifetime, the distribution of NO$_2$ is strongly heterogeneous with highest abundances in industrialized regions. Substantial changes in its concentrations have occurred over the past decade (Richter et al., 2005).

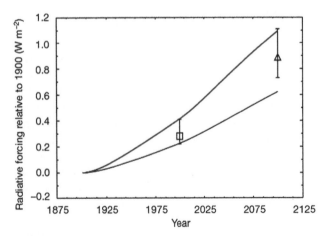

FIGURE 12.27 Indirect radiative forcing due to ozone on plants affecting the CO_2 uptake for 'high' (red) and 'low' (blue) plant sensitivity to ozone (from Sitch et al., 2007). Direct radiative forcing due to ozone is included as bars.

A first estimate of RF due to NO_2 changes gives a value of 0.04 W m^{-2} (Kvalevåg and Myhre, 2007) and is less than 2% of the total RF of the WMGG.

12.5.2. Direct Aerosol Effect

In the AR4 IPCC report (Forster et al., 2007) the total direct aerosol effect was quantified with a best estimate of −0.5 W m^{-2}, together with a rather large uncertainty range from −0.9 to −0.1 W m^{-2}. The large uncertainty range results from significant differences between estimates from global aerosol models (see also discussion in Section 12.4.3). Using a combination of models and satellite observations Myhre (2009) show that the RF due to the direct effect of aerosols is around −0.3 W m^{-2}, and that the uncertainty range can be halved compared to the estimate by IPCC (Forster et al., 2007). In the global aerosol inter-comparison study AEROCOM, nine models used identical emissions for calculations of RF (Schulz et al., 2006); they obtained a mean of −0.2 W m^{-2} and an uncertainty range of 0.2 W m^{-2}. Schulz et al. (2006) also summarized some earlier model studies, which gave similar results. Studies based on observations estimate a much stronger RF for the total direct aerosol effect, with values between −0.8 and −0.5 W m^{-2} (Bellouin et al., 2005; Chung et al., 2005). This section presents advances since the AR4 report in global aerosol models development, and in our understanding of the differences between model estimates and observational based studies.

12.5.2.1. Recent Progress in Estimates of the RF

Bellouin et al. (2008) have updated the simulation from Bellouin et al. (2005) using new satellite data and performed additional sensitivity simulations to investigate the difference between model calculations and methods based on observations. The latest version of the MODIS satellite retrieval, named Collection 5, is implemented instead of the MODIS Collection 4 used in Bellouin et al. (2005). This decreased the RF from −0.8 to −0.65 W m^{-2}. It was further shown that the treatment of clouds was a key cause to the difference between models (Bellouin et al., 2008). It was also shown that assumption of zero RF in cloudy region used in Bellouin et al. (2005) is questionable. In Schulz et al. (2006) the RF in cloudy region varied substantially among the global aerosol models. Schulz et al. (2006) considered only sulfate, primary BC, and OC from fossil and biomass burning as anthropogenic aerosols. Anthropogenic nitrate, secondary organic carbon (SOA), and dust were not included and these three aerosol components are all estimated to have a negative RF (Forster et al., 2007). The diversity of the RF estimates due to nitrate is large, from around −0.2 to −0.02 W m^{-2} (Adams et al., 2001; Bauer et al., 2007; Jacobson, 2001; Liao and Seinfeld, 2005; Myhre et al., 2006, 2009). The formation of nitrate particles is dependent on other aerosol components and precursors in

a complex way and is likely the main reason for the large range in the RF due to nitrate. The formation of nitrate in the fine mode depends on the excess of ammonia and it is thus important to obtain a sufficiently low amount of sulfate to allow for the formation of ammonium nitrate. Observations show an important land–sea contrast, where fine mode nitrate dominates over land and coarse mode nitrate dominates over ocean (Putaud et al., 2004; Quinn and Bates, 2005). The size of the nitrate particles is important since coarse mode nitrate particles give a weak RF while fine mode nitrate generate a strong RF.

SOA particles were included in the Schulz et al. (2006) study, but only a natural component was considered, while the anthropogenic was not taken into contribution. Several global aerosol modelling studies of SOA have been performed (Chung and Seinfeld, 2002; Henze and Seinfeld, 2006; Hoyle et al., 2007; Tsigaridis and Kanakidou, 2003). The results show that SOA constitute a significant fraction of the total organic matter and even dominate in certain regions. Observations also show a dominance of SOA, when compared to primary organic matter in industrialized regions (Crosier et al., 2007; Gelencser et al., 2007).

As shown in Schulz et al. (2006) the spread in RF due to the direct aerosol effect among the models is substantial even though the emissions are similar. The diversity among the global aerosol models is found to be more related to the parameterizations of transport, removal chemistry, and aerosol microphysics than to the aerosol emissions used in the models (Textor et al., 2007). Substantial differences in RF were also found using a single model, with various meteorological input data (Liu et al., 2007).

12.5.2.2. Atmospheric Absorption by Carbonaceous Aerosols

Ramanathan and Carmichael (2008) estimate that the direct aerosol effect of BC leads to an RF at the TOA of 0.9 W m^{-2} and a reduced radiation reaching the surface of 1.7 W m^{-2}. The estimate by Ramanathan and Carmichael (2008) is based on a method using a combination of observations, satellite retrievals, and models. It is important to note that the RF due to BC is a result of fossil fuel as well as biomass burning changes. The emission of BC from biomass burning is also associated with emissions of organic matter and the net effect of these emissions is an RF close to zero (Forster et al., 2007; Schulz et al., 2006). The RF due to BC is particularly strong in Asia (Ramanathan and Carmichael, 2008) and contributes to an enhanced solar heating rate in the atmosphere (Ramanathan et al., 2007).

The state of mixing of BC particles is important (Fuller et al., 1999; Haywood and Shine, 1995). The aerosol optical properties of BC have been quantified and better constrained by observations showing that aged BC is often coated with other aerosol components; while more fresh BC particles are often not mixed with other aerosol components (externally mixed) (Bond and Bergstrom, 2006; Bond et al., 2006). Suggestions that OC can absorb at short wavelengths in the solar spectrum (Jacobson, 1999; Dinar et al., 2008; Barnard et al., 2008; Sun et al., 2007), has not been quantified in terms of solar absorption or RF.

12.5.3. Semidirect Effects of Aerosols

The degree of absorption versus scattering by aerosols depends on type and mixing state. Aerosol species that are efficient absorbers of solar radiation in the atmosphere are mainly BC, but also, to some extent, mineral dust and some organic compounds. An absorbing aerosol (e.g., soot) layer, as typically encountered in the boundary layer, may therefore lead to a substantial radiative heating in this layer and a corresponding reduction in the solar radiation reaching the surface (Kaufman et al., 2002; Ramanathan et al., 2001). The term *semidirect aerosol effect* was first introduced by Hansen et al. (1997). They investigated the effects of radiative forcing in a GCM and found that the large-scale cloud cover was sensitive to the amount of absorbing aerosols in the lowest model layers. A reduction in the low-level cloud cover corresponds to a positive radiative forcing at the TOA, and thereby a warming of the Earth–atmosphere system.

However, the concept of absorbing aerosols influencing clouds had already been introduced by Grassl (1975). In this study, absorbing aerosols within cloud droplets and in cloudy air were found to reduce cloud albedo and heat the cloud layer. Observational support for absorbing aerosols influencing cloud cover in the Tropics was presented by Ackerman et al. (2000), as a minimum in trade cumulus cloud coverage was found in regions and seasons where clouds were typically embedded in dark haze. Ackerman et al. (2000) also presented large eddy-model (LES) simulations showing that absorbing aerosols may reduce trade cumulus cloud coverage by decreasing the boundary layer convection that drives stratocumulus formation.

This triggered a number of model studies investigating the semidirect aerosol effect on a global-scale. Some studies (Lohmann and Feichter, 2001; Jacobson, 2002; Cook and Highwood, 2004) found that absorbing aerosols decreased low-level cloud cover, in agreement with Hansen et al. (1997) and Ackerman et al. (2000), thereby reducing reflection of solar radiation and having a warming effect on the climate. Others, such as Menon et al. (2002) and Penner et al. (2003), found the opposite. In Menon et al. (2002) absorbing aerosols increased large-scale cloud cover. Penner et al. (2003) pointed out that the injection height of the absorbing aerosols is crucial for their effects on clouds,

possibly explaining the apparently contradicting results arising from different GCM studies. The importance of the vertical structure of the absorbing aerosols for the semi-direct aerosol effect was also pointed out by Johnson et al. (2004). In LES model simulations of marine stratocumulus, they found absorbing aerosols located in the boundary layer to increase cloudiness, while the opposite was true for aerosol absorption above the clouds.

Stuber et al. (2011) performed idealized simulations with absorbing aerosols at different levels in the troposphere using two GCMs to investigate the robustness of the responses. Inclusion of absorbing aerosols in the lower part of the troposphere led to a surface warming, whereas absorbing aerosols in the middle and higher troposphere caused a surface cooling in both models. The GCM responses of absorbing aerosols clearly show changes in the local cloud distribution and water content (liquid and frozen) but also change in the instability of the troposphere. The climate sensitivity of absorbing aerosols was mostly higher than for a doubling of CO_2. These idealized experiments show that a realistic absorbing aerosol distribution may impact the clouds and tropospheric and surface temperature in a complex way.

Further observational support for the semidirect aerosol effect was given by Koren et al. (2004), based on satellite observations from the MODIS instrument. Focusing on the Amazon basin they found cloud cover to be anti-correlated with smoke optical depth, and attributed this to absorption within the smoke layer, which stabilized the boundary layer and reduced the moisture fluxes from the surface. In Denman et al. (2007), the semidirect aerosol effect was summarized to correspond to a negligible or slightly positive radiative forcing at the TOA.

In terms of radiative fluxes at the TOA, the semidirect aerosol effect has been reported to be dominated by a more conventional aerosol effect on clouds, where aerosols acting as cloud condensation nuclei (CCN) increase cloud lifetime and coverage by delaying precipitation release (i.e., the cloud lifetime effect, see Section 12.6.4). However, a recent study by Koren et al. (2008) shed new light on the roles of these competing effects. Based on a theoretical framework and satellite observations from the MODIS instrument, they reported that, while the cloud lifetime effect dominates the semidirect aerosol effect in regimes with relatively low AODs, the opposite is true for larger AODs. AOD was in this study taken as a proxy for both CCN and the aerosol potential to absorb solar energy. The physical reason for this transition was that, while polluted clouds are relatively insensitive to further increases in CCN, there is no corresponding saturation effect for aerosol absorption, which increases steadily with aerosol loading. The study of Koren et al. (2008) uses satellite data over Amazon, but a similar pattern for the relationship between AOD and cloud cover is found in many regions in a global study using MODIS data (Myhre et al., 2007b). Koren et al. (2008) challenged the modelling community to test or incorporate this new concept in cloud-resolving or global models.

12.5.4. Aerosol Indirect Effects

More than three decades ago, Twomey (1974) stated the first theory on how anthropogenic aerosols may influence climate through their impact on clouds. According to this hypothesis, often called the first aerosol indirect effect, an increase in atmospheric pollution will lead to an increase in cloud albedo, everything else being equal. Since then, a number of hypotheses on how anthropogenic aerosols may influence climate through clouds have been published. While some of the hypotheses have been studied extensively using models, satellite data or laboratory experiments, other hypotheses are new and controversial and have not yet been sufficiently tested. Current suggestions for possible aerosol indirect effects are:

1) The first indirect effect or cloud albedo effect: In warm clouds, anthropogenic aerosols increase the cloud albedo by acting as CCN, thereby increasing cloud droplet number concentrations and decreasing cloud droplet sizes in the case of constant liquid water content (LWC) (Twomey, 1977; see also Section 12.2.1.2).

2) The second indirect effect or cloud lifetime effect: The anthropogenic decrease in cloud droplet sizes causes a less efficient precipitation production, which could lead to increased cloud lifetimes or increased cloud horizontal and vertical extension (Albrecht, 1989).

3) The cloud glaciation effect: Ice formation in clouds at temperatures above approximately −35°C occurs by the aid of so-called ice nuclei (IN) (i.e., heterogeneous freezing). Such IN are typically insoluble particles with crystalline structure. Candidates are mineral dust, biological particles, and soot (Pruppacher and Klett, 1997). Soot particles are largely of anthropogenic origin, so anthropogenic activity may have introduced additional IN into the atmosphere. Such an increase in IN would lead to an anthropogenic increase in freezing and cloud glaciation (e.g., Lohmann, 2002).

4) The de-activation effect: A relatively new idea is that anthropogenic sulfur coatings can potentially de-activate IN (Girard et al., 2004), or alternatively make them less efficient in mixed phase clouds (Storelvmo et al., 2008b; Hoose et al., 2008).

5) The thermodynamic effect: As large droplets freeze more readily than small droplets (Pruppacher and Klett, 1997), the anthropogenic decrease in cloud droplet radius is expected to counteract freezing (e.g., Rosenfeld and Woodley, 2000).

6) Aerosol indirect effects on cirrus: The formation of pure ice clouds (i.e., cirrus) by homogeneous freezing of super-cooled aerosols may occur at temperatures below approximately −35°C. Although this freezing process can take place without the aid of IN, the presence of IN may delay or entirely prevent homogenous freezing, as heterogeneous freezing processes may deplete water vapour to the extent that the high super saturation required for the onset of homogeneous freezing are not reached (e.g., DeMott et al., 1997).

Figure 12.28 shows a schematic overview of how the above effects influence precipitation, cloud microphysical properties, and radiation. Except from the deactivation effect, all of the above effects were discussed in IPCC AR4 (Denman et al., 2007) in light of recent publications. Rather than repeating what was already thoroughly

FIGURE 12.28 Schematic of aerosol indirect effects on warm and mixed-phase clouds. *(Source: Hoose et al., 2008.)*

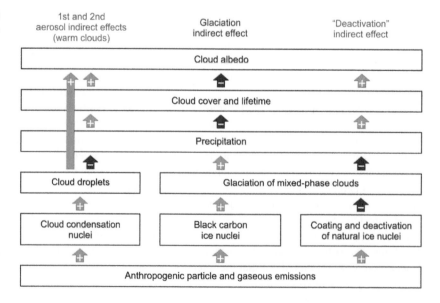

discussed in IPCC AR4, we here focus on aerosol indirect effects, which have recently received particular attention from the scientific community. The following sections review recent publications on aerosol indirect effects, and are further divided into subsections on warm and cold clouds, respectively. In Section 12.5.4.2, a discussion of the recent results and suggestions of future studies required for increased understanding of aerosol indirect effects, and ultimately the climate system, is given.

12.5.4.1. Aerosol Indirect Effects Associated with Warm Clouds

As mentioned above, the aerosol indirect effect on warm clouds is typically divided further into the cloud albedo effect and the cloud lifetime effect. Both effects act to increase cloud albedo and cool the Earth–atmosphere system by increasing the amount of solar radiation reflected back to space. In IPCC AR4, the model estimates of radiative forcing (i.e., the perturbation to the global energy balance of the Earth–atmosphere system) associated with the cloud albedo effect range from -0.22 W m^{-2} to -1.85 W m^{-2}, with a best estimate of -0.7 W m^{-2} (Forster et al., 2007). Low estimates are associated with model simulations using empirical relationships between aerosol mass and cloud droplet number concentration (CDNC) fitted to satellite data to estimate the cloud albedo effect (ranging from -0.22 W m^{-2} to -0.5 W m^{-2}). A recent study estimating the cloud albedo effect from satellite data alone (Quaas et al., 2008) gives an even lower estimate of -0.2 ± 0.1 W m^{-2}, suggesting that global models are exaggerating the effect compared to satellite measurements. Furthermore, based on ground-based remote sensing at the Southern Great Planes, Kim et al. (2008) reported that while the first aerosol indirect effect could be observed in adiabatic clouds, it was not readily observed in subadiabatic cases. A possible explanation is that entrainment and mixing processes attenuate the effect. However, high uncertainties are associated not only with model estimates, but also with satellite retrievals and in-situ measurements.

McComiskey and Feingold (2008) recently demonstrated how measurements of ACI differ substantially among various observational platforms, leading to radiative forcing that differ by several W m^{-2} for the same anthropogenic aerosol perturbation in an idealized cloud case. Storelvmo et al. (2009) analysed the cloud albedo effect in the IPCC AR4 models used for transient simulations of future climate. They found that about 1.3 W m^{-2} of the 2 W m^{-2} spread in estimated SW forcing for the year 2000 can be explained by the different empirical relationships between aerosol mass and CDNC used to calculate the aerosol indirect effect in the models.

The range of model estimates of the total anthropogenic aerosol effect on warm clouds (i.e., both the direct and indirect aerosol effects) range from -0.2 W m^{-2} to -2.3 W m^{-2} in IPCC AR4 (note that this range is based on a slightly different set of model studies than that of the cloud albedo effect) (Denman et al., 2007). The parameterization of auto conversion (i.e., warm-phase precipitation formation) is crucial when calculating the aerosol lifetime effect in the models. Many different parameterizations of auto conversion are available in the literature, yielding precipitation rates that differ by orders of magnitude for the same cloud LWC and CDNC (e.g., Sotiropoulou et al., 2007). Consequently, an increase in CDNC caused by an anthropogenic aerosol perturbation can lead to substantially different precipitation responses, depending on the auto-conversion scheme. Most, if not all, GCM estimates of the cloud lifetime effect predict increased cloud thicknesses and/or lifetimes as precipitation release is reduced in response to anthropogenic aerosol perturbations.

Furthermore, it has been suggested that in the presence of giant CCN (GCCN; typically large natural particles such as sea salt and dust), the effect of increased CCN concentrations on precipitation release is significantly reduced (Feingold et al., 1999; Zhang et al., 2006; Cheng et al., 2007; Posselt and Lohmann, 2008). The GCCN initiate coalescence and precipitation, and thereby have the ability to trigger precipitation in otherwise non-precipitating clouds. To date, most GCM estimates of the aerosol lifetime effect have restricted themselves to warm stratiform clouds, despite several publications (e.g., Rosenfeld, 1999, based on satellite data; Wang, 2005, based on cloud-resolving modelling) suggesting a pronounced aerosol effect on convective clouds. Menon and Rotstayn (2006) provided the first global radiative forcing estimate of aerosol effects on (warm) convective clouds, based on two GCMs (CSIRO and GISS). Whereas the CSIRO simulations suggested an enhanced total aerosol indirect effect when aerosol influences on convective clouds were included, the opposite was true for GISS simulations. Lohmann (2008) included aerosol effects on both warm and cold stratiform and convective clouds, and found a slight reduction in the total aerosol indirect effect due to aerosol influence on convective clouds.

12.5.4.2. Aerosol Indirect Effects Associated with Cold Clouds

Similar to aerosol effects on convective clouds, aerosol effects on cold clouds have only very recently been included in global climate models. The first global estimate of the cloud glaciation effect was provided by Lohmann (2002), suggesting a reduced aerosol indirect effect under the extreme assumptions that IN are present in unlimited amounts compared to an atmosphere containing no IN at all. A more elaborate study of the cloud glaciation effect was presented by Lohmann and Diehl (2006), concluding that anthropogenic activity may lead to increased heterogeneous freezing and thereby influence cloud lifetime and precipitation processes, depending on the concentration and nature of the natural IN present. Mineral dust and certain primary biogenic aerosol particles (PBAPs) constitute natural IN in the atmosphere. While the impact of including realistic dust mineralogy was tested in a GCM by Hoose et al. (2008), no GCM has so far included PBAPs. Although realistic natural and anthropogenic IN concentrations are required for simulations of aerosol effects on mixed-phase clouds, the interaction of such particles with mixed-phase cloud microphysical processes such as the Bergeron—Findeisen (BF) process is as important in this context. Such processes have typically been treated in a very simplistic manner in GCMs. Recent in-cloud observations from the high alpine research station Jungfraujoch (Verheggen et al. 2007) shed new light on the BF process and its interaction with aerosols. Further guidance on the BF process was given in terms of a theoretical framework presented in Korolev (2007); the implementation of this theoretical framework in a global model was found to have a strong impact on the simulated aerosol indirect effect (Storelvmo et al., 2008a).

Girard et al. (2004) were the first to test the hypothesis of IN deactivation by anthropogenic sulfate in a modelling framework, using a single column model to study arctic haze events. Assuming that sulfur coatings deactivated the IN entirely, they found pronounced impacts on precipitation rates and cloud radiative properties. A partial deactivation of IN as a result of sulfate coatings was also found in recent modelling studies: in the ECHAM5 GCM (Hoose et al., 2008) and in the CAM-Oslo GCM (Storelvmo et al., 2008b). Rather than assuming that a sulfate coating deactivates the IN completely, both studies showed a shift of the IN from contact freezing nuclei to immersion freezing nuclei. In contact freezing mode, IN initiate freezing when colliding with cloud droplets. This freezing mechanism becomes efficient at relatively warm subzero temperatures, but requires insoluble IN particles. Coatings typically render IN slightly soluble, allowing them to activate to form cloud droplets. Thereafter they may initiate freezing of the droplet from within (i.e., immersion freezing), but this freezing mechanism becomes efficient at lower temperatures. Although model studies have so far focused on de-activation due to sulfate coatings, other species may have the same coating effect. The fact that the efficiency of IN, such as mineral dust, may be reduced by physiochemical transformation of other aerosol species has recently been discussed in Baker and Peter (2008). In summary, the de-activation effect counteracts the

cloud glaciation effect, and which effect dominates in any given case depends on the concentration of natural and anthropogenic IN, and the amount of anthropogenic material available for IN coating.

In a recent study, Penner et al. (2009) have shown that the effect on cirrus clouds of emission of anthropogenic aerosols, for instance, from aircraft, could have significant impact and reverse the sign of radiative forcing. Note that the results are preliminary, and further studies need to be done before conclusions can be made.

12.5.4.3. Aerosol Indirect Effects Associated with Various Cloud Types

A recent and comprehensive report on aerosol pollution impact on precipitation, produced by the International Aerosol Precipitation Science Assessment Group (IAP-SAG) (Levin and Cotton, 2007), concluded that despite the progress made in recent years, pollution effects on precipitation are not understood, neither on the scale of a local storm nor on the global-scale. Obtaining observational evidence for pollution affecting precipitation release and cloud lifetimes is challenging, and recent modelling results render not only the magnitude but also the sign of this effect as highly uncertain. However, findings in recent publications suggest that the effect of increased aerosol concentrations on convective precipitation is to a large extent determined by environmental parameters such as the unperturbed aerosol concentration or the relative humidity.

Figure 12.29 gives an overview of the majority of global estimates of the aerosol indirect effect. Evidently, the spread in estimates has not decreased with time. On the contrary, estimates span a larger range than ever, much due to the new aerosol indirect effects recently taken into account.

12.5.5. Radiative Forcing Summary

Figure 12.30 and Table 12.1 show estimates of the RF for seven different types of climate forcing mechanisms over the industrial era up to 2007. Only anthropogenic atmospheric compounds are included. For anthropogenic climate forcing on surface and natural climate forcing, we refer to Forster et al. (2007). The WMGG dominates in terms of RF and consists of CO_2, CH_4, N_2O, and halocarbons. We group ozone, NO_2, and stratospheric water vapour into a common class of short-lived gases (SL-G). These gases have quite variable lifetimes that are too short for them to be well-mixed in the atmosphere. Estimates for five groups of aerosol effects are included in Figure 12.30. RF due to the direct aerosol effect and the cloud albedo effect were estimated earlier, but separate estimates of the semidirect aerosol effect and the cloud lifetime effect

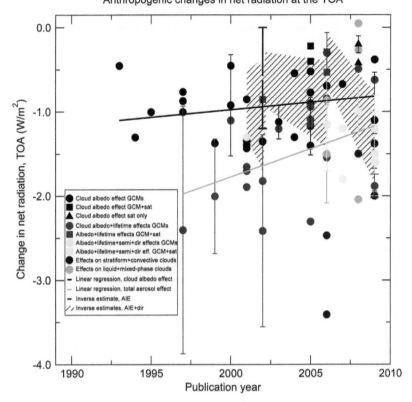

Published estimates of the aerosol indirect effect
Anthropogenic changes in net radiation at the TOA

FIGURE 12.29   Model and satellite estimates of the aerosol indirect effects over the last two decades. Blue represents estimates of the first AIE from GCMs (circles), GCMs combined with satellite measurements (triangles), and satellite only (square). Red represents estimates of both the first and second AIE from GCMs (circles) and GCMs combined with satellite estimates (triangles). The yellow circle represents an estimate of the first and second aerosol indirect effects, the direct, and semi-direct effects. Green circles represent estimates of the first and second indirect aerosol effects on mixed-phase clouds, and black circles represent the first and second AIE on both stratiform and mixed-phase clouds.

were not provided earlier (Forster et al., 2007; Hansen et al., 2005). We also discuss here an estimate for indirect aerosol effects of mixed phase clouds.

12.6. CONTRIBUTIONS TO TROPOSPHERIC CHANGES FROM THE TRANSPORT SECTOR AND FOR DIFFERENT REGIONS

Several recent studies on the impact of emissions from the transport sectors, in particular ship and aircraft have been performed (Eyring et al., 2010; Hoor et al., 2009; Collins et al., 2009; Dalsøren et al., 2009a). Both sectors differ from land-based sectors. A typical example of the differences is illustrated by emission from ship traffic. Ocean-going ship types (e.g., bulk, oil transport, container ship) have extremely different ship tracks, with different environmental impact, and very different growth rates (Dalsøren et al., 2009a). Furthermore, the climate impact from the different transport sectors is different, which makes it necessary to treat sectors individually.

Ships and aircraft have so far not been affected by international policy measures to reduce emissions. Understanding and quantifying the impact of ship emissions is

particularly challenging since the associated composition changes lead to large positive and negative global climate forcings.

Recent regional differences in NO_x emissions play a significant role in the atmospheric oxidation process and on the impact on climate compounds such as O_3 and CH_4 and oxidants such as OH. For the period 1996—2004 substantial reductions of NO_2 were found over Europe and the United States, corresponding to regional emission decreases during this period, while emissions continued to increase in developing countries and accelerated in some transition economy regions, as demonstrated in Figure 12.31 (Akimoto, 2003).

There has been a large increase in NO_x emissions since pre-industrial time, which have led to significant increase in the atmospheric nitrogen burden. Tsigaridis et al. (2006) estimate global increase by a factor of 4. Galloway et al. (2004) calculated an increase in the nitrogen oxide deposition from 12.8 Tg(N) year^{-1} in 1860 to 45.8 Tg(N) year^{-1} in the early 1990s. Using emissions valid for 1992 and a critical load varying between different natural ecosystems, Bouwman et al. (2002) estimated that sulfur and nitrogen deposition exceeded critical loads over 7% of the region covered by natural vegetation (see also Fowler et al., 2009). Changes in modelled NO_2 concentrations

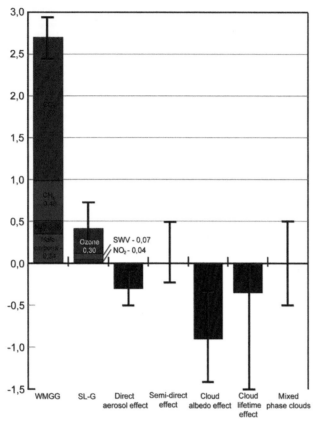

FIGURE 12.30 Radiative forcing estimates of atmospheric compounds from the pre-industrial period 1750 to 2007. RF estimates of WMGG are similar to the one reported in Forster et al. (2007) except that it has been taken into account that growth in RF due to CO_2 is 0.03 W m^{-2} yr^{-1} between 2005 and 2007. Estimates for ozone and stratospheric water vapour are similar to the value given in Forster et al. (2007), whereas the estimate of NO_2 of 0.04 W m^{-2} is from Kvalevåg and Myhre (2007). The estimate of the direct aerosol effect is from Myhre (2009). The RF for the cloud albedo effect is calculated based on Forster et al. (2007), with the modification that no model estimates have been excluded based on the number of aerosol species treated in the various models. Additionally, estimates published after Forster et al. (2007) have also been taken into account. While Forster et al. (2007) reported the median value of the estimates considered, we here report the mean. However, the mean and median values differ by only 2%. As few model studies report estimates of the cloud lifetime effect only; the estimate for cloud lifetime effect is calculated by subtracting the mean RF for the cloud albedo effect from the mean RF for both the albedo and lifetime effect. The latter is calculated based on studies reported in Denman et al. (2007), in addition to recently published estimates. This way of calculating the RF for the cloud lifetime effect relies on the assumption that the total (albedo and lifetime) effect equals the linear sum of the two separate effects to a good approximation (e.g., Kristjánsson, 2002). The few RF estimates for aerosol effects on mixed-phase clouds available in the literature suggest a potentially large effect, but so far model studies are inconclusive in terms of the sign of this RF. Hence, we refrain from giving an RF estimate here, but rather give an uncertainty range from −0.5 W m^{-2} to 0.5 W m^{-2}. The semi-direct effect is estimated based on a combination of published results of realistic BC atmospheric distributions and idealized experiments in Stuber et al. (2011). The absorption AOD included in the idealized experiments in Stuber et al. (2011) is 0.05, which is an order of magnitude larger than the total (anthropogenic and natural) absorption AOD from global aerosol models (Kinne et al., 2006). All the RF from the idealized semi-direct

TABLE 12.1 Summary of Radiative Forcing (RF) for Well-Mixed and Short-Lived Greenhouse Gases, and the Contribution from Different Aerosol Effects

Radiative Forcing Mechanism	Best Estimate	Range
WMGG	2.69	±0.27
CO_2	1.72	±0.17
CH_4	0.48	±0.05
N_2O	0.16	±0.02
Halocarbons	0.34	±0.03
SL-GHG	0.41	0.25, 0.73
Ozone	0.30	0.15, 0.60
Stratospheric water vapor	0.07	±0.05
NO_2	0.04	±0.02
Direct aerosol effect	−0.30	±0.20
Semidirect effect	NA	−0.25, +0.50
Cloud albedo effect	−0.90	−1.4, −0.3
Cloud lifetime effect	−0.35	−1.5,0
Mixed phase clouds	NA	−0.5, +0.5

(the uncertainty range is included)

during the 1990s qualitatively reproduce the different changes in regional emissions (Granier et al., 2003). Large decreases are found over Europe and Russia, smaller decreases are identified over western North America, and large increases in populated or industrialized areas in Asia and parts of Latin America, which is in line with reported emission trends in Figure 12.31. For further discussions on chemical interactions of pollution emissions, in particular the regional impact of compounds such as NO_x, CO, NMVOC, and SO_2 we refer to the accompanying papers by Monks et al. (2009) and Fowler et al. (2009).

12.6.1. Composition Change Due to Emission from the Transport Sectors

During recent years a strong focus has been on the atmospheric impact from the transport sectors. Several ongoing

aerosol effect experiments are within the range −0.25 to +0.5 W m^{-2} if the values are divided by 10 (Stuber et al., 2011). The factor of 10 is adopted since the idealized experiments are performed with an absorption AOD of this magnitude larger than global aerosol models predicts for current condition.

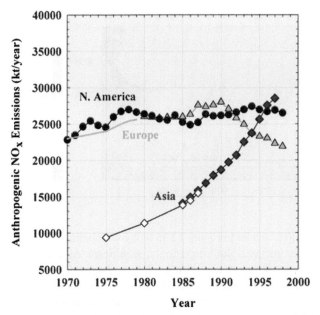

FIGURE 12.31 Changes in anthropogenic NO$_x$ emissions in North America (United States and Canada), Europe (including Russia and the Middle East), and Asia (east, South East, and South Asia). *(Source: Akimoto, 2003. Reprinted with permission from AAAS.)*

EU projects (ACCENT, QUANTIFY, ATTICA, ECATS) investigate the effects of aviation, shipping, road traffic, and other land-based transport on the atmosphere (Eyring et al., 2007a; Hoor et al., 2009; Eyring et al., 2010; Dalsøren et al., 2007; Dalsøren et al., 2009a; Lee et al., 2010). The focus in these projects is primarily on regional- to global-scales.

12.6.1.1. Studies of Current Impact

The European Integrated Project QUANTIFY has investigated the impact of the various transport sectors on atmospheric composition and climate. A comprehensive model activity has evaluated the impact of transport emissions on ozone, methane, and the resulting radiative forcing. Figure 12.32 shows the combined effect of shipping, aviation, and road traffic on column ozone. Perturbations are most pronounced during the summer months in the Northern Hemisphere, with local perturbations of 4% above the background levels, although individual model increases can be larger.

Gas and particle emissions from ocean-going ships are a significant and growing contribution to the total emissions from the transportation sector. The EU financed ATTICA project has published a detailed review on the current state

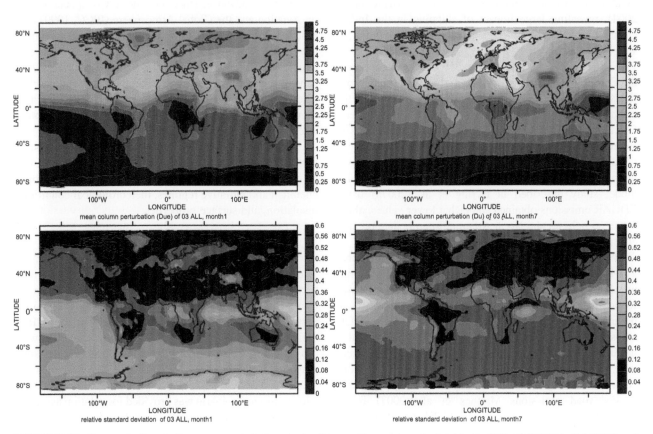

FIGURE 12.32 Top panels: Mean column ozone perturbations due to all types of transport, as simulated by the TM4, LMDzINCA, OsloCTM2, and p-TOMCAT models (ensemble mean) for January (left) and July (right). Bottom panels: Corresponding relative standard deviations for January (left) and July (right). *(Source: Hoor et al., 2009.)*

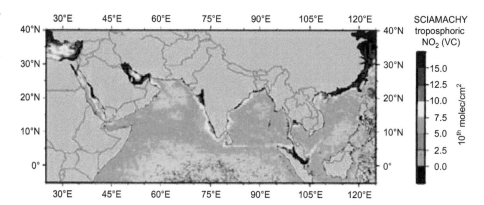

FIGURE 12.33 NO$_x$ signature of shipping in the Indian Ocean as observed by satellite. *(Source: Richter et al., 2004 © Copyright 2004 American Geophysical Union. Reproduced/modified by permission of American Geophysical Union.)*

of research concerning the effects of shipping emissions on atmospheric chemistry and climate (Eyring et al., 2010). Nearly 70% of ship emissions occur within 400 km of coastlines, causing AQ problems through the formation of ground-level ozone, sulfur compounds, and particulate matter in coastal areas. Ozone and aerosol precursors emitted to the atmospheres from ships are transported over several hundreds of kilometres, contributing to AQ problems further inland. Figure 12.33 shows satellite measurements of NO$_2$ columns retrieved by the SCIAMACHY instrument over marine areas (Richter et al., 2004), revealing large impact of shipping in the Indian Ocean, Eastern Mediterranean, and coastlines in general, as well as further inland.

Figure 12.34 shows global ozone enhancements from the OsloCTM2 model for year 2000 due to ship emissions (Dalsøren et al., 2007). The signal clearly follows the conventional ship tracks, but also affects continental areas. Calculations of ozone enhancements from ship emissions by Endresen et al. (2003) give maximum ozone perturbations of 12 ppb in the marine boundary layer during summer over the northern Atlantic and Pacific regions, while the multi-model study described in Eyring et al. (2007a) using EDGAR emissions obtained somewhat lower values, about 5 ppb—6 ppb, for the North Atlantic. In an update on ship emissions and their environmental impact, Dalsøren et al. (2009b) isolated the impacts of major ship types and ports. The large and increasing contribution from container ships was highlighted. In agreement with earlier studies, it was found that ship emissions contribute significantly to pollutants such as ozone, sulfate, nitrate, BC, and OC aerosols. In another study, Dalsøren et al. (2009a) show that in regions where acidification is still of some concern (e.g., northwest America and Scandinavia) ship emissions contribute heavily to wet deposition of nitrate.

For road emissions Matthes et al. (2007) found maximum surface ozone increases of 12% in northern mid-latitudes during July. New estimates of the impact of road traffic emissions on ozone have been calculated by Hoor et al. (2009) using an ensemble of six global models. The result is shown in Figure 12.35 along with impacts from the other sectors. Not surprisingly, significant enhancements due to shipping and road traffic are largely confined to marine and continental areas, respectively. However, the ship signals also affect coastal areas, and the impact of road traffic is not confined to the maximum emissions source regions in Europe and North America. Significant increases in ozone due to the aviation sector are confined to the NH, where they are zonally mixed and, during summer, reach a maximum at high latitudes.

Scientific advances since the 1999 IPCC Special Report on Aviation and the Global Atmosphere (IPCC, 1999) have reduced key uncertainties, yet the basic conclusions remain the same. The ATTICA project has recently delivered a review of the current impact of aviation (Lee et al., 2010). Much progress has been made in the past 10 years on characterizing aviation emissions, although major uncertainties remain, especially for particle emissions, and the combined effect on ozone and methane through NO$_x$ emissions.

The spatial variability in the effects of the transport sectors is large, not only because of the uneven distribution of emissions, but also because of differences in ambient chemical and radiative regimes, meteorological conditions, insolation, and surface properties.

12.6.1.2. Studies on Future Trends

The model groups involved in the QUANTIFY project have performed detailed calculations of future impact from the transport sectors. The studies focus on the time evolution between 2000 and 2050, based on the different SRES scenarios. QUANTIFY project emission scenarios were provided for different transport sectors (road traffic, shipping, and aviation) based on the IPCC SRES scenarios A1, A2, B1, and B2 (Nakicenovic et al., 2000). In addition, a low-emission scenario was provided for aviation assuming the implementation of all available technology. The assumption is that emissions of ozone precursors from road traffic will drop to negligible amounts during this century, due to emission reduction technology. At the same time

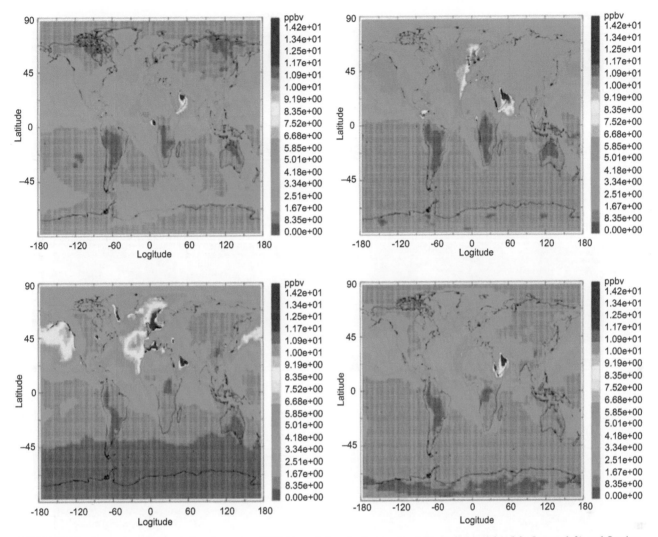

FIGURE 12.34 Ozone change at the surface due to year 2000 ship emissions for January (top left), April (top right), July (bottom left), and October (bottom right). *(Source: Dalsøren et al., 2007 © Copyright 2007 American Geophysical Union. Reproduced/modified by permission of American Geophysical Union.)*

emissions from shipping and aviation are expected to increase during coming decades, since improving technology will not be able to compensate for the increased traffic.

The climate impact from shipping is complex, since current emission trends lead to a global average negative radiative forcing (see Section 12.6.2.1). Future climate impact will, to a large extent, be determined by trends in emissions and on implementation of emission regulations. It should be recognized that environmental impact (on climate and AQ) from ships is local and regional in nature due to contributions from short-lived compounds such as sulfate aerosols. Dalsøren et al. (2007) have studied local effects of ship emissions in coastal regions and in the Arctic from future seaborne traffic.

Collins et al. (2009) have calculated the impact from ship emissions on atmospheric chemistry for the different land-based emissions in 2000 and 2030, ACCENT

PhotoComp, presented in Section 12.4.2.1 (Figure 12.36). The three plots show the change in the tropospheric ozone burden, production, and methane lifetime due to ship emission. Shipping has the largest effect when using the 2030 MFR land-based scenario, a smaller impact when using the 2000 or 2030 CLE, and an even smaller effect when using the high NO_x SRES A2 scenario for the land-based emissions, illustrating the non-linearity of ozone chemistry with lower effects of emissions when added to a higher background. The non-linearity is stronger when looking at the change in the methane lifetime.

Collins et al. (2009) have also performed calculations showing the effect of sulfur emissions, according to which the increase in sulfate aerosols and sulfur deposition due to the SO_2 from shipping will offset 75% of the benefits in AQ that would be expected from land-based emission controls under the SRES A2 scenario.

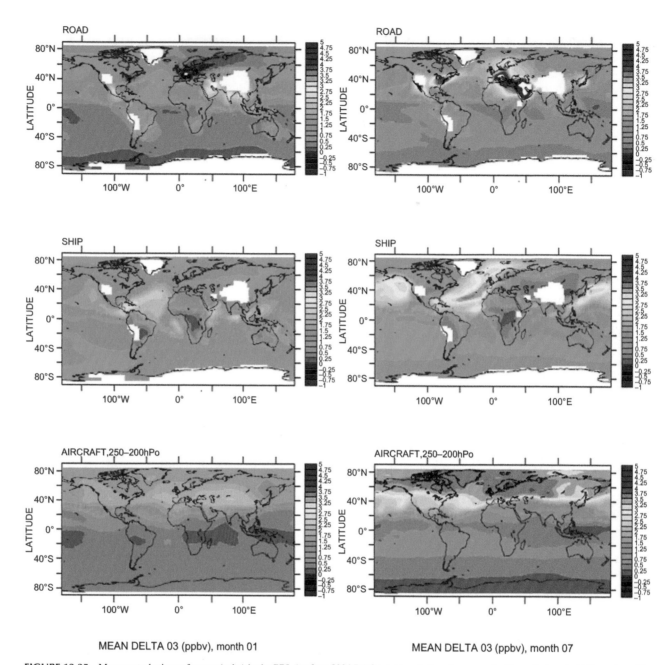

MEAN DELTA 03 (ppbv), month 01

MEAN DELTA 03 (ppbv), month 07

FIGURE 12.35 Mean perturbations of ozone (ppbv) in the PBL (surface-800 hPa) for road emissions (upper row), ship emissions (middle row), and in the upper troposphere (200–250 hPa) for aircraft emissions (lower row) during January (left column) and July (right column). *(Source: Hoor et al., 2009.)*

12.6.2. Climate Impact from the Transport Sectors

12.6.2.1. Impact from the Different Sectors

In general, emissions from the transport sectors increase ozone and reduce methane. Ozone changes from the road sector result in stronger forcing because of the more effective vertical mixing over land that enhances the ozone change at higher altitudes where the forcing efficiency is higher.

In a recent study, Fuglestvedt et al. (2008) show that the net RF due to ship traffic from pre-industrial times to present is negative, opposite to what is found for other sectors. Aerosols are the main reason for the negative RF from shipping, but atmospheric reactive gases (e.g., methane) contribute to the negative RF. Figure 12.37 depicts the combined RF from ozone changes and changes in methane. Ship emissions stand out with a combined negative effect. The reason for this is that NO_x emissions from ship traffic occur over pristine areas where OH enhances efficiently,

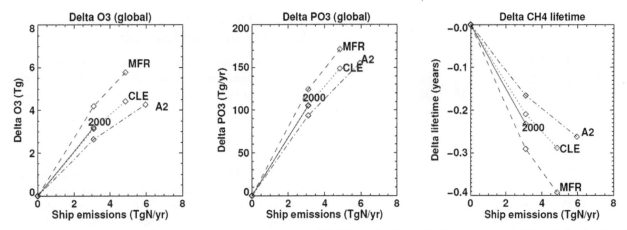

FIGURE 12.36 Change in tropospheric ozone burden, production, and methane lifetime for changes in ship emissions according to the four land-based emission scenarios. *(Source: Collins et al., 2009. www.borntraeger-cramer.de)*

causing a strong reduction in the CH_4 lifetime and concentration (Dalsøren et al., 2009a; Fuglestvedt et al., 2008).

To investigate future impact, Fuglestvedt et al. (2008) have integrated radiative forcing for a current emission over different time horizons in order to compare the effect of different transport sectors and chemical components.

Figure 12.38 shows the integrated radiative forcing due to the various components and sectors with a time horizon of 100 years, as used in the Kyoto Protocol. According to Fuglestvedt et al. (2008) CO_2 is by far the most important substance on this timescale, with the largest contribution coming from road transport. The second largest positive forcing comes from tropospheric ozone, again with a dominating contribution from road transport. Changes in methane and sulfate lead to a negative radiative forcing, with the largest contributions coming from the shipping sector. On the chosen timescale of 100 years the contributions from the short-lived and intermediate perturbations (ozone, aerosols, and methane) become significantly

smaller compared with CO_2. Of the 100-year integrated net radiative forcing from total anthropogenic emissions, the net radiative forcing from transport amounts to 16%.

12.6.2.2. Comparison with Other Sectors

Recent studies have improved our understanding of the contribution from various emission sectors and regions to RF from ozone and aerosol changes. Berntsen et al. (2006), Fuglestvedt et al. (2008), and Unger et al. (2008) estimated the contribution to ozone radiative forcing. Unger et al. (2008) evaluated the RF due to ozone for the year 2030 from six emission sectors, as indicated in Figure 12.39, with the largest contribution from the transportation and biomass burning sectors.

Koch et al. (2007) calculate the radiative forcing for the direct aerosol from six different sectors. The impacts of present-day aerosols emitted from particular regions and from particular sectors using the Goddard Institute for Space

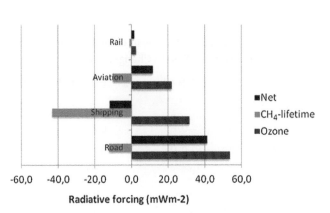

FIGURE 12.37 Radiative forcing due to ozone, impact of changes in CH_4 lifetime, and the corresponding net radiative forcing for four transport sectors from pre-industrial to present, as calculated by Fuglestvedt et al. (2008). *(Source: Copyright 2008 National Academy of Sciences, U.S.A.)*

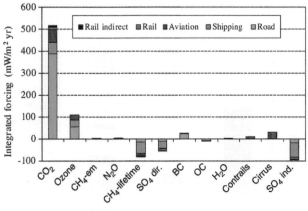

FIGURE 12.38 Integrated radiative forcing of current emissions, by substance and transport subsector, over different time horizons. Integrated global mean RF (mWm^2yr) due to 2000 transport emissions, time horizon H = 100 years. *(Source: Fuglestvedt et al., 2008. Copyright 2008 National Academy of Sciences, U.S.A.)*

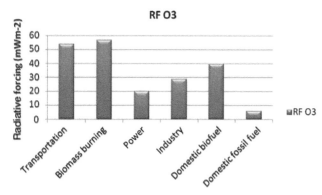

FIGURE 12.39 Radiative forcing due to ozone changes in 2030 from different sectors, as calculated by Unger et al. (2008). *(Source: Copyright 2008 American Geophysical Union. Reproduced/modified by permission of American Geophysical Union.)*

Studies (GISS) GCM were studied. The conclusion from the work was that South East Asia exports over two-thirds of its emitted BC and sulfate burden over the NH. While Africa has the largest biomass burning emissions, South America generates a larger (about 20% of the global carbonaceous) aerosol burden; about half of this burden is exported and dominates the carbonaceous aerosol load in the SH. The resulting calculated direct anthropogenic radiative forcings are -0.29, -0.06, and 0.24 W m^{-2} calculated for sulfate, organic, and BC, respectively. Figure 12.40 shows the percentage radiative forcing from various sectors for BC, OC, and sulfate. Two of the largest BC radiative forcings are from residential (0.09 W m^{-2}) and transport (0.06 W m^{-2}) sectors, making these potential targets to counter global warming suggested by Koch et al. (2007). Most anthropogenic sulfate comes from the power and industry sectors, and these sectors are responsible for the large negative aerosol forcings over the central NH. Further studies have been performed by Shindell et al. (2008a,b), examining AQ and radiative forcing due to emissions reductions by economic sector for different regions.

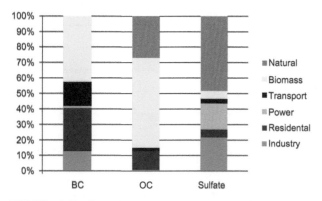

FIGURE 12.40 Sectoral contributions to aerosol forcings shown as percentages relative to total TOA forcing for each species. *(Source: Results taken from Koch et al., 2007 © Copyright 2007 American Geophysical Union. Reproduced/modified by permission of American Geophysical Union.)*

12.6.3. The Impact of Large Emission Increases in South East Asia

South East Asia is where emission of pollutants have increased significantly, particularly during the last decade, and where it is likely that future emissions will continue to grow fast (Isaksen et al., 2009). These emissions are at such a high level that they have the potential to markedly increase transport of ozone to the US west coast (Zhang et al., 2009), and reduce global methane levels through enhanced OH levels (Dalsøren et al., 2009b). In the global model study by Dalsøren et al. (2009b) perturbations in 2000 and 2006, and future (2020) perturbations of OH and methane lifetimes are calculated. The adopted NO$_x$ emissions and results for the global average changes in OH and methane lifetime are given. For details of the adopted emissions in the model studies, see Dalsøren et al. (2009a). The high emission cases are based on the rapid growth in energy use after 2000, and assumption of future significant growth (see also Isaksen et al., 2009). Note that due to the methane response time of approximately 10 to 12 years to perturbation changes, there will be a lag in the methane response to lifetime changes.

In agreement with former studies it is found that global anthropogenic emission changes have resulted in small changes in OH and in methane lifetime from 1980 to 2000 (Table 12.2). As shown in Figure 12.31 emission of NO$_x$, a main OH precursor, has decreased significantly during this time period in Europe. The model studies therefore indicate a significant cancellation of the increase in South East Asia from the overall global emission.

Comparison for 2000 with satellite measurements show that the model is able to reproduce retrieved tropospheric NO$_2$ columns over South East Asia. In the high emission case the 2006 NO$_x$ emissions were fitted to the 2000–2006 increase in energy consumption in China. This case shows better agreement with the satellite retrieved 2000–2006

TABLE 12.2 Estimated Changes in Global Average OH and Methane Lifetime Due to Changes in NO$_x$ Emissions

	NO$_x$	OH Change (%)	CH$_4$ Lifetime (%)
1980	4.8		
2000	8.2	−0.1	0.22
2006 ref	11.5	−0.15	0.25
2006 high	12.7	0.9	−1.1
2020 ref	15	0.25	−0.5
2020 high	22.3	2.75	−2.7

2000 changes are relative to 1980, and changes after 2000 are relative to 2000.

tropospheric NO_2 column trend over central and eastern China than the reference case. The satellite comparison thereby highlights an issue noted in other studies: The question addressed whether current emission inventories underestimate the recent increase in NO_x emissions in South East Asia. The development of global OH and methane due to emission changes in South East Asia after 2000 is also dependent on the balance between changes in NO_x and CO emissions. From 2000 to 2006 the sign of OH and methane lifetime changes are different in the reference case and the high NO_x emission case. Though the high NO_x emission case is an upper estimate of OH changes it is better than the reference scenario in reproducing measured NO_2 trends. From 2000 to 2020 the contributions from South East Asia are moderate in the base scenario but much larger than the small overall global emission impact from 1980 to 2000. In the 2020 high South East Asian NO_x emission case the impact on global OH and methane is large. It is of comparable absolute magnitude to the international ship traffic (Endresen et al., 2003; Dalsøren et al., 2007; Dalsøren et al., 2009a). The OH increase in South East Asia is 4.7% in the 2020 high NO_x case. Such a large change would have a significant effect on the oxidation of regional pollutants with short and intermediate lifetimes.

Zhang et al. (2009) compiled a detailed anthropogenic emission inventory for Asia for the spring 2006 period of INTEX-B but, similar to Dalsøren et al. (2009b), used a higher estimate for Asian NO_x emissions by constraining them to satellite observations. Zhang et al. (2009) found that Asian pollution enhanced surface-ozone concentrations by 5–7 ppbv over western North America in spring 2006. The 2000–2006 rise in Asian anthropogenic emissions increased this influence by 1–2 ppbv.

12.6.4. Impact on the Arctic (Arctic Haze)

Human-induced climate–chemistry interactions could be of particular importance in the Arctic. The Arctic has warmed more rapidly during recent decades than most parts of the globe, and it is projected to warm even more during the twenty-first century. During the past 30 years, the extent of Arctic sea-ice during the summer has decreased by about 25% (Stroeve et al., 2007), and the seasonal melt area of the Greenland ice sheet has increased rapidly, by ~7% per decade since 1979. While much of the Arctic warming originates from the same factors underlying global warming, studies suggest that both tropospheric aerosols and ozone play an important role in the Arctic.

12.6.4.1. Observational Data on Trends

Measurements of pollutants in the Arctic are relatively limited compared to more populated mid-latitude regions. Long-term (20–30 years) surface observations are available for ozone, sulfate, BC, and dust, though only from a few stations. Data from Alert, Canada show no appreciable trends during the 1980s, followed by decreases in many aerosols concentrations (sulfate, nitrate, sea-salt) during the 1990s (Sirois and Barrie, 1999). Data from both Alert and Barrow, Alaska show likewise decreasing BC during the 1990s, but suggest that concentrations began to increase again around 2000 (Sharma et al., 2006). The trends in individual aerosol species are consistent with observations of the overall light scattering by aerosols, which peaks in spring as inflow of pollutants is large and near-surface temperature inversions inhibit deposition, leading to the seasonal aerosol build-up known as Arctic Haze (Iversen and Joranger, 1995). Measurements of the Arctic Haze began in the late 1970s: they show a maximum scattering during the 1980s and 1990s, followed by decreases through around 2000 (Quinn et al., 2007).

Ozone data from Barrow show a slightly increasing trend from 1975–1995, but they are not statistically significant (Oltmans et al., 1998). In contrast, measurements of CO, the only ozone precursor (other than methane) with long-term Arctic records, show decreases at northern high latitudes during the 1990s (Novelli et al., 1998). Only few long-term data on pollutants are available for the Eastern Hemisphere side of the Arctic.

On much longer timescales, the recent analysis of BC concentrations from a Greenland ice core has allowed reconstruction of BC deposition since the late eighteenth century (McConnell et al., 2007). These suggest that emissions in the main source regions, North America, Europe, and more recently (and to a lesser extent) Asia, were extremely large in the early 1900s, declined thereafter through the middle of the century, and increased again during recent decades.

12.6.4.2. Climate Impact of Chemically and Radiatively Active Short-Lived Species in the Arctic

As emissions within the Arctic are relatively small, most Arctic pollution comes from distant sources. The relative importance of pollutants located within the Arctic as compared with those at lower latitudes is not clear, but remote radiative forcing appears to play a large role (Shindell et al., 2007). Direct radiative forcing due to changes in pollutants located within the Arctic can also be substantial, however, with seasonal values as large as 1 W m^{-2} (Quinn et al., 2008). The contribution of various species to the total pre-industrial to present-day Arctic surface temperature trends shows that BC, ozone, and methane have together caused roughly as much warming as CO_2, while reflective aerosols have caused a cooling of comparable magnitude as seen in Figure 12.41. Shortwave radiation is primarily perturbed via absorption by ozone and BC and reflection by sulfate, OC, and nitrate.

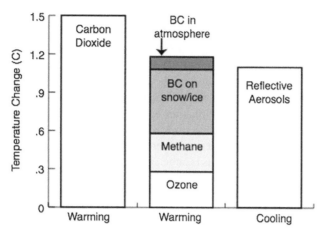

FIGURE 12.41 Estimates of the contribution of particular species to pre-industrial to present-day Arctic (60° to 90°N) surface temperature trends. Values are based on the assessment of modelling and observations of Quinn et al. (2008), and do not include aerosol indirect effects. Reflective aerosols include sulfate and organic carbon.

Longwave fluxes are affected by ozone (and other greenhouse gases) and by aerosols. While aerosol forcing is usually dominated by shortwave impacts, the longwave effects play a larger role at high latitudes during polar darkness when observations show a distinct effect of Arctic Haze on thermal radiation (Ritter et al., 2005). Hence under Arctic conditions, aerosols can sometimes have similar impacts as those to greenhouse gases.

Indirect effects of aerosols can also be large in the Arctic. One prominent indirect effect is that of BC deposited onto snow or ice surfaces, darkening them and reducing their albedo (Warren and Wiscombe, 1980; Clarke and Noone, 1985; Jacobson, 2004). Another is the aerosol indirect effect on clouds, which can lead to very large radiative forcings (>3 W m^{-2}) by increasing cloud cover under conditions of low insolation, which leads to a net warming effect due to absorption of outgoing longwave radiation (Garrett and Zhao, 2006; Lubin and Vogelmann, 2006). There are limited understanding of any of these indirect processes. For example, aerosol indirect effects on clouds are not well-characterized for ice phase clouds (Quinn et al., 2008).

Forcing by Arctic aerosols has a strong seasonal variation for both shortwave and longwave components, while forcing by ozone appears to maximize during boreal spring when Arctic ozone abundances are comparatively large and sunlight is relatively abundant (Quinn et al., 2008).

12.7. IMPACT ON TROPOSPHERIC COMPOSITION FROM CLIMATE CHANGE AND CHANGES IN STRATOSPHERIC COMPOSITION

Changes in climate have the potential to affect tropospheric composition and the distribution in several ways. Changes in

circulation will alter the transport of pollutants, and in particular if stagnant conditions during high pressure situations become more and less frequent, conditions will be more and less favourable for the build-up of high pollution levels (e.g., Royal Society, 2008; Matsueda et al., 2009). Furthermore, higher temperatures will increase water vapour with enhanced OH production, and lightning activities could be enhanced in a warmer climate leading to enhanced NO$_x$ production. Changes in surface characteristics (dryness, wind speed) will affect the exchange of components between the biosphere and the atmosphere (see Fowler et al., 2009). Stratospheric/tropospheric exchange of, in particular, ozone is affected by climate change, and also by changes in ozone depletion in the stratosphere. For instance, the large stratospheric ozone depletion in the 1980s and 1990s had an impact on stratosphere—troposphere exchange and on ozone transport to the troposphere (e.g., van Noije et al., 2004).

12.7.1. Impact of Climate Change on Future Tropospheric Composition

12.7.1.1. Ozone and Its Precursors

Several modelling studies have investigated how future climate change may impact tropospheric composition, ozone in particular. An early study (Thompson et al., 1989) suggested that increases in water vapour, associated with warmer temperatures, would increase the flux through the Reaction (R12.2) (O(^{1}D) + H$_2$O → 2OH), enhancing the loss of ozone and production of the OH. Subsequent work (e.g., Zeng and Pyle, 2003; Collins et al., 2003; Sudo et al., 2003) additionally found that climate change led to an increased flux of ozone from the stratosphere to the troposphere. More recent results from climate models, with detailed stratospheric dynamics have confirmed that a more vigorous Brewer—Dobson circulation (the large-scale circulation that exerts a major control on stratosphere—troposphere exchange (STE)) is a robust feature of models (Butchart et al., 2006). Stevenson et al. (2006) presented results from ten models that simulated the impact of climate change between 2000 and 2030 on ozone. The main climate feedback processes affecting ozone in these models appeared to be the two discussed above. Some of the models were dominated by the negative water vapour feedback, while the positive STE feedback was more prominent in others.

Changes in lightning (and associated NO$_x$ production) in connection with climate change have been studied. Most model studies show an increase in lightning as climate warms (Schumann and Huntrieser, 2007), with increases of up to 60% per degree K of global mean surface warming (Lamarque et al., 2005), although other studies have shown no global trend, just a geographical shift from tropics to mid-latitudes (Stevenson et al., 2005), or decreases (Jacobson and Streets, 2009). Other studies show similar discrepancies, with no clear conclusions on climate change and lightning activity. Observations show no relationship between surface temperature and lightning in the tropics, but a strong positive correlation over NH land areas (Reeve and Toumi, 1999). Some studies suggest that increases in lightning may be the dominant climate-related driver of future upper tropospheric O_3 increases (Wu et al., 2008a). Wild (2007) demonstrated that changes in lightning NO_x emissions had a large influence over the tropospheric ozone burden.

Biogenic VOC emissions are a function of temperature, photosynthetically available radiation (PAR), leaf area index (LAI), and plant species (Guenther et al., 1995, 2006). Several global chemistry models account for some or all of these dependencies (e.g., Hauglustaine et al., 2005; Stevenson et al., 2006; Wu et al., 2008a). Most models show an increase in tropospheric O_3 as VOC emissions rise, particularly in regions with relatively high NO_x levels; however, in low NO_x regions, O_3 often decreases. VOC degradation chemistry influences ozone's sensitivity to VOC emissions, in particular whether isoprene nitrate formation represents a temporary or terminal sink for NO_x (Giacopelli et al., 2005; Wu et al., 2008b). Horowitz et al. (2007) present observational evidence that isoprene nitrates do permanently remove NO_x, and in models that reflect this finding, the surface O_3 response to rising biogenic VOC emissions quite rapidly saturates (Wu et al., 2008b), and can even result in O_3 reductions. For example, Wu et al. (2008b) find that future surface O_3 levels in the southeast United States are reduced as isoprene emissions rise. One study by Arneth et al. (2007) links biogenic VOC emissions also to CO_2 levels (Arneth et al., 2007) and land-use change. The process is not well-known, and has generally not yet been considered in global models.

Changes in the biosphere associated with climate change may have other important effects. Soil NO_x emissions, produced from bacterial activity, are likely to increase as the land warms (Granier et al., 2003; Hauglustaine et al., 2005; Liao et al., 2006; Wu et al., 2008a). Changes in the hydrological cycle, and in particular soil moisture, may significantly affect stomatal opening and hence dry deposition fluxes of ozone and other gases (e.g., Solberg et al., 2008). Enhanced deposition will occur under wetter conditions and vice versa. Increases in CO_2 levels may reduce stomatal opening and hence ozone deposition (Sanderson et al., 2007). Furthermore, higher temperatures will increase evaporative anthropogenic VOC emissions (e.g., Rubin et al., 2006).

Trends in cloud optical depth, aerosol loading and the solar cycle in the tropics (Chandra et al., 1999; Fusco and Logan, 2003) might modulate tropospheric ozone to some

extent. Climate-related changes in temperature, humidity, meteorology, and land use/vegetation can also affect ozone. Fusco and Logan (2003) estimate a 3% ozone increase over Western Europe and 1% decline over much of North America due to 1970–1995 temperature increase and decrease respectively. The overall change in tropospheric ozone due to climate change since the pre-industrial period was positive in 5 of 6 studies in a model assessment (Gauss et al., 2006). However, model studies based on future scenarios taking into account climate changes (Hauglustaine et al., 2005; Stevenson et al., 2006; Dentener et al., 2005; Dentener et al., 2006c) report a significantly reduced global tropospheric burden due to the effects of climate change, partly as a result of enhanced OH formation. It should, however, be noted that the signs of the changes vary in different regions of the troposphere both for past and future changes.

More prolonged and frequent occurrence of heatwaves of the type experienced over Western Europe in August 2003 is expected to occur in a future warmer climate. Solberg et al. (2008) applied meteorological data in a global-scale Chemical Transport Model (the Oslo CTM2), covering the heatwave period, and showed that under such extreme weather situations large ozone levels build up in the planetary boundary layer with current emissions for extended time periods (in excess of 100 ppb for several days). If the 2003 heatwave is taken as a proxy for more frequent severe weather situations during future warmer climates, the effect of measures to control regional pollution could be strongly reduced.

12.7.1.2. Methane

The natural source of methane from wetlands is climate-sensitive (Gedney et al., 2004; Shindell et al., 2004). Changes in OH concentrations directly affect the rate of methane oxidation. Higher levels of water vapour lead to higher levels of OH (Reaction (R12.2)). The reaction of methane with OH is strongly temperature dependent, and proceeds faster at higher temperatures. Several model studies have noted the negative feedback of climate on methane through these mechanisms (e.g., Johnson et al., 2001; Stevenson et al., 2005). All ten of the models in Stevenson et al. (2006) showed a reduction in the methane lifetime in a warmer climate, with an average reduction of 4% for a global mean surface warming of around 0.7 K.

Thawing of permafrost represents a potential significant source of methane to the atmosphere, and may constitute a positive feedback mechanism in the climate–chemistry system (Osterkamp, 2005; Lawrence and Slater, 2005; Walter et al., 2006a,b). With strongly enhancing temperatures in the Arctic, permafrost degradation has been observed to occur rapidly (Jorgensen et al., 2006), and methane has been observed to be bubbling from melting lakes at an increasing rate (Walter et al., 2006a,b). Large amounts of organic carbon are stored in the permafrost, which is partly converted to methane and emitted to the atmosphere, following the thawing of the permafrost (Walter et al., 2006a,b; Zimov et al., 2006). There is a non-linear response in atmospheric methane concentrations to enhanced emissions, currently giving an additional 40% increase in atmospheric concentrations (positive feedback) (Prather and Ehhalt, 2001). The non-linear feedback is likely to be much larger with strong enhancements in methane concentrations, since a larger fraction of the OH loss is through the methane reaction. Methane emissions over periods of decades to several centuries in connection with permafrost thawing (Walter et al., 2006a,b), combined with a methane lifetime of the order of 10 years, makes the impact of methane permafrost emissions depend critically on the time horizon of permafrost thawing. Similar and rapid impact on methane can occur if large amounts of methane trapped as methane hydrates under Arctic sea-ice (Buffett and Archer, 2004) are released following ocean temperature increases and extensive sea-ice melting. Although the contributions from the described processes are uncertain, observations of emission from melting permafrost lakes nevertheless show increase, although at a small rate compared with other methane sources (Walter et al., 2006a,b). Sea-ice melting in the Arctic was recently found to be faster than predicted (Stroeve et al., 2007).

12.7.1.3. Aerosol

A few studies have considered the effects of climate change on aerosol concentrations. Higher levels of oxidants tend to increase secondary aerosol formation rates (Stevenson et al., 2005; Unger et al., 2006; Tsigaridis and Kanakidou, 2007). Changes in the hydrological cycle have strong regional influences on wet removal processes, generally increasing wet deposition (Liao et al., 2006; Racherla and Adams, 2006). The thermodynamic equilibrium of some aerosol species may be shifted by changes in climate (Liao et al., 2006). Some sources of aerosols, such as biomass burning, may increase as climate warms (Jacob and Winner, 2009). Following the A2 scenario, global aerosol burdens are reduced by 20% (2100−2000) in the study of Liao et al. (2006), and by 2−18% (2050−1990) in that of Racherla and Adams (2006), mainly due to enhanced deposition.

Feedback mechanisms between the physical climate state (wind, land cover, ice cover, vegetation distribution, and temperature), and natural emissions lead to changes in aerosol abundances (dust loads, sea salt, and biogenic organic aerosols) (Mahowald and Luo, 2003; Jones et al., 2007; Tsigaridis and Kanakidou, 2007). The quantification of such changes will require considerable research efforts on relevant parameterizations in global models, in order to define credible scenarios including such feedbacks.

Large uncertainties still exist in terms of natural emissions, for example, DMS from the oceans and hydrocarbons from land vegetation, and their response to climate change. There are also large uncertainties in the contribution of biomass burning, which occurs both naturally and due to human activity and is characterized by large inter-annual variations, thus complicating trend analyses.

12.7.1.4. Effects of Climate Change on Arctic Composition

Climate change can influence Arctic pollutants in many ways. Changes in the transport of pollutants to the Arctic region could occur because enhanced polar warming would reduce the gradient between the Arctic and lower latitude pollutant source regions, facilitating transport which is currently limited by the polar front (constant potential temperature surfaces which intersect the Earth's surface and form a dome over the polar region) (Stohl, 2006). Additionally, climate models generally project an enhancement of the strength of NH mid-latitude westerly winds associated with the large climate variability patterns of the North Atlantic Oscillation and Northern Annular Mode (Miller et al., 2006), which would lead to enhanced transport of pollutants to the Arctic (Eckhardt et al., 2003; Duncan and Bey, 2004; Sharma et al., 2006).

Emissions of ozone precursors in and near the Arctic may also increase as boreal regions warm and forest fire frequency increases. Release of methane from enhanced boreal wetland productivity is an additional possible consequence of Arctic warming, whose effects would however be global given methane's long lifetime. The retreat of Arctic sea-ice cover during springtime is also likely to alter the ozone-depleting halogen chemistry that takes place in the marine boundary layer (Simpson et al., 2007).

The relative importance of climate-induced changes on Arctic pollutants versus anthropogenic emission changes will clearly depend on future emissions. A shift of pollutant emissions from industrialized nations to developing nations at lower latitudes is already taking place. This may have a greater impact on pollutant transport to the Arctic than the climate-induced changes discussed above as pollution transport from lower latitude Asian sources is substantially less efficient per unit emission than transport from Europe (Stohl, 2006; Shindell et al., 2008a,b). Furthermore, substantial increases in local emissions are possible under a scenario where shipping in the Arctic ocean increases substantially as sea-ice cover retreats (Granier et al., 2006; Dalsøren et al., 2007). Hence much work remains to better understand climate—chemistry interactions in the Arctic and to project their role in future Arctic climate change.

12.7.2. Impact of Stratospheric Changes on Tropospheric Composition

The stratosphere can affect tropospheric ozone by direct downwards transport of stratospheric high-ozone air, and by changes in the stratospheric ozone column affecting the penetration of UV radiation, and hence tropospheric photolysis rates. Changes in the stratosphere can also affect tropospheric circulation patterns particularly in the polar regions such as Antarctic circumpolar westerly winds (e.g., Perlwitz et al., 2008), and the Arctic oscillation (Hess and Lamarque, 2007), and hence indirectly affect tropospheric ozone distributions.

12.7.2.1. Contributions of Stratosphere–Troposphere Exchange to the Tropospheric Ozone Budget

Irreversible exchange of air between the stratosphere and troposphere occurs largely in synoptic scale events. Global estimates have been made using correlations between N_2O and O_3 (e.g., Murphy and Fahey, 1994), by 3D chemistry models (e.g., Roelofs and Lelieveld, 1995; Jöckel et al., 2006), and by combining satellite ozone observations with meteorological analyses (e.g., Olsen et al., 2003). These typically give fluxes in the range 450–600 $Tg(O_3)$ yr^{-1}. However there are some discrepancies in the definition of the flux. Some studies quote a flux through a fixed pressure level, others through a PV surface. Many tropospheric chemistry models use the 150 ppbv ozone surface as a definition of the tropopause.

Stevenson et al. (2006) assessed the ozone budgets of 26 tropospheric chemistry models. They inferred the STE flux in the models across the 150 ppbv ozone surface by assuming a balance between the stratospheric input, net chemical production, and dry deposition of ozone. The results gave a mean STE ozone flux of 552 Tg yr^{-1} with a range of 151–930 Tg yr^{-1}. This compares to a mean tropospheric chemical production of 5110 Tg yr^{-1}. Hence STE contributes about 10% to the total tropospheric ozone budget. This does not necessarily mean that STE contributes 10% to the tropospheric burden since the ozone is injected into the upper troposphere where the ozone lifetime is longer. By tagging stratospheric ozone in their model simulation, Sudo and Akimoto (2007) calculated that STE contributed 23% to the tropospheric ozone burden. Variations in STE account for 25% of the variance in the tropospheric ozone burden between models (Wild, 2007).

12.7.2.2. Trends in STE

The exchange of mass between the stratosphere and troposphere depends on the Brewer–Dobson circulation in the stratosphere (upwelling in the tropics, downwelling in the extra-tropics), which is driven by extra-tropical wave forcing. A study by Butchart and Scaife (2001) found that the rate of global mass exchange between the stratosphere and troposphere will increase by 3% per decade in future. They put this down to an increased wave forcing from more vigorous extra-tropical planetary waves in the warmer climate. The overall trend of the Brewer–Dobson circulation is a consistent feature of coupled CCMs. Butchart et al. (2006) estimated an average 2% per decade increase in the tropical upwelling in 11 models over the twenty-first century.

The ozone concentrations in the lower stratosphere depend on chemistry and transport, and could be strongly influenced by climate change. Zeng and Pyle (2003) found a substantial increase in lower stratospheric ozone, with a future climate in their model with a top at 36 km. Another study, using a model with an 85 km top, found that while lower temperatures increased ozone in the upper stratosphere, this reduced the lower stratospheric ozone as less UV could penetrate to the lower levels.

12.7.2.3. Trends in Tropospheric Ozone

Collins et al. (2003), Sudo et al. (2003), and Zeng and Pyle (2003) found that STE of ozone by 2100 would increase by 40%, 130%, and 80% respectively. These models were tropospheric chemistry models with tops between 36 and 40 km. The three studies investigated changes in both circulation and lower stratospheric ozone. In all these cases the change in lower stratospheric ozone had a large effect on STE. Increased water vapour in a warmer climate is expected to increase destruction of tropospheric ozone, but in the above three studies increased STE partially or completely offset the increased destruction. In the Stevenson et al. (2006) model inter-comparison, models had varying methods of accounting for STE of ozone. These gave STE increases with 2030s climate of 0–19%.

Recent trends in tropospheric ozone may be at least partly due to changes in the lower stratosphere. Ordóñez et al. (2007) have shown that a 16% per decade trend in lowermost stratospheric ozone over Europe (1992–2004) correlates well with a 5–10% per decade trend in lower tropospheric background ozone (Figure 12.42).

Stratospheric ozone depletion and its subsequent recovery will affect lower stratospheric concentrations and hence influx of ozone into the troposphere. Ozone depletion may have offset two thirds of the tropospheric ozone burden increase from pre-industrial times to the present day, according to Shindell et al. (2006). They also suggest that of their expected 124% increase in STE in future, only 33% is due to climate change, the rest is due to ozone recovery.

12.7.2.4. Impact of UV Changes

Since chemical processes in the troposphere are initiated by solar UV-B penetration into the troposphere, changes in

FIGURE 12.42 12-month running means of the normalized monthly anomalies with respect to the 1992—2004 mean annual cycle of lower free tropospheric ozone (blue) at (top) Jungfraujoch and (bottom) Zugspitze, and 150 hPa (~13.6 km altitude) ozone averaged for Payerne and Hohenpeissenberg (red). (*Source: Ordóñez et al., 2007 © Copyright 2007 American Geophysical Union. Reproduced/modified by permission of American Geophysical Union.*)

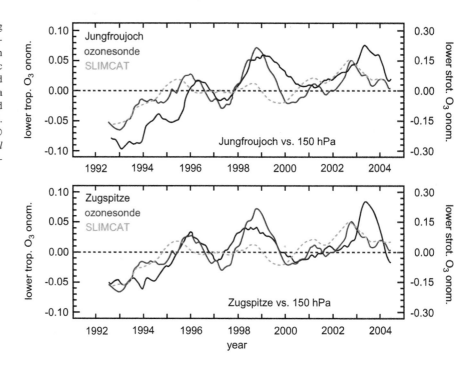

stratospheric ozone could affect the impact on the tropospheric oxidation. Available measurements show that surface UV radiation levels have generally increased with a geographical pattern similar to the observed reductions in stratospheric ozone. In addition surface UV levels will be strongly influenced by cloud cover, local ground albedo, and the atmospheric aerosol content. UV increases lead to an enhanced OH production, which reduces the lifetime of methane and influences ozone production and loss rates. The impact on ozone is more complex (Isaksen et al., 2005). While decreased ozone columns enhanced surface ozone in polluted regions, surface ozone decreased in pristine background regions with low NO_x. Collins et al. (2009) showed that during summer with very high NO_x levels, titration lowered ozone production. Shindell et al. (2006) calculated that ozone depletion since 1979 had increased OH levels by 3.5%. This would correspond to a 5% decrease in methane concentrations (using the factor 1.4 feedback adjustment, from Prather and Ehhalt (2001), and assuming initial reduction in methane lifetime is the same as the global OH increase). Conversely ozone recovery would be expected to increase methane and hence tropospheric ozone.

12.8. CROSS CUTTING ISSUES (POLICY RELATIONS, INTEGRATION)

The 1992 UN Framework Convention on Climate Change (UNFCCC) states that policies and measures to address a human induced climate change shall stabilize atmospheric concentrations of greenhouse gases (GHGs) "at a level that would prevent dangerous anthropogenic interference with the climate system" (Art. 2), and that the measures should be "comprehensive" and "cost-effective" (Art. 3.3). Article 2 forms the basis for determining the total reductions in emissions. However, due to incomplete understanding of the climate system it is difficult to determine the appropriate level of mitigation. One major issue is the possibility that natural fluctuations such as the solar forcing have contributed significantly to the observed global temperature increase. One way to assure comprehensiveness and cost-effectiveness, as stated in Article 3.3, is to allow flexibility with respect to which species should be mitigated. This requires an emission metric whereby the global impact of emissions of different gases or aerosols with different atmospheric lifetimes and different radiative properties can be compared and weighted.

In the 1997 Kyoto Protocol, the target is formulated in terms of 'CO₂ equivalents' using the Global Warming Potential (GWP) metric concept. The principle of comprehensiveness and cost effectiveness are made operative as the aggregate anthropogenic carbon dioxide equivalent emissions of six specified GHGs or groups of GHGs: carbon dioxide (CO_2), methane (CH_4), nitrous oxide (N_2O), hydrofluorocarbons (HFCs), perfluorocarbons (PFCs), and sulfur hexafluoride (SF_6) (Art. 3.1, Annex A). Note that apart from being used in binding agreements, emission metrics can also be used to form a basis for common understanding of the climate impact that can be used for example when new technology standards are developed.

While a number of interactions between atmospheric chemistry and climate are now recognized, quantifying these interactions and their relative importance is more difficult and many uncertainties remain. Part of the difficulty in fully quantifying the relative importance of different climate–chemistry interactions is that the composition of the atmosphere and the climate processes are strongly coupled, thus a change in the emissions or concentration of one gas or particle can feed back on the atmospheric composition and climate in multiple ways. Furthermore, the large variability both in time and space in the secondary compounds such as ozone and sulfate and organic particles make the quantification of the climate–chemistry interactions and adopting suitable metric for policy measures highly complex.

12.8.1. Climatic Response to Solar Forcing: Overview of Theories

Variations in solar activity may give rise to changes in Earth's atmosphere both in terms of energy as well as its composition. Furthermore, some hypotheses suggest that cloud formation and the creation of CCN may be linked indirectly to the level of solar activity through the action of galactic cosmic rays.

The notion that changes in the Sun may affect weather on Earth is old and can be traced back to early discussions in the late seventeenth century (Benestad, 2002). It was established in 1896 (Kristian Birkeland) that Aurora was connected to the solar activity, and in more modern times an unequivocal response has been found in the upper atmosphere (Labitzke, 1987; Labitzke and van Loon, 1988; Haigh, 1994, 2003; Salby and Callagan, 2000, 2004; Shindell et al., 1999, 2001). But no definite and convincing evidence has yet been found suggesting a strong link for the troposphere and Earth's surface climate, although changes in the temperature structure in the stratosphere may affect the planetary wave dynamics and hence the advection of heat.

There have been many speculations over the past centuries about a solar connection to variations in quantities such as temperature, precipitation, and sea level pressure, and attempts to find a solar link. Friis-Christensen and Lassen (1991) presented an analysis suggesting a strong association between the warming over the twentieth century in the NH and the solar cycle length, but Laut (2003) and Damon and Laut (2004) reported errors in their analysis and argued that their conclusions were invalid. Subsequent analysis by Benestad (2005) also failed to reproduce their results. Furthermore, it is not clear how the temperature on Earth would be sensitive to the solar cycle length and exactly which physical processes would be involved.

Analysis by Crook and Gray (2005a,b) and Lean and Rind (2008) found a solar cycle signature in climatic variables, although these were weak compared to the internal variability. Furthermore, simulations with general circulation models (GCMs) also suggest a solar response, albeit modest (Cubash et al., 1997; Meehl et al., 2003). The level of solar activity is estimated to be high for this millennium (Solanki et al., 2004), however, direct modern instrument measurements indicate that the level of solar activity has not increased significantly over the last 30–50 years (Richardson et al., 2002; Lockwood and Frölich, 2007; Benestad, 2005).

Traditional studies have relied on empirical evidence and sunspot observations. More recently, additional indices as well as more physical variables, such as the total solar irradiance (TSI) and magnetic indices (Lockwood, 2002), have been included in similar analyses. The level of solar output is, however, still uncertain (Lean, 2006), and although several reconstructions of past variations have been published (Lean et al., 1995; Lean, 2004; Wang et al., 2005), there are still holes in our knowledge of past variations in solar forcings. Forster et al. (2007) estimated a radiative forcing due to an increase in TSI (1750–2005) of +0.12 (0.06–0.18) W m^{-2} (90% confidence interval).

There are three main explanations for how changes in the Sun may affect Earth's climate: (i) direct effect from TSI; (ii) stratospheric response (Haigh, 1994, 2003); and (iii) modification of clouds through shielding of galactic cosmic rays (GCR; Dickinson, 1975; Carslaw et al., 2002; Marsh and Svensmark, 2000).

Climatic response involving the stratosphere is believed to involve chemical reactions in the stratosphere (e.g., ozone), and the fact that solar UV varies strongly over a solar cycle compared to the total solar irradiance strongly suggests a mechanism involving both atmospheric chemistry and a response in the atmospheric dynamics (e.g., wave propagation). However, the importance of this mechanism depends on whether the variations in the stratosphere have an influence on the lower troposphere.

While the direct TSI effect and stratospheric mechanisms have been investigated in several modelling studies, the effect of GCR has been difficult to test in the GCMs. One problem with GCR and clouds is that the physical mechanisms are poorly known, and neither is it firmly established whether the GCR mechanism takes place in Earth's atmosphere. There have been several studies questioning the hypothesis due to a lack of empirical basis (Kristjánsson et al., 2008; Sloan and Wolfendale, 2008; Erlykin et al., 2009), but also some suggesting an effect of GCRs on clouds (Harrison and Stephenson, 2006; Knudsen and Riisager, 2009).

However, the fact that clouds may be affected by either GCR or the atmospheric electric field does not necessarily imply the mechanism proposed by Marsh and Svensmark (2000), where GCR is alleged to produce higher low-cloud coverage, higher planetary albedo, and hence lower temperatures. Pierce and Adams (2009) attempted to model the effect of GCR on the low clouds, but found that the changes from the cloud condensation nuclei (CCN) due to changes in the GCR were two orders of magnitude too small to account for the observed variations in the cloud properties. For instance, while Knudsen and Riisager (2009) reported high correlation between ^{18}O records and the Earth's magnetic dipole moment and concluded that the tropical precipitation has been influenced by variations in the geomagnetic field, they also argued that their findings gives some support to the link between GCR particles, cloud formation, and climate. However, if GCRs result in an increase in the planetary albedo and hence a cooler climate, it is unclear how this could lead to higher rainfall in the tropics. On the other hand, the rainfall in the region of their study is strongly linked to the Monsoon system, and a latitudinal shift in the position of the system can easily translate into large increases or decreases locally. There have also been other claims of strong solar-terrestrial links even over the most recent decades (e.g., Scafetta and West, 2005), but these are controversial and have been disputed (Lean, 2006; Benestad and Schmidt, 2009). Furthermore, a lack of trend in the GCR since 1952 suggests that GCR is not responsible for the most recent global warming (Benestad, 2005), and recent trends in the stratosphere further disagree with the notion of recently enhanced levels of solar activity (IPCC, 2007a).

12.8.2. Metrics

12.8.2.1. Definition of a Metric

In general the 'CO$_2$ equivalent' emissions of any component is not linked exclusively to the GWP metric, but can be derived for any metric M by

$$Em(CO_2\text{-}eq) = Em_i \cdot M_i \qquad (12.1)$$

where M_i is the metric value for component i and Em_i is the emissions of component i. The metric value (M_i) will not ensure CO$_2$-equivalence in all aspects of climate change, but only with respect to the definition of the metric (O'Neill, 2000). For example the climate impact of emissions of 1 kg of methane (with GWP$_{100}$ = 25) is only equal to that of 25 kg emission of CO$_2$ in that the integrated radiative forcing over 100 years after the emissions are equal. The more the physical properties (e.g., atmospheric lifetime) of component i deviates from that of CO$_2$ (e.g., as for NO$_x$ and aerosols) the larger differences can be expected for other climate impacts (e.g., on other time-scales) than the impacts that defines the metric.

The Kyoto Protocol excludes emissions of short-lived species including all types of aerosols, or their precursors, despite the fact that they are believed to contribute significantly to climate change. This could weaken efforts to mitigate climate change by weakening the comprehensive approach embodied in the UNFCCC; indeed, the US administration has cited the absence of BC and tropospheric ozone from the Protocol as one reason why they have not become signatories. The absence could also lead to a distortion (for example, away from economically more efficient measures) of priorities in emission reductions.

It is important to stress that the choices of a given metric do not define the policy — they are tools that enable implementation of the policy. However, different policy frameworks require different metric concepts (e.g., Manne and Richels, 2001; Fuglestvedt et al., 2003; Shine et al., 2007; IPCC, 2009). For example, the Global Temperature Potential (GTP; Shine et al., 2005) is more consistent with a climate policy with a long-term temperature target (e.g., EU 2°C target) than the GWP metric used in the Kyoto Protocol.

12.8.2.2. Special Challenges for Metrics of Chemically Active Short-Lived Species

Inclusion of chemically active short-lived components such as aerosols and ozone precursors, together with long-lived greenhouse gases, in a climate policy using a common metric concept gives rise to several challenges. There are significant uncertainties in our quantitative understanding of the climate impacts for these species. This includes quantitative and even qualitative understanding of the many chemical and physical indirect effects. As an example of indirect effects, the current definition of GWP$_{100}$ for methane that is in operational use in the Kyoto Protocol, includes indirect effects on ozone and stratospheric water vapour (IPCC, 2001a; Forster et al., 2007), but there may be further indirect effects of ozone on the carbon cycle as discussed in Section 12.5.1.3 (Sitch et al., 2007). In addition, and maybe more importantly, there are also more fundamental methodological issues caused by the very wide range of lifetimes for the chemically active species compared to the long-lived greenhouse gases including CO$_2$. An emission pulse of a short-lived component will naturally give a climate response on a much shorter time-scale. The climate response is dominated by the response time of the mixed layer of the ocean, that is, 5–10 years (e.g., Hartmann, 1996).

Emissions of short-lived components will not lead to globally or hemispheric scale well-mixed increases. A global metric value for each component as indicated in Equation (12.1) (and applied in the form of GWPs in the Kyoto Protocol) may not be applicable to short-lived components, and metric values that are regionally dependent and possibly also time (e.g., seasonally) dependent should be used. However, policymakers may decide that metric values that depend on region and time are too difficult to handle in a negotiation process and instead accept the inaccuracies that are embedded in a global metric value if inclusion of short-lived species is deemed to be important.

Radiative forcing is the basis for the GWP metric based on the assumption that other climate impacts such as global mean temperature increase is scaling linearly with radiative forcing, that is, that the efficacy (cf. Section 12.5) is equal for all forcing mechanisms. For most of the short-lived components there are indications that the efficacy is relatively close to 1.0 (Hansen et al., 2005; Forster et al., 2007), there are indications that the efficacy can be significantly enhanced for forcing mechanisms, that is, spatially correlated with regions of strong feedbacks, or if species specific processes are involved. In particular will BC aerosols that are deposited on snow trigger a strong snow albedo feedback with efficacy estimates in the range from 1.8 to 3.5 (Hansen and Nazarenko, 2004; Flanner et al., 2007). If indirect cloud effects of aerosols, except the cloud albedo effect (e.g., the cloud lifetime and semidirect effects) are not considered as forcing mechanisms but rather as feedbacks (as in Forster et al., 2007) the net effect of these emissions may have a quite different efficacy than the LLGHGs.

Chemically active components and aerosols frequently cause negative radiative forcing and thus a cooling. For some components such as SO_2 and OC aerosols there will be a net cooling, while for others such as NO_x, the warming (ozone) and cooling (decreased methane through increased OH and formation of nitrate aerosols) effects can be of similar magnitude. In the case of NO_x the warming and cooling may cancel on a global and time integrated level, but since the methane perturbation is much more long-lived, the cooling will affect both hemispheres while the warming will mainly be confined to the hemisphere where NO_x is emitted. To what extent these cooling effects should be included in the metric values and thus in a policy aimed at reducing global warming is basically a political questions. For instance, whether or not to include SO_2 with a negative metric value is in many ways similar to the discussion about the use of geo-engineering to mitigate global warming (Crutzen, 2006; Rasch et al., 2008; Latham et al., 2008).

Emissions of chemically active short-lived species often lead to other non-climate environmental effects such as degradation of AQ, acid deposition, and eutrophication. To the extent that these effects do not influence the climate parameter used to calculate the metric value, these effects should not be included in the metric value, but will, of course, in the end affect the policy development.

As discussed above there are significant uncertainties in our quantitative understanding of the climate impact of the chemically active short-lived species. A metric that is to be used in binding agreements needs to be relatively transparent and robust; robust in the sense that the scientific understanding must be sufficiently mature so that the metric value is not expected to change significantly with

every new study. The many indirect effects are probably the main source of uncertainty (e.g., indirect aerosol effects, Section 12.5.4).

12.8.2.3. Examples of Published Metrics

Metrics have been developed by economists based on different kinds of optimization frameworks such as cost-benefit analysis (e.g., Kandlikar, 1995) and cost-effectiveness (e.g., Manne and Richels, 2001). Common for these metrics is that mitigation cost estimates and discount rates are needed and, in the case of cost benefit analysis, also damage costs (could be positive or negative) of climate change are needed. These cost estimates are highly uncertain, and the determination of an appropriate discount rate is very controversial.

Purely physical metrics have also been developed, the Global Warming Potential (GWP) being the most widely known and used through its inclusion in the Kyoto Protocol. The AGWP(H) (Absolute GWP) of a component i is defined as the integrated radiative forcing over a given time horizon (H) following a pulse emission of 1 kg of i.

$$AGWP_i(H) = \int_0^H RF_i(t) \cdot dt \qquad (12.2)$$

where RF_i is the radiative forcing following a pulse emission of 1 kg of i at $t = 0$. The GWP_i is defined as the ratio of $AGWP_i$ and the AGWP for CO_2 which is used as a reference gas.

If component i is decaying with a constant lifetime then

$$AGWP_i(H) = a_i \int_0^H C_i(t)dt = a_i \int_0^H C_0 \cdot e^{-t/\tau} \cdot dt \qquad (12.3)$$

where a_i is the radiative efficiency (RF per unit concentration change) and $C_i(t)$ is the concentration perturbation following a pulse emission of 1 kg of i, C_0 is the initial concentration perturbation at $t = 0$ and τ is the lifetime of component i.

A second purely physical metric, the GTP has more recently been proposed (Shine et al., 2005). The AGTP(H) (absolute GTP) is defined as the temperature perturbation (global annual mean) at time H following a pulse emission at time $t = 0$ of 1 kg of component i. As for the GWP the GTP is defined as the ratio of $AGTP_i$ and the AGTP for CO_2. To derive values for the GTP metric knowledge of the temporal response of the climate system to a perturbation is needed. Several methods have been proposed and used (Shine et al., 2005; Shine et al., 2007; Boucher and Reddy, 2008; Berntsen and Fuglestvedt, 2008), using both simple analytical climate models and impulse response functions fitted to AOGCMs. Consensus about what is the appropriate method has not yet been established.

Figure 12.43 shows examples of published and derived metric values for NO_x and BC particles (see Fuglestvedt et al., 2010). For NO_x it is the indirect effects through ozone formation (positive RF) and decrease in methane lifetime (negative RF) that is included. Due to the integrating nature of the GWP concept the strong, but short-term, positive RF caused by ozone production has a strong impact on the GWP even with a 100-year time horizon. For the GTP metric, even after 20 years most of the warming of the ozone has disappeared and the more long-term cooling of methane reductions dominates giving negative GTP values.

Tol et al. (2008) show that the purely physical metrics GWP and GTP are special cases that can be derived through a number of simplifications from a cost-benefit analysis (GWP) and a cost-effective analysis (GTP).

Although there are significant uncertainties and it can be discussed if the GWP and GTP concepts are appropriate measures of the climate impacts due to emissions short-lived chemically active components, the metric values can be readily calculated based on results from CTM of GCMs with chemistry. Fuglestvedt et al. (2010) give an overview of metric values for both the GWP and GTP for ozone precursors, SO_2, and carbonaceous aerosols with some information about regional differences.

12.8.3. Future Directions for Climate—Chemistry Research

12.8.3.1. The Role of Model—Observation and Model—Model Comparisons

Studies of future climate—chemistry interactions rely heavily on how models can reproduce distribution and changes in the chemical active greenhouse gases. It has been demonstrated that there are significant uncertainties with the models, which should be demonstrated by a couple of examples: The global tropospheric concentration of OH, which is a key component in tropospheric oxidation, shows large spread in the comparison given in Figure 12.14 (Shindell et al., 2006). Likewise, aerosol burdens in state-of-the-art models, given in Figure 12.23 (the AEROCOM experiment), reveal large uncertainties in model estimates.

One of the key first steps in quantifying climate—chemistry interactions and perform assessments of coupled climate—chemistry models is to improve the modelling tools through extensive comparisons with observations. Some past efforts have already been made towards assessing CCMs, including Austin et al. (2003) and Eyring et al. (2005a,b, 2006).

There are a number of ongoing activities around the world to study various aspects of climate—chemistry interactions, in the troposphere and stratosphere. The international Aerosol model inter-Comparison (AERO-COM), the ACCENT model inter-comparison, the assessment of the atmospheric impact of the transport sector in the EU funded project ATTICA, and the Task Force on Hemispheric Transport of Atmospheric Pollutants (TF HTAP) focus on climate relevant modelling studies. Under AEROCOM, global tropospheric aerosol models were compared and tested against satellite, lidar, and Sun photometer measurements. The ACCENT-MIP effort previously focused on coordinating and comparing IPCC scenarios (Nakicenovic et al., 2000), contrasting the climate in 2030 versus 2000 across a suite of tropospheric chemistry—climate models, with a goal towards capturing how climate change might affect AQ (gas species only).

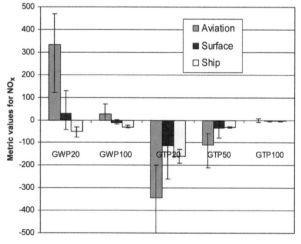
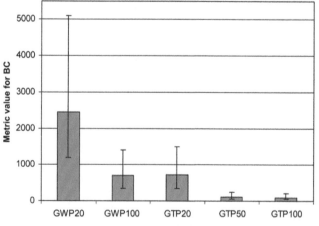

FIGURE 12.43 Overview of published and derived metric values for NO_x and BC emissions (*Source: based on Fuglestvedt et al., 2010. Reprinted from Fuglestvedt et al., 2010; with permission from Elsevier.*). The metrics shown are GWPs with 20- and 100-year time horizons, and GTPs with 20-, 50-, and 100-year time horizons. The error bars indicate the full range of values across regions of emissions and published studies. For NO_x it is distinguished between emissions from aviation, shipping, and surface land sources.

This effort is extended to encompass the activities of the TF HTAP. The activities focus on understanding and quantifying Northern Hemispheric transport of gaseous and particulate air pollutants and their precursors from source to receptor regions.

A major new international effort towards analysing climate–chemistry models is being coordinated, called the Atmospheric Chemistry and Climate (AC&C) Initiative. The focus of AC&C is on identifying science issues, providing analysis tools, coordinating and integrating activities, and furthermore to provide input on climate–chemistry interactions to the IPCC process.

12.8.3.2. Other Issues in Improved Modelling of Chemistry–Climate Interactions

As noted already, advancing the quantification of climate–chemistry interactions will also require improvement in the fundamental building blocks of our understanding of tropospheric and stratospheric chemical and physical processes. These include improvements in characterizing the rates of gas phase, heterogeneous, and photolytic processes; process studies in the atmosphere to test our understanding of the chemical processes; and incorporation of these processes in a realistic way in global climate–chemistry models.

Long-term monitoring and observation is a prerequisite to better understand and quantify the key atmospheric processes affecting climate–chemistry interactions. These observations are also a key to determining the scientific basis for understanding how and why changes are occurring. Although significant amounts of data for large-scale model validation and process studies have been made available during the last decade, in particular satellite observations, the actual needs for observing the Earth's system are expanding and require more complex experiments covering a wide range of instruments and observed species on a global spatial and temporal scale. The value of an integrated global observing system is recognized worldwide as the only way to produce systematic, coherent data, capable of addressing climate related issues on a global-scale. The different platforms used for data collection play complementary roles in the establishment of long-term monitoring capability of global coverage for process and global changes studies, and remain essential to constrain models and predict the future of our planet with reduced uncertainties.

12.9. SUMMARY AND CONCLUSIONS

Potentially significant contributions to the climate impact are provided by compounds such as CO_2, CH_4, O_3, particles, and cirrus clouds. For the chemically active gases, processes in the atmosphere are important, with large spatial and temporal variations. The climate–chemistry interactions are therefore characterized by significant regional differences with regions such as South East Asia being a future key region due to significant increases in energy use and pollution emission (Akimoto, 2003; Kupiainen and Klimont, 2007; Amann et al., 2008; Klimont et al., 2009; Isaksen et al., 2009). Likewise, ship and air traffic represent important sectors because of significant increases in emissions in recent years (Eyring et al., 2005a,b, 2007a; Dalsøren et al., 2009a; Lee et al., 2010). The relative contributions to the emissions from various sectors are expected to change significantly over the next few decades due to differences in mitigation options and costs.

Our ability to predict the extensive interaction between climate and chemistry has improved significantly over the last decade due to an extensive increase in observational data and computer resources, which has led to improved large-scale modelling capability for tropospheric distribution and trends. Both CTMs and CCMs are available (Eyring et al., 2005a,b; Bousquet et al., 2006; Gauss et al., 2006; Stevenson et al., 2006) with more extensive chemical processes included and higher resolutions, and with observations available for model evaluation and long-term model analysis, in particular satellite observations (Gerasopoulos et al., 2005; Richter et al., 2005; Laj et al., 2009). Extensive multimodel comparisons and studies of emission impact have been performed in projects such as ACCENT, AEROCOM, and QUANTIFY during the last few years. This demonstrates our ability to model man-made impact on atmospheric composition of chemically active climate compounds (Shindell et al., 2006; Stevenson et al., 2006; Textor et al., 2006, Textor et al., 2007; Hoor et al., 2009; Fiore et al., 2009). These model comparisons show, however, that there are significant model-to-model differences.

New satellite observations have provided data on the global distribution of tropospheric column burden of ozone and methane, as well as a number of ozone precursors (NO_2, CO, and HCHO) (Laj et al., 2009). Over the last two decades tropospheric ozone trends from surface observations give a mixed signal depending on regional trends in emissions of ozone precursors and possibly an influence of a reversed trend in the influx of stratospheric ozone (Ordóñez et al., 2007). Shindell et al. (2006) report that a significant fraction of the increase in tropospheric ozone since pre-industrial time may have been reduced due to stratospheric ozone depletion. Referring to changes since the 1970s, Ordóñez et al. (2007) came to the same conclusion. Models driven by changes in emissions are capable of capturing some but not all of the features of the observed trends in tropospheric ozone. A key point here is the difference in emissions from different emission sectors, and a number of studies have focused specifically on attributing changes in atmospheric concentrations to specific emission sectors. For instance, it is possible that the large increase in emissions during the last decade (Eyring et al., 2010; Dalsøren et al., 2009a), from ocean-going ships, may have led to increases in observed ozone at coastal stations in Europe and the western US.

OH is the key compound in the tropospheric oxidation process (Monks et al., 2009). Long-term trend in global OH is significant for methane trend. Observations show that methane concentrations levelled off during the last decade, although there was a significant increase between 2007 and 2008. It is debated if the levelling off in methane is primarily caused by a levelling off in the emissions or if the primary cause is an increase in global OH levels, possibly caused by an increase in the NO/CO ratio in global emissions (Dalsøren and Isaksen, 2006). Although there is a general understanding of the OH distribution, there are significant uncertainties connected to OH trends and to processes affecting distribution and trends (Cariolle et al., 2008). NO_x emissions from ship traffic represent a particularly important source for atmospheric OH (Eyring et al., 2010; Hoor et al., 2009; Dalsøren et al., 2009a). Surface emissions of NO_x from ships in pristine areas efficiently increase OH, thereby reducing methane lifetime and atmospheric methane. It has been shown that the efficient methane reduction together with aerosol emissions give a negative radiative forcing for current ship traffic. Future long-term impact will depend on mitigation measures adopted to reduce emissions (reduction of sulfur and NO_x emissions).

For aerosols a number of satellite retrievals of bulk aerosol properties backed up by surface measurements and in-situ campaigns have improved the understanding of the global distribution of aerosols significantly. Extensive model validation and inter-comparisons within the AEROCOM project have revealed significant inter-model differences and discrepancies compared to observations. Differences are mainly originating from differences in model constructions (Textor et al., 2006; Schulz et al., 2006; Textor et al., 2007). Anthropogenic aerosols' ability to alter the radiative budget has been singled out as the most uncertain contributor to the total anthropogenic forcing of the Earth—atmosphere system (Forster et al., 2007). Several recent studies improve our understanding of the RF from different climate compounds. In a new study, combining models and satellite data, Myhre (2009) estimates the RF due to the direct effect of aerosols to be around -0.3 W m^{-2}, with halved uncertainty range compared to the estimates by IPCC (Forster et al., 2007).

Although the cloud albedo effect has been studied extensively for several decades, it is still characterized by a low level of understanding. A number of studies on aerosol indirect effects have been published since IPCC

AR4, showing that the scientific community is currently making an effort to increase the knowledge and reduce uncertainties associated with aerosol indirect effects. Continued use of satellite data combined with models, and continuous improvements of cloud parameterizations, may reduce the uncertainties in model estimates (Anderson et al., 2003) of the aerosol indirect effect. The comparison of the spread in estimates of indirect aerosol effects in Figure 12.29 shows that the uncertainties have not decreased with time. In fact, estimates span a larger range than ever, much due to the new aerosol indirect effects recently taken into account.

A number of climate processes that influence the levels of methane, tropospheric ozone, and the oxidation capacity have been identified through recent studies and to some extent quantified through sensitivity experiments in models. These processes include: change in OH from temperature and humidity perturbations, change in tropospheric ozone through STE exchange, and changes in stratospheric ozone; enhancements in methane emissions from melting permafrost, changes in NO_x emissions from lightning, and change in biogenic VOC and NO_x emissions from changes in surface characteristics (humidity, temperature) (Zeng and Pyle, 2003; Collins et al., 2003; Isaksen et al., 2005; Horowitz et al., 2007; Sitch et al., 2007; Walter et al., 2006a,b; Wu et al., 2008b).

Models have been used for sensitivity studies and for studies of changes in tropospheric and surface ozone for 2030, 2050, and 2100 due to both emission changes and climate change (Prather et al., 2003; Stevenson et al., 2006; Royal Society, 2008). Differences in assumptions about improvements in control of precursor emissions have significant impacts on future tropospheric ozone concentrations over the 30-year period from 2000 to 2030. An increase between 10 ppb and 14 ppb in annual zonal-mean ozone at 30°N is calculated for the unregulated SRES A2 case, and a decrease of up to 6 ppb for the B2 + MFR (maximum feasible reduction) case. Changes in climate are found to slightly reduce the amount of tropospheric ozone, through enhanced OH formation. Model studies of OH concentrations in 2030 and 2050 indicate that global OH concentrations will increase slightly due to emission changes, but that this increase will be more than compensated by effects of climate change. Higher temperatures give more water vapour and enhanced OH production.

Changes in chemically active greenhouse gases and aerosols have contributed significantly to the overall radiative forcing since pre-industrial times. Two considerations are important for estimates of future climate impact from anthropogenic emissions: the short chemical lifetime of most of the important chemically active climate compounds as compared to CO_2 and the possibility for mitigation options to reduce emissions. It is likely that in the future the relative contribution from the chemically active climate compounds become smaller compared to the contribution from CO_2. A consequence of the large differences in the lifetimes of the climate compounds is that future effects are strongly dependent on the scenarios adopted for emissions, and it is not obvious what type of metric should be adopted (Shine et al., 2005; Shine et al., 2007; Berntsen and Fuglestvedt, 2008; Boucher and Reddy, 2008; Fuglestvedt et al., 2010). Continued growth in emissions, for instance, favours impact from the short-lived compounds.

Long-term monitoring and observation is a prerequisite to better understand and quantify the key atmospheric processes affecting climate–chemistry interactions. These observations are also a key to building the scientific basis for understanding how and why changes are occurring. Although significant amounts of data for large-scale model validation and process studies have been made available during the last decade, in particular satellite observations (see Laj et al., 2009), the actual needs for observing the Earth's system are expanding and require more complex experiments covering a wide range of instruments and observed species on a global spatial and temporal scale. A key question is the issue of the natural variability of composition on different scales associated with the climate variability, which indicate that longer, and possibly ensemble simulations, may be required to extract the signal of anthropogenic climate change on composition from that related to natural climate variability.

ACKNOWLEDGEMENTS

Significant parts of the work reported in this article are based on studies performed in the EU projects ACCENT, ATTICA, QUANTIFY, EUCAARI, and HYMN. We are thankful to Dr. Corinna Hoose for making Figure 12.28 available and to Dr. Keith Shine for valuable discussions and for his contribution to the article. We are thankful to the ACCENT project office for their support in the preparation of the article.

Climate–Chemistry Interaction: Future Tropospheric Ozone and Aerosols

Wei-Chyung Wang,[a] Jen-Ping Chen,[b] Ivar S. A. Isaksen,[c] I-Chun Tsai,[b] Kevin Noone[d], and Kendal McGuffie[e]

[a]Atmospheric Sciences Research Center, State University of New York, Albany, USA, [b]Department of Atmospheric Sciences, National Taiwan University, Taiwan, China, [c]Department of Geosciences, University of Oslo, Oslo, Norway, [d]ITM, Stockholm University, Stockholm, Sweden, [e]School of Physics and Advanced Materials, University of Technology, Sydney, Australia

Chapter Outline

13.1. ATMOSPHERIC COMPOSITION, CHEMISTRY, AND CLIMATE

13.1.1. Background

Fossil fuel and biomass burning emissions increase the atmospheric concentrations of primary climate compounds: CO_2, black carbon (BC), and primary organic aerosols (POA) and of the chemically-active compounds: CH_4, tropospheric O_3, sulfate, and secondary organic aerosols (SOA). Such emissions are also important sources for tropospheric NO_x, volatile organic compounds (VOCs), and CO. Anthropogenic additions of CO_2 and these chemically

The Future of the World's Climate. DOI: 10.1016/B978-0-12-386917-3.00013-0

active, radiatively important gases increase the natural greenhouse effect and cause a warming of the troposphere—surface climate system (e.g., IPCC, 2007a).

The climate implications of chemically-active compounds can be grouped into: (a) the primary climate gas CH_4, which is emitted from different natural and anthropogenic surface sources and which, through atmospheric oxidation processes, leads to enhanced abundance of itself, and of O_3, stratospheric H_2O, and CO_2; and (b) secondary climate compounds: O_3, sulfate aerosols, and SOA, which are formed by chemical processes in the atmosphere. Direct emissions of BC warm both the atmosphere, through increasing the absorption of solar radiation, and the surface by decreasing the snow albedo. On the other hand, sulfate aerosols and POA reflect solar radiation and cause a direct cooling effect.

Emissions of pollutants such as NO_x, CO, VOCs, and SO_2 change the concentrations of secondary climate compounds since they affect the atmospheric oxidation processes. In addition, these pollutants are likely to be affected by climate change through changes in natural emissions, in the oxidation, and deposition processes, and in the meteorology (temperature, atmospheric stability, convection, and cloudiness). NO_x and VOCs are also precursors of SOA and nitrate aerosols whose climate effect on a global scale is similar to, but smaller than, the effect of sulfate aerosols. Changes in sulfate aerosols and SOA cause a direct cooling effect by scattering solar radiation and thus increasing the Earth's albedo. In addition, these aerosols induce indirect climate

effects by acting as cloud condensation nuclei (CCN), thereby changing both the cloud albedo (the first indirect effect) and the cloud lifetime and precipitation (the second indirect effect). At present, the aerosol—cloud microphysical interaction is recognized as one of the largest uncertainties in future climate changes (IPCC, 2007a).

Atmospheric O_3 plays an important role in the atmospheric radiation balance. The radiative forcing from O_3 is sensitive to changes in its vertical distribution. A decrease in stratospheric O_3 acts to provide two opposite climate effects for the troposphere—surface climate system: a *warming* effect due to increased available solar radiation for absorption and a *cooling* effect due to decreased downward longwave radiation; the net effect will depend on the location and the time of year. On the other hand, an increase in tropospheric O_3 can warm the troposphere—surface climate system through increases in absorption of both the solar and longwave radiation. Consequently, climate—chemistry interactions are important for understanding the present and future climate effects of changing atmospheric chemical composition due to anthropogenic activities.

13.1.2. Anthropogenic Activity and Climate Changes

As listed in Table 13.1, fossil fuel and biomass burning are the main anthropogenic sources of the climate compounds. With the possible exception of CO_2 and BC, all compounds are of relevance for climate—chemistry two-way

TABLE 13.1 Atmospheric Compounds with Climate Implications

Compound	Source	Character	Key Compounds Affected	Climate Effect
CO_2	Fossil fuel burning	GHG	–	Warming
CH_4	Fossil fuel burning	GHG (chemistry)	OH, O_3, H_2O	Warming
NO; NO_2	Fossil fuel burning Biomass burning	GHG (chemistry)	OH, O_3, CH_4,	Warming/cooling
CO; VOC	Fossil fuel burning Biomass burning	GHG (chemistry)	OH, O_3, SOA	Warming
SO_2	Fossil fuel burning	Particles (chemistry)	Sulfate aerosol	Cooling
Organic carbon (OC)	Fossil fuel burning Biomass burning	Particles (chemistry)	SOA	Cooling
Black carbon (BC)	Fossil fuel burning Biomass burning	Particles	–	Warming

The compounds (except, arguably, CO_2 and black carbon (BC)) involved in climate—chemistry interaction. They are emitted either from anthropogenic activities and controlled by atmospheric chemistry or are pollutants that affect climate compounds O_3, CH_4, sulfate aerosols, and secondary organic carbon (SOA). OH is a major compound in the tropospheric O_3 chemistry and interacts with CH_4 in a non-linear way. GHG refers to greenhouse gases. Emission of NO_x (leading to the formation of NO_2) gives warming through increased O_3, and cooling through reduced CH_4.

TABLE 13.2 Radiative Forcing of Climate Between 1750 and 2005

Compound	Radiative Forcing
CO_2	1.66 (±0.17)
CH_4	0.48 (±0.05)
N_2O	0.16 (±0.02)
Halocarbons	0.34 (±0.03)
Stratospheric O_3	−0.05 (±0.10)
Tropospheric O_3	0.35 (−0.1, +0.3)
Sulfate direct	−0.4 (±0.20)
Sulfate first indirect	−0.7 (−1.1, +0.4)
Organic carbon direct	−0.5 (±0.5)

The radiative forcing (W m^{-2}) caused by the concentration changes between 1750 and 2005 (IPCC, 2007a). The numbers in parenthesis are the uncertainties.

interaction and are likely to be affected by future climate change, although internal carbon cycling exists, as discussed by Dickinson (2012, this volume). Table 13.2 shows the radiative forcing attributed to changes in atmospheric composition between 1750 (marking the beginning of the industrial era) and 2005. Both the magnitudes and uncertainties are given; for the latter, detailed discussion has been provided in IPCC (2007a).

Tropospheric O_3 and sulfate aerosols produce substantial radiative forcing. For example, the increase in tropospheric O_3 provides a warming of 0.35 W m^{-2}, which is about 3/4 of the forcing due to the increase of CH_4. While the CH_4 change is spatially homogeneous because of its decadal lifetime, tropospheric O_3, with its much shorter lifetime, is dominated by the pattern of emissions of its precursors, which is distinctly regional. The estimated direct radiative forcing from sulfate is a cooling of about 0.4 W m^{-2}, which, when combined with the sulfate first indirect (cloud albedo) effect that is estimated to be −0.7 W m^{-2}, offsets almost 2/3 of the 1.66 W m^{-2} CO_2 warming effect. Large uncertainties exist in the sulfate indirect effect due to inadequate understanding of sources and sinks of sulfate (and other aerosol particles) and the associated aerosol—cloud microphysics. The direct effect of organic carbon (OC) is a cooling of about −0.5 W m^{-2}, but with large uncertainties.

13.1.3. Climate—Chemistry Interaction: Regional-Scales

Our current understanding of future climate—chemistry interaction is evolving. To illustrate the importance and

complexity of predicting climate inclusive of fully coupled chemistry, we focus on tropospheric ozone and secondary aerosols (sulfate and organics) within the context of the primary climate compounds: CO_2, CH_4, dust, BC, and primary organics. Of the primary compounds, CH_4 is of particular importance since its climatic impact is strongly enhanced through atmospheric chemical interaction involving tropospheric O_3 and stratospheric H_2O.

Two factors make tropospheric O_3 and secondary aerosols (sulfate and SOA) distinctly different from the long-lived climate gases CO_2, CH_4, N_2O, and halocarbons. Firstly, they are secondary compounds and formed by chemical reactions in the atmosphere (Isaksen et al., 2009; Monks et al., 2009). Their abundance is strongly influenced by physical and chemical processes in the atmosphere, as well as the intensity of solar radiation. Furthermore, their distributions depend on the atmospheric oxidation capacity and on the emission of precursor pollutants such as NO_x, CO, VOCs, and SO_2, which vary strongly with time and in space, and are also affected by regional climate condition. Secondly, changes in atmospheric O_3 and secondary aerosol distributions show large spatial and temporal variations, a consequence of their short chemical lifetimes. Tropospheric lifetimes of O_3 and aerosols are days to weeks whereas, in the lower stratosphere, O_3 lifetime is of the order of months. Abundances and radiative forcing from changes of these compounds are, therefore, highly non-uniform. Studies of their distribution and changes, and the associated climate impact, therefore, require further investigation on regional-scales (cf. Evans et al., 2012, this volume).

Isaksen et al. (2009; reproduced as Isaksen et al., 2012, this volume) provided a comprehensive review of observational and modelling studies of atmospheric composition change with emphasis on climate—chemistry interactions on a global scale. This chapter addresses the effect of climate—chemistry interaction associated with tropospheric O_3 and sulfate aerosols on a *regional* basis. The three focus regions in this chapter: East Asia, North America, and Western Europe, were selected because, during the last few decades, changes in the surface emission patterns of SO_2 and O_3 precursors have evolved differently and are likely to vary differently in the future. These regions also have different regional climate characteristics, in particular the atmospheric circulation and precipitation, which affect atmospheric oxidation, transport and removal processes.

East Asia is influenced by the monsoon with heavy precipitation in summer and cold (and dry) surges in winter, both affected by the strength of the monsoonal flows and the amount of water vapour transported. Central and northern North America are influenced by mid-latitude cyclones. The North American monsoon system develops in early July; prevailing winds over the Gulf of California undergo a seasonal reversal, from northerly in winter to southerly in summer, bringing a pronounced increase in rainfall over southwest North America. Variations in the atmospheric circulation influence the European climate on an inter-annual timescale, providing meteorological settings for extreme weather. A recent example is the central European heatwave in the summer of 2003, which demonstrated the potential for ozone–climate interactions in a future warmer climate (Solberg et al., 2008).

13.1.4. Focus of This Chapter

In this chapter, we will describe the relevant physical and chemical processes affecting tropospheric ozone and sulfate aerosols, and their coupling to climate. While simulations from global chemical transport models (CTMs) are used to illustrate the spatial and temporal distribution of these compounds, a climate model with interactive chemistry is adopted to examine the issues relevant for climate–chemistry interactions affecting the regional concentrations of these compounds in the present-day climate and during future climate changes.

To provide a broader perspective, Section 13.2 presents a summary of the current understanding of the climatically-important chemical compounds. CTMs, in particular the Oslo-CTM2 (Isaksen and Hov, 1987; Berntsen and Isaksen, 1997, 1999; Isaksen et al., 2005), are used, together with results from a coupled chemistry–climate model (Liao et al., 2009), to facilitate the presentation. Sections 13.3 and 13.4 address the climate–chemistry interaction concerning, respectively, tropospheric O_3 and sulfate aerosols. To illustrate the relevant issues, we analyse simulations from a global climate–chemistry model (GCCM), which is a climate model with interactive O_3 chemistry (Wong et al., 2004) and sulfur chemistry (Tsai et al., 2010). Specifically, we discuss the processes and parameters describing global mean and regional interactions. Discussions on the mitigation for climate and air quality are given in Section 13.5. Conclusions and the outlook for future climates are presented in Section 13.6.

13.2. CLIMATICALLY-IMPORTANT CHEMICAL COMPOUNDS

13.2.1. Tropospheric Ozone

13.2.1.1. Observational Analyses

Based on analyses at a few stations in the Northern Hemisphere (Vingarzan, 2004), it is found that surface O_3 has approximately doubled since the late nineteenth century and early twentieth century. For the last two to three decades, the change identified from observations has not been uniform in time. In general, the increases are larger from the 1970s to mid-1980s than in recent years; for the latter period, the signal is mixed. Measurements at some background stations have shown that O_3 concentrations have consistently increased since the 1950s. European stations showed an increase in surface O_3 until approximately the year 2000, and remained constant afterwards. However, on the west coast of the USA, the concentration continued to increase after 2000. The upward trends in O_3 in Europe and the USA before the 1990s were caused by increasing emissions of O_3 precursors (CO, NO_x, and VOCs). During the 1990s, the cause was less clear. Ordonez et al. (2007) argued that the increase in the USA can be attributed, to a large degree, to increased downward O_3 fluxes from the stratosphere. The downward trend in O_3 over Europe after 2000 is generally considered to have been caused by decreased precursor emissions. Dalsøren et al. (2009) suggest that the upward trend in O_3 on the west coast of the USA, despite control on land-based precursor emissions, was probably due to increases in pollution transport from Asia and from ship emissions, both of which have shown significant growth during the last decade. Below, we demonstrate similar developments for sulfate particles. These differences in precursors and the enhancement of secondary climate compounds underline the importance of considering regional and sector-wise distributions and changes in attempting future climate prediction.

Tropospheric composition derived from satellite measurements was first assessed with the Total Ozone Mapping Spectrometer (TOMS) instrument (Krueger, 1989). Although the instrument was basically designed to observe the total O_3 column, it was also used to retrieve tropospheric O_3. A lengthy time series has been constructed for tropospheric O_3, based on data taken by TOMS together with other datasets (Fishman and Brackett, 1997; Ziemke et al., 2001). The Global Ozone Monitoring Experiment (Burrows et al., 1999) was the first campaign in which the satellite instruments had adequate spectral resolution to observe O_3 and some of its precursors such as NO_2. Satellite observations of chemically active compounds O_3, CH_4, CO, and NO_2 have provided information on the global O_3 distribution for validating CTMs. Analyses of satellite observations have also improved knowledge about key chemical and physical processes in the troposphere, thereby improving the capability to reproduce and assess the impact of changes in anthropogenic emissions (Richter et al., 2004; Schoeberl et al., 2007). Such studies include the identification of key source regions, the quantification of source strengths and the assessment of changes and trends over past decades.

Harries et al. (2001) used changes in observed outgoing radiation at the top-of-the-atmosphere (TOA) within the spectral regions where O_3 absorbs to demonstrate its contribution to the greenhouse effect. The clear-sky total radiative forcing from upper tropospheric O_3 over the oceans has recently been estimated by Worden et al. (2008) to be 0.48 ± 0.14 W m^{-2}.

13.2.1.2. Modelling Studies

A characteristic feature of tropospheric O_3 (the same for sulfate and SOA) and its radiative forcing is the large hemispheric asymmetry in their spatial distributions (Isaksen et al., 2009; reproduced as Isaksen et al., 2012, this volume). Figure 13.1 depicts the comparison of mid-tropospheric O_3 distribution between model estimates and satellite results from the tropospheric emission spectrometer; both model and observations indicate a large regional contrast and a pronounced hemispheric asymmetry. Further comparison also shows that the observed O_3 column distribution can be well-represented by model calculations. Enhanced tropospheric O_3 is found over polluted continents in the Northern Hemisphere. Also, some distinct transport of O_3 from the North American continent into the North Atlantic is identified.

To illustrate the strong impact of changes in precursor emissions on regional O_3, Figure 13.2 shows the changes of surface O_3 resulting from changes in the precursor emissions (NO_x, VOCs, and CO) between the two periods, 1990–1992 and 1999–2001. Reductions or moderate changes occurred over the USA and Europe, reflecting control measures implemented to reduce precursor emissions during the decade. In contrast, significant O_3 increases were estimated over Southeast Asia caused by continued emission increases in the last decade. Since the column O_3 changes shown correspond to the year 2000, the increases underestimate the current increases over regions such as Southeast Asia where the precursor emissions have increased significantly during the last decade. In addition to large regional differences in the O_3 changes, large seasonal differences in O_3 were estimated, particularly at high northern latitudes, indicating regional and seasonal differences in oxidation capacity.

The evolution of tropospheric O_3 due to anthropogenic activities in the future and the associated radiative forcing have been thoroughly investigated (e.g., Stevenson et al., 2000; Hauglustain et al., 2005; Liao et al., 2006; Zeng et al., 2008). In general, the surface precursors dominate and cause an increase in tropospheric column O_3. More discussion will be given in Section 13.3.3. Finally, Unger et al. (2008) have researched the contribution to radiative

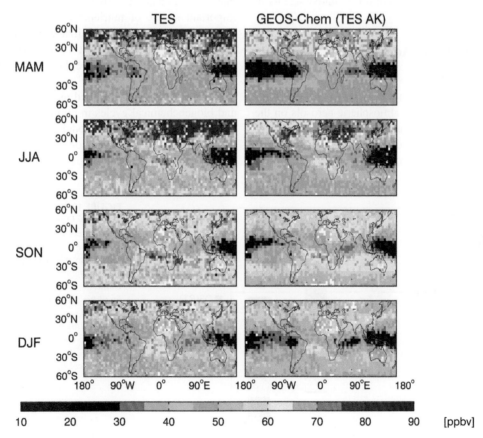

FIGURE 13.1 Tropospheric ozone distributions in unit of ppbv from TES at 500 hPa for the different seasons of 2006 — March-April-May (MAM), June-July-August (JJA), September-October-November (SON), and December-January-February (DJF) — compared with the GEOS-Chem ozone simulation. The purple colour represents relatively low values and the red colour represents relatively high values. White areas indicate lack of data meeting the retrieval quality criteria. *(Source: Zhang et al., 2010.)*

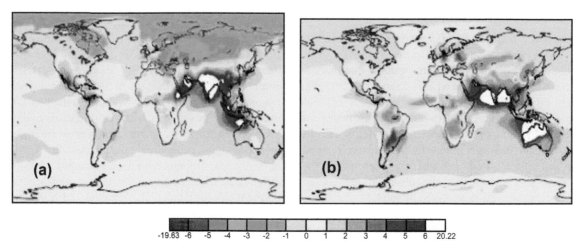

FIGURE 13.2 Calculated increase of surface O_3 (ppbv) as a result of emission changes between two periods (1990–1992 and 1999–2001) simulated by Oslo-CTM2: (a) April and (b) October.

forcing due to emissions of air pollutants from individual sectors between 2000 and 2030. As illustrated in Figure 13.3, the total O_3 forcing, which is dominated by biomass burning, transportation, and domestic biofuel consumption, is nevertheless small (cf. radiative agents in Table 13.2).

13.2.2. Tropospheric Aerosols

13.2.2.1. Anthropogenic Sources

Perhaps one of the earliest and most pervasive couplings between human activities and atmospheric aerosols on the large-scale is the nexus of anthropogenic land use for agriculture and animal husbandry and atmospheric mineral dust. Dust plays an important role in global biogeochemical

cycles (Falkowski et al., 1998; Mahowald et al., 2009) and has important climatic effects (Ramanathan et al., 2001; Stith et al., 2009). Furthermore, it plays an important role in planetary-scale teleconnections, an example being the fact that dust transported from the Saharan region is an important source of nutrients for the Amazon rainforest (Ben-Ami et al., 2010). Mulitza et al. (2010) show that there is a clear increase in dust flux in Africa at the onset of commercial agriculture in the Sahel region at the beginning of the nineteenth century.

Dynamically coupling land-use changes with aerosols and biogeochemical cycles in climate models is a topic at the cutting edge of climate model development. Mahowald et al. (2011) show that inclusion of dust and other aerosols has statistically significant impacts on regional climate and

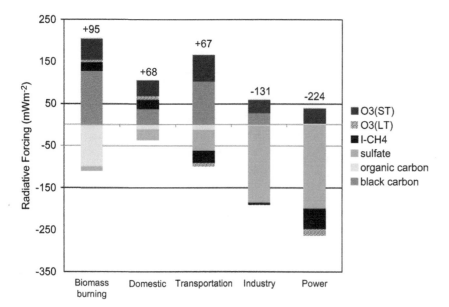

FIGURE 13.3 Annual global mean radiative forcing by O_3 (ST (short term)), sulfate, BC, OC, indirect CH_4 (I-CH_4), and O_3 (LT (long term)) due to individual emission sectors at 2030 in an A1B atmosphere (mW m^{-2}). BC and OC values are from Koch et al. (2007). Domestic sector includes both biofuel and fossil fuel. The net sum over all components for each sector is shown above each bar. *(Source: Unger et al., 2008 © Copyright 2008 American Geophysical Union. Reproduced/ modified by permission of American Geophysical Union.)*

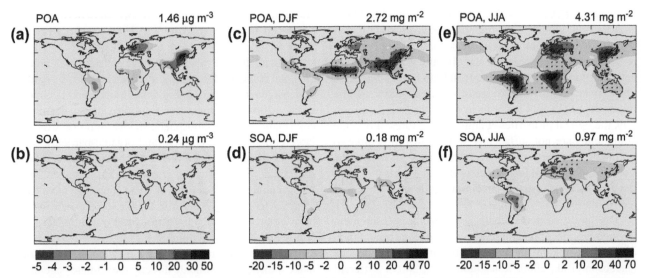

FIGURE 13.4 Predicted changes in Primary Organic Aerosol (POA) and Secondary Organic Aerosol (SOA) between 2000 and 2100. (a) and (b) show mean surface-layer concentrations ($\mu g\ m^{-3}$); (c) through (f) show column burdens (mg m^{-2}). *(Source: Liao et al., 2009 © Copyright 2009 American Geophysical Union. Reproduced/modified by permission of American Geophysical Union.)*

biogeochemistry. These effects are through the perturbations to the ocean nitrogen cycle and primary productivity due to altered iron inputs from desert dust deposition. Dust is a powerful mechanism and vehicle for processes that link terrestrial vegetation, oceanic biology, atmospheric chemistry and physics, and biogeochemical cycling (cf. Dickinson, 2012, this volume; Sen Gupta and McNeil, 2012, this volume).

Organic aerosols (OAs) are also particularly sensitive to feedbacks between changes in climate and associated changes in atmospheric chemistry. Figure 13.4 (from Liao et al., 2009) shows predicted changes in seasonal POA and SOA concentrations between 2000 and 2100. Figures 13.4a and 13.4b show changes in annual mean surface-layer concentrations (units of $\mu g\ m^{-3}$).

Figures 13.4c and 13.4e show column burdens (units of mg m^{-2}) for POA, and Figures 13.4d and 13.4f show column burdens for SOA. The global SOA burden is 13% higher in 2100 for the fully coupled calculations compared with the non-coupled simulations. This result illustrates a very important interaction between changes in climate and changes in chemistry and thermodynamics. Higher temperatures in the coupled simulations lead to a 92% increase in the emissions of monoterpenes and other VOC in 2100. However, the higher temperatures also shift the gas–particle partitioning of these compounds strongly towards the gaseous phase. Similar results were obtained for nitrate compounds by Pye et al. (2009).

An important feedback loop exists between aerosols and precipitation. Recent results show the feedback between aerosols and precipitation to be important. If the non-modelled interactions between aerosols and precipitation are included in terms of how aerosols influence cloud droplet

spectra (and thereby the coagulation/coalescence process), as well as aerosol effects on ice formation (a critical precipitation formation process particularly in tropical convective clouds), the processes linking aerosols to precipitation must be considered a very important area for future climate research (Lohmann et al., 2010).

Heald et al. (2008) investigated the response of SOA concentrations to changes in climate, emissions, and land use. They used the Community Atmosphere Model (CAM3) driven with the SRES A1B and A2 scenarios (see Evans et al., 2012, this volume). Their calculations for the A1B scenario predict that the SOA burden increases by 36% in 2100, roughly 90% of which is biogenic in origin. They predict increases in biogenic SOA production for all regions. The production rate of anthropogenic SOA is predicted to decrease in Europe and North America, but to increase in other regions — particularly in Asia, South America, and Africa. Anthropogenic SOA increases are greater using the A2 scenario, although projected agricultural expansion and urbanisation reduces the biogenic SOA source strength in South America, Sub-Saharan Africa, China, and the United States. Heald et al. (2008) caution that anthropogenic SOA is thought to be underestimated in current climate models.

Heald et al. (2008) also point out that the present-day burden of SOA is comparable to the current sulfate burden. Sulfur emissions are predicted to decrease by more than 50% in the A1B scenario, with corresponding decreases in the burden of atmospheric sulfate. This decrease in sulfate aerosol would possibly be compensated by increases in SOA (such as the 36% increase predicted in the Heald et al. (2008) study) and by projected increases in POA emissions (Chung and Seinfeld, 2002). However, a global-scale tradeoff between sulfate and organic aerosol mean mass burdens does not imply a corresponding compensation in the climatic effects of these aerosols. Most of the decrease in sulfate burden will take place over industrialized areas, whereas most of the increase in OAs will occur in tropical regions. Additionally, the optical properties of complex mixtures of organic compounds are not sufficiently well-known to be able to accurately predict what effect they may have on the Earth's radiative balance, much less what effects they may have on cloud properties, precipitation development, and ice formation.

OAs are a particular challenge from the measurement perspective, since 'organics' at any given location and time could be a mixture of hundreds of different compounds (Jacobson et al., 2000; Kanakidou et al., 2005). Different mixtures of compounds will lead to different properties for the particles, for example, refractive index, chemical reactions and reactivity, and thermodynamic properties. Identifying and quantifying all of the compounds present in OAs will remain a major measurement challenge for some time to come. Even more challenging will be the development of methods for incorporating new knowledge about OAs into climate models. Linking the kind of observational information available at a single-particle level to the scale of a global model is a formidable challenge. Fuzzi et al. (2006) suggest that a potential way forward in this regard is to use molecular functional groups as a way of reducing the number of parameters necessarily needing to be incorporated in global models, while still retaining a sufficient amount of information about the important properties of OAs (Isaksen et al., 2009).

13.2.2.2. Observational Analyses

Deep ice-core drillings allow the identification of pre-industrial concentration levels of key aerosol components and their changes during the industrial period. For example, sulfate concentration levels in ice cores suggest a massive increase in aerosol loading after 1900. Decreases were observed after 1980 in some major industrial regions (e.g., the USA and later in Europe) due to the SO_2 abatement policies in relation to power plants. Specifically, an increase from pre-industrial levels by a factor of about 4 in 1980 was found for Greenland and the Canadian Arctic (Legrand et al., 1997). Ice cores in the Alps show larger increases due to the proximity of European sources. Preunkert et al. (2001) measured an increase in sulfates from 80 pg g^{-1} (10^{-12} g g^{-1}) in the pre-industrial period to 860 pg g^{-1} in 1980, and subsequently a decrease to 600 pg g^{-1} in the 1990s. A similar development is also seen in remote locations; for example, Quinn et al. (2007) reported trends in the Arctic: peak

concentrations of sulfate aerosols at Alert, Canada decreased from 0.8 (μgS m^{-3}) in 1982 to 0.3 μgS m^{-3} in 2004.

Ice-core records also show that emissions of mineral dust can change dramatically in different climate regimes, at least over glacial—interglacial transitions (Delmonte et al., 2004). Biomass burning provides a significant contribution to the atmospheric oxidation process through large emissions of CO, NO$_x$, and organic particles. Inter-annual variability in the biomass source could be significant for climate—chemistry interactions. For instance, large inter-annual variations in biomass burning have been observed (van der Werf et al., 2006), some of which have been related to climate disturbances such as large El Niño—Southern Oscillation events.

Although the strong regional differences in aerosols make it difficult to evaluate directly the climate—chemistry interaction, attempts have been made by using a variety of observations to study the implications to atmospheric compositions during the periods of short-term climate anomalies. For example, as expected from the atmospheric increase in sulfate aerosols, global temperatures decreased significantly following the 1991 eruption of Mount Pinatubo. Observations show that tropospheric water vapour decreased after the eruption (Soden et al., 2002), in agreement with the response seen in climate model simulations (MS). Additionally, CH$_4$ emissions from wetlands were reduced during this time (Walter et al., 2001).

Quaas et al. (2008) conducted a satellite-based estimate of the direct aerosol radiative forcing and found that the TOA all-sky forcing is about -0.9 ± 0.4 W m^{-2}. They indicated that the range of aerosol forcing values inferred from satellites agrees with the range of values calculated by forward modelling[1], implying large uncertainties associated with the use of satellite measurements to infer aerosol forcing. The study also estimated the indirect effect of aerosols on cloud albedo (the first indirect effect), calculating a value of -0.2 ± 0.1 W m^{-2}. Although the uncertainty range could be larger than the range given, the results indicate that current global climate models may overestimate the cloud albedo effect. Although there are limitations in the forcing estimate using satellite observations, an advantage is that such data have better global coverage than those from ground and aircraft measurements. It should be pointed out that radiative forcing estimates based on satellite data cannot separate the natural and anthropogenic contributions.

Zerefos et al. (2009) examined the trend of the impact due to aerosol abundances during the last two decades. Figure 13.5 shows significant different impacts on the

---

1. Forward modelling provides a numerical solution to a physical process where no direct analytical solution exists.

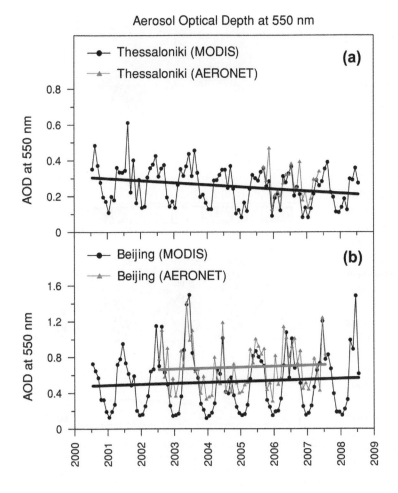

FIGURE 13.5 Time series of aerosol optical depth (AOD) at 550 nm (1 nanometre = 10^{-9} metre) at Thessaloniki from MODIS satellite data (circles) and from the AERONET data database (triangles). (b) Same as (a) but for Beijing. *(Source: Zerefos et al., 2009.)*

penetration of solar radiation since 2000 over Europe, where air pollution abatement strategies have been implemented during this period, than over China, where pollution emissions have continued to increase. After a reduction in solar ultra-violet band A (UV–A) radiation over Thessaloniki, Greece until the early 1990s, there was an increase afterwards. In contrast, over Beijing, China, where air quality measures were taken later, no reversal in overall solar radiation was seen.

Major technology developments are expected to lead to a shift in emissions and aerosol distributions, with a larger fraction over the next 20–30 years emitted in regions such as India and China (Cofala et al., 2007), a tendency already observed during the 1990s (Akimoto, 2003). The inadequate global distribution of aerosol monitoring sites has made satellite-retrieved aerosol distributions a major source of information for global coverage. Calibration and precision of the satellite-retrieved aerosol optical thickness have improved, in particular over land and sea since the late 1990s with the launch of a number of satellite-borne instruments (Laj et al., 2009). It is anticipated that further significant progress will be made in the coming years.

13.2.2.3. Modelling Studies

SOA represents an important, and in some regions the major fraction of, the burden of OAs (Kanakidou et al., 2005). There has been considerable change in the composition and magnitude of emissions from anthropogenic activities. Changes in SOA production and burden since pre-industrial times were first examined with a global model by Kanakidou et al. (2005). Subsequent studies involved a range of SOA precursors. A recent estimate of the SOA burden between the pre-industrial period and the present was made by Hoyle et al. (2009) who found a 60% increase from about 44 Tg to 70 Tg. The increases are greatest over industrialized areas, the majority of the increases being found to be caused by POA from fossil fuel and biofuel combustions.

The formation of secondary aerosols such as sulfate is often extremely non-linear due to the dependence on cloud distribution and OH gas chemistry in the troposphere. In addition, anthropogenically-induced secondary aerosols have a non-uniform distribution due to their short lifetime. Therefore, future aerosol distributions will be strongly linked to their regional emission patterns. Assessing the impact on climate is therefore not straightforward. It requires detailed knowledge of how the abundances and trends of atmospheric aerosols are affected by atmospheric physical and chemical processes.

Sulfate is a major anthropogenic aerosol component in the atmosphere (IPCC, 2007a). Since sulfate aerosols are short-lived in the troposphere and sulfate loading (the total sulfur mass in the atmosphere) is nearly proportional to the emissions of its precursor SO_2 (Andreae et al., 2005), the significant increase since pre-industrial time reflects the increasing anthropogenic SO_2 emission from fossil fuel combustion. The anthropogenic fraction of global sulfur emissions is estimated to clearly dominate over the natural emissions. The global anthropogenic SO_2 emissions were estimated to be in the range 66.8 TgS–92.3 TgS per year in the 1990s, while the natural SO_2 sources include 20 TgS per year of DMS (dimethyl sulfide) over oceans (Rotstayn and Penner, 2001; Verma et al., 2007) and volcanic eruption of 8 TgS per year in 1990 (Rodhe, 1999).

Over recent decades, increases in sulfate burden and the contribution to surface cooling show large regional differences due to different implementations of control measures to reduce SO_2 emissions. For example, Figure 13.6 illustrates the large regional difference in tropospheric sulfate burden between 1985 and 1996 as simulated from the Oslo-CTM2. Sulfate was reduced significantly over Europe over this period reflecting the large reduction in SO_2 emission. The reduction in sulfate over the USA was less pronounced since most of the reduction in emission of SO_2 occurred prior to 1985. On the other hand, Southeast Asia experienced a significant increase in SO_2 emission throughout the decade. This increase in emission continued during the last decade. The corresponding radiative forcing from sulfate changes in the same time period is striking, with strong regional effects. Over Europe and to a lesser degree over the USA, where sulfate levels decreased during the time period, a positive forcing is calculated due to less reflection of solar radiation. Southeast Asia on the other hand during the same time period experienced a significant negative radiative forcing due to the enhanced sulfate levels from large increases in SO_2 emissions. Analysis of Figure 13.6 clearly demonstrates the large potential for regional climate impact from short-lived climate compounds.

Studies of regional differences in sulfate perturbations have been conducted by using observed temperature trends together with the spatial pattern of temperature over the Northern Hemisphere (Santer et al., 1996; Kaufmann and Stern, 2002; Stott et al., 2006). In addition, a major modelling effort together with observations, the AM project (Schulz et al., 2006), was undertaken to study the global mass distributions for the major aerosol components. Nine models with detailed aerosol modules participated in order to simulate instantaneous direct radiative forcing due to anthropogenic aerosols. In the study, the aerosol effect

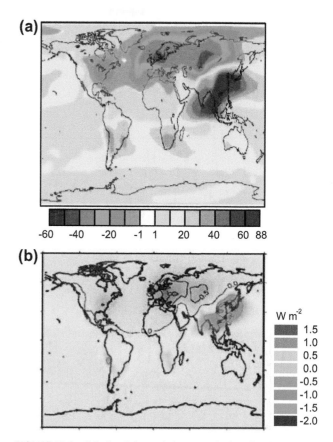

FIGURE 13.6 Calculated change in (a) tropospheric sulfate loading (%), and (b) the corresponding radiative forcing (W m^{-2}) between 1985 and 1996 from changes in emission of the sulfate precursor SO_2. The values are yearly average changes as simulated by Oslo-CTM2.

was derived from the difference between two sets of MS with prescribed aerosol emissions, one for the present-day and one for the pre-industrial condition. The contribution to radiative forcing was estimated for the three major anthropogenic aerosol compounds: sulfate, OA, and BC. Figure 13.7a shows the ensemble mean of the increase in optical depth since pre-industrial time from the models that participated in the AeroCom study. The large aerosol loadings found in North America, Western Europe, equatorial West Africa, and the areas of East and Southeast Asia, as well as the Indian subcontinent, are attributed to different types of aerosols. Ensemble estimates of aerosol radiative forcing from the AeroCom project (Figure 13.7b) indicate that there are large negative forcings over the industrial regions of the USA, Europe, and East Asia. However, the uncertainties in the modelling study of aerosol enhancement and of radiative forcing due to the direct aerosol effect are large, as reflected in the large standard deviations among the models shown in Figure 13.7c. The comparisons demonstrate that there are significant differences in the predictions of sulfate distribution and in the comparisons with observations. Due to their short atmospheric lifetime,

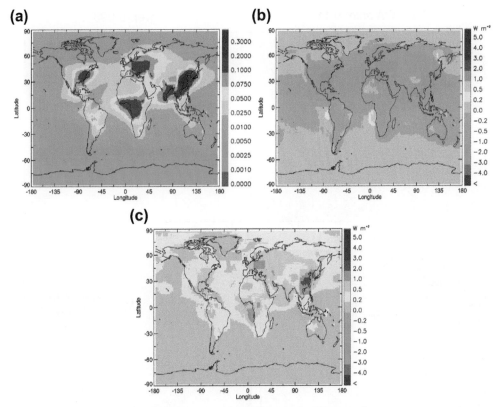

FIGURE 13.7 Ensemble mean increase of (a) total AOD, (b) mean direct radiative forcing (W m^{-2}), and (c) standard deviation of direct radiative forcing (W m^{-2}) due to anthropogenic contribution since the pre-industrial period. *(Source: Schulz et al., 2006.)*

global particle distribution can be linked to the major source regions. From the distribution patterns of the different aerosol components, including sulfate, considerable local variability in the aerosol composition can be inferred. As is the case for all climate predictions, it is not recommended to base conclusions on estimates from individual models but to use a combination of several models to provide a mean estimate (e.g., Evans et al., 2011).

In a recent study, Hoyle et al. (2009) found that the change in SOA burden caused a radiative forcing of −0.09 W m^{-2} when SOA was partitioned in organic and sulfate aerosols. This gives a radiative forcing of SOA that is substantially stronger than the best estimate for POA in the Fourth IPCC Assessment (IPCC, 2007a).

The sources for aerosols and their contribution to radiative forcing vary significantly with sectors and with aerosol types. A study of the contribution to total aerosol forcing from different sectors was conducted by Koch et al. (2007) who used a GCM to calculate the direct radiative forcing for present-day aerosols emitted from particular regions and from different sectors. Figure 13.8 shows the individual contribution to the total radiative forcing from various sectors from Koch et al. (2007). They found that two of the largest radiative forcings for BC are from

residential and transport sectors, making these, as Koch et al. (2007) suggested, potential targets to counter global warming. Most anthropogenic sulfate aerosols come from the power and industry sectors, and they account for the large negative aerosol forcing over the mid-latitudes of the Northern Hemisphere. There is also strong absorption (positive radiative forcing) over Southeast Asia.

In addition to the direct climatic effect, aerosols can serve as CCN to initiate liquid cloud droplets and regulate cloud droplet number and size, and hence cloud radiative properties and precipitation. Most aerosol–cloud inter-action studies that involve warm clouds (Kogan, 1991; Flossmann, 1998; Ghan et al., 1997, 2001; Lohmann et al., 1999; Menon et al., 2002) indicate consistent response in cloud properties: more aerosols result in more but smaller cloud droplets and thus enhanced cloud albedo (Twomey, 1977; Coakley et al., 1987; Ackerman et al., 2000; Cheng et al., 2007), known as the first aerosol indirect effect. In addition, the smaller cloud droplets would reduce the efficiency of cloud droplet coagulation and raindrop formation and thus usually prolong the cloud lifetime (Albrecht, 1989; Rosenfeld, 2000), known as the second aerosol indirect effect. Moreover, the reduced raindrop formation would lead to less or less intense precipitation.

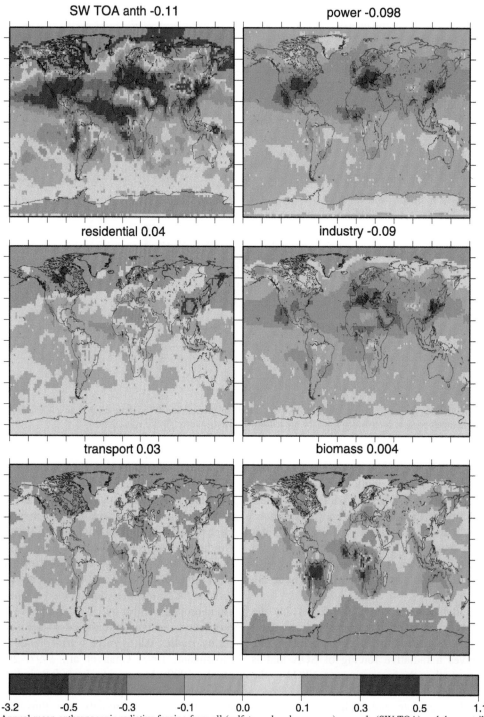

FIGURE 13.8 Annual mean anthropogenic radiative forcing from all (sulfate and carbonaceous) aerosols (SW-TOA) and the contributions from each sector. Global mean contribution is given for each sector. Units are W m^{-2}. *(Source: Koch et al., 2007 © Copyright 2007 American Geophysical Union. Reproduced/modified by permission of American Geophysical Union.)*

13.2.3. Coupling Changes of Chemistry and Climate

Formation of the secondary chemical compounds O_3, sulfate, and SOA depends on several other chemical constituents. In addition to their precursors, H_2O_2 and radicals like OH, HO_2 are important in the oxidation processes, which are extremely non-linear. This non-linearity is important for O_3 formation, as well as for sulfate and SOA-forming processes and thereby the cloud—chemistry interaction. The distribution of O_3 is also important for the formation of SOA. In modelling the chemical processes in the atmosphere, it is common to develop and apply extensive chemical schemes of primary and secondary compounds to represent the tropospheric O_3 chemistry, a task that is far from straightforward. A consequence of the complexity of atmospheric chemistry is that studies of chemical composition, composition changes, and estimates of climate impact are extremely uncertain.

One of the major advances since the publication of the IPCC AR4 has been the improvement of the treatment of aerosols and clouds and the development of climate models with the ability to simulate the interactions and feedbacks between changes in climate, chemistry, and atmospheric aerosols. A major impetus for this development has been to provide more accurate attribution of global environmental changes (such as changes in surface temperature, precipitation, or air quality) to their underlying causes.

Levy et al. (2008) examined the sensitivity of calculations for late twenty-first century climate to projected changes in short-lived air pollutants. Using variations of the A1B emissions scenario (in which SO_2 emissions decrease by 65% and BC emissions double), they calculated that changes in aerosol properties account for a significant fraction of the total warming predicted for 2100. The net global radiative forcing due to short-lived species is negligible in 2030 (relative to 2000). By 2050, short-lived species account for approximately 19% of the positive change in radiative forcing while, by 2100, decreasing sulfate and increasing BC concentrations account for about 26% of the global radiative forcing and 36% of the change in the Northern Hemisphere. In Levy et al.'s (2008) calculations, the contributions of O_3 and OAs to changes in radiative forcing are modest and tend to offset each other. Most of these changes were controlled by emission changes at the end of the century. Importantly, Levy et al. (2008) found that in their calculations the regional patterns of surface warming did not follow the regional patterns in respect to changes in short-lived species. This lack of correlation may be due to the model's climate response, and not significantly influenced by changes in aerosol properties. The calculations did not include the indirect effects of aerosols on clouds, nor was there any feedback between changes in climate and changes in aerosols and O_3.

The differences in predictions between models that do and do not include interactive coupling of climate changes, emissions, atmospheric composition, and chemistry have been examined by Liao et al. (2006, 2009) who used SRES A2 scenario, and by Pye et al. (2009) who adopted the A1B scenario. The three studies investigate different aspects of aerosol—chemistry—climate interactions.

Liao et al. (2009) use the Goddard Institute of Space Studies (GISS) general circulation model II with online simulation of sulfate, nitrate, ammonium,

BC, primary OC, secondary OC, and tropospheric O_3—NO_x—hydrocarbon chemistry, driven by the SRES A2 scenario. They use five different model experiments to examine the effect of chemistry—aerosol—climate coupling on the predictions of future climate. Figure 13.9 shows the seasonal equilibrium temperature response (the difference between years 2000 and 2100) when all species are treated interactively (Figures 13.9a and 13.9b), when only greenhouse gases (GHGs) are considered (Figures 13.9c and 13.9d), and when the effects of O_3 and aerosols (panels E and F) are included. When all species are considered, the equilibrium global mean surface air temperature increase over the century is +6.1 K during December—February (DJF) and +5.8 K during June—August (JJA). Most of this increase is due to GHGs, illustrated in panels C and D. Aerosols and O_3 contribute +0.82 K in DJF and +0.85 K in JJA to this increase. The contributions of O_3 and aerosols to the overall warming are very inhomogeneous both spatially and seasonally. In the Liao et al. (2009) calculations, O_3 and aerosols lead to a pronounced warming in the Arctic region, especially in the Northern Hemisphere winter season. Ozone and aerosols lead to regional cooling in South and Southeast Asia in DJF, while in JJA the model predicts warming in east Asia, but cooling over the Indian subcontinent and over the southern Saharan region.

The climate sensitivity in the model used for these calculations is 0.8 K m^2 W^{-1}, which is similar to many of the models included in the IPCC AR4. Andreae et al. (2005) point out that, if the present-day aerosol forcing is underestimated, the climate sensitivity to a doubling of CO_2 would also be underestimated (cf. Harvey, 2012, this volume). They caution that incomplete consideration of aerosols in current models may have led to an underestimation of the true climate sensitivity. The estimate of global mean net aerosol forcing given in the IPCC AR4 is −1.2 W m^{-2}, with a probable range between −0.6 and −2.4 (Forster et al., 2007). A more recent estimate using historical data derives a value of −1.31 ± 0.52 W m^{-2}, and concludes that decreasing concentrations of sulfate and increasing concentrations of BC aerosols have substantially contributed to the rapid Arctic warming during the past 30 years (Shindell and Faluvegi, 2009). The fact that the climate sensitivity in the Liao et al. (2009) calculations (with a more comprehensive treatment of aerosols compared to most of the models included in the IPCC AR4) is similar to previous model estimates is perhaps somewhat comforting in this regard, but insufficient to put to rest the issue of climate sensitivity in current models. While global mean surface temperature has been a major concern, other changes are of equal (if not greater) importance to society. Figure 13.10 (also from Liao et al., 2009) shows changes between 2000—2100 in the seasonal equilibrium precipitation rate (left panels) and surface-layer aerosol concentrations (right panels).

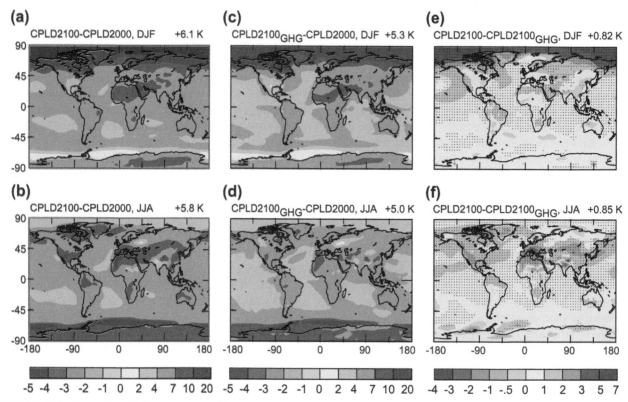

FIGURE 13.9 Seasonal equilibrium temperature differences (K) to changes in all species: (a) and (b) GHGs, ozone, aerosols; (c) and (d) only GHGs; (e) and (f) ozone and aerosols (change between present day and 2100. DJF — December—February; JJA — June—August. (*Source: Liao et al., 2009.* © *Copyright 2009 American Geophysical Union. Reproduced/modified by permission of American Geophysical Union.*)

In these calculations, precipitation is seen to increase in the Inter-Tropical Convergence Zone (ITCZ) in both seasons. The global mean increase in precipitation in the fully coupled simulation is 0.34 mm d^{-1} for DJF and 0.31 mm d^{-1} for JJA, roughly a 10% increase. These increases in precipitation are driven primarily by the large warming at the surface and in the lower troposphere, and are consistent with previous calculations (Meehl et al., 2007). An interesting aspect of these (and of other) calculations is that precipitation changes are driven by changes in energetics and dynamics, not by changes in cloud properties (Ming et al., 2010). Precipitation processes have been shown in many investigations to be sensitive to aerosol—cloud interactions, not only to the influence of dynamics and energetics (e.g., Andreae et al., 2004; Ferek et al., 2000; Rosenfeld, 2000). Presumably, incorporating improved descriptions of how aerosols affect precipitation formation would alter these results; however, it is impossible to predict at this time how the predictions of future climate may change.

Also shown in Figure 13.10 are results for calculations of seasonal changes in surface-layer aerosol concentrations (panels C and D). The largest changes are seen over East Asia, over areas of substantial biomass burning and over Europe. These increases are a result of a positive feedback

loop: reduced convection and precipitation lead to decreased wet deposition of aerosols; this leads to higher concentrations, which further influences convection and precipitation.

13.3. CLIMATE—CHEMISTRY INTERACTION OF TROPOSPHERIC OZONE

13.3.1. The Role of Ozone As a Climatically Active Compound

Atmospheric ozone has been recognized as a climatically active gas (Wang and Isaksen, 1995) and its change will affect atmospheric temperature, as well as the solar and longwave radiation reaching the surface. In particular, changes in absolute O_3 densities in the lower stratosphere and upper troposphere lead to the most pronounced impact on surface temperature (Wang and Sze, 1980; Lacis et al., 1990; Forster and Shine, 1997). Climate also affects O_3 because its formation depends on temperature, moisture, clouds, and winds (for transport), as well as the presence of other chemical compounds. This section provides a broader perspective on tropospheric O_3, covering the relevant chemical and climate processes. The importance of climate—chemistry interactions in affecting tropospheric O_3

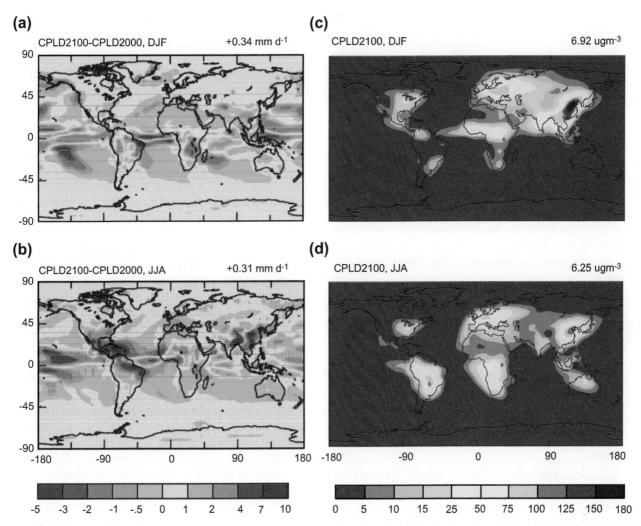

FIGURE 13.10 Changes between 2000 and 2100 in seasonal mean equilibrium precipitation rate (mm d-1) for (c) December–February (DJF), and (b) and June–August (JJA). Surface-layer aerosol concentrations ($\mu g\ m^{-3}$) for (a) DJF, and (d) JJA. *(Source: Liao et al., 2009 © Copyright 2009 American Geophysical Union. Reproduced/modified by permission of American Geophysical Union.)*

concentration and the feedback effect to climate is illustrated by analysing simulations from a global climate model with interactive chemistry within the context of global warming.

13.3.2. Ozone Chemistry

The primary processes leading to tropospheric O_3 production proceed via a chemical reaction (R) involving nitrogen oxides (NO_x) and hydrogen radicals OH and HO_2.

The formation of OH, initiated via:

$$O_3 + h\nu \rightarrow O(^1D) + O_2 \qquad (R13.1)$$

and followed by:

$$O(^1D) + H_2O \rightarrow OH + OH \qquad (R13.2)$$

are the key initial steps of the oxidation process in the troposphere, where $h\nu$ represents solar UV–B flux. Consequently, the UV–B flux and the distributions of O_3

and H_2O are important for the formation of tropospheric OH. In addition, other gases affect OH; for example, increasing atmospheric CH_4 can decrease OH in the troposphere, leading to O_3 increases.

Ozone production in the troposphere occurs almost exclusively through the reaction:

$$NO + HO_2 \rightarrow NO_2 + OH \qquad (R13.3)$$

or similar reactions of NO with peroxy radicals, followed by Reactions (R13.4) and (R13.5). The sources for HO_2 include oxidation of VOCs initiated by reactions of the type given by Reaction (R13.7) and Reaction (R13.8) of OH with CO. Subsequently, O_3 is produced through the reaction sequence:

$$NO_2 + h\nu \rightarrow NO + O \qquad (R13.4)$$

$$O + O_2 + M \rightarrow O_3 + M \qquad (R13.5)$$

An important O_3 loss path in the free troposphere is through:

$$O_3 + HO_2 \rightarrow OH + 2O_2 \qquad (R13.6)$$

Although there are a large number of reactions affecting the chemical composition in the atmosphere, as a first-order approximation, there is a competition in the free troposphere between O_3 production represented by Reaction (R13.3) and O_3 loss represented by Reaction (R13.6). Therefore, the availability of NO_x ($NO + NO_2$) in the free troposphere is a major requirement for a net O_3 production. In a NO_x-rich region, the chemical processes give a net O_3 production through Reaction (R13.3) while, in a NO_x poor region, O_3 loss dominates mainly through Reaction (R13.6) (Isaksen et al., 1978).

OH represents the dominant CH_4 loss through the reaction:

$$CH_4 + OH \rightarrow CH_3 + H_2O \qquad (R13.7)$$

This reaction is also an important loss for OH in the troposphere, thereby representing a positive feedback for the OH–CH_4 chemical interaction. In the present atmosphere, the feedback of 40% reported in IPCC (2001a) is still the current best estimate. The feedback is highly significant for estimating the climate impact of CH_4 emission (Isaksen et al., 2009).

The main loss of atmospheric OH occurs via the reaction:

$$CO + OH + O_2 \rightarrow CO_2 + HO_2 \qquad (R13.8)$$

with Reaction (R13.3) being a major source reaction of OH in regions affected by the emission of pollutants such as NO_x and CO, which could have a major indirect impact on CH_4 through their control of OH. The anthropogenic NO_x and CO surface emissions are concentrated mainly in the Northern Hemisphere.

Increases of atmospheric CH_4 can lead to increased H_2O in the stratosphere. The formation occurs mainly through Reaction (R13.7), followed by reactions where the CH_3 radical is rapidly oxidized to OH or HO_2. H_2O is then formed through the reaction:

$$OH + HO_2 \rightarrow H_2O + O_2 \qquad (R13.9)$$

As a result, one molecule of CH_4 is oxidized in the stratosphere to form two H_2O molecules (except for a small fraction of H_2).

The reactions discussed above are the main reactions in the O_3 forming process. A key factor in limiting O_3 production in the troposphere is the efficiency of NO_x removal. NO_2 has a lifetime of only a few hours to days in the troposphere. Thus, there are large spatial and temporal variations in the NO_x distribution in the troposphere and thereby in O_3 formation.

13.3.3. Ozone–Climate Coupling

Tropospheric O_3 is closely coupled to the climate system, with its concentration link to parameters (e.g., temperature, moisture, and clouds) and processes (e.g., transport and dry deposition). Temperature affects chemical reaction rates for O_3 production and loss through the rate constants: Reactions (R13.1) and (R13.4) are positively related to temperature due to the temperature-dependence of the absorption cross-section and quantum yield; on the other hand, increases in temperature decrease the reaction rates of Reactions (R13.3) and (R13.5). Overall, the net effect of temperature increase is to yield more O_3. The reaction rate for CH_4 oxidation (Reaction (R13.7)) increases with temperature.

Increases in moisture associated with a warming climate increase the odd hydrogen (e.g., OH) through the Reaction (R13.2), resulting in a significant chemical O_3 sink (Reaction (R13.1)). On the other hand, if OH increases, then, through Reactions (R13.7) and (R13.8), this results in an enhancement of the oxidation of CH_4 and CO respectively and produces more peroxy radicals (i.e., HO_2). Subsequently, HO_2 converts NO back to NO_2 (Reaction (R13.3)), which contributes to the production of O_3 through the photolysis of NO_2 (Reactions (R13.4) and (R13.5)). The net effect of increases in moisture is to decrease O_3. Climate change might also be accompanied by significant changes in cloud cover and cloud lifetime, which may affect tropospheric photochemistry (Mao et al., 2003) and, in particular, O_3. A rise in cloud fraction decreases the amount of actinic flux available for photochemical reactions (Reactions (R13.1) and (R13.4)). However, the effects of cloud changes on O_3 and aqueous phase chemistry are less certain.

Studies have shown that there is a potential for significant biosphere–atmosphere interaction affecting O_3 distribution in a warmer climate (Fowler et al., 2009). The changes in surface characteristics due to climate change can influence dry deposition of O_3. Removal of O_3 at the surface is most effective over vegetated surfaces. Surface O_3 deposition could affect its abundance strongly in the lower troposphere. This represents a potentially important process linking O_3 abundance to climate change. Solberg et al. (2008) showed that a drier and warmer climate could yield significantly enhanced tropospheric O_3 levels. Sitch et al. (2007) argued that O_3 could have a significant indirect positive effect on climate through interaction with CO_2 surface uptake. Furthermore, climate change is likely to change precipitation and the frequency and intensity of rainfall events (IPCC, 2007a), which affect O_3 concentration. Climate change also influences anthropogenic surface emissions of O_3 precursors (e.g., from fossil fuel and biomass burning; Fowler et al., 2009) and climate-sensitive natural emissions (e.g., N_2O and NO_x from soils; NO_x from lightning; CH_4 from wetlands; hydrocarbons from vegetation; various gases from the oceans; and particles by physical processes).

The impact on O_3 of changes in dynamics and atmospheric stability in a warmer climate has been investigated by Solberg et al. (2008), who used an observed extreme high pressure situation over Western Europe in August 2003 as an analogue. The study used a CTM to simulate the chemical evolution during such an episode, and the results reveal, in line with the observations, a strong O_3 increase and a degradation of air quality due to changes in the dynamics.

Therefore, climate changes, such as changes in temperature, water vapour, dynamics, and interaction with vegetation, may affect O_3. Due to the strong variation in space and time of the climate states and surface emissions of O_3 precursor gases, the O_3 responses will also exhibit different regional characteristics.

Many coupling mechanisms between climate and O_3 are understood, but the net effect of climate change on O_3 and vice versa have significant uncertainty, in particular in the troposphere. Atmospheric O_3 climate–chemistry interaction has been studied using a vertical 1D model to consider temperature change (e.g., Wang and Sze, 1980) but, in recent years, 3D global models (Liao et al., 2006; Unger et al., 2006; Zeng et al., 2008) have also been employed, including a CTM with prescribed meteorology such as reanalysis data, and an interactive climate–chemistry model with consistently simulated meteorology. The net impact on O_3 is a result of two large but compensating effects: a temperature increase leads to more O_3 production, but an increase of water vapour in the troposphere causes more O_3 destruction via Reaction (R13.2). These studies indicate that the change in total chemical destruction is larger than total chemical production, thus causing a decrease in net chemical production. The predicted climate changes predominantly decrease tropospheric column O_3, while the anticipated increases in surface emissions of O_3 precursors dominate the column O_3 increase. The results imply that physical climate changes play a minor role relative to changes in tropospheric O_3 caused by surface emissions of precursors.

13.3.4. Effect of Ozone–Climate Interaction

13.3.4.1. Model Study Setup

Because the climate states and surface emissions of O_3 precursor gases exhibit distinctively different regional characteristics, the present analyses focus on changes in regional tropospheric column O_3 and surface O_3 due to future changes in global climate and surface emissions, and the subsequent feedback effect of O_3 changes to climate.

To illustrate the effects, we used the findings from Tanaka (2009) who conducted Global Climate–Chemistry Model

(GCCM) (Wong et al., 2004) simulations to specifically examine the responses of tropospheric O_3 to (a) climate changes due to increasing GHGs and (b) the future surface emissions of O_3 precursor gases. Three MS were conducted: MS1 and MS2 use respectively the 1990 and 2100 levels of GHG concentrations and surface emissions of O_3 precursors (CO, NO_x, and NMVOCs [non-methane volatile organic compounds]); MS3 uses the 2100 GHG concentrations but 1990 surface emissions; and for 2100, the A2 scenario of IPCC (2007a) was utilized. Increases of surface emissions of O_3 precursors between 1990 and 2100 occurred mainly over mid-latitude land areas where the three regions − North American, Western Europe, and East Asia − are located. Large increases in CO and NMVOCs are also identified over India and Saudi Arabia. The contribution of lightning to NO_x remains unchanged but aircraft emissions are considered. In the simulations, the sea surface temperature (SST) is prescribed based on the 1990−1999 monthly mean values for the MS1 case while, for MS2 and MS3, the SST is increased by 2 K uniformly over the oceans to approximate anticipated global warming (but cf. Sen Gupta and McNeil, 2012, this volume). The climate responses of the land surface and atmosphere to increases of GHG, and the coupling between climate and chemistry, are explicitly simulated by the model (cf. Pitman and de Noblet-Ducoudré, 2012, this volume).

Differences between MS1 and MS2 reveal tropospheric O_3 changes due to both increased surface emissions and climate warming, while differences between MS1 and MS3 provide changes attributed exclusively to climate warming. Comparisons of these two sets of values provide indications of the relative role of climate warming versus increased surface emissions in affecting tropospheric O_3. In addition to the global mean, we examine results for three regions (land only): East Asia, North America, and Western Europe. The model-simulated O_3 changes between 1990 (MS1) and 2100 (MS2 and MS3) are then used for calculating the radiative forcing. The results indicate that the temperatures on the surface and throughout the troposphere are increased due to increases of GHG, with a peak warming of 3.5 K found at the 300 hPa level over the mid-latitudes of both hemispheres. The column moisture change shows a more distinctive zonal characteristic with a much larger increase at low latitudes. Total cloud cover change is small. The feedback to climate from O_3 changes for both climate warming and the combined global warming and surface emissions is small.

13.3.4.2. Impact on Column Ozone

As shown in Figure 13.11b, the effect of global warming (MS3-MS1) yields a small decrease in column O_3 over most of the areas around the globe, resulting in a 1 DU (Dobson Unit) decrease in the annual and global mean column O_3. The seasonal contrast of the global column O_3

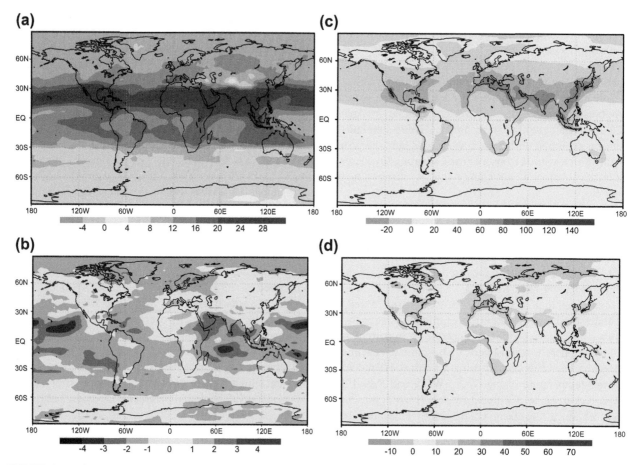

FIGURE 13.11 Simulated O₃ changes in: (a) and (b) tropospheric column amount (DU), and (c) and (d) near surface (ppbv), due to combined increases in surface precursor emissions and GHG (a,c), and increases of GHG alone (b,d). The changes are the differences between 1990 and 2100 simulated with the global climate-chemistry model (GCCM) using the corresponding GHG and surface emissions of O₃ precursors. (*Source: Tanaka, 2009.*)

change is small: 1.1 DU in winter and −0.9 DU in summer. The simulations suggest that there exists a good correlation between the pattern of column O_3 decrease and the pattern of column moisture increase. For example, O_3 decreases of 3 DU are seen over one area of the eastern Pacific Ocean, where increases of tropospheric column water vapour are also large. This feature is consistent with the notion that more water vapour causes more O_3 destruction, as discussed in Sections 13.3.1 and 13.3.2. However, it is difficult to relate the O_3 change to changes of temperature and clouds, although Mao et al. (2003) have shown that low and middle level clouds are more important than high clouds in affecting the O_3 chemistry.

When both the climate warming and surface emission of O_3 precursors are considered, the annual and global mean column O_3 change is calculated to be 13.7 DU which is within the range 13.2−19.5 DU estimated by other model studies (e.g., Stevenson et al., 2000; Hauglustain et al., 2005; Liao et al., 2006; Zeng et al., 2008). The value is an order of magnitude larger than the value of 1 DU caused by

the effect of climate warming alone. Large increases are seen over the areas between 30°S and 30°N. Again, the seasonal contrast in the column O_3 changes is small. Comparison between the two sets of simulations (i.e. MS2-MS1: changing both GHG and surface emissions, Figure 13.11a; and MS3-MS1: changing only GHGs and not surface emissions, Figure 13.11b), clearly indicate that, in the future, changes in surface emission of O_3 precursors will dominate the column O_3 changes while the effect of global warming is small. This feature was also found by Stevenson et al. (2000) and Hauglustain et al. (2005) and, to a lesser extent, by Liao et al. (2006).

Table 13.3 summarizes the changes in annual mean surface emissions of O_3 precursors, key climate parameters, and tropospheric column O_3 between 1990 and 2100. Values for both combined climate warming and increased surface emissions and climate warming only are listed for the global average and for the three individual regions. For the case of combined effects, increases of 17.0, 18.1, and 16.1 DU respectively are found over East Asia, North

TABLE 13.3 Changes in Surface Emissions of O_3 Precursors, and the Climate Parameters due to the Greenhouse Warming, and the Associated Tropospheric O_3 Column

Parameter	Globe	East Asia	North America	Western Europe
CO (10^6 TgCO km^{-2} yr^{-1})	1.54 (0.0)	9.77 (0.0)	7.06 (0.0)	9.63 (0.0)
NO$_x$ (10^6 TgN km^{-2} yr^{-1})	0.13 (0.0)	0.60 (0.0)	0.91 (0.0)	1.06 (0.0)
NMVOCs (10^6 TgC km^{-2} yr^{-1})	0.42 (0.0)	2.23 (0.0)	1.56 (0.0)	2.36 (0.0)
T_s (K)	2.34 (2.29)	2.55 (2.50)	2.48 (2.37)	2.60 (2.42)
T_m (K)	2.89 (2.86)	2.89 (2.80)	2.88 (2.82)	2.93 (2.64)
Q_m (mm)	4.64 (4.57)	4.88 (3.85)	3.59 (2.87)	2.80 (2.46)
C (%)	−0.13 (−0.05)	−0.53 (−2.03)	−0.05 (−0.47)	−1.35 (−3.21)
Column O_3 (DU)	13.7 (−1.00)	17.0 (−0.79)	18.1 (−0.81)	16.1 (−0.71)

GCCM-simulated annual mean changes in tropospheric column O_3 between 1990 and 2100. Two sets of values are given, corresponding respectively to the combined effect of increases in surface emissions of O_3 precursors CO, NO$_x$, NMVOCs, and GHG, and (in parentheses) increase of GHG alone. The responses over three regions: East Asia, North America, and Western Europe are given in order to highlight the regional contrast. Changes in key climate parameters of surface air temperature (T_s), tropospheric mean temperature (T_m), tropospheric column water vapour (Q_m), and total cloudiness (C) are also given.

America, and Western Europe. When only climate warming is considered, the tropospheric column O_3 is decreased but the effect is small for the global mean and the individual regions, more than a factor of 10−20 smaller than when both factors are considered. These results imply that changes in surface emissions of O_3 precursors dominate the tropospheric column O_3 changes over the globe as well as for the three regions, while climate warming plays a relatively minor role.

It is interesting to note that, although the differences in temperature changes among the three regions are small, the differences in the changes in moisture and clouds are quite substantial, in particular between East Asia and Western Europe. In addition, since the O_3 change is small in the case of climate warming alone, the feedback effect of O_3 change on climate can be revealed by comparing the changes in climate parameters in Table 13.4. For example, the global mean surface temperature caused by tropospheric O_3 increase is 0.05 K (2.34 K−2.29 K) with a larger warming of 0.18 K (2.60 K−2.42 K) for Western Europe. The tropospheric mean temperature and column water vapour are also increased, again showing a larger effect over Western Europe. The effect on cloud, however, is less clear, mainly because of the large inter-annual variability seen from the MS, as discussed below.

One important aspect in the GCCM study of tropospheric O_3 changes is the model inter-annual variability — the model 'noise' inherent in the variation of model-simulated tropospheric O_3. This aspect is important when studying the signal of changes in tropospheric O_3 due to climate changes. For example, we examined the last 5-years' means and standard deviations of tropospheric O_3 and several key tropospheric climate parameters simulated in the MS1 case.

On a global mean basis, the inter-annual variability of tropospheric O_3 column of 0.05 DU is small when compared with the column mean of 29.5 DU. However, the variability is much larger on regional-scales. For example, for the three regions, the respective values are 0.37 versus 42.1 DU for East Asia; 0.20 versus 38.0 DU for North America; and 0.27 versus 35.2 for Western Europe. The causes for the inter-annual variability of tropospheric O_3 can be traced to the variability of tropospheric clouds and water vapour and, to a less extent, the surface temperature over the three regions. Consequently, considering the magnitude of a decrease of 1 DU in tropospheric column O_3 due to climate warming only, the regional inter-annual variability can be significant. This aspect needs to be carefully addressed when studying the tropospheric O_3 response to climate changes and its feedback effect to climate. The use of ensemble runs and longer MS is necessary in order to ascertain the robustness of these results.

13.3.4.3. Impact on Surface Ozone

The geographical distribution of changes in annual mean surface O_3 concentration in response to combined climate warming and increased surface emissions, and climate warming alone are also studied. When both factors are included (Figure 13.11c), the change in global and annual mean surface O_3 is simulated to be 22.1 ppbv, with much larger increases within 0−60°N. For the case of climate warming alone (Figure 13.11d), instead of the small decrease in column O_3 presented above, the annual and global mean surface O_3 is increased by 4.6 ppbv, with more pronounced increases over low and middle latitude land areas of both hemispheres. The results further indicate that,

TABLE 13.4 Global Climate–Chemistry Model Simulated Sulfate Aerosols and Radiative Forcing

Parameter	Globe	East Asia	North America	Western Europe
SO$_2$ emission	213 (237/197)	1120 (1307/1036)	1049 (1102/1028)	1861 (2299/1433)
SO$_2$ loading	1.17 (1.20/1.20)	2.90 (3.28/2.83)	3.09 (3.70/2.50)	4.34 (4.87/3.96)
Sulfate loading	2.30 (1.93/2.95)	5.89 (4.65/8.37)	5.11 (3.07/7.78)	7.79 (4.95/10.9)
Direct radiative forcing	−0.24 (−0.13/−0.43)	−1.82 (−1.49/−2.12)	−1.39 (−0.96/−1.92)	−2.57 (−1.93/−4.04)
(Direct+first indirect) radiative forcing	−1.85 (−1.24/−2.23)	−4.49 (−3.83/−4.60)	−5.15 (−4.25/−7.16)	−5.97 (−3.38/−7.84)

Surface SO$_2$ emissions (mg m^{-2} yr^{-1}) for 1985 and GCCM-simulated column SO$_2$ (mg m^{-2}), column sulfate (mg m^{-2}), and radiative forcing (radiative forcing; W m^{-2}) for the globe and three regions: East Asia, North America, and Western Europe. Values correspond to annual means with December–January–February/June–July–August values in parentheses.

during June–July–August, much larger increases, ~100 ppbv, occur over East Asia, North America, and Western Europe. In particular, the effect of climate warming on surface O$_3$ concentrations (MS3-MS1) could account for up to 50% of the increases in total ground-level O$_3$ concentrations (MS2-MS1) over the eastern United States, indicating that climate warming plays a much more important role. Although high ground-level O$_3$ concentration is usually considered a local air quality issue, the model simulations here show that surface O$_3$ concentrations could exceed 75 ppbv over many areas of North America, Western Europe, and East Asia in the summer months, caused by climate warming.

In a warmer climate, there is a tendency for an increased frequency of extremes such as heatwaves (IPCC, 2007a), which may have implications for surface O$_3$. The notion that heatwaves could create higher surface O$_3$ was investigated by Solberg et al. (2008) for the case of August 12, 2003 in Europe. The study was conducted by using the Oslo-CTM2 with ECMWF meteorological data to simulate O$_3$. The results indicate that the surface O$_3$ levels were found to exceed 100 ppbv over extended areas of Europe related to several competing factors. Increases in surface temperature increase the surface O$_3$ chemical production, which is larger than its chemical destruction, and more surface O$_3$ is destroyed due to increased humidity. When static stability is greater than zero, the 1000 hPa–850 hPa atmosphere is stable, resulting in more near-surface O$_3$.

To further address this aspect, we analysed the 1946–2006 NCEP/NCAR reanalysis data of surface specific humidity, 1000 hPa–850 hPa column water vapour and 1000 hPa–850 hPa mean static stability, which are all closely associated with the surface temperature and relevant to near-surface O$_3$. We used the 3 K surface temperature anomaly (Frich et al., 2002) over the study period for each month to define the heatwave regions, and then examined the characteristics of the three parameters during the high surface warming events such as those during the 2003 June–July–August and the 2006 July heatwaves over Europe. The analyses suggest that the three parameters were all high during the heatwave episodes when compared with other periods without heatwaves. In a way, meteorological conditions during heatwaves tend to provide a favourable setting for higher near-surface O$_3$. The study also shows that changes in these climate indicators are similar to the ones occurring during the heatwaves (e.g., 2003 JJA and 2006 July) over Europe. Thus, climate warming increases the frequency of the heatwave events that may imply a higher surface O$_3$ concentration.

13.3.4.4. Impact on Radiative Forcing

To calculate the radiative forcing requires the information of the vertical distributions of temperature, water vapour, clouds, and O$_3$ for the reference case, MS1, and O$_3$ changes from the other two cases: combined global warming and increased surface emissions (MS3-MS1) and global warming alone (MS2-MS1). For the case MS3-MS1, the O$_3$ decreases throughout the atmosphere, except in the areas of the mid-to-upper troposphere in the 30°S–30°N zones where the O$_3$ increases are related to a warmer temperature and a decrease of cloud cover. The tropospheric O$_3$ decrease in the lower troposphere within 30°S–30°N is consistent with a moisture increase. Nevertheless, comparison of the O$_3$ vertical distribution between the two cases indicates clearly the dominant effect of surface emissions of O$_3$ precursors when the O$_3$ increases throughout the troposphere, in particular in

(a)

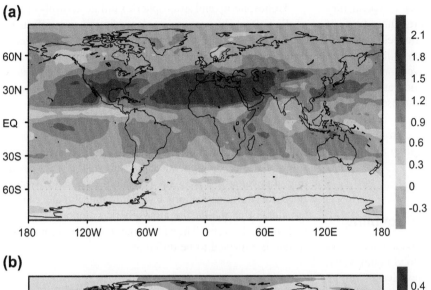

(b)

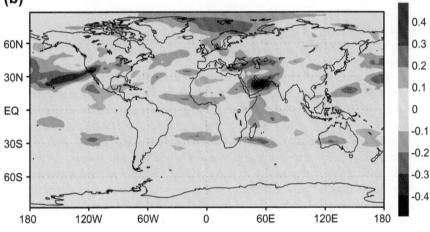

FIGURE 13.12 July radiative forcing (W m^{-2}) due to O$_3$ changes caused by: (a) combined increases of surface precursor emissions and GHG, and (b) increases of GHG alone for the period between 1990 and 2100 simulated with the GCCM using the corresponding GHG and surface emissions of O$_3$ precursors. *(Source: Tanaka, 2009.)*

mid-latitudes over the Northern Hemisphere where the major sources of surface emissions are located. The increases also reach the upper troposphere and spread to high latitudes.

Figure 13.12 shows the July instantaneous radiative forcing due to O$_3$ changes for both cases (MS2-MS1: both GHG and surface emission change, Figure 13.12a; and MS3-MS1: only GHGs change, Figure 13.12b). In MS3-MS1, the radiative forcing shows small values of warming/cooling over different regions with a negligible global mean value. The averaged January and July radiative forcing is calculated to be −0.084 W m^{-2} (corresponding to a tropospheric column mean O$_3$ change of −1 DU), which is within the range of values of +0.004 and −0.21 W m^{-2} from other model studies (Stevenson et al., 2000; Hauglustain et al., 2005; Liao et al., 2006). On the other hand, Figure 13.12a indicates increases over the whole globe with much larger values found at mid-latitudes over the Northern Hemisphere, a similar pattern being found for January. The global mean January and July radiative forcings are calculated to be 0.66 and 0.93 W m^{-2} respectively.

When both the global warming and surface emissions are considered, the present model calculated a radiative forcing of 0.79 W m^{-2} and 14.1 DU tropospheric O$_3$ increase, which are respectively within the range of other model studies referenced in Section 13.2.1.2. The separate shortwave and longwave forcings contribute 42% and 58% of the total radiative forcing, respectively, which are also comparable to the study of Liao et al. (2006) with 39% for the shortwave and 61% for the longwave forcing. For the global warming only case, both values for O$_3$ change and radiative forcing are small.

13.4. CLIMATE–CHEMISTRY INTERACTION OF TROPOSPHERIC SULFATE AEROSOLS

13.4.1. The Role of Sulfate Aerosols as a Climatically-Active Compound

Sulfate aerosols, either naturally produced or anthropogenerated, have been studied extensively. The sulfur cycle involving production, chemical conversion, transport, and removal of sulfate aerosol exemplifies a typical life-cycle of trace chemicals in the atmosphere. Other types of aerosols may also experience a similar lifecycle, although without strong chemical conversions (such as mineral dust and BC), while other chemical conversions are too complicated and not fully understood (such as OAs). We use sulfate as an example to demonstrate aerosol climate–chemistry interactions, with an added note that, unlike mineral dust and BC aerosols, which have strong absorption of solar radiation, sulfate aerosols mainly reflect solar radiation.

13.4.2. Sulfate Aerosol–Climate Coupling

The important climate factors in sulfate production and loss processes are temperature, cloud cover, cloud liquid water, and precipitation. Temperature has a positive effect on sulfate gaseous-phase production, but a negative effect on sulfate aqueous-phase production. Cloud cover and cloud liquid water have positive effects on sulfate aqueous phase production and negative effects on gaseous-phase sulfate production due to blocking of actinic flux. In most cases, cloud cover is positively related to cloud liquid water and precipitation, so the effects on gaseous-phase versus aqueous-phase reactions tend to partly offset each other. Precipitation also increases the wet deposition of SO_2 and thus has a negative effect on sulfate production efficiency. Therefore, the composite effects of temperature, cloud cover, cloud liquid water, and precipitation can either increase or decrease sulfate production. Indirectly, these climate parameters also influence sulfate precursors in the atmosphere. H_2O_2 is one of the most important precursors of sulfate and the H_2O_2 pathway accounts for more than 50% of total sulfate production. Note that H_2O_2 concentration is determined by gaseous-phase reactions involving hydroxyl radicals (OH and HO_2) and pollutants such as NO_x, VOC, and CO. DMS is an important natural precursor of SO_2, and its emissions from oceans are correlated positively with surface wind but negatively with SST (see Rice and Henderson-Sellers, 2012, this volume).

Atmospheric sulfate loading and sulfate aerosol radiative forcing have been simulated by a large number of CCMs and CTMs since the 1990s. The simulated global sulfate loading for present-day conditions has varied from about 0.24 TgS to 0.71 TgS, with a sulfate lifetime ranging from 2.0 to 5.7 days (cf. Tsai et al., 2010). The adopted SO_2 emissions were in the range 64 TgS to 81 TgS per year (Chin et al., 2000; Chuang et al., 2002; Iversen and Seland, 2002; Liao and Seinfeld, 2005; Lucas and Akimoto, 2007; Manktelow et al., 2007; Unger et al., 2006; Verma et al., 2007). The effects of changes in future anthropogenic SO_2 emissions have also been reported in some of these studies and in recent investigations of geo-engineering.

Tropospheric (and stratospheric) sulfate aerosols induce cooling effects by scattering solar radiation, thus offsetting the warming effects from the GHG. The cooling effect is particularly large in the regions of large anthropogenic SO_2 emissions such as East Asia, North America, and Western Europe. Climate characteristics differ significantly between these regions, which can result in different production (such as clean-air and in-cloud oxidation of SO_2) and removal (such as wet scavenging) processes that affect their chemical lifetime. Most of the previous studies have investigated sulfate aerosols on a global scale. Here again, we focus on three key regions: East Asia, North America, and Western Europe. The three regions have different climate characteristics, in particular the atmospheric components of the hydrological cycle. Therefore, the effect of climate on sulfate is expected to be different.

13.4.3. Effect of Climate–Chemistry Interactions

Tsai et al. (2010) incorporated the sulfur chemistry scheme (Berglen et al., 2004) into the GCCM (Wong et al., 2004) to investigate the sulfur cycle and the radiative forcing of sulfate aerosol. Four new species, DMS, SO_2, methane sulfonate (MSA), and sulfate, are added with related processes: emissions of SO_2, DMS, and sulfate; dry and wet deposition of SO_2, MSA, and sulfate; and gaseous- and aqueous-phase chemical reactions. In Tsai et al. (2010), the 1985 anthropogenic SO_2 emission was used to simulate the sulfate aerosol and the associated direct and indirect (cloud albedo) radiative forcing in comparison with pre-industrial sulfur emission conditions. The simulation results indicate that the 1985 anthropogenic emission tripled the global SO_2 and sulfate loadings from their natural values of 0.16 TgS and 0.10 TgS, respectively. The sulfate aerosol produces a direct radiative forcing of -0.43 W m^{-2} for clear-sky and -0.24 W m^{-2} for all-sky conditions, and a significant first indirect forcing of -1.85 W m^{-2}.

The model-simulated geographical distribution of annual mean, near-surface sulfate aerosol is shown in Figure 13.13. Because of its short atmospheric lifetime, sulfate is closely correlated with the pattern of surface SO_2 emissions, in particular over the three emission centres of East Asia, North America, and Western Europe. The spatial coverage of near-surface sulfate aerosol is broader than the SO_2 source regions, reflecting the effect of horizontal transport and the time needed for chemical transformation. The simulated vertical distribution of sulfate further highlights the characteristics of large concentrations at mid-latitudes in the lower troposphere in the Northern Hemisphere with decreases in concentration vertically. A large seasonal contrast in the tropospheric column sulfate aerosol is also found in the simulations.

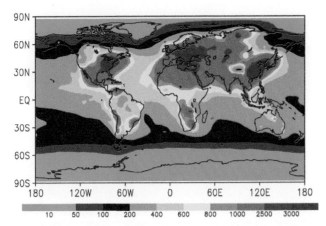

FIGURE 13.13 Annual mean near-surface sulfate aerosol concentration (pptv) simulated with the GCCM using the 1985 emission scenario. *(Source: cf. Tsai et al., 2010.)*

Below, we present an analysis of the GCCM 1985 simulation to investigate the effect of climate—chemistry interactions on a *regional*-scale. Specifically, we examine the correspondence among the surface SO_2 emission, SO_2 atmospheric loading, and the sulfate aerosol loading both globally and over the three regions. The difference in the radiative forcing and removal processes among the three regions will be used as an indication of the effect of climate—chemistry interactions on sulfate aerosols.

13.4.3.1. Sulfate Loading and Radiative Forcing

Table 13.4 lists the values for the globe and the three regions of the relevant parameters, including the surface SO_2 emission, the simulated atmospheric SO_2, and sulfate loadings, and the direct and first indirect radiative forcing caused by the sulfate aerosols. For the globe and all three regions, the surface SO_2 emission is larger in the winter (December—January—February) than in summer (June—July—August). The region of Western Europe is unique in two aspects: large seasonal contrast with the emissions for both seasons larger than either season for the other two regions. The atmospheric SO_2 loading and its seasonal variations are generally in line with the surface SO_2 emissions, although a small deviation is seen in the annual values between East Asia (2.90/1120) and North America (3.09/1049).

However, the simulated sulfate shows different characteristics, especially in respect of seasonal responses. For the annual mean values, larger surface SO_2 emission leads to larger sulfate loading. Firstly, for all regions, although the surface SO_2 emission is smaller in the summer, the sulfate loading is actually larger, thus implying a different climate effect on the sulfate loading. For example, over North America, the (boreal) winter and summer surface SO_2 emission is nearly the same; however, the (boreal)

summer-simulated sulfate loading is much higher, by more than a factor of two, indicating faster oxidation processes in the summer. This seasonal contrast is even more pronounced for East Asia and Western Europe, that is, larger sulfate is associated with lower surface SO_2 emission in summer, a reflection of the difference in chemical transformation, as well as in transport and removal processes. For example, because of a stronger vertical transport, the removal of sulfur (especially SO_2) by surface dry deposition is weaker in the summer for all three regions. As for the wet deposition, it is weaker in the summer for Western Europe and North America due to the precipitation seasonality mentioned earlier.

The global, annual mean radiative forcings are calculated to be -0.24 W m^{-2} for the direct effect, and -1.85 W m^{-2} for the combined direct and first indirect effect, with much larger values during summer. The radiative forcing is also much larger on a regional basis with its magnitude proportional to the sulfate loading. Western Europe has the highest sulfate loading and the annual mean total cooling can be as large as -5.97 W m^{-2}, with the summer cooling (-7.84 W m^{-2}) more than a factor of 2 larger than the winter cooling (-3.38 W m^{-2}) despite the fact that the summer SO_2 emission is a factor of 1.60 smaller. This feature is also identified for East Asia and North America although the seasonal contrast is less. These results suggest that the radiative forcing of sulfate aerosol is sensitive to the climate regime.

Similar to tropospheric O_3, the model internal variability is quite substantial, and the magnitude of its inter-annual variability is larger than the sulfate direct climate effect. However, the first indirect effect signal is found to be significantly larger than the model inter-annual variability so that the regional climate responses to sulfate can be estimated. As indicated in Tsai et al. (2010), the surface temperature over central Europe, eastern North America, and East Asia could be decreased by 0.64 K, 0.14 K, and 0.53 K, respectively, leading to circulation changes. For example, as pointed out by Tsai et al. (2010), the cooling over East Asia decreases the land—ocean surface temperature contrast in the summer, resulting in a weaker East Asia summer monsoon (i.e., a decrease in the northward moisture transport and the precipitation over East Asia). On the other hand, the winter monsoon could be strengthened due to the land—surface cooling caused by sulfate aerosols. These changes are similar to the findings by Huang et al. (2007) who used a regional climate—chemistry model and by Liu et al. (2009) who conducted global model simulations with assimilated aerosols. These characteristics are very similar to the climate effect of volcanic eruptions documented in Peng et al. (2010) and those invoked as part of the explanation of catastrophic climate change in Belcher and Mander (2012, this volume). The dynamic mechanisms

TABLE 13.5 Global Climate–Chemistry Model Simulated Sulfur Removal Processes and Lifetime

	Dry Deposition	Wet Deposition	Transport	Lifetime
Global	0.41 (0.43/0.37)	0.51 (0.52/0.50)	0.08 (0.05/0.12)	2.72 (1.30/2.87)
East Asia	0.68 (0.78/0.52)	0.48 (0.46/0.48)	−0.160 (−0.24/−0.01)	2.46 (1.85/2.04)
North America	0.63 (0.71/0.51)	0.29 (0.30/0.27)	0.08 (−0.01/0.22)	2.81 (1.90/3.60)
Western Europe	0.46 (0.59/0.51)	0.38 (0.51/0.35)	0.16 (−0.11/0.13)	2.41 (1.69/3.84)

GCCM-simulated tropospheric sulfur removal efficiency (as a fraction of SO_2 emission) and chemical lifetime (days). The positive value indicates removal from the regions. The three regions are the East Asia, North America, and Western Europe. Annual values with, in parentheses, December–January–February/June–July–August values.

in response to sulfate radiative forcing certainly differ among different regions, and more careful analyses are required to study the cause–effect relationship.

13.4.3.2. Efficiency of Sulfur Removal Processes

For a given region, the emitted atmospheric sulfur, including SO_2 and its sulfate product, are removed by dry and wet depositions, as well as being transported into and out of the region by wind. Therefore, the relative importance of the three processes can be used to compare and categorize the individual regions in regard to the effect of climate–chemistry interaction. Table 13.5 compares the efficiency of the three processes as well as the resulting chemical lifetime of the atmospheric sulfur.

On a global and annual mean basis, dry and wet depositions remove 41% and 51% of sulfur respectively from the atmosphere, while about 8% is transported to the stratosphere; the lifetime is calculated to be 2.72 days. The seasonal contrast can mainly be attributed to differences in the dry deposition and transport, which yield a factor of 2 differences in lifetime, being longer in the (boreal) summer. On the other hand, for the three regions, the dry deposition dominates, particularly over East Asia where this removal mechanism accounts for 68%. The wet deposition exhibits a strong regional difference, ranging from 29% for North America to 48% for East Asia. It is also interesting to note that for East Asia the total deposition is larger than the emissions, with 16% of sulfur being transported into the region; whereas North America and Western Europe export sulfur. [The situation probably changed in the past decade due to changes in geographical distribution of energy consumptions.] The regional lifetime of sulfur in the atmosphere ranges from 2.81 days in North America to 2.41 days in Western Europe.

13.4.4. Predicting Future Aerosol Impact on Climate

13.4.4.1. Aerosol Distribution

Estimating future aerosol properties and processes is critically dependent on the assumptions made regarding anthropogenic sources. Three important anthropogenic sources — oxides of sulfur and nitrogen (producing sulfate and nitrate aerosol) and BC — are important because near-term mitigation efforts will likely be aimed at these species. Other sources, such as aerosols produced by fires or by vegetation can be a combination of natural and anthropogenic sources. Finally, volatile species such as nitrate and many organic compounds may partition differently between the condensed and gaseous phases in a warmer world.

Table 13.6 shows the source strengths, mass loading, lifetime, and resultant optical depth of five major aerosol types. The data (other than for nitrate) are based on results from 16 models from the Aerosol Comparisons between Observations and Models (AeroCom) project (Kinne et al., 2005; Textor et al., 2006).

It is important to examine how aerosol properties have changed in the past. Horowitz (2006) estimated that tropospheric sulfate burdens have increased by a factor of 3 and BC burdens by a factor of 6 since pre-industrial times. Figure 13.14 shows comparisons of the spatial distributions of sulfate and BC between pre-industrial (Figures 13.14a and 13.14c) and the year 2000 (Figures 13.14b and 13.14d). In the pre-industrial period, sulfate concentrations were highest over the oceans (reflecting the marine DMS source) while BC concentrations were highest over land, reflecting combustion and fires from terrestrial locations. Sulfate concentrations in 2000 were highest over the

TABLE 13.6 Characteristics of Major Atmospheric Aerosols

Aerosol Type	Source (Tg yr[-1])	Mass Loading (Tg)	Lifetime (days)	Anthrop. Fraction[5] (%)
Sulfate[1]	190 (100–230)	2.0 (0.9–2.7)	4.1 (2.6–5.4)	61
Black Carbon[1]	11 (8–20)	0.2 (0.05–0.5)	6.5 (5.3–15)	n/a
Organics[1]	100 (50–140)	1.8 (0.5–2.6)	6.2 (4.3–11)	51
Dust[1]	1600 (700–4000)	20 (5–30)	4.0 (1.3–7)	22
Sea salt[1]	6000 (2000–120,000)	6 (3–13)	0.4 (0.03–1.1)	0
Nitrate (non-dust)[2,3]	21.3[4]	0.3 (0.04–0.63)	4.5	n/a

Source strength, mass loading, lifetime, and optical depth for five major aerosol types, given as median values and ranges
(Sources: 1. Chin et al., 2009; 2. Tsigaridis et al., 2006; 3. Liao et al., 2004; 4. gas-to-particle conversion from nitric acid; 5. anthropogenic fraction estimated from supplementary material from Ramanathan et al., 2001.)

industrialized areas of North America, Europe, and Asia. BC concentrations show a similar spatial distribution, with additional maxima over areas of substantial biomass burning.

13.4.4.2. Model Development

The spatial distribution of these species has important ramifications in terms of mitigation strategies. Sulfate aerosols tend to have a cooling effect on the Earth's

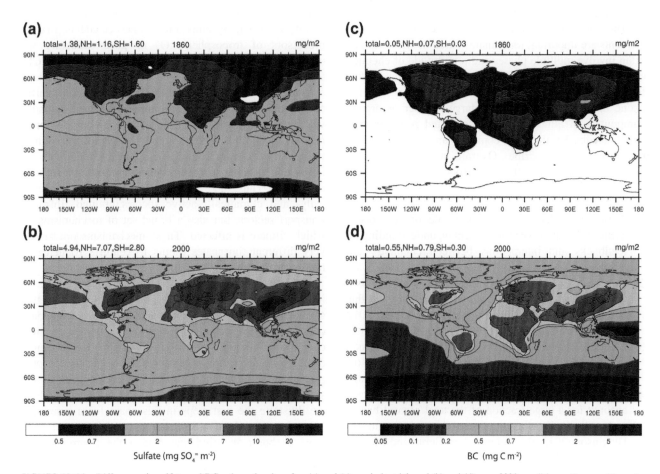

FIGURE 13.14 Differences in sulfate and BC column burdens for: (a) and (c) pre-industrial, and (b) and (d) year 2000 conditions. *(Source: Horowitz, 2006 © Copyright 2006 American Geophysical Union. Reproduced/modified by permission of American Geophysical Union.)*

climate. BC aerosols can lead to a surface cooling by absorbing incoming solar radiation, preventing it from reaching and warming the surface. This same absorption tends to heat the atmosphere, leading to a net warming of the climate system. Additionally, BC aerosols deposited on highly reflective surfaces such as snow or ice can lead to increased surface warming and potentially more rapid melting of the snow or ice (Clarke and Noone, 1985; Lau et al., 2010). Legislation to control the emissions of BC and sulfate precursors can have very different climatic effects (Arneth et al., 2009).

When aerosols (having sulfur as a major chemical component) were established as having a significant effect on the Earth's radiation balance (Charlson et al., 1992), the sulfur cycle was also added to the biogeochemical cycles of direct climate relevance. Currently, an active carbon cycle, dynamic vegetation, and some atmospheric chemistry have been incorporated into state-of-the-art climate models. Despite all of these advances, aerosols and clouds remain the largest uncertainties in current climate models in terms of constraining the Earth's energy balance. The importance of these processes is evident by the inclusion of a new chapter in the forthcoming Fifth IPCC Assessment (AR5): Clouds and Aerosols (Chapter 7 of the Working Group I contribution).

The interactions between the cycles of the major elements are important for climate. These interactions very often involve multiple phases and a multitude of chemical components (e.g., solid particles from vegetation or dust, liquid droplets in clouds, gas-phase components such as ozone or nitrous oxide, a very large number of organic compounds in atmospheric aerosols) and result in non-linear feedbacks that are complex and challenging to understand and describe. Overall, a number of key processes need to be better understood in order to predict the behaviour of aerosols in a future climate:

1. A proper understanding of the role of aerosols in a future climate is very dependent on understanding the feedbacks and interactions between chemistry and climate.
2. Uncertainty in current aerosol radiative forcing leads to uncertainty in estimates of climate sensitivity and thus our ability to predict future climate (cf. Harvey, 2012, this volume).
3. Land-use change and desertification may drive significant changes in the cycle of dust in the atmosphere (cf. Pitman and de Noblet-Ducoudré (2012, this volume).
4. Biogenic emissions of aerosols and aerosol precursors account for a large fraction of the total aerosol burden and will likely change in a changing climate (cf. Dickinson, 2012, this volume; Sen Gupta and McNeil, 2012, this volume).
5. SOA burdens may increase in a warmer climate by about the same amount as sulfate aerosols decrease;

however, due to the different spatial and seasonal emission patterns, these effects will likely not simply offset each other.
6. Changes in aerosol burdens and properties will influence not only the radiative balance but also precipitation patterns and atmospheric dynamics at least regionally (cf. Evans et al., 2012, this volume).
7. Reductions in sulfate and increases in BC aerosols may lead to substantial additional warming by 2100.

Aerosols influence not only climate, but also human health and environmental acidification over land and oceans. For this reason, mitigation policies concerning air quality and climate need to be considered together (cf. Taplin, 2012, this volume).

13.5. MITIGATION POLICIES FOR CLIMATE AND AIR QUALITY

13.5.1. Mitigation Studies from the Transport Sector

There are four main mechanisms by which emissions from transport affect climate: (i) by emission of direct GHGs, mainly CO_2; (ii) by emission of indirect GHGs, that is, precursors of tropospheric O_3 or gases affecting the oxidation capacity of the atmosphere, such as NO_x, CO, and VOC; (iii) by the direct effect of emission of aerosols or aerosol precursors, in particular BC and OC, and sulfur compounds; and (iv) by the indirect effect of aerosols, which trigger changes in the distribution and properties of clouds. In addition, water vapour from aircraft engines triggers formation of contrails and cirrus clouds, with the latter effect being strongly dependent on atmospheric conditions. Thus, there is not only a broad mix of chemically- and climatically-active species emitted from the transport sectors, but also a broad set of mechanisms by which climate is affected. These mechanisms operate on very different timescales (hours to centuries) and cause both warming and cooling.

To illustrate the chain of events from emissions to impact and damage caused by human activity we give a general perspective in Figure 13.15 through a standard cause-and-effect chain from emissions through physical changes in climate to ecological, health, or economic damage. In general, moving down the chain, the adopted parameters become more relevant to society but, at the same time, they become more difficult to quantify. Here, we illustrate the first steps in this chain for a simplified example, with all quantities taken to be global averages. It must be borne in mind that neither the figure, nor the simplified example, attempt to represent the many complex feedbacks that could exist in the climate system; changes in any of the boxes in the figure have the potential to impact

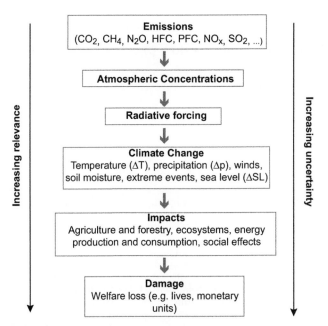

Emissions
(CO_2, CH_4, N_2O, HFC, PFC, NO_x, SO_2, ...)

↓

Atmospheric Concentrations

↓

Radiative forcing

↓

Climate Change
Temperature (ΔT), precipitation (Δp), winds, soil moisture, extreme events, sea level (ΔSL)

↓

Impacts
Agriculture and forestry, ecosystems, energy production and consumption, social effects

↓

Damage
Welfare loss (e.g. lives, monetary units)

Increasing relevance ←→ *Increasing uncertainty*

FIGURE 13.15 Possible chain of events of the potential effect on climate and society of anthropogenic emissions. Note the correspondence of increasing relevance and increasing uncertainty. *(Source: Fuglestvedt et al., 2003.)*

on the processes in all the other boxes (for example, changes in temperature can impact on CO_2 emissions and on precursors of other climate gases or the rate of reactions that affect CH_4 and O_3).

Transportation contributes many different gases and aerosols to global emissions that can have an impact on climate, either directly or indirectly (e.g., Eyring et al., 2007; Fuglestvedt et al., 2008). In 2000, the global transport sector was the largest source of anthropogenic emissions of oxides of nitrogen (NO_x) (estimated to be 37% of the total anthropogenic emissions of NO_x) and was also a major contributor of fossil fuel CO_2 (21%), VOC (19%), CO (18%), and BC (14%) (Fuglestvedt et al., 2008).

Gases and aerosols emitted from the transport sector have distinctly different characteristics, which influence climate directly and indirectly via distinct chemical and physical processes. They have a wide range in atmospheric lifetimes, which influences the overall concentrations and spatial distributions and their abilities to affect climate. Many of the non-CO_2 emissions from the transport sector, however, are short-lived substances, not currently covered by the Kyoto Protocol. Of particular interest are the future emissions from aircraft and shipping which inject directly the short-lived species into the remote and pristine atmosphere, and the mitigation options for these sub-sectors. The main emissions from aircraft occur in the upper troposphere and lower stratosphere with significant impact on atmospheric composition and radiative forcing. Ship emissions occur mostly in

pristine surface background regions with a resulting strong impact, through the increase in OH, affecting the climate impact of O_3 and CH_4. Emissions from aircraft and ship are not included in the Kyoto Protocol, but significant reductions are possible, since there are large differences in future emission scenarios for the climate component precursors NO_x and SO_2, even for a moderate time horizon (2030).

In adopting measures to mitigate climate change, we need to define tools that allow emissions to be placed on some kind of common scale in terms of their impact on climate. They should have a number of possible uses such as agreements on emission trading schemes; consideration of potential tradeoffs between changes in emissions resulting from technological or operational developments; and/or comparing the impact of different environmental impacts of transport activities. Such metrics represent a challenge — dealing with a combination of long- and short-lived climate compounds in the transport sector. A second difficulty is present for some transport-related emissions. Since the values of metrics (unlike the gases included in the Kyoto Protocol) depend on where and when the emissions are introduced into the atmosphere, both the regional distribution and, for aircraft, the distributions as a function of altitude, are important. Currently, metrics such as Global Warming Potentials (GWPs) and Global Temperature Change Potentials (GTPs) are being recommended (Fuglestvedt et al., 2003). For example, it has been shown that there are large variations in GWP and GTP values for NO_x, mainly due to the dependence on emission location, but also because of inter-model differences and differences in experimental design. A significant uncertainty in the presented metric values reflects the current, preliminary state of understanding.

13.5.2. Tropospheric Aerosols

The studies discussed so far in this chapter have been aimed mainly at investigating how more comprehensive treatments of aerosols, aerosol precursors, and atmospheric chemistry can influence the results of climate model predictions. They illustrate that aerosols are an important factor in determining the nature of the future climate of the Earth. Perhaps more importantly, these studies point to a number of important feedback processes that have implications for mitigation strategies: the feedback between changing aerosol loading and composition and precipitation; the feedback to increasing temperatures; the emission of OAs and their precursors; and the thermodynamic partitioning of components between the gaseous and condensed phases. Monks et al. (2009) provide a very comprehensive review of the current state-of-the-science in our understanding of these kinds of feedbacks and processes.

TABLE 13.7 Global Model Simulated Possible Effects of Aerosol Pollution Reductions on Future Climate

Aerosol Burden (Tg)	GHG	GHG + IP	GHG + DT	GHG + AE
Sulfate (S)	+4	−25	+5	−32
Black carbon	+6	−3	−12	−13
Organics	+6	+1	−4	−1
Dust	+7	−3	−1	−1
Sea salt	+2	+2	+2	+3
Effect				
Temperature increase (%)	1.20 (+8)	1.89 (+13)	1.39 (+10)	2.18 (+15)
Precipitation (%)	0.07 (+2)	0.13 (+4)	0.08 (+3)	0.15 (+5)
Climate sensitivity (K W^{-1} m^2)	0.78	0.84	0.84	0.82
Hydrological sensitivity (% K^{-1})	1.96	2.24	1.86	2.36

Changes between 2000 to 2030 in aerosol burdens (top, in %) and effects (bottom) for different model simulations. Adapted from Kloster et al. (2010). Here, IP refers to the industrial and powerplant sectors, DT refers to the domestic and transport sectors, and AE to the MFR aerosol scenario.

Looking into the future, a recent study examined more specifically possible effects of aerosol pollution reductions on future climate response. Kloster et al. (2010) used the ECHAM5-HAM model, that is, ECHAM5 (Roeckner et al. 2003) of the Max Planck Institute for Meteorology extended by a microphysical aerosol model HAM (Stier et al. 2005), driven by two different sets of assumptions: (a) future emissions that would result from current emission control legislation applied to future economic development (CLE), and (b) the lowest emission levels that could be achieved with the most advanced control technologies currently available. Emissions of anthropogenic SO$_2$, BC, and OC are taken from the aerosol emission inventory of the International Institute for Applied System Analysis (IIASA), which considers two possible futures: current legislation (CLE) and MFR: the former accounting for decided control legislations while MFR assumes a full implementation of today's most advanced technologies (Cofala et al. 2007). Future emissions of GHGs were taken from the SRES B2 scenario. Kloster et al. (2010) use a number of model runs to attempt to flesh out the effects of different mitigation pathways and emission sectors. In their GHG simulation, only GHG concentrations change while aerosol and other precursor emissions remain at their 2000 levels. In their AE simulation, only aerosol and precursor emissions change, whilst GHG concentrations remain at 2000 levels. The GHG+AE simulation assumes that aerosol emissions follow the maximum feasible reduction (MFR) path, while GHG concentrations follow the B2 scenario. In the GHG+DT simulation, GHGs follow the B2 scenario, emissions from the domestic and transport sectors follow the MFR pathway, and all other sectors follow the CLE

pathway. Finally, the GHG+IP (GHGs with aerosols abated only in the industrial and powerplant sectors) simulation has GHGs following the B2 scenario, emissions from the industry and power sectors follow the MFR pathway, while all other sectors follow the CLE pathway. These simulations are compared with a 'control' simulation that uses present-day GHG concentrations, aerosols, and aerosol precursors. The results from Kloster et al.'s (2010) simulations for a number of selected compounds and effects are summarized in Table 13.7.

Burdens for all of the compounds increase between 2% to 7% in the GHG simulation, driven by reduced removal rates and longer aerosol lifetimes — consistent with the results in the other model simulations discussed above. Changes in deposition are inhomogeneous across regions and are not linearly related to precipitation rate or aerosol burdens; precipitation frequency and intensity are also important in determining wet removal. When emissions from the industry and power sectors are reduced dramatically, the sulfate burden decreases by 25%, while the BC burden only decreases by 3%. Other species show only small changes in this simulation. When emissions from the domestic and traffic sectors are substantially reduced, the BC burden decreases by 12%, while the sulfate burden increases by 5%. When all anthropogenic aerosol and precursor emissions follow the MFR pathway, sulfate burdens decrease by 32% and BC burdens decrease by 13%, while other components show only small changes.

In Kloster et al.'s (2010) calculations, an increase in GHG concentrations alone leads to an increase in the equilibrium temperature response of 1.2 K (+8%) by 2030. The MFR in anthropogenic aerosol and precursor

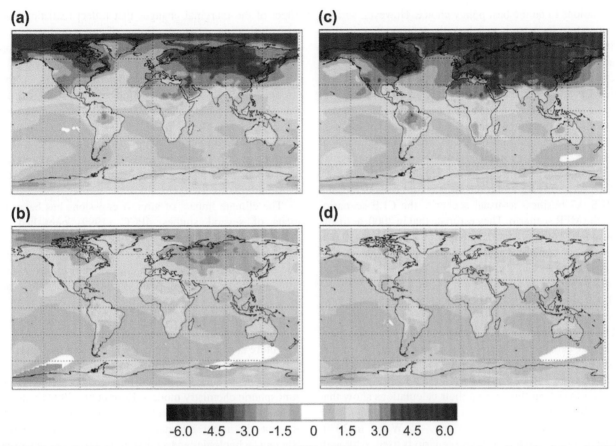

FIGURE 13.16 Equilibrium surface temperature response (K) between 2000 and 2030 as compared with their control case for the simulations: (a) GHG-IP (GHGs with aerosols abated only in the industrial and powerplant sectors); (b) GHG+DT (GHGs with aerosols abated only in the domestic and transport sectors); (c) GHG+AE (aerosols according to the MFR scenario); and (d) GHG alone. *(Source: Kloster et al., 2010. With kind permission from Springer Science+Business Media.)*

emissions would lead to an increase in temperature of 2.18 K (+15%). The global annual average precipitation rate increases in all cases. It should be kept in mind, however, that this global average increase masks substantial heterogeneity in spatial distribution and intensity. Precipitation is predicted to decrease in some areas (mainly over the oceans in the Southern Hemisphere mid-latitudes) and increase in others (mainly in the ITCZ region).

Figure 13.16 shows the spatial variability in equilibrium surface temperature increase between 2000 and 2030 for four different simulations from Kloster et al. (2010). Figure 13.16d shows results when GHG concentrations increase but with emissions of anthropogenic aerosols and precursors held at their year 2000 levels. The global average temperature increase in this case is 1.2 K. Figure 13.16a shows the simulation in which anthropogenic emissions from the industry and power sectors are substantially reduced (31% decrease in sulfate, 17% decrease in BC, and 9% decrease in particulate organic matter (POM) source strengths). Warming in the Northern Hemisphere increases substantially, with the global average temperature increasing by 1.89 K. Figure 13.16b is for the

case in which anthropogenic emissions from the domestic and transport sectors are substantially reduced (2% decrease in sulfate, 23% decrease in BC, and 12% decrease in POM source strengths). The overall warming (+1.39 K) is substantially less than in the GHG+IP case, in part because the BC burden has been lowered substantially more than the sulfate burden (which actually increases slightly in this case). Finally, Figure 13.16c shows the case in which the emissions of all anthropogenic aerosols and precursors are substantially reduced (GHG+AE). The spatial pattern of the warming in this case is similar to the GHG+IP simulation, but the magnitude is larger: a global average increase of +2.18 K. Similar to previous studies, Kloster et al. (2010) conclude that the precipitation response and hydrological sensitivity differs strongly for GHG and aerosol forcings, and that changes in aerosol burdens are not independent of changes in climate.

If the only criterion for determining which mitigation strategy to choose was minimizing the annual average surface temperature increase, then using the maximum feasible control technology on BC emissions and applying CLE to sulfur emissions would appear from these simulations to be the best policy choice. However, surface temperature increase is not the only appropriate criterion. Much of the impetus for past legislation to reduce sulfur and NO_x emissions has been driven by the effects that aerosols have on acidification and human health. These tradeoffs are illustrated in a study by Saikawa et al. (2009) which examines the effects of future aerosol concentrations in China on premature mortality and on radiative forcing.

Saikawa et al. (2009) use the MOZART-2 chemical transport model, combined with several emissions scenario. The CLE scenario is used for the year 2000. Three scenarios for emissions in 2030 are considered: the SRES A2 business-as-usual scenario, the CLE scenario, and the MFR scenario. They estimate that in 2000, aerosols caused 470,000 premature deaths in China and 30,000 premature deaths in countries downwind. Aggressive emissions reductions following the MFR pathway could reduce the number of deaths in China to 240,000 and 10,000 downwind by 2030. Under the high emissions scenario, premature deaths would increase by 2030 to 720,000 in China and 40,000 downwind. They also calculate that aerosols have a very substantial impact on the radiative balance for China. As the negative radiative forcing from sulfates and OC is larger than the positive forcing by BC in this region, their calculations show that aerosols account for a net direct radiative forcing of -74 W m^{-2} in 2000, and between -15 and -97 W m^{-2} in 2030, depending on which emissions scenario is used. Although they do not calculate the corresponding temperature response, clearly radiative forcings of this magnitude will lead to substantial differences in temperature, and the associated aerosol burdens will substantially perturb the hydrological cycle in the region. Their conclusion is that further efforts are necessary in China to reduce GHG emissions.

13.5.3. Tropospheric Ozone

As mentioned above, the transport sector contributes to the pollution of the atmosphere by means of several climate change agents not covered by the Kyoto Protocol, most notably tropospheric O_3 driven by NO_x, CO, and VOC. From a 'Kyoto Protocol perspective', the full climate impact of transport, especially the negative contributions from the shipping sub-sector, is not captured. Despite their large contribution to SO_2 and NO_x emissions, the Kyoto Protocol does not cover emissions from international aviation and shipping.

There are a number of non-linear relationships in the climate system, which makes it difficult to estimate the complete climate impact of humanity's activity. Because non-linear processes control concentrations of some substances (e.g., O_3 and OH), the radiative forcing of gases such as ozone and methane cannot be scaled to obtain the effect of the marginal changes that reflect realistic short-term mitigation measures. Some of this non-linearity is accounted for in model frameworks using results from more sophisticated and complex studies.

A climate system component such as O_3 has effects that strongly depends on the location of emissions of, for example, NOx (Dalsøren et al., 2009), as well as on CO and VOC emissions. The radiative forcing for O_3 can, to some extent, be scaled with global emissions if the geographical distribution of the emissions is assumed to remain unchanged. This represents an approximation, since the distribution is likely to change in the future.

The climate impact of aircraft emissions has been the focus of several studies. IPCC (1999) published an assessment of the impact of aviation on climate, which was later updated by Sausen et al. (2005) and more recently by Lee et al. (2009). Extensive studies of impacts from ship emissions have been published during the last five years (e.g., Dalsøren and Isaksen, 2006; Eyring et al., 2007; Dalsøren et al., 2009). Some of the studies of aircraft and ship impact include future climate impact evaluation.

Future emissions from aviation and shipping impact the atmospheric chemical composition (Hoor et al., 2009) and have been estimated using an ensemble of six different atmospheric chemistry models. Hoor et al. (2009) consider an optimistic emission scenario (B1), which takes into account the rapid introduction of clean and resource-efficient technologies and a mitigation option for the aircraft sector (B1 Advisory Council for Aeronautics Research in Europe (ACARE)), assuming likely further technological improvements. Results from sensitivity simulations, where emissions from each of the transport sectors were reduced by 5%, show that emissions from both aircraft and shipping will have a larger impact on atmospheric ozone and OH in the near future (2025; B1) and for a longer time horizons (2050; B1) compared to the current impact (2000). However, the ozone and OH impact from aircraft can be reduced substantially by 2050 if the technological improvements considered in the B1 ACARE are achieved.

Model studies have demonstrated strong regional and sector-wise differences in emissions of the precursors of secondary climate compounds and their subsequent impact on radiative forcing. Studies of the impact of emissions from the transport sector have been extensive and the results show strong sectoral differences. The difference in the regional impact of anthropogenic emission of NO_x is particularly large. Emissions of NO_x lead to increased concentrations of OH, thus reducing CH_4 concentration through enhanced CH_4 loss (Reaction (R13.7)). Hoor et al. (2009) showed that, while the pattern and the impact from land-based transport emissions do not differ much from other land-based emissions, the impact of NO_x emission from ships and aircraft is particularly pronounced, since it often occurs in remote, pristine regions with little impact from other sources (Eyring et al., 2007; Hoor et al., 2009). The result is that the reduction in CH_4, particularly in the case of ship emissions, causes a significant, negative radiative forcing.

Since the precursors and atmospheric composition of the chemical compounds are extremely different for the different transport sub-sectors, and their climate impacts are different, it is necessary to take this into consideration when climate−chemistry studies are performed, mitigation action recommended, and the policy implications addressed for the implementation of new regulations (Fuglestvedt et al., 2009). Of the short-lived species, tropospheric ozone as well as methane and BC are key contributors to global warming.

Quantifying the combined impact of chemical species on Earth's radiative forcing is complex, particularly the impact of those with an atmospheric lifetime of less than two months, since they tend to be poorly mixed and concentrated close to their sources. This uneven distribution, combined with physical and chemical heterogeneities in the atmosphere, means that the impact of short-lived species on radiative forcing can vary by more than a factor of ten depending on location. This situation is further complicated by non-linear chemical reactions between short-lived species in polluted areas.

13.6. FUTURE STUDY OF CLIMATE−CHEMISTRY INTERACTION

13.6.1. Extending Current Case Studies

This chapter has highlighted the importance of climate−chemistry interactions in affecting the atmospheric concentration of tropospheric O_3 and secondary aerosols (e.g., sulfate, organics) and their subsequent feedback to future climates especially at a regional-scale. Furthermore, studies have shown that future global warming due to increases in atmospheric GHGs could have a significant impact on the near-surface O_3, but less effect on free tropospheric O_3. The near-surface impact would be particularly large in regions where O_3 precursor emissions are large (North America, East Asia, Western Europe). One outstanding issue that requires further study would be whether a higher frequency of heatwaves with more stable and moist meteorological conditions and with different surface characteristics, such as those that occurred in the summer of 2003 over Europe, would trigger enhanced surface O_3.

The impact of changes in climate−chemistry interactions on the *regional* secondary aerosols is found to be sensitive to location and to season. It is dominated by the removal processes such as dry and wet deposition, and to some extent by the local stability. The current situation is that anthropogenic SO_2 emissions decrease at northern mid-latitudes but increase at low latitudes. Therefore, one issue that needs to be further addressed is how future patterns of SO_2 emission change regional sulfate aerosol loading and radiative forcing. For example, in the IPCC SRES scenarios, from 1985 to 2100, although the global anthropogenic SO_2 emission could be decreased by 9.7% (6.7 TgS), there could be significant regional differences. Over oceans, extensive future ship activity would lead to an increase in emissions of 1.3 TgS; increases are also expected over India, the Arabian Peninsula, and southern Africa. In contrast, there would be significant reductions over North America (4.07 TgS, −46%) and Western Europe (7.89 TgS, −63%), with less reduction over East Asia (0.66 TgS, −8.0%). Given the different regional climate characteristics like dry and wet removals, cloud distribution and solar incidence, this shift in the SO_2 emission pattern may yield a different spatial distribution in sulfate aerosol loading and in radiative forcing. As shown in Table 13.5, dry deposition removal efficiency in the tropics (e.g., East Asia), for the 1985 case study, is comparable to similar removal in the North American region, but the wet deposition in East Asia is much more efficient than in the North American region.

The challenges facing humanity have been characterized by a number of 'planetary boundaries', which include climate change, land use, and atmospheric aerosols (Rockström et al., 2009a; 2009b). These boundaries (including loss of biodiversity, ocean acidification, and nitrogen and phosphorus use) are tightly coupled and it would be unwise to develop strategies without being cognizant of how the associated actions would impact on the other areas. On the other hand, it is exceedingly difficult to conceive and negotiate international legislation of all of these issues taken together. Recent studies do indicate, however, that we are at the point, both in terms of our process-level understanding and our model tools, where the scientific community should be able to connect two of the governance frameworks − air pollution and climate − in a unified Earth system policy (Rice and Henderson-Sellers, 2012, this volume).

13.6.2. Climate−Chemistry 'Known Unknowns'

CTMs and coupled climate−chemistry models are becoming useful tools for estimates of mitigation action to reduce the impact of anthropogenic emissions on climate. It is therefore important to emphasize that there should be a strong focus on continued development and evaluation of such models. In addition to improving our understanding of the effects of changes in the concentration of tropospheric O_3 and secondary aerosols on regional climate, these studies also provide input as references to the impacts of primary climate compounds CO_2, CH_4, BC, and primary aerosols.

We have seen in this chapter that some of the tools necessary to produce results that are directly relevant for decision support in both climate and air quality arenas are already available. There are a number of limiting factors even in these state-of-the-art models that need to be addressed in order to make even more progress.

13.6.2.1. Feedbacks between Aerosols and Precipitation

Model simulations by Liao et al. (2009) have demonstrated an important feedback loop between aerosol concentrations (and composition) and precipitation. This feedback in their model was due largely to the influence of aerosols on the energy balance of the surface and atmosphere, the water balance at the surface, and associated changes in convection. Liao et al. (2009) did not include explicit descriptions of how changing aerosol microphysics and chemistry influence the cloud properties directly, including precipitation development, although these effects have been shown to be important (Andreae and Rosenfeld, 2008; Andreae et al., 2004; Chen et al., 2010; Rosenfeld et al., 2008), and improved parameterizations of aerosol—cloud interactions are clearly necessary for improved predictions of these effects. Quaas et al. (2009) compared simulations of aerosol indirect effects from ten different climate models with satellite observations using three different instruments. They found positive correlations between cloud droplet number and optical depth, with models overestimating the relationship over land. All models overestimate the relationship between cloud optical depth and liquid water path compared with satellite observations. They conclude that climate model parameterizations need to be improved in order to more properly account for the indirect effects of aerosols.

13.6.2.2. Feedbacks between Temperature, Chemical Equilibria, Organic Aerosols, and Land Use

Biogenic emissions of OAs and their precursors are extremely temperature-dependent, as well as being dependent on the type of vegetation, surface hydrological conditions, and nutrient availability. Vegetation, hydrology, nutrient cycling (Dickinson, 2012, this volume), and even surface temperature are connected to anthropogenic land-use change. Once in the atmosphere, the partitioning of volatile compounds like nitrate and many organic species will depend heavily on the ambient temperature. While multiphase, multicomponent systems are very complex, studies such as the ones described in this chapter show that many of the component tools are already in place to begin to address these feedbacks (Barth et al., 2005). Some 'low-hanging fruit' in terms of candidates for near-term improvements include treatments of fire

(e.g., Spracklen et al., 2009), or how urbanisation and megacities influence atmospheric chemistry (Butler and Lawrence, 2009); see also Cleugh and Grimmond (2012, this volume) and Pitman and de Noblet-Ducoudré (2012, this volume).

13.6.2.3. Feedbacks between Aerosols and Monsoon Circulations

Recent investigations have shown that aerosols may have a substantial influence on the Asian monsoon circulation (Lau and Kim, 2006; Lau et al., 2008; Ramanathan and Carmichael, 2008; Ramanathan et al., 2005). Lau and Kim (2006) showed that absorbing aerosols over the Indo-Gangetic plain near the foothills of the Himalaya act as an extra heat source aloft, enhancing the incipient monsoon circulation. The same aerosols lead to a surface cooling over central India, shifting rainfall to the Himalayan region. This 'elevated heat pump' effect leads to the monsoon rain beginning earlier in May—June in northern India and the southern Tibetan plateau, and to an increase in monsoon rainfall over all of India in July—August, with a corresponding reduction in rainfall over the Indian Ocean. While it is generally accepted that aerosols influence the Asian monsoon, there is still a great deal of uncertainty in the physical processes underlying the effects, and in their interactions (Kuhlmann and Quaas, 2010; Evans et al., 2012, this volume).

13.6.2.4. Glimpsing the Hazy Future

Fleshing out the causal links between increasing aerosol burdens and negative human health effects will require developing a good deal of new process-level knowledge. While there is overwhelming evidence for links between aerosols and health (e.g., Pope and Dockery, 2006), the causal mechanisms still need to be identified.

More work is needed to create enhanced opportunities for co-creation of new knowledge between modellers and experimentalists. Process model calculations and experimental field campaigns often share similar scales, as do satellite observations and climate models. While there is a very large body of information in all of these different areas, substantial methodological hurdles remain in terms of effective and accurate ways to compare observations and model results across spatial and temporal scales. Something as simple as strongly encouraging modellers and experimenters to collaborate from the outset of planning new activities in each other's domains may be a way of building the necessary new methodologies (cf. Rice and Henderson-Sellers, 2012, this volume).

The indirect effects of aerosols on clouds have consistently been seen as the largest uncertainty in our understanding of the Earth's radiative balance in all of the IPCC assessment reports (Schwartz et al., 2010). From the point of view of potential impacts, how aerosols affect precipitation is clearly the most important area in which we need better process-level understanding, supported by observations, and used to develop better parameterizations of these processes in climate models. In addition to the precipitation issue, the degree to which changing aerosol concentrations and properties will influence optically thin clouds in the upper troposphere is of particular importance. Depending on their optical thickness, upper tropospheric clouds can either warm or cool the planet. Aerosol—cloud interactions in these kinds of clouds are currently very poorly understood, due in part to the difficulty of obtaining in-situ observations. Understanding the aerosol—cloud interactions that determine the properties of these clouds will become even more important if geo-engineering schemes such as injecting aerosols into the stratosphere are to be properly assessed.

13.6.3. Atmospheric Chemistry and Future Climate

The issues discussed in the previous section dealt with suggestions for areas in which current climate—chemistry models could be improved. While GCCM are an absolutely necessary tool to have in any decision-support toolbox for mitigation and adaptation, we should also ask the question whether we need additional new tools to complement the ones we currently possess.

Doherty et al. (2009) give a number of recommendations for future climate research. Inter alia, they recommend that the framework for future climate research and observations should be redesigned with the specific goal of producing the information needed for decisions about impacts, adaptation, and mitigation. Firstly, vulnerability to climate change must be defined and mapped out. The next step would be to use this vulnerability 'map' to focus research strategies for, and to prioritize among, the urgent scientific issues in climate research. At the intersection of vulnerability and science issues are research and observational strategies specifically aimed at improving the predictability and understanding of impacts, adaptive capacity, and societal and ecosystem vulnerabilities. Starting with vulnerability does not mean that urgent scientific questions will be ignored. Rather, it ensures that research investments addressing these urgent scientific issues are carried out in a way that more effectively enables the results to be used as decision support for global environmental issues.

State-of-the-art climate—chemistry models are expensive in terms of the time and resources needed to develop, maintain, and run them. One simpler model framework for looking at how different decisions can be taken about control strategies for BC emissions is given in Bond (2007), who uses a simple two-box model to investigate the amount of forcing produced by burning 1 kg of various different fuels, comparing these results with the forcing from the CO_2 emitted from burning 1 kg of the same fuel. Bond (2007) concludes from this simple model that there is a high probability that aerosols emitted from diesel engines and biofuels used for cooking warm the climate. Simple models of this kind should be seen as complements to complex climate models, not as replacements. In addition, a truly useful decision-support toolbox must contain more than just different global climate models, however comprehensive they may grow to be. We need to work on techniques that enable us to engage in a large number of rapid 'what if' simulations with stakeholders focusing on the kinds of results that are possible from different adaptation and mitigation decisions, not simply continue to develop the most comprehensive models.

ACKNOWLEDGEMENTS

This research was supported by a grant (to SUNYA) from the Office of Science (BER), U. S. Department of Energy; K. J. Noone was supported by FORMAS grant #214-2009-409 (MACCII).

Learning Lessons

Climate prediction methods are built on process understanding established from data from the present and the past. Undertaking hindcasts of earlier periods and past epochs also tests prediction skill. Extreme and catastrophic events in Earth's climate history illuminate the future of climate thresholds and sensitivity to disturbances.

Human activities might fundamentally alter the dynamics of the climate system, switching it out of its recent mode of roughly 100,000-year glacial cycles and back into an earlier (Pliocene-like) state in which Greenland lives up to its name and remains un-glaciated, West Antarctica only sporadically has an ice sheet, East Antarctica holds less ice, and sea levels are in the long term around 25 metres higher.

Lenton, 2012, this volume

Chapter 14, ***Records from the Past, Lessons for the Future: What the Palaeorecord Implies about Mechanisms of Global Change*** by Sandy Harrison and Patrick Bartlein, examines the palaeorecord, which documents variations in climate on multiple time- and space-scales. At one end of the scale, there is the gradual cooling over the past 70 million years that resulted in the Earth shifting from a warm, ice-free state to a predominantly cold, glaciated state. At the other extreme, the Dansgaard—Oeschger cycles are rapid shifts between warm and cold states that occurred in some cases within decades, and were most marked in regions bordering the North Atlantic. These very different types of changes can be explained in terms of insolation forcing and the differentially lagged response of components of the Earth system. Biophysical and biogeochemical feedbacks associated with changes in the hydrosphere, as well as the marine and terrestrial biosphere, are responsible for amplification of initial climate changes and can result in extremely rapid climate changes. Although the past does not provide direct analogues for potential future climate changes, it does provide insight into the mechanisms of climate change of similar magnitude to, and at time- and space-scales congruent with, the changes that might occur in response to anthropogenic forcing. Palaeo-environmental records of the response of physical and biological systems to past climate changes provide targets for the evaluation of the Earth system models used to project potential future climate change. How the wealth of information available from the palaeorecord should be interpreted, and how these records illuminate the complex mechanisms of climate change and interactions between components of the Earth system, provides a framework both for future developments in climate science and for understanding and predicting future climate change.

Chapter 15, ***Modelling the Past and Future Interglacials in Response to Astronomical and Greenhouse Gas Forcing*** by André Berger and Qiuzhen Yin, shows that the greenhouse gas concentrations play a dominant role in the variations of the annual temperature averaged over the globe and over the southern high latitudes, whereas insolation plays a dominant role in the northern high latitudes. Modelling climate at the peaks of the interglacials of the last 800,000 years in response to insolation and greenhouse gas forcings uses an Earth system model of intermediate complexity: LOVECLIM. The interglacial marine isotope stage 11, assumed to be a good analogue of our interglacial, is warm only because of its high greenhouse gas concentrations, its insolation contributing to a cooling. The warmest interglacials are stages 9 and 5, and their greenhouse gas concentrations and insolation reinforce each other. Marine isotope stage 19, which has the same greenhouse gas concentrations and insolation pattern as marine isotope stage 1, appears to be the best analogue of our interglacial. In response to insolation, the annual mean temperatures averaged over the globe and over southern high latitudes are highly linearly correlated with obliquity. However, precession becomes important in the temperature of the northern high latitudes. During local winters, the response over the polar oceans, although the available energy is small, is larger than during the local summers due to the summer-remnant effect. Finally, the sensitivity to doubled CO_2 is the highest for the coolest interglacial.

Chapter 16, ***Catastrophe: Extraterrestrial Impacts, Massive Volcanism, and the Biosphere*** by Claire Belcher and Luke Mander, reviews two intervals of major environmental change in the geological past that are associated with mass extinction of life. The first occurred at the transition from the Triassic to the Jurassic Period (Tr—J) about 200 million years ago (Ma) and the second at the end of the Cretaceous Period (K—Pg) about 65 Ma. The K—Pg event highlights the possible environmental and biological consequences of the impact of an extraterrestrial body with the Earth, and the Tr—J highlights the possible environmental and biological consequences of massive volcanism. Such cataclysmic events appear to have greatly influenced Earth's climate such that the resulting climatic changes contributed to extinctions on a global scale. Understanding the environmental and climatic effects of each of these processes serves to enhance more complete appreciation of the potential risk facing life on Earth when such cataclysmic events occur in Earth's future.

Be careful what you wish for!—what if you get it?
…moving into somewhat uncharted intellectual territory is why Climatic Change got launched in the first place 35 years ago, and why we will continue to experiment with new ways to grow and serve the science and policy communities.
We must always be flexible to new situations, and to be willing to adjust our practices to the evolving state of the intellectual endeavors we work so hard to honestly report to the broad interdisciplinary readership

Stephen H. Schneider, Editorial 2010

Records from the Past, Lessons for the Future: What the Palaeorecord Implies about Mechanisms of Global Change

Sandy P. Harrison[a] and Pat Bartlein[b]

[a]School of Biological Sciences, Macquarie University, Sydney, Australia, [b]Department of Geography, University of Oregon, Oregon, USA

Chapter Outline

14.1. TIMESCALES OF CLIMATE CHANGE, THEIR CAUSATION, AND DETECTION

The climate of a region is most commonly expressed in terms of the 30-year average of key variables such as seasonal temperature, precipitation, and sea-level pressure. In this sense, climate could be considered as average weather conditions. However, this view does not work well with the perspective offered by the study of past climates, palaeoclimatology, which emphasizes climate variability on multiple timescales ranging from inter-annual to multimillennial. In this light, climate is not 'average weather', but rather the state of the Earth system at the particular timescale.

Over the past 70 million years (Figure 14.1a), $\delta^{18}O$ records (which can be regarded as a gross index of global temperature) show the Earth shifting from a warm, ice-free state to a predominantly cold state (Zachos et al., 2001). The long-term trend towards cooler conditions during the Cenozoic is marked by more rapid transitions, for example, around 35 Ma. On multimillennial timescales (Figure 14.1b), transitions between cold (glacial) and warmer (interglacial) climates show periodicity, but also variability in the amplitude of the transitions between warm and cold states, including a gradual increase in the amplitude of the temperature shifts after ~3 Ma (Lisiecki and Raymo, 2007). Both the periodicity and the variability are predictable consequences of the inherent variation in the Earth's energy budget caused by the superimposition of changes in the Earth's orbit on the configuration of the oceans and continents (Berger and Yin, 2012, this volume). Glacial intervals are characterized by pronounced, high-amplitude, millennial-scale variability (Figure 14.1c). These Dansgaard–Oeschger (D–O) cycles (Dansgaard

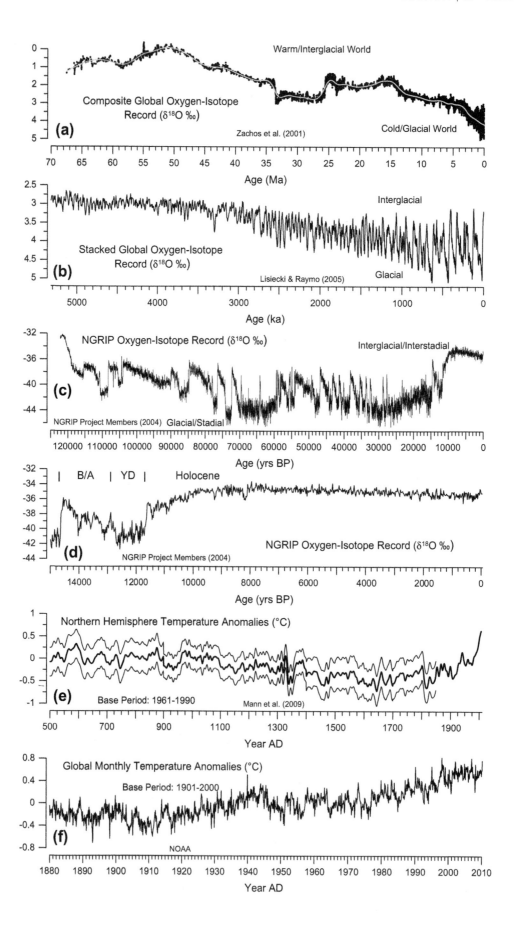

et al., 1984) appear to be related to the behaviour of the meridional overturning circulation (MOC). The most recent of these events is the Younger Dryas climate reversal (Figure 14.1d). Similar millennial-scale variability, including Younger Dryas-like climate reversals, is a regular feature of glacial intervals and is not unique to the last 100,000 years (Martrat et al., 2007; Cheng et al., 2009).

Millennial-scale variability has also been recognized during interglacial intervals (see e.g., Marchant and Hooghiemstra, 2004; Booth et al., 2005; Rohling and Palike, 2005) but is generally of lower amplitude than during glacial intervals. The climate record from Greenland (Figure 14.1e) shows a progressive, but small, cooling trend during the recent (Holocene) interglacial interval. Climate variability is also evident on centennial timescales (Figure 14.1e) although, again, over the last two millennia, this is superimposed on a longer term cooling trend (Mann et al., 2009) that culminates in the so-called Little Ice Age and is followed by the comparatively rapid warming characteristic of the industrial era (post 1750 A.D.). Finally, in addition to the strong imprint of the seasonal cycle, the observational record of the last century (Figure 14.1f) shows variability at both inter-annual and inter-decadal timescales (cf. Latif and Park, 2012, this volume).

Thus, climate variability has considerable structure on a wide range of timescales. The available records show several styles of variability: periodicity, resulting from astronomical or 'orbital' forcing on multimillennial and seasonal timescales; progressive changes, such as the long-term cooling through the Cenozoic in response to changes in land—sea configuration and atmospheric composition (Zachos et al., 2001; Fletcher et al., 2007), or the more recent cooling trend of the last two millennia; and rapid climate shifts, for example, following re-organisation of the coupled atmospheric—oceanic circulation during the D—O cycles (Bond et al., 1993; Kageyama et al., 2010). At each timescale, the climate system is characterized by changes in both frequency and magnitude of its variations. However, the magnitude of climate shifts is not a simple function of the timescale. Large and rapid changes in climate occur both on multimillennial and much shorter timescales (Figure 14.1).

On any timescale, global climate can be seen to be continuously varying, never dwelling long at any one value and frequently crossing the mean, thereby further rendering the definition of climate as a long-term mean unworkable when considering the history of changes in the Earth system. However, over some timespans, such as the past million years (Figure 14.1b) and over the interval between 80,000 and 10,000 years ago (Figure 14.1c), the variations of climate remain within a 'corridor' of values, indicating that the climate system is not completely non-stationary, in the sense of having continuously varying statistical properties.

Variations in climate influence many other aspects of the Earth system, including atmospheric composition, the hydrological cycle, and marine and terrestrial biology. Records of these changes are preserved in natural archives (see Bradley, 1999): for example, variations in atmospheric composition are recorded by air trapped in bubbles in the slowly accumulating ice sheets, changes in the hydrological cycles are recorded by lake shorelines or fluvial deposits, while changes in vegetation cover are recorded through pollen and plant macrofossils trapped in anoxic lake or bog sediments. Provided that such records can be unambiguously dated, they can be used as 'sensors' of climate and environmental change. (We eschew the term 'proxy', used as an abbreviation for 'climate proxy', preferring to use the term 'sensor' because any environmental archive is a palimpsest of both direct and indirect climate and other influences — and the primary goal of palaeo-environmental research is to disentangle the multiple factors influencing the record at any one time.)

Since environmental conditions change as climate changes, it is unsurprising that a large number of environmental archives show similar scales and types of variability (Figure 14.2). As an example, Figure 14.2 shows the variability in the incidence of wildfire on different timescales. Wildfire is controlled both by the occurrence of suitable weather conditions (e.g., convective activity controls the frequency of lightning ignitions, while seasonal drought controls the dryness of fuels) and by the nature of the vegetation cover, and hence the availability of fuel to burn, which in itself is determined by climate (Dwyer et al., 2000; Harrison et al., 2010). Sedimentary charcoal records document variability in fire regimes over millions of years (Figure 14.2a) through multimillennial (Figure 14.2b), millennial (Figure 14.2c), centennial (Figure 14.2d), decadal (Figure 14.2e) to annual (Figure 14.2f) timescales, and show progressive, quasi-periodic and abrupt changes in fire regimes. Furthermore, as with more direct indicators of

---

FIGURE 14.1   Variability in temperature on multiple timescales. The record for the last 65 Myr (a) is based on a global compilation of ^{18}O isotope records on benthic foraminifera (Zachos et al., 2001), and that for last 5500 kyr (b) on 57 ^{18}O isotope records (Lisiecki and Raymo, 2007). The $\delta^{18}O$ record from the NGRIP ice core (c; North Greenland Ice Core Project Members, 2004) is based on 50-year mean values over the past 125 kyr. An expansion of the last part of this record (from 15 ka onwards) is also shown (d). The decadally-resolved record for the past 1500 years (e; from 500 AD onwards) is based on a synthesis of instrumental and historical documentary records with palaeo-data including tree-ring reconstructions (Mann et al., 2009). The data for the last century (f) is from the NOAA historical climatology (*http://www.ncdc.noaa.gov/ghcnm*). The curves illustrate that climate is always varying and has no particular average value (but often varies within a particular corridor). The curves also show the rich set of trends (a, b, c, d, e, and f), periodic and quasi-periodic (b and c) variations, and abrupt changes (a and c), that both require explanation and provide 'natural experiments' with which to test models.

FIGURE 14.2 Variability in global fire on multiple timescales, as an illustration of the variability of a particular set of environmental subsystems and processes (i.e., disturbance of terrestrial ecosystems by fire). The record for the past billion years (1 Gyr) (a) is a qualitative index of global fire based on discontinuous sedimentary charcoal records (Bowman et al., 2009). The record for the past 80 kyr (b) is a global composite of 30 sedimentary charcoal records (Daniau et al., 2010), that for the past 21 kyr (c) is a global composite of ~700 sedimentary charcoal records (Daniau et al., submitted) and that for the past 2 kyr is a global composite of ~400 sedimentary charcoal records (d; Marlon et al., 2008). Global area burnt over the twentieth century (e) is estimated by combining data from tree-ring, historical, and remotely-sensed sources (Mouillot and Field, 2005), while global area burnt from 1997 to 2006 (f) is derived from satellite-based remote sensing (GFED v3.1, Giglio et al., 2010). Like palaeoclimatic records in Figure 14.1, the palaeofire records show continuous variability, as well as similarly recurring patterns of variability and abrupt changes.

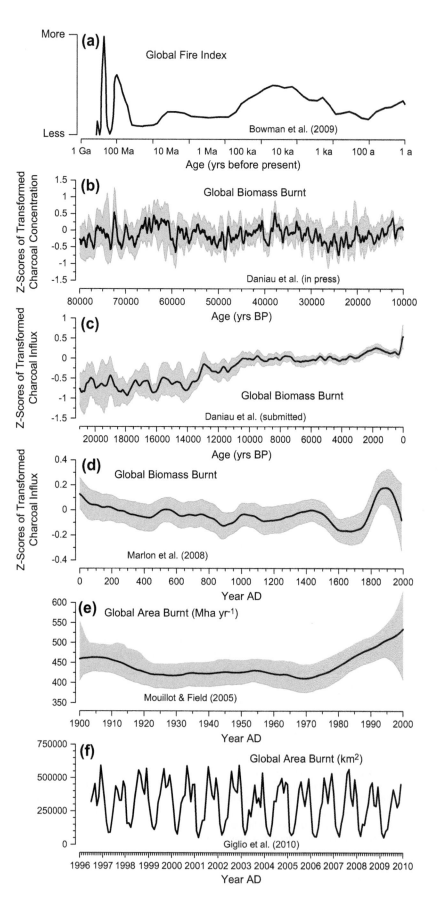

climate, the magnitude of the changes in fire regime is not necessarily related to the timescale or the frequency of variability.

Climatic variations on all timescales evoke responses in the terrestrial and marine biota that range from evolution and disappearance of species and genera on the longest of timescales, to wholesale re-organisations of the biosphere accompanying glacial—interglacial variations, to impacts on the phenology and growth of terrestrial plants and the distribution of short-lived or migratory organisms on the inter-annual timescale (Dickinson, 2012, this volume; Sen Gupta and McNeil, 2012, this volume). Re-organisations of climate and the biosphere on orbital timescales provide the most useful guide to the likely response of environmental systems related to climate during the coming centuries — not through exact analogy, but by revealing the mechanism and ways in which the climate system responds to changes in external forcing comparable in magnitude to those underway at present.

14.1.1. The Climate System and Timescales of Variability

The number of basic components of the climate system (atmosphere, ocean, biosphere, cryosphere) is small, but the number of variables that describe those components (e.g., regional seasonal temperatures, carbon stocks) is

enormous. The many variables describing the climate system fall into one of three categories (Figure 14.3; after Saltzman, 2002): (i) those that describe the external forcing of the system (*boundary conditions*; e.g., insolation, volcanic aerosols, and a broad set of 'geodynamic' variables, including the configuration, topography and bathymetry of continents and ocean basins); (ii) those that describe the more slowly varying aspects of the system (*slow-response variables*; e.g., the ice sheets, bedrock, and mantle that are deformable by ice sheets; sea level and the temperature and salinity of the deep ocean, and the long-term state of its overturning and horizontal circulation; and the slowly varying reservoirs involved in biogeochemical cycling that ultimately determine atmospheric composition); and (iii) those that describe the internal variables that are ordinarily thought of as weather, but also include rapidly varying biophysical and biogeochemical processes (*fast-response variables*; e.g., precipitation, evapotranspiration, respiration, seasonally varying sea-ice extent, soil-moisture content, vegetation properties, the temperature, depth, and other physical and biological characteristics of the mixed layer of the ocean). A fourth category of variables, *subsystem variables*, describes the state and function of the many environmental subsystems that are governed by climate, and which, depending on context, include many of the fast-response variables. For example, evapotranspiration can be considered both as a fast-response variable that is

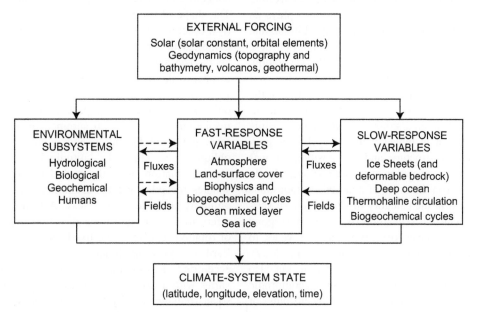

FIGURE 14.3 The climate system. The variables that describe the external forcing of the system (or boundary conditions) directly or indirectly influence the slow-response, fast-response, and environmental-subsystem variables; these, in turn, influence each other and determine the state of the climate system as a function of longitude, latitude, elevation, and time. The arrows labelled 'fields' indicate that one set of variables influences the other through patterns of atmospheric circulation, moisture, and heat, while those labelled 'fluxes' indicate that one set of variables influences another through the transfer of mass and energy. The dashed arrows indicate that the influence of the fast-response variable on environmental subsystem variables is currently unidirectional in climate models, but that eventually environmental subsystem variables will interact with the fast-response variables as climate models develop (after Saltzman, 2002: Bartlein and Hostetler, 2005). Many of the individual variables listed here leave evidence (i.e., palaeoclimatic data) of their variations over time.

a component of the surface water and energy balance, and as a subsystem variable when describing hydrological systems on scales from watershed to continents. Human activities such as fossil-fuel burning are often thought of as 'external' variables, but are more realistically considered as an environmental subsystem that both influences and is influenced by climate (i.e., as part of a 'coupled Earth system', e.g., Rice and Henderson-Sellers, 2012, this volume).

Many environmental systems respond to variations of climate. These systems are also characterized by a large number of variables, including some that play a role in the interaction and feedback between the atmosphere and the surface (and might therefore be thought of as fast-response variables), and some that are dependent on climate but do not feed back to the climate system except in limited ways. Human social and economic systems can also be considered as another environmental subsystem that responds to climate, but ultimately feeds back to it, as is illustrated by the current focus on the mitigation of anthropogenic climate change (Metz et al., 2007) and by the new approaches for developing emissions scenarios that include input from integrated assessment models of the coupled natural and human environment (Moss et al., 2008; Rice and Henderson-Sellers, 2012, this volume).

Some variables are not easily categorized. Vegetation plays a key role in the instantaneous coupling of the atmospheric boundary layer and land surface by controlling the exchanges of energy and moisture, and also plays an important role in biogeochemical cycles (e.g., Pitman and de Noblet-Ducoudré, 2012, this volume). The rates of these fast exchanges depend on the structure of the vegetation, and the states of the atmosphere and underlying soil (including atmospheric humidity, wind, net radiation at the surface and soil-moisture availability) that together influence plant physiology (e.g., stomatal conductance), while the role of vegetation in biogeochemical cycling is governed by vegetation structure and composition. It was formerly thought that vegetation structure responds slowly to climate changes (on the order of hundreds to thousands of years), placing it in the category of slow-response variables. It is now clear that vegetation structure can respond rapidly to climate changes over timespans of years to decades (Tinner and Lotter, 2001; Shuman et al., 2009). Soils are dependent on climate and vegetation, but also have strongly expressed geological and geomorphic controls (e.g., Harvey, 2012, this volume). Key attributes of the soil such as water-holding capacity may be dominated by parent material (as in arenaceous soils), and so this might be regarded as a boundary condition; in other situations water-holding capacity is dominantly controlled by soil morphology, and hence acts like a slow-response variable. The particular category a variable falls into is thus largely dependent on context, location, and scale.

14.1.2. Insolation Variations

The temporal-scales of variability in the climate system depend on the nature of the external forcing of the climate system and the response time of the internal components of the climate system to this forcing. On orbital timescales (Figure 14.1b and 14.1c), the ultimate cause of natural climate variability is changes in the latitudinal and seasonal distribution of incoming solar radiation (insolation) as a result of changes in the Earth's orbit (Berger, 1978, 1981, 1988; Berger and Yin, 2012, this volume). These insolation variations govern the principal mode of climate variability over the past several million years—the variations between glacial and interglacial states—and these variations therefore provide a perspective on the behaviour of the climate system to changes in its external controls, and on the resulting re-organisation of the terrestrial and marine biospheres and biogeochemical cycles, and on the mechanisms by which the climate system amplifies the external forcing.

Earth's orbit can be described by three parameters (Figure 14.4a): the shape of the orbit about the Sun (eccentricity), the tilt of the Earth's axis relative to the Sun (obliquity), and the time of the year when the Earth is closest to the Sun (precession). Strictly speaking, obliquity is an 'astronomical' parameter, not a characteristic of the Earth's orbit about the Sun, but we include it here in the collection of 'orbital' parameters that influence insolation. The shape of the Earth's orbit varies from nearly circular (low eccentricity of 0.005) to elliptical (high eccentricity of 0.058), with periodicities of roughly 100,000 and 400,000 years. The angle of the Earth's axial tilt (obliquity) with respect to the plane of the orbit varies between 22.1° and 24.5° with a periodicity of around 41,000 years. As obliquity increases, the amplitude of the seasonal cycle of insolation increases, with summers in both hemispheres receiving more insolation and winters less. Changes in obliquity also affect the latitudinal variations of annual mean insolation, with the high latitudes of both hemispheres receiving greater insolation during times of high obliquity and the tropics less. Astronomical precession refers to the change in the direction of the Earth's axis of rotation relative to the fixed stars (i.e., the 'wobble' of Earth's axis) while climatic precession, which depends on eccentricity and the time of year of perihelion (and hence varies with astronomical precession), governs the average solar irradiance on any given day (Loutre, 2009). Climatic precession varies with periodicities of 23,000 and 19,000 years (Berger, 1978; Crucifix et al., 2009), and leads to opposing variations in insolation between the Northern and Southern Hemispheres. From the perspective of one of the hemispheres, when the axis is aligned towards the Sun during the time when the Earth is closest to the Sun (perihelion), the seasonal difference in radiation receipt is

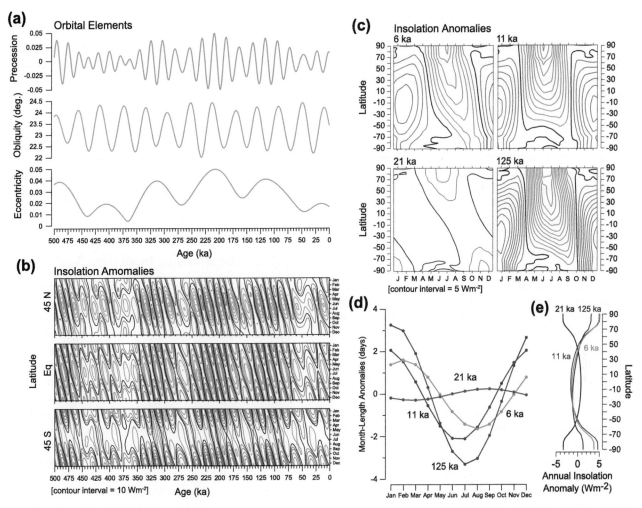

FIGURE 14.4 Effects of changes in orbital parameters on insolation. (a) Changes in climatic precession, obliquity, and eccentricity over the past 500 kyr (Berger, 1978). These changes in orbital configuration give rise to complex changes (b) in seasonal insolation through time (here represented as plots of mid-monthly insolation in W m⁻² at 45°N, the equator, and 45°S over the past 500 kyr). The difference (relative to present) in seasonal insolation by latitude (in W m⁻²) is shown for four iconic time periods (c): the mid-Holocene (6 ka), the early Holocene (11 ka), the Last Glacial Maximum (LGM; 21 ka), and the peak of the last interglacial (125 ka). The impact of changes in orbital configuration on (d) 'month length' anomalies (Kutzbach and Gallimore, 1988), or the difference in the time taken by the Earth to sweep through one-twelfth of its orbit, and (e) the latitudinal distribution of insolation is also shown for these iconic time periods. In panels (b), (c), and (d), months are labelled using their conventional names; strictly speaking 'months' are defined by the angular position of Earth relative to that at the vernal equinox. The first month of the year (labelled 'J' or 'Jan' is the interval between the time of Earth's angular location at −81° and −50° (before the vernal equinox). Along with the annual cycle, these insolation variations are the only controls of climate variations that vary in a strictly periodic, or cyclical, way.

enhanced in that hemisphere (the hemisphere experiencing 'summer' conditions) and reduced in the other hemisphere.

The superimposition of variations in these three parameters gives rise to remarkably complex patterns in the seasonal and latitudinal distribution of insolation through time (Figure 14.4b). Precessional variations produce the most obvious pattern in the variations of the annual cycle of insolation over time at all latitudes: the repeating pattern of positive and negative insolation anomalies at the precessional timescale. These translate into changes in seasonality within a single hemisphere, and to opposition in the sign of the anomaly between hemispheres during any

particular month and opposition of the magnitude of the anomaly during any particular season (i.e., positive summer anomalies in the north are accompanied by negative summer anomalies in the south). The amplitude of the precessional maxima and minima is related to eccentricity: times of high eccentricity produce larger differences between insolation maxima and minima (e.g., between 250,000 and 175,000 years ago), while times of low eccentricity produce smaller differences (e.g., between 450,000 and 350,000 years ago). Obliquity variations modify these variations by amplifying the summer insolation maxima, the winter minima, and seasonal contrast when obliquity is high (such as during the maximum of the last precessional cycle, ~10,000 years ago), and damping it when low (as ~30,000 years ago). The precessional maxima and minima can also be seen to progress through the year, with the extremes occurring progressively later in the year during any given cycle.

The net effect of the variations in the different orbital elements can be seen in latitude by month anomalies in insolation at particular times (Figure 14.4c). At 6000 years ago, perihelion occurred at the end of the northern summer, while obliquity was slightly greater and eccentricity about the same as present (see Berger and Yin, 2012, this volume). Summer insolation was roughly 6% greater than present and winter 8% less in the mid-latitudes to high latitudes of the Northern Hemisphere, while in the Southern Hemisphere the anomalies were reversed. At 21,000 years ago, all three of the orbital elements were close to their present values, and so the insolation anomalies relative to present were small. The previous interglacial period, ~125,000 years ago, was characterized by perihelion during the northern summer (and at the boreal summer solstice at 127,000 years ago, Berger and Yin, 2012, this volume), high obliquity and high eccentricity. Consequently, the insolation anomalies were greater than those at 6000 years ago, and the seasonality of insolation in both hemispheres was high relative to both 6000 years ago and to present.

Eccentricity of the orbit also leads to differences in the length of the summer and winter seasons (Figure 14.4d), because the Earth moves more rapidly along its orbital track near perihelion and less rapidly near aphelion (Joussaume and Braconnot, 1997; Braconnot et al., 2008; Berger et al., 2010). This effect can oppose that of precession. For example, at both 6000 and 125,000 years ago, the effect of the climatic precession-related maximum in the Northern Hemisphere insolation is partly compensated by a reduction in the length of the summer months relative to today (Berger et al., 2010; Berger and Yin, 2012, this volume).

The time course of insolation at a specific latitude differs from that at adjacent latitudes, even in the same hemisphere. Consequently, while it is often convenient to refer to situations like 'Northern Hemisphere summer insolation maxima' or 'interglacials,' the regional and temporal expression of these situations vary. For example, the interglacials of the last million or so years are all different in terms of the particular sequence and spatial pattern of insolation variations that caused them, although some may be broadly similar (Berger and Yin, 2012, this volume). This fact severely limits the potential for constructing analogies between them, for example, specific sequences of variations in greenhouse gases or global ice volume (e.g., Ruddiman, 2008). There is no reason to expect that the sequence of events in one interglacial should be exactly similar to that in another, and indeed there is ample evidence that this is not the case.

Insolation variations can potentially be expressed as any number of sinusoidal curves; for example, one could produce a curve of April mid-month insolation anomalies at a particular latitude, or of the February–October difference in insolation integrated over a specific hemisphere. This situation creates a danger of finding spurious explanations for a particular palaeoclimatic time series, in the same way that any curve can be represented by a Fourier series. Consequently, the search for explanations of climatic variations in terms of a specific record of insolation forcing must be based on an underlying mechanistic or conceptual model that specifies why that particular linkage should occur and, furthermore, allows testing of explicit hypotheses about how it occurs.

14.1.3. Implications of Insolation Variations

The temporal variations in insolation are gradual, producing smoothly varying shifts in climate on orbital timescales and giving rise to alternations between globally cold, glacial states and globally warm, interglacial states. These changes in insolation affect other elements of the climate system: large ice sheets grow during cold states and decay during warm states, while changes in land–sea geography consequent on the growth and decay of these ice sheets affects ocean circulation. Vegetation responds to changes in global temperature, with increases in the area of forests during warm periods and decreases during cold periods. Likewise, major changes in the carbon cycle (see below) occur on orbital timescales.

However, each of these elements of the climate system has an inherent timescale of response to the initial orbital forcing. Ice sheets take many millennia to build up and decay, changes in ocean circulation take centuries to millennia, and vegetation migration takes decades to centuries. The timescale of interest determines whether each of these elements has to be considered as a dependent or independent variable in the climate system. On timescales of 10^4 to 10^6 years, ice sheets are dependent variables in the climate system, with their build-up and decay controlled by orbitally driven variations in insolation. At shorter timescales (10^3 to 10^4 years), ice sheets are independent variables that have an important impact on atmospheric circulation and global temperature. Similarly, at millennial timescales (10^3 years) vegetation changes are driven by climate changes, but on shorter timescales (10^2 to 10^3) changes in vegetation distribution affect climate through changing albedo and other land-surface characteristics, as well as through changing emissions of climatically important trace gases and aerosols (Arneth et al., 2010).

Changes in insolation provide the explanation for the first-order climate variation evident on orbital timescales. However, analysis of the response of the climate system to insolation forcing provides numerous insights into the functioning of the Earth system, which under some circumstances responds to orbital forcing in extremely non-linear ways. The insolation variations provide what might be thought of as a continuous experiment with the global energy balance involving manipulations of all aspects of the incoming radiation at the top of the atmosphere, including weak variations of the annual average insolation integrated over the whole planet, as well as the large perturbations of its latitudinal and seasonal distribution. The transmission of these perturbations through the climate system can be used to understand the general linkages among different pathways of energy flow in the Earth−atmosphere energy balance, and to estimate the sensitivity of climate to such perturbations.

Although the insolation variations can be regarded as the 'pacemaker' of the glacial−interglacial variations of climate (Imbrie et al., 1984), they fail to fully explain those variations in several important ways: (i) the variations in annual insolation are small and insufficient by themselves to generate the glacial−interglacial variations; (ii) the main variations in insolation occur on the precessional timescale (Figure 14.4a and 14.4b), while the principal variations of global climate occur variously on the obliquity and eccentricity timescales (Figure 14.1b), and (iii) the expression of these timescales of variation in terms of global climate can change relatively abruptly (as for example around 1 million years ago, when 41-kyr variations gave way to 100-kyr variations). Often referred to as the 100-kyr or 41-kyr problems (Raymo and Nisancioglu, 2003; Lisiecki and Raymo, 2007), several ideas have been advanced that emphasize either variations in the mode of transmission of the 'signals' of the three orbital elements through the climate system (Imbrie et al., 1992; Ruddiman, 2006), or to specific features of one of the elements (e.g., obliquity and summer insolation, Huybers and Wunch, 2005), but as yet no consensus has emerged. Nevertheless, although the explanation is still incomplete, the empirical link between insolation and glacial−interglacial variations of climate attests to the role of mechanisms internal to the climate system that amplify the effects of externally forced perturbations of the energy balance. There is strong evidence that this amplification involves the effects of the ice sheets themselves on planetary albedo (along with that of changing vegetation distributions), global biogeochemical cycles (in particular those of the long-lived greenhouse gases: carbon dioxide and methane), and physically and/or biologically mediated changes in the dust and aerosol content of the atmosphere − all of which substantially modify the Earth−atmosphere energy balance (Hansen et al., 1984; Forster et al., 2007).

Several proposed explanations for the amplification of climate change on orbital timescales rest on a combination of geophysical and biological mechanisms. Some biological mechanisms operate at the level of the physiology of individual organisms; but large-scale re-organisations of the biosphere also involve geographical range shifts. Variations in climate on orbital timescales have been a constantly present feature in the evolution of life on Earth, and so it should not be surprising that species have developed strategies for dealing with large-scale re-organisations of the biosphere. Species persist for periods that are typically several orders of magnitude longer that the timescales of insolation variations (Bartlein and Prentice, 1989; Bennett, 2004). Because of the conservatism of species' environmental niches, range boundary shifts are a near-universal feature of species' responses to climate change (Huntley and Webb, 1989; Davis et al., 2005). The responses of species are individualistic, but species generally are capable of migration, allowing major biogeographical re-organisations to take place along with re-organisations of climate.

14.1.4. Co-variation of Climate and Biogeochemical Cycles Over the Past 800 kyr

The co-variation among components that describe the general state of the climate system, as represented, for example, by global ice volume and elements of biogeochemical cycles, can be seen through the perspective provided by polar ice cores (Figure 14.5). The longest and most comprehensively analysed record to date is the EPICA Dome C (EDC) record from Antarctica (Jouzel et al., 2007a; Barbante et al., 2010), which spans the past 800,000 years. In addition to chemical and physical measures (such as dust concentration), ice core records provide samples of the actual atmosphere at different times, and thereby (subject to synchronization of records derived from trapped air and records derived from the ice itself) provide an internally consistent record of climate and biogeochemical cycles. Atmospheric composition in terms of long-lived greenhouse gases (carbon dioxide, methane, nitrous oxide), dust and local temperature indicators (Figure 14.5) all show the signatures of insolation forcing (represented here by the often-used index of July insolation at 65°N), in common with the oxygen isotope record of global ice volume from marine sediments. Individual precessional peaks are evident to a greater (e.g., methane) or lesser (e.g., carbon dioxide, but see Ahn and Brook, 2008) extent and are modulated by the 100-kyr eccentricity cycle. Interglacials are characterized by high levels of greenhouse gases, relatively warm Antarctic conditions, and low dust concentrations, while glacials show the opposite features.

The specific contributions of both the external (insolation) and internal (the ice sheets, sea-ice, land cover, greenhouse gases, and dust) drivers to variations in the net radiative forcing over the past 800,000 years has been

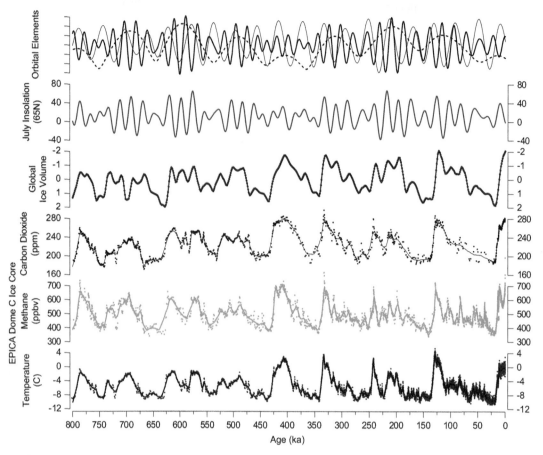

FIGURE 14.5   Co-variation of climate and biogeochemical cycles over the past 800 kyr. Changes in precession (black solid line), obliquity (grey line), and eccentricity (dotted line), and the resultant July insolation at 65°N are compared with global ice volume (Martinson et al., 1987), changes in CO_2 (Lüthi et al., 2008), and CH_4 (Loulergue et al., 2008) from the EPICA Dome C (EDC) ice core and with inferred temperature from the deuterium isotope record from EPICA (Jouzel et al., 2007a). The imprint of the insolation variations is clearly expressed in temperature and greenhouse gas records.

estimated by Köhler et al. (2010) using a combination of observations from the EDC records and other palaeo-climatic data and model-based interpretations of these records. 'Radiative forcing' is a way to express the impact of a heterogeneous set of potential controls (and feedbacks) on the global energy balance, and in turn on global average temperatures (e.g., Forster et al., 2007; Arneth et al., 2010). The analysis by Köhler et al. (2010) suggests that the ice sheets make the largest contribution to the overall radiative forcing variations over the past 800,000 years, followed by greenhouse gases and sea-ice, and dust and vegetation, with the feedbacks all much larger than the direct effects of insolation (see figure 7 in Köhler et al., 2010). They did not directly estimate the impacts of additional feedbacks from water vapour, clouds, and related changes in lapse rates (which are highly uncertain and model dependent), but these are roughly comparable to the others, and again exceed the direct effects of insolation. The feedbacks are all positive in the sense of driving the climate system towards colder conditions when Northern Hemisphere insolation

decreases, and towards warmer conditions when it increases. Although the specific mechanisms, pathways, and spatial variations in feedbacks remain to be disclosed, the overall potential of the climate system to amplify relatively weak changes in radiative forcing is clear (cf. Harvey, 2012, this volume).

On long timescales (e.g., over the full span of the EDC ice core), insolation, ice volume, and greenhouse gases co-vary. On shorter timescales, such as the interval since the Last Glacial Maximum (LGM, Figure 14.6), they are quasi-independent drivers of climate changes because of inherent time lags in their response to inso-lation changes. What is considered an internal component of the climate system and what is an external control is not arbitrary, but depends on the specific timescale of interest and the response time of a particular component relative to that timescale.

The LGM-to-present interval has been a major focus of palaeoclimate investigations because it provides a range of 'natural experiments' allowing differentiation of the role

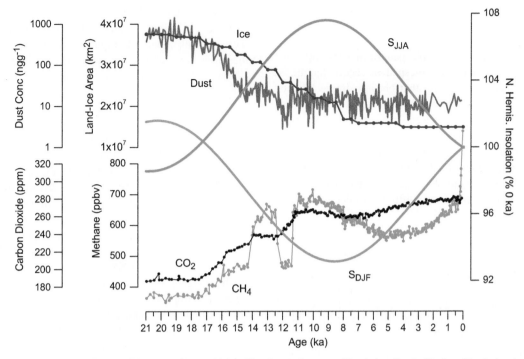

FIGURE 14.6 Changes in boundary conditions over the past 21 kyr. The changes in seasonal insolation (for the Northern Hemisphere) are compared with changes in global ice volume from ICE-5G (Peltier, 2004), and changes in CO_2 (Lüthi et al., 2008), CH_4 (Loulergue et al., 2008), and dust (Lambert et al., 2008) from the EDC ice core. The curves jointly illustrate the natural experiments that provide targets for palaeoclimatic simulations. At 21 ka, the distribution of insolation is close to that at present, but there were large ice sheets, high concentrations of dust, and low concentrations of the greenhouse gases, carbon dioxide and methane, providing an experiment that shows the impact of these 'glacial-age' boundary-condition settings. At 6 ka, most of the boundary conditions were close to their pre-industrial values, except for insolation, providing an experiment involving a perturbation of the Earth–atmosphere energy balance.

of individual drivers on regional climates. At the LGM, and over the first part of deglaciation (i.e., to ~18 ka), insolation levels were near their present ones, but ice sheets were larger than today, greenhouse gas concentrations were lower, and dust loadings higher. In contrast, at 6 ka, the ice sheets were nearly at their present extents, and the atmospheric constituents were close to their 'pre-industrial' (~1750 AD) values.

14.1.5. The Hierarchy of Climatic Variations and the Explanation of Palaeoclimatic Records

Climatic variations occur within a hierarchy of controls and responses, which begin at the top level with the external controls of climate, proceed through global, hemispheric, continental, and regional-scales, and end with the variations of individual climate variables at specific locations at the bottom level (Bartlein, 1997). Responses at any one level of the hierarchy become the controls of variations of the components at lower levels. With the exception of the annual cycle, there is a general tendency for the variations of components at higher levels in the hierarchy to show more long-term variability while those at lower levels experience more short-term variability.

The existence of this hierarchy also has implications for attempts to explain the variations at a particular place, or to interpret the 'signal' encoded in a particular palaeoclimatic record. For example, although climatic variations at a place are ultimately governed by global-scale controls, a specific palaeoclimatic record generally cannot be representative of the general state of the global system. This situation arises because the intermediate controls and responses have the potential of reinforcing, cancelling, or even reversing the longer-term, larger-scale trends. Gradual changes in large-scale controls may sometimes produce abrupt local changes when atmospheric circulation is re-organised. Conversely, abrupt changes in the large-scale circulation may produce warming in some regions and cooling or no change in others, as can be seen in the spatial anomaly patterns of year-to-year variations in climate. Consequently, while it may be difficult or even impossible to ascribe a particular climate variation at a place to a specific configuration of higher-level controls, shorter-term variations at lower levels are strongly conditioned by the particular state of the system at higher levels. Therefore, any discussion of the timescales of climatic variability should explicitly acknowledge the spatial-scale or extent of the system being discussed.

14.1.6. Cycles and Spurious Periodicity: A Warning

Rather than displaying a simple pattern of variability that increases as a function of timespan or record length, as would be typical of a system that obeyed simple 'scaling laws' (Kantz and Schreiber, 2004), the climate system instead has some preferred temporal-scales of variability that reflect the nature of the external forcing of the climate system (e.g., insolation on the orbital timescale), or the internal time constants of the components of the climate system itself — like the slow build-up and decay of ice sheets, or the inter-annual variability associated with the El

Niño–Southern Oscillation (ENSO; Saltzman, 2002; Latif and Park, 2012, this volume). The most common way of describing this apparently organised variability in climatic time series is to refer to the variations as *cycles*.

Several factors can conspire to predispose researchers to see cycles in time series when in fact none exist, and then to invoke some kind of regular cyclical mechanisms, either external (Sun, moon, planets) or internal (oscillatory 'climate-modes') to account for them. Firstly, there *are* cycles in palaeoclimatic time series, in the form of the variations of ice volume and many other variables that occur in response to the orbitally driven variations of insolation or the quasi-periodicities in tropical Pacific ocean–atmosphere interactions (ENSO) that are related to a specific physical mechanism (the propagation of Kelvin waves across the Pacific). Secondly, spectral analysis (Jenkins and Watts, 1968), which was developed for statistical signal processing, has been used successfully to detect the imprint of the orbital variations in many palaeoclimatic time series. Thirdly, there is a tendency for palaeoclimatic time series to vary between general limits; this, coupled with our tendency to seek order in variable data, can lead to the perception of periodic variations when none exist. Fourthly, quasi-periodic variations in time series can arise from simple short-term memory or persistence (as in autoregressive moving-average models; Box and Jenkins, 1976). Fifthly, some common data-analysis tools or procedures can impart spurious periodicity. These last two sources of apparent periodicity are particularly problematic because failure to recognize them can lead to incorrect inferences about causality.

Figure 14.7 shows three examples of time series that feature quasi-periodic variations, each generated by filtering or transforming a series of normally distributed random numbers in an intrinsically aperiodic way. (In each panel, the random numbers are plotted in grey in the background on an arbitrary scale, and the generated series in black.) The top curve shows the output of a second-order auto-regressive or AR(2) model in which the current value of the time series depends on the previous two values, plus the random input (plotted in grey). In this particular example, the parameters of the model are those that apply to the well-known Wölfer sunspot series, which is well-described by an AR(2) model (Box and Jenkins, 1976). The resulting series clearly shows a rough 10 time-step oscillation, and further displays a change in amplitude and regularity of this oscillation midway through the series that would likely provoke additional comment if this were a real palaeoclimatic time series.

The middle series shows a 'difference stationary' time series generated using an integrated autoregressive model (ARI(1,1); Box and Jenkins, 1976), that was linearly detrended. (Long-period trends in time series are of two types: 'difference stationary' series that can be detrended

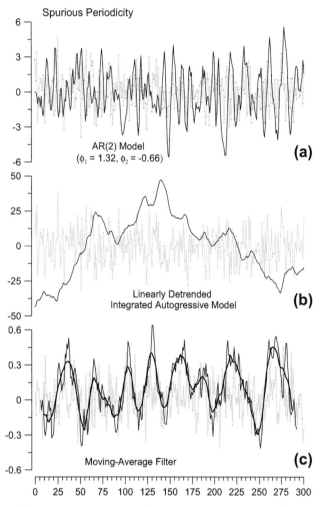

FIGURE 14.7 Illustrations of spurious periodicity. Series plotted in grey are the white-noise time series used to generate the simulations in black; (a) shows a second-order autoregressive process (AR(2)); (b) shows an improperly detrended integrated autoregressive model; (c) shows running mean-filtered white noise (thin black line represents the output from a 15-term running mean, thick black line from a further 10-term running mean of the data represented by the thin line). The curves show how apparent 'cycles' can arise in time series even when no underlying cyclical mechanism exists.

by taking the first differences between observations, and 'trend stationary' series that can be detrended by fitting a straight-line curve to the data: Nelson and Kang, 1981). The example series shows a broad cycle with a wavelength close to the record length and, in this particular realization, happens to also show lower-amplitude, higher-frequency variations, with a period around 20 time-steps. Again, although generated by an aperiodic process, the resulting series would likely be viewed as cyclic. Palaeoclimatic time series are frequently detrended as part of preliminary steps in data analysis.

Probably the most frequently applied method for generating apparently periodic variations in a time series when none really exist is illustrated by the bottom panel of Figure 14.7, which demonstrates the Slutsky—Yule effect (von Storch and Zwiers, 2001). The thin black line in Figure 14.7c is the result of applying a 15-term running mean to a white-noise time series, while the smoother solid line is simply the first series further smoothed by a 10-term running mean. The resulting series are clearly periodic, and it would be hard not to apply the term 'cycle' in describing or explaining these data if they were real. Slutsky's theorem (Jenkins and Watts, 1968, p. 297) shows that by repeated application of summing (as in the running mean) or differencing filters, white noise can be reduced to a sine wave. In practice, the smoothing produced by the running mean can occur naturally in palaeoclimatic records that integrate environmental conditions over time, but most often the smoothing occurs during data analysis.

All three series appear superficially periodic yet were generated by relatively simple processes, without invoking any kind of mechanism that could genuinely be called cyclical or periodic. Because there are multiple ways of generating series with apparent periodicity, some of which are related directly to commonly applied data-analysis procedures, a relatively high standard should exist for declaring a particular series cyclical and therefore explicable using the class of oscillatory or cyclical physical mechanisms. The premature declaration of a series as being cyclical may limit the search for other, possibly better, mechanistic explanations.

14.2. REGIONAL RESPONSES TO MILLENNIAL-SCALE FORCING

Faced with the richness and complexity of the palaeo-record, one strategy adopted by palaeoclimatologists to analyse key aspects of the record has been to study 'snapshots' of the state of the world corresponding to iconic periods with well-defined boundary conditions. An alternative strategy, the analysis of time-dependent changes, has always been the primary focus for the analysis of palaeo-environmental observations, but modelling

of the transient behaviour of the climate system has only recently become feasible with the advent of fast climate models (see e.g., Timm and Timmermann, 2007; Liu et al., 2009). Here we discuss the two most heavily studied intervals of the recent geological record (the LGM and the mid-Holocene) before considering some generalizations that have emerged from two decades of focusing on these intervals.

14.2.1. The Last Glacial Maximum

The LGM (~21,000 years ago; 26.5 ka—20 ka according to Clark et al., 2009) has been a focus for modelling experiments since the early days of palaeoclimate modelling (Williams et al., 1974; Gates, 1976; COHMAP, 1988; Kutzbach et al., 1993; Braconnot et al., 2007a, 2007b, and references therein; Otto-Bleisner et al., 2009). This is, in part, because it represents a time when most of the climate drivers (or boundary conditions) were radically different from today and, in part, because of the wealth of palaeo-environmental data that has been assembled to document regional climate conditions during this period (Figure 14.8).

At the LGM, palaeo-environmental data show colder (Figures 14.8a and 14.8b) and drier conditions in most of the northern extra-tropics. Vegetation records (Figure 14.8c) from Europe, Eurasia, and Alaska indicate a landscape dominated by treeless vegetation, with a significant expansion of graminoid and forb grassland and xerophytic shrubland in northern and central Eurasia (Prentice et al., 2011). Plant macrofossil data, however, indicate that trees persisted in local refugial situations (Willis and Whittaker, 2000; Willis et al., 2000; McLachlan et al., 2005). Forests were present south of the ice sheet in North America. Sedimentary charcoal records indicate a reduction in the amount of biomass burning (Figure 14.8d), partly as a result of a temperature-dependent reduction of fuel loads and partly because colder, drier climates give rise to a reduction in convection and hence lightning ignition (Power et al., 2008). The absence of forest cover, and the generally drier-than-present conditions, gave rise to increased deflation of surface material by winds and hence atmospheric dust loadings were between 2 and 5 times higher than today in the northern extra-tropics (Figure 14.8e; Harrison et al., 2001; Kohfeld and Harrison, 2001). Sea surface temperatures (SSTs) in the Northern Hemisphere were considerably lower than today (Figure 14.8b). The strongest annual mean cooling (up to −10°C) occurred in the mid-latitude North Atlantic, with more pronounced cooling in the eastern than in the western Atlantic (MARGO Project Members, 2009).

In contrast, both vegetation (Figure 14.8c) and lake data (Figure 14.8e) indicate wetter-than-present conditions in the American southwest and in western China, and the

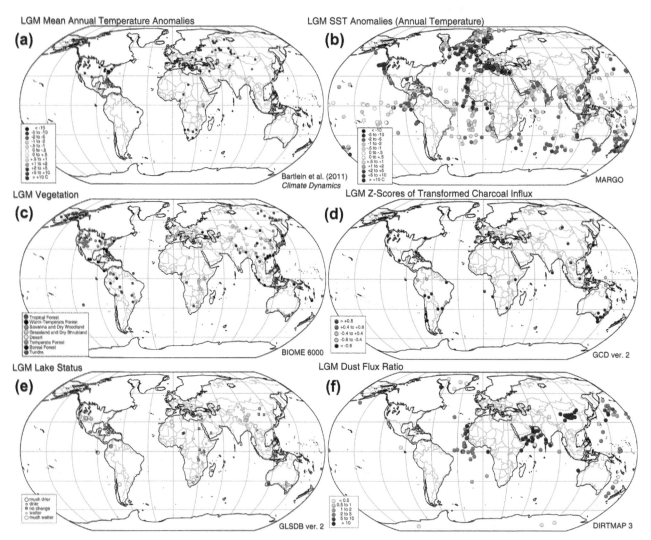

FIGURE 14.8 The world at the LGM (21 ka) as shown by palaeo-environmental data: (a) changes in mean annual temperature (MAT) compared to present-day reconstructed from pollen and plant macrofossil data (Bartlein et al., 2011); (b) changes in mean annual sea surface temperature (SST) compared to present day reconstructed from biological and geochemical climate proxies (MARGO Project Members, 2009); (c) vegetation reconstructions from the BIOME 6000 project (Prentice et al., 2000; Bigelow et al., 2003; Pickett et al., 2004; Marchant et al., 2009; and unpublished data), reclassified using the megabiome scheme described by Harrison and Prentice (2003); (d) changes in biomass burning compared to the long-term average between 21 and 0.25 ka, expressed as, from the Global Palaeofire Working Group Database (Version 2, Daniau et al., submitted); (e) changes in lake status compared to present from the Global Lake Status Database (Kohfeld and Harrison, 2000); (f) changes in dust deposition compared to present from the DIRTMAP database (v2: Kohfeld and Harrison, 2001).

lake data (though not vegetation records) suggest that the region around the Mediterranean Sea was also wetter. Charcoal records from China indicate increased fire (Figure 14.8d).

During the LGM, the tropics were colder and drier than today. Terrestrial records (Figure 14.8a) indicate an average cooling at sea level across the tropics of 2.5−3°C. This average hides considerable regional differentiation, with circum-Pacific regions experiencing relatively little cooling (1−2°C) and stronger cooling (5−6°C) in Central and tropical South America (Farrera et al., 1999; Bartlein et al., 2010). Marine records (MARGO Project Members, 2009)

show an average cooling in the tropics of 3−4°C, but again considerable regional differentiation in the strength of the change (Figure 14.8b). Cooling at high elevations, as shown by the lowering of snowline equilibrium lines, was larger than the cooling registered at sea level (Mark et al., 2005). Thus, tropical lapse rates must have been steeper than today. There was regional differentiation in the degree of high-elevation cooling, with sites in the northern Andes, Central America, and Papua New Guinea showing larger snowline depressions than sites in the Himalayas, the southern Andes, and eastern Africa. The charcoal records (Power et al., 2008) show less fire over most regions of the

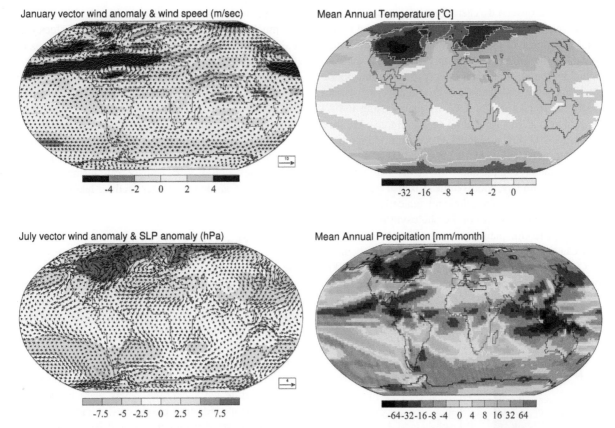

FIGURE 14.9 The world at the LGM (21 ka) as shown by ensemble-averaged differences between 21 ka and a pre-industrial control from coupled ocean—atmosphere climate model simulations from the Palaeoclimate Modelling Intercomparison Project (PMIP2, *http://pmip2.lsce.ipsl.fr*): January 500 hPa wind vector and wind speed (m s^{-1}) anomalies, mean annual temperature anomalies (°C), July sea-level pressure (SLP, hPa), surface-wind vector anomalies, and mean annual precipitation anomalies (mm per month). Comparison of this figure with Figure 14.8 illustrates one of the challenges of comparing palaeoclimatic simulations with observations: the climate model output must be expressed in the same terms as the palaeoclimatic data using 'forward' models, or the palaeoclimatic data must be interpreted in terms of specific climate variables using 'inverse' modelling procedures.

tropics, although some sites in South East Asia and Papua New Guinea (where the regional cooling appears to have been small) show more fire than today (Figure 14.8d).

Regional climate changes in the southern extra-tropics (Figures 14.8a, 14.8b) were less pronounced than those registered in the northern extra-tropics, close to the ice sheets. Nevertheless, both vegetation (Figure 14.8c) and lake (Figure 14.8e) data show colder and drier climates than today, while charcoal records indicate less fire (Figure 14.8d). Dust deposition records indicate that the decrease in vegetation cover and the increased aridity led to an increase in dust erosion and transport in the southern extra-tropics (Figure 14.8f), although this appears to have been less marked than the increase in the Northern Hemisphere. Marine data from the Southern Ocean show a northward shift of the polar front to about 45°S during the LGM, and sea surface temperatures during austral summer up to 2—6°C cooler than today (Figure 14.8b).

The changes in regional climates shown by these various palaeo-environmental sensors can be explained in

terms of the glacial-age boundary conditions. Climate model experiments (Figure 14.9) show that the presence of large ice sheets, increased sea-ice cover, and low greenhouse gases led to globally colder and drier conditions (see Braconnot et al., 2007a). The wet conditions in American southwest, in the region around the Mediterranean Sea, and in western China are a specifically predicted consequence of the southern displacement of the Northern Hemisphere Westerlies as a consequence of the presence of the large, mountain-like mass of the Laurentide Ice Sheet (Kutzbach et al., 1993). Model simulations also show that changes in the tropics are more muted than those in the extra-tropics (see Braconnot et al., 2007a). The depression of tropical snowline is a consequence of the lowering of tropical sea surface temperatures, which is responsible for a drier atmosphere and, therefore higher lapse rates (Kageyama et al., 2005). This first-order effect is amplified by a weakening of the Asian monsoon, which led to a further increase in lapse rates in the northern tropics and around the western Pacific.

Although the observed vegetation changes at the LGM (relative to the Holocene) can be partially explained by the differential response of plant functional types to the change in climate at the LGM, vegetation distribution was also influenced by the physiological effects of low atmospheric CO_2 concentration on plant growth (Figure 14.10), and especially the impact of low CO_2 on plants using the C_3 photosynthetic pathway (including nearly all trees: Prentice and Harrison, 2009). Under low CO_2 concentrations, transpiration per unit leaf area is increased as a result of increased stomatal conductance. In plants using the C_3 photosynthetic pathway, photosynthesis is also reduced due to reduced substrate concentration and reduced competition by O_2 for carboxylation sites of the photosynthetic enzyme Rubisco. As a result, plants using the C_4 photosynthesis pathway can compete with C_3 plants more effectively under low CO_2 concentrations (Bond and Midgley, 2000). Atmospheric CO_2 concentration is a limiting factor for C_3

photosynthesis even at modern values (>380 ppmv) and was much more strongly limiting at glacial values (Polley et al., 1993; Guiot et al., 2001; Harrison and Sanchez Goñi, 2010). Modelling experiments have shown that the restricted forest cover during the LGM can only be reproduced accurately when physiologically mediated CO_2 effects on plant competition are taken into account (Harrison and Prentice, 2003). Since the nature of the vegetation cover influences the partitioning of precipitation into evapotranspiration and runoff, the influence of low CO_2 values could also impact on surface hydrology, including the amount of water feeding lakes and rivers at the LGM.

14.2.2. The Mid-Holocene

The mid-Holocene ($\sim$6000 years ago, 6 ka) was selected for detailed study because of its capacity to illustrate nearly pure effects of the insolation anomaly centred on 10,000 years ago. Although some high northern latitude records show a temperature maximum close to the peak of the insolation anomaly (Kaufman et al., 2004), most northern mid-latitude regions experienced a delayed temperature maximum due to the persistent regional cooling effect of the Laurentide ice sheet. Ice sheets were nearly at their present extents and atmospheric constituents were close to their pre-industrial values by the mid-Holocene, while the seasonal and latitudinal distribution of insolation was still substantially different from today, with increased summer insolation and enhanced seasonal contrast in the Northern Hemisphere and reduced summer insolation and decreased seasonal contrast in the Southern Hemisphere.

There is considerable asymmetry in the regional climate changes at high northern latitudes (Figures 14.11a, 14.11b), implying modulation of the direct insolation forcing through atmospheric circulation. Vegetation records (Figure 14.11c) show a northward extension of the Arctic tree line in Europe and Eurasia, as a consequence of increased warmth during the growing season (Prentice et al., 2000; Bigelow et al., 2003). Northern temperate forest zones were also shifted northwards, with displacements of even greater magnitude than that shown by the Arctic tree line. Warmer-than-present winters as well as summers are required to explain all of the observed northward shifts in the temperate forest zones. However, vegetation records from much of northern Canada and Alaska show no discernible northward shift (Edwards et al., 2000; Williams et al., 2000) and the tree line in eastern Canada was shifted southward compared to today (Richard, 1995; Williams et al., 2000). The southward expansion of shrub tundra and boreal woodlands was accompanied by an increase in fire in eastern Canada and the limited amount of data from northern Europe suggests that northward expansion of forests led to a reduction in fire (Figure 14.11d; Power et al., 2008).

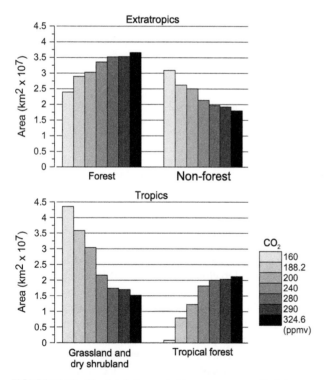

FIGURE 14.10 Simulated changes in biome area as a result of changes in CO_2 on the competition between C_3 and C_4 plants under modern climatic conditions. Changes in extra-tropical regions are summarized by changes in total forest versus total non-forest biomes; changes in the tropics by changes in the area of tropical forest compared to grassland/dry shrubland. In these simulations with the BIOME4 equilibrium biogeography—biochemistry model (Kaplan et al., 2003), atmospheric CO_2 levels have been systematically varied between 180 and 'modern' levels of 324.6 ppmv but climate was kept constant. The simulations show that the direct physiological impact of CO_2 can have major impacts on vegetation distribution *(Source: redrawn from Harrison and Prentice, 2003; Harrison and Sanchez Goñi, 2010).*

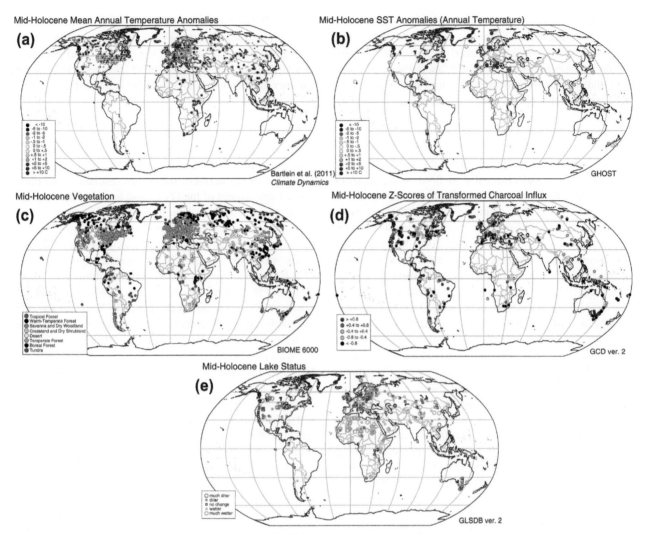

FIGURE 14.11 The world during the mid-Holocene (6 ka) as shown by palaeo-environmental data: (a) changes in mean annual temperature (MAT) compared to present-day, reconstructed from pollen and plant macrofossil data (Bartlein et al., 2010); (b) changes in mean annual SST compared to present-day, reconstructed from biological and geochemical climate proxies (Leduc et al., 2010); (c) vegetation reconstructions from the BIOME 6000 project (Prentice et al., 2000; Bigelow et al., 2003; Pickett et al., 2004; Marchant et al., 2009; and unpublished data, reclassified using the megabiome scheme described by Harrison and Prentice, 2003); (d) changes in biomass burning compared to the long-term average between 21−0.25 ka, expressed as z-cores, from the Global Palaeofire Working Group Database (Version 2, Daniau et al., submitted); (e) changes in lake status compared to present-day from the Global Lake Status Database (Kohfeld and Harrison, 2000). For most of the palaeoclimatic data types, the mid-Holocene is richer than that for the LGM, but the record of SSTs is sparser and confined mainly to oceanic regions with high sedimentation rates.

Temperate deciduous forests extended southward into the Mediterranean zone in Europe at 6 ka (Figure 14.11c), indicating summers wetter than today. Lake data (Figure 14.11e) also show wetter conditions at that time, and a progressive increase in aridity thereafter (Yu and Harrison, 1995). Charcoal records indicate a reduction of fire in lowland parts of the Mediterranean region, although sites at higher elevations tend to show increased biomass burning (Power et al., 2008).

The most pronounced changes in regional climate were registered in the region dominated by the Afro−Asian monsoon. Enhanced monsoons extended forest biomes inland in China, and Sahelian vegetation northward into the

Sahara (Figure 14.11c). Lake data (Figure 14.11e) and aeolian data (Kohfeld and Harrison, 2003) also show an expansion of the region influenced by monsoon precipitation in both Africa and Asia. The African tropical rainforest was also reduced in extent (Figure 14.11c), consistent with a northward shift of the inter-tropical convergence zone (ITCZ) and a more seasonal climate in the equatorial zone. Vegetation and lake data show wetter conditions in the American southwest and Central America, consequent on the expansion of the North American monsoon, accompanied by expansion of steppe vegetation and aeolian activity in interior North America; this observed duality is explicitly predicted by climate modelling and attributed to enhanced subsidence around the monsoon core region (Harrison et al., 2003). The charcoal record shows that changes in fire regimes were highly heterogeneous. This finding is consistent with our understanding that increased precipitation can lead to increased fuel loads and hence more fire in fuel-limited systems while, at sites where fuel is not limiting, increased precipitation suppresses fire (Van der Werf et al., 2008; Daniau et al., submitted).

The monsoon systems of the Southern Hemisphere were generally weaker during the mid-Holocene than today. Vegetation, lake, and charcoal data from South America suggest drier-than-present conditions (Markgraf, 1989; Mayle and Power, 2008; Marchant et al., 2009). Charcoal data indicate reduced fire in the southern tropics of Africa and northern Australia (Power et al., 2008; Mooney et al., 2010), consistent with reduced monsoons. However, some geomorphic evidence from northwestern Australia and the continental interior suggests that the Australian monsoon may have been stronger than at present (Shulmeister, 1999; Wyrwoll and Miller, 2001; Lynch et al., 2007).

Vegetation data from the southern extra-tropics (Figure 14.11c) show comparatively little change from present (Jolly et al., 1998; Marchant et al., 2009). Vegetation data from southern Australia show changes in plant-available moisture during the mid-Holocene, but adjacent regions show opposite signals in the direction of the inferred changes with the southernmost part of the region and Tasmania somewhat wetter than today and sites lying further north and along the east coast showing somewhat drier conditions (Pickett et al., 2004). Mooney et al. (2010) have identified a similar opposition in the regional changes in fire and suggest that this is consistent with shifts in atmospheric circulation. In contrast, lake data (Figure 14.11e) show wetter conditions uniformly across the southern extra-tropics, although this may reflect the comparative paucity of records from these regions.

The mid-Holocene has not been a major focus for the synthesis of marine records, in part because low sedimentation rates make it difficult to identify changes through the Holocene. In general, and in contrast to the marked changes in regional climates shown over the continents, ocean surface temperatures were similar to today over much of the world (Figure 14.11b). Records from the North Atlantic suggest that the mean annual surface ocean temperature was slightly higher than today, although late summer and autumn temperatures at high northern latitudes may have been slightly cooler (Kim et al., 2004; Leduc et al., 2010). Data from the tropical Pacific and Indian Oceans suggest that ocean temperatures there were slightly lower than today (Rimbu et al., 2004; Stott et al., 2004; Lorenz et al., 2006), possibly as a result of reduced ENSO variability (Tudhope et al., 2001). While these regional signals are consistent with changes in adjacent land areas, there is a lack of information from critical oceanic regions.

The first-order features of regional climate shown by these reconstructions can be explained as a consequence of known changes in forcing during the mid-Holocene. Orbitally-induced enhancement of Northern Hemisphere summer insolation at 6 ka resulted in increased heating over the Northern Hemisphere continents (Figure 14.12), deepening the thermal lows over the land and thus intensifying the flux of moisture from the tropical ocean to the continents (Kutzbach and Street-Perrott, 1985; COHMAP, 1988; Joussaume et al., 1999; Braconnot et al., 2007a). Increased heating over the northern subtropics resulted in the northward displacement of the ITCZ, and hence of the monsoon front, leading to drier conditions in the equatorial zone. Monsoon expansion is most pronounced in Africa and Asia because of their large continental area; the expansion of the North American monsoon is correspondingly more muted. Mid-Holocene aridity in the mid-continent of North America is caused by enhanced subsidence over the continental interior that is dynamically linked to the orbitally-induced enhancement of the summer monsoon in the American southwest (Harrison et al., 2003).

Warmer-than-present summers at high northern latitudes, as indicated by the northward shift of the Arctic treeline, are also a consequence of increased Northern Hemisphere summer insolation and the prolongation of warm conditions into the autumn caused by warmer oceans and less extensive sea-ice (Wohlfahrt et al., 2004). The pronounced longitudinal asymmetry in this warming appears to reflect the nature of ocean circulation in the Arctic, which results in sea-ice transport away from the northern coast of Siberia and towards Canada and which therefore amplifies the impact of orbitally-induced sea-ice reduction on the high latitudes of Eurasia while minimizing its impact on the high latitudes of Canada. The observed shifts in Northern Hemisphere temperate forests attest to warmer-than-present winters during the mid-Holocene, a signal opposite to the direct consequence of orbital forcing which would tend in the direction of colder winters (Kutzbach and Guetter, 1986). The paradox of warm mid-Holocene winters has been extensively discussed for Europe (see e.g., Prentice et al., 1998 and references therein), where it is generally attributed to stronger westerly flow around a strengthened Icelandic Low allowing warm air to penetrate further into the continental interior than is the case today.

In general, model simulations show a reduction in the Southern Hemisphere monsoons and comparatively little change in the southern extra-tropics (Braconnot et al., 2007a). However, some simulations have shown an enhancement of the Australasian monsoon during the mid-Holocene (see e.g., Liu et al., 2004; Marshall and Lynch, 2006). Reduction of the Southern Hemisphere monsoons is a direct response to changes in orbital forcing during the mid-Holocene. However, analyses of a large ensemble of mid-Holocene climate simulations (Zhao and Harrison,

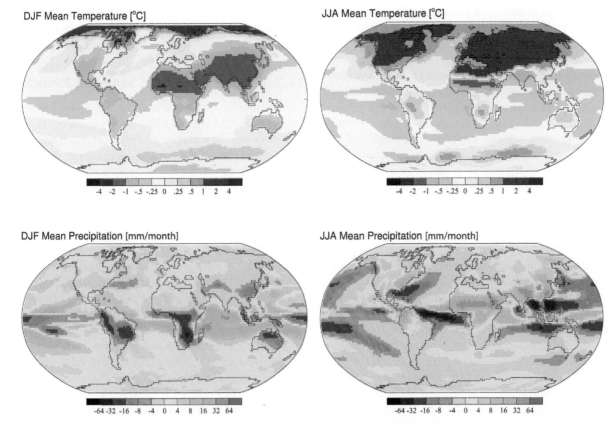

FIGURE 14.12   The world during the mid-Holocene (6 ka) as shown by ensemble averaged differences between 6 ka and a pre-industrial control from coupled ocean—atmosphere climate model simulations from the Palaeoclimate Modelling Intercomparison Project (PMIP2, *http://pmip2.lsce.ipsl.fr*): boreal winter (December, January, February, DJF) and boreal summer (June, July, August, JJA) mean temperature (°C) and precipitation (mm per month) anomalies.

in press) suggest that ocean feedbacks weaken the impact of orbital forcing in the Southern Hemisphere leading to a smaller decrease in monsoon rainfall than might otherwise be expected. In the case of the Australian monsoon, local changes in SSTs that generate a low-pressure cell over the Indian Ocean can create increased precipitation (compared to present) in northern and central Australia depending on the exact location of the low-pressure cell (Zhao and Harrison, in press). Observed differences in the timing of maximum monsoon expansion during the Holocene between different parts of the Afro—Asian monsoon region can be explained parsimoniously by two facts: (i) that the seasonal timing of the summer monsoon onset differs among subregions; and (ii) that the Holocene timing of the peak insolation anomaly is different for different months of the year (Marzin and Braconnot, 2009).

14.2.3. Consistency of Spatial Responses in Warm and Cold Climates

The palaeorecord shows that there are common features of the response to climate forcing in both cold and warm

climates, and that these features can be explained relatively simply. Most prominent among these is the comparatively muted response of the tropics to changes in forcing. At the LGM, for example, ice core (Stenni et al., 2001; Jouzel et al., 2003; Masson-Delmotte et al., 2005), sea-ice (Gersonde et al., 2005), permafrost (Renssen and Vandenberghe, 2003), and vegetation data (Prentice et al., 2000; Bigelow et al., 2003; Bartlein et al., 2010) all show major changes in temperature in the higher latitudes (>10°C) while both terrestrial (Farrera et al., 1999) and marine (Ballantyne et al., 2005; MARGO Project Members, 2009) data show comparatively small changes in temperature (<4°C) in the tropics. Climate changes during the glacial, associated with D—O cycles, are also stronger in the extra-tropics than in the tropics (Hessler et al., 2010; Harrison and Sanchez Goñi, 2010). Polar amplification (Masson-Delmotte et al., 2006) of temperature changes is also a feature of warm climates, including the mid-Holocene, the Last Interglacial, and the mid-Pliocene (CAPE-Last Interglacial Project Members, 2006; Jansen et al., 2007; Miller et al., 2010), and of projected future climates (Holland and Bitz, 2003; Masson-Delmotte et al., 2006; Meehl et al., 2007). Polar amplification of temperature changes is, at least in part, a predictable consequence of feedbacks associated with changes in sea-ice, snow cover, and vegetation, and is reproduced in climate models.

Palaeo-environmental data show larger changes of land than of ocean temperatures in both cold and warm climate intervals, and this difference is observed in both tropical (e.g., Figure 14.13) and extra-tropical regions. Climate model projections of the response to greenhouse gas forcing consistently show that temperatures over land increase more rapidly than over sea, with a ratio in the range 1.36–1.84 independent of the simulated global mean temperature change (Sutton et al., 2007; Crook et al., 2011). The difference between land and ocean warming appears to be associated with land-surface feedbacks (Joshi et al., 2008). Analyses of palaeoclimate simulations (Lainé et al., 2009) suggest that this 'land/sea warming ratio' appears to be remarkably consistent through time.

Although glacial–interglacial temperature changes in the tropics are comparatively muted, there are large changes in precipitation associated with the waxing and waning of the monsoons (e.g., Dupont et al., 2000; Weldeab et al., 2007; Cai et al., 2010; Revel et al., 2010). The palaeorecord emphasizes the fact that climate changes in mid- to high latitudes are dominated by shifts in temperature, whereas changes in precipitation are the dominant influence in tropical latitudes. The debate about polar amplification has, to some extent, obscured the importance of the large variations in tropical precipitation.

14.2.4. Different Spatial Scales of Response

The broad-scale patterns shown by palaeo-environmental data at the LGM and mid-Holocene can be explained as a consequence of changes in surface energy balance and atmospheric circulation due to large-scale changes in climate forcing. However, mapped patterns of environmental variables always show some heterogeneity: either individual sites that show a signal different from that registered at nearby sites, or small-scale regional patterns that run counter to the more zonal pattern of climate change. This heterogeneity is frequently thought of as noise or attributed to errors in the interpretation of the palaeo-records or to uncertainties in dating. However, this is not necessarily the case.

The characteristics of the sensor itself influence its sensitivity to climate change. For example, the response of the level of a lake to increased precipitation is determined by the relative size and shape of the lake and of its catchment (Harrison et al., 2002). The same change in precipitation is expected to produce different changes in lake level and area in nearby lakes that are otherwise

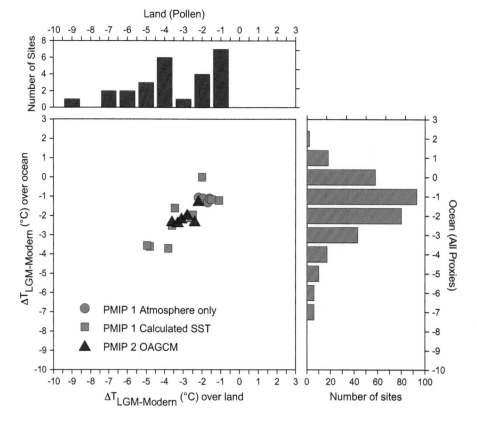

FIGURE 14.13 Comparison of simulated and observed changes in tropical temperatures at the LGM (21 ka) relative to present. The inferred temperature changes from pollen are from Ferrara et al. (1999), the inferred changes in SST are from the MARGO data set (MARGO Project Members, 2009). Zonally-averaged temperatures from the models are shown in grey for atmosphere-only simulations from PMIP1 and in colour for coupled ocean–atmosphere simulations from PMIP2 (Pinot et al., 1999; Kohfeld and Harrison, 2000). The data clearly illustrate the greater temperature anomaly over land relative to that over the ocean.

identical, for example, if the size of the catchment differs. Similarly, plants growing close to the limit of their range are expected to be more sensitive to a climate change of a given magnitude than the same plants growing in the middle of their range (Bartlein et al., 2010). A decrease in precipitation could lead to either an increase or a decrease in fire depending on whether the change leads to a decrease in the amount of fuel available or an increase in the dryness of the fuel (Van der Werf et al., 2008; Daniau et al., submitted.). Changes in the oxygen isotopic composition of ice reflect local temperatures but can also be influenced by changes in the seasonal timing of the precipitation that produced the ice (Werner et al., 2000; see also Lee et al., 2008). Changes in the composition of foraminiferal assemblages reflect the growth temperature of the assemblage, but may not directly reflect SST if changes in the structure of the mixed layer have caused changes in the depth at which the organisms live (Morey et al., 2005) or if other environmental factors influence productivity change (Siccha et al., 2009).

Spatial heterogeneity can also reflect the fact that contiguous areas of climate space are not necessarily contiguous in geographical space. This is most obvious in areas of complex terrain, where large spatial variations of climate occur in a limited geographical area. Physiography exerts an additional influence on climate through its modulation of the large-scale atmospheric circulation and results in the occurrence of distinctly different climate regimes in close proximity. The modern climate of the Yellowstone area in the western USA, for example, is characterized by a mosaic of summer- and winter-dominated rainfall areas (Figure 14.14). During the mid-Holocene, the vegetation record shows that sites that lie today in summer-dominated rainfall areas showed conditions wetter than present while sites from winter-dominated rainfall areas showed conditions drier than present. The apparently contradictory patterns of climate change in closely adjacent sites shown by the palaeorecords from this region (Figure 14.14) are nevertheless a predictable consequence of orbitally-induced changes in the North American monsoon. Sites fed by summer rains today received more rainfall with the amplified monsoon, while sites that are now summer-dry were drier still because of increased evaporation (Shafer et al., 2005).

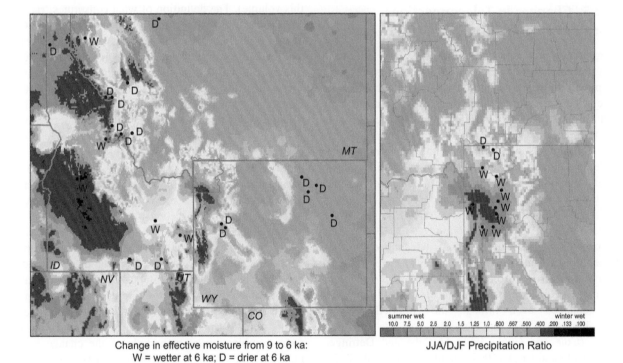

Change in effective moisture from 9 to 6 ka:
W = wetter at 6 ka; D = drier at 6 ka

JJA/DJF Precipitation Ratio

FIGURE 14.14 Relationship between modern climate patterns, as mediated by topography, and mid-Holocene climate changes in the northern Rocky Mountains, USA. The left panel shows modern precipitation seasonality (ratio of total precipitation in June–July–August, JJA, to December–January–February, DFJ), where green indicates a summer maximum and blue a winter maximum. The overlain letters indicate the change in moisture, as reconstructed from pollen data, between the early (~9 ka) and mid-Holocene (MH, ~6 ka) where W equals sites that have become wetter since the early Holocene and D equals sites that have become drier. The right panel is a blow-up of the region around Yellowstone National Park. The direction of the change in moisture between the MH and today is predicted by the modern seasonality. Sites that today are in the summer-dry/winter-wet region (blue) were drier in the early Holocene, while those in the summer-wet/winter-dry region (green) were wetter, paralleling the climate changes that made the summer-dry regions drier in the early Holocene and the summer-wet regions wetter *(Source: see Harrison et al., 2003. Redrawn from Shafer et al., 2005.)*

Physiography also has an impact at larger regional-scales. For example, lake records from the peri-Baltic region show drier conditions during the mid-Holocene (Figure 14.11e), and vegetation records show a more northerly penetration of warmth-demanding forests along the coast than in the interior of Sweden (Figure 14.11c). The Baltic was larger than today during the mid-Holocene, and the presence of this water body would have led to warmer winters (thus favouring northward penetration of temperate forest) and somewhat cooler summers with a circulation regime that would have favoured blocking of the Westerlies and hence increased aridity (Yu and Harrison, 1995). Vegetation records from northeastern North America (Figure 14.11c) show that Hudson Bay also had a localized impact on seasonal climates (Prentice et al., 2000; Bartlein et al., 2010). However, just as climate varies on multiple temporal-scales, so there is variability on multiple spatial-scales, reflecting the modulation of large-scale controls by a hierarchy of regional and local influences. The strength of the overall climate changes may determine the coherency of the spatial response: larger climate changes (e.g., LGM climates) may be sufficient to override topographic and physiographic influences.

14.2.5. Changes in Teleconnections/Short-Term Variability

Interpretation of inter-annual/inter-decadal variability from palaeo-environmental records is complicated by the complexity of the climatic and environmental controls on these sensors. One interpretative approach that has been used is to link the observed variability in the recent past to variations in modes of atmospheric circulation (e.g., the North Atlantic Oscillation; ENSO; the Southern Annular Mode) and then to interpret longer-term variability shown in an individual palaeorecord as evidence for the changing strength of these modes (see e.g., Bradbury et al., 1993; Hammarlund et al., 2002; Donders et al., 2005; Björck et al., 2006; Quigley et al., 2010). This approach assumes that observed teleconnections are stable through time. However, analyses of meteorological records have shown that the strength of correlations between specific modes and local climate variables have changed even through the twentieth century (e.g., Figure 14.15) and certainly over the historical period (Cole and Cook, 1998; McCabe and Dettinger, 1998). Cole and Cook (1998; see also Cook et al., 2000) have shown that the modern correlation between ENSO and droughts in the southwestern USA was not present in the early part of the nineteenth century. Climate model simulations of the mid-Holocene response to orbital forcing are characterized by changes in the spatial patterns of teleconnections associated with the Arctic Oscillation (Otto-Bliesner et al., 2003; Figure 14.16) and

somewhat weaker ENSO teleconnections than seen in simulations of the present day (Otto-Bliesner et al., 2003; Figure 14.16), reinforcing the idea that such linkages are time-varying. Thus, while short-term variability in the region directly influenced by a specific mode may reflect changes in that mode, the use of this approach to interpret variability in more distant regions, teleconnected today, will lead to spurious conclusions. Furthermore, it obscures the work that is required: namely the use of networks of palaeo-environmental records to reconstruct modes of climate variability through time, paralleling the approach used with, for example, documentary evidence to reconstruct circulation patterns during the historical period (e.g., Luterbacher et al., 2002; Brázdil et al., 2005; Luterbacher et al., 2010).

14.3. RAPID CLIMATE CHANGES

A common feature of palaeoclimatic time series that appears on many different timescales (Figure 14.1) is the frequent occurrence of rapid or abrupt changes in the level or variability of the time series. These are also features of great importance for future climates (e.g., Lenton, 2012, this volume). The definition of what constitutes 'rapid' or 'abrupt' is somewhat arbitrary (Alley et al., 2002; Clark et al., 2008), but in general includes the idea of a change that occurs over several decades (or longer, on longer timescales), and that persists for an interval several times longer than the time taken for the change. The National Research Council (US) Committee on Abrupt Climate Change (Alley et al., 2002), in a definition adopted by the IPCC (Meehl et al., 2007), define *abrupt climate change* as one that takes place more rapidly than the underlying forcing, pointing out that this kind of behaviour can only occur when the climate system crosses a critical threshold defining the limit between two different climate states. This definition provides a theoretical basis for understanding abrupt climate changes (see e.g., Kageyama et al., 2010) but the rapidity of the climate change is a function of the temporal-scale of the specific forcing involved and the definition is thus difficult to apply in the case of palaeo-records where the nature of the forcing is *a priori* unknown. For practical reasons, many authors therefore identify abrupt climate changes in geological records in terms of some combination of magnitude of the change and the rapidity with which it is accomplished (see e.g., Martrat et al., 2004; Clark et al., 2008; Belcher and Mander, 2012, this volume). Although rapid transitions are inevitable in any time series that show the kind of short- or long-memory behaviour that climate time series do, there is a particular combination of mechanism and spatial pattern of response that involves the Atlantic Meridional Overturning Circulation (AMOC; also referred to as the thermohaline

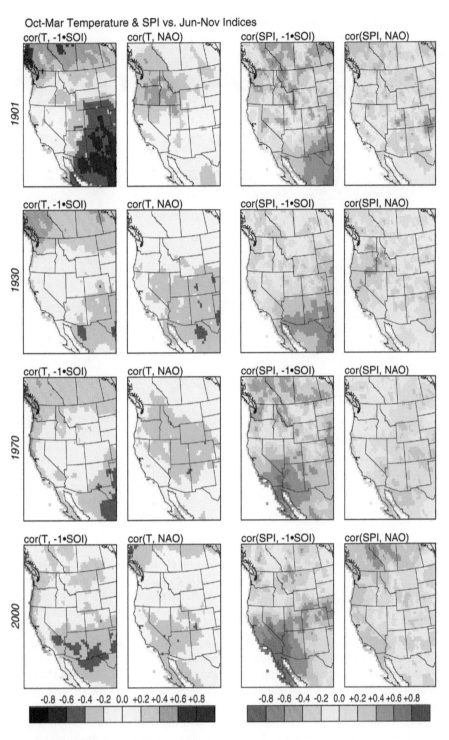

CRU Data: CRU TS 2.0 [http://www.cru.uea.ac.uk/]
Image: Dept Geography, Univ. Oregon [http://geography.uoregon.edu/envchange/]

FIGURE 14.15 Maps of the correlations between temperature (left two columns) and the standardized precipitation index (SPI, right two columns) with two climate-mode indices, the Southern Oscillation Index (SOI) and the North Atlantic Oscillation (NAO) over western North America. The maps show correlations for four periods during the twentieth century (from top to bottom: 1901–1915, 1915–1945, 1955–1985, 1985–2000). The climate data are from the CRU TS 2.0 data set (Mitchell and Jones, 2005), the SOI was obtained at *http://www.cpc.noaa.gov/data/indices/* and the NAO index was obtained at *http://www.esrl.noaa.gov/psd/data/climateindices/list/*. Although the sign of the teleconnections (correlations) remain constant in the 'core' regions of the teleconnections (i.e., western Canada and the southwestern United States) the magnitudes and spatial patterns of the correlations vary, and in many regions correlations change sign over the twentieth century.

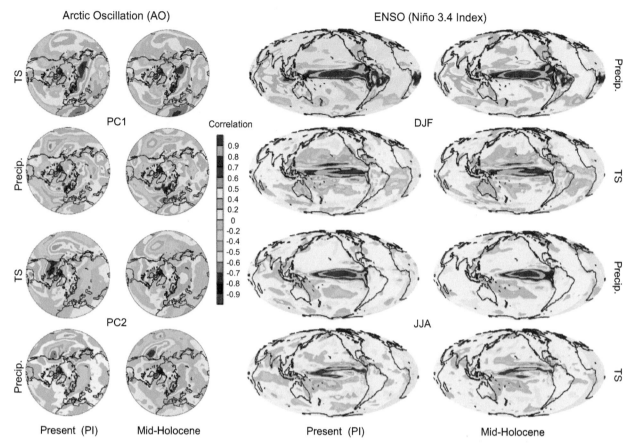

FIGURE 14.16   Correlations between two climate-mode indices, the Arctic Oscillation (left panels) and the El Niño–Southern Oscillation (ENSO), as represented by the Niño 3.4 index, and temperature (TS) and precipitation (Precip.) in pre-industrial (PI) and mid-Holocene simulations with the NCAR CCSM coupled ocean–atmosphere climate model (Otto-Bliesner et al., 2003). The PI is conventionally used for the modern baseline (control) simulation to avoid the impact of twentieth century climate change. As was the case for the regional teleconnection patterns shown in Figure 14.15, global teleconnection patterns vary between the mid-Holocene and pre-industrial (PI).

circulation, THC) that recurs frequently enough to warrant discussion as a distinct class of climate variability.

14.3.1. Examples of Rapid/Abrupt Climate Changes

On the longest of timescales (Figures 14.1a and 14.1b), climatic time series can be seen to experience changes in level (such as that around 34 Ma, accompanying the glaciations of Antarctica) or variability (such as that at 2.6 Ma, when the oxygen isotopic variations on the 41-kyr obliquity timescale became amplified and the repeated Northern Hemisphere glaciations began, or about 1 Ma, when 100-kyr variations became prominent). These transitions, which are rapid from the perspective of the particular timescale at which they appear, tend not to recur.

In contrast, on the millennial timescale over the past glacial interval (~80,000 to 11,700 years ago), there is a series of 20 instances of abrupt warming (Figure 14.1c), followed by more gradual cooling, referred to as D–O

'cycles' or Greenland Interstadials/Stadials (GI/GS) (Steffensen et al., 2008; Sanchez-Goñi and Harrison, 2010; Wolff et al., 2010). Although spaced roughly 1500 years apart, the specific length and form of the variations vary too much for the series to be regarded as periodic. Embedded in these variations are series of six occurrences of extreme cooling, known as Heinrich events or stadials. The individual warming events involved temperature increases on the order of 10°C in Greenland, and were accompanied by changes in climate, and in terrestrial and marine ecosystems, on a global scale.

During the last deglaciation (Figure 14.1d), there was an abrupt warming in the Northern Hemisphere ~14.7 ka, known as the Bølling–Allerød (BA) interstadial, that was followed by gradual cooling into the Younger Dryas chronozone (YDC). In the Southern Hemisphere, which had been warming gradually since the LGM (not evident in Figure 14.1), the Antarctic Cold Reversal — an episode of cooler temperatures — began around the time of the BA, and lasted through the YDC. The apparent opposition in deglacial temperature trends in the two hemispheres is apparent in polar ice-core records throughout the last glacial, and has been referred to as the 'bipolar see-saw'. However, as will be discussed further below, the majority of the variability in temperature throughout the last glacial and into the deglaciation has been coherent between the hemispheres (Shakun and Carlson, 2010).

The abrupt changes apparent during the last glacial period (between 80,000 and 11,700 years ago) are also apparent in higher resolution marine and terrestrial records that span the past four glacial–interglacial cycles (Martrat et al., 2007), and so are a pervasive feature of past climatic variability. The particular abrupt changes of the last deglaciation are also present in some previous glacial terminations (i.e., Termination III, ~250,000 years ago), and so these abrupt changes are also not unique to the last glacial–interglacial cycle (Cheng et al., 2009).

During the Holocene (i.e., since 11,700 years ago), there are also examples of abrupt changes, some of these related to deglaciation (such as the '8.2 ka event' visible in Figure 14.1d), but others related to the effects of insolation on mid-continental aridity and the strength of the monsoon (reviewed in Cook et al., 2008; see also Williams et al., 2010). In general, these tend to be regional in extent, rather than global. On the centennial and decadal timescales over the past 1000 years, abrupt changes in the form of multi-centennial droughts are also evident in palaeoclimatic records (Cook et al., 2008, 2010), related to ocean–atmosphere interactions.

14.3.2. Characteristics of Dansgaard–Oeschger (D–O) Cycles

The nature of the D–O events can be explored using time series of terrestrial and marine records, and mapped syntheses of data similar to those reviewed above (Vogelsang et al., 2001; Voelker, 2002; Harrison and Sanchez Goñi, 2010). A typical terrestrial (pollen) record that spans the past glacial interval and is of high enough resolution to reveal the structure of the D–O cycles is that from Lago Grande di Monticchio, Italy (Allen et al., 2000; Fletcher et al., 2010; Figure 14.17). The record shows alternations between grassland and steppe vegetation during the cold (GS) part of one D–O cycle, and forest, with temperate elements during the warm (GI) part, reflecting variation in climate between cool/dry and warm/moist conditions (Allen et al., 2000; Fletcher et al., 2010). This pattern is superimposed on longer-term climatic variations related to orbital timescale variations in global ice volume, with more steppic vegetation during cool stages (i.e., Marine Isotopic Stages 4, MIS 4, ~74,000 to 59,000 years ago, and MIS 2, ~27,500 to 15,000 years ago), and more forested vegetation during warm stages

(i.e., MIS 3, ~59,000 to 27,500 years ago) (see also Berger and Yin, 2012, this volume).

These variations, which involve biome-level variations in vegetation (as opposed to more modest changes in species abundance), are registered in terrestrial records globally (Figure 14.17). In the northern extra-tropics, in particular, the difference in vegetation between a GI and GS consists of one or more 'steps' along a vegetation continuum between forest and steppe, or between tropical, warm, temperate, or boreal forest — changes in vegetation that encompass much of the total range of vegetation change between full glacial and interglacial conditions. Furthermore, the rapidity of the change in vegetation seen in these records attests to the rapid response of vegetation to the underlying change in climate.

The different components of the climate system adjust to, or participate in, these rapid climate changes with very little delay. The characteristic response of different variables can be seen by superimposing segments of (appropriately detrended) individual time series, aligning them relative to key times (such as the times of the GIs and GSs), a procedure known as superposed epoch analysis. If there is a consistent response, then this emerges as a distinct pattern in the average of the superimposed series; if not, the individual events cancel one another out, and no pattern appears. Confidence limits for the average series can be calculated using a Monte Carlo approach.

The NGRIP Greenland oxygen isotope record (Figure 14.18a), an index of regional temperature (North Greenland Ice Core Project Members, 2004; Svensson et al., 2008), shows the characteristic saw-tooth pattern of an individual D–O cycle, and all of the other records show distinctive responses associated with the occurrence of abrupt warming or cooling. The responses of all time series to the events are non-linear (linear responses would appear as inverted mirror images). The individual responses to abrupt warming can be categorized as *rapid* with no appreciable lag, such as those for methane (Loulergue et al., 2008; Figure 14.18b), dust (Lambert et al., 2008; Figure 14.18e), and biomass burning (Daniau et al., 2010; Figure 14.18f), all influenced by the hydrological status of the land surface, the nature of the vegetation cover and vegetation productivity, and *progressive* such as those for CO_2 (Ahn and Brook, 2008; Figure 14.18c) and N_2O (Figure 14.18d), which apply to atmospheric constituents with longer lifetimes. The responses to rapid cooling are similarly mixed, with those for CH_4 and dust again abrupt and rapid with no lag, and those for CO_2 and N_2O more gradual. There is an initial decrease of biomass burning in response to cooling, followed by a gradual recovery.

The superposed epoch analysis curves show that all parts of the climate system examined here co-vary with the

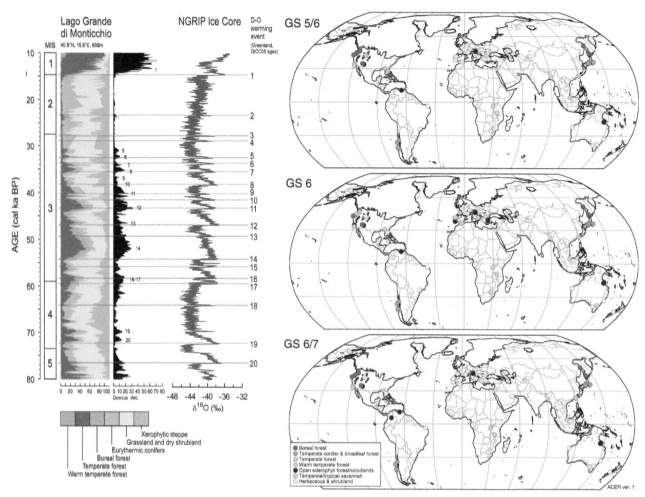

FIGURE 14.17 Vegetation changes associated with Dansgaard–Oeschger (D–O) cycles during the last glacial. The record from the Lago Grande di Monticchio, Italy *(Source: redrawn from Fletcher et al., 2010)* shows oscillations between grassland or steppe during colder intervals and forest during warm intervals. These changes can be related to the D–O cycles as registered by changes in $\delta^{18}O$ in the NGRIP ice core *(Source: redrawn from Wolff et al., 2010)*. The individually numbered D–O warming events are also shown. The maps show the biome registered at individual sites during a single D–O cycle (D–O 6): Greenland Stadial (GS) 6/7 through Greenland Interstadial (GI) 6 and into GS 5/6 *(Source: redrawn from Harrison and Sanchez-Goñi, 2010)*.

D–O events. Furthermore, the feedbacks from these co-variations are all positive, reinforcing either the warming or cooling recorded by the ice-core records. A general observation could be made that the response to the D–O changes is rapid, especially for those processes where changes in hydrology, productivity, or biophysics are involved, and are somewhat longer for carbon- and nitrogen-cycle related responses.

14.3.3. Mechanisms for D–O Cycles

The mechanism most often invoked for the generation of D–O variability involves variations in the strength of the AMOC, and the changes in ocean heat transport associated with that circulation (Rial et al., 2004; Delworth et al., 2008). The AMOC is the vertical component of the circulation of the North Atlantic that is driven by

increases in the density (through decreases in temperature and increases in salinity) of surface water in the North Atlantic and its consequent sinking (Kuhlbrodt et al., 2007; Latif and Park, 2012, this volume). Heat transported by both the AMOC, and by atmospheric circulation, into the North Atlantic ultimately warms the whole of the Northern Hemisphere and adjacent southern tropics (Stouffer et al., 2006; Pitman and Stouffer, 2006). Because the AMOC is driven in part by increasing salinity of the North Atlantic (as water evaporates from the warm, northward-flowing Gulf Stream), the intensity of the circulation can be diminished by increasing the flow of freshwater into the North Atlantic from increases in precipitation relative to evaporation, melting sea and land ice, and from river flow into the Arctic basin and North Atlantic. Both modelling studies and palaeoclimatic observations suggest that the changes in

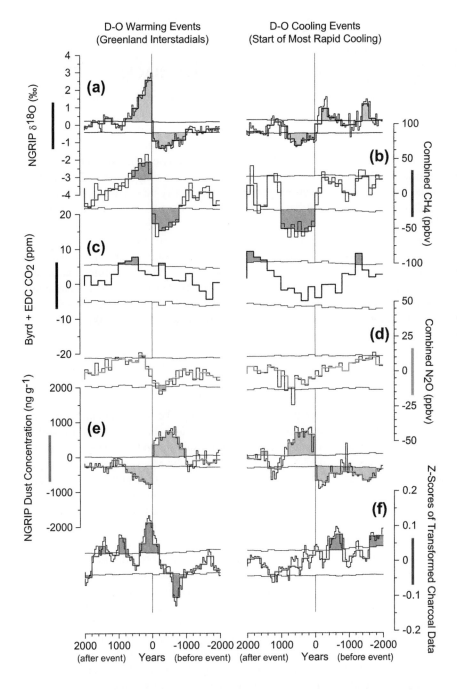

FIGURE 14.18 Superposed epoch analysis (SEA) of ice-core and biomass-burning records over the interval 80 ka to 10 ka. This analysis shows the consistent response of a time series to the repeated occurrence of the abrupt warming and cooling events that define the D−O cycles during the last glacial (MIS 4, 3, and 2). Shading indicates significant patterns in the response of the time series to the events of abrupt warming and rapid cooling. (a) The Greenland oxygen isotope record, an index of regional temperature, shows the characteristic saw-tooth pattern of an individual D−O cycle; (b−f) Distinctive responses of greenhouse gases (CO_2, CH_4, N_2O), dust, and fire associated with the occurrence of abrupt warming or cooling. BYRD: Byrd station, Antarctica; EDC: EPICA Dome C; NGRIP: North Greenland Ice Core Project. Redrawn from Arneth et al. (2010), see text for data sources. The curves show that all of the components of the climate system represented by these variables respond or co-vary with the abrupt changes recorded in the Greenland oxygen isotope record, with variables that describe terrestrial ecosystems or surface hydrology (dust, CH_4, biomass burning) responding relatively rapidly.

intensity can be abrupt, involving both threshold exceedance and hysteresis (Ganopolski and Rahmstorf, 2001; Rind et al., 2001; Roche et al., 2004; Braun et al., 2005; Rahmstorf et al., 2005; Clement and Peterson, 2008). In a typical progression through a GS and GI pair, the AMOC gradually slows or collapses producing the cold stadial (or Heinrich event) and then abruptly resumes, producing the rapid warming. However, the magnitude of the variations in greenhouse gases and dust accompanying the abrupt warming or cooling (Figures 14.18a−14.18e), indicates that the radiative

forcing by these controls of the energy balance, when further amplified by, for example, water vapour, cloud, and land-cover changes, must contribute significantly to the amplitude of the D−O variations.

14.3.4. Spatial Patterns of D−O Cycles

As might be expected, the D−O variations are strongly expressed in the circum-North Atlantic region, but Figure 14.17b suggests that the variations are expressed globally to one extent or another. There are three potential mechanisms for 'transmission' of the North Atlantic-focused climatic variations: (i) transmission by the AMOC itself, (ii) transmission via atmospheric circulation changes, and (iii) transmission via changes in atmospheric composition and its influence on radiative forcing; or some combination of these mechanisms. Transmission by changes in the AMOC circulation is consistent with the idea of a 'bipolar see-saw' in which the high-latitude climates of both hemispheres vary in opposition with one another. When the AMOC circulation shuts down, or diminishes in intensity, the heat formerly transported to the high northern latitudes would remain in the tropics (and presumably also in the Southern Hemisphere). This mechanism has been invoked to explain the apparent inverse correlation between millennial-scale variations in the Arctic and Antarctic ice cores (EPICA Community Members, 2006). Transmission by atmospheric circulation changes has been proposed, motivated by the large impact that the North American ice sheet and North Atlantic sea-ice have on atmospheric circulation of the Northern Hemisphere, including that of the monsoon regions (Clark et al., 2002). Although plausible changes in the Laurentide Ice Sheet that accompany Heinrich events evoke a large-scale response in an atmospheric model, it is not clear whether the same responses would occur for smaller variations in the ice sheets or in a coupled atmosphere—ocean model. The potential of the third mechanism, changes in atmospheric composition, can be seen in the superposed epoch analysis (SEA) (Figure 14.18), where the effects of the changes in greenhouse gases and dust accompanying both the warming and cooling events would be almost instantaneous (on this timescale) around the globe.

Although the polar ice core records support the idea of opposition (and the AMOC circulation mechanism), the latitudinal extents of the opposition cannot be ascertained from those records alone. Analyses of networks of palaeoclimatic time series suggest, however, that the latitudinal extent of the opposition is indeed restricted. Shakun and Carlson (2010) analysed 104 high-resolution records (primarily marine, but including some terrestrial records) spanning the last deglaciation and found that over 60% of the total variability among the records had a coherent global pattern that was similar in sign, while a hemispherically contrasting pattern explained only 11% of the total variability. The time series describing the expression of the first pattern could be related to the general trend of both CO_2 concentration and sea level over the interval, while that for the second pattern resembled an indicator of AMOC strength. Clark et al. (2007) examined 39 records spanning MIS 3 and found a similar ordering of modes, with the 'northern' (i.e., general global) mode again more important than a hemispherically contrasting one. These results support the transmission of the D—O variations by the atmospheric circulation or atmospheric composition

mechanisms, but do not preclude a role for AMOC changes in their origin.

14.4. BIOSPHERE FEEDBACKS

Palaeo-environmental records document large-scale changes in vegetation, surface hydrology, and other land-surface properties in response to changes in external forcing. These changes tracked changes in climate with no discernible lag, even in response to rapid climate changes (Arneth et al., 2010; Harrison and Sanchez Goñi, 2010). Such readjustments necessarily affected the surface-water and surface-energy balances, leading to feedbacks to climate at a regional scale. Biophysical feedbacks have been identified as playing an important role in, for example, maintaining differences in the ratio of land/sea responses to climate forcing (Lainé et al., 2009), amplification of high latitude climate changes (Foley et al., 1994; Ganopolski et al., 1998; Jahn et al., 2005), and monsoon enhancement (Kutzbach et al., 1996; Schurgers et al., 2007; Dallmeyer et al., 2010; Dekker et al., 2010).

High latitude climate warming causes expansion of forest at the expense of tundra vegetation. In addition to being darker, trees shelter the surface, thus reducing the contribution of snow cover to surface albedo and leading to increased surface warming. The immediate effect of these changes in albedo (and associated changes in surface roughness and evapotranspiration rates) is large, potentially doubling the orbitally induced warming; but perhaps a more important consequence of such biosphere feedbacks is on the persistence of the influence of the climate forcing. Model-based studies have shown that vegetation feedback leads to year-round warming (de Noblet et al., 1996; Ganopolski et al., 1998; Wohlfahrt et al., 2004), effectively reversing the impact of direct orbital forcing on winter temperatures. Model-based studies of the Northern Hemisphere monsoons have also shown that land-surface feedbacks affect precipitation seasonality, prolonging the monsoon season into the autumn when the direct effect on insolation forcing is waning (Broström et al., 1998; Braconnot et al., 1999; Irizarry-Ortiz et al., 2003; Dallmeyer et al., 2010). Thus, biospheric feedback provides a mechanism for producing counterintuitive responses to orbital forcing.

Biophysical feedbacks have also been implicated in the generation of abrupt responses to gradual changes in insolation. Studies with simplified climate models, for example, have suggested a role for vegetation—climate feedback in hastening the apparent collapse of the northern African monsoon around 5.4 ka (e.g., Claussen et al., 1999, Renssen et al., 2003), although this mechanism is not supported by more recent simulations with coupled atmosphere—ocean—vegetation models (Liu et al., 2006, 2007; Braconnot et al., 2007b).

Vegetation-controlled changes in atmospheric dust concentrations have also been implicated in rapid climate change, through both biophysical and biogeochemical mechanisms. During Heinrich Stadials and the last deglaciation, for example, atmospheric dust concentration decreased rapidly (deMenocal et al., 2000; Peck et al., 2004; Mulitza et al., 2008), implying a rapid decay of both its net radiative cooling effect (Claquin et al., 2003) and the aeolian supply of iron to the Southern Ocean, which may have increased marine export production and thus contributed to keeping CO_2 low during cold climate phases (Bopp et al., 2003). Analyses of records of marine export production records during the last interglacial and glacial periods show that iron fertilization cannot explain the initial glacial drawdown in CO_2 but could be responsible for the further 15–20 ppmv lowering that occurred during the glacial maximum (Kohfeld et al., 2005).

Carbon cycle feedbacks involving the terrestrial biosphere are potentially important on various timescales. A pervasive *negative* feedback results from the CO_2 fertilization effect on terrestrial primary production. This effect is needed to explain the 300–700 PgC increase of terrestrial carbon storage from the LGM to the Holocene, as indicated by $\delta^{13}C$ in marine benthic foraminiferal tests (Bird et al., 1996; Prentice and Harrison, 2009). This storage increase opposed the CO_2 rise, by taking additional carbon out of the atmosphere. Rapid terrestrial biosphere growth has also been invoked to account for the observed (transient) dip in atmospheric CO_2 concentration that culminated around 8000 years ago, after the initial deglacial rise (Joos et al., 2004). On the other hand, warming alone is expected to reduce terrestrial carbon storage (a positive feedback: Denman et al., 2007; Dickinson, 2012,

this volume). This mechanism has been invoked to explain the observed dip in CO_2 concentration during the Little Ice Age (Figure 14.19: Joos and Prentice, 2004; Friedlingstein and Prentice, 2010). Furthermore, the oceanic $CaCO_3$ compensation mechanism may partially explain the steady (about 20 ppmv) rise in atmospheric CO_2 that took place during the Holocene, after 8 ka (Joos et al., 2004). This mechanism involves the slow (multimillennial-scale) response of marine $CaCO_3$ sedimentation to the extraction of CO_2 from the ocean–atmosphere system as the deglaciation progressed.

Modelling of carbon cycle changes over these timescales has generally neglected the role of peatlands, yet these are a significant and dynamic carbon store (Figure 14.19). Northern peatlands cover around 4 million km^2, contain an estimated 545 PgC (Yu et al., 2010), and also contribute as much as 5–20% of total contemporary CH_4 emissions (Zhuang et al., 2004; Denman et al., 2007; Harvey, 2012, this volume). Peatland growth has been estimated to produce a long-term carbon sink of 0.07–0.1 PgC per year (Dean and Gorham, 1998; Yu et al., 2010). There was little or no peatland in the high northern latitudes during the LGM (Figure 14.19). New peatland growth over large areas began during the deglaciation; almost half of the modern area of peatlands accumulated before 8 ka (Figure 14.19; MacDonald et al., 2006; Gorham et al., 2007; Yu et al., 2010). It has been estimated that peatland growth prior to 8 ka sequestered about 100 PgC, potentially contributing to the observed dip in CO_2 and also to the peak in atmospheric CH_4 in the early Holocene (Yu et al., 2010). Peatland productivity and decomposition are highly sensitive to temperature changes; hence, carbon sequestration in peatlands can vary on decadal to millennial timescales

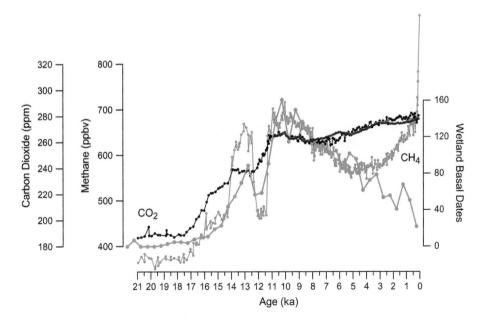

FIGURE 14.19 Relationship between peatland growth and carbon cycle indicators since the LGM (~21 ka). Basal radiocarbon dates on northern peatlands (blue dots, Harrison, unpublished data) provide an index for the timing of peatland initiation and spread. The CO_2 (black: Lüthi et al., 2008) and CH_4 (green: Loulergue et al., 2008) records are from Antarctic ice cores. Simulated changes in CO_2 during the Holocene (purple) from Joos et al. (2004).

(Singarayer et al., 2011). Increased accumulation rates are registered during the early to mid-Holocene thermal maximum (e.g., Kaufman et al., 2004), for example, and substantially reduced accumulation occurred during the cold intervals of the Younger Dryas and the Little Ice Age (e.g., Mauquoy et al., 2002; Charman et al., in review). The impact of changes in peatland accumulation on climate is uncertain because of the competing influences of changes in C sequestration and CH_4 emission and the timescales over which these operate. Frolking et al. (2006) have suggested that the initial impact of northern peatland growth is net warming that peaks after about 50 years after peatland initiation but remains positive for the next several hundred to several thousand years, depending on the rate of carbon sequestration, although in the longer term this would change and peatlands would have an increasing net cooling impact. However, these calculations were made without taking into account the impact of climate variability on peatland growth, carbon uptake, and CH_4 emissions.

Analyses of inter-hemispheric gradients in ice core CH_4 concentrations and carbon isotope composition ($\delta^{13}C$-CH_4) have been interpreted as suggesting that changes in wetland emissions drove glacial—interglacial changes in CH_4 concentration (Chappellaz et al., 1993, 1997; Schaefer et al., 2006; Harvey, 2012, this volume). However, published simulations to date using simple formulations of wetland extent and emissions have been unable to reduce wetland sources sufficiently to effect the observed changes in atmospheric CH_4 concentration (Kaplan, 2002; Valdes et al., 2005). It has been speculated that rapid shifts in atmospheric CH_4 during the last glacial (associated with D—O events) and after the deglaciation were caused by changes in Northern Hemisphere wetland emissions (Brook et al., 2000; Dallenbach et al., 2000; Korhola et al., 2010), although this finding is not consistent with recent modelling experiments (Singarayer et al., 2011).

The large-scale changes in vegetation distribution and productivity, and in fire regimes, shown on both glacial—interglacial and millennial timescales could potentially have had significant impacts on other biogeochemical cycles, and most particularly on the release of atmospheric trace gases and aerosol precursors. Model simulations show a substantial reduction (about 40%) in non-methane biogenic volatile organic compound emissions at the LGM, resulting in an increased atmospheric sink for CH_4 and, thus, potentially helping to explain the reduction of methane during glacial intervals (Valdes et al., 2005). However, this result could be negated by the finding that isoprene emission is enhanced at low CO_2 concentration (Arneth et al., 2010). N_2O variations during warm periods have also been tentatively attributed to changes in vegetation (e.g., Flückiger et al., 2002; Spahni et al., 2005). Thonicke et al. (2005) suggested that biomass burning was reduced by only 25% globally but with a marked increase

in fire in the equatorial zone. The overall reduction in pyrogenic emissions was insufficient to explain observed changes in atmospheric composition. However, Thonicke et al. (2005) argued that enhanced NO_x emissions as a result of increased burning in the tropics could have increased the oxidizing capacity of the atmosphere and helped to explain the observed low atmospheric CH_4 concentrations during the glacial. This hypothesis has not been examined quantitatively. There remain large uncertainties regarding the control of atmospheric CH_4 and N_2O.

Models used to project the emissions of greenhouse gases consistent with stabilization of climate at different levels usually assume a quasi-linear behaviour of the climate system that does not produce abrupt changes, even if feedbacks are considered (House et al., 2008). However, this modelling approach would not fully account for the observational record of the past (Alley et al., 2003; Jansen et al., 2007), which shows that periods of stability or gradual change (e.g., due to orbital forcing) have been interrupted by rapid state transitions, with large-scale warming and cooling events linked to atmospheric circulation shifts that in some cases took less than a decade to complete (Steffensen et al., 2008). Abrupt transitions and events have occurred during periods when changes in external drivers were much more gradual. Furthermore, rapid changes have commonly been associated with changes in atmospheric composition, including changes in atmospheric concentrations of greenhouse gases (CO_2, CH_4, N_2O) and aerosols (e.g., black carbon, mineral dust) that indicate changes in the land biosphere and/or the circulation and biogeochemistry of the oceans, with the potential to reinforce the climate change. For example, concentrations of these three greenhouse gases closely tracked polar temperatures during the global warming intervals that initiated the Holocene and last interglacial periods (EPICA Community Members, 2004). The change in CO_2, especially, was large enough to contribute substantially to the subsequent maintenance of warm conditions, implying a positive feedback that presumably contributed to the rapidity of the warming (Jansen et al., 2007). Ice core records show that atmospheric composition has tracked changes in climate at least over the past 800,000 years (EPICA Community Members, 2004; Spahni et al., 2005; Lambert et al., 2008; Loulergue et al., 2008; Lüthi et al., 2008), with the phasing of changes in individual greenhouse gases modulated by differences in the temporal and spatial patterning of biospheric feedbacks (see e.g., Flückiger et al., 2002).

14.5. LESSONS FROM THE PAST FOR THE STUDY OF CLIMATE CHANGES

The palaeorecord is rich in information, providing opportunities for exploring the mechanisms of climate changes and cautioning against simplistic or single-factor explanations for such changes.

Climate changes impact on all aspects of the environment, which provides a wealth of different possible indicators or sensors. However, each of these sensors responds to a different set of climate controls and on different timescales. The hierarchical nature of the climate system means that changes in these climate controls can be brought about in many different ways. Thus, individual sensors may apparently display congruent or non-congruent responses depending on how the overarching controls impact on the direct controls of the individual sensors. The inherent complexity of the climate system makes it important to develop a mechanistic understanding of the direct controls on individual sensors in order to be able to interpret the records in terms of climate change.

The hierarchical nature of the climate system also makes it clear that simple, single-factor climate explanations for changes in palaeorecords are inherently likely to be wrong. This point applies equally to geochemical and biological sensors. By exploiting the fact that individual sensors have different controls, and that no one sensor is more closely tied to climate than any other, we should be able to exploit multiple sensors to reconstruct a more complete picture of the nature of climate change at any time. As a corollary, forcing multiple sensors to reconstruct the same climate variable (e.g., July temperature) involves both a loss of explanatory power and an incomplete understanding of how the climate system works.

The hierarchical nature of climate is also apparent in terms of spatial patterns. Large-scale controls can be registered quasi-globally, but regional atmospheric circulation patterns and local factors (e.g., the presence of water bodies, complex topography) modulate the global signal. Processes affecting the sensor, and that themselves are influenced by climate, may further modulate the global signal. As a consequence of the mediation of large-scale controls by local features and processes, climate changes display spatial patterns at multiple scales. Furthermore, simple patterns in climate space may not map into simple patterns in geographical space. The ice-core records of well-mixed greenhouse gases are an important exception, but most records (including ice-core records of temperature) can only represent local or regional climate signals. Consequently, iconic or 'golden spike' records do not provide an adequate description of past climate changes. Such records should not be extrapolated to continental, hemispheric, or global-scales. Large-scale data syntheses frequently disprove simplistic interpretations based on one or a few sites. Large-scale, large magnitude climate changes may produce a homogeneous response, but spatial heterogeneity is much more common.

A further consequence of the non-stationarity and hierarchical nature of climate is that teleconnection patterns must vary through time. This can be verified both in historical observations and palaeoclimate simulations. The interpretation of palaeorecords in terms of an apparent modern linkage to some distant phenomenon, in the absence of a mechanistic relationship, is therefore not justified.

Many authors have identified multiple cycles or periodicities in palaeoclimate time series. However, this does not make them real. Apparent cyclicity can be generated by simple time series models or by data-analytic techniques. While this does not preclude the existence of real cyclicity in palaeorecords, the only periodicities that can be explained mechanistically are those tied to orbital forcing and the seasonal cycle. Identification of periodicities does not provide insights into the mechanisms of climate change, although understanding of the mechanisms helps to distinguish between real and spurious climate cycles.

We have shown that orbital forcing, and concomitant re-organisation of the atmospheric and oceanic circulations, explains many aspects of the temporal and spatial variability of climate during the late Quaternary. Nevertheless, much remains unclear about the detailed mechanisms by which these initial forcings are translated into the observed patterns of spatial and temporal variability. While it is clear that orbital variations drive climate variability on glacial–interglacial timescales, for example, the precise cause of glacial initiation and the explanation of the strength of the 100 kyr cycle remain obscure. Similarly, D–O cycles are linked to changes in the AMOC but such changes fail to explain why this signal is quasi-global or the differences among D–O cycles and between D–O cycles and Heinrich events. Biosphere feedbacks are likely to play a role here, both in amplifying the impact of relatively small changes in forcing and, through their control of greenhouse gas concentrations, in translating a localized forcing into a global response. Biophysical feedbacks associated with the terrestrial biosphere, and carbon cycle feedbacks associated with both the marine and terrestrial biosphere, are now being taken into account in modelling. However, the exploration of how changes in biogeochemical cycles interact with one another, and how these changes impact on atmospheric chemistry, offers many more possibilities for explaining observed climate patterns. The development of Earth system models that allow the complexity of biogeochemical cycles to be taken into consideration is a research priority and will likely shed considerable light on the mechanisms of past climate changes.

The complexity of the climate system, the multiple inter-linkages between different components, the existence of feedbacks, and the high degree of spatial and temporal variability, all militate against explaining past climate changes from observations alone. The use of a hierarchy of mechanistic models in conjunction with appropriate

Mid-Holocene

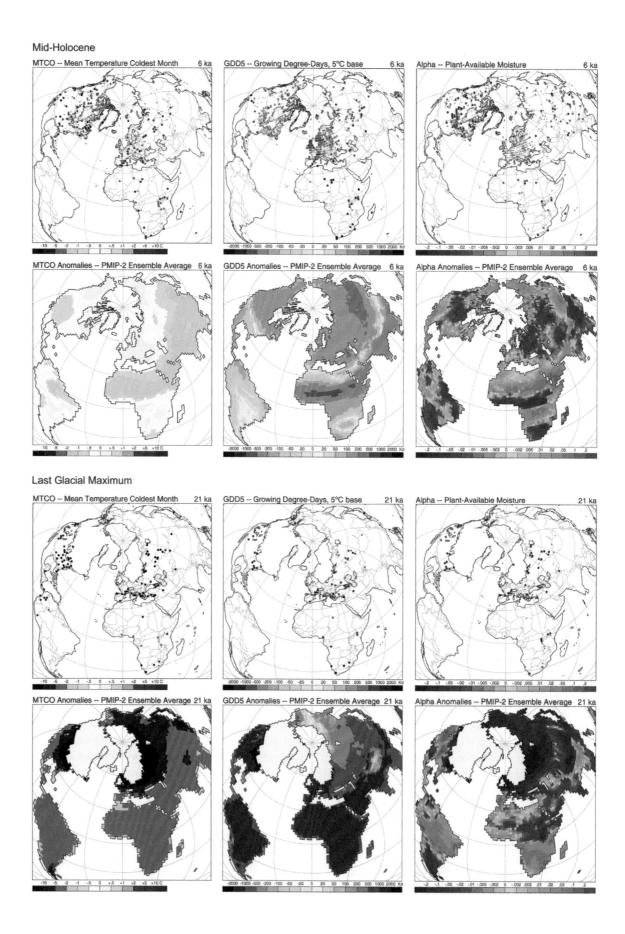

large-scale syntheses of data is the best way forward. Although this approach is often advocated, it has rarely been practised.

14.6. LESSONS FROM THE PAST FOR FUTURE CLIMATES

The palaeoclimate record shows climate variability well outside the range seen during the twentieth century, or indeed during the last two millennia. This means that the recent period does not provide an adequate sampling of how climate behaves. Changes in the controls of climate in recent decades, and those expected during the twenty-first century, exceed those of the last two millennia but are comparable in magnitude to changes seen on longer palaeoclimate timescales. Thus, on the one hand, past climate provides insights into mechanisms that cannot be studied in the historical past and, on the other hand, includes changes that are comparable in magnitude (though not cause) to future climate changes.

Changes in forcing that drive natural climate variability, and give rise to natural climate cycles, will not offset the impacts of anthropogenically induced climate changes. Ongoing changes in orbital parameters, for example, would not be expected to trigger the next ice age for 50 kyr (Berger and Loutre, 2002). Cyclicity on non-orbital timescales is not supported by the palaeoclimate record, and thus cannot be invoked as a potential mechanism to offset anthropogenically induced changes.

Nevertheless, the palaeoclimate record does show instances of short-term and rapid climate changes, both during cold-climate and warm-climate states. Evidence suggests that the MOC is weakening and will continue to do so, but it is unlikely that the MOC will collapse during the twenty-first century. Thus, the D–O cycles do not provide a guide to what might happen in the future. However, the palaeoclimate record shows that rapid re-organisations of the climate system trigger adjustments in biogeochemical cycles on timescales of decades to centuries. Thus, any mechanisms that generate rapid climate changes in the future will evoke a range of feedbacks that could substantially amplify the initial climate change, as they have in the past. Furthermore, these feedbacks have the potential to translate regional forcings into quasi-global responses.

Feedbacks play an important role in generating some of the large-scale patterns seen in simulations of future climate, most noticeably the fact that the high latitudes warm considerably more than other regions. The palaeo-climate record shows that the amplitude of climate variations at high latitudes was greater than that at low latitudes during both the mid-Holocene and the LGM. This suggests that polar amplification is a pervasive feature of the climate system. In both future and palaeoclimate simulations, this high-latitude warming is triggered by feedbacks associated with changes in the extent of sea-ice and biophysical changes in land-surface properties. However, the palaeorecord indicates that further amplification of the signal is likely through changes in the carbon cycle, in particular through changes in methane emissions from peatlands.

The ice-core record shows that atmospheric CO_2 has varied over a narrowly defined range (~180–280 ppmv) on glacial–interglacial timescales. The terrestrial biosphere takes up CO_2 during warming intervals, and thus opposes the observed trend, so the mechanism must lie in changes in the ocean carbon cycle. Although the details remain enigmatic, the asynchronous responses of the terrestrial biosphere and $CaCO_3$ compensation in the ocean broadly explain Holocene changes in CO_2. Furthermore, palaeo-observations in conjunction with biogeochemical modelling experiments place limits on the possible impact of ocean fertilization on glacial–interglacial timescales. While our incomplete understanding of the carbon cycle on palaeotimescales suggest a need to exercise caution when interpreting projections of future changes, the palaeorecord supports model results that indicate that afforestation/reforestation on land and iron fertilization of the ocean have a strictly limited role to play in mitigating future changes in CO_2 (House et al., 2008). Furthermore, $CaCO_3$ compensation operates on multimillennial timescales and so will not mitigate anthropogenic changes in CO_2 on centennial timescales.

The palaeorecord provides insights into the mechanisms of climate change that have direct relevance of our understanding of likely future climate changes. Palaeo-climate data, however, also have an additional role to play through the evaluation of state-of-the-art climate and Earth system models. This is best done through confronting model simulations with well-documented global-scale reconstructions of climate or environmental data (see e.g., Figure 14.20); such comparisons can provide a quantitative assessment of individual model performance, discrimination between models, and

FIGURE 14.20 Comparison of reconstructed and simulated mean temperature of the coldest month (MTCO), accumulated temperature above 5°C during the growing season (GDD5), and plant-available moisture as measured by the ratio between actual and potential evapotranspiration (alpha) at 6 ka and 21 ka. The reconstructions are from Bartlein et al. (2011) and the simulations are an ensemble average of the coupled ocean–atmosphere simulations runs in PMIP2 (Braconnot et al., 2007a). The ensemble-average simulations show anomaly patterns that are much smoother than the reconstructions. The reconstructions include spatial variations of climate that are not realizable in the simulations because of the coarse resolution of the models and the averaging across models.

diagnosis of the sources of model error. The Palaeo-climate Modelling Intercomparison Project (PMIP) is coordinating the systematic use of the palaeorecord for climate model evaluation at an international level. This evaluation should inform model development or improvement by individual modelling groups and identify high-impact priorities for data gathering and synthesis.

ACKNOWLEDGEMENTS

We thank Pat McDowell and Colin Prentice for providing support during the initial drafting of this paper; our colleagues in COHMAP and PMIP for stimulating discussions over several decades; Kenji Izumi for help with model output; and André Berger, John Dodson, and Colin Prentice for helpful comments on the manuscript.

Chapter 15

Modelling the Past and Future Interglacials in Response to Astronomical and Greenhouse Gas Forcing

André Berger and Qiuzhen Yin

Université catholique de Louvain Georges Lemaître Center for Earth and Climate Research (formely Institut d'Astronomie et de Géophysique G. Lemaître) SST/ELI/ELIC Louvain-la-neuve, Belgium

15.1. INTRODUCTION: INTERGLACIALS AND WARM CLIMATE

As we are presently in an interglacial (the Holocene) that is predicted to be exceptionally long (Crutzen and Stoermer, 2000; Berger and Loutre, 2002), past interglacials are particularly relevant to better understand our warm climate and its future (Harrison and Bartlein, 2012, this volume). Indeed, they provide a quite complete view of the range and underlying physics of natural warm climate variability (e.g., Tzedakis et al., 2009). They also provide insights into climate processes and feedbacks during warm intervals that are characterized by different combinations of climate forcings such as insolation, greenhouse gases (GHG), and

ice sheets (Harvey, 2012, this volume). It is therefore expected that investigating the climate response to these different forcings during the past interglacials will help to improve our estimate of the sensitivity of the climate system, a key point for the prediction of future climate change.

The succession of glacial–interglacial cycles is actually a prominent feature of the last three million years. From 3 to 1 Ma BP (million years before present), climate variations were characterized by a 41 ka (ka = 1000 years) quasi-cyclicity (Ruddiman et al., 1986a). After the Mid-Pleistocene Transition, at about 900 ka BP, this cyclicity became progressively 100 ka, a transition that was successfully simulated by Berger et al. (1999). In addition,

The Future of the World's Climate. DOI: 10.1016/B978-0-12-386917-3.00015-4

during the last one million years, marine and ice-core records show clearly that the amplitude of the glacial—interglacial cycles has significantly increased from around 430 ka BP (at the Mid-Brunhes Event, MBE) (Figure 15.1). Consequently, the pre-MBE interglacials seem cooler — at least in Antarctica (Jouzel et al., 2007) — and probably more glaciated than the post-MBE ones.

It, therefore, appears interesting to analyse the difference between the characteristics of the interglacials before and after MBE (Yin and Berger, 2010, 2011), as a first step towards understanding the reason(s) for such a difference. Along the same lines, it is also important to quantify the climate response to the two primary forcings, insolation (incoming solar radiation) and GHG, of all nine individual

interglacials of the past 800 ka (Figure 15.1). During the last decade, efforts have been made to model the most recent interglacials: the Holocene (e.g., Ganopolski et al., 1998; Braconnot et al., 2000; Crucifix et al., 2002), the Eemian (e.g., Montoya et al., 1998; Kubatzki et al., 2000; Crucifix and Loutre, 2002; Otto-Bliesner et al., 2006) and the other post-MBE interglacials (Loutre et al., 2007). These efforts continue to be made, in particular, through international networks like the Palaeoclimate Modelling Intercomparison Project (PMIP, see Harrison and Bartlein, 2012, this volume). The research presented in this chapter is part of this global effort and extends these lines of research over the whole upper Pleistocene and Holocene. Firstly, the model and the strategy for the experiment

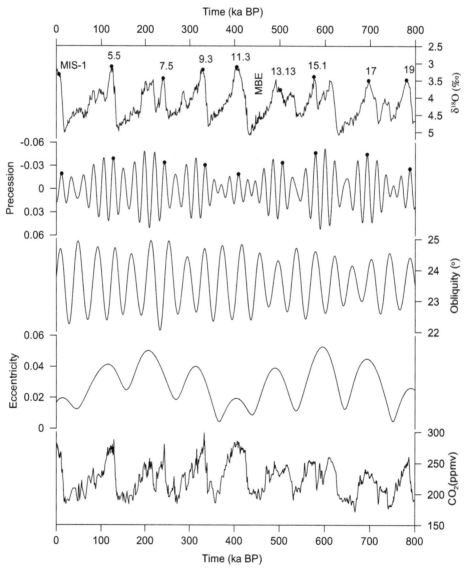

FIGURE 15.1 Benthic $\delta^{18}O$ (Lisiecki and Raymo, 2005), astronomical parameters (Berger, 1978) and CO_2 concentration (Luthi et al., 2008) of the last 800 ka. The points on the $\delta^{18}O$ curve indicate the selected peaks of the marine isotope stages, and those on the precession curve indicate the dates of NH summer at perihelion.

design used for simulating the climate of the last nine interglacial peaks are introduced in Section 15.2. Discussions on the astronomical parameters and insolation of the nine interglacials are given in Sections 15.3 and 15.4. In Section 15.5, the differences between the pre-MBE and post-MBE interglacials are discussed. In Section 15.6, the individual and combined contributions of insolation and GHG are analysed. In Section 15.7, discussions focus on the future of our current interglacial, and a summary is given in the final section.

15.2. MODEL AND EXPERIMENTS USED FOR SIMULATING THE LAST NINE INTERGLACIALS

The model we used to simulate the last nine interglacials is LOVECLIM[1]. It includes representations of the atmosphere, the ocean and sea-ice, the land surface (including vegetation), the ice sheets, and the carbon cycle. Its most recent version is documented in detail in Goosse et al. (2010). It is a three-dimensional Earth system model of intermediate complexity, that is, with coarser spatial resolution and simpler physical processes than that of state-of-the-art General Circulation Models (GCMs) (Claussen et al., 2002). Due to its reduced complexity, LOVECLIM can be used in millennial-scale transient climate simulations or large ensembles for which higher resolution coupled GCMs are too expensive. However, due again to its reduced complexity, LOVECLIM cannot be expected to reproduce all the observations with the same skill and the same level of detail as a GCM. Goosse et al. (2010) show that the most serious biases are mainly located at low latitudes with an overestimation of the temperature there, a too symmetric distribution of precipitation between the two hemispheres, and an overestimation of precipitation and vegetation cover in the subtropics. In addition, the atmospheric circulation is too weak and the ocean heat uptake observed over the last decade is overestimated.

In this study, the atmosphere (ECBilt), the ocean and sea-ice (CLIO), and the terrestrial biosphere (VECODE) components are interactively coupled, but the ice-sheet and the carbon-cycle components are kept to their present values. ECBilt is a quasi-geostrophic potential vorticity atmospheric model with 3 levels and a T21 horizontal resolution (Opsteegh et al., 1998). CLIO consists of an ocean general circulation model coupled to a comprehensive thermodynamic–dynamic sea-ice model (Goosse and Fichefet, 1999). Its horizontal resolution is 3° by 3°, and there are 20 levels in the ocean. VECODE is a reduced-

form model of vegetation dynamics and of the terrestrial carbon cycle (Brovkin et al., 1997). It computes the evolution of the vegetation cover described as a fractional distribution of desert, tree, and grassland at the same resolution as that of ECBilt. All the experiments start with the same initial conditions, which come from a quasi-equilibrium run of several thousands of years' duration, corresponding to the forcing applied in 1500 A.D. The simulations are 1000-years long, and the last 100-year average climatology is analysed. Various versions of ECBilt-CLIO, ECBilt-CLIO-VECODE, and LOVECLIM have been used in a large number of climate studies of the past, present, and future.

Such a model of intermediate complexity allows a large number of snapshot experiments to be undertaken in a reasonable period of time and computer cost. A snapshot experiment provides an instantaneous picture of the climate in equilibrium with the forcings and the boundary conditions. It has advantages and limitations. Its limitation comes from the long memory of the climate system, which implies that climate results not only from the external conditions prevailing at a single time but also from its past history (Harvey, 2012, this volume). Its main advantage is that it allows, rather easily, the use of atmosphere–ocean coupled models which otherwise remain costly to run in a transient mode. A large series of snapshot simulations could therefore be performed to intercompare the interglacials of the last 800 ka, a complementary approach to the transient simulation done by, for example, Ganopolski and Calov (2010).

The strategy for selecting insolation and GHG forcings for each of the last nine interglacials is the same as in Yin and Berger (2010). Firstly, the interglacials were identified by their peaks in the marine isotopic records $\delta^{18}O$ (Figure 15.1). In such palaeoclimate records, the glacials are labelled by even numbers and the interglacials by odd numbers. Over the last 800 ka, the interglacials are identified as Marine Isotopic Stages (MIS) 1 to 19 and some of them are subdivided in substages (e.g., MIS-13 with its MIS-13.3, MIS-13.13, and MIS-13.11). The $\delta^{18}O$ peaks can differ slightly in amplitude and in time from one record to another and, within a given record, the selection of the peaks is not necessarily straightforward. For MIS-19, MIS-17, and MIS-15.1, the peak is unique and well-defined. For MIS-13.1, MIS-11.3, MIS-9.3, MIS-7.5, and MIS-5.5, the different peaks extend over periods of time ranging from 2 to 10 ka. The most difficult problem is raised by MIS-15 and MIS-7. In both cases, two main peaks of about equivalent magnitude are present and separated by about 30 to 40 ka. In MIS-15, in addition to MIS-15.1, there is a single peak at 610 ka BP (MIS-15.3). In MIS-7, in addition to MIS-7.5, there is a broad peak covering 17 ka with maxima at 217 ka BP (MIS-7.3) and at 200 ka BP (MIS-7.1). The reason we first selected MIS-7.5 and MIS-15.1 for these two interglacials is because they

---

1. LOch-Vecode-Ecbilt-CLio-agIsm Model: More information about the model and a complete list of references is available at http://www.astr.ucl.ac.be/index.php?page=LOVECLIM%40Description

correspond to absolute $\delta^{18}O$ minima. Our selection is also based on the assumption that the astronomical forcing is partly responsible for the interglacials and leads to about 100 ka between them (Imbrie et al., 1993). However, MIS-15.3 and MIS-7.3 deserve further attention.

Secondly, following the hypothesis that an interglacial is caused by a maximum insolation in the Northern Hemisphere (NH) during its summer (Kukla et al., 1981), the astronomical parameters (Berger, 1978) were taken at the dates when NH summer occurs at perihelion just preceding the $\delta^{18}O$ peaks. Table 15.1 shows that these dates correspond quite well to the peaks of the $\delta^{18}O$-curve if we accept that the response time of the climate system to the astronomically-induced insolation is a few thousands of years. This lag between NH summer at perihelion and the peak of the interglacials means also that warm climates occur when NH autumn is close to perihelion. This phase agreement between the $\delta^{18}O$ proxy data and the astronomical parameters could have been expected from an astronomically tuned record, but there are two exceptions. One is during MIS-17 where NH summer at perihelion occurred 3 ka later than the $\delta^{18}O$ peak (Figure 15.2). The other, more serious, one is during MIS-13.1. At that time, NH summer occurred at perihelion 506 ka ago, a date more coherent to MIS-13.13 (501 ka BP) of SPECMAP (Imbrie et al., 1984) than to MIS-13.11 (491 ka BP) of Lisiecki and Raymo (2005). Indeed just before MIS-13.11, 495 ka ago, NH summer occurred at aphelion, a situation associated more naturally with a cold period than with a warm one. This 'anomalous' relationship between insolation and the interglacial peaks at MIS-13 and MIS-17, if not found in other proxies and/or climate simulations, will have to be further examined by testing the accuracy of the timescales

and/or of the intensity of the peaks. On the other hand, if Lisiecki and Raymo (2005) data are confirmed, it is the straightforward relationship between climate and the astronomical forcing that will have to be revisited and/or complemented. MIS-13.13 was first selected because we were primarily interested in a sensitivity analysis to the astronomical forcing corresponding to NH summer at perihelion, in this case occurring at 506 ka BP, but further analyses involve the whole MIS-13.1.

This selection of the interglacial peaks provides a more homogenous ensemble of experiments than if we had selected the dates of the $\delta^{18}O$ peaks themselves. Consequently, our work can be considered as a sensitivity analysis to the astronomical forcing under the hypothesis that NH summer at perihelion drives the interglacial climates.

Thirdly, for the GHG forcing, CO_2 (Luthi et al., 2008), CH_4 (Loulergue et al., 2008), and N_2O (Schilt et al., 2010) were taken into account in our simulations. In order to maximize the response of the climate model, their concentrations were taken at the CO_2 peaks just preceding the $\delta^{18}O$ ones (Yin and Berger, 2010). To test the robustness of our results, we have also used an average over a few thousands of years around the peaks of CO_2, CH_4, and N_2O just preceding the $\delta^{18}O$ peaks or even the maximum concentration of each gas (supplementary materials of Yin and Berger, 2010). Such sensitivity experiments to the GHG concentrations show that our conclusions are not affected by the strategy used.

Finally, to isolate the pure contributions of insolation and of GHG, the factor separation technique by Stein and Alpert (1993) was used (see also Alpert and Sholokhman, 2011). This method allows the identification of the individual contribution of each factor to any climatic variable, as well as

TABLE 15.1 Astronomical Parameters and CO_2 Equivalent (CO_2e) Concentration of the Experiments f_{11} for the Past Nine Interglacials Indicated by Marine Isotope Stage (MIS; Shackleton, 1967)

| MIS | Dates of $\delta^{18}O$ Peaks (ka BP) | Orbital Parameters | | | CO_2e (ppmv) | June 65°N Insolation (Wm^{-2}) |
		Dates (ka BP)	Eccentricity	Obliquity		
1	6	12	0.019608	24.152	264	529
5.5	123	127	0.039378	24.040	284	549
7.5	239	242	0.034033	23.262	276	528
9.3	329	334	0.031539	24.239	300	543
11.3	405	409	0.019322	23.781	286	522
13.13	501	506	0.034046	23.377	237	531
15.1	575	579	0.047152	24.17	248	560
17	696	693	0.045154	23.438	234	544
19	780	788	0.026196	24.003	265	533

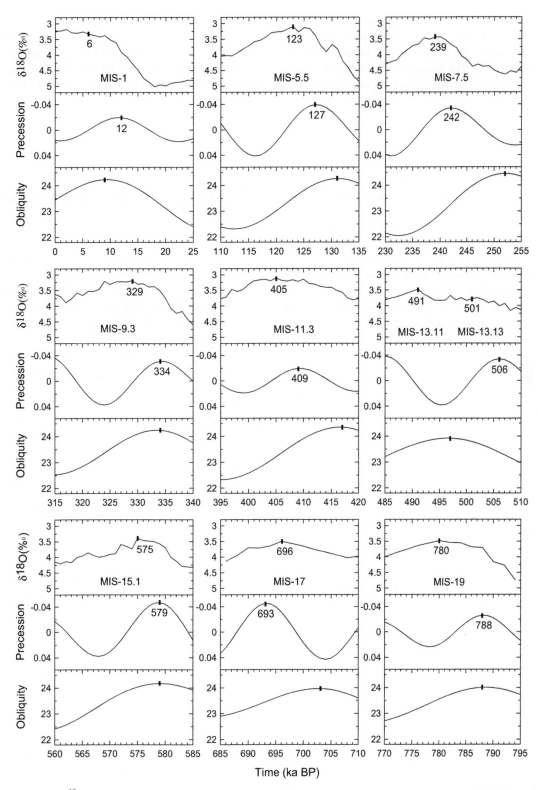

FIGURE 15.2 Benthic δ^{18}O (Lisiecki and Raymo, 2005), precession, and obliquity (Berger, 1978) for the last nine interglacials. The points identify δ^{18}O minima, NH summer at perihelion, and obliquity maxima. The dates of NH at perihelion (on the precession curve), δ^{18}O minima, and corresponding marine isotope stages (MIS) are indicated.

their synergistic effects. It has been used in many modelling studies to quantify the relative importance of specific processes and their interactions as, for example, for the Holocene (Berger, 2001; Claussen et al., 2002) and for the Eemian (Crucifix and Loutre, 2002). In all these studies, the factor separation method has been applied to components of the climate system such as the ocean, the atmosphere, and the vegetation; whereas, in this chapter, it is applied to the two climate forcings, insolation and GHG. This requires four experiments to be made that are labelled 00, 01, 10, and 11 for any climatic variable, f. In the reference experiment (leading to f_{00}), the averages of the astronomical parameters and of the GHG concentrations of the nine interglacials are used in order to make the intercomparison between the interglacials easier. In the control experiments (leading to f_{11}), both insolation and GHG are taken at their interglacial levels. In the other two simulations, only one factor is allowed to change at a time from its reference to its interglacial level (insolation in f_{10} and GHG in f_{01}). The pure contributions of insolation and of GHG to any climate variable are symbolized by $\widehat{f}_{10}$ and $\widehat{f}_{01}$, respectively, and the contribution of their interactions (synergism) is symbolized by $\widehat{f}_{11}$. According to Stein and Alpert (1993), we have:

$$\widehat{f}_{10} = f_{10} - f_{00} \qquad (15.1)$$

$$\widehat{f}_{01} = f_{01} - f_{00} \qquad (15.2)$$

$$\widehat{f}_{11} = f_{11} - (f_{10} + f_{01}) + f_{00}$$
$$= f_{11} - f_{00} - \widehat{f}_{10} - \widehat{f}_{01} \qquad (15.3)$$

In our experiments, the synergism between GHG and insolation is the additional change in a given climatic parameter that is obtained when GHG and insolation are both altered from their reference levels to their control ones, as compared to the sum of their pure contributions. It is actually equivalent to:

$$(f_{11} - f_{10}) - (f_{01} - f_{00}) \qquad (15.4)$$

or to:

$$(f_{11} - f_{01}) - (f_{10} - f_{00}) \qquad (15.5)$$

These differences are those between the sensitivity of the climate system to the change of one given factor (GHG in Equation (15.4) and insolation in Equation (15.5)) from its reference level to its control level when the other factor is kept at its control value and the sensitivity when the latter is kept at its reference value. In Equations (15.4) and (15.5), the first term is a measure of the traditional way to estimate the sensitivity to a parameter change, whereas the second is the pure contribution as defined in the Stein–Alpert method. It is clear that when the synergism is about zero, the

two techniques for calculating the sensitivity to a given parameter are equivalent (i.e., almost independent of the value of the other), which appears to be the case in this study.

Therefore, according to the factor separation method, each interglacial needs three experiments to be made, f_{11}, f_{10}, and f_{01}, which with the common reference experiment f_{00}, leads to a total of 28 experiments. $\widehat{f}_{10}$ and $\widehat{f}_{01}$ are primarily analysed in this chapter, the synergism being relatively small, as shown in Figure 15.11.

15.3. PRECESSION AND OBLIQUITY DURING THE INTERGLACIALS

The occurrence of glacials was associated with NH summer at aphelion, a large eccentricity, and a low obliquity by Milankovitch (1941). For the interglacials, Kukla et al. (1981) and Berger et al. (1981) tentatively associated them with NH summer at perihelion, a large eccentricity, and a large obliquity. Such simultaneous occurrences guarantee maximized energy received by the NH during its summer. A large eccentricity minimizes the Earth–Sun distance at perihelion and a large obliquity increases the energy received in high northern latitudes during their summer. The existence of such a large energy is assumed to lead to an interglacial peak after a few thousand years, as shown in Figure 15.2, that also provides the difference in phase between precession and obliquity at all interglacials with identifiable characteristics.

Firstly, for the peaks of MIS-1, MIS-5.5, MIS-9.3, MIS-15.1, and MIS-19, the precession minima and the obliquity maxima are more or less in phase and lead the $\delta^{18}O$ minima by about 5 ka (for MIS-1 the obliquity maximum lags behind the precession minimum by 3 ka; for MIS-5.5 it precedes the precession minimum by 3 ka; and for the other three, they are in phase). Three of these interglacials occur before MBE and three after, making a criterion based only on the phasing relationship between precession and obliquity difficult to use to distinguish between 'warmer' (after MBE) and 'cooler' (before MBE) interglacials.

Secondly, at the peaks of MIS-7.5, MIS-11.3, and MIS-17, obliquity maxima precede the precession minima by about 9 ka, making precession and obliquity almost in opposite phase. With its large eccentricity, MIS-17 is much more insolated than MIS-7.5 and MIS-11.3.

Finally, MIS-13.1 is the most puzzling. At MIS-13.11, obliquity and precession maxima are almost in phase and lead the $\delta^{18}O$ minima by about 5 ka, a situation that challenges the relationship between precession and climate, the interglacial being here associated with NH summer at aphelion (or SH summer at perihelion). For MIS-13.13, the precession minimum (at 506 ka BP) occurs 9 ka before the obliquity maximum, a situation similar to MIS-1, but with a larger phase shift.

This analysis of the precession—obliquity phase relationship leads to about all possible combinations and no straightforward conclusion can be drawn. Although it is difficult to see any difference in the astronomical elements of the interglacials before and after the MBE, it happens, however, that averaged over the five post-MBE interglacials, the values of eccentricity (0.028776) and of obliquity (23.90°) are significantly different from those calculated over the pre-MBE interglacials (respectively 0.038137 and 23.75°). This is related to both the fact that the 400-ka cycle of eccentricity is much stronger after the MBE than before, and to the 1.3 million years cycle in the amplitude modulation of obliquity (Berger et al., 1998).

15.4. LATITUDINAL AND SEASONAL DISTRIBUTION OF INSOLATION

As the climate model employed here (LOVECLIM) is actually forced by the latitudinal and seasonal distribution of insolation (Berger et al., 1993), the comparison of the insolation between the interglacials is worthy of discussion.

Firstly, as expected, there is a strong coherency between the insolation patterns at all the selected dates when compared to present-day, a coherency that comes from the insolation being calculated for NH summer at perihelion (see figure S2 in Yin and Berger, 2010). It also originates from the fact that the seasonal cycle is described in terms of the true longitude of the Sun (Berger, 1978).

Secondly, using NH summer at perihelion leads to the immediate conclusion that the insolation at the selected dates depends only on obliquity and eccentricity. Therefore, their values explain the differences between the different insolation patterns, which in Figure 15.3 are the deviations from the average of the last nine interglacials, but also lead to the following grouping:

(i) MIS-1, MIS-11.3, and MIS-19 are fairly similar with a low value of eccentricity around 0.02 (although for MIS-19 the eccentricity is slightly larger) and a moderate value of obliquity at around 24°. The higher obliquity of MIS-1 and MIS-19 counteracts the negative anomalies of the northern polar latitudes and reinforces the positive anomalies of the southern ones. The lower obliquity of MIS-11 does the reverse.

(ii) MIS-7.5 and MIS-13.13 are very similar with a moderate value of eccentricity around 0.034 and a low obliquity of about 23.3°. In both cases, the insolation pattern shows a tripole signature: the hemispheres are under-insolated during their local summers with a deep minimum over the poles, and over-insolated during their local winters with a maximum over the mid-latitudes, a signature of a lower obliquity.

(iii) MIS-9.3 is an interglacial with a unique insolation pattern that is the opposite to MIS-7.5 and MIS-13.13. Its eccentricity is moderate, but its obliquity is the largest (24.24°).

(iv) MIS-5.5, MIS-15.1, and MIS-17 have in common the largest value of eccentricity (of about 0.04 and a large value of obliquity, except for MIS-17 for which obliquity is low). This explains that (a) the maximum negative anomaly is centred at the South Pole during austral summer for MIS-17, whereas it is centred in the tropical/equatorial regions for the other two interglacials, and (b) the maximum positive anomaly is centred at the North Pole for MIS-5.5 and MIS-15, whereas it is at the Equator for MIS-17.

Thirdly, MIS-5.5, MIS-9.3, and MIS-15.1 are definitely the most highly insolated over the northern high latitudes during boreal summer. This could have been expected from the analysis of the astronomical elements themselves: MIS-1, MIS-5.5, MIS-9.3, MIS-15.1, and MIS-19 show a good correlation between precession minimum and obliquity maximum (Figure 15.2). However, the insolation is not large at MIS-1 and MIS-19 because of a low eccentricity, which is also the case for MIS-11.3. This low value is associated with the 400 ka cycle of eccentricity (Berger and Loutre, 2003). It was also at the basis of taking MIS-11 as a potentially good analogue for MIS-1 and at the origin of the prediction of an exceptionally long Holocene interglacial (Berger and Loutre, 2002). However, as the simulation will confirm (Section 15.6), MIS-19 appears to be an even better analogue of MIS-1 than MIS-11.

15.5. MODELLING THE GHG AND INSOLATION CONTRIBUTIONS TO THE DIFFERENCE BETWEEN PRE- AND POST-MBE INTERGLACIALS

Despite no systematic difference appearing between the insolation patterns of the individual interglacials of pre- and post-MBE, a clear difference is apparent when their averages are considered. This difference was quite unexpected (Figure 15.4a): NH summer is significantly under-insolated and SH summer overinsolated for post-MBE interglacials than for pre-MBE ones. This is mainly due to precession, but as here the longitude of the perihelion is fixed, the difference in precession results strictly from a difference in eccentricity. A smaller post-MBE eccentricity value leads to a larger distance at perihelion and therefore to a lower insolation during the NH summer all over the Earth. At the same time, however, the influence of a greater obliquity is felt because the high polar latitudes are less under-insolated than expected from a smaller eccentricity.

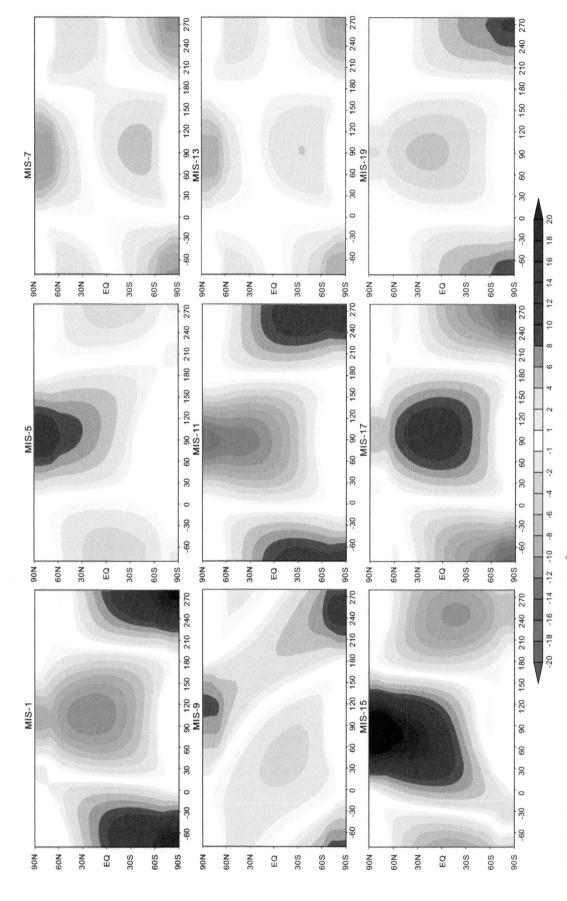

FIGURE 15.3   Latitudinal and seasonal distribution of insolation (Wm^{-2}) for the interglacials of the last 800 ka (deviation from the reference, which is the average of the 9 interglacials). The horizontal axis indicates the true longitude of the Sun from the beginning to the end of the year (0° and 180° are for the spring and autumn equinoxes; 90° and 270° are for the summer and winter solstices). Insolation is calculated from the long-term variations of eccentricity, precession, and obliquity (*Source: Berger, 1978.*)

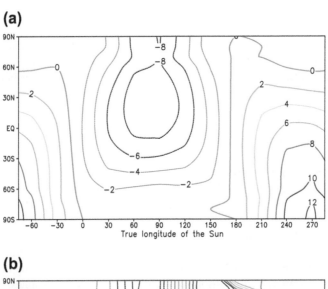

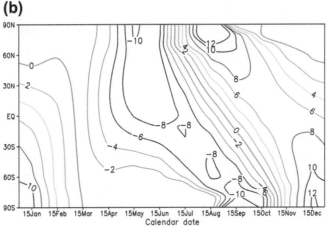

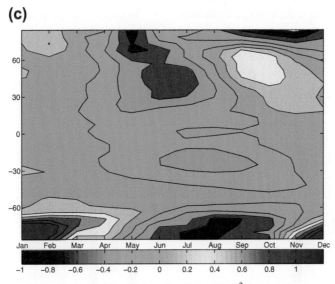

FIGURE 15.4 (a) Difference in the seasonal and latitudinal distribution of insolation (Wm^{-2}) between the average of the post-MBE interglacials and of the pre-MBE interglacials. The horizontal axis gives the true longitude of the Sun (as in Figure 15.3). (b) As (a), but the horizontal axis is calendar date. (c) Difference in the seasonal and latitudinal distribution of the insolation-induced temperature (°C) between the average of the post-MBE and pre-MBE interglacials.

The concentration in CO_2 equivalent characterizing the pre-MBE interglacials varies between 234 and 265 ppmv; for the post-MBE it varies between 264 and 300 ppmv. There is, on average, a difference of 36 ppmv between the post- and the pre-MBE interglacials.

Figure 15.5a shows the simulated global annual mean temperature for each of the last 9 interglacials (see Section 15.6.4) compared to their average, which is 16.2°C. The average temperature of the post-MBE interglacials (MIS-1 to MIS-11) is higher than that of the pre-MBE ones (MIS-13 to MIS-19) by 0.33°C. The post-MBE interglacials, except MIS-7, are all above their average value, with MIS-9 being the warmest and MIS-7, the coolest. The pre-MBE temperatures are all below the average except MIS-19 with MIS-13 being the coolest. As will be discussed later, the insolation patterns and CO_2 concentrations are well-imprinted in the climate of all the individual interglacials. Although the model used here is only forced by insolation and GHG, the amplitude of the annual mean temperature change between pre- and post-MBE interglacials is in reasonable agreement with the amplitude change of the marine $\delta^{18}O$ records (see Section 15.6.4.3). It is worth noting that, if an extra ice sheet is introduced in the interglacials before the MBE (as the $\delta^{18}O$ records might suggest), it will reinforce the results by further cooling the system.

At the seasonal and global scales, the post-MBE interglacials are on average 0.49°C warmer than the pre-MBE ones in December−January−February (DJF, Figure 15.5b), but only 0.12°C in June−July−August (JJA, Figure 15.5c). Moreover, in DJF, all post-MBE interglacials, except MIS-7, are well above the average, and all the pre-MBE ones, except MIS-19, are well below the average. Definitely, the warming on an annual average is mainly due to the

warming in DJF, which fits very well with the insolation pattern of Figure 15.4a. In JJA, the difference between the pre- and the post-MBE interglacials is much smaller and only MIS-5, MIS-9, and MIS-15 are above the average. Analysing the differences between the hemispheres during the same local seasons shows that the post-MBE interglacials are on average significantly warmer than the pre-MBE ones in the winter hemispheres (0.44°C for NH in DJF and 0.25°C for SH in JJA). For the summer hemispheres, the situation is more complex. The SH summer (DJF) is 0.54°C warmer during the post-MBE interglacials, but there is no evident difference between the post-MBE and the pre-MBE interglacials for the NH summer (JJA). In summary, the response to the astronomical and GHG forcings is, in general, more coherent in DJF. The climate of boreal winter, globally and in the Southern Hemisphere (SH) in particular, is definitely warmer during the post-MBE interglacials than during the pre-MBE ones, as a direct response to the astronomical forcing.

At the regional-scale, the post-MBE interglacials are annually warmer than the pre-MBE ones over almost the whole Earth, with the largest warming reaching 4°C over the high latitudes in both hemispheres. There is a slight cooling of less than 1°C over a small area north of eastern Asia and over western Australia. Seasonally (Figure 15.6), the post-MBE interglacials are on average warmer than the pre-MBE ones over the whole Earth in boreal winter, spring, and autumn, and cooler mostly over the continents, but warmer over the Southern Ocean in boreal summer (austral winter) due to the SH summer-remnant effect (see Section 15.6).

To understand the relative role of insolation and GHG in generating an average climate warmer over the post-MBE

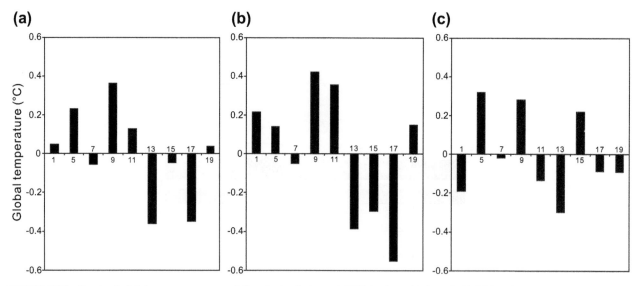

FIGURE 15.5 Simulated global mean temperatures (°C) under insolation and GHG forcings. (a) Annual, (b) DJF, and (c) JJA. Deviation from the respective averages (which are 16.2°C, 13.6°C, and 19.3°C for the annual mean, for DJF, and for JJA respectively) for the nine interglacials.

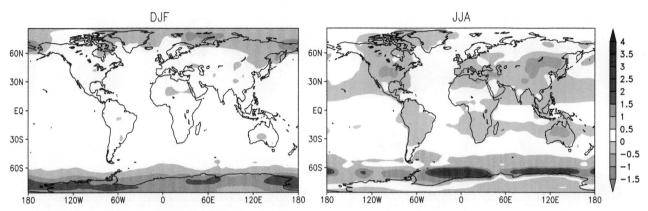

FIGURE 15.6 Geographical distribution of the simulated DJF and JJA temperature differences (°C) between the average of the post-MBE and pre-MBE interglacials, due to both the insolation and GHGs.

interglacials than over the pre-MBE ones, the factor separation method was used first for the insolation and GHG forcings averaged over the post-MBE and over the pre-MBE interglacials. The results show that the higher global annual mean temperature of the post-MBE interglacials is almost entirely due to its higher average GHG concentration (as might have been expected). However, this is not the case at the seasonal and hemispheric scales, where the insolation signature becomes highly significant in the response to the differential impact shown in Figure 15.4a. As the model is run in calendar days, the same pattern as in Figure 15.4a can be drawn along the calendar days (Figure 15.4b) instead of using the true longitude of the Sun, as in Figure 15.4a. The difference between Figures 15.4a and 15.4b is related to the variable length of the seasons: boreal summer/austral winter lasts longer and austral spring is shorter during post-MBE interglacials than during pre-MBE ones. As shown by Figure 15.4c, compared to the pre-MBE interglacials, insolation is responsible for the global Earth being warmer over the post-MBE interglacials during boreal winter, reinforcing the warming expected from the higher GHG concentrations. During boreal summer, the insolation deficit cools the Earth (Figure 15.4c), counteracting the global GHG warming.

If we now analyse the NH, insolation plays a very minor role in winter, but negative insolation anomalies largely cool the Earth in summer, leading to a slight annual cooling. It is different for the SH. Insolation warms the SH significantly during its summer, but cools the SH during its winter, leading to a slight annual warming over SH. The insolation change is therefore responsible for an increase in the difference between the NH and SH summers in agreement with the pattern of Figure 15.4a.

The relative importance of insolation and GHG on the amplitude difference between the pre-MBE and the post-MBE interglacials has been investigated further through the factor separation analysis for each individual

interglacials in Yin and Berger (2011). It is shown that the MBE is clearly seen in the simulated global annual mean temperature and in the temperature of the southern high latitudes in relationship to the dominant effect of GHG there. However, the dominant effect of insolation on the simulated temperature of the northern high latitudes prevents the MBE being obvious there. These results are confirmed by proxy data: MBE is clearly expressed in the marine $\delta^{18}O$ data (e.g., Lisiecki and Raymo, 2005) — a global proxy — and in the Antarctica temperature (Jouzel et al., 2007), but it is not present in some regional proxy records from the northern high latitudes (e.g., in the data of Ruddiman et al., 1986b).

15.6. GHG AND INSOLATION CONTRIBUTIONS TO THE INDIVIDUAL INTERGLACIAL CLIMATES

15.6.1. The Reference Climate

In the factor separation analysis, f_{11}, f_{10}, and f_{01} of each interglacial are compared to the reference experiment f_{00} for obtaining the combined effect of insolation and GHG, their pure contributions, and their synergism. To better understand the reference climate, we first compare it with the simulated pre-industrial climate that is described in Yin et al. (2008). The CO_2e concentration of f_{00} is 14 ppmv lower than the pre-industrial one (266 vs. 280 ppmv). Its eccentricity is larger (0.032936 vs. 0.016724) as well as its obliquity (23.829° vs. 23.446°). Its NH summer occurs at perihelion, a situation opposite to that in the pre-industrial period. Such a difference in the astronomical parameters leads to much more insolation received over the Earth during boreal summer and much less insolation during boreal winter in f_{00} than in the pre-industrial case. As a result, the f_{00} climate is 0.2°C warmer than pre-industrial for the global annual mean temperature. It is 3.3°C warmer during boreal summer (Figure 15.7a) and 2.5°C cooler

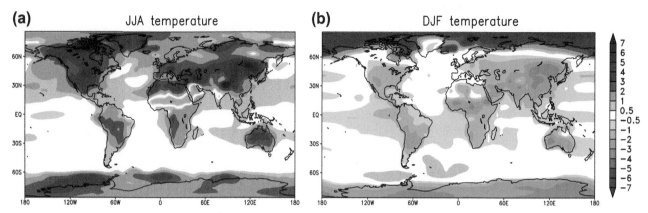

FIGURE 15.7   The differences in surface temperature (°C) between the reference experiment and the pre-industrial one in (a) JJA and (b) DJF.

during boreal winter (Figure 15.7b). During boreal summer, the lower CO_2e concentration reduces the warming due to insolation, and, during boreal winter, it strengthens the insolation-induced cooling. At the regional-scale, in boreal summer, f_{00} is warmer than pre-industrial over almost the whole Earth with the largest warming, by up to 5°C, over all the continents and the Southern Ocean (where it is austral summer). In boreal winter, f_{00} is much cooler over all the continents, but is warmer by up to 1°C over the Southern Ocean and by up to 5°C over the Arctic and its surrounding lands (the summer insolation remnant effect). Such a large difference must be kept in mind when using present-day climate as a reference.

15.6.2. Pure Contribution of GHG

As expected from the greenhouse effect, the temperature response to GHG follows very well the variation of the CO_2e concentration, with an almost perfectly linear relationship between the two (Figure 15.8a). MIS-9 has the highest CO_2e (300 ppmv) and the highest GHG-induced temperature. MIS-17 has the lowest CO_2e (234 ppmv) and the lowest GHG-induced temperature, 0.61°C below MIS-9. MIS-11 having a CO_2e concentration closer to MIS-5 than to MIS-1, partly explains why it is not as good an analogue of MIS-1 as MIS-19. The equilibrium climate sensitivity of LOVECLIM to a doubling of CO_2 at the

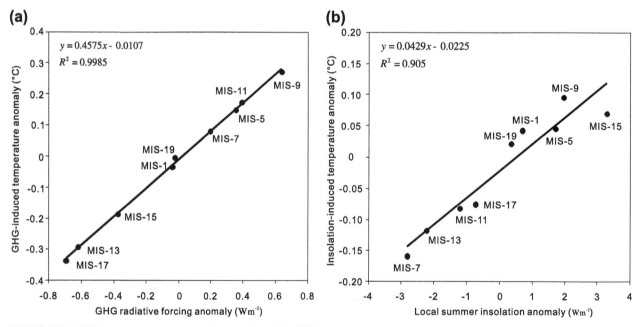

FIGURE 15.8   (a) Linear regression between the anomalies of the GHG-induced global annual mean surface temperature and of the GHG radiative forcing. (b) Linear regression between the anomalies of insolation-induced global annual mean surface temperature and of the average insolation calculated over the NH during boreal summer and over SH during austral summer. All anomalies are deviations from the reference case. Points show individual MISs.

pre-industrial condition is 1.6°C. Being at the lower end of the range obtained by other GCMs (Randall et al., 2007; Harvey, 2012, this volume), this low sensitivity might be responsible for our model underestimating the temperature response to GHG.

At the regional-scale, the largest impact of GHG on temperature is, for all the interglacials, over the high latitudes both annually and seasonally. This is a typical feature seen in many greenhouse experiments performed with coupled atmosphere—ocean models (e.g., Manabe and Stouffer, 1980; Boer et al., 1992; Murphy and Mitchell, 1995; Stouffer and Wetherald, 2007). The large impact of GHG on the high latitudes temperature is mainly due to the positive snow and sea-ice albedo—temperature feedback and to the difference between the lapse rates of low and high latitudes. In low latitudes, moist convection, indeed, tends to adjust the static stability towards the moist adiabatic lapse rate and this remains so after perturbation. Since the moist adiabatic lapse rate reduces with increasing temperature, it is expected that the CO_2-induced warming is larger in the upper troposphere than at the surface. This explains why moist convection is responsible for the small value of the warming in low latitudes. It is not the case in high latitudes, where more stable lapse rates prevent vertical mixing and favour larger surface warming. Annually (Figure 15.9), MIS-9 is warmer than the reference climate by up to 2.5°C over the Southern Ocean and 1.5°C over the Arctic. MIS-17 is cooler than the reference climate by up to 2.5°C over the Southern Ocean and 2°C over the Arctic. During boreal summer (Figure 15.9), the largest impact is over the Southern Ocean (where it is winter), with a warming of up to 4°C at MIS-9 and a cooling of up to 4°C

at MIS-17. The impact of GHG over the Antarctic continent is less and it is quite small over the Arctic Ocean. During boreal winter (Figure 15.9), the sensitivity to GHG change over the northern high latitudes, including the Arctic and its surrounding lands increases, but it decreases over the southern high latitudes (where it is summer). At MIS-9, the maximum warming reaches 2.5°C over the Arctic and 2°C over Antarctica and, at MIS-17, the maximum cooling reaches 2.5°C over both these regions. The larger sensitivity to GHG change is therefore definitely in high latitudes and, over the polar oceans, it is larger during the local winter than during the local summer.

This seasonal behaviour of the GHG-induced temperature change over the polar oceans is consistent with earlier CO_2-increase simulation results (e.g., Manabe and Stouffer, 1980), where the thermal insulation of sea-ice and the large heat capacity of the ocean both play an important role. Figure 15.10a shows the seasonal march of the difference due to GHG between MIS-9 (taken as an example of high-CO_2 interglacial) and the reference experiment in the surface heat balance components. In boreal summer, the shortwave radiation changes because the sea-ice—snow—vegetation albedo induced by the GHG changes. The strong energy excess is transferred directly downward to the sea-ice—ocean system, enhancing the melting of ice and warming the upper ocean. As a consequence, it does not contribute to the summer warming of the model atmosphere, which explains the very small GHG-induced surface temperature change in summer. The additional heat received by the ocean delays the formation of sea-ice and reduces its thickness during the following winter. Such a reduction of the thermal insulation effect of

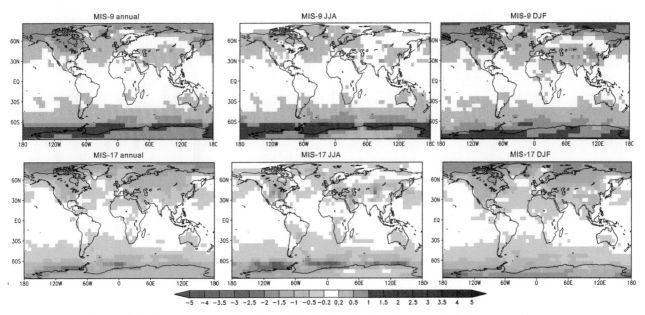

FIGURE 15.9 Pure impact of GHG on the annual, JJA, and DJF surface temperature (°C) for MIS-9 and MIS-17.

the sea-ice leads to an increase in the sensible and latent heat fluxes from the surface into the atmosphere (Figure 15.10a). This transfer of the summer surface energy excess to the ocean and its release in winter explains why the GHG-induced warming is more pronounced in winter than in summer (a phenomenon named here the summer-remnant effect). A similar annual march of the heat balance anomaly also occurs at the surface of the Southern Ocean, which explains why its warming is also larger during its local winter than during its local summer. In the low GHG experiments, such as MIS-17, the reverse happens with, relative to the reference experiment, a cooling of the high latitudes during their local winter larger than during their local summer.

15.6.3. Pure Contribution of Insolation

15.6.3.1. Annual Global Mean Temperature and Obliquity

As shown in Figure 15.11, the response of the global annual mean temperature to insolation alone is, for each interglacial, significantly different from the reference climate. The insolation contributes so as to warm the interglacials MIS-9, MIS-15, MIS-5, MIS-1, MIS-19, but to cool MIS-17, MIS-11, MIS-13, and MIS-7. It is not straightforward to explain such a response of the annual mean temperature to the insolation distribution, which varies only with seasons and latitudes. On the one hand, the difference in the insolation pattern between the interglacials (Figure 15.3) is the most important over the summer hemispheres. On the other hand, the simulated global annual mean temperature is highly correlated ($R^2 = 0.99$) with the average of the local summer temperatures (the JJA temperatures averaged over

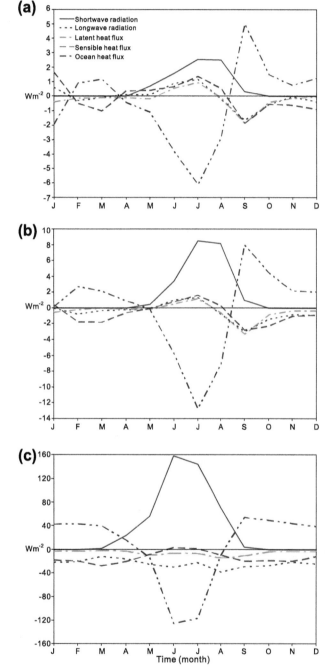

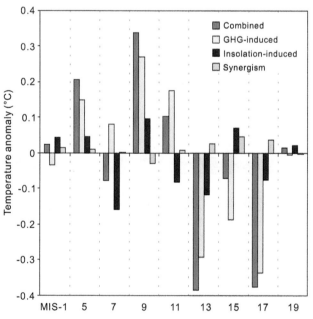

FIGURE 15.10 (a) Annual march of the GHG-induced difference between MIS-9 and the reference for the heat balance components at the Arctic surface (averaged over a region between 82°N and 90°N). (b) As (a), but for insolation-induced difference. (c) Annual march of the heat balance components at the Arctic surface of the reference experiment. Positive means gained by the surface.

FIGURE 15.11 Individual and combined effects of GHG and insolation on the global annual mean temperature (°C) (departure from the reference experiment) at each interglacial, as well as their synergism.

the NH plus the DJF temperatures averaged over the SH). This average summer temperature is, in turn, highly correlated with the average between the insolation of the NH during boreal summer and the insolation of the SH during austral summer. The insolation-induced warm and cool interglacial groups correspond therefore to respectively positive and negative anomalies in the average insolation of the two summer hemispheres (Figure 15.8b) and, consequently, to obliquity. This relationship between the annual global mean temperature and obliquity comes from the influence of obliquity on both the annual and seasonal insolations. Firstly, the signal of obliquity in daily insolation increases polewards (Berger et al., 1993) and an increase of obliquity increases the insolation over the latitudes of the summer hemisphere with a maximum at the poles (Berger and Loutre, 1994). Secondly, a high obliquity leads to a high annual mean irradiation in high latitudes of both hemispheres (Berger et al., 2010). Although this is compensated by a decrease in low latitudes, the strong feedbacks in high latitudes contribute to a net positive

impact of obliquity on the global annual temperatures. Thirdly, the irradiation received over one hemisphere during an astronomical season is only a function of obliquity (this explains why in Figure 15.8b, the horizontal axis can also be labelled in terms of obliquity with a higher obliquity being associated with a larger insolation). To better understand the response of the global mean temperature to the insolation forcing, the regional temperature anomalies must be investigated.

15.6.3.2. Annual Temperature at the Regional-Scale

At the regional-scale (Figure 15.12a), the most important difference between the annual mean surface temperatures of the nine interglacials is over the polar regions. Accordingly, the nine interglacials can be divided into three groups, the first two corresponding to the warm insolation-induced interglacials and the third to the cool ones. This is in pretty close relationship with the insolation grouping of

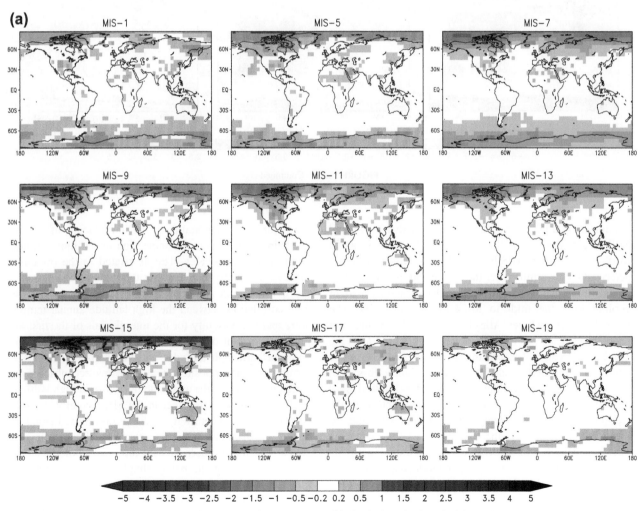

FIGURE 15.12 Pure impact of insolation on the surface temperature (°C) for the last nine interglacials. (a) Annual, (b) JJA, and (c) DJF.

(b)

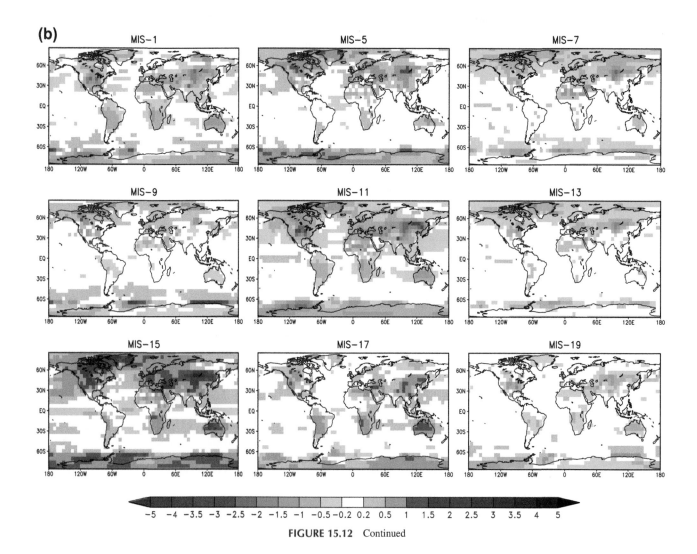

FIGURE 15.12   Continued

Section 15.4, at least if we take into account that MIS-9 and MIS-17 are more difficult to categorize. The first group is made up of MIS-5, MIS-9, and MIS-15, which are characterized by a warming over the Arctic and its surrounding lands, Antarctica, and the Southern Ocean, and by a slight cooling over the mid- to low-latitude continents. The second group includes MIS-1 and MIS-19, which are characterized by a warming over Antarctica and the Southern Ocean but a cooling over the Arctic. As the cooling in northern high latitudes is slightly less than the warming in southern high latitudes, MIS-1 and MIS-19 belong to the warm insolation-induced interglacials. Although the pattern of temperature change is similar for these two interglacials, the magnitude of the change is larger at MIS-1 than at MIS-19 in agreement with the insolation distribution. Finally, the interglacials of the third group (MIS-7, MIS-13, and MIS-17) are characterized by a cooling over the high latitudes in both hemispheres. The largest cooling of up to 2°C occurs at MIS-7 and MIS-13.

The temperature anomalies of the interglacials of the first and third groups have the same sign in the high latitudes of the two hemispheres. This is related to the obliquity-induced symmetry between the total annual insolation of the northern and southern high latitudes, explaining the high correlation between the high latitude annual temperature and the obliquity for the interglacials of the first and third groups. For the interglacials of the second group, the cool northern high latitudes and the warm southern ones result from the opposite effect of a low eccentricity and a high obliquity on the annual temperature in high latitudes of the NH and a joint effect in the SH. Actually, MIS-11, with its lowest eccentricity and low obliquity, belongs to the second group as far as the astronomical parameters are concerned, its Southern Ocean, although cooler than the reference experiment, being simply less cool (warmer) than the Arctic regions. This definitely makes MIS-11 and MIS-19 analogues for MIS-1 from the astronomical point of view.

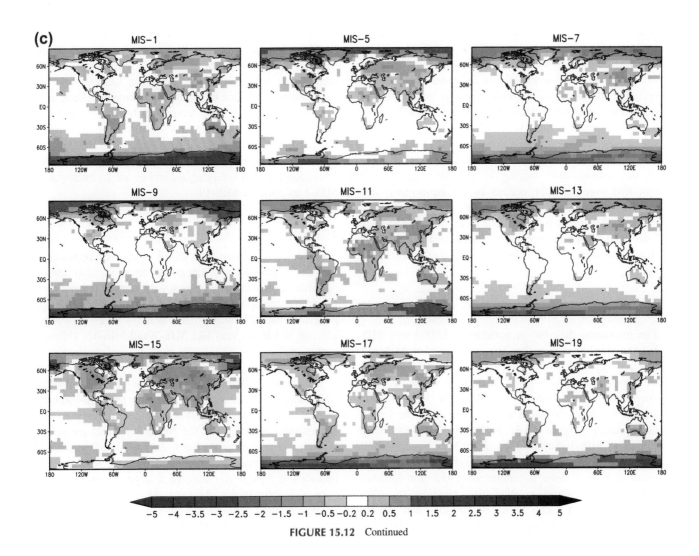

FIGURE 15.12   Continued

To explain these regional features at the annual-scale, the response to insolation at the seasonal-scale must be analysed.

15.6.3.3. The Anomalous High Latitude Local Winter Warming

The simulated seasonal and regional temperatures (Figures 15.12b and 15.12c) follow well the seasonal and latitudinal insolation distributions, except over the polar oceans of the winter hemisphere. To analyse this problem, MIS-7 and MIS-13 with their similar insolation patterns, are selected. The hemispheres being under-insolated during their local summers and over-insolated during their local winters, there is, in boreal summer, a cooling of up to 2°C over the northern mid- to high-latitudes (Figure 15.12c). In austral summer (Figure 15.12b), there is a cooling of up to 3°C over Antarctica and to a lesser degree over the Southern Ocean, and a warming of up to 1°C over northern Africa and central, southern, and

eastern Asia. However, although there is a small increase in the high latitudes of the winter hemispheres, there is a large decrease in temperature both over the Arctic region in DJF (which is even larger than the temperature decrease in boreal summer) and over the Southern Ocean in JJA.

The insolation pattern of MIS-9 is almost the opposite of MIS-7 and MIS-13: over-insolated over the local summer hemispheres with a maximum centred at the poles and under-insolated over the local winter hemispheres with a minimum insolation centred over the mid-latitudes, a signature of a larger obliquity. Such an insolation pattern leads to a temperature response: the opposite to what happens in MIS-7 and MIS-13. In particular, and inconsistent with the insolation distribution, there is an anomalous and large local winter warming at MIS-9 over the Arctic regions and over the Southern Ocean. To understand these anomalous local winter warmings/coolings, we analyse the insolation-induced heat balance at the surface of the Arctic and Southern Ocean. The results for the Arctic at MIS-9 (Figure 15.10b) are very similar to the summer-remnant effect related to the GHG change (Figure 15.10a and Section 15.6.2), but with a larger amplitude. Given the small amount of energy available in winter over the high latitudes, the summer-remnant effect is of the utmost importance in this determination of the pure contribution of insolation during winter. At MIS-7 and MIS-13 the opposite occurs, leading to a significant cooling (relative to the reference experiment) in local winters. The seasonal impact of the astronomical forcing is therefore significantly modified over the polar oceans by the atmosphere—sea-ice—ocean interactions, which is in line with the findings of earlier studies made for the glacial inception (Khodri et al., 2001) and for the Holocene (e.g., Ganopolski et al., 1998; Braconnot et al., 2000; Crucifix et al., 2002; Otto et al., 2009).

15.6.3.4. MIS-1 Analogues

The insolation distribution patterns of MIS-1 and MIS-19 are similar except that the amplitude at MIS-1 is larger. According to the phasing between the astronomical parameters, MIS-19 is suggested to be a better astronomical analogue than MIS-11 for the Holocene and its future evolution. This is in line with the insolation pattern (Figure 15.3) with MIS-19 being more similar to MIS-1 than MIS-11. Both MIS-1 and MIS-19 are under-insolated over the whole globe during boreal summer with a deep minimum over the low latitudes around the summer solstice (as mentioned earlier, a signature of a much lower eccentricity and slightly larger obliquity, as compared to the reference). Both are over-insolated during boreal winter, with a maximum over the South Pole. This insolation distribution leads to a cooling over all the continents in boreal summer (Figure 15.12c), and a warming over the whole Earth, except the Arctic, in boreal winter (Figure 15.12d). These can be considered as a direct impact of insolation on temperature. However, the warming over the Southern Ocean in austral winter, although insolation is slightly reduced, and the cooling over the Arctic in boreal winter, although insolation is slightly increased, are due to the summer-remnant effect.

The insolation pattern at MIS-11 is similar to MIS-1 except that, in MIS-1, the minimum of insolation during boreal summer is located in the northern mid-latitudes. Such a similarity exists also with MIS-19, but the magnitude of the anomaly is much smaller at MIS-19. This leads, in boreal summer, to a large cooling over the whole globe in particular over all the continents, but to a slight warming over the equatorial Pacific, the equatorial Atlantic, and northern Africa due to a decrease of precipitation over these low latitudes. Over the Southern Ocean, the large global cooling during the austral winter is

dominating the remnant effect of the austral summer leading to a deep cooling there, unlike at MIS-1 and MIS-19. In boreal winter, there is a warming over all the continents, but a cooling over the Arctic and its surrounding lands due to the summer-remnant effect, as in MIS-1 and MIS-19. This seasonal behaviour leads to an annual mean temperature below the reference everywhere on Earth, except over northern Africa, southern, and eastern Asia, and explains why MIS-11 belongs to the cool insolation-induced interglacials and is not an analogue of MIS-1 as good as MIS-19.

15.6.3.5. MIS-17, MIS-5, and MIS-15

The insolation pattern of MIS-17 is just the opposite of MIS-1, leading to a temperature response also almost opposite to MIS-1. As the boreal winter cooling anomaly is larger than the boreal summer warming, the annual average pattern of MIS-17 means that it belongs to the cool interglacials and looks very like MIS-7 and MIS-13, although with a smaller cooling over the northern and southern high latitudes.

MIS-5 and MIS-15 have a similar insolation pattern with a larger amplitude at MIS-15. They are over-insolated over almost the whole Earth during boreal summer with a maximum centred over the North Pole, and are under-insolated over the whole Earth during boreal winter, the insolation minimum being centred over the southern low latitudes, a typical signature of a larger eccentricity and even a slightly larger obliquity. As a direct response to insolation, their temperature changes are very similar. In boreal summer (Figure 15.12c), there is a large warming over the whole Earth, in particular over the continents, the warming being larger at MIS-15 than at MIS-5 in response to the insolation forcing. However, there is a cooling over northern Africa, the equatorial Pacific, and the equatorial Atlantic. This cooling results from increased precipitation over these regions following the northward shift of the inter-tropical convergence zone. In boreal winter (Figure 15.12b), there is a significant warming over the high-latitude oceans and a cooling over all the continents (except over Antarctica at MIS-5). As the whole Earth is under-insolated during boreal winter, the warming over the northern high latitude oceans in this season is due to the local summer-remnant effect (a situation opposite to what happens at MIS-7 and MIS-13). However, during austral winter, the temperature over the Southern Ocean responds to the global warming induced by high insolation over the whole Earth and does not reflect the local summer-remnant effect. This is because the change in local summer insolation over the southern high latitudes is too small to influence the following local winter climate, especially when compared to the large warming occurring over the whole Earth during austral winter.

15.6.4. Combined Effect of Insolation and GHG

15.6.4.1. Annual and Seasonal Global Mean Temperature

In response to the combined effect of insolation and GHG (Figures 15.11 and 15.13), MIS-9 is the warmest interglacial, followed in decreasing order by MIS-5, MIS-11, MIS-1, MIS-19 (all warmer than the reference), and MIS-15, MIS-7, MIS-17, and MIS-13 (all cooler than the reference). The difference between MIS-9 and MIS-13 is 0.72°C, which might be underestimated due to the low sensitivity of our model. As a palaeoclimatic contribution to better understanding of the sensitivity of the climate system to a doubled CO_2 concentration (Hansen et al., 2008; Harvey, 2012, this volume), an additional equilibrium simulation was made for each of the last nine interglacials (Yin and Berger, 2011). Within the range of the interglacial variability with the CO_2e concentration going from 234 to 300 ppmv, LOVECLIM's climate sensitivity is shown to generally decrease with

increasing temperature: MIS-9 has the lowest sensitivity and MIS-13 the highest. The sensitivity at MIS-5 is 10% lower than for the pre-industrial period.

The simulated temperature change is very different between the seasons, in particular over the continents. During boreal summer (Figure 15.5c), MIS-5, MIS-9, and MIS-15 are the three warmest and MIS-11, MIS-1, and MIS-13 the three coolest. During this season, although the NH of MIS-13 is, on average, almost similar to the NH of MIS-1 and even slightly warmer than MIS-11, its SH is much cooler than during any of the other interglacials (Figure 15.5c). This is in line with the hypothesis proposed by Guo et al. (2009) based on a collection of proxy data that the boreal summer climate of the two hemispheres during MIS-13 would be characterized by an asymmetry with a mild NH but a cold southern one. With a cold boreal winter everywhere, MIS-13 is globally the coolest of the last nine interglacials. During boreal winter (Figure 15.5b), MIS-9, MIS-11, and MIS-1 are the three warmest and MIS-15, MIS-13, and MIS-17 the three coolest. The

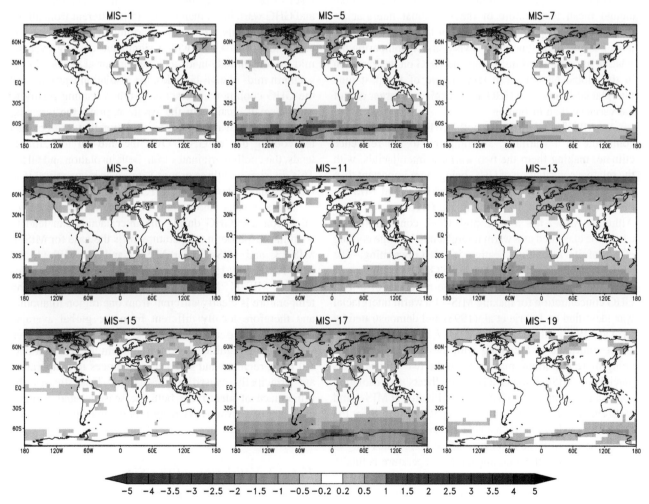

FIGURE 15.13 Combined effect of insolation and GHG on the annual mean surface temperature (°C) for the last nine interglacials.

difference between the seasons is mainly caused by precession (eccentricity here due to the selection of NH summer occurring at perihelion). MIS-15, which has the largest eccentricity, has the largest seasonal contrast with large warming in boreal summer and large cooling in boreal winter. MIS-5 is similar to MIS-15 but with a reduced amplitude. MIS-11 with the smallest eccentricity has a largely reduced seasonal contrast. For the interglacials, MIS-7, MIS-9, and MIS-13, with an eccentricity value similar to the reference experiment, the change in the seasonal contrast is small. Because of the damping effect of the oceans, the seasonal contrast in the SH is much smaller than in the NH, with the largest at MIS-1, MIS-11, and MIS-19. The regional and seasonal difference in the temperature change stresses the importance of identifying both the seasonal and regional character of a given proxy record for its proper interpretation.

15.6.4.2. Relative Importance of GHG and Insolation During Each Individual Interglacial

The combined impact of insolation and GHG on the global annual mean temperature of the individual interglacials (Figure 15.11) is primarily the sum of their pure contributions because their synergism is very small. Comparing the relative importance of insolation and GHG on the variation in the global annual mean temperature of the past nine interglacials shows that GHG alone explains a larger part of the variance than insolation.

At MIS-5 and MIS-9, both GHG and insolation contribute to a warming, as compared to the reference climate, making them the two warmest interglacials, with the relative contribution of GHG being more important than that of insolation (Figure 15.11). At MIS-7 and MIS-11, insolation contributes to a cooling and GHG to a warming. Although MIS-7 has a quite high CO_2e concentration, it cannot be classified as a warm interglacial, the large cooling effect of its insolation beating its GHG warming. The opposite happens at MIS-11, its large GHG warming overwhelming the insolation cooling. The necessity of a high GHG concentration for making MIS-11 a warm interglacial was identified by Imbrie et al. (1993) and demonstrated by Li et al. (1998). At MIS-13 and MIS-17, both insolation and GHG contribute to a cooling with the GHG contribution being more important, making them the two coolest interglacials. At MIS-15, the GHG cooling 'beats' the insolation warming, leading to a global cooling. Finally, at MIS-1 and MIS-19, as their CO_2e is only slightly lower than the reference level, the GHG leads to a very small cooling. As the insolation of these two interglacials contributes only to a slight warming, the total impact on temperature is finally not very different from the reference case.

At the regional-scale and for the annual mean temperature, the nine interglacials can be divided into four groups

(Figure 15.13). Three interglacials of the third group of Section 15.6.3 (pure insolation contribution), MIS-7, MIS-13, and MIS-17, are the three coolest interglacials with the most important cooling over the mid- to high-latitudes in both hemispheres. The reason for their cooling is, however, different. At MIS-7, the cooling is due to insolation (Figure 15.12a) and is everywhere larger than the warming due to its GHG, whereas the cooling at MIS-13 and MIS-17 is due to both insolation and GHG. MIS-5 and MIS-9 belong to a group characterized by a large warming in the mid- to high-latitudes, the opposite of what occurs at MIS-7, MIS-13, and MIS-17. Both insolation and a low GHG concentration contribute to this regional behaviour. MIS-1, MIS-11, and MIS-19 belong to the same group characterized by a significant warming in the southern high latitudes but a cooling in the northern high latitudes. These three interglacials correspond to a minimum of the 400 ka cycle in the eccentricity. For MIS-1 and MIS-19, this regional temperature pattern is almost entirely caused by insolation, their GHG concentrations being more or less equivalent to the reference value. For MIS-11, the temperature pattern results from the competition between its insolation cooling and GHG warming. Finally, MIS-15 is different from all the other interglacials, with a large warming in the Arctic and its surrounding lands, and with a significant cooling over the mid- to low-latitude lands and the equatorial oceans. In the southern mid- to high-latitudes, the insolation warming and the GHG cooling counteract each other, leading finally to a very small change. In the Arctic regions, the warming results from the warming due to insolation that is larger than the cooling due to GHG. Over the mid- to low-latitude lands, the cooling originates from both insolation and GHG. This division into three groups is more or less applicable at the seasonal-scale (Figures 15.12b and 15.12c).

These results show that different combinations of insolation and GHG during the interglacials can lead to a similar response in temperature, as is the case for MIS-7, MIS-13, and MIS-17, for MIS-1, and for MIS-11 and MIS-19 (Figure 15.13). Moreover, for some interglacials (MIS-1, MIS-11, MIS-15, and MIS-19), the annual mean temperature is clearly different from one region to another and therefore locally different from the global average. In addition, the annual response pattern of any given interglacial is significantly different from the seasonal one. All these regional and seasonal differences in temperature are principally caused by insolation. This stresses the importance of identifying both the seasonal and regional character of a given proxy record for using its interpretation correctly.

15.6.4.3. Validation of the Simulations

To validate our modelling results, a comparison between them and proxy data and a comparison between them and the simulation results of earlier studies are given in this section. Harrison and Bartlein (2012, this volume) offer a detailed survey of palaeoclimate data.

The first validation is on the relative magnitude of the simulated interglacials. Although the model is forced only by insolation and GHG, the relative magnitude of the simulated interglacials is in reasonable agreement with proxy data. The simulated global annual mean temperature is significantly correlated with the benthic $\delta^{18}O$ records (R = 0.79, Figure 15.14a) that are assumed to partly reflect the global ice volume and, in turn, the global temperature (as in Yin and Berger, 2010). The deep-sea records show that the post-MBE interglacials are on average warmer than the pre-MBE ones. They also show that MIS-7 is the coolest interglacial over the last 430 ka, and MIS-13 is among the coolest over the last one million years: these are consistent with the modelling results presented here. On a more regional-scale, the simulated Antarctica temperature is also significantly correlated with the Antarctica ice core temperature record (Jouzel et al., 2007) (R = 0.77, Figure 15.14b). Regarding the MBE event, as already discussed in Section 15.5, our simulation shows that the MBE is mainly a feature of the global annual mean temperature and of the temperature of the southern high latitudes, but is not clearly seen in the northern high latitude temperature record due to the different dominant effects of insolation and of GHG at the regional-scale. This result is confirmed by proxy data (see Section 15.5) and provides an explanation for them.

The second validation is offered for the last interglacial MIS-5 because it is often considered as an analogue for our future warming (in addition to MIS-11) and also because there are more available proxy records, as well as simulations, for this recent interglacial. In our simulations, the global annual mean temperatures of MIS-5 and MIS-11 are higher than the simulated pre-industrial climate by 0.41°C and 0.3°C and are higher than the simulated present-day climate by 0.14°C and 0.04°C. Compared to Kubatzki et al. (2000), who used CLIMBER-2 with a climate sensitivity to doubling CO_2 of 3°C, MIS-5 relative to the pre-industrial period is underestimated in our model due to its low sensitivity. However, the main seasonal and regional features of MIS-5, which have either been simulated by other climate models or been reconstructed from proxy records, are well-reproduced in our simulation. Figure 15.15 shows that, compared to the pre-industrial period, the simulated MIS-5 is characterized by: (i) a significant warming over all the continents during boreal summer and a significant cooling over all the continents during boreal winter leading to a largely increased seasonal contrast over the NH land and a reduced seasonal contrast in the SH. As MIS-5 has a CO_2 concentration similar to pre-industrial, this seasonal contrast is mainly due to the much higher (lower) insolation received by the whole Earth during boreal summer (winter) at MIS-5 (during this interglacial NH summer (winter) occurs at perihelion (aphelion), exactly the opposite of the pre-industrial situation); (ii) a warming over the Arctic region through the whole year due to a higher insolation over this region in summer and the summer-remnant effect in winter; (iii) a warmer Southern Ocean and an annually warmer Antarctica (as explained in Section 15.6.3.5); (iv) a large increase of precipitation over all the northern monsoon regions due to the intensification of the monsoon related to the high NH summer insolation; (v) a large increase of the tree fraction over northern Africa, the Middle East, central, southern, and eastern Asia, as well as a large part of North America and northern Eurasia. These features have also

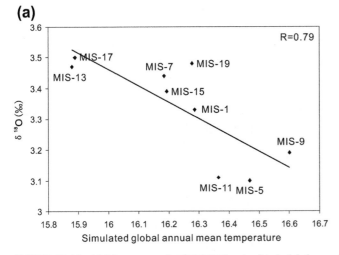

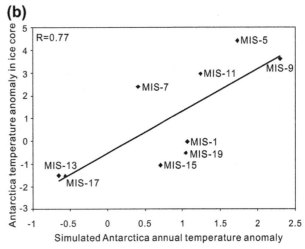

FIGURE 15.14 (a) Linear regression between the simulated global annual mean temperature (°C) and the benthic $\delta^{18}O$ at the peak of the nine interglacials. (b) Linear regression between the simulated annual mean temperature anomaly for Antarctica and that recorded in the EPICA ice core *(Source: Jouzel et al., 2007.)*

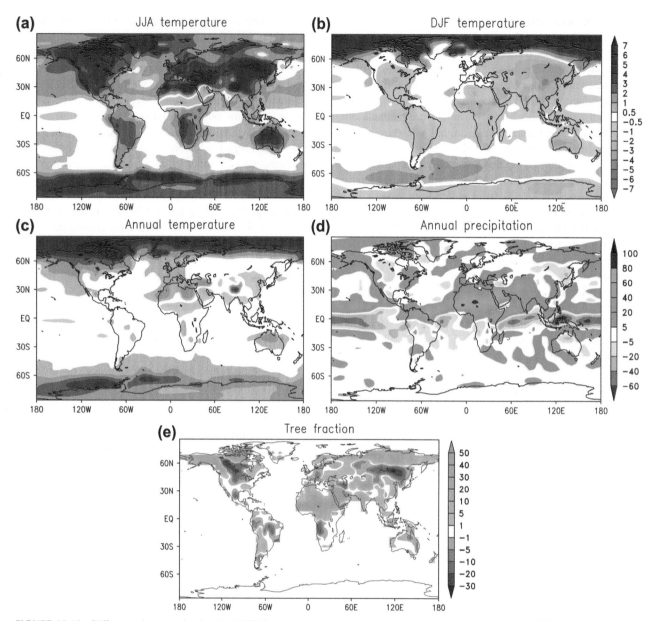

FIGURE 15.15 Differences between the simulated MIS-5 and pre-industrial climate in (a) JJA surface temperature, (b) DJF surface temperature, (c) annual mean surface temperature (°C), (d) annual mean precipitation (cm per year), and (e) simulated tree fraction (%).

been simulated by other climate models of different complexity (e.g., Harrison et al., 1995; Montoya et al., 1998; Kubatzki et al., 2000; Otto-Bliesner et al., 2006) and are in agreement with proxy data. For example, Lozhkin and Anderson (1995) show a warmer summer temperature in the Arctic than today and a polarward expansion of boreal forest. A warming of about 2°C over Antarctica and over the Southern Ocean is shown by Vimeux et al. (2002) and Pahnke et al. (2003), respectively.

The reasonable agreement between our simulated results, proxy data, and other model results give credence to our result, but further intercomparisons are currently being undertaken. Moreover, qualitatively, our simulated climatic response to GHG and feedbacks such as the summer-remnant effect are similar to the results of other models.

15.7. FUTURE OF OUR INTERGLACIAL

15.7.1. Future Insolation and Analogues for the Holocene

In 1972, a group of palaeoclimatologists gathered at a working conference on how and when the present interglacial will end (Kukla et al., 1972). Based on geological records available at that time, the length of each of the last two interglacials was assumed to be approximately 10 ka. As it is about the length of our Holocene warm interval to date and assuming a common duration of the interglacials, they concluded, "it is likely that the present-day warm epoch will terminate relatively soon if man does not intervene" (Kukla et al., 1972, p. 267).

Only limited attempts have been made to tentatively predict the future of our interglacial on a quantitative basis. Most of the early simulations (Imbrie and Imbrie, 1980; Berger et al., 1991) were predicting a continuation of the cooling that started at the peak of the Holocene some 6000 years ago, a cold interval around 25 ka AP (after present-day), and a glaciation around 55 ka AP. These projections into an ice age were based on statistical rules or simple models not including any CO_2 forcing. They implicitly assumed a GHG value equal to the average of the last glacial–interglacial cycles (~225 ppmv). On the other hand, the simple ice-sheet model of Oerlemans and Van der Veen (1984) predicted a long interglacial of 50 ka, followed by a first glacial maximum at ~ 65 ka AP. This result was in line with Ledley (1995), who found that an ice age is unlikely to begin in the next 70 ka, a statement based on a significant relationship that

she found between the observed rate of ice volume change and the summer solstice radiation. Other experiments were more oriented towards a modelling approach including the possible effects of anthropogenically-increased CO_2 on the dynamics of the ice-age cycles (Saltzman et al., 1993; Loutre, 1995). Such a possible intervention by human activities on the length of the present interglacial had begun to be investigated in the 1990s, based on the understanding that the insolation change over the next several thousand years is quite unusual (Berger and Loutre, 1996).

The insolation of the next 100 ka is indeed characterized by the small amplitude of its variations (Figure 15.16), much smaller than during MIS-5. For example, this amplitude at 65°N in June will be less than 25 Wm^{-2} over the next 25 ka, whereas it was 120 Wm^{-2} between 125 and 115 ka BP. Consequently, from the point of view of insolation, the Eemian can hardly be taken as an analogue for the next thousands of years, as pointed out earlier in this chapter. Such a future weakly varying insolation is really exceptional. Comparing the variations in the insolation at the NH summer solstice over the last 1 million years shows that the interval 405–340 ka BP, which overlaps MIS-11, is the most recent astronomical analogue for the period extending from 5 ka BP to 60 ka AP (Berger and Loutre,

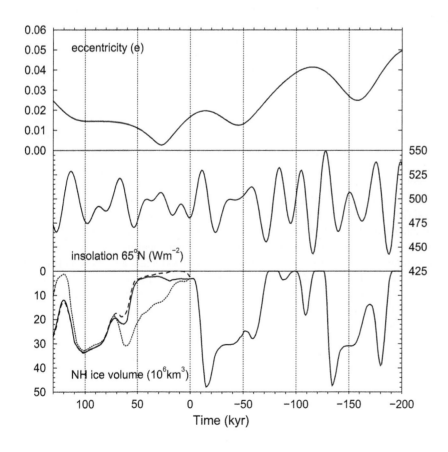

FIGURE 15.16 Long-term variations over the period from 200 ka BP to 130 ka AP of eccentricity (e), June insolation at 65°N (Wm^{-2}) (Berger, 1978), and NH ice volume (10^6 km³) (Berger and Loutre, 2002). In the panel Ice Volume: the solid line is the ice volume simulated by using the Vostok CO_2 concentration for the past (Petit et al., 1999) and a scenario reproducing the last 120 ka for the future; the dotted line gives the future ice volume if the future CO_2 would remain constant and equal to 210 ppmv; the dashed line is for the future ice volume under a scenario where the CO_2 concentration would reach 750 ppmv within the next two centuries and return to the 'natural Vostok level' 1000 years later.

2003; Loutre and Berger, 2003; Loutre, 2003). However, the precise chronological alignment of these interglacials has different implications over the natural length of the current interglacial. Berger and his colleagues favoured a synchronization of MIS-1 and MIS-11 by using the precessional variations based on the hypothesis that the variations in NH summer insolation at northern high latitudes are well-correlated to changes in global ice volume (Crucifix and Berger, 2006). This alignment, which differs from the synchronization of the obliquity signal used by EPICA (2004), is actually confirmed by the vegetation trends observed by Tzedakis (2010) in MIS-1 and MIS-11. MIS-19, also associated with a low eccentricity of the 400 ka cycle, is the next analogue of MIS-1. That MIS-19 may be an even better analogue is discussed in Section 15.6.4.3, but it must be stressed that, in this section, only the peaks of the interglacials are considered.

15.7.2. Modelling the Future of Holocene

To analyse the impact of this eccentricity related low amplitude of the solar radiation change, experiments have been made with the LLN (Louvain-la-Neuve) 2D sectorial climate model (Gallée et al., 1991) forced with both the insolation and the CO_2 variations over the next 100 ka, over MIS-11, and over the last 1 Ma. This Earth Model of Intermediate Complexity (EMIC) has a sensitivity to a doubling of CO_2 of 2.2°C, well within the range of the general circulation models used by IPCC (Randall et al., 2007). As there is no CO_2 cycle yet coupled to the LLN model, CO_2 had to be considered as a forcing although, at the geological timescale, it is definitely a feedback (cf. Harvey, 2012, this volume).

As a consequence, many CO_2 scenarios were constructed for the next 100 ka (Loutre and Berger, 2000). Constant CO_2 concentrations of 210, 250, and 290 ppmv were first assumed in agreement with their glacial–interglacial values. CO_2 was then allowed to vary according to the Vostok reconstruction (Petit et al., 1999) assuming a future CO_2 evolution equivalent to that which occurred at either MIS-5 or MIS-11. Most of these natural (i.e., without anthropogenic effects) scenarios, and many others, lead to an exceptionally long interglacial lasting from 5 ka BP to 50 ka AP (Figure 15.16, full line in the ice volume panel), the next glacial maximum being reached at 100 ka AP. The only exceptions happen when CO_2 is less than 220 ppmv (Figure 15.16 dotted line in the ice volume panel) that leads to the early entrance into glaciation predicted in the 1980s and in agreement with Ruddiman's hypothesis (see below). This kind of result was confirmed by Vettoretti and Peltier (2004) who, using the Canadian climate general circulation model, showed that under the present-day insolation regime and pre-industrial CO_2 concentration, no glacial inception is possible.

Further comparisons of our likely future climate with similar situations of the past could give credibility to our result. Using the Petit et al. (1999) CO_2 reconstruction from Vostok leads to a simulated MIS-11 interglacial 50 ka long, very similar to that obtained for the future. It is significant that such a predicted, particularly long, MIS-11 interglacial was later observed in the EPICA core (2004) and more recently confirmed by the palynological record of Lake Baikal (Prokopenko et al., 2010). In order to understand the reason for such a long interglacial, sensitivity experiments have been conducted. In particular, the Vostok CO_2 timescale was shifted forward and backward by 10 ka. Making the CO_2 'younger' confirmed the very long interglacial, but making it 'older' reduced its length considerably. Actually, during the decreasing phase of insolation, between 410 and 395 ka BP, a sustained high value of CO_2 creates a very long interglacial (50 ka) whereas a decreasing CO_2 leads to a much shorter interglacial (20 ka). It is therefore the high value of CO_2 that counteracts the impact of the decreasing insolation on the ice volume. This is exactly what is happening between 400 and 350 ka BP and over the next 50 ka, during which only the lowest CO_2 scenario can shorten the interglacial. This confirms the potential of MIS-11 for being, at least partly, one of the analogues for the future of our Holocene interglacial. As far as MIS-19 is concerned, the 2D LLN simulation of the last 1 Ma (Berger et al., 1999) shows that it was also a very long interglacial. However, in this simulation, the CO_2 scenario used sustains a moderate value of CO_2 for a sufficiently long period of time to prevent the declining insolation from shortening the interglacial. Given the robustness of the simulated ice volume over the next 130 ka, it is worth testing the sensitivity of this natural evolution to the anthropogenic CO_2 increase.

Similarly to the IPCC scenarios, Berger and Loutre (2002) assumed that the CO_2 concentration would increase from the unperturbed level to 750 or 550 ppmv over the next 200 years, would decrease to 300 ppmv over the following 450 or 300 years, would reach linearly the 1 ka AP concentration of the natural 'Vostok' scenario, and would follow this scenario thereafter. The responses of the NH ice sheets to these scenarios are very different. In the 550 ppmv experiment, a slight melting of Greenland ice occurs only over the next 1000 years and there is little or no difference with the results from the natural scenario based on the Vostok data. For 750 ppmv, the impact is far more pronounced. The Greenland ice sheet melts completely between roughly 8 and 15 ka AP and reaches its natural size again at 40 ka AP (Figure 15.16 dashed line in ice volume panel). These results seem to indicate that, under very small insolation variations, there is a threshold value of CO_2 above which the Greenland ice sheet disappears in the LLN 2D model and that the climate system would take ~50 ka to assimilate the impacts of human activities triggered during the early centuries of the third millenium. Such a possible melting of the Greenland ice sheet is similar to the result of Huybrechts and De Wolde (1999), who showed that, even if GHG concentrations were stabilized by the early twenty-second century, a Greenland meltdown is irreversible for CO_2e concentrations more than twice the present-day value. Archer and Ganopolski (2005) predicted that a carbon release from fossil fuels or methane hydrate deposits of 5000 GtC could prevent glaciation for the next 500 ka, that is, after two 400 ka cycles of the eccentricity minima. This is because, in their model CLIMBER-2, a deep minimum in summer insolation is required to nucleate an ice sheet when the CO_2 concentration is raised. More recently, Charbit et al. (2008), using a climate ice-sheet model, showed also that above an emission of 3000 GtC, the simulated melting of Greenland appears irreversible, while below 2500 GtC Greenland experiences only a partial melting, followed by a regrowth phase.

15.7.3. Ruddiman Early Anthropogenic Hypothesis

The question of when human activities started to influence the natural evolution of climate was raised by Ruddiman in 2003. He claimed that humans began modifying GHG concentrations thousands of years before the industrial era, with the first clearance of forests emitting CO_2 and the intensification of rice agriculture emitting CH_4. These elevated GHG concentrations have countered the natural (astronomical) cooling trend and prevented global climate from entering into glaciations. His hypothesis that the increase in GHG concentrations during the Holocene was not natural was based on the alignment of MIS-11 and MIS-1 by using precessional variations (Ruddiman, 2005), as done by Berger and his colleagues. This alignment led him to assume that ice sheets would have started forming in the northern polar regions in the absence of anthropogenic interference. Such an early entrance into glaciations is not seen in Berger and Loutre (2002), stressing that extrapolations based on statistical rules are not necessarily confirmed by modelling experiments. However, simulations with climate models forced by GHG concentration changes (Ruddiman et al., 2005) support Ruddiman's earlier hypothesis, although experiments by Crucifix et al. (2005) and Berger and Loutre (2006) showed that a glacial inception would have required a CO_2 lower than 230 ppmv or a decrease since 7 ka BP at a much faster rate than during MIS-7. Moreover, the simulation of Singarayer et al. (2011) indicates that the late Holocene increase in the methane concentration results from natural changes in the Earth's astronomical configuration with no early agricultural source required to account for such an increase over the last 5000 years.

15.8. PROBING FUTURE ASTRO-CLIMATES

To better understand our current interglacial and its future, we have investigated the response of the climate system to insolation and GHG during the interglacials of the past 800,000 years. The following results and conclusions might provide ideas for future research and also for better understanding of the future of this interglacial, the Holocene.

If we consider only the peaks of the interglacials, MIS-19 and MIS-1, having the same CO_2e concentration and a similar insolation distribution pattern, their temperature response is similar and MIS-19 can be considered a good analogue of the Holocene interglacial. The difficulty with MIS-11, which was generally considered a good analogue of MIS-1 (Berger and Loutre, 2003), comes from its higher CO_2e concentration and a slightly lower obliquity, which leads to a cooler insolation-induced SH during its local summer and much cooler during its local winter. If we look now for analogues of the whole Holocene and its future, it must be stressed that the next minimum of eccentricity of the 400-ka cycle is approaching. As the amplitude of the climatic precession parameter is modulated by the eccentricity, it is the main forcing of the daily insolation all over the Earth, so that the seasonal and latitudinal distribution of insolation is not going to vary much over the next tens of thousands of years. This will allow the other forcing feedbacks (e.g., GHGs) to play an even more important role. With CO_2 at interglacial levels, and even larger under human influence, our interglacial is therefore predicted to be exceptionally long-lasting. The same happened during MIS-11. Being associated with the last deep minimum of eccentricity, it is the most recent astronomical analogue of MIS-1 and its future, and the existence of a rather high CO_2 concentration during the decreasing phase of insolation around 400 ka BP at the origin of its long duration, as confirmed by the EPICA record. MIS-19, also associated with a very low eccentricity, is the next best astronomical analogue. According to the sensitivity experiments of Berger et al. (1999), moderate values of CO_2 sustained sufficiently long might have led to an interglacial even much longer than MIS-11 and MIS-1.

The interglacials MIS-9 and MIS-5 are the warmest over the last 800 ka and, as such, are analogues for our CO_2-induced future warm interglacial, although their astronomical forcings are largely different from MIS-1 and its future. MIS-9 is the warmest and MIS-5, which is generally assumed to be a good analogue for the future warmth of our interglacial, is slightly warmer than the simulated present-day climate. The warm climates of MIS-9 and MIS-5 are due to a warming contribution from both insolation and GHG, but MIS-11 is a warm interglacial only because of its high GHG concentrations, its insolation contributing to a cooling. Its possible analogy with MIS-1, although not perfect, might confirm the hypothesis of

Ruddiman (2003, 2005) claiming that a high GHG concentration (human-induced) has already prevented an early end of our Holocene interglacial climate and so an early entrance into a glaciation (cf. Lenton, 2012, this volume).

From the regional point of view, the annual mean temperature of the warm interglacials (MIS-1 to 11) and of MIS-19 increases more in the SH than in the NH. This is also the case for boreal winter, except for MIS-7, but MIS-9, the warmest, is now followed by MIS-11, MIS-1, MIS-19, and MIS-5, the mildest. The situation for boreal summer is quite different as MIS-1, MIS-7, MIS-11, and MIS-19 are cooler than the average, with only MIS-9, MIS-5, and MIS-15 being warmer.

Over the polar oceans, the local summer response to either GHG or insolation has a remnant effect on the local winter climate through the buffering effect of the oceans on sea-ice formation, cover, and thickness. This remnant effect is of the utmost importance for the high latitude winter climate: on the one hand, there is a larger climate response to GHG during winter than during summer and, on the other hand, there is a large impact of insolation during winter although the insolation forcing is very weak during this season. In terms of response to insolation, there are, however, three exceptions. During MIS-11, the large local winter cooling over the Southern Ocean results from the very deep cooling of the whole Earth during this season that overtakes the local over-insolated summer-remnant effect. During MIS-5 and MIS-15, the insolation change during austral summer over the Southern Ocean is too small to trigger a significant remnant effect there and the global warming during the whole boreal summer drives the simulated warming local to the Southern Ocean.

The overall variation in the global annual mean temperature of the interglacials is explained more by GHG than by insolation. However, the diversity of the climate patterns among the interglacials is due to the different relative importance of insolation and GHG. For example, the cool interglacials MIS-13 and MIS-17 are due to the cooling effects of both insolation and GHG, whereas the cool MIS-7 is due to the cooling effect of its insolation that beats its GHG warming and the cool MIS-15 is due to the reverse situation. At the regional-scale, different combinations of insolation and GHG can lead to a similar response, for example, at MIS-7, MIS-13, and MIS-17; at MIS-5 and MIS-9; and at MIS-1, MIS-11, and MIS-19. MIS-15 is the only interglacial that does not have any analogue.

The pure contribution of GHG on temperature change is quite straightforward, with a linear relationship between the global annual mean temperature and the GHG concentrations (the greenhouse effect). This means that the warmest GHG-induced interglacial is MIS-9 and the coolest MIS-17. In line with earlier CO_2-increase studies, the largest impact of GHG is over the high latitudes and, over the polar oceans, it is larger during the local winter than during the local summer.

Insolation plays a significant role on the global annual mean temperature, which is highly and positively correlated with obliquity. The warmest insolation-induced interglacial is MIS-9 and the coolest is MIS-7. In the annual mean temperature, the most important signal is over the high latitudes. For all interglacials, except MIS-1 and MIS-19, the insolation-induced temperature anomalies in high latitudes of both hemispheres have the same sign. This is because a high obliquity leads to a high annual mean irradiation in the high latitudes of both hemispheres and the strong feedbacks in high latitudes contribute to the high positive correlation of obliquity with the global annual mean temperatures. At the seasonal-scale, the spatial response of temperature follows well the seasonal and latitudinal distribution of insolation, except over the polar oceans of the winter hemisphere.

The post-MBE interglacials are warmer than the pre-MBE ones, as expected from their higher GHG concentration. However, insolation plays a significant role at the seasonal and hemispheric-scales. In the explanation of the warmer post-MBE interglacials, (i) boreal winter − or equivalently austral summer − is a key season; and (ii) the SH plays a more important role than the NH as it warms significantly during both seasons. We suggest that this might be considered when trying to understand the underlying causes of the higher CO_2 concentration during the interglacials after the MBE.

Finally, most of the natural (astronomical) scenarios for future CO_2 concentrations lead to an exceptionally long interglacial, unless the CO_2 concentration falls below 220 ppmv. Below this value, entrance into glaciations would agree with the hypothesis of Ruddiman. According to different scenarios for future anthropogenic CO_2 emissions, there seems to be a threshold in the LLN model, above which the Greenland ice sheet would melt totally over the next 5000 years.

ACKNOWLEDGEMENTS

This work is supported by the European Research Council Advanced Grant EMIS (No. 227348 of the Programme 'Ideas'). Q.Z.Y. is supported by the Belgian National Fund for Scientific Research (F.R.S.-FNRS). Access to computer facilities was made easier through sponsorship from S. A. Electrabel, Belgium.

Chapter 16

Catastrophe: Extraterrestrial Impacts, Massive Volcanism, and the Biosphere

Claire M. Belcher[a,b] and Luke Mander[c]

[a]School of Geosciences, University of Edinburgh, Edinburgh, UK, [b]BRE Centre for Fire Safety Engineering, University of Edinburgh, Edinburgh, UK, [c]Department of Plant Biology, University of Illinois at Urbana–Champaign, Urbana, USA

16.1. INTRODUCTION: WHAT IS A CLIMATE CATASTROPHE?

The geological record reveals that throughout the Phanerozoic Eon (~542 million years ago (Ma) to present day; Gradstein et al., 2004) there have been numerous radical and sudden changes to the Earth's environment. There are examples of global cooling, such as the transition from the Eocene to the Oligocene Epoch (33.5 Ma–34 Ma; Pearson et al., 2009), and examples of rapid global warming, such as at the Palaeocene–Eocene Thermal Maximum (PETM; ~55 Ma; Zachos et al., 2008) and the transition from the Triassic to the Jurassic Period (~200 Ma; McElwain et al., 1999). There are also examples of rapid environmental change related to the impact of an extraterrestrial body, such as that which occurred at the end of the Cretaceous

The Future of the World's Climate. DOI: 10.1016/B978-0-12-386917-3.00016-6

Period (~65 Ma; Hildebrand, 1993). The geological record thus represents a vast store of natural experiments that can help researchers investigate the dynamics of climate change over a wide range of timescales. This can provide a better understanding of the natural variability of the Earth's geophysical environment and its influence on the evolution of life.

In this chapter, we focus on two intervals of major environmental change in the geological past. The first occurred at the transition from the Triassic to the Jurassic Period (Tr–J) and the second at the end of the Cretaceous Period (K–Pg). There is evidence for the impact of an extraterrestrial body and the eruption of large volumes of basalt lava during both these intervals of time, and there is also evidence of rising atmospheric CO_2 and consequent global warming during both intervals. The Tr–J and K–Pg also both represent episodes of mass extinction among terrestrial and marine life. We aim to use the Tr–J to highlight the possible environmental and biological consequences of massive volcanism, and the K–Pg to highlight the possible environmental and biological consequences of the impact of an extraterrestrial body. We do not intend to provide definitive taxon-by-taxon accounts of each extinction event, nor do we claim that either extraterrestrial impact or massive volcanism were solely responsible for environmental and biotic change during each time interval. Rather, we aim to highlight the possible environmental effects of each process and then attempt to illustrate how these reconcile with the observed evidence of environmental and biotic change at the Tr–J and K–Pg.

16.2. MASSIVE VOLCANISM: CASE STUDY OF THE TRIASSIC–JURASSIC (TR–J) EVENT

16.2.1. Introduction

Earth history is punctuated by several episodes of massive continental flood basalt volcanism. These result in the outpouring of large volumes of basalt, which are preserved as Large Igneous Provinces (LIPS) that can cover millions of square kilometres (Wignall, 2001). There is a first-order correlation between the emplacement of LIPs and episodes of major palaeoclimatic change and mass extinction. For example, the Siberian basalts roughly coincide with mass extinction at the end of the Permian Period (~251 Ma; Kamo et al., 2003), the basalts of the Central Atlantic Magmatic Province (CAMP) coincide with mass extinction at the Tr–J (Marzoli et al., 1999), and the main eruptive phase of the Deccan basalts of India coincides with mass extinction at the end of the K–Pg (~65 Ma; e.g., Courtillot et al., 1986; Self et al., 2006). The Karoo basalts of South Africa coincide with a minor extinction at the transition from the Pliensbachian to the

Toarcian Epoch (~186.3 Ma–179.3 Ma; Pálfy and Smith 2000; Bambach, 2006), and the North Atlantic basalts coincide with a minor extinction during the PETM (Storey et al., 2007; Harrison and Bartlein, 2012, this volume).

Despite this first-order correlation, it is often difficult to demonstrate a causal relationship between massive volcanism and palaeoclimatic change and mass extinction in the geological record (e.g., Twitchett, 2006). One approach to this problem is to correlate the lava flows of LIPs to records of environmental change and extinction preserved in individual marine and terrestrial rock successions using a combination of palaeobiological, geochemical, and lithological data. Although the Tr–J is often labelled as the least studied of the Phanerozoic mass extinctions (e.g., Hallam 2002; Tanner et al., 2004; Bambach, 2006), much recent work has been undertaken on the Tr–J that broadly follows this approach (Hesselbo et al., 2007) and this has led to significant improvement in understanding the nature and timing of CAMP volcanism relative to Tr–J environmental and biotic change.

In order to review this body of work it is necessary to provide the following: (i) a formal definition of the Tr–J (Section 16.2.2); (ii) an assessment of the age of the CAMP basalts; (iii) an overview of the pattern of Tr–J environmental change and mass extinction at this time; (iv) a discussion that explores whether or not the observed biotic and environmental changes at the Tr–J can be linked with any degree of confidence to CAMP volcanism (Section 16.2.6).

16.2.2. A Definition of the Triassic–Jurassic Boundary

This review is primarily concerned with the time period spanning the Rhaetian (uppermost Triassic) to Hettangian (lowermost Jurassic) stages. The Rhaetian currently lacks a Global Stratotype Section and Point (GSSP). The only candidate GSSP at the time of writing is located at Steinbergkogel near Hallstatt, Austria (Krystyn et al., 2007). The base of the Rhaetian in this GSSP proposal is defined by the first appearance of the conodont *Misikella posthernsteinii* (Krystyn et al., 2007). This GSSP proposal is likely to be accepted by the Subcommission on Triassic Stratigraphy (McRoberts, 2009) and is adopted as a working definition of the Rhaetian in this review. Based on palaeomagnetic studies, the Rhaetian Stage appears to be around 9 Ma in duration (Hüsing et al., 2011).

The GSSP for the Hettangian, and therefore the Triassic–Jurassic (Tr–J) boundary, is located at Kuhjoch in the northern Calcareous Alps, Austria (von Hillebrandt et al., 2007; Morton, 2008). The boundary is defined by the first appearance of the ammonite *Psiloceras spelae* (von Hillebrandt et al., 2007). The GSSP for the Sinemurian Stage, and therefore the top of the Hettangian, is located in a coastal cliff north of the village of East Quantoxhead in Somerset, southwest UK (Bloos and Page, 2002). The Hettangian–Sinemurian boundary is defined by the first appearance of the ammonite genera *Vermiceras* (*V. quantoxense*, *V. palmeri*) and *Metophioceras* (*M. sp. indet. 1*) (Bloos and Page, 2002). Based on cyclostratigraphy, the Hettangian Stage appears to be 1.8 Ma in duration (Ruhl et al., 2010).

The most recent age estimate of the Tr–J boundary is $201.31 \pm 0.18/0.38/0.43$ Ma (Schoene et al., 2010). This age is based on $^{206}Pb/^{238}U$ geochronology from volcanic ash beds within a marine Tr–J boundary section in the Pucara basin, Peru (Schoene et al., 2010).

16.2.3. Break-Up of Pangaea and Massive Volcanism at the Tr–J Transition

The break–up of the supercontinent Pangaea began around 230 Ma with the rifting of the southeastern part of North America, which led to the opening of the Atlantic Ocean (Schlische et al., 2002). The opening of this ocean resulted in the emplacement of the CAMP basalts, which are preserved in northern South America, eastern North America, North Africa, and Europe (e.g., McHone, 1996, 2003). The CAMP covers $\sim 1 \times 10^7$ km^2 (Nomade et al., 2007) and has a preserved magmatic volume of $\sim 1 \times 10^6$ km^3 (McHone, 2003). The original magmatic volume is estimated to be greater than 2.5×10^6 km^3 (McHone, 2003), which is comparable to voluminous LIPs, such as the Siberian and Deccan basalts (Nomade et al., 2007). Exceptionally large basalt lava flow successions 100 m to 300 m in thickness are preserved in basins in Morocco, North America, and South America (Nomade et al., 2007). The CAMP basalts were emplaced primarily as dikes, sills, and lava flows, and emplacement proceeded in a series of discrete pulses that were diachronous across the proto-Atlantic Ocean (Deenen et al., 2010). A palaeogeographical map of the Earth at the Tr–J and the geographical extent of CAMP basalts is shown in Figure 16.1.

There are over 100 published radiometric ages for the CAMP basalts, which have been generated using $^{206}Pb/^{238}U$ and $^{40}Ar/^{39}Ar$ geochronology (Hesselbo et al., 2007). Although several studies have noted that $^{40}Ar/^{39}Ar$ ages are systematically 0.3% to 1.0% younger than $^{206}Pb/^{238}U$ ages of the same rocks (Renne et al., 1998; Min et al., 2000; Villeneuve et al., 2000), and are also $\sim 1.0\%$ younger than astronomically calibrated stratigraphic ages (Schaltegger et al., 2008), recent calibration of the $^{40}Ar/^{39}Ar$ technique (Kuiper et al., 2008) and new $^{206}Pb/^{238}U$ studies have permitted a more precise dating of the CAMP basalts. $^{40}Ar/^{39}Ar$ dating of CAMP basalts has yielded a mean age of 199.0 ± 2.4 Ma (Marzoli et al., 1999; Nomade et al., 2007). Application of the astronomical recalibration of the $^{40}Ar/^{39}Ar$ dating standard to this date

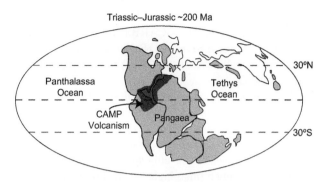

Triassic–Jurassic ~200 Ma

FIGURE 16.1 Schematic palaeogeographical cartoon showing the position of the Earth's continents at the Triassic–Jurassic transition. Map modified from Scotese Paleomap Project 2000 (*http://www.scotese.com/Jurassic.htm*) and Ziegler (1990). Palaeolatitudes from Kent and Tauxe (2005). Dark grey shading highlights the extent of basalts from the Central Atlantic Magmatic Province (CAMP) (extent of CAMP basalts from Hesselbo et al., 2002).

requires an increase in age of 0.65% (Kuiper et al., 2008), resulting in a mean age of 200.3 ± 2.4 Ma for the CAMP (Deenen et al., 2010). This recalibrated $^{40}Ar/^{39}Ar$ mean age is similar, within error, to the $^{206}Pb/^{238}U$ age of the large North Mountain CAMP basalt flow in the Bay of Fundy, Canada, which was calculated as 201.38 ± 0.02 Ma (Schoene et al., 2010). Such precise dating is essential in order to firmly establish the sequence of events in the geological past, and is essential when comparing the timing of geological events such as volcanic eruptions with biological events such as extinction.

Geochronology indicates that CAMP volcanism occurred roughly synchronously with the Tr–J boundary (the first appearance of *P. spelae*; von Hillebrandt et al., 2007). The opening of the Atlantic Ocean and the emplacement of one of the most voluminous LIPs of the Phanerozoic thus form the large-scale plate tectonic backdrop for the transition from the Rhaetian to the Hettangian Stage.

16.2.4. The Earth's Physical Environment at the Triassic–Jurassic Transition

The Earth's physical environment at the Tr–J has been subject to extensive scientific investigation using a combination of numerical models and a suite of recently published geochemical and biological proxy records. Much effort has been focused on evaluating the composition of the Earth's atmosphere at the Tr–J, and on understanding the carbon cycle at this time.

16.2.4.1. The Earth's Atmosphere and Global Temperatures at the Triassic–Jurassic Transition

Models of biogeochemical cycling over Phanerozoic time provide estimates of atmospheric composition at the Tr–J.

Several models have been published, including the most recent version of the GEOCARBSULF model (Berner, 2006, 2009), the COPSE model (Bergman et al., 2004), and the model of Falkowski et al. (2005) that only deals with atmospheric O_2.

Both GEOCARBSULF and COPSE agree that atmospheric CO_2 was considerably higher at the Tr–J than a pre-industrial present day value of 280 parts per million by volume (ppmv) (e.g., Bergman et al., 2004). GEOCARBSULF suggests that atmospheric CO_2 at this time was between 4 and 5 times higher than present day, around 1120–1400 ppmv (Berner, 2006), and the COPSE model suggests atmospheric CO_2 levels approximately 3 times higher than present day, around 800 ppmv–850 ppmv (Bergman et al., 2004).

Models of atmospheric O_2 are constrained within a 'fire window', defined by the presence or absence of fossil charcoal, which is produced naturally by wildfire. The upper limit of the 'fire window' is 35% O_2, above which runaway wildfire would engulf the Earth's vegetation, and the lower limit of the 'fire window' is 15% O_2, below which there is insufficient O_2 to support fire (Belcher and McElwain, 2008). Charcoal is preserved throughout the Rhaetian and Hettangian (e.g., Belcher and McElwain, 2008; Marynowski and Simoneit, 2009; Belcher et al., 2010), and so atmospheric O_2 is unlikely to have been as low as 12% at the Tr–J, as suggested by the model of Falkowski et al. (2005). The most recent revision of the GEOCARBSULF model indicates that atmospheric O_2 may have been around 15% at the Tr–J (Berner, 2009), and the COPSE model suggests that Tr–J atmospheric O_2 levels were 20% (Bergman et al., 2004), slightly lower than the present day value of 21%.

The GEOCARBSULF model inputs data every ten million years (Berner, 2006), and forcings in the COPSE model are updated every million years (Bergman et al., 2004). Both models are therefore unable to detect more rapid episodic changes in atmospheric CO_2 concentration. In contrast, palaeoproxies of ancient atmospheric CO_2 can be well-constrained stratigraphically and can detect changes in atmospheric CO_2 at a higher temporal resolution. The stable carbon isotopic composition ($\delta^{13}C$) of pedogenic (soil) carbonates is an established proxy for atmospheric CO_2 (e.g., Cerling, 1992; Royer et al., 2001; Breecker et al., 2010). Pedogenic carbonates incorporate carbon from two sources: atmospheric CO_2 that diffuses into the soil directly, and in situ CO_2 from biological respiration (e.g., Royer et al., 2001). The $\delta^{13}C$ of these two sources is distinct and pedogenic carbonates form in isotopic equilibrium with soil CO_2 (Cerling et al., 1991). As a result, the concentration of CO_2 in the atmosphere can be inferred if the concentration of soil CO_2, the $\delta^{13}C$ of the two sources, and the $\delta^{13}C$ of soil carbonate are known (Cerling, 1991). The concentration of soil CO_2 in ancient

palaeosols is assumed to be similar to values recorded from modern soils, which have long been estimated to range from around 9000 ppmv in highly productive soils to around 3000 ppmv in less productive soils (see review in Royer et al., 2001). The $\delta^{13}C$ of atmospheric CO_2 is typically derived from marine carbonates, assuming a constant fractionation and ocean–atmosphere equilibrium (e.g., Veizer et al., 1999), and the $\delta^{13}C$ of in-situ CO_2 from biological respiration closely corresponds to the $\delta^{13}C$ of soil organic matter (Cerling, 1992; Royer et al., 2001). Measurements of the $\delta^{13}C$ of pedogenic carbonate nodules are used as a proxy for the $\delta^{13}C$ of soil CO_2. This pedogenic carbonate CO_2 proxy indicates that atmospheric CO_2 rose by ~1020 ppmv from the upper Triassic (Carnian and Norian) into the lower Jurassic (Hettangian) (Tanner et al., 2001; Beerling, 2002), and that CO_2 values fluctuated between ~1500 and ~3000 ppmv in the Rhaetian (Cleveland et al., 2008).

A detailed Tr–J atmospheric CO_2 reconstruction from the Newark Basin, North America, indicates that Rhaetian CO_2 levels were ~2000 ppmv and increased to ~4400 ppmv in the interval spanning the Tr–J boundary (Schaller et al., 2011). Atmospheric CO_2 levels were elevated for at least 500 ka in the section investigated by Schaller et al. (2011). It has been suggested that the concentration of soil CO_2 in modern soils is on average just ~2500 ppmv (Breecker et al., 2010), which is less than the value used in most palaeo-atmospheric CO_2 reconstructions based on palaeosols (e.g., Cleveland et al., 2008; Schaller et al., 2011). These results imply that atmospheric CO_2 values across the Tr–J from existing palaeosol studies may be too high, and could be reduced by as much as half (Royer, 2010).

Another proxy for atmospheric CO_2 exploits the observation that the numbers of stomata on the surface of a leaf are inversely related to ambient CO_2 concentration during leaf growth (Woodward, 1987; Dickinson, 2012, this volume). This relationship has been applied to fossil leaves and has been used to reconstruct CO_2 concentrations from the Quaternary (e.g., Beerling et al., 1993; McElwain et al., 1995), the Neogene (e.g., van der Burgh et al., 1993; Kürschner et al., 1996), and as far back in time as the Palaeozoic (e.g., McElwain and Chaloner, 1995). Analyses of the stomatal numbers of Ginkgoalean and Cycadalean leaves preserved in Greenland and Sweden indicate a fourfold rise in CO_2 across the Tr–J, from ~600 ppmv to ~2100 ppmv–2400 ppmv (McElwain et al., 1999). This transient rise in CO_2 began in the Late Rhaetian, in strata that correlate to rocks that predate the first appearance of *P. spelae*, and lasted approximately until the middle Hettangian (McElwain et al., 1999; Hesselbo et al., 2002; McElwain et al., 2007). The CO_2 values from this palaeobotanical record are similar to palaeosol estimates of CO_2 if a ~2500 ppmv soil CO_2 value is assumed (Breecker et al., 2010). Analyses of the stomatal numbers of the leaves of the seed-fern *Lepidopteris* and the Ginkgoalean *Ginkgoites* from Germany indicate that atmospheric CO_2 rose during the Rhaetian and reached a maximum of between 1650 and 2750 ppmv during the latest Rheatian (Bonis et al., 2010). A key difference between the stomatal CO_2 records from Greenland and Sweden (McElwain et al., 1999) and Germany (Bonis et al., 2010), is that the German records indicate that atmospheric CO_2 levels in the middle Rhaetian were almost as high as in the latest Rhaetian (Bonis et al., 2010), whereas the records from Greenland and Sweden indicate much lower middle Rhaetian values (McElwain et al., 1999). One possible explanation for this discrepancy is that the high CO_2 values recorded in the middle Rhaetian of Germany (Bonis et al., 2010) represent a transient pulse of CO_2 that occurred within the 2 m–15 m sampling gaps in Greenland and Sweden.

Global temperatures are thought to have co-varied with atmospheric CO_2 for 450 Ma owing to the enhanced greenhouse effect (Royer et al., 2001), and the rise in atmospheric CO_2 across the Tr–J is thought to have led to a 3–4°C rise in the mean global surface air temperature through such greenhouse warming of the Earth (McElwain et al., 1999). These temperature estimates are supported by GCMs, which indicate that this rise in CO_2 may have raised the average surface air temperature by 6.1°C globally and 8.0°C on land, compared to 4.5°C over the oceans (Huynh and Poulsen, 2005). Globally averaged mean annual ocean temperatures may have risen by 5°C, with the greatest warming occurring in the surface ocean: 5.4°C at 10 m depth to 4.5°C at 100 m (Huynh and Poulsen, 2005). Oxygen isotopes from diagenetically unaltered oyster shells from the UK indicate a local warming of up to 10°C at the Tr–J (Korte et al., 2009). There are currently no published Tr–J temperature records based on the TEX_{86} palaeothermometer (Schouten et al., 2002) with which to evaluate palaeotemperatures from models or oxygen isotopes. Warming at the Tr–J is independently supported by palaeobotanical records, which indicate poleward migration of plants at the Tr–J in eastern Greenland (McElwain et al., 2007).

16.2.4.2. The Triassic–Jurassic Carbon Cycle

Proxy records of atmospheric CO_2 indicate that the atmospheric carbon reservoir was perturbed at the Tr–J (McElwain et al., 1999; Beerling and Berner, 2002; Hesselbo et al., 2002). Carbon isotope records indicate that a perturbation of the Tr–J carbon cycle also affected shallow-ocean and terrestrial carbon reservoirs (McRoberts et al., 1997; Pálfy et al., 2001; Ward et al., 2001; Hesselbo et al., 2002; Guex et al., 2004; Galli et al., 2007; Kuerschner et al., 2007; Williford et al., 2007; Ruhl et al., 2009). This perturbation is typically represented by more

than one isotopic excursion, and the carbon isotope pattern that appears to be consistent in Tr–J boundary sections around the world is an abrupt 'initial' negative carbon isotope excursion, closely followed by a 'main' negative carbon isotope excursion that may represent longer-term ^{12}C enrichment into the Sinemurian (see Hesselbo et al., 2002, 2007; Ruhl et al., 2010; Whiteside et al., 2010). These two negative excursions are separated by a pronounced positive excursion (see Hesselbo et al., 2007; Williford et al., 2007). Most Tr–J carbon isotope records from marine sections are based on bulk organic matter or bulk carbonate. However, $\delta^{13}C$ records from diagenetically unaltered skeletal low-Mg calcite (from the oyster *Liostrea hisingeri*) follow carbon isotope trends established from bulk organic matter samples, which indicates that carbon isotope records based on bulk organic matter capture the isotopic characteristic of the local water bodies (Korte et al., 2009). The carbon isotope composition of long-chain *n*-alkanes, derived from epicuticular plant waxes, also follows the pattern from bulk organic matter, confirming strong transient ^{13}C depletion of the atmosphere during the Tr–J (Ruhl, 2010; Whiteside et al., 2010).

The 'initial' carbon isotope excursion has received particular attention owing to its magnitude and geologically brief duration (e.g., Hesselbo et al., 2002; Ruhl et al., 2010). In bulk organic matter samples, the 'initial' negative carbon isotope excursion is represented by a rapid 2‰–3‰ deviation (see discussion in Beerling and Berner, 2002). In the *n*-alkane record however, the magnitude of the 'initial' carbon isotope excursion is considerably larger, and is represented by a 5‰–6‰ negative excursion from Rhaetian base values of about −29‰ (Ruhl, 2010). The total duration of the 'initial' carbon isotope excursion is estimated to be ~20–40 thousand years (ka), based on cyclostratigraphy (Ruhl et al., 2010), which implies strong $\delta^{13}C$ depletion of the atmosphere within 5 ka–10 ka (Deenen et al., 2010).

Pronounced carbon isotope excursions are recognized in Tr–J boundary sections that are geographically widespread and record different sea-level and thermal subsidence histories (e.g., Hesselbo et al., 2007). As a result, the Tr–J carbon isotope excursions are thought not to result from changing proportions of organic components of varied isotopic composition, nor from diagenentic effects (see also Hesselbo et al., 2002). Additionally, a detailed characterization of sedimentary organic matter during the 'initial' carbon isotope excursion at the Tr–J GSSP at Kuhjoch, Austria (von Hillebrandt et al., 2007; Morton, 2008), revealed only minor changes in kerogen type, which was mainly of terrestrial origin (Ruhl et al., 2010). This supports the view that changes in the $\delta^{13}C$ of organic matter at this time are genuine and represent a real disturbance to the global carbon cycle (Ruhl et al., 2010).

Input of a large volume of isotopically light carbon is required to explain both the rise in atmospheric CO_2 across the Tr—J and the patterns evident in the carbon isotope record. Identifying the source of this carbon is important. At the global-scale, the CAMP eruptions and the rise in atmospheric CO_2 at the Tr—J began synchronously (at least within the errors of $^{206}Pb/^{238}U$ and $^{40}Ar/^{39}Ar$ geochronology) and, as a result of this tight correlation, it is thought that volcanic degassing from the CAMP basalts provided at least some of the CO_2 input to the atmosphere at this time (e.g., McElwain et al., 1999; Beerling and Berner, 2002). The reconstruction of Tr—J atmospheric CO_2 values from pedogenic carbonates in North America (Schaller et al., 2011) provides persuasive evidence of the link between CAMP volcanism and CO_2. Atmospheric CO_2 levels increased from ~2000 ppmv to ~4400 ppmv following the eruption of the Orange Mountain Basalt in the Newark Basin, and declined to pre-eruption levels over the following 300 ka (Schaller et al., 2011), a pattern that is repeated for the two succeeding lava flows: the Preakness Basalt and the Hook Mountain Basalt (Schaller et al., 2011).

Carbon cycle simulations indicate that a total 8000—9000 GtC may have been degassed as CO_2 during the CAMP basaltic eruptions (Beerling and Berner, 2002), but it is thought that this is insufficient to entirely account for the rise in atmospheric CO_2 observed in proxy records and the magnitude of the observed 'initial' negative carbon isotope excursion. Consequently, it is thought that an additional 5000 GtC was released as CH_4, which is isotopically lighter ($\delta^{13}C$ values around -60%) (Beerling and Berner, 2002; see also Hesselbo et al., 2002). Dissociation of CH_4 from destabilized gas hydrates, which would rapidly oxidize to CO_2 in the atmosphere and ocean (Zachos et al., 2008; Harvey, 2012, this volume), is a possible source of this carbon, as has been suggested for the PETM (Dickens et al., 1995) and the Toarcian oceanic anoxic event in the Jurassic (e.g., Kemp et al., 2005). Combustion of subsurface organic-rich rocks by CAMP basalts (van de Schootbrugge et al., 2009) and contact metamorphism (Schaller et al., 2011) have also been proposed as potential sources of isotopically light carbon at the Tr—J.

Natural variations in atmospheric CH_4 concentration over Phanerozoic time have not yet featured widely in discussions of Tr—J atmospheric change and the Tr—J carbon cycle. The most important sources of atmospheric CH_4 in the present day are continental wetlands (IPCC, 2007a). By scaling a wetland CH_4 emission estimate for the middle Pliocene (3.6 Ma—2.6 Ma) by the relative rate of coal basin deposition in the geological past, Beerling et al. (2009) suggest that input of CH_4 to the atmosphere has fluctuated dramatically over the past 400 Ma. Although the estimates of atmospheric CH_4 are of low temporal resolution and exclude short bursts of CH_4 input from geospheric reservoirs such as gas hydrates and thermal decomposition of organic matter (Beerling et al., 2009), the model results

indicate that the global wetland CH_4 flux rose from ~100 ppbv in the Late Triassic to ~700 ppbv—1000 ppbv in the Early Jurassic (Beerling et al., 2009). Although the results of this model indicate that changes in atmospheric CH_4 from continental wetlands are not as important as CO_2 in terms of their effect on ancient climates (Beerling et al., 2009), the seven-to-tenfold rise in CH_4 emissions from wetlands at the Tr—J may be significant in the context of Tr—J carbon cycle changes.

16.2.5. Mass Extinction and Biotic Changes at the Triassic—Jurassic Transition

Compilations of stratigraphic ranges of organisms through geological time indicate that the Tr—J witnessed one of the major mass extinction episodes of the Phanerozoic (e.g., Newell, 1963; Raup and Sepkoski, 1982; Sepkoski, 1993; Benton, 1995; Bambach et al., 2004). Mass extinctions are defined as "any substantial increase in the amount of extinction (lineage termination) suffered by more than one geographically wide-spread higher taxon during a relatively short interval of geologic time, resulting in an at least temporary decline in their standing diversity" (Sepkoski, 1986, p. 278). Such compilations indicate that 23% of marine families and 22% of terrestrial families became extinct at the Tr—J (Benton, 1995). Estimates of Tr—J extinction rates from databases of marine invertebrates are varied. Using the Paleobiology Database (*http://www.pbdb.org*), Kiessling et al. (2007) reported a Rhaetian extinction rate of 41% among benthic marine invertebrate genera that crossed the Norian—Rhaetian boundary. Using the same database but including all known metazoans (except the tetrapods), Alroy et al. (2008) reported a genus-level extinction rate of 63% for a combined Norian—Rhaetian time bin. Bambach (2006) analysed the database of Sepkoski (2002) (with all marine taxa included) and reported a genus-level extinction rate of 47% in a combined Late Norian—Rhaetian time bin. Reduced origination rates of organisms were also a factor in lowering global marine diversity at the Tr—J (Bambach et al., 2004; Kiessling et al., 2007). Compilations of stratigraphic ranges of land plants through geological time do not indicate elevated global extinction rates among plants at the Tr—J (Niklas et al., 1980, 1985; Niklas and Tiffney, 1994).

Global databases provide a broad perspective on biotic events at the Tr—J and reveal the gross nature and pattern of extinction among different groups of organisms at this time. The Ceratitina, a group of ammonites, and the conodonts suffered complete extinction during the Late Triassic (e.g., Benton, 1993; Hallam, 2002). Calcareous sponges were almost eliminated at the Tr—J, with just two families surviving and 16 suffering extinction in the Rhaetian (Benton, 1993; Hallam, 2002), while spiriferid and terebratulid brachiopods suffered more than 67%

genus-level extinction during the Late Triassic (Bambach, 2006). Five bivalve families suffered extinction in the Rhaetian and 52 families range through into the Hettangian (Benton, 1993). Reefs and reef-building organisms were profoundly disturbed at the Tr—J (Stanley, 1988; Kiessling et al., 2007, 2009), leading to a pronounced depression in coral reef building in the Hettangian and Sinemurian stages (e.g., Kiessling et al., 2009). Lathuilière and Marchal (2009) report that 60% of Rhaetian coral genera do not range into the Hettangian, although this estimate falls to 53% when Rhaetian singletons are removed from the analysis (Kiessling et al., 2009). Certain groups of organisms appear to have passed through the Tr—J largely unscathed. Among echinoderms, all 34 families of echinoid present in the Late Triassic range into the Jurassic and just one family of crinoids, the Somphocrinidae, suffered extinction in the Rhaetian (Hallam, 2002). Similarly, all 35 gastropod families that are present in the Rhaetian range into the Hettangian (Hallam, 2002). Eleven reptile families suffered extinction in the Late Triassic (Benton, 1993) and, among higher plant taxa, only the Peltaspermaceae, a family of seed-ferns, were lost from the Earth's biota at the Tr—J (e.g., McElwain and Punyasena, 2007).

16.2.6. Relationship Between CAMP Volcanism and Biotic Change at the Tr—J

The stratigraphic relationship between CAMP volcanism and biotic turnover, which occurred in the Late Rhaetian prior to the Tr—J boundary (the first appearance of *P. spelae*; von Hillebrandt et al., 2007), has received much recent scientific attention. It is clear that the emplacement of CAMP basalts took place roughly at the same time as biotic turnover (e.g., Deenen et al., 2010; Schoene et al., 2010) and continued into the Hettangian. Geochronology, palaeomagnetics, and trace-element ratios have been used to provide information on the relative timing of volcanic pulses within the CAMP province (e.g., Deenen et al., 2010; Schoene et al., 2010) and have resulted in precise stratigraphic correlation between individual basalt flows and biotic turnover in marine sediments.

The first pulse of CAMP volcanism occurred in Morocco (Deenen et al., 2010), although there is some evidence from Lanthanum—Ytterbium versus Thorium—Yttrium ratios that the Taoudenni dikes in Mali, West Africa, may predate the oldest CAMP basalts in Morocco (Verati et al., 2005; see also Deenen, 2010, p. 114). Four pulses of CAMP volcanism occur in the Late Rhaetian, before the first appearance of *P. spelae*. These are the Lower and Intermediate Units in Morocco, the Orange Mountain Basalt in North America (Deenen et al., 2010), and the North Mountain Basalt in Canada (Schoene et al., 2010). The Lower Unit and the North Mountain Basalt were emplaced around the same time as the 'initial' carbon isotope excursion (Deenen et al., 2010; Schoene et al., 2010). The Intermediate Unit and Orange Mountain Basalt correspond to one another, are considered to be coeval (Deenen et al., 2010), and were emplaced just prior to the onset of the 'main' carbon isotope excursion (Deenen et al., 2010). These four basalt flows are tightly correlated to marine extinction and biotic turnover (e.g., Deenen et al., 2010; Schoene et al., 2010), and were emplaced over ~20 ka (Deenen et al., 2010). Other CAMP basalt flows in North America, the Preakness Basalt, and the Hook Mountain Basalt are younger and do not correspond to any stratigraphic levels containing evidence of biotic extinction.

The tight stratigraphic correlation between CAMP volcanism and biotic turnover at the Tr—J has led to the suggestion that "the case for a dominant volcanic *deus ex machina* [at the Tr—J] now looks incontestable" (Hesselbo et al., 2007, p. 1). Environmental changes resulting from CO_2 and SO_2 emissions from CAMP volcanism are thought to have been primary causes of the Tr—J mass extinction and biotic turnover. Certain traits apparently elevated extinction risk among particular organisms at the Tr—J, and such studies of extinction selectivity on global and local-scales can be used to test the biological effect of proposed stress mechanisms (e.g., Knoll et al., 1996, 2007). These can then narrow down the possible causal mechanisms for mass extinction at the Tr—J (e.g., Kiessling and Simpson, 2011).

16.2.6.1. Atmospheric CO_2: Rising Temperatures and Plant Extinction

The observed fourfold rise in atmospheric CO_2 across the Tr—J and associated 3—4°C greenhouse warming of the Earth is thought to have had deleterious effects on plant life (McElwain et al., 1999). For plants growing in a warm Late Triassic climate (summer temperatures ~30°C) in Greenland and Sweden, additional CO_2-induced warming may have raised noon leaf temperatures for canopy and open-habitat taxa above the observed heat limit for CO_2 uptake in modern day tropical taxa, which ranges from 45—52°C owing to enzyme denaturing (McElwain et al., 1999). Small and/or highly dissected leaves dissipate heat more effectively than large and/or entire-margined leaves, and the selective replacement of taxa with large entire-margined leaves (canopy Ginkgoales) by taxa with highly dissected leaves (e.g., *Czekanowskia* and *Baiera*) across the Tr—J in Greenland and Sweden may be an expression of rising temperatures at this time (McElwain et al., 1999). Such greenhouse warming has also been invoked to explain the ~17% genus-level extinction among plants in Jameson Land, eastern Greenland, and may also explain a ~85% decline in standing species richness in this region (McElwain et al., 2007). Such high local to regional turnover rates among plant genera/species is at odds with the picture of static plant diversity among higher taxa in global databases (e.g., Niklas et al., 1980, 1985; Niklas and Tiffney, 1994). One explanation for this difference may be that single-section or regional studies record only local to regional events that are outweighed in global compilations by data from regions where diversity remained static (Wing, 2004). A second explanation is that inconsistent taxonomic practice masks diversity decline and extinction in global compilations of plant taxa (Wing, 2004).

These estimates of plant extinction rates and species richness are derived from investigations of plant mega-fossils (mostly leaves), and are typically much higher than estimates of diversity loss based on sporomorphs (pollen and spores). A regional sporomorph diversity loss of ~60% is characteristic of the Tr–J in the Newark Basin, North America (Fowell and Olsen, 1993; Olsen et al., 2002), but existing sporomorph records spanning the Tr–J in Europe provide little evidence of such catastrophic diversity loss and instead show a pattern of gradual compositional change from the Rhaetian to the Hettangian (Kuerschner et al., 2007; Bonis et al., 2009). There is also little evidence for abrupt diversity loss in sporomorph records from East Greenland, despite strong evidence of plant extinction and palaeo-ecological change in the megafossil record in this region (Mander et al., 2010). This discrepancy has led some authors to question the impact of Tr–J environmental change on plant life at this time (e.g., Bonis et al., 2009).

The discrepancy between megafossil and sporomorph records of Tr–J plant diversity in East Greenland may be explained by the poor representation of ginkgos and repro-ductively specialized (insect pollinated) plants such as cycads, bennettites, and the seed-fern *Lepidopteris* in the sporomorph record (Mander et al., 2010). Additionally, the limited taxonomic resolution of the Tr–J sporomorph record masks plant extinctions at this time. For example, the extinction of the Peltaspermaceae at the Tr–J (e.g., McElwain and Punyasena, 2007) is masked in the sporomorph record because the pollen produced by this plant family (*Cycadopites*; Townrow, 1960) was also produced by several other plant groups including cycads, bennettites, and ginkgos (Mander et al., 2010; Mander, in press). In order to establish whether or not diversity loss and extinction among reproductively specialized plants at the Tr–J is characteristic of other regions, future studies of Tr–J plant life should aim to incorporate megafossils, which census this group of plants more accurately by virtue of greater taxonomic resolution (Mander et al., 2010; Mander, in press).

16.2.6.2. Atmospheric CO_2: A Biocalcification Crisis at the Tr–J?

Rising atmospheric CO_2 has also been invoked to explain the apparent temporary suppression of carbonate sedimentation at the Tr–J and the extinction patterns among shelly marine invertebrates. It has been postulated that high atmospheric CO_2 led to undersaturation of seawater with respect to aragonite leading to a biocalcification crisis at the Tr–J (Hautmann, 2004). This idea is supported by the replace-ment of aragonite by calcite in the shells of epifaunal bivalves across the Tr–J (Hautmann, 2004), and the selec-tive extinction of organisms with aragonitic or high-Mg calcitic skeletons and little physiological control of bio-calcification (Hautmann et al., 2008). Additionally, numer-ical models of the carbon cycle suggest that CO_2 and SO_2 were released from CAMP volcanism (together with possible feedbacks from CH_4 input from gas hydrates) in sufficient quantities to have resulted in carbonate under-saturation of the ocean (Berner and Beerling, 2007). However, such a scenario is highly unlikely, as it is necessary that the Tr–J oceans were close to undersaturation at the point of CO_2 emission, and that estimates of the amount of volcanic gas released during CAMP eruptions approach "the very upper limits of plausibility" (Hesselbo et al., 2007, p. 8).

Indeed, recognizing the effects of Tr–J ocean acidifi-cation in the fossil record is not straightforward. The thickness of bivalve shells across the Tr–J in the UK show a temporary twofold increase, rather than a decrease as might be expected as organisms struggled to calcify in suboptimal conditions (Mander et al., 2008), and an anal-ysis of extinction selectivity at the Tr–J based on the Palaeobiology Database found that an aragonitic skeletal mineralogy represented a selective disadvantage only for bivalves, and there was no evidence of selectivity for skeletal mineralogy when all organisms were considered (Kiessling et al., 2007; Hautmann et al., 2008). However, it has been demonstrated that some taxa increase, rather than decrease, their calcification rates under conditions of reduced pH in the oceans (Wood et al., 2008), suggesting that the twofold rise in bivalve shell thickness in southwest UK may indeed be a response to oceanic pH reduction (cf. Mander et al., 2008). Kiessling and Simpson (2011) concluded that the evidence for ocean acidification was strong at the Tr–J and the clear selection against hyper-calcifying sponges and corals and physiologically unbuf-fered organisms at this time is persuasive. However, as noted by Kiessling and Simpson (2011), tropical taxa were more affected than non-tropical taxa, and the earliest Jurassic reefs and reef corals were situated in deeper water and in mid-latitudes, suggesting that these animals were escaping excess heat rather than reduced pH. Much further work is required to test the extent to which ocean acidifi-cation contributed to biotic turnover at the Tr–J and this should prove a fruitful avenue of future research. An additional concern is that the terrestrial CO_2 proxy records (e.g., McElwain et al., 1999) remain poorly correlated to marine Tr–J boundary sections (e.g., Bonis et al., 2010).

Improved stratigraphic correlation of the Tr—J transition in eastern Greenland to other Tr—J boundary sections, using techniques such as palynostratigraphy, palaeomagnetism, and geochronology, would significantly enhance understanding of the relative timing of CO_2 change and biotic turnover at the Tr—J.

16.2.6.3. Volcanism: Atmospheric Pollution and Global Cooling

There is growing evidence that the effects of rising CO_2 are not sufficient to explain all abiotic and biotic patterns that have been described at the Tr—J, and that oxygen isotopes indicate that a relatively cool interval (7—14°C) immediately followed the 'initial' carbon isotope excursion in the UK before temperatures rose by ~10°C during the 'main' carbon isotope excursion (Korte et al., 2009). Additionally, some workers have suggested that rapid sea-level fluctuations at the Tr—J indicate that global cooling and glaciation were closely associated with the Tr—J mass extinction (Guex et al., 2004; Schoene et al., 2010), while others have suggested that abrupt patterns of Tr—J floral change and plant diversity loss are inconsistent with the gradual pattern of vegetation change that might be expected from CO_2-induced global warming (e.g., McElwain et al., 2009). Indeed, of Tr—J vegetation change in northern and central Europe, van de Schootbrugge et al. (2009, p. 589) state: "the terrestrial vegetation shift is so severe and wide ranging that it is unlikely to have been triggered by greenhouse warming alone". Could CAMP volcanism also explain abrupt patterns of Tr—J vegetation change and possibly global cooling?

CAMP basalts contain an average of 460 ppmv of sulfur (Gottfried et al., 1991), and it is thought that CAMP volcanism released ~2300 Gt of SO_2 to the atmosphere (McHone, 2003; van de Schootbrugge et al., 2009). SO_2 emitted by volcanic activity quickly forms H_2SO_4 and in modern systems these molecules are washed out of the atmosphere in a matter of days to months (van de Schootbrugge et al., 2009). However, pulses of CAMP volcanism over ~20 ka (Deenen et al., 2010) are thought to have led to a build-up of aerosols in the atmosphere that would have persisted for much longer than implied by observations of modern volcanic eruptions (e.g., Tanner et al., 2004; van de Schootbrugge et al., 2009). In central and northern Europe, an abrupt switch from gymnosperm forests to transient fern-dominated vegetation coincides with an enrichment of Polycyclic Aromatic Hydrocarbons (PAHs), providing evidence for tight stratigraphic correlation between volcanic pollution and abrupt vegetation change in this region (van de Schootbrugge et al., 2009). Additionally, sporomorphs from this fern-dominated vegetation at the Tr—J are markedly darker than sporomorphs from older Rhaetian sediments of younger

Hettangian sediments (van de Schootbrugge et al., 2009). This pronounced darkening is apparently unconnected to palynological processing methods or thermal history and has been interpreted as evidence of soil acidification from sulfuric acid deposition during CAMP eruptions, which may also have contributed to biotic change in this region (van de Schootbrugge et al., 2009). The lack of an SO_2 proxy record renders the precise connection between biotic turnover at the Tr—J and atmospheric pollution from CAMP somewhat enigmatic, and the precise mechanisms by which volcanic pollution such as SO_2 and PAHs would damage plants at the Tr—J are currently unclear (e.g., McElwain et al., 2009). Nevertheless, this example highlights that volcanic pollutants may have played a direct and significant role in Tr—J biotic turnover.

H_2SO_4 aerosol droplets increase the opacity of the atmosphere, which leads to a reduction in the amount of solar radiation reaching the Earth from the Sun (Sigurdsson, 1990). This is cited as a mechanism by which SO_2 emissions from CAMP may have led to a cooling of the Earth at the Tr—J (e.g., Guex et al., 2004; Schoene et al., 2010). However, the relationship of cooling to biotic change at the Tr—J is also rather obscure. For example, oxygen isotope data indicate cooling after the 'initial' isotope excursion in the UK (Korte et al., 2009), which therefore post-dates the onset of palaeo-ecological change in the benthic marine ecosystem in the same area (e.g., Mander et al., 2008), and cooling does not appear a viable explanation for the selective extinction of large entire-margined leaves in Greenland and Sweden (McElwain et al., 1999). Cooling may have indirectly influenced biotic change at the Tr—J by lowering sea level through glaciation, and marine extinction in the Late Rhaetian occurs, in part, during an almost global phase of marine regression (e.g., Hallam and Wignall, 1999; Guex et al., 2004; Hesselbo et al., 2004; Mander et al., 2008). Sea-level fall is thought to result in extinction of marine organisms through shelf habitat loss (e.g., Hallam and Wignall, 1999) but, as noted by Jablonksi (1985) sea-level fall may also create new shelf habitat around oceanic islands, and sea-level fall also fails to explain the extinction of terrestrial biota at the Tr—J (e.g., McElwain et al., 2007).

16.2.7. Summary

The Tr—J represents an interval of major natural environmental and climatic change and of mass extinction. There is tight stratigraphic—temporal correlation between the eruption of CAMP basalts, proxy records of environmental change that are consistent with the possible effects of massive volcanism, and palaeobiological records of extinction at the Tr—J. A summary of environmental and biotic from the terrestrial Tr—J transition in eastern Greenland is shown in Figure 16.2. A summary

Terrestrial Tr—J Transition in East Greenland

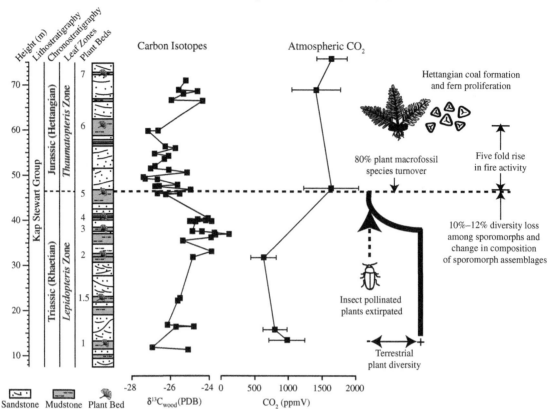

FIGURE 16.2 Environmental and biotic changes recorded from the terrestrial Tr—J transition in East Greenland. Schematic sedimentary log of a Tr—J boundary section at Astartekløft adapted from Belcher et al. (2010). The Tr—J boundary at Astartekløft is approximated by the first occurrence of the pollen morphospecies *Cerebropollenites thiergartii* following recommendations by Kuerschner et al. (2007) (see also Belcher et al., 2010; Mander et al., 2010), and is shown as a dashed horizontal line. *(Source: Carbon isotope data and stratigraphic scheme from Hesselbo et al., 2002. Atmospheric CO_2 record from McElwain et al., 1999. 80% plant megafossil species turnover from Harris, 1937 and McElwain et al., 2007. Demise of insect pollinated plants from McElwain et al., 2007. Hettangian coal formation and fern proliferation from McElwain et al., 2007. Schematic terrestrial plant diversity from McElwain et al., 2009. Fire activity from Belcher et al., 2010. Sporomorph diversity loss and compositional change from Mander et al., 2010.)*

of environmental and biotic changes from the marine Tr—J transition in southern UK is shown in Figure 16.3. Much attention has been paid to atmospheric CO_2 levels at the Tr—J, and proxy records from soil carbonates and stomata indicate that atmospheric CO_2 was elevated at this time. However, there is an urgent need for the development of proxies of other proposed agents of palaeo-environmental and biotic change at the Tr—J, such as CH_4 and volcanic gases such as SO_2 (see also Wang et al., 2012, this volume).

16.3. EXTRATERRESTRIAL IMPACTS: CASE STUDY OF THE END-CRETACEOUS EVENTS

The K—Pg is the most recent of the 'big five' biological crises of the Phanerozoic and has long been recognized by palaeontologists as a time when many groups of organisms

became extinct. This event is probably one of the most studied of intervals of Earth history, not least because it famously saw the end of the reign of the dinosaurs. This has meant that the K—Pg event has enjoyed considerable media attention, providing people the world over with an introduction to the Earth sciences. Consequently the K—Pg event has been one of the most hotly debated topics in the Earth sciences for around three decades. Raup (1989, p. 181) remarked of the K—Pg that "several hundred research papers have been published on the subject since 1980, and commentaries number in the thousands. The literature has thus become unmanageably large, so that a comprehensive review is impossible"; many years on and the K—Pg literature is still expanding towards the impenetrable. However, such intensive study has in no small part formed the majority of our understanding of the potential effects of extraterrestrial impacts on the Earth. In this section we review the physical

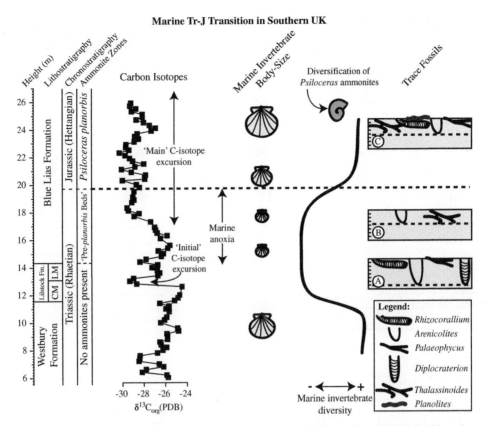

FIGURE 16.3 Environmental and biotic changes recorded from the marine Tr—J transition in southern UK. The Tr—J boundary at St Audrie's Bay is approximated by the first occurrence of the ammonite *Psiloceras planorbis* following Warrington et al. (1994), and is shown as a dashed horizontal line. Carbon isotope data and stratigraphic scheme from Hesselbo et al. (2002). Schematic marine invertebrate body-size and diversity from Mander et al. (2008) and Mander and Twitchett (2008). Diversification of *Psiloceras* ammonites from Warrington et al. (1994). Trace fossils from Barras and Twitchett (2007), A = Langport Member, B = Pre-*planorbis* Beds, C = *Psiloceras planorbis* Zone. Note reduction in maximum burrow depth in Pre-*planorbis* Beds and *P. planorbis* Zone. Lithostratigraphic abbreviations at St Audrie's Bay: CM = Cotham Member; LM = Langport Member.

characteristics of the Cretaceous—Palaeogene boundary in relation to a K—Pg impact and in Section 16.4, we consider the effects of the K—Pg impact on the climate system and the resulting extinction mechanisms.

16.3.1. A Definition of the Cretaceous—Palaeogene Boundary

The Cretaceous—Palaeogene boundary 65.5 ± 0.3 Ma marks the base of the Danian Stage in the Palaeocene Epoch, of the Palaeogene Period, which comprise part of the 'Tertiary' Subera of the Cenozoic Era (Molina et al., 2006). The GSSP for the K—Pg boundary is located at El Kef in Tunisia and is defined as the base of the boundary clay (Molina et al., 2006). The GSSP lies in the upper Maastrichtian to Palaeocene El Haria Formations, which is dominated by marls deposited in a marine environment. The boundary clay itself is 0.5 m thick and marked at its base with a 1 mm—3 mm thick rust-red coloured (iron rich layer), which is the signature of the boundary event(s). This

red layer can be traced globally, is considered to be isochronous, and is preserved in both marine and continental sections (Molina et al., 2006).

16.3.2. Impact at the End of the Cretaceous

Until the 1980s there was little physical evidence available to explain why so much of life on Earth suffered so greatly at the end of the Cretaceous. In 1980 Alvarez et al. (1980) reported an abundance of the element iridium in K—Pg sections from Italy, Denmark, and New Zealand. The iridium was shown to peak in abundance in the red clay that marks the K—Pg boundary (Figure 16.4). Platinum metals are depleted in Earth's crust and consequently such abundances are considered to be indicative of influxes of extraterrestrial material. In 1991, an 180 km diameter circular structure was identified on the Yucatan Peninsula in Mexico (Hildebrand et al., 1991). This crater was confirmed as K—Pg age by radiometric dating techniques using the $^{40}Ar/^{39}Ar$ record of minerals in the melt rock

The Signature of the End Cretaceous Event

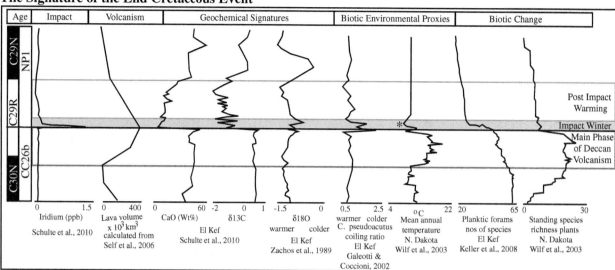

Record from the Western Interior of North America

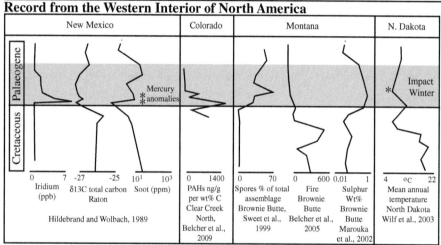

FIGURE 16.4 The global signature of the K–Pg event and the record from the western interior of North America. Data sources are shown in figure. We show evidence of extraterrestrial impact (Iridium ppb), evidence for significant volcanism (lava volume), geochemical data, and biotic changes observed in the marine realm at El Kef (the GSSP for the K–Pg) and the terrestrial plant record of biotic change from N. Dakota. It can be seen that disruptions occur following the peak in Iridium. We also show the iridium anomaly, carbon isotope anomaly, soot abundance, PAHs, spore spike, fire, sulfur, and the mean annual temperature estimated from plants all for the western interior of North America. ** shows occurrence of Mercury anomalies, * highlights minimum temperature after the impact, highlighting a cooling trend in the plant record (impact winter). C30N, C29R, and so on, are divisions of the geomagnetic polarity timescale, and CC26b and NP1 are divisions of the biostratigraphic timescale.

recovered in drill core samples from within the crater (Swisher et al., 1992). The Chicxulub crater, named after the nearby town Chicxulub Puerto, is now generally accepted to be the mark of an extraterrestrial impact event at the end Cretaceous (see Figure 16.5 for the position of the crater) although there is mounting evidence that multiple extraterrestrial impacts may have been involved in this event (Jolley et al., 2010).

16.3.3. Deccan and Other Volcanism

The latest Cretaceous was a period of intense volcanic activity that saw the eruption of huge volumes of basalts

in the Deccan province of India (Courtillot et al., 1986; Self et al., 2006) (Figure 16.3). ^{40}Ar/^{39}Ar dates of the Deccan lava flows suggest that the main outpouring occurred between 68 Ma–64 Ma but was centred around 66 Ma, with a distinct peak between 67 Ma–65 Ma in the first half of Chron 29R (a division of the geomagnetic polarity timescale) (Self et al., 2006) (see Figure 16.4). It is suggested that this main Deccan phase lasted around 1 million years emplacing 3.65×10^5 km^3 during its peak. Moreover, large-scale explosive volcanism occurred from the Albian through to the Maastrichtian in the Ochotsk–Chukotka volcanogenic belt in northeast Russia (Akinin and Miller, 2011).

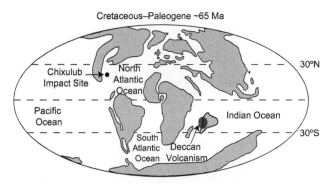

FIGURE 16.5 Schematic palaeogeographical cartoon showing the position of the Earth's continents at the Cretaceous–Palaeogene boundary. Map modified from Scotese Paleomap Project 2000 (*http://www.scotese.com/K/t.htm*). Position of Chixulub impact site from Scotese Paleomap Project 2000 (*http://www.scotese.com/K/t.htm*). Dark grey shading highlights extent of basalts from the Deccan volcanic province (extent of Deccan volcanism from Chenet et al., 2009).

A simplistic single cause is unlikely to explain all the extinctions seen during a major mass extinction event (such as the K–Pg) therefore recognition and future research into multiple simultaneous cause scenarios will better enhance our understanding of major mass extinction events. We note here, however, that the goal of this chapter is to use the K–Pg event as a case study towards highlighting the potential environmental and climate perturbations associated with a large body extraterrestrial impact. Therefore we will not debate the consequences of K–Pg aged volcanism on Earth's ecosystems.

16.3.4. Mass Extinction and Biotic Changes at the Cretaceous–Palaeogene Boundary

An apparently abrupt and global-scale turnover of Earth's biota has long been recognized to have occurred at the end of the Cretaceous. Raup and Boyajian (1988) estimated that 60%–80% of all living species became extinct at the boundary. The K–Pg event caused the least severe loss of taxonomic diversity but led to the second most ecologically severe biotic crisis of the big five Phanerozoic extinction events (McGhee et al., 2004). Several major animal groups are known to have disappeared at this time: non-avian dinosaurs, marine and flying reptiles, ammonites, and rudists (Fastovsky and Sheehan, 2005).

In the ocean foraminifera, terebratulid brachiopods, bivalve mollusks, ammonites, and marine reptiles all experienced greater than 60% genus-level extinction. The most marked extinctions occurred in calcareous nanno-plankton, that show a 90% loss of species at the K–Pg boundary (Bown, 2005) and in ammonites and rudists that were extirpated forever from the oceans (Schulte et al., 2010). These extinctions appear to be globally synchronous suggesting instantaneous extinction. Planktic foraminifera

that were abundant and diverse in the late Cretaceous also reveal a significant decline. The record at El Kef reveals the disappearance of 68.7% of species at the K–Pg boundary (Arenillas et al., 2000) (Figure 16.4). A similar pattern is recorded at sites across the globe. The extinction of corals appears to have been moderate compared to other marine invertebrates, with 30%–45% of genera became extinct at this time (Kiessling and Baron-Szabo, 2004).

Not all marine biota suffered extinction, but many groups underwent changes in composition for a time. Benthic foraminifera did not suffer mass extinction, but reveal drastic restructuring of their communities (Peryt et al., 2002). The dinoflagellate group of phytoplankton appear to indicate large fluctuations in abundance and composition in the Tethyan region (Galeotti and Coccioni, 2002; Galeotti et al., 2004).

The K–Pg event had a more profound ecological impact on the continents (McGhee et al., 2004), where it brought to an end the dinosaur-dominated ecosystems, by eradicating the non-avian dinosaurs and flying reptiles (Fastovsky and Sheehan, 2005). The majority of Earth's dominant megafauna was lost as part of the K–Pg events. However, turtles/tortoises, champosaurs, crocodilians, neornithine birds, fishes, most mammals, lizards and amphibians, snakes, and amphisbaenians survived relatively unscathed (Robertson et al., 2004). Insect family diversity also appears not to have been drastically affected by the end-Cretaceous extinction, although specialized insect feeding habits appear to have been lost (Labandeira et al., 2002).

Land plants were profoundly regionally disturbed at the K–Pg boundary. The plant megafossil record of North America reveals 70%–90% extinction at a number of sites (Johnson et al., 1989; Wilf and Johnson, 2004). Many Palaeocene floras in this region bear an imprint of ecological trauma where all dominant Cretaceous species are lost and diminishing richness is observed after the K–Pg event (Wilf et al., 2003). Palynological records from the same area reveal a lower extinction rate of ~30%–35% (Wilf and Johnson, 2004). The difference in extinction rates between plant mega and microfossils is probably the result of lower taxonomic resolution of palynomorphs (Mander et al., 2010).

An abundance of fern spores occurs in/just above the K–Pg rock layers (Tschudy et al., 1984; Vadja et al., 2001; Hotton, 2002) (Figure 16.4). This along with severe changes in the abundance and diversity of pollen and spores are recognized at locations across the globe (North America, Europe, Japan, India, New Zealand). These fern rich assemblages are interpreted as an early successional flora that is thought to demonstrate the first phases of plant recovery following major disturbance to plant ecosystems following the asteroid impact at Chicxulub. The story however, is far from simple as there is no record of extinction in Arctic or Antarctic floras (Spicer, 1989; Spicer and Herman, 2010) and far greater floral turnovers can be recognized in the early Maastrichtian in Asia, perhaps suggesting impact was not the only driving mechanism of floral turnovers at this time. It is clear, nonetheless, that land plants in many regions of the world suffered ecological disturbance and that extinctions occurred regionally in association with the K–Pg events.

16.4. THE POTENTIAL OF THE K–PG IMPACT TO CAUSE ENVIRONMENTAL CHANGE

Alvarez (1983, p. 637) suggested that the K–Pg impact released the equivalent of 100 million megatons of energy. He wrote "the worst nuclear scenario I have ever heard considered is when all 50,000 bombs that we [America] and the Russians own go off pretty much at the same time. The energy in that case would be less than what we got in the [K–Pg] asteroid impact by a factor of ~10^{-4}". The tremendous quantity of energy deposited on the Earth by the K–Pg impact is sufficient to change the global terrestrial environment far more and far faster than any other known terrestrial process (Hildebrand, 1993).

16.4.1. K–Pg Ground Zero

The immediate area around the impact would have been obliterated within seconds, with the impact generating a plume of incandescent (impact fireball) material in excess of 10,000°C (Pierazzo et al., 1998). This impact fireball, containing partially melted target rock and projectile material, is likely to have devastated life up to 1500 km from the crater as it expanded upwards and around the impact site. On escaping the atmosphere, the impact fireball would have cooled and collapsed causing the particles entrained in the plume to decelerate. Many of these particles would have fallen back through the atmosphere. As each particle hits the atmosphere, kinetic energy is lost due to the drag imposed by the atmosphere, causing heat to be radiated at infrared wavelengths and heating the surrounding air (Goldin and Melosh, 2009). By this means, a thermal effect may have been felt across the globe.

Thick impact ejecta deposits (see Kring, 2007, and references therein) buried the area around the crater. Shock waves and earthquakes up to magnitude 10 on the Richter-scale likely shook the ground following the impact (one minute after impact, according to Conway Morris, 1998) accompanied by an air blast of winds in excess of 1000 km hr^{-1} (Kring, 2007). The air blast potentially extended to a radius of 900 km–1800 km from the crater (Toon et al., 1997). Shock waves sent tsunamis across the Gulf of Mexico with wave heights estimated up to 300 m (Bourgeois et al., 1988) ripping up the seafloor to a depth

of 500 m (Smit, 1999). It is without doubt that Earth's biota would have faced extreme conditions within 1000 km radius of the K–Pg 'ground zero'.

16.4.2. Global Effects

16.4.2.1. Thermal Radiation

One of the most significant environmental consequences of the K–Pg extraterrestrial impact is the delivery of large volumes of thermal radiation across the globe. The re-entering ejecta from the impact would have produced a global pulse of thermal radiation (Melosh et al., 1990; Hildebrand, 1993; Kring and Durda, 2002), although this probably varied regionally and diminished with distance from the impact site (Hildebrand, 1993). This has been suggested to have been sufficient to have ignited wildfires locally, if not globally (e.g., Melosh et al. 1990; Kring and Durda 2002), although observational palaeontological data have been used to both support and dispute this idea. Wolbach et al. (1985) reported an enrichment of soot in the K–Pg boundary impact rocks (Figure 16.4). Enhanced abundances of PAHs (which can be produced in forest fires) have been reported in K–Pg boundary impact rocks in Denmark, Italy, and New Zealand (Venkatesan and Dahl, 1989). In contrast, the record of fossil charcoal from K–Pg boundary impact rocks has provided arguments against the wildfire hypothesis (Figure 16.4). The total amount of charred plant material in and directly above the K–Pg boundary rock layers is lower than the average amounts recorded in either the Cretaceous or Palaeogene strata. This pattern is consistent across the western interior of North America from sites near to the impact in the USA and north into Canada (Belcher et al. 2003, 2005). Moreover, it is now known that the K–Pg soot and PAHs reveal a signature consistent with hydrocarbon combustion at the impact site (Belcher et al., 2005, 2009; Harvey et al., 2008) (see also PAH abundance curve on Figure 16.4). The lack of evidence for wildfires, along with the high abundance of non-charred plant remains found in non-marine K–Pg boundary impact rocks, implies that the thermal radiation from the K–Pg impact produced ground temperatures of the order of 325°C with a short-lived peak of no more than 545°C (Belcher et al., 2003). These require that the thermal radiation delivered by the impact cannot have exceeded 19 kW m^{-2} at any point and not more than 6 kW m^{-2} for more than a few hours (Belcher et al., 2005).

More recently Goldin and Melosh (2009) modelled the atmospheric re-entry of the K–Pg asteroid impact melt spherules arriving at distal sites. They concluded that the thermal flux to the ground surface was a maximum of 19 kW m^{-2} but not more than 10 kW m^{-2} for over 20 minutes. These figures are in good agreement with observed palaeontological data (absence of fires). This implies that surface temperatures did not exceed 325°C (for more than an hour) and that large wildfires themselves were not responsible for the extinctions seen at this time. These estimates do not eliminate the role of relatively high temperatures (it is possible to roast a turkey in an oven at 200°C) in playing a role in the K–Pg extinctions. It is clear that significant environmental perturbations associated with an intense heat blast would be expected following the K–Pg impact event even in the absence of extensive K–Pg wildfires. The effect of such a heat blast on Earth's organisms would have depended on their location, size, habitat, and lifecycle stage.

This initial heat blast is likely to have lasted of the order of hours and likely had its most profound influence on terrestrial ecosystems. Robertson et al. (2004) highlight that water is opaque to incident thermal radiation. The heat pulse would have been dissipated by the top few micrometres of any water body as latent heat of vapourization. Therefore, thermal radiation is unable to account for the major extinctions seen in the marine realm.

It is also possible to speculate that thick cloud layers may also have shielded parts of the ground from severe levels of thermal radiation (Melosh et al., 1990) since cloud layers are typically thick at the thermal equator and at high latitudes. Such an effect could possibly have been more pronounced during the late Cretaceous and studies have postulated the existence of a permanent polar cloud cap (Spicer and Herman, 2010), which might help explain some of the regional nature of the extinctions (e.g., land plants, see Section 16.4.3).

16.4.2.2. Climatic Influence – Impact Winter

Following the heat blast, the asteroid impact is likely to have influenced the Earth's climate in several ways and on differing timescales. Alvarez et al. (1980) proposed that the dominant cause of the extinctions was probably a globe-encircling dust cloud that blocked out sunlight leaving the world in cold and darkness for several years. This original estimate was revised to 3–6 months of darkness based on models of dust settling by Toon et al. (1982), but see also the review by Starley Thompson and Stephen Schneider (Thompson and Schneider, 1986). The period of global darkness is estimated to have cooled continental interiors to below freezing. Pope (2002) argued against the ability of impact dust to cause global darkness claiming that the amount of silicate dust thrown into the atmosphere by the impact was insufficient. However, if this is combined with impact-generated sulfate aerosol and soot, then cooling might be expected. The impact hit an anhydrite rich rock sequence releasing 10,000 Gt of SO_2 instantaneously (O'Keefe and Ahrens 1989). This would have been supplemented with sulfate aerosols from Deccan volcanism (of the order of 0.0023 Gt per year and totalling ~225 Gt SO_2 in total being delivered

to the atmosphere prior to and at the K–Pg boundary – estimated from Chenet et al., 2009) and from explosive and persistent volcanism in the Ochotsk–Chukotka volcanogenic belt (injecting considerable sulfate aerosols into the polar stratosphere) (Spicer and Herman, 2010; Akinin and Miller, 2011). Large sulfate aerosol loadings along with soot from vapourization of hydrocarbons in the target rocks (Belcher et al., 2009) and the impact dust ought to have been enough to create global darkness and global cooling.

The impact winter hypothesis can be backed up by several observations of major historical volcanic eruptions. Violent volcanic eruptions may influence short-term global climate in two ways that may both be related to the effects of the K–Pg impact: sulfate aerosols and albedo changes (Wang et al., 2012, this volume).

Large eruptive columns deliver large volumes of SO_2 to the stratosphere (i.e., similar to the K–Pg impact fireball). This SO_2 will combine with other elements in the atmosphere to form sulfuric acid and sulfates, which in combination block out sunlight (de Jong Boers, 1995). The Laki fissure eruption in Iceland created a persistent dry fog composed of around 0.28 Gt of sulfate aerosols for at least 2 months in the summer of 1783 (Stothers, 1996). This is correlated with the coldest or one of the coldest winters ever recorded in numerous localities across Europe, eastern North America, and Japan. The climatic effects of the Laki eruption are believed to have influenced global climate for up to three years (Stothers 1996). The largest historical eruption is that of Tambora in Indonesia in 1815, being estimated to have caused a peak stratospheric loading of 0.01 Gt of sulfate aerosols (Oppenheimer, 2003). The effects of this eruption on climate are evidenced by the following year that is known as 'the year without a summer' which was exceptionally cold, as much as 2.5°C lower than normal, particularly in the northeastern parts of America and Canada. For example, on the 6th of June snow fell in New York while 30 cm of snow fell near Quebec City. Such adverse conditions are known to have killed forests in Connecticut and crops across northeast America (Oppenheimer, 2003). Europe also experienced a cold summer causing crop failures on a wide scale. The amounts of aerosols ejected in both the Laki and Tambora eruptions are evidenced by the amount of acid fallout observed in ice cores from Greenland (Clausen and Hammer, 1988). These examples highlight the effects that known volcanic eruptions have had on climate, where these effects were felt under the influence of 0.01 Gt–0.28 Gt of sulfate aerosols 35,000 times less than that estimated to have been delivered to the atmosphere from the Chicxulub impact on a similar timescale (although more than the estimated 0.0023 Gt per year estimated from Deccan volcanism). This highlights the influence that sulfate aerosols are likely to have had on Earth's climate following the impact.

The second mechanism that may influence short time changes in climate is the influence that the impact had on the albedo. We consider here for the first time the previously neglected climatic effects of covering a whole continent (North America), if not much of the land mass of the globe, in impact ejecta. The impact ejecta would have led to a surface with a higher reflectivity, so that greater proportions of incoming radiation would be reflected back to space, leading to surface cooling.

Jones et al. (2007) modelled the effects of the Yellowstone super-volcano by simulating the effects of placing an ash blanket across much of North America, as is similar for the K–Pg ejecta and fireball layers. The proposed combination of deforestation and increased albedo led to a local surface cooling exceeding 5°C throughout North America and influenced climate worldwide. Such a large volcanic ash blanket would be likely to remain a significant climate forcing for at least 10–50 years. Therefore the K–Pg ejecta blanket coupled with the large volume of sulfur ejected into the atmosphere probably imposed significant climatic cooling, potentially for half a century following the initial blast of thermal radiation.

16.4.2.3. Climatic Influence – CO_2 Effects

The impact at Chicxulub, into a rock sequence dominated by carbonate rocks, would have led to shock devolatilization of CO_2 in vast amounts (O'Keefe and Ahrens, 1989). A ~1 million Gt CO_2 pulse (O'Keefe and Ahrens, 1989) was released which probably remained resident in the atmosphere on a timescale of 10^3 to 10^5 years (Berner et al., 1983). This quantity of CO_2 is ~50 times the current atmospheric level and would have been capable of producing greenhouse warming by as much as ~15°C (Hildebrand, 1993), following the initial period of cold and darkness. Again, additions of volcanogenic CO_2 (of the order of 0.01 Gt per year and totalling at least 468 Gt prior to and at the K–Pg boundary from the Deccan – estimated from Chenet et al., 2009) from volcanism could have exacerbated the effects. Water vapour and CH_4 (both greenhouse gases) would also have been added to the atmosphere (Kring, 2007). The residence times of gases such as CO_2 are greater than that of dust and sulfate aerosols. Therefore, the greenhouse warming is likely to have persisted after an initial period of global cooling during which aerosol cooling overwhelmed GHG warming.

Based on estimates of the CO_2 released from the impact, Pierazzo et al. (1998) estimate the global warming to be on average 1–1.5°C. Beerling et al. (2002), estimated atmospheric CO_2 levels increased to at least 2300 ppmv (from ~350 ppmv) within 10,000 years of the K–Pg event, leading to a ~7.5°C warming. A warming trend is also reflected in $\delta^{18}O$ data from planktic and benthic foraminfera in the North Pacific, that imply a 3°C warming of the

ocean following the K–Pg boundary (Zachos et al., 1989) (Figure 16.4). We note that this warming trend is not apparent in mean annual temperatures as estimated from fossil leaves (Wilf et al., 2003; Wilf and Johnson, 2004) (Figure 16.4); this most likely being due to a lack of leaf fossils available since no leaf fossils were available in this study between 65.4 Ma and 65 Ma, thereby likely missing the warm period suggested by the $\delta^{18}O$ record (Zachos et al., 1989).

16.4.2.4. Sulfates and Acid Rain

Hildebrand and Boynton (1989) and Brett (1992) suggested that acid rain could have been produced as a consequence of the impact at Chicxulub. The sediments at Chicxulub include a significant fraction (up to 50%) of gypsum and anhydrite evaporates (Hildebrand, 1993). If devolatization of the sulfates occurred on impact, this would release 10,000 Gt of SO_2 (Hildebrand, 1993). Furthermore, the enormous amount of radiant energy in the rising impact fireball would pass through the atmosphere and fix nitrogen, generating huge amounts of nitrogen oxides (Alvarez, 1983). On reaction with water in the atmosphere, the impact-generated SO_2 and nitrogen oxides would have produced acid rain that may have begun shortly after the impact (again with possible additions from volcanism). It is suggested that of the order of 1×10^{15} mol of nitric acid rain could be produced by a Chicxulub-sized impact and 6.1×10^{15} mol of sulfuric acid rain (Kring, 2007). The acid rain is not likely to have been sufficient to acidify ocean basins (D'Hondt et al., 1994), but was likely able to influence freshwater and terrestrial ecosystems.

16.4.2.5. Ozone and UV Radiation

Large amounts of Cl and Br may have been released during vapourization of the asteroid and the target rocks (Kring, 2007). An impact of a 1 km diameter asteroid (ten times smaller than the K–Pg asteroid) is estimated to increase upper atmospheric Cl and Br densities by >130 times the normal amount (Pierazzo et al., 2010). Both of these gases destroy stratospheric ozone. Several years of stratospheric ozone depletion are estimated for a 1 km diameter sized impact (Pierazzo et al., 2010) altering the UV irradiance delivered to the Earth's surface. It is likely that levels of UV-B radiation, dangerous to living organisms, would be delivered to the Earth's surface (Pierazzo et al., 2010). Therefore, high levels of UV-B may have been able to reach the Earth's surface if the dust and sulfate aerosols from the impact cleared before stratospheric ozone levels were able to recover.

The impact may also have caused a build-up of ozone in the troposphere (Kawaragi et al., 2009; Kikuchi and Vanneste, 2010). This is firstly thought to have resulted in an intense global warming of ~2–5°C for several years due to a radiative forcing potential of ~+7 kW m^{-2} (Kawaragi et al., 2009). Secondly, ozone is toxic and at ground level would have been potentially life-threatening to standing (non-dormant or seed-based) plant and animal life for several years following the K–Pg impact (Kikuchi and Vanneste, 2010).

16.4.3. Extinction Mechanisms and Biotic Change at the K–Pg Boundary

The extinctions at the K–Pg boundary were selective in nature and can be used to aid our understanding of the kill mechanisms proposed. We discuss here the potential environmental perturbations beyond K–Pg ground zero and how the biotic changes observed in the fossil record might support these hypotheses.

16.4.3.1. Thermal Radiation

Thermal radiation is unlikely to have been a significant influence in the oceanic realm. Therefore, we consider here the evidence in support of, and the effects of, thermal radiation on terrestrial ecosystems. Robertson et al. (2004) (using calculations from Melosh et al., 1990) suggested that the atmosphere itself would have largely been transparent to the thermal radiation, such that air temperatures at points distant to the impact would only have been elevated by ~10°C. However, the surfaces of standing plants or animals would have quickly absorbed the intense thermal radiation coming from the sky. Land animals would not have been able to breathe without searing their respiratory membranes unless they could shelter from the pulse of thermal radiation, which would have been rapidly absorbed by their surficial tissues (e.g., skin) (Robertson et al., 2004). Goldin and Melosh (2009) suggest that radiation intensity of 10 kW m^{-2} would result in a 50% probability of death in a 100 second exposure, such that the thermal pulse would have been lethal to thin-skinned animals and that it may have caused some dermal damage to those with thick skin.

A significantly higher probability of survival existed for creatures that either lived in soils, used burrows, or bathed or swam in water — or at least those that had the potential to shelter from the thermal radiation using the above means (Robertson et al., 2004). This is consistent with the survival of turtles/tortoises, champosaurs, crocodilians, fishes, most mammals, lizards, amphibians, and snakes, which likely had the ability to shelter in water, in burrows, or in caves or under rocks. However, the survival of birds is harder to explain. Robertson et al. (2004) note that semi-aquatic behaviour and sheltering underground are widespread in bird groups today that may be related to Cretaceous ancestor survivors. It is likely that this sheltering strategy extends to terrestrial invertebrates also (e.g., insects), which most likely had populations that lived in water or burrows

or that had eggs or pupae underground, thus allowing them to survive the thermal stress (Robertson et al., 2004). Large land animals, for example dinosaurs, are suggested as lacking the ability to shelter adequately from such a pulse of thermal radiation.

The record of terrestrial plants also seems likely to support the idea that thermal radiation played a role in the extinctions seen at the time. Plants provide an excellent tool towards understanding environmental stress, because they are essentially static and therefore cannot move to escape high levels of thermal radiation. Severe changes in the abundance and diversity of plants are recognized at locations across the globe (North America, Europe, Japan, India, New Zealand) (Spicer, 1989). Floral extinction in North America reaches 75% in the south yet is as low as 24% in the north (Spicer, 1989), such that even on this continental-scale large variations in extinction can be observed. Wolfe and Upchurch (1986, 1987) suggest a steady recovery occurring over 1.5 million years in southern US sites. In contrast, the timespan for the perturbation seen in the Canadian floras is probably not >5000–10,000 years, with the interval occupied by opportunistic floras representing ~100 years, which is succeeded by Taxodiaceous–Cupressaceous swamps and raised mires and the return to canopied swamp-forest within 10 cm–40 cm above the K–Pg boundary (Sweet, 2001). Plants appear to have been affected less in northern North America as compared to the south. This might suggest that a more significant thermal effect was felt in the south closer to the impact site. This would accord with the idea that the thermal radiation was concentrated by the impact site and likely decreased with increasing distance (Hildebrand, 1993).

Ground temperatures following the impact are estimated to have been of the order of a couple of hundred degrees centigrade for a few hours. While such temperatures are incapable of directly igniting vegetation, they most likely did dry out and damage plants. Temperatures in excess of 55°C can cause irreversible tissue damage to plants leaves (Kolb and Robberecht, 1996, and references therein). However, the temperatures estimated from the K–Pg impact would probably not have damaged woody organs, as evidenced from measured cambium temperatures and tree survival in forest fires (Haase and Sackett, 1998). We note that many forest fires are more than twice as hot as the ground temperatures estimated at the K–Pg; therefore, the effects of the K–Pg thermal radiation are likely to have been less severe than those imposed by modern forest fires (see also Harrison and Bartlein, 2012, this volume). Plant also have roots, seeds, and vegetative parts that even in the face of significant biomass destruction are capable of regrowing again once conditions become favourable. Regrowth of herbaceous stems within the first 1 m above the K–Pg boundary (Wolfe and Upchurch, 1987;

Beerling et al., 2001) and the dominance of floral components typical of disturbed conditions is consistent with the idea that plants were able to regrow fairly rapidly from soil seed banks and/or via vegetative regrowth following initial destruction. This may suggest that the land plant extinctions observed by Johnson et al. (1989) and Wilf and Johnson (2004) may more likely represent regional extirpations, due to secondary longer-term Palaeocene environmental changes than direct K–Pg extinctions.

16.4.3.2. Extinction Mechanisms – Impact Winter

The K–Pg event greatly affected marine organisms living in the water column and it seems that primary production was wiped out (Zachos et al., 1989) (see Figure 16.4, CaO and $\delta^{13}C$ records). A 90% loss of calcareous nannoplankton species at the K–Pg boundary (Bown, 2005) and its apparent global synchroneity would imply a huge shock to the ocean system. Calcareous nannoplankton (marine phytoplankton) require light for their energy. Twelve nannofossil species are known to have survived the K–Pg events, all of which are potentially opportunist species that were restricted to neritic and high-latitude environments during the late Cretaceous (Bown, 2005). These might likely have been better adapted to lower light levels than their oceanic counterparts. These opportunist taxa extended into oceanic habitats following the K–Pg events in the place of the previously dominant specialist species (Bown, 2005). The sudden decline of these photosynthetic organisms is good evidence in support of the impact winter scenario, where the survivors were those able to temporarily endure low light levels and cooler temperatures. This is also consistent with laboratory tests on modern species subjected to prolonged darkness that reveal survivorship of only 2–8 weeks in light levels similar to those estimated in the K–Pg impact winter (Griffis and Chapman, 1988). This is significantly less than the minimum 3 month estimated duration of the impact winter.

The major disruption to marine primary productivity is reflected in the collapse of the surface to deep-water $\delta^{13}C$ gradient, which highlights a likely switching off of the ocean's biological pump (Schulte et al., 2010). The ocean would have seen exceedingly low productivity during the interval of darkness resulting from the K–Pg events (D'Hondt et al., 1998). Phytoplankton typically double on timescales of hours to days implying that once the dust and soot cleared from the atmosphere, the remaining survivors would have re-established, boosting ocean productivity (D'Hondt et al., 1998). D'Hondt et al. (1998) suggest that marine biological production returned to levels sufficient enough to support abundant small zooplankton after a few thousand years. The $\delta^{13}C$ record of these organisms stabilizes within a few hundred thousand years although it

probably took ~3 million years for the flux of organics to the deep ocean to recover, indicating that Earth's biogeochemical cycles were perturbed for a significant duration (D'Hondt et al., 1998).

The idea that primary productivity collapsed in ocean surface waters is supported by the record of benthic foraminifera that indicate a strong decrease in the food supply to the seafloor coincident with the K–Pg boundary (Alegret et al., 2004). Benthic foraminiferal assemblages in the latest Cretaceous were diverse containing a mixture of infaunal and epifaunal morphogroups indicative of mesotrophic conditions. Danian assemblages show a strong decrease in diversity and a drastic drop in infaunal morphogroups, which is interpreted to indicate a dramatic decrease in the food supply to the ocean floor (Alegret et al., 2004). This is believed to be related to a collapse in the food chain due to the mass extinction of primary producers. A similar observation is made in Danian benthic mollusc shelf ecosystems that show an increase in starvation-resistant non-planktotrophic deposit feeders and chemosymbionts with inactive lifestyles (Aberhan et al., 2007).

The decline in photosynthetic organisms in surface waters is matched by the coral record where photosymbiotic corals appear to have been significantly more affected by the K–Pg events than non-photosymbiotic forms (Kiessling and Baron-Szabo, 2004).

The marine record also reveals evidence for a cooling event in ocean waters. The coiling ratio (i.e., the number of sinistral versus dextrally coiled individuals) is suggested as a proxy for water temperature in both planktic and benthic foraminifera (Galeotti and Coccioni, 2002). A shift in the coiling ratio of *Cibicidoides pseudoactus* in Tunisian K–Pg sections is consistent with cooler waters (Figure 16.4) and co-occurs with an invasion of boreal dinoflagelette species and a brief cold spike shortly following the K–Pg event as revealed from oxygen isotope ($\delta^{18}O$) (Figure 16.4). This is interpreted as a response to a short-term cooling event in ocean bottom waters, suggesting an invasion of the boreal bioprovince into western Tethys immediately following the K–Pg events (Galeotti and Coccioni, 2002; Galeotti et al., 2004). An invasion of boreal species versus Tethyan species of dinocyst has also been observed in the earliest Danian (just after the K–Pg event) (Brinkhuis et al., 1998) and is also interpreted to represent a post-K–Pg boundary cooling event (Brinkhuis et al., 1998; Galeotti et al., 2004). The oceanic record of events seems to support the idea of an impact winter.

In the terrestrial realm, these creatures able to shelter (this time from the cold), enter a period of dormancy (hibernation), or feed on detritus might have had the best chance of survival. Of the aquatic terrestrial fauna, those able to feed on plant detritus (the base of the food chain) would probably have had a sufficient food source (e.g., amphibians, fish, turtles), to ensure their survival, which would then have provided a necessary food source for crocodiles and larger fish (Retallack, 2004). The survival of large ectotherms such as champsosaurs and crocodiles may suggest that low ground temperatures cannot have persisted for a significant duration or else that refugia outside frozen continental interiors existed to support such communities. Mammals are likely to have survived the dark and cold by burrowing or sheltering. Insectivorous, omnivorous, and herbivore—frugivores mammals appear also to have survived (Retallack, 2004), suggesting that they were able to modify their feeding patterns and perhaps survive by eating dead-decaying plant matter and surviving invertebrates.

Mean annual temperatures as estimated from fossil leaves (Wilf et al., 2003; Wilf and Johnson, 2004) also reveal a general cooling trend across the K—Pg boundary reaching minimum temperatures shortly after the impact (see * on Figure 16.4). Although this cooling trend does appear to begin ahead of the impact, the record of terrestrial plants in North America reveals that broadleaved evergreen vegetation suffered more than broadleaved deciduous vegetation such that, within conifer communities, evergreen species became extinct whereas deciduous species survived (Spicer, 1989). This has been linked to a palaeolatitudinal gradient where Wolfe and Upchurch (1986) suggested that the latest Cretaceous forests north of 60° palaeolatitude were deciduous, giving plant communities in these areas the survival advantage over their southern North American evergreen counterparts. However, the response in the Southern Hemisphere appears to be less severe where both the mid-latitudes and high latitudes suggest little difference in susceptibility between evergreen and deciduous plants and indeed no widespread ecological disruption associated with the K—Pg events (Spicer, 1989).

It is likely that the initial heat blast would have dried and caused considerable defoliation of plants (at least across North America). In general, it appears that deciduousness (the plants' pre-adaptation for dormancy) is a key life strategy of the K—Pg plant survivors and that this strategy might aid survival not only following a heat pulse but also a subsequent dark and cold impact winter. Spicer (1989) suggests that late Cretaceous high latitude coastal vegetation was well-adapted to freezing conditions and long periods without sunlight, but that such conditions were rarely experienced by mid-latitude coastal vegetation. This might help to explain the higher extinction rates observed among thermophilous broad-leaved evergreen taxa in the middle northern latitudes and the survival of deciduous northern taxa, as these plants were likely better pre-adapted to darkness, cold, or both. It seems likely that plants with an ability to enter a dormant phase and regrow from rhizomes and soil seed banks would be favoured in re-establishment once the heat and then the cold and darkness had passed.

Seasonality might be able to explain the difference observed in Northern versus Southern Hemisphere extinctions. If the impact occurred in the Northern Hemisphere's late-summer, evergreen plants in this area might be expected to suffer more than deciduous plants, since deciduous forms would already be preparing for their normal period of dormancy (Spicer, 1989). If this is the case, then the Southern Hemisphere would be experiencing late winter and, assuming the impact created no fatal chill, plants might continue to survive relatively unaffected. An alternative hypothesis is that there was no global impact winter and that sunlight-blocking dust did not extend across the Southern Hemisphere.

Historical volcanic eruptions support the idea that an impact winter would impact plant communities. The eruption of Tambora in Indonesia in the spring of 1815 caused snow to fall in summer the following year in New York and Quebec. Widespread frosts occurred across the USA and Europe as a result of the climatic effects of this eruption, leading to widespread crop death and significant damage to forests (Oppenheimer, 2003). Tambora probably had a lesser impact on global temperatures than the K—Pg impact, yet its effects were felt across the globe, highlighting the likely ability of an impact winter to significantly affect Earth's ecosystems (cf. Thompson and Schneider, 1986).

Terrestrial ecosystems appear to have recovered fairly rapidly following the K—Pg events. The atmospheric $\delta^{13}C$ excursion (e.g., see Figure 16.4) from terrestrial records appears to have lasted ~130 ka (Arens and Jahren, 2000) which is similar in duration to the time taken for planktic and benthic $\delta^{13}C$ records to stabilize in the ocean, but contrasts to the three million years taken for the surface—deep ocean $\delta^{13}C$ gradient to re-establish (D'Hondt et al., 1998). Lomax et al. (2001) suggest that terrestrial primary productivity recovered from the consequences of the impact winter within a decade of the impact, which is considerably more rapid than the time taken for ocean plankton productivity to re-establish (e.g., D'Hondt et al., 1998). Global terrestrial carbon storage in vegetation biomass, however, is estimated to have taken of the order of 60—80 years to recover (Lomax et al., 2001). Continued depression of plant species richness continued in hard-hit areas well into the Tertiary, taking of the order of 1—2 million years (Wolfe and Upchurch, 1987). Beerling et al. (2001) emphasize that recovery of ecosystems biodiversity following episodes of extinction operate on long, multi-million-year timescales.

16.4.3.3. Biotic Influences — CO₂ Effects

An extraterrestrial impact into carbonate rocks is suggested to directly cause a significant global warming event. For the K—Pg event, this is estimated to be ~7.5°C within 10,000 years (Beerling et al., 2002). Zachos et al. (1989) suggested that primary productivity remained low in the early Palaeocene oceans due to climate and oceanic instability. Rapid extinction and shutdown of ocean primary productivity most likely led to a two or threefold increase in atmospheric CO_2 over several oceanic mixing cycles due to the breakdown of organic flux to the deep sea, which is a major sink of CO_2 from the atmosphere (Zachos et al., 1989; D'Hondt et al., 1998). The deep-sea carbon sink is estimated to have taken ~3 million years to re-establish, based on restoration of the surface to deep-water $\delta^{13}C$ gradient (D'Hondt et al., 1998). Therefore, the cessation of oceanic primary productivity slowed the CO_2 drawdown, probably providing a positive feedback to any impact-induced global warming. This idea is supported by evidence of a gradual warming of ~3°C in the ocean during the initial 0.5 million years of the Palaeocene according to $\delta^{18}O$ records from planktic and benthic foraminifera and nannoplankton (Zachos et al., 1989).

Such global warming during the earliest Palaeocene would have undoubtedly affected ecosystems already stressed by the immediate K—Pg aftermath (heat, cold, darkness, acid rain, etc.) (Beerling et al., 2002). It seems likely that the imposed global warming would have hindered ecosystem recovery and forced further evolutionary change in earliest Palaeocene ecosystems, although it is noted that increased atmospheric CO_2 levels may have aided the initial rebound of terrestrial primary productivity by stimulating ecosystem photosynthetic productivity (Lomax et al., 2001).

16.4.3.4. Sulfates and Acid Rain

The levels of acid rain estimated to have been produced by the K—Pg impact are considered to be insufficient to have acidified the ocean (D'Hondt et al., 1994). Acid rain is, however, likely to have influenced terrestrial ecosystems, particularly plants. Prinn and Fegley (1987) suggested that an acid rain pulse would produce geochemical anomalies by leaching cations from subaerial environments. This leaching model is supported by Retallack (1996, 2004) who reported acid leaching of the boundary claystone (ejecta layer) and palaeosols. Hildebrand and Boynton (1989) found mercury anomalies in the K—Pg boundary sediments that they suggested are a result of leaching by acid rains (see *s marked on Figure 16.4). Maruoka et al. (2002) studied sulfur isotopes and carbon contents across the K—Pg boundary in the western interior of North America (see Figure 16.4), suggesting that the low ratio of organic C to non-organic S in the K—Pg layers coupled with sulfur isotopic data supports the idea of a high input of sulfate into freshwater wetlands delivered either from melted ejecta and/or acid rain.

There is good evidence from modern observations that acid rain damages plants. Relative sensitivity to acid rain

has been ranked as herbaceous dicots > woody dicots > monocots > conifers (Percy, 1991). Percy (1986) suggested that evergreen trees were more susceptible to acid treatment than deciduous forest trees. Therefore, it seems likely that acid rain may have impacted more on evergreen communities than deciduous forms that would have inbuilt pre-adaptation to acid-cold induced dormancy.

16.4.3.5. Ozone and UV Radiation

Stratospheric ozone depletion leading to increased levels of UV-B radiation (Pierazzo et al., 2010) as a result of the K—Pg impact may be supported by the large extinction seen in marine phytoplankton. Increased UV-B levels associated with the hole in the ozone layer above Antarctica have been shown to inhibit phytoplankton activity (Smith et al., 1992). Increased UV-B might also be expected to adversely affect plant life, where studies reveal decreases in plant height and shoot mass as well as a reduction in foliage area (Caldwell et al., 2003). However, taking into account the relatively good evidence for a heavily-dust laden atmosphere with increased sulfate aerosols, both of which would create cloud haze across the atmosphere and probably an impact winter, then the capacity for high levels of UV-B to be delivered at ground level seems unlikely in the direct aftermath of the K—Pg events. If the dust and sulfate aerosol clouds cleared more rapidly than stratospheric ozone regenerated, then increased UV-B radiation levels may have influenced the recovery of ecosystems or the survival of those flora and fauna that had been spared the affects of heat, cold, and darkness.

Stratospheric ozone may have been depleted by the impact. However, there is some evidence to suggest that tropospheric or ground-level ozone could have been increased (Kawaragi et al., 2009; Kikuchi and Vanneste, 2010). Ozone is a powerful respiratory irritant (Kikuchi and Vanneste, 2010) and is suggested to have varied in abundance throughout the day. Land animals capable of entering a torpor state during the period of peak ozone levels may have been capable of reducing their respiration rate significantly such that they could avoid damage to their respiratory tract (Kikuchi and Vanneste, 2010). Many small mammals today are capable of hibernation or torpor, or have nocturnal habits. It seems likely that mammals at the K—Pg might have adopted similar habits, aiding their survival/protection from high levels of ozone. Semi-aquatic and aquatic animals would have been able to avoid high levels of ozone as the half-life of ozone in water is short such that semi-aquatic animals could have sheltered in water during the daytime peak of ozone. These mechanisms would appear to be able to explain the survival of amphibians, turtles, and crocodiles.

Plants are also susceptible to ozone damage; for example, in a study of 35 trees exposed to ozone all decreased their biomass (Pye, 1988). Ferns have been found to be relatively insensitive to ozone (Renfro, 1989), which would correlate with their ability to regrow rapidly following the K–Pg events as documented by the K–Pg fern spike (Tschudy et al., 1984; Vadja et al., 2001; Hotton, 2002). It may be that ozone toxicity contributed to some of the extinctions and disruptions seen in the terrestrial realm following the K–Pg impact.

16.4.4. Concluding Remarks on the K–Pg Event

Hildebrand (1993, p. 78) commented, "Remarkable hypotheses require extraordinary proof". He was referring to the doubters of the Alvarez et al. (1980) impact hypothesis as a mechanism for the K–Pg extinctions. It seems now that the globally recognized signature of the K–Pg event, the discovery of the Chicxulub crater, all of which are closely matched to the timing and patterns of the end of the Cretaceous biotic turnover, suggests that this remarkable hypothesis has now found its extraordinary proof. Extraordinary proof is not easy to find in the geological record where time stretches back over millions of years. Indeed, Luis Alvarez (1983, p. 641) calculated that "if the asteroid had nothing to do with dinosaur extinction… then the probability that this happened by luck was about 1.5×10^{-4}."

16.5. COMPARISON OF THE TR–J AND K–PG EVENTS

Here, we briefly summarize the similarities and differences between the biotic changes that occurred during the Tr–J and K–Pg mass extinctions. We first consider terrestrial plants and then consider marine organisms. There were no angiosperms (flowering plants) at the Tr–J, whereas by the Cretaceous the Earth's flora was taking on an increasingly modern aspect following the evolution and rise to ecological dominance of this plant group (e.g., Crane and Lidgard, 1989). This makes comparison between the two time periods rather problematic. The rarity of megafossil records that span the Tr–J boundary interval also makes comparison difficult, especially with the more extensive but varied K–Pg plant records. Nonetheless, there are some intriguing similarities in the patterns of vegetation change as recorded by plant megafossils at the Tr–J in East Greenland and the K–Pg in America.

During both intervals of biotic change, there is evidence for extremely high local/regional extinction or diversity loss at low taxonomic levels such as species or genus, despite negligible extinction rates among higher plant taxa such as families (e.g., Wing 2004). For example, the megafossil record preserves an extinction of 70%–90% across the K–Pg in America (Johnson et al., 1989; Wilf and

Johnson, 2004) and an 85% decline in standing species richness across the Tr–J in East Greenland (Harris, 1937; McElwain et al., 2007). In both cases it appears that full recovery of vegetation diversity took place over millions of years (Wolfe and Upchurch, 1986, 1987; McElwain et al., 2007). Additionally, at the Tr–J in East Greenland and the K–Pg in America there was a loss of plants that dominated the pre-event vegetation (e.g., Wilf et al., 2003; McElwain et al., 2007), and insect pollinated plants appear to have been at great risk of extinction during both intervals (McElwain and Punyasena, 2007). It is unclear, however, whether these similarities reflect an underlying similarity in the way that plants respond to major environmental upheaval (e.g., McElwain et al., 2007), or whether they have arisen by chance.

There is a strong contrast in the groups of marine organisms selected against during the Tr–J and the K–Pg events. At the Tr–J, there is evidence of strong selection against taxa that invest heavily in the production of a calcium carbonate skeleton and/or are physiologically unbuffered, such as corals and hypercalcifying sponges (e.g., Kiessling and Simpson, 2011). In contrast, physiologically unbuffered organisms were significantly less affected at the K–Pg than other groups (Kiessling and Simpson, 2011), while marine organisms that were direct consumers of photosynthetic products rather than detritus suffered greatly at the K–Pg (e.g., Sheehan et al., 1996; Bambach 2006). These different patterns of extinction selectivity strongly suggest that the primary mechanism of extinction was different at the Tr–J and K–Pg. The fossil record of marine invertebrates is consistent with the idea that hypercapnia (see Knoll et al., 2007) and/or ocean acidification played some role in biotic events at the Tr–J (e.g., Hautmann et al., 2008; Kiessling and Simpson, 2011), but suggests that these factors were perhaps less important at the K–Pg. The selection against marine photosynthetic primary producers (e.g., Bown, 2005; Kiessling and Baron-Szabo, 2004) and selective survival of detritus-feeders at the K–Pg (e.g., Sheehan et al., 1996) is consistent with an impact winter kill mechanism. Of course, this does not mean that other mechanisms of extinction did not operate during the Tr–J and K–Pg. Distinguishing heat escape from pH escape at the Tr–J, for example, is of particular importance to determine the relative importance of global warming versus ocean acidification in driving the extinction of marine organisms at this time (see Kiessling and Simpson, 2011).

It is also noteworthy that patterns of extinction selectivity at the K–Pg in the ocean and on the land are similar. For example, terrestrial primary productivity (plants) was severely impacted at many localities around the globe (see Section 16.3.4), which most probably propagated down the food chain. As in the ocean, terrestrial detritus-based food chains were much less affected (Schulte et al., 2010).

In general, organisms that were non-photosynthetic, starvation-resistant, sessile, or detritus-feeders appear to have been K—Pg survivors (Retallack, 2004). It is of crucial importance that future research attempts to tease apart the relative roles of short-term changes such as thermal radiation, impact winter, and acid rain, from longer-term environmental changes such as increased CO_2 and global warming at the K—Pg.

16.6. 'DEEP-TIME' CONTEXT FOR ANTHROPOGENIC ENVIRONMENTAL AND CLIMATE CHANGE

The Tr—J and K—Pg events offer a view of how the Earth's environment and ecosystems have been influenced by vast volcanic eruptions and a large body extraterrestrial collision in the geological past. Certain aspects of climate and environmental change that occurred as part of the Tr—J and K—Pg events are in some way analogous to changes that are occurring on our planet today, or are anticipated in the near future (cf. Lenton 2012, this volume). That caution should be exercised when using data from these two intervals of time to understand the possible future climate of our planet needs no belabouring, not least because Earth's biological and physical systems were very different at the Tr—J and K—Pg compared to the present day. Nonetheless, in the following paragraphs we attempt to outline how the Tr—J and K—Pg events serve as reference points, particularly for estimates of CO_2 rise and global warming, which can be used to place contemporary global change into the context of Earth's geological history.

The Tr—J and K—Pg events and anthropogenic climate change can be placed on a conceptual temporal continuum. At one end of this continuum are processes that occur instantaneously such as the impact of an extraterrestrial body, and at the other are processes that take place over millions or billions of years such as the assembly and break-up of continents (Figure 16.6). Global atmospheric CO_2 levels have risen by ~100 ppmv since the onset of the western Industrial Revolution (IPCC, 2007a). Exceptional evidence comes from the monthly mean carbon dioxide levels measured at the Mauna Loa observatory in Hawaii since 1958 (e.g., see *http://celebrating200years.noaa.gov/datasets/mauna/image3b.html*). This rise in CO_2 corresponds with the global land—ocean temperature anomaly, which shows that global mean annual temperatures have risen by ~0.6°C as part of a consistent warming trend from 1880 to the present day (Hansen et al., 2006; Sen Gupta and McNeil, 2012, this volume). In some areas this warming trend has been amplified, particularly at high latitudes where some arctic regions have seen increases of 2.1°C since 1951 (Hansen et al., 2006). Future estimates (e.g., IPCC, 2007a) suggest that if anthropogenic emissions

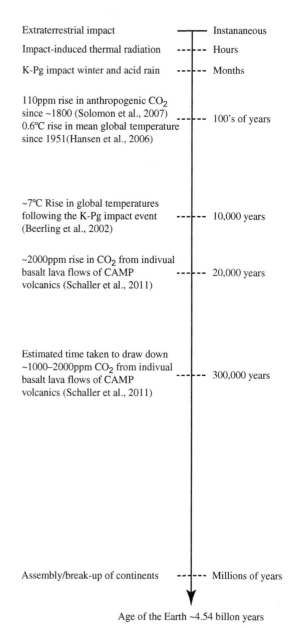

FIGURE 16.6 Conceptual temporal continuum showing anthropogenic CO_2 rise and global warming and selected processes that occurred as part of the Tr—J and K—Pg events. Processes that occur over short timescales are shown toward the top of the diagram and processes that occur over long timescales are shown at the bottom of the diagram.

continue at a similar level then the Earth may experience a global warming of between 1.5 and 6°C (cf. Harvey, 2012, this volume). This anthropogenic rise in CO_2 and associated greenhouse warming has occurred on a timescale of hundreds of years (Figure 16.6). How does this compare to changes in atmospheric CO_2 and global temperatures at the Tr—J and K—Pg?

Changes in atmospheric CO_2 at the Tr—J have been reasonably well-studied (e.g., McElwain et al., 1999; Cleveland et al., 2008; Schaller et al., 2011; Bonis et al.,

2010). Estimates of CO_2 change across the Tr–J range from the stomatal proxy record of McElwain et al. (1999) that indicates a rise from ~600 ppmv in the Rhaetian to ~2100–2400 ppmv in the Tr–J boundary interval, to the soil carbonate proxy record of Schaller et al. (2011) that indicates a rise from ~2000 ppmv in the Rhaetian to ~4400 ppmv in the Tr–J boundary interval. The CO_2 reconstruction by Schaller et al. (2011) is particularly well-constrained stratigraphically, and indicates that each of the three major basalt lava flows of CAMP volcanism in the Newark Basin increased atmospheric CO_2 by ~2000 ppmv (either by direct outgassing or by contact metamorphism), and that each rise probably occurred within 20 ka (Schaller et al., 2011). There has been less work on the K–Pg atmospheric CO_2 record, but there is some evidence of a rise in CO_2 at the K–Pg boundary (e.g., Pierazzo et al., 1998; Beerling et al., 2002). Atmospheric CO_2 apparently increased from 350–500 ppmv in the Late Cretaceous–Early Palaeocene to at least 2300 ppmv within 10 ka of the K–Pg boundary (Beerling et al., 2002). Additionally, the shutdown of the oceanic biological pump at the K–Pg (Zachos et al., 1989; D'Hondt et al., 1998) probably slowed the rate of CO_2 drawdown in the early Palaeocene. Changes in CO_2 at the Tr–J and K–Pg clearly occurred over substantially longer timescales and were of greater magnitude than the current anthropogenic rise in CO_2 (Figure 16.6). The ~300 ka taken to drawdown ~1000–2000 ppmv of CO_2 from individual CAMP basalt lava flows at the Tr–J (Schaller et al., 2011; Figure 16.6) provides empirical evidence of the amount of time taken for CO_2 to be removed from the atmosphere by natural processes such as silicate weathering (e.g., Schaller et al., 2011).

Estimates for the increase in mean global surface air temperatures across the Tr–J range from 3–4°C (McElwain et al., 1999) to ~6.1°C (Huynh and Poulsen, 2005). Oxygen isotope studies, however, suggest warming of local water masses in the UK of up to 10°C (Korte et al., 2009), highlighting that estimates of temperature change on a global-scale at the Tr–J may smooth out much spatial and environmental variation. The K–Pg event is estimated to have resulted in a ~1.5–7.5°C rise in global temperatures (Pierazzo et al., 1998; Beerling et al., 2002; Figure 16.6), and the oxygen isotope record from planktic and benthic foraminifera suggests water-mass warming of the order of 3°C (Zachos et al., 1989). Estimates of mean annual temperature using leaf margin analysis from fossil leaves, however, reveal little increase in temperatures in North America (Wilf et al., 2003), and this reinforces the view that major global change can create extremely heterogeneous patterns of temperature change across the Earth.

16.7. FUTURE CLIMATE CATASTROPHES

By outlining the Tr–J and K–Pg events, we have attempted to highlight how Earth's geological history can be used to augment our understanding of the processes that have shaped the history of life on Earth, and also emphasize how such knowledge can be used to help understand the processes that will shape the face of our planet in the future (cf. Lovelock, 2009). Both events reveal that climatic and environmental changes led to significant reorganisation and mass extinction of the Earth's biota. In both cases, the time taken to perturb the environment is substantially less than the time taken for the Earth to reach a new equilibrium (see Figure 16.6). Earth's deep geological past provides us with empirical data on the behaviour of the Earth's physical environment and biota under conditions outside the range that is measurable over historical timescales. Such data can play a valuable role in refining numerical models of the Earth's climate and more complete Earth system models (e.g., Rice and Henderson-Sellers, 2012, this volume), and can provide information on the resilience of ecosystems in the face of global change. Such knowledge will allow us to adapt our lifestyles and ecosystem management practices to make better use of our planet's changing resources.

ACKNOWLEDGEMENTS

Robert A. Spicer and Wolfgang Kiessling are thanked for constructive comments that improved this chapter. CMB thanks Margaret E. Collinson and Andrew C. Scott for inspiring her to study K–Pg wildfires. CMB and LM thank Jennifer C. McElwain and Wolfram M. Kürschner for extensive discussion of the Tr–J.

CMB acknowledges the following funding: Natural Environment Research Council Studentship NER/S/A/2001/06342; CASE support in association with the Royal Botanic Gardens, Kew; European Union Marie Curie Intra-European Fellowship FILE-PIEF-GA-2009-253780. CMB and LM acknowledge funding through a European Union Marie Curie Excellence Grant MEXT-CT-2006-042531.

Understanding the Unknowns

Future prediction is very difficult. This section returns attention to the characteristics of the Anthropocene introduced in Section I. To be successful in projection of likely, and even of less likely, futures it is necessary to understand complex systems behaviour and human interaction in, and interference with, biogeochemical systems.

We should all keep looking for civil discourse, it is the only way we'll get science back into the right place for our unwritten social contract with societal decision-makers: to give them what they need from us — expert risk assessment — so they can do their job for which they were elected — normative risk management. But these days that is so broken by special interest campaign contributions, false media 'balance', lies and spin, that I am fearful our social contract is past the Humpty Dumpty stage. But we all should keep trying…. What other reasonable choice is there if civil dialogue and rational discourse is to become the rule again.

Stephen H. Schneider, in an email to Paul Gross, 13 April, 2010

Chapter 17, **Future Climate Surprises** by Timothy Lenton, addresses the surprises that future climate change may have in store for us. Firstly, it considers how to categorize climate surprises, based on concepts from dynamical systems theory. Existing definitions of tipping elements and tipping points are reviewed, adding the case of noise-induced transitions. The climate subsystems that could undergo surprise changes are discussed in three broad categories: the melting of large masses of ice, changes in atmospheric and ocean circulation, and loss of biomes. Then, the focus switches to how science can help societies cope with climate surprises, starting with how to assess the risk they pose, by combining information on their likelihood (under a given scenario) and their impacts. The prospects for removing the element of surprise — by achieving early warning of approaching thresholds — are considered in detail. Finally, the available response and recovery strategies are assessed, assuming societies may be faced with unwelcome climate surprises.

Chapter 18, **Future Climate: One Vital Component of Trans-disciplinary Earth System Science** by Martin Rice and Ann Henderson-Sellers, places climate into the 'bigger picture' that goes by many names including Gaia and Earth system science. The future climate is just one component of a larger, complex system including people and their practices as well as the physical and biogeochemical aspects of the Earth. Recognizing that climate has been used frequently as a way of illustrating emergent Earth system behaviour, how future climates are integral to the whole-of-Earth system is explored. That future climatic and Earth system challenges will have to be communicated clearly and effectively to the public, who demand or refuse policy direction changes, is undeniable. Choice of frameworks and illustration of aspects of narrative by metaphor and even cartooning is in keeping with the aim of James Lovelock, the creator of the Gaia hypothesis around which this chapter is organized. Indeed, readily accessible modes of sharing understanding support the clear communications goal of Stephen Schneider to whom this book is dedicated.

"Every citizen in a democracy is capable of joining in those decisions because in the end they are value judgments based on common sense, plus an awareness of the risks and benefits of alternative strategies.

M[m]any people feel they can't say anything, that they must be ignorant because they can't understand the details. So they just punt. They kick the decision over to others who supposedly 'know better.' What I call the 'one fax-one-vote' system comes into play, whereby special interests shout loud enough to confuse nearly all lay people. That is how special interests manage to gain equal credibility in the public arena for what really is not a very credible position."

Stephen H. Schneider, AAAS, 1997

Chapter 17

Future Climate Surprises

Tim Lenton

College of Life and Environmental Sciences, University of Exeter, Exeter, UK

Chapter Outline

17.1. INTRODUCTION: PROBING FUTURE CLIMATES

Fifteen years ago, when the first edition of this book was published (Henderson-Sellers, 1995), much scientific attention was directed at one source of climate surprise: changes in the Atlantic thermohaline circulation (THC; Peng, 1995). In the intervening time, scientists have identified many more systems that could produce climate surprises (Lenton et al., 2008). Improvements to the observational and palaeorecords have reinforced the view that climate can change abruptly at large scales. Furthermore, recent, striking developments in the climate system have added to the concern that human-induced climate change is unlikely to involve a smooth and entirely predictable transition into the future. The record minimum area coverage of Arctic sea-ice in September 2007 drew widespread attention, as has the accelerating loss of mass from the Greenland and West Antarctic Ice Sheets (Pritchard et al., 2009; Rignot et al., 2008). Droughts have afflicted the Amazon rainforest (Phillips et al., 2009) and a massive insect outbreak has struck Canada's boreal forest (Kurz et al., 2008b). These large-scale components of the Earth system are among those that have been identified as potential 'tipping elements' — climate subsystems that could exhibit a 'tipping point' where a small change in forcing (in particular, global temperature change) causes a qualitative change in their future state (Lenton et al., 2008). The resulting transition may be either abrupt or irreversible or, in the worst cases, both. In IPCC terms, such changes are referred to as 'large-scale discontinuities' (Smith et al., 2009). Should they occur, they would surely qualify as dangerous climate changes (Schellnhuber et al., 2006). However, not all are equally dangerous. While, for the most part, the impacts are clearly damaging and large, there is at least

The Future of the World's Climate. DOI: 10.1016/B978-0-12-386917-3.00017-8

one case (greening of the Sahel) where the climate surprise could be a pleasant one.

While the terminology of 'climate surprises' implies their predictability is limited (there is some irreducible uncertainty), those discussed herein are not completely unpredictable. The trigger of any tipping-point change is likely to be a combination of natural variability on top of an underlying forcing due to human activities. Hence, one can only talk in terms of probabilities of passing particular tipping points. However, recent expert elicitation has obtained some useful information on these probabilities for different future warming scenarios (Kriegler et al., 2009). The probabilities are imprecise but, *even with the most conservative assumptions*, they indicate that in a 4°C warmer world it is more likely than not that at least one of five large-scale thresholds will be passed. The key message from recent studies is that large climate surprises now appear significantly closer, in terms of global temperature change, than they did in earlier assessments (Smith et al., 2009).

This chapter has the following aims: Firstly, it addresses how to categorize climate surprises, reviewing existing definitions of tipping elements and tipping points, and adding the case of noise-induced transitions. Next, the list of potential policy-relevant tipping elements (Lenton et al., 2008) is revisited (and slightly revised) considering them in three categories: the melting of large masses of ice, changes in atmospheric and ocean circulation, and loss of biomes. Then, the discussion turns to how science can help societies cope with climate surprises, starting with how to assess the risk they pose, and then examining in detail the prospects for early warning of them. Finally, the available response and recovery strategies are considered for societies faced with unwelcome climate surprises.

17.2. DEFINING CLIMATE SURPRISES

The nature of surprise is that it is unexpected, and abrupt. So, 'climate surprises' is taken here to refer to events where there is a stochastic (i.e., random) component driving them, as well as a deterministic one, and where the resulting changes are unexpectedly large, relative to the factors driving them. Surprises can be pleasant or unpleasant, but in the case of anthropogenic climate change it is usually assumed that changes are for the worst. Abrupt and unpredicted changes are seen as particularly undesirable, because they are most difficult to adapt to. In the following subsections, climate surprises are subdivided into those where a system is forced past a 'tipping point' and those where internal variability (noise) triggers a transition. The reason, as we will see later, is that there is some prospect of predicting 'tipping points' (because they have a deterministic component), whereas purely noise-induced

transitions are completely unpredictable surprises. However, the subdivision is over-idealized, and in reality a mixture of both factors is likely to be at work in future climate change surprises.

17.2.1. Tipping Points and Noise-Induced Transitions

In colloquial terms, the phrase 'tipping point' captures the notion that "little things can make a big difference" (Gladwell, 2000, p. 1). In other words, at a particular moment in time, a small change can have large, long-term consequences for a system. To apply the term usefully to the climate (or in any other scientific context), it is important to be precise about what qualifies as a tipping point, and about the class of systems that can undergo such change. To this end, the term 'tipping element' has been introduced (Lenton et al., 2008) to describe large-scale subsystems (or components) of the Earth system that can be switched — under certain circumstances — into a qualitatively different state by small perturbations. In this context, the tipping point is the corresponding critical point — in forcing and a feature of the system — at which the future state of the system is qualitatively altered. For a system to possess a tipping point, there must be strong positive feedback in its internal dynamics (see Harvey, 2012, this volume).

To formalize the notion of a tipping element further (Lenton et al., 2008), it is important to define a spatial-scale; here, only components of the Earth system associated with a specific region or collection of regions, which are at least subcontinental in scale (length-scale of order ~1000 km), are considered. Then, for such a system to qualify as a tipping element, it must be possible to identify a single control parameter (ρ), for which there exists a critical control value (ρ_{crit}), from which a small perturbation ($\delta\rho > 0$) leads to a qualitative change in a crucial feature of the system (ΔF) after some observation time ($T > 0$). In this definition (Lenton et al., 2008), the critical threshold (ρ_{crit}) is the tipping point, beyond which a qualitative change occurs — this change may occur immediately after the cause or much later.

Many scientists intuitively take 'tipping point' to be synonymous with a 'bifurcation point' in the equilibrium solutions of a system, as schematically illustrated in Figure 17.1a. This implies that passing a tipping point necessarily carries some irreversibility. However, other classes of non-linear transition can meet the definition above, and one schematic example is given in Figure 17.1b. Again this shows the (time-independent) equilibrium solutions of a system, but here they are continuous (there is no bifurcation) and, therefore, the transition is reversible.

In reality, the existence or not of a tipping point should be considered in a time-dependent fashion, and there could be several other possible types of tipping elements (for the

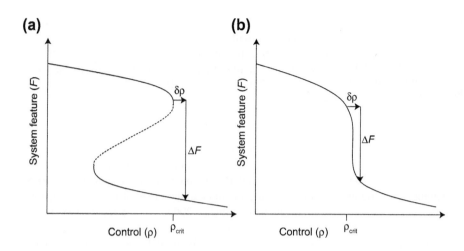

(a)

(b)

FIGURE 17.1 Two types of tipping point. The schematics show the time-independent equilibrium solutions of a system: (a) a system with bi-stability passing a bifurcation point, (b) a mono-stable system exhibiting highly non-linear change. These cases meet the definition (Lenton et al., 2008) of a tipping element passing a tipping point (ρ_{crit}), where a small change in control ($\delta\rho$) results in a large change in a system feature (ΔF).

mathematical details, see the supplementary information of Lenton et al., 2008). Theoretically, one can construct elements that react infinitely slowly to tipping, yet do this in an entirely irreversible fashion. Also, recent work has identified examples of rate-dependent tipping: where a system undergoes a large and rapid change, but only when the rate at which it is forced exceeds a critical value (Levermann and Born, 2007; Sebastian Wieczorek, personal communication, 2010).

A different class of climate surprise are noise-induced transitions, as illustrated in Figure 17.2. In such cases, internal variability causes a system to leave its current state (or attractor) and transition to a different state (or attractor). This does not require a bifurcation point to be passed, potentially it can occur without any change in forcing (the control parameter, ρ). However, it does require the co-existence of multiple states under a given forcing, which implies that the underlying system possesses bifurcation-type tipping points. As illustrated in Figure 17.2b, one can think of noise as pushing a system out of a valley (one stable steady state), up to the top of a hill (an unstable

steady state), and it then rolling down the other side into a different valley (some new stable state). One might call perching on the top of the hill (at the unstable steady state) a 'tipping point' and define it in terms of the corresponding value of the system feature (F_{crit}). However, the value of F_{crit} is a function of ρ (Figure 17.2a), so is not as well-defined as ρ_{crit}. Thus, noise-induced transitions do not strictly fit the tipping point definition given above (Lenton et al., 2008), but they are related to it. Clearly, they are a kind of surprise that could occur in the climate system and should be considered, an example being abrupt monsoon transitions (Levermann et al., 2009).

In general, noise-induced transitions become more likely to occur the closer one is to a bifurcation point. Thus, given that the climate system has its own internally-generated noise (familiar to us as the weather), we can expect that, if it is approaching a bifurcation point, it will leave its present state before the bifurcation point is reached (Kleinen et al., 2003). This means that some future climate surprises could involve a mixture of the idealized mechanisms shown in Figure 17.1a and Figure 17.2.

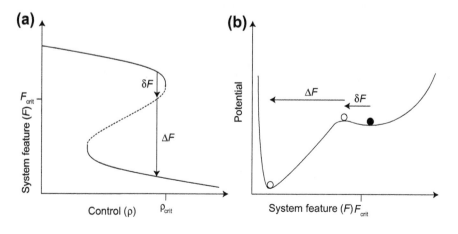

(a)

(b)

FIGURE 17.2 Noise-induced transition. The schematics show a system with bistability undergoing a noise-induced transition between states: (a) time-independent equilibrium solutions (for comparison with Figure 17.1), (b) representation of the transition within the system potential (y-axis of (a) has become x-axis of (b)). In (b), the wells represent stable steady states, the hilltop represents an unstable steady state, and the ball represents the actual state of the system (filled black is the initial state). In this case, a small perturbation due to noise (δF), with no change in control (ρ), results in a large change in a system feature (ΔF). This suggests an alternative class of tipping point (F_{crit}).

17.2.2. Policy-Relevant Tipping Elements

The above definitions are quite general and could conceivably be applied at any point in Earth's climate history. However, the focus here is on future climate surprises, so we need to narrow them down somewhat. Previous work (Lenton et al., 2008) has defined a subset of 'policy-relevant' tipping elements by adding the following conditions to our tipping element definition:

(i) Human activities are interfering with the system such that decisions taken within a 'political time horizon' (T_P ~ 100 years) can determine whether the tipping point (ρ_{crit}) is crossed.

(ii) The time to observe a qualitative change (including the time to trigger it) lies within an 'ethical time horizon' (T_E ~ 1000 years).

(iii) A significant number of people care about the fate of the system because either it contributes significantly to the overall mode of operation of the Earth system, it contributes significantly to human welfare, or it has great value in itself as a unique feature of the biosphere.

This definition focuses on the consequences of decisions enacted within this century that could lead to large changes within this millennium. For a system to meet the definition of a tipping element, there needs to be some theoretical basis for expecting it to exhibit a critical threshold at a subcontinental-scale and/or past evidence of threshold behaviour. To identify the subset of policy-relevant tipping elements, the conditions given above were evaluated. For (i), the 'accessible neighbourhood' of climate out to 2100 was defined, by considering the range of IPCC Special Report on Emissions Scenarios (SRES) climate forcing factors and the range of resulting projected climate changes (IPCC, 2007a). To evaluate (ii), model projections and palaeo data were used, taking into account known shortcomings of the models. To evaluate (iii) inevitably involves some subjective judgments. Figure 17.3 shows the resulting map of the potential policy-relevant tipping elements in the climate system, updated somewhat from the one originally introduced (Lenton et al., 2008).

Before getting into the details of specific tipping elements, it is worth pausing to consider whether those identified on the map (Figure 17.3) include all the systems

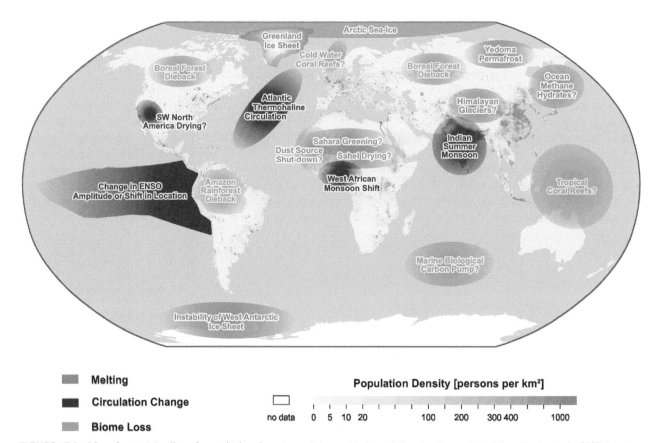

FIGURE 17.3 Map of potential policy-relevant tipping elements overlain on global population density — adjusted from Lenton et al. (2008) based on contents of this chapter. Question marks indicate systems whose status as policy-relevant tipping elements is particularly uncertain. *(Source: Figure by Veronika Huber, Martin Wodinski, Timothy M. Lenton, and Hans-Joachim Schellnhuber.)*

that might undergo noise-induced transitions in the future and, if not, how they might be included? Conceivably there could be systems that do not reach a bifurcation point (Figure 17.1a) due to anthropogenic climate change, but can still be knocked out of their present state by a relatively high degree of internal variability (Figure 17.2). The Holocene climate is generally characterized as having relatively low internal variability (compared to, for example, the ice age climate; cf. Berger and Yin, 2012, this volume). Consequently, it tends to be assumed that the 'signal' of anthropogenic climate change will be large compared to the 'noise' of internal variability. However, when one goes down to the subcontinental-scale, variability is much greater than in the global mean. (Furthermore, internal variability may itself vary with changes in the mean climate state.) To include the possibility of noise-induced transitions, one could broaden the definition out somewhat, by offering an alternative to condition (i):

> (i-alt.) Internal variability within the 'political time horizon' ($T_P \sim 100$ years) could be sufficient to push the system past an unstable state (F_{crit}) into a new basin of attraction.

Strictly speaking, this criterion has not been applied in coming up with the following list of potential climate surprises. However, both anthropogenic forcing and internal climate variability could play a role in tipping several of the systems. In the following sections, potential tipping elements are subdivided into those involving melting of large masses of ice, those involving changes in the circulation of the atmosphere and ocean, and those involving the loss of unique biomes.

17.3. MELTING OF LARGE MASSES OF ICE

The concept of a threshold is intuitively obvious when thinking about ice melting to liquid water — an example of a first-order phase transition. However, that happens on a relatively small-scale. For major masses of ice on Earth to qualify as climate tipping elements, they must exhibit a large-scale threshold due to strong positive feedbacks in their internal dynamics, coupled to the climate.

17.3.1. Arctic Sea-Ice

The summer minimum area cover of Arctic sea-ice has declined markedly in recent decades, most strikingly in 2007. Observations have fallen below all IPCC model projections, despite the models having been in agreement with the observations in the 1970s (Stroeve et al., 2007). Winter sea-ice is also declining in area (though less rapidly), with a loss of 1.5 million km^2 of multiyear ice coverage in the past decade (Nghiem et al., 2007). There is also an overall, progressive thinning of the ice cap,

with observations showing a decrease of mean winter multiyear ice thickness from 3.6 m to 1.9 m over the past three decades (Kwok and Rothrock, 2009). The observed decline in sea-ice is consistent with acceleration due to the ice–albedo positive feedback, as exposure of the dark ocean surface causes increased absorption of solar radiation. This is warming the upper ocean and contributing significantly to melting on the bottom of the sea-ice. Over 1979 to 2007, 85% of the Arctic region has received an increase in solar heat input at the surface, with an increase of 5% per year in some regions, including the Beaufort Sea (Perovich et al., 2007). In situ measurements in this region show that there was a three times greater bottom ice melt in 2007, compared to earlier years (with relatively little change in surface melt) (Perovich et al., 2008). Other factors contributing to record ice loss include patterns of atmospheric circulation (Maslanik et al., 2007; Rigor and Wallace, 2004) and ocean circulation (Nghiem et al., 2007), which have exported multiyear ice out of the Arctic basin through the Fram Strait, reductions in summertime cloud cover (Kay et al., 2008), and increased input of ocean heat from the Pacific (Shimada et al., 2006; Woodgate et al., 2006).

Further warming 'in the pipeline' raises the possibility that the Arctic may already be committed to a qualitative change in which the ocean becomes largely ice-free in summer (e.g., Harvey, 2012, this volume). The year that the North Pole becomes seasonally ice-free will likely be seen as a 'tipping point' by non-experts. While politically important (cf. Taplin, 2012, this volume), it is unlikely that such a transition involves an irreversible bifurcation (as in Figure 17.1a) (Eisenman and Wettlaufer, 2009). Summer sea-ice quickly recovers once the climate turns cold again because of a stabilizing feedback related to the ice growth rate (Notz, 2009). Yet it may still qualify as a tipping element because, as the ice cap gets thinner, it becomes prone to larger fluctuations in area, which can be triggered by relatively small changes in forcing (Holland et al., 2006). Also, it is conceivable that loss of summer ice would involve additional dynamical feedbacks that lead to a qualitative change in atmosphere or ocean circulation and heat transports. If so, the impacts are likely to be felt further afield, for example in Europe. Loss of winter (i.e., year-round) ice is more likely to represent a bifurcation (Figure 17.1a), where the system can switch rapidly and irreversibly from one state (with seasonal ice) to another (without any) (Eisenman and Wettlaufer, 2009). However, the threshold for year-round ice loss requires around 13°C warming at the North Pole (Winton, 2006). Whether this could occur this century depends on anthropogenic emissions and the uncertain strength of polar amplification of warming.

17.3.2. Greenland Ice Sheet (GIS)

The Greenland ice sheet (GIS) is currently losing mass at a rate that has been accelerating (Rignot et al., 2008). In summer 2007, there was an unprecedented increase in surface melt, mostly south of 70°N and also up the west side of Greenland, due to an up to 50-day longer melt season than average with an earlier start (Mote, 2007). This is part of a longer-term trend of increasing melt extent since the 1970s. Recent observations show that seasonal surface melt has led to accelerated glacier flow (Joughin et al., 2008; van de Wal et al., 2008). The surface mass balance of the GIS is still positive (there is more incoming snowfall than melt at the surface, on an annual average), but the overall mass balance of the GIS is negative due to an increased loss flux from calving of glaciers that outweighs the positive surface mass balance. The margins of the GIS are thinning at all latitudes (Pritchard et al., 2009), and the rapid retreat of calving glaciers terminating in the ocean, most notably Jakobshavn Isbrae, is probably linked to warming ocean waters (Holland et al., 2008).

The GIS will be committed to irreversible meltdown if the surface mass balance goes negative, most notably because as the altitude of the surface declines, it gets warmer (a positive feedback). Initial assessments put the temperature threshold for this to occur at around 3°C of regional warming, based on a positive-degree-days model for the surface mass balance (Huybrechts and De Wolde, 1999; Cogley, 2012, this volume). Results from an expert elicitation concur that if global warming exceeds 4°C, there is a high probability of passing the threshold (Kriegler et al., 2009). An alternative surface energy balance model predicts a more distant threshold at 8°C regional warming or ~6°C global warming (J. Bamber, personal communication, 2010). However, recent work suggests the threshold could be much closer at 1.3°C−2.3°C global warming above pre-industrial (A. Robinson and A. Ganopolski, personal communication, 2010). The actual threshold for massive GIS shrinkage must lie before the surface mass balance goes negative. A more nuanced possibility, which is emerging from some coupled climate−ice sheet model studies, is that there could be multiple stable states for GIS volume, and, hence, multiple tipping points (Ridley et al., 2009). Passing a first tipping point where the GIS retreats onto land could lead to ~15% loss of the ice sheet and about 1 m of global sea-level rise. As for the rate at which this could occur, an upper limit is that the GIS could contribute around 50 cm to global sea-level rise this century (Pfeffer et al., 2008).

17.3.3. West Antarctic Ice Sheet (WAIS)

The West Antarctic Ice Sheet (WAIS) is also losing mass at present, and some parts, particularly those draining into the

Amundsen Sea, are thinning rapidly (Pritchard et al., 2009). While air temperatures have recently been shown to be warming across West Antarctica (Steig et al., 2009), the shrinkage of the WAIS is more sensitive to the intrusion of warming ocean waters and the collapse of floating ice shelves that buttress the main ice sheet. The WAIS may be vulnerable to large-scale collapse, due to retreat of the grounding line where the ice sheet is pinned to the bedrock below sea level (Mercer, 1978; Weertman, 1974). Having been strongly questioned when it was first introduced, the paradigm of a potential abrupt collapse of the WAIS has recently gained new momentum (Vaughan, 2008). Recent theory has confirmed the potential for multiple stable states of the grounding line and, hence, bifurcation-type tipping points (Figure 17.1a) (Schoof, 2007). Also, new palaeo data has shown that the WAIS collapsed repeatedly during the ~3°C warmer world of the early Pliocene (5 Ma−3 Ma) (Naish et al., 2009). Modelling supports this and suggests further collapses during some (but not all) of the more recent Pleistocene interglacials (Pollard and DeConto, 2009). Furthermore, East Antarctic ice cores show anomalous spikes of warmth (above present) during all of the past four interglacial intervals, which might be explained by repeated WAIS collapse (Holden et al., 2010). Data from the last (Eemian) interglacial suggest the up-to-2°C-warmer world of the time may have had peaks of sea level up to 9 m higher than present and rates of sea-level rise of 1.6 m ± 1.0 m per century (Rohling et al., 2008). To achieve such rates of sea-level rise probably required rapid grounding line retreat of the WAIS, and possibly parts of the periphery of the East Antarctic Ice Sheet (EAIS) that are also grounded below sea level. Current models put the threshold for WAIS collapse when the surrounding ocean warms by ~5°C (Pollard and DeConto, 2009), and expert elicitation concurs that if global warming exceeds 4°C, it is more likely than not that the WAIS will collapse (Kriegler et al., 2009).

17.3.4. Yedoma Permafrost

Continuous permafrost is the perennially frozen soil, which currently covers ~10.5 million km^2 of the Arctic land surfaces but is melting rapidly in some regions. This area could be reduced to as little as 1.0 million km^2 by the year 2100, which would represent a qualitative change in state (Lawrence and Slater, 2005; Harvey, 2012, this volume). However, permafrost did not make the original shortlist of tipping elements (Lenton et al., 2008) because of a lack of evidence for a large-scale threshold for permafrost melt. Instead, in future projections the local threshold of freezing temperatures is exceeded at different times in different localities. Yet more recent work has suggested that at least one large area of permafrost could exhibit coherent threshold behaviour. The frozen loess (windblown organic material) of northeastern Siberia

(150°E−168°E and 63°N−70°N), also known as yedoma, is deep (~25m) and has an extremely high carbon content (2%−5%); thus, it may contain ~500 PgC (Zimov et al., 2006). Recent studies have shown the potential for this regional frozen carbon store to undergo self-sustaining collapse, due to an internally-generated source of heat released by biochemical decomposition of the carbon, triggering further melting in a runaway positive feedback (Khvorostyanov et al., 2008a, 2008b). Once underway, this process could release 2.0 PgC−2.8 PgC per year (mostly as CO_2, but with some methane) over about a century, removing ~75% of the initial carbon stock. The collapse would be irreversible in the sense that removing the forcing would not stop it continuing. To pass the tipping point requires an estimated >9°C of regional warming (Khvorostyanov et al., 2008a). However, this is a region already experiencing strongly amplified warming, partly linked to shrinkage of the Arctic sea-ice (Lawrence et al., 2008). During August−October 2007, Arctic land temperatures jumped around 3°C above the mean for the preceding 30 years (from analysis of the HadCRUT data). Thus, the yedoma tipping point may be accessible this century under high emissions scenarios as discussed in Harvey (2012, this volume).

17.3.5. Ocean Methane Hydrates

Recent model estimates suggest that up to 2000 PgC are stored as methane hydrates beneath the ocean floor (Archer et al., 2009). As the deep ocean warms, heat diffuses into the sediment layer and may destabilize this reservoir of frozen methane. Bubbles associated with the melting of methane may trigger submarine landslides (Kayen and Lee, 1991). This finding raises the concern that the destabilization of methane hydrates could result in an abrupt massive release of methane into the atmosphere. If this scenario were plausible, methane hydrates would clearly qualify as a policy-relevant tipping element. However, palaeoclimatic evidence makes this scenario very unlikely (Archer, 2007). Instead, the most likely impact of a melting hydrate reservoir is a long-term chronic methane source (Archer et al., 2009; Harvey, 2012, this volume). An additional warming of 0.4°C−0.5°C from the hydrate response to fossil fuel CO_2 release is estimated, persisting over several millennia (Archer et al., 2009). This estimate is, however, subject to large uncertainties in particular with regard to the magnitude of temperature forcing required to trigger the destabilization of methane hydrates. Even in a ~1.5°C global warming scenario, ~2°C warmer 'heat bubbles' may persist at depth in the ocean for many centuries (Schewe et al., 2010). In summary, a qualitative change in this Earth subsystem is unlikely to occur on a policy-relevant timescale (as defined by Lenton et al., 2008). Yet, methane hydrates can be considered a slow and,

for societal purposes, irreversible tipping element in the global carbon cycle.

17.3.6. Himalayan Glaciers

It has been suggested that the Hindu−Kush−Himalaya−Tibetan (HKHT) glaciers should be added to the list of tipping elements because much of their mass could be lost this century (Ramanathan and Feng, 2008). The loss of mountain glaciers (Cogley, 2012, this volume) involves a positive feedback whereby dust accumulation lowers the surface albedo, thus accelerating melting (Oerlemans et al., 2009). Also, where snow or ice disappears altogether, the further lowering of albedo (ice−albedo feedback) amplifies warming (Pepin and Lundquist, 2008). However, it needs to be examined whether such positive feedbacks will cause HKHT mass loss to exhibit strong non-linearity in response to warming, and therefore qualify as a climate surprise or tipping element.

17.4. CHANGES IN ATMOSPHERIC AND OCEANIC CIRCULATION

The circulations of the ocean and atmosphere, coupled together and to the land surface, can exhibit different dynamical stable states and modes of variability, with potential thresholds between them (cf. Latif and Park, 2012, this volume). They can also be particularly sensitive to gradients of forcing as these are usually what drive the circulations in the first place. Monsoons are a seasonal example, initiated by more rapid heating of the land than ocean, which causes warm air to rise over the continent, creating a pressure gradient that sucks in moist air from over the ocean, which then rises, its water condenses, and rain falls, releasing latent heat that reinforces the circulation (Levermann et al., 2009).

17.4.1. Indian Summer Monsoon (ISM)

The Indian Summer Monsoon (ISM) system is already being influenced by aerosol and greenhouse gas forcing. Palaeorecords indicate its volatility, with flips on and off of monsoonal rainfall linked to climate changes in the North Atlantic (Burns et al., 2003; Goswami et al., 2006; Gupta et al., 2003). Greenhouse warming, which is stronger over Northern Hemisphere land than over the Indian Ocean, would on its own be expected to strengthen the monsoonal circulation. However, the observational record shows declines in ISM rainfall, which have been linked to an 'atmospheric brown cloud' (ABC) haze created by a mixture of black carbon (soot) and sulfate aerosols (Ramanathan et al., 2005). The ABC haze is more concentrated over the continent than over the ocean to the south, and it causes more sunlight to be absorbed in the atmosphere and less heating at the surface. Hence, it tends to weaken the monsoonal circulation (Meehl et al., 2008; Ramanathan and Carmichael, 2008). In simple models, there is a tipping point for the regional planetary albedo (reflectivity) over the continent which, if exceeded, causes the ISM to collapse altogether (Levermann et al., 2009; Zickfeld et al., 2005). The real picture is likely to be more complex with the potential for switches in the strength and location of the monsoonal rains. Increasing aerosol forcing could further weaken the monsoon but, if then removed, greenhouse warming could trigger a stronger monsoon, producing a climatic 'roller-coaster ride' for hundreds of millions of people (Zickfeld et al., 2005).

17.4.2. El Niño–Southern Oscillation (ENSO)

The El Niño–Southern Oscillation (ENSO) phenomenon is the most significant natural mode of coupled ocean–atmosphere variability in the climate system. Over the past century, warming has been greater in the eastern than the western equatorial Pacific and this has been linked to El Niño events becoming more severe (e.g., in 1983 and 1998). Recently, a changing pattern of El Niño has been noted towards 'Modiki' events where the warm pool shifts from the West to the middle (rather than the East) of the equatorial Pacific (Ashok and Yamagata, 2009; Yeh et al., 2009). In future projections, the first coupled model studies predicted a shift from current ENSO variability to more persistent or frequent El Niño conditions. Now that numerous models have been inter-compared, there is no consistent trend in frequency. However, in response to a stabilized 3°C–6°C warmer climate, the most realistic models simulate increased El Niño amplitude (with no change in frequency) (Guilyardi, 2006). Also, a shift towards Modiki events has recently been forecast (Yeh et al., 2009). Furthermore, palaeo data indicate different ENSO regimes under different climates of the past. Despite large persisting uncertainties, the probability of tipping point behaviour, in the sense that ENSO either vanishes or becomes overly strong, is estimated to be rather low during the twenty-first century (Latif and Keenlyside, 2009, Latif and Park, 2012, this volume). The mechanisms and time-scale of any transition are unclear, but a gradual increase in El Niño amplitude and/or a shift in location is consistent with the recent observational record and would, nevertheless, have severe impacts in many regions.

17.4.3. Atlantic Thermohaline Circulation (THC)

The archetypal example of a climate surprise is a reorganisation of the Atlantic THC, which is prone to collapse when sufficient freshwater enters the North Atlantic to halt density-driven deep water (NADW) formation there (Peng,

1995; Stommel, 1961). Modelling that minimizes artefacts arising from numerical diffusion shows that a hysteresis-type response to freshwater perturbations is a characteristic, robust feature of the THC (Hofmann and Rahmstorf, 2009). However, the shutdown of the THC may actually be one of the more distant tipping points. Expert elicitation suggests that THC collapse becomes as likely as not with >4°C warming this century (Kriegler et al., 2009). The IPCC (2007a) views the threshold as even more remote, but recent analysis suggests the AR4 models are systematically biased towards a stable THC (Drijfhout et al., 2010). Although a collapse of the THC may be one of the more distant tipping points, a weakening of the THC this century is robustly predicted (IPCC, 2007a). This, in turn, will have similar, though smaller, effects as a total collapse. A potential tipping point that occurs in some models is a switch of the subpolar gyre in which deep convection and NADW formation shuts off in the Labrador Sea region (to the west of Greenland) and convection switches to only occurring in the Greenland–Iceland–Norwegian Seas (to the east of Greenland) (Born and Levermann, 2010; Levermann and Born, 2007). This would have dynamic effects on sea level, increasing it down the eastern seaboard of the USA by around 25 cm in the regions of Boston, New York, and Washington DC (in addition to the global steric effects of ocean warming) (Yin et al., 2009).

17.4.4. West African Monsoon (WAM) and Sahel-Sahara

Past intervals of severe drought in West Africa have been linked to weakening of the THC (Chang et al., 2008; Shanahan et al., 2009). This event seems to trigger a phenomenon known as the Atlantic Niño (by analogy with El Niño events in the equatorial Pacific), involving reduced stratification and warming of the sea surface in the Gulf of Guinea. This disrupts the West African Monsoon (WAM), which is usually enhanced by the development of a 'cold tongue' in the eastern equatorial Atlantic that increases the temperature contrast between the Gulf of Guinea and the land to the north. In a typical year, there is also a northward 'jump' of the monsoon into the Sahel in July, which corresponds to a rapid decrease in coastal rainfall and the establishment of the West African Westerly Jet in the atmosphere (Hagos and Cook, 2007). The jump is due to a tipping point in atmospheric dynamics: when the east/west wind changes sharply in the north/south direction, this instability causes the northward perturbation of an air parcel to generate additional northward flow (a strong positive feedback).

It is not clear in which direction the WAM might shift in the future. The more benign alternative is a greening of the Sahel-Sahara, by a mechanism that can be related to observations of the seasonal northward shift of the WAM (although this would restrict dust-borne nutrient supply to the tropical Atlantic and the Amazon rainforest (Washington et al., 2009). The more dangerous option is a southward shift of the WAM. If ocean temperatures change such that the West African Westerly Jet fails to form or is weakened below the tipping point needed to create inertial instability, then the rains may fail to move into the continental interior, drying the Sahel. Recent simulations suggest a tipping point if the THC weakens below ~8 Sv, causing the subsurface North Brazil Current to reverse and an abrupt warming in the Gulf of Guinea (a persistent Atlantic Niño state). The WAM then shifts such that there is a large reduction in rainfall in the Sahel and an increase in the Gulf of Guinea and coastal regions (Chang et al., 2008). Such a transition is forecast in one of only three IPCC (2007a) models that produces a realistic present climate in this region (Cook and Vizy, 2006). However, the other two models give conflicting responses: in one the Sahel gets markedly wetter despite a collapse of the WAM and in the other there is little net change.

17.4.5. Southwest North America (SWNA)

Increased humidity in a warmer world causes increased moisture divergence, changing global atmospheric circulation — including poleward expansion of the Hadley cells and the subtropical dry zones (Held and Soden, 2006; Lu et al., 2007), a development that tends to strongly reduce runoff in these regions (Milly et al., 2005). One area that may be particularly affected is southwest North America (SWNA), defined as all land in the region 125°W—95°W and 25°N—40°N. Aridity in this domain is robustly predicted to intensify and persist in future and a transition is probably already underway: to something which has been described as "…unlike any climate state we have seen in the instrumental record" (Seager et al., 2007, p. 1183). Recently, increased SWNA aridity has been linked to the potential for increased flooding in the Great Plains (Cook et al., 2008). The key driver is model-projected relatively higher summer warming over land than over ocean (analogous to what drives seasonal monsoons). In simulations of future dynamics, an increased contrast between the continental low and the North Atlantic subtropical high strengthens the Great Plains low-level jet, which transports moisture from the Caribbean to the upper Great Plains, triggering flooding there but starving SWNA of moisture (Cook et al., 2008). However, it is unclear whether there is strong non-linearity in response to warming, and, therefore, whether drying of SWNA qualifies as a climate surprise or tipping element.

17.5. LOSS OF BIOMES

At the ecosystem-scale there are probably many potential thresholds related to climate change, the most obvious of which is the 'extinction' (disappearance) of a unique type of ecosystem, because it has nowhere to retreat to (for example the Fynbos on the southern tip of Africa, or ecosystems already high in mountains). Such changes are clearly a concern to policymakers, but it is not clear that they involve a dynamical threshold within the relevant ecosystems. Also, here the focus is on the larger-scale of what on land are called biomes. Biome tipping points can come about due to local biophysical feedbacks that exist between the land surface and climate (Claussen et al., 1999), with the Amazon rainforest and boreal forest being leading candidates (Lenton et al., 2008). Although marine extinctions are clear in the geologic record following climate 'catastrophes' (e.g., Belcher and Mander, 2012, this volume), relatively little attention has been directed at the search for tipping elements in marine 'biomes' (or 'biogeographical provinces').

17.5.1. Amazon Rainforest

A severe drought occurred from July to October in 2005 in the western and southern parts of the Amazon basin, which led the Brazilian government to declare a state of emergency. Despite initial 'greening up' of large areas of forest (Saleska et al., 2007), the 2005 drought made the Amazon region a significant episodic carbon source, when otherwise it has been a carbon sink (Phillips et al., 2009). The 2005 drought has been linked to unusually warm sea surface temperatures in the North Atlantic (Cox et al., 2008). However, lengthening of the Amazon dry season is also part of a wider trend in seasonality, associated with weakening of the zonal tropical Pacific atmospheric circulation as attributed to anthropogenic greenhouse gas forcing (Vecchi et al., 2006). The trend of a lengthening dry season is forecast to continue, with one model predicting that the 2005 drought will be the norm by 2025 (Cox et al., 2008). If drying continues, several model studies have shown the potential for significant dieback of up to ~70% of the Amazon rainforest by late this century, and its replacement by savanna and caatinga (mixed shrubland and grassland) (Cook and Vizy, 2008).

There are positive feedbacks related to the ways in which rainforests store and recycle water to the atmosphere, which could accelerate the Amazon demise: increased (decreased) forest cover leads to increased (decreased) precipitation and vice versa (Betts et al., 2004). Such a feedback opens the possibility of bistability and tipping point behaviour. Ecosystem disturbance processes such as increased fire frequency and pest infestation could also amplify a transition initially driven by drought. Experts suggest Amazon dieback is more likely than not if global warming exceeds 4°C (Kriegler et al., 2009). However, the Amazon rainforest may lag climate forcing significantly and, hence, it may be committed to some dieback long before it is apparent. In one model, committed dieback begins at 1°C global warming above pre-industrial, even though it does not begin to be observed until global warming approaches 4°C (Jones et al., 2009). The existence and extent of forecast Amazon dieback depends on the choice of climate model (Salazar et al., 2007; Scholze et al., 2006) because not all global climate model projections give a regional, seasonal drying trend in the Amazon. A recent analysis based on 19 GCMs has indicated that many models tend to underestimate current rainfall and that, although dry-season water stress is likely to increase over the twenty-first century, the rainfall regime of east Amazonia is likely to shift in the direction of seasonal forests, rather than savanna (Malhi et al., 2009). Dieback is generally less sensitive to the choice of vegetation model, but the direct effect of CO_2 increasing the water use efficiency of vegetation can have a strong effect of tending to shift the dieback threshold further away (P. M. Cox, personal communication, 2010).

17.5.2. Boreal Forest

The boreal forest in western Canada is currently suffering from an invasion of mountain pine beetle that has caused widespread tree mortality (Kurz et al., 2008b). This has turned the nation's forests from a carbon sink to a carbon source (Kurz et al., 2008a). Fire frequencies have also been increasing across the boreal forest zone. In the future, widespread dieback has been predicted in at least one model, when regional temperatures reach around 7°C above present, corresponding to around 3°C global warming. Expert elicitation concurs that above 4°C global warming dieback becomes more likely than not (supplementary information of Kriegler et al., 2009). The causal mechanisms include increasingly warm summers becoming too hot for the currently dominant tree species, increased vulnerability to disease, decreased reproduction rates and more frequent fires causing significantly higher mortality. The forest would be replaced over large areas by open woodlands or grasslands, which would in turn amplify summer warming and drying and increase fire frequency, producing a potentially strong positive feedback. (A warmer future climate may also enable northward expansion of the boreal forest into tundra regions (Scholze et al., 2006; Sitch et al., 2008) but this is not forecast to involve threshold behaviour.)

17.5.3. Coral Reefs

Coral bleaching events, linked to ocean warming, have become much more widespread and detrimental in recent decades, and marine biologists are talking about being at a 'point of no return' for tropical coral reefs (Veron et al., 2009). Ocean acidification, as a direct consequence of rising atmospheric CO_2 emissions, may also contribute to threshold-like changes in coral reef ecosystems (Riebesell et al., 2009; Sen Gupta and McNeil, 2012, this volume). Cold-water corals that grow down to 3000 m depth will be particularly vulnerable to acidification. They will be first affected as the saturation horizon of aragonite (a crystalline form of calcium carbonate) shallows because of ocean acidification. Once bathed in corrosive waters and undersaturated in aragonite, the skeletons and shells may dissolve and the reefs collapse. It has been estimated that with unabated CO_2 emissions, 70% of the presently known deep-sea coral reef locations will be in corrosive waters by the end of this century (Guinotte et al., 2006). However, whether large areas will reach a threshold together (and thus qualify as a tipping element), such as the Great Barrier Reef, or the cold-water coral reef systems extending from northern Norway to the west coast of Africa, warrants further research.

17.6. COPING WITH CLIMATE SURPRISES

Having detailed several potential climate surprises, the overriding question becomes 'How should (climate) policy respond?' In human endeavours, the prospect of having to deal with unpleasant surprises — high-impact, relatively low-probability events, including a strong element of unpredictability — is not new. Think of earthquakes or hurricanes making landfall. Systems exist (albeit flawed ones) for dealing with such events, and they hinge around a risk management approach (e.g., Taplin, 2012, this volume). Although these are relatively short-timescale 'events', some of the risk management principles may be usefully mapped over to climate tipping points. Risk, in the formal sense, is the product of the likelihood (or probability) of something happening and its (negative) impact. So a meaningful risk assessment of tipping elements would demand careful assessment of the likelihood of passing various tipping points (under different forcing scenarios), as well as the associated impacts.

17.6.1. Risk Assessment

There already exists some information about the likelihood of passing different tipping points as a function of global temperature change. Results of a workshop and literature review cover eight tipping elements (Lenton and Schellnhuber, 2007; Lenton et al., 2008), and a process of expert elicitation considered six of these under three different future climate trajectories, and involved the elicitation of imprecise probability statements from 52 experts (Kriegler et al., 2009). Useful results were obtained for five tipping elements: the Greenland Ice Sheet, West Antarctic Ice Sheet, ENSO, Amazon rainforest, and the THC. The imprecise probability statements were then formally combined to give lower-bound probabilities. These reveal that the likelihood of passing at least one of five tipping points rises from >16% under a midrange (2°C–4°C) global warming corridor to >56% (i.e., more likely than not) under a high-warming (>4°C) corridor. In Figure 17.4, the 'burning embers' diagram of Lenton and Schellnhuber (2007) is updated to summarize likelihoods as a function of global warming, based on the expert elicitation results and recent literature.

There also exists some information on the impacts of tipping the different elements, but the gaps are larger and this area needs more detailed work. A recent study has articulated the implications of four different tipping point scenarios for the insurance sector (Lenton et al., 2009a), considering Amazon rainforest dieback, Indian Summer Monsoon disruption (coupled with melt of HKHT glaciers), a shift to a more arid climate in southwest North America (including loss of mountain snowpacks), and high-end sea-level rise from melting ice sheets with additional regional sea-level rise along the northeastern seaboard of the USA related to weakening of the THC. However, tipping point impacts will depend on human responses and are thus a more epistemologically contested area than assigning likelihoods to events. The resulting ambiguity needs to be reduced if risk assessment is to be usefully pursued (Stirling, 2003).

With these caveats, let us offer an initial 'straw man' illustration of how a tipping point risk assessment might look (Table 17.1). Here a scenario of *partial* mitigation of greenhouse gas emissions is assumed, leading to roughly 3°C global warming by 2100. The focus is on tipping elements from the original shortlist (Lenton et al., 2008) where a threshold can be meaningfully linked to global temperature change (thus excluding the Indian Summer Monsoon). Likelihoods and relative impacts are assessed on a five-point scale: low, low–medium, medium, medium–high, and high. Information on likelihood is taken from

a review of the literature (Lenton and Schellnhuber, 2007; Lenton et al., 2008) and, where available, expert elicitation (Kriegler et al., 2009). Impacts are considered in relative terms based on an initial subjective judgment (noting that most tipping point impacts, if placed on an absolute scale compared to other climate eventualities, would be high). Impacts depend on timescale and here the full 'ethical time horizon' of 1000 years is considered (Lenton et al., 2008), assuming minimal discounting of impacts on future generations. Likelihood and impacts are simply multiplied together to give a measure of risk, and a ranking emerges (Table 17.1).

What this simple tabulation readily illustrates are some familiar dilemmas for the would-be risk manager: relatively high-impact, low-probability events, such as WAM collapse, come out with a similar risk to relatively over-impact, high-probability events, such as Arctic summer sea-ice loss. However, what stand out are the *high-impact, high-probability scenarios* as a priority for risk management effort, in this case the melting of the Greenland ice sheet, followed by West Antarctic Ice Sheet collapse. These risks will be best managed by restricting the extent of future warming of these systems and their surrounding ocean. Of course, this exercise would be better conducted with a wider team of experts and relevant stakeholders to get a more scientifically credible and socially legitimate assessment (cf. Taplin, 2012, this volume). It is simply hoped that the ranking in Table 17.1 encourages some thought and activity in this area.

17.6.2. Removing the Element of Surprise

Faced with the risk of unpleasant climate surprises, perhaps the most useful information that science could provide to help societies cope is some early warning of an approaching tipping point. There are several degrees of early warning, from simply identifying possible threats, to

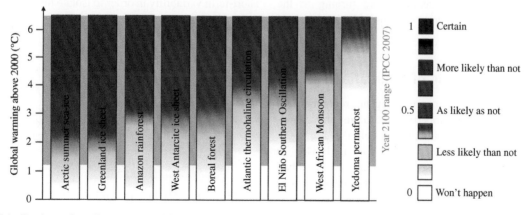

FIGURE 17.4 Burning embers diagram summarizing current information on the likelihood of tipping different elements under different degrees of global warming. *(Source: Updated from Lenton and Schellnhuber, 2007, based on expert elicitation results (Kriegler et al., 2009) and recent literature discussed therein.)*

TABLE 17.1 A Simple 'Straw Man' Example of Tipping Element Risk Assessment

Tipping Element	Likelihood of Passing a Tipping Point (by 2100)	Relative Impact** of Change in State (by 3000)	Risk Score (likelihood × impact)	Risk Ranking
Arctic summer sea-ice	High	Low	3	4
Greenland ice sheet	Medium–High*	High	7.5	1 (highest)
West Antarctic Ice Sheet	Medium*	High	6	2
Atlantic THC	Low*	Medium–High	2.5	6
ENSO	Low*	Medium–High	2.5	6
West African monsoon	Low	High	3	4
Amazon rainforest	Medium*	Medium	4	3
Boreal forest	Low	Low–Medium	1.5	8 (lowest)

*Likelihoods informed by expert elicitation.
**Initial judgment of relative impacts is the subjective assessment of the author.

being able to forecast that a tipping point is imminent. For other high-impact events, such as hurricanes or tsunamis, there are already quite sophisticated early-warning systems in place, which climate policy could potentially learn from.

Where a potential tipping point threat has been convincingly identified, the challenge becomes 'Can any early warning signs be detected before the threshold is breached?' Here the answer depends critically on the nature of the underlying tipping phenomenon. Bifurcation-type tipping points (Figure 17.1a) offer the best prospects for early warning. In contrast, purely noise-induced transitions (Figure 17.2) are fundamentally unpredictable (Ditlevsen and Johnsen, 2010; Hastings and Wysham, 2010). Non-bifurcation type tipping points (Figure 17.1b) present an intermediate case; their response is expected to resemble bifurcation-type behaviour to a certain degree. The prospects for early warning of an approaching bifurcation are now considered in more detail, before turning to the question of how to characterize systems with high levels of noise.

17.6.3. Early Warning of Bifurcations

Physical systems that are approaching bifurcation points show a nearly universal property of becoming more sluggish in response to a perturbation (Scheffer et al., 2009; Wiesenfeld and McNamara, 1986; Wissel, 1984). To visualize this, picture the present state of a system as a ball in a curved potential well (attractor) that is being nudged around by some stochastic noise process, such as weather (Figure 17.5). The ball continually tends to roll back towards the bottom of the well — its lowest potential state — and the rate at which it rolls back is determined by the

curvature of the potential well. As the system is forced towards a bifurcation point, the potential well becomes flatter. Hence, the ball will roll back ever more sluggishly. At the bifurcation point, the potential becomes flat and the ball is destined to roll off into some other state (alternative potential well). Mathematically speaking, the leading eigenvalue, which characterizes the rates of change around the present equilibrium state, tends to zero as the bifurcation point is approached.

So, for those tipping elements that exhibit true bifurcation points (e.g., Figure 17.1a), an observed slowing down in timeseries data could provide a basis for early warning. This should be manifested as an increasing autocorrelation in the time-series data (in simple terms, each data point becomes more like the surrounding ones). Following this rationale, a method of examining the decay rate to perturbations using a simple lag-1 autocorrelation function (ACF) was developed, averaging over short-term variability in order to isolate the dynamics of the

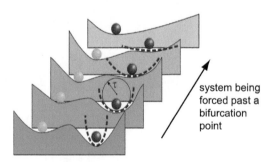

FIGURE 17.5 Schematic representation of a system being forced past a bifurcation point. The system's response time to small perturbations, τ, is related to the growing radius of the potential well. *(Source: Figure by Hermann Held, from Lenton et al., 2008. Copyright 2008 National Academy of Science, U.S.A.)*

longest immanent timescale of a system (Held and Kleinen, 2004). The approach was subsequently modified by using detrended fluctuation analysis (DFA) to assess the proximity to a threshold from the power law exponent describing correlations in the timeseries data (Livina and Lenton, 2007). At a critical threshold, the data become highly correlated across short and middle-range timescales and the time series behaves as a random walk with uncorrelated steps. Both methods need to span a sufficient time interval of data to capture what can be a very slow decay rate, and they suffer the problem that rapid forcing of a system could alter the dynamics and override any slowing down.

Model tests have shown that both early warning methods work in principle, in simple (Dakos et al., 2008), intermediate complexity (Held and Kleinen, 2004; Livina and Lenton, 2007), and fully three-dimensional (Lenton et al., 2009b) models. The challenge is to get the methods to work in practice, in the complex and noisy climate system. Initial tests found that the ending of the last ice age recorded in the ice core data is detected as a critical transition using the DFA method (Livina and Lenton, 2007). Subsequent work showed increasing autocorrelation in eight palaeoclimate timeseries approaching transitions, using the ACF method (Dakos et al., 2008).

In Figure 17.6, existing DFA analysis of a Greenland ice core record (Livina and Lenton, 2007) is compared with the ACF method (with or without detrending). Here, the data come from the GISP2 ice core and are the $\delta^{18}O$ stable water isotope record, which is a proxy for past air temperature at the ice-core site (and can also be influenced by changing water source temperatures and snowfall seasonality). Increasing $\delta^{18}O$ corresponds to warming. The data in this case are sparse and unevenly spaced (this is typical of many palaeo data records, e.g., Harrison and Bartlein, 2012, this volume) and are not interpolated, since this can introduce a high degree of correlation and overwrite any potential early warning signal. Both approaches detect critical behaviour during the last deglaciation (the indicators approach or exceed a critical value of 1). However, using only the more sparse data prior to the transition (to the left of the dotted vertical line in Figure 17.6, top panel) the upwards trends, indicative of slowing down, are rather weak (to the left of the dotted vertical lines in Figure 17.6, middle and bottom panels).

Recent work has emphasized that to get a reliable signal of approaching bifurcation, one should monitor changes in variance, as well as autocorrelation in the data (Ditlevsen and Johnsen, 2010). As a threshold is approached and the potential becomes flatter (Figure 17.5), one intuitively expects the variance to go up (i.e., the ball to make greater departures from the local stable state). If at the same time the system is slowing down, one must be

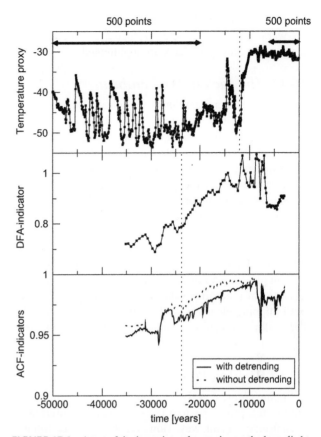

FIGURE 17.6 A test of tipping point early warning methods applied to the GISP2 Greenland ice core $\delta^{18}O$ proxy record of palaeotemperature, following Livina and Lenton (2007) with the addition of ACF analysis. A sliding window of 500 data points is used, and the data are non-uniform, as indicated in the top panel. Results are plotted in the middle of the sliding windows. The vertical dotted line in the bottom two panels indicates where the analyses include the 500 points before the transition into the Holocene marked in the top panel. The ACF method was applied with and without the detrending method described in Dakos et al. (2008).

careful to choose a long enough time window to accurately sample the variance. The resulting changes in variance may provide a statistically more robust and earlier warning signal than changes in autocorrelation (Ditlevsen and Johnsen, 2010). However, in the climate system, which contains many sources of inertia or 'memory', the two properties can be expected to change together. Other potential early warning indicators are being explored for ecological tipping points and could potentially be applied to climate. These include increasing skewness of responses (Biggs et al., 2009; Guttal and Jayaprakash, 2008), spatial variance, and spatial skewnesses (Guttal and Jayaprakash, 2009).

17.6.4. Limitations on Early Warning

It is encouraging that there is some theoretical scope for early warning of an approaching bifurcation, but there are considerable practical limitations on whether an effective early-warning system could be deployed for specific systems. Here three problems are highlighted.

17.6.4.1. The Lack of Data Problem

A key consideration is 'What is the longest internal time-scale of the system in question?' It is changes in this that the bifurcation early warning method is trying to detect. In the case of the ocean circulation or ice sheet dynamics, these timescales are long (in the thousands of years). Therefore, one needs a long and relatively high-resolution (palaeo) timeseries record for the system in question, in order to get an accurate picture of its natural state of variability from which to detect changes. Often such records are lacking. However, some potential tipping points have much faster dynamics and relatively little internal memory, for example the monsoons. For such systems, existing observational timeseries data may be sufficient. Also, for specific tipping elements, such as the THC, there may be other leading indicators of vulnerability that are deducible from observational data (Drijfhout et al., 2010).

17.6.4.2. The Lag Problem

If a tipping element is forced slowly (keeping it in quasi-equilibrium), proximity to a threshold may be inferred in a model-independent way. However, humans are forcing the climate system relatively rapidly, so inherently 'slow' tipping elements, such as ice sheets and the THC, will be well out of equilibrium with the forcing (e.g., Harvey, 2012, this volume). This means that a dynamical model simulating transient behaviour will also be needed to establish proximity to a threshold.

17.6.4.3. The Noise Problem

Noise (or internal variability) in a system may be such that it does not allow the detection of any trend towards slowing down. For example, in the case of the THC there are now several years of direct observations showing that its strength exhibits high internal variability (Latif and Park, 2012, this volume). Where internal variability is high, a tipping element could exit its present state well before a bifurcation point is reached (Figure 17.2a). Hence, for a method of anticipating a threshold to be useful, the time it takes to find out its proximity to a threshold must be shorter than the time in which noise would be expected to cause the system to change state — the 'mean first exit time' (Kleinen et al., 2003). A sophisticated early warning system should take account of the noise level for a particular tipping element and adjust its estimates accordingly (Thompson and Sieber, 2010).

In systems with a high level of noise, flickering between states (i.e., noise-induced transitions in both directions) may occur prior to a more permanent transition (Bakke

et al., 2009). We now turn to consider whether, in such noisy systems, one can detect bifurcations.

17.6.5. Bifurcations in Noisy Systems

Although individual noise-induced transitions in the climate system are inherently unpredictable, if a system is experiencing a relatively high level of noise and sampling several different states (or modes of operation), then in principle one can deduce how many states it has, as well as their relative stability or instability. To do so successfully requires a sufficiently long timeseries that all available states are being sampled. If the number of states and/or the stability of states changes over time then in principle this can also be detected. However, to do so requires a long window that is moving through an even longer timeseries. These ideas are at the heart of a recently developed method called 'potential analysis' (Livina et al., 2010, 2011). While the 'slowing down' method described above assumes that a system subject to noise is contained within one potential well, and looks for signals that the corresponding state is becoming unstable, 'potential analysis' assumes that a system is sampling a number of potential wells (e.g., Figure 17.2b) and tries to deduce the number of wells and their stability properties.

Potential analysis assumes that the dynamics of a chosen climate variable can be described by a stochastic differential equation (i.e., one that includes a noise term), with an underlying potential (i.e., series of wells and hill-tops) whose shape can be described by a polynomial equation. The chosen stochastic differential equation has a corresponding Fokker–Planck equation, describing the probability density function; crucially, this has a stationary solution that depends only on the underlying potential function and the noise level. This gives a one-to-one correspondence between the potential and the stationary probability density of the system (the potential is directly proportional to minus the logarithm of the probability density, scaled by the square of the noise level). This allows the underlying potential to be reconstructed, given the probability density of a stretch of timeseries data and an estimate of the noise level. For the mathematics, see Livina et al. (2011).

The method is illustrated in Figure 17.7. It starts by transforming a window of timeseries data into a histogram of the data. Next, this is converted into an empirical probability density of the data using a standard Gaussian kernel estimator. If the system has a single, stable state, the resulting distribution should have a single mode with smooth sides. Any deviations from this immediately provide a visual clue as to the existence of other states. In the example, the probability density has a distinct 'shoulder' suggesting the existence of a second state. Next, the probability density is inverted and log-transformed.

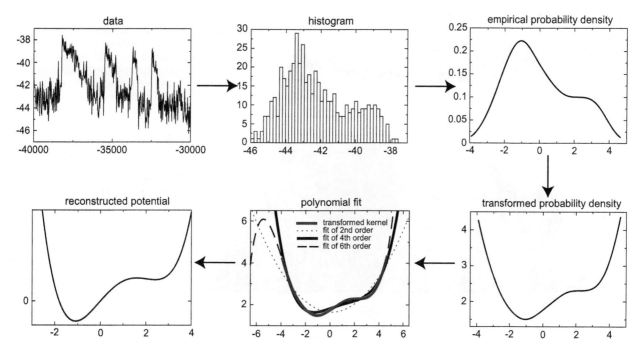

FIGURE 17.7 Illustration of the method of potential analysis for a window of timeseries data (here NGRIP δ^{18}O data 40–30 ka BP). The steps are: obtain histogram of the data; convert into empirical probability density (using a standard Gaussian kernel estimator); transform this distribution (take natural logarithm, invert, and scale by noise level); least-squares fit the transformed distribution with polynomial functions of increasing even order and select the highest order before encountering a negative leading coefficient; accurately determine the coefficients using the Unscented Kalman Filter; and reconstruct the potential.

The method then attempts to least-square fit the transformed distribution with polynomial functions of increasing, even order (starting with second order, i.e., a quadratic equation). At some point, the least-square fit returns a negative leading coefficient, which is not physically reasonable (this happens for a sixth-order fit in the example). So, the polynomial of highest degree before this is encountered is taken as the most appropriate representation of the probability density of the timeseries (fourth-order in the example in Figure 17.7).

The number of states in the system is then determined from the number of inflection points in the fitted polynomial potential (and the results of this are plotted in Figure 17.8, using a colour scheme). The simplest potential has a single state with no inflection points. Each pair of inflection points corresponds to an additional state. This approach picks up real minima (wells) in the potential, or just flattening of the potential. The latter, importantly, are degenerate states corresponding to bifurcation points. In a final step, having determined its order, the coefficients of the potential can be accurately estimated using the Unscented Kalman Filter (UKF; Kwasniok and Lohmann, 2009). For this to work well, the noise level must be accurately estimated, which can be done separately using a wavelet denoising routine that separates the signal into the potential and the noise (Livina et al., 2011). The end result is a reconstructed potential.

17.6.6. Application to Past Abrupt Climate Changes

In principle, the method of potential analysis just described can detect bifurcations in a noisy climate system. Their timing can never be precisely tied down, because the method always has to work with a relatively long time window of data (typically of order 1000 data points). However, by moving a sliding window through a long timeseries, one should be able to detect changes in the number of states over time.

To test this, potential analysis has been applied to a classic case study of abrupt climate change in Earth's recent past: the Dansgaard–Oeschger events (or 'D–O events' for short) (Livina et al., 2010). These were rapid climate changes that occurred repeatedly during the last ice age, were concentrated in the North Atlantic region but had widespread effects, and are recorded most strikingly in Greenland ice cores. Figures 17.8a and 17.8c show two such ice core records: the GRIP (Dansgaard et al., 1993) and NGRIP (NGRIP, 2004) δ^{18}O stable water isotope records, which are a proxy for past air temperature at the ice-core sites, which are 325 km apart. The D–O events are the abrupt increases in δ^{18}O (corresponding to warming).

What do these D–O events reveal about the nature of past climate surprises? The precise mechanism for what

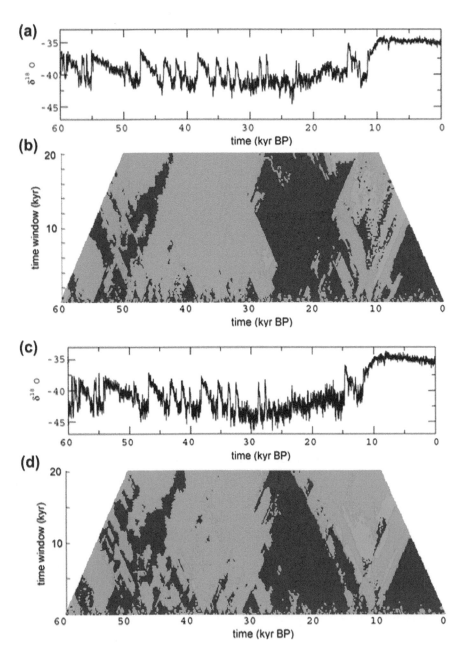

FIGURE 17.8 Ice-core $\delta^{18}O$ proxy records of palaeotemperature over the past 60 ka at two different sites 325 km apart in Greenland: (a) GRIP (Dansgaard et al., 1993) and (c) NGRIP (NGRIP, 2004); both at 20-year resolution and on the most recent GICC05 timescale (Svensson et al., 2008). Contour plots of the number of detected states in the records are a function of time and sliding window length (results plotted at the midpoints of the sliding windows), for (b) GRIP and (d) NGRIP, where; red = 1 state, green = 2, cyan = 3, purple = 4. This shows the loss of a second climate state (green-to-red transition across a wide range of window lengths) around 25 ka BP. *(Source: Livina et al., 2010.)*

was going on in the climate system continues to be debated (Colin de Verdière, 2006), although most studies assign a key role to changes in the Atlantic THC, coupled to changes in sea-ice cover. Regardless of the underlying mechanism, recent work has shown that the repeated transitions from cold 'stadial' to warm 'interstadial' states can be well-described by a model of purely noise-induced transitions (Ditlevsen et al., 2005) (Figure 17.2). Recent analysis suggests that the cold stadial state remains stable during the last ice age and does not experience a bifurcation (Ditlevsen and Johnsen, 2010). However, when looking across the interval 70 ka−20 ka BP, the warm state is

characterized as being only marginally stable (Kwasniok and Lohmann, 2009).

On examining this more closely using the method of potential analysis (Figure 17.8b and 17.8d), the ice core records are best characterized as having two states from 60 ka BP to about 25 ka BP, but only one during the depths of the Last Glacial Maximum (LGM) (Livina et al., 2010). It is inferred that a bifurcation occurred (Figure 17.1a) sometime prior to 25 ka BP, in which the warm interstadial state became unstable, and later it disappeared altogether. Figure 17.9 shows this bifurcation occurring in reconstructed potentials from the NGRIP data. It is reflected in

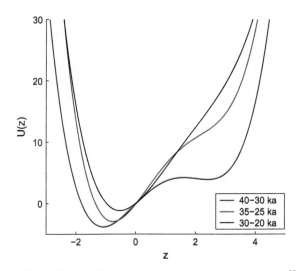

FIGURE 17.9 Potential curves reconstructed from the NGRIP $\delta^{18}O$ palaeotemperature proxy data: for the intervals 40 ka—30 ka (blue line, as in Figure 17.7), 35 ka—25 ka (red line), and 30 ka—20 ka (black line). Here 'z' represents anomalies in NGRIP $\delta^{18}O$ and U(z) is the derived potential, which shows the steady states in the system and describes the rate and direction of response of 'z' to deviations from steady state (according to dz/dt = −dU/dz). The results show a bifurcation in which the warm interstadial climate state becomes degenerate (unstable) and then disappears, at some time prior to 25 ka. *(Source: Livina et al., 2011. With kind permission from Springer Science + Business Media)*

the original records as the warm events becoming progressively shorter, until they are very short-lived indeed, and then they cease through the LGM (Figures 17.8a and 17.8c). In the case of a bifurcation such as this, which reduces the number of system states, detection (the switch from green to red in Figures 17.8b and 17.8d) typically occurs after the potential has become degenerate and a bifurcation has occurred (Figure 17.9). That is because a system with a second state that has lost its stability, but not disappeared, altogether gives a 'shoulder' in the probability density and is still best described by a fourth-order polynomial (e.g., the reconstructed potential for NGRIP 35 ka—25 ka BP in Figure 17.9). However, this means that, conversely, in a system that is gaining an extra state, the method may pick this up before that state becomes stable (i.e., bifurcation occurs) and, hence, provide some early warning that a new climate state is appearing. Currently palaeo and observational timeseries data are being analysed to search for such cases where potential analysis indicates a new climate state is appearing (Livina et al., 2011).

The simple model of D—O events just described does not account for all the features in the ice core records, in particular, the gradual cooling that typically follows an abrupt warming (an asymmetrical response) (Figures 17.8a and 17.8c). Furthermore, it does not account for variations in the noise level (i.e., internal variability) between the cold and the warm states (it appears to be considerably greater in the cold state). Hence, future work can be expected to produce

more advanced models. Still, to summarize our present understanding, the archetypal example of past abrupt climate changes illustrates that the climate system has undergone both noise-induced transitions and bifurcations in the past. These two types of climate surprise have very different implications for whether individual events are predictable.

17.7. FUTURE CLIMATE: SURPRISES, RESPONSES, AND RECOVERY STRATEGIES

Assuming science could provide society with reliable early warning of an approaching tipping point, and that this information was viewed as sufficiently credible, salient, and legitimate to warrant action, the question becomes, 'What could (or should) we do about it?' Obvious courses of action are to try and avoid reaching the tipping point, or to try and 'pre-adapt' (build resilience) to better cope with the changes that are due to occur. Which is feasible or most appropriate will depend somewhat on the tipping element in question and the forcing factors responsible, but let us start with some fairly general observations.

The natural response to warning of the prospect of a high-impact event is to try to avoid it. However, our options for effective intervention ('steering') in the climate system are actually somewhat limited. That is because there are several sources of inertia in the Earth system — one might liken heading towards a climate tipping point as like the Titanic heading towards the fateful iceberg: it is a big ship and it is hard to change its trajectory quickly. An analogous problem of avoiding an approaching tipping point in an ecological system (a fishery) shows that once there is a reliable early warning of an approaching tipping point, it is probably too late for slow intervention methods to avoid it (Biggs et al., 2009).

17.7.1. Mitigation

The conventional avoidance strategy is to mitigate the emissions of greenhouse gases, especially carbon dioxide, but this is a slow intervention. The climate is responding to the concentration of these agents (and the resulting radiative forcing), not their emissions (Harvey, 2012, this volume). The prime mitigation target, CO_2, is a very long-lived pollutant and the current policy challenge is still framed as stopping it rising and holding a stabilization level, not bringing its concentration down. Even if radiative forcing is stabilized, the climate will continue to warm for some considerable time because heat will still be entering the system (primarily the deep ocean) and slow positive feedbacks will be operating (such as loss of methane from hydrates, e.g., Harvey, 2012, this volume).

Potentially, radiative forcing could be reduced by mitigating emissions of short-lived greenhouse gases (e.g., tropospheric ozone) and black carbon (BC) aerosols, whose concentrations will decline fairly rapidly. Ozone in the lower atmosphere currently contributes 20% as much as CO_2 to global warming. Rigorous global implementation of air pollution regulations and available technologies could drive down emissions of ozone precursors (especially carbon monoxide and nitrogen oxides) quickly, producing a climate response within decades (Molina et al., 2009). BC is estimated to be the second or third largest global warming agent, although large uncertainties about its exact radiative forcing exist. Due to its deposition on snow and resulting positive feedbacks, BC may be responsible for as much as ~0.5°C–1.4°C of the 1.9°C warming observed in the Arctic from 1890 to 2007 (Shindell and Faluvegi, 2009) and for approximately 0.6°C of the 1°C warming in the Tibetan Himalayas since the 1950 (Ramanathan and Carmichael, 2008). Halving BC emissions could be achieved by 2030 with full application of existing technologies (Cofala et al., 2007). Lowering BC emissions would have the 'double benefit' of reducing global warming and raising the temperature level at which tipping points involving melting of ice and snow are triggered (Hansen and Nazarenko, 2004). A further 'fast-action' strategy to consider is phasing down the production and consumption of hydrofluorocarbon (HFCs) with high global warming potential (Molina et al., 2009).

The bulk of mitigation will always be a relatively slow way of turning the climate ship; the inertia in society when it comes to replacing energy and built infrastructure has not even been mentioned. The steering might be speeded up by starting to actively remove CO_2 (or other positive radiative forcing agents) from the atmosphere (Lenton and Vaughan, 2009). However, it will still take timescales of half a century before CO_2 concentration can be stabilized and a century before temperature stops rising (Lenton, 2010). A comprehensive analysis of the complex control problem involved is given elsewhere (Schellnhuber et al., 2009), based on the limitation of cumulative CO_2 emissions worldwide.

17.7.2. Geo-engineering

So, are there any faster avoidance strategies available? This raises the controversial topic of geo-engineering the amount of sunlight absorbed by the Earth, with the aim of deliberately counteracting the positive radiative forcing coming primarily from accumulated greenhouse gases (Lenton and Vaughan, 2009). Some proponents of this type of geo-engineering are arguing that it should be developed as a potential emergency response to the prospect of an approaching climate tipping point. For their argument to hold up, one would need to be confident that such intervention could work fast enough to avoid the transgression of a threshold. The currently much-discussed method of injecting aerosols (probably of sulfate) globally into the stratosphere (Crutzen, 2006) could conceivably rebalance

radiative fluxes at the tropopause, and could begin to cool the climate within a year, if it mimicked the Mount Pinatubo volcanic eruption. However, it would take several years of repeated injection to feel the full climate cooling effect (as the temperature of the ocean mixed layer adjusted) (Wigley, 2006). To this must be added the time to develop the technology and the infrastructure needed to deploy it, which is probably at least a decade at present. So, we are probably talking about 10–20 years for this type of intervention to take effect. In principle, this could be helpful in avoiding an approaching temperature threshold in a relatively sluggish system such as the Greenland ice sheet. Perhaps a tailor-made regional version of radiation management — 'regio-engineering' (H. J. Schellnhuber, personal communication, 2010) — would be the preferable option under these circumstances.

However, geo-engineering is not going to be much use for fast response systems or ones where the threshold is not clearly linked to global temperature, such as the west African or Indian monsoons. Indeed, past volcanic aerosol injections are known to have slowed down the hydrological cycle and reduced rainfall in such regions (Robock et al., 2008; Trenberth and Dai, 2007) (and the theory behind this is robust). So, aerosol geo-engineering interventions might pose a greater risk to some tipping elements than the reduction in risk they achieved for others. Once again, a careful risk management approach is needed to evaluate costs, benefits, and likelihoods.

17.7.3. Rational Responses?

In a rational world, any early warning would be useful, even if it turns out to be impossible to avoid an approaching tipping point, because it would give societies some time to prepare themselves. Effective adaptation can certainly happen faster than mitigation can alter the climate trajectory, and some types of adaptation can probably happen faster than geo-engineering could alter the climate trajectory. However, these may be types of adaptation, such as mass migration, that carry their own considerable risks of triggering undesirable social 'tipping points', that is, conflicts rather than co-operative responses. Research on 'social tipping elements' is urgently needed to better anticipate this type of dynamics (Schellnhuber, 2009).

Furthermore, there is the problem that humans are not perfectly rational actors. Hence, receiving an alarm signal carries the danger of triggering maladaptive responses, especially when our fallible, internal, human methods of risk assessment are at play. Such maladaptive responses to early warning of approaching climate tipping points cannot be ruled out (Travis, 2010). Indeed, despite the known incidence of hurricane landfalls, there has been a demographic trend of people moving *to* Florida. Ironically, US citizens have also historically been leaving the central

Great Plains and moving to the southwest, which is one of the regions highlighted as undergoing a transition to a new climate state less able to support agriculture and people.

17.7.4. Recovery Prospects

If societies experience climate surprises, and the corresponding tipping elements change state, is there any prospect for recovering their original state? Again the answer will depend on the system in question, and the timescale. The definition of a tipping element includes systems that can exhibit reversible or irreversible changes in state (Figure 17.1). A reversible transition means that, if the forcing is returned below the tipping point, the system will recover its original state (either abruptly or gradually). An irreversible transition means that it will not — it takes a larger change in forcing to recover (and hence there is some hysteresis in the trajectory of the system in phase space). An example of a transition that should be reversible in principle is the loss of Arctic summer sea-ice; whereas transitions that could exhibit some irreversibility are the loss of large ice sheets, changes in the Atlantic THC and/or the loss of major biomes.

Even if a transition is reversible in principle, it does not mean that the changes will be reversible in practice. A key problem (discussed above) is that it is rather difficult to reduce radiative forcing of the climate system, unless one indulges in geo-engineering, and the climate further lags the forcing (the problem of 'committed climate change'). In other words, global warming is hard to reverse. Even loss of Arctic summer sea-ice, which seems like a highly reversible system because the ice regrows each winter, has some longer-term 'memory'. It cannot be completely recovered in one season because some of the ice that has been lost consisted of thick, multiyear strata that take several winters to accumulate.

Irreversible transitions vary in what is needed to recover from them, and even with respect to whether recovery is possible at all. The 'strongest' form of irreversibility is extinction — the loss of species and, hence, genetic information and diversity (e.g., Belcher and Mander, 2012, this volume). Extinctions would surely accompany major biome transitions, such as dieback of the Amazon rainforest or boreal forests, should they occur. Thus, although something resembling the current rainforest or boreal forest might eventually grow back, they would never be the same. Ice sheets, such as those on Greenland and West Antarctica, if lost, might eventually be recovered with appropriate forcing, but it would take timescales of tens of thousands of years to regrow a major ice sheet. Alternatively, human activities might fundamentally alter the dynamics of the climate system, switching it out of its recent mode of roughly 100,000-year glacial cycles and back into an earlier

(Pliocene-like) state in which Greenland lives up to its name and remains un-glaciated, West Antarctica only sporadically has an ice sheet, East Antarctica holds less ice, and sea levels are in the long term around 25 metres higher (Rohling et al., 2009).

17.8. CONCLUSION: GAPS IN KNOWLEDGE

Existing work has probably not identified all possible future surprises in the climate system, so more research is definitely warranted to systematically search for them. The latest methods for detecting threshold behaviour could be applied to a wide range of palaeo data, as well as to the instrumental record, and to the output of existing climate model runs, such as from the IPCC Assessment Reports. A useful theoretical starting point would be to try to identify all the potentially strong positive feedbacks in components of the Earth system that could be manifested at large spatial-scales — because these are a necessary condition for tipping point behaviour. Historically, climate science has been good at identifying global-scale positive feedbacks on temperature. However, the focus here is on intermediate (but still large) spatial-scales, and about feedbacks that are internal to the dynamics of a part of the system, and sometimes have little or no effect on global temperature (although they may be triggered by it changing). In recent years, at least one such feedback has been discovered (the potential for runaway breakdown of yedoma permafrost), and others have been better formalized (the multiple stable states of ice sheet grounding lines). Once identified, such feedbacks should then be included in Earth system models, and the phase space of the models systematically searched for multiple stable states at the regional-scale and other signs of strong non-linearity. There are recent examples of the successful detection of multiple states in complex models, for example in the Amazon basin (Oyama and Nobre, 2003). Further effort is encouraged in this area, although, with state-of-the-art models, it will challenge current computing resources.

ACKNOWLEDGEMENTS

I thank John Schellnhuber for first encouraging me to work on climate tipping points and for input to this chapter. Veronika Huber and Martin Wodinski produced Figure 17.3, with input from John, and I. Hermann Held originally produced Figure 17.5. Valerie Livina performed the analyses in, and provided, Figures 17.6, 17.8, and 17.9, and helped with Figure 17.7. Anders Levermann, Veronika Huber, John Schellnhuber, and an anonymous referee provided helpful reviews.

Chapter 18

Future Climate: One Vital Component of Trans-disciplinary Earth System Science

Martin Rice and Ann Henderson-Sellers

Department of Environment and Geography, Macquarie University, Sydney, Australia

18.1. GAIA AND EARTH SYSTEM SCIENCE

This chapter examines climate futures in the context of Earth system (ES) study. Section 18.1 describes how, over the past half century, attempts to describe the way in which the complete ES may work have frequently used climate to illustrate processes and, importantly, to demonstrate the emergence of new characteristics. Schematics and models of Gaian worlds have enabled ES thinking to move from a climate focus to more holistic views. Increasingly, thinking of the ES in terms of climate futures is being replaced by a new systems approach to the integrated study of the Earth that goes beyond traditional disciplinary boundaries (e.g., Figure 18.1). This trans-disciplinary Earth system science (ESS) is discussed in Section 18.3. In Section 18.2, we consider the feature that separates Earth from all other planets as far as we know: people. Human transformation of the ES is unprecedented (Lubchenko,

1998; Steffen et al., 2004; Leemans et al., 2009). This section considers humans in the ES, with climate as an exemplar; in particular, the generation of integrated ES knowledge and what impact (if any) it has on global efforts to tackle planetary challenges, such as anthropogenic climate change.

18.1.1. Earth: An Integrated System

The notion of Earth as a living, integrated system is not new. In a lecture to the Royal Society of Edinburgh over 200 years ago, geologist James Hutton described the Earth as "a compound system of things, forming together one whole living world... the matter of this active world is perpetually moved, in that salutary circulation by which provision is so wisely made for the growth and prosperity of plants, and for the life and comfort of its various animals" (Hutton, 1795, Vol. 2, p. 560).

The Future of the World's Climate. DOI: 10.1016/B978-0-12-386917-3.00018-X

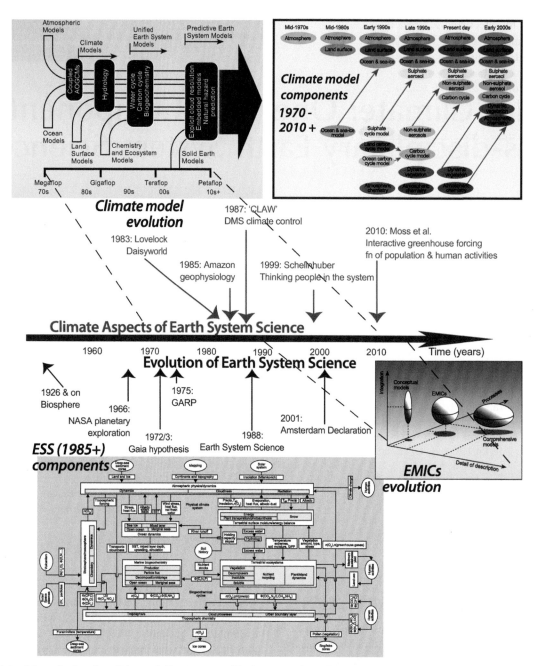

FIGURE 18.1   Schematic of a selected history of climate aspects of Earth system science (ESS). Below the timeline (in blue) describes milestones in the development of Earth system (ES) science from the definition of the 'biosphere', through the development of the Gaia hypothesis to the Amsterdam declaration, with the recognition of the Earth as a holistic system capable of self-regulation. Above the timeline (in red) are climate illustrative examples of how the complete ES works and demonstrations of how processes can work (e.g., Daisyworld in 1983, dimethyl sulfide (DMS) control, and human transformation of the ES). Inset boxed diagrams (*clockwise from the top left*) show the evolution of models from atmospheric models to predictive Earth system models (*Source: McGuffie and Henderson-Sellers, 2005*); the gradually increasing component set in such models (*Source: NOAA, 2010*); the evolution of Earth Models of Intermediate Complexity (EMICs) (*Source: McGuffie and Henderson-Sellers, 2005*); and a version of the Bretherton diagram, emphasizing humans as an integral component of the ES (*Source: Schellnhuber, 1999. Reprinted by permission from Macmillan Publishers Ltd.*).

Revisiting this notion many decades later, James Lovelock explained, "the search for life on Mars led to another look at Hutton's super-organism" (Lovelock, 1986, p. 25). Lovelock (1987a) postulated that the study of the solar system and planets with different atmospheres could advance our understanding of the Earth's climate. Indeed, Carl Sagan's work in 1961 on the climatic state of Venus prompted James Hansen of NASA to investigate the role of sulfate aerosols on the Earth's climate (Weart, 2008). In 1966, NASA designed experiments to detect life on other planets (Figure 18.1). Hitchcock and Lovelock (1967) believed these experiments would fail due to their 'geocentric approach'. An alternative method was proposed, based on advances in infrared astronomy and compositional analysis of planetary atmospheres. This allowed a new systems view of the possibility of life on Mars and Venus. The basic premise was that if life is a global entity then its presence could be detected by a change in the chemical make-up of the planet's atmosphere. This approach led Hitchcock and Lovelock to investigate, with the articulation of the Gaia hypothesis, how the Earth and the nature of the system "could hold so unstable an atmosphere in a steady state that was even more remarkably just right for life" (Lovelock, 1987b, p. 13).

18.1.2. The Gaia Hypothesis

The Gaia hypothesis (Lovelock, 1972; Lovelock and Margulis, 1973) postulates: "the climate and chemical composition of the Earth's surface environment is, and has been, regulated in a state tolerable for the biota" (Lovelock, 1989, p. 215). The Gaia hypothesis (Figure 18.1) — named after the Greek goddess of Earth by the author William Golding (Lovelock, 2009) — was criticized for not fully factoring in evolution by natural selection and, in particular, competition between organisms (e.g., Dawkins, 1982). In response, Lovelock contended that, "In no way is this [Gaia] theory a contradiction of Darwin's great vision. It is an extension to it to include the largest living organism in the Solar System, the Earth itself" (Lovelock, 1986, p. 25). The Daisyworld model (Figure 18.2) (Lovelock, 1983; Watson and Lovelock, 1983) was developed to illustrate how Gaia may work (Kump and Lovelock, 1995). It also provides an initial 'mathematical framework' for understanding self-regulation (Lenton, 1998).

Daisyworld is an imaginary planet, similar in many respects to Earth, on which grow only daisies. The daisies have an abundance of nutrients and water. Their ability to spread across the planetary surface depends only on temperature, and the relationship is parabolic, with minimum, optimum, and maximum temperatures for growth. The climate system is correspondingly simple. There are no clouds, and no greenhouse gases [GHGs]. The planetary energy balance is a function only of solar insolation, albedo and surface temperature, and planetary albedo depends on the areal coverage of the soil (which is grey) by black and white daisies.

(Kump and Lovelock, 1995, p. 539)
(Reprinted from Kump and Lovelock, 1995; with permission from Elsevier.)

The Gaia hypothesis has evolved over time, generating further research to test its robustness and advance the

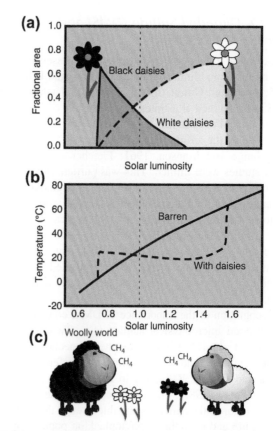

FIGURE 18.2 Daisyworld and Woollyworld are both extremely simplified depictions of planetary systems: the former due to Lovelock (1979) and the latter described in outline by Schellnhuber (1999). On Daisyworld there are only two life forms: white and black daisies. (a) Fraction of planet Daisyworld covered by different daisies as solar luminosity increases over time; (b) how the global mean planetary temperature on Daisyworld is 'controlled' by the daisies; (c) schematic of some of the inhabitants of Woollyworld: populated by sheep that graze, reflect solar radiation, and emit the greenhouse gas methane (CH_4). *(Source: Parts (a) and (b) after McGuffie and Henderson-Sellers, 2005.)*

notion of a holistic ES. For example, "When introduced, this [Gaia] hypothesis was contrary to conventional wisdom that life adapted to planetary conditions as it and they evolved in their separate ways. We now know that the hypothesis as originally stated was wrong because it is not life alone but the whole ES that does the regulating" (Lovelock, 2009, p. 166). In a paper presented at the United Nations University in Tokyo on 25 September 1992, Lovelock explained that, although contentious, the Gaia hypothesis has generated many experiments (Lovelock, 1993). This section describes how — through these experiments — researchers have attempted to elucidate how the ES works by using climate as an illustrative example of how processes and feedbacks can operate homeostasis. Indicative of this thinking is the investigation of the role of algae in the ocean and its control of Earth's climate through the dimethyl sulfide (DMS) process.

18.1.2.1. Dimethyl Sulfide (DMS) and Climate Regulation

DMS, central to numerous atmospheric processes, plays an important role in climate regulation (Kump and Lovelock, 1995; Ayers and Gillett, 2000). How an understanding of DMS was attained illustrates the evolution of integrated ESs thinking. For instance, Ayers and Gillett (2000) explain that, despite the effort of a small group of researchers (e.g., Junge and Manson, 1961; Fletcher, 1962), initial sulfur studies were limited. This was particularly because the source of aerosol sulfur in regions far removed from volcanoes or anthropogenic emissions of sulfur dioxide had yet to be determined, making it problematical to balance global sulfur budgets. The solution came "with the suggestion by Lovelock et al. (1972) that DMS was the 'missing' biogenic source of sulfur needed to balance the global atmospheric sulfur budget" (Ayers and Gillett, 2000, p. 276).

Recognition of the importance of DMS, combined with earlier cloud microphysical studies (e.g., Twomey, 1977; Twomey et al., 1984) that made the connection between droplet numbers and cloud radiative transfer properties (Ayers and Gillett, 2000), led to the CLAW[1] hypothesis (Charlson et al., 1987). This hypothesis postulates, "biological regulation of the climate is possible through the effects of temperature and sunlight on phytoplankton population and dimethyl sulfide production" (Charlson et al., 1987, p. 665). In other words, DMS emissions from the oceans are influenced by climate and climate (through the impact of cloud albedo on the radiation budget) is affected by Cloud Condensation Nuclei (CCN) emanating from DMS emissions, "making climate and DMS emissions interdependent and closing a feedback loop" (Ayers and Gillett, 2000, p. 276).

Looking at the elucidation of DMS as part of the whole planet's chemistry and its importance to climate regulation, it seems that the systems approach advocated by Lovelock and others was an important framework. For example, Lenton (1998) argues that the Gaia hypothesis was used to make predictions, such as "marine organisms would make volatile compounds that can transfer essential elements from the ocean to the land. The discovery that dimethyl sulfide and methyl iodide are the major atmospheric carriers of the sulfur and iodine cycles, respectively, support this suggestion." (Lenton, 1998, p. 440). Another early example of ES framing using climate as an illustrator is seen in research on vegetation and climate interactions.

18.1.2.2. Vegetation and Climate Interactions

When large changes were recognized as occurring in tropical rainforests (e.g., Salati and Vose, 1984), tests

were conducted to try to determine their climatic impact (e.g., Henderson-Sellers and Gornitz, 1984). Fundamental aspects of this research included the use of stable water isotopes to track hydrological changes (e.g., Salati et al., 1979; McGuffie and Henderson-Sellers, 2004) and model simulations of tropical deforestation that helped elucidate the importance of an accurate representation of vegetation in global climate modelling (e.g., Dickinson and Henderson-Sellers, 1988; Henderson-Sellers et al., 2008).

Tropical deforestation simulations indicated a "sensitivity of the local climate to the removal of tropical forest.... Moreover, the scale of moisture convergence changes, and possibly also cloud and convection changes, is such that there is a possibility that nonlocal climatic impacts may also occur" (Zhang et al., 1996, p. 1516). Further studies (e.g., Zhang et al., 2001) found that tropical deforestation can impact large-scale atmospheric circulation. This supported previous Global Climate Model (GCM) studies (e.g., Sud et al., 1988) and suggested that land-use change (e.g., tropical deforestation) may affect projections of future climate (cf. Pitman and de Noblet-Ducoudré, 2012, this volume). However, research in Amazonia had yet to be studied in an interdisciplinary manner (Dickinson, 1987), a central tenet of an ESs approach.

Although it was not clear how deforestation might threaten interdependent (homeostatic) systems because "our scientific framework is yet inadequate to make such judgments" (Dickinson, 1987, p. 1), and well before detailed disciplinary research of the 1990s−2000s, research scientists joined an international conference on 'Climatic, Biotic, and Human Interactions in the Humid Tropics with Emphasis on the Vegetation and Climatic Interactions in Amazonia' in Brazil in 1985. This meeting brought together some of the world's top scientists to examine critical processes linking climate and vegetation in the tropics. The humid tropics were chosen as the focus because they were deemed of fundamental importance to the global climate. The urgent need to carefully analyse land-use change and climate in the humid tropics was combined with a desire to communicate research findings clearly (Figure 18.3).

Tropical forests are vulnerable to anthropogenic climate change through disturbances in precipitation and temperature (e.g., Lewis et al., 2011) and the compounding effects of tropical deforestation and greenhouse warming on climate have been investigated for some time (e.g., Zhang et al., 2001; Fearnside, 2011). There are many synergies operating among local people's survival, climate, vegetation, and land-use change in the humid tropics. For example, as Fearnside (2011) notes, "Because half of the dry weight of the trees in a tropical forest is carbon, either deforestation or forest die-off releases this carbon in the

---

1. CLAW is an acronym from the authors of the first paper viz: Charlson, Lovelock, Andreae, and Warren.

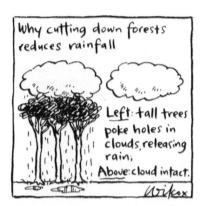

FIGURE 18.3 Forest moisture recycling increases precipitation in the Amazon, that is, why removing trees reduces rainfall. A Cathy Wilcox cartoon (first published on 4 March 2005 on the front page of *The Sydney Morning Herald*, Australia) illustrating a geophysiological discovery made by tracking and modelling stable water isotopes. *(Source: Reproduced by permission of Cathy Wilcox, SMH.)*

form of greenhouse gases such as carbon dioxide (CO_2) and methane (CH_4), whether the trees are burned or simply left to rot" (Fearnside, 2011, p. 1283).

Gradually, as tropical forests became a key part of climate change research and policy debate, simulations became more like 'Gaian-type experiments' in which researchers attempted to describe how the ES works by using disturbances to the tropical forests' climate as an exemplar (e.g., Henderson-Sellers et al., 1988). An integrated systems approach (big picture perspective) evolved through the lens of ESS. This understanding prompted the concept of teleconnections and tipping points resulting from tropical deforestation in Amazonia, Africa, and South East Asia, as discussed by Lenton (2012, this volume). Nobre (2011, personal communication) made the following comments:

Prompted by a need to create a scientific framework to better understand these complex processes, the workshop on Vegetation and Climatic Interactions in Amazonia in 1985 helped advance an integrated Earth systems approach. The Conference recommendations evolved into central Large-Scale Biosphere Atmosphere Experiment in Amazonia (LBA) themes of understanding the Amazon as a regional entity of the Earth system and of studying how climate and land cover changes can alter its physical, chemical and biological functioning.

C. Nobre, personal communication, 2011

Lovelock's Gaia hypothesis advanced understanding that a planet with abundant life will have an atmosphere with 'thermodynamic disequilibrium' and that "Earth is habitable because of complex linkages and feedbacks between the atmosphere, oceans, land, and biosphere", which helped shape ESS (Lawton, 2001, p. 1965). The remainder of this section focuses on the genesis and evolution of ESS.

18.1.3. Earth System Science

ESS is the study of the Earth and its response to anthropogenic change (Pitman, 2005). It has evolved from the visionary scientists (e.g., Vernadsky – biosphere, 1926; Lovelock – Gaia, 1979) who "stood back from the minutiae of classical scientific investigation and looked at the Earth as a single integrated system. While Lovelock's (1979) ideas were initially viewed unfavourably within the scientific community, they now form the key building block of the Earth system science revolution" (Pitman, 2005, p. 140).

In the mid-1980s, a group of geoscientists called for the need for the organisation of "an integrated understanding of how the various elements of the Earth system, including climate, hydrology, biospheric processes, and human activities, interact to produce current and possible future Earth system conditions" (Goward and Williams, 1997, p. 887). In response, NASA initiated a new mission for its Earth science programme in 1989 to examine the Earth from space as an interconnected, holistic system (Figure 18.1). NASA's current ESS programme derived from the 1988 report led by Francis Bretherton on 'Earth System Science: A Closer View' (NASA, 1988). The basis for this report arose from the NASA Advisory Council's establishment of the Earth system science Committee in 1983, together with awareness of other initiatives, including those of the World Climate Research Programme (WCRP), the Intergovernmental Panel on Climate Change (IPCC), and the International Geosphere–Biosphere Programme (IGBP) (Stewart, 2005). One example of a bigger picture outcome arising from NASA's ESS programme was the deployment of NASA's Landsat Thematic Mapper instrument in the early 1980s and the subsequent advances in acquiring satellite images specifically to assess human impacts within the ES (Goward and Williams, 1997).

This growing awareness of the human transformation of the ES urged ES thinkers to consider options for promoting sustainable stewardship of our planet. This, in turn, acted as a catalyst for Earth system analysis (ESA) to arise as a useful tool to explore the kind of world we have, the kind we want, and what we must do to get there (Schellnhuber, 1999).

18.1.3.1. Earth System Analysis (ESA)

ESA is a "trans-discipline striving to perceive the big picture, to ask and answer the genuine systems questions, and to identify the prime pathways toward global sustainability" (Schellnhuber et al., 2005, p. 14). ESA is complementary to sustainability science (Biermann, 2007), which investigates relationships among environment, development, and knowledge to help identify and respond to emerging societal challenges (Clark et al., 2005; Clark 2007; Carpenter et al., 2009). In this chapter we take trans-disciplinary to "include the disciplines but goes further than multi-disciplinarity to include all validated constructs of knowledge and their worldviews and methods of inquiry" (Brown et al., 2010, p. 4).

The need for science and technology to underpin sustainable development in the Anthropocene[2] was acknowledged at the World Summit on Sustainable Development in Johannesburg in 2002; how to actually achieve this was not (Clark et al., 2005). Exploring options for promoting efforts towards sustainable ES management was, therefore, the central goal of the 91st Dahlem Workshop, held in May 2003, on 'Earth System Analysis for Sustainability' (Schellnhuber et al., 2005). The workshop's scientific foci (e.g., geosphere—biosphere interactions on Earth, understanding of the Earth's functioning during the Late Quaternary, human transformation of the ES, and global sustainability) used Gaia as an 'integrating factor and unifying metaphor' (Clark et al., 2005). Furthermore, by bringing together scientists from various disciplines (i.e., geologists, physicists, ES scientists, and policy and governance experts), with a range of skills and perspectives, we contend that this workshop helped advance interdisciplinary ES research and encouraged the nascent big picture Earth perspective. Clark et al. (2005) recognized "the emergence of a new scientific paradigm that is driven by unprecedented planetary-scale challenges, operationalized by trans-disciplinary centennial-scale agendas, and developed by multiple-scale co-production based on a new contract between science and society" (Clark et al., 2005, p. 24). Within the framework of ESA, the Anthropocene can be considered as the "latest step on the grand co-evolutionary ladder" that, driven by global industrialization, "may in turn provoke a transition to an even higher form of worldwide socio-political organisation" (Lenton et al., 2004, p. 913) (Figure 18.4). This notion of a new contract between science and society is explored in Section 18.3 where we investigate a trans-disciplinary ES lens as a means of enabling humans to perceive themselves as able to have a positive influence on future climates.

Schellnhuber (1999) draws an analogy between the beginnings of holistic ESA at the end of the twentieth century with the extraordinarily profound Copernican revolution of the fifteenth century. To support this comparison, he constructs a quasi-mathematical representation of this new view of the Earth's 'system' (E) in the form of pseudo-vector (i.e., multidimensional) relationships combining the traditional Earth science with all aspects of human influence. For the 'old view' of Earth, Schellnhuber follows GARP (1975) reiterating that the

whole is some complex function of the atmosphere (a), the biosphere (b), the cryosphere (c), and, no doubt, many geophysical/geochemical subsystems (and letters). This well-established view is, however, now augmented by the human system, itself a (vector) combination of the recently recognized Anthropocene and Schellnhuber's (1999) novel addition: the self-awareness and still more importantly self-control factor exerted by humans on the ES.

Schellnhuber (1999) uses an apt analogy comparing the effect of sheep on a planet's biosphere with that of humans on Earth. The sheep (like Lovelock's white daisies) reflect solar radiation but they also emit methane (a greenhouse gas) and, by overgrazing, cause soil degradation (Figure 18.2c). They are mindless geo-modifiers while people, Schellnhuber asserts, now have the ability to create interpretive models (Earth System Models of Intermediate Complexity (EMICs) are an example in his view) and use these for prediction and hence as the basis for policies to intentionally modify the planet's environment (Figure 18.1). Arguably, Schellnhuber (1999) is the first to formalize the growing realization that the holistic Earth comprises a planet plus the blind actions of its biota (including people), *plus* the emerging self-aware actions of these people as they come to understand this system and attempt to actively manage its future.

Considering the human role at the same level as the biophysical role in understanding the ES led to some "interesting debate within the literature... and is considered a revolution by some biophysical scientists, and wrong by others" (Pitman, 2005, pp. 139—140). For example, Johnson et al. (1997) consider that there is a risk of ESS becoming "diluted by generalities without gaining meaningful insight into the fundamental physical processes which govern the system" (Johnson et al., 1997, p. 688). Lovelock (2009) explains, however, that this new discipline (ESS) has arisen from the geoscience community as a result of discontent with conventional geology and its inability to process increasingly sophisticated ES knowledge. The need to develop a new scientific framework to handle increasingly complex challenges was strengthened and underlined in the 2001 Amsterdam Declaration on Global Change (Box 18.1).

18.1.3.2. The Amsterdam Declaration

The Amsterdam declaration describes the ES as a single, self-regulating system consisting of physical, chemical, and human components (Steffen et al., 2004; Lovelock, 2009). The declaration led to the establishment of the Earth System Science Partnership (ESSP) by the four global environmental change (GEC) research programmes (DIVERSITAS: an international programme on biodiversity science; IGBP: International Geosphere—Biosphere Programme; IHDP: International Human Dimensions

---

2. Human activities manifest as a geological force, see e.g., Crutzen (2002).

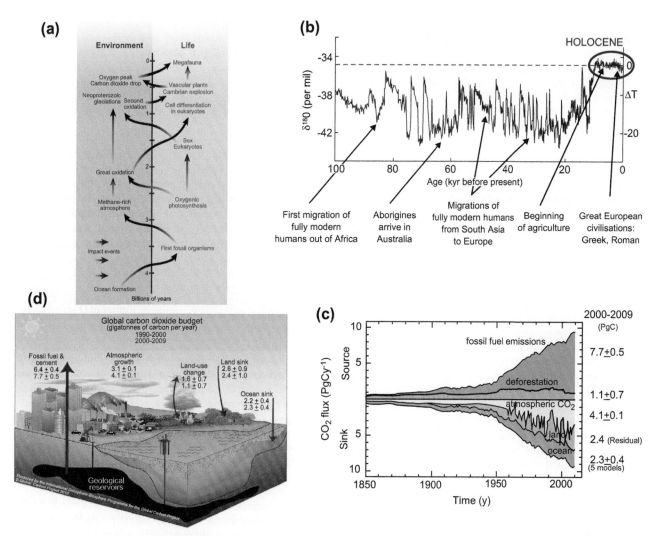

FIGURE 18.4 Life interactions with Earth's 4.5 billion-year planetary history. (*Clockwise from the top left*): (a) from the Earth's 'co-evolutionary ladder' *(Source: Lenton et al., 2004. Reprinted by permission from Macmillan Publishers Ltd.)*; (b) through human history to the Holocene (last 10,000 years) *(Source: Rockström et al., 2009a, adapted from Young and Steffen, 2009)*; (c) and (d) to the Anthropocene (generally taken to be the last 200 years) in which humans became a geological-scale force (depicted here by human perturbation of the global carbon budget) *(Source: GCP, 2010)*.

Programme on global environmental change; and the WCRP: World Climate Research Programme). The 2001 Amsterdam declaration acknowledged the underpinning scientific enterprise of the GEC research programmes and "identified the need for a new partnership to respond to the challenges associated with greater integration of emerging scientific knowledge of the Earth system and the need for timely information for decision makers, hence the birth of the ESSP" (Leemans et al., 2009, p. 6).

18.1.3.3. Integrated ESS Research

This subsection introduces a set of integrated research projects as illustrative of the importance of integrated ESS to deeper and more complete global comprehension.

The ESSP joint projects relate to GEC and sustainability on the carbon cycle (Global Carbon Project); food systems (Global Environmental Change and Food Systems); water systems (Global Water System Project); and more recently on human health (Global Environmental Change and Human Health). These four projects (Box 18.2) not only synthesize fundamental process-related research and knowledge from the core projects of individual GEC research programmes, but also conduct research at global, regional, and local scales that integrates findings into knowledge products. One of the strengths of this type of ES research is that it helps bridge the boundaries of social and natural sciences. For example, over the past decade, these joint projects have developed their own methodologies and approaches to build the scientific infrastructure

BOX 18.1 The Amsterdam Declaration on Global Change

The scientific communities of four international global change research programmes — the International Geosphere–Biosphere Programme (IGBP), the International Human Dimensions Programme on Global Environmental Change (IHDP), the World Climate Research Programme (WCRP), and the international biodiversity programme DIVERSITAS — recognize that, in addition to the threat of significant climate change, there is growing concern over the ever-increasing human modification of other aspects of the global environment and the consequent implications for human well-being. Basic goods and services supplied by the planetary life support system, such as food, water, clean air, and an environment conducive to human health, are being affected increasingly by global change.

Research carried out over the past decade under the auspices of the four programmes to address these concerns has shown that:

- The Earth system (ES) behaves as a single, self-regulating system comprised of physical, chemical, biological, and human components. The interactions and feedbacks between the component parts are complex and exhibit multiscale temporal and spatial variability. The understanding of the natural dynamics of the ES has advanced greatly in recent years and provides a sound basis for evaluating the effects and consequences of human-driven change.

- Human activities are significantly influencing Earth's environment in many ways in addition to greenhouse gas emissions and climate change. Anthropogenic changes to Earth's land surface, oceans, coasts, and atmosphere and to biological diversity, the water cycle, and biogeochemical cycles are clearly identifiable beyond natural variability. They are equal to some of the great forces of nature in their extent and impact. Many are accelerating. Global change is real and is happening *now*.

- Global change cannot be understood in terms of a simple cause–effect paradigm. Human-driven changes cause multiple effects that cascade through the ES in complex ways. These effects interact with each other and with local- and regional-scale changes in multidimensional patterns that are difficult to understand and even more difficult to predict. Surprises abound.

- ES dynamics are characterized by critical thresholds and abrupt changes. Human activities could inadvertently trigger such changes with severe consequences for Earth's environment and inhabitants. The ES has operated in different states over the last half million years, with abrupt transitions (a decade or less) sometimes occurring between them. Human activities have the potential to switch the ES to alternative modes of operation that may prove irreversible and less hospitable to humans and other life. The probability of a human-driven abrupt change in Earth's environment has yet to be quantified but is not negligible.

- In terms of some key environmental parameters, the ES has moved well outside the range of the natural variability exhibited over at least the last half million years. The nature of changes now occurring simultaneously in the ES, their magnitudes, and rates of change are unprecedented. The Earth is currently operating in a no-analogue state.

On this basis, the international global change programmes urge governments, public, and private institutions and people of the world to agree that:

- An ethical framework for global stewardship and strategies for ES management are urgently needed. The accelerating human transformation of the Earth's environment is not sustainable. Therefore, the business-as-usual way of dealing with the ES is not an option. It has to be replaced — as soon as possible — by deliberate strategies of good management that sustain the Earth's environment while meeting social and economic development objectives.

- A new system of global environmental science is required. This is beginning to evolve from complementary approaches of the international global change research programmes and needs strengthening and further development. It will draw strongly on the existing and expanding disciplinary base of global change science; integrate across disciplines, environment, and development issues and the natural and social sciences; collaborate across national boundaries on the basis of shared and secure infrastructure; intensify efforts to enable the full involvement of developing country scientists; and employ the complementary strengths of nations and regions to build an efficient international system of global environmental science.

The global change programmes are committed to working closely with other sectors of society and across all nations and cultures to meet the challenge of a changing Earth. New partnerships are forming among university, industrial, and governmental research institutions. Dialogues are increasing between the scientific community and policy-makers at a number of levels. Action is required to formalize, consolidate, and strengthen the initiatives being developed. The common goal must be to develop the essential knowledge base needed to respond effectively and quickly to the great challenge of global change.

(Source: Reprinted from Leemans et al., 2009; with permission from Elsevier.)

that allows for a more integrated approach (Leemans et al., 2009). Schmidt and Moyer (2008) contend that these projects provide an outlet for a new generation of inter-disciplinary and trans-disciplinary researchers needed to

tackle today's global environmental change and sustainability challenges.

The importance of such an integrated ESS approach was recently reiterated by the president of the American

---

BOX 18.2 ESSP Joint Projects
Global Carbon Project (GCP)

The GCP was established in 2001 in recognition of the enormous scientific challenge and fundamentally critical nature of the carbon cycle for Earth sustainability. The scientific goal of the project is to develop a complete picture of the global carbon cycle, including both its biophysical and human dimensions, together with the interactions and feedbacks between them.

Global Environmental Change and Food Systems (GECAFS)

The GECAFS goal is to determine strategies to cope with the impacts of global environmental change on food systems and to assess the environmental and socio-economic consequences of adaptive responses aimed at improving food security.

Global Water System Project (GWSP)

The GWSP aims to address the central research question: How are humans changing the global water cycle, the associated biogeochemical cycles, and the biological components of the global water system and what are the social feedbacks arising from these changes?

Global Environmental Change and Human Health (GECHH)

GECHH is the fourth joint project within the ESSP. It is being developed as a logical complement to the three ongoing ESSP projects, as changes in each of those three systems (food, carbon, water) influence, via diverse pathways, human well-being, and health.

(Source: Reprinted from Leemans et al., 2009; with permission from Elsevier.)

Association for the Advancement of Science who stated that, "Its [ESSP's] initiatives relating to the carbon cycle, food security, water, and human health in the context of global environmental change will provide essential new understanding as society steers to a future that diminishes risk for future human well-being and life all across our planet" (McCarthy, 2009, p. 1665).

18.1.4. Advances in Earth System Science

Recent advances in ESS include the recognition of an integrated ES being disturbed by humans not acting as mindless geo-modifiers, but who (unlike sheep) have the capacity to orchestrate planetary welfare.

18.1.4.1. Carbon Cycle and Carbon Budgets

Strenuous efforts to estimate how the Earth might change over the next century revealed that our understanding of ESs was incomplete (Falkowski et al., 2000). For example,

Tans et al. (1990) contended that the annual atmospheric CO_2 budget could not be balanced. Falkowski et al. (2000), building on previous insights from ES thinkers described in previous sections (e.g., Lovelock, 1979; Schellnhuber, 1999), add that a systems approach is required as our knowledge is inadequate to elucidate the interactions between the components of the ES and the relationships between the carbon cycle and other biogeochemical and climatological processes.

The overarching goal of the carbon cycle research community in responding to the climate change challenge is "to understand the role of the natural and managed carbon cycle in the dynamics of the climate system. That requires quantifying the effect of human activities on the carbon cycle; determining the response of natural systems to these disturbances; projecting future behaviour of carbon pools and fluxes; and exploring pathways to atmospheric stabilization through the management of the carbon—climate—human system" (Canadell et al., 2010, p. 301). Carbon science has now developed an annual update (coordinated by the GCP) on the global carbon budget and its trends and analysis (e.g., Canadell et al., 2007; Le Quéré et al., 2009; Friedlingstein et al., 2010). In the recently published 2009 global carbon budget (Friedlingstein et al., 2010), ES scientists explain that the annual growth rate of atmospheric CO_2 was 1.6 parts per million by volume (ppmv) in 2009, bringing the atmospheric CO_2 concentration to 387 ppmv by the end of 2009. This is 39% above the atmospheric burden at the start of the Industrial Revolution (about 280 ppmv in 1750). Friedlingstein et al. (2010) underline the extent of this disturbance by noting that the current concentration is the highest during at least the past 2 million years (see Harrison and Bartlein, 2012, this volume). This synthesis of carbon cycle research, and in particular its implications for the world's future climate, has prompted significant interest from policy-makers and the media. For example,

As the experience with the global carbon budgets (http://www. globalcarbonproject.org/carbonbudget) has shown, integrative and synthetic science products that are released and updated regularly, and have a direct connection with the policy process, can be very effective in influencing key policy discussions.

Leemans et al., 2009, p. 11

The Global Carbon Project (Box 18.2) (launched in concert with the Amsterdam declaration in 2001 — Box 18.1) has helped carbon cycle science become mainstream; particularly through its regular annual trends and analysis of the CO_2 budget (and soon the CH_4 budget). This has been achieved by embarking on a major communications effort (e.g., press releases and policy briefs — 'knowledge products'), making the science accessible and of relevance to society. The entire global CO_2 budget has become one more key indicator of the evolution of climate change combining with sea-level rise and the annual mean global temperature increase. Furthermore, by integrating data on physical quantities of CO_2 with information on the relative source of emissions (fossil fuel types), growth in population and per capita wealth, the carbon research community can now explain some of the underlying drivers of change consistent with the emerging global picture. This knowledge should aid global governance and inform equity issues in climate negotiations. The GCP is also the first group of scientists to identify and publish data about the possibility of inefficiencies of CO_2 sinks, with a slow decline over the last 60 years. Despite being contested by some, this challenge remains at the forefront of future pathways for CO_2 stabilization and is attracting a great deal of scientific and policy interest that would not be there without the GCP.

The carbon science community has launched a REgional Carbon Cycle Assessment and Processes (RECCAP): a global coordination effort among researchers and institutions (GCP, 2010). The objective of RECCAP is to "establish the mean carbon balance and change over the period 1990–2009 for all subcontinents and ocean basins. The global coverage will provide, for the first time, opportunities to link regional budgets with the global carbon budget. Regional details on or insights into processes driving fluxes have not, to date, been incorporated into efforts addressing the global carbon budget (Canadell et al., 2007; Le Quéré et al., 2009). The consistency checks between the sum of regional fluxes and the global budget will be a unique measure of the level of confidence there is in scaling carbon budgets up and down" (Canadell et al., 2011, p. 81). Another important area of ES research that has advanced in recent years is our understanding of the global water system.

18.1.4.2. Water System and Cycling

According to model predictions, the most "significant manifestation of climate change for humans and the environment is an intensification of the global water cycle" (Asrar et al., 2001, p. 1313). The development of advanced technology-derived data, statistical analysis, and models has assisted the ESS community to contribute significantly to a comprehensive global water assessment (Vörösmarty, 2002). The extent of the world's rivers' crisis (Vörösmarty et al., 2010) was the first integrated ESS study to simultaneously incorporate the effects of, for instance, pollution, dam building, agricultural runoff, the conversion of wetlands, and the introduction of exotic species on the health of the world's rivers (Vörösmarty et al., 2010). Indicative of ESs thinking, the importance of water is not limited to human society in these analyses. For instance, ecosystems and aquatic and terrestrial biodiversity cannot exist without adequate quantity and

quality of water. One of the key findings of the Vörösmarty et al. (2010) study is the inherent competition between human water security and that of biodiversity. Similar to advances in carbon cycle research, we contend that the generation of water system knowledge would not have been possible without an integrated ES approach. Another example of progress created and encouraged through ESS can be found in the advancement of scenarios of future climates.

18.1.4.3. Coherent Future Scenarios for Climate

Advances in the science and observation of climate change are providing a clearer understanding of the "inherent variability of Earth's climate system and its likely response to human and natural influences" (Moss et al., 2010, p. 747; see Figure 18.1). Projections of future climates also depend on climate sensitivity (see Harvey, 2012, this volume); itself a function of fast and slow feedbacks in future climates. However, significant uncertainties exist not only in the responses to climate change but also in future forcings. Recognizing the need for scenarios of the future in comprehending the potential consequences of different human response options, interdisciplinary climate researchers responded by developing "a new coordinated parallel process for developing scenarios, with 'representative concentration pathways' (RCPs) that [will] provide a framework for modelling in the next stages of scenario-based research" (Moss et al., 2010, p. 747). This new scenarios approach differed from the 'storylines' originally employed by the IPCC (IPCC, 1992) because it integrated postulated social response strategies with radiative forcing used in numerical climate predictions (Hibbard et al., 2007).

The approach, which was led by the climate research community (particularly IGBP– Analysis, Integration and Modelling of the Earth System (AIMES) and WCRP Working Group on Coupled Modelling (WGCM)) was ultimately a "direct and collaborative influence on the IPCC process" (Leemans et al., 2010, p. 1215) when the strategy was adopted by the 28th IPCC plenary meeting in Budapest, Hungary in April 2008. However, this development is by no means complete. Indeed, Hibbard et al. (2009) comment, "There are opportunities to develop a networked suite or system of regional to global models through collaborations, integrative research that can provide insight into the coupled human–environmental system that is meaningful for resource managers, decision makers and policy communities" (Hibbard et al., 2009).

The potential for 'climate surprises', which include the melting of large masses of ice (e.g., Arctic sea-ice and the Greenland ice sheet), changes in atmospheric and oceanic circulation (e.g., El Niño–Southern Oscillation and the

Indian Summer Monsoon), and — complementing vegetation and climate interaction research discussed earlier in this section — the loss of biomes (e.g., the Amazon rainforest) has been recognized for a few years (e.g., Lenton et al., 2008). However, Lenton (2012, this volume) ponders the prospect of people acting to try to remove or at least reduce the element of surprise. Section 18.3, therefore, describes trans-disciplinary ESS as a means to help assist embryonic planetary management conjectures. Such a trans-disciplinary perspective can be found in a new 'planetary boundaries' concept that — building on recent advances in ESS (examples of which are outlined earlier in this section) — has been proposed to help establish parameters also termed 'sustainability guardrails' in which humanity can operate safely.

18.1.4.4. Planetary Boundaries

Rockström et al. (2009a, 2009b) have proposed a new approach to global sustainability, based on a 'Planetary Boundaries' concept that sets parameters in which human society may continue successfully. The original idea grew from the notion that integrating resilience theory and ESS could have profound implications for the way we approach global sustainability in the Anthropocene. The discussions building the concept started in 2007 and continued in 2008, through a series of scientific meetings at Tällberg fora in Sweden (J. Rockström, personal communication, 2011).

The planetary boundaries concept is based on the notion that "transgressing one or more planetary boundaries is deleterious and may be catastrophic due to the risk of crossing thresholds that will trigger non-linear, abrupt environmental change within continental- to planetary-scale systems" (Rockström et al., 2009a). Nine planetary boundaries have been identified (Figure 18.5): (i) climate change (which is the focus of this book); (ii) ocean acidification (e.g., Sen Gupta and McNeil, 2012, this volume); (iii) stratospheric ozone; (iv) biogeochemical nitrogen and phosphorus cycles; (v) global freshwater use; (vi) land-system change; (vii) biodiversity loss; (viii) chemical pollution; and (ix) atmospheric aerosol loading (e.g., Wang et al., 2012, this volume). Rockström et al. (2009a, 2009b) contend that these planetary boundaries define the 'playing field' outside which humanity should not transgress. The planetary boundaries concept builds on recognition of human transformation of the ES (e.g., Crutzen, 2002; Steffen et al., 2007) and the notion of 'non-linear' transitions of the system (Schellnhuber, 2002; Lenton et al., 2008). Rockström et al. (2009a, 2009b) indicate that humanity, as mindless geo-modifiers, has already 'transgressed' three boundaries: the rate of biodiversity loss; the rate of transformation of the nitrogen cycle; and climate change.

Following the Rockström's presentation of the planetary boundaries concept during a major pre-UNFCCC

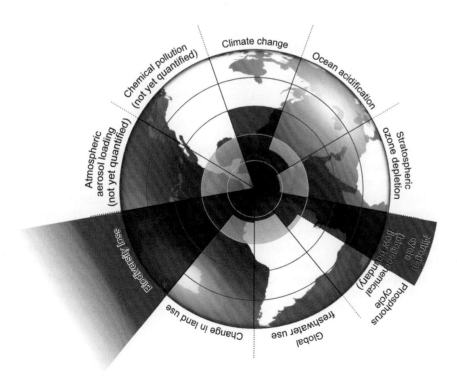

Figure 18.5 Nine 'planetary boundaries': (i) climate change; (ii) ocean acidification; (iii) stratospheric ozone depletion; (iv) biogeochemical flow boundary — nitrogen and phosphorus cycles; (v) global freshwater use; (vi) changes in land use; (vii) biodiversity loss; (viii) chemical pollution; and (ix) atmospheric aerosol loading. The inner green shading marks humanity's 'safe operating space'. The red wedges indicate that the boundaries of three critical systems have been transgressed; that is, climate change, biodiversity loss, and human interference with the nitrogen cycle. The atmospheric aerosol loading and chemical pollution boundaries are not yet quantified. (*Source: Rockström et al., 2009b. Reprinted by permission from Macmillan Publishers Ltd.*)

(United Nations Framework Convention on Climate Change) Conference of Parties (COP)15 science congress in Copenhagen in March 2009, we asked him if this new idea was created intentionally to coincide with the UNFCCC COP15 in December 2009. His response was as follows:

The planetary boundaries analysis was not in any way triggered by or tactically initiated with COP15 in mind. Although once we approached March 2009, and the large Copenhagen climate conference (co-organised by several PB [planetary boundaries] co-authors — Katherine Richardson, John Schellnhuber, Will Steffen), we realized that the PB analysis should be considered by and thus influence the UNFCCC negotiations, and that it therefore was important to aim for a publication before COP15 (in the end it came out in September 2009). My thinking here was that the planetary boundaries analysis should influence the COP meeting in two ways. First, that the compelling and comprehensive insight of the risk of catastrophic regime shifts, not only triggered by the climate system but certainly closely linked to the climate system, should generate an even stronger realization among the climate negotiators that a global legally binding agreement on emission cuts is necessary. Secondly, and even more importantly, I hoped (and we hoped as co-authors) that the PB paper would clearly show that it is impossible to stabilize the climate by only addressing the climate agenda. Instead, meeting the 2° guardrail will require a global agreement for sustainable development, addressing all interacting planetary boundaries that influence the climate system, including global freshwater use, land management, interference with the N and P cycles, stewardship of ecosystems and biodiversity, reduction of air polluting aerosols, and securing the continued reduction of ozone depleting substances.

J. Rockström, personal communication, 2011

The planetary boundaries concept, building on the integrated, big picture perspective of ESS (as described in this section) is, arguably, the first concrete attempt to help identify and inform response options for tackling global challenges.

This section has described how ES thinkers — by using climate processes as illustrative examples (e.g., dimethyl sulfide and climate regulation) — advanced the notion of the Earth as an integrated system. This big picture perspective supported the genesis and evolution of ESS: generating new data pertaining to the world's environmental challenges and analysis of the many globally significant systems' crises (Zhao and Running, 2010). The publication of the IPCC Fourth Assessment Report (AR4) and the Nobel Peace Prize award generated great expectations of a global political response (at UNFCCC COP15 in 2009). Section 18.2 describes humans in the ES and the political and social responses (or lack of) to one of the most urgent ES challenges: anthropogenic climate change.

18.2. HUMANS IN THE EARTH SYSTEM

Anthropogenic climate change has become political. It is one of the most prominent ES challenges of our time. The research community has generated a wealth of knowledge on a changing climate system and how humans are the dominant driving force. The consequences of these disturbances are already evident in cities (Cleugh and Grimmond, 2012, this volume), in regional-scale shifts (Evans et al., 2012, this volume), and in environmental indicators including glaciers (Cogley, 2012, this volume) and the world's oceans (Latif and Park, 2012, this volume).

The IPCC synthesizes knowledge for political consideration at the global level, through the UNFCCC. Climate science was on a 'high' in 2007, with the publication of the IPCC AR4 and the joint Nobel Peace Prize award to Al Gore and the IPCC. At the political level, the UNFCCC Bali roadmap also marked significant progress in 2007. It seemed that the world was finally attempting to begin to seriously tackle anthropogenic climate change. The situation, however, dramatically changed in 2009, leading up to the UNFCCC Conference of the Parties (COP15) in Copenhagen with journalistic coverage of alleged IPCC errors and issues of possible research misconduct. The outcome of the UNFCCC COP15 (the Copenhagen 'Accord') is considered weak: a failure of global governance. Moreover, the UNFCCC COP16 in Cancun in December 2010 merely formalized the Copenhagen 'Accord'. This section discusses a 'social tipping point' and how climate discourse has flipped from a 'virtuous' to a 'vicious' cycle. We also examine the notion of an 'integrity paradox' concerning the issue of policy prescription or people's ponzi[3]; whereby we fail to differentiate between the quantification of climate risk (the role of the research community) and the development of responses to these risks (the role of community decision-making) and, in so doing, create 'Gaian governance monkeys'.

18.2.1. Climate Change and the Gaian Governance Monkeys

Anthropogenic climate change is real and accelerating; it is wealth creating and it is media titillating. This climate change arises because of abuse of uncosted and under-valued environmental services and is generationally postponed, affecting our grandchildren and beyond. The governance response is through a convention (the United Nations Framework Convention on Climate Change — UNFCCC) that is unique in that it was initiated by scientists

---

3. Ponzi schemes are fraudulent investment operations rewarding investors not from any true earnings but from their own funds or subscriptions by subsequent investors.

and NGOs (non-governmental organizations), not by nation states. The debacle of the UNFCCC's COP15 in Copenhagen in December 2009 was not caused by challenges to the fact of global warming, nor to the need for action, but by the inability to reach politically acceptable agreements: by a failure of global governance. It seems now that, faced by an almost universal preference for obfuscation and denial (Hamilton, 2010), the challenge of climate futures is not about improved quality assurance or more careful management of climate change's massive data deluge (although this is undoubtedly a daunting task — Table 18.1), it is whether researchers, funders, policymakers, and the public clearly differentiate between climate change risk research — which requires a meritocracy — as opposed to the response to this risk, which as a social issue demands democratic (or other) community decision-making. The communities in which we live continue to support the truth of anthropogenic climate change but whether our governance processes enable us to prioritize climate change response to really dangerous outcomes that cannot currently be ruled out with less than a 10% chance remains unknown.

Climate policy is now a choice between a bad or a very bad future (Stern, 2006; Garnaut, 2008). In 1989, the American Geophysical Society sponsored a Chapman Conference on Gaia (Schneider and Boston, 1991). During one of the discussion sessions, one of us coined the idea of the Three Gaian Monkeys: those who cannot see Gaia, those who hear nothing that convinces them of the existence of Gaia, and those who will not discuss Gaia. Nowadays, the three wise monkeys' proverb can be just as

FIGURE 18.6 An analogy is drawn between the scientific response to the Gaia hypothesis in the 1980s, caricatured in terms of the three Gaian monkeys, and today's reaction to anthropogenic climate change in which there are three global warming governance monkeys: see no change despite (pictured) exponential increases; listen to neither research nor research assessments (such as the IPCC); and unwilling or unable to discuss constructive responses to change (for example in the UNFCCC Conference of the Parties).

aptly employed to illustrate those in the public (and in policy) who pretend to fail to recognize the changes in our climate, refuse to hear about the 'unequivocal' fact of warming, and will not discuss responses to the anthropogenic climate crisis (Figure 18.6). Wise determination of the most beneficial path under all likely future climates involves clear research and urgent action arising from decisions made by an adequately informed public.

18.2.2. Social Tipping Points in Climate Change: 2007 to 2010

Tipping points and positive feedback are two very well-known climate phenomena (e.g., McGuffie and

TABLE 18.1 Climate Data Deluge

Unit	Size	Meaning/Derivation/Climate Example
Bit (b)	1 or 0	A binary digit
Byte (B)	8 bits	Enough bits for a single letter or number
Kilobyte (kB)	1000 B or 2^{10} bytes	Greek for thousand (1 page is ~2 kB)
Megabyte (MB)	1000 KB or 2^{20} bytes	Large in Greek (MP3 song; a chapter of 'A Climate Modelling Primer'[4] as a PDF)
Gigabyte (GB)	1000 MB or 2^{30} bytes	Giant in Greek (typical movie ~2 GB)
Terabyte (TB)	1000 GB or 2^{40} bytes	Monster in Greek (CMIP3 for AR4 ~35 TB)
Petabyte (PB)	1000 TB or 2^{50} bytes	CMIP5 created for AR5 ~1 PB in 2011
Exabyte (EB)	1000 PB or 2^{60} bytes	Humanity created ~150 EB in 2005; world's storage capacity in 2010 ~500 EB
Zettabyte (ZB)	1000 EB or 2^{70} bytes	All information in existence 2010 ~1.2 ZB
Yottabyte (YB)	1000 ZB or 2^{80} bytes	Roughly 35 years of SKA data archiving

[4]*McGuffie and Henderson-Sellers (2005).*
(Source: © The Economist Newspaper Limited, London, 5/3/2010)

Henderson-Sellers, 2005; Harvey, 2012, this volume; Lenton, 2012, this volume). In everyday life, positive feedback is usually a good thing — say for example from your boss on your performance (Henderson-Sellers, 2010a) — but, in systems theory, positive feedbacks exaggerate any disturbance. Positive feedback magnifies changes and can lead to the crossing of thresholds beyond which lies a new state from which return is not achievable by removing (or reversing) the original disturbance (Lenton et al., 2008).

Climate change entered a different regime in 2007 with the publication of the IPCC's AR4 and the joint award of the Nobel Peace Prize to Al Gore and the IPCC. People no longer asked 'whether' human activities are changing the climate but the more urgent questions of 'how fast?' 'with what impacts?' and 'demanding what responses?' Between 2005 and 2007, the World Climate Research Programme (WCRP) assembled the largest ever collection of coherent climate change simulations: the Coupled Model Inter-comparison Project 3 (CMIP3). When the Intergovernmental Panel on Climate Change (IPCC) Fourth Assessment Report (IPCC, 2007a) Working Group 1 (WG 1) component was signed off in January 2007, this dataset had been downloaded by more than 1200 researchers (over 337 terabytes (TB) of data — Table 18.1) and used in more than 250 peer-reviewed papers (CMIP, 2010). Ironically, despite this heroic data management effort, the IPCC WG 1 'news' came from a single zero-dimensional energy balance model run on a laptop.

In late 2007, a virtuous cycle reinforced the public's recognition of the need for urgent action to mitigate change. Positive feedback (Figure 18.7a), including in the media, recognized climate change to be a risk management problem to be solved by all nations (UNFCCC, 2007). The mass media reversed (from virtuous to vicious) the direction of its positive feedback on anthropogenic climate change late in 2009 (Figure 18.7b). This time, the character of public perception of anthropogenic climate change was horribly transformed by media coverage of an unimportant but high-profile error in Working Group Two (WG 2) IPCC AR4 and selected contents of the Climatic Research Unit of the UK's University of East Anglia (CRU UEA) emails. The press pushed public perception past a social tipping point in December 2009. The weak Copenhagen Accord (UNFCCC, 2009) and the reduced pressure for climate mitigation legislation felt by world leaders are symptoms of the new social state resulting from crossing this irrevocable social threshold (but cf. Leiserowitz et al., 2010a). Forty years ago, Albert Crewe said, "It is up to the scientific community to point out where they can help… government cannot be expected to seek our advice, because they are much more accustomed to solving problems by new legislation…" and, "Perhaps better solutions exist… (but) until we can make ourselves heard… problems are in

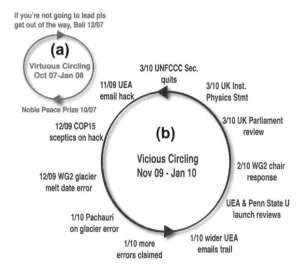

FIGURE 18.7 Positive feedback as a social phenomenon reinforced by and through the mass media and greatly affecting the perception of climate change information. Public understanding and political will were reinforced through media coverage of the issue of anthropogenic climate change in two strong processes of positive feedback in 2007 and 2009. (a) Virtuous circle — positive feedback strengthens public belief in the reality of climate change and, thus, the political will to respond to the threat at COP13 in Bali in December 2007; and (b) vicious circle — a media clamour of positive feedback strengthens public interest in IPCC errors and in claims (later refuted) of abuse of the processes of IPCC and, thus, greatly reduces the political will to respond to the global warming threat at COP15 in Copenhagen in December 2009. *(Source: Henderson-Sellers, 2010c.)*

danger of being grossly underestimated" (Crewe, 2007, pp. 25—30). Today, the challenge of anthropogenic climate change remains the inability of science to be clearly understood and the search for clarity in public understanding seems to be alternately aided and thwarted by the mass media.

18.2.3. Research Requires a Meritocracy; Decisions Demand Democracy

The rules that govern IPCC (to which they sadly failed to adhere themselves during the AR4) include making policy-relevant but no policy-prescriptive statements. The tension between the IPCC prescription: "When risks cannot be well quantified, it is the job of policy to make decisions.… Scientists must make it clear where our job stops and the job of policy begins" (Solomon, 2007) and the good governance requirement: "The ultimate policy-maker is the public. Unless the public is provided with unfiltered scientific information that accurately reflects the views of the scientific community, policy-making is likely to suffer" (Hansen, 2005) has not been, and may never be, resolved. Great care must be exercised to avoid both false positive (colloquially 'crying wolf') as well as false negative (i.e., waiting too long) forecast errors (Schneider and Mastrandrea, 2010). There is also a need to establish the appropriateness of scepticism in scientific enquiry while avoiding negative terminology for those who seem deaf and blind (Figure 18.6) to research-based truths (e.g., O'Neill and Boykoff, 2010).

In the unfolding global tragedy of our planetary commons, painfully depicted in Copenhagen in December 2009, national and international leaders are risking the future of the Earth by policy failure; the mass media highlights ethically bankrupt behaviours but fails to demand alternatives; and the assessment instrument of the UNFCCC, the IPCC, has metamorphosed from a useful policy tool into one that, at best, encourages no action and, at worst, justifies inadequate responses (Henderson-Sellers, 2010b). In December 2007, the director of the WCRP advised the heads of the two UN agencies that jointly sponsor the IPCC to close it because it had "completed its task of assessing the unique challenge that faces humanity" (e.g., IPCC, 2007d). Her recommendation, for an immediate UN action agenda on global warming mitigation and adaptation to inevitable climate change, was ignored then, although by December 2009 the IPCC may have wished it had been accepted. By early 2010, the defence of climate science was insipid as governments and their agencies strove to distance themselves from the tarnished IPCC (e.g., *Nature*, 2010; Netherlands Environmental Assessment Agency, 2010).

Future climates are inevitably a matter of conjecture. We cannot know them, so we try to predict their character. Research publication and peer review is an essential part of this predictive process. Schultz (2010) shows that 79% of all atmospheric science journals and virtually all journals in the field of climate reject more than 30% of submitted papers. Whether rejection of around one third of all submissions is desirable or inadequate may not matter since rejection rate is not an important predictor of the publication's importance (half-life of papers, as measured by citations). Whether peer review is now a handicap or a benefit to the sharing of climate change information is debatable, but it is demonstrably unfair (Aldhous, 2010; Russell, 2010). While the principle that experts evaluate research is widely agreed (Fuller, 2002), peer-review implementation has changed over time (e.g., blind peer reviews began in journals only in the 1950s to 1960s) and is disputed (only 8% of members of the Scientific Research Society agreed that "peer review works well as it is"; Chubin and Hackett, 1990). Worse still, Freedom of Information legislation and other even less appropriate laws (such as on tax malfeasance) are being abused in attempts to access climate data and other records (AAAS, 2010; Adam, 2010).

The flaws inherent in the IPCC increasingly render it unhelpful (e.g., Hulme et al., 2010). The top ten challenges

for the IPCC are: its linear structure (first 'science', then 'impacts' and, finally, 'mitigation'); peer review (poorly defined and facing an information avalanche: CMIP5 ~1 PB by 2011; Knutti, 2010); protracted gestation (fifth assessment not due until 2014 even though the fourth was out-of-date in late 2006); mandated incapacity to make policy statements; no-preference display of results (failure to 'out' bad models); model inter-comparison paradox (gradual community-wide performance improvement masks fundamental failures such as non-conservation: Henderson-Sellers et al., 2008); cost of participation vying with national and laboratory kudos; consensus requirement manifested as fear of highlighting shortcomings and failures (in models and observations); unknown(able) fatness of the probability distribution function tail; and poor handling of published errors and a few "awful emails". Climate prediction has languished near the top of local 'highs' in predictive skill for years (Greene, 2006) and, while there are possible routes to improving the current models, few if any groups seem poised to pursue these, perhaps because of the massive burden of merely participating in IPCC assessments.

At the moment when other global sustainability challenges are seeking to establish IPCC-like assessments (e.g., Larigauderie and Mooney, 2010), the facts of climate change are well-established (e.g., Garnaut, 2008). IPCC's failings are well-known among participants (Doherty et al., 2009): "adding complexity to models, when some basic elements are not working right (e.g., the hydrological cycle), is not sound science"; "until and unless major (climate) oscillations can be predicted to the extent that they are predictable, regional climate is not a well defined problem. It may never be. If that is the case, then climate science must say so" (Henderson-Sellers, 2008). Research managers and funders should not continue to seek the ill-achieved assurance of consensus as a reason for action (cf. McKibben, 2010).

18.2.4. Integrity Paradox: Policy Prescription or People's Ponzi

The public knows climate change is already occurring and its impacts are bad, but allows wrong behaviour to persist, accepting media titillation and subsidized fossil fuel resources (e.g., Obama, 2010; Oreskes and Conway, 2010). The reality that policy is failing to address is that we must limit all future fossil fuel use to less than the amount already consumed: 0.5 trillion tonnes of carbon (Allen et al., 2009). Questioning research integrity is just the latest manifestation of the public's fear of acting on well-known facts and public failure to move to minimize risks about which there is virtual certainty (Ereaut and Segnit, 2006).

In the second half of the seventeenth century, Blaise Pascal debated with himself about the existence of God (e.g., Rescher, 1985). His conclusion was that, despite recognized uncertainty, his best risk reduction strategy was belief. The same Precautionary Principle was embedded in the UNFCCC in 1992 (Schellnhuber, 2010). Despite these guiding ideas, and since at least 1637 (the Tulip Scandal), societies have behaved in highly aberrant ways in a variety of 'Ponzi' schemes. Climate change, which trades the benefits of cheap fossil fuel-derived energy now against future environmental degradation and energy depletion and conflict over scarce remaining energy resources, is rapidly replacing subprime loans as the bursting greed bubble (*The Economist*, 2010).

In the space of a couple of months from late November 2009 to mid-February 2010, two well-known features of the climate system — positive feedback and passing an irrevocable threshold (Figure 18.7) — transformed climate policy from an action agenda (UNFCCC, 2007) to an ineffectual and non-binding statement (UNFCCC, 2009). In this period, both emission of GHGs and their impacts accelerated (Friedlingstein et al., 2010). Media coverage of investigations into anthropogenic climate change greatly exacerbated public desire: in 2007 for action and in 2009–2010 against action (Leiserowitz et al., 2010a, 2010b).

Climate change research has been found to have integrity by inquiries in the UK, the Netherlands, and the USA (UK Parliament, 2010; Oxburgh, 2010; AAAS, 2010; Netherlands Environmental Assessment Agency, 2010; Russell, 2010). While democracy must uphold law (e.g., Act on CO_2 advertisements banned in the UK in March 2010), the current pretence that climate change or any other research is democratic (e.g., Royal Society UK and Australian Academy, 2010) is a fabrication. The right of individuals, bodies, and groups to make statements cannot be denied but equally their opinions must not override the truth of our current understanding of science (Anderegg et al., 2010). This was most painfully demonstrated by the UK Institute of Physics' humiliating withdrawal of its 2010 submission to the UK parliamentary climate emails inquiry (Adam, 2010). Leaders need to uphold truth: not with weak praise (*Nature*, 2010), not by imposing additional quality assurance or further review (e.g., Netherlands Environmental Assessment Agency, 2010), and certainly not by becoming part of the audience titillation trap (cf. Mooney, 2010).

18.2.5. Gaian Governance

Future climates are unknown, but we do know a great deal about how they occur and which forces prevail over different time periods, as this book demonstrates. Dangerous climate change is so named because it negatively affects some community or system. Anthropogenic climate change seems very likely to negatively impact people (who cause it) as well as a wide range of natural and human systems. However, for the first time in the Earth's evolution, people have the opportunity to be more than a sheep-like cause of climate change (Figure 18.2) by actively managing aspects of homeostasis. One of the most challenging aspects of today's climate change is that we seem to be failing to clearly differentiate between climate change risk quantification (which is best undertaken by researchers) and the response to the identified risk (which, as a social issue, demands community decision-making). Conflating these two separate aspects of future climates carries with it the danger of turning both researchers and the general public into bad governance monkeys (Figure 18.6).

People are smarter than sheep (cf. Figure 18.2) but the slowness with which humanity is grappling the climate challenge may render most solutions unobtainable (Anderson and Bows, 2011). The Cancun Agreement (UNFCCC, 2010) while praised by some delegates to UNFCCC COP16 did little more than formalize the Copenhagen Accord (UNFCCC, 2009). The main agreement elements are outlined in Box 18.3. A (potential) positive outcome of the Cancun Agreement is a financial pledge from industrialized countries to support climate change action in developing countries. It remains to be seen, however, if (new) funds materialize and are used to support climate adaptation efforts in developing nations. From a scientific (and ethical) standpoint, despite the praiseworthy goals of the Cancun Agreement, even if fully implemented and rigorously monitored, it cannot save the Earth from human-induced climate change of at least $+2°C$ and probably higher (e.g., Betts et al., 2011): thresholds agreed as 'dangerous interference'. Irrespective of the Earth system scientific community describing the potential threats posed by anthropogenic climate change (e.g., IPCC, 2007a; Richardson et al., 2009), Cancun is yet another example of failed global governance.

This section has considered the steps that we, people, must take to become more than climate-modifying sheep. As expounded in Section 18.1, human behaviour is now a massive component of Gaian Earth management. For example, we illustrated that mass media reporting of climate change switched from virtuous (reinforcing a 'good' message) in 2007, when Al Gore and the IPCC were awarded the Nobel Peace Prize, to vicious (escalating a 'bad' message) in late 2009/early 2010, with journalistic coverage of the claims of errors in the IPCC Fourth Assessment Report (AR4) and other issues of alleged research misconduct. We have witnessed a 'social tipping point'. The outcomes of the UNFCCC COP15 in Copenhagen and the COP16 in Cancun indicate a systemic failure in global governance. In response, in Section 18.3 we propose that

BOX 18.3 The Cancun Agreement (main objectives)

1. Establish clear objectives for reducing human-generated greenhouse gas emissions over time to keep the global average temperature rise below 2°

2. Encourage the participation of all countries in reducing these emissions, in accordance with each country's different responsibilities and capabilities to do so

3. Ensure the international transparency of the actions taken by countries and ensure that global progress towards the long-term goal is reviewed in a timely way

4. Mobilize the development and transfer of clean technology to boost efforts to address climate change, getting it to the right place at the right time and for the best effect

5. Mobilize and provide scaled-up funds in the short and long term to enable developing countries to take greater and effective action

6. Assist the particularly vulnerable people in the world to adapt to the inevitable impacts of climate change

7. Protect the world's forests, which are a major repository of carbon

8. Build up global capacity, especially in developing countries, to meet the overall challenge

9. Establish effective institutions and systems that will ensure these objectives are implemented successfully

(Source: From Cancun Agreement UNFCCC, 2010.)

trans-disciplinary ESS can aid future climate management. If people can indeed understand ESS and become an active and intelligent force within it, homeostasis may yet be possible.

18.3. TRANS-DISCIPLINARY EARTH SYSTEM SCIENCE

Section 18.1 described how ES thinkers use climate processes as exemplars of how the ES works through Gaian-type experiments (e.g., DMS and tropical deforestation). Arising from this research, ESS has advanced our understanding of a holistic system (Figure 18.8) under pressure from human transformation.

18.3.1. Creating a Social Contract with Society

Despite scientific evidence describing disturbance of the ES on an unprecedented scale (e.g., IPCC, 2007a; Richardson et al., 2009), humanity continues to mindlessly geo-modify the planet. In Section 18.2, the systemic failure of global governance to tackle one of the most urgent ES challenges of our time — anthropogenic climate change — was discussed. There is, therefore, a critical need to develop a framework for a comprehensive, long-term 'environmental stewardship' that supplants "our own personal self-interest and market forces in deciding how resources are used" (Australian Academy of Science, 2010, p. 71). We argue that science should revisit a call by Lubchenko (1998) to uphold a social contract with society. "This contract represents a commitment on the part of all scientists to devote their energies and talents to the most pressing problems of the day, in proportion to their importance, in exchange for public funding. The new and unmet needs of society include more comprehensive information,

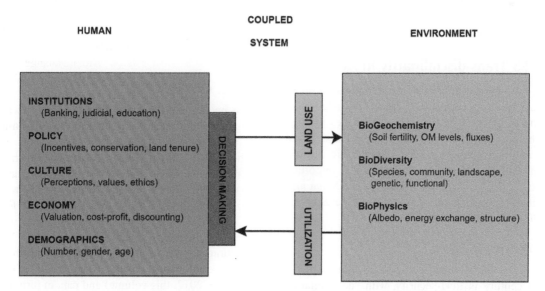

FIGURE 18.8 The interconnectedness of the ES is illustrated by the two-way interactions between the coupled human (*left*) and environmental (*right*) subsystems. The human components include institutions, policy, culture, economy, and demographics. Although 'culture' is listed separately here, it can be considered a crosscutting component of human actions (Proctor, 1998). The environmental aspects of the system include biogeochemistry, biodiversity, and biophysics. Coupling is illustrated here in terms of land use and decision-making regarding its utilization (see Pitman and de Noblet-Ducoudré, 2012, this volume). *(Source: Nobre et al., 2010; © American Meteorological Society. Reprinted with permission.)*

understanding, and technologies for society to move toward a more sustainable biosphere — one which is ecologically sound, economically feasible, and socially just. New fundamental research, faster and more effective transmission of new and existing knowledge to policy- and decision-makers, and better communication of this knowledge to the public will all be required to meet this challenge" (Lubchenko, 1998, p. 491). This section proposes that trans-disciplinary ESS (i.e., engagement of academics and non-academics to provide solutions to 'real world' problems) can help contribute to this social contract with society.

For trans-disciplinary research (TDR) to be effective, there needs to be a combination of factors, including the co-design of a vision, a common language, understanding the tradeoffs of different options and collaborative learning (Cronin, 2008; Jäger, 2008). Tackling complex problems, such as climate change, collectively requires transcending boundaries both 'horizontally', that is, across disciplines, and 'vertically', that is, across society, including scientists, researchers, policy-makers, practitioners, and the public (Klein, 2004). TDR must produce knowledge that is both scientifically credible and of relevance to future planetary management. ES researchers must inform global governance, and actors and stakeholders in society must be part of the co-production and uptake of this knowledge (Diggelmann et al., 2001) not sightless, deaf, and voiceless 'monkeys' (Figure 18.6). This requires a new type of research capacity and institutional responsiveness with active engagement between academics and non-academics, where "science becomes part of the societal process" (Cronin, 2008, p. 10). The ESS community has already started to engage in this social contract exploration. In this section, we explore this development and look hopefully into its future.

18.3.2. ESS Trans-disciplinarity in Action

This section reviews examples of ESS trans-disciplinary research and outreach, selected because they attempt to tackle some of the most pressing ES challenges through engagement of academics (including natural and social scientists) and non-academics. The examples are illustrative not exhaustive: there are other equally important examples of ESS-type trans-disciplinary research that also merit attention such as integrated coastal zone management (Harvey, 2006; Lange et al., 2010).

18.3.2.1. Food Security

One of the most pressing and complex challenges of the twenty-first century is food security, with "accumulating evidence that the food security and livelihoods of hundreds of millions of people who depend on small-scale agriculture are under significant threat from climate change" (CCAFS, 2009, p. 5). Food security is "when all people, at

all times, have physical and economic access to sufficient, safe, and nutritious food to meet their dietary needs and food preferences for an active and healthy life" (World Food Summit, 1996). The complexity of the challenge of climate change and food security will most likely be exacerbated by adaptive responses to climate change, many of which have negative consequences for food security. At the same time, measures taken to increase food security may further perturb anthropogenic climate change (CCAFS, 2009).

Ziervogel and Ericksen (2010) argue that adapting food systems both to enhance food security for the poor and vulnerable and to prevent future detrimental impacts from climate change requires an analysis framework beyond the traditional focus of agricultural production. Examples of this approach can be found in 'Food Security and Global Environmental Change' (Ingram et al., 2010), a major synthesis of the current state of knowledge and thinking on the relationships between GEC and food security. In this compendium, Holmes et al. (2010) explain that stakeholder dialogues can assist in raising awareness of, for example, climate change issues within the policy and other participating communities. Proactive discussion can promote awareness of the policy and resource management issues within the climate science community and the process can help establish a network of stakeholders (including natural and social scientists) willing to work collaboratively. Furthermore, the inclusion of wide-ranging stakeholders in Ingram et al. (2010), facilitated by GECAFS, illustrates how a TDR approach provides a more attractive foundation within which to engage broad stakeholder participation. Another example of a commitment from science to enter a social contract with society can be found in an ongoing activity in Asia on extreme weather events and their potential impact on coastal cities.

18.3.2.2. Cities at Risk

Asia is undergoing rapid population and economic growth in large coastal cities that are at high risk from sea-level rise and other climate change-induced weather extremes (Nicholls et al., 2007; Alexander and Tebaldi, 2012, this volume). Asia's densely populated deltas and mega-deltas and other low-lying coastal urban areas have been identified in the IPCC Fourth Assessment Report as "key societal hotspots of coastal vulnerability" with many millions of people potentially affected (IPCC, 2007b). Population growth in coastal areas compounds the climatic disturbance of urban development per se (Cleugh and Grimmond, 2012, this volume) and can, in turn, lead to an increased potential for loss of life and property. In recent years, there has been severe flooding particularly when high tides were combined with storm surges and high river flows (McGranahan et al., 2007). Despite these urgent threats, local governments and the international development community have yet to factor in the implications of climate change and sea-level rise on rapidly growing coastal populations and infrastructure. Part of this reticence arises because forecasting future extreme weather events is difficult (e.g., Alexander and Tebaldi, 2012, this volume). There is, therefore, a need to enhance efforts on risk and vulnerability assessment, awareness raising, and integration of science into planning and policy for the potentially affected areas (APN, 2009).

An international project on 'Cities at Risk: Developing Adaptive Capacity for Climate Change in Asia's Coastal Mega Cities' was launched in 2008, partnered by the ESS community, which included the global change SysTem for Analysis, Research, and Training (START), the World Climate Research Programme (WCRP), and the Asia–Pacific Network for Global Change Research (APN). The Cities at Risk workshop in Bangkok in 2009 brought together ES scientists, urban planners and officials, and representatives of disaster management and development agencies to review scientific findings and projections regarding climate-related risks, including sea-level rise, extreme climate events, intensification of storms, and storm surges, for Asia's coastal megacities. The workshop spelled out the need to resolve the disconnect between the differing geographies and timescales used by the scientific and planning/policy communities, as well as the importance of effective governance implemented at the systemic level in order to place adaptation strategies more centrally in the policy arena (APN, 2009).

In order to foster a trans-disciplinary approach to examine potential vulnerabilities and instil a dynamic process of adaptation to change, in 2008 START and the World Bank carried out a set of 'visioning the future exercises' in several Asian cities. This exercise engaged and consolidated a network of urban planners/officials and representatives of civil organisations to foster strategies for resilience building that deal with multiple stresses, including climate change. This was followed by an APN-START intensive training session to prepare city teams to assess natural and social vulnerabilities to climate variability and change.

Hassan Virji, personal communication, 2011

The example of cities at risk underlines the challenge of making predictions (in this case of changes in weather extremes) effectively for a community (here urban planners and local government officials) well outside the normal 'customer base' of climate models. This problem is widespread in ESS, as discussed in the next section.

18.3.2.3. Earth System Prediction

Nobre et al. (2010) contend that, in consideration of rapid human transformation of the ES, there is a need for considerable advances in the development of a monitoring

and prediction system that integrates physical, biogeochemical, and social processes. This is of fundamental importance if researchers are to provide mitigation and adaptation options for a changing environment. Many active biogeochemical feedback systems display significant non-linear behaviour (see Dickinson, 2012, this volume). Changes to the system can be prompted by both natural and human activities. Nobre et al. (2010) further argue that society would be well advised to strive to avoid abrupt tipping points in the ES (Steffen et al., 2004; Lenton et al., 2008; Rockström et al., 2009a, 2009b). For instance, as illustrated in Section 18.1, the ecosystem of the Amazon basin plays a critical role in relation to regional-to-global teleconnections with weather and climate (Zhang et al., 2001). Furthermore, climate model simulations indicate that changes in the Amazon basin biosphere can influence surface temperature and precipitation in North America, Africa, and the Himalayas. These system links have implications for the African and Asian monsoons (Nobre et al., 1991; Gedney and Valdes, 2000; Werth and Avissar, 2002; Nobre and Borma, 2009). The recognition of ES integration is prompting calls for extensions of ES prediction models to include impacts on society, with a particular focus on water, food, health, and air quality (Nobre et al., 2010). While commenting that "to date, no fully coupled model exists", Nobre et al. (2010, p. 1391) emphasize the urgent need to develop ES models that integrate human interactions such as socio-economics and land-use changes (see Pitman and de Noblet-Ducoudré, 2012, this volume).

Initial attempts from the ESS community, in partnership with other key stakeholders, to address societal challenges (e.g., climate change and food security) have been described in this section. Whether (and how) people may become more than a woolly world sheep-like force by using the data and tools of ESS to begin holistic planetary management is the topic of the next section, which describes some of the opportunities and challenges of future trans-disciplinary ESS.

18.3.3. Future Climates: Exploiting Trans-disciplinary Earth System Science

The climate change boundary was debated at the global level at the UNFCCC COP in Denmark in 2009 and in Mexico in 2010. A '2°C guardrail' was noted in Copenhagen (UNFCCC, 2009) and supported in Cancun (UNFCCC, 2010). Key findings from the Climate Change Congress Synthesis Report (Richardson et al., 2009) imply that, even if we do manage to avoid global warming above 2°C, it is virtually certain that humanity will experience extreme effects of climate change including sea-level rise, ocean temperature increases, glacier melt, Arctic sea-ice melt, and ocean acidification. A recent study on 'Four degrees and beyond: the potential for a global temperature increase of four degrees and its implications' (New et al., 2011) offers a much bleaker outlook. New et al. (2011) contend that a continued rise in greenhouse gas emissions over the past ten years (Canadell et al., 2010), coupled with the political impasse at reaching a global emissions reduction agreement, means that achieving this 2°C target is extremely difficult. Humanity might well face global temperature rises of 3 or 4°C within this century (New et al., 2011).

Human activities, especially the emissions of greenhouse gases, are already overwhelming the long-term effects of natural causes of climatic change (e.g., cycles in received solar energy; (Berger and Yin, 2012, this volume) and changes in volcanic and meteoritic activity (Belcher and Mander, 2012, this volume)). Furthermore, the future of the world's climate includes inertia in the system because CO_2 added to the atmosphere requires on the order of 100,000 years to be fully removed (Harvey, 2012, this volume). Schellnhuber (2011), in an interview in *Der Spiegel* on 23 March 2011, states that, with the prospect of the Earth becoming 4 to 6°C warmer by 2200, the world is on the precipice of having to deal with 'the unavoidability of the improbable'. The possibility of humanity and the rest of the ES facing 'the risk of unpleasant surprises' that future climates may 'have in store for us' is discussed by Lenton (2012, this volume). Lenton proposes, however, that science can help future planetary management by establishing some early warning of encroaching tipping points (similar to the planetary boundaries concept). We contend that trans-disciplinary ESS will have an important role in managing this risk. However, as Section 18.2 explains, one of the key challenges of climate change (and other ES challenges) is the need for a clear demarcation between risk quantification (the role of researchers) and responses to the identified risk or tipping point (the role of society that requires community decision-making).

The future climate of the world is one of the most important scientific and communication challenges of our time (Henderson-Sellers, 2012, this volume). The public's support of political efforts to tackle ES challenges, such as anthropogenic climate change, "hinges on robust and transparent information from the scientific community" (Le Quéré et al., 2010, p. 1). Sadly, existing structures that support the science of climate change, according to Le Quéré et al. (2010), do not have the capacity to support political efforts to limit global warming (see also Taplin, 2012, this volume). Section 18.2 describes the 'top ten' challenges for the IPCC, including its protracted gestation and mandated incapacity to make policy statements. The ES research community can and is helping resolve this issue (e.g., the annual update of the carbon budget and its trends, facilitated by the GCP; e.g., Friedlingstein et al., 2010). There has been a proposal to bolster this effort with

a call for the establishment of an International Carbon Office that would "provide full GHG balance at a regional and global level, and to respond quickly to other needs for information as they emerge" (Le Quéré et al., 2010, p. 297).

In line with the lifelong work of Stephen H. Schneider, to whose memory this book is dedicated, Section 18.2 depicts the public as the 'ultimate policy-maker' and illustrates the need for the provision of reliable and accurate scientific information to the public (e.g., Hansen, 2005; Solomon, 2007). Furthermore, "as the debate around climate change is now demonstrating, we cannot rely upon democratically elected governments to take the right decisions, from a scientific point of view. If you want to change the prevailing do nothing attitudes of the majority of the world's governments towards the enormous issues that confront us, you must first change the mindset of the human race at large" (Cribb, 2011). In response, one nation – Australia – established a 'Climate Commission' on 10 February 2011. The Climate Commission is an independent body with a mandate to provide reliable information on climate change to help inform a national debate (Climate Commission, 2011). The commission reflects a trans-disciplinary perspective, with members from science communications, business, public policy, economics, and the ESS community (including one of the planetary boundary authors and a key figure in the articulation of the Amsterdam declaration and creation of the ESSP).

ES may have originated in the geosciences but the complexity of the challenges humanity faces demands a trans-disciplinary approach, as illustrated in this chapter. Furthermore, the inclusion of, for example, behavioural and organisational sciences in ES research could help foster an understanding of the social, attitudinal, behavioural, and management changes required to tackle and resolve ES challenges identified by research in the geophysical sciences (Hartel and Pearman, 2010). "The need for further and on-going [trans]-multi-disciplinary and international research is both necessary and pressing. Moreover, it is an ethical and practical responsibility that individuals in all scientific persuasions cannot afford to shirk" (Hartel and Pearman, 2010, p. 16). A significant challenge of future ESS research is, therefore, to bring together different disciplines (and their approaches, theories, and methodologies) from different fields and to collectively ensure the joint framing, design, execution, and application of ES research. This research effort should include new educational opportunities to train the next generation of transdisciplinary ES researchers in developing as well as in developed countries. The ESS community could also explore new technologies (e.g., web-based platforms) to explain ESS and engender community participation in response to ES challenges, including anthropogenic climate change.

This chapter contends that Gaia, using climate processes as an exemplar, made a founding contribution to ESS. Increased recognition of the Earth as an integrated system has helped advance understanding of the unprecedented scale of the (mindless geo-modifying) human transformation of the ES (cf. Figure 18.2). In response to the recognition that people are a geological force, the cause of the Anthropocene, big picture ES thinkers have helped identify what we, people, must consider as the necessary steps to become more than climate-modifying sheep. In other words, human behaviour is now a massive component of Gaian Earth management. The outcomes of the UNFCCC COP15 in Copenhagen and the COP16 in Cancun indicate a systemic failure in global governance, arguably exacerbated when mass media reporting of climate change passed social tipping points to create virtuous and vicious reinforcement. This chapter has shown that climate change is too narrow a framing of the challenges we now face. To move forwards we need a trans-disciplinary ES approach that can contribute to holistic planetary management.

ACKNOWLEDGEMENTS

The authors thank Janos Bogardi, Josep (Pep) Canadell, Michelle Carnegie, John Ingram, Rik Leemans, Carlos Nobre, Johan Rockström and Hassan Virji for valuable comments.

AAAS (2010), Statement of the AAAS board of directors concerning the Virginia attorney general's investigation of Prof. Michael Mann's work while on the faculty of University of Virginia, *http://www.aaas. org/news/releases/2010/media/0518board_statement_cuccinelli.pdf.* (Accessed 26/7/10.)

Abbot, D. S., P. I. Palmer, R. V. Martin, K. V. Chance, D. H. Jacob, and A. Guenther (2003), Seasonal and inter-annual variability of North American isoprene emissions as determined by formaldehyde column measurements from space, *Geophys. Res. Lett., 30*(17), 1886, doi:10.1029/2003GL017336.

ABC 'AM' (2010), The Murray—Darling Basin Authority has bowed to public pressure and ordered a new study into the likely social and economic impacts of the proposed management plan for the river system, 18 October, 2010, *http://www.abc.net.au/am/content/ 2010/s3040716.htm.* (Accessed 21/10/10.)

ABC Radio National (2010), Climate change and behavioural change: What will it take? *All in the Mind,* 4 September.

Aberhan, M., S. Weidemeyer, W. Kiessling, R. A, Scasso, and F. A. Medina (2007), Faunal evidence for reduced productivity and uncoordinated recover in Southern Hemisphere Cretaceous-Paleogene boundary sections, *Geology, 35,* 227—230, doi:10.1130/G23197A.1.

Abram, N. J., M. K. Gagan, J. E. Cole, W. S. Hantoro, and M. Mudelsee (2008), Recent intensification of tropical climate variability in the Indian Ocean, *Nat. Geosci.,* doi:10.1038/ngeo357.

Abramowitz, G. (2010), Model independence in multi-model ensemble prediction, *Aust. Meteorol. Oceanogr. J., 59,* 3—6.

Abramowitz, G., and H. Gupta (2008), Toward a model space and model independence metric, *Geophys. Res. Lett., 35.*

Ackerman, A. S., O. B. Toon, J. P. Taylor, D. Johnson, P. V. Hobbs, and R. J. Ferek (2000), Effects of aerosols on cloud albedo: Evaluation of Twomey's parameterization of cloud susceptibility using measurements of ship tracks, *J. Atmos. Sci., 57,* 2684—2695.

Adam, D. (2010), Climate emails inquiry: Energy consultant linked to physics body's submission, *The Guardian,* 5 March, *www.guardian. co.uk/environment/2010/mar/05/climate-emails-institute-of-physics-submission.* (Accessed 26/7/10.)

Adams, P. J., J. H. Seinfeld, D. Koch, L. Mickley, and D. Jacob (2001), General circulation model assessment of direct radiative forcing by the sulfate—nitrate—ammonium—water inorganic aerosol system, *J. Geophys. Res., 106*(D1), 1097—1111.

Adger, W. N., N. Arnell, and E. Tompkins (2005), Successful adaptation to climate change across scales, *Global Environ. Chang., 15,* 77—86.

Aguilar, E., I. Auer, M. Brunet, T. C. Peterson, and J. Wieringa (2003), Guidelines on climate metadata and homogenization, *WCDMP-no. 53, WMOTD No. 1186,* World Meteorological Organization, Geneva.

Aguilar, E., T. C. Peterson, P. Ramírez Obando, R. Frutos, J. A. Retana, M. Solera, I. González Santos, R. M. Araujo, A. Rosa Santos, V. E. Valle, M. Brunet India, L. Aguilar, L. Álvarez, M. Bautista, C. Castañón, L. Herrera, E. Ruano, J. J. Siani, F. Obed, F., G. I. Hernández Oviedo, J. E. Salgado, J. L. Vásquez, M. Baca, M. Gutiérrez, C. Centella, J. Espinosa, D. Martínez, B. Olmedo, C. E. Ojeda Espinoza, M. Haylock, R. Núñez, H. Benavides, and R. Mayorga (2005), Changes in precipitation and temperature extremes in Central America and Northern South America, 1961—2003, *J. Geophys. Res., 110,* D23107, doi:10.1029/2005JD006119.

Aguilar E., A. A. Barry, and M. Brunet (2009), Changes in temperature and precipitation extremes in western Central Africa, Guinea Conakry, and Zimbabwe, 1955—2006, *J. Geophys. Res.-Atmos., 114,* D02115.

Ahn, J., and E. J. Brook (2007), Atmospheric CO_2 and climate from 65 to 30 ka B.P, *Geophys. Res. Lett., 34,* L10703, doi:10.1029/ 2007GL029551.

Ahn, J., and E. J. Brook (2008), Atmospheric CO_2 and climate on millennial time scales during the Last Glacial Period, *Science, 322*(5898), 83—85, doi:10.1126/science.1160832.

Ainsworth, E. A., and S. P. Long (2005), What have we learned from 15 years of free-air CO_2 enrichment (FACE)? A meta-analytic review of the responses of photosynthesis, canopy properties and plant production to rising CO_2, *New Phytol., 165,* 351—372.

Akbari, H., and S. Konopacki (2004), Energy effects of heat-island reduction strategies in Toronto, Canada, *Energy, 29,* 191—210.

Akbari, H., R. Levinson, and L. Rainer (2005), Monitoring the energy-use effects of cool roofs on California commercial buildings, *Energ. Buildings, 37,* 1007—1016.

Akbari, H., S. Menon, and A. Rosenfeld (2009), Global cooling: Increasing world-wide urban albedos to offset CO_2, *Climatic Change, 94,* 275—286, doi:10.1007/s10584-008-9515-9.

Akimoto, H. (2003), Global air quality and pollution, *Science, 302*(5651), 1716—1719, doi:10.1126/science.1092666.

Akinin, V. V., and E. L. Miller (2011), Evolution of calc-alkaline magmas of the Okhotsk-Chukotka volcanic belt, *Petrologiya, 19,* 237—277, (in press)(in Russian).

Albrecht, B. A. (1989), Aerosols, cloud microphysics, and fractional cloudiness, *Science, 245,* 1227—1230.

Aldhous, P. (2010), Paper trail: Inside the stem cell wars, *New Sci.,* 2764, *http://www.newscientist.com/article/mg20627643.700-paper-trail-inside-the-stem-cell-wars.html.* (Accessed 26/7/10.)

Alegret, L., M. A. Kaminski, and E. Molina (2004), Paleoenvironmental recovery after the Cretaceous/Paleogene boundary crisis: Evidence from the marine Bidart section (SW France), *Palaios, 19,* 574—586.

Alexander, L. V. (2009), Extreme Measures: Mechanisms driving changes in climate extremes in Australia, PhD thesis, Monash University, Melbourne, Australia.

Alexander, L. V. (2011), Climate Science: Extreme heat rooted in dry soils, *Nat. Geosci., 4*(1), 12—13, doi:10.1038/ngeo1045.

Alexander, L. V., and J. M. Arblaster (2009), Assessing trends in observed and modelled climate extremes over Australia in relation to future projections, *Int. J. Climatol., 29*, 417–435, doi:10.1002/joc.1730.

Alexander, L., and C. Tebaldi (2012), Climate and weather extremes: Observations, modelling, and projections, in *The Future of the World's Climate*, edited by A. Henderson-Sellers and K. McGuffie, pp. 253–258, Elsevier, Amsterdam.

Alexander, L. V., S. F. B. Tett, and T. Jonsson (2005), Recent observed changes in severe storms over the United Kingdom and Iceland, *Geophys. Res. Lett., 32*, L13704, doi:10.1029/2005GL022371.

Alexander, L. V., X. Zhang, T. C. Peterson, J. Caesar, B. Gleason, A. M. G. Klein Tank, M. Haylock, D. Collins, B. Trewin, F. Rahim, A. Tagipour, R. Kumar Kolli, J. V. Revadekar, G. Griffiths, L. Vincent, D. B. Stephenson, J. Burn, E. Aguilar, M. Brunet, M. Taylor, M. New, P. Zhai, M. Rusticucci, and J. Luis Vazquez Aguirre (2006), Global observed changes in daily climate extremes of temperature and precipitation, *J. Geophys. Res., 111*, D05109, doi:10.1029/2005JD006290.

Alexander, L. V., P. Hope, D. Collins, B. Trewin, A. Lynch, and N. Nicholls (2007), Trends in Australia's climate means and extremes: A global context, *Aust. Meteorol. Mag., 56*, 1–18.

Alexander, L. V., N. Tapper, X. Zhang, A. Lynch, H. J. Fowler, and C. Tebaldi (2009a), Climate extremes: Progress and future directions, *Int. J. Climatol., 29*, 317–319, doi:10.1002/joc.1861.

Alexander L. V., P. Uotila, and N. Nicholls (2009b), The influence of sea surface temperature variability on global temperature and precipitation extremes, *J. Geophys. Res.-Atmos., 114*, D18116, doi:10.1029/2009JD012301.

Alexander L. V., P. Uotila, N. Nicholls, and A. Lynch (2010), A new daily pressure dataset for Australia and its application to the assessment of changes in synoptic patterns during the last century, *J. Climate, 23*(5), 1111–1126.

Alexander L. V., X. L. Wang, H. Wan, and B. Trewin (2011), Significant decline in storminess over south-east Australia since the late 19th century, *Aust. Meteorol. Oceanogr. J., 61*, 23–30.

Alexander, M., and C. Pendland (1996), Variability in a mixed layer ocean-model driven by stochastic atmospheric forcing, *J. Climate, 9*, 2424.

Alexandersson, H., H. Tuomenvirta, T. Schmith, and K. Iden (2000), Trends of storms in northwest Europe derived from an updated pressure data set, *Clim. Res., 14*, 71–73.

Allan, R., S. Tett, and L. V. Alexander (2009), Fluctuations in autumn-winter severe storms over the British Isles: 1920 to present, *Int. J. Climatol.*, doi:10.1002/joc.1765.

Allen, A., D. Milenic, and P. Sikora (2003), Shallow gravel aquifers and the urban 'heat island' effect: A source of low enthalpy geothermal energy, *Geothermics, 32*, 569–578.

Allen, C. D., A. K. Macalady, H. Chenchouni, D. Bachelet, N. McDowell, M. Vennetier, T. Kitzberger, A. Rigling, D. D. Breshears, E. H. Hogg, P. Gonzalez, R. Fensham, Z. Zhang, J. Castro, N. Demidova, J.-H. Lim, G. Allard, S. W. Running, A. Semerci, and N. Cobb (2010), A global overview of drought and heat-induced tree mortality reveals emerging climate change risks for forests, *For. Ecol. Manage., 259*, 660–684.

Allen, J. R. M., W. A. Watts, and B. Huntley (2000), Weichselian paly-nostratigraphy, palaeovegetation and palaeoenvironment; The record from Lago Grande di Monticchio, southern Italy, *Quatern. Int., 73–74*, 91–110, doi:10.1016/S1040-6182(00)00067-7.

Allen, L., F. Lindberg, and C. S. B. Grimmond (2010), Global to city scale urban anthropogenic heat flux: Model and variability, *Int. J. Climatol., 31*, doi:10.1002/joc.2210.

Allen, M. R., D. J. Frame, C. Huntingford, C. D. Jones, J. A. Lowe, M. Meinshausen, and N. Meinshausen (2009), Warming caused by cumulative carbon emissions towards the trillionth tonne, *Nature, 458*, 1163–1166, doi:10.1038/nature08019, *http://www.nature.com/nature/journal/v458/n7242/full/nature08019.html.* (Accessed 26/7/10.)

Allen, R. J., and S. C. Sherwood (2008), Warming maximum in the tropical upper troposphere deduced from thermal wind observations, *Nat. Geosci., 65*, 399–403.

Alley, R. B., J. Marotzke, W. D. Nordhaus, J. T. Overpeck, D. M. Peteet, R. A. Pielke Jr., R. T. Pierrehumbert, P. B. Rhines, T. F. Stocker, L. D. Talley, and J. M. Wallace (2002), Abrupt climate change: Inevitable surprises, US National Research Council Report, Washington, DC.

Alley, R. B., J. Marotzke, W. D. Nordhaus, J. T. Overpeck, D. M. Peteet, R. A. Pielke, R. T. Pierrehumbert, P. B. Rhines, T. F. Stocker, L. D. Talley, and J. M. Wallace (2003), Abrupt climate change, *Science, 299*(5615), 2005–2010, doi:10.1126/science.1081056.

Allison, I., R. B. Alley, H. A. Fricker, R. H. Thomas, and R. C. Warner (2009a), Ice sheet mass balance and sea level, *Antarct. Sci., 21*(05), 413–426.

Allison, I., N. L. Bindoff, R. A. Bindschadler, P. M. Cox, N. de Noblet, M. H. England, J. E. Francis, N. Gruber, A. M. Haywood, D. J. Karoly, G. Kaser, C. Le Quéré, T. M. Lenton, M. E. Mann, B. I. McNeil, A. J. Pitman, S. Rahmstorf, E. Rignot, H. J. Schellnhuber, S. H. Schneider, S. C. Sherwood, R. C. J. Somerville, K. Steffen, E. J. Steig, M. Visbeck, and A. J. Weaver (2009), The Copenhagen Diagnosis, 2009: Updating the world on the Latest Climate Science. The University of New South Wales Climate Change Research Centre (CCRC), Sydney, Australia, 60pp.

Alory, G., S. Wijffels, and G. Meyers (2007), Observed temperature trends in the Indian Ocean over 1960–1999 and associated mechanisms, *Geophys. Res. Lett., 34*, L02606, doi:10.1029/2006GL028 044.

Alpert, P., and T. Sholokhman (Eds.) (2011), *Factor Separation in the Atmosphere, Applications and Future Prospects*, Cambridge University Press.

Alpert, P., P. Kishcha, Y. J. Kaufman, and R. Schwarzbard (2005), Global dimming or local dimming?: Effect of urbanization on sunlight availability, *Geophys. Res. Lett., 32*, L17802, doi:10.1029/2005GL023320.

Alroy, J., and 34 others (2008), Phanerozoic trends in the global diversity of marine invertebrates, *Science, 321*, 97–100.

Alvarez, L. W. (1983), Experimental evidence that an asteroid impact led to the extinction of many species 65 million years ago, *Proc. Natl. Acad. Sci. USA, 80*, 627–642.

Alvarez, L. W., W. Alvarez, F. Asaro, and H. V. Michel (1980), Extra-terrestrial cause for the Cretaceous-Tertiary extinction, *Science, 208*, 1095–1108.

Alvarez-Garcia, F., M. Latif, and A. Biastoch (2008), On multidecadal and quasi-decadal North Atlantic variability, *J. Climate, 21*, 3433–3452.

Amann, M., J. Cofala, C. Heyes, Z. Klimont, R. Mechler, M. Posch, and W. Schöpp (2004), *The RAINS Model, Documentation of the Model Approach Prepared for the RAINS Review*, International Institute for Applied Systems Analysis (IIASA), Laxenburg, Austria.

Amann, M., I. Bertok, J. Cofala, C. Heyes, Z. Klimont, P. Rafaj, W. Schöpp, and F. Wagner (2008), National emission ceilings for 2020 based on the 2008 Climate and Energy Package, *NEC Scenario Analysis Report 6*, Final report to the European Commission, *www.iiasa.ac.at/rains/reports/NEC6 final110708.pdf.*

Ammann, C. M., F. Joos, D. S. Schimel, B. L. Otto-Bliesner, and R. A. Tomas (2007), Solar influence on climate during the past millennium: Results from transient simulations with the NCAR Climate System Model, *Proc. Natl. Acad. Sci. USA, 104,* 3713–3718.

Anderegg W. R. L., 2010, Stephen H. Schneider: In Memoriam, *Thought and Action: Magazine of the Higher Education Association (NEA),* Fall 2010, 33–34.

Anderegg, W. R. L., J. W. Prall, J. Harold, and S. H. Schneider (2010), Expert credibility in climate change, *Proc. Natl. Acad. Sci. USA, 107,* 12107–12109. doi:10.1073/pnas.1003187107, *www.pnas.org/cgi/doi/10.1073/pnas.1003187107.* (Accessed 26/7/10.)

Anderson, K., and A. Bows (2011), Beyond 'dangerous' climate change: Emission scenarios for a new world, *Philos. Trans. R. Soc. London Ser. A, 369,* 20–44, doi:10.1098/rsta.2010.0290.

Anderson, T. L., S. E. Charlson, S. E. Schwartz, R. Knutti, O. Boucher, H. Rodhe, and J. Heintzenberg (2003), Climate forcing by a aerosols — A hazy picture, *Science, 300,* 1103–1104.

Andreadis, K. M., and D. P. Lettenmaier (2006), Trends in 20th century drought over the continental United States, *Geophys. Res. Lett., 33*(10), L10403.

Andreadis, K. M., E. A. Clark, A. W. Wood, A. F. Hamlet, and D. P. Lettenmaier (2005), 20th Century drought in the conterminous United States, *J. Hydromet., 6,* 985–1001.

Andreae, M. O., and D. Rosenfeld (2008), Aerosol-cloud-precipitation interactions: 1. The nature and sources of cloud-active aerosols, *Earth-Sci. Rev., 89,* 13–41.

Andreae, M. O., D. Rosenfeld, P. Artaxo, A. A. Costa, G. P. Frank, K. M. Longo, and M. A. F. Silva Dias (2004), Smoking rain clouds over the Amazon, *Science, 303*(27 February), 1337–1342.

Andreae, M. O., C. D. Jones, and P. M. Cox (2005), Strong present-day aerosol cooling implies a hot future, *Nature, 435,* 1187–1190.

Andrews, E. D., R. C. Antweiler, P. J. Neiman, and F. M. Ralph (2004), Influence of ENSO on flood frequency along the California coast, *J. Climate, 17,* 337–348.

Annan, J. D., and J. C. Hargreaves (2006), Using multiple observationally-based constraints to estimate climate sensitivity, *Geophys. Res. Lett., 33,* L06704.

Anthes, R. (1978), Height of planetary boundary-layer and production of circulation in a sea breeze model, *J. Atmos. Sci., 35*(7), 1231–1239.

Antonov, J. I., S. Levitus, and T. P. Boyer (2002), Steric sea level variations during 1957–1994: Importance of salinity, *J. Geophys. Res., 107*(C12), 8013.

Aragão, L. E., Y. Malhi, R. M. Roman-Cuesta, S. Saatchi, L. O. Anderson, and Y. E. Shimabukuro (2007), Spatial patterns and fire response of recent Amazonian droughts, *Geophys. Res. Lett., 34,* L07701, doi:10.1029/2006GL028946.

Arblaster, J. M., and G. A. Meehl (2006), Contributions of external forcings to Southern Annular Mode trends, *J. Climate, 19*(12), 2896–2905, doi:10.1175/JCLI3774.1.

Archer, D. (2007), Methane hydrate stability and anthropogenic climate change, *Biogeosciences, 4,* 521–544.

Archer, D., and A. Ganopolski (2005), A movable trigger: Fossil fuel CO_2 and the onset of the next glaciation, *Geochemistry, Geophysics, Geosystems, Research Letter, 6*(3), Q05003, doi:10.1029/2004GC000891.

Archer, D., B. Buffet, and V. Brovkin (2009), Ocean methane hydrates as a slow tipping point in the global carbon cycle, *Proc. Natl. Acad. Sci. USA, 106,* 20596–20601.

Archer, D., M. Eby, V. Brovkin, A. Ridgwell, L. Cao, U. Mikolajewicz, K. Caldeira, K. Matsumoto, G. Munhoven, A. Montenegro, and

K. Tokos (2009), Atmospheric lifetime of fossil fuel carbon dioxide, *Ann. Rev. Earth Planet. Sci., 37,* 117–134.

Arendt, A. A., K. A. Echelmeyer, W. D. Harrison, C. S. Lingle, S. Zirnheld, V. Valentine, B. Ritchie, and M. Druckenmiller (2006), Updated estimates of glacier volume changes in the Western Chugach Mountains, Alaska, and a comparison of regional extrapolation methods, *J. Geophys. Res., 111,* F03019, doi:10.1029/2005JF000436.

Arendt, A. A., S. B. Luthcke, C. F. Larsen, W. Abdalati, W. B. Krabill, and M. J. Beedle (2008), Validation of high-resolution GRACE mascon estimates of glacier mass change in the St. Elias Mountains, Alaska, USA, using aircraft laser altimetry, *J. Glaciol., 54*(188), 778–787.

Arenillas, I., J. A. Arz, E. Molina, and C. Dupuis (2000), An independent test of planktic foraminiferal turnover across the Cretaceous/Paleogene (K/P) boundary at El Kef, Tunisia: Catastrophic mass extinction and possible survivorship, *Micropaleontology, 46,* 31–49.

Arens, N. C., and A. H. Jahren (2000), Carbon isotopic excursion in atmospheric CO_2 at the Cretaceous-Tertiary boundary: Evidence from terrestrial sediments, *Palaios, 15,* 314–322.

Arnao, B. M. (1998), Glaciers of Peru, in *Satellite Image Atlas of Glaciers of the World: Glaciers of South America,* Professional Paper 1386-I, edited by R. S. Williams, Jr. and J. G. Ferrigno, pp. 51–79, U.S. Geological Survey, Reston, VA.

Arndt, D. S., M. O. Baringer, M. R. Johnson (Eds.) (2010), State of the climate in 2009, *Bull. Am. Meteorol. Soc., 91*(7), S1–S224, (see also NOAA, *http://www1.ncdc.noaa.gov/pub/data/cmb/bams-sotc/2009/bams-sotc-2009-brochure-lo-rez.pdf* [Accessed 22/10/10]).

Arneth, A., P. A. Miller, M. Scholze, T. Hickler, G. Schurgers, B. Smith, and I. C. Prentice (2007), CO_2 inhibition of global terrestrial isoprene emissions: Potential implications for atmospheric chemistry, *Geophys. Res. Lett., 34,* L18813, doi:10.1029/2007GL030615.

Arneth, A., N. Unger, M. Kulmala, and M. O. Andreae (2009), Clean the air, heat the planet?, *Science, 326*(5953), 672–673.

Arneth, A., S. P. Harrison, S. Zaehle, K. Tsigaridis, S. Menon, P. J. Bartlein, J. Feichter, A. Korhola, M. Kulmala, D. O'Donnell, G. Schurgers, S. Sorvari, and T. Vesala (2010), Terrestrial biogeochemical feedbacks in the climate system, *Nat. Geosci., 3,* 525–532, doi:10.1038/ngeo905.

Arnfield, A. J. (2003), Two decades of urban climate research: A review of turbulence, exchanges of energy and water, and the urban heat island, *Int. J. Clim., 23,* 1–26.

Arnfield A. J., and C. S. B. Grimmond (1998), An urban canyon energy budget model and its application to urban storage heat flux modelling, *Energ. Buildings, 27,* 61–68.

Arora, K., and G. J. Boer (2010), Uncertainties in the 20th century carbon budget associated with land use change, *Glob. Change Biolog., 16,* 3327–3348, doi:10.1111/j.1365-2486.2010.02202.x.

Arora, V. (2002), Modeling vegetation as a dynamic component in soil-vegetation-atmosphere transfer schemes and hydrological models, *Rev. Geophys., 40,* 3.1–3.26.

Ashcroft, M., L. Chisholm, and K. French (2008), The effect of exposure on landscape scale soil surface temperatures and species distribution models, *Landscape Ecol., 23*(2), 211–225, doi:10.1007/s10980-007-9181-8.

Ashfaq, M., C. B. Skinner, and N. S. Diffenbaugh (2010), Influence of SST biases in future climate projections, *Clim. Dynam.,* doi:10.1007/s00382-010-0875-2.

Ashok, K., and T. Yamagata (2009), Climate change: The El Niño with a difference, *Nature, 461*(7263), 481–484.

Ashok, K., S. K. Behera, S. A. Rao, H. Weng, and T. Yamagata (2007), El Niño Modoki and its possible teleconnection, *J. Geophys. Res., 112,* 27, doi:200710.1029/2006JC003798.

Asia-Pacific Network for Global Change Research (APN) (2009), Cities at Risk: Developing Adaptive Capacity for Climate Change in Asia's Coastal Mega Cities, CAPaBLE Project, *http://www.apn.gr.jp/new APN/resources/projectBulletinOutputs/finalProjectReports/2008/CBA 2008-06NSY-Fuchs-FinalReport.pdf.* (Accessed 27/11/10.)

Asrar, G., J. A. Kaye, and P. Morel (2001), NASA research strategy for Earth system science: Climate component, *Bull. Am. Meteorol. Soc., 82*(7), 1309−1329.

Austin, J., D. Shindell, S. R. Beagley, C. Brühl, M. Dameris, E. Manzini, T. Nagashima, P. Newman, S. Pawson, G. Pitari, E. Rozanov, C. Schnadt, and T. G. Shepherd (2003), Uncertainties and assessments of chemistry-climate models of the stratosphere, *Atmos. Chem. Phys., 3,* 1−27.

Australian Academy of Science (2010), *To Live Within Earth's Limits: An Australian Plan to Develop a Science of the Whole Earth System,* edited by R. M. Gifford, et al., Canberra.

Avissar, R., and D. Werth (2005), Global hydroclimatological teleconnections resulting from tropical deforestation, *J. Hydrometeor., 6,* 134−145.

Aw, J., and M. J. Kleeman (2003), Evaluating the first-order effect of intra-annual temperature variability on urban air pollution, *J. Geophys. Res., 108,* 4365, doi:10.1029/2002JD002688.

Ayers, G. P., and J. M. Cainey (2007), The CLAW hypothesis: A review of the major developments, *Environ. Chem., 4*(6), 366−374.

Ayers, G. P., and R. W. Gillett (2000), DMS and its oxidation products in the remote marine atmosphere: Implications for climate and atmospheric chemistry, *J. Sea Res., 43,* 275−286.

Azar, C., and S. H. Schneider (2002), Are the economic costs of stabilizing the atmosphere prohibitive?, *Ecol. Econ., 42,* 73−80.

Bader, J., and M. Latif (2003), The impact of decadal-scale Indian Ocean sea surface temperature anomalies on Sahelian rainfall and the North Atlantic Oscillation, *Geophys. Res. Lett., 30*(22), 2169, doi:10.1029/ 2003GL018426.

Bahr, D. B., M. F. Meier, and S. D. Peckham (1997), The theoretical basis of volume-area scaling, *J. Geophys. Res., 102*(B9), 20355−20362.

Bahr, D. B., M. Dyurgerov, and M. F. Meier (2009), Sea-level rise from glaciers and ice caps: A lower bound, *Geophys. Res. Lett., 36,* L03501, doi:10.1029/ 2008GL036309.

Baik, J. J., Y. H. Kim, and H. Y. Chun (2001), Dry and moist convection forced by an urban heat island, *J. Appl. Meteorol., 40,* 1462−1475.

Baines, T. (2010), Cancun conclusions, *Climate Spectator,* 14 December, *http://www.climatespectator.com.au/commentary/cancun-conclusions.* (Accessed 30/3/11.)

Baird, A. J., L. R. Belyea, X. Comas, A. S. Reeve, and L. D. Slater (2009), Carbon cycling in Northern Peatlands, *Geoph. Monog. Series, 184,* 299.

Baker, M. B., and T. Peter (2008), Small-scale cloud processes and climate, *Nature, 451,* 299−300.

Bakke, J., O. Lie, E. Heegaard, T. Dokken, G. H. Haug, H. H. Birks, P. Dulski, and T. Nilsen (2009), Rapid oceanic and atmospheric changes during the Younger Dryas cold period, *Nat. Geosci, 2,* 202−205.

Baklanov, A., C. S. B. Grimmond, A. Mahura, and M. Athanassiadou (Eds.) (2009), *Urbanization of Meteorological and Air Quality Models,* Springer-Verlag.

Bala, G., K. Caldeira, M. Wickett, T. J. Phillips, D. B. Lobell, C. Delire, and A. Mirin (2007), Combined climate and carbon-cycle effects of large-scale deforestation, *Proc. Natl. Acad. Sci. USA,* 106, 6550−6555, doi:10.1073/pnas.0608998104.

Baldocchi, D. (2008), Turner Review No. 15, 'Breathing' of the terrestrial biosphere: Lessons learned from a global network of carbon dioxide flux measurement systems, *Aust. J. Bot., 56,* 1−26, doi:10.1071/ BT07151.

Ball, J. T., I. E. Woodrow, and J. A. Berry (1987), A model predicting stomatal conductance and its contribution to the control of photosynthesis under different environmental conditions, in *Progress in Photosynthesis Research,* Volume IV, edited by J. Biggins, pp. 221−224, Martinus, Nijhoff, Dordrecht, Netherlands.

Ballantyne, A. P., M. Lavine, T. J. Crowley, J. Liu, and P. B. Baker (2005), Meta-analysis of tropical surface temperatures during the Last Glacial Maximum, *Geophys. Res. Lett., 32*(5), doi:10.1029/2004GL021217.

Ballester, J., F. Giorgi, and X. Rodo (2010), Changes in European temperature extremes can be predicted from changes in pdf central statistics: A letter, *Climatic Change, 98*(1−2), 277−284.

Balshi, M. S., A. D. McGuire, P. Duffy, M. Flannigan, D. W. Kicklighter, and J. Melillo (2009), Vulnerability of carbon storage in North American boreal forests to wildfires during the 21st century, *Glob. Change Biol., 15,* 1491−1510.

Bambach, R. K. (2006), Phanerozoic biodiversity mass extinctions, *Annu. Rev. Earth Pl. Sc., 34,* 127−155.

Bambach, R. K., A. H. Knoll, and S. Wang (2004), Origination, extinction, and mass depletions of marine diversity, *Paleobiology, 30,* 522−542.

Bamber, J. L., R. L. Layberry, and S. Gogineni (2001), A new ice thickness and bed dataset for the Greenland ice sheet, 1. Measurement, data reduction, and errors, *J. Geophys. Res., 106*(D24), 33773−33780.

Bamber, J. L., S. Ekholm, and W. B. Krabill (2001), A new, high-resolution digital elevation model of Greenland fully validated with airborne altimeter data, *J. Geophys. Res., 106,* 6733−6745.

Bambrick, H., K. Dear, and I. Hanigan (2009), Deaths in a sunburnt country: Differential effects of temperature on mortality in a 'hot' country and implications for climate change adaptation, IOP Conference Series: Earth and Environmental Science, 6.

Banta, R. M., R. K. Newsom, J. K. Lundquist, Y. L. Pichugina, R. L. Coulter, and L. Mahrt (2002), Nocturnal low-level jet characteristics over Kansas during CASES-99, *Bound.-Lay. Meteorol., 105,* 221−252.

Barbante, C., H. Fischer, V. Masson-Delmotte, C. Waelbroeck, and E. W. Wolff (2010), Climate of the last million years: New insights from EPICA and other records, *Quaternary Sci. Rev., 29*(1−2), 1−7, doi:10.1016/j.quascirev.2009.11.025.

Barlow, J., and C. A. Peres (2008), Fire-mediated dieback and compositional cascade in an Amazonian forest, *Philos. Trans. R. Soc. London Ser. B, 363,* 1787−1794.

Barlow, J. F., I. N. Harman, and S. E. Belcher (2004), Scalar fluxes from urban street canyons, Part I: Laboratory simulation, *Bound.-Lay. Meteorol., 113,* 369−385.

Barlow, M., S. Nigam, and E. H. Berbery (2001), ENSO, Pacific Decadal Variability, and US summertime precipitation, drought, and stream flow, *J. Climate, 14,* 2105−2128.

Barnard, J. C., R. Volkamer, and E. I. Kassianov (2008), Estimation of the mass absorption cross section of the organic carbon component of aerosols in the Mexico City metropolitan area, *Atmos. Chem. Phys., 8,* 6665−6679.

Barras, C. G., and R. J. Twitchett (2007), Response of marine infauna to Triassic-Jurassic environmental change: Ichnological data from Southern England, *Palaeogeogr. Palaeocl., 244,* 223−241.

Barreiro, M., G. Philander, R. Pacanowski, and A. Fedorov (2006), Simulations of warm tropical conditions with application to middle Pliocene atmospheres, *Clim. Dynam., 26*, 349–365.

Bärring, L., and H. von Storch (2004), Scandinavian storminess since about 1800, *Geophys. Res. Lett., 31*, L20202, doi:10.1029/2004GL020441.

Barriopedro, D., E. M. Fischer, J. Luterbacher, R. M. Trigo, and R. García-Herrera (2011), The hot summer of 2010: Redrawing the temperature record map of Europe, *Science, 332*(6026), 220–224, doi:10.1126/science.1201224.

Barron, A. R., D. W. Purves, and L. O. Hedin (2011), Facultative nitrogen fixation by canopy legumes in a lowland tropical forest, *Oecologia, 165*, 511–520, doi:10.107/s00442-010-1838-3.

Barsugli, J. J., and D. S. Battisti (1998), The basic effects of atmosphere-ocean thermal coupling on midlatitude variability, *J. Atmos. Sci., 55*, 477–93.

Barth, M., et al. (2005), Coupling between land ecosystems and the atmospheric hydrologic cycle through biogenic aerosol pathways, *Bull. Am. Meteorol. Soc., 86*(12), 1738–1742.

Bartholy, J., and R. Pongracz (2007), Regional analysis of extreme temperature and precipitation indices for the Carpathian Basin from 1946 to 2001. *Glob. Planet. Change, 57*(1–2), 83–95.

Bartlein, P. J. (1997), Past environmental changes: Characteristic features of Quaternary climate variations, in *Past and Future Rapid Environmental Changes: The Spatial and Evolutionary Responses of Terrestrial Biota*, edited by B. Huntley, et al., Springer-Verlag, Berlin.

Bartlein, P. J., and S. W. Hostetler (2004), Modeling paleoclimates, in *The Quaternary Period in the United States*, edited by A. Gillespie et al., pp. 563–582, Elsevier.

Bartlein, P. J., and I. C. Prentice (1989), Orbital variations, climate and paleoecology, *Trends Ecol. Evol., 4*(7), 195–199, doi:10.1016/0169-5347(89)90072-4.

Bartlein, P. J., S. P. Harrison, S. Brewer, S. Connor, B. Davis, K. Gajewski, J. Guiot, T. Harrison-Prentice, A. Henderson, O. Peyron, I. Prentice, M. Scholze, H. Seppä, B. Shuman, S. Sugita, R. Thompson, A. Viau, J. Williams, and H. Wu (2010), Pollen-based continental climate reconstructions at 6 and 21 ka: A global synthesis, *Clim. Dynam.*, doi:10.1007/s00382-010-0904-1.

Bassirirad, H. (2000), Kinetics of nutrient uptake by roots: Response to global change, *New Phytol., 147*, 155–169.

Bates, B. C., P. Hope, B. Ryan, I. Smith, and S. Charles (2008), Key findings from the Indian Ocean Climate Initiative and their impact on policy development in Australia, *Climatic Change*, doi:10.1007/s10584-007-9390-9.

Bauer, S. E., D. Koch, N. Unger, S. M. Metzger, D. T. Shindell, and D. G. Streets (2007), Nitrate aerosols today and in 2030: A global simulation including aerosols and tropospheric ozone, *Atmos. Chem. Phys., 7*(19), 5043–5059.

Beaugrand, G. (2004), The North Sea regime shift: Evidence, causes, mechanisms and consequences, *Prog. Oceanogr., 60*, 245–262.

Beaugrand, G., M. Edwards, K. Brander, C. Luczak, and F. Ibañez (2008), Causes and projections of abrupt climate-driven ecosystem shifts in the North Atlantic, *Ecol. Lett., 11*, 1157–1168.

Beaugrand, G., M. Edwards, and L. Legendre (2010), Marine biodiversity, ecosystem functioning, and carbon cycles, *Proc. Natl. Acad. Sci. USA, 107*, 10120–10124.

Beauvais, L. (1984), Evolution and diversification of Jurassic Scleractinia, *Palaeontographica Americana, 54*, 219–224.

Beerling, D. J. (2002), CO_2 and the end-Triassic mass extinction, *Nature, 415*, 386–387.

Beerling, D. J., and R. A. Berner (2002), Biogeochemical constraints on the Triassic-Jurassic boundary carbon cycle event, *Global Biogeochem. Cycles, 16*, 101–113.

Beerling, D. J., W. G. Chaloner, B. Huntley, J. A. Pearson, and M. J. Tooley (1993), Stomatal density responds to the glacial cycle of environmental change, *Proc. Roy. Soc. B-Biol. Sci., 251*, 133–138.

Beerling, D. J., et al. (2001), Evidence for the recovery of terrestrial ecosystems ahead of marine primary production following a biotic crisis at the Cretaceous-Tertiary boundary, *J. Geol. Soc. London, 158*, 737–740.

Beerling, D. J., B. H. Lomax, D. L. Royer, G. R. Upchurch, Jr., and L. R. Kump (2002), An atmospheric pCO_2 reconstruction across the Cretaceous-Tertiary boundary from leaf megafossils, *Proc. Natl. Acad. Sci. USA, 99*, 7836–7840.

Beerling, D. J., R. A. Berner, F. T. Mackenzie, M. B. Harfoot, and J. A. Pyle (2009), Methane and the CH_4-related greenhouse effect over the past 400 million years, *Am. J. Sci., 309*, 97–113.

Behling, H. (1995), A high-resolution Holocene pollen record from Lago Do Pires, Se Brazil — Vegetation, climate and fire history, *J. Paleolimnol., 14*(3), 253–268, doi:10.1007/BF00682427.

Behrenfeld, M. J., R. T. O'Malley, D. A. Siegel, C. R. McClain, J. L. Sarmiento, G. C. Feldman, A. J. Milligan, P. G. Falkowski, R. M. Letelier, and E. S. Boss (2006), Climate-driven trends in contemporary ocean productivity, *Nature, 444*(7120), 752–755, doi:10.1038/nature05317.

Beirle, S., U. Platt, R. von Glasow, M. Wenig, and T. Wagner (2004), Estimate of nitrogen oxide emissions from shipping by satellite remote sensing, *Geophys. Res. Lett., 31*(18), L18102, doi:10.1029/2004GL020312.

Beirle, S., N. Spichtinger, A. Stohl, K. L. Cummins, T. Turner, D. Boccippio, O. R. Cooper, M. Wenig, M. Grzegorski, U. Platt, and T. Wagner (2006), Estimating the NO_x produced by lightning from GOME and NLDN data: A case study in the Gulf of Mexico, *Atmos. Chem. Phys., 6*, 1075–1089.

Belcher, C. M., and L. Mander (2012), Catastrophe: Extraterrestrial impacts, massive volcanism, and the biosphere, in *The Future of the World's Climate*, edited by A. Henderson-Sellers and K. McGuffie, pp. 463–485, Elsevier, Amsterdam.

Belcher, C. M., and J. C. McElwain (2008), Limits for combustion in low O_2 redefine paleoatmospheric predictions for the Mesozoic, *Science, 321*, 1197–1200.

Belcher, C. M., M. E. Collinson, A. R. Sweet, A. R Hildebrand, and A. C. Scott (2003), "Fireball passes and nothing burns" — The role of thermal radiation in the K-T event: Evidence from the charcoal record of North America, *Geology, 31*, 1061–1064.

Belcher, C. M., M. E. Collinson, and A. C. Scott (2005), Constraints on the thermal power released from the Chicxulub impactor: New evidence from multi-method charcoal analysis, *J. Geol. Soc. London, 162*, 591–602.

Belcher, C. M., P. Finch, M. E. Collinson, A. C. Scott, and N. V. Grassineau (2009), Geochemical evidence for combustion of hydro-carbons during the K-T impact event, *Proc. Natl. Acad. Sci. USA, 6*, 4112–4117.

Belcher C. M., L. Mander, G. Rein, F. X. Jervis, M. Haworth, S. P. Hesselbo, I. J. Glasspool, and J. C. McElwain (2010), Increased fire activity at the Triassic/Jurassic boundary in Greenland due to climate-driven floral change, *Nat. Geosci., 3*, 26−429.

Beljaars, A. C. M., P. Viterbo, M. J. Miller, and A. K. Betts (1996), The anomalous rainfall over the United States during July 1993: Sensitivity to land surface parameterization and soil moisture anomalies, *Mon. Weather Rev., 124*, 362−383.

Bell, J. D., J. E. Johnson, and A. J. Hobday (Eds.) (2011), *Vulnerability of Fisheries and Aquaculture in the Tropical Pacific to Climate Change*, Secretariat of the Pacific Community, Noumea, New Caledonia, (in press).

Bellouin, N., O. Boucher, J. Haywood, and M. S. Reddy (2005), Global estimate of aerosol direct radiative forcing from satellite measurements, *Nature, 438*(7071), 1138−1141.

Bellouin, N., A. Jones, J. Haywood, and S. A. Christopher (2008), Updated estimate of aerosol direct radiative forcing from satellite observations and comparison against the Hadley Centre Climate Model, *J. Geophys. Res., 113*(D10), D10205.

Ben-Ami, Y., I. Koren, Y. Rudich, P. Artaxo, S. T. Martin, and M. O. Andreae (2010), Transport of North African dust from the Bodélé depression to the Amazon Basin: A case study, *Atmos. Chem. Phys., 10*(16), 7533−7544.

Bender, F. A.-M., A. M. L. Ekman, and H. Rodhe (2010), Response to the eruption of Mount Pinatubo in relation to climate sensitivity in the CMIP3 models, *Clim. Dynam., 35*, 875−886.

Benestad, R. E. (2002), *Solar Activity and Earth's Climate*, Praxis-Springer, Berlin and Heidelberg.

Benestad, R. E. (2005), A review of the solar cycle length estimates, *Geophys. Res. Lett., 32*, doi:10.1029/2005GL023621.

Benestad, R. E., and G. A. Schmidt (2009), Solar trends and global warming, *J. Geophys. Res., 114*, D14101, doi:10.1029/2008JD011639.

Bengtsson, L., V. A. Semenov, and O. M. Johannessen (2004), The early twentieth-century warming in the Arctic − A possible mechanism, *J. Climate, 17*, 4045−4057.

Bengtsson, L., K. I. Hodges, and E. Roekner (2006), Storm tracks and climate change, *J. Climate, 19*, 3518−3543.

Benito, G., J. Corominas, and J. M. Moreno (2005), Impacts on natural hazards of climate origin, in *A Preliminary General Assessment of the Impacts in Spain Due to the Effects of Climate Change*, edited by J. M. Moreno, pp. 505−559, Ministry of Environment, Spain.

Bennett, K. D. (2004), Continuing the debate on the role of Quaternary environmental change for macroevolution, *Philos. Trans. R. Soc. London Ser. B, 359*(1442), 295−303, doi:10.1098/rstb.2003.1395.

Benson, D. R., L.-H. Young, S.-H. Lee, T. L. Campos, D. C. Rogers, and J. Jensen (2008), The effects of airmass history on new particle formation in the free troposphere: Case studies, *Atmos. Chem. Phys., 8*, 3015−3024.

Benton, M. J. (1993), *The Fossil Record 2*, Chapman and Hall, London.

Benton, M. J. (1995), Diversification and extinction in the history of life, *Science, 268*, 52−58.

Bentz, B. J., J. Regniere, and C. J. Fettig (2010), Climate change and bark beetles of the western United States and Canada: Direct and indirect effects, *Bioscience, 60*, 602−613.

Bergamaschi, P., C. Frankenberg, J. F. Meirink, M. Krol, F. Dentener, T. Wagner, U. Platt, J. O. Kaplan, S. Körner, M. Heimann, E. J. Dlugokencky, and A. Goede (2007), Satellite chartography of atmospheric methane from SCIAMACHY on board ENVISAT: 2. Evaluation based on inverse model simulations, *J. Geophys. Res., 112*, D02304, doi:10.1029/2006JD007268.

Berger, A. (1978), Long-term variations of daily insolation and Quaternary climatic changes, *J. Atmos. Sci., 35*(12), 2361−2367, doi:10.1016/0033-5894(78)90064-9.

Berger, A. (Ed.) (1981), *Climatic Variations and Variability: Facts and Theories*, D. Riedel Publishing Company, Dordrecht, Holland.

Berger, A. (1988), Milankovitch theory and climate, *Rev. Geophys., 26*(4), 624−657, doi:10.1029/RG026i004p00624.

Berger, A. (2001), The role of CO_2, sea-level and vegetation during the Milankovitch-forced glacial-interglacial cycles, in *Geosphere-Biosphere Interactions and Climate*, edited by L. O. Bengtsson and U. C. Hammer, pp. 119−146, Cambridge University Press, New York.

Berger, A., and M. F. Loutre (1994), Long-term variations of the astronomical seasons, in *Topics in Atmospheric and Interstellar Physics and Chemistry*, edited by C. F. Boutron, pp. 33−61, Les Editions de Physique, Les Ulis, France.

Berger, A., and M. F. Loutre (1996), Modelling the climate response to astronomical and CO_2 forcings, *C. R. Acad. Sci. Paris, t.323*(IIa), 1−16.

Berger, A., and M. F. Loutre (2002), An exceptionally long interglacial ahead?, *Science, 297*(5585), 1287−1288, doi:10.1126/science.1076120.

Berger, A., and M. F. Loutre (2003), Climate 400,000 years ago, a key to the future?, in *Earth Climate and Orbital Eccentricity: The Marine Isotope Stage 11 question, Geophysical Monograph*, vol. 137, edited by A. Droxler, L. Burckle, and A. Poore, pp. 17−26, AGU, Washington, DC.

Berger, A., and M. F. Loutre (2006), De la théorie astronomique au réchauffement global, in *L'Homme Face au Climat*, edited by E. Bard, pp. 15−33, Odile Jacob, Paris.

Berger, A., and Q. Yin (2011), Modelling the interglacials of the last one million years, in *Climate change at the eve of the second decade of the century. Inferences from paleoclimates and regional aspects, Milankovitch 130th Anniversary Symposium*, edited by A. Berger, F. Mesinger, and D. Sijacki, (in press).

Berger, A., and Q. Z. Yin (2012), Modelling the past and future interglacials in response to astronomical and greenhouse gas forcing, in *The Future of the World's Climate*, edited by A. Henderson-Sellers and K. McGuffie, pp. 437−462, Elsevier, Amsterdam.

Berger, A., J. Guiot, G. Kukla, and P. Pestiaux (1981), Long-term variations of monthly insolation as related to climatic changes, *Geologischen Rundschau, 70*, 748−758.

Berger, A., H. Gallée, and J. L. Mélice (1991), The Earth's future climate at the astronomical time scale, in *Future Climate Change and Radioactive Waste Disposal, NIREX Safety Series NSS/R257*, edited by Cl. Goodess and J. Palutikof, pp. 148−165.

Berger, A., J. L. Melice, and L. Hinnov (1991), A strategy for frequency spectra of Quaternary climate records, *Clim. Dynam., 5*, 227−240, doi:10.1016/0277-3791(93)90013-C.

Berger, A., M. F. Loutre, and Ch. Tricot (1993), Insolation and Earth's orbital periods, *J. Geophys. Res., 98*(D6), 10341−10362.

Berger, A., M. F. Loutre, and J. L. Mélice (1998), Instability of the astronomical periods over the last and next millions of years, *Paleoclimate Data and Modelling, 2*(4), 239−280

Berger, A., X. S. Li, and M. F. Loutre (1999), Modelling Northern Hemisphere ice volume over the last 3 Ma., *Quaternary Science Reviews, 18*(1), 1−11.

Berger, A., M. F. Loutre, and Q. Z. Yin (2010), Total irradiation during any time interval of the year using elliptic integrals, *Quaternary Sci. Rev., 29*, 1968−1982, doi:10.1016/j.quascirev.2010.05.007.

Bergin, M. S., J. J. West, T. J. Keating, and A. G. Russell (2005), Regional atmospheric pollution and transboundary air quality management, *Annu. Rev. Env. Resour., 30*, 1−37, doi:10.1146/annurev.energy. 30.050504.144138.

Berglen, T. F., T. K. Berntsen, I. S. A. Isaksen, and J. K. Sundet (2004), A global model of the coupled sulfur/oxidant chemistry in the troposphere: The sulfur cycle, *J. Geophys. Res., 109*, D19310, doi:10.1029/2003JD003948.

Bergman, N. M., T. M. Lenton, and A. J. Watson (2004), COPSE: A new model of biogeochemical cycling over Phanerozoic time, *Am. J. Sci., 304*, 397−437.

Berner, R. A. (2006), GEOCARBSULF: A combined model for Phanerozoic atmospheric O_2 and CO_2, *Geochim. Cosmochim. Acta, 70*, 5653−5664.

Berner, R. A. (2009), Phanerozoic atmospheric oxygen: New results using the GEOCARBSULF model, *Am. J. Sci., 309*, 603−606.

Berner, R. A., and D. J. Beerling (2007), Volcanic degassing necessary to produce a $CaCO_3$ undersaturated ocean at the Triassic-Jurassic boundary, *Palaeogeogr. Palaeocl., 244*, 368−373.

Berner, R. A., A. C. Lasaga, and R. M. Garrels (1983), The carbonate-silicate geochemical cycle and its effect on atmospheric carbon dioxide over the past 100 million years, *J. Am. Sci., 283*, 641−683.

Berntsen, T. K., and J. S. Fuglestvedt (2008), Global temperature responses to current emissions from the transport sectors, *Proc. Natl. Acad. Sci. USA, 105*(49), 19154−19159.

Berntsen, T. K., and I. S. A. Isaksen (1997), A global three-dimensional chemical transport model for the troposphere: 1. Model description and CO and ozone results, *J. Geophys. Res., 102*, 21239−21280.

Berntsen, T. K., and I. S. A. Isaksen (1999), The effect of lightning and convection on changes in tropospheric ozone due to NO_x emission from aircraft, *Tellus B, 51*, 766−788.

Berntsen, T. K., I. S. A. Isaksen, G. Myhre, J. S. Fuglestvedt, F. Stordal, T. A. Larsen, R. S. Freckleton, and K. P. Shine (1997), Effects of anthropogenic emissions on tropospheric ozone and its radiative forcing, *J. Geophys. Res., 102*(D23), 28101−28126, doi:10.1029/97JD02226.

Berntsen, T. K., J. Fuglestvedt, G. Myhre, F. Stordal, and T. F. Berglen (2006), Abatement of greenhouse gases: Does location matter?, *Climatic Change, 74*(4), 377−411.

Berresheim, H., C. Plass-Dülmer, T. Elste, N. Mihalopoulos, and F. Rohrer (2003), OH in the coastal boundary layer of Crete during MINOS: Measurements and relationship with ozone photolysis, *Atmos. Chem. Phys., 3*, 639−649.

Berrio, J. C., H. Hooghiemstra, H. Behling, P. Botero, and K. Van der Borg (2002), Late-Quaternary savanna history of the Colombian Llanos Orientales from Lagunas Chenevo and Mozambique: A transect synthesis, *Holocene, 12*(1), 35−48, doi:10.1191/0959683602hl518rp.

Bertram, T. H., A. Heckel, A. Richter, J. P. Burrows, and R. C. Cohen (2005), Satellite measurements of daily variations in soil NO_x emissions, *Geophys. Res. Lett., 32*(24).

Best, M. J. (2005), Representing urban areas within operational numerical weather prediction models, *Bound.-Lay. Meteorol., 114*, 91−109.

Best, M. J. (2006), Progress towards better weather forecasts for city dwellers: From short range to climate change, *Theor. Appl. Climatol., 84*, 47−55, doi:10.1007/s00704-005-0143-2.

Best, M. J., C. S. B. Grimmond, and M. G. Villani (2006), Evaluation of the urban tile in MOSES using surface energy balance observations, *Bound.-Lay. Meteorol., 118*, 503−525, doi:10.1007/s10546-005-9025-5.

Betts, A. K. (2004), Understanding hydrometeorology using global models, *Bull. Am. Meteorol. Soc., 85*(11), 1673−1688.

Betts, A. K., J. H. Ball, A. C. M. Beljaars, M. J. Miller, and P. A. Viterbo (1996), The land surface-atmosphere interaction: A review based on observational and global modeling perspectives, *J. Geophys. Res, 101*, 7209−7225.

Betts, R. A. (2000), Offset of the potential carbon sink from boreal forestation by decreases in surface albedo, *Nature, 408*, 187−190.

Betts, R. A. (2001), Biogeophysical impacts of land use on the present-day climate: near-surface temperature change and radiative forcing. *Atmos. Sci. Lett.*, doi: 10.1006/asle.2001.0023.

Betts, R. A., P. M. Cox, S. E. Lee, and F. I. Woodward (1997), Contrasting physiological and structural vegetation feedbacks in climate change simulations, *Nature, 387*, 796−799.

Betts, R. A., P. N. Cox, M. Collins, P. P. Harris, C. Huntingford, and C. D. Jones (2004), The role of ecosystem-atmosphere interactions in simulated Amazonian precipitation decrease and forest dieback under global climate warming, *Theor. Appl. Climatol., 78*, 157−175.

Betts, R. A., O. Boucher, M. Collins, P. M. Cox, P. D. Falloon, N. Gedney, D. L. Hemming, C. Huntingford, C. D. Jones, D. M. H. Sexton, and M. J. Webb (2007), Projected increase in continental runoff due to plant responses to increasing carbon dioxide, *Nature, 448*, 1037−1042.

Betts, R. A., M. Collins, D. L. Hemming, C. D. Jones, J. A. Lowe, and M. G. Sanderson (2011), When could global warming reach 4°C?, *Philos. Trans. R. Soc. London Ser. A, 369*, 67−84, doi:10.1098/rsta.2010.0292.

Bian, H., M. Chin, J. Rodriguez, H. Yu, J. E. Penner, and S. Strahan (2009), Sensitivity of aerosol optical thickness and aerosol direct radiative effect to relative humidity, *Atmos. Chem. Phys., 9*, 2375−2386.

Biastoch, A., C. W. Böning, J. Getzlaff, J.-M. Molines, and G. Madec (2008), Causes of interannual−decadal variability in the meridional overturning circulation of the midlatitude North Atlantic Ocean, *J. Climate, 21*(24), 6599−6615, doi:10.1175/2008JCLI2404.1.

Biastoch, A., C. W. Böning, F. U. Schwarzkopf, and J. R. E. Lutjeharms (2009), Increase in Agulhas leakage due to poleward shift of Southern Hemisphere westerlies, *Nature, 462*(7272), 495−498.

Biermann, F. (2007), 'Earth system governance' as a crosscutting theme of global change research, *Glob. Environ. Change, 17*, 326−337.

Bigelow, N. H., L. B. Brubaker, M. E. Edwards, S. P. Harrison I. C. Prentice, P. M. Anderson, A. A. Andreev, P. J. Bartlein, T. R. Christensen, W. Cramer, J. O. Kaplan, A. V. Lozhkin, N. V. Matveyeva, D. F. Murray, A. D. McGuire, V. Y. Razzhivin, J. C. Ritchie, B. Smith, D. A. Walker, K. Gajewski, V. Wolf, B. H. Holmqvist, Y. Igarashi, K. Kremenetskii, A. Paus, M. F. J. Pisaric, and V. S. Volkova (2003), Climate change and Arctic ecosystems: 1. Vegetation changes north of 55 degrees N between the Last Glacial Maximum, mid-Holocene, and present, *J. Geophys. Res.-Atmos., 108*(D19), doi:10.1029/2002JD002558.

Biggs, R., S. R. Carpenter, and W. A. Brock (2009), Turning back from the brink: Detecting an impending regime shift in time to avert it, *Proc. Natl. Acad. Sci. USA, 106*(3), 826−831.

Bijl, P. K., A. J. P. Houben, S. Schouten, S. M. Bohaty, A. Sluijs, G.-J. Reichart, J. S. S. Damsté, and H. Brinkhuis (2010), Transient middle Eocene atmospheric CO_2 and temperature variations, *Science, 330*, 819−821.

Bindoff, N. L., J. Willebrand, V. Artale, A, Cazenave, J. Gregory, S. Gulev, K. Hanawa, C. Le Quéré, S. Levitus, Y. Nojiri, C. K. Shum, L. D. Talley, and A. Unnikrishnan (2007), Observations: Oceanic Climate Change and Sea Level. In *Climate Change*

2007: The Physical Science Basis. Contribution of Working Group I to the Fourth Assessment Report of the Intergovernmental Panel on Climate Change, edited by S. Solomon, D. Qin, M. Manning, Z. Chen, M. Marquis, K. B. Averyt, M. Tignor and H. L. Miller, Cambridge University Press, Cambridge, United Kingdom and New York, NY, USA.

Bintanja, R. (1998), The contribution of snowdrift sublimation to the surface mass balance of Antarctica, *Ann. Glaciol., 27*, 251−259.

Bird, M. I., J. Lloyd, and G. D. Farquhar (1996), Terrestrial carbon-storage from the Last Glacial Maximum to the present, *Chemosphere, 33*(9), 1675−1685, doi:10.1016/0045-6535(96)00187-7.

Birkeland, J. (2002), *Design for Sustainability: A Sourcebook of Integrated Ecological Solutions*, Earthscan, London.

Bjerknes, J. (1964), Atlantic air-sea interaction, *Adv. Geophys., 10*, 1−82.

Bjerknes, J. (1969), Atmospheric teleconnections from the equatorial Pacific, *Mon. Weather Rev., 97*(3), 163−172.

Björck, S., T. Rittenour, P. Rosén, Z. França, P. Möller, I. Snowball, S. Wastegård, O. Bennike, and B. Kromer (2006), A Holocene lacustrine record in the central North Atlantic: Proxies for volcanic activity, short-term NAO mode variability, and long-term precipitation changes, *Quaternary Sci. Rev., 25*(1−2), 9−32, doi:10.1016/j.quascirev.2005.08.008.

Black, E., M. Blackburn, G. Harrison, and J. Methven (2004), Factors contributing to the Summer 2003 European heatwave, *Weather, 59*, 217−223.

Blanchard, D., and R. Lopez (1985), Spatial patterns of convection in South Florida, *Mon. Weather Rev., 113*(8), 1282−1299.

Bloom, A. J., M. Burger, J. S. R Asensio, and A. B. Cousins (2010), Carbon dioxide enrichment inhibits nitrate assimilation in wheat and Arabidopsis, *Science, 328*, 899−903.

Bloos, G., and K. N. Page (2002), Global stratotype section and point for base of the Sinemurian Stage (lower Jurassic), *Episodes, 25*, 22−28.

Bloss, W. J., J. D. Lee, D. E. Heard, R. A. Salmon, S. J.-B. Bauguitte, H. K. Roscoe, and A. E. Jones (2007), Observations of OH and HO_2 radicals in coastal Antarctica, *Atmos. Chem. Phys., 7*, 4171−4185.

Boden, T., and G. Marland (2010), *Global CO_2 Emissions from Fossil-Fuel Burning, Cement Manufacture and Gas Flaring: 1751−2007*, Carbon Dioxide Information Analysis Center, Oak Ridge National Laboratory, Oak Ridge, TN, *http://cdiac.esd.ornl.gov/*.

Boer, G. J. (2010), Decadal potential predictability of 21st century climate, *Clim. Dynam., 36*(5−6), 1119−1133, doi:10.1007/s00382-010-0747-9.

Boer, G. J., and S. J. Lambert (2008), Multi-model decadal potential predictability of precipitation and temperature, *Geophys. Res. Lett., 35*, L05706, doi:10.1029/2008GL033234.

Boer, G. J., N. A. McFarlane, M. Lazare (1992), Greenhouse gas−induced climate change simulated with the CCC second-generation general circulation model. *J. Climate, 5*, 1045−1077.

Boersma, K. F., H. J. Eskes, E. W. Meijer, and H. M. Kelder (2005), Estimates of lightning NO_x production from GOME satellite observations, *Atmos. Chem. Phys., 5*, 2311−2331.

Boisvenue, C., and S. W. Running (2006), Impacts of climate change on natural forest productivity − Evidence since the middle of the 20th century, *Glob. Change Biol., 12*, 862−882.

Bolin, B. (2007), *A History of the Science and Politics of Climate Change: The Role of the Intergovernmental Panel on Climate Change*, Cambridge University Press, Cambridge, UK.

Bonan, G. B. (1995), Land-atmosphere CO_2 exchange simulated by a land surface process model coupled to an atmospheric general circulation model, *J. Geophys. Res., 100*, 2817−2831.

Bonan, G. B. (1996), A land surface model (LSM version 1.0) for ecological, hydrological, and atmospheric studies: Technical description and user's guide, NCAR/TN-417+STR NCAR Technical Note, Climate and Global Dynamics Division, National Center for Atmospheric Research, Boulder, Colorado.

Bonan, G. B. (1997), Effects of land use on the climate of the United States, *Climatic Change, 37*, 449−486.

Bonan, G. B. (1999), Frost followed the plow: Impacts of deforestation on the climate of the United States, *Ecol. Appl., 9*, 1305−1315.

Bonan, G. B. (2001), Observational evidence for reduction of daily maximum temperature by croplands in the midwest United States, *J. Climate, 14*, 2430−2442.

Bonan, G. B. (2002), *Ecological Climatology*, Cambridge University Press.

Bonan, G. B. and S. Levis (2010), Quantifying carbon-nitrogen feedbacks in the community land model (CLM4), *Geophys. Res. Lett., 37*, L07401, doi:10.1029/2010GL042430.

Bonan, G. B., P. J. Lawrence, K. W. Oleson, and S. Levis (2011), Improving canopy processes in the community land model (CLM4) using global flux fields empirically from FLUXNET data, *J. Geophys. Res.- Biogeo.*, (in press).

Bond, G., W. Broecker, S. Johnsen, J. McManus, L. Labeyrie, J. Jouzel, and G. Bonani (1993), Correlations between climate records from North Atlantic sediments and Greenland ice, *Nature, 365*(6442), 143−147.

Bond, T. C. (2007), Can warming particles enter global climate discussions?, *Environ. Res. Lett., 2*(4), 045030, doi:10.1088/1748-9326/2/4/045030.

Bond, T. C., and R. W. Bergstrom (2006), Light absorption by carbonaceous particles: An investigative review, *Aerosol Sci. Tech., 40*(1), 27−67.

Bond, T. C., D. G. Streets, K. F. Yarber, S. M. Nelson, J. H. Woo, and Z. Klimont (2004), A technology-based global inventory of black and organic carbon emissions from combustion, *J. Geophys. Res., 109*, D14203, doi:10.1029/2003JD003697.

Bond, T. C., G. Habib, and R. W. Bergstrom (2006), Limitations in the enhancement of visible light absorption due to mixing state, *J. Geophys. Res., 111*(D20), D20211.

Bond, T. C., E. Bhardwaj, R. Dong, R. Jogani, S. Jung, C. Roden, D. G. Streets, and N. M. Trautmann (2007), Historical emissions of black and organic carbon aerosol from energy-related combustion, 1850−2000, *Global Biogeochem. Cycles, 21*, GB2018, doi:10.1029/2006GB002840.

Bond, W. J., and G. F. Midgley (2000), A proposed CO_2-controlled mechanism of woody plant invasion in grasslands and savannas, *Glob. Change Biol., 6*(8), 865−869.

Böning, C. W., A. Dispert, M. Visbeck, S. Rintoul, and F. U. Schwarzkopf (2008), The response of the Antarctic Circumpolar Current to recent climate change, *Nat. Geosci., 1*, 864−869, doi:10.1038/ngeo362.

Böning, C., R. Timmermann, S. Danilov, and J. Schröter (2010), On the representation of transport variability of the Antarctic Circumpolar Current in GRACE gravity solutions and numerical ocean model simulations, In F. Flechtner, T. Gruber, A. Güntner, M. Mandea, M. Rothacher, T. Schöne, and J. Wickert (Eds) *Satellite Geodesy and Earth System Science*, Springer-Verlag, Berlin, Heidelberg, Part 2, 187−199, doi: 10.1007/978-3-642-10228-8_15.

Bonis, N. R., W. M. Kürschner, and L. Krystyn (2009), A detailed palynological study of the Triassic-Jurassic transition in key sections of the Eiberg Basin (Northern Calcareous Alps, Austria), *Rev. Palaeobot. Palyno., 156*, 376–400.

Bonis, N. R., J. H. A. van Konijnenburg-van Cittert, and W. M. Kürschner (2010), Changing CO_2 conditions during the end-Triassic inferred from stomatal frequency analysis on *Lepidopteris ottonis* (Goeppert) Schimper and *Ginkgoites taeniatus* (Braun) Harris, *Palaeogeogr. Palaeocl., 290*, 151–159.

Bonner, W. (1968), Climatology of the low level jet, *Mon. Weather Rev., 96*(12), 833–850.

Bony, S., R. Colman, V. M. Kattsov, R. P. Allan, C. S. Bretherton, J.-L. Dufresne, A. Hall, S. Hallegatte, M. M. Holland, W. Ingram, D. A. Randall, B. J. Soden, G. Tselioudis, and M. J. Webb (2006), How well do we understand and evaluate climate change feedback processes? *J. Climate, 19*, 3445–3482.

Booth, R. K., S. T. Jackson, S. L. Forman, J. E. Kutzbach, E. A. Bettis, J. Kreig, and D. K. Wright (2005), A severe centennial-scale drought in mid-continental North America 4200 years ago and apparent global linkages, *Holocene, 15*, 321–328.

Bopp, L., C. Le Quéré, M. Heimann, and A. C. Manning (2002), Climate-induced oceanic oxygen fluxes: Implications for the contemporary carbon budget, *Global Biogeochem. Cycles, 16*, doi:10.1029/2001GB001445.

Bopp, L., K. E. Kohfeld, C. Le Quéré, and O. Aumont (2003), Dust impact on marine biota and atmospheric CO_2 during glacial periods, *Paleoceanography, 18*(2), doi:10.1029/2002PA000810.

Bopp, L., O. Aumont, P. Cadule, S. Alvain, and M. Gehlen (2005), Response of diatoms distribution to global warming and potential implications: A global model study, *Geophys. Res. Lett., 32*, L19606, doi:10.1029/2005GL02653.

Born, A., and A. Levermann (2010), The 8.2 ka event: Abrupt transition of the subpolar gyre toward a modern North Atlantic circulation, *Geochem. Geophys. Geosyst., 11*(6), Q06011.

Bornstein, R., and Q. Lin (2000), Urban heat islands and summertime convective thunderstorms in Atlanta: Three case studies, *Atmos. Environ., 34*, 507–516.

Bossel, H. (1998), *Earth at a Crossroads: Paths to a Sustainable Future.* Cambridge University Press, Cambridge, UK.

Boucher, O., and M. S. Reddy (2008), Climate trade-off between black carbon and carbon dioxide emissions, *Energ. Policy, 36*, 193–200.

Boucher, O., G. Myhre, and A. Myhre (2004), Direct human influence of irrigation on atmospheric water vapour and climate, *Clim. Dynam., 22*(6–7), 597–603, doi:10.1007/s00382-004-0402-4.

Bougamont, M., J. L. Bamber, J. K. Ridley, R. M. Gladstone, W. Greuell, E. Hanna, A. J. Payne, and I. Rutt (2007), Impact of model physics on estimating the surface mass balance of the Greenland Ice Sheet, *Geophys. Res. Lett., 34*(17), L17501, doi:10.1029/2007GL030700.

Bounoua, L., R. DeFries, G. J. Collatz, P. J. Sellers, and H. Khan (2002), Effects of land cover conversion on surface climate, *Climatic Change, 52*, 29–64.

Bounoua, L., F. G. Hall, P. J. Sellers, A. Kumar, G. J. Collatz, C. J. Tucker, and M. L. Imhoff (2010), Quantifying the negative feedback of vegetation to greenhouse warming: A modeling approach, *Geophys. Res. Lett., 37*, L23701, doi:10.1029/2010GL045338.

Bourgeois, J., T. A. Hansen, P. L. Wiberg, and E. G. Kauffmann (1988), A tsunami deposit at the Cretaceous-Tertiary boundary in Texas, *Science, 241*, 567–570.

Bousquet, P., D. A. Hauglustaine, P. Peylin, C. Carouge, and P. Ciais (2005), Two decades of OH variability as inferred by an inversion of atmospheric transport and chemistry of methyl chloroform, *Atmos. Chem. Phys., 5*, 2635–2656.

Bousquet, P., P. Ciais, J. B. Miller, E. J. Dlugokencky, D. A. Hauglustaine, C. Prigent, G. R. Van der Werf, P. Peylin, E. G. Brunke, C. Carouge, R. L. Langenfelds, J. Lathiere, F. Papa, M. Ramonet, M. Schmidt, L. P. Steele, S. C. Tyler, and J. White (2006), Contribution of anthropogenic and natural sources to atmospheric methane variability, *Nature, 443*, 439–443.

Bouwman, A. F., D. P. von Duuren, R. G. Derwent, and M. Posch (2002), A global analysis of acidification and eutrophication of terrestrial ecosystems, *Water Air Soil Poll., 141*, 349–382.

Bovensmann, H., J. P. Burrows, M. Buchwitz, J. Frerick, S. Noël, V. V. Rozanov, K. V. Chance, and A. P. H. Goede (1999), SCIAMACHY: Mission objectives and measurement modes, *J. Atmos. Sci., 56*(2), 127–150.

Bowler, D. E., L. Buyung-Ali, T. M. Knight, and A. S. Pullin (2010), Urban greening to cool towns and cities: A systematic review of the empirical evidence, *Landscape Urban Plan., 97*, 147–155.

Bowman, D. M. J. S., J. K. Balch, P. Artaxo, W. J. Bond, J. M. Carlson, M. A. Cochrane, C. M. D'Antonio, R. S. DeFries, J. C. Doyle, S. P. Harrison, F. H. Johnston, J. E. Keeley, M. A. Krawchuk, C. A. Kull, J. B. Marston, M. A. Moritz, I. C. Prentice, C. I. Roos, A. C. Scott, T. W. Swetnam, G. R. van der Werf, and S. J. Pyne (2009), Fire in the Earth system, *Science, 324*(5926), 481–484.

Bown, P. (2005), Selective calcareous nannoplankton survivorship at the Cretaceous-Tertiary boundary, *Geology, 33*, 653–656.

Box, G. E. P., and G. Jenkins (1976), *Time Series Analysis: Forecasting and Control*, Holden-Day, San Francisco, CA.

Box, J. E., and A. Rinke (2003), Evaluation of Greenland Ice Sheet surface climate in the HIRHAM regional climate model, *J. Climate, 16*, 1302–1319.

Box, J. E., D. H. Bromwich, and L.-S. Bai (2004), Greenland ice sheet surface mass balance 1991–2000: Application of Polar MM5 mesoscale model and in situ data, *J. Geophys. Res., 109*, D16105, doi:10.1029/2003JD004451.

Boyce, D. G., M. R. Lewis, and B. Worm (2010), Global phytoplankton decline over the past century, *Nature, 466*(7306), 591–596, doi:10.1038/nature09268.

Boyce, D. G., M. R. Lewis, and B. Worm (2011), Boyce et al. reply, *Nature, 472*(7342), E8–E9, doi:10.1038/nature09953.

Boyer, T. P., S. Levitus, J. I. Antonov, R. A. Locarnini, and H. E. Garcia (2005), Linear trends in salinity for the world ocean, 1955–1998, *Geophys. Res. Lett, 32*, L01604, doi:10.1029/2004GL021791.

Boyle, J. S. (1998), Evaluation of the annual cycle of precipitation over the United States in GCMs: AMIP simulations, *J. Climate, 11*, 1041–1055.

Braconnot, P., S. Joussaume, O. Marti, and N. de Noblet (1999), Synergistic feedbacks from ocean and vegetation on the African monsoon response to mid-Holocene insolation, *Geophys. Res. Lett., 26*(16), 2481–2484, doi:10.1029/1999GL006047.

Braconnot, P., O. Marti, S. Joussaume, and Y. Leclainche (2000), Ocean feedback in response to 6 kyr BP insolation, *J. Climate, 13*, 1537–1553.

Braconnot, P., B. Otto-Bliesner, S. P. Harrison, S. Joussaume, J. Y. Peterchmitt, A. Abe-Ouchi, M. Crucifix, E. Driesschaert, T. Fichefet, C. D. Hewitt, M. Kageyama, A. Kitoh, A. Lainé, M. F. Loutre,

O. Marti, U. Merkel, G. Ramstein, P. Valdes, S. L. Weber, Y. Yu, and Y. Zhao (2007a), Results of PMIP2 coupled simulations of the mid-Holocene and Last Glacial Maximum: 1. Experiments and large-scale features, *Clim. Past, 3*(2), 261–277, doi:10.5194/cp-3-261-2007.

Braconnot, P., B. Otto-Bliesner, S. P. Harrison, S. Joussaume, J. Y. Peterchmitt, A. Abe-Ouchi, M. Crucifix, E. Driesschaert, T. Fichefet, C. D. Hewitt, M. Kageyama, A. Kitoh, M. F. Loutre, O. Marti, U. Merkel, G. Ramstein, P. Valdes, L. Weber, Y. Yu, and Y. Zhao (2007b), Results of PMIP2 coupled simulations of the mid-Holocene and Last Glacial Maximum: 2. Feedbacks with emphasis on the location of the ITCZ and mid- and high latitudes heat budget, *Clim. Past, 3*(2), 279–296, doi:10.5194/cp-3-279-2007.

Braconnot, P., C. Marzin, L. Grégoire, E. Mosquet, and O. Marti (2008), Monsoon response to changes in Earth's orbital parameters: Comparisons between simulations of the Eemian and of the Holocene, *Clim. Past, 4*(2), 281–294, doi:10.5194/cp-4-281-2008.

Bradbury, J. P., W. E. Dean, and R. Y. Anderson (1993), Holocene climatic and limnologic history of the North-Central United States as recorded in the varved sediments of Elk Lake, Minnesota: A synthesis, *Geol. S. Am. S., (276)*, 309.

Bradley, R. S. (1999), *Paleoclimatology: Reconstructing Climates of the Quaternary*, Academic Press, San Diego, California; London.

Brandefelt, J., and E. Källén (2004), The response of the Southern Hemisphere atmospheric circulation to an enhanced greenhouse gas forcing, *J. Climate, 17*(22), 4425–4442, doi:10.1175/3221.1.

Brandt, P., V. Hormann, A. Körtzinger, M. Visbeck, G. Krahmann, L. Stramma, R. Lumpkin, and C. Schmid (2010), Changes in the ventilation of the oxygen minimum zone of the tropical North Atlantic, *J. Phys. Oceanogr., 40*(8), 1784–1801, doi:10.1175/2010JPO4301.1.

Braun, H., M. Christl, S. Rahmstorf, A. Ganopolski, A. Mangini, C. Kubatzki, K. Roth, and B. Kromer (2005), Possible solar origin of the 1470-year glacial climate cycle demonstrated in a coupled model, *Nature, 438*(7065), 208–211, doi:10.1038/nature04121.

Brázdil, R., C. Pfister, H. Wanner, H. V. Storch, and J. Luterbacher (2005), Historical climatology in Europe – The state of the art, *Climatic Change, 70*(3), 363–430, doi:10.1007/s10584-005-5924-1.

Breda, N., and V. Badeau (2008), Forest tree responses to extreme drought and some biotic events: Towards a selection according to hazard tolerance?, *C. R. Geosci., 340*, 651–662.

Breecker, D. O., Z. D. Sharp, and L. D. McFadden (2010), Atmospheric CO_2 concentrations during ancient greenhouse climates were similar to those predicted for A.D. 2100, *Proc. Natl. Acad. Sci. USA, 107*, 576–580.

Breshears, D. D., N. S. Cobb, P. M. Rich, K. P. Price, C. D. Allen, R. G. Balice, W. H. Romme, J. H. Kastens, M. L. Floyd, J. Belnap, J. J. Anderson, O. B. Myers, and C. W. Meyer (2005), Regional vegetation die-off in response to global-change-type drought, *Proc. Natl. Acad. Sci. USA, 102*, 15144–15148.

Brett, R. (1992), The Cretaceous-Tertiary extinction: A lethal mechanism involving anhydrite target rocks, *Geochem. Cosmochim. Acta, 56*, 3603–3606.

Brinkhuis, H., J. P. Bujak, J. Smit, G. J. M. Versteegh, and H. Visscher (1998), Dinoflagellete-based sea surface temperature reconstructions across the Cretaceous-Tertiary boundary, *Palaeogeogr. Palaeocl., 14*, 67–83.

Britter, R. E., and S. R. Hanna (2003), Flow and dispersion in urban areas, *Annu. Rev. Fluid Mech., 35*, 469–496.

Brohan, P., J. J. Kennedy, I. Harris, S. F. B. Tett, and P. D. Jones (2006), Uncertainty estimates in regional and global observed temperature changes: A new dataset from 1850, *J. Geophys. Res., 111*, D12106, doi:10.1029/2005JD006548.

Brook, E. J., S. Harder, J. Severinghaus, E. J. Steig, and C. M. Sucher (2000), On the origin and timing of rapid changes in atmospheric methane during the Last Glacial Period, *Global Biogeochem. Cycles, 14*(2), 559–572, doi:10.1029/1999GB001182.

Brooks, H. E., and N. Dotzek. (2008), The spatial distribution of severe convective storms and an analysis of their secular changes, in *Climate Extremes and Society*, edited by R. Murnane, and H. Diaz, pp. 35–53, Cambridge University Press.

Brooks, H. E., C. A. Doswell, and M. P. Kay (2003), Climatological estimates of local daily tornado probability for the United States, *Weather Forecast, 18*, 626–640.

Broström, A., M. Coe, S. P. Harrison, R. Gallimore, J. E. Kutzbach, J. Foley, I. C. Prentice, and P. Behling (1998), Land surface feedbacks and palaeomonsoons in northern Africa, *Geophys. Res. Lett., 25*, 3615–3618.

Brovkin, V., A. Ganapolski, and Y. Svirezhev (1997), A continuous climate-vegetation classification for use in climate-biosphere studies, *Ecol. Modell., 101*, 251–261.

Brown, J., J. Harper, and N. Humphrey (2010), Cirque glacier sensitivity to 21st century warming: Sperry Glacier, Rocky Mountains, USA, *Glob. Planet. Change, 74*(2), 91–98.

Brown, S. J., J. Caesar, and C. A. T. Ferro (2008), Global changes in extreme daily temperature since 1950, *J. Geophys. Res.-Atmos., 113*(D5).

Brown, V. A., P. M. Deane, J. A. Harris, and J. Y. Russell (2010), Towards a just and sustainable future, in *Tackling Wicked Problems: Through the Transdisciplinary Imagination*, edited by V. A. Brown, P. M. Harris, and J. Y. Russelal, Earthscan, London.

Brunner, D., J. Staehelin, H. L. Rogers, M. O. Kohler, J. A. Pyle, D. A. Hauglustaine, L. Jourdain, T. K. Berntsen, M. Gauss, I. S. A. Isaksen, E. Meijer, P. van Velthoven, G. Pitari, E. Mancini, V. Grewe, and R. Sausen (2005), An evaluation of the performance of chemistry transport models: 2. Detailed comparison with two selected campaigns, *Atmos. Chem. Phys., 5*, 107–129.

Bryden, H. L., H. R. Longworth, and S. A. Cunningham (2005), Slowing of the Atlantic meridional overturning circulation at 25°N, *Nature, 438*(7068), 655–657.

Buchwitz, M., R. de Beek, J. P. Burrows, H. Bovensmann, T. Warneke, J. Notholt, J. F. Meirink, A. P. H. Goede, P. Bergamaschi, S. Korner, M. Heimann, and A. Schulz (2005), Atmospheric methane and carbon dioxide from SCIAMACHY satellite data: Initial comparison with chemistry and transport models, *Atmos. Chem. Phys., 5*, 941–962.

Buckley, T. N., and M. A. Adams (2011), An analytical model of non-photorespiratory CO_2 release in the light and dark in leaves of C_3 species based on stoichiometric flux balance, *Plant Cell Environ., 34*, 89–112, doi:10.111/j.1365-3040.2010.02228.x.

Budyko, M. I. (1968), On the origin of ice ages, *Meteorologiya i Gidrologiya, 11*, 3–12.

Buffett, B., and D. Archer (2004), Global inventory of methane clathrate: Sensitivity to changes in the deep ocean, *Earth Planet. Sci. Lett., 227*, 185–199.

Bührs, T. (2008), Climate change policy and New Zealand's 'national interest', *Polit. Sci., 60*(1), 61–72.

Bunce, J. A. (2008), Acclimation of photosynthesis to temperature in Arabidopsis thaliana and Brassica oleracea, *Photosynthetica, 46*, 517–524.

Burke, E. J., S. J. Brown, and N. Christidis (2006), Modeling the recent evolution of global drought and projections for the twenty-first century with the Hadley centre climate model, *J. Hydrometeorol.*, 7(5), 1113−1125.

Burns, S. J., D. Fleitmann, A. Matter, J. Kramers, and A. A. Al-Subbary (2003), Indian Ocean climate and an absolute chronology over Dansgaard/Oeschger events 9 to 13, *Science*, 301, 1365−1367.

Burrows, J. P., M. Weber, M. Buchwitz, V. Rozanov, A. Landstätter-Weißenmayer, A. Richter, R. Debeek, R. Hoogen, K. Bramstedt, K.-U. Eichmann, M. Eisinger, and D. Perner (1999), The global ozone monitoring experiment (GOME): Mission concept and first scientific results, *J. Atmos. Sci.*, 56(2), 151−175.

Buser, C., H. Kunsch, and C. Schar (2010), Bayesian multi-model projections of climate: Generalization and application to ENSEMBLES results, *Clim. Res.*, 44(2−3), 227−241, doi:10.3354/cr00895.

Butchart, N., and A. A. Scaife (2001), Removal of chlorofluorocarbons by increased mass exchange between the stratosphere and troposphere in a changing climate, *Nature*, 410(6830), 799−802.

Butchart, N., et al. (2006), Simulations of anthropogenic change in the strength of the Brewer−Dobson circulation, *Clim. Dynam.*, 27, 727−741.

Butkovskaya, N., A. Kukui, and G. Le Bras (2007), HNO_3 forming channel of the HO_2 + NO reaction as a function of pressure and temperature in the ranges of 72−600 Torr and 223−323 KJ, *Phys. Chem. A*, 111(37), 9047−9053, doi:10.1021/jp074117m.

Butler, T. M., and M. G. Lawrence (2009), The influence of megacities on global atmospheric chemistry: A modelling study, *Environ. Chem.*, 6(3), 219−225.

Butler, T. M., P. J. Rayner, I. Simmonds, and M. G. Lawrence (2005), Simultaneous mass balance inverse modeling of methane and carbon monoxide, *J. Geophys. Res.*, 110, D21310, doi:10.1029/2005JD006071.

Caesar, J., L. V. Alexander, and R. Vose (2006), Large-scale changes in observed daily maximum and minimum temperatures, 1946−2000, *J. Geophys. Res.*, 111 (D5), D05101, doi:10.1029/2005/JD006280.

Caesar, J., L. V. Alexander, B. Trewin, K. Tse-ring, L. Sorany, V. Vuniyawaya, N. Keosavang, A. Shimana, M. M. Htay, J. Karmacharya, D. A. Jayasinghearachchi, J. Sakkamart, E. Soares, L. T. Hung, L. T. Thoung, C. T. Hue, N. T. T. Dung, P. V. Hung, H. D. Cuong, N. M. Cuong, and S. Sirabaha (2011), Changes in temperature and precipitation extremes over the Indo-Pacific region from 1971 to 2005, *Int. J. Climatol.*, 31, 791−801, doi:0.1002/joc.2118.

Cai, W. (2006), Antarctic ozone depletion causes an intensification of the Southern Ocean super-gyre circulation, *Geophys. Res. Lett.*, 33(3), L03712.

Cai, W., and T. Cowan (2006), SAM and regional rainfall in IPCC AR4 models: Can anthropogenic forcing account for southwest Western Australian winter rainfall reduction?, *Geophys. Res. Lett.*, 33, L24708, doi:10.1029/2006GL028037.

Cai, W., and T. Cowan (2007), Trends in Southern Hemisphere circulation in IPCC AR4 models over 1950−99: Ozone depletion versus greenhouse forcing, *J. Climate*, 20(4), 681−693, doi:10.1175/JCLI4028.1.

Cai, W., G. Shi, and Y. Li (2005), Multidecadal fluctuations of winter rainfall over southwest Western Australia simulated in the CSIRO Mark 3 coupled model, *Geophys. Res. Lett.*, 32, L12701, doi:10.1029/2005GL022712.

Cai, W., G. Shi, T. Cowan, D. Bi, and J. Ribbe (2005), The response of the Southern Annular Mode, the East Australian Current, and the Southern mid-latitude ocean circulation to global warming, *Geophys. Res. Lett.*, 32(23), 1−4, doi:10.1029/2005GL024701.

Cai, W., T. Cowan, A. Sullivan (2009), Recent unprecedented skewness towards positive Indian Ocean Dipole occurrences and its impact on Australian rainfall, *Geophys. Res. Lett.*, 36, L11705, doi:10.1029/2009GL037604.

Cai, W., T. Cowan, S. Godfrey, and S. Wijffels (2010), Simulations of processes associated with the fast warming rate of the Southern Midlatitude Ocean, *J. Climate*, 23(1), 197−206.

Cai, Y. J., H. Cheng, Z. S. An, R. L. Edwards, X. F. Wang, L. C. Tan, and J. Wang (2010), Large variations of oxygen isotopes in precipitation over South-Central Tibet during Marine Isotope Stage 5, *Geology*, 38(3), 243−246, doi:10.1126/science.1177840.

Caldeira, K., and P. B. Duffy (2000), The role of the Southern Ocean in uptake and storage of anthropogenic carbon dioxide, *Science*, 287(5453), 620.

Caldeira, K., and M. E. Wickett (2003), Anthropogenic carbon and ocean pH, *Nature*, 425, 365−365.

Caldwell, M. M., et al. (2003), Terrestrial ecosystems, increased solar ultraviolet radiation and interaction with other climatic change factors, UNEP special issue, *Photochem. Photobiol. S.*, 2, 29−38.

Callies, J., E. Corpaccioli, M. Eisinger, A. Hahne, and A. Lefebvre (2000), GOME-2-MetOp's second generation sensor for operational ozone monitoring, *ESA Bulletin*, 102.

Camargo, S. J., K. A. Emanuel, and A. H. Sobel (2007), Use of a genesis potential index to diagnose ENSO effects on tropical cyclone genesis, *J. Climate*, 20, 4819−4834.

Came, R. E., J. M. Eiler, J. Veizer, K. Azmy, U. Brand, and C. R. Weidman (2007), Coupling of surface temperatures and atmospheric CO_2 concentrations during the Palaeozoic era, *Nature*, 449, 198−201.

Campbell, G. S. (1977), *An Introduction to Environmental Biophysics*, Springer-Verlag, New York.

Canadell, J., R. B. Jackson, J. R. Ehleringer, H. A. Mooney, O. E. Sala, and E.-D. Schulze (1996), Maximum rooting depth of vegetation types at the global scale, *Oecologia*, 108, 583−595.

Canadell, J. G., C. Le Quéré, M. R. Raupach, C. B. Field, E. T. Buitenhuis, P. Ciais, T. J. Conway, N. P. Gillett, R. A. Houghton, and G. Marland (2007), Contributions to accelerating atmospheric CO_2 growth from economic activity, carbon intensity, and efficiency of natural sinks, *Proc. Natl. Acad. Sci. USA*, 104(47), 18866−18870, doi:10.1073/pnas.0702737104.

Canadell, J. G., D. E. Pataki, R. Gifford, R. A. Houghton, Y. Luo, M. R. Raupach, P. Smith, and W. Steffen (2007), Saturation of the terrestrial carbon sink, in *Terrestrial Ecosystems in a Changing World*, The IGBP Series, edited by J. G. Canadell et al., pp. 59−78, Springer-Verlag, Berlin, and Heidelberg.

Canadell, J. G., C. Philippe, S. Dhakal, H. Dolman, P. Friedlingstein, K. R. Gurney, A. Held, R. B. Jackson, C. Le Quéré, E. L. Malone, D. S. Ojima, A. Patwardhan, G. P. Peters, and M. R. Raupach (2010), Interactions of the carbon cycle, human activity, and the climate system: A research portfolio, *Curr. Opin. Environ. Sustain.*, 2, 301−311, doi:10.1016/j.cosust.2010.08.003.

Canadell, J. G., P. Ciais, J. Gurney, C. Le Quéré, S. Piao, M. R. Raupach, and C. L. Sabine (2011), An international effort to quantify regional carbon fluxes, *Eos Trans. AGU*, 92(10), 81−88.

Cao, L., and A. Jain (2005), An Earth system model of intermediate complexity: Simulation of the role of ocean mixing parameterizations and climate change in estimated uptake for natural and bomb radiocarbon and anthropogenic CO$_2$, *J. Geophys. Res., 110,* C09002, doi:10.1029/2005JC002919.

Cao, L., G. Bala, K. Caldeira, R. Nemani, and G. Ban-Weiss (2010), Importance of carbon dioxide physiological forcing to future climate change, *Proc. Natl. Acad. Sci. USA, 107,* 9513–9518.

Cao, Z. H. (2008), Severe hail frequency over Ontario, Canada: Recent trend and variability, *Geophys. Res. Lett., 35,* L14803.

CAPE Last Interglacial Project Members (2006), Last Interglacial Arctic warmth confirms polar amplification of climate change, *Quaternary Sci. Rev., 25,* 1383–1400, doi:10.1016/j.quascirev. 2006.01.033.

Carey, L. D., and S. A. Rutledge (2000), The relationship between precipitation and lightning in tropical island convection: A C-band polarimetric study, *Mon. Weather Rev., 128,* 2687.

Cariolle, D., and J.-J. Morcrette (2006), A linearized approach to the radiative budget of the stratosphere: Influence of the ozone distribution, *Geophys. Res. Lett., 33,* L05806, doi:10.1029/ 2005GL025597.

Cariolle, D., M. J. Evans, M. P. Chipperfield, N. Butkovskaya, A. Kukui, and G. Le Bras (2008), Impact of the new HNO$_3$-forming channel of the HO$_2$ + NO reaction on tropospheric HNO$_3$, NO$_x$, HO$_2$ and ozone, *Atmos. Chem. Phys., 8,* 4061–4068.

Carnicer, J., M. Coll, M. Ninyerola, X. Pons, G. Sánchez, and J. Peñuelas (2011), Widespread crown condition decline, food web disruption, and amplified tree mortality with increased climate change-type drought, *Proc. Natl. Acad. Sci. USA, 108,* 1474–1478.

Carpenter, S. R., H. A. Mooney, J. Agard, D. Capistrano, R. S. DeFries, S. Díaz, T. Dietz, A. K. Duraiappah, A. Oteng-Yeboah, H. M. Pereira, C. Perrings, W. Reid, J. Sarukhan, R. J. Scholes, and A. Whyte (2009), Science for managing ecosystem services: Beyond the Millennium Ecosystem Assessment, *Proc. Natl. Acad. Sci. USA, 106*(5), 1305–1312, doi:10.1073/pnas.0808772106.

Carroll, L. (1865), *Alice's Adventures in Wonderland,* Macmillan and Co., London, UK.

Carslaw, K. S., R. G. Harrison, and J. Kirkby (2002), Cosmic rays, clouds, and climate, *Science, 298,* 1732–1737.

Carslaw, K. S., O. Boucher, D. V. Spracklen, G. W. Mann, J. G. L. Rae, S. Woodward, and M. Kulmala (2009), Atmospheric aerosols in the Earth system: A review of interactions and feedbacks, *Atmos. Chem. Phys. Disc., 9*(3), 11087–11183.

Carslaw, K. S., O. Boucher, D. V. Spracklen, G. W. Mann, J. G. L. Rae, S. Woodward, and M. Kulmala (2010), A review of natural aerosol interactions and feedbacks within the Earth system, *Atmos. Chem. Phys., 10,* 1701–1737.

Catton, W. R., Jr., and R. E. Dunlap (1978), Environmental sociology: A new paradigm, *Am. Sociol., 13,* 41–49.

CCAFS (2009), *Climate Change, Agriculture and Food Security,* A CGIAR Challenge Program, The Alliance of the CGIAR Centers and ESSP, Rome and Paris.

CDKN (2010), Climate and Development Knowledge Network (CDKN), *http://www.odi.org.uk/work/projects/details.asp?id=2202&;title= climate-development-knowledge-network-cdkn.* (Accessed 9/11/10.)

Cerling, T. E. (1991), Carbon dioxide in the atmosphere: Evidence from Cenozoic and Mesozoic paleosols, *Am. J. Sci., 291,* 377–400.

Cerling, T. E. (1992), Use of carbon isotopes in paleosols as an indicator of the PCO$_2$ of the paleoatmosphere, *Global Biogeochem. Cycles, 6,* 307–314.

Cerling, T. E., D. K. Solomon, J. Quade, and J. R. Bowman (1991), On the isotopic composition of carbon in soil carbon dioxide, *Geochim. Cosmochim. Acta, 55,* 3403–3405.

Chan, F., J. A. Barth, J. Lubchenco, A. Kirincich, H. Weeks, W. T. Peterson, and B. A. Menge (2008), Emergence of anoxia in the California Current large marine ecosystem, *Science, 319*(5865), 920.

Chance, K., P. I. Palmer, R. J. D. Spurr, R. V. Martin, T. P. Kurosu, and D. J. Jacob (2000), Satellite observations of formaldehyde over North America from GOME, *Geophys. Res. Lett., 27*(21), 3461–3464.

Chandra, S., J. R. Ziemke, and R. W. Stewart (1999), An 11-year solar cycle in tropospheric ozone from TOMS measurements, *Geophys. Res. Lett., 26*(2), 185–188.

Chang, P., R. Zhang, W. Hazeleger, C. Wen, X. Wan, L. Ji, R. J. Haarsma, W.-P. Breugem, and H. Seidel (2008), Oceanic link between abrupt change in the North Atlantic Ocean and the African monsoon, *Nat. Geosci., 1,* 444–448.

Changnon, S. A., Jr. (1981), METROMEX: A review and summary, *Meteor. Mon., 18,* 181.

Chapin, F. S., III (1995), New cog in the nitrogen cycle, *Nature, 377,* 199–200.

Chapin, F. S., III, P. M. Vitousek, and K. van Cleve (1986), The nature of nutrient limitation in plant communities, *Am. Nat., 127,* 48–58.

Chapman, C. R., and D. Morrison (1994), Impacts on the Earth by asteroids and comets: Assessing the hazard, *Nature, 367,* 33–40.

Chappellaz, J. A., I. Y. Fung, and A. M. Thompson (1993), The atmospheric CH$_4$ increase since the Last Glacial Maximum: 1. Source estimates, *Tellus B, 45*(3), 228, doi:10.1034/j.1600-0889.1993.t01-2-00002.x.

Chappellaz, J., T. Blunier, S. Kints, A. Dallenbach, J. M. Barnola, J. Schwander, D. Raynaud, and B. Stauffer (1997), Changes in the atmospheric CH$_4$ gradient between Greenland and Antarctica during the Holocene, *J. Geophys. Res.-Atmos., 102*(D13), 15987–15997, doi:10.1029/97JD01017.

Charbit, S., D. Paillard, and G. Ramstein (2008), Amount of CO$_2$ emissions irreversibly leading to total melting of Greenland, *Geophys. Res. Letters, 35,* L12503, 5, doi:10.1029/2008GL033472.

Charles, S., B. Bates, I. Smith, and J. Hughes (2004), Statistical downscaling of daily precipitation from observed and modelled atmospheric fields, *Hydrol. Process., 18*(8), 1373–1394.

Charlson, R. J., J. E. Lovelock, M. O. Andreae, and S. G. Warren (1987), Oceanic phytoplankton, atmospheric sulfur, cloud albedo and climate, *Nature, 326,* 655–661.

Charlson, R. J., S. E. Schwartz, J. M. Hales, R. D. Cess, J. A. Coakley, J. E. Hansen, and D. J. Hofmann (1992), Climate forcing by anthropogenic aerosols, *Science, 255,* 423–430.

Charman, D. J., D. W. Beilman, M. Blaauw, R. K. Booth, S. Brewer, F. M. Chambers, J. A. Christen, A. Gallego-Sala, S. P. Harrison, P. D. M. Hughes, S. T. Jackson, A. Korhola, D. Mauquoy, F. J. G. Mitchell, I. C. Prentice, M. van der Linden, F. de Vleeschouwer, Z. C. Yu, J. Alm, I. E. Bauer, Y. M. C. Corish, M. Garneau, V. Hohl, Y. Huang, E. Karofeld, G. Le Roux, R. Moschen, J. E. Nichols, T. Nieminen, G. M. McDonald, N. R. Phadtare, N. Rausch, W. Shotyk, U. Sillasoo, G. T. Swindles, E. S. Tuittila, L. Ukonmaanaho, M. Väliranta, S. van Bellen, B. van Geel, D. H. Vitt, and Y. Zhao (2011), Climate-driven changes in peatland carbon accumulation during the last millennium, (in review).

Chase, T. N., R. A. Pielke, T. G. F. Kittel, R. Nemani, and S. W. Running (2000), Simulated impacts of historical land cover changes on global climate in northern winter, *Clim. Dynam., 16,* 93–105.

Chaves, M. M. (1991), Effects of water deficits on carbon assimilation, *J. Exp. Bot., 42*, 1–16.

Chavez, F. P., M. Messié, and J. T. Pennington (2011), Marine primary production in relation to climate variability and change, *Annu. Rev. Marine. Sci., 3*(1), 227–260, doi:10.1146/annurev.marine.010908.163917.

Chen, C.-T., and T. Knutson (2008), On the verification and comparison of extreme rainfall indices from climate models, *J. Climate, 21*, 1605–1621.

Chen, F., H. Kusaka, R. Bornstein, J. Ching, C. S. B. Grimmond, S. Grossman-Clarke, T. Loridan, K. W. Manning, A. Martilli, S. Miao, D. Sailor, F. P. Salamanca, H. Taha, M. Tewari, X. Wang, A. A. Wyszogrodzki, and C. Zhang (2011), The integrated WRF/urban modelling system: Development, evaluation, and applications to urban environmental problems, *Int. J. Climatol., 31*, 273–288, doi:10.1002/joc.2158.

Chen, H., R. E. Dickinson, Y. Dai, and L. Zhou (2010), Sensitivity of simulated terrestrial carbon assimilation and canopy transpiration to different stomatal conductance and carbon assimilation schemes, *Clim. Dynam.*, doi:10.1007/s00382-010-0741-2.

Chen, J., and A. Ohmura (1990), Estimation of Alpine glacier water resources and their change since the 1870s, *Int. Assoc. Hydrolog. Sci. Pub., 193*, 127–135.

Chen, S., J. Price, W. Zhao, M. Donelan, and E. Walsh (2007), The CBLAST-hurricane program and the next-generation fully coupled atmosphere-wave-ocean models for hurricane research and prediction, *Bull. Am. Meteorol. Soc., 88*(3), 311–317, doi:10.1175/BAMS-88-3-311.

Chen, S. T., P. S. Yu, and Y. H. Tang (2010), Statistical downscaling of daily precipitation using support vector machines and multivariate analysis, *J. Hydrol., 385*(1–4), 13–22.

Chen, T. H., A. Henderson-Sellers, P. C. D. Milly, A. J. Pitman, A. C. M. Beljaars, J. Polcher, F. Abramopoulos, A. Boone, S. Chang, F. Chen, Y. Dai, C. E. Desborough, R. E. Dickinson, L. Dumenil, M. Ek, J. R. Garratt, N. Gedney, Y. M. Gusev, J. Kim, R. Koster, E. A. Kowalczyk, K. Laval, J. Lean, D. Lettenmaier, X. Liang, J.-F. Mahfouf, H.-T. Mengelkamp, K. Mitchell, O. N. Nasonova, J. Noilhan, A. Robock, C. Rosenzweig, J. Schaake, C. A. Schlosser, J.-P. Schulz, A. B. Shmakin, D. L. Verseghy, P. Wetzel, E. F. Wood, Y. Xue, Z.-L. Yang, and Q. Zeng (1997), Cabauw experimental results from the project for intercomparison of land-surface parameterization schemes, *J. Climate, 10*, 1144–1215.

Chen, W.-T., A. Nenes, H. Liao, P. J. Adams, J.-L. F. Li, and J. H. Seinfeld (2010), Global climate response to anthropogenic aerosol indirect effects: Present day and year 2100, *J. Geophys. Res., 115*(D12), D12207.

Chen, Y. H., and R. G. Prinn (2006), Estimation of atmospheric methane emissions between 1996 and 2001 using a three-dimensional global chemical transport model, *J. Geophys. Res., 111*, doi:10.1029/2005JD006058.

Chenet, A.-L., et al. (2009), Determination of rapid Deccan eruptions across the Cretaceous-Tertiary boundary using paleomagnetic secular variation: 2. Constraints from analysis of eight new sections and synthesis for a 3500-m-thick composite section, *J. Geophys. Res., 114*, doi:10.1029/2008JB005644.

Cheng, C.-T., W.-C. Wang, and J.-P. Chen (2007), A modeling study of aerosol impacts on cloud microphysics and radiative properties, *Q. J. Roy. Meteor. Soc., 133*, 283–297.

Cheng, H., R. L. Edwards, W. S. Broecker, G. H. Denton, X. Kong, Y. Wang, R. Zhang, and X. Wang (2009), Ice Age terminations, *Science, 326*(5950), 248–252, doi:10.1126/science.1177840.

Chiew, F., J. Teng, J. Vaze, D. Post, J. Perraud, D. Kirono, and N. Viney (2009), Estimating climate change impact on runoff across southeast Australia: Method, results, and implications of the modelling method, *Water Resour. Res., 45*, doi:10.1029/2008WR007338.

Chiew, F., D. Kirono, D. Kent, A. Frost, S. Charles, B. Timbal, K. Nguyen, and G. Fu (2010), Comparison of runoff modelled using rainfall from different downscaling methods for historical and future climates, *J. Hydrol., 387*(1–2), 10–23, doi:10.1016/j.jhydrol.2010.03.025.

Chin, M., and D. J. Jacob (1996), Anthropogenic and natural contributions to tropospheric sulfate: A global model analysis, *J. Geophys. Res., 101*(D13), 18691–18699, doi:199610.1029/96JD01222.

Chin, M., D. L. Savoie, B. J. Huebert, A. R. Bandy, D. C. Thornton, T. S. Bates, P. K. Quinn, E. S. Saltzman, and W. J. De Bruyn (2000), Atmospheric sulfur cycle simulated in the global model GOCART: Comparison with field observations and regional budgets, *J. Geophys. Res., 105*, 24689–24712.

Chin, M., R. A. Kahn, and S. E. Schwartz (2009), CCSP 2009: Atmospheric aerosol properties and climate impacts, a report by the U.S. Climate Change Science Program and the Subcommittee on Global Change Research, National Aeronautics and Space Administration, Washington, DC.

Christen, A., and R. Vogt (2004), Energy and radiation balance of a central European city, *Int. J. Climatol., 24*(11), 1395–1421, doi:10.1002/joc.1074.

Christensen, J. H., and O. Christensen (2007), A summary of the PRUDENCE Model projections of changes in European climate by the end of this century, *Climatic Change, 81*, 7–30, doi:10.1007/s10584-006-9210-7.

Christensen, J. H., T. R. Carter, M. Rummukainen, and G. Amanatidis (2007), Evaluating the performance and utility of regional climate models: The PRUDENCE Project, *Climatic Change, 81*(Supplement 1), 1–6, doi:10.1007/s10584-006-9211-6.

Christensen, J. H., E. Kjellstrom, F. Giorgi, G. Lenderink, and M. Rummukainen (2010), Weight assignment in regional climate models, *Clim. Res., 44*(2–3), 179–194, doi:10.3354/cr00916.

Christidis, N., P. A. Stott, S. Brown, G. Hegerl, and J. Caesar (2005), Detection of changes in temperature extremes during the second half of the 20th century, *Geophys. Res. Lett., 32*, L20716, doi:10.1029/2005GL023885.

Christidis, N., P. A. Stott, F. W. Zwiers, H. Shiogama, and T. Nozawa, (2010), Probabilistic estimates of recent changes in temperature forced by human activity: A multi-scale attribution analysis, *Clim. Dynam., 34*, 1139–1156, doi:10.1007/s00382-009-0615-7.

Choi, G., D. Collins, G. Y. Ren, B. Trewin, M. Baldi, Y. Fukuda, M. Afzaal, T. Pianmana, P. Gomboluudev, P. T. T. Huong, N. Lias, W. T. Kwon, K. O. Boo, Y. M. Cha, and Y. Q. Zhou (2009), Changes in means and extreme events of temperature and precipitation in the Asia-Pacific Network region, 1955–2007, *Int. J. Climatol., 29*(13), 1956–1975.

Chuang, C. C., J. E. Penner, J. M. Prospero, K. E. Grant, G. H. Rau, and K. Kawamoto (2002), Cloud susceptibility and the first aerosol indirect forcing: Sensitivity to black carbon and aerosol concentrations, *J. Geophys. Res., 107*, 4564–4587, doi:10.1029/2000JD000215.

Chubin, D. E., and E. J. Hackett (1990), *Peerless Science: Peer Review and U.S. Science Policy,* State University of New York Press, Albany, NY.

Chuck, A., T. Tyrrell, I. J. Totterdell, and P. M. Holligan (2005), The oceanic response to carbon emissions over the next century: Investigations using three ocean carbon cycle models, *Tellus B, 57,* 70–86.

Chung, C. E., V. Ramanathan, D. Kim, and I. A. Podgorny (2005), Global anthropogenic aerosol direct forcing derived from satellite and ground-based observations, *J. Geophys. Res., 110*(D24), D24207.

Chung, S. H., and J. H. Seinfeld (2002), Global distribution and climate forcing of carbonaceous aerosols, *J. Geophys Res.-Atmos., 107*(D19), 4407.

Church, J. A., and N. J. White (2006), A 20th century acceleration in global sea-level rise, *Geophys. Res. Lett., 33,* L01602.

Church, J. A., J. M. Gregory, P. Huybrechts, M. Kuhn, K. Lambeck, M. T. Nhuan, D. Qin, and P. L. Woodworth (2001), Changes in sea level, in *Climate Change 2001: The Scientific Basis, contribution of Working Group I to the Third Assessment Report (TAR) of the Intergovernmental Panel on Climate Change,* edited by J. T. Houghton et al., pp. 639–693, Cambridge University Press, Cambridge, UK.

Churkina, G. (2008), Modeling the carbon cycle of urban systems, *Ecol. model., 216,* 107–113.

Chylek, P., U. Lohmann, M. Dubey, M. Mishchenko, R. Kahn, and A. Ohmura (2007), Limits on climate sensitivity derived from recent satellite and surface observations, *J. Geophys. Res., 112,* D24S04, doi:10.1029/2007JD008740.

Ciais, P., M. Reichstein, N. Viovy, A. Granier, J. Ogée, V. Allard, M. Aubinet, N. Buchmann, C. Bernhofer, A. Carrara, F. Chevallier, N. De Noblet, A. D. Friend, P. Friedlingstein, T. Grünwald, B. Heinesch, P. Keronen, A. Knohl, G. Krinner, D. Loustau, G. Manca, G. Matteucci, F. Miglietta, J. M. Ourcival, D. Papale, K. Pilegaard, S. Rambal, G. Seufert, J. F. Soussana, M. J. Sanz, E. D. Schulze, T. Vesala, and R. Valentini (2005), Europe-wide reduction in primary productivity caused by the heat and drought in 2003, *Nature, 437,* 529–533.

Ciais, P., J. G. Canadell, S. Luyssaert, F. Chevallier, A. Shvidenko, Z. Poussi, M. Jonas, P. Peylin, A. W. King, E.-D. Schulze, S. Piao, C. Rödenbeck, W. Peters, and F.-M. Bréon (2010), Can we reconcile atmospheric estimates of the Northern terrestrial carbon sink with land-based accounting?, *Curr. Opin. Environ. Sustain., 2,* 225–230, doi:10.1016/j.cosust.2010.06.008.

CIF (2010), MDB Role, Climate Investment Funds, *http://www.climateinvestmentfunds.org/cif/MDB-Role.* (Accessed 9/11/10.)

Claquin, T., C. Roelandt, K. E. Kohfeld, S. P. Harrison, I. Tegen, I. C. Prentice, Y. Balkanski, G. Bergametti, G. Hansson, N. Mahowald, H. Rodhe, and M. Schulz (2003), Radiative forcing of climate by Ice-Age atmospheric dust, *Clim. Dynam., 20*(2–3), 193–202, doi:10.1007/s00382-002-0269-1.

Clark, P. U., N. G. Pisias, T. F. Stocker, and A. J. Weaver (2002), The role of the thermohaline circulation in abrupt climate change, *Nature, 415*(6874), 863–869, doi:10.1038/415863a.

Clark, P. U., S. W. Hostetler, N. G. Pisias, A. Schmittner, and K. J. Meissner (2007), Mechanisms for a ca.7-kyr climate and sea-level oscillation during Marine Isotope Stage 3, *AGU Monogr.,* 209–246.

Clark, P. U., A. J. Weaver, E. Brook, E. R. Cook, T. L. Delworth, and K. Steffen (2008), Abrupt climate change, a report by the U.S. Climate Change Science Program and the Subcommittee on Global Change Research, U.S. Geological Survey, Reston, VA.

Clark, P. U., A. S. Dyke, J. D. Shakun, A. E. Carlson, J. Clark, B. Wohlfarth, J. X. Mitrovica, S. W. Hostetler, and A. M. McCabe (2009), The Last Glacial Maximum, *Science, 325*(5941), 710–714, doi:10.1126/science.1172873.

Clark, W. C. (2007), Sustainability science: A room of its own, *Proc. Natl. Acad. Sci. USA, 104,* 1737–1738.

Clark, W. C., P. J. Crutzen, and H. J. Schellnhuber (2005), Science for global sustainability: Toward a new paradigm, Center for International Development, *Working Paper No. 120,* Harvard University, Cambridge, MA.

Clarke, A. D., and K. J. Noone (1985), Soot in the Arctic snowpack: A cause for perturbations in radiative transfer, *Atmos. Environ., 19,* 2045–2053.

Clausen, H. B., and C. U. Hammer (1988), The Laki and Tambora eruptions as revealed in Greenland ice cores from 11 locations, *Ann. Glaciol., 10,* 16–22.

Claussen, M., C. Kubatzki, V. Brovkin, A. Ganopolski, P. Hoelzmann, and H.-J. Pachur (1999), Simulation of an abrupt change in Saharan vegetation in the mid-Holocene, *Geophys. Res. Lett., 26*(14), 2037–2040, doi:10.1029/1999GL900494.

Claussen, M., L. A. Mysak, A. J. Weaver, M. Crucifix, T. Fichefet, M.-F. Loutre, S. L. Weber, J. Alcamo, V. A. Alexeev, A. Berger, R. Calov, A. Ganopolski, H. Goosse, G. Lohman, F. Lunkeit, I. I. Mokhov, V. Petoukhov, P. Stone, and Z. Wang (2002), Earth system models of intermediate complexity: Closing the gap in the spectrum of climate system models, *Clim. Dynam., 18,* 579–586.

Clement, A. C., and L. C. Peterson (2008), Mechanisms of abrupt climate change of the last glacial period, *Rev. Geophys.,* 46, RG4002, doi:10.1029/2006RG000204.

Clerbaux, C., A. Boynard, L. Clarisse, M. George, J. Hadji-Lazaro, H. Herbin, D. Hurtmans, M. Pommier, A. Razavi, S. Turquety, C. Wespes, and P.-F. Coheur (2009), Monitoring of atmospheric composition using the thermal infrared IASI/MetOp sounder, *Atmos. Chem. Phys., 9,* 6041–6054.

Cleugh, H. A. (1990), Development and evaluation of a suburban evaporation model: A study of surface and atmospheric controls on the suburban evaporation regime, Ph.D. thesis, University of British Columbia, Vancouver.

Cleugh, H. A. (1995), Urban climates, in *Future Climates of the World: A Modelling Perspective, World Survey of Climatology,* vol. 16, edited by A. Henderson-Sellers, pp. 477–514, Elsevier, Amsterdam.

Cleugh, H. A., and C. S. B. Grimmond (2012), Urban climates and global climate change, in *The Future of the World's Climate,* edited by A. Henderson-Sellers, and K. McGuffie, pp. 47–76, Elsevier, Amsterdam.

Cleugh, H. A., and D. E. Hughes (2002), The impact of shelter in crop microclimates and water use: A synthesis of results from wind tunnel and field measurements, *Aust. J. Exp. Agri., 42,* 679–701.

Cleugh, H. A., and T. R. Oke (1986), Suburban-rural energy balance comparisons in summer for Vancouver, BC, *Bound.-Lay. Meteorol., 36,* 351–369.

Cleugh, H. A., E. Bui, D. Simon, J. Xu, and V. G. Mitchell (2005), The impact of suburban design on water use and microclimate. Proc. MODSIM05, 10–14 Dec 2005, Melbourne, Australia. *http://www.mssanz.org.au/modsim05/papers/cleugh_2.pdf.*

Cleveland, D. M., L. C. Nordt, S. I. Dworkin, and S. C. Atchley (2008), Pedogenic carbonate isotopes as evidence for extreme climatic events preceding the Triassic-Jurassic boundary: Implications for the biotic crisis?, *Geol. Soc. Am. Bull., 120,* 1408−1415.

Climate Commission (Australia) (2011), *http://climatecommission.gov.au.* (Accessed 28/02/11.)

CLIVAR, (2009), 2009: International Clivar Project Office, report of the 12th Session of the JSC/CLIVAR Working Group on Coupled Modelling (WGCM), International CLIVAR Project Office, CLIVAR Publication Series No. 136. (Not peer reviewed.)

CMIP (2010), Coupled Model Intercomparison Project − Overview, *http://cmip-pcmdi.llnl.gov.* (Accessed 12/8/10.)

Coakley, J. A., R. L. Bernstein, Jr., and P. A. Durkee (1987), Effect of ship-stack effluents on cloud reflectivity, *Science, 237,* 1020−1022.

Cobb, K., C. Charles, H. Cheng, and R. Edwards (2003), El Niño/Southern Oscillation and tropical Pacific climate during the last millennium, *Nature, 424*(6946), 271−276, doi:10.1038/nature01779.

Coceal O., A. Dobre, and T. G. Thomas (2007), Unsteady dynamics and organized structures from DNS over an idealized building canopy, *Int. J. Climatol., 27,* 1943−1953.

Cofala, J., M. Amann, Z. Klimont, K. Kupiainen, and L. Höglund-Isaksson (2007), Scenarios of global anthropogenic emissions of air pollutants and methane until 2030, *Atmos. Environ., 41,* 8486−8499, doi:10.1016/j.atmosenv.2007.07.010.

Cogley, J. G. (2003), *GGHYDRO* − Global Hydrographic Data, Release 2.3, Trent Technical Note 2003-1, Department of Geography, Trent University, Peterborough, Ont., *http://www.trentu.ca/geography/glaciology.*

Cogley, J. G. (2005), Mass and energy balances of glaciers and ice sheets, in *Encyclopaedia of Hydrological Sciences,* vol. 4, edited by M. G. Anderson, pp. 2555−2573, Wiley, New York.

Cogley, J. G. (2008), Measured rates of glacier shrinkage, *Geophys. Res. Abstr., 10,* EGU2008.A-11595, paper presented at Annual General Assembly, European Geosciences Union, Vienna, 14 April 2008, *http://www.cosis.net/abstracts/EGU2008/11595/EGU2008-A-11595.pdf.*

Cogley, J. G. (2009a), A more complete version of the World Glacier Inventory, *Ann. Glaciol., 50*(53), 32−38.

Cogley, J. G. (2009b), Geodetic and direct mass-balance measurements: Comparison and joint analysis, *Ann. Glaciol., 50*(50), 96−100.

Cogley, J. G. (2011a), Himalayan glaciers now and in 2035, in *Encyclopaedia of Snow, Ice and Glaciers,* edited by V. P. Singh, et al., Springer Verlag, Part 8, pp. 520−523.

Cogley, J. G. (2011b), Present and future states of Himalaya and Karakoram glaciers, *Ann. Glaciol., 52*(59), 69−73.

Cogley, J. G. (2012), The future of the world's glaciers, in *The Future of the World's Climate,* edited by A. Henderson-Sellers and K. McGuffie, pp. 197−222, Elsevier, Amsterdam.

Cogley, J. G., and M. S. McIntyre (2003), Hess altitudes and related morphological estimators of glacier equilibrium lines, *Arct. Antarct. Alp. Res., 35*(4), 482−488.

Cogley, J. G., J. S. Kargel, G. Kaser, and C. J. van der Veen (2010), Tracking the source of glacier misinformation, *Science, 327*(5965), 522.

Cogley, J. G., R. Hock, L. A. Rasmussen. A. A. Arendt, A. Bauder, R. J. Braithwaite, P. Jansson, G. Kaser, M. Möller, L. Nicholson, and M. Zemp (2011), *Glossary of Glacier Mass Balance and Related Terms,* IHP-VII Technical Documents in Hydrology No. 86/IACS Contribution No. 2, UNESCO-IHP, Paris.

Cohen, S. (2010), From observer to extension agent-using research experiences to enable proactive response to climate change, *Climatic Change, 100,* 131−135.

COHMAP Members (1988), Climatic changes of the last 18,000 years: Observations and model simulations, *Science, 241,* 1043−1052, doi:10.1126/science.241.4869.1043.

Cole, J. E., and E. R. Cook (1998), The changing relationship between ENSO variability and moisture balance in the Continental United States, *Geophys. Res. Lett., 25*(24), 4529−4532, doi:10.1029/1998GL900145.

Coles, S. (2001), *An Introduction to Statistical Modeling of Extreme Values,* Springer Series in Statistics, Springer-Verlag, London.

Coley, P. D. (1998), Possible effects of climate change on plant/herbivore interactions in moist tropical forests, *Climatic Change, 39,* 455−472.

Colin de Verdière, A., Ben Jelloul, M., and Sévellec, F. (2006), Bifurcation structure of thermohaline millennial oscillations, *J. Climate, 19,* 5777−5795.

Collatz, G. J., J. T. Ball, C. Grivet, and J. A. Berry (1991), Physiological and environmental regulation of stomatal conductance, photosynthesis, and transpiration: A model that includes a laminar boundary layer, *Agr. Forest Meteorol., 54,* 107−136.

Collatz, G. J., M. Ribas-Carbo, and J. A. Berry (1992), A coupled photosynthesis-stomatal conductance model for leaves of C4 plants, *Aust. J. Plant Physiol., 19,* 519−538.

Collatz, G. J., L. Bounoua, S. O. Los, D. A. Randall, I. Y. Fung, and P. J. Sellers (2000), A mechanism for the influence of vegetation on the response of the diurnal temperature range to changing climate, *Geophys. Res. Lett., 27*(20), 3381−3384.

Colle, B., J. Wolfe, W. Steenburgh, D. Kingsmill, J. Cox, and J. Shafer (2005), High-resolution simulations and microphysical validation of an orographic precipitation event over the Wasatch Mountains during IPEX IOP3, *Mon. Weather Rev., 133*(10), 2947−2971.

Collier, C. G. (2006), The impact of urban areas on weather, *Q. J. Roy. Meteor. Soc., 132,* 1−25.

Collins, M. (2007), Ensembles and probabilities: A new era in the prediction of climate change, *Philos. Trans. R. Soc. London Ser. A, 365*(1857), 1957−1970, doi:10.1098/rsta.2007.2068.

Collins, W. J., D. S. Stevenson, C. E. Johnson, and R. G. Derwent (2000), The European regional ozone distribution and its links with the global scale for the years 1992 and 2015, *Atmos. Environ., 34,* 255−267.

Collins, M., M. Botzet, A. Carril, H. Drange, A. Jouzeau, M. Latif, O. H. Ottera, S. Masina, H. Pohlmann, A. Sorteberg, R. Sutton, and L. Terray (2006), Inter-annual to decadal climate predictability in the North Atlantic: A multimodel-ensemble study, *J. Climate, 19,* 1195−1203.

Collins, M., et al. (2010), The impact of global warming on the tropical Pacific Ocean and El Niño, *Nat. Geosci., 3*(6), 391−397, doi:10.1038/ngeo868.

Collins, W. J., R. G. Derwent, B. Garnier, C. E. Johnson, M. G. Sanderson, and D. S. Stevenson (2003), The effect of stratosphere−troposphere exchange on the future tropospheric ozone trend, *J. Geophys. Res., 108,* doi:10.1029/2002JD002617.

Collins, W. J., M. G. Sanderson, and C. E. Johnson (2009), Impact of increasing ship emissions on air quality and deposition over Europe by 2030, *Meteorologische Zeitschrift, 18,* 25−39.

Commonwealth Scientific and Industrial Research Organisation (CSIRO) (2010), Climate variability and change in south-eastern Australia: A synthesis of findings from Phase 1 of the South Eastern Australian Climate Initiative (SEACI).

Conway Morris, S. (1998), The evolution and diversity of ancient ecosystems: A review, *Philos. Trans. R. Soc. London Ser. B, 353,* 327−345.

Cook, E. R., B. M. Buckley, R. D. D'Arrigo, and M. J. Peterson (2000), Warm-season temperatures since 1600 BC reconstructed from Tasmanian tree rings and their relationship to large-scale sea surface temperature anomalies, *Clim. Dynam., 16*(2), 79–91, doi:10.1007/s003820050006.

Cook, E. R., P. J. Bartlein, N. S. Diffenbaugh, R. Seager, B. N. Shuman, R. S. Webb, and J. W. Williams (2008), Hydrological variability and change, in Abrupt climate change, a report by the US Climate Change Science Program and the Subcommittee on Global Change Research, edited by P. U. Clark et al., *Synthesis and Assessment Product 3.4.,* US Climate Change Research Program.

Cook, E. R., K. J. Anchukaitis, B. M. Buckley, R. D. D'Arrigo, G. C. Jacoby, and W. E. Wright (2010), Asian monsoon failure and megadrought during the last millennium, *Science, 328*(5977), 486–489.

Cook, J., and E. J. Highwood (2004), Climate response to tropospheric absorbing aerosols in an intermediate general-circulation model, *Q. J. Roy. Meteor. Soc., 130*(596), 175–191.

Cook, K. H., and E. K. Vizy (2006), Coupled model simulations of the West African monsoon system: Twentieth- and twenty-first-century simulations, *J. Climate, 19*(15), 3681–3703.

Cook, K. H., and E. K. Vizy (2008), Effects of twenty-first-century climate change on the Amazon Rain Forest, *J. Climate, 21*, 542–560.

Cook, K. H., E. K. Vizy, Z. S. Launer, and C. M. Patricola (2008), Springtime intensification of the Great Plains low-level jet and Midwest precipitation in GCM simulations of the twenty-first century, *J. Climate, 21*, 6321–6340.

Cooley, D., D. Nychka, and P. Naveau (2007), Bayesian spatial modelling of extreme precipitation return levels, *J. Am. Stat. Assoc., 102*(479), 824–840.

Cooley, D., R. A. Davis, and P. Naveau (2010), The pairwise beta distribution: A flexible parametric multivariate model for extremes, *J. Multivariate Anal., 101*(9), 2103–2117.

Copenhagen Accord (2009), Decision -/CP.15, 18 December, *http://unfccc.int/files/meetings/cop_15/application/pdf/cop15_cph_auv.pdf.* (Accessed 9/11/10.)

Corney, S. P., J. J. Katzfey, J. L. McGregor, M. R. Grose, J. C. Bennett, C. J. White, G. K. Holz, S. M. Gaynor, and N. L. Bindoff (2010), Climate futures for Tasmania: Climate modelling, Technical report, Antarctic Climate and Ecosystems Cooperative Research Centre, Hobart.

Corti, S., F. Molteni, and T. N. Palmer (1999), Signature of recent climate change in frequencies of natural atmospheric circulation regimes, *Nature, 398*(6730), 799–802, doi:10.1038/19745.

Costa, M. H., and J. A. Foley (2000), Combined effects of deforestation and doubled atmospheric CO_2 concentrations on the climate of Amazonia, *J. Climate, 13*, 18–34.

Cotgrove, S., and A. Duff (1980), Environmentalism, middle-class radicalism and politics, *Sociol. Rev., 28*(2), 333–351.

Courtillot, V., J. Besse, D. Vandamme, J. J. Jaeger, and R. Montigny (1986), Deccan trap volcanism as a cause of biologic extinctions at the Cretaceous-Tertiary boundary, *C. R. Acad. Sci. Paris II, 303*, 863–868.

Coutts A. M., J. Beringer, and N. J. Tapper (2007a), Characteristics influencing the variability of urban CO_2 fluxes in Melbourne, Australia, *Atmos. Environ., 41*, 51–62.

Coutts, A. M., J. Beringer, and N. J. Tapper (2007b), Impact of increasing urban density on local climate: Spatial and temporal variations in the surface energy balance in Melbourne, Australia, *J. Appl. Meteorol. Clim., 47*, 477–493.

Covey, C., K. M. Achuta Rao, U. Cubasch, P. Jones, S. J. Lambert, M. E. Mann, T. J. Phillips, and K. E. Taylor (2003), An overview of results from the Coupled Model Intercomparison Project, *Glob. Planet. Change, 37*(1–2), 103–133, doi:10.1016/S0921-8181(02)00193-5.

Cox, P., and C. Jones (2008), Illuminating the modern dance of climate and CO_2, *Science, 321*, 1642–1643.

Cox, P. M., R. A. Betts, C. D. Jones, S. A. Spall, and I. J. Totterdell (2000), Acceleration of global warming due to carbon-cycle feedbacks in a coupled climate model, *Nature, 408*, 184–187.

Cox, P. M., R. A. Betts, M. Collins, P. P. Harris, C. Huntingford, and C. D. Jones (2004), Amazonian forest dieback under climate–carbon cycle projections for the 21st century, *Theor. Appl. Climatol., 78*, 137–156.

Cox, P. M., P. P. Harris, C. Huntingford, R. A. Betts, M. Collins, C. D. Jones, T. E. Jupp, J. A. Marengo, and C. A. Nobre (2008), Increasing risk of Amazonian drought due to decreasing aerosol pollution, *Nature, 453*, 212–215.

Craine, J. M., N. Fierer, and K. K. McLauchian (2010), Widespread coupling between the rate and temperature sensitivity of organic matter decay, *Nat. Geosci., 3*, 854–857, doi:10.1038/ngeo1009.

Crane, P. R., and S. Lidgard, (1989), Angiosperm diversification and paleolatitudinal gradients in Cretaceous floristic diversity, *Science, 246*, 675–678.

Cravatte, S., T. Delcroix, D. Zhang, M. Mcphaden, and J. Leloup (2009), Observed freshening and warming of the Western Pacific Warm Pool, *Clim. Dynam., 33*(4), 565–589, doi:10.1007/s00382-009-0526-7.

Crawford B., C. S. B. Grimmond, and A. Christen (2011), Five years of carbon dioxide fluxes measurements in a highly vegetated suburban area, *Atmos. Environ., 45*, 896–905, doi:10.1016/j.atmosenv.2010.11.017.

Cressie, N. (1995), *Statistics for Spatial Data*, Revised edition, Wiley, New York.

Crewe, A. (1967) (and July 2007), 'Science and the war on...', Head Argonne National Lab., *Phys. Today, 20*(10), 5–30.

Cribb, J. (2011), The case of open science, ABC Radio (Australia), *Ockham's Razor,* 9 January 2011, *http://www.abc.net.au/rn/ockhamsrazor/stories/2011/3089284.htm.* (Accessed 10/01/11.)

Cronin, K. (2008), Transdisciplinary Research (TDR) and Sustainability: Overview report prepared for the Ministry of Research, Science and Technology (MoRST), New Zealand.

Crook, J. A., P. M. Forster, and N. Stuber (2011), Spatial patterns of modeled climate feedback and contributions to temperature response and polar amplification, *J. Climate, 24*, 3575–3592. doi:10.1175/2011JCLI3863.1.

Crook, S. A., and L. J. Gray (2005a), Characterization of the 11-year solar signal using a multiple regression analysis of the ERA-40 dataset, *J. Climate, 18*, 996–1015.

Crosier, J., J. D. Allan, H. Coe, K. N. Bower, P. Formenti, and P. I. Williams (2007), Chemical composition of summertime aerosol in the Po Valley (Italy), northern Adriatic and Black Sea, *Q. J. Roy. Meteor. Soc., 133*, 61–75.

Crosman, E., and J. Horel (2010), Sea and lake breezes: A review of numerical studies, *Bound.-Lay. Meteorol., 137*(1), 1–29, doi:10.1007/s10546-010-9517-9.

Crous, K. Y., P. B. Reich, M. D. Hunter, and D. S. Ellsworth (2010), Maintenance of leaf N controls the photosynthetic CO_2 response of grassland species exposed to 9 years of free-air CO_2 enrichment,

Glob. Change Bio., 16, 2076–2088, doi:10.1111/j.1365-2486.2009.02058x.

CRU (2005), STARDEX: Downscaling climate extremes, Climate Research Unit, University of East Anglia, UK, *http://www.cru.uea.ac.uk/projects/stardex/reports/STARDEX_FINAL_REPORT.pdf.*

Crucifix, M., and A. Berger (2006), How long will our interglacial be?, *EOS, Transactions, AGU, 87*(35), 352–353.

Crucifix, M., and M. F. Loutre (2002), Transient simulations over the last interglacial period (126–115 kyr BP): Feedback and forcing analysis, *Clim. Dynam., 19*, 417–433.

Crucifix, M., M. F. Loutre, P. Tulkens, T. Fichefet, and A. Berger (2002), Climate evolution during the Holocene: A study with an Earth system model of intermediate complexity, *Clim. Dynam., 19*, 43–60.

Crucifix, M., M. F. Loutre, and A. Berger (2005), Commentary on 'The Anthropogenic Greenhouse Era began thousands of years ago', *Climatic Change, 69*, 419–426.

Crucifix, M., M. Claussen, G. Ganssen, J. Guiot, Z. Guo, T. Kiefer, M.-F. Loutre, D.-D. Rousseau, and E. Wolff (2009), Climate change: From the geological past to the uncertain future – A symposium honouring André Berger, *Clim. Past, 5*, 707–711.

Crueger, T., E. Roeckner, T. Raddatz, R. Schnur, and P. Wetzel (2008), Ocean dynamics determine the response of oceanic CO_2 uptake to climate change, *Clim. Dynam., 31*, 151–168.

Crutzen, P. J. (2002), Geology of mankind, *Nature, 415*, 23.

Crutzen, P. J. (2006), Albedo enhancement by stratospheric sulfur injections: A contribution to resolve a policy dilemma?, *Climatic Change, 77*, 211–220, doi:10.1007/s10584-006-9101-y.

Crutzen, P. J., and E. S. Stoermer (2000), The Anthropocene, *IGBP Newsletter, 41*, 17–18.

Cruz, R. V., H. Harasawa, M. Lal, S. Wu, Y. Anokhin, B. Punsalmaa, Y. Honda, M. Jafari, C. Li, and N. Huu Ninh (2007), Asia, in *Climate Change 2007: Impacts, Adaptation and Vulnerability, contribution of Working Group II to the Fourth Assessment Report (FAR) of the Intergovernmental Panel on Climate Change*, edited by M. L. Parry et al., pp. 469–506, Cambridge University Press, Cambridge, UK.

CSIRO (2010), Climate variability and change in south-eastern Australia: A synthesis of findings from phase 1 of the South Eastern Australian Climate Initiative (SEACI).

Cubash, U., R. Voss, G. C. Hegerl, J. Waszkewitz, and T. J. Crowley (1997), Simulation of the influence of solar radiation variations on the global climate with an ocean-atmosphere general circulation model, *Clim. Dynam., 13*, 757–767.

Cuffey, K. M., and W. S. B. Paterson (2010), *The Physics of Glaciers*, 4th edition, Academic Press, Amsterdam.

Cullen, N. J., T. Mölg, G. Kaser, K. Hussein, K. Steffen, and D. R. Hardy (2006), Kilimanjaro glaciers: Recent areal extent from satellite data and new interpretation of observed 20th century retreat rates, *Geophys. Res. Lett., 33*, L16502, doi:10.1029/2006GL027084.

Cunderlik, J. M., and T. B. M. J. Ouarda (2009), Trends in the timing and magnitude of floods in Canada, *J. Hydrol., 375*, 471–480.

Cunningham, S. A., T. Kanzow, D. Rayner, M. O. Baringer, W. E. Johns, J. Marotzke, H. R. Longworth, E. M. Grant, J. J. M. Hirschi, L. M. Beal, C. S. Meinen, and H. L. Bryden (2007), Temporal variability of the Atlantic meridional overturning circulation at 26.5°N, *Science, 317*(5840), 935–938, doi:10.1126/science.1141304.

Curry, C. L. (2007), Modeling the soil consumption of atmospheric methane at the global scale, *Global Biogeochem. Cycles, 21*, GB4012, doi:10.1029/ 2006GB002818.

Curry, R. G., M. S. McCartney, and T. M. Joyce (1998), Oceanic transport of subpolar climate signals to mid-depth subtropical waters, *Nature, 391*(6667), 575–577.

Curry, R., and C. Mauritzen (2005), Dilution of the northern North Atlantic in recent decades, *Science, 308*, 1772–1774.

D'Hondt, S., M. E. Q. Pilson, H. Sigurdsson, A. K. Hanson, Jr., and S. Carey (1994), Surface-water acidification and extinction at the Cretaceous-Tertiary boundary, *Geology, 22*, 983–986.

D'Hondt, S., P. Donaghay, J. C. Zachos, D. Luttenberg, and M. Lindinger (1998), Organic carbon fluxes and ecological recovery from the Cretaceous-Tertiary mass extinction, *Science, 282*, 276–279.

D'Ippoliti, P. Michelozzi, and C. Marino (2010), The impact of heat waves on mortality in 9 European cities: Results from the EuroHEAT project, *Environ. Health-UK, 9*, 37.

d'Orgeville, M., and W. R. Peltier (2007), On the Pacific Decadal Oscillation and the Atlantic Multidecadal Oscillation: Might they be related? *Geophys. Res. Lett., 34*, L23705, doi:10.1029/2007GL031584.

da Costa, A. C. L., D. Galbraith, S. Almeida, B. T. T. Portela, M. da Costa, J. de A. Silva, A. P. Braga, P. H. L. de Gonçalves, A. A. de Oliveira, R. Fisher, O. L. Phillips, D. B. Metcalfe, P. Levy, and P. Meir (2010), Effect of 7 yr of experimental drought on vegetation dynamics and biomass storage of an eastern Amazonian rainforest, *New Phytol., 187*, 579–591.

Dai, A. (2006), Recent climatology, variability, and trends in global surface humidity, *J. Climate, 19*, 3589–3606.

Dai, A. (2010), Drought under global warming: A review, *Wiley Interdiscip. Rev.: Clim. Change, 2*(1), 45–65, doi:10.1002/wcc.81.

Dai, A., K. E. Trenberth, and T. Qian (2004), A global data set of Palmer Drought Severity Index for 1870–2002: Relationship with soil moisture and effects of surface warming, *J. Hydrometeorol., 5*, 1117–1130.

Dai, Y., R. E. Dickinson, and Y.-P. Wang (2004), A two-big-leaf model for canopy temperature, photosynthesis, and stomatal conductance, *J. Climate, 17*, 2281–2299.

Dakos, V., M. Scheffer, E. H. van Nes, V. Brovkin, V. Petoukhov, and H. Held (2008), Slowing down as an early warning signal for abrupt climate change, *Proc. Natl. Acad. Sci. USA, 105*(38), 14308–14312.

Dällenbach, A., T. Blunier, J. Flückiger, B. Stauffer, J. Chappellaz, and D. Raynaud (2000), Changes in the atmospheric CH_4 gradient between Greenland and Antarctica during the Last Glacial and the transition to the Holocene, *Geophys. Res. Lett., 27*(7), 1005–1008, doi:10.1029/1999GL010873.

Dallmeyer, A., M. Claussen, and J. Otto (2010), Contribution of oceanic and vegetation feedbacks to Holocene climate change in monsoonal Asia, *Clim. Past, 6*(2), 195–218.

Dalsøren, S. B., and I. S. A. Isaksen (2006), CTM study of changes in tropospheric hydroxyl distribution 1990–2001 and its impact on methane, *Geophys. Res. Lett., 33*(7), L23811, doi:10.1029/2006GL027295.

Dalsøren, S. B., Ø. Endresen, I. S. A. Isaksen, G. Gravir, and E. Sørgård (2007), Environmental impacts of the expected increase in sea transportation, with a particular focus on oil and gas scenarios for Norway and northwest Russia, *J. Geophys. Res., 112*, D02310, doi:10.1029/2005JD006927.

Dalsøren, S. B., M. S. Eide, Ø. Endresen, A. Mjelde, G. Gravir, and I. S. A. Isaksen (2009a), Update on emissions and environmental impacts from the international fleet of ships: The contribution from major ship types and ports, *Atmos. Chem. Phys., 9*, 2171–2194.

Dalsøren, S. B., I. S. A. Isaksen, L. Li, and A. Richter (2009b), Effect of emission changes in Southeast Asia on global hydroxyl and methane lifetime, *Tellus B, 61*(4), 588–601, doi:10.1111/j.1600-0889.2009.00429.x.5184.

Damoah, R., N. Spichtinger, R. Servranckx, M. Fromm, E. W. Eloranta, I. A. Razenkov, P. James, M. Shulski, C. Forster, and A. Stohl (2006), A case study of pyro-convection using transport model and remote sensing data, *Atmos. Chem. Phys., 6*, 173–185.

Damon, P. E., and P. Laut (2004), Pattern of strange errors plagues solar activity and terrestrial climate data, *Eos Trans. AGU, 85*(39), 370.

Daniau, A. L., S. P. Harrison, and P. J. Bartlein (2010), Fire regimes during the Last Glacial, *Quaternary Sci. Rev., 29*(21–22), 2918–2930.

Daniau, A. L., P. J. Bartlein, S. P. Harrison, I. C. Prentice, S. Brewer, P. Friedlingstein, T. I. Harrison-Prentice, J. Inoue, J. R. Marlon, S. D. Mooney, M. J. Power, J. Stevenson, W. Tinner, and M. J. A. Andrič, H. Behling, M. Black, O. Blarquez, K. J. Brown, C. Carcaillet, E. Colhoun, D. Colombaroli, B. A. S. Davis, D. D'Costa, J. Dodson, L. Dupont, Z. Eshetu, D. G. Gavin, A. Genries, T. Gebru, S. Haberle, D. J. Hallett, S. Horn, G. Hope, F. Katamura, L. Kennedy, P. Kershaw, S. Krivonogov, C. Long, D. Magri, E. Marinova, G. M. McKenzie, P. I. Moreno, P. Moss, F. H. Neumann, E. Norström, C. Paitre, D. Rius, N. Roberts, G. Robinson, N. Sasaki, L. Scott, H. Takahara, V. Terwilliger, F. Thevenon, R. B. Turner, V. G. Valsecchi, B. Vannière, M. Walsh, N. Williams, and Y. Zhang (2011), Predictability of biomass burning in response to climate changes, (Submitted).

Dansgaard, W., S. J. Johnsen, H. B. Clausen, D. Dahl-Jensen, N. Gundestrup, C. U. Hammer, and H. Oeschger (1984), North Atlantic climatic oscillations revealed by deep Greenland ice cores, in *Climatic Processes and Climate Sensitivity – Geophysical Monograph*, edited by J. E. Hansen, et al., pp. 288–298, Maurice Ewing Series 29.

Dansgaard, W., et al. (1993), Evidence for general instability of past climate from a 250-ka ice-core record, *Nature, 364*(6434), 218–220.

Daschkeit, A., and P. Mahrenholz (2010), Adaptation policy in Germany and multi-level governance, paper presented at International Climate Change Adaptation Conference, Climate Adaptation Futures: Preparing for the unavoidable impacts of climate change, National Climate Change Adaptation Research Facility and the CSIRO Climate Adaptation Flagship, 29 June –1 July, Gold Coast.

Daufresne, M., K. Lengfellner, and U. Sommer (2009), Global warming benefits the small in aquatic ecosystems, *Proc. Natl. Acad. Sci. USA, 106*, 12788–12793.

Davidson, E. A., and I. A. Janssens (2006), Temperature sensitivity of soil carbon decomposition and feedbacks to climate change, *Nature, 440*, 165–173, doi:10.1038/nature04514.

Davin, E. L., and N. de Noblet-Ducoudré (2010), Climatic impact of global-scale deforestation: Radiative versus non-radiative processes, *J. Climate, 23*, 97–112, doi:10.1175/2009JCLI3102.1.

Davin, E. L., N. de Noblet-Ducoudré, and P. Friedlingstein (2007), Impact of land-cover change on surface climate: Relevance of the radiative forcing concept, *Geophys. Res. Lett., 34*, L13702, doi:10.1029/2007GL029678.

Davis, C., W. Wang, S. S. Chen, Y. S. Chen, K. Corbosiero, M. DeMaria, J. Dudhia, G. Holland, J. Klemp, J. Michalakes, H. Reeves, R. Rotunno, C. Snyder, and Q. N. Xiao (2008), Prediction of landfalling hurricanes with the Advanced Hurricane WRF model, *Mon. Weather Rev., 136*(6), 1990–2005, doi:10.1175/2007MWR2085.1.

Davis, M. B., R. G. Shaw, and J. R. Etterson (2005) Evolutionary responses to changing climate, *Ecology, 286*, 1704–1714.

Davis, R. E. (1976), Predictability of sea surface temperature and sea level pressure anomalies over the North Pacific Ocean, *J. Phys. Oceanogr., 6*, 249–266.

Dawkins, R. (1982), *The Extended Phenotype: The Long Reach of the Gene*, Oxford University Press.

Dawson, J. P., P. J. Adams, and S. N. Pandis (2007), Sensitivity of PM2.5 to climate in the Eastern US: A modeling case study, *Atmos. Chem. Phys., 7*, 4295–4309.

Dean, W. E., and E. Gorham (1998), Magnitude and significance of carbon burial in lakes, reservoirs, and peatlands. *Geology, 26*(6), 535–538, doi: 10.1130/0091—7613(1998)026<0535:MASOCB>2.3.CO;2.

de Boyer Montegut, C., J. Mignot, A. Lazar, and S. Cravatte (2007), Control of salinity on the mixed layer depth in the world ocean: 1. General description, *J. Geophys. Res., 112*(C6), C06011.

de Bruin, K. C., R. B. Dellink and R. S. J. Tol (2009), AD-DICE: an implementation of adaptation in the DICE model, *Climatic Change, 95*(1–2), 63–81, doi:10.1007/s10584-008-9535-5.

de Jong Boers, B. (1995), Mount Tambora in 1815: A volcanic eruption in Indonesia and its aftermath, *Indonesia, 60*, 37–60.

de Noblet, N., I. C. Prentice, S. Joussaume, D. Texier, A. Botta, and A. Haxeltine (1996), Possible role of atmosphere-biosphere interactions in triggering the last glaciation. *Geophys. Res. Lett., 23*(22), 3191–3194.

de Pury D. G. G., and G. D. Farquhar (1997), Simple scaling of photosynthesis from leaves to canopies without the errors of big-leaf models, *Plant Cell Environ., 20*(5), 537–557.

De Ridder, K., and H. Gallée (1998), Land surface-induced regional climate change in Southern Israel, *J. Appl. Meteorol., 37, 1470–1485*.

de Rosnay, P., J. Polcher, K. Laval, and M. Sabre (2003), Integrated parameterization of irrigation in the land surface model ORCHIDEE, Validation over Indian Peninsula, *Geophys. Res. Lett., 30*, 1986, doi:10.1029/2003GL018024.

De Smedt, I., J.-F. Müller, T. Stavrakou, R. van der A, H. Eskes, and M. Van Roozendael (2008), Twelve years of global observation of formaldehyde in the troposphere using GOME and SCIAMACHY sensors, *Atmos. Chem. Phys., 8*, 4947–4963.

De'ath, G., J. M. Lough, and K. E. Fabricius (2009), Declining coral calcification on the Great Barrier Reef, *Science, 323*(5910), 116–119, doi:10.1126/science.1165283.

DeBeer, C. M., and M. J. Sharp (2007), Recent glacier retreat within the southern Canadian Cordillera, *Ann. Glaciol., 46*, 215–221.

Decker, E. H., S. Elliot, F. A. Smith, D. R. Blake, and F. S. Rowland (2000), Energy and material flow through the urban ecosystem, *Annu. Rev. Energ. Env., 25*, 685–740.

Deenen, M. H. L. (2010), A new chronology for the late Triassic to early Jurassic, *Geologica Utraiectina, 323*, 215.

Deenen, M. H. L., M. Ruhl, N. R. Bonis, W. Krijgsman, W. M. Kuerschner, M. Reitsma, and M. J. van Bergen (2010), A new chronology for the end-Triassic mass extinction, *Earth Planet. Sci. Lett., 291*, 113–125.

DeFries, R. S., C. B. Field, I. Fung, C. O. Justice, S. Los, P. A. Matson, E. Matthews, H. A. Mooney, C. A. Potter, K. Prentice, P. J. Sellers, J. R. G. Townshend, C. J. Tucker, S. L. Ustin, and P. M. Vitousek (1995), Mapping the land surface for global atmosphere-biosphere models: Toward continuous distributions of vegetation's functional properties, *J. Geophys. Res., 100*, 20867–20882.

Defries, R. S., L. Bounoua, and G. Collatz (2002), Human modification of the landscape and surface climate in the next fifty years, *Glob. Change Biol., 8*, 438–458.

Dekker, S. C., H. J. de Boer, V. Brovkin, K. Fraedrich, M. J. Wassen, and M. Rietkerk (2010), Biogeophysical feedbacks trigger shifts in the modelled vegetation-atmosphere system at multiple scales, *Biogeosciences, 7*, 1237–1245.

Delgado, J. M., H. Apel, and B. Merz (2009), Flood trends and variability in the Mekong River, *Hydrol. Earth Sys. Sci., 14*, 407–418.

Della-Marta, P. M., M. R. Haylock, J. Luterbacher, and H. Wanner (2007), Doubled length of Western European summer heat waves since 1880, *J. Geophys. Res.-Atmos., 112*, D15103, doi:10.1029/2007JD008510.

Delmonte, B., I. Basile-Doelsch, J.-R. Petit, V. Maggi, M. Revel-Rolland, A. Michard, E. Jagoutz, and F. Grousset (2004), Comparing the Epica and Vostok dust records during the last 220,000 years: Stratigraphical correlation and provenance in glacial periods, *Earth-Sci. Rev., 66*, 63–87.

Delpierre, N., K. Soudani, C. François, B. Köstner, J.-Y. Pontailler, E. Nikinmaa, L. Misson, M. Aubinet, C. Bernhofer, A. Granier, T. Grünwald, B. Heinesch, B. Longdoz, J.-M. Ourcival, S. Rambal, T. Vesala, and E. Dufrêne (2009), Exceptional carbon uptake in European forests during the warm spring of 2007: A data-model analysis, *Glob. Change Biol., 15*, 1455–1474.

Delworth, T. L., and R. J. Greatbatch (2000), Multidecadal thermohaline circulation variability driven by atmospheric surface flux forcing, *J. Climate, 13*, 1481–1495.

Delworth, T. L., and M. E. Mann (2000), Observed and simulated multidecadal variability in the Northern Hemisphere, *Clim. Dynam., 16*, 661–676, doi:10.1007/s003820000075.

Delworth, T. L., S. Manabe, and R. J. Stouffer (1993), Inter-decadal variations of the thermohaline circulation in a coupled ocean-atmosphere model, *J. Climate, 6*, 1993–2011.

Delworth, T. L., P. U. Clark, M. Holland, W. E. Johns, T. Kuhlbrodt, J. Lynch-Stieglitz, C. Morrill, R. Seager, A. J. Weaver, and R. Zhang (2008), The potential for abrupt change in the Atlantic Meridional Overturning Circulation, in Abrupt climate change, a report by the US Climate Change Science Program and the Subcommittee on Global Change Research, edited by P.U. Clark et al., pp. 258–359, *Synthesis and Assessment Product 3.4*, US Climate Change Research Program.

deMenocal, P., J. Ortiz, T. Guilderson, J. Adkins, M. Sarnthein, L. Baker, and M. Yarusinsky (2000), Abrupt onset and termination of the African Humid Period: Rapid climate responses to gradual insolation forcing, *Quaternary Sci. Rev, 19*(1–5), 347–361, doi:10.1006/qres.2001.2261.

DeMott, P. J., D. C. Rogers, and S. M. Kreidenweis (1997), The susceptibility of ice formation in upper tropospheric clouds to insoluble aerosol components, *J. Geophys. Res., 102*, 19575–19584.

Denman, K. L., G. Brasseur, A. Chidthaisong, P. Ciais, P. M. Cox, R. E. Dickinson, D. Hauglustaine, C. Heinze, E. Holland, D. Jacob, U. Lohmann, S. Ramachandran, P. L. da Silva Dias, S. C. Wofsy, and X. Zhang (2007), Couplings between changes in the climate system and biogeochemistry, in *Climate Change 2007: The Physical Science Basis, contribution of Working Group I to the Fourth Assessment Report (FAR) of the Intergovernmental Panel on Climate Change*, edited by S. Solomon, D. Quin, M. Manning, M. Marquis, K. Averyt, M. M. B. Tignor, H. L. Miller, and Z. Chen, Cambridge University Press, Cambridge, UK, and New York, NY, 499–587.

Dentener, F., D. Stevenson, J. Cofala, R. Mechler, M. Amann, P. Bergamaschi, F. Raes, and R. Derwent (2005), The impact of air pollutant and methane emission controls on tropospheric ozone and radiative forcing: CTM calculations for the period 1990–2030, *Atmos. Chem. Phys., 5*, 1731–1755.

Dentener, F., J. Drevet, J. F. Lamarque, I. Bey, B. Eickhout, A. M. Fiore, D. Hauglustaine, L. W. Horowitz, M. Krol, U. C. Kulshrestha, M. Lawrence, C. Galy-Lacaux, S. Rast, D. Shindell, D. Stevenson, T. Van Noije, C. Atherton, N. Bell, D. Bergman, T. Butler, J. Cofala, B. Collins, R. Doherty, K. Ellingsen, J. Galloway, M. Gauss, V. Montanaro, J. F. Müller, G. Pitari, J. Rodriguez, M. Sanderson, F. Solmon, S. Strahan, M. Schultz, K. Sudo, S. Szopa, and O. Wild (2006a), Nitrogen and sulfur deposition on regional and global scales: A multimodel evaluation, *Global Biogeochem. Cycles, 20*, GB4003, doi:10.1029/2005GB002672.

Dentener, F., D. Stevenson, K. Ellingsen, T. van Noije, M. Schultz, M. Amann, C. Atherton, N. Bell, D. Bergmann, I. Bey, L. Bouwman, T. Butler, J. Cofala, B. Collins, J. Drevet, R. Doherty, B. Eickhout, H. Eskes, A. Fiore, M. Gauss, D. Hauglustaine, L. Horowitz, I. S. A. Isaksen, B. Josse, M. Lawrence, M. Krol, J. F. Lamarque, V. Montanaro, J. F. Müller, V. H. Peuch, G. Pitari, J. Pyle, S. Rast, J. Rodriguez, M. Sanderson, N. H. Savage, D. Shindell, S. Strahan, S. Szopa, K. Sudo, R. Van Dingenen, O. Wild, and G. Zeng (2006b), Global atmospheric environment for the next generation, *Environ. Sci. Technol., 40*, 3586–3594.

Dentener, F., S. Kinne, T. Bond, O. Boucher, J. Cofala, S. Generoso, P. Ginoux, S. Gong, J. J. Hoelzemann, A. Ito, L. Marelli, J. E. Penner, J.-P. Putaud, C. Textor, M. Schulz, G. R. van der Werf, and J. Wilson (2006c), Emissions of primary aerosol and precursor gases in the years 2000 and 1750 prescribed data-sets for AeroCom, *Atmos. Chem. Phys., 6*, 4321–4344.

Deo, R. C., J. I. Syktus, C. A. McAlpine, P. J. Lawrence, H. A. McGowan, and S. R. Phinn (2009), Impact of historical land-cover change on daily indices of climate extremes including droughts in eastern Australia, *Geophys. Res. Lett., 36*, L08705, doi:10.1029/2009GL037666.

Déqué, M., and J. P. Piedlievre (1995), High resolution climate simulation over Europe, *Clim. Dynam., 11*, 321–339.

Déqué, M., and S. Somot (2010), Weighted frequency distributions express modelling uncertainties in the ENSEMBLES regional climate experiments, *Clim. Res., 44*(2–3), 195–209, doi:10.3354/cr00866.

Déqué, M., D. P. Rowell, and D. Luthi (2007), An intercomparison of regional climate simulations for Europe: Assessing uncertainties in model projections, *Climatic Change, 81*, 53–70.

Derwent, R. G., P. G. Simmonds, S. O'Doherty, D. S. Stevenson, W. J. Collins, M. G. Sanderson, C. E. Johnson, F. Dentener, J. Cofala, R. Mechler, and M. Amann (2006), External influences on Europe's air quality: Baseline methane, carbon monoxide and ozone from 1990 to 2030 at Mace Head, Ireland, *Atmos. Environ., 40*, 844–855.

Desai, A. R. (2010), Climatic and phenological controls on coherent regional interannual variability of carbon dioxide flux in a heterogeneous landscape, *J. Geophys. Res.*, *115*, G00J02, doi:10.1029/2010JG001423.

Deser, C., and M. L. Blackmon (1993), Surface climate variations over the North Atlantic Ocean during winter: 1900–1989, *J. Climate, 6*, 1743–1753.

Deser, C., and J. Wallace (1987), El-Nino events and their relation to the Southern Oscillation – 1925–1986, *J. Geophys. Res.-Oceans, 92*(C13), 14189–14196.

Deser, C., and J. M. Wallace (1990), Large-scale atmospheric circulation features of warm and cold episodes in the tropical pacific, *J. Climate, 3*, 1254–1281.

Deser, C., J. Walsh, and M. Timlin (2000), Arctic sea-ice variability in the context of recent atmospheric circulation trends, *J. Climate, 13*, 617–633.

Deser, C., A. S. Phillips, and M. A. Alexander (2010), Twentieth century tropical sea surface temperature trends revisited, *Geophys. Res. Lett., 37*(10), L10701, doi:10.1029/2010GL043321.

Dessler, A. E., and S. M. Davis (2010), Trends in tropospheric humidity from reanalysis systems, *J. Geophys. Res., 115*, D19127, doi:10.1029/2010JD014192.

Dessler, A. E., and S. Wong (2009), Estimates of the water vapour climate feedback during El Niño-Southern Oscillation, *J. Climate, 22*, 6404–6412.

Deutsch, C., S. Emerson, and L. Thompson (2006), Physical-biological interactions in North Pacific oxygen variability, *J. Geophys. Res., 111*(C9), C09S90.

Dewar, R. C. (1996), The correlation between plant growth and intercepted radiation: An interpretation in terms of optimal plant nitrogen content, *Ann. Bot., 78*, 125–136.

DFID (2010), *New climate network to link developing countries and drive adaptation and mitigation policy development*, Press Release 17 March, UK National Archives, *http://www.preventionweb.net/english/professional/news/v.php?id=13214*.

Dickens, G. R., J. R. O'Neil, D. K. Rea, and R. M. Owen (1995), Dissociation of oceanic methane hydrate as a cause of the carbon isotope excursion at the end of the Paleocene, *Paleoceanography, 10*, 965–971.

Dickinson, R. E. (1975), Solar variability and the lower atmosphere, *Bull. Am. Meteorol. Soc., 56*, 1240–1248.

Dickinson, R. E. (1987), *The Geophysiology of Amazonia: Vegetation and Climate Interactions*, John Wiley & Sons, New York.

Dickinson, R. E. (2011), *Coupled Atmospheric Circulation Models to Biophysical, Biochemical, and Biological Processes at the Land Surface*, edited by L. Donner et al., Cambridge University Press.

Dickinson, R. E. (2012), Interaction between future climate and terrestrial carbon and nitrogen, in *The Future of the World's Climate*, edited by A. Henderson-Sellers and K. McGuffie, pp. 289–308, Elsevier, Amsterdam.

Dickinson, R. E., and A. Henderson-Sellers (1988), Modelling tropical deforestation: A study of GCM land-surface parameterizations, *Q. J. Roy. Meteor. Soc., 114*(b), 439–462.

Dickinson, R. E., A. Henderson-Sellers, P. J. Kennedy, and M. F. Wilson (1986), Biosphere- Atmosphere Transfer Scheme (BATS) for the NCAR Community Climate Model, *NCAR Technical Note TN-275 + STR*.

Dickinson, R. E., A. Henderson-Sellers, C. Rosenzweig, and P. J. Sellers (1991), Evapotranspiration models with canopy resistance for use in climate models: A review, *Agr. Forest Meteorol., 54*, 373–388.

Dickinson, R. E., et al. (2002), Nitrogen controls on climate model evapotranspiration, *J. Climate, 15*, 278–295.

Dickson, A. G., and F. J. Millero (1987), A Comparison of the Equilibrium-Constants for the Dissociation of Carbonic-Acid in Seawater Media. *Deep-Sea Research Part a-Oceanographic Research Papers, 34*, 1733–1743.

Dieleman, W. I. J., et al. (2010), Soil [N] modulates soil C cycling in CO_2-fumigated tree stands: A meta-analysis, *Plant Cell Environ., 33*, 2001–2011, doi:10.1111/j.1365-3040.2010.02201.x.

Diffenbaugh, N. S., J. S. Pal, F. Giorgi, and X. Gao, (2007), Heat stress intensification in the Mediterranean climate change hotspot, *Geophys. Res. Lett., 34*, L11706.

Diggelmann, H., G. H. Hadorn, R. Kaufmann-Hayoz, J. R. Randegger, and S. Smoliner (2001), The Swiss Transdisciplinarity Award, in *Transdisciplinarity: Joint Problem Solving among Science, Technology and Society*, edited by T. K. Klein, W. Grossenbacher-Mansuy, R. Haberli, A. Bill, R. W. Scholz, and M. Welti, M. Birkhauser Verlag, Basel, Switzerland.

Dinar, E., A. Abo Riziq, C. Spindler, C. Erlick, G. Kiss, and Y. Rudich (2008), The complex refractive index of atmospheric and model humic-like substances (HULIS) retrieved by a cavity ring down aerosol spectrometer (CRD-AS), *Faraday Discuss., 137*, 279–295.

DiNezio, P. N., A. C. Clement, G. A. Vecchi, B. J. Soden, B. P. Kirtman, and S.-K. Lee (2009), Climate response of the equatorial Pacific to global warming, *J. Climate, 22*(18), 4873–4892.

Ding, T., W. Qian, and Z. Yan (2010), Changes in hot days and heat waves in China during 1961–2007, *Int. J. Climatol, 30*(10), 1452–1462.

Ditlevsen, P. D., and S. J. Johnsen (2010), Tipping points: Early warning and wishful thinking, *Geophys. Res. Lett.*, doi:10.1029/2010GL044486.

Ditlevsen, P. D., M. S. Kristensen, and K. K. Andersen (2005), The recurrence time of Dansgaard–Oeschger events and limits on the possible periodic component, *J. Climate, 18*(14), 2594–2603.

Dodman, D. (2009), Blaming cities for climate change? An analysis of urban greenhouse gas emissions inventories, *Environ. Urban., 21*(1), 185–201, doi:10.1177/0956247809103016.

Doherty, S. J, S. Bojinski, A. Henderson-Sellers, K. Noone, D. Goodrich, N. L. Bindoff, J. Church, K. A. Hibbard, T. R. Karl, L. Kajfez-Bogataj, A. H. Lynch, P. J. Mason, D. E. Parker, C. Prentice, V. Ramaswamy, R. W. Saunders, A. J. Simmons, M. Stafford Smith, K. Steffen, T. F. Stocker, P. W. Thorne, K. Trenberth, M. M. Verstraete, and F. W. Zwiers (2009), Lessons learned from IPCC: Developments needed to understand and predict climate change for adaptation, *Bull. Am. Meteorol. Soc., 90*, 497–513, doi:10.1175/2008BAMS2643.1.

Domingues, C. M., J. A. Church, N. J. White, P. J. Gleckler, S. E. Wijffels, P. M. Barker, and J. R. Dunn (2008), Improved estimates of upper-ocean warming and multi-decadal sea-level rise, *Nature, 453*(7198), 1090–1093.

Dommenget, D. (2009), The ocean's role in continental climate variability and change, *J. Climate, 22*(18), 4939–4952.

Dommenget, D., and M. Latif (2008), Generation of hyper climate mode, *Geophys. Res. Lett., 35*, L02706, doi:10.1029/2007GL031087.

Donders, T. H., F. Wagner, D. L. Dilcher, and H. Visscher (2005), Mid-to-late-Holocene El Niño–Southern Oscillation dynamics reflected in the subtropical terrestrial realm, *Proc. Natl. Acad. Sci. USA, 102*(31), 10904–10908.

Doney, S. C., V. J. Fabry, R. A. Feely, and J. A. Kleypas (2009), Ocean acidification: The other CO_2 problem, *Annu. Rev. Marine Sci., 1*, 169–192.

Dorrepaal, E., S. Toet, R. S. P. van Logstestijn, E. Swart, M. J. van de Weg, T. V. Callaghan, and R. Aerts (2009), Carbon respiration from subsurface peat accelerated by climate warming in the subarctic, *Nature, 460,* 616–620.

Doswell, C. A., H. E. Brooks, and M. P. Kay (2005), Climatological estimates of daily local non-tornadic severe thunderstorm probability for the United States, *Weather Forecast., 20,* 577–595.

Douglas, E. M., A. Beltrán-Przekurat, D. Niyogi, R. A. Pielke, Sr., and C. J. Vörösmarty (2009), The impact of agricultural intensification and irrigation on land–atmosphere interactions and Indian monsoon precipitation – A mesoscale modeling perspective, *Glob. Planet. Change, 67,* 117–128, doi:10.1016/j.gloplacha.2008.12.007.

Doutriaux-Boucher, M., M. J. Webb, J. M. Gregory, and O. Boucher (2009), Carbon dioxide induced stomatal closure increases radiative forcing via a rapid reduction in low cloud, *Geophys. Res. Lett., 36,* L02703, doi:10.1029/2008GL036273.

Dovers, S. (2005), *Environment and Sustainability Policy: Creation, Implementation and Evaluation,* Federation Press, Sydney.

Downes, S. M., N. L. Bindoff, and S. R. Rintoul (2010), Changes in the subduction of Southern Ocean water masses at the end of the 21st century in eight IPCC models, *J. Climate, 23*(24), 6526–6541.

Drijfhout, S., S. Weber, and E. van der Swaluw (2010), The stability of the MOC as diagnosed from model projections for pre-industrial, present and future climates, *Clim. Dynam.,* 1–12.

Drinkwater, K. F. (2006), The regime shift of the 1920s and 1930s in the North Atlantic, *Progr. Oceanogr., 68,* 134–151.

Drinkwater, K. F., G. Beaugrand, M. Kaeriyama, S. Kim, G. Ottersen, R. I. Perry, H.-O. Pörtner, J. J. Polovina, and A. Takasuka (2010), On the processes linking climate to ecosystem changes, *J. Marine Syst., 79*(3–4), 374–388, doi:10.1016/j.jmarsys.2008.12.014.

Drought Policy Review Expert Social Panel (2008), It's about people: Changing perspectives on dryness, a report to government by an expert social panel, *http://www.daff.gov.au/agriculture-food/drought/national_review_of_drought_policy/social_assessment/dryness-report.* (Accessed 21/10/10.)

Drummond, J. R., and G. S. Mand (1996), The measurements of pollution in the troposphere (MOPITT) instrument: Overall performance and calibration requirements, *J. Atmos. Oceanic Technol., 13*(2), 314–320.

Ducharne, A., K. Laval, and J. Polcher (1998), Sensitivity of the hydrological cycle to the parameterization of soil hydrology in a GCM, *Clim. Dynam., 14,* 307–327.

Dudhia, J. (1989), Numerical study of convection observed during the winter monsoon experiment using a mesoscale two-dimensional model, *J. Atmos. Sci., 46*(20), 3077–3107.

Dudhia, J. (1993), A nonhydrostatic version of the Penn State–NCAR mesoscale model: Validation tests and simulation of an Atlantic cyclone and cold front, *Mon. Weather Rev., 121,* 1493–1513.

Dufek, A. S., and T. Ambrizzi (2008), Precipitation variability in Sao Paulo State, Brazil, *Theor. Appl. Climatol., 93*(3–4), 167–178.

Duncan, B. N., and I. Bey (2004), A modeling study of the export pathways of pollution from Europe: Seasonal and interannual variations (1987–1997), *J. Geophys. Res., 109,* D08301, doi:10.1029/2003JD004079.

Dunlap, R. E. (2008), The New Environmental Paradigm scale: From marginality to worldwide use, *J. Environ. Educ., 40*(1), 3–18.

Dunlap, R. E., and K. D. Van Liere (1978), The new environmental paradigm, *J. Environ. Educ., 9,* 10–19.

Dunlap, R. E., K. D. Van Liere, A. Mertig, and R. E. Jones (2000), Measuring endorsement of the New Ecological Paradigm: A revised NEP scale, *J. Soc. Issues, 56,* 425–442.

Dupont, L. M., S. Jahns, F. Marret, and S. Ning (2000), Vegetation change in equatorial West Africa: Time-slices for the last 150 ka, *Palaeogeogr. Palaeoclimatol. Palaeoecol., 155*(1–2), 95–122, doi: 10.1016/S0031-0182(99)00095-4.

Durack, P. J., and S. E. Wijffels (2010), Fifty-year trends in global ocean salinities and their relationship to broad-scale warming, *J. Climate, 23*(16), 4342–4362.

Durante, F., and G. Salvadori (2010), On the construction of multivariate extreme value models via copulas, *Environmetrics, 21*(2), 143–161.

Durre, I., J. M. Wallace, and D. P. Lettenmaier (2000), Dependence of extreme daily maximum temperatures on antecedent soil moisture in the contiguous United States during summer, *J. Climate, 13*(14), 2641–2651.

Dutton, E. G., D. W. Nelson, R. S. Stone, D. Longenecker, G. Carbaugh, J. M. Harris, and J. Wendell (2006), Decadal variations in surface solar irradiance as observed in a globally remote network, *J. Geophys. Res., 111,* doi:10.1029/2005JD006901.

Dwyer, E., S. Pinnock, J. M. Gregoire, and J. M. C. Pereira (2000), Global spatial and temporal distribution of vegetation fire as determined from satellite observations, *Int. J. Remote Sens., 21*(6–7), 1289–1302.

Dyer, G. (2008), Climate Wars, Random House Canada.

Dyson, F. W., A. S. Eddington, and C. R. Davidson (1920), A determination of the deflection of light by the sun's gravitational field, from observations made at the solar eclipse of May 29, 1919, *Phil. Trans. Roy. Soc. A, 220*(571–581), 291–333, doi:10.1098/rsta.1920.0009.

Dyurgerov, M. B. (2010), Reanalysis of glacier changes: From the IGY to the IPY, 1960–2008, *Materialy Glyatsiologicheskikh Issledovanij, 108,* 5–116.

Dyurgerov, M. B., and M. F. Meier (2005), Glaciers and the changing Earth system: A 2004 snapshot, *Occasional Paper 58,* Institute of Arctic and Alpine Research, University of Colorado, Boulder, CO.

Easterling, D. R., G. A. Meehl, C. Parmesan, S. A. Changnon, T. R. Karl, and L. O. Mearns (2000), Climate extremes: Observations, modeling, and impacts, *Science, 289,* 2068–2074.

Easterling, D. R., L. V. Alexander, A. Mokssit, and V. Detemmerman (2003), CCl/CLIVAR workshop to develop priority climate indices, *Bull. Am. Meteorol. Soc., 84,* 1403–1407.

Easterling, D. R., D. M. Anderson, S. J. Cohen, W. J. Gutowski, G. J. Holland, K. E. Kunkel, T. C. Peterson, R. S. Pulwarty, R. J. Stouffer, and M. F. Wehner (2008), Measures to improve our understanding of weather and climate extremes, in *Weather and Climate Extremes in a Changing Climate. Regions of focus: North America, Hawaii, Caribbean, and US Pacific Islands,* edited by T. R. Karl et al., pp. 117–126, A Report by the U.S. Climate Change Science Program and the Subcommittee on Global Change Research, Washington, DC.

ECA (Economics of Climate Adaptation) (2009), *Shaping Climate Resilient Development: A Framework for Decision-Making,* ECA Working Group, London.

Eckert, C., and M. Latif (1997), Predictability limits of ENSO: The role of stochastic forcing, *J. Climate, 10,* 1488–1504.

Eckhardt, S., A. Stohl, S. Beirle, N. Spichtinger, P. James, C. Forster, C. Junker, T. Wagner, U. Platt, and S. G. Jennings (2003), The North Atlantic oscillation controls air pollution transport to the Arctic, *Atmos. Chem. Phys., 3,* 1769–1778.

Economist, The (2010), The clouds of unknowing: The science of climate change, 24 March, *http://www.economist.com/displaystory.cfm?story_id=15719298.* (Accessed 26/7/10.)

Eden, C., and R. J. Greatbatch (2003), A damped decadal oscillation in the North Atlantic Ocean climate system, *J. Climate, 16*, 4043−4060.

Eden, C., and T. Jung (2001), North Atlantic inter-decadal variability: Oceanic response to the North Atlantic Oscillation (1865−1997), *J. Climate, 14*, 676−691.

Eden, C., and J. Willebrand (2001), Mechanism of inter-annual to decadal variability of the North Atlantic Circulation, *J. Climate, 14*, 2266−2280.

Editorial Board of Climatic Change (2010), Stephen H. Schneider 1945−2010, *Climatic Change, 102*, 1−7, doi: 10.1007/s10584-010-9934-2.

Edwards, M., and A. J. Richardson (2004), Impact of climate change on marine pelagic phenology and trophic mismatch, *Nature, 430*(7002), 881−884, doi:10.1038/nature02808.

Edwards, M. E., P. M. Anderson, L. B. Brubaker, T. A. Ager, A. A. Andreev, N. H. Bigelow, L. C. Cwynar, W. R. Eisner, S. P. Harrison, F. S. Hu, D. Jolly, A. V. Lozhkin, G. M. MacDonald, C. J. Mock, J. C. Ritchie, A. V. Sher, R. W. Spear, J. W. Williams, and G. Yu (2000), Pollen-based biomes for Beringia 18,000, 6000 and 0 C-14 yr B.P, *J. Biogeogr., 27*(3), 521−554, doi:10.1046/j.1365−2699.2000.00426.x.

Ehleringer, J. R., T. E. Cerling, and B. R. Helliker (1997), C_4 photosynthesis, atmospheric CO_2, and climate, *Oceologia, 112*, 285−299.

Ehrlich, P. R. (2010), Stephen Schneider (1945−2010) retrospective, *Science, 329*(13), 776.

Eichelmann, H., et al. (2005), Adjustment of leaf photosynthesis to shade in a natural canopy: reallocation of nitrogen, *Plant Cell Environ., 28*, 389−401.

Einstein, A. (1915), Die Feldgleichungen der Gravitation (The Field Equations of Gravitation), *Königlich Preussische Akademie der Wissenschaften*, 844−847.

Eisenman, I., and J. S. Wettlaufer (2009), Nonlinear threshold behavior during the loss of Arctic sea ice, *Proc. Natl. Acad. Sci. USA, 106*(1), 28−32.

Ek, M., and L. Mahrt (1994), Daytime evolution of relative-humidity at the boundary-layer top, *Mon. Weather Rev., 122*(12), 2709−2721.

Elsner, J. B., J. P. Kossin, and T. H. Jagger (2008), The increasing intensity of the strongest tropical cyclones, *Nature, 455*(7209), 92−95.

Emanuel, K. (2005), Increasing destructiveness of tropical cyclones over the past 30 years, *Nature, 436*, 686−688, doi:10.1038/nature03906.

Emanuel, K. (2007), Comment on "Sea surface temperatures and tropical cyclones in the Atlantic basin" by Patrick J. Michaels, Paul C. Knappenberger, and Robert E. Davis. *Geophys. Res. Lett., 34*, doi:10.1029/2006GL026942.

Emanuel, K., R. Sundararajan, and J. Williams (2008), Hurricanes and global warming: Results from down-scaling IPCC AR4 simulations, *Bull. Am. Meteorol. Soc., 89*, 347−367, doi:10.1175/BAMS-89-3-347.

Emanuel, K., K. Oouchi, M. Satoh, T. Hirofumi, and Y. Yamada (2010), Comparison of explicitly simulated and downscaled tropical cyclone activity in a high-resolution global climate model, *J. Adv. Model. Earth Syst., 2*, doi:10.3894/JAMES.2010.2.9.

Emanuel, K. A. (1988), The maximum intensity of tropical cyclones, *J. Atmos. Sci., 45*, 1143−1155.

EMEP (2004), *EMEP Assessment Part I European Perspective*, edited by G. Lövblad, L. Tarrasón, K. Tørseth, and S. Dutchak, Oslo.

EMEP (2006), Transboundary acidification, eutrophication and ground level ozone in Europe since 1990 to 2004, *EMEP Status Report 1/06* to support the Review of Gothenburg Protocol.

Emmerson, K. M., N. Carslaw, D. C. Carslaw, J. D. Lee, G. McFiggans, W. J. Bloss, T. Gravestock, D. E. Heard, J. Hopkins, T. Ingham, M. J. Pilling, S. C. Smith, M. Jacob, and P. S. Monks (2007), Free radical modelling studies during the UK TORCH Campaign in Summer 2003, *Atmos. Chem. Phys., 7*, 167−181.

Endresen, Ø., E. Sørgård, J. K. Sundet, S. B. Dalsøren, I. S. A. Isaksen, T. F. Berglen, and G. Gravir (2003), Emission from international sea transportation and environmental impact, *J. Geophys. Res., 108*(D17), 4560, doi:10.1029/2002JD002898.

Engelbrecht, F. A., J. L. McGregor, and C. J. Engelbrecht (2009), Dynamics of the Conformal-Cubic Atmospheric Model projected climate-change signal over southern Africa, *Int. J. Climatol.*, 1013−1033, doi:10.1002/joc., 29.

Environics Research Group (2010), National climate change adaptation benchmark survey, prepared for Natural Resources Canada, Ottawa, *http://epe.lac-bac.gc.ca/100/200/301/pwgsc-tpsgc/por-ef/natural_resources/2010/075-08/index.html*. (Accessed 9/9/10.)

EPA (2003), Environmental Protection Agency: National Air Quality and Emissions Trends Report, 2003 Special Studies Edition, *http://www.epa.gov/air/airtrends/aqtrnd03*.

EPICA (2004), Eight glacial cycles from an Antarctic ice core, *Nature, 429*(6992), 623−628, doi:10.1038/nature02599.

EPICA Community Members, C. Barbante, J. M. Barnola, S. Becagli, J. Beer, M. Bigler, C. Boutron, T. Blunier, E. Castellano, and O. Cattani (2006), One-to-one coupling of glacial climate variability in Greenland and Antarctica, *Nature, 444*(7116), 195−198, doi:10.1038/nature05301.

Ereaut G., and N. Segnit (2006), *Warm Words: How Are We Telling the Climate Story and Can We Tell It Better?*, Institute for Public Policy Research, UK.

Erlykin, A. D., T. Sloan, and A. W. Wolfendale (2009), Solar activity and the mean global temperature, *Environ. Res. Lett., 4*, doi:10.1088/1748-9326/4/1/014006,014006.

Esser, G., J. Kattge, and A. Sakalli (2011), Feedback of carbon and nitrogen cycles enhances carbon sequestration in the terrestrial biosphere, *Glob. Change Bio., 17*, 819−842, doi:10.1111/j.1365-2486.2010.02261.x.

Estoque, M. (1961), A theoretical investigation of the sea breeze, *Q. J. Roy. Meteor. Soc., 87*(372), 136−146.

Ettema, J., M. R. van den Broeke, E. van Meijgaard, W. J. van de Berg, J. L. Bamber, J. E. Box, and R. C. Bales (2009), Higher surface mass balance of the Greenland Ice Sheet revealed by high-resolution climate modelling, *Geophys. Res. Lett., 36*, L12501, doi:10.1029/2009GL038110.

EUR-Lex (2009), Report from the Commission to the Council, the European Parliament, the European Economic and Social Committee and the Committee of the Regions on the application and effectiveness of the Directive on Strategic Environmental Assessment (Directive 2001/42/EC), *http://eur-lex.europa.eu/LexUriServ/LexUriServ.do?uri=CELEX:52009DC0469:EN:NOT*. (Accessed 1/11/10.)

Evans, J. P. (2008), Changes in water vapour transport and the production of precipitation in the Eastern Fertile Crescent as a result of global warming, *J. Hydrometeorol., 9*(6), 1390−1401, doi:10.1175/2008JHM998.1.

Evans, J. P. (2009), 21st century climate change in the Middle East, *Climatic Change, 92*(3−4), 417−432.

Evans, J. P. (2010), Global warming impact on the dominant precipitation processes in the Middle East, *Theor. Appl. Climatol., 99*(3−4), 389−402.

Evans, J. P., and M. F. McCabe (2010), Regional climate simulation over Australia's Murray—Darling basin: A multitemporal assessment, *J. Geophys. Res., 115*, D14114, doi:10.1029/2010JD013816.

Evans, J. R., and H. Poorter (2001), Photosynthetic acclimation of plants to growth irradiance: The relative importance of specific leaf area and nitrogen partitioning in maximizing carbon gain, *Plant Cell Environ., 24*, 755—767.

Evans, J. P., and S. Schreider (2002), Hydrological impacts of climate change on inflows to Perth, Australia, *Climatic Change, 55*(3), 361—393.

Evans, J. P., and R. B. Smith (2006), Water vapor transport and the production of precipitation in the Eastern fertile crescent, *J. Hydrometeorol., 7*(6), 1295—1307.

Evans, J. P., R. B. Smith, and R. J. Oglesby (2004), Middle East climate simulation and dominant precipitation processes, *Int. J. Climatol., 24*(13), 1671—1694.

Evans, J. P., R. J. Oglesby, and W. M. Lapenta (2005), Time series analysis of regional climate model performance, *J. Geophys. Res., 110*(4), 1—23.

Evans, J. P., J. L. McGregor, and K. McGuffie (2012), Future regional climates, in *The Future of the World's Climate*, edited by A. Henderson-Sellers, and K. McGuffie, pp. 489—507, Elsevier, Amsterdam.

Eyring, V., N. R. P. Harris, M. Rex, T. G. Shepherd, D. W. Fahey, G. T. Amanatidis, J. Austin, M. P. Chipperfield, M. Dameris, P. M. F. Forster, A. Gettelman, H. F. Graf, T. Nagashima, P. A. Newman, S. Pawson, M. J. Prather, J. A. Pyle, R. J. Salawitch, B. D. Santer, and D. W. Waugh (2005a), A strategy for process-oriented validation of coupled chemistry-climate models, *Bull. Am. Meteorol. Soc., 86*, 1117—1133.

Eyring, V., H. W. Köhler, J. van Aardenne, and A. Lauer (2005b), Emissions from international shipping: 1. The last 50 years, *J. Geophys. Res., 110*, D17305, doi:10.1029/2004JD005619.

Eyring, V., N. Butchart, D. W. Waugh, H. Akiyoshi, J. Austin, S. Bekki, G. E. Bodeker, B. A. Boville, C. Brühl, M. P. Chipperfield, E. Cordero, M. Dameris, M. Deushi, V. E. Fioletov, S. M. Frith, R. R. Garcia, A. Gettelman, M. A. Giorgetta, V. Grewe, L. Jourdain, D. E. Kinnison, E. Mancini, E. Manzini, M. Marchand, D. R. Marsh, T. Nagashima, P. A. Newman, J. E. Nielsen, S. Pawson, G. Pitari, D. A. Plummer, E. Rozanov, M. Schraner, T. G. Shepherd, K. Shibata, R. S. Stolarski, H. Struthers, W. Tian, and M. Yoshiki (2006), Assessment of temperature, trace species and ozone in chemistry-climate model simulations of the recent past, *J. Geophys. Res., 111*, D22308, doi:10.1029/2006JD007327.

Eyring, V., D. S. Stevenson, A. Lauer, F. J. Dentener, T. Butler, W. J. Collins, K. Ellingsen, M. Gauss, D. A. Hauglustaine, I. S. A. Isaksen, M. G. Lawrence, A. Richter, J. M. Rodriguez, M. Sanderson, S. E. Strahan, K. Sudo, S. Szopa, T. P. C. van Noije, and O. Wild (2007a), Multi-model simulations of the impact of international shipping on atmospheric chemistry and climate in 2000 and 2030, *Atmos. Chem. Phys., 7*, 757—780.

Eyring, V., D. W. Waugh, G. E. Bodeker, E. Cordero, H. Akiyoshi, J. Austin, S. R. Beagley, B. Boville, P. Braesicke, C. Brühl, N. Butchart, M. P. Chipperfield, M. Dameris, R. Deckert, M. Deushi, S. M. Frith, R. R. Garcia, A. Gettelman, M. Giorgetta, D. E. Kinnison, E. Mancini, E. Manzini, D. R. Marsh, S. Matthes, T/Nagashima, P. A. Newman, J. E. Nielsen, S. Pawson, G. Pitari, D. A. Plummer, E. Rozanov, M. Schraner, J. F. Scinocca, K. Semeniuk, T. G. Shepherd, K. Shibata, B. Steil, R. Stolarski,

W. Tian, and M. Yoshiki (2007b), Multimodel projections of stratospheric ozone in the 21st century, *J. Geophys. Res., 112*, D16303, doi:10.1029/2006JD008332.

Eyring, V., I. S. A. Isaksen, T. Berntsen, W. J. Collins, J. J. Corbett, Ø. Endresen, R. G. Grainger, J. Moldanova, H. Schlager, and D. S. Stevenson (2010), Transport impacts on atmosphere and climate: Shipping, *Atmos. Environ., 44*, 4735—4771, doi:10.1016/j.atmosenv.2009.04.059.

Fabry, V. J., B. A. Seibel, R. A. Feely, and J. C. Orr (2008), Impacts of ocean acidification on marine fauna and ecosystem processes, *ICES J. Mar. Sci.: J. Conseil, 65*(3), 414—432, doi:10.1093/icesjms/fsn048.

Falkowski, P. G., R. T. Barber, and V. Smetacek (1998), Biogeochemical controls and feedbacks on ocean primary production, *Science, 281*(5374), 200—206.

Falkowski, P., R. J. Scholes, E. Boyle, J. G. Canadell, D. Canfield, J. Elser, N. Gruber, K. Hibbard, P. Hoegberg, S. Linder, F. T. Mackenzie, B. Moore, III, T. Pedersen, Y. Rosenthal, S. Seitzinger, B. Smetacek, and W. Steffen (2000), The global carbon cycle: A test of our knowledge of Earth as a system, *Science, 290*, 291—296.

Falkowski, P. G., M. E. Katz, A. J. Milligan, K. Fennel, B. S. Cramer, M. P. Aubry, R. A. Berner, M. J. Novacek, and W. M. Zapol (2005), The rise of oxygen over the past 205 million years and the evolution of large placental mammals, *Science, 309*, 2202—2204.

Farquhar, G. D. (1989), Models of integrated photosynthesis of cells and leaves, *Philos. T. Roy. Soc. B, 323*(1216), 357—386.

Farquhar, G. D., S. von Caemmerer, and J. A. Berry (1980), A biochemical model of photosynthetic CO_2 assimilation in leaves of C_3 species, *Planta, 149*, 78—90.

Farrera, I., S. P. Harrison, I. C. Prentice, G. Ramstein, J. Guiot, P. J. Bartlein, R. Bonnefille, M. Bush, W. Cramer, U. von Grafenstein, K. Holmgren, H. Hooghiemstra, G. Hope, D. Jolly, S. E. Lauritzen, Y. Ono, S. Pinot, M. Stute, and G. Yu (1999), Tropical climates at the Last Glacial Maximum: A new synthesis of terrestrial palaeoclimate data: 1. Vegetation, lake levels and geochemistry, *Clim. Dynam., 15*(11), 823—856, doi:10.1007/s003820050317.

Fastovsky, D. E., and P. M. Sheehan (2005), The extinction of the dinosaurs in North America, *GSA Today, 15*, 4—10.

Fearnside, P. M. (2011), Global warming: How much of a threat to tropical forests?, in *Survival and Sustainability, Environmental Earth Sciences*, edited by H. Gökçekus, U. Türker, and J. W. La Moreaux, pp. 1283—1292, Springer-Verlag, Berlin, doi:10.1007/978-3-540-95991-5_120.

Feddema, J. J., K. W. Oleson, G. B. Bonan, L. O. Mearns, L. E. Buja, G. A. Meehl, and W. M. Washington (2005), The importance of land-cover change in simulating future climates, *Science, 310*, 1674—1678, doi:10.1126/science.1118160.

Feddes, R. A., H. Hoff, M. Bruen, T. Dawson, P. de Rosnay, P. Dirmeyer, R. B. Jackson, P. Kabat, A. Kleidon, A. Lilly, and A. J. Pitman (2001), Modeling root water uptake in hydrological and climate models, *Bull. Am. Meteorol. Soc.*, 2797—2809.

Fedorov, A. V., and S. G. Philander (2000), Is El Niño changing?, *Science, 288*(5473), 1997—2002, doi:10.1126/science.288.5473.1997.

Fedorov, A. V., R. C. Pacanowski, S. G. Philander, and G. Boccaletti (2004), The effect of salinity on the wind-driven circulation and the thermal structure of the upper ocean, *J. Phys. Oceanogr., 34*, 1949—1966.

Fedorov, A. V., P. S. Dekens, M. McCarthy, A. C. Ravelo, P. B. deMenocal, M. Barreiro, R. C. Pacanowski, and S. G. Philander (2006),

The Pliocene paradox (mechanisms for a permanent El Niño), *Science, 312,* 1485–1489.

Feeley, K., S. J. Wright, M. N. N. Supardi, A. R. Kassim, and S. J. Davies (2007), Decelerating growth in tropical forest trees, *Ecol. Lett., 10,* 461–469.

Feely, R. A., R. Wanninkhof, T. Takahashi, and P. Tans (1999), Influence of El Niño on the equatorial Pacific contribution to atmospheric CO_2 accumulation, *Nature, 398*(6728), 597–601, doi:10.1038/19273.

Feely, R. A., J. Boutin, C. E. Cosca, Y. Dandonneau, J. Etcheto, H. Y. Inoue, M. Ishii, C. L. Quéré, D. J. Mackey, M. McPhaden, N. Metzl, A. Poisson, and R. Wanninkhof (2002), Seasonal and interannual variability of CO_2 in the equatorial Pacific, *Deep-Sea Res. Pt. II, 49*(13–14), 2443–2469, doi:10.1016/S0967-0645(02)00044-9.

Feely, R. A., T. Takahashi, R. Wanninkhof, M. J. McPhaden, C. E. Cosca, S. C. Sutherland, and M.-E. Carr (2006), Decadal variability of the air-sea CO_2 fluxes in the equatorial Pacific Ocean, *J. Geophys. Res., 111*(C8), C08S90.

Feingold, G., W. R. Cotton, S. Kreidenweis, and J. T. Davis (1999), The impact of giant cloud condensation nuclei on drizzle formation in stratocumulus: Implications for cloud radiative properties, *J. Atmos. Sci., 56,* 4100–4117.

Feng, Y. (2009), K-Model-A continuous model of soil organic carbon dynamics: Theory, *Soil Sci., 174,* 482–493, doi:10.1097/SS.obo13e3181bb0f80.

Ferek, R. J., et al. (2000), Drizzle suppression in ship tracks, *J. Atmos. Sci., 57*(16), 2707–2728.

Fettweis, X., J.-P. van Ypersele, H. Gallée, H. Lefebre, and W. Lefebvre (2007), The 1979–2005 Greenland ice melt extent from passive microwave data using an improved version of the melt retrieval XPGR algorithm, *Geophys. Res. Lett., 34,* L05502.

Field, C. (1983), Allocation leaf nitrogen for the maximization of carbon gain: Leaf age as a control on the allocation program, *Oecologia, 56,* 341–347.

Field C., R. Jackson, and H. Mooney (1995), Stomatal responses to increased CO_2: Implications from the plant to the global scale, *Plant Cell Environ., 18,* 1214–1255, doi:10.1111/j.1365-3040.1995.tb00630.x.

Field, C. B., M. J. Behrenfeld, J. T. Randerson, and P. Falkowski (1998), Primary production of the biosphere: Integrating terrestrial and oceanic components, *Science, 281*(5374), 237–240.

Findell, K. L., and E. A. B. Eltahir (2003), Atmospheric controls on soil moisture–boundary layer interactions, Part II: Feedbacks within the continental United States, *J. Hydrometeorol., 4,* 570–583.

Findell, K. L., T. R. Knutson, and P. C. D. Milly (2006), Weak simulated extratropical responses to complete tropical deforestation, *J. Climate, 19,* 2835–2850.

Findell, K. L., E. Shevliakova, P. C. D. Milly, and R. J. Stouffer (2007), Modeled impact of anthropogenic land-cover change on climate, *J. Climate, 20,* 3621–3634, doi:10.1175/JCLI4185.1.

Findell, K. L., A. J. Pitman, M. H. England, and P. Pegion (2009), Regional and global impacts of land-cover change and sea surface temperature anomalies, *J. Climate, 22,* 3248–3269, doi:10.1175/2008JCLI2580.1.

Finnigan, J. J. (2000), Turbulence in plant canopies, *Annu. Rev. Fluid Mech., 32,* 519–571.

Finnigan, J. J., M. R. Raupach, and H. A. Cleugh (1994), The impact of vegetation on the physical environment of cities, in *A Vision for a Greener City: The Role of Vegetation in Urban Environments, Proceedings,* edited by M. A. Scheltema, 1994 Greening Australia

Conference, Fremantle, Western Australia, 4–6 October 1994, pp. 23–37, Greening Australia Ltd., Canberra.

Finzi, A. C., et al. (2006), Progressive nitrogen limitation of ecosystem processes under elevated CO_2 in a warm-temperate forest, *Ecology, 87*(1), 15–25.

Fiore, A. M., et al. (2009), Multimodel estimates of intercontinental source-receptor relationships for ozone pollution, *J. Geophys. Res., 114,* D04301, doi:10.1029/2008JD010816.

Fischer, E. M., S. I. Seneviratne, D. Luthi, and C. Schar (2007), Contribution of land-atmosphere coupling to recent European summer heat waves, *Geophys. Res. Lett., 34,* L06707.

Fischer, E., S. Seneviratne, P. Vidale, D. Lüthi, and C. Schär (2007), Soil moisture-atmosphere interactions during the 2003 European summer heatwave, *J. Climate, 20*(20), 5081–5099.

Fisher, J. B., S. Y. Malhi, R. A. Fisher, C. Huntingford, and S.-Y. Tan (2010), Carbon cost of plant nitrogen acquisition: A mechanistic, globally applicable model of plant nitrogen uptake, retranslocation, and fixation, *Global Biogeochem. Cycles, 24,* GB1014, doi:10.1029/2009GB003621.

Fishman, J., and V. G. Brackett (1997), The climatological distribution of tropospheric ozone derived from satellite measurements using version 7 Total Ozone Mapping Spectrometer and Stratospheric Aerosol and Gas Experiment data sets, *J. Geophys. Res., 102,* 19275–19278.

Fitzjarrald, D. R., and R. K. Sakai (2010), LBA-ECO CD-03 Flux-Meteorological Data, km 77 Pasture Site, Para, Brazil: 2000–2005, data set, available online from Oak Ridge National Laboratory Distributed Active Archive Center, Oak Ridge, Tennessee, *http://daac.ornl.gov,* doi:10.3334/ORNLDAAC/962.

Flanner, M. G., C. S. Zender, J. T. Randerson, and P. J. Rasch (2007), Present-day climate forcing and response from black carbon in snow, *J. Geophys. Res., 112,* D11202, doi:10.1029/2006JD008003.

Fletcher, N. H. (1962), *The Physics of Rainclouds,* Cambridge University Press, Cambridge, UK.

Fletcher, S. E. M., N. Gruber, A. R. Jacobson, M. Gloor, S. C. Doney, S. Dutkiewicz, M. Gerber, F. Follows, F. Joos, and K. Lindsay (2007), Inverse estimates of the oceanic sources and sinks of natural CO_2 and the implied oceanic carbon transport *Global Biogeochem. Cycles, 21,* GB1010, doi:10.1029/2006GB002751.

Fletcher, W. J., M. F. Sánchez Goñi, J. R. M. Allen, R. Cheddadi, N. Combourieu-Nebout, B. Huntley, I. Lawson, L. Londeix, D. Magri, V. Margari, U. C. Müller, F. Naughton, E. Novenko, K. Roucoux, and P. C. Tzedakis (2010), Millennial-scale variability during the last glacial in vegetation records from Europe, *Quaternary Sci. Rev., 29*(21–22), 2839–2864, doi:10.1016/j.quascirev.2010.06.031.

Flossmann, A. I. (1998), Interaction of aerosol particles and clouds, *J. Atmos. Sci., 55,* 879–887.

Flückiger, J., E. Monnin, B. Stauffer, J. Schwander, T. F. Stocker, J. Chappellaz, D. Raynaud, and J.-M. Barnola (2002), High-resolution Holocene N_2O ice core record and its relationship with CH_4 and CO_2, *Global Biogeochem. Cycles, 16*(1), 1010, doi:10.1029/2001GB001417.

Foley, A. (2010), Uncertainty in regional climate modelling: A review, *Prog. Phys. Geogr., 34*(5), 647–670, doi:10.1177/0309133310375654.

Foley, J. A., J. E. Kutzbach, M. T. Coe, and S. Levis (1994), Feedbacks between climate and boreal forests during the Holocene Epoch, *Nature, 371*(6492), 52–54, doi:10.1038/371052a0.

Folkins, I., and R. Chatfield (2000), Impact of acetone on ozone production and OH in the upper troposphere at high NO_x, *J. Geophys. Res., 105*(D9), 11585–11600.

Folland, C. K., T. N. Palmer, and D. E. Parker (1986), Sahel rainfall and worldwide sea temperatures, 1901−1985, *Nature, 320*, 602−607.

Folland, C. K., T. R. Karl, N. Nicholls, B. S. Nyenzi, D. E. Parker, and K. Y. Vinnikov (1992), Observed climate variability and change, in *Climate change 1992*, edited by J. T. Houghton et al., Cambridge University Press, Cambridge, UK.

Folland, C. K., T. R. Karl, J. R. Christy, R. A. Clarke, G. V. Gruza, J. Jouzel, M. E. Mann, J. Oerlemans, M. J. Salinger, and S. W. Wang (2001), Observed climate variability and change, in *Climate Change 2001: The Scientific Basis, contribution of Working Group I to the Third Assessment Report (TAR) of the Intergovernmental Panel on Climate Change (IPCC)*, Cambridge University Press, Cambridge, UK, and New York, NY.

Forest, C. E., P. H. Stone, and A. P. Sokolov (2008), Constraining climate model parameters from observed 20th century changes, *Tellus A, 60*, 911−920.

Forster, P. M. D., and K. P. Shine (1997), Radiative forcing and temperature trends from stratospheric ozone changes, *J. Geophys. Res., 102*, 10841−10857.

Forster, P. M. D., V. Ramaswamy, P. Artaxo, T. Berntsen, R. Betts, D. W. Fahey, J. Haywood, J. Lean, D. C. Lowe, G. Myhre, J. Nganga., R. Prinn, G. Raga, M. Schulz, R. van Dorland, G. Bodeker, O. Boucher, W. D. Collins, T. J. Conway, E. Dlugokencky, J. W. Elkins, D. Etheridge, P. Foukal, P. Fraser, M. Geller, F. Joos, C. D. Keeling, S. Kinne, K. Lassey, U. Lohmann, A. C. Manning, S. Montzka, D. Oram, K. O'Shaughnessy, S. Piper, G.-K. Plattner, M. Ponater, N. Ramankutty, G. Reid, D. Rind, K. Rosenlof, R. Sausen, D. Schwarzkopf, S. K. Solanki, G. Stenchikov, N. Stuber, T. Takemura, C. Textor, R. Wang, R. Weiss, and T. Whorf (2007), Changes in atmospheric constituents and in radiative forcing, in *Climate Change 2007: The Physical Science Basis. Contribution of Working Group I to the Fourth Assessment Report of the Intergovernmental Panel on Climate Change*, edited by S. Solomon, D. Qin, M. Manning, Z. Chen, M. Marquis, K. B. Avery, M. Tignor, and H. L. Miller, Cambridge University Press, Cambridge, UK.

Fowell, S. J., and P. E. Olsen (1993), Time calibration of Triassic/Jurassic microfloral turnover, eastern North America, *Tectonophysics, 222*, 361−369.

Fowler, H. J., and M. Ekstrom (2009), Multi-model ensemble estimates of climate change impacts on UK seasonal precipitation extremes, *Int. J. Climatol., 29*, 385−416.

Fowler, H. J., M. Ekstrom, S. Blenkinsop, and A. P. Smith (2007), Estimating change in extreme European precipitation using a multi-model ensemble, *J. Geophys. Res.-Atmos., 112*, D18104, doi:10.1029/2007JD008619.

Fowler, H., S. Blenkinsop, and C. Tebaldi (2007a), Linking climate change modelling to impacts studies: Recent advances in downscaling techniques for hydrological modelling, *Int. J. Climatol., 27*(12), 1547−1578.

Fowler, H., M. Ekström, S. Blenkinsop, and A. Smith (2007b), Estimating change in extreme European precipitation using a multimodel ensemble, *J. Geophys. Res., 112*(18).

Fowler, D., K. Pilegaard, M. A. Sutton, P. Ambus, M. Raivonen, J. Duyzer, D. Simpson, H. Fagerli, J. K. Schjoerring, A. Neftel, J. Burkhardt, U. Daemmgen, J. Neirynck, E. Personne, R. Wichink-Kruit, K. Butterbach-Bahl, C. Flechard, J. P. Tuovinen, M. Coyle, S. Fuzzi, G. Gerosa, C. Granier, B. Loubet, N. Altimir, L. Gruenhage, C. Ammann, S. Cieslik, E. Paoletti, T. N. Mikkelsen, H. Ro-Poulsen, P. Cellier, J. N. Cape, I. S. A. Isaksen, L. Horvath, F. Loreto, U. Niinemets, P. I. Palmer, J. Rinne, P. Laj, M. Maione, P. Misztal, P. Monks, E. Nemitz, D. Nilsson, S. Pryor, M. W. Gallagher, T. Vesala, U. Skiba, N. Brueggemann, S. Zechmeister-Boltenstern, J. Williams, C. O'Dowd, M. C. Facchini, G. de Leeuw, A. Flossman, N. Chaumerliac, and J. W. Erisman (2009), Atmospheric composition change: Ecosystems atmosphere interactions, *Atmos. Environ., 43*, 5138−5192.

Fox-Rabinovitz, M. S., L. L. Takacs, R. C. Govindaraju, and M. J. Suarez, (2001), A variable-resolution stretched-grid general circulation model: Regional climate simulation, *Mon. Weather Rev., 129*, 453−469.

Fox-Rabinovitz, M. S., J. Côté, B. Dugas, M. Déqué, and J. L. McGregor (2006), Variable resolution general circulation models: Stretched-grid model intercomparison project (SGMIP), *J. Geophys. Res., 111*, D16104, doi:10.1029/2005JD006520.

Frank, D. C., et al. (2010), Ensemble reconstruction constraints on the global carbon cycle sensitivity to climate, *Nature, 463*, 527−530, doi:10.1038/nature8769.

Frank, W. (1977), Structure and energetics of tropical cyclone, 1. Storm structure, *Mon. Weather Rev., 105*(9), 1119−1135.

Frankenberg, C., J. F. Meirink, M. van Weele, U. Platt, and T. Wagner (2005), Assessing methane emissions from global space-borne observations, *Science, 308*(5724), 1010−1014.

Frankenberg, C., J. F. Meirink, P. Bergamaschi, A. P. H. Goede, M. Heimann, S. Körner, U. Platt, M. van Weele, and T. Wagner (2006), Satellite chartography of atmospheric methane from SCIAMACHY on board ENVISAT: Analysis of the years 2003 and 2004, *J. Geophys. Res., 111*, D07303, doi:10.1029/2005JD006235.

Frankenberg, C., P. Bergamaschi, A. Butz, S. Houweling, J. F. Meirink, J. Notholt, A. K. Petersen, H. Schrijver, T. Warneke, and I. Aben (2008), Tropical methane emissions: A revised view from SCIA-MACHY onboard ENVISAT, *Geophys. Res. Lett. 35*, L15811, doi:10.1029/2008GL034300.

Frankignoul, C., and K. Hasselmann (1977), Stochastic climate models, part II: Application to sea surface temperature anomalies and thermocline variability, *Tellus, 29*, 284−305.

Frankignoul, C., and R. W. Reynolds (1983), Testing a dynamical model for midlatitude sea surface temperature anomalies, *J. Phys. Oceanogr., 13*, 1131−1145.

Frankignoul, C., P. Müller, and E. Zorita (1997), A simple model of the decadal response of the ocean to stochastic wind forcing, *J. Phys. Oceanogr., 27*, 1533−1546.

Franklin, C., G. J. Holland, and P. May (2005), Sensitivity of tropical cyclone rainbands to ice-phase microphysics, *Mon. Weather Rev., 133*, 2473−2493.

Franklin, O. (2007), Optimal nitrogen allocation controls tree responses to elevated CO_2, *New Phytol., 174*, 811−822, doi:10.1111/j.1469-8137.2007.02063.x.

Frederiksen, J. S., and C. S. Frederiksen (2007), Interdecadal changes in Southern Hemisphere winter storm track modes, *Tellus A, 59*, 599−617.

Frei, E., J. Bodin, and G. R. Walther (2010), Plant species' range shifts in mountainous areas—all uphill from here?, *Bot. Helv.*, 1−12.

Frich, P., L. Alexander, P. Della-Marta, B. Gleason, M. Haylock, A. Klein Tank, and T. Peterson (2002), Observed coherent changes in climatic extremes during the second half of the twentieth century, *Clim. Res., 19*(3), 193−212.

Friederichs, P. (2010), Statistical downscaling of extreme precipitation events using extreme value theory, *Extremes, 13*(2), 109−132, doi:10.1007/s10687-010-0107-5.

Friedlingstein, P. (2008), A steep road to climate stabilization, *Nature, 451*(7176), 297–298, doi:10.1038/nature06593.

Friedlingstein, P., and I. C. Prentice (2010), Carbon-climate feedbacks: A review of model and observation based estimates, *Curr. Opin. Environ. Sustain., 2*(4), 251–257, doi:10.1016/j.cosust.2010.06.002.

Friedlingstein, P., P. Cox, R. Betts, L. Bopp, W. von Bloh, V. Brovkin, P. Cadule, S. Doney, M. Eby, I. Fung, G. Bala, J. John, C. Jones, F. Joos, T. Kato, M. Kawamiya, W. Knorr, K. Lindsay, H. D. Matthews, T. Raddatz, P. Rayner, C. Reick, E. Roeckner, K.-G. Schnitzler, R. Schnur, K. Strassmann, A. J. Weaver, C. Yoshikawa, and N. Zeng (2006), Climate–carbon cycle feedback analysis: Results from the C4MIP model intercomparison, *J. Climate, 19*(14), 3337–3353.

Friedlingstein, P., R. A. Houghton, G. Marland, J. Hackler, T. J. Conway, J. G. Canadell, M. R. Raupach, P. Ciais, and C. Le Quéré (2010), Update on CO₂ emissions, *Nat. Geosci., 3*, 811–812, doi:10.1038/ngeo1022.

Friend, A. D. (2001), Modelling canopy CO₂ fluxes: are 'big-leaf' simplifications justified?, *Global Ecol. Biogeogr., 10*, 603–619.

Friend, A. D., et al. (2007), FLUXNET and modeling the global carbon cycle, *Glob. Change Biol., 13*, 610–633, doi:10.1111/j.1365-2486.2006.01223.x.

Friend, A. D., R. J. Geider, M. J. Behrenfeld, and C. J. Still (2009), Photosynthesis in global-scale models, in *Photosynthesis in silico: Understanding Complexity from Molecules to Ecosystems*, edited by A. Laisk et al., pp. 465–497, Springer.

Friis-Christensen, E., and K. Lassen (1991), Length of the solar cycle: An indicator of solar activity closely associated with climate, *Science, 254*, 698–700.

Frolking, S., N. Roulet, and J. Fuglestvedt (2006), How northern peat-lands influence the Earth's radiative budget: Sustained methane emission versus sustained carbon sequestration, *J. Geophys. Res., 111*, G01008, doi:10.1029/2005JD006588.

Fu, C., S. Wang, Z. Xiong, W. J. Gutowski, Jr., D.-K. Lee, J. L. McGregor, Y. Sato, H. Kato, J.-W. Kim, and M.-S. Suh (2005), Regional climate model intercomparison project for Asia, *Bull. Am. Meteorol. Soc., 86*, 257–266.

Fu, L. L., E. Christensen, C. Yamarone, Jr., M. Lefebvre, Y. Menard, M. Dorrer, and P. Escudier (1994), TOPEX/POSEIDON mission overview, *J. Geophys. Res., 99*(12) 24369–24381.

Fu, T. M., D. J. Jacob, P. I. Palmer, K. Chance, Y. X. X. Wang, B. Barletta, D. R. Blake, J. C. Stanton, and M. J. Pilling (2007), Space-based formaldehyde measurements as constraints on volatile organic compound emissions in east and south Asia and implications for ozone, *J. Geophys. Res.-Atmos., 112*(D6), doi:10.1029/2006JD007853.

Fudeyasu, H., Y. Wang, M. Satoh, T. Nasuno, H. Miura, and W. Yanase (2008), The global cloud-system-resolving model NICAM success-fully simulated the lifecycles of two real tropical cyclones, *Geophy. Res. Lett., 35*, doi:10.1029/2008GL036003.

Fuglestvedt, J. S., T. Berntsen, O. Godal, R. Sausen, K. P. Shine, and T. Skodvin (2003), Metrics of climate change: Assessing radiative forcing and emission indices, *Climatic Change, 58*(3), 267–331.

Fuglestvedt, J. S., T. K. Berntsen, G. Myhre, K. Rypdal, and R. B. Skeie (2008), Climate forcing from the transport sectors, *Proc. Natl. Acad. Sci. USA, 105*(2), 454–458.

Fuglestvedt, J. S., T. K. Berntsen, V. Eyring, I. S. A. Isaksen, D. S. Lee, and R. Sausen (2009), Shipping emissions: From cooling to warming of climate-and reducing impacts on health, how should shipping emissions' ability to cool or warm the climate be addressed?, *Environ. Sci. Tech., 43*, 9057–9062.

Fuglestvedt, J. S., K. P. Shine, J. Cook, T. Berntsen, D. S. Lee, A. Stenke, R. B. Skeie, G. J. M. Velders, and I. A. Waitz (2010), Transport impacts on atmosphere and climate: Metrics, *Atmos. Environ., 44*(37), 4648–4677, doi:10.1016/j.atmosenv.2009.04.044.

Fujita, K., A. Sakai T. Nuimura, S. Yamaguchi, and R. R. Sharma (2009), Recent changes in Imja Glacial Lake and its damming moraine in the Nepal Himalaya revealed by in situ surveys and multi-temporal ASTER imagery, *Environ. Res. Lett., 4*(4), 045205.

Fuller, K. A., W. C. Malm, and S. M. Kreidenweis (1999), Effects of mixing on extinction by carbonaceous particles, *J. Geophys. Res., 104*(D13), 15941–15954.

Fuller, S. (2002), *Knowledge Management Foundations*, Butterworth-Heinemann, Boston, MA.

Fung, I. P., E. Matthews, J. Lerner, and G. Russell (1983), Three-dimensional tracer model study of atmospheric CO₂: Response to seasonal exchanges with the terrestrial biosphere, *J. Geophys. Res., 88*, 1281–1294.

Furrer, E. M., and R. W. Katz (2008), Improving the simulation of extreme precipitation events by stochastic weather generators, *Water Resour. Res., 44*(12), W12439.

Furrer, E. M., R. W. Katz, M. D. Walter, and R. Furrer (2010), Statistical modeling of hot spells and heat waves, *Clim. Res., 43*(3), 191–205.

Fusco, A. C., and J. A. Logan (2003), Analysis of 1970–1995 trends in tropospheric ozone at Northern Hemisphere midlatitudes with the GEOS-CHEM model, *J. Geophys. Res., 108*, 4449, doi:10.1029/2002JD002742.

Fuzzi, S., et al. (2006), Critical assessment of the current state of scientific knowledge, terminology, and research needs concerning the role of organic aerosols in the atmosphere, climate, and global change, *Atmos. Chem. Phys., 6*(7), 2017–2038.

Fyfe, J. C., and O. A. Saenko (2006), Simulated changes in the extra-tropical Southern Hemisphere winds and currents, *Geophys. Res. Lett., 33*, L06701, doi:10.1029/2005GL025332.

Fyfe, J. C., G. J. Boer, and G. M. Flato (1999), The Arctic and Antarctic Oscillations and their projected changes under global warming, *Geophys. Res. Lett., 26*(11), 1601–1604.

Galbraith, D., et al. (2010), Multiple mechanisms of Amazonian forest biomass losses in three dynamic global vegetation models under climate change, *New Phytol., 187*, 647–665, doi:10.1111/j.1469-8137.2010.03350.x.

Galeotti, S., and R. Coccioni (2002), Changes in coiling direction of *Cibcidoides pseudoacutus* (Nakkady) across the Cretaceous-Tertiary boundary of Tunisia: Palaecological and biostratigraphic implica-tions, *Palaeogeogr. Palaeocl., 178*, 197–210.

Galeotti, S., H. Brinkhuis, and M. Huber (2004), Records of post-Cretaceous-Tertiary boundary millennial-scale cooling from the western Tethys: A smoking gun for the impact-winter hypothesis?, *Geology, 32*, 529–532.

Gallée, H., J. P. van Ypersele, T. Fichefet, C. Tricot, and A. Berger (1991), Simulation of the last glacial cycle by a coupled, sectorially averaged climate-ice sheet model, I. The climate model, *J. Geophys. Res., 96*(D7), 13,139–13,161.

Galli, M. T., F. Jadoul, S. M. Bernasconi, S. Cirilli, and H. Weissert (2007), Stratigraphy and palaeoenvironmental analysis of the Triassic-Jurassic transition in the western Southern Alps (Northern Italy), *Palaeogeogr. Palaeocl., 244*, 52–70.

Gallo, K. P., T. W. Owen, D. R. Easterling, and P. F. Jamason (1999), Temperature trend of the U.S. historical climatology network based on satellite-designated land use/land cover, *J. Climate, 12*, 1344−1348.

Galloway, J. N., F. J. Dentener, D. G. Capone, E. W. Boyer, R. W. Howarth, S. P. Seitzinger, G. P. Asner, C. Cleveland, P. Green, E. Holland, D. M. Karl, A. F. Michaels, J. H. Porter, A. Townsend, and C. Vörösmarty (2004), Nitrogen cycles: Past, present and future, *Biogeochemistry, 70*(2), 153−226.

Ganachaud, A., A. Sen Gupta, J. C. Orr, S. E. Wiffels, K. Ridgway, M. A. Herner, C. Maes, C. Steinberg, A. Tribollet, B. Qiu, and J. Kruger (2011), Observed and expected changes to the tropical Pacific Ocean, in *Vulnerability of Fisheries and Aquaculture in the Tropical Pacific to Climate Change,* edited by J. D. Bell, J. E. Johnson, and A. J. Hobday, Secretariat of the Pacific Community, Noumea, New Caledonia.

Ganopolski, A., and R. Calov (2010), Simulation of glacial cycles with an Earth system model, in *Climate Change at the Eve of the Second Decade of the Century. Inferences from Paleoclimates and Regional aspects,* edited by A. Berger, F. Mesinger, and D. Sijacki, Milankovitch 130th anniversary symposium, Belgrade, 2009.

Ganopolski, A., and S. Rahmstorf (2001), Rapid changes of glacial climate simulated in a coupled climate model, *Nature, 409*(6817), 153−158, doi:10.1038/35051500.

Ganopolski, A., C. Kubatzki, M. Claussen, V. Brovkin, and V. Petoukhov (1998), The influence of vegetation-atmosphere-ocean interaction on climate during the mid-Holocene, *Science, 280*(5371), 1916−1919, doi:10.1126/science.280.5371.1916.

Gao X. J., Z. C. Zhao, and F. Giorgi (2002), Changes of extreme events in regional climate simulate over East Asia, *Adv. Atmos. Sci., 19*(5), 927−942.

Gao X. J., Y. Luo, W. T. Lin, Z. C. Zhao, and F. Giorgi (2003), Simulation of effects of land use change on climate in China by a regional climate model, *Adv. Atmos. Sci., 20*(4), 583−592.

Gao, X. J., J. S. Pal, and F. Giorgi (2006), Projected changes in mean and extreme precipitation over the Mediterranean region from a high resolution double nested RCM simulation, *Geophys. Res. Lett., 33*(3), L03706.

Garnaut, R. (2008), *The Garnaut Climate Change Review,* Australian National University, Canberra.

GARP (1975), The physical basis of climate and climate modelling, *GARP Publication Series 16,* ICSU/WMO, Geneva.

Garrett, T. J., and C. Zhao (2006), Increased Arctic cloud longwave emissivity associated with pollution from mid-latitudes, *Nature, 440*, 787−789.

Gates, W. L. (1976), The numerical simulation of Ice-Age climate with a global general circulation model, *J. Atmos. Sci., 33*(10), 1844−1873.

Gattuso, J. P., M. Frankignoulle, I. Bourge, S. Romaine, and R. W. Buddemeier (1998), Effect of calcium carbonate saturation of seawater on coral calcification, *Glob. Planet. Change, 18*(1−2), 37−46.

Gauss, M., G. Myhre, I. S. A. Isaksen, V. Grewe, G. Pitari, O. Wild, W. J. Collins, F. J. Dentener, K. Ellingsen, L. K. Gohar, D. A. Hauglustaine, D. Iachetti, J.-F. Lamarque, E. Mancini, L. J. Mickley, M. J. Prather, J. A. Pyle, M. G. Sanderson, K. P. Shine, D. S. Stevenson, K. Sudo, S. Szopa, and G. Zeng (2006), Radiative forcing since preindustrial times due to ozone change in the troposphere and the lower stratosphere, *Atmos. Chem. Phys., 6*, 575−599.

Gazal, R., M. A. White, R. Gillies, E. Rodemakers, E. Sparrow, and L. Gordon (2008), GLOBE students, teachers, and scientists demonstrate variable differences between urban and rural leaf phenology, *Glob. Change Biol., 14*, 1568−1580, doi:10.1111/j.1365-2486.2008.01602.x.

Gedney, N., and P. J. Valdes (2000), The effect of Amazonian deforestation on the Northern Hemisphere circulation and climate, *Geophys. Res. Lett., 27*, 12753−12758.

Gedney, N., P. M. Cox, and C. Huntingford (2004), Climate feedback from wetland methane emissions, *Geophys. Res. Lett., 31*, L20503, doi:10.1029/2004GL020919.

Gedney, N., P. M. Cox, R. A. Betts, O. Boucher, C. Huntingford, and P. A. Stott (2006), Detection of a direct carbon dioxide effect in continental river runoff records. *Nature, 439*, 835−838.

Gelencser, A., B. May, D. Simpson, A. Sanchez-Ochoa, A. Kasper-Giebl, H. Puxbaum, A. Caseiro, C. Pio, and M. Legrand (2007), Source apportionment of PM2.5 organic aerosol over Europe: Primary/ secondary, natural/anthropogenic, and fossil/biogenic origin, *J. Geophys. Res., 112*(D23), D23S04.

Geng, Q., and M. Sugi (2001), Variability of the North Atlantic cyclone activity in winter analyzed from NCEP-NCAR reanalysis data, *J. Climate, 14*, 3863−3873.

Gentry, M., and G. Lackmann (2010), Sensitivity of simulated tropical cyclone structure and intensity to horizontal resolution, *Mon. Weather Rev., 138*(3), 688−704, doi:10.1175/2009MWR2976.1.

Gerasopoulos, E., G. Kouvarakis, M. Vrekoussis, M. Kanakidou, and N. Mihalopoulos (2005), Ozone variability in the marine boundary layer of the eastern Mediterranean based on 7-year observations, *J. Geophys. Res., 110*, D15309, doi:10.1029/2005JD005991.

Gerber, S., L. O. Hedin, M. Oppenheimer, S. W. Pacala, and E. Shevliakova (2010), Nitrogen cycling and feedback in a global dynamic land model, *Global Biogeochem. Cycles, 24*, GB1001, doi:10.1029/2008GB003336.

Gero, A., and A. J. Pitman (2006), The impact of land-cover change on a simulated storm event in the Sydney Basin, *J. Appl. Meteorol., 45*, 283−300.

Gershunov, A., and T. Barnett (1998), Inter-decadal modulation of ENSO teleconnections, *Bull. Am. Meteorol. Soc., 79*, 2715−2725.

Gersonde, R., X. Crosta, A. Abelmann, and L. Armand (2005), Sea-surface temperature and sea ice distribution of the southern ocean at the EPILOG Last Glacial Maximum − A circum-Antarctic view based on siliceous microfossil records, *Quaternary Sci. Rev., 24*, 869−896, doi:10.1016/j.quascirev.2004.07.015.

Gettelman, A., and Q. Fu (2008), Observed and simulated upper-tropospheric water vapor feedback, *J. Climate, 21*, 3282−3289.

Ghan, S. J., L. R. Leung, R. C. Easter, and H. Abdul-Razzak (1997), Prediction of droplet number in a general circulation model, *J. Geophys. Res., 102*, 21777−21794.

Ghan, S. J., R. Easter, E. Chapman, H. Abdul-Razzak,Y. Zhang, L. R. Leung, N. Laulainen, R. D. Saylor, and R. Zaveri (2001), A physically-based estimate of radiative forcing by anthropogenic sulfate aerosol, *J. Geophys. Res., 106*(D6), 5279−5294, doi:10.1029/ 2000JD900503.

GHG Protocol (The Greenhouse Gas Protocol) (2011), *The Greenhouse Gas Protocol Initiative,* World Resources Institute and World Business Council for Sustainable Development, *http://www. ghgprotocol.org/about-ghgp.* (Accessed 29/8/11.).

Ghosh, S. (2010), SVM-PGSL coupled approach for statistical downscaling to predict rainfall from GCM output, *J. Geophys. Res. − Atmos., 115*, doi:10.1029/2009JD013548.

Giacopelli, P., K. Ford, C. Espada, and P. B. Shepson (2005), Comparison of the measured and simulated isoprene nitrate distributions above a forest canopy, *J. Geophys. Res., 110*, D01304, doi:10.1029/2004JD005123.

Giannini, A., R. Saravanan, and P. Chang (2003), Oceanic forcing of Sahel rainfall on inter-annual to inter-decadal timescales, *Science, 302*, 1027−1030.

Gifford, R. (2008), Psychology's essential role in alleviating the impacts of climate change, *Can. Psychol., 49*(4), 273−280.

Giglio, L., J. T. Randerson, G. R. Van Der Werf, P. S. Kasibhatla, G. J. Collatz, D. C. Morton, and R. S. Defries (2010), Assessing variability and long-term trends in burned area by merging multiple satellite fire products, *Biogeosciences, 7*(3), 1171−1186.

Gilgen, H., M. Wild, and A. Ohmura (1998), Means and trends of shortwave irradiance at the surface estimated from GEBA, *J. Climate, 11*, 2042−2061.

Gille, S. T. (2008), Decadal-scale temperature trends in the Southern Hemisphere ocean, *J. Climate, 21*(18), 4749, doi:10.1175/2008JCLI2131.1.

Gillett, N. P., and D. W. J. Thompson (2003), Simulation of recent Southern Hemisphere climate change, *Science, 302*(5643), 273−275.

Gillett, N. P., H. F. Graf, and T. J. Osborn (2003), Climate change and the North Atlantic Oscillation, *Geoph. Monog. Series, 134*, 193−210.

Gillett, N. P., A. J. Weaver, F. W. Zwiers, and M. D. Flannigan (2004), Detecting the effect of climate change on Canadian forest fires, *Geophys. Res. Lett., 31*, L18211, doi:10.1029/2004GL020876.

Gillett, N. P., T. D. Kell, and P. D. Jones (2006), Regional climate impacts of the Southern Annular Mode, *Geophys. Res. Lett., 33*, 4, doi:200610.1029/2006GL027721.

Gillet, N. P., P. A. Stott, and B. D. Santer (2008), Attribution of cyclogenesis region sea surface temperature change to anthropogenic influence, *Geophys. Res. Lett., 35*(9), L09707.

Giorgi, F. (1990), Simulation of regional climate using a limited area model nested in a general circulation model, *J. Climate, 3*, 941−963.

Giorgi, F., and L. O. Mearns (1991), Approaches to the simulation of regional climate change: A review, *Rev. Geophys., 29*, 191−216.

Giorgi, F., and L. O. Mearns (1999), Introduction to special section: Regional climate modelling revisited, *J. Geophys. Res.-Atmos., 104*(D6), 6335−6352.

Giorgi, F., and L. O. Mearns, (2002), Calculation of average, uncertainty range and reliability of regional climate changes from AOGCM simulations via the 'reliability ensemble averaging' (REA) method, *J. Climate, 15*(10), 1141−1158.

Girard, E., J.-P. Blanchet, and Y. Dubois (2004), Effects of arctic sulphuric acid aerosols on wintertime low-level atmospheric ice crystals, humidity and temperature at Alert, Nunavut, *Atmos. Res., 73*, 131−148.

Giorgi, F., C. Jones, and G. R. Asrar (2009), Addressing climate information needs at the regional level: The CORDEX framework, *WMO Bulletin, 58*(3), 175−183.

Gladwell, M. (2000), *How Little Things Can Make a Big Difference*, Little Brown, Boston.

Glahn, H. R., and D. A. Lowry (1972), The use of model output statistics (MOS) in objective weather forecasting, *J. Appl. Meteorol., 11*, 1203−1211.

Gleckler, P. J., K. E. Taylor, and C. Doutriaux (2008), Performance metrics for climate models, *J. Geophys. Res., 113*, D06104, doi:10.1029/2007JD008972.

Glickman, T. (Ed.) (2000), *Glossary of Meteorology*, Second Edition, American Meteorological Society, Boston, MA.

Global Carbon Project (2010), 10 Years of Advancing Knowledge on the Global Carbon Cycle and its Management, *http://www.global carbonproject.org/global/pdf/GCP_10years_high_res.pdf.* (Accessed 12/02/11.)

Gohm, A., G. Zangl, and G. Mayr (2004), South foehn in the Wipp Valley on 24 October 1999 (MAP IOP 10): Verification of high-resolution numerical simulations with observations, *Mon. Weather Rev., 132*(1), 78−102.

Goldenberg, S. B., C. W. Landsea, A. M. Mestas-Nuñez, and W. M. Gray (2001), The recent increase in Atlantic hurricane activity: Causes and implications, *Science, 293*, 474−479.

Goldin, T. J., and H. J. Melosh (2009), Self-shielding of thermal radiation by Chicxulub impact ejecta: Firestorm of fizzle?, *Geology, 37*, 1135−1138.

Goody, R. M. (1952). A statistical model for water-vapour absorption, *Q. J. Roy. Meteor. Soc., 78*, 165−169.

Goody, R. M. (2002), Observing and thinking about the atmosphere, *Annu. Rev. Energ. Env., 27*, 1−20, doi:10.1146/annurev.energy.27.122001.083412.

Goosse, H., and T. Fichefet (1999), Importance of ice-ocean interactions for the global ocean circulation: A model study, *J. Geophys. Res., 104*(C10), 23,337−23,355.

Goosse, H., V. Brovkin, T. Fichefet, R. Haarsma, J. Jongma, A. Huybrechts, A. Mouchet, F. Selten, P.-Y. Barriat, J.-M. Campin, E. Deleersnijder, E. Driesschaert, H. Goelzer, I. Janssens, M.-F. Loutre, M. A. Morales Maqueda, T. Opsteegh, P.-P. Mathieu, G. Munhoven, E. Petterson, H. Renssen, D. M. Roche, M. Schaeffer, C. Severijns, B. Tartinville, A. Timmermann, and N. Weber (2010), Description of the Earth system model of intermediate complexity LOVECLIM version 1.2, *Geoscientific Model Development, 3*, 603−633.

Görgen, K., A. H. Lynch, A. G. Marshall, and J. Beringer (2006), Impact of abrupt land cover changes by savanna fire on northern Australian climate, *J. Geophys. Res., 111*, D19106, doi:10.1029/2005JD006860.

Gorham, E., C. Lehman, A. Dyke, J. Janssens, and L. Dyke (2007), Temporal and spatial aspects of peatland initiation following deglaciation in North America, *Quaternary Sci. Rev., 26*(3−4), 300−311, doi:10.1016/j.quascirev.2006.08.008.

Goswami, B. N., M. S. Madhusoodanan, C. P. Neema, and D. Sengupta (2006), A physical mechanism for North Atlantic SST influence on the Indian summer monsoon, *Geophys. Res. Lett., 33*(2), L02706.

Goto-Azuma, K., and R. M. Koerner (2001), Ice core studies of anthropogenic sulfate and nitrate trends in the Arctic, *J. Geophys. Res., 106*(D5), 4959−4969.

Govindasamy, B., P. B. Duffy, and K. Caldeira (2001), Land use changes and northern hemisphere cooling, *Geophys. Res. Lett., 28*, 291−294.

Goward, N., and D. L. Williams (1997), Landsat and Earth systems science: Development of terrestrial monitoring, *Photogramm. Eng. Rem. S., 6*(7), 887−900.

Gradstein, F. M., J. G. Ogg, and A. G. Smith (2004), *A Geologic Timescale*, Cambridge University Press.

Granier, C., U. Niemeier, J. F. Müller, J. Olivier, A. Richter, H. Nuess, and J. Burrows (2003), Variation of the atmospheric composition over the 1990−2000 period, *POET Report 6*, EU project EVK2-1999-00011.

Granier, C., U. Niemeier, J. H. Jungclaus, L. Emmons, P. Hess, J.-F. Lamarque, S. Walters, and G. P. Brasseur (2006), Ozone pollution from future ship traffic in the Arctic northern passages, *Geophys. Res. Lett. 33*, L13807, doi:10.1029/2006GL026180.

Grassl, H. (1975), Albedo reduction and radiative heating of clouds by absorbing aerosol particles, *Contrib. Atmos. Phys.*, *48*, 199–210.

Greene, A., L. Goddard, and U. Lall (2006), Probabilistic multimodel regional temperature change projections, *J. Climate, 19*(17), 4326–4343.

Greene, M. T. (2006), Looking for a general for some modern major models, *Endeavour, 30*(2), 55–59.

Gregory, J. M., and P. M. Forster (2008), Transient climate response estimated from radiative forcing and observed temperature change, *J. Geophys. Res., 113*, D23105, doi:10.1029/2008JD010405.

Gregory, J., and M. Webb (2008), Tropospheric adjustment induces a cloud component to CO_2 forcing, *J. Climate, 21*, 58–71.

Gregory, J. M., W. J. Ingram, M. A. Palmer, G. S. Jones, P. A. Stott, R. B. Thorpe, J. A. Lowe, T. C. Johns, and K. D. Williams (2004), A new method for diagnosing radiative forcing and climate sensitivity, *Geophys. Res. Lett., 31*, L03205, doi:10.1029/2003GL018747.

Gregory, J. M., C. D. Jones, P. Cadule, and P. Friedlingstein (2009), Quantifying carbon cycle feedbacks, *J. Climate, 22*, 5232–5250.

Grell, G. A., J. Dudhia, and D. R. Stauffer (1994), A description of the fifth-generation Penn State/NCAR mesoscale model (MM5), *NCAR Tech. Note NCAR/TN-398STR*, 117.

Greve, R., and H. Blatter (2009), *Dynamics of Ice Sheets and Glaciers*, Springer, Dordrecht.

Griffies, S. M., and E. Tziperman (1995), A linear thermohaline oscillator driven by stochastic atmospheric forcing, *J. Climate, 8*, 2440–2453.

Griffis, K., and D. J. Chapman (1988), Survival of phytoplankton under prolonged darkness: Implications for the Cretaceous-Tertiary boundary darkness hypothesis, *Palaeogeogr. Palaeocl., 67*, 305–314.

Griffiths, A., N. Haigh, and J. Rassias (2007), A framework for understanding governance systems and climate change: The case of Australia, *Euro. Manag. J., 25*(6), 415–427.

Grimm, A. M. (2010), Interannual climate variability in South America: Impacts on seasonal precipitation, extreme events, and possible effects of climate change, *Stoch. Env. Res. Risk A., 25*(4), 537–554, doi:10.1007/s00477-010-0420-1.

Grimmond, C. S. B. (1992), The suburban energy balance: Methodological considerations and results for a mid-latitude west coast city under winter and spring conditions, *Int. J. Climatol., 12*, 481–497.

Grimmond, C. S. B. (2006), Progress in measuring and observing the urban atmosphere, *Theor. Appl. Climatol., 84*, 3–22.

Grimmond, C. S. B. (2007), Urbanization and global environmental change: Local effects of urban warming, *Geogr. J., 173*, 83–88.

Grimmond, C. S. B., and T. R. Oke (1986), Urban water balance II: Results from a suburb of Vancouver, BC, *Water Resour. Res., 22*, 1404–1412.

Grimmond, C. S. B., and T. R. Oke (1991), An evapotranspiration-interception model for urban areas, *Water Resour. Res., 27*, 1739–1755.

Grimmond, C. S. B., and T. R. Oke (1995), Comparison of heat fluxes from summertime observations in the suburbs of four North American Cities, *J. Appl. Meteor., 34*, 873–889.

Grimmond, C. S. B., and T. R. Oke (1999a), Aerodynamic properties of urban areas derived from analysis of surface form, *J. Appl. Meteorol., 38*, 1262–1292.

Grimmond, C. S. B., and T. R. Oke (1999b), Heat storage in urban areas: Observations and evaluation of a simple model, *J. Appl. Meteorol., 38*, 922–940.

Grimmond, C. S. B., and T. R. Oke (1999c), Rates of evaporation in urban areas, in *Impacts of Urban Growth on Surface and Ground Waters,*

Publication 259, pp. 235–243, International Association of Hydrological Sciences.

Grimmond, C. S. B., and T. R. Oke (2002), Turbulent fluxes in urban areas: Observations and a local-scale urban meteorological parameterization scheme (LUMPS), *J. Appl. Meteorol., 41*, 792–810.

Grimmond, C. S. B., T. R. Oke, and D. G. Steyn (1986), Urban water balance I: A model for daily totals, *Water Resour. Res., 22*, 1397–1403.

Grimmond, C. S. B., T. R. Oke, and H. A. Cleugh (1993), The role of "rural" in comparisons of observed suburban and rural flux differences, *IAHS Pub. No. 212*, 165–174.

Grimmond, C. S. B., C. Souch, and M. D. Hubble (1996), Influence of tree cover on summertime surface energy balance fluxes, San Gabriel Valley, Los Angeles, *Clim. Res., 6*, 45–57.

Grimmond, C. S. B., H.-B. Su, B. Offerle, B. Crawford, S. Scott, S. Zhong, and C. Clements (2004), Variability of sensible heat fluxes in a suburban area of Oklahoma City, American Meteorological Society Symposium on Planning, Nowcasting and Forecasting in the Urban Zone, Eighth Symposium on Integrated Observing and Assimilation Systems for Atmosphere, Oceans, and Land Surface, *http://ams.confex.com/ams/84Annual/techprogram/paper_67542.htm.*

Grimmond, C. S. B., J. A. Salmond, T. R. Oke, B. Offerle, and A. Lemonsu (2004a), Flux and turbulence measurements at a densely built-up site in Marseille: Heat, mass (water and carbon dioxide), and momentum, *J. Geophys. Res.-Atmos., 109*, D24.

Grimmond, C. S. B., W. Kuttler, S. Lindqvist, and M. Roth (2007), Special issue: Urban climatology ICUC6, *Int. J. Climatol., 27*(14), 1847–1848, doi:10.1002/joc.1638.

Grimmond, C. S. B., M. Best, J. Barlow, A. J. Arnfield, J.-J. Baik, S. Belcher, M. Bruse, I. Calmet, F. Chen, P. Clark, A. Dandou, E. Erell, K. Fortuniak, R. Hamdi, M. Kanda, T. Kawai, H. Kondo, S. Krayenhoff, S. H. Lee, S.-B. Limor, A. Martilli, V. Masson, S. Miao, G. Mills, R. Moriwaki, K. Oleson, A. Porson, U. Sievers, M. Tombrou, J. Voogt, and T. Williamson (2009), Urban surface energy balance models: Model characteristics & methodology for a comparison study, in *Urbanization of Meteorological and Air Quality Models*, edited by A. Baklanov, C. S. B. Grimmond, A. Mahura, and M. Athanassiadou, pp. 97–123, Springer-Verlag.

Grimmond, C. S. B., M. Blackett, M. L. Gouvea, T. Loridan, D. Young, M. J. Best, M. Hendry, J. Barlow, S. E. Belcher, S. I. Bohnenstengel, J.-J. Baik, S.-H. Lee, Y.-H. Ryu, I. Calmet, F. Chen, K. Oleson, A. Dandou, M. Tombrou, K. Fortuniak, R. Hamdi, T. Kawai, Y. Kawamoto, H. Kondo, E. S. Krayenhoff, A. Martilli, F. Salamanca, V. Masson, G. Pigeon, S. Miao, A. Porson, L. Shashua-Bar, G.-J. Steeneveld, J. Voogt, and N. Zhang (2010), The international urban energy balance models comparison project: First results from phase 1, *J. Appl. Meteorol. Clim.*, doi:10.1175/2010JAMC2354.1

Grimmond, C. S. B., M. Roth, T. R. Oke, Y. C. Au, M. Best, R. Betts, G. Carmichael, H. Cleugh, W. Dabberdt, R. Emmanuel, E. Freitas, K. Fortuniak, S. Hanna, P. Klein, L. S. Kalkstein, C. H. Liu, A. Nickson, D. Pearlmutter, D. Sailor, and J. Voogt (2010a), Climate and more sustainable cities: Climate information for improved planning & management of cities (producers/capabilities perspective), *Procedia Environmental Sciences, 1*, 247–274, doi:10.1016/j.proenv.2010.09.016.

Grimmond, C. S. B., M. Blackett, M. Best, J. Barlow, J.-J. Baik, S. Belcher, S. I. Bohnenstengel, I. Calmet, F. Chen, A. Dandou, K. Fortuniak, M. L. Gouvea, R. Hamdi, M. Hendry, T. Kawai, Y. Kawamoto, H. Kondo, E. S. Krayenhoff, S. H. Lee, T. Loridan, A. Martilli,

V. Masson, S. Miao, K. Oleson, G. Pigeon, A. Porson, Y. H. Ryu, F. Salamanca, G. J. Steeneveld, M. Tombrou, J. Voogt, D. Young, and N. Zhang (2010b), The International Urban Energy Balance Models Comparison Project: First results from Phase 1, *J. Appl. Meteorol. Clim., 49*, 1268–1292, doi:10.1175/2010JAMC2354.1.

Grimmond C. S. B., M. Blackett, M. J. Best, J.-J. Baik, S. E. Belcher, J. Beringer, S. I. Bohnenstengel, I. Calmet, F. Chen, A. Coutts, A. Dandou, K. Fortuniak, M. L. Gouvea, R. Hamdi, M. Hendry, M. Kanda, T. Kawai, Y. Kawamoto, H. Kondo, E. S. Krayenhoff, S.-H. Lee, T. Loridan, A. Martilli, V. Masson S. Miao, K. Oleson, R. Ooka, G. Pigeon, A. Porson, Y.-H. Ryu, F. Salamanca, G.-J. Steeneveld, M. Tombrou, J. A. Voogt, D. Young, and N. Zhang (2011), Initial results from phase 2 of the International Urban Energy Balance Comparison Project, *Int. J. Climatol., 31*, 244–272, doi:10.1002/joc.2227.

Grinsted, A., J. C. Moore, and S. Jevrejeva (2009), Reconstructing sea level from paleo and projected temperatures 200 to 2100 AD, *Clim. Dynam., 34*(4), 461–472, doi:10.1007/s00382-008-0507-2.

Grize, L., A. Huss, O. Thommen, C. Schindler, and C. Braun-Fabrlander (2005), Heat wave 2003 and mortality in Switzerland, *Swiss Med. Wkly., 135*, 200–205.

Groisman, P. Y., and E. Y. Rankova (2001), Precipitation trends over the Russian permafrost-free zone: Removing the artifacts of pre-processing, *Int. J. Climatol., 21*, 657–678.

Groisman, P. Y., R. W. Knight, T. R. Karl, D. R. Easterling, B. Sun, and J. H. Lawrimore (2004), Contemporary changes of the hydrological cycle over the contiguous United States: trends derived from in situ observations, *J. Hydrometeorol., 5*, 64–85.

Groisman., P. Y., R. W. Knight, D. R. Easterling, T. R. Karl, G. C. Hegerl, and V. A. N. Razuvaev (2005), Trends in intense precipitation in the climate record, *J. Climate, 18*, 1326–1350.

Grossman-Clarke, S., J. A. Zehnder, T. Loridan, and C. S. B. Grimmond (2010), Contribution of land use changes to near-surface air temperatures during recent summer extreme heat events in the Phoenix metropolitan area, *J. Appl. Meteorol. Climatol., 49*, 1649–1664.

Grubb, M., T. L. Brewer, M. Sato, R. Heilmayr, and D. Fazekas (2009), Climate policy and industrial competitiveness: Ten insights from Europe on the EU Emissions Trading System, Climate & Energy Paper Series, German Marshall Fund of the United States, 3 August 2009, *http://climatestrategies.org/component/reports/category/61/204.html*.

Gruber, N., S. C. Doney, S. R. Emerson, D. Gilbert, T. Kobayashi, A. Körtzinger, G. C. Johnson, K. S. Johnson, S. C. Riser, and O. Ulloa (2007), The ARGO-Oxygen Program – A white paper to promote the addition of oxygen sensors to the international ARGO float program.

Gruber, S., M. Hoelzle, and W. Haeberli (2004), Permafrost thaw and destabilization of Alpine rock walls in the hot summer of 2003, *Geophys. Res. Lett., 31*, L13504.

Gu, D., and S. G. H. Philander (1997), Inter-decadal climate fluctuations that depend on exchanges between the tropics and extratropics, *Science, 275*, 805–807, doi:10.1126/science.275.5301.805.

Gu, L., S. G. Pallardy, K. Tu, B. E. Law, and S. D. Wullschleger (2010), Reliable estimation of biochemical parameters from C_3 leaf photosynthesis-intercellular carbon dioxide response curves, *Plant Cell Environ., 33*, 1852–1874 doi:10.111/j.1365-3040.2010.02192.x.

Guan, L., and H. Kawamura (2003), SST availabilities of satellite infrared and microwave measurements, *J. Oceanogr., 59*(2), 201–209.

Guenther, A., C. N. Hewitt, D. Erickson, R. Fall, C. Geron, T. Graedel, P. Harley, L. Klinger, M. Lerdau, W. A. Mckay, T. Pierce, B. Scholes, R. Steinbrecher, R. Tallamraju, J. Taylor, and P. A. Zimmerman (1995), Global-model of natural volatile organic-compound emissions, *J. Geophys. Res.-Atmos., 100*(D5), 8873–8892.

Guenther, A., T. Karl, P. Harley, C. Wiedinmyer, P. I. Palmer, and C. Geron (2006), Estimates of global terrestrial isoprene emissions using MEGAN (Model of Emissions of Gases and Aerosols from Nature), *Atmos. Chem. Phys., 6*, 3181–3210.

Guex, J., A. Bartolini, V. Atudorei, and D. Taylor (2004), High-resolution ammonite and carbon isotope stratigraphy across the Triassic-Jurassic boundary at New York Canyon (Nevada), *Earth Planet. Sci. Lett., 225*, 29–41.

Guilyardi, E. (2006), El Niño–mean state–seasonal cycle interactions in a multi-model ensemble, *Clim. Dynam., 26*(4), 329–348.

Guinotte, J. M., J. Orr, S. Cairns, A. Freiwald, L. Morgan, and R. George (2006), Will human-induced changes in seawater chemistry alter the distribution of deep-sea scleractinian corals?, *Front. Ecol. Environ., 4*, 141–146.

Guiot, J., I. C. Prentice, C. Peng, D. Jolly, F. Laarif, and B. Smith (2001), Reconstructing and modelling past changes in terrestrial primary production, in *Terrestrial Global Productivity*, edited by J. Roy et al., Academic Press.

Gunderson, C. A., K. H. O'Hara, C. M. Campion, A. V. Walker, and N. T. Edwards (2010), Thermal plasticity of photosynthesis: The role of acclimation in forest responses to a warming climate, *Glob. Change Bio., 16*, 2272–2286, doi:10.1111/j.1365-2486.2009.02090.x.

Guo, Z.T., A. Berger, Q. Z. Yin, and L. Qin (2009), Strong asymmetry of hemispheric climates during MIS-13 inferred from correlating China loess and Antarctica ice records, *Clim. Past, 5*, 21–31.

Gupta, A. K., D. M. Anderson, and J. T. Overpeck (2003), Abrupt changes in the Asian southwest monsoon during the Holocene and their links to the North Atlantic Ocean, *Nature, 431*, 354–357.

Gurney, K. R., D. Baker, and S. Denning (2008), Interannual variations in continental-scale net carbon exchange and sensitivity to observing networks estimated from atmospheric CO_2 inversions for the period 1980 to 2005, *Global Biogeochem. Cycles, 22*, GB3025, doi:10.1029/2007GB003082.

Gutschick, V. P. (1982), Energetics of microbial fixation dinitrogen, *Adv. Biochem. Eng. Biot., 21*, 109–167, doi:10.1007/3-540-11019-4_7.

Guttal, V., and C. Jayaprakash (2008), Changing skewness: An early warning signal of regime shifts in ecosystems, *Ecol. Lett., 11*, 450–460.

Guttal, V., and C. Jayaprakash (2009), Spatial variance and spatial skewness: Leading indicators of regime shifts in spatial ecological systems, *Theor. Ecol., 2*, 3–12.

Haase, S., and S. S. Sackett (1998), Effects of prescribed fire in giant sequoia-mixed conifer stands in Sequoia and Kings Canyon National Parks, in Fire in ecosystem management: Shifting the paradigm from suppression to prescription, edited by T. L. Priden, and L. A. Brennan, *Tall Timbers Fire Ecology Conf. Proc. 20*, Tall Timbers Research Station, Tallahassee, Florida.

Haeberli, W., and M. Hoelzle (1995), Application of inventory data for estimating characteristics of and regional climate-change effects on mountain glaciers: A pilot study with the European Alps, *Ann. Glaciol., 21*, 206–212.

Hagedorn, R., F. Doblas-Reyes, and T. Palmer (2005), The rationale behind the success of multi-model ensembles in seasonal forecasting, 1. Basic concept, *Tellus A, 57*(3), 219–233.

Hagos, S. M., and K. H. Cook (2007), Dynamics of the West African Monsoon Jump, *J. Climate, 20*, 5264–5284.

Haigh, J. D. (1994), The role of stratospheric ozone in modulating the solar radiative forcing of climate, *Nature, 370*, 544–546.

Haigh, J. D. (2003), The effects of solar variability on the Earth's climate, *Philos. T. Roy. Soc. A, 361*, 95–111.

Häkkinen, S., and P. B. Rhines (2004), Decline of subpolar North Atlantic circulation during the 1990s, *Science, 304*, 555–559.

Hale, R. C., K. P. Gallo, and T. R. Loveland (2008), Influences of specific land use/land cover conversions on climatological normals of near-surface temperature, *J. Geophys. Res., 113*, D14113, doi:10.1029/2007JD009548.

Hall, A. (2004), The role of surface albedo feedback in climate, *J. Climate, 17*(7), 1550–1568.

Hall, A., and S. Manabe (1997), Can local, linear stochastic theory explain sea surface temperature and salinity variability?, *Clim. Dynam., 13*, 167–180.

Hall, A., and M. Visbeck (2002), Synchronous variability in the Southern Hemisphere atmosphere, sea-ice, and ocean resulting from the annular mode, *J. Climate, 13*, 3043–3057.

Hall, N. L., and R. Taplin (2007), Solar festivals and climate bills: Comparing NGO climate change campaigns in the UK and Australia, *Voluntas: International Journal of Voluntary and Nonprofit Organizations, 18*(4), 317–338.

Hall, N. L., R. Taplin, and W. Goldstein (2010), Empowerment of individuals and realization of community agency: Applying action research to climate change responses in Australia, *Action Res., 8*, 71–91.

Hallam, A. (2002), How catastrophic was the end-Triassic mass extinction?, *Lethaia, 35*, 147–157.

Hallam, A., and P. B. Wignall (1999), Mass extinctions and sea-level changes, *Earth Sci. Rev., 48*, 217–250.

Hamilton, C. (2010), *Requiem for a Species: Why We Resist the Truth about Climate Change*, Allen & Unwin, Sydney, Australia.

Hammarlund, D., L. Barnekow, H. J. B. Birks, B. R. Buchardt, and T. W. D. Edwards (2002), Holocene changes in atmospheric circulation recorded in the oxygen-isotope stratigraphy of lacustrine carbonates from northern Sweden, *Holocene, 12*(3), 339–351, doi:10.1191/0959683602hl548rp.

Han, W., G. A. Meehl, and A. Hu (2006), Interpretation of tropical thermocline cooling in the Indian and Pacific Oceans during recent decades, *Geophys. Res. Lett, 33*, 1961–2000.

Hannachi, A., I. Jolliffe, and D. Stephenson (2007), Empirical orthogonal functions and related techniques in atmospheric science: A review, *Int. J. Climatol., 27*(9), 1119–1152, doi:10.1002/joc.1499.

Hansen, J. (2005), Is there still time to avoid 'dangerous anthropogenic interference' with global climate?, A Tribute to Charles David Keeling, presentation at American Geophysical Union, 6 December, San Francisco, USA.

Hansen, J. (2009), *Storms of My Grandchildren: The Truth about the Coming Climate Catastrophe and Our Last Chance to Save Humanity*, Bloomsbury, USA.

Hansen, J., and L. Nazarenko (2004), Soot climate forcing via snow and ice albedos, *Proc. Natl. Acad. Sci. USA, 101*, 423–428.

Hansen, J., A. Lacis, D. Rind, G. Russell, P. Stone, I. Fung, R. Ruedy, and J. Lerner (1984), Climate sensitivity: Analysis of feedback mechanisms — Geophysical monograph, in *Climatic Processes and Climate Sensitivity*, edited by J. E. Hansen and T. Takahash, pp. 130–163, Maurice Ewing Series 29.

Hansen, J., M. Sato, and R. Ruedy (1997), Radiative forcing and climate response, *J. Geophys. Res., 102*, 6831–6864.

Hansen, J., M. Sato, R. Ruedy, L. Nazarenko, A. Lacis, G. A. Schmidt, G. Russell, I. Aleinov, M. Bauer, S. Bauer, N. Bell, B. Cairns, V. Canuto, M. Chandler, Y. Cheng, A. Del Genio, G. Faluvegi, E. Fleming, A. Friend, T. Hall, C. Jackman, M. Kelley, N. Kiang, D. Koch, J. Lean, J. Lerner, K. Lo, S. Menon, R. Miller, P. Minnis, T. Novakov, V. Oinas, J. Perlwitz, J. Perlwitz, D. Rind, A. Romanou, D. Shindell, P. Stone, S. Sun, N. Tausnev, D. Thresher, B. Wielicki, T. Wong, M. Yao, and S. Zhang (2005), Efficacy of climate forcings, *J. Geophys. Res.-Atmos., 110*, D18104.

Hansen, J., M. Sato, R. Ruedy, K. Lo, D. W. Lea, and M. Median-Elizade (2006), Global temperature change, *Proc. Natl. Acad. Sci. USA, 103*, 14288–14293.

Hansen, J., M. Sato, P. Kharecha, D. Beerling, R. Berner, V. Masson-Delmotte, M. Pagani, M. Raymo, D. L. Royer, and J. C. Zachos (2008), Target atmospheric CO_2: Where should humanity aim?, *Open Atmos. Sci. J., 2*, 217–231.

Harding, S. (2006), *Animate Earth: Science, intuition and Gaia*, Green Books, Dartington, UK.

Hare, S. R., and N. J. Mantua (2000), Empirical evidence for North Pacific regime shifts in 1977 and 1989, *Progr. Oceanogr., 47*(2–4), 103–145.

Harley, P. C., and T. D. Sharkey (1991), An improved model of C_3 photosynthesis at high CO_2: Reversed O_2 sensitivity explained by lack of glycerate reentry into the choloroplast, *Photosynth. Res., 27*, 169–178.

Harley, P. C., F. Loreto, G. D. Marco, and T. D. Sharkey (1992), Theoretical considerations when estimation the mesophyll conductance to CO_2 flux by analysis of the response of photosynthesis to CO_2, *Plant Physiol., 98*, 1429–1436.

Harman I. N., J. F. Barlow, and S. E. Belcher (2004), Scalar fluxes from urban street canyons, part II: Model, *Bound.-Lay. Meteorol., 113*, 387–410.

Harries, J. E., H. E. Brindley, P. J. Sagoo, and R. J. Bantges (2001), Increases in greenhouse forcing inferred from the outgoing longwave radiation spectra of the Earth in 1970 and 1997, *Nature, 410*, 355–357.

Harrison, G. R., and D. B. Stephenson (2006), Empirical evidence for a nonlinear effect of galactic cosmic rays on clouds, *Proc. Roy. Soc. A, 462*(2068), 1221–1233, doi:10.1098/rspa.2005.1628.

Harrison, S. P., and P. J. Bartlein (2012), Records from the past, lessons for the future: What the palaeorecord implies about mechanisms of global change, in *The Future of the World's Climate*, edited by A. Henderson-Sellers and K. McGuffie, pp. 403–436, Elsevier, Amsterdam.

Harrison, S. P., and M. F. Sánchez Goñi (2010), Global patterns of vegetation response to millennial-scale variability and rapid climate change during the last glacial period, *Quaternary Sci. Rev., 29*(21–22), 2957–2980, doi:10.1016/j.quascirev.2010.07.016.

Harrison S. P., J. E. Kutzbach, I. C. Prentice, P. J. Behling, and M. T. Sykes (1995), The response of Northern Hemisphere extratropical climate and vegetation to orbitally induced changes in insolation during the last interglacial, *Quat. Res., 43*, 174–184.

Harrison, S. P., K. E. Kohfeld, C. Roelandt, and T. Claquin (2001), The role of dust in climate changes today, at the Last Glacial Maximum and in the future, *Earth-Sci. Rev., 54*(1–3), 43–80, doi:10.1016/S0012-8252(01)00041-1.

Harrison, S. P., and A. I. Prentice (2003), Climate and CO_2 controls on global vegetation distribution at the Last Glacial Maximum: Analysis based on palaeovegetation data, biome modelling and palaeoclimate simulations, *Glob. Change Biol.*, *9*(7), 983–1004, doi:10.1046/j.1365-2486.2003.00640.x.

Harrison, S. P., G. Yu, and J. Vassiljev (2002), Climate changes during the Holocene recorded by lakes from Europe, in *Climate Development and History of the North Atlantic Realm*, edited by G. Wefer et al., pp. 191–204, Springer-Verlag, Berlin/Heidelberg.

Harrison, S. P., J. E. Kutzbach, Z. Liu, P. J. Bartlein, B. Otto-Bliesner, D. Muhs, I. C. Prentice, and R. S. Thompson (2003), Mid-Holocene climates of the Americas: A dynamical response to changed seasonality, *Clim. Dynam.*, *20*(7–8), 663–688, doi:10.1007/s00382-002-0300-6.

Harrison, S. P., J. R. Marlon, and P. J. Bartlein (2010), Fire in the Earth system, in *Changing Climates, Earth Systems and Society*, edited by J. Dodson, pp. 21–48, Springer-Verlag.

Hartel, C. E. J., and G. Pearman (2010), Understanding and responding to the climate change issue: Towards a whole-of-science research agenda, *J. Manag. Org.*, *16*, 16–47.

Hartmann, D. L. (1996), *Global Physical Climatology*, Academic Press.

Harvey, L. D. D. (2001), A quasi-one-dimensional coupled climate–carbon cycle model, Part II: The carbon cycle component, *J. Geophys. Res.*, *106*, 22355–22372.

Harvey, L. D. D. (2010a), *Energy and the New Reality, Volume 1: Energy Efficiency and the Demand for Energy Services*, Earthscan, London, UK.

Harvey, L. D. D. (2010b), *Energy and the New Reality, Volume 2: Carbon-Free Energy Supply*, Earthscan, London, UK.

Harvey, L. D. D. (2011), Climate and climate-system modelling, in *Environmental Modelling: Finding Simplicity in Complexity*, second ed., edited by J. Wainwright, and M. Mulligan, Wiley-Blackwell (in press).

Harvey, L. D. D. (2012), Fast and slow feedbacks in future climates, in *The Future of the World's Climate*, edited by A. Henderson-Sellers and K. McGuffie, pp. 99–139, Elsevier, Amsterdam.

Harvey, L. D. D., and R. K. Kaufmann (2002), Simultaneously constraining climate sensitivity and aerosol radiative forcing, *J. Climate*, *15*, 2837–2861.

Harvey, L. D. D., and Z. Huang (1995), Evaluation of the potential impact of methane clathrate destabilization on future global warming, *J. Geophys. Res.*, *100*, 2905–2926.

Harvey, L. D. D., and Z. Huang (2001), A quasi-one-dimensional coupled climate–carbon cycle model, part I: Description and behaviour of the climate component, *J. Geophys. Res.*, *106*, 22339–22353.

Harvey, L. D. D., J. Gregory, M. Hoffert, A. Jain, M. Lal, R. Leemans, S. Raper, T. Wigley, and J. de Wolde (1997), *An Introduction to Simple Climate Models Used in the IPCC Second Assessment Report*, *Intergovernmental Panel on Climate Change, Technical Paper No. 2*, Intergovernmental Panel on Climate Change, World Meteorological Organization, Geneva.

Harvey, M. V., S. C. Brassell, C. M. Belcher, and A. Montanari (2008), Combustion of fossil organic matter at the Cretaceous–Paleogene (K–P) boundary, *Geology*, *36*, 335–358.

Harvey, N. (Ed.) (2006), Global change and integrated coastal management: The Asia Pacific Region, in *Coastal Systems and Continental Margins*, vol. 10, 2006, Springer, Dordrecht, doi:10.1007/1-4020-3628-0.

Hasler, N., D. Werth, and R. Avissar (2009), Effects of tropical deforestation on global hydroclimate: A multimodel ensemble analysis, *J. Climate*, *22*, 1124–1141, doi:10.1175/2008JCLI2157.1.

Hasselmann, K. (1976), Stochastic climate models, Part I: Theory, *Tellus*, *28*, 473–485.

Hassol S. J. (2008), Improving how scientists communicate about climate change, *EOS Trans. AGU*, *89*(11), March 11, 2008, 106–107 and Reply to Comments on 'Improving how scientists communicate about climate change', *EOS Trans. AGU*, *89*(33), August 12, 2008, 304.

Hastings, A., and D. B. Wysham (2010), Regime shifts in ecological systems can occur with no warning, *Ecol. Lett.*, *13*(4), 464–472.

Hatzianastassiou, N., C. Matsoukas, A. Fotiadi, K. G. Pavlakis, E. Drakakis, D. Hatzidimitriou, and I. Vardavas (2005), Global distribution of Earth's surface shortwave radiation budget, *Atmos. Chem. Phys.*, *5*, 2847–2867.

Hauglustaine, D. A., and G. P. Brasseur (2001), Evolution of tropospheric ozone under anthropogenic activities and associated radiative forcing of climate, *J. Geophys. Res.-Atmos.*, *106*(D23), 32337–32360, doi:10.1029/2001JD900175.

Hauglustaine, D. A., J. Lathière, S. Szopa, and G. A. Folberth (2005), Future tropospheric ozone simulated with a climate–chemistry–biosphere model, *Geophys. Res. Lett.*, *32*, L24807, doi:10.1029/2005GL024031.

Hautmann, M. (2004), Effect of end-Triassic CO_2 maximum on carbonate sedimentation and marine mass extinction, *Facies*, *50*, 257–261.

Hautmann, M., F. Stiller, C. Huawei, and S. Jingeng (2008), Extinction-recovery pattern of level-bottom faunas across the Triassic-Jurassic boundary in Tibet: Implications for potential killing mechanisms, *Palaios*, *23*, 711–718.

Hawkins, E., and S. Rowan (2009), The potential to narrow uncertainty in regional climate predictions, *Bull. Am. Meteorol. Soc.*, *90*, 1095–1107. doi:10.1175/2009BAMS2607.1.

Hawkins, E., and R. Sutton (2009), The potential to narrow uncertainty in regional climate predictions, *Bull. Am. Meteorol. Soc.*, *90*, 1095–1107.

Haxeltine, A., and I. C. Prentice (1996), A general model for the light-use efficiency of primary production, *Funct. Ecol.*, *10*, 551–561.

Hayakawa, Y. S., T. Oguchi, and Z. Lin (2008), Comparison of new and existing global digital elevation models: ASTER G-DEM and SRTM-3, *Geophys. Res. Lett.*, *35*(17), L17404, doi:10.1029/2008GL035036.

Hayes, S. P., L. J. Mangum, J. Picaut, A. Sumi, and K. Takeuchi (1991), TOGA-TAO: A moored array for real-time measurements in the tropical Pacific Ocean, *Bull. Am. Meteorol. Soc.*, *72*(3), 339–347.

Haylock, M. R., and C. M. Goodess (2004), Interannual variability of extreme European winter rainfall and links with mean large-scale circulation, *Int. J. Climatol.*, *24*, 759–776.

Haylock, M. R., T. Peterson, J. R. Abreu de Sousa, L. M. Alves, T. Ambrizzi, J. Baez, J. I. Barbosa, V. R. Barros, M. A. Berlato, M. Bidegain, G. Coronel, V. Corradi, A. M. Grimm, R. Jaildo dos Anjos, D. Karoly, J. A. Marengo, M. B. Marino, P. R. Meira, G. C. Miranda, L. Molion, D. F. Moncunill, D. Nechet, G. Ontaneda, J. Quintana, E. Ramirez, E. Rebello, M. Rusticucci, J. L. Santos, I. T. Varillas, J. G. Villanueva, L. Vincent, and M. Yumiko (2006), Trends in total and extreme South America rainfall 1960–2000 and links with sea surface temperature, *J. Climate*, *19*, 1490–1512.

Haylock, M., G. Cawley, C. Harpham, R. Wilby, and C. Goodess (2006), Downscaling heavy precipitation over the United Kingdom: A comparison of dynamical and statistical methods and their future scenarios, *Int. J. Climatol.*, *26*(10), 1397–1415, doi:10.1002/joc.1318.

Haywood, J. M., and K. P. Shine (1995), The effect of anthropogenic sulfate and soot aerosol on the clear-sky planetary radiation budget, *Geophys. Res. Lett.*, *22*(5), 603–606.

Heald, C. L., et al. (2008), Predicted change in global secondary organic aerosol concentrations in response to future climate, emissions, and land use change, *J. Geophys. Res., 113*(D5), D05211.

Heck, P., D. Luthi, H. Wernli, and C. Schar (2001), Climate impacts of European-scale anthropogenic vegetation changes: A sensitivity study using a regional climate model, *J. Geophys. Res., 106*, 7817–7836.

Hegerl, G. C., K. Hasselmann, U. Cubasch, J. F. B. Mitchell, E. Roeckner, R. Voss, and J. Waszkewitz (1997), Multi-fingerprint detection and attribution analysis of greenhouse gas, greenhouse gas-plus-aerosol, and solar forced climate change, *Clim. Dynam., 13*, 613–634.

Hegerl, G. C., F. Zwiers, S. Kharin, and P. Stott, (2004), Detectability of anthropogenic changes in temperature and precipitation extremes, *J. Climate, 17*, 3683–3700.

Hegerl, G. C., O. Hoegh-Guldberg, G. Casassa, M. P. Hoerling, R. S. Kovats, C. Parmesan, D. W. Pierce, and P. A. Stott, (2010), Good practice guidance paper on detection and attribution related to anthropogenic climate change, in *Meeting Report of the Intergovernmental Panel on Climate Change Expert Meeting on Detection and Attribution of Anthropogenic Climate Change*, edited by T. F. Stocker, C. B. Field, D. Qin, V. Barros, G.-K. Plattner, M. Tignor, P. M. Midgley, and K. L. Ebi, IPCC Working Group I Technical Support Unit, University of Bern, Bern, Switzerland.

Heim Jr, R. R. (2002), A review of twentieth-century drought indices used in the United States, *Bull. Am. Meteorol. Soc., 19*(21), 5686–5699.

Hein, R., P. J. Crutzen, and M. Heimann (1997), An inverse modeling approach to investigate the global atmospheric methane cycle, *Global Biogeochem. Cycles, 11*(1), 43–76.

Heinze, C. (2004), Simulating oceanic $CaCO_3$ export production in the greenhouse, *Geophys. Res. Lett., 31*(4).

Held, H., and T. Kleinen (2004), Detection of climate system bifurcations by degenerate fingerprinting, *Geophys. Res. Lett., 31*, L23207.

Held, I. M., and B. J. Soden (2006), Robust responses of the hydrological cycle to global warming, *J. Climate, 19*, 5686–5699.

Helfter C., D. Famulari, G. J. Phillips, J. F. Barlow, C. R. Wood, C. S. B. Grimmond, and E. Nemitz (2011), Controls of carbon dioxide concentrations and fluxes above central London, *Atmos. Chem. Phys., 11*, 1913–1928, doi:10.5194/acp-11-1913-2011.

Helm, K. P., N. L. Bindoff, and J. A. Church (2010), Changes in the global hydrological-cycle inferred from ocean salinity, *Geophys. Res. Lett., 37*(18), L18701.

Helsen, M. M., M. R. van den Broeke, R. S. W. van de Wal, W. J. van de Berg, E. van Meijgaard, C. H. Davis, Y. H. Li, and I. Goodwin (2008), Elevation changes in Antarctica mainly determined by accumulation variability, *Science, 320*(5883), 1626–1629.

Henderson-Sellers, A. (1995a), *Future Climates of the World: A Modelling Perspective, Volume 16, World Survey of Climatology*, Elsevier Science, Amsterdam.

Henderson-Sellers, A. (1995b), Climates of the future, in *Future Climates of the World*, edited by A. Henderson-Sellers, pp. 1–18, Elsevier, Amsterdam.

Henderson-Sellers, A. (2008), The IPCC report: What the lead authors really think?, Talking Point Invited Article, *Environ. Res. Lett., http://environmentalresearchweb.org/cws/article/opinion/35820*. (Accessed 26/7/10.)

Henderson-Sellers, A. (2010a), How seriously are we taking climate change? Monitoring climate change communication, in *Climate Alert Climate Change Monitoring and Strategy*, edited by Y. Yu, and A. Henderson-Sellers, pp. 28–65, Sydney University Press, Australia.

Henderson-Sellers, A. (2010b), Climatic change: Communication changes over this journal's first 'century', *Climatic Change, 100*, 215–227, doi:10.1007/s10584-010-9814-9, *http://www.springerlink.com/openurl.asp?genre=article&;id=doi:10.1007/s10584-010-9814-9*. (Accessed 26/7/10.)

Henderson-Sellers, A. (2010c), Research integrity's burning fuse: Climate truth before change explodes, Invited presentation at the Second World Conference on Research Integrity, Singapore, 21–24 July, *http://www.wcri2010.org*. (Accessed 4/4/11.)

Henderson-Sellers, A. (2012), Seeing further: The futurology of climate, in *The Future of the World's Climate*, edited by A. Henderson-Sellers and K. McGuffie, pp. 3–25, Elsevier, Amsterdam.

Henderson-Sellers, A., and V. Gornitz (1984), Possible climatic impacts of land cover transformations, with particular emphasis on tropical deforestation, *Climatic Change, 6*, 231–257.

Henderson-Sellers, A., R. E. Dickinson, and M. F. Wilson (1988), Tropical deforestation: Important processes for climate models, *Climatic Change, 13*, 43–67.

Henderson-Sellers, A., R. E. Dickinson, T. B. Durbidge, P. J. Kennedy, K. McGuffie, and A. J. Pitman (1993), Tropical deforestation: Modelling local- to regional-scale climate change, *J. Geophys. Res., 98*, 7289–7315.

Henderson-Sellers, A., H. Zhang, G. Berz, K. Emanuel, W. Gray, C. Landsea, G. Holland, J. Lighthill, S.-L. Shieh, P. Webster, and K. McGuffie (1998), Tropical Cyclones and Global Climate Change: A Post-IPCC Assessment. *Bull. Amer. Meteor. Soc., 79*, 19–38.

Henderson-Sellers, A., P. Irannejad, and K. McGuffie (2008), Future desertification and climate change: The need for land-surface system evaluation improvement, *Glob. Planet. Change, 64*, 129–138.

Hendon, H. H., D. W. J. Thompson, and M. C. Wheeler (2007), Australian rainfall and surface temperature variations associated with the Southern Hemisphere Annular Mode, *J. Climate, 20*(11), 2452–2467, doi:10.1175/JCLI4134.1.

Hennessy, K. J., R. Fawcett, D. G. C. Kirono, F. S. Mpelasoka, D. Jones, J. M. Bathols, P. H. Whetton, M. Stafford Smith, M. Howden, C. D. Mitchell, and N. Plummer (2008), *An assessment of the impact of climate change on the nature and frequency of exceptional climatic events*, CSIRO Marine and Atmospheric Research, Aspendale.

Henson, S. A., J. P. Dunne, and J. L. Sarmiento (2009), Decadal variability in North Atlantic phytoplankton blooms, *J. Geophys. Res., 114*(C4), C04013.

Henson, S. A., J. L. Sarmiento, J. P. Dunne, L. Bopp, I. Lima, S. C. Doney, J. John, and C. Beaulieu (2010), Detection of anthropogenic climate change in satellite records of ocean chlorophyll and productivity, *Biogeosciences, 7*(2), 621–640, doi:10.5194/bg-7-621-2010.

Henze, D. K., and J. H. Seinfeld (2006), Global secondary organic aerosol from isoprene oxidation, *Geophys. Res. Lett., 33*(9), L09812.

Herrmann, A. D., M. E. Patzkowsky, and D. Pollard (2004), The impact of paleogeography, pCO_2, poleward ocean heat transport and sea level change on global cooling during the late Ordovician, *Palaeogeogr. Palaeocl., 206*, 59–74.

Herterich K., and K. Hasselmann (1987), Extraction of mixed layer advection velocities, diffusion coefficients, feedback factors, and atmospheric forcing parameters from the statistical analysis of the North Pacific SST anomaly fields, *J. Phys. Oceanogr., 17,* 2145–2156.

Hertig, E., and J. Jacobeit (2008), Downscaling future climate change: Temperature scenarios for the Mediterranean area, *Glob. Planet. Change, 63*(2–3), 127–131, doi:10.1016/j.gloplacha.2007.09.003.

Hess, P. G., and J.-F. Lamarque (2007), Ozone source attribution and its modulation by the Arctic oscillation during the spring months, *J. Geophys. Res., 112,* D11303, doi:10.1029/2006JD007557.

Hesselbo, S. P., D. R. Gröcke, H. C. Jenkyns, C. J. Bjerrum, P. Farrimond, H. S. M. Bell, and O. R. Green (2000), Massive dissociation of gas hydrate during a Jurassic oceanic anoxic event, *Nature, 406,* 392–395.

Hesselbo, S. P., S. A. Robinson, F. Surlyk, and S. Piasecki (2002), Terrestrial and marine extinction at the Triassic-Jurassic boundary synchronized with major carbon cycle perturbation: A link to initiation of massive volcanism?, *Geology, 30,* 251–254.

Hesselbo, S. P., S. A. Robinson, and F. Surlyk (2004), Sea-level change and facies development across potential Triassic/Jurassic boundary horizons, SW Britain, *J. Geol. Soc. London, 161,* 365–379.

Hesselbo, S. P., C. A. McRoberts, and J. Pálfy (2007), Triassic-Jurassic boundary events: Problems, progress, possibilities, *Palaeogeogr. Palaeocl., 244,* 1–10.

Hessler, I., L. Dupont, R. Bonnefille, H. Behling, C. Gonzalez, K. F. Helmens, H. Hooghiemstra, J. Lebamba, M. P. Ledru, A. M. Lezine, J. Maley, F. Marret, and A. Vincens (2010), Millennial-scale changes in vegetation records from tropical Africa and South America during the last glacial, *Quaternary Sci. Rev., 29*(21–22), 2882–2899, doi:10.1016/j.quascirev.2009.11.029.

Hewitt, C. D., (2005), The ENSEMBLES Project: Providing ensemble-based predictions of climate changes and their impacts, *EGGS newsletter, 13,* 22–25.

Hibbard, K. H., G. A. Meehl, P. Cox, and P. Friedlingstein (2007), A strategy for climate change stabilization experiments, *Eos Trans. AGU, 88*(20), 217, 219, 221.

Hibbard, K. H., G. A. Meehl, N. Nakicenovic, S. Rose, J.-F. Lamarque, D. van Vuuren, J. Edmonds, T. Janetos, A. Thomson, S. Smith, G. Hurtt, S. Frolking, and L. Chini (2009), The handshake between IAMs and ESMs: IPCC AR5 as a catalyst, IAV, Brazil, Nov 4, *http://www.ccst. inpe.br/iavbr/files/AOGCM_IAM_Brazil.pdf.* (Accessed 7/4/11.)

Hidalgo, J., V. Masson, A. Baklanov, G. Pigeon, and L. Gimenoa (2008), Advances in urban climate modeling, in Trends and directions in climate research, *Ann. N.Y. Acad. Sci., 1146,* 354–374, doi:10.1196/annals.1446.015.

Hiddink, J. G., and R. Ter Hofstede (2008), Climate induced increases in species richness of marine fishes, *Glob. Change Biol., 14*(3), 453–460, doi:10.1111/j.1365-2486.2007.01518.x.

Hijioka, Y., Y. Matsuoka, H. Nishimoto, M. Masui, and M. Kainuma (2008), Global GHG emissions scenarios under GHG concentration stabilization targets. *Journal of Global Environmental Engineering, 13,* 97–108.

Hildebrand, A. R. (1993), The Cretaceous/Tertiary boundary impact (or the dinosaurs didn't have a chance), *J. Roy. Astron. Soc. Can., 87,* 77–118.

Hildebrand, A. R., and W. V. Boynton (1989), Hg anomalies at the K/T boundary − Evidence for acid-rain, *Meteoritics, 24,* 277–278.

Hildebrand, A. R., and W. S. Wolbach (1989), *XX LPSC abstracts,* p. 414, Houston, Texas, Lunar and Planetary Science Institute.

Hildebrand, A. R., et al. (1991), The Chicxulub crater: A possible Cretaceous-Tertiary boundary impact crater on the Yucatan Peninsula, Mexico, *Geology, 19,* 867–871.

Hill, K., S. Rintoul, R. Coleman, and K. Ridgway (2008), Wind forced low frequency variability of the East Australia Current, *Geophys. Res. Lett., 35*(8), L08602.

Hirabayashi, Y., P. Döll, and S. Kanae (2010), Global-scale modeling of glacier mass balances for water resources assessments: Glacier mass changes between 1948 and 2006, *J. Hydrol., 390*(3–4), 245–256.

Hirschi, M., S. Seneviratne, V. Alexandrov, F. Boberg, C. Boroneant, O. Christensen, H. Formayer, B. Orlowsky, and P. Stepanek (2011), Observational evidence for soil-moisture impact on hot extremes in southeastern Europe, *Nat. Geosci., 4,* 17–21, doi:10.1038/ngeo1032.

Hitchcock, D. R., and J. E. Lovelock (1967), Life detection by atmospheric analysis, *Icarus, 7*(1–3), 149–159, doi:10.1016/0019-1035(67)90059-0.

Hock, R. (1999), A distributed temperature-index ice- and snow-melt model including potential direct solar radiation, *J. Glaciol., 45*(149), 101–111.

Hock, R. (2003), Temperature index melt modelling in mountain areas, *J. Hydrol., 282,* 104–115.

Hock, R., M. de Woul, V. Radić, and M. B. Dyurgerov (2009), Mountain glaciers and ice caps around Antarctica make a large sea-level rise contribution, *Geophys. Res. Lett., 36,* L07501, doi:10.1029/2008GL037020.

Hoffert, M. I., and C. Covey (1992), Deriving global climate sensitivity from paleoclimate reconstructions, *Nature, 360,* 573–576.

Hofmann, M., and S. Rahmstorf (2009), On the stability of the Atlantic Meridional Overturning Circulation, *Proc. Natl. Acad. Sci. USA, 106,* 20584–20589.

Hofstra, N., M. Haylock, M. New, P. Jones, and C. Frei (2008), Comparison of six methods for the interpolation of daily, European climate data, *J. Geophys. Res.-Atmos., 113,* D21110.

Hofstra, N., M. New, and C. McSweeney (2010), The influence of interpolation and station network density on the distributions and trends of climate variables in gridded daily data, *Clim. Dynam., 35,* 841–858.

Hogrefe, C., B. Lynn, K. Civerolo, J.-Y. Ku, J. Rosenthal, C. Rosenzweig, R. Goldberg, S. Gaffin, K. Knowlton, and P. L. Kinney (2004), Simulating changes in regional air pollution over the eastern United States due to changes in global and regional climate and emissions, *J. Geophys. Res., 109,* D22301, doi:10.1029/2004JD004690.

Hohenegger, C., P. Brockhaus, C. Bretherton, and C. Schär (2009), The soil moisture-precipitation feedback in simulations with explicit and parameterized convection, *J. Climate, 22,* 5003–5020.

Holden, P. B., N. R. Edwards, E. W. Wolff, N. J. Lang, J. S. Singarayer, P. J. Valdes, and T. F. Stocker (2010), Interhemispheric coupling, the West Antarctic Ice Sheet and warm Antarctic interglacials, *Clim. Past, 6*(4), 431–443.

Holdren, J. (2010), Climate change science and sanity: Stephen Schneider's contributions to both, keynote address at Public Memorial Service, 12 December 2010, Stanford, Palo Alto, USA, *http://woods.stanford.edu/woods/steve-schneider-memorial.html.* (Accessed 3/5/11.)

Holland, G. J. (1997), The maximum potential intensity of tropical cyclones, *J. Atmos. Sci., 54,* 2519–2541.

Holland, D. M., and A. Jenkins (1999), Modeling thermodynamic ice-ocean interactions at the base of an ice shelf, *J. Phys. Oceanogr., 29*(8), 1787–1800.

Holland, D. M., R. H. Thomas, B. De Young, M. H. Ribergaard, and B. Lyberth (2008), Acceleration of Jakobshavn Isbræ triggered by warm subsurface ocean waters, *Nat. Geosci., 1*(10), 659–664.

Holland, G. J. (2009), Predicting El Niño's impacts, *Science, 325(5936)*, 47, doi:10.1126/science.1176515.

Holland, G. J., and P. J. Webster (2007), Heightened tropical cyclone activity in the North Atlantic: Natural variability or climate trend?, AMS Forum: Climate Change Manifested by Changes in Weather (87. AMS Meeting), San Antonio, TX (USA), 13–18 Jan 2007, 365, 2695–2716.

Holland, M. M., and C. M. Bitz (2003), Polar amplification of climate change in coupled models, *Clim. Dynam., 21(3–4)*, 221–232, doi:10.1007/s00382-003-0332-6.

Holland, M. M., C. M. Bitz, and B. Tremblay (2006), Future abrupt reductions in the summer Arctic sea ice, *Geophys. Res. Lett., 33*, L23503.

Holliday, N. P., S. L. Hughes, S. Bacon, A. Beszczynska-Möller, B. Hansen, A. Lavin, H. Loeng, K. A. Mork, S. Osterhus, T. Sherwin, et al. (2008), Reversal of the 1960s to 1990s freshening trend in the northeast North Atlantic and Nordic Seas, *Geophys. Res. Lett., 35*, L03614.

Holmes, J., G. Bammer, J. Young, M. Sax, and B. Stewart (2010), The science-policy interface, in *Food Security and Global Environmental Change*, edited by J. Ingram, P. Ericksen, and D. Liverman, Earthscan, UK.

Hoor, P., J. Borken-Kleefeld, D. Caro, O. Dessens, Ø. Endresen, M. Gauss, V. Grewe, D. Hauglustaine, I. S. A. Isaksen, P. Jöckel, J. Lelieveld, E. Meijer, D. Olivie, M. Prather, C. Schnadt Poberaj, J. Staehelin, Q. Tang, J. van Aardenne, P. van Velthoven, and R. Sausen (2009), The impact of traffic emissions on atmospheric ozone and OH: Results from QUANTIFY, *Atmos. Chem. Phys., 9*, 3113–3136, doi:10.5194/acp-9-3113-2009.

Hoose, C., U. Lohmann, R. Erdin, and I. Tegen (2008), Global influence of dust mineralogical composition on heterogeneous ice nucleation, *Environ. Res. Lett., 3*, 025003, doi:10.1088/1748-9326/3/2/025003.

Hope, P. K. (2006), Projected future changes in synoptic systems influencing southwest Western Australia, *Clim. Dynam., 26*, 751–764.

Hope, P. K., W. Drosdowsky, and N. Nicholls (2006), Shifts in the synoptic systems influencing southwest Western Australia, *Clim. Dynam., 26*, 751–764.

Horowitz, L. W. (2006), Past, present, and future concentrations of tropospheric ozone and aerosols: Methodology, ozone evaluation, and sensitivity to aerosol wet removal, *J. Geophys. Res., 111*(D22), D22211.

Horowitz, L. W., A. M. Fiore, G. P. Milly, R. C. Cohen, A. Perring, P. J. Wooldridge, P. G. Hess, L. K. Emmons, and J.-F. Lamarque (2007), Observational constraints on the chemistry of isoprene nitrates over the Eastern United States, *J. Geophys. Res., 112*, D12S08, doi:10.1029/2006JD007747.

Horton, E. B., C. K. Folland, and D. E. Parker (2001), The changing incidence of extremes in worldwide and Central England temperatures to the end of the twentieth century, *Climatic Change, 50*, 267–295.

Horton, R., C. Herweijer, C. Rosenzweig, J. Liu, V. Gornitz, and A. C. Ruane (2008), Sea level rise projections for current generation CGCMs based on the semi-empirical method, *Geophys. Res. Lett., 35*(2), L02715.

Hosoda, S., T. Suga, N. Shikama, and K. Mizuno (2009), Global surface layer salinity change detected by ARGO and its implication for hydrological cycle intensification, *J. Oceanogr., 65*(4), 579–586.

Hotton, C. (2002), Palynology of the Cretaceous-Tertiary boundary in Central Montana: Evidence for extraterrestrial impact as a cause of the terminal Cretaceous extinctions, in The Hell Creek formation of the Cretaceous-Tertiary boundary in the Northern Great Plains, *Geol. Soc. Am. Spec. Pap. 361*, edited by J. D. Hartman et al., pp. 473–501.

Houghton, R. A. (2003), Revised estimates of the annual net flux of carbon to the atmosphere from changes in land use and land management 1850–2000, *Tellus B, 55(2)*, 378–390.

Houlton, B. Z., Y.-P. Wang, P. M. Vitousek, and C. B. Field (2008), A unifying framework for dinitrogen fixation in the terrestrial biosphere, *Nature, 454*, 327–331, doi:10.1038/nature07028.

House, J. I., C. Huntingford, W. Knorr, S. E. Cornell, P. M. Cox, G. R. Harris, C. D. Jones, J. A. Lowe, and I. C. Prentice (2008), What do recent advances in quantifying climate and carbon cycle uncertainties mean for climate policy?, *Env. Res. Lett., 3*(4), doi:10.1088/1748–9326/3/4/044002.

Houweling, S., T. Rockmann, I. Aben, F. Keppler, M. Krol, J. F. Meirink, E. J. Dlugokencky, and C. Frankenberg (2006), Atmospheric constraints on global emissions of methane from plants, *Geophys. Res. Lett., 33*, ISI:000239991400002.

Hoyle, C. R., T. Berntsen, G. Myhre, and I. S. A. Isaksen (2007), Secondary organic aerosol in the global aerosol-chemical transport model Oslo CTM2, *Atmos. Chem. Phys., 7*(21), 5675–5694.

Hoyle, C. R., G. Myhre, T. Berntsen, and I. S. A. Isaksen (2009), Anthropogenic influence on SOA and the resulting radiative forcing, *Atmos. Chem. Phys., 9*, 2715–2728

Hoyle, G. (2010), *2011: Living in the Future*, (first published in 1972 by Heinemann, London, UK), Darling and Company.

Hoyos, C. D., P. A. Agudelo, P. J. Webster, and J. A. Curry (2006), Deconvolution of the factors contributing to the increase in global hurricane intensity, *Science, 312(5770)*, 94–97, doi:10.1126/science.1123560.

Hu, F. S., P. E. Higuera, J. E. Walsh, W. L. Chapman, P. A. Duffy, L. B. Brubaker, and M. L. Chipman (2010), Tundra burning in Alaska: Linkage to climatic change and sea ice retreat, *J. Geophys. Res., 115*, G04002, doi:10.1029/2009/JG001270.

Huang, Y. J., H. Akbari, H. Taha, and A. H. Rosenfeld (1987), The potential of vegetation in reducing summer cooling loads in residential buildings, *J. Clim. Appl. Meteorol., 26*, 1103–1116.

Huang, Y., W. L. Chameides, and R. E. Dickinson (2007), Direct and indirect effects of anthropogenic aerosols on regional precipitation over East Asia, *J. Geophys. Res., 112*, D03212, doi:10.1029/2006JD007114.

Hulme, M. (2009), *Why We Disagree about Climate Change: Understanding Controversy, Inaction and Opportunity*, Cambridge University Press, Cambridge, UK.

Hulme, M., and M. Mahoney (2010), Climate change: What do we know about the IPCC?, *Prog. Phys. Geo. 24*, 591–599.

Hulme, M., E. Zorita, J. Price, and J. R. Christy (2010), IPCC: Cherish it, tweak it or scrap it?, *Nature, 463*, 730–732.

Hungate, B. A., P. D. Stiling, P. Dijkstra, D. W. Johnson, M. E. Ketterer, G. J. Hymus, C. R. Hinkle, and B. G. Drake (2004), CO_2 elicits long-term decline in nitrogen fixation, *Science, 304*, 1291, doi:10.1126/science.1095549.

Huntington, T. G. (2008), CO_2-induced suppression of transpiration cannot explain increasing runoff, *Hydrol. Process., 22*, 311–314, doi:10.1002/hyp.6925.

Huntley, B., and T. Webb, III (1989), Migration: Species' response to climatic variations caused by changes in the Earth's orbit, *J. Biogeogr., 16*(1), 5–19.

Hurrell, J. W. (1995), Decadal trends in the North Atlantic Oscillation: Regional temperatures and precipitation, *Science, 269*, 676–679.

Hurrell, J., Y. Kushnir, and M. Visbeck (2001), Climate — The North Atlantic Oscillation, *Science, 291*(5504), 603—605.

Hurrell, J. W., Y. Kushnir, M. Visbeck, and G. Ottersen (2003), An overview of the North Atlantic Oscillation, in *The North Atlantic Oscillation: Climate Significance and Environmental Impact*, edited by J. W. Hurrell et al., pp. 1—35, *Geophys. Monog. Series, 134*.

Hurtt, G. C., S. Frolking, M. G. Fearon, B. Moore III, E. Shevliakova, S. Malyshev, S. W. Pacala, and R. A. Houghton (2006), The underpinnings of land-use history: Three centuries of global gridded land-use transitions, wood harvest activity, and resulting secondary lands, *Glob. Change Biol., 12*, 1208—1229.

Hüsing, S. K., M. H. L. Deenen, J. Koopmans, J. Krijsman, and W. M. Kürschner (2011), Magnetostratigraphic dating of the Rhaetian GSSP at Steinbergkogel (upper Triassic, Austria): Implications for the late Triassic Time Scale, *Earth Planet. Sci. Lett., 302*, 203—216.

Huss, M., R. Hock, A. Bauder, and M. Funk (2010a), 100-year mass changes in the Swiss Alps linked to the Atlantic Multidecadal Oscillation, *Geophys. Res. Lett., 37*(10), L10501.

Huss, M., R. Hock, A. Bauder, and M. Funk (2010b), Reply to the comment of Leclercq, et al. on 100-year mass changes in the Swiss Alps linked to the Atlantic Multidecadal Oscillation, *The Cryosphere Discuss., 4*, 2587—2592.

Hussain, M., and B. E. Lee (1980), A wind tunnel study of the mean pressure forces acting on large groups of low-rise buildings, *J. Wind. Eng. Ind. Aero., 6*, 207—225.

Hutton, J. (1795), *Theory of the Earth with Proofs and Illustrations*, vols. I—II, Creech, Edinburgh.

Hüve, K., I. Bichele, B. Rasulov, and Ü. Ninemets (2011), When it is too hot for photosynthesis: Heat-induced instability of photosynthesis in relation to respiratory burst, cell permeability changes, and H_2O_2 formation, *Plant Cell Environ., 34*, 113—126, doi:10.111/j.1365-3040.2010.02229.x.

Huybers, P., and C. Wunsch (2005), Obliquity pacing of the late-Pleistocene glacial terminations, *Nature, 434*(7032), 491—494, doi:10.1038/nature03401.

Huybrechts, P., and J. De Wolde (1999), The dynamic response of the Greenland and Antarctic Ice Sheets to multiple-century climatic warming, *J. Climate, 12*, 2169—2188.

Huynh, T. T., and C. J. Poulsen (2005), Rising atmospheric CO_2 as a possible trigger for the end-Triassic mass extinction, *Palaeogeogr. Palaeoclim., 217*, 223—242.

IAC (2010), Climate change assessments: Review of the processes and procedures of the IPCC, InterAcademy Council, Amsterdam, *http://reviewipcc.interacademycouncil.net/report.html*. (Accessed 9/9/10.)

Iati, I. (2008), The potential of civil society in climate change adaptation strategies, *Pol. Sci., 60*(1), 19—30.

IDS (2011), Adaptation screening tools for development cooperation: Piloting ORCHID and other approaches, *http://www.ids.ac.uk/climatechange/orchid*. (Accessed 31/3/11.)

Iglesias-Rodriguez, M. D., P. R. Halloran, R. E. M. Rickaby, I. R. Hall, E. Colmenero-Hidalgo, J. R. Gittins, D. R. H. Green, T. Tyrrell, S. J. Gibbs, P. von Dassow, E. Rehm, E. V. Armbrust, and K. P. Boessenkool (2008), Phytoplankton Calcification in a High-CO2 World, *Science, 320*(5874), 336—340.

Ihara, C., Y. Kushnir, and M. A. Cane (2008), Warming trend of the Indian Ocean SST and Indian Ocean dipole from 1880 to 2004, *J. Climate, 21*, 2035—2046.

Imbrie, J., and J. Z. Imbrie (1980), Modelling the climatic response to orbital variations, *Science, 207*(29), 943—953.

Imbrie, J., J. D. Hays, D. G. Martinson, A. McIntyre, A. C. Mix, J. J. Morley, N. G. Pisias, W. L. Prell, and N. J. Shackleton (1984), The orbital theory of Pleistocene climate: Support from a revised chronology of the marine _δ18O record, in *Milankovitch and Climate, Part 1*, edited by A. L. Berger et al., pp. 269—305, D. Reidel, Dordrecht.

Imbrie, J., E. A. Boyle, S. C. Clemens, A. Duffy, W. R. Howard, G. Kukla, J. Kutzbach, D. G. Martinson, A. McIntyre, A. C. Mix, B. Molfino, J. J. Morley, L. C. Peterson, N. G. Pisias, W. L. Prell, M. E. Raymo, N. J. Shackleton, and J. R. Toggweiler (1992), On the structure and origin of major glaciation cycles: 1. Linear responses to Milankovitch forcing, *Paleoceanography, 7*(6), 701—738, doi:10.1029/92PA02855.

Imbrie, J., A. Berger, E. A. Boyle, S. C. Clemens, A. Duffy, W. R. Howard, G. Kukla, J. Kutzbach, D. G. Martinson, A. McIntyre, A. C. Mix, B. Molfino, J. J. Morley, L. C. Peterson, N. G. Pisias, W. L. Prell, M. E. Raymo, N. J. Shackleton, and J. R. Toggweiler (1993), On the structure and origin of major glaciation cycles. 2. The 100,000-year cycle, *Paleoceanography, 8*(6), 699—735.

Imhoff, M. L., B. Lahouari, R. DeFries, W. T. Lawrence, D. Stutzer, C. J. Tucker, and T. Ricketts, (2004), The consequences of urban land transformation on net primary productivity in the United States, *Remote Sens. Environ., 89*, 434—443.

Ingram, J., P. Ericksen, and D. Liverman. (Eds.) (2010), *Food Security and Global Environmental Change*, Earthscan, UK.

IPCC (1990a), *Climate Change: The IPCC Scientific Assessment. Report prepared for Intergovernmental Panel on Climate Change by Working Group I*, edited by J. T. Houghton, G. J. Jenkins, and J. J. Ephraums, Cambridge University Press, Cambridge, UK.

IPCC (1990b), *Climate Change: The IPCC Impacts Assessment. Report prepared for Intergovernmental Panel on Climate Change by Working Group II*, edited by W. J. McG. Tegart, G. W. Sheldon, and D. C. Griffiths, Australian Government Publishing Service, Canberra, Australia, and Cambridge University Press, Cambridge, UK.

IPCC (1990c), *Climate Change: The IPCC Response Strategies. Report prepared for Intergovernmental Panel on Climate Change by Working Group III*, Cambridge University Press, Cambridge, UK.

IPCC (1992), *Climate Change 1992: The Supplementary Report to the IPCC Scientific Assessment Report prepared for Intergovernmental Panel on Climate Change by Working Group I combined with Supporting Scientific Material*, J.T. Houghton, B.A. Callander and S.K. Varney (Eds.), Cambridge University Press, Cambridge, UK.

IPCC (1995a), *Climate Change 1995: The Science of Climate Change. Contribution of Working Group I to the Second Assessment Report of the Intergovernmental Panel on Climate Change*, edited by J. T. Houghton, L. G. Meira Filho, B. A. Callander, N. Harris, A. Kattenberg, and K. Maskell, Cambridge University Press, Cambridge, UK.

IPCC (1995b), *Climate Change 1995: Impacts, Adaptations and Mitigation of Climate Change: Scientific-Technical Analyses Contribution of Working Group II to the Second Assessment Report of the Intergovernmental Panel on Climate Change*, edited by R. T. Watson, M. C. Zinyowera, and R. H. Moss, Cambridge University Press, Cambridge, UK.

IPCC (1995c), *Climate Change 1995: Economic and Social Dimensions of Climate Change. Contribution of Working Group III to the Second Assessment Report of the Intergovernmental Panel on Climate Change*, edited by J. P. Bruce, H. Lee, and E. F. Haites, Cambridge University Press, Cambridge, UK.

IPCC (1999), *Aviation And The Global Atmosphere: A Special Report of IPCC Working Groups I and III*, J. E. Penner, D. H. Lister, D. J.

Griggs, D. J. Dokken, M. McFarland (Eds.), Cambridge University Press, Cambridge, UK.

IPCC (2000), *Emissions Scenarios: A Special Report of Working Group III of the IPCC*, N. Nakicenovic and R. Swart (Eds.), Cambridge University Press, Cambridge. UK.

IPCC (2001a), *Climate Change 2001: The Scientific Basis. Contribution of Working Group I to the Third Assessment Report of the Intergovernmental Panel on Climate Change*, edited by J. T. Houghton, Y. Ding, D. J. Griggs, M. Noguer, P. J. van der Linden, X. Dai, K. Maskell, and C. A. Johnson, Cambridge University Press, Cambridge, UK.

IPCC (2001b), *Climate Change 2001: Impacts, Adaptation and Vulnerability. Contribution of Working Group II to the Third Assessment Report of the Intergovernmental Panel on Climate Change*, edited by J. J. McCarthy, O. F. Canziani, N. A. Leary, D. J. Dokken, and K. S. White, Cambridge University Press, Cambridge, UK.

IPCC (2001c), *Climate Change 2001: Mitigation. Contribution of Working Group III of the Intergovernmental Panel on Climate Change*, edited by B. Metz, O. Davidson, R. Swart, and J. Pan, Cambridge University Press, Cambridge, UK.

IPCC (2001d), *Climate Change 2001: Synthesis Report. Contribution of Working Groups I, II, and III to the Third Assessment Report of the Intergovernmental Panel on Climate Change*, edited by Watson, R.T. and the Core Writing Team, Cambridge University Press, Cambridge, UK.

IPCC/TEAP (2005), *Safeguarding the Ozone Layer and the Global Climate System: Issues Related to Hydrofluorocarbons and Perfluorocarbons*, B. Metz, L. Kuijpers, S. Solomon, S. O. Andersen, O. Davidson, J. Pons, D. de Jager, T. Kestin, M. Manning, and L. Meyer (Eds), Cambridge University Press, Cambridge, UK.

IPCC (2007a), *Climate Change 2007: The Physical Science Basis. Contribution of Working Group I to the Fourth Assessment Report of the Intergovernmental Panel on Climate Change*, edited by S. Solomon, D. Qin, M. Manning, Z. Chen, M. Marquis, K. B. Avery, M. Tignor and H. L. Miller, Cambridge University Press, Cambridge, UK and New York, USA.

IPCC (2007b), *Climate Change 2007: Impacts, Adaptation and Vulnerability. Contribution of Working Group II to the Fourth Assessment Report of the Intergovernmental Panel on Climate Change*, edited by M. L. Parry, O. F. Canziani, J. P. Palutikof, P. J. van der Linden, and C. E. Hanson, Cambridge University Press, Cambridge, UK.

IPCC (2007c), *Climate Change 2007: Mitigation of Climate Change. Contribution of Working Group III to the Fourth Assessment Report of the Intergovernmental Panel on Climate Change*, edited by B. Metz, O. R. Davidson, P. R. Bosch, R. Dave, and L. A. Meyer, Cambridge University Press, Cambridge, UK.

IPCC (2007d), *Climate Change 2007: Synthesis Report. Contribution of Working Groups I, II, and III to the Fourth Assessment Report of the Intergovernmental Panel on Climate Change*, edited by R. K. Pachauri and A. Reisinger, IPCC, Geneva, Switzerland.

IPCC (2009), *Meeting Report of the Expert Meeting on the Science of Alternative Metrics*, Plattner, G-K., Stocker, T.F., Midgley, P., Tignor, M. (Eds.), IPCC Working Group I Technical Support Unit. University of Bern, Bern, Switzerland.

IPCC (2010), *Working groups/task force*, Intergovernmental Panel on Climate Change (IPCC), Geneva, *www.ipcc.ch/working_groups/ working_groups.htm*. (Accessed 27/9/10.)

Irizarry-Ortiz, M. M., G. Wang, and E. A. B. Eltahir (2003), Role of the biosphere in the mid-Holocene climate of West Africa, *J. Geophys. Res., 108*(D2), 4042, doi:10.1029/2001JD000989.

Isaksen, I. S. A. and Ø. Hov (1987), Calculation of trends in the tropospheric concentration of O_3, OH, CO, CH_4 and NO_x, *Tellus B, 39*, 271−285.

Isaksen, I. S. A., E. Hesstvedt, and O. Hov (1978), A chemical model for urban plumes: Test for ozone and particulate sulfur formation in the St. Louis urban plume, *Atmos. Environ., 12*, 599−604.

Isaksen, I. S. A., C. Zerefos, K. Kourtidis, C. Meleti, S. B. Dalsøren, J. K. Sundet, P. Zanis, and D. Balis (2005), Tropospheric ozone changes at unpolluted and semipolluted regions induced by stratospheric ozone changes, *J. Geophys. Res., 110*, DO2302, doi:10.1029/ 2004JD004618.

Isaksen, I. S. A., S. B. Dalsøren, L. Li, and W.-C. Wang (2009), Introduction to special section on East Asia Climate and Environment, *Tellus, 61*(4), 583−589, doi:10.1111/j.1600-0889.2009.00432.x.

Isaksen, I. S. A., C. Granier, G. Myhre, T. K. Berntsen, S. B. Dalsøren, M. Gauss, Z. Klimont, R. Benestad, P. Bousquet, W. Collins, T. Cox, V. Eyring, D. Fowler, S. Fuzzi, P. Jockel, P. Laj, U. Lohmann, M. Maione, P. Monks, A. S. H. Prevot, F. Raes, A. Richter, B. Rognerud, M. Schulz, D. Shindell, D. S. Stevenson, T. Storelvmo, W. -C. Wang, M. van Weele, M. Wild, and D. Wuebbles (2009), Atmospheric composition change: Climate−chemistry interactions, *Atmos. Environ., 43*, 5138−5192.

Ise, T., A. L. Dunn, S. C. Wofsy, and P. R. Moorcroft (2008), High sensitivity of peat decomposition to climate change through watertable feedback, *Nat. Geosci., 1*, 763−766.

Ishii, M., and M. Kimoto (2009), Reevaluation of historical ocean heat content variations with time-varying XBT and MBT depth bias corrections, *J. Oceanogr, 65*(3), 287−299, doi:10.1007/s10872-009-0027-7.

ISO (2006), Greenhouse gases *ISO 14064:2006, http://www.iso.org/iso/ catalogue_detail?csnumber=38381*. (Accessed 9/11/10.)

Iversen, C. M., and R. J. Norby (2008), Nitrogen limitation in a sweetgum plantation: Implications for carbon allocation and storage, *Can. J. Forest Res., 38*, 1021−1032, doi:10.1139/X07-213.

Iversen, T., and E. Joranger (1995), Arctic air pollution and large scale atmospheric flows, *Atmos. Environ., 19*, 2099−2108.

Iversen, T., and Ø. Seland (2002), A scheme for process-tagged SO_4 and BC aerosols in NCAR CCM3: Validation and sensitivity to cloud processes, *J. Geophys. Res., 107*(D24), 4751, doi:10.1029/2001JD000885.

Jablonski, D. (1985), Marine regressions and mass extinctions: A test using modern biota, in *Phanerozoic Diversity Patterns*, edited by J. W. Valentine, pp. 335−354, Princeton University Press, Princeton, NJ.

Jackson, R. B., J. Canadell, J. R. Ehleringer, H. A. Mooney, O. E. Sala, and E. D. Schulze (1996), A global analysis of root distributions for terrestrial biomes, *Oecologia, 108*, 389−411.

Jacob, D. J., and. D. A. Winner (2009), Effect of climate change on air quality, *Atmos. Environ., 43*, 51−63.

Jacobson, M. C., H. C. Hansson, K. J. Noone, and R. J. Charlson (2000), Organic atmospheric aerosols: Review and state of the science, *Rev. Geophys., 38*(2), 267−294.

Jacobson, M. Z. (1999), Isolating nitrated and aromatic aerosols and nitrated aromatic gases as sources of ultraviolet light absorption, *J. Geophys. Res., 104*(D3), 3527−3542.

Jacobson, M. Z. (2001), Global direct radiative forcing due to multicomponent anthropogenic and natural aerosols, *J. Geophys. Res., 106*(D2), 1551−1568.

Jacobson, M. Z. (2002), Control of fossil fuel particulate black carbon and organic matter, possibly the most effective method of slowing global warming, *J. Geophys. Res.*, *107*, doi:10.1029/2001JD001376.

Jacobson, M. Z. (2004), Climate response of fossil fuel and biofuel soot, accounting for soot's feedback to snow and sea ice albedo and emissivity, *J. Geophys. Res.*, *109*, D21201, doi:10.1029/2004JD004945.

Jacobson, M. Z., and D. G. Streets (2009), Influence of future anthropogenic emissions on climate, natural emissions, and air quality, *J. Geophys. Res.*, *114*, D08118, doi:10.1029/2008JD011476.

Jaegle, L., R. V. Martin, K. Chance, L. Steinberger, T. P. Kurosu, D. J. Jacob, A. I. Modi, V. Yoboue, L. Sigha-Nkamdjou, and C. Galy-Lacaux (2004), Satellite mapping of rain induced nitric oxide emissions from soils, *J. Geophys. Res.-Atmos.*, *109*(D21).

Jäger, J. (2008), Foreword, in *Handbook of Transdisciplinary Research*, edited by G. H. Hadorn, H., Hoffman-Riem, S., Biber-Klemm, W., Grossenbacher-Mansuy, D., Joye, C., Pohl, U., Wiesmann, and E., Zemp, Springer, USA.

Jahn, A., M. Claussen, A. Ganopolski, and V. Brovkin (2005), Quantifying the effect of vegetation dynamics on the climate of the Last Glacial Maximum, *Clim. Past*, *1*, 1−7.

Jain, A., X. Yang, H. Kheshgi, A. D. McGuire, W. Post, and D. Kicklighter (2009), Nitrogen attenuation of terrestrial carbon cycle response to global environmental factors, *Global Biogeochem. Cycles*, *23*, GB4028, doi:10.1029/2009GB003519.

Jansen, E., J. Overpeck, K. R. Briffa, J. C. Duplessy, F. Joos, V. Masson-Delmotte, D. Olago, B. Otto-Bliesner, W. R. Peltier, S. Rahmstorf, R. Ramesh, D. Raynaud, D. Rind, O. Solomina, R. Villalba, and D. Zhang (2007), Palaeoclimate, in *Climate Change 2007: The Physical Science Basis, contribution of Working Group I to the Fourth Assessment Report (FAR) of the Intergovernmental Panel on Climate Change (IPCC)*, edited by S. Solomon, et al., pp. 434−497.

Järvi L., C. S. B. Grimmond, and A. Christen (2011), The surface urban energy and water balance scheme (SUEWS): Evaluation in Vancouver and Los Angeles, *J. Hydrol.*, (in review).

Jarvis, P. G. (1976), The interpretation of the variations in leaf water potential and stomatal conductance found in canopies in the field, *Philos. T. Roy. Soc. B*, *273*, 593−610.

Jarvis, P., and S. Linder (2000), Constraints to growth of boreal forests, *Nature*, *405*, 904−905.

Jauregui E., and E. Luyando (1998), Long-term association between pan evaporation and the urban heat island in Mexico City, *Atmosfera*, *11*, 45−60.

Jeffrey, S. W., R. F. C. Mantoura, and S. W. Wright (1997), *Phytoplankton Pigments in Oceanography: Guidelines to Modern Methods*, Unesco.

Jenkins, G., and D. Watts (1968), *Spectral Analysis and Its Applications*, Holden-Day.

Jenkinson, D. S., and K. Coleman (2008), The turnover of organic carbon in subsoils, 2. Modelling carbon turnover, *Euro. J. Soil Sci.*, *59*, 400−413, doi:10.1111/j.1365-2389.2008.01026x.

Jevrejeva, S., J. C. Moore, and A. Grinsted (2010), How will sea level respond to changes in natural and anthropogenic forcings by 2100?, *Geophys. Res. Lett.*, *37*(7), L07703.

Ji, R., M. Edwards, D. L. Mackas, J. A. Runge, and A. C. Thomas (2010), Marine plankton phenology and life history in a changing climate: Current research and future directions, *J. Plankton Res.*, *32*(10), 1355.

Jiang, T., Z. W. Kundzewicz, and B. Su (2008), Changes in monthly precipitation and flood hazard in the Yangtze River Basin, China, *Int. J. Climatol.*, *28*, 1471−1481.

Jin, F.-F., J. D. Neelin, and M. Ghil (1994), El Niño on the Devil's Staircase: Annual subharmonic steps to chaos, *Science*, *264*, 70−72.

Jin, M., R. E. Dickinson, and D. L. Zhang (2005), The footprint of urban areas on global climate as characterized by MODIS, *J. Climate*, *18*, 1551−1565.

Jin, X., and N. Gruber (2003), Offsetting the radiative benefit of ocean iron fertilization by enhancing N_2O emissions, *Geophys. Res. Lett.*, *30*(24), 2249.

Jöckel, P., H. Tost, A. Pozzer, C. Brühl, J. Buchholz, L. Ganzeveld, P. Hoor, A. Kerkweg, M. G. Lawrence, R. Sander, B. Steil, G. Stiller, M. Tanarhte, D. Taraborrelli, J. van Aardenne, and J. Lelieveld (2006), The atmospheric chemistry general circulation model ECHAM5/MESSy1: Consistent simulation of ozone from the surface to the mesosphere, *Atmos. Chem. Phys.*, *6*, 5067−5104.

Jóhannesson, T., C. Raymond, and E. Waddington (1989), Time-scale for adjustment of glaciers to changes in mass balance, *J. Glaciol.*, *35*(121), 355−369.

Johnson, B. T., K. P. Shine, and P. M. Forster (2004), The semi-direct aerosol effect: Impact of absorbing aerosols on marine stratocumulus, *Q. J. Roy. Meteor. Soc.*, *130*, 1407−1422.

Johnson, C., D. Stevenson, W. Collins, and R. Derwent (2001), Role of climate feedback on methane and ozone studied with a coupled ocean-atmosphere-chemistry model, *Geophys. Res. Lett.*, *28*(9), 1723−1726.

Johnson, D. R., M. Ruzek, and M. Kalb (1997), What is Earth system science?, Proceedings of the 1997 International Geoscience and Remote Sensing Symposium, Singapore, August 4−8, pp. 688−691.

Johnson F., and A. Sharma (2009), Assessing future droughts in Australia − A nesting model to correct for long-term persistence in general circulation model precipitation simulations, in: *18th IMACS World Congress and MODSIM09*, vol. 1−6, edited by R. S. Anderson, et al., p. 297, Modelling and Simulation Society of Australia, Cairns.

Johnson, G., and S. Doney (2006), Recent western South Atlantic bottom water warming, *Geophys. Res. Lett.*, *33*(14), L14614.

Johnson, G., S. Mecking, B. Sloyan, and S. Wijffels (2007), Recent bottom water warming in the Pacific Ocean, *J. Climate*, *20*(21), 5365−5375, doi:10.1175/2007JCLI1879.1.

Johnson, G., S. Purkey, and J. Bullister (2008), Warming and freshening in the abyssal Southeastern Indian Ocean, *J. Climate*, *21*(20), 5351−5363.

Johnson, K. R., D. J. Nichols, M. Attrep, Jr., and C. J. Orth (1989), High-resolution leaf-fossil record spanning the Cretaceous/Tertiary boundary, *Nature*, *340*, 708−711.

Jolley, D., I. Gilmour, E. Gurov, S. P. Kelley, and J. Watson (2010), Two large meteorite impacts at the K/Pg boundary, *Geology*, *38*, 835−838.

Jolly, D., S. P. Harrison, B. Damnati, and R. Bonnefille (1998), Simulated climate and biomes of Africa during the Late Quaternary: Comparison with pollen and lake status data, *Quaternary Sci. Rev.*, *17*(6−7), 629−657, doi:10.1016/S0277-3791(98)00015-8.

Jones, A., J. M. Haywood, and O. Boucher (2007), Aerosol forcing, climate response and climate sensitivity in the Hadley Centre Climate Model, *J. Geophys. Res.*, *112*, D20211, doi:10.1029/2007JD008688.

Jones, C., J. Lowe, S. Liddicoat, and R. Betts (2009), Committed ecosystem change due to climate change, *Nat. Geosci.*, *2*, 484−487.

Jones, C., J. Lowe, S. Liddicoat, and R. Betts (2009), Committed terrestrial ecosystem changes due to climate change, *Nat. Geosci.*, *2*, 484−487.

Jones, G. S., P. A. Stott, and N. Christidis (2008), Human contribution to rapidly increasing frequency of very warm Northern Hemisphere summers, *J. Geophys. Res., 113*, D02109.

Jones, M. T., R. S. J. Sparks, and P. J. Valdes (2007), The climatic impact of supervolcanic ash blankets, *Clim. Dynam., 29*, 553–564.

Jones, P. D., and M. E. Mann (2004), Climate over past millennia, *Rev. Geophys., 42*, RG2002, doi:10.1029/2003RG000143.

Jones, P. D., and A. Moberg (2003), Hemispheric and large-scale surface air temperature variations: Extensive revisions and an update to 2001, *J. Climate, 16*(2), 206–223.

Jones, P. D., and T. M. L. Wigley (2010), Estimation of global temperature trends: What's important and what isn't, *Climatic Change, 100*, 59–69, doi:10.1007/s10584-010-9836-3.

Jones, R. G., J. M. Murphy, and M. Noguer (1995), Simulation of climate change over Europe using a nested regional-climate model, 1: Assessment of control climate, including sensitivity to location of lateral boundaries, *Q. J. Roy. Meteor. Soc., 121*, 1413–1449.

Jonson, J. E., D. Simpson, H. Fagerli, and S. Solberg (2006), Can we explain the trends in European ozone levels?, *Atmos. Chem. Phys., 6*, 51–66.

Joos, F. P., and I. C. Prentice (2004), A paleo-perspective on changes in atmospheric CO_2 and climate, in *The Global Carbon Cycle*, edited by C. B. Field and M. R. Raupach, pp. 165–186, Island Press, Washington, DC.

Joos, F., G. K. Plattner, T. F. Stocker, A. Körtzinger, and D. W. R. Wallace (2003), Trends in marine dissolved oxygen: Implications for ocean circulation changes and the carbon budget, *Eos Trans. AGU, 84*(21), 197.

Joos, F. P., S. Gerber, I. C. Prentice, B. L. Otto-Bliesner, and P. J. Valdes (2004), Transient simulations of Holocene atmospheric carbon dioxide and terrestrial carbon since the Last Glacial Maximum, *Global Biogeochem. Cycles, 18*(2), GB2002, doi:10.1029/2003GB002156.

Jorgenson, M. T., Y. L. Shur, and E. R. Pullman (2006), Abrupt increase in permafrost degradation in Arctic Alaska, *Geophys. Res. Lett., 33*(354), L02503, doi:10.1029/2005GL024960.

Joshi, M., K. P. Shine, M. Ponater, N. Stuber, R. Sausen, and L. Li (2003), A comparison of climate response to different radiative forcings in three general circulation models: Towards an improved metric of climate change, *Clim. Dynam., 20*(7–8), 843–854.

Joshi, M. M., J. M. Gregory, M. J. Webb, D. M. H. Sexton, and T. C. Johns (2008), Mechanisms for the land/sea warming contrast exhibited by simulations of climate change, *Clim. Dynam., 30*(5), 455–465, doi:10.1007/s00382-007-0306-1.

Joughin, I., S. B. Das, M. A. King, B. E. Smith, I. M. Howat, and T. Moon (2008), Seasonal speedup along the western flank of the Greenland Ice Sheet, *Science, 320*, 781–783.

Joussaume, S., and P. Braconnot (1997), Sensitivity of paleoclimate simulation results to season definitions, *J. Geophys. Res., 102*(D2), 1943–1956.

Joussaume, S., K. E. Taylor, P. Braconnot, J. F. B. Mitchell, J. E. Kutzbach, S. P. Harrison, I. C. Prentice, A. J. Broccoli, A. Abe-Ouchi, P. J. Bartlein, C. Bonfils, B. Dong, J. Guiot, K. Herterich, C. D. Hewitt, D. Jolly, J. W. Kim, A. Kislov, A. Kitoh, M. Loutre, V. Masson, B. McAvaney, N. McFarlane, N. de Noblet, W. R. Peltier, J. Y. Peterschmitt, D. Pollard, D. Rind, J. F. Royer, M. E. Schlesinger, J. Syktus, S. Thompson, P. Valdes, G. Vettoretti, R. S. Webb, and U. Wyputta (1999), Monsoon changes for 6000 years ago: Results of 18 simulations from the Paleoclimate Modeling Intercomparison Project (PMIP), *Geophys. Res. Lett., 26*(7), 859–862, doi:10.1029/1999GL900126.

Jouzel, J., F. Vimeux, N. Caillon, G. Delaygue, G. Hoffmann, V. Masson-Delmotte, and F. Parrenin (2003), Magnitude of isotope/temperature scaling for interpretation of central Antarctic ice cores, *J. Geophys. Res., 108*(D12), 4361, doi:10.1029/2002JD002677.

Jouzel, J., V. Masson-Delmotte, O. Cattani, G. Dreyfus, S. Falourd, G. Hoffmann, B. Minster, J. Nouet, J. M. Barnola, J. Chappellaz, H. Fischer, J. C. Gallet, S. Johnsen, M. Leuenberger, L. Loulergue, D. Luethi, H. Oerter, F. Parrenin, G. Raisbeck, D. Raynaud, A. Schilt, J. Schwander, E. Selmo, R. Souchez, R. Spahni, B. Stauffer, J. P. Steffensen, B. Stenni, T. F. Stocker, J. L. Tison, M. Werner, and E. W. Wolff (2007a), Orbital and millennial Antarctic climate variability over the past 800,000 years, *Science, 317*(5839), 793–796.

Jouzel, J., M. Stievenard, S. J. Johnsen, A. Landais, V. Masson-Delmotte, A. Sveinbjornsdottir, F. Vimeux, U. von Grafenstein, and J. W. White (2007b), The GRIP deuterium-excess record, *Quaternary Sci. Rev., 26*(1–2), 1–17, doi:10.1016/j.quascirev.2006.07.015.

Juday, G. P., V. Barber, P. Duffy, H. Linderholm, S. Rupp, S. Sparrow, E. Vaganov, and J. Yarie (2005), Forests, land management, and agriculture, in *Arctic Climate Impact Assessment, Scientific Report*, pp. 781–862, Cambridge University Press, Cambridge, UK, and New York, NY.

Judt, F., and S. Chen (2010), Convectively generated potential vorticity in rainbands and formation of the secondary eyewall in Hurricane Rita of 2005, *J. Atmos. Sci., 67*(11), 3581–3599, doi:10.1175/2010JAS3471.1.

June, T., J. R. Evans, and G. D. Farquhar (2004), A simple new equation for the reversible temperature dependence of photosynthetic transport: a study on soybean leaf, *Funct. Plant Biol., 31*, 275–283, doi:10.1071/FP03250.

Jungclaus, J. H., S. J. Lorenz, C. Timmreck, C. H. Reick, V. Brovkin, K. Six, J. Segschneider, M. A. Giorgetta, T. J. Crowley, J. Pongratz, N. A. Krivova, L. E. Vieira, S. K. Solanki, D. Klocke, M. Botzet, M. Esch, V. Gayler, H. Haak, T. J. Raddatz, E. Roeckner, R. Schnur, H. Widmann, M. Claussen, B. Stevens, and J. Marotzke (2010), Climate and carbon-cycle variability over the last millennium, *Clim. Past, 6*, 723–737, doi:10.5194/cp-6-723-2010.

Junge, C. E., and J. E. Manson (1961), Stratospheric aerosol studies, *J. Geophys. Res., 66*, 2163–2182.

Kageyama, M., S. P. Harrison, and A. Abe-Ouchi (2005), The depression of tropical snowlines at the Last Glacial Maximum: What can we learn from climate model experiments?, *Quatern. Int., 138–139*, 202–219.

Kageyama, M., A. Paul, D. M. Roche, C. J. Van Meerbeeck (2010), Modelling glacial climatic millennial-scale variability related to changes in the Atlantic Meridional Overturning Circulation: A review, *Quaternary Sci. Rev.*, doi:10.1016/j.quascirev.2010.05.029.

Kallache, M., M. Vrac, P. Naveau, and P.-A. Michelangeli (2011), Nonstationary probabilistic downscaling of extreme precipitation, *J. Geophys. Res., 116*, D05113, doi:10.1029/2010JD014892.

Kallis, G. (2008), Droughts, *Annu. Rev. Env. Resour., 33*, 85–118.

Kalnay, E. (2003), *Atmospheric modelling, Data Assimilation and Predictability*, Cambridge University Press, New York.

Kalnay, E., M. Kanamitsu, R. Kistler, W. Collins, D. Deaven, L. Gandin, M. Iredell, S. Saha, G. White, J. Woollen, Y. Zhu, M. Chelliah, W. Ebisuzaki, W. Higgins, J. Janowiak, K. C. Mo, C. Ropelewski, J. Wang, A. Leetmaa, R. Reynolds, R. Jenne, and D. Joseph (1996),

The NCEP/NCAR 40-Year Reanalysis Project, *Bull. Am. Meteorol. Soc., 77*(3), 437–471.

Kamo, S. L., G. K. Czamanske, Y. Amelin, V. A. Fedorenko, D. W. Davis, and V. R. Trofimov (2003), Rapid eruption of Siberian flood-volcanic rocks and evidence for coincidence with the Permian-Triassic boundary and mass extinction at 251 Ma, *Earth Planet. Sci. Lett., 214*, 75–91.

Kanakidou, M., J. H. Seinfeld, S. N. Pandis, I. Barnes, F. J. Dentener, M. C. Facchini, R. van Dingenen, B. Ervens, A. Nenes, C. J. Nielsen, E. Swietlicki, J. P. Putaud, Y. Balkanski, S. Fuzzi, J. Horth, G. K. Moortgat, R. Winterhalter, C. E. L. Myhre, K. Tsigaridis, E. Vignati, E. G. Stephanou, and J. Wilson (2005), Organic aerosol and global climate modelling: A review, *Atmos. Chem. Phys., 5*, 1053–1123.

Kanamaru, H., and M. Kanamitsu (2007), Scale-selective bias correction in a downscaling of global analysis using a regional model, *Mon. Weather Rev., 135*, 334–350.

Kanda, M. (2006), Progress in the scale modelling of urban climate: Review, *Theor. Appl. Climatol., 84*, 23–33, doi:10.1007/s00704-005-0141-4.

Kandlikar, M. (1995), The relative role of trace gas emissions in greenhouse abatement policies, *Energ. Policy, 23*(10), 879–883.

Kantz, H., and T. Schreiber (2004), *Nonlinear Time Series Analysis*, 2nd edition, Cambridge University Press, Cambridge, UK.

Kaplan, J. O. (2002), Wetlands at the Last Glacial Maximum: Distribution and methane emissions, *Geophys. Res. Lett., 29*(6), 1079.

Kaplan, J. O., N. H. Bigelow, I. C. Prentice, S. P. Harrison, P. J. Bartlein, T. R. Christensen, W. Cramer, N. V. Matveyeva, A. D. McGuire, D. F. Murray, V. Y. Razzhivin, B. Smith, D. A. Walker, P. M. Anderson, A. A. Andreev, L. B. Brubaker, M. E. Edwards, and A. V. Lozhkin (2003), Climate change and Arctic ecosystems: 2. Modeling, paleodata-model comparisons, and future projections, *J. Geophys. Res.-Atmos., 108*(D19).

Karl, T. R., and R. W. Knight (1997), The 1995 Chicago heat wave: How likely is a recurrence?, *Bull. Am. Meteorol. Soc., 78*, 1107–1119.

Karl, T. R., H. F. Diaz, and G. Kukla (1988), Urbanization: Its detection and effect in the United States climate record, *J. Climate, 1*, 1099–1123.

Karlsdóttir, S., I. S. A. Isaksen, G. Myhre, and T. K. Berntsen (2000), Trend analysis of O_3 and CO in the period 1980–1996: A three-dimensional model study, *J. Geophys. Res., 105*(D23), 28907–28934.

Karoly, D. J., and Q. G. Wu (2005), Detection of regional surface temperature trends, *J. Climate, 21*, 4337–4343.

Kaser, G., J. G. Cogley, M. B. Dyurgerov, M. F. Meier, and A. Ohmura (2006), Mass balance of glaciers and ice caps: Consensus estimates for 1961–2004, *Geophys. Res. Lett., 33*, L19501, doi:10.1029/2006GL027511.

Kasischke, E. S., and M. R. Turetsky (2006), Recent changes in the fire regime across the North American boreal region – Spatial and temporal patterns of burning across Canada and Alaska, *Geophys. Res. Lett., 33*, L09703, doi:10.1029/2006GL025677.

Kasischke, E. S., D. L. Verbyla, T. S. Rupp, A. D. McGuire, K. A. Murphy, R. Jandt, J. L. Barnes, E. E. Hoy, P. A. Duffy, M. Calef, and M. R. Turetsky (2010), Alaska's changing fire regime – Implications for the vulnerability of its boreal forests, *Can. J. Forest Res., 40*, 1313–1324.

Kaster-Klein, P., and M. W. Rotach (2004), Mean flow and turbulence characteristics in an urban roughness sublayer, *Bound.-Lay. Meteorol., 111*, 55–84.

Katz, R. W. (1999), Extreme value theory for precipitation: Sensitivity analysis for climate change, *Adv. Water Resour., 23*, 133–139.

Katz, R. W., and B. G. Brown (1992), Extreme events in a changing climate: Variability is more important than averages, *Climatic Change, 21*, 289–302.

Katzfey, J. (1995), Simulation of extreme New Zealand precipitation events, 1. Sensitivity to orography and resolution, *Mon. Weather Rev., 123*(3), 737–754.

Kaufman, D. S., T. A. Ager, N. J. Anderson, P. M. Anderson, J. T. Andrews, P. J. Bartlein, L. B. Brubaker, L. L. Coats, L. C. Cwynar, M. L. Duvall, A. S. Dyke, M. E. Edwards, W. R. Eisner, K. Gajewski, A. Geirsdottir, F. S. Hu, A. E. Jennings, M. R. Kaplan, M. N. Kerwin, A. V. Lozhkin, G. M. MacDonald, G. H. Miller, C. J. Mock, W. W. Oswald, B. L. Otto-Bliesner, D. F. Porinchu, K. Ruhland, J. P. Smol, E. J. Steig, and B. B. Wolfe (2004), Holocene thermal maximum in the Western Arctic (0–180 degrees W), *Quaternary Sci. Rev., 23*(5–6), 529–560, doi:10.1016/j.quascirev.2003.09.007.

Kaufman, Y. J., D. Tanre, and O. Boucher (2002), A satellite view of aerosols in the climate system, *Nature, 419*(6903), 215–223.

Kaufmann, R. K., and D. I. Stern (2002), Cointegration analysis of hemispheric temperature relations, *J. Geophys. Res., 107*(D2), 4012, doi:10.1029/2000JD000174.

Kawaragi, K., et al. (2009), Direct measurements of chemical composition of shock-induced gases from calcite: An intense global warming after the Chicxulub impact due to the indirect greenhouse effect of carbon monoxide, *Earth Planet. Sci. Lett., 282*, 56–64, doi:10.1016/j.epsl.2009.02.037.

Kawase, H., T. Yoshikane, M. Hara, F. Kimura, T. Yasunari, B. Ailikun, H. Ueda, and T. Inoue (2009), Intermodel variability of future changes in the Baiu rainband estimated by the pseudo global warming downscaling method, *J. Geophys. Res., 114*, D24110, doi:10.1029/2009JD011803.

Kay, J. E., T. L'Ecuyer, A. Gettelman, G. Stephens, and C. O'Dell (2008), The contribution of cloud and radiation anomalies to the 2007 Arctic sea ice extent minimum, *Geophys. Res. Lett., 35*, L08503.

Kaya, Y., and K. Yokobori (1997), *Environment, Energy, and Economy: Strategies for Sustainability*, UN University Press, Tokyo.

Kaya, Y., and K. Yokobori (2007), *Environment, Energy, and Economy: Strategies for Sustainability*, UN University Press, Tokyo.

Kayen, R. E., and H. J. Lee (1991), Pleistocene slope instability of gas hydrate-laden sediment of Beaufort Sea margin, *Mar. Geotechnol., 10*, 125–141.

Keating, T. J., J. J. West, and A. E. Farrell (2004), Prospects for international management of inter-continental air pollution transport, in *Intercontinental Transport of Air Pollution*, edited by A. Stohl, pp. 295–320, Springer, Berlin.

Keeling, R. F., and H. E. Garcia (2002), The change in oceanic O_2 inventory associated with recent global warming, *Proc. Natl. Acad. Sci. USA, 99*(12), 7848.

Keeling, R. F., A. Körtzinger, and N. Gruber (2010), Ocean deoxygenation in a warming world, *Annu. Rev. Marine Sci., 2*(1), 199–229, doi:10.1146/annurev.marine.010908.163855.

Keenan, T., S. Sabate, and C. Gracia (2010), The importance of mesophyll conductance in regulation forest ecosystem productivity during drought periods, *Glob. Change Bio., 16*, 1019–1034, doi:10.1111/j.1365-2486.2009.02017.x.

Keenan, T., P. May, G. Holland, S. Rutledge, R. Carbone, J. Wilson, M. Moncrieff, A. Crook, T. Takahashi, N. Tapper, M. Platt, J. Hacker, S. Sekelsky, K. Saito, and K. Gage (2000), The Maritime

Continent Thunderstorm Experiment (MCTEX): Overview and some results, *Bull. Amer. Meteor. Soc., 81*, 2433–2455.

Keenlyside, N. S., and J. Ba (2010), Prospects for decadal climate prediction, *Wiley Interdiscip. Rev.: Clim. Change, 1*, doi:10.1002/wcc.69.

Keller, G., T. Adatter, S. Gardin, A. Bartolini, and S. Bajpai, (2008), Main Deccan volcanism phase ends near the K-T boundary: Evidence from the Krishna-Godavari basin, SE India, *Earth Planet. Sci. Lett., 268*, 293–311.

Kennedy, M., D. Mrofka, and C. von der Borch (2008), Snowball Earth termination by destabilization of equatorial permafrost methane clathrate, *Nature, 453*, 642–645.

Kennett, J. P., K. G. Cannariato, I. L. Hendy, and R. J. Behl (2000), Carbon isotope evidence for methane hydrate instability during Quaternary interstadials, *Science, 288*, 128–133.

Kent, D. V., and L. Tauxe (2005), Corrected late Triassic latitudes for continents adjacent to the North Atlantic, *Science, 307*, 240–244.

Kenyon, J., and G. C. Hegerl (2008), Influence of modes of climate variability on global temperature extremes, *J. Climate, 21*, 3872–3889.

Kenyon, J., and G. C. Hegerl (2010), Influence of modes of climate variability on global precipitation extremes, *J. Climate, 23*, 6248–6262.

Kerr, R. A. (2000), A North Atlantic climate pacemaker for the centuries, *Science, 288*(5473), 1984–1985, doi:10.1126/science.288.5473.1984.

Kerr, R. A. (2005), Global climate change: The Atlantic conveyor may have slowed, but don't panic yet, *Science, 310*(5753), 1403.

Kerr, R. A. (2006), Yes, it's been getting warmer in here since the CO_2 began to rise, *Science, 312*, 1854.

Kershaw, A. P., J. S. Clark, A. M. Gill, and D. M. D'Costa (2002), A history of fire in Australia, in *Flammable Australia; The Fire Regimes and Biodiversity of a Continent*, edited by R. Bradstock, et al., Cambridge University Press, Cambridge, UK.

Key, R. M., A. Kozyr, C. L. Sabine, K. Lee, R. Wanninkhof, J. L. Bullister, R. A. Feely, F. J. Millero, C. Mordy, and T. H. Peng (2004), A global ocean carbon climatology: Results from Global Data Analysis Project (GLODAP). *Global Biogeochem. Cycles, 18*(4), GB4031, doi:10.1029/2004GB002247.

Khalil, A. F., H. H. Kwon, and U. Lall (2010), Predictive downscaling based on non-homogeneous hidden Markov models, *Hydrolog. Sci. J., 55*, 333–350.

Kharin, V. V., and F. W. Zwiers (2005), Estimating extremes in transient climate change simulations, *J. Climate, 18*, 1156–1173.

Kharin, V. V., F. Zwiers, X. Zhang, and G. C. Hegerl (2007), Changes in temperature and precipitation extremes in the IPCC Ensemble of Global Coupled Model Simulations, *J. Climate, 20*, 1419–1444.

Khodri, M., Y. Leclainche, G. Ramstein, P. Braconnot, O. Marti, and E. Cortijo (2001), Simulating the amplification of orbital forcing by ocean feedbacks in the last glaciations, *Nature, 410*(6828), 570–574.

Khvorostyanov, D. V., G. Krinner, P. Ciais, M. Heimann, and S. A. Zimov (2008), Vulnerability of permafrost carbon to global warming, 1. Model description and role of heat generated by organic matter decomposition, *Tellus B, 60*, 250–264, doi:10.1111/j.1600-0889.2007.00333.x.

Khvorostyanov, D. V., P. Ciais, G. Krinner, and S. A. Zimov (2008a), Vulnerability of East Siberia's frozen carbon stores to future warming, *Geophys. Res. Lett., 35*, L10703, doi:10.1029/2008GL033639.

Khvorostyanov, D. V., P. Ciais, G. Krinner, S. A. Zimov, Ch. Corrado, and G. Guggenberger (2008b), Vulnerability of permafrost carbon to global warming. Part II: sensitivity of permafrost carbon stock to global warming, *Tellus, 60B*, 265–275.

Kida, H., T. Koide, H. Sasaki, and M. Chiba (1991), A new approach for coupling a limited area model to a GCM for regional climate simulations, *J. Meteorol Soc. Jpn., 69*, 723–728.

Kiessling, W., and R. C. Baron-Szabo (2004), Extinction and recovery patterns of scleractinian corals at the Cretaceous-Tertiary boundary, *Palaeogeogr. Palaeoclim., 214*, 195–223.

Kiessling, W., and C. Simpson (2011), On the potential for ocean acidification to be a general cause of ancient reef crisis, *Glob. Change Biol., 17*, 56–67.

Kiessling, W., M. Aberhan, B. Brenneis, and P. J. Wagner (2007), Extinction trajectories of benthic organisms across the Triassic/Jurassic boundary, *Palaeogeogr. Palaeoclim., 244*, 201–222.

Kiessling, W., W. Roniewicz, L. Villier, P. Léonide, and U. Struck (2009), An early Hettangian coral reef in southern France: Implications for the end-Triassic reef crisis, *Palaios, 24*, 657–671.

Kiktev, D., D. M. H. Sexton, L. V. Alexander, and C. K. Folland (2003), Comparison of modeled and observed trends in indicators of daily climate extremes, *J. Climate, 16*, 3560–3571.

Kiktev, D., J. Caesar, L. V. Alexander, H. Shiogama, and M. Collier (2007), Comparison of observed and multimodelled trends in annual extremes of temperature and precipitation, *Geophys. Res. Lett., 34*, L10702, doi:10.1029/2007GL029539.

Kikuchi, R., and M. Vanneste (2010), A theoretical exercise in the modelling of ground-level ozone resulting from the K-T asteroid impact: Its possible link with the extinction selectivity of terrestrial vertebrates, *Palaeogeogr. Palaeoclim., 288*, 14–23.

Kiladis, G. N., and H. F. Diaz (1989), Global climatic anomalies associated with extremes in Southern Oscillation, *J. Climate, 2*, 1069–1090.

Kilsby, C. G., P. D. Jones, and A. Burton (2007), A daily weather generator for use in climate change studies, *Environ. Modell. Softw., 22*(12), 1705–1719.

Kim, B.-G., M. A. Miller, S. E. Schwartz, Y. Liu, and Q. Min (2008), The role of adiabaticity in the aerosol first indirect effect, *J. Geophys. Res., 113*, doi:10.1029/2007JD008961.

Kim, E., and S. Hong (2010), Impact of air-sea interaction on East Asian summer monsoon climate in WRF, *J. Geophys. Res. – Atmos., 115*, doi:10.1029/2009JD013253.

Kim, J. H., N. Rimbu, S. J. Lorenz, G. Lohmann, S. I. Nam, S. Schouten, C. Rühlemann, and R. R. Schneider (2004), North Pacific and North Atlantic sea-surface temperature variability during the Holocene, *Quaternary Sci. Rev., 23*(20–22), 2141–2154, doi:10.1016/j.quascirev.2004.08.010.

Kim, S. (2010), Modelling of precipitation downscaling using MLP-NNM and SVM-NNM approach, *Disaster Adv., 3*(4), 13–23.

Kim, S. W., A. Heckel, S. A. McKeen, G. J. Frost, E.-Y. Hsie, M. K. Trainer, A. Richter, J. P. Burrows, S. E. Peckham, and G. A. Grell (2006), Satellite-observed US power plant NO_x emission reductions and their impact on air quality, *Geophys. Res. Lett., 33*(22).

Kinne, S., M. Schulz, C. Textor, S. Guibert, Y. Balkanski, S. E. Bauer, T. Berntsen, T. F. Berglen, O. Boucher, M. Chin, W. Collins, F. Dentener, T. Diehl, R. Easter, J. Feichter, D. Fillmore, S. Ghan, P. Ginoux, S. Gong, A. Grini, J. Hendricks, M. Herzog, L. Horowitz, I. Isaksen, T. Iversen, A. Kirkevag, S. Kloster, D. Koch, J. E. Kristjansson, M. Krol, A. Lauer, J. F. Lamarque, G. Lesins, X. Liu, U. Lohmann, V. Montanaro, G. Myhre, J. E. Penner, G. Pitari, S. Reddy, O. Seland, P. Stier, T. Takemura, X. Tie (2005), An

AeroCom initial assessment: Optical properties in aerosol component modules of global models, *Atmos. Chem. Phys., 5*, 8285–8330.

Kinne, S., M. Schulz, C. Textor, S. Guibert, Y. Balkanski, S. E. Bauer, T. Berntsen, T. F. Berglen, O. Boucher, M. Chin, W. Collins, F. Dentener, T. Diehl, R. Easter, J. Feichter, D. Fillmore, S. Ghan, P. Ginoux, S. Gong, A. Grini, J. Hendricks, M. Herzog, L. Horowitz, I. S. A. Isaksen, T. Iversen, A. Kirkevåg, S. Kloster, D. Koch, J. E. Kristjansson, M. Krol, A. Lauer, J. F. Lamarque, G. Lesins, X. Liu, U. Lohmann, V. Montanaro, G. Myhre, J. E. Penner, G. Pitari, S. Reddy, O. Seland, P. Stier, T. Takemura, and X. Tie (2006), An AeroCom initial assessment – optical properties in aerosol component modules of global models, *Atmos. Chem. Phys., 6*, 1815–1834.

Kirchner, J. W. (1989), The Gaia hypothesis: Can it be tested?, *Rev. Geophys., 27*(2), 223–235.

Kirono, D. G. C., and D. M. Kent (2010), Assessment of rainfall and potential evaporation from global climate models and its implications for Australian regional drought projection, *Int. J. Climatol.*, doi:10.1002/joc.2165.

Kjellstrom, E., and F. Giorgi (2010), Regional climate model evaluation and weighting, *Clim. Res., 44*(2–3), 117–119, doi:10.3354/cr00976.

Klaas, C., and D. E. Archer (2002), Association of sinking organic matter with various types of mineral ballast in the deep sea: Implications for the rain ratio, *Global Biogeochem. Cycles, 16*(4), 1116.

Klatt, O., O. Boebel, and E. Fahrbach (2007), A profiling float's sense of ice, *J. Atmos. Oceanic Technol., 24*(7), 1301–1308.

Kleber, M., P. S. Nico, A. Plante, T. Filleys, M. Kramer, C. Swanson, and P. Sollins (2011), Old and stable soil organic matter is not necessarily chemically recalcitrant: implications for modeling concepts and temperature sensitivity, *Glob. Change Bio., 17*, 1097–1107, doi:10.1111/j.1365-2486.2010.02278.x.

Kleeman, M. J. (2007), A preliminary assessment of the sensitivity of air quality in California to global change, *Climatic Change, 87*, S273–S292.

Kleffmann, J., T. Gavriloaiei, A. Hofzumahaus, F. Holland, R. Koppmann, L. Rupp, E. Schlosser, M. Siese, and A. Wahner (2005), Daytime formation of nitrous acid: A major source of OH radicals in a forest, *Geophys. Res. Lett., 32*, L05818, doi:10.1029/2005GL022524.

Kleidon, A., and M. Heimann (2000), Assessing the role of deep rooted vegetation in the climate system with model simulations: Mechanism, comparison to observations, and implications for Amazonian deforestation, *Clim. Dynam., 16*, 183–199.

Klein Goldewijk, K. (2001), Estimating global land use change over the past 300 years: The HYDE Database, *Global Biogeochem. Cy., 15*, 417–434.

Klein Tank, A. M. G., and G. P. Können (2003), Trends in indices of daily temperature and precipitation extremes in Europe, 1946–1999, *J. Climate, 16*, 3665–3680.

Klein Tank, A. M. G., J. B. Wijngaard, G. P. Können, R. Böhm, G. Demarée, A. Gocheva, M. Mileta, S. Pashiardis, L. Hejkrlik, C. Kern-Hansen, R. Heino, P. Bessemoulin, G. Müller-Westmeier, M. Tzanakou, S. Szalai, T. Pálsdóttir, D. Fitzgerald, S. Rubin, M. Capaldo, M. Maugeri, A. Leitass, A. Bukantis, R. Aberfeld, A. V. F. van Engelen, E. Forland, M. Mietus, F. Coelho, C. Mares, V. Razuvaev, E. Nieplova, T. Cegnar, J. Antonio López, B. Dahlström, A. Moberg, W. Kirchhofer, A. Ceylan, O. Pachaliuk, L. V. Alexander, and P. Petrovic (2002), Daily surface air temperature and precipitation dataset 1901–1999 for European Climate Assessment (ECA), *Int. J. Climatol., 22*, 1441–1453.

Klein Tank, A. M. G., T. C. Peterson, D. A. Quadir, S. Dorji, Z. Xukai, T. Hongyu, K. Santhosh, U. R. Joshi, A. K. Jaswal, R. K. Kolli, A. Sikder, N. R. Deshpande, J. Revadekar, K. Yeleuova, S. Vandasheva, M. Faleyeva, P. Gomboluudev, K. P. Budhathoki, A. Hussain, M. Afzaal, L. Chandrapala, H. Anvar, D. Amanmurad, V. S. Asanova, P. D. Jones, M. G. New, and T. Spektorman (2006), Changes in daily temperature and precipitation extremes in Central and South Asia, *J. Geophys. Res., 111*, D16105, doi:10.1029/2005JD006316.

Klein Tank, A. M. G., F. W. Zwiers, and X. Zhang (2009), Guidelines on analysis of extremes in a changing climate in support of informed decisions for adaptation, Climate Data and Monitoring, *WCDMP-72, WMO-TD 1500.*

Klein, J. T. (2004), Prospects for transdisciplinarity, *Futures, 36*, 512–526.

Kleinen, T., H. Held, and G. Petschel-Held (2003), The potential role of spectral properties in detecting thresholds in the Earth system: Application to the thermohaline circulation, *Ocean Dynam., 53*, 53–63.

Kleypas, J. A., R. W. Buddemeier, D. Archer, J. P. Gattuso, C. Langdon, and B. N. Opdyke (1999), Geochemical consequences of increased atmospheric carbon dioxide on coral reefs, *Science, 284*, 118–120.

Klimont, Z., and D. G. Streets (2007), Emissions inventories and projections for assessing hemispheric or intercontinental transport, in Hemispheric transport of air pollution 2007, edited by T. Keating, and A. Zuber, *Atmospheric Pollution Studies No. 16, ECE/EB.AIR/94*, United Nations, Geneva.

Klimont, Z., J. Cofala, J. Xing, W. Wei, C. Zhang, S. Wang, J. Kejun, P. Bhandari, R. Mathura, P. Purohit, P. Rafaj, A. Chambers, M. Amann, and J. Hao (2009), Projections of SO_2, NO_x, and carbonaceous aerosols emissions in Asia., *Tellus B, 61*(4), doi:10.1111/j.1600-0889.2009.00428.x.

Kloster, S., F. Dentener, J. Feichter, F. Raes, U. Lohmann, E. Roeckner, and I. Fischer-Bruns (2010), A GCM study of future climate response to aerosol pollution reductions, *Clim. Dynam., 34*(7), 1177–1194.

Knapp, A. K., et al. (2008), Shrub encroachment in North American grasslands: Shifts in growth form dominance rapidly alters control of ecosystem carbon inputs, *Glob. Change Bio., 14*, 615–623, doi:10.111/j.1365-2486.2007.01512x.

Knight, J. R., R. J. Allan, C. K. Folland, M. Vellinga, and M. E. Mann (2005), A signature of persistent natural thermohaline circulation cycles in observed climate, *Geophys. Res. Lett., 32*(20), L20708, doi:10.1029/2005GL024233.

Knoll, A. H., R. K. Bambach, J. P. Grotzinger, and D. Canfield (1996), Comparative Earth history and late-Permian mass extinction, *Science, 273*, 452–457.

Knoll, A. H., R. K. Bambach, J. L. Payne, S. Pruss, and W. W. Fischer (2007), Paleophysiology and the end-Permian mass extinction, *Earth Planet. Sci. Lett., 256*, 295–313.

Knorr, W., T. Kaminski, M. Scholze, N. Gobron, B. Pinty, R. Giering, and P.-P. Mathieu (2010), Carbon cycle data assimilation with a generic phenology model, *J. Geophys. Res., 115*, G04017, doi:10.1029/2009JG001119.

Knudsen, M. F., and P. Riisager (2009), Is there a link between Earth's magnetic field and low-latitude precipitation?, *Geology, 37*(1), 71–74, doi:10.1130/G25238A.1.

Knutson, T. R., J. J. Sirutis, S. T. Garner, I. M. Held, and R. E. Tuleya, (2007), Simulation of the recent multidecadal increase of Atlantic Hurricane activity using an 18-km-Grid Regional Model, *Bull. Am. Meteorol. Soc., 88*, 1549−1565, doi:10.1175/BAMS-88-10-1549.

Knutson, T. R., J. J. Sirutis, S. T. Garner, G. A. Vecchi, and I. M. Held (2008), Simulated reduction in Atlantic hurricane frequency under twenty-first-century warming conditions, *Nat. Geosci., 1*, 359−364, doi:10.1038/ngeo202.

Knutson, T. R., J. L. McBride, J. Chan, K. Emanuel, G. Holland, C. Landsea, I. Held, J. P. Kossin, A. K. Srivastava, and M. Sugi (2010), Tropical cyclones and climate change, *Nat. Geosci., 3*, 157−163, doi:10.1038/ngeo779.

Knutti, R. (2010), The end of model democracy? (An editorial comment), *Climatic Change, 102*, 395−404, doi:10.1007/s10584-010-9800-2.

Knutti, R., G. A. Meehl, M. R. Allen, and D. A. Stainforth (2006), Constraining climate sensitivity from the seasonal cycle in surface temperature, *J. Climate, 19*, 4224−4233.

Knutti, R., G. Abramowitz, M. Collins, V. Eyring, P. J. Gleckler, B. Hewitson, and L. O. Mearns (2010), Good practice guidance paper on assessing and combining multi model climate projections, in *Meeting Report of the Intergovernmental Panel on Climate Change Expert Meeting on Assessing and Combining Multi Model Climate Projections*, edited by T. F. Stocker, D. Qin, G.-K. Plattner, M. Tignor, and P. M. Midgley, IPCC Working Group I Technical Support Unit, University of Bern, Bern, Switzerland.

Koch, D., T. C. Bond, D. G. Streets, N. Unger, and G. R. van der Werf (2007), Global impacts of aerosols from particular source regions and sectors, *J. Geophys. Res., 112*, D02205, doi:10.1029/2005JD007024.

Kogan, Y. L. (1991), The simulation of a convective cloud in a 3-D model with explicit microphysics: 1. Model description and sensitivity experiments, *J. Atmos. Sci., 48*, 1160−1189.

Kohfeld, K. E., and S. P. Harrison (2000), How well can we simulate past climates? Evaluating the models using global palaeoenvironmental datasets, *Quaternary Sci. Rev., 19*(1−5), 321−346.

Kohfeld, K. E., and S. P. Harrison (2001), DIRTMAP: The geological record of dust, *Earth-Sci. Rev., 54*(1−3), 81−114.

Kohfeld, K. E., and S. P. Harrison (2003), Glacial-interglacial changes in dust deposition on the Chinese Loess Plateau, *Quaternary Sci. Rev., 22*(18−19), 1859−1878.

Kohfeld, K. E., C. Le Quéré, S. P. Harrison, and R. F. Anderson (2005), Role of marine biology in glacial-interglacial CO_2 cycles, *Science, 308*, 74−78, doi:10.1126/science.1105375.

Köhler, P., H. Fischer, and J. Schmitt (2010), Atmospheric $\delta^{13}CO_2$ and its relation to pCO_2 and deep ocean $\delta^{13}C$ during the Late Pleistocene, *Paleoceanography, 25*, PA1213.

Kolb, P. F., and R. Robberecht (1996), High temperature and drought stress effects on survival of *Pinus ponderosa* seedlings, *Tree Physiol., 16*, 665−672.

Konovalov, I. B., M. Beekmann, A. Richter, and J. P. Burrows (2006), Inverse modelling of the spatial distribution of NO_x emissions on a continental scale using satellite data, *Atmos. Chem. Phys., 6*, 1747−1770.

Kopp, R. E., F. J. Simons, J. X. Mitrovica, A. C. Maloof, and M. Oppenheimer (2009), Probabilistic assessment of sea level during the last interglacial stage, *Nature, 462,* 863−868.

Koppe, C., R. Kovats, G. Jendritzky, and B. Menne (2004), Heat-waves: Impacts and responses, Technical report, World Health Organization Regional Office for Europe, Copenhagen, Sweden, Health and Global Environmental Change Series No. 2.

Koren, I., Y. J. Kaufman, L. A. Remer, and J. V. Martins (2004), Measurement of the effect of Amazon smoke on inhibition of cloud formation, *Science, 303*, 1342−1345.

Koren, I., J. V. Martins, L. A. Remer, and H. Afargan (2008), Smoke invigoration versus inhibition of clouds over the Amazon, *Science, 321*, 946−949.

Korhola, A., M. Ruppel, H. Sieppa, M. Valiranta, T. Virtanen and J. Weckstrom (2010), The importance of northern peatland expansion to the Late-Holocene rise of atmospheric methane, *Quaternary Sci. Rev., 29*, 611−617, doi:10.1016/j.quascirev.2009.12.010.

Korolev, A. (2007), Limitations of the Wegener-Bergeron-Findeisen mechanism in the evolution of mixed-phase clouds, *J. Atmos. Sci., 64*, 3372−3375.

Korte, C., S. P. Hesselbo, H. C. Jenkyns, R. E. M. Rickaby, and C. Spötl (2009), Palaeoenvironmental significance of carbon and oxygen isotope stratigraphy of marine Triassic-Jurassic boundary sections in SW Britain, *J. Geol. Soc. London, 166*, 431−445.

Kossin, J. P., K. R. Knapp, D. J. Vimont, R. J. Murnane, and B. A. Harper (2007), A globally consistent reanalysis of hurricane variability and trends, *Geophys. Res. Lett., 34*(4), L04815.

Koster, R. D., M. J. Suarez, and M. Heiser (2000), Variance and predictability of precipitation at seasonal-to-interannual timescales, *J. Hydrometeorol., 1*, 26−46.

Koster, R. D., M. J. Suarez, W. Higgins, and H. M. Van den Dool (2003), Observational evidence that soil moisture variations affect precipitation, *Geophys. Res. Lett., 30*(5), 1241, doi:10.1029/2002GL016571.

Koster, R. D., P. A. Dirmeyer, Z. C. Guo, G. Bonan, E. Chan, P. Cox, C. T. Gordon, S. Kanae, E. Kowalczyk, D. Lawrence, P. Liu, C.-H. Lu, S. Malyshev, B. J. McAvaney, K. Mitchell, D. Mocko, T. Oki, K. W. Oleson, A. J. Pitman, Y. C. Sud, C. M. Taylor, D. Verseghy, R. Vasic, Y. Xue, and T. Yamada (2004), Regions of strong coupling between soil moisture and precipitation, *Science, 305*, 1138−1140.

Koster R. D., Z. Guo, P. A. Dirmeyer, G. Bonan, E. Chan, P. Cox, H. Davies, C. T. Gordon, S. Kanae, E. Kowalczyk, D. Lawrence, P. Liu, C.-H. Lu, S. Malyshev, B. J. McAvaney, K. Mitchell, D. Mocko, T. Oki, K. W. Oleson, A. J. Pitman, Y. C. Sud, C. M. Taylor, D. Verseghy, R. Vasic, Y. Xue, and T. Yamada (2006), GLACE: The global land-atmosphere coupling experiment, part I: Overview, *J. Hydrometeorol., 7*, 590−610.

Kostopoulou, E., K. Tolika, I. Tegoulias, C. Giannakopoulos, S. Somot, C. Anagnostopoulou, and P. Maheras (2009), Evaluation of a regional climate model using in situ temperature observations over the Balkan Peninsula, *Tellus A, 61*(3), 357−370.

Kotamarthi, V. R., D. J. Wuebbles, and R. A. Reck (1999), Effects of non-methane hydrocarbons on lower stratospheric and upper tropospheric 2-D zonal average model chemical climatology, *J. Geophys. Res., 104*, 21537−21547.

Kovats, R., S. Hajat, and P. Wilkinson (2004), Contrasting patterns of mortality and hospital admissions during hot weather and heat waves in Greater London, United Kingdom, *Occup. Environ. Med., 61*, 893−898.

Krayenhoff, S., and J. Voogt (2010), Impacts of urban albedo increase on local air temperature at daily−annual time scales: Model results and synthesis of previous work, *J. Appl. Meteorol. Clim., 49*, 1634−1648.

Kriegler, E., J. W. Hall, H. Held, R. Dawson, and H. J. Schellnhuber (2009), Imprecise probability assessment of tipping points in the climate system, *Proc. Natl. Acad. Sci. USA, 106*(13), 5041–5046.

Kring, D. A. (2007), The Chicxulub impact event and its environmental consequences at the Cretaceous-Tertiary boundary, *Palaeogeogr. Palaeoclim., 225*, 4–21.

Kring, D. A., and D. D. Durda (2002), Trajectories and distribution of material ejected from the Chicxulub impact crater: Implications for post impact wildfires, *J. Geophys. Res., 107*, 6–22.

Kristjánsson, J. E. (2002), Studies of the aerosol indirect effect from sulfate and black carbon aerosols, *J. Geophys. Res., 107*, doi:10.1029/2001D000887.

Kristjánsson, J. E., C. Stjern, F. Stordal, A. M. Fjaeraa, G. Myhre, and K. Jonasson (2008), Cosmic rays and clouds – a reassessment using MODIS data, *Atmos. Chem. Phys., 8*, 7373–7387.

Krueger, A. J. (1989), The global distribution of total ozone – Toms satellite measurements, *Planet. Space Sci., 37*(12), 1555–1565.

Krystyn, L., H. Bouquerel, W. M. Kuerschner, S. Richoz, and Y. Gallet (2007), Proposal for a candidate GSSP for the base of the Rhaetian Stage, in *The Global Triassic, New Mexico Museum of Natural History and Science Bulletin, 41*, edited by S. G. Lucas and J. A. Spielmann, pp. 189–199.

Kubatzki, C., M. Montoya, S. Rahmstorf, A. Ganopolski, and M. Claussen (2000), Comparison of the last interglacial climate simulated by a coupled global model of intermediate complexity and AOGCM, *Climate Dynamics, 16*, 799–814.

Kuerschner, W. M., N. R. Bonis, and L. Krystyn (2007), Carbon-isotope stratigraphy and palynostratigraphy of the Triassic-Jurassic transition in the Tiefengraben section – Northern Calcareous Alps (Austria), *Palaeogeogr. Palaeoclim., 244*, 257–280.

Kuhlbrodt, T., A. Griesel, M. Montoya, A. Levermann, M. Hofmann, and S. Rahmstorf (2007), On the driving processes of the Atlantic Meridional Overturning Circulation, *Rev. Geophys., 45*(2), RG2001.

Kuhlmann, J., and J. Quaas (2010), How can aerosols affect the Asian summer monsoon? Assessment during three consecutive pre-monsoon seasons from CALIPSO satellite data, *Atmos. Chem. Phys., 10*(10), 4673–4688.

Kuiper, K. F., A. Deino, F. J. Hilgen, W. Krijgsman, P. R. Renne, and J. R. Wijbrans (2008), Synchronising rock clocks of Earth history, *Science, 320*, 500–504.

Kukla, G. J., R. K. Matthews, and M. J. Mitchell (1972), Present interglacial: How and when will it end?, *Quaternary Res., 2*(3), 261–269.

Kukla, G., A. Berger, R. Lotti, and J. Brown (1981), Orbital signature of interglacials, *Nature, 290*, 295–300.

Kull, O., and B. Kruijt (1998), Leaf photosynthetic light response: A mechanistic model for scaling photosynthesis to leaves and canopies, *Funct. Ecol., 12*, 767–777.

Kull, O., and B. Kruijt (1999), Acclimation of photosynthesis to light: A mechanistic approach, *Funct. Ecol., 13*, 24–36.

Kulmala, M., A. Reissell, M. Sipila, B. Bonn, T. M. Ruuskanen, K. E. J. Lehtinen, V.-M. Kerminen, and J. Strom (2006), Deep convective clouds as aerosol production engines: Role of insoluble organics, *J. Geophys. Res., 111*, D17202, doi:10.1029/2005JD006963.

Kumar, R., J. Dudhia, and S. Bhowmik (2010), Evaluation of physics options of the Weather Research and Forecasting (WRF) model to simulate high impact heavy rainfall events over Indian Monsoon region, *Geofizika, 27*(2), 101–125.

Kump, L. R., and D. Pollard (2008), Amplification of Cretaceous warmth by biological cloud feedbacks, *Science, 320*, 195.

Kump, L. R., and J. E. Lovelock (1995), The geophysiology of climate, in *Future Climates of the World*, edited by A. Henderson-Sellers, vol. 16, pp. 537–553, Elsevier, Amsterdam.

Kundzewicz, Z. W. (2007), Freshwater resources and their management, in *Impacts, Adaptation and Vulnerability, contribution of Working Group II to the Fourth Assessment Report (FAR) of the Intergovernmental Panel on Climate Change (IPCC)*, edited by M. L. Parry, et al., pp. 173–210, Cambridge University Press, Cambridge, UK and New York, NY.

Kunkel, K. E., D. R. Easterling, K. Redmond, and K. H. Hubbard (2003), Temporal variations of extreme precipitation events in the United States: 1895–2000, *Geophys. Res. Lett., 30*(17), 1900, doi:10.1029/2003GL018052.

Kunkel, K. E. (2008), Observed changes in weather and climate extremes, in *Weather and Climate Extremes in a Changing Climate, Regions of focus: North America, Hawaii, Caribbean, and US Pacific Islands*.

Kupiainen, K., and Z. Klimont (2007), Primary emissions of fine carbonaceous particles in Europe, *Atmos. Environ., 41*(10), 2156–2170, doi:10.1016/j.atmosenv.2006.10.066.

Kurbis, K., M. Mudelsee, G. Tetzlaff, and R. Brazdil (2009), Trends in extremes of temperature, dew point, and precipitation from long instrumental series from Central Europe, *Theor. Appl. Climatol., 98*(1–2), 187–195.

Kuroda, Y., K. Yamazaki, and K. Shibata (2008), Role of ozone in the solar cycle modulation of the North Atlantic Oscillation, *J. Geophys. Res., 113*, D14122, doi:10.1029/2007JD009336.

Kürschner, W. M., J. van der Burgh, H. Visscher, and D. L. Dilcher (1996), Oak leaves as biosensors of late Neogene and early Pleistocene paleoatmospheric CO_2 concentrations, *Mar. Micropaleontol., 27*, 299–312.

Kurz, W. A., C. C. Dymond, G. Stinson, G. J. Rampley, E. T. Neilson, A. L. Carroll, T. Ebata, and L. Safranyik (2008), Mountain pine beetle and forest carbon feedback to climate change, *Nature, 452*, 987–990.

Kurz, W. A., G. Stinson, G. J. Rampley, C. C. Dymond, and E. T. Neilson (2008a), Risk of natural disturbances makes future contribution of Canada's forests to the global carbon cycle highly uncertain, *Proc. Natl. Acad. Sci. USA, 105*(5), 1551–1555.

Kurz, W. A., C. C. Dymond, G. Stinson, G. J. Rampley, E. T. Neilson, A. L. Carroll, T. Ebata, and L. Safranyik (2008b), Mountain pine beetle and forest carbon feedback to climate change, *Nature, 452*, 987–990.

Kutzbach, J. E., and R. G. Gallimore (1988), Sensitivity of a coupled atmosphere/mixed layer ocean model to changes in orbital forcing at 9000 years B.P, *J. Geophys. Res., 93*(D1), 803–821.

Kutzbach, J. E., and P. J. Guetter (1986), Influence of changing orbital parameters and surface boundary conditions on climate simulations for the past 18,000 years, *J. Atmos. Sci., 43*(16), 1726–1759.

Kutzbach, J. E., and F. A. Street-Perrott (1985), Milankovitch forcing of fluctuations in the level of tropical lakes from 18 to 0 kyr B.P, *Nature, 317*(6033), 130–134, doi:10.1038/317130a0.

Kutzbach, J. E., P. J. Guetter, P. J. Behling, R. Selin (1993), Simulated climatic changes: Results of the COHMAP climate-model experiments, in *Global Climates Since the Last Glacial Maximum*, edited by H. E. Wright, Jr., et al., University of Minnesota Press, Minneapolis.

Kutzbach, J., G. Bonan, J. Foley, and S. P. Harrison (1996), Vegetation and soil feedbacks on the response of the African monsoon to orbital forcing in the Early to Middle Holocene, *Nature, 384*(6610), 623–626.

Kuzyakov, Y. (2002), Review: Factors affecting rhizosphere priming effects, *J. Plant Nutr. Soil Sc., 165*, 382–396.

Kvalevåg, M. M., and G. Myhre (2007), Human impact on direct and diffuse solar radiation during the industrial era, *J. Climate, 20*(19), 4874–4883.

Kvenvolden, K. A., and B. W. Rogers (2005), Gaia's breath – Global methane exhalations, *Mar. Petrol. Geol., 22*, 579–590.

Kwasniok, F., and G. Lohmann (2009), Deriving dynamical models from paleoclimatic records: Application to glacial millennial-scale climate variability, *Phys. Rev. E, 80*(6), 066104.

Kwok, R., and D. A. Rothrock (2009), Decline in Arctic sea ice thickness from submarine and ICESat records: 1958–2008, *Geophys. Res. Lett., 36*, L15501.

Kwon, E. Y., F. Primeau, and J. L. Sarmiento (2009), The impact of remineralization depth on the air-sea carbon balance, *Nat. Geosci., 2*, 630–635.

Kysely J., J. Picek, and R. Beranova (2010), Estimating extremes in climate change simulations using the peaks-over-threshold method with a non-stationary threshold, *Glob. Planet. Change, 72*(1–2), 55–68.

Labandeira, C. C., K. R. Johnson, and P. Wilf (2002), Impact of the terminal Cretaceous event on plant-insect associations, *Proc. Natl. Acad. Sci. USA, 99*, 2061–2066.

Labitzke, K. (1987), Sunspots, the QBO, and the stratospheric temperature in the North Polar region, *Geophys. Res. Lett., 14*(5), 535–537.

Labitzke, K., and H. van Loon (1988), Association between the 11-year solar cycle, the QBO, and the atmosphere: 1. The troposphere and stratosphere on the Northern Hemisphere winter, *J. Atmos. Terr. Phys., 50*, 197–206.

Lacis, A. A., D. J. Wuebbles, and J. A. Logan (1990), Radiative forcing of climate by changes in the vertical distribution of ozone, *J. Geophys. Res., 95*, 9971–9981, doi:10.1029/90JD00092.

Lacis, A. A., G. A. Schmidt, D. Rind, and R. A. Ruedy (2010), Atmospheric CO_2: Principal control knob governing Earth's temperature, *Science, 330*, 356–359.

Lagerloef, G., J. Boutin, J. Carton, Y. Chao, T. Delcroix, J. Font, J. Lilly, N. Reul, R. Schmitt, S. Riser, et al. (2009), Resolving the global surface salinity field and variations by blending satellite and in situ observations, Community White Paper, OceanObs '09 Conference, 21 September, Venice, Italy, *https://abstracts.congrex.com/scripts/jmevent/abstracts/FCXNL-09A02a-1727018%20-1-cwp2b06001.pdf*.

Lainé, A., M. Kageyama, P. Braconnot, and R. Alkama (2009), Impact of greenhouse as concentration changes on surface energetics in IPSL-CM4: Regional warming patterns, land-sea warming ratios, and glacial-interglacial differences, *J. Climate, 22*(17), 4621–4635.

Laj, P., J. Klausen, M. Bilde, C. Plass-Dülmer, G. Pappalardo, C. Clerbaux, U. Baltensperger, J. Hjorth, D. Simpson, S. Reimann, P.-F. Coheur, A. Richter, M. De Mazière, Y. Rudich, G. McFiggans, K. Torseth, A. Wiedensohler, S. Morin, M. Schulz, J. Allan, J.-L. Attié, I. Barnes, W. Birmilli, P. Cammas, J. Dommen, H.-P. Dorn, D. Fowler, J.-S. Fuzzi, M. Glasius, C. Granier, M. Hermann, I. S. A. Isaksen, S. Kinne, I. Koren, F. Madonna, M. Maione, A. Massling, O. Moehler, L. Mona, P. Monks, D. Müller, T. Müller, J. Orphal, V.-H. Peuch,

F. Stratmann, D. Tanré, G. Tyndall, A. A. Riziq, M. Van Roozendael, P. Villani, B. Wehner, H. Wex, and A. A. Zardini (2009), Measuring atmospheric composition change, *Atmos. Environ., 43*, 5352–5415, doi:10.1016/j.atmosenv.2009.08.020.

Lamarque, J.-F., P. Hess, L. Emmons, L. Buja, W. Washington, and C. Granier (2005), Tropospheric ozone evolution between 1890 and 1990, *J. Geophys. Res., 110*, D08304, doi:10.1029/2004JD005537.

Lambert, F., B. Delmonte, J. R. Petit, M. Bigler, P. R. Kaufmann, M. A. Hutterli, T. F. Stocker, U. Ruth, J. P. Steffensen, and V. Maggi (2008), Dust-climate couplings over the past 800,000 years from the EPICA Dome C ice core, *Nature, 452*(7187), 616–619, doi:10.1038/nature06763.

Landman, W., M. Kgatuke, M. Mbedzi, A. Beraki, A. Bartman, and A. du Piesanie (2009), Performance comparison of some dynamical and empirical downscaling methods for South Africa from a seasonal climate modelling perspective, *Int. J. Climatol., 29*(11), 1535–1549, doi:10.1002/joc.1766.

Landsea, C. W. (2007), Counting Atlantic tropical cyclones back to 1900, *EOS Trans., 88*(18), 197–202.

Langdon, C., T. Takahashi, C. Sweeney, D. Chipman, J. Goddard, F. Marubini, H. Aceves, H. Barnett, and M. J. Atkinson (2000), Effect of calcium carbonate saturation state on the calcification rate of an experimental coral reef, *Global Biogeochem. Cycles, 14*(2), 639–654.

Lange, M., B. Burkhard, S. Garthe, K. Gee, A. Kannen, H. Lenhart, and W. Windhorst (2010), Analyzing coastal and marine changes: Offshore wind farming as a case study, Zukunft Küste – Coastal Futures Synthesis Report, *LOICZ Research & Studies No. 36*, GKSS Research Center, Geesthacht.

Langner, J., R. Bergstrom, and V. Foltescu (2005), Impact of climate change on surface ozone and deposition of sulphur and nitrogen in Europe, *Atmos. Environ., 39*, 1129–1141.

Lantuit, H., P. P. Overduin, N. Couture, S. Wetterich, F. Aré, D. Atkinson, J. Brown, G. Cherkashov, D. Drozdov, D. Lawrence Forbes, A. Graves-Gaylord, M. Grigoriev, H.-W. Hubberten, J. Jordan, T. Jorgenson, R. S. Ødegård, S. Ogorodov, W.-H. Pollard, V. Rachold, S. Sedenko, S. Solomon, F. Steenhuisen, I. Streletskaya, and A. Vasiliev (2011), The Arctic coastal dynamics database: A new classification scheme and statistics on Arctic permafrost coastlines, *Estuar. Coast.*, doi:10.1007/s12237-010-9362-6.

Larigauderie, A., and H. A. Mooney (2010), The intergovernmental science-policy platform on biodiversity and ecosystem services: Moving a step closer to an IPCC-like mechanism for biodiversity, *Curr. Opin. Environ. Sustain., 2*, 1–6, doi:10.1016/j.cosust.2010.02.006.

Latham, J., P. Rasch, C.-C. Chen, L. Kettles, A. Gadian, A. Gettelman, H. Morrison, K. Bower, and T. Choularton (2008), Global temperature stabilization via controlled albedo enhancement of low-level maritime clouds, *Philos. T. Roy. Soc. A*, doi:10.1098/rsta.2008.0137.

Lathuilière, B., and D. Marchal (2009), Extinction, survival and recovery of corals from the Triassic to middle Jurassic time, *Terra Nova, 21*, 57–66.

Latif, M. (2010), Uncertainty in climate change projections, *J. Geochem. Explor.*, (Special Issue: Geochemical Cycling), doi:10.1016/j.gexplo.2010.09.011.

Latif, M., and T. P. Barnett (1994), Causes of decadal climate variability over the North Pacific and North America, *Science, 266*, 634–637.

Latif, M., and N. S. Keenlyside (2009), El Niño/Southern Oscillation response to global warming, *Proc. Natl. Acad. Sci. USA, 106*, 20578–20583.

Latif, M., and N. S. Keenlyside (2011), A perspective on decadal climate variability and predictability, *Deep-Sea Res. Pt. II, 58 (17-18)*, 1880–1894. doi:10.1016/j.dsr2.2010.10.066.

Latif, M. and W. Park (2012), Climatic variability on decadal to century timescales, in *The Future of the World's Climate*, edited by A. Henderson-Sellers and K. McGuffie, pp. 167–195, Elsevier, Amsterdam.

Latif, M., E. Roeckner, M. Botzet, M. Esch, H. Haak, S. Hagemann, J. Jungclaus, S. Legutke, S. Marsland, U. Mikolajewicz, and J. Mitchell (2004), Reconstructing, monitoring, and predicting multidecadal-scale changes in the North Atlantic thermohaline circulation with sea surface temperature, *J. Climate, 17*, 1605–1614.

Latif, M., C. Böning, J. Willebrand, A. Biastoch, J. Dengg, N. Keenlyside, U. Schweckendiek, and G. Madec (2006), Is the thermohaline circulation changing?, *J. Climate, 19*(18), 4631–4637.

Latif, M., M. Collins, H. Pohlmann, and N. Keenlyside (2006a), A review of predictability studies of the Atlantic sector climate on decadal time scales, *J. Climate, 19*, 5971–5987.

Latif, M., C. Böning, J. Willebrand, A. Biastoch, J. Dengg, N. Keenlyside, U. Schweckendiek, and G. Madec (2006b), Is the thermohaline circulation changing?, *J. Climate, 19*, 4631–4637.

Latif, M., W. Park, N. Keenlyside, and H. Ding (2009), Internal and external North Atlantic sector variability in the Kiel Climate Model, *Meteorol. Z., 18*, 433–443.

Lau, K.-M., and K.-M. Kim (2006), Observational relationships between aerosol and Asian monsoon rainfall, and circulation, *Geophys. Res. Lett., 33*, L21810, doi:10.1029/2006GL027546.

Lau, K., and H. Wu (2001), Principal modes of rainfall-SST variability of the Asian summer monsoon: A reassessment of the monsoon-ENSO relationship, *J. Climate, 14*(13), 2880–2895.

Lau, K.-M., et al. (2008), The joint aerosol-monsoon experiment: a new challenge for monsoon climate research, *Bull. Am. Meteorol. Soc., 89*, 1–5, doi:10.1175/BAMS-89-3-369.

Lau, K.-M., M.-K. Kim, K.-M. Kim, and W.-S. Lee (2010), Enhanced surface warming and accelerated snow melt in the Himalayas and Tibetan Plateau induced by absorbing aerosols, *Environ. Res. Lett., 5*(2), 025204.

Laurance, W. F. (2000), Mega-development trends in the Amazon: Implications for global change, *Environ. Monit. Assess., 61*, 113–122.

Laut, P. (2003), Solar activity and terrestrial climate: An analysis of some purported correlations, *J. Atmos. Sol.-Terr. Phys., 65*, 801–812.

Law, R. M., R. J. Matear, and R. J. Francey (2008), Comment on "Saturation of the Southern Ocean CO_2 sink due to recent climate change," *Science, 319*(5863), 570a, doi:10.1126/science.1149077.

Lawrence, D. M., and A. G. Slater (2005), A projection of severe near-surface permafrost degradation during the 21st century, *Geophys. Res. Lett., 32*, L24401, doi:10.1029/2005GL025080.

Lawrence, D. M., A. G. Slater, R. A. Tomas, M. M. Holland, and C. Deser (2008), Accelerated Arctic land warming and permafrost degradation during rapid sea ice loss, *Geophys. Res. Lett., 35*, L11506.

Lawrence, P. J., and T. N. Chase (2010), Investigating the climate impacts of global land cover change in the community climate system model, *Int. J. Climatol., 30*, 2066–2087, doi: 10.1002/joc.2061.

Laws, E. A., P. G. Falkowski, W. O. Smith, H. Ducklow, and J. J. McCarthy (2000), Temperature effects on export production in the open ocean, *Global Biogeochem. Cycles, 14*(4), 1231–1246.

Lawton, J. (2001), Earth system science, *Science, 292*(5524), 1965, doi:10.1126/science.292.5524.1965.

Le Quéré, C., C. Rödenbeck, E. T. Buitenhuis, T. J. Conway, R. Langenfelds, A. Gomez, C. Labuschagne, M. Ramonet, T. Nakazawa, N. Metzl, N. Gillett, and M. Heimann (2007), Saturation of the Southern Ocean CO_2 sink due to recent climate change, *Science, 316*(5832), 1735–1738.

Le Quéré, C., C. Rodenbeck, E. T. Buitenhuis, T. J. Conway, R. Langenfelds, A. Gomez, C. Labuschagne, M. Romonet, T. Nakazawa, N. Metzl, N. P. Gillett, and M. Heimann (2008), Response to comments on "Saturation of the Southern Ocean CO_2 sink due to recent climate change," *Science, 319*(5863), 570c, doi:10.1126/science.1147315.

Le Quéré, C., M. R. Raupach, J. G. Canadell, G. Marland, L. Bopp, P. Ciais, T. J. Conway, S. C. Doney, R. A. Feely, P. Foster, P. Friedlingstein, K. Gurney, R. A. Houghton, J. I. House, C. Huntingford, P. E. Levy, M. R. Lomas, N. Metzl, J. P. Ometto, G. P. Peters, I. C. Prentice, J. T. Randerson, S. W. Running, J. L. Sarmiento, U. Schuster, S. Sitch, T. Takahashi, N. Viovy, G. R. van der Werf, and F. I. Woodward (2009), Trends in the sources and sinks of carbon dioxide, *Nat. Geosci., 2*(12), 831–836, doi:10.1038/ngeo689.

Le Quéré, C., J. G. Canadell, P. Ciais, S. Dhakal, A. Patwardhan, M. R. Raupach, and O. R. Young (2010), An International Carbon Office to assist policy-based science, *Curr. Opin. Environ. Sustain., 2*, 297–300, doi:10.1016/j.cosust.2010.06.010.

Le Quéré, C., T. Takahashi, E. T. Buitenhuis, C. Rödenbeck, and S. C. Sutherland (2010), Impact of climate change and variability on the global oceanic sink of CO_2, *Global Biogeochem. Cycles, 24*, GB4007, doi:10.1029/2009GB003599.

Le Treut, H., R. Somerville, U. Cubasch, Y. Ding, C. Mauritzen, A. Mokssit, T. W. Peterson, and M. Prather (2007), Historical overview of climate change, in *Climate Change 2007: The Physical Science Basis, contribution of Working Group 1 to the Fourth Assessment Report (FAR) of the Intergovernmental Panel on Climate Change (IPCC)*, edited by S. Solomon, et al., pp. 93–127, Cambridge University Press, Cambridge, UK and New York, NY.

Le Verrier, U. (1859), Lettre de M. Le Verrier à M. Faye sur la théorie de Mercure et sur le mouvement du périhélie de cette planète, *Comptes rendus hebdomadaires des séances de l'Académie des sciences (Paris), 49*, 379–383.

Lea, D. W. (2004), The 100,000-yr cycle in tropical SST, greenhouse forcing, and climate sensitivity, *J. Climate, 17*, 2170–2179.

Leakey, A. D. B., E. A. Ainsworth, C. J. Bernacchi, A. Rogers, S. P. Long, and D. R. Ort (2009), Elevated CO_2 effects on plant carbon, nitrogen, and water relations: Six important lessons from FACE, *J. Exp. Bot., 60*, 2859–2876, doi:10.1093/jxb/erp096.

Lean, J. (2004), Solar irradiance reconstruction, IGBP PAGES/World Data Center for Paleoclimatology, *2004-035*, GBP PAGES/WDCA.

Lean, J. L. (2006), Comment on 'Estimated solar contribution to the global surface warming using the ACRIM TSI satellite composite' by N. Scafetta, B. J. West, *Geophys. Res. Lett., 33*, L15701, doi:10.1029/2005GL025342.

Lean, J., J. Beer, and R. Bradley (1995), Reconstruction of solar irradiance since 1610–Implications for climate-change, *Geophys. Res. Lett., 22*, 3195–3198.

Lean, J. L., and D. H. Rind (2008), How natural and anthropogenic influences alter global and regional surface temperatures: 1889 to 2006, *Geophys. Res. Lett., 35*, L18701, doi:10.1029/2008GL034864.

Leclercq, P. W., R. S. W. van de Wal, and J. Oerlemans (2010), Comment on "100-year mass changes in the Swiss Alps linked to the Atlantic Multidecadal Oscillation" by Matthias Huss, et al., *The Cryosphere Discuss., 4*, 2475–2481.

Leclercq, P. W., J. Oerlemans, and J. G. Cogley (2011), Estimating the glacier contribution to sea-level rise for the period 1800–2005, *Surv. Geophys., 32*(4), doi:10.1007/s10712-011-9121-7, *http://www.springerlink.com/content/3404k8273781233m/fulltext.pdf.*

Ledley, T. S. (1995), Summer solstice solar radiation, the 100-kyr ice age cycle, and the next ice age, *Geophys. Res. Lett., 22*(20), 2745–2748.

Leduc, M., and R. Laprise (2009), Regional climate model sensitivity to domain size, *Clim. Dynam., 32*(6), 833–854.

Leduc, G., R. Schneider, J. H. Kim, and G. Lohmann (2010), Holocene and Eemian sea surface temperature trends as revealed by alkenone and Mg/Ca paleothermometry, *Quaternary Sci. Rev., 29*(7–8), 989–1004.

Lee, D. S., D. W. Fahey, P. M. Forster, P. J. Newton, R. C. N. Wit, L. L. Lim, B. Owen, and R. Sausen (2009), Aviation and global climate change in the 21st century, *Atmos. Environ., 43*, 3520–3537, doi:10.1016/j.atmosenv.2009.04.024.

Lee, D. S., G. Pitari, V. Grewe, K. Gierens, J. E. Penner, A. Petzold, M. Prather, U. Schumann, A. Bais, T. Berntsen, D. Iachetti, L. L. Lim, and R. Sausen (2010), Transport impacts on atmosphere and climate: Aviation, *Atmos. Environ., 44*, 4678–4742.

Lee, J.-E., I. Fung, D. J. DePaolo, and B. Otto-Bliesner (2008), Water isotopes during the Last Glacial Maximum: New general circulation model calculations, *J. Geophys. Res., 113*, D19109, doi:10.1029/2008JD009859.

Lee, T., and M. J. McPhaden (2008), Decadal phase change in large-scale sea level and winds in the Indo-Pacific region at the end of the twentieth century, *Geophys. Res. Lett., 35*, L01605, doi:10.1029/2007GL032419.

Lee, T., and M. J. McPhaden (2010), Increasing intensity of El Niño in the central-equatorial Pacific, *Geophy. Res. Lett., 37*, L14603, doi:10.1029/2010GL044007.

Leemans, R., G. Asrar, J. G. Canadell, J. Ingram A. Larigauderie, H. Mooney, C. Nobre, A. Patwardhan, M. Rice, S. Seitzinger, H. Virji, C. Vörösmarthy, and O. R. Young (2009), Developing a common strategy for integrative global change research and outreach: The Earth System Science Partnership (ESSP), *Curr. Opin. Environ. Sustain., 1*, 1–10, doi:10.1016/j.cosust.2009.07.013.

Leemans, R., M. Rice, A. Henderson-Sellers, and K. Noone (2010), Research agenda and policy input of the Earth System Science Partnership for coping with global environmental change, in *Coping with Global Environmental Change, Disasters and Security: Threats, Challenges, Vulnerabilities and Risks*, edited by H. G. Brauch, et al., Hexagon Series on Human and Environmental Security and Peace, vol. 4, pp. 1205–1220, Springer, Berlin, Heidelberg, and New York.

Lefèvre, N., A. J. Watson, A. Olsen, A. F. Ríos, F. F. Pérez, and T. Johannessen (2004), A decrease in the sink for atmospheric CO_2 in the North Atlantic, *Geophys. Res. Lett., 31*, L07306, doi:10.1029/2003GL018957.

Lefohn, A. S., J. D. Husar, and R. B. Husar (1999), Estimating historical anthropogenic global sulfur emission patterns for the period 1850–1990, *Atmos. Environ., 33*, 3435–3444.

Leggett, J., W. J. Pepper, and R. J. Swart (1992), Emissions scenarios for the IPCC: An update, in *Climate Change 1992: The Supplementary Report to the IPCC Scientific Assessment*, edited by J. T. Houghton, B. A. Callander, and S. K. Varney, Cambridge University Press, Cambridge, UK, 62–95.

Legrand, M., C. Hammer, M. De Angelis, J. Savarino, R. Delmas, H. Clausen, and S. J. Johnsen (1997), Sulfur-containing species (methanesulfonate and SO_4) over the last climatic cycle in the Greenland Ice Core Project (Central Greenland) ice core, *J. Geophys. Res., 102*(C12), 26663–26679.

Lehodey, P., M. Bertignac, J. Hampton, A. Lewis, and J. Picaut (1997), El Niño Southern Oscillation and tuna in the western Pacific, *Nature, 389*(6652), 715–718, doi:10.1038/39575.

Lehodey, P., I. Senina, J. Sibert, L. Bopp, B. Calmettes, J. Hampton, and R. Murtugudde (2010), Preliminary forecasts of Pacific bigeye tuna population trends under the A2 IPCC scenario, *Prog. Oceanogr., 86*(1–2), 302–315, doi:10.1016/j.pocean.2010.04.021.

Leiserowitz, A., E. Maibach, and C. Roser-Renouf (2010a), Climate change in the American Mind: Americans' global warming beliefs and attitudes in January 2010, Yale Project on Climate Change, Yale University, and George Mason University, New Haven, CT, *http://environment.yale.edu/uploads/AmericansGlobalWarmingBeliefs2010.pdf.* (Accessed 26/7/10.)

Leiserowitz, A., E. Maibach, and C. Roser-Renouf (2010b), Global warming's six Americas, June 2010, Yale Project on Climate Change, Yale University, and George Mason University, New Haven, CT, *http://environment.yale.edu/climate/files/SixAmericasJune2010.pdf.* (Accessed 18/8/10.)

Lelieveld, J., and F. J. Dentener (2000), What controls tropospheric ozone?, *J. Geophys. Res., 105*, 3531–3551.

Lelieveld, J., W. Peters, F. J. Dentener, and M. C. Krol (2002), Stability of tropospheric hydroxyl chemistry, *J. Geophys. Res., 107*(D23), 4715.

Lelieveld, J., J. van Aardenne, H. Fischer, M. de Reus, J. Williams, and P. Winkler (2004), Increasing ozone over the Atlantic ocean, *Science, 304*, 1483–1487.

Lelieveld, J., T. M. Butler, J. N. Crowley, T. J. Dillon, H. Fischer, L. Ganzeveld, H. Harder, M. G. Lawrence, M. Martinez, D. Taraborrelli, and J. Williams (2008), Atmospheric oxidation capacity sustained by a tropical forest, *Nature, 452*(7188), 737–740.

Leloup, J., M. Lengaigne, and J. P. Boulanger (2007), Twentieth century ENSO characteristics in the IPCC database, *Clim. Dynam., 30*, 277–291.

Lemke, P., E. W. Trinkl, and K. Hasselmann (1980), Stochastic dynamic analysis of sea-ice variability, *J. Phys. Oceanogr., 10*, 2100–2120.

Lemke, P., J. Ren, R. B. Alley, I. Allison, J. Carrasco, G. Flato, Y. Fujii, G. Kaser, P. Mote, R. H. Thomas, and T. Zhang (2007), Observations: Changes in snow, ice and frozen ground, in *Climate Change 2007: The Physical Science Basis, contribution of Working Group I to the Fourth Assessment Report (FAR) of the Intergovernmental Panel on Climate Change*, edited by S. Solomon, et al., pp. 337–383, Cambridge University Press, Cambridge, UK.

Lemonsu, A., and V. Masson (2002), Simulation of a summer urban breeze over Paris, *Bound.-Lay. Meteorol., 104*, 463–490.

Lemonsu A., S. Belair, J. Mailhot, M. Benjamin, F. Chagnon, G. Morneau, B. Harvey, J. Voogt, and M. Jean (2008), Overview and first results of the Montreal urban snow experiment 2005, *J. Appl. Meteorol. Clim., 47*(1), 59–75.

Lenton, A., and R. J. Matear (2007), Role of the Southern Annular Mode (SAM) in Southern Ocean CO_2 uptake, *Global Biogeochem. Cycles, 21*(2).

Lenton, A., F. Codron, L. Bopp, N. Metzl, P. Cadule, A. Tagliabue, and J. Le Sommer (2009), Stratospheric ozone depletion reduces ocean carbon uptake and enhances ocean acidification, *Geophys. Res. Lett., 36*(12), L12606, doi:10.1029/2009GL038227.

Lenton, T. M. (1998), Gaia and natural selection, *Nature, 394*, 439–447.

Lenton, T. M. (2009), Tipping points in the Earth system, Earth System Modelling Group, University of East Anglia, Norwich, *http://researchpages.net/ESMG/people/tim-lenton/tipping-points/*. (Accessed 9/11/10.)

Lenton, T. M. (2010), The potential for land-based biological CO_2 removal to lower future atmospheric CO_2 concentration, *Carbon Manag., 1*(1), 145–160.

Lenton, T. M. (2012), Future climate surprises, in *The Future of the World's Climate*, edited by A. Henderson-Sellers and K. McGuffie, pp. 489–507, Elsevier, Amsterdam.

Lenton, T. M., and H. J. Schellnhuber (2007), Tipping the scales, *Nature Reports Clim. Change, 1*, 97–98.

Lenton, T. M., and N. E. Vaughan (2009), The radiative forcing potential of different climate geoengineering options, *Atmos. Chem. Phys., 9*, 5539–5561.

Lenton, T. M., H. J. Schellnhuber, and E. Szathmary (2004), Climbing the co-evolution ladder, *Nature, 431*, 913, doi:10.1038/431913a.

Lenton T. M., H. Held, E. Kriegler, J. W. Hall, W. Lucht, S. Rahmstorf, and H. J. Schellnhuber (2008), Tipping elements in the Earth's climate system, *Proc. Natl. Acad. Sci. USA, 105*(6), 1786–1793.

Lenton, T. M., A. Footitt, and A. Dlugolecki (2009a), Major tipping points in the Earth's climate system and consequences for the insurance sector, report, Tyndall Centre for Climate Change Research.

Lenton, T. M., R. J. Myerscough, R. Marsh, V. N. Livina, A. R. Price, S. J. Cox, and GENIEteam (2009b), Using GENIE to study a tipping point in the climate system, *Philos. T. Roy. Soc. A, 367*(1890), 871–884.

Levelt, P. F., G. H. J. van den Oord, M. R. Dobber, A. Mälkki, H. Visser, J. de Vries, P. Stammes, J. O. V. Lundell, and H. Saari (2006), Science objectives of the ozone monitoring instrument, *IEEE Trans. Geosci. Remote Sens., 44*(5), 1199–1208.

Levin, K., and R. Bradley (2010), Comparability of annex I emission reduction pledges, World Resources Institute Working Paper, February, *http://pdf.wri.org/working_papers/comparability_of_annex1_emission_reduction_pledges_2010-02-01.pdf*. (Accessed 9/10/11.)

Levermann, A., and A. Born (2007), Bistability of the Atlantic subpolar gyre in a coarse-resolution climate model, *Geophys. Res. Lett., 34*, L24605.

Levermann, A., J. Schewe, V. Petoukhov, and H. Held (2009), Basic mechanism for abrupt monsoon transitions, *Proc. Natl. Acad. Sci. USA*.

Levin, Z., and W. Cotton (2007), Aerosol pollution impact on precipitation: A scientific review, Report from the WMO/IUGG international Aerosol Precipitation Science Assessment Group (IAPSAG), World Meteorological Organization, Geneva, Switzerland.

Levitus, S., J. Antonov, and T. Boyer (2005), Warming of the world ocean, 1955–2003, *Geophys. Res. Lett., 32*(2), L02604.

Levitus, S., J. I. Antonov, T. P. Boyer, R. A. Locarnini, H. E. Garcia, and A. V. Mishonov (2009), Global ocean heat content 1955–2008 in light of recently revealed instrumentation problems, *Geophys. Res. Lett., 36*, L07608.

Levy, H., II, M. D. Schwarzkopf, L. Horowitz, V. Ramaswamy, and K. L. Findell (2008), Strong sensitivity of late 21st century climate to projected changes in short-lived air pollutants, *J. Geophys. Res., 113*(D6), D06102.

Lewis, S. L., P. M. Brando, O. L. Phillips, G. M. F. van der Heijden, and D. Nepstad (2011), The 2010 Amazon Drought, *Science, 331*, 554, doi:10.1126/science.1200807.

Li, J., and Y. Chen (1998), Barrier jets during TAMEX, *Mon. Weather Rev., 126*(4), 959–971.

Li, X. S., A. Berger, and M. F. Loutre (1998), CO_2 and Northern Hemisphere ice volume variations over the middle and late Quaternary, *Clim. Dynam., 14*(7–8), 537–544.

Liao, H., and J. H. Seinfeld (2005), Global impacts of gas-phase chemistry-aerosol interactions on direct radiative forcing by anthropogenic aerosols and ozone, *J. Geophys. Res., 110*, D18208.

Liao, H., J. H. Seinfeld, P. J. Adams, and L. J. Mickley (2004), Global radiative forcing of coupled tropospheric ozone and aerosols in a unified general circulation model, *J. Geophys. Res., 109*, doi:10.1029/2003JD004456.

Liao, H., W.-T. Chen, and J. H. Seinfeld (2006), Role of climate change in global predictions of future tropospheric ozone and aerosols, *J. Geophys. Res., 111*, D12304, doi:10.1029/2005JD006852.

Liao, H., Y. Zhang, W.-T. Chen, F. Raes, and J. H. Seinfeld (2009), Effect of chemistry-aerosol-climate coupling on predictions of future climate and future levels of tropospheric ozone and aerosols, *J. Geophys. Res., 114*(D10), D10306.

Liepert, B. G. (2002), Observed reductions of surface solar radiation at sites in the United States and worldwide from 1961 to 1990, *Geophys. Res. Lett., 29*, doi:10.1029/2002GL014910.

Lietzke, B., and R. Vogt (2009), Part I: Energy in the urban system, from Inventory of current state of empirical and modeling knowledge of energy, water and carbon sinks, sources and fluxes, Report on EU Framework 7 Project D.2.1, edited by C. S. B. Grimmond, Ref. 211345_001_PU_KCL, *http://www.bridge-fp7.eu/*.

Lindzen, R. S., M.-D. Chou, and A. Y. Hou (2001), Does the Earth have an adaptive infrared iris?, *Bull. Am. Meteorol. Soc., 82*(3), 417–432, doi:10.1175/1520-0477(2001)082<0417:DTEHAA>2.3.CO;2.

Ling, S. D., C. R. Johnson, K. Ridgway, A. J. Hobday, and M. Haddon (2009), Climate-driven range extension of a sea urchin: Inferring future trends by analysis of recent population dynamics, *Glob. Change Biol., 15*(3), 719–731, doi:10.1111/j.1365-2486.2008.01734.x.

Lisiecki, L. E., and M. E. Raymo (2005), A Pliocene-Pleistocene stack of 57 globally distributed benthic delta ^{18}O records, *Paleoceanography, 20*(1), PA1003, doi:10.1029/2004PA001071.

Lisiecki, L. E., and M. E. Raymo (2007), Plio-Pleistocene climate evolution: Trends and transitions in glacial cycle dynamics, *Quaternary Sci. Rev., 26*(1–2), 56–69, doi:10.1016/j.quascirev.2006.09.005.

Liu, X. H., J. E. Penner, B. Y. Das, D. Bergmann, J. M. Rodriguez, S. Strahan, M. Wang, and Y. Feng (2007), Uncertainties in global aerosol simulations: Assessment using three meteorological data sets, *J. Geophys. Res., 112*(D11).

Liu, Y., J. Sun, and B. Yang (2009), The effects of black carbon and sulfate aerosols in China regions on East Asia monsoons, *Tellus B*, *61*, 642–656.

Liu, Z., S. P. Harrison, J. Kutzbach, and B. Otto-Bliesner (2004), Global monsoons in the mid-Holocene and oceanic feedback, *Clim. Dynam.*, *22*(2–3), 157–182, doi:10.1007/s00382-003-0372-y.

Liu, Z., Y. Wang, R. Gallimore, M. Notaro, and I. C. Prentice (2006), On the cause of abrupt vegetation collapse in North Africa during the Holocene: Climate variability vs. vegetation feedback, *Geophys. Res. Lett.*, *33*, L22709, doi:10.1029/2006GL028062.

Liu, Z., B. L. Otto-Bliesner, F. He, E. C. Brady, R. Tomas, P. U. Clark, A. E. Carlson, J. Lynch-Stieglitz, W. Curry, E. Brook, D. Erickson, R. Jacob, J. Kutzbach, and J. Cheng (2009), Transient simulation of last deglaciation with a new mechanism for Bølling-Allerød warming, *Science*, 325 (5938), 310–314, doi:10.1126/science.1171041.

Liverman, D. (2010), Bridging the science policy interface, paper presented at International Climate Change Adaptation Conference, Climate Adaptation Futures: Preparing for the unavoidable impacts of climate change, National Climate Change Adaptation Research Facility and the CSIRO Climate Adaptation Flagship, 29 June–1 July, Gold Coast, *http://www.nccarf.edu.au/conference2010/wp-content/uploads/Diana-Liverman.pdf*.

Livina, V. N., and T. M. Lenton (2007), A modified method for detecting incipient bifurcations in a dynamical system, *Geophys. Res. Lett.*, *34*, L03712.

Livina, V. N., F. Kwasniok, and T. M. Lenton (2010), Potential analysis reveals changing number of climate states during the last 60 ka, *Clim. Past*, *6*(1), 77–82.

Livina, V. N., F. Kwasniok, G. Lohmann, J. W. Kantelhardt, Y. I. Sapronov, and T. M. Lenton (2011), Changing climate states: From Pliocene to present, *Clim. Dynam.*, (in press).

Lliboutry, L., B. M. Arnao, A. Pautre, and B. Schneider (1977), Glaciological problems set by the control of dangerous lakes in Cordillera Blanca, Peru, I. Historical failures of morainic dams, their causes and prevention, *J. Glaciol.*, *18*(79), 239–254.

Lloyd-Hughes, B., and M. A. Saunders (2002), A drought climatology for Europe, *Int. J. Climatol.*, *22*, 1571–1592.

Lloyd, J., and G. D. Farquhar (2008), Effects of rising temperature and [CO_2] on the physiology of tropical forest trees, *Philos. T. Roy. Soc. B*, *363*, 1811–1817, doi:10.1098/rstb.2007.0032.

Lockwood, M. (2002), Long-term variations in the open solar flux and possible links to Earth's climate, in *Proceedings of the SOHO 11 Symposium on From Solar Min to Max: Half a Solar Cycle with SoHO, 11–15 March 2002, Davos, Switzerland, a symposium dedicated to Roger M. Bonnet*, edited by A. Wilson, *ESA SP-508*, pp. 507–522, ESA Publications Division, Noordwijk.

Lockwood, M., and C. Frölich (2007), Recent oppositely directed trends in solar climate forcings and the global mean surface air temperature, *Proc. Roy. Soc. A*, doi:10.1098/rspa.2007.1880.

Logan, J. A. (1998), An analysis of ozonesonde data for the troposphere: Recommendations for testing 3-D models and development of a gridded climatology for tropospheric ozone, *J. Geophys. Res.*, *10*(D13), 16115–16149.

Lohmann, K., and M. Latif (2005), Tropical Pacific Decadal Variability and the subtropical–tropical cells, *J. Climate*, *18*, 5163–5178.

Lohmann, U. (2002), Possible aerosol effects on ice clouds via contact nucleation, *J. Atmos. Sci.*, *59*, 647–656.

Lohmann, U. (2008), Global anthropogenic aerosol effects on convective clouds in ECHAM5-HAM, *Atmos. Chem. Phys.*, *8*, 2115–2131.

Lohmann, U., and K. Diehl (2006), Sensitivity studies of the importance of dust ice nuclei for the indirect aerosol effect on stratiform mixed-phase clouds, *J. Atmos. Sci.*, *63*, 968–982.

Lohmann, U., and J. Feichter (2001), Can the direct and semi-direct aerosol effect compete with the indirect effect on a global scale, *Geophys. Res. Lett.*, *28*(1), 159–161.

Lohmann, U., J. Feichter, C. C. Chuang, and J. E. Penner (1999), Predicting the number of cloud droplets in the ECHAM GCM, *J. Geophys. Res.*, *104*, 9169–1998.

Lohmann, U., L. Rotstayn, T. Storelvmo, A. Jones, S. Menon, J. Quaas, A. M. L. Ekman, D. Koch, and R. Ruedy (2010), Total aerosol effect: radiative forcing or radiative flux perturbation?, *Atmos. Chem. Phys.*, *10*(7), 3235–3246.

Lomax, B., D. Beering, G. Upchurch, Jr., and B. Otto-Bliesner (2001), Rapid (10-yr) recovery of terrestrial productivity in a simulation study of the terminal Cretaceous impact event, *Earth Planet. Sci. Lett.*, *192*, 137–144.

López-Urrutia, Á., E. San Martin, R. P. Harris, and X. Irigoien (2006), Scaling the metabolic balance of the oceans, *Proc. Natl. Acad. Sci. USA*, *103*, 8739–8744.

Lorenz, E. N. (1964), The problem of deducing the climate from the governing equations, *Tellus*, *16*, 1–11.

Lorenz, S. J., J.-H. Kim, N. Rimbu, R. R. Schneider, and G. Lohmann (2006), Orbitally driven insolation forcing on Holocene climate trends: Evidence from alkenone data and climate modeling, *Paleoceanography*, *21*(1), PA1002.

Loridan, T, and C. S. B. Grimmond (2011), Characterization of energy flux partitioning in urban environments: Links with surface seasonal properties, *J. Appl. Meteorol. Clim.*, (submitted).

Loughnan, M. E., N. Nicholls, and N. J. Tapper (2010), The effects of summer temperature, age, and socioeconomic circumstance on acute myocardial infarction admissions in Melbourne, Australia, *Int. J. Health Geogr.*, *9*, 41.

Loulergue, L., A. Schilt, R. Spahni, V. Masson-Delmotte, T. Blunier, B. Lemieux, J. M. Barnola, D. Raynaud, T. F. Stocker, and J. Chappellaz (2008), Orbital and millennial-scale features of atmospheric CH_4 over the past 800,000 years, *Nature*, *453*(7193), 383–386, doi:10.1038/nature06950.

Loutre, M. F. (1995), Greenland ice sheet over the next 5000 years, *Geophys. Res. Lett.*, *22*(7), 783–786.

Loutre, M. F. (2003), Clues from MIS-11 to predict the future climate – a modelling point of view, *Earth Planet. Sci. Lett.*, *212*, 213–224, doi:10.1016/S0012-821X(03)00235-8.

Loutre, M.-F. (2009), Eccentricity, in *Encyclopedia of Paleoclimatology and Ancient Environments*, edited by V. Gornitz, pp. 825–826, Springer.

Loutre, M. F., and A. Berger (2000), Future climatic changes: Are we entering an exceptionally long interglacial?, *Climatic Change, 46*, 61–90.

Loutre, M. F., and A. Berger (2003), Marine Isotope Stage 11 as an analogue for the present interglacial, *Global Planet. Change, 36*, 209–217, doi:10.1016/S0921-8181(02)00186-8.

Loutre, M. F., A. Berger, M. Crucifix, S. Desprat, and M. F. Sanchez-Goñi (2007), Interglacials simulated by the LLN 2-D NH and MoBidiC climate models, in *The Climate of Past Interglacials, Developments in Quaternary Science*, edited by F. Sirocko, et al., pp. 547–561, Elsevier, Amsterdam.

Loveland, T. R., B. C. Reed, J. F. Brown, O. Ohlen, Z. Zhu, L. Yang, and J. W. Merchant (2000), Development of a global land cover characteristics database and IGBP DISCover from 1 km AVHRR data, *Int. J. Remote Sens., 21,* 1303–1330, doi:10.1080/014311600210191.

Lovelock, J. E. (1972), Gaia as seen through the atmosphere, *Atmos. Environ., 6* (8), 579–580, doi:10.1016/0004-6981(72)90076-5.

Lovelock, J. E. (1979), *Gaia: A New Look at Life on Earth,* Oxford University Press, Oxford, UK.

Lovelock, J. E. (1983), Daisy World: A cybernetic proof of the Gaia hypothesis, *CoEvol. Quart., 38,* 66–72.

Lovelock, J. E. (1986), Gaia: The world as a living organism, *New Sci.,* 18 December, 25–28.

Lovelock, J. E. (1987a), *Gaia: A New Look at Life on Earth,* Oxford University Press, Oxford and New York.

Lovelock, J. E. (1987b), Geophysiology: A new look at Earth science, in *The Geophysiology of Amazonia: Vegetation and Climate Interactions,* edited by R. E. Dickinson, pp. 11–23, John Wiley & Sons, New York.

Lovelock, J. E. (1989), Geophysiology, the science of Gaia, *Rev. Geophys., 27*(2), 215–222.

Lovelock, J. E. (1993), The evolving Gaia theory, a paper presented in September 1992 at The United Nations University, Tokyo, Japan, *http://unu.edu/unupress/lecture1.html,* (Accessed 27/11/10.)

Lovelock, J. E. (2006), *The Revenge of Gaia: Why the Earth is Fighting Back – And How We Can Still Save Humanity,* Allen Lane, London, Santa Barbara, California, Allen Lane.

Lovelock, J. E. (2009), *The Vanishing Face of Gaia: A Final Warning,* Penguin Books, London.

Lovelock, J. E. (2010), *The Vanishing Face of Gaia: A Final Warning,* Basic Books, New York.

Lovelock J. E., and L. Margulis (1974), Atmospheric homeostasis by and for the biosphere: The Gaia hypothesis, *Tellus,* 26, 1–10.

Lovelock, J. E., and L. Margulis (1973), Atmospheric homeostasis by and for the biosphere: The Gaia hypothesis, *Tellus, 26*(1974), 1–10.

Lovenduski, N. S., and N. Gruber (2005), Impact of the Southern Annular Mode on Southern Ocean circulation and biology, *Geophys. Res. Lett., 32,* 4, doi:200510.1029/2005GL022727.

Low, P. S. (2010), Copenhagen Accord and National Communications for Non-Annex I Parties, Asia-Pacific Human Development Network, Climate Change Discussion, *http://www2.undprcc.lk/ext/HDRU/files/climet_change/global_negotiations/APHDNet_Pak_Sum_Low_contribution_SubTheme_7_4_July_2010.pdf.* (Accessed 9/11/10.)

Lowe, I. (2010), US climate scientist Stephen Schneider dies at 65 – Australian tributes, on 'Rapid round up', *Australia Science Media Centre,* 20 July 2010, *http://www.aussmc.org/2010/07/rapid-roundup-us-climate-scientist-stephen-schneider-dies-at-65-australian-tributes/.* (Accessed 5/5/11.)

Lowry, W. (1998), Urban effects on precipitation amount, *Prog. Phys. Geog., 22,* 477–520.

Lozhkin, A. V., and P. M. Anderson (1995), The last Interglaciation in Northeast Siberia, *Quaternary Res., 43,* 147–158.

Lozier, M. S., S. J. Leadbetter, R. G. Williams, V. Roussenov, M. S. C. Reed, and N. J. Moore (2008), The spatial pattern and mechanisms of heat content change in the North Atlantic, *Science, 319,* 800–803.

Lu, J., and T. Delworth (2005), Oceanic forcing of the late 20th century Sahel drought, *Geophys. Res. Lett., 32,* L22706, doi:10.1029/2005GL023316.

Lu, J., G. A. Vecchi, and T. Reichler (2007), Expansion of the Hadley cell under global warming, *Geophys. Res. Lett., 34,* L06805.

Lubchenko, J. (1998), Entering the century of the environment: A new social contract for science, *Science, 279*(5350), 491–497, doi:10.1126/science.279.5350.491.

Lubin, D., and A. M. Vogelmann (2006), A climatologically significant aerosol longwave indirect effect in the Arctic, *Nature, 439,* 453–456.

Lucas, D. D., and A. H. Akimoto (2007), Contributions of anthropogenic and natural sources of sulfur to SO_2, $H_2SO_4(g)$ and nano-particles formation, *Atmos. Chem. Phys., 7,* 7679–7721.

Lukac, M., C. Calfapietra, A. Lagomarsino, and F. Loreto (2010), Global climate change and tree nutrition: effects of elevated CO_2 and temperature, *Tree Physiol., 30,* 1209–1220, doi:10.1093/treephys/tpq040.

Luke, C. M., and P. A. Cox (2011), Soil carbon and climate change: From the Jenkinson effect to the compost-bomb instability, *Euro. J. Soil Sci., 62,* 5–12, doi:10.1111/j.136-2389-2010.01312.x.

Luo, Y., L. M. Rothstein, and R.-H. Zhang (2009), Response of Pacific subtropical-tropical thermocline water pathways and transports to global warming, *Geophys. Res. Lett., 36*(4), L04601, doi:10.1029/2008GL036705.

Luo, Y., et al. (2011), Coordinated approaches to quantify long-term ecosystem dynamics in response to global change, *Glob. Change Bio., 17,* 843–854, doi:10.1111/j.1365-2486.2010.02265.x.

Luterbacher, J., E. Xoplaki, D. Dietrich, R. Rickli, J. Jacobeit, C. Beck, D. Gyalistras, C. Schmutz, and H. Wanner (2002), Reconstruction of sea level pressure fields over the eastern North Atlantic and Europe back to 1500, *Clim. Dynam., 18*(7), 545–561, doi:10.1007/s00382-001-0196-6.

Luterbacher, J., S. Koenig, J. Franke, G. van der Schrier, E. Zorita, A. Moberg, J. Jacobeit, P. Della-Marta, M. Küttel, E. Xoplaki, D. Wheeler, T. Rutishauser, M. Stössel, H. Wanner, R. Brázdil, P. Dobrovolný, D. Camuffo, C. Bertolin, A. van Engelen, F. Gonzalez-Rouco, R. Wilson, C. Pfister, D. Limanówka, Ø. Nordli, L. Leijonhufvud, J. Söderberg, R. Allan, M. Barriendos, R. Glaser, D. Riemann, Z. Hao, and C. Zerefos (2010), Circulation dynamics and its influence on European and Mediterranean January–April climate over the past half millennium: Results and insights from instrumental data, documentary evidence and coupled climate models, *Climatic Change, 101*(1), 201–234.

Lüthi, D., M. Le Floch, B. Bereiter, T. Blunier, J. M. Barnola, U. Siegenthaler, D. Raynaud, J. Jouzel, H. Fischer, K. Kawamura, and T. F. Stocker (2008), High-resolution carbon dioxide concentration record 650,000–800,000 years before present, *Nature, 453*(7193), 379–382, doi:10.1038/nature06949.

Lyman, J. M., S. A. Good, V. V. Gouretski, M. Ishii, G. C. Johnson, M. D. Palmer, D. M. Smith, and J. K. Willis (2010), Robust warming of the global upper ocean, *Nature, 465*(7296), 334–337, doi:10.1038/nature09043.

Lynch, A., P. Uotila, and J. J. Cassano (2006), Changes in synoptic weather patterns in the polar regions in the 20th and 21st centuries, Part 2, Antarctic, *Int. J. Climatol., 26,* 1181–1199.

Lynch, A. H., J. Beringer, P. Kershaw, A. Marshall, S. Mooney, N. Tapper, C. Turney, and S. Van Der Kaars (2007), Using the paleorecord to evaluate climate and fire interactions in Australia, *Annu. Rev. Earth Planet. Sci., 35,* 215–240.

Lyons, T. J. (2002), Clouds prefer native vegetation, *Meteor. Atmos. Phy., 80,* 131–140.

Lyons T. J., P. Schwerdtferger, J. M. Hacker, I. J. Foster, R. C. G. Smith, and X. Huang (1993), Land atmosphere interaction in a semiarid region: The bunny fence experiment, *Bull. Am. Meteorol. Soc., 74,* 1327–1334.

Lyons, T. J., R. C. G. Smith, and H. Xinmei (1996), The impact of clearing for agriculture on the surface energy balance, *Int. J. Climatol., 16,* 551–558.

Lythe, M. B., and D. G. Vaughan (2001), BEDMAP: A new ice thickness and subglacial topographic model of Antarctica, *J. Geophys. Res., 106*(B6), 11335–11351.

Maayar, M. E., N. Ramankutty, and C. Kucharik (2006), Modeling global and regional net primary production under elevated atmospheric CO_2: On a potential source of uncertainty, *Earth Interact., 10,* 1–20.

MacDonald, G. M., D. W. Bielman, K. V. Kremenetski, Y. Sheng, L. C. Smith, and A. A. Velichko (2006), Rapid early development of circumarctic peatlands and atmospheric CH_4 and CO_2 variations, *Science, 314*(5797), 285–288, doi:10.1126/science.1131722.

Mackas, D. L. (2011), Does blending of chlorophyll data bias temporal trend?, *Nature, 472*(7342), E4–E5, doi:10.1038/nature09951.

Mackas, D. L., S. Batten, and M. Trudel (2007), Effects on zooplankton of a warmer ocean: Recent evidence from the Northeast Pacific, *Prog. Oceanogr., 75*(2), 223–252, doi:10.1016/j.pocean.2007.08.010.

Mackey, D. J., J. E. O'Sullivan, and R. J. Watson (2002), Iron in the western Pacific: A riverine or hydrothermal source for iron in the equatorial undercurrent?, *Deep-Sea Res. Pt I, 49*(5), 877–893.

Madden, R. A., and P. R. Julian (1971), Detection of a 40–50 day oscillation in the zonal wind in the tropical Pacific, *J. Atmos. Sci., 28,* 702–708.

Mahecha, M. D., et al. (2010a), Comparing observations and process-based simulations of biosphere-atmosphere exchanges on multiple time-scales, *J. Geophys. Res., 115,* G02003, doi:10.129/2009JG001016.

Mahecha, M. D., et al. (2010b), Global convergence in the temperature sensitivity of respiration at ecosystem level, *Science, 329,* 838–840, doi:10.1126/science.1189587.

Mahmood, R., A. I. Quintanar, G. Conner, R. Leeper, S. Dobler, R. A. Pielke, Sr., A. Beltran-Przekurat, K. G. Hubbard, D. Niyogi, G. Bonan, P. Lawrence, T. Chase, R. McNider, Y. Wu, C. McAlpine, R. Deo, A. Etter, S. Gameda, B. Qian, A. Carleton, J. O. Adegoke, S. Vezhapparambu, S. Asefi, U. S. Nair, E. Sertel, D. R. Legates, R. Hale, O. W. Frauenfeld, A. Watts, M. Shepherd, C. Mitra, V. G. Anantharaj, S. Fall, H.-I. Chang, R. Lund, A. Treviño, P. Blanken, J. Du, and J. Syktus (2010), Impacts of land use/land-cover change on climate and future research priorities, *Bull. Am. Meteorol. Soc., 91,* 37–46.

Mahowald, N. M., and C. Luo (2003), A less dusty future?, *Geophys. Res. Lett., 30*(17), 1903, doi:10.1029/2003GL017880.

Mahowald, N. M., et al. (2009), Atmospheric iron deposition: Global distribution, variability, and human perturbations, *Annu. Rev. Marine Sci., 1*(1), 245–278.

Mahowald, N., K. Lindsay, D. Rothenberg, S. C. Doney, J. K. Moore, P. Thornton, J. T. Randerson, and C. D. Jones (2011), Desert dust and anthropogenic aerosol interactions in the Community Climate System Model coupled-carbon-climate model, *Biogeosciences, 8*(2), 387–414.

Maisch, M., and W. Haeberli (1982), Interpretation geometrischer Parameter von Spaetglazialgletschern im Gebiet Mittelbuenden, Schweizer Alpen, *Physische Geographie, 1,* 111–126.

Malcolm, G. M., J. C. López-Gutiérrez, R. T. Koide, and D. M. Eissenstat (2008), Acclimation to temperature and temperature sensitivity of metabolism by ectomycorrhizal fungi, *Glob. Change Biol., 14,* 1169–1180.

Malhi, Y. (2010), The carbon balance of tropical forest regions, 1990–2005, *Curr. Opin. Environ. Sustain., 2,* 237–244, doi:10.1016/j.cosust.2010.08.002.

Malhi, Y., L. E. O. Aragão, D. Galbraith, C. Huntingford, R. Fisher, P. Zelazowski, S. Sitch, C. McSweeney, and P. Meir (2009), Exploring the likelihood and mechanism of a climate-change-induced dieback of the Amazon rainforest, *Proc. Natl. Acad. Sci. USA, 106,* 20610–20615.

Manabe, S., and R. J. Stouffer (1980), Sensitivity of a global climate model to an increase of CO_2 concentration in the atmosphere, *J. Geophys. Res., 85,* 5529–5554.

Mander, L. (2011), Taxonomic resolution of the Triassic-Jurassic sporomorph record in East Greenland, *J. Micropalaeontol.,* (in press).

Mander, L., and R. J. Twitchett (2008), Quality of the Triassic-Jurassic bivalve fossil record in northwest Europe, *Palaeontology, 51,* 1213–1223.

Mander, L., R. J. Twitchett, and M. J. Benton (2008), Palaeoecology of the late Triassic extinction event in the SW UK, *J. Geol. Soc. London, 165,* 319–332.

Mander, L., W. M. Kürschner, and J. C. McElwain (2010), An explanation for conflicting records of Triassic-Jurassic plant diversity, *Proc. Natl. Acad. Sci. USA, 107,* 15351–15356.

Manktelow, P. T., G. W. Mann, K. S. Carslaw, D. V. Spracklen, and M. P. Chipperfield (2007), Regional and global trends in sulfate aerosol since the 1980s, *Geophys. Res. Lett., 34,* L14803, doi:10.11029/12006GL028668.

Mann, M. E., R. S. Bradley, and M. K. Hughes (1999), Northern Hemisphere temperatures during the past millennium: Inferences, uncertainties, and limitations, *Geophys. Res. Lett., 26,* 759–762, doi:10.1029/1999GL900070.

Mann, M. E., K. A. Emanual, G. J. Holland, and P. J. Webster (2007), Atlantic tropical cyclones revisited, *EOS Trans., 88,* 349–350.

Mann, M. E., Z. Zhang, S. Rutherford, R. S. Bradley, M. K. Hughes, D. Shindell, C. Ammann, G. Faluvegi, and F. Ni (2009), Global signatures and dynamical origins of the Little Ice Age and Medieval Climate Anomaly, *Science, 326,* 1256–1260, doi:10.1126/science.1177303

Manne, A. S., and R. G. Richels (2001), An alternative approach to establishing trade offs among greenhouse gases, *Nature, 410,* 675–677.

Manton, M. J., P. M. Della-Marta, M. R. Haylock, K. J. Hennessy, N. Nicholls, L. E. Chambers, D. A. Collins, G. Daw, A. Finet, D. Gunawan, K. Inape, H. Isobe, T. S. Kestin, P. Lefale, C. H. Leyu, T. Lwin, L. Maitrepierre, N. Ouprasitwong, C. M. Page, J. Pahalad, N. Plummer, M. J. Salinger, R. Suppiah, V. L. Tran, B. Trewin, I. Tibig, and D. Yee (2001), Trends in extreme daily rainfall and temperature in Southeast Asia and the South Pacific: 1961–1998, *Int. J. Climatol., 21,* 269–284.

Mantua, N. J., and S. R. Hare (2002), The Pacific Decadal Oscillation, *J. Oceanogr., 58*(1), 35–44.

Mantua, N. J., S. R. Hare, Y. Zhang, J. M. Wallace, and R. C. Francis (1997), A Pacific decadal climate oscillation with impacts on salmon, *Bull. Am. Meteorol. Soc., 78,* 1069–1079.

Manzoni, S., G. Katul, P. A. Fay, H. Wayne Polley, and A. Porporato (2011), Modeling the vegetation–atmosphere carbon dioxide and water vapor interactions along a controlled CO_2 gradient, *Ecol. Model., 222*(3), 653–665, doi:10.1016/j.ecolmodel.2010.10.016.

Mao, H., W.-C. Wang, X.-Z. Liang, and R. W. Talbot (2003), Global and seasonal variations of O_3 and NO_2 photodissociation rate coefficients, *J. Geophys. Res., 108*(D7), 4216, doi:10.1029/2002JD002760.

Maraun, D., F. Wetterhall, A. M. Ireson, R. E. Chandler, E. J. Kendon, M. Widmann, S. Brienen, H. W. Rust, T. Sauter, M. Themessl,

V. K. C. Venema, K. P. Chun, C. M. Goodess, R. G. Jones, C. Onof, M. Vrac, and I. Thiele-Eich (2010), Precipitation downscaling under climate change: Recent developments to bridge the gap between dynamical models and the end user, *Rev. Geophys., 48*, doi:10.1029/2009RG000314.

Marbaix, P., H. Gallee, O. Brasseur, and J. van Ypersele (2003), Lateral boundary conditions in regional climate models: A detailed study of the relaxation procedure, *Mon. Weather Rev., 131*(3), 461–479.

Marchant, R., and H. Hooghiemstra (2004), Rapid environmental change in African and South American tropics around 4000 years before present: A review, *Earth-Sci. Rev., 66*, 217–260.

Marchant, R., A. Cleef, S. P. Harrison, H. Hooghiemstra, V. Markgraf, J. van Boxel, T. Ager, L. Almeida, R. Anderson, C. Baied, H. Behling, J. C. Berrio, R. Burbridge, S. Bjorck, R. Byrne, M. Bush, J. Duivenvoorden, J. Flenley, P. De Oliveira, B. van Geel, K. Graf, W. D. Gosling, S. Harbele, T. van der Hammen, B. Hansen, S. Horn, P. Kuhry, M. P. Ledru, F. Mayle, B. Leyden, S. Lozano-Garcia, A. M. Melief, P. Moreno, N. T. Moar, A. Prieto, G. van Reenen, M. Salgado-Labouriau, F. Schabitz, E. J. Schreve-Brinkman, and M. Wille (2009), Pollen-based biome reconstructions for Latin America at 0, 6000 and 18 000 radiocarbon years ago, *Clim. Past, 5*(4), 725–767, doi:10.5194/cp-5-725-2009.

MARGO Project Members (2009), Constraints on the magnitude and patterns of ocean cooling at the Last Glacial Maximum, *Nat. Geosci., 2*(2), 127–132, doi:10.1038/ngeo411.

Marengo, J. A., M. Rusticucci, O. Penalba, and M. Renom (2009), An intercomparison of observed and simulated extreme rainfall and temperature events during the last half of the twentieth century: part 2: historical trends, *Climatic Change, 98*(3–4), 509–529.

Mariotte, E. (1681), *Essays de phisique, ou Mémoires pour servir à la science des choses naturelles*, A Paris: Chez Estienne Michalle 1679 [–1681], 4 parts in 2 vols.

Mark, B. G., S. P. Harrison, A. Spessa, M. New, D. J. A. Evans, and K. F. Helmens (2005), Tropical snowline changes at the Last Glacial Maximum: A global assessment, *Quatern. Int., 138*, 168–201.

Markgraf, V. (1989), Palaeoclimates in Central and South America since 18,000 B.P based on pollen and lake-level records, *Quaternary Sci. Rev., 8*(1), 1–24.

Marks, K. (2010), On the frontline of climate change, *The Independent*, 19 August 2010, *http://www.independent.co.uk/environment/climate-change/on-the-frontline-of-climate-change-2056322.html.* (Accessed 22/10/10.)

Marlon, J. R., P. J. Bartlein, C. Carcaillet, D. G. Gavin, S. P. Harrison, P. E. Higuera, F. Joos, M. J. Power, and I. C. Prentice (2008), Climate and human influences on global biomass burning over the past two millennia, *Nat. Geosci., 1*(10), 697–702.

Marros, S., and D. Spano (2009), Part III: Carbon, from Inventory of current state of empirical and modeling knowledge of energy, water and carbon sinks, sources and fluxes, Report on EU Framework 7 Project D.2.1, edited by C. S. B. Grimmond, Ref. 211345_001_PU_KCL, *http://www.bridge-fp7.eu/.*

Marsh, N. D., and H. Svensmark (2000), Low cloud properties influenced by cosmic rays, *Phys. Rev. Lett., 85*(23), 5004–5007.

Marshall, A. G., and A. H. Lynch (2006), Time-slice analysis of the Australian summer monsoon during the Late Quaternary using the Fast Ocean Atmosphere Model, *J. Quatern. Sci., 21*(7), 789–801.

Martilli, A. (2007), Current research and future challenges in urban mesoscale modelling, *Int. J. Climatol., 27*, 1909–1918.

Martin, J. H. (1990), Glacial-interglacial CO_2 change: The iron hypothesis, *Paleoceanography, 5*(1), 1–13.

Martin, R. V., A. M. Fiore, and A. Van Donkelaar (2004), Space-based diagnosis of surface ozone sensitivity to anthropogenic emissions, *Geophys. Res. Lett., 31*, L06120, doi:10.1029/2004GL019416.

Martin, R. V., B. Sauvage, I. Folkins, C. E. Sioris, C. Boone, P. Bernath, and J. Ziemke (2007), Space-based constraints on the production of nitric oxide by lightning, *J. Geophys. Res.-Atmos., 112*, D09309, doi:10.1029/2006JD007831.

Martinez, E., D. Antoine, F. D'Ortenzio, and B. Gentili (2009), Climate-driven basin-scale decadal oscillations of oceanic phytoplankton, *Science, 326*, 1253–1256.

Martinson, D. G., N. G. Pisias, J. D. Hays, J. Imbrie, T. C. Moore, Jr., and N. J. Shackleton (1987), Age dating and the orbital theory of the ice ages: Development of a high-resolution 0 to 300,000-year chronostratigraphy, *Quatern. Res., 27*(1), 1–29.

Martrat, B., J. O. Grimalt, C. Lopez-Martinez, I. Cacho, F. J. Sierro, J. A. Flores, R. Zahn, M. Canals, J. H. Curtis, and D. A. Hodell (2004), Abrupt temperature changes in the western Mediterranean over the past 250,000 years, *Science, 306*(5702), 1762–1765, doi:10.1126/science.1101706.

Martrat, B., J. O. Grimalt, N. J. Shackleton, L. De Abreu, M. A. Hutterli, and T. F. Stocker (2007), Four climate cycles of recurring deep and surface water destabilizations on the Iberian margin, *Science, 317*(5837), 502–507, doi:10.1126/science.1139994.

Maruoka, T., C. Koeberl, J. Newton, I. Gilmour, and B. F. Bohor (2002), Sulfur isotopic compositions across terrestrial Cretaceous-Tertiary boundary successions, in Catastrophic events and mass extinctions: Impacts and beyond, edited by C. Koeberl and K. G. MacLeod, *Geol. Soc. Am. Spec. Pap. 356*, Boulder, Colorado, pp. 337–344.

Marynowski, L., and B. R. T. Simoneit (2009), Widespread upper Triassic to lower Jurassic wildfire records from Poland: Evidence from charcoal and pyrolytic polycyclic aromatic hydrocarbons, *Palaios, 24*, 785–798.

Marzin, C., and P. Braconnot (2009), Variations of Indian and African monsoons induced by insolation changes at 6 and 9.5 kyr B.P, *Clim. Dynam., 33*(2), 215–231, doi:10.1007/s00382-009-0538-3.

Marzoli, A., P. R. Renne, E. M. Piccirillo, M. Ernesto, G. Bellieni, and A. De Min (1999), Extensive 200-million-year-old continental flood basalts of the Central Atlantic Magmatic Province, *Science, 284*, 616–618.

Maslanik, J., S. Drobot, C. Fowler, W. Emery, and R. Barry (2007), On the Arctic climate paradox and the continuing role of atmospheric circulation in affecting sea ice conditions, *Geophys. Res. Lett., 34*, L03711.

Masson, V. (2000), A physically-based scheme for the urban energy budget in atmospheric models, *Bound-Layer Meteor., 94*, 357–397.

Masson, V. (2006), Urban surface modelling and the meso-scale impact of cities, *Theor. Appl. Climatol., 84*, 35–45, doi:10.1007/s00704-005-0142-3.

Masson V., C. S. B. Grimmond, and T. R. Oke (2002), Evaluation of the Town Energy Balance (TEB) scheme with direct measurements from dry districts in two cities, *J. Appl. Meteorol., 41*, 1011–1126.

Masson, V., L. Gomes, G. Pigeon, C. Liousse, V. Pont, J.-P. Lagouarde, J. Voogt, J. Salmond, T. R. Oke, J. Hidalgo, D. Legain,

O. Garrouste, C. Lac, O. Connan, X. Briottet, S. Lachérade, and P. Tulet (2008), The canopy and aerosol particles interactions in Toulouse urban layer (CAPITOUL) experiment, *Meteorol. Atmos. Phys., 102*, 135−157.

Masson-Delmotte, V., J. Jouzel, A. Landais, M. Stievenard, S. J. Johnsen, J. W. C. White, M. Werner, A. Sveinbjornsdottir, and K. Fuhrer (2005), Atmospheric science: GRIP deuterium excess reveals rapid and orbital-scale changes in Greenland moisture origin, *Science, 309*(5731), 118−121, doi:10.1126/science.1108575.

Masson-Delmotte, V., M. Kageyama, P. Braconnot, S. Charbit, G. Krinner, C. Ritz, E. Guilyardi, J. Jouzel, A. Abe-Ouche, M. Crucifix, R. M. Gladstone, C. D. Hewitt, A. Kitoh, A. N. LeGrande, O. Marti, U. Merkel, T. Motoi, R. Ohgaito, B. Otto-Bliesner, W. R. Peltier, I. Ross, P. J. Valdes, G. Vettoretti, S. L. Weber, F. Wolk, and Y. Yu (2006), Past and future polar amplification of climate change: Climate model intercomparisons and ice-core constraints, *Clim. Dynam., 26*(5), 513−529, doi:10.1007/s00382-005-0081-9.

Mastrandrea, M. D., and S. H. Schneider (2010), *Preparing for Climate Change*, MIT Press, Cambridge, MA.

Mastrandrea, D. M., N. E. Heller, T. L. Root, and S. H. Schneider (2010), Bridging the gap: linking climate-impacts research with adaptation planning and management, *Climatic Change, 100*, 87−101, doi:10.1007/s10584-010-9827-4.

Matear, R. J., and A. C. Hirst (1999), Climate change feedback on the future oceanic CO_2 uptake, *Tellus B, 51*(3), 722−733.

Matear, R. J., and A. C. Hirst (2003), Long-term changes in dissolved oxygen concentrations in the ocean caused by protracted global warming, *Global Biogeochem. Cycles, 17*(4), 1125, doi:10.1029/2002GB001997.

Matear, R. J., A. C. Hirst, and B. I. McNeil (2000), Changes in dissolved oxygen in the Southern Ocean with climate change, *Geochem. Geophys. Geosy., 1*(11).

Matear, R. J., Y. P. Wang, and A. Lenton (2010), Land and ocean nutrient and carbon cycle interactions, *Curr. Opin. Environ. Sustain., 2*(4), 258−263.

Matese A., B. Gioli, F. P. Vaccari, A. Zaldei, and F. Miglietta (2009), CO_2 emissions of the city center of Firenze, Italy: Measurement, evaluation and source partitioning, *J. Appl. Meteorol. Clim., 48*, 1940−1947.

Mathew, S., A. Henderson-Sellers, and R. Taplin (2011), Prioritizing climatic change adaptation at local government levels, in *The Economic, Social and Political Elements of Climate Change*, edited by W. Leal Filho, pp. 733−751, Springer Verlag, Berlin.

Matsueda, M., R. Mizuta, and S. Kusunoki (2009), Future change in wintertime atmospheric blocking simulated using a 20-km-mesh atmospheric global circulation model, *J. Geophys. Res., 114*, D12114, doi:10.1029/2009JD011919.

Matthes, S., V. Grewe, R. Sausen, and G.-J. Roelofs (2007), Global impact of road traffic emissions on tropospheric ozone, *Atmos. Chem. Phys., 6*, 1075−1089.

Matthews, E. (1997), Global litter production, pools, and turnover times: Estimates from measurement data and regression models, *J. Geophys. Res., 102*, 18771−18800.

Matthews, H. D., A. J. Weaver, K. J. Meissner, N. P. Gillett, and M. Eby (2004), Natural and anthropogenic climate change: Incorporating historical land cover change, vegetation dynamics, and the global carbon cycle, *Clim. Dynam., 22*, 461−479, doi:10.1007/s00382-004-0392-2.

Matthews, H. D., and D. W. Keith (2007), Carbon-cycle feedbacks increase the likelihood of a warmer future, *Geophys. Res. Lett., 34*, L09702, doi:10.1029/2006GL028685.

Matthews, H. D., M. Eby, T. Ewen, P. Friedlingstein, and B. J. Hawkins (2007), What determines the magnitude of carbon-cycle climate feedbacks? *Global Biogeochem. Cycles, 21*, GB2012, doi:10.1029/2006GB002733.

Matulla, C., W. Schöner, H. Alexandersson, H. von Storch, and X. L. Wang (2008), European storminess: Late 19th century to present, *Clim. Dynam., 31*, 1125−1130.

Mauquoy, D., T. Engelkes, M. H. M. Groot, F. Markesteijn, M. G. Oudejans, J. van der Plicht, and B. van Geel (2002), High-resolution records of Late-Holocene climate change and carbon accumulation in two North-West European ombrotrophic peat bogs, *Palaeogeogr. Palaeoclimatol. Palaeoecol., 186*(3−4), 275−310, doi:10.1016/S0031-0182(02)00513-8.

Maurer, E. P., and H. G. Hidalgo (2008), Utility of daily vs. monthly large-scale climate data: An intercomparison of two statistical downscaling methods, *Hydrol. Earth Sys. Sci., 12*(2), 551−563.

Mayle, F. E., and M. J. Power (2008), Impact of a drier early-mid-Holocene climate upon Amazonian forests, *Philos. Trans. Roy. Soc. B, 363*(1498), 1829−1838, doi:10.1098/rstb.2007.0019.

Mayr, G., L. Armi, A. Gohm, G. Zangl, D. Durran, C. Flamant, S. Gabersek, S. Mobbs, A. Ross, and M. Weissmann (2007), Gap flows: Results from the Mesoscale Alpine Programme, *Q. J. Roy. Meteor. Soc., 133*(625), 881−896, doi:10.1002/qj.66.

McAlpine, C. A., J. Syktus, J. G. Ryan, R. C. Deo, P. J. Lawrence, G. M. McKeon, H. A. McGowan, and S. R. Phinn (2009), A continent under stress: Interactions, feedbacks, and risks associated with impact of modified land cover on Australia's climate, *Glob. Change Biol., 15*, 1−18, doi:10.1111/j.1365-2486.2009.01939.x.

McBride, J. L., and N. Nicholls (1983), Seasonal relationships between Australian rainfall and the Southern Oscillation, *Mon. Weather Rev., 111*, 1998−2004.

McCabe, G. J., and M. D. Dettinger (1999), Decadal variations in the strength of ENSO teleconnections with precipitation in the western United States, *Int. J. Climatol., 19*(13), 1399−1410.

McCarthy, J. (2009), AAAS presidential address: Reflections on our planet and its life, origins, and futures, *Science, 326*, 1665.

McCarthy, M. P., P. W. Thorne, and H. A. Titchner (2009), An analysis of tropospheric humidity trends from radiosondes, *J. Climate, 22*, 5820−5838.

McCarthy, M. P., M. J. Best, and R. A. Betts (2010), Climate change in cities due to global warming and urban effects, *Geophys. Res. Lett., 37*, L09705, doi:10.1029/2010GL042845.

McClain, C. R. (2009), A decade of satellite ocean color observations*, *Annu. Rev. Marine. Sci., 1*(1), 19−42, doi:10.1146/annurev.marine.010908.163650.

McComiskey, A., and G. Feingold (2008), Quantifying error in the radiative forcing of the first aerosol indirect effect, *Geophys. Res. Lett., 35*, doi:10.1029/2007GL032667.

McConnell, J., R. Edwards, G. L. Kok, M. G. Flanner, C. S. Zender, E. S. Saltzman, J. R. Banta, D. R. Pasteris, M. M. Carter, and J. D. W. Kahl (2007), 20th-century industrial black carbon emissions altered Arctic climate forcing, *Science, 317*, 1381−1384, doi:10.1126/science.1144856.

McDonald, R. E., D. G. Bleaken, D. R. Cresswell, V. D. Pope, and C. A. Senior (2005), Tropical storms: representation and diagnosis in climate models and the impacts of climate change. *Climate Dynamics, 25,* 19–36.

McDowell, N., W. T. Pockman, C. D. Allen, D. D. Breshears, N. Cobb, T. Kolb, J. Plaut, J. Sperry, A. West, D. G. Williams, and E. A. Yepez (2008), Mechanisms of plant survival and mortality during drought: Why do some plants survive while others succumb to drought? *New Phytol., 178,* 719–739.

McElwain, J. C., and W. G. Chaloner (1995), Stomatal density and index of fossil plants track atmospheric carbon dioxide in the Palaeozoic, *Ann. Bot.-London, 76,* 389–395.

McElwain, J. C., and S. W. Punyasena (2007), Mass extinction events and the plant fossil record, *Trends Ecol. Evol., 22,* 548–557.

McElwain, J. C., F. J. G. Mitchell, and M. B. Jones (1995), Relationship of stomatal density and index of *Salix cinerea* to atmospheric carbon dioxide concentrations in the Holocene, *Holocene, 5,* 216–219.

McElwain, J. C., D. J. Beerling, and F. I. Woodward (1999), Fossil plants and global warming at the Triassic-Jurassic boundary, *Science, 285,* 1386–1390.

McElwain, J. C., M. E. Popa, S. P. Hesselbo, M. Haworth, and F. Surlyk (2007), Macroecological responses of terrestrial vegetation to climatic and atmospheric change across the Triassic/Jurassic boundary in East Greenland, *Paleobiology, 33,* 547–573.

McElwain, J. C., P. J. Wagner, and S. P. Hesselbo (2009), Fossil plant relative abundances indicate sudden loss of late Triassic biodiversity in East Greenland, *Science, 324,* 1554–1556.

McFarland, J. W., R. W. Ruess, K. Kielland, K. Pregitzer, R. Hendrick, and M. Allen (2010), Cross-ecosystem comparisons of in situ plant uptake of amino acid-N and NH_4^+, *Ecosystems, 13,* 177–193, doi:10.1007/s10021-009-9309-6.

McGhee, Jr., G. T., P. M. Sheehan, D. J. Bottjer, and M. L. Droser (2004), Ecological ranking of Phanorozoic biodiversity crises: Ecological and taxonomic severities are decoupled, *Palaeogeogr. Palaeoclim., 211,* 289–297.

McGranahan, G., D. Balk, and B. Anderson (2007), The rising tide: Assessing the risks of climate change and human settlements in low elevation coastal zones, *Environ. Urban., 19,* 17–37.

McGregor, J. L. (1997), Regional climate modelling, *Meteorol. Atmos. Phys., 63,* 105–117.

McGregor, J. L., and K. Walsh (1993), Nested simulations of perpetual January climate over the Australian region, *J. Geophys. Res., 98,* 23283–23290.

McGregor, J. L., and M. R. Dix (2008), An updated description of the Conformal-Cubic Atmospheric Model, in *High Resolution Simulation of the Atmosphere and Ocean*, edited by K. Hamilton, and W. Ohfuchi, pp. 51–76, Springer.

McGuffie, K., and A. Henderson-Sellers (2004), Stable water isotope characterization of human and natural impacts on land-atmosphere exchanges in the Amazon basin, *J. Geophys. Res.-Atmos., 109,* D17104, doi:10.1029/2003JD004388.

McGuffie, K., and A. Henderson-Sellers (2005), *A Climate Modelling Primer*, 3rd edition, John Wiley & Sons, UK.

McGuffie, K., A. Henderson-Sellers, H. Zhang, T. B. Durbridge, and A. J. Pitman (1995), Global climate sensitivity to tropical deforestation, *Global Planet. Change, 10,* 97–128.

McGuire, A. D., et al. (2009), Sensitivity of the carbon cycle in the Arctic to climate change, *Ecolog. Monogr., 79*(4), 523–555.

McGuire, A. D., R. W. Macdonald, E. A. G. Schuur, J. W. Harden, P. Kuhry, D. J. Hayes, T. R. Christensen, and M. Heimann (2010), The carbon budget of the northern cryosphere region, *Curr. Opin. Environ. Sustain., 2,* 231–236, doi:10.1016/j.cosust.2010.05.003.

McHone, J. G. (1996), Broad-terrane Jurassic flood basalts across northeastern North America, *Geology, 24,* 319–322.

McHone, J. G. (2003), Volatile emissions from Central Atlantic Magmatic Province basalts; Mass assumptions and environmental consequences, in *The Central Atlantic Magmatic Province; Insights from Fragments of Pangea: Geophysical Monograph*, edited by W. E. Hames, et al., pp. 241–254, American Geophysical Society, Washington, DC.

McKibben, W. (2010), *Eaarth: Making a Life on a Tough New Planet*, Times Books, New York.

McLachlan, J. S., J. S. Clark, and P. S. Manos (2005), Molecular indicators of tree migration capacity under rapid climate change, *Ecology, 86*(8), 2088–2098.

McManus, J. F., R. Francois, J.-M. Gherardi, L. D. Keigwin, and S. Brown-Leger (2004), Collapse and rapid resumption of Atlantic meridional circulation linked to deglacial climate changes, *Nature, 428*(6985), 834–837, doi:10.1038/nature02494.

McNeil, B. I., and R. J. Matear (2007), Climate change feedbacks on future oceanic acidification, *Tellus B, 59*(2), 191–198.

McNeil, B. I., and R. J. Matear (2008), Southern Ocean acidification: A tipping point at 450-ppm atmospheric CO_2, *Proc. Natl. Acad. Sci. USA, 105*(48), 18860.

McNeil, B. I., R. J. Matear, R. M. Key, J. L. Bullister, and J. L. Sarmiento (2003), Anthropogenic CO_2 uptake by the ocean based on the global chlorofluorocarbon data set, *Science, 299*(5604), 235.

McPhaden, M. J., and D. Zhang (2002), Slowdown of the meridional overturning circulation in the upper Pacific Ocean, *Nature, 415,* 603–608.

McPhaden, M., and D. Zhang (2004), Pacific Ocean circulation rebounds, *Geophys. Res. Lett., 31*(18), L1830, doi:10.1029/2004GL020727.

McQuatters-Gollop, A., et al. (2011), Is there a decline in marine phytoplankton?, *Nature, 472*(7342), E6–E7, doi:10.1038/nature09950.

McRoberts, C. A. (2009), From the secretary ICS Subcommission on Triassic Stratigraphy: Minutes of the business meeting of the STS, Bad Goisern, Austria, October 1, 2008, *Albertiana, 37,* 4–5.

McRoberts, C. A., H. Furrer, and D. S. Jones (1997), Palaeoenvironmental interpretation of a Triassic-Jurassic boundary section from western Austria based on palaeoecological and geochemical data, *Palaeogeogr. Palaeoclim., 136,* 79–95.

Mearns, L. O., I. Brogardi, F. Giorgi, I. Matyasovszky, and M. Palecki (1999), Comparison of climate change scenarios generated from regional climate model experiments and statistical downscaling, *J. Geophys. Res., 104*(D6), 6603–6621.

Mearns, L. O., W. Gutowski, R. Jones, R. Leung, S. McGinnis, A. Nunes, and Y. Qian (2009), A regional climate change assessment program for North America, *EOS Trans., 90,* 311.

Medlyn, B. E., et al. (2011), Reconciling the optimal and empirical approaches to modelling stomatal conductance, *Glob. Change Bio., 17,* doi:10.1111/j.1365.2486.2010.02375x.

Meehl, G. A., and C. Tebaldi (2004), More intense, more frequent, and longer lasting heat waves in the 21st century, *Science, 305,* 994–997.

Meehl, G. A., W. M. Washington, T. M. L. Wigley, J. M. Arblaster, and A. Dai (2003), Solar and greenhouse gas forcing and climate response in the twentieth century, *J. Climate,* 16426–16444.

Meehl, G. A., C. Tebaldi, and D. Nychka (2004), Changes in frost days in simulations of 21st century climate, *Clim. Dynam., 23,* 495–511.

Meehl, G. A., J. M. Arblaster, and C. Tebaldi (2005), Understanding future patterns of precipitation extremes in climate model simulations, *Geophys. Res. Lett., 32*, L18719, doi:10.1029/2005GL023680.

Meehl, G. A., T. F. Stocker, W. D. Collins, P. Friedlingstein, A. T. Gaye, J. M. Gregory, A. Kitoh, J. M. Murphy, A. Noda, S. C. B. Raper, I. G. Watterson, A. J. Weaver and Z.-C. Zhao (2007), Global climate projections, in *Climate Change 2007: The Physical Science Basis, contribution of Working Group I to the Fourth Assessment Report (FAR) of the Intergovernmental Panel on Climate Change*, edited by S. Soloman, D. Qin, M. Manning, M. Marquis, K. Averyt, M. M. B. Tignor, H. L. Miller, and Z. Chen, pp. 747–845, Cambridge University Press, Cambridge, UK.

Meehl, G. A, J. M. Arblaster, and C. Tebaldi (2007a), Contributions of natural and anthropogenic forcings to changes in temperature extremes over the US, *Geophys. Res. Lett., 34*, L19709, doi:10.1029/2007GL030948.

Meehl, G. A., C. Tebaldi, H. Teng, and T. C. Peterson (2007b), Current and future U.S. weather extremes and El Niño, *Geophys. Res. Lett., 34*, L20704, doi:10.1029/2007GL031027.

Meehl, G. A., J. M. Arblaster, and W. D. Collins (2008), Effects of black carbon aerosols on the Indian monsoon, *J. Climate, 21*, 2869–2882.

Meehl, G. A., J. M. Arblaster, K. Matthes, F. Sassi, and H. van Loon (2009), Amplifying the Pacific climate system response to a small 11-year solar cycle forcing, *Science, 325*, 1114–1118.

Meehl, G. A., L. Goddard, J. Murphy, R. J. Stouffer, G. Boer, G. Danabasoglu, K. Dixon, M. A. Giorgetta, A. Greene, E. Hawkins, G. Hegerl, D. Karoly, N. Keenlyside, M. Kimoto, B. Kirtman, A. Navarra, R. Pulwarty, D. Smith, D. Stammer, and T. Stockdale (2009), Decadal prediction: Can it be skilful? *Bull. Am. Meteorol. Soc., 90*, 1467–1485.

Meehl, G. A., C. Tebaldi, G. Walton, D. Easterling, and L. McDaniel (2009), The relative increase of record high maximum temperatures compared to record low minimum temperatures in the U.S., *Geophys. Res. Lett., 36*, L23701, doi:10.1029/2009GL040736.

Mei, R., and G. Wang (2010), Rain follows logging in the Amazon? Results from CAM3–CLM3, *Clim. Dynam., 34*, 983–996, doi:10.1007/s00382-009-0592-x.

Meier, M. F., and A. Post (1987), Fast tidewater glaciers, *J. Geophys. Res., 92B*, 9051–9058.

Meier, M. F., D. B. Bahr, M. B. Dyurgerov, and W. T. Pfeffer (2005), Comment on "The potential for sea level rise: New estimates from glacier and ice cap area and volume distribution" by S. C. B. Raper, and R. J. Braithwaite, *Geophys. Res. Lett., 32*, L17501, doi:10.1029/2005GL023319.

Meier M. F., M. B. Dyurgerov, U. K. Rick, S. O'Neel, W. T. Pfeffer, R. S. Anderson, S. P. Anderson, and A. F. Glazovsky (2007), Glaciers dominate eustatic sea-level rise in the 21st century, *Science, 317*(5841), 1064–1067.

Melosh, H. J., N. M. Schneider, K. J. Zahnle, and D. Latham (1990), Ignition of global wildfires at the Cretaceous/Tertiary boundary, *Nature, 343*, 251–254.

Menary, M. B., W. Park, K. Lohmann, M. Vellinga, M. Palmer, M. Latif, and J. Jungclaus (2011), A multimodel comparison of centennial Atlantic Meridional Overturning Circulation variability, *J. Climate*, (in prep.).

Menéndez, M., and P. L. Woodworth (2010), Changes in extreme high water levels based on a quasi-global tide-gauge data set, *J. Geophys. Res., 115*, C10011, doi:10.1029/2009JC005997.

Menne, M. J., and C. N. Williams (2005), Detection of undocumented change points: On the use of multiple test statistics and composite reference series, *J. Climate, 18*, 4271–4286.

Menon, S., and L. Rotstayn (2006), The radiative influence of aerosol effects on liquid-phase cumulus and stratiform clouds based on sensitivity studies with two climate models, *Clim. Dynam., 27*, 345–356.

Menon, S., A. D. Del Genio, D. Koch, and G. Teslioudis (2002), GCM simulations of aerosol indirect effect: Sensitivity to cloud parameterization and aerosol burden, *J. Atmos. Sci., 59*, 692–713.

Menon, S., J. Hansen, L. Nazarenko, and Y. Luo (2002), Climate effects of black carbon aerosols in China and India, *Science, 297*, 2250–2252.

Menon, S., H. Akbari, S. Mahanama, I. Sednev, and R. Levinson (2010), Radiative forcing and temperature response to changes in urban albedos and associated CO_2 offsets, *Environ. Res. Lett., 5*, doi:10.1088/1748-9326/5/1/014005.

Menviel, L., A. Timmermann, O. E. Timm, and A. Mouchet (2011), Deconstructing the Last Glacial termination: The role of millennial and orbital-scale forcings, *Quaternary Sci. Rev.*, doi:10.1016/j.quascirev.2011.02.005.

Mercado, L. M., N. Bellouin, S. Sitch, O. Boucher, C. Huntingford, M. Wild, and P. M. Cox (2009), Impact of changes in diffuse radiation on the global land carbon sink, *Nature, 458*, 1014–1018.

Mercer, J. H. (1978), West Antarctic Ice Sheet and CO_2 greenhouse effect: A threat of disaster, *Nature, 271*, 321–325.

Meredith, M. P., and A. M. Hogg (2006), Circumpolar response of Southern Ocean eddy activity to a change in the Southern Annular Mode, *Geophys. Res. Lett., 33*(16), L16608.

Meredith, M. P., P. L. Woodworth, C. W. Hughes, and V. Stepanov (2004), Changes in the ocean transport through Drake Passage during the 1980s and 1990s, forced by changes in the Southern Annular Mode, *Geophys. Res. Lett., 31*(21), L21305.

Merryfield, W. J., and G. J. Boer (2005), Variability of upper Pacific Ocean overturning in a coupled climate model, *J. Climate, 18*, 666–683.

Metz, B., O. R. Davidson, P. R. Bosch, R. Dave, and L. A. Meyer (Eds.) (2007), *Contribution of Working Group III to the Fourth Assessment Report (FAR) of the Intergovernmental Panel on Climate Change (IPCC)*, Cambridge University Press, Cambridge, UK, and New York, NY.

Michaud, L. M. (2001), Total energy equation method for calculating hurricane intensity, *Meteorol. Atmos. Phys., 78*, 35–43.

Mickley, L. J., P. P. Murti, D. J. Jacob, J. A. Logan, D. Rind, and D. Koch (1999), Radiative forcing from tropospheric ozone calculated with a unified chemistry-climate model, *J. Geophys. Res., 104*, 30153–30172.

Miguez-Macho, G., G. Stenchikov, and A. Robock (2004), Spectral nudging to eliminate the effects of domain position and geometry in regional climate model simulations, *J. Geophys. Res., 109*(D13), doi:10.1029/2003JD004495.

Mikolajewicz, U., and E. Maier-Reimer (1990), Internal secular variability in an ocean general circulation model, *Clim. Dynam., 4*, 145–156.

Milankovitch, M. M. (1941), Kanon der Erdbestrahlung und seine Anwendung auf des Eizeitenproblem, *R. Serbian Acad. Spec. Publ., 132, Sect. Math. Nat. Sci., 633*, (Canon of Insolation and the ice-age problem, English translation by Israel Program for Scientific Translation, Jerusalem,1969).

Milbrath, L. W. (1984), *Environmentalists: Vanguard for a New Society*, State University of New York Press, New York.

Milbrath, L. W. (1989), *Envisioning a Sustainable Society: Learning Our Way Out*, State University of New York Press, New York.

Miller, G. H., R. B. Alley, J. Brigham-Grette, J. J. Fitzpatrick, L. Polyak, M. C. Serreze, and J. W. C. White (2010), Arctic amplification: Can the past constrain the future?, *Quaternary Sci. Rev., 29*(15−16), 1779−1790, doi:10.1016/j.quascirev.2010.02.008.

Miller, R. L., G. A. Schmidt, and D. T. Shindell (2006), Forced variations of annular modes in the 20th century Intergovernmental Panel on Climate Change Fourth Assessment Report (FAR) models, *J. Geophys. Res., 111*, D18101, doi:10.1029/ 2005JD006323.

Miller, S., M. Goulden, and H. R. da Rocha (2009), LBA-ECO CD-04 Meteorological and Flux Data, km 83 Tower Site, Tapajos National Forest, data set, available online from Oak Ridge National Laboratory Distributed Active Archive Center, Oak Ridge, Tennessee, *http://daac.ornl.gov*, doi:10.3334/ORNLDAAC/946.

Mills, G. (2004), The urban canopy layer heat island, *http://www.urban-climate.org/UHI_Canopy.pdf*, (Accessed 29/09/2010.)

Mills, G. (2006), Progress toward sustainable settlements: A role for urban climatology, *Theor. Appl. Climatol., 84*, 69−76.

Mills, G. (2007), Cities as agents of global change, *Int. J. Climatol., 27*, 1849−1857, doi:10.1002/joc.1604.

Mills, G., H. A. Cleugh, R. Emmanuel, W. Endlicher, E. Erell, G. McGranahan, E. Ng, A. Nickson, J. Rosenthal, and K. Steemer (2010), Climate information for improved planning and management of mega cities (needs perspective), *Procedia Environmental Sciences, 1*, 228−246.

Milly, P. C. D., K. A. Dunne, and A. V. Vecchia (2005), Global pattern of trends in streamflow and water availability in a changing climate, *Nature, 438*, 347−350.

Min, K., R. Mundil, P. R. Renne, and K. R. Ludwig (2000), A test for systematic errors in 40Ar/39Ar geochronology through comparison with U-Pb analysis of a 1.1 Ga rhyolite, *Geochim. Cosmochim. Acta, 64*, 73−98.

Min, S.-K., X. Zhang, F. W. Zwiers, P. Friederichs, and A. Hense (2009), Signal detectability in extreme precipitation changes assessed from twentieth-century climate simulations, *Clim. Dynam., 32*, 95−111.

Min, S.-K., X. Zhang, F. W. Zwiers, and G. C. Hegerl (2011), Human contribution to more-intense precipitation extremes, *Nature, 470*, 378−381, doi:10.1038/nature09763.

Ming, Y., V. Ramaswamy, and G. Persad (2010), Two opposing effects of absorbing aerosols on global-mean precipitation, *Geophys. Res. Lett., 37*(13), L13701.

Minschwaner, K., and A. E. Dessler (2004), Water vapor feedback in the tropical upper troposphere: Model results and observations, *J. Climate, 17*, 1272−1282.

Minschwaner, K., A. E. Dessler, and P. Sawaengphokhai (2006), Multi-model analysis of the water vapor feedback in the tropical upper troposphere, *J. Climate, 19*, 5455−5463.

Mishchenko, M. I., I. V. Geogdzhayev, W. B. Rossow, B. Cairns, B. E. Carlson, A. A. Lacis, L. Liu, and L. D. Travis (2007), Long-term satellite record reveals likely recent aerosol trend, *Science, 315*, 1543.

Mitchell, T. D., and P. D. Jones (2005), An improved method of constructing a database of monthly climate observations and associated high-resolution grids, *Int. J. Climatol., 25*, 693−712.

Mitchell, V. G., R. G. Mein, and T. A. McMahon (2001), Modelling the urban water cycle, *Environ. Modell. Softw., 16*, 615−629.

Mitchell, V. G., H. A. Cleugh, C. S. B. Grimmond, and J. Xu (2008), Linking urban water balance and energy balance models to analyse urban design options, *Hydrol. Process., 22*, 2891−2900, doi:10.1002/hyp.6868.

Mitrovica, J. X., N. Gomez, and P. U. Clark (2009), The sea-level fingerprint of West Antarctic collapse, *Science, 323*(5915), 753, doi:10.1126/science.1166510.

Moberg, A., P. D. Jones, D. Lister, A. Walther, M. Brunet, J. Jacobeit, L. V. Alexander, P. M. Della-Marta, J. Luterbacher, P. Yiou, D. L. Chen, A. Tank, O. Saladie, J. Sigro, E. Aguilar, H. Alexandersson, C. Almarza, I. Auer, M. Barriendos, M. Begert, H. Bergstrom, R. Bohm, C. J. Butler, J. Caesar, A. Drebs, D. Founda, F. W. Gerstengarbe, G. Micela, M. Maugeri, H. Osterle, K. Pandzic, M. Petrakis, L. Srnec, R. Tolasz, H. Tuomenvirta, P. C. Werner, H. Linderholm, A. Philipp, H. Wanner, and E. Xoplaki (2006), Indices for daily temperature and precipitation extremes in Europe analyzed for the period 1901−2000, *J. Geophys. Res.-Atmos., 111*(D22), D22106.

Mokssit, A. (2003), Development of priority climate indices for Africa: A CCI/CLIVAR workshop of the World Meteorological Organization, in *Mediterranean Climate: Variability and Trends*, edited by H. J. Bolle, pp. 116−123, Springer, Berlin.

Molina, E., et al. (2006), The global boundary stratotype section and point for the base of the Danian Stage (Paleocene, Paleogene, "Tertiary", Cenozoic) at El Kef, Tunisia − Original definition and revision, *Episodes, 29*, 263−273.

Molina, M., D. Zaelke, K. M. Sarma, S. O. Andersen, V. Ramanathan, and D. Kaniaru (2009), Reducing abrupt climate change risk using the Montreal Protocol and other regulatory actions to complement cuts in CO_2 emissions, *Proc. Natl. Acad. Sci. USA, 106*, 20616−20621.

Monks, P. S., C. Granier, S. Fuzzi, A. Stohl, M. Williams, H. Akimoto, M. Amann, A. Baklanov, U. Baltensperger, I. Bey, N. Blake, R. S. Blake, K. Carslaw, O. R. Cooper, F. Dentener, D. Fowler, E. Fragkou, G. Frost, S. Generoso, P. Ginoux, V. Grewe, A. Guenther, H. C. Hansson, S. Henne, J. Hjorth, A. Hofzumahaus, H. Huntrieser, I. S. A. Isaksen, M. E. Jenkin, J. Kaiser, M. Kanakidou, Z. Klimont, M. Kulmala, P. Laj, M. G. Lawrence, J. D. Lee, C. Liousse, M. Maione, G. McFiggans, A. Metzger, A. Mieville, N. Moussiopoulos, J. J. Orlando, C. O'Dowd, P. I. Palmer, D. Parrish, A. Petzold, U. Platt, U. Pöschl, A. S. H. Prévôt, C. E. Reeves, S. Reiman, Y. Rudich, K. Sellegri, R. Steinbrecher, D. Simpson, H. ten Brink, J. Theloke, G. van der Werf, R. Vautard, V. Vestreng, C. Vlachokostas, and R. von-Glasow (2009), Atmospheric composition change − Global and regional air quality, *Atmos. Environ., 43*, 5268−5350, doi:10.1016/j.atmosenv.2009.08.021.

Montgomery, M., R. Smith, and S. Nguyen (2010), Sensitivity of tropical-cyclone models to the surface drag coefficient, *Q. J. Roy. Meteor. Soc., 136*(653), 1945−1953, doi:10.1002/qj.702.

Montoya, M., T. J. Crowley, and H. von Storch (1998), Temperature at the last interglacial simulated by means of a coupled general circulation model, *Paleoceanography, 13*, 170−177.

Mooney, C. (2010), The climate trap: This account of last year's 'climategate scandal' inadvertently plays the sceptics game (review of *The Climate Files* by Fred Pearce), *New Sci., 2767*, 42.

Mooney, S. D., S. P. Harrison, P. J. Bartlein, A. L. Daniau, J. Stevenson, K. C. Brownlie, S. Buckman, M. Cupper, J. Luly, M. Black, E. Colhoun, D. D'Costa, J. Dodson, S. Haberle, G. S. Hope, P. Kershaw, C. Kenyon, M. McKenzie, and N. Williams (2010), Late-Quaternary fire regimes of Australasia, *Quaternary Sci. Rev., 30*(1−2), 28−46, doi:10.1016/j.quascirev.2010.10.010.

Morán, X. A. G., Á. López-Urrutia, A. Calvo-Díaz, and W. K. W. Li (2010), Increasing importance of small phytoplankton in a warmer ocean, *Glob. Change Biol., 16*, 1137−1144.

Morey, A. E., A. C. Mix, and N. G. Pisias (2005), Planktonic foraminiferal assemblages preserved in surface sediments correspond to multiple environment variables, *Quaternary Sci. Rev., 24*, 925−950, doi:10.1016/j.quascirev.2003.09.011.

Morisette, J. T., A. D. Richardson, A. K. Knapp, J. I. Fisher, E. A. Graham, J. Abatzoglou, B. E. Wilson, D. D. Breshears, G. M. Henebry, J. M. Hanes, and L. Liang (2009), Tracking the rhythm of the seasons in the face of global change: phenological research in the 21st century, *Frontiers in Ecology and the Environment, 7*, 253−260, doi:10.1890/070217.

Moriwaki, R., and M. Kanda (2004), Seasonal and diurnal fluxes of radiation, heat, water vapor and CO_2 over a suburban area, *J. Appl. Meteorol., 43*, 1700−1710.

Morrison, M., and D. Hatton MacDonald (2010), Economic valuation of environmental benefits in the Murray-Darling basin, report prepared for the Murray-Darling Basin Authority, October 2010, *http://www. mdba.gov.au/files/bp-kid/1282-MDBA-NMV-Report-Morrison-and-Hatton-MacDonald-20Sep2010.pdf.* (Accessed 8/11/10.)

Mortland, M. M. (1955), Absorption of ammonia by clays and muck, *Soil Sci., 80*, 11−18.

Morton, N. (2008), Details of voting on proposed GSSP and ASSP for the base of the Hettangian Stage and Jurassic System, *International Subcommission on Jurassic Stratigraphy Newsletter, 35*, 74.

Moss, R., M. Babiker, S. Brinkman, E. Calvo, T. Carter, J. Edmonds, I. Elgizouli, S. Emori, L. Erda, K. Hibbard, R. Jones, M. Kainuma, J. Kelleher, J. F. Lamarque, M. Manning, B. Matthews, J. Meehl, L. Meyer, J. Mitchell, N. Nakicenovic, B. O'Neill, R. Pichs, K. Riahi, S. Rose, P. Runci, R. Stouffer, D. van Vuuren, J. Weyant, T. Wilbanks, J. P. van Ypersele, and M. Zurek (2008), *Towards New Scenarios for Analysis of Emissions, Climate Change, Impacts, and Response Strategies*, Intergovernmental Panel on Climate Change (IPCC), Geneva.

Moss, R. H., J. A. Edmonds, K. A. Hibbard, M. R. Manning, S. K. Rose, D. P. van Vuuren, T. R. Carter, S. Emori, M. Kainuma, T. Kram, G. A. Meehl, J. F. B. Mitchell, N. Nakicenovic, K. Riahi, S. J. Smith, R. J. Stouffer, A. M. Thomson, J. P. Weyant, and T. J. Wilbanks (2010), The next generation of scenarios for climate change research and assessment, *Nature, 463*(7282), 747−756, doi:10.1038/nature08823.

Moss, T., M. Babiker, S. Brinkman, E. Calvo, T. Carter, J. Edmonds, I. Elgizouli, S. Emori, L. Erda, K. Hibbard, R. Jones, M. Kainuma, J. Kelleher, J.-F. Lamarque, M. Manning, B. Matthews, G. Meehl, L. Meyer, J. Mitchell, N. Nakicenovic, B. O'Neill, T. Pichs, K. Riahi, S. Rose, P. Runci, R. Stouffer, D. van Vuuren, J. Weyant, T. Wilbanks, J. P. van Ypersele, and M. Zurek (2008), Towards new scenarios for analysis of emissions, climate change, impacts, and response strategies, Intergovernmental Panel on Climate Change (IPCC), Geneva.

Mote, T. L. (2007), Greenland surface melt trends 1973−2007: Evidence of a large increase in 2007, *Geophys. Res. Lett., 34*, L22507.

Mouillot, F., and C. B. Field (2005), Fire history and the global carbon budget: a $1° × 1°$ fire history reconstruction for the 20th century, *Glob. Change Biol., 11*(3), 398−420.

Moussiopoulos, N., and I. S. A. Isaksen (Eds.) (2006), Proceedings of the workshop on model benchmarking and quality assurance, 29/30 May 2006, Greece, Thessaloniki, *http://www.accent-network.org/farcry_*

accent/download.cfm?DownloadFile=4C5CBBCB-BCDC-BAD1-A23 A2B54239B5545.

Moy, A. D., W. R. Howard, S. G. Bray, and T. W. Trull (2009), Reduced calcification in modern Southern Ocean planktonic foraminifera, *Nat. Geosci., 2*(4), 276−280, doi:10.1038/ngeo460.

Mulitza, S., M. Prange, J.-B. Stuut, M. Zabel, T. von Dobeneck, A. C. Itambi, J. Nizou, M. Schulz, and G. Wefer (2008), Sahel megadroughts triggered by glacial slowdowns of Atlantic Meridional Overturning, *Paleoceanography, 23*(4), PA4206.

Mulitza, S., et al. (2010), Increase in African dust flux at the onset of commercial agriculture in the Sahel region, *Nature, 466*(7303), 226−228.

Muller, J. F., and T. Stavrakou (2005), Inversion of CO and NO_x emissions using the adjoint of the IMAGES model, *Atmos. Chem. Phys., 5*, 1157−1186.

Munich Re (2009), Heavy losses due to severe weather in the first six months of 2009, Press release, 27 July 2009, *http://www.munichre. com/en/press/press_releases/2009/2009_07_27_press_release.aspx.*

Murphy, D. M., and D. W. Fahey (1994), An estimate of the flux of stratospheric reactive nitrogen and ozone into the troposphere, *J. Geophys. Res., 99*(D3), 5325−5332.

Murphy, D. M., S. Solomon, R. W. Portmann, K. H. Rosenlof, P. M. Forster, and T. Wong (2009), An observationally-based energy balance for the Earth since 1950, *J. Geophys. Res., 114*, D17107, doi:10.1029/2009JD012105.

Murphy, J. (1999), An evaluation of statistical and dynamical techniques for downscaling local climate, *J. Climate, 12*(8), 2256−2284.

Murphy, J. M., and J. F. B. Mitchell (1995), Transient response of the Hadley Centre coupled ocean-atmosphere model to increasing carbon dioxide. Part II: spatial and temporal structure of response, *J. Climate, 8*, 57−80.

Murray−Darling Basin Authority (MDBA) (2010a), *Guide to the Proposed Basin Plan- Overview, Volume 1, http://download.mdba.gov.au/Guide_ to_the_Basin_Plan_Volume_1_web.pdf.* (Accessed 11/10/10.)

Murray−Darling Basin Authority (MDBA) (2010b), *Guide to the Proposed Basin Plan- Technical Background, Volume 2, http://download.mdba.gov.au/Guide-to-proposed-BP-vol2-0-12.pdf.* (Accessed 22/10/10.)

Murtugudde, R., J. Beauchamp, C. R. McClain, M. Lewis, and A. J. Busalacchi (2002), Effects of penetrative radiation on the upper tropical ocean circulation, *J. Climate, 15*(5), 470−486.

Myhre, G. (2009), Consistency between satellite-derived and modeled estimates of the direct aerosol effect, *Science, 325*(5937), 187−190, doi:10.1126/science.1174461.

Myhre, G., A. Grini, and S. Metzger (2006), Modelling of nitrate and ammonium-containing aerosols in presence of sea salt, *Atmos. Chem. Phys., 6*, 4809−4821.

Myhre, G., J. S. Nilsen, L. Gulstad, K. P. Shine, B. Rognerud, and I. S. A. Isaksen (2007a), Radiative forcing due to stratospheric water vapour from CH_4 oxidation, *Geophys. Res. Lett., 34*(1), L01807, doi:10.1029/2006GL027472.

Myhre, G., F. Stordal, M. Johnsrud, Y. J. Kaufman, D. Rosenfeld, T. Storelvmo, J. E. Kristjansson, T. K. Berntsen, A. Myhre, and I. S. A. Isaksen (2007b), Aerosol-cloud interaction inferred from MODIS satellite data and global aerosol models, *Atmos. Chem. Phys., 7*(12), 3081−3101.

Myhre, G., T. F. Berglen, M. Johnsrud, C. R. Hoyle, T. K. Berntsen, S. A. Christopher, D. W. Fahey, I. S. A. Isaksen, T. A. Jones, R. A. Kahn, N. Loeb, P. Quinn, L. Remer, J. P. Schwarz, and K. E. Yttri

(2009), Modelled radiative forcing of the direct aerosol effect with multi-observation evaluation, *Atmos. Chem. Phys., 9*, 1365–1392.

Naish, T., et al. (2009), Obliquity-paced Pliocene West Antarctic Ice Sheet oscillations, *Nature, 458*(7236), 322–328.

Nakicenovic, N., J. Alcamo, G. Davis, B. de Vries, J. Fenhann, S. Gaffin, K. Gregory, A. Grübler, T. Y. Jung, T. Kram, E. L. La Rovere, L. Michaelis, S. Mori, T. Morita, W. Pepper, H. Pitcher, L. Price, K. Raihi, A. Roehrl, H.-H. Rogner, A. Sankovski, M. Schlesinger, P. Shukla, S. Smith, R. Swart, S. van Rooijen, N. Victor, and Z. Dadi (2000), *Emission Scenarios, Special report of the Intergovernmental Panel on Climate Change (IPCC)*, edited by N. Nakicenovic and S. Swart, Cambridge University Press, Cambridge, UK.

Nakicenovic, N., O. Davidson, G. Davis, A. Grubler, T. Kram, E. L. L. Rovere, B. Metz, T. Morita, W. Pepper, H. Pitcher, A. Sankovski, P. Shukla, R. Swart, R. Watson, and Z. Dadi (2000), *Emissions scenarios: A special report of Working Group III of the Intergovernmental Panel on Climate Change* (Summary for Policy Makers), IPCC, 27.

Narisma, G. T., and A. J. Pitman (2003), The impact of 200-years land-cover change on the Australian near-surface climate, *J. Hydrometeorol., 4*, 424–436.

Narisma, G. T., and A. J. Pitman (2004), The effect of including biospheric feedbacks on the impact of land-cover change over Australia, *Earth Interact., 8*, doi:10.1175/1087-3562(2004)008.

National Aeronautics and Space Administration (1988), Earth system science: A closer view, report of the Earth Systems Sciences Committee to the NASA Advisory Committee National Aeronautics and Space Administration, Washington, DC.

National Climate Change Adaptation Research Facility (NCCARF) (2010), 2010 Climate Adaptation Futures Conference. *www.nccarf.edu.au/conference2010*, (Accessed 25/10/10.)

National Oceanic and Atmospheric Administration (2010), Geophysical Fluid Dynamics Laboratory – Climate Modeling, *http://www.gfdl.noaa.gov/climate-modeling*. (Accessed 27/11/10.)

National Research Council (2010), *When Weather Matters: Science and Services to Meet Critical Societal Needs*, National Academies Press, Washington, DC, *http://www.nap.edu/catalog.php?record_id=12888*.

Natural England (2010), Climate change adaptation indicators for the natural environment, *Report No. NECR038*, Sheffield, *http://naturalengland.etraderstores.com/NaturalEnglandShop/NECR038*. (Accessed 9/9/10.)

Nature (2010), Editorial, Climate of fear, *Nature, 464*(7286), 141, doi:10.1038/464141a.

NCCARF (2010), The National Climate Change Adaptation Research Facility, *http://www.nccarf.edu.au*. (Accessed 9/11/10.)

Neelin, D. J. (2011), Climate Change and Climate Modelling, Cambridge University Press, Cambridge, 282pp.

Neelin, J. D., D. S. Battisti, A. C. Hirst, F.-F. Jin, Y. Wakata, T. Yamagata, and S. E. Zebiak (1998), ENSO theory, *J. Geophys. Res., 103*(C7), 14261–14290, doi:10.1029/97JC03424.

Neil, K. L., L. Landrum, and J. Wu (2010), Effects of urbanization on flowering phenology in the metropolitan Phoenix region of USA: Findings from herbarium records, *J. Arid Environ., 74*, 440–444.

Nelson, C. R., and H. Kang (1981), Spurious periodicity in inappropriately detrended time series, *Econometrica, 49*(3), 741–751.

Nelson, D. (2009), Bangladesh to host centre for climate adaptation knowhow, Science and Development Network, 11 September, *http://www.scidev.net/en/news/bangladesh-to-host-centre-for-climate-adaptation-k.html*. (Accessed 31/3/11.)

Nepstad, D. C., I. M. Tohver, D. Ray, P. Moutinho, and G. Cardinot (2007), Mortality of large trees and lianas following experimental drought in an Amazon forest, *Ecology, 88*, 2259–2269.

Netherlands Environmental Assessment Agency (2010), Assessing an IPCC assessment: An analysis of statements on projected regional impacts in the 2007 report, [see also Q. Schiermier (2010), Few fishy facts found in climate report, Dutch investigation supports key warnings from the IPCC's most recent assessment, *Nature, 466*(7303), doi: 10.1038/466170a].

New, M., M. Hulme, and P. D. Jones (2000), Representing twentieth-century space-time climate variability, 2.: Development of 1901–96 monthly grids of terrestrial surface climate, *J. Climate, 13*, 2217–2238.

New, M., B. Hewitson, D. B. Stephenson, A. Tsiga, A. Kruger, A. Manhique, B. Gomez, C. A. S. Coelho, D. N. Masisi, E. Kululanga, E. Mbambalala, F. Adesina, H. Saleh, J. Kanyanga, J. Adosi, L. Bulane, L. Fortunata, M. L. Mdoka, and R. Lajoie (2006), Evidence of trends in daily climate extremes over Southern and West Africa, *J. Geophys. Res., 111*, D14102, doi:10.1029/2005JD006289.

New, M., A. Lopez, S. Dessai, and R. Wilby (2007), Challenges in using probabilistic climate change information for impact assessments: An example from the water sector, *Philos. Trans. R. Soc. London Ser. A, 365*(1857), 2117–2131, doi:10.1098/rsta.2007.2080.

New, M., D. Liverman, H. Schroder, and K. Anderson (2011), Four degrees and beyond: The potential for a global temperature increase of four degrees and its implications, *Proc. Trans. Roy. Soc. A, 369*, 6–19, doi:10.1098/rsta.2010.0303.

Newell, N. D. (1963), Crises in the history of life, *Sci. Am., 208*, 76–92.

Newton, I. (1687), Philosophiæ Naturalis Principia Mathematica, S. Pepys, London, UK.

Newton, P. C. D., et al. (2010), The rate of progression and stability of progressive nitrogen limitation at elevated atmospheric CO_2 in a grazed grassland over 11 years of free air CO_2 enrichment, *Plant Soil, 336*, 433–441, doi:10.107/s11104-010-0493-0.

Newton, T., T. R. Oke, C. S. B. Grimmond, and M. Roth (2007), The suburban energy balance in Miami, Florida, *Geogr. Ann., 89A*(4), 331–347.

Nghiem, S. V., I. G. Rigor, D. K. Perovich, P. Clemente-Colon, J. W. Weatherly, and G. Neumann (2007), Rapid reduction of Arctic perennial sea ice, *Geophys. Res. Lett., 34*, L19504.

NGRIP (2004), High-resolution record of Northern Hemisphere climate extending into the last interglacial period, *Nature, 431*(7005), 147–151.

Nguyen, K. C., and J. L. McGregor (2009), Modelling the Asian summer monsoon using CCAM, *Clim. Dynam., 32*, 219–236.

Nicholls, N. (1996), Long-term climate monitoring and extreme events, in *Long-Term Climate Monitoring by the Global Climate Observing System*, edited by T. R. Karl, Kluwer, Dordrecht.

Nicholls, N. (2004), The changing nature of Australian droughts, *Climatic Change, 63*, 323–336.

Nicholls, N., and L. V. Alexander (2007), Has the climate become more variable or extreme? Progress 1992–2006, *Progr. Phys. Geogr., 31*(1), 77–87.

Nicholls, N., and M. Manton (2005), APN workshops on daily extremes across East Asia and the West Pacific, *Asia-Pacific Network for Global Change Research Newsletter, 11*, 4–5.

Nicholls, N., G. V. Gruza, J. Jouzel, T. R. Karl, L. A. Ogallo, and D. E. Parker (1995), Observed climate variability and change, in *Climate Change 1995*, edited by J. T. Houghton, et al., Cambridge University Press, Cambridge, UK.

Nicholls, N., W. Drosdowsky, and B. Lavery (1997), Australian rainfall variability and change, *Weather, 52,* 66—72.

Nicholls, N., H.-J. Baek, A. Gosai, L. E. Chambers, Y. Choi, D. Collins, P. M. Della-Marta, G. M. Griffiths, M. R. Haylock, R. Lata, L. Maitrepierre, M. J. Manton, H. Nakamigawa, N. Ouprasitwong, D. Solofa, D. T. Thuy, L. Tibig, B. Trewin, K. Vediapan, and P. Zhai (2005), The El Niño—Southern Oscillation and daily temperature extremes in East Asia and the West Pacific, *Geophys. Res. Lett., 32,* L16714, doi:10.1029/2005GL022621.

Nicholls, N., C. Skinner, M. Loughnan, and N. Tapper (2008), A simple heat alert system for Melbourne, Australia, *Int. J. Biometeorol., 52,* 375—384, doi:10.1007/s00484-007-0132-5.

Nicholls, R. J., P. P. Wong, V. R. Burkett, J. O. Codignotto, J. E. Hay, R. F. McLean, S. Ragoonaden, and C. D. Woodroffe (2007), Coastal systems and low-lying areas, in *Climate Change 2007: Impacts, Adaptation and Vulnerability, contribution of Working Group II to the Fourth Assessment Report (FAR) of the Intergovernmental Panel on Climate Change,* edited by M. L. Parry, O. F. Canziani, J. P. Palutikof, P. J. Van der Linden, and C. E. Hanson, pp. 315—356, Cambridge University Press, Cambridge, UK.

Niklas, K. J., and B. H. Tiffney (1994), The quantification of plant biodiversity through time, *Philos. Trans. R. Soc. London Ser. B, 345,* 35—44.

Niklas, K. J., B. H. Tiffney, and A. H. Knoll (1980), Apparent changes in diversity of fossil plants, *Evol. Biol., 12,* 1—89.

Niklas, K. J., B. H. Tiffney, and A. H. Knoll (1985), Patterns in vascular land plant evolution diversification: An analysis at the species level, in *Phanerozoic Diversity Patterns: Profiles in Macroevolution* , edited by J. W. Valentine, pp. 97—128, Princeton University Press, Princeton, NJ.

Nine News (2010), Murray plan 'could cost 10,000 jobs', 15 October, 2010, *http://news.ninemsn.com.au/national/8106430/science-still-needs-to-be-tested.* (Accessed 21/10/10.)

NOAA (2007), Mauna Loa Carbon Dioxide Record, *http://celebrating200years.noaa.gov/datasets/mauna/image3b.html.*

Noble, I. (2010), Financing adaptation: International transfers and global geopolitics, Panel Presentation, International Climate Change Adaptation Conference, Climate Adaptation Futures: Preparing for the unavoidable impacts of climate change, National Climate Change Adaptation Research Facility and the CSIRO Climate Adaptation Flagship, 29 June—1 July, Gold Coast.

Nobre, C. A., and L. Borma (2009), "Tipping points" of the Amazon forest, *Curr. Opin. Environ. Sustain., 1,* 28—36.

Nobre, C. A., P. Sellers, and J. Shukla (1991), Amazonian deforestation and regional climate change, *J. Climate, 4,* 957—988.

Nobre, C., G. P. Brasseur, M. A. Shapiro, M. Lahsen, G. Brunet, A. J. Busalacchi, K. Hibbard, S. Seitzinger, K. Noone, and J. P. Ometto (2010), Addressing the complexity of the Earth system, *Bull. Am. Meteorol. Soc., 91,* 1389—1396, doi:10.1175/2010BAMS3012.1.

Nogaj, N., S. Parey, and D. Dacunha-Castelle (2007), Non-stationary extreme models and a climatic application, *Nonlinear Proc. Geoph., 14*(3), 305—316.

Nomade, S., et al. (2007), Chronology of the Central Atlantic Magmatic Province: Implications for the Central Atlantic rifting processes and the Triassic-Jurassic biotic crises, *Palaeogeogr. Palaeoclim., 244,* 326—344.

Norby, R. J., J. M. Warren, C. M. Iversen, B. E. Medlyn, and R. E. McMurtrie (2010), CO_2 enhancement of forest productivity constrained by limited nitrogen availability, *Proc. Natl. Acad. Sci. USA, 107,* 19368—19373.

Norris, J. R., and M. Wild (2007), Trends in aerosol radiative effects over Europe inferred from observed cloud cover, solar "dimming" and solar "brightening," *J. Geophys. Res., 112,* D08214, doi:10.1029/2006JD007794.

North Greenland Ice Core Project Members (2004), High-resolution record of Northern Hemisphere climate extending into the last interglacial period, *Nature, 431*(7005), 147—151, doi:10.1038/nature02805.

Notaro, M., S. Vavrus, and Z. Liu (2007), Global vegetation and climate change due to future increases in CO_2 as projected by a fully coupled model with dynamic vegetation, *J. Climate, 20,* 70—90.

Notz, D. (2009), The future of ice sheets and sea ice: Between reversible retreat and unstoppable loss, *Proc. Natl. Acad. Sci. USA, 106,* 20590—20595.

Novelli, P. C., K. A. Masarie, and P. M. Lang (1998), Distributions and recent changes of carbon monoxide in the lower troposphere, *J. Geophys. Res., 103*(D15), 19015—19034.

Novelli, P. C., K. A. Masarie, P. M. Lang, B. D. Hall, R. C. Myers, and J. W. Elkins (2003), Reanalysis of tropospheric CO trends: Effects of the 1997—1998 wildfires, *J. Geophys. Res., 108*(D15), 4464, doi:10.1029/2002JD003031.

NRC (2009), *America's Climate Choices,* National Research Council, National Academy of Sciences, Washington, DC, *http://americasclimatechoices.org.* (Accessed 9/9/10.)

O'ishi, R., and A. Abe-Ouchi (2009), Influence of dynamic vegetation on climate change arising from increasing CO_2, *Clim. Dynam., 33,* 645—663.

O'ishi, R., A. Abe-Ouchi, I. C. Prentice, and S. Sitch (2009), Vegetation dynamics and plant CO_2 responses as positive feedbacks in a greenhouse world, *Geophys. Res. Lett., 36,* L11706, doi:10.1029/2009GL038217.

O'Keefe, J. D., and T. J. Ahrens (1989), Impact production of CO_2 by the Cretaceous-Tertiary extinction bolide and the resultant heating of the Earth, *Nature, 338,* 247—249.

O'Neil, S., and M. Boycoff (2010), Climate denier, skeptic or contrarian?, *Proc. Natl. Acad. Sci. USA, 107*(39), doi:10.1073/pnas.1010507107, online only.

O'Neill, B. C. (2000), The jury is still out on global warming potentials, *Climatic Change, 44,* 427—443.

O'Neill, S. J., and M. Boykoff (2010), Climate denier, skeptic or contrarian?, *Proc. Natl. Acad. Sci. USA, 107*(39), E151, doi:10.1073/pnas.1010507107.

O'Riordan, T., and H. Voisey (1998), *The Transition to Sustainability: The Politics of Agenda 21 in Europe,* Earthscan, London.

Obama, B. (2010), President Obama's Oval Office Address on BP Oil Spill & Energy, US Presidential address, 15 June, *http://www.youtube.com/watch?v=Gh76oepKFc8&feature=related.* (Accessed 26/7/10.)

OECD (2009), Policy Guidance on Integrating Climate Change into Development Cooperation, 28—29 May, Paris.

Oechel, W. C., G. L. Vourlitis, S. J. Hastings, R. C. Zulueta, L. Hinzman, and D. Kane (2000), Acclimation of ecosystem CO_2 exchange in the Alaskan Arctic in response to decadal climate warming, *Nature, 406,* 978—981.

Oerlemans, J. (2001), *Glaciers and Climate Change,* A. A. Balkema, Lisse.

Oerlemans, J., and B. K. Reichert (2000), Relating glacier mass balance to meteorological data using a seasonal sensitivity characteristic, *J. Glaciol., 46*(152), 1—6.

Oerlemans, J., and C. J. Van der Veen (1984), *Ice Sheets and Climate,* Reidel Publishing, Dordrecht.

Oerlemans, J., R. H. Giesen, and M. R. van den Broeke (2009), Retreating alpine glaciers: Increased melt rates due to accumulation of dust (Vadret da Morteratsch, Switzerland), *J. Glaciol., 55*, 729−736.

Offerle, B., C. S. B. Grimmond, and K. Fortuniak (2005a), Heat storage and anthropogenic heat flux in relation to the energy balance of a central European city centre, *Int. J. Climatol., 25*, 1405−1419.

Offerle, B., P. Jonsson, I. Eliasson, and C. S. B. Grimmond (2005b), Urban modification of the surface energy balance in the west African Sahel: Ouagadougou, Burkina Faso, *J. Climate, 18*, 3983−3995.

Offerle, B., C. S. B. Grimmond, K. Fortuniak, and W. Pawlak (2006a), Intraurban differences of surface energy fluxes in a central European city, *J. Appl. Meteorol. Clim., 45*, 125−136.

Offerle, B., C. S. B. Grimmond, K. Fortuniak, K. Kłysik, and T. R. Oke (2006b). Temporal variations in heat fluxes over a central European city centre, *Theor. Appl. Climatol., 84*, 103−116.

Ohara, T., H. Akimoto, J. Kurokawa, N. Horii, K. Yamaji, X. Yan, and T. Hayasaka (2007), An Asian emission inventory of anthropogenic emission sources for the period 1980−2020, *Atmos. Chem. Phys., 7*, 4419−4444.

Ohmura, A. (2001), Physical basis for the temperature-based melt index method, *J. Appl. Meteorol., 40*, 753−761.

Ohmura, A. (2004), Cryosphere during the twentieth century, *Geophys. Monog. Series, 150*, 239−257.

Ohmura, A., and H. Lang (1989), Secular variation of global radiation over Europe, in *Current Problems in Atmospheric Radiation*, edited by J. Lenoble, and J. F. Geleyn, pp. 98−301, Deepak, Hampton, VA.

Ohmura, A., E. G. Dutton, B. Forgan, C. Frohlich, H. Gilgen, H. Hegner, A. Heimo, G. Konig-Langlo, B. McArthur, G. Muller, R. Philipona, R. Pinker, C. H. Whitlock, K. Dehne, and M. Wild (1998), Baseline Surface Radiation Network, a new precision radiometry for climate research, *Bull. Am. Meteorol. Soc., 79*, 2115−2136.

Oke, T. R. (1973), City size and the urban heat island, *Atmos. Environ., 7*, 769−779.

Oke, T. R. (1976), The distinction between canopy and boundary-layer urban heat islands, *Atmosphere, 14*, 268−277.

Oke, T. R. (1979), Advectively-assisted evapotranspiration from irrigated urban vegetation, *Bound.-Lay. Meteorol., 17*, 163−173.

Oke, T. R. (1981), Canyon geometry and the nocturnal urban heat island: Comparison of scale model and field observations, *J. Climatol., 1*, 237−254.

Oke, T. R. (1982), The energetic basis of the urban heat island, *Q. J. Roy. Meteor. Soc., 108*, 1−24.

Oke, T. R. (1984), Methods in urban climatology. In W. Kirchofer, A. Ohmura, and H. Wanner (Editors), *Applied Climatology (Zurcher Geog. Schriften, 14)*. Eidgenossiche Technische Hochschule Geographische Institut, Zurich, p. 19−29.

Oke, T. R. (1987), *Boundary Layer Climates*, 2nd ed., Methuen, London.

Oke, T. R. (1988), The surface energy budgets of urban areas, *Prog. Phys. Geog., 12*, 471−508.

Oke, T. R. (1989), The micrometeorology of the urban forest, *Philos. Trans. R. Soc. London Ser. B, 324*, 335−349.

Oke, T. R. (1994), *Global Change and Urban Climates*. Proc. 13th Intern. Congr. Biometeor., 12 − 18 Sept. 1993, Calgary, Canada, pp. 123−134.

Oke, T. R., and H. A. Cleugh (1986), Urban heat storage derived as energy balance residuals, *Bound.-Lay. Meteorol., 39*, 233−245.

Oke, T. R., and J. H. McCaughey (1983), Suburban-rural energy balance comparisons for Vancouver, BC: An extreme case?, *Bound.-Lay. Meteorol., 26*, 337−354.

Oke, T. R., G. Zeuner, and E. Jauregui (1992), The surface energy balance in Mexico City, *Atmos. Environ. B-Urb., 26*(4), 433−444.

Oke, T. R., R. A. Spronken-Smith, E. Jauregui, and C. S. B. Grimmond (1999), The energy balance of central Mexico City during the dry season, *Atmos. Environ., 33*, 3919−3930.

Oleson, K. W., G. B. Bonan, S. Levis, and M. Vertenstein (2004), Effects of land use change on North American climate: Impact of surface datasets and model biogeophysics, *Clim. Dynam., 23*, 117−132, doi:10.1007/s00382-004-0426-9.

Oleson, K. W., G. B. Bonan, J. Feddema, M. Vertenstein, and C. S. B. Grimmond (2008), An urban parameterization for a global climate model, part I: Formulation and evaluation for two cities, *J. Appl. Meteorol. Clim., 47*, 1038−1060, doi:10.1175/2007JAMC1597.1.

Oleson, K. W., et al. (2010), Technical description of version 4.0 of the community land model (CLM), NCAR/TN-478+STR, NCAR Technical Note, Climate and Global Dynamics Division, National Center for Atmospheric Research.

Oleson, K. W., G. B. Bonan, and J. Feddema (2010), The effects of white roofs on urban temperature in a global climate model, *Geophys. Res. Lett., 37*, L03701, doi:10.1029/2009GL042194.

Olivier J., J. Peters, C. Granier, G. Pétron, J. F. Müller, and S. Wallens (2003), Present and future surface emissions of atmospheric compounds, POET Report 2, *EU Project EVK-1999−00011.*

Olsen, M. (1965), *The Logic of Collective Action*, John Hopkins University Press, Baltimore.

Olsen, M. A., A. R. Douglass, and M. R. Schoeberl (2003), A comparison of Northern and Southern Hemisphere cross-tropopause ozone flux, *Geophys. Res. Lett., 30*(7), 1412, doi:10.1029/2002GL016538.

Olsen, P. E., et al. (2002), Ascent of dinosaurs linked to an iridium anomaly at the Triassic/Jurassic boundary, *Science, 296*, 1305−1307.

Oltmans, S. J., A. S. Lefohn, H. E. Scheel, J. M. Harris, H. Levy, II, I. E. Galbally, E.-G. Brunke, C. P. Meyer, J. A. Lathrop, B. J. Johnson, D. S. Shadwick, E. Cuevas, F. J. Schmidlin, D. W. Tarasick, H. Claude, J. B. Kerr, O. Uchino, and V. Mohnen (1998), Trends of ozone in the troposphere, *Geophys. Res. Lett., 25*, 139−142.

Oltmans, S. J., A. S. Lefohn, J. M. Harris, I. Galbally, H. E. Scheel, G. Bodeker, E. Brunke, H. Claude, D. Tarasick, B. J. Johnson, P. Simmonds, D. Shadwick, K. Anlauf, K. Hayden, F. Schmidlin, T. Fujimoto, K. Akagi, C. Meyer, S. Nichol, J. Davies, A. Redondas, and E. Cuevas (2006), Long-term changes in tropospheric ozone, *Atmos. Environ., 40*, 3156−3173.

Omta, A. W., J. Bruggeman, S. A. L. M. Kooijman, and H. A. Dijkstra (2006), Biological carbon pump revisited: Feedback mechanisms between climate and the Redfield ratio, *Geophys. Res. Lett., 33*, L14613, doi:10.1029/2006GL026213.

Oppenheimer, C. (2003), Climatic, environmental and human consequences of the largest known historic eruption: Tambora volcano (Indonesia) 1815, *Phys. Geogr., 27*, 230−259.

Oppenheimer, M. (2010), Michael Oppenheimer, quoted in Stephen H. Schneider, climate change expert, dies at 65, obituary by T. R. Shapiro, *Washington Post*, 20 July, 2010, *www.washingtonpost.com/wp-dyn/content/article/2010/07/19/AR2010071905108.html.* (Accessed 24/10/10.)

Opsteegh, J. D., R. J. Haarsma, F. M. Selten, and A. Kattenberg (1998), ECBILT: A dynamic alternative to mixed boundary conditions in ocean models, *Tellus, 50*(A), 348−367.

Ordóñez, C., D. Brunner, J. Staehelin, P. Hadjinicolaou, J. A. Pyle, M. Jonas, H. Wernli, and A. S. H. Prévôt (2007), Strong influence of lowermost stratospheric ozone on lower tropospheric background ozone changes over Europe, *Geophys. Res. Lett., 34*, L07805, doi:10.1029/2006GL029113.

Oreskes, N., and E. M. Conway (2010), *Merchants of Doubt: How a Handful of Scientists Obscured the Truth on Issues from Tobacco Smoke to Global Warming*, Bloomsbury Press, New York.

Orlowsky, B., and S. I. Seneviratne (2011), Global changes in extremes events: Regional and seasonal dimension, *Climatic Change*, available on-line, doi:10.1007/s10584-011-0122-9.

Orr, J. C. (2002), Final OCMIP-2 Report.

Orr, J. C., V. J. Fabry, O. Aumont, L. Bopp, S. C. Doney, R. A. Feely, A. Gnanadesikan, N. Gruber, A. Ishida, F. Joos, R. M. Key, K. Lindsay, E. Maier-Reimer, R. Matear, P. Monfray, A. Mouchet, R. G. Najjar, G.-K. Plattner, K. B. Rodgers, C. L. Sabine, J. L. Sarmiento, R. Schlitzer, R. D. Slater, I. J. Totterdell, M.-F. Weirig, Y. Yamanaka, and A. Yool (2005), Anthropogenic ocean acidification over the twenty-first century and its impact on calcifying organisms, *Nature, 437*(7059), 681−686, doi:10.1038/nature04095.

Orville, R. E., G. Huffines, J. Nielsen-Gammon, R. Y. Zhang, B. Ely, S. Steiger, S. Phillips, S. Allen, and W. Read (2001), Enhancement of cloud-to-ground lightning over Houston, Texas, *Geophys. Res. Lett., 28*, 2597−2600.

Oschlies, A., K. G. Schulz, U. Riebesell, and A. Schmittner (2008), Simulated 21st century's increase in oceanic suboxia by CO_2-enhanced biotic carbon export, *Global Biogeochem. Cycles, 22*(4), GB4008.

Osterkamp, T. E. (2005), The recent warming of permafrost in Alaska, *Glob. Planet. Change, 49*, 187−202.

Ostrom, E. (2009a), A polycentric approach for coping with climate change, *The World Bank Policy Research Working Paper 5095*, Washington, DC, *http://www-wds.worldbank.org/servlet/WDSContentServer/WDSP/IB/ 2009/10/26/000158349_20091026142624/Rendered/PDF/WPS5095. pdf.* (Accessed 9/11/10.)

Ostrom, E. (2009b), A general framework for analyzing sustainability of socio-ecological systems, *Science, 325*, 419−422.

Ostrom, E. (2010), Polycentric systems for coping with collective action and global environmental change, *Global Environ. Chang., 20*(4), 550−557.

Ottera, O. H., M. Bentsen, H. Drange, and L. Suo (2010), External forcing as a metronome for Atlantic Multidecadal Variability, *Nat. Geosci., 3*, 688−694, doi:10.1038/ngeo955.

Otto J., T. Raddatz, M. Claussen, V. Brovkin, and V. Gayler (2009), Separation of atmosphere-ocean-vegetation feedbacks and synergies for mid-Holocene climate, *Geophys. Res. Lett., 36*, L09701, doi:10.1029/2009GL037482.

Otto-Bliesner, B. L., E. C. Brady, S. I. Shin, Z. Liu, and C. Shields (2003), Modeling El-Niño and its tropical teleconnections during the last glacial-interglacial cycle, *Geophys. Res. Lett., 30*(23), 2198.

Otto-Bliesner, B. L., S. J. Marshall, J. T. Overpeck, G. H. Miller, A. Hu, and CAPE Last Interglacial Project members (2006), Retreat in the last interglaciation simulating Arctic climate warmth and icefield, *Science, 311*, 1751−1753.

Otto-Bliesner, B., R. Schneider, E. Brady, M. Kucera, A. Abe-Ouchi, E. Bard, P. Braconnot, M. Crucifix, C. Hewitt, M. Kageyama, O. Marti, A. Paul, A. Rosell-Melé, C. Waelbroeck, S. Weber, M. Weinelt, and Y. Yu (2009), A comparison of PMIP2 model simulations and the MARGO proxy reconstruction for tropical sea surface temperatures at Last Glacial Maximum, *Clim. Dynam., 32*(6), 799−815, doi:10.1007/s00382-008-0509-0.

Overland, J., and N. Bond (1995), Observations and scale analysis of coastal wind jets, *Mon. Weather Rev., 123*(10), 2934−2941.

Ow, L. F., D. Whitehead, A. S. Walcroft, and M. H. Turnbull (2010), Seasonal variation in foliar carbon exchange in *Pinus radiate* and *Populus deltoides:* Respiration acclimates fully to changes in temperature but photosynthesis does not, *Glob. Change Biol., 16*, 288−302.

Oxburgh, R. (2010), Lord Oxburgh Scientific Assessment Panel, Second CRU inquiry reports, April, *http://www.realclimate.org/ index.php/archives/2010/04/second-cru-inquiry-reports.* (Accessed 26/7/10.)

Oxburgh, R., H. Davies, K. Emanuel, L. Graumlich, D. Hand, H. Huppert, and M. Kelly (2010), Report of the International Panel set up by the University of East Anglia to examine the research of the Climatic Research Unit, University of East Anglia and the Royal Society, 19 April, *www.uea.ac.uk/mac/comm/media/press/CRUstatements/SAP.* (Accessed 28/3/11.)

Oyama, M. D., and C. A. Nobre (2003), A new climate-vegetation equilibrium state for tropical South America, *Geophys. Res. Lett., 30*(23), 2199.

Pachauri, R. K. (2010a), Statement by the Chair of the Intergovernmental Panel on Climate Change (IPCC) at the opening session of the 16th Conference of the Parties, Cancun, November 29, *http://unfccc.int/ files/meetings/cop_16/media/application/pdf/101129_cop16_oc_ rpac.pdf.* (Accessed 28/3/11.)

Pachauri, R. K. (2010b), Opening plenary presentation, International Climate Change Adaptation Conference, Climate Adaptation Futures: Preparing for the unavoidable impacts of climate change, National Climate Change Adaptation Research Facility and the CSIRO Climate Adaptation Flagship, 29 June−1 July, Gold Coast.

Paeth, H., K. Born, R. Girmes, R. Podzun, and D. Jacob (2009), Regional climate change in tropical and northern Africa due to greenhouse forcing and land use changes, *J. Climate, 22*, 114−132, doi:10.1175/ 2008JCLI2390.1.

Pagani, M., K. Caldeira, D. Archer, and J. C. Zachos (2006), An ancient carbon mystery, *Science, 314*, 1556−1557.

Pagani, M., Z. Liu, J. LaRiviere, and A. C. Ravelo (2010), High Earth-system climate sensitivity determined from Pliocene carbon dioxide concentrations, *Nat. Geosci., 3*, 27−30.

Page, C. M., N. Nicholls, N. Plummer, B. C. Trewin, M. J. Manton, L. V. Alexander, L. E. Chambers, Y. Choi, D. A. Collins, A. Gosai, P. Della-Marta, M. R. Haylock, K. Inape, V. Laurent, L. Maitrepierre, E. E. P. Makmur, H. Nakamigawa, N. Ouprasitwong, S. McGree, J. Pahalad, M. J. Salinger, L. Tibig, T. D. Tran, K. Vediapan, and P. Zhai (2004), Data rescue in the southeast Asia and South Pacific region: Challenges and Opportunities, *Bull. Am. Meteorol. Soc., 85*, 1483−1489.

Page, S. E., F. Siegert, J. O. Rieley, H.-D. V. Boehm, A. Jaya, and S. Limin (2002), The amount of carbon released from peat and forest fires in Indonesia during 1997, *Nature, 420*, 61−65.

Page, S. E., R. A. J. Wüst, D. Weiss, J. O. Rieley, W. Shotyk, and S. H. Limin (2004), A record of Late Pleistocene and Holocene carbon accumulation and climate change from an equatorial pear bog (Kalimantan, Indonesia): Implications for past, present and future carbon dynamics, *J. Quaternary Sci.,* *19,* 625−635.

Pahnke, K., R. Zahn, H. Elderfield, and M. Schulz (2003), 340,000-year centennial-scale marine record of southern hemisphere climatic oscillation, *Science, 301,* 948−952.

Pal, J. S., and E. A. B. Eltahir (2008), Pathways relating soil moisture conditions to future summer rainfall within a model of the land−atmosphere system, *J. Climate, 14,* 1227−1242.

Palatella, L., M. Miglietta, P. Paradisi, and P. Lionello (2010), Climate change assessment for Mediterranean agricultural areas by statistical downscaling, *Nat. Hazard. Earth Sys., 10*(7), 1647−1661, doi:10.5194/nhess-10-1647-2010.

Pálfy, J., and P. L. Smith (2000), Synchrony between early Jurassic extinction, oceanic anoxic event and the Karoo-Ferrar flood basalt volcanism, *Geology, 28,* 747−750.

Pálfy, J., A. Demeny, J. Haas, M. Htenyi, M. J. Orchard, and I. Veto (2001), Carbon isotope anomaly at the Triassic-Jurassic boundary from a marine section in Hungary, *Geology, 29,* 1047−1050.

Pall, P., T. Aina, D. A. Stone, P. A. Stott, T. Nozawa, A. G. J. Hilberts, D. Lohmann, and M. R. Allen (2011), Anthropogenic greenhouse gas contribution to flood risk in England and Wales in autumn 2000, *Nature, 470,* 382−385.

Pallé, E., P. Montanes-Rodriguez, P. R. Goode, S. E. Koonin, M. Wild, and S. Casadio (2005), A multi-data comparison of shortwave climate forcing changes, *Geophys. Res. Lett., 32*(21), L21702, doi:10.1029/2005GL023847.

Palmer, P. I., D. J. Jacob, A. M. Fiore, R. V. Martin, K. Chance, and T. P. Kurosu (2003), Mapping isoprene emissions over North America using formaldehyde column observations from space, *J. Geophys. Res.-Atmos., 108*(D6).

Palmer, P. I., D. S. Abbot, T. M. Fu, D. J. Jacob, K. Chance, T. P. Kurosu, A. Guenther, C. Wiedinmyer, J. C. Stanton, M. J. Pilling, S. N. Pressley, B. Lamb, and A. L. Sumner (2006), Quantifying the seasonal and interannual variability of North American isoprene emissions using satellite observations of the formaldehyde column, *J. Geophys. Res.-Atmos., 111*(D12), doi:10.1029/2005JD006689.

Palmer, W. C. (1965), Meteorological drought, *Report 45,* US Weather Bureau, Washington, DC.

Pan, L. L., J. C. Wei, D. E. Kinnison, R. R. Garcia, D. J. Wuebbles, and G. P. Brasseur (2007), A set of diagnostics for evaluating chemistry-climate models in the extratropical tropopause region, *J. Geophys. Res., 112,* D09316, doi:10.1029/2006JD007792.

Park, J., and D. L. Royer (2011), Geologic constraints on the glacial amplification of Phanerozoic climate sensitivity, *Am. J. Sci., 311,* 1−26.

Park, W., and M. Latif (2008), Multidecadal and multicentennial variability of the meridional overturning circulation, *Geophys. Res. Lett., 35,* L22703, doi:10.1029/2008GL035779.

Park, W., and M. Latif (2010), Pacific and Atlantic Multidecadal Variability in the Kiel Climate Model, *Geophys. Res. Lett., 37,* L24702, doi:10.1029/2010GL045560.

Park, W., and M. Latif (2011), Atlantic Meridional Overturning Circulation response to idealized solar forcing, *Clim. Dynam.,* (Submitted).

Parker, D. E., T. P. Legg, and C. K. Folland (1992), A new daily Central England temperature series, 1772−1991, *Int. J. Climatol., 12,* 317−342.

Parrish, D. D., D. B. Millet, and A. H. Goldstein (2008), Increasing ozone concentrations in marine boundary layer air flow at the west coasts of North America and Europe, *Atmos. Chem. Phys. Discuss., 8,* 13847−13901.

Pataki, D. E., R. J. Alig, A. S. Fung, N. E. Golubiewski, C. A. Kennedy, E. G. McPherson, D. J. Nowak, R. V. Pouyat, and P. Romero Lankao (2006), Urban ecosystems and the North American carbon cycle, *Glob. Change Bio., 12,* 2092−2102.

Pataki, D. E., P. C. Emmi, C. B. Forster, J. I. Mills, E. R. Pardyjak, T. R. Peterson, J. D. Thompson, and E. Dudley-Murphy (2009), An integrated approach to improving fossil fuel emissions scenarios with urban ecosystem studies, *Ecol. Complex., 6*(1), 1− 14, doi:10.1016/j.ecocom.2008.09.003.

Patz, J. A., D. Campbell-Lendrum, T. Holloway, and J. A. Foley (2005), Impact of regional climate change on human health, *Nature, 438,* 310−317.

Patzek, T. W., and G. D. Croft (2010), A global coal production forecast with multi-Hubbert cycle analysis, *Energ. Policy, 35,* 3109−3122.

Pearce, F. (2010), *The Climate Files: The Battle for the Truth about Global Warming,* Guardian Books, London, UK.

Pearlmutter, D., P. Berlinera, and E. Shaviv (2007), Urban climatology in arid regions: Current research in the Negev desert, *Int. J. Climatol., 27,* 1875−1885, doi:10.1002/joc.1523.

Pearson, P. N., G. L. Foster, and B. S. Wade (2009), Atmospheric carbon dioxide through the Eocene-Oligocene climate transition, *Nature, 461,* 1110−1113.

Peck, J. A., R. R. Green, T. Shanahan, J. W. King, J. T. Overpeck, and C. A. Scholz (2004), A magnetic mineral record of Late-Quaternary tropical climate variability from Lake Bosumtwi, Ghana, *Palaeogeogr. Palaeoclimatol. Palaeocol., 215,* 37−57, doi:10.1016/j.palaeo.2004.08.003.

Pelejero, C., E. Calvo, and O. Hoegh-Guldberg (2010), Paleo-perspectives on ocean acidification, *Trends Ecol. Evol., 25*(6), 332−344, doi:10.1016/j.tree.2010.02.002.

Peltier, W. R. (2004), Global glacial isostasy and the surface of the Ice-Age Earth: The ICE-5G (VM2) model and GRACE, *Annu. Rev. Plant Biol., 32,* 111−149.

Penalba, O. C., and F. A. Robeldo (2010), Spatial and temporal variability of the frequency of extreme daily rainfall regime of the La Plata Basin during the 20th century, *Climatic Change, 98*(3−4), 531−550.

Pendall, E., S. del Grosso, J. Y. King, D. R. LeCain, D. G. Milchunas, J. A. Morgan, A. R. Mosier, D. S. Ojima, W. A. Parton, P. P. Tans, and J. W. C. White (2003), Elevated atmospheric CO_2 effects and soil water feedbacks on soil respiration components in a Colorado grassland, *Global Biogeochem. Cycles, 17,* 1046, doi:10.1029/2001GB001821.

Peng, T. H. (1995), Future climate surprises, in *Future Climates of the World: A Modelling Perspective,* edited by A. Henderson-Sellers, pp. 517−535, Elsevier.

Peng, Y.-B., C.-M. Shen, W.-C. Wang, and Y. Xu (2010), Response of summer precipitation over Eastern China to large volcanic eruptions, *J. Climate, 23,* 818−824, doi:10.1175/2009JCL. I2950.1.

Penner, J. E., D. J. Bergmann, J. J. Walton, D. Kinnison, M. J. Prather, D. Rotman, C. Price, K. E. Pickering, and S. Baughcum (1998), An evaluation of upper tropospheric NO_x with two models, *J. Geophys. Res., 103,* 22097−22113.

Penner, J. E., S. Y. Zhang, and C. C. Chuang (2003), Soot and smoke aerosols may not warm climate, *J. Geophys. Res., 108*(21), 4657, doi:10.1029/2003JD003409.

Penner, J. E., Y. Chen, M. Wang, and X. Liu (2009), Possible influence of anthropogenic aerosols on cirrus clouds and anthropogenic forcing, *Atmos. Chem. Phys., 9*, 879−896.

Pepin, N. C., and J. D. Lundquist (2008), Temperature trends at high elevations: Patterns across the globe, *Geophys. Res. Lett., 35*(14), L14701.

Percy, K. (1986), The effects of simulated acid rain on germinative capacity, growth and morphology of forest tree seedlings, *New. Phytol., 104*, 473−484.

Percy, K. (1991), Effects of acid rain on forest vegetation: Morphological and non-mensurational growth effects, in Effects of acid rain on forest resources, Proceedings of a conference held in Ste. Foy, Quebec Forestry Canada, Ottawa, pp. 97−110.

Perkins, S. E. and A. J. Pitman, (2009), Do weak AR4 models bias projections of future climate changes over Australia?, *Climatic Change, 93*(3−4), 527−58.

Perkins, S. E., A. J. Pitman, N. J. Holbrook, and J. McAneney (2007), Evaluation of the AR4 climate models' simulated daily maximum temperature, minimum temperature, and precipitation over Australia using probability density functions, *J. Climate, 20*, 4356−4376.

Perkins, S. E., A. J. Pitman, and S. A. Sisson (2009), Smaller projected increases in 20-year temperature returns over Australia in skill-selected climate models, *Geophys. Res. Lett., 36*, L06710, doi:10.1029/2009GL037293.

Perlwitz, J., S. Pawson, R. L. Fogt, J. E. Nielsen, and W. D. Neff (2008), Impact of stratospheric ozone hole recovery on Antarctic climate, *Geophys. Res. Lett., 35*, L08714, doi:10.1029/2008GL033317.

Perovich, D. K., B. Light, H. Eicken, K. F. Jones, K. Runciman, and S. V. Nghiem (2007), Increasing solar heating of the Arctic Ocean and adjacent seas, 1979−2005: Attribution and role in the ice-albedo feedback, *Geophys. Res. Lett., 34*, L19505.

Perovich, D. K., J. A. Richter-Menge, K. F. Jones, and B. Light (2008), Sunlight, water, and ice: Extreme Arctic sea ice melt during the summer of 2007, *Geophys. Res. Lett., 35*, L11501.

Perry, A. L., P. J. Low, J. R. Ellis, and J. D. Reynolds (2005), Climate change and distribution shifts in marine fishes, *Science, 308*(5730), 1912−1915, doi:10.1126/science.1111322.

Peryt, D., L. Alegret, and E. Molina (2002), The Cretaceous/Paleogene (K/P) boundary at Ain Settara, Tunisia: Restructuring of benthic foraminiferal assemblages, *Terra Nova, 14*, 101−107.

Peterson, T. C., and M. J. Manton (2008), Monitoring changes in climate extremes: A tale of international collaboration, *Bull. Am. Meteorol. Soc., 89*, 1266−1271, doi:10.1175/2008BAMS2501.1.

Peterson, T. C., D. R. Easterling, T. R. Karl, P. Groisman, N. Nicholls, N. Plummer, S. Torok, I. Auer, R. Boehm, D. Gullett, L. Vincent, R. Heino, H. Tuomenvirta, O. Mestre, T. Szentimrey, J. Salinger, E. J. Forland, I. Hanssen-Bauer, H. Alexandersson, P. Jones, and D. Parker (1998), Homogeneity adjustments of in situ atmospheric climate data: A review, *Int. J. Climatol., 18*, 1493−1517.

Peterson, T. C., M. A. Taylor, R. Demeritte, D. L. Duncombe, S. Burton, F. Thompson, A. Porter, M. Mercedes, E. Villegas, R. Semexant Fils, A. M. G. Klein-Tank, R. Warner, A. Joyette, W. Mills, L. V. Alexander, and B. Gleason (2002), Recent changes in climate extremes in the Caribbean region, *J. Geophys. Res., 107*(D21), 4601, doi:10.1029/2002JD002251.

Peterson, T. C., X. B. Zhang, M. Brunet-India, and J. L. Vazquez-Aguirre, (2008), Changes in North American extremes derived from daily weather data, *J. Geophys. Res.-Atmos., 113*(D7), D07113.

Petit, J. R., J. Jouzel, D. Raynaud, N. I. Barkov, J. M. Barnola, I. Basile, M. Bender, J. Chapellaz, M. Davis, G. Delaaygue, M. Delmotte, V. M. Kotlyakov, M. Legrand, V. Y. Lipenkov, C. Lorius, L. Pépin, C. Ritz, E. Saltzman, and M. Stievenard (1999), Climate and atmospheric history of the past 420,000 years from Vostok ice core, Antarctica, *Nature, 399*(6735), 429−436.

Petrescu, A. M. R., L. P. H. van Beek, J. van Huissteden, C. Prigent, T. Sachs, C. A. R. Corradi, F. J. W. Parmentier, and A. J. Dolman (2010), Modeling regional to global CH4 emissions of boreal and arctic wetlands, *Global Biogeochem. Cycles, 24*, GB4009, doi:10.1029/2009GB003610.

Petron, G., C. Granier, B. Khattatov, V. Yudin, J. F. Lamarque, L. Emmons, J. Gille, and D. P. Edwards (2004), Monthly CO surface sources inventory based on the 2000−2001 MOPITT satellite data, *Geophys. Res. Lett., 31*, L21107, doi:10.1029/2004GL020560.

Petrow, T., and B. Merz (2009), Trends in flood magnitude, frequency and seasonality in Germany in the period 1951−2002, *J. Hydrol., 371*(1−4), 129−141.

Pew Center on Global Climate Change, 2010, Summary of COP 16 and CMP 6, December, *http://www.pewclimate.org/docUploads/cancun-climate-conference-cop16-summary.pdf.* (Accessed 30/3/11.)

Pfeffer, W. T., J. T. Harper, and S. O'Neel (2008), Kinematic constraints on glacier contributions to 21st-century sea-level rise, *Science, 321*(5894), 1340−1343.

Phillips, O. L., L. E. O. C. Aragão, S. L. Lewis, J. B. Fisher, J. Lloyd, G. López-González, Y. Malhi, A. Monteagudo, J. Peacock, C. A. Quesada, G. van der Heijden, S. Almeida, I. Amaral, L. Arroyo, G. Aymard, T. R. Baker, O. Bánki, L. Blanc, D. Bonal, P. Brando, J. Chave, Á. C. A. de Oliveira, N. D. Cardozo, C. I. Czimczik, T. R. Feldpausch, M. A. Freitas, E. Gloor, N. Higuchi, E. Jiménez, G. Lloyd, P. Meir, C. Mendoza, A. Morel, D. A. Neill, D. Nepstad, S. Patiño, M. C. Peñuela, A. Prieto, F. Ramírez, M. Schwarz, J. Silva, M. Silveira, A. S. Thomas, H. ter Steege, J. Stropp, R. Vásquez, P. Zelazowski, E. A. Dávila, S. Andelman, A. Andrade, K.-J. Chao, T. Erwin, A. Di Fiore, E. Honorio C., H. Keeling, T. J. Killeen, W. F. Laurance, A. Peña Cruz, N. C. A. Pitman, P. Núñez Vargas, H. Ramírez-Angulo, A. Rudas, R. Salamão, N. Silva, J. Terborgh, and A. Torres-Lezama (2009), Drought sensitivity of the Amazon rainforest, *Science, 323*(5919), 1344−1347.

Phillips, R. P., A. C. Finzi, and E. S. Bernhardt (2011), Enhanced root exudation induces microbial feedbacks to N cycling in a pine forest under long-term CO$_2$ fumigation, *Ecol. Lett., 14*, 187−194, doi:10.1111/j.1461-0248.2010.01570.x.

Piani, C., D. Frame, D. Stainforth, and M. Allen (2005), Constraints on climate change from a multi-thousand member ensemble of simulations, *Geophys. Res. Lett., 32*(23), doi:10.1029/2005GL024452.

Piao, S., P. Friedlingstein, P. Ciais, L. Zhou, and A. Chen (2006), Effect of climate and CO$_2$ changes on the greening of the Northern Hemisphere over the past two decades, *Geophys. Res. Lett., 33*, L23402, doi:10.1029/2006GL028205.

Picaut, J., F. Masia, and Y. du Penhoat (1997), An advective-reflective conceptual model for the oscillatory nature of the ENSO, *Science, 277*(5326), 663−666, doi:10.1126/science.277.5326.663.

Pickett, E. J., S. P. Harrison, G. Hope, K. Harle, J. R. Dodson, A. P. Kershaw, I. C. Prentice, J. Backhouse, E. A. Colhoun, D. D'Costa, J. Flenley, J. Grindrod, S. Haberle, C. Hassell, C. Kenyon, M. Macphail, H. Martin, A. H. Martin, M. McKenzie,

J. C. Newsome, D. Penny, J. Powell, J. I. Raine, W. Southern, J. Stevenson, J. P. Sutra, I. Thomas, S. van der Kaars, and J. Ward (2004), Pollen-based reconstructions of biome distributions for Australia, Southeast Asia and the Pacific (SEAPAC region) at 0, 6000 and 18,000 C-14 yr B.P, *J. Biogeogr., 31*(9), 1381—1444.

Pielke, R., Jr. (2010), Blog entry: He's baaack, 17 May, 2010, *http:// rogerpielkejr.blogspot.com/2010/05/hes-baaack.html.* (Accessed 21/10/10.)

Pielke, R. A., Sr., R. Avissar, M. Raupach, A. J. Dolman, X. Zeng, and A. S. Denning (1998), Interactions between the atmosphere and terrestrial ecosystems: Influence on weather and climate, *Glob. Change Biol., 4,* 461—475.

Pielke, R. A., Sr., G. Marland, R. A. Betts, T. N. Chase, J. L. Eastman, J. O. Niles, D. Niyogi, and S. W. Running (2002), The influence of land-use change and landscape dynamics on the climate system relevance to climate change policy beyond the radiative effect of greenhouse gases, *Philos. Trans. R. Soc. London Ser. A, 360,* 1705—1719.

Pierazzo, E., D. A. Kring, and H. J. Melosh (1998), Hydrocode simulation of the Chicxulub impact event and the production of climatically active gases, *J. Geophys. Res., 103,* 28607—28625.

Pierazzo, E., R. R. Garcia, D. E. Kinnison, D. R. Marsh, J. Lee-Taylor, and P. J. Crutzen (2010), Ozone perturbation from medium-size asteroid impacts in the ocean, *Earth Planet. Sci. Lett.*, doi:10.1016/j.epsl.2010.08.036.

Pierce, D. W., T. P. Barnett, and U. Mikolajewicz (1995), Competing roles of heat and fresh water fluxes in forcing thermohaline oscillations, *J. Phys. Oceanogr., 25,* 2046—2064.

Pierce, D. W., T. P. Barnett, and M. Latif (2000), Connections between the Pacific Ocean tropics and midlatitudes on decadal timescales, *J. Climate, 13,* 1173—1194.

Pierce, J. R., and P. J. Adams (2009), Can cosmic rays affect cloud condensation nuclei by altering new particle formation rates?, *Geophys. Res. Lett., 36,* L09820, doi:10.1029/2009GL037946.

Pierson, T. C., R. J. Janda, J. C. Thouret, and C. A. Borrero (1990), Perturbation and melting of snow and ice by the 13 November 1985 eruption of Nevado del Ruiz, Colombia, and consequent mobilization, flow and deposition of lahars, *J. Volcanol. Geoth. Res., 41*(1—4), 17—66.

Pigeon G., A. Lemonsu, C. S. B. Grimmond, P. Durand, O. Thouron, and V. Masson (2007), Divergence of turbulent fluxes in the surface layer: Case of a coastal city, *Bound.-Lay. Meteorol., 124,* 269—290.

Pinker, R. T., B. Zhang, and E. G. Dutton (2005), Do satellites detect trends in surface solar radiation?, *Science, 308,* 850—854.

Pinot, S., G. Ramstein, S. P. Harrison, I. C. Prentice, J. Guiot, S. Joussaume, M. Stute, and PMIP participating groups (1999), Tropical palaeoclimates at the Last Glacial Maximum: comparison of Paleoclimate Modeling Intercomparison Project (PMIP) simulations and paleodata. *Climate Dynamics, 15*(11), 857—874.

Piringer, M., C. S. B. Grimmond, S. M. Joffre, P. Mestayer, D. R. Middleton, M. W. Rotach, A. Baklanov, K. D. Ridder, J. Ferreira, E. Guilloteau, A. Karppinen, A. Martilli, V. Masson, and M. Tombrou (2002), Investigating the surface energy balance in urban areas — Recent advances and future needs, *Water Air Soil Pollut. Focus, 2,* 1—16.

Piringer, M., S. M. Joffre, A. Baklanov, A. Christen, M. Deserti, K. De Ridder, S. Emeis, P. Mestayer, M. Tombrou, D. Middleton, K. Baumann-Stanzer, A. Dandou, A. Karppinen, and J. Burzynski (2007), The surface energy balance and the mixing height in urban areas — Activities and recommendations of COST-Action 715, *Bound.-Lay. Meteorol., 124,* 3—24.

Pisaric, M. F. J., S. K. Carye, S. V. Kokelj, and D. Youngblut (2007), Anomalous 20th century tree growth, Mackenzie Delta, Northwest Territories, Canada, *Geophys. Res. Lett., 34,* L05714, doi:10.1029/2006GL029139.

Pison, I., P. Bousquet, F. Chevallier, S. Szopa, and D. Hauglustaine (2009), Multi-species inversion of CH_4, CO and H_2 emissions from surface measurements, *Atmos. Chem. Phys., 9,* 5281—5297.

Pitman, A. J. (2003), The evolution of, and revolution in, land surface schemes designed for climate models, *Int. J. Climatol., 23,* 479—510.

Pitman, A. J. (2005), On the role of geography in Earth system science, *Geoforum, 36,* 137—148.

Pitman A. J., and G. T. Narisma (2005), The role of land surface processes in regional climate change: A case study of future land cover change over south-western Australia, *Meteor. Atmos. Phys., 89,* 235—249.

Pitman, A. J., and N. de Noblet-Ducoudré (2012), Human effects on climate through land-use-induced land-cover change, in *The Future of the World's Climate*, edited by A. Henderson-Sellers and K. McGuffie, pp. 77—95, Elsevier, Amsterdam.

Pitman, A. J., and R. J. Stouffer (2006), Abrupt change in climate and climate models, *Hydrol. Earth Syst. Sc., 10*(6), 903—912.

Pitman, A. J., G. T. Narisma, R. Pielke, and N. J. Holbrook (2004), The impact of land-cover change on the climate of south west Western Australia, *J. Geophys. Res., 109,* D18109, doi:10.1029/2003JD004347.

Pitman, A. J., N. de Noblet-Ducoudré, F. T. Cruz, E. L. Davin, G. B. Bonan, V. Brovkin, M. Claussen, C. Delire, L. Ganzeveld, V. Gayler, B. J. J. M. van den Hurk, P. J. Lawrence, M. K. van der Molen, C. Müller, C. H. Reick, S. I. Seneviratne, B. J. Strengers, and A. voldoire (2009), Uncertainties in climate responses to past land cover change: First results from the LUCID intercomparison study, *Geophys. Res. Lett., 36,* L14814, doi:10.1029/2009GL039076.

Plattner, G. K., F. Joos, T. F. Stocker, and O. Marchal (2001), Feedback mechanisms and sensitivities of ocean carbon uptake under global warming, *Tellus B, 53*(5), 564—592.

Plattner, G.-K., R. Knutti, F. Joos, T. F. Stocker, W. von Bloh, V. Brovkin, D. Cameron, E. Driesschaert, S. Dutkiewicz, M. Eby, N. R. Edwards, T. Fichefet, J. C. Hargreaves, C. D. Jones, M. F. Loutre, H. D. Matthews, A. Mouchet, S. A. Müller, S. Nawrath, A. Price, A. Sokolov, K. M. Strassmann, and A. J. Weaver (2008), Long-term climate commitments projected with climate—carbon cycle models, *J. Climate, 21,* 2721—2751.

Polcher, J. (1995), Sensitivity of tropical convection to land surface processes, *J. Atmos. Sci., 52,* 3143—3161.

Polcher, J., and K. Laval (1994), The impact of African and Amazonian deforestation on tropical climate, *J. Hydrol., 15,* 389—405.

Pollard, D., and R. M. DeConto (2009), Modelling West Antarctic Ice Sheet growth and collapse through the past five million years, *Nature, 458,* 329—332.

Polley, H. W., H. B. Johnson, B. D. Marinot, and H. S. Mayeux (1993), Increase in C_3 plant water-use efficiency and biomass over glacial to present CO_2 concentrations, *Nature, 361*(6407), 61—64, doi:10.1038/361061a0.

Polovina, J. J., E. A. Howell, and M. Abecassis (2008), Ocean's least productive waters are expanding, *Geophys. Res. Lett., 35,* L03618, doi:10.1029/2007GL031745.

Pongratz, J., C. H. Reick, T. Raddatz, and M. Claussen (2008), A reconstruction of global agricultural areas and land cover for the last millennium, *Global Biogeochem. Cy.*, 22, GB3018, doi:10.1029/2007GB003153.

Pongratz, J., C. H. Reick, T. Raddatz, and M. Claussen (2010), Biogeophysical versus biogeochemical climate response to historical anthropogenic land cover change, *Geophys. Res. Lett.*, 37, L08702, doi:10.1029/2010GL043010.

Pope, III, C. A., and D. W. Dockery (2006), Health effects of fine particulate air pollution: Lines that connect, *JAPCA J. Air Waste Ma.*, 56, 709–742.

Pope, K. O. (2002), Impact dust not the cause of the Cretaceous-Tertiary mass extinction, *Geology*, 30, 99–102.

Popper, K. (1959), *The Logic of Scientific Discovery*, (translation of *Logik der Forschung*) (revised 1968), Hutchinson, London.

Posselt, R., and U. Lohmann (2008), Influence of Giant CCN on warm rain processes in the ECHAM5 GCM, *Atmos. Chem. Phys.*, 8, 3769–3788.

Post, W., et al. (2009), Terrestrial biological sequestration: Science for enhancement and implementation, in *Carbon Sequestration and Its Role in the Global Carbon Cycle*, edited by B. McPherson and E. T. Sundquist, pp. 183, 350, American Geophysical Union.

Pouillet, C. S. M. (1838), Mémoire sur la chaleur solaire, sur les pouvoirs rayonnants et absorbants de l'air atmosphérique et sur la température de l'espace, *Comptes Rendus de l'Académie des Sciences*, 7, 24–65.

Poulter, B., F. Hattermann, E. Hawkins, S. Zaehles, S. Sitch, N. R-. Coupe, U. Heyder, and W. Cramer (2010), Robust dynamics of Amazon dieback to climate change with perturbed ecosystem model parameters, *Glob. Change Bio.*, 16, 2476–2495, doi:10.1111/j.1365-2486.2009.02157.x.

Poumadere, M., C. Mays, S. Le Mer, and R. Blong (2005), The 2003 heat wave in France: Dangerous climate change here and now, *Risk Anal.*, 25, 1483–1494.

Power, S., F. Tseitkin, S. Torok, B. Lavery, R. Dahni, and B. McAvaney (1998), Australian temperature, Australian rainfall, and the Southern Oscillation, 1910–1992: Coherent variability and recent changes, *Aust. Meteorol. Mag.*, 47(2), 85–101.

Power, S., M. Haylock, R. Colman, and X. Wang (2006), The predictability of interdecadal changes in ENSO activity and ENSO teleconnections, *J. Climate*, 19(19), 4755–4771.

Power, M. J., J. Marlon, N. Ortiz, P. J. Bartlein, S. P. Harrison, F. E. Mayle, A. Ballouche, R. H. W. Bradshaw, C. Carcaillet, C. Cordova, S. Mooney, P. I. Moreno, I. C. Prentice, K. Thonicke, W. Tinner, C. Whitlock, Y. Zhang, Y. Zhao, A. A. Ali, R. S. Anderson, R. Beer, H. Behling, C. Briles, K. J. Brown, A. Brunelle, M. Bush, P. Camill, G. Q. Chu, J. Clark, D. Colombaroli, S. Connor, A. L. Daniau, M. Daniels, J. Dodson, E. Doughty, M. E. Edwards, W. Finsinger, D. Foster, J. Frechette, M. J. Gaillard, D. G. Gavin, E. Gobet, S. Haberle, D. J. Hallett, P. Higuera, G. Hope, S. Horn, J. Inoue, P. Kaltenrieder, L. Kennedy, Z. C. Kong, C. Larsen, C. J. Long, J. Lynch, E. A. Lynch, M. McGlone, S. Meeks, S. Mensing, G. Meyer, T. Minckley, J. Mohr, D. M. Nelson, J. New, R. Newnham, R. Noti, W. Oswald, J. Pierce, P. J. H. Richard, C. Rowe, M. F. S. Goni, B. N. Shuman, H. Takahara, J. Toney, C. Turney, D. H. Urrego-Sanchez, C. Umbanhowar, M. Vandergoes, B. Vanniere, E. Vescovi, M. Walsh, X. Wang, N. Williams, J. Wilmshurst, and J. H. Zhang (2008), Changes in fire regimes since the Last Glacial Maximum: An assessment based on a global synthesis and analysis of charcoal data, *Clim. Dynam.*, 30(7–8), 887–907, doi:10.1007/s00382-007-0334-x.

Prather, M., and D. Ehhalt (2001), Atmospheric chemistry and greenhouse gases, Chapter 4, in *Climate Change 2001: The Scientific Basis*, edited by J. T. Houghton, et al., pp. 239–287, Cambridge University Press, Cambridge, UK.

Prather, M., M. Gauss, T. Berntsen, I. S. A. Isaksen, J. Sundet, I. Bey, G. Brasseur, F. Dentener, R. Derwent, D. Stevenson, L. Grenfell, D. Hauglustaine, L. Horowitz, D. Jacob, L. Mickley, M. Lawrence, R. von Kuhlmann, J.-F. Muller, G. Pitari, H. Rogers, M. Johnson, J. Pyle, K. Law, M. van Weele, and O. Wild (2003), Fresh air in the 21st century?, *Geophys. Res. Lett.*, 30(2), 1100, doi:10.1029/2002GL016285.

Prell, W., and J. Kutzbach (1987), Monsoon variability over the past 150,000 years, *J. Geophys. Res.-Atmos.*, 92(D7), 8411–8425.

Prentice, I. C. (2001), The carbon cycle and atmospheric carbon dioxide, in *Climate Change 2001: The Scientific Basis, contribution of Working Group I to the IPCC Third Assessment Report (TAR)*, 218.

Prentice, I. C., and S. P. Harrison (2009), Ecosystem effects of CO_2 concentration: Evidence from past climates, *Clim. Past*, 5(3), 297–307.

Prentice, I. C., S. P. Harrison, D. Jolly, and J. Guiot (1998), The climate and biomes of Europe at 6000 yr B.P: Comparison of model simulations and pollen-based reconstructions, *Quaternary Sci. Rev.*, 17(6–7), 659–668, doi:10.1016/S0277-3791(98)00016-X.

Prentice, I. C., D. Jolly, and Biome 6000 Participants (2000), Mid-Holocene and glacial-maximum vegetation geography of the northern continents and Africa, *J. Biogeogr.*, 27(3), 507–519.

Prentice, I. C., G. D. Farquhar, M. J. R. Fasham, M. L. Goulden, M. Heimann, V. J. Jaramillo, H. S. Kheshgi, C. Le Quéré, R. J. Sholes, and D. W. R. Wallace (2001), The carbon cycle and atmospheric carbon dioxide, in *Climate Change 2001: The Scientific Basis, contribution of Working Group I to the Third Assessment Report (TAR) of the Intergovernmental Panel on Climate Change (IPCC)*, edited by J. T. Houghton, Y. Ding, D. J. Griggs, M. Noguer, P. J. van der Linden, X. Dai, K. Maskell, and C. A. Johnson, pp. 183–237, Cambridge University Press, Cambridge, UK.

Prentice, I. C., S. P. Harrison, and P. J. Bartlein (2011), Tropical forests, ice ages and the carbon cycle, *New Phytol.*, 189, 988–998.

Preunkert, S., M. Legrand, and D. Wagenbach (2001), Sulfate trends in a Col du Dôme (French Alps) ice core: A record of anthropogenic sulfate levels in the European midtroposphere over the twentieth century, *J. Geophys. Res.*, 106, 31991–32004.

Prinn, R. G., and B. Fegley, Jr. (1987), Bolide impacts, acid rain, and biospheric traumas at the Cretaceous-Tertiary boundary, *Earth Planet. Sci. Lett.*, 83, 1–15.

Pritchard, H. D., R. J. Arthern, D. G. Vaughan, and L. A. Edwards (2009), Extensive dynamic thinning on the margins of the Greenland and Antarctic Ice Sheets, *Nature*.

Proctor, J. D. (1998), The meaning of global environmental change: Retheorizing culture in human dimensions research, *Glob. Environ. Change*, 8, 227–248.

Productivity Commission (2009), Economic impacts of drought, report by Productivity Commission for DAFF, *http://www.daff.gov.au/agriculture-food/drought/national_review_of_drought_policy/economic_assessment*.

Prokopenko, A. A., E. V. Bezrukova, G. K. Khursevich, E. P. Solotchina, M. I. Kuzmin, and P. E. Tarasov (2010), Climate in continental interior Asia during the longest interglacial of the past 500,000 years:

The new MIS-11 records from Lake Baikal, SE Siberia, *Clim. Past, 6*, 31–48.

Pruppacher, H. R., and J. D. Klett (1997), *Microphysics of Clouds and Precipitation*, Second edition, Kluwer Academic, Norwell, MA.

Pryor, S. C., J. A. Howe, and K. E. Kunkel (2009), How spatially coherent and statistically robust are temporal changes in extreme precipitation in the contiguous USA?, *Int. J. Climatol., 29*(1), 31–45.

Putaud, J. P., F. Raes, R. Van Dingenen, E. Bruggemann, M. C. Facchini, S. Decesari, S. Fuzzi, R. Gehrig, C. Hüglin, P. Laj, G. Lorbeer, W. Maenhaut, N. Mihalopoulos, K. Müller, X. Querol, S. Rodriguez, J. Schneider, G. Spindler, H. ten Brinkj, K. Tørseth and A. Wiedensohler (2004), European aerosol phenomenology-2: Chemical characteristics of particulate matter at kerbside, urban, rural and background sites in Europe, *Atmos. Environ., 38*(16), 2579–2595.

Pye, J. M. (1988), Impact of ozone on the growth and yield of trees: A review, *J. Environ. Qual., 17*, 347–360.

Pye, H. O. T., H. Liao, S. Wu, L. J. Mickley, D. J. Jacob, D. K. Henze, and J. H. Seinfeld (2009), Effect of changes in climate and emissions on future sulfate-nitrate-ammonium aerosol levels in the United States, *J. Geophys. Res., 114*(D1), D01205.

Qian, H., R. Joseph, and N. Zing (2008), Response of the terrestrial carbon cycle to the El Niño-Southern Oscillation, *Tellus B, 60*, 537–550.

Qian, H., R. Joseph, and N. Zeng (2009), Enhanced terrestrial carbon uptake in the northern high latitudes in the 21st century from the coupled carbon cycle climate model intercomparison project model projections, *Glob. Change Bio., 16*, 641–656, doi:10.1111/j.1365-2486.2009.01989.x.

Qian, J., A. Robertson, and V. Moron (2010), Interactions among ENSO, the monsoon, and diurnal cycle in rainfall variability over Java, Indonesia, *J. Atmos. Sci., 67*(11), 3509–3524, doi:10.1175/2010JAS3348.1.

Qu, X., and A. Hall, (2006), Assessing snow albedo feedback in simulated climate change, *J. Climate, 19*, 2617–2630, doi:10.1175/JCLI3750.1.

Quaas J., O. Boucher, N. Bellouin, and S. Kinne (2008), Satellite-based estimate of the direct and indirect climate forcing, *J. Geophys. Res.*, doi:10.1029/2007JD008962.

Quaas, J., Y. Ming, S. Menon, T. Takemura, M. Wang, J. E. Penner, A. Gettelman, U. Lohmann, N. Bellouin, O. Boucher, A. M. Sayer, G. E. Thomas, A. McComiskey, G. Feingold, C. Hoose, J. E. Kristjansson, X. Liu, Y. Balkanski, L. J. Donner, P. A. Ginoux, P. Stier, B. Grandey, J. Feichter, I. Sednev, S. E. Bauer, D. Koch, R. G. Grainger, A. Kirkevag, T. Iversen, O. Seland, R. Easter, S. J. Ghan, P. J. Rasch, H. Morrison, J. F. Lamarque, M. J. Iacono, S. Kinne, and M. Schulz (2009), Aerosol indirect effects – general circulation model intercomparison and evaluation with satellite data, *Atmos. Chem. Phys., 9*(22), 8697–8717.

Qui, J. (2009), Tundra's burning, *Nature, 461*, 34–36.

Quigley, M. C., T. Horton, J. C. Hellstrom, M. L. Cupper, and M. Sandiford (2010), Holocene climate change in arid Australia from speleothem and alluvial records, *Holocene, 20*(7), 1093–1104, doi:10.1177/0959683610369508.

Quinn, P. K., and T. S. Bates (2005), Regional aerosol properties: Comparisons of boundary layer measurements from ACE 1, ACE 2, aerosols99, INDOEX, ACE Asia, TARFOX, and NEAQS, *J. Geophys. Res., 110* (D14), D14202.

Quinn, P. K., G. Shaw, E. Andrews, E. G. Dutton, T. Ruoho-Airola, and S. T. Gong (2007), Arctic haze: Current trends and knowledge gaps, *Tellus B, 59*, 99–114.

Quinn, P. K., T. S. Bates, E. Baum, N. Doubleday, A. M. Fiore, M. Flanner, A. Fridlind, T. J. Garrett, D. Koch, S. Menon, D. Shindell, A. Stohl, and S. G. Warren (2008), Shortlived pollutants in the Arctic: Their climate impact and possible mitigation strategies, *Atmos. Chem. Phys., 8*, 1723–1735.

Racherla, P. N., and P. J. Adams (2006), Sensitivity of global tropospheric ozone and fine particulate matter concentrations to climate change, *J. Geophys. Res., 111*, D24103, doi:10.1029/2005JD006939.

Radić, V., and R. Hock (2010), Regional and global volumes of glaciers derived from statistical upscaling of glacier inventory data, *J. Geophys. Res., 115*, F01010, doi:10.1029/2009JF001373.

Radić, V., and R. Hock (2011), Regionally differentiated contribution of mountain glaciers and ice caps to future sea-level rise, *Nat. Geosci., 4*(2), 90–94.

Radić, V., R. Hock, and J. Oerlemans (2008), Analysis of scaling methods in deriving future volume evolutions of valley glaciers, *J. Glaciol., 54*(187), 601–612.

Rafelski, L. E., S. C. Piper, and R. F. Kelling (2009), Climate effects on atmospheric carbon dioxide over the last century, *Tellus, 61B*, 718–731.

Raffa, K. F., B. H. Aukema, B. J. Bentz, A. L. Carroll, J. A. Hicke, M. G. Turner, and W. H. Romme (2008), Cross-scale drivers of natural disturbances prone to anthropogenic amplification: The dynamics of bark beetle eruptions, *BioScience, 58*, 501–517.

Rahimzadeh, F., A. Asgari, and E. Fattahi (2009), Variability of extreme temperature and precipitation in Iran during recent decades, *Int. J. Climatol., 29*, 329–343.

Rahmstorf, S. (2007), A semi-empirical approach to projecting future sea-level rise, *Science, 315*(5810), 368–370, doi:10.1126/science.1135456.

Rahmstorf, S. (2010), A new view on sea level rise, *Nat. Clim. Change, 4*, 44–45, doi:10.1038/climate.2010.29.

Rahmstorf, S., M. Crucifix, A. Ganopolski, H. Goosse, I. V. Kamenkovich, R. Knutti, G. Lohmann, R. Marsh, L. A. Mysak, Z. Wang, and A. J. Weaver (2005), Thermohaline circulation hysteresis: A model intercomparison, *Geophys. Res. Lett., 32*, L23605, doi:10.1029/2005GL023655.

Rahmstorf, S., A. Cazenave, J. A. Church, J. E. Hansen, R. F. Keeling, D. E. Parker, and R. C. J. Somerville (2007), Recent climate observations compared to projections, *Science, 316*(5825), 709, doi:10.1126/science.1136843.

Ramanathan, V., and G. Carmichael (2008), Global and regional climate changes due to black carbon, *Nat. Geosci., 1*(4), 221–227.

Ramanathan, V., and Y. Feng (2008), On avoiding dangerous anthropogenic interference with the climate system: Formidable challenges ahead, *Proc. Natl. Acad. Sci. USA, 105*(38), 14245–14250.

Ramanathan, V., L. Callis, R. Cess, J. Hansen, I. S. A. Isaksen, W. Kuhn, A. Lacis, F. Luther, J. Mahlman, R. Reck, and M. Schlesinger (1987), Climate-chemical interactions and effects of changing atmospheric trace gases, *Rev. Geophys. Space Phys., 25*, 1441–1482.

Ramanathan, V., P. J. Crutzen, J. T. Kiehl, and D. Rosenfeld (2001), Atmosphere – Aerosols, climate, and the hydrological cycle, *Science, 294*(5549), 2119–2124.

Ramanathan, V., C. Chung, D. Kim, T. Bettge, L. Buja, J. T. Kiehl, W. M. Washington, Q. Fu, D. R. Sikka, and M. Wild (2005), Atmospheric brown clouds: Impacts on South Asian climate and hydrological cycle, *Proc. Natl. Acad. Sci. USA, 102*(15), 5326–5333.

Ramanathan, V., M. V. Ramana, G. Roberts, D. Kim, C. Corrigan, M. V. Ramana, G. Roberts, D. Kim, C. Corrigan, C. Chung, and

D. Winker (2007), Warming trends in Asia amplified by brown cloud solar absorption, *Nature, 448*(7153) 575−578.

Ramankutty, N., and J. A. Foley (1999), Estimating historical changes in global land cover: Croplands from 1700 to 1992, *Global Biogeochem. Cy., 13,* 997−1027.

Ramaswamy, V., and M. M. Bowen (1994), Effect of changes in radiatively active species upon the lower stratospheric temperatures, *J. Geophys. Res., 99*(D9), 18909−18921.

Rampino, M. R. (2010), Mass extinctions of life and catastrophic flood basalt volcanism, *Proc. Natl. Acad. Sci. USA, 107,* 6555−6556.

Randall, D. A., R. A. Wood, S. Bony, R. Coleman, T. Fichefet, J. Fyfe, V. Kattsov, A. Pitman, J. Shukla, J. Srinivasan, R. J. Stouffer, A. Sumi, and K. E. Taylor (2007), Climate models and their evaluation, in *Climate Change 2007: The Physical Science Basis, contribution of Working Group I to the Fourth Assessment Report (FAR) of the Intergovernmental Panel on Climate Change,* edited by S. Solomon, et al., Cambridge University Press, Cambridge, UK.

Randel, W. J., F. Wu, H. Vomel, G. E. Nedoluha, and P. Forster (2006), Decreases in stratospheric water vapor after 2001: Links to changes in the tropical tropopause and the Brewer-Dobson circulation, *J. Geophys. Res., 111,* D12312, doi:10.1029/2005JD006744.

Randerson, J. T., et al. (2009), Systematic assessment of terrestrial biogeochemistry in coupled climate-carbon models, *Glob. Change Bio., 15,* 2462−2484.

Rao, S., K. Riahi, K. Kupiainen, and Z. Klimont (2005), Long-term scenarios for black and organic carbon emissions, *Environ. Sci., 2*(2−3), 205−216.

Raper, S. C. B., and R. J. Braithwaite (2005a), Reply to comment by M. F. Meier, et al. on "The potential for sea level rise: New estimates from glacier and ice cap area and volume distributions," *Geophys. Res. Lett., 32,* L17502, doi:10.1029/2005GL023460.

Raper, S. C. B., and R. J. Braithwaite (2005b), The potential for sea level rise: New estimates from glacier and ice cap area and volume distributions, *Geophys. Res. Lett., 32,* L05502, doi:10.1029/2004GL021981.

Raper, S. C. B., and R. J. Braithwaite (2006), Low sea level rise projections from mountain glaciers and icecaps under global warming, *Nature, 439,* 311−313.

Rappin, E., D. Nolan, and K. Emanuel (2010), Thermodynamic control of tropical cyclogenesis in environments of radiative-convective equilibrium with shear, *Q. J. Roy. Meteor. Soc., 136*(653), 1954−1971, doi:10.1002/qj.706.

Rasch, P. J., P. J. Crutzen, and D. B. Coleman (2008), Exploring the geoengineering of climate using stratospheric sulfate aerosols: The role of particle size, *Geophys. Res. Lett., 35,* L02809, doi:10.1029/2007GL032179.

Rasmusson, E. M., and T. H. Carpenter (1982), Variations in tropical sea surface temperature and surface wind fields associated with the Southern Oscillation/El Ñino, *Mon. Weather Rev., 110,* 354−384.

Rasool, S. I., and S. H. Schneider (1971), Atmospheric carbon dioxide and aerosols: Effects of large increases on global climate, *Science, 173,* 3992.

Raup, D. M. (1989), The case for extraterrestrial causes of extinction, *Philos. Trans. R. Soc. London Ser. B, 325,* 421−435.

Raup, D. M., and G. E. Boyajian (1988), Patterns of generic extinction in the fossil record, *Paleobiology, 14,* 109−125.

Raup, D. M., and J. J. Sepkoski (1982), Mass extinctions in the marine fossil record, *Science, 215,* 1501−1503.

Raup, B., A. Racoviteanu, S. J. S. Khalsa, C. Helm, R. Armstrong, and Y. Arnaud (2007), The GLIMS geospatial glacier database: A new tool for studying glacier change, *Glob. Planet. Change, 56*(1−2), 101−110.

Raupach, M. R., and J. G. Canadell (2010), Carbon and the Anthropocene, *Curr. Opin. Environ. Sustain.,* doi:10.1016/j.cosust.2010.04.003.

Rauscher, S. A., E. Coppola, and C. Piani (2010), Resolution effects on regional climate model simulations of seasonal precipitation over Europe, *Clim. Dynam., 35*(4), 685−711.

Raven, J., K. Caldeira, H. Elderfield, O. Hoegh-Guldberg, P. Liss, U. Riebesell, J. Shepherd, C. Turley, and A. Watson (2005), Ocean acidification due to increasing atmospheric carbon dioxide, The Royal Society, Science Policy Section.

Ray, D. K., U. S. Nair, R. M. Welch, Q. Han, J. Zeng, T. Su, T. Kikuchi, and T. J. Lyons (2003), Effects of land use in southwest Australia: 1. Observations of cumulus cloudiness and energy fluxes, *J. Geophys. Res., 108,* 4414, doi:10.1029/2002JD002654.

Raymo, M. E., and K. Nisancioglu (2003), The 41 kyr world: Milankovitch's other unsolved mystery, *Paleoceanography, 18*(1), 1011.

Rayner, P. J. (2010), The current state of carbon-cycle data assimilation, *Curr. Opin. Environ. Sustain., 2,* 289−296, doi:10.1016/j.cosust.2010.05.005.

Rayner, N. A., D. E. Parker, E. B. Horton, C. K. Folland, L. V. Alexander, D. P. Rowell, E. C. Kent, and A. Kaplan (2003), Global analyses of sea surface temperature, sea ice, and night marine air temperature since the late nineteenth century, *J. Geophys. Res., 108*(D14), 4407, doi:10.1029/ 2002JD002670.

Rayner, N. A., P. Brohan, D. E. Parker, C. K. Folland, J. J. Kennedy, M. Vanicek, T. J. Ansell, and S. F. B. Tett (2006), Improved analyses of changes and uncertainties in sea surface temperature measured in situ since the mid-nineteenth century: The HadSST2 dataset, *J. Climate, 19*(3), 446−469, doi:10.1175/JCLI3637.1.

Re, M., and V. Ricardo Barros (2009), Extreme rainfalls in SE South America, *Climatic Change, 96*(1−2), 119−136.

Reagan, M. T., and G. J. Moridis (2007), Oceanic gas hydrate instability and dissociation under climate change scenarios, *Geophys. Res. Lett., 34,* L22709, doi:10.1029/2007GL031671.

Reagan, M. T., and G. J. Moridis (2009), Large-scale simulation of methane hydrate dissociation along the West Spitsbergen margin, *Geophys. Res. Lett., 36,* L23612, doi:10.1029/2009GL041332.

Rees, H. G., and D. N. Collins (2006), Regional differences in response of flow in glacier-fed Himalayan rivers to climatic warming, *Hydrol. Process., 20,* 2157−2169.

Reeve, N., and R. Toumi (1999), Lightning activity as an indicator of climate change, *Q. J. Roy. Meteor. Soc., 125*(555), 893−903.

Reich, P. B., B. A. Hungate, and Y. Luo (2006), Carbon-nitrogen interactions in terrestrial ecosystems in response to rising atmospheric carbon dioxide, *Annu. Rev. Ecol. Evol. Syst., 37,* 611−636, doi:10.1146/annurev.ecolsys.37.091305.110039.

Reichert, B. K., L. Bengtsson, and J. Oerlemans (2002), Recent glacier retreat exceeds internal variability, *J. Climate, 15*(21), 3069−3081.

Reichstein, M., J.-A. Subke, A. C. Angeli, and J. D. Tenhunen (2005), Does the temperature sensitivity of decomposition of soil organic matter depend upon water content, soil horizon, or incubation time?, *Glob. Change Bio., 11,* 1754−1767, doi:10.1111/j.1365-2486.2005.01010.x.

Reichstein, M., P. Ciais, D. Papale, R. Valentini, S. Running, N. Viovy, W. Cramer, A. Granier, J. Ogée, V. Allard, M. Aubinet, C. Bernhofer,

N. Buchmann, A. Carrara, T. Grünwald, M. Heimann, B. Heinesch, A. Knohl, W. Kutsch, D. Loustau, G. Manca, G. Matteucci, F. Miglietta, J. M. Ourcival, K. Pilegaard, J. Pumpanen, S. Rambal, S. Schaphoff, G. Seufert, J.-F. Soussana, M.-J. Sanz, T. Vesala, and M. Zhao (2007), Reduction of ecosystem productivity and respiration during the European summer 2003 climate anomaly: A joint flux tower, remote sensing and modelling analysis, *Glob. Change Biol., 13,* 634–651.

Renfro, J. R. (1989), Evaluating the effects of ozone on plants of Great Smoky Mountains National Park, *Park Sci., 9,* 22–23.

Renne, P. R., C. C. Swisher, III, A. L. Deino, D. B. Karner, T. Owens, and D. J. DePaolo (1998), Intercalibration of standards, absolute ages and uncertainties in 40Ar/39Ar dating, *Chem. Geol., 145,* 117–152.

Renssen, H., and J. Vandenberghe (2003), Investigation of the relationship between permafrost distribution in NW Europe and extensive winter sea-ice cover in the North Atlantic Ocean during the cold phases of the Last Glaciation, *Quaternary Sci. Rev., 22*(2–4), 209–223, doi:10.1016/S0277-3791(02)00190-7.

Renssen, H., V. Brovkin, T. Fichefet, and H. Goosse (2003), Holocene climate instability during the termination of the African Humid Period, *Geophys. Res. Lett., 30*(4), 1184.

Renwick, J. A., J. J. Katzfey, K. C. Nguyen, and J. L. McGregor (1998), Regional model simulations of New Zealand climate, *J. Geophys. Res., 103*(D6), 5973–5982, doi:10.1029/97JD02939.

Rescher, N. (1985), *Pascal's Wager: An Essay on Practical Reasoning in Philosophical Theology,* University of Notre Dame Press, Notre Dame.

Retallack, G. J. (1996), Acid trauma at the Cretaceous-Tertiary boundary in Eastern Montana, *GSA Today, 6,* 1–7.

Retallack, G. J. (2004), End-Cretaceous acid rain as a selective extinction mechanism between birds and dinosaurs, in *Feathered Dragons: Studies on the Transition from Dinosaurs to Birds,* edited by P. J. Currie, et al., pp. 35–64, Indiana University Press, Bloomington and Indianapolis.

Reth, S., W. Graf, M. Reichstein, and J. C. Munch (2009), Sustained stimulation of soil respiration after 10 years of experimental warming, *Environ. Res. Lett., 4,* 024005, doi:10.1088/1748-9326/4/2/024005.

Revel, M., E. Ducassou, F. E. Grousset, S. M. Bernasconi, S. Migeon, S. Revillon, J. Mascle, A. Murat, S. Zaragosi, and D. Bosch (2010), 100,000 years of African monsoon variability recorded in sediments of the Nile margin, *Quaternary Sci. Rev., 29*(11–12), 1342–1362, doi:10.1016/S0277-3791(02)00190-7.

Reynolds, R. W., N. A. Rayner, T. M. Smith, D. C. Stokes, and W. Wang (2002), An improved in situ and satellite SST analysis for climate, *J. Climate, 15*(13), 1609–1625.

Riahi, K., A. Grübler, and N. Nakicenovic (2006), Scenarios of long-term socio-economic and environmental development under climate stabilization, *Technol. Forecast. Soc., 74*(7), 887–935.

Riahi, K., A. Gruebler, and N. Nakicenovic (2007), Scenarios of long-term socio-economic and environmental development under climate stabilization. *Technological Forecasting and Social Change, 74, 7,* 887–935.

Rial, J. A., R. A. Pielke, Sr., M. Benniston, M. Claussen, J. Canadell, P. Cox, J. Held, N. de Noblet-Ducoudré, R. Prinn, J. F. Reynolds, and J. D. Salas (2004), Nonlinearities, feedbacks and critical thresholds within the Earth's climate system, *Climatic Change, 65,* 11–38.

Rice, M. and A. Henderson-Sellers (2012), Future climate: One vital component of trans-disciplinary Earth system science, in *The Future of the World's Climate,* edited by A. Henderson-Sellers and K. McGuffie, pp. 509–529, Elsevier, Amsterdam.

Richard, P. J. H. (1995), Le couvert végétal du Québec-Labrador il y a 6000 ans B.P: Essai, *Geogr. Phys. Quatern., 49,* 117–140.

Richardson, A. J., and E. S. Poloczanska (2008), Under-resourced, under threat, *Science, 320*(5881), 1294–1295, doi:10.1126/science.1156129.

Richardson, I. G., E. W. Cliver, and H. V. Cane (2002), Long-term trends in interplanetary magnetic field strength and solar wind structure during the twentieth century, *J. Geophys. Res., 107,* doi:10.1029/2001JA000507.

Richardson, K., W. Steffen, H. J. Schellnhuber, J. Alcamo, T. Barker, D. Kammen, R. Leemans, D. Liverman, M. Monasinghe, and B. Osman-Elasha (2009), Climate change: Global risks, challenges and decisions, Synthesis Report, University of Copenhagen, Denmark.

Richardson, S. D., and J. M. Reynolds (2000), An overview of glacial hazards in the Himalayas, *Quatern. Int., 65/66,* 31–47.

Richter, A., V. Eyring, J. P. Burrows, H. Bovensmann, A. Lauer, B. Sierk, and P. J. Crutzen (2004), Satellite measurements of NO$_2$ from international shipping emissions, *Geophys. Res. Lett., 31,* L23110, doi:10.1029/2004GL020822.

Richter, A., J. P. Burrows, H. Nüß, C. Granier, and U. Niemeier (2005), Increase in tropospheric nitrogen dioxide over China observed from space, *Nature, 437,* 129–132.

Ridgwell, A., J. S. Singarayer, A. H. Hetherington, and P. J. Valdes (2009), Tackling regional climate change by leaf albedo bio-geoengineering, *Curr. Biol., 19,* 146–150, doi:10.1016/j.cub.2008.12.025.

Ridley, J., J. Gregory, P. Huybrechts, and J. Lowe (2009), Thresholds for irreversible decline of the Greenland ice sheet, *Clim. Dynam.,* 1–9.

Riebesell, U., K. G. Schulz, R. G. J. Bellerby, M. Botros, P. Fritsche, M. Meyerhofer, C. Neill, G. Nondal, A. Oschlies, J. Wohlers, and E. Zollner (2007), Enhanced biological carbon consumption in a high CO$_2$ ocean, *Nature, 450*(7169), 545–548, doi:10.1038/nature06267.

Riebesell, U., A. Körtzinger, and A. Oschlies (2009), Sensitivities of marine carbon fluxes to ocean change, *Proc. Natl. Acad. Sci. USA, 106*(49), 20602–20609.

Rignot, E. (2006), Changes in ice dynamics and mass balance of the Antarctic ice sheet, *Philos. Trans. R. Soc. London Ser. A, 364*(1844), 1637–1655, doi:10.1098/rsta.2006.1793.

Rignot, E., and S. S. Jacobs (2002), Rapid bottom melting widespread near Antarctic Ice Sheet grounding lines, *Science, 296,* 2020–2023.

Rignot, E., J. L. Bamber, M. R. van den Broeke, C. Davis, Y. H. Li, W. J. van de Berg, and E. van Meijgaard (2008), Recent Antarctic ice mass loss from radar interferometry and regional climate modelling, *Nat. Geosci., 1*(2), 106–110, doi:10.1038/ngeo102.

Rignot, E., J. E. Box, E. Burgess, and E. Hanna (2008), Mass balance of the Greenland ice sheet from 1958 to 2007, *Geophys. Res. Lett., 35,* L20502.

Rignot, E., M. Koppes, and I. Velicogna (2010), Rapid submarine melting of the calving faces of West Greenland glaciers, *Nat. Geosci., 3,* 187–191.

Rignot, E., I. Velicogna, M. R. van den Broeke, A. Monaghan, and J. Lenaerts (2011), Acceleration of the contribution of the Greenland and Antarctic Ice Sheets to sea level rise, Geophys. Res. Lett., 38, L05503, doi:10.1029/2011GL046583.

Rigor, I. G., and J. M. Wallace (2004), Variations in the age of Arctic sea-ice and summer sea-ice extent, *Geophys. Res. Lett., 31*, L09401.

Rimbu, N., G. Lohmann, S. J. Lorenz, J. H. Kim, and R. R. Schneider (2004), Holocene climate variability as derived from alkenone sea surface temperature and coupled ocean-atmosphere model experiments, *Clim. Dynam., 23*(2), 215–227, doi:10.1007/s00382-004-0435-8.

Rind, D., P. DeMenocal, G. Russell, S. Sheth, D. Collins, G. Schmidt, and J. Teller (2001), Effects of glacial meltwater in the GISS coupled atmosphere-ocean model: 1. North Atlantic deep water response, *J. Geophys. Res. D, 106*(D21), 27335–27353, doi:10.1029/2000JD000070.

Rinke, A., R. Gerdes, K. Dethloff, T. Kandlbinder, M. Karcher, F. Kauker, S. Frickenhaus, C. Köberle, and W. Hiller (2003), A case study of the anomalous Arctic sea ice conditions during 1990: Insights from coupled and uncoupled regional climate model simulations, *J. Geophys. Res., 108*(D9), 4275, doi:10.1029/2002JD003146.

Rinke, A., W. Maslowski, K. Dethloff, and J. Clement (2006), Influence of sea ice on the atmosphere: A study with an Arctic atmospheric regional climate model, *J. Geophys. Res., 111*, D16103, doi:10.1029/2005JD006957.

Rintoul, S. R. (2007), Rapid freshening of Antarctic Bottom Water formed in the Indian and Pacific Oceans, *Geophys. Res. Lett., 34*(6), L06606.

Risbey, J., M. Pook, P. McIntosh, M. Wheeler, and H. Hendon (2009), On the remote drivers of rainfall variability in Australia, *Mon. Weather Rev., 137*(10), 3233–3253.

Riser, S. C., and K. S. Johnson (2008), Net production of oxygen in the subtropical ocean, *Nature, 451*(7176), 323–325, doi:10.1038/nature06441.

Risk Management Solutions (RMS), Inc. (2003), Central Europe flooding, August 2002, event report, from *http://www.rms.com/Publications/Central Europe Floods Whitepaper_final.pdf.*

Ritter, C., J. Notholt, J. Fisher, and C. Rathke (2005), Direct thermal radiative forcing of tropospheric aerosol in the Arctic measured by ground based infrared spectrometry, *Geophys. Res. Lett., 32*, doi:10.1029/2005GL024331.

Roberts, S. M., T. R. Oke, C. S. B. Grimmond, and J. A. Voogt (2006), Comparison of four methods to estimate urban heat storage, *J. Appl. Meteorol. Clim., 45*, 1766–1781.

Robertson, D. S., M. C. McKenna, O. B. Toon, S. Hope, and J. A. Lillegraven (2004), Survival in the first hours of the Cenozoic, *Geol. Soc. Am. Bull., 116*, 760–768.

Robeson, S. M. (2005), Statistical climatology, in *Encyclopedia of World Climatology*, edited by J. E. Oliver, pp. 687–694, Springer, New York.

Robinson, D. A., K. F. Dewey, and R. R. Heim (1993), Global snow cover monitoring: An update, *Bull. Am. Meteorol. Soc., 74*(9), 1689–1696.

Robock, A., K. Y. Vinnikov, G. Srinivasan, J. K. Entin, S. E. Hollinger, N. A. Speranskaya, S. Liu, and A. Namkhai, (2000), The global soil moisture data bank, *Bull. Am. Meteorol. Soc., 81*(6), 1281–1299.

Robock, A., L. Oman, and G. L. Stenchikov (2008), Regional climate responses to geoengineering with tropical and Arctic SO_2 injections, *J. Geophys. Res.-Atmos., 113*, D16101.

Roche, D., D. Paillard, A. Ganopolski, and G. Hoffmann (2004), Oceanic oxygen-18 at the present day and LGM: Equilibrium simulations with a coupled climate model of intermediate complexity, *Earth Planet. Sci. Lett., 218*, 317–330.

Rockström, J., et al. (2009a), A safe operating space for humanity, *Nature, 461*(24 September), 472–475.

Rockström, J., et al. (2009b), Planetary boundaries: Exploring the safe operating space for humanity, *Ecol. Soc., 14*(2), 32.

Rockström, J., W. Steffen, K. Noone, A. Persson, F. S. Chapin, III, E. Lambin, T. Lenton, M. Scheffer, H. J. Schellnhuber, B. Nykvist, C. A. de Wit, T. Hughes, S. van der Leeuw, H. Rodhe, S. Sorlin, P. K. Snyder, R. Costanza, U. Svedin, M. Falkenmark, L. Karlberg, R. W. Corell, V. J. Fabry, J. Hansen, B. Walker, D. Liverman, K. Richardson, P. Crutzen, and J. Foley (2009a), Planetary boundaries: Exploring the safe operating space for humanity, *Ecol. Soc., 14*(2), 32, *www.ecologyandsociety.org/vol14/iss2/art32.* (Accessed 27/11/10.)

Rockström, J., W. Steffen, K. Noone, A. Persson, F. S. Chapin, III, E. Lambin, T. Lenton, M. Scheffer, H. J. Schellnhuber, B. Nykvist, C. A. de Wit, T. Hughes, S. van der Leeuw, H. Rodhe, S. Sorlin, P. K. Snyder, R. Costanza, U. Svedin, M. Falkenmark, L. Karlberg, R. W. Corell, V. J. Fabry, J. Hansen, B. Walker, D. Liverman, K. Richardson, P. Crutzen, and J. Foley (2009b), Planetary boundaries: Exploring the safe operating space for humanity, *Nature, 461*, 472–475.

Rodgers, K. B., P. Friederichs, and M. Latif (2004), Tropical Pacific Decadal Variability and its relation to decadal modulations of ENSO, *J. Climate, 17*, 3761–3774.

Rodhe, H. (1999), Human impact on the atmospheric sulfur balance, *Tellus, 51*, 110–122.

Rodó, X., E. Baert, and F. A. Comín (1997), Variations in seasonal rainfall in southern Europe during the present century: Relationships with the North Atlantic Oscillation and the El Niño–Southern Oscillation, *Clim. Dynam., 13*(4), 275–284, doi: 10.1007/s003820050165.

Roe, G. (2005), Orographic precipitation, *Ann. Rev. Earth Planet. Sci., 33*, 645–671.

Roeckner, E., G. Baeuml, L. Bonventura, R. Brokopf, M. Esch, M. Giorgetta, S. Hagemann, I. Kirchner, L. Kornblueh, E. Manizini, A. Rhodin, U. Schlese, U. Schulzweida, and A. Tompkins (2003), The atmospheric general circulation model ECHAM5: 1. Model description, *Report 349*, Max Planck Institute for Meteorology, Hamburg, Germany, available from: *http://www.mpimet.mpg.de.*

Roelofs, G.-J., and J. Lelieveld (1995), Distribution and budget of O_3 in the troposphere calculated with a chemistry general circulation model, *J. Geophys. Res., 100*(D10), 20983–20998.

Roemmich, D., J. Gilson, R. Davis, P. Sutton, S. Wijffels, and S. Riser (2007), Decadal spinup of the South Pacific subtropical gyre, *J. Phys. Oceanogr., 37*(2), 162–173.

Rogers, A., E. A. Ainsworth, and A. D. B. Leakey (2009), Will elevated carbon dioxide concentrations amplify the benefits of nitrogen fixation in legumes?, *Plant Physiol., 151*, 1009–1016.

Rohling, E. J., and H. Palike (2005), Centennial-scale climate cooling with a sudden cold event around 8200 years ago, *Nature, 434*(7036), 975–979.

Rohling, E. J., K. Grant, C. Hemleben, M. Siddall, B. A. A. Hoogakker, M. Bolshaw, and M. Kucera (2008), High rates of sea-level rise during the last interglacial period, *Nat. Geosci., 1*, 38–42.

Rohling, E. J., K. Grant, M. Bolshaw, A. P. Roberts, M. Siddall, C. Hemleben, and M. Kucera (2009), Antarctic temperature and global sea level closely coupled over the past five glacial cycles, *Nat. Geosci., 2,* 500–504.

Root, T. L., and S. H. Schneider (1995), Ecology and climate: Research strategies and implications, *Science, 269,* 334–341.

Root, T. L., and S. H. Schneider (2006), Conservation and climate change: The challenges ahead, *Conservation Biology, 20,* 706–708.

Ropelewski, C. F., and M. S. Halpert (1986), North American precipitation and temperature patterns associated with the El Ñino/Southern Oscillation (ENSO), *Mon. Weather Rev., 114,* 2352–2362.

Rosen, R. D. (1999), The global energy cycle, in *Global Energy and Water Cycles,* edited by K. A. Browning, and R. J. Gurney, pp. 1–9, Cambridge University Press.

Rosenfeld, D. (1999), TRMM observed first direct evidence of smoke from forest fires inhibiting rainfall, *Geophys. Res. Lett., 26,* 3105–3108.

Rosenfeld, D. (2000), Suppression of rain and snow by urban and industrial air pollution, *Science 287,* 1793–1796.

Rosenfeld, D., and W. L. Woodley (2000), Deep convective clouds with sustained supercooled liquid water down to $-37.5°C$, *Nature, 405,* 440–442.

Rosenfeld, D., U. Lohmann, G. B. Raga, C. D. O'Dowd, M. Kulmala, S. Fuzzi, A. Reissell, and M. O. Andreae (2008), Flood or drought: How do aerosols affect precipitation?, *Science, 321*(5894), 1309–1313.

Rosenzweig, C. (2007), Assessment of observed changes and responses in natural and managed systems, in *Impacts, Adaptation and Vulnerability, contribution of Working Group II to the Fourth Assessment Report (FAR) of the Intergovernmental Panel on Climate Change (IPCC),* edited by M. L. Parry, et al., pp. 79–131, Cambridge University Press, Cambridge, UK, and New York, NY.

Rosenzweig, C., D. Karoly, M. Vicarelli, P. Neofotis, Q. Wu, G. Casassa, A. Menzel, T. L. Root, N. Estrella, B. Seguin, P. Tryjanowski, C. Liu, S. Rawlins, and A. Imeson (2008), Attributing physical and biological impacts to anthropogenic climate change, *Nature, 453*(7193), 353–357.

Rotach, M. W., R. Vogt, C. Bernhofer, E. Batchvarova, A. Christen, A. Clappier, B. Feddersen, S.-E. Gryning, G. Martucci, H. Mayer, V. Mitev, T. R. Oke, E. Parlow, H. Richner, M. Roth, Y. Roulet, D. Ruffieux, J. A. Salmond, M. Schatzmann, and J. A. Voogt (2005), BUBBLE — an urban boundary layer meteorology project, *Theor. Appl. Climatol., 81,* 231–261.

Roth, M. (2000), Review of atmospheric turbulence over cities, *Q. J. Roy. Meteor. Soc., 126,* 941–990.

Roth, M. (2007), Review of urban climate research in (sub)tropical regions. *Int. J. Climatol., 27,* 1859–1873, doi:10.1002/joc.1591.

Rotstayn, L. D., and U. Lohmann (2002), Tropical rainfall trends and the indirect aerosol effect, *J. Climate, 15,* 2103–2116.

Rotstayn, L. D., and J. E. Penner (2001), Indirect aerosol forcing, quasi-forcing, and climate response, *J. Climate, 14,* 2960–2975.

Rotstayn, L. D., W. Cai, M. R. Dix, G. D. Farquar, Y. Feng, P. Ginoux, M. Herzog, A. Ito, J. E. Penner, M. L. Roderick, and M. Wang (2007), Have Australian rainfall and cloudiness increased due to the remote effects of the Asian anthropogenic aerosols?, *J. Geophys. Res.-Atmos., 112,* D09202, doi:10.1029/2006JD007712.

Rotunno, R. (1983), On the linear-theory of the land and sea breeze, *J. Atmos. Sci., 40*(8), 1999–2009.

Royal Society (2008), Ground-level ozone in the 21st century: Future trends, impacts and policy implications, *Science Policy Report,* 15/08, *http://royalsociety.org/WorkArea/DownloadAsset.aspx?id=5484.*

Royer, D. L. (2006), CO_2-forced climate thresholds during the Phanerozoic, *Geochim. Cosmochim. Acta, 70,* 5665–5675.

Royer, D. L. (2010), Fossil soils constrain ancient climate sensitivity, *Proc. Natl. Acad. Sci. USA, 107,* 517–518.

Royer, D. L., R. A. Berner, and D. J. Beerling (2001), Phanerozoic atmospheric CO_2 change: Evaluating geochemical and paleobiological approaches, *Earth Sci. Rev., 54,* 349–392.

Royer, D. L., R. A. Berner, and J. Park (2007), Climate sensitivity constrained by CO_2 concentrations over the past 420 million years, *Nature, 446,* 530–532.

Rozoff, C., W. Cotton, and J. Adegoke (2003), Simulation of St. Louis, Missouri, land use impacts on thunderstorms, *J. Appl. Meteorol., 42,* 716–738.

Rubin, J. I., A. J. Kean, R. A. Harley, D. B. Millet, and A. H. Goldstein (2006), Temperature dependence of volatile organic compound evaporative emissions from motor vehicles, *J. Geophys. Res., 111,* D03305, doi:10.1029/2005JD006458.

Ruddiman, W. F. (2003), The Anthropogenic greenhouse era began thousands of years ago, *Climatic Change, 61*(3), 261–293.

Ruddiman, W. F. (2005), Cold climate during the closest Stage 11 analog to recent Millennia, *Quarternary Sci. Rev., 24*(10–11), 1111–1121.

Ruddiman, W. F. (2006), What is the timing of orbital-scale monsoon changes?, *Quaternary Sci. Rev., 25*(7–8), 657–658, doi:10.1016/j.quascirev.2006.02.004.

Ruddiman, W. F. (2008), The challenge of modeling interglacial CO_2 and CH_4 trends, *Quaternary Sci. Rev., 27*(5–6), 445–448, doi:10.1016/j.quascirev.2007.11.007.

Ruddiman, W. F., M. Raymo, and A. McYntyre (1986a), Matuyama 41,000-year cycles: North Atlantic Ocean and Northern Hemisphere ice sheets, *Earth Planet. Sci. Lett., 80,* 117–129.

Ruddiman, W. F., N. J. Shackleton, and A. McIntyre (1986b), North Atlantic sea-surface temperatures for the last 1.1 million years, in *North Atlantic Palaeoceanography. Geological Society Special Publication No. 21,* edited by C. P. Summerhayes and N. J. Shackleton, pp. 155–173.

Ruddiman, W. F., S. J. Vavrus, and J. E. Kutzbach (2005), A test of the overdue-glaciation hypothesis, *Quaternary Sci. Rev., 24,* 1–10.

Ruhl, M. (2010), Carbon cycle changes during the Triassic-Jurassic transition, Laboratory of Paleobotany and Palynology, *Contribution Series 28,* Utrecht University, The Netherlands.

Ruhl, M., W. M. Kürschner, and L. Krystyn (2009), Triassic-Jurassic organic carbon isotope stratigraphy of key sections in the western Tethys realm (Austria), *Earth Planet. Sci. Lett., 281,* 169–187.

Ruhl, M., M. H. L. Deenen, H. A. Abels, N. R. Bonis, W. Krijgsman, and W. M. Kürschner (2010), Astronomical constraints on the duration of the early Jurassic Hettangian stage and recovery rates following the end-Triassic mass extinction (St Audrie's Bay/East Quantoxhead, UK), *Earth Planet. Sci. Lett., 295,* 262–276.

Rummukainen, M. (2010), State-of-the-art with regional climate models, *Wiley Interdiscip. Rev.: Clim. Change, 1*(1), 82–96.

Rummukainen, M., J. Raisanen, B. Bringfelt, A. Ullerstig, A. Omstedt, U. Willen, U. Hansson, and C. Jones (2001), A regional climate model for Northern Europe: Model description and results from the downscaling of two GCM control simulations, *Clim. Dynam., 17*(5–6), 339–359.

Running, S. W. (2006), Is global warming causing more, larger wildfires? *Science, 313,* 927–928.

Russell, C. (2010), Stephen Schneider: Climate communicator, remembering an esteemed scientist's contributions to the media over three decades, *Columbia Journal. Rev.*, The Observatory, 20 July, 2010, *http://www.cjr.org/the_observatory/stephen_schneider_climate_comm. php.* (Accessed 23/10/10.)

Russell, M. (2010), The independent climate change e-mails review, 7 July, http://www.cce-review.org/pdf/FINALREPORT.pdf, www.cce-review. org/pdf/FINAL, http://www.cce-review.org/pdf/FINALREPORT.pdf REPORT.pdf. (Accessed 26/7/10.)

Rutter, A. J., K. A. Kershaw, P. C. Robins, and A. J. Morton (1972), A predictive model of rainfall interception in forest, *Agr. Forest Meteorol., 9,* 857–861.

Rütting, T., T. J. Clough, C. Müller, M. Lieffering, and P. C. D. Newton (2010), Ten years of elevated atmospheric carbon dioxide alters soil nitrogen transformations in a sheep-grazed pasture, *Glob. Change Biol., 16,* 2530–2542.

Rykaczewski, R. R., and J. P. Dunne (2011), A measured look at ocean chlorophyll trends, *Nature, 472*(7342), E5–E6, doi:10.1038/nature09952.

Sabine C. L., R. A. Feely, N. Gruber, R. M. Key, K. Lee, J. L. Bullister, R. Wanninkhof, C. S. Wong, D. W. R. Wallace, B. Tilbrook, F. J. Millero, T-H. Peng, A. Kozyr, T. Ono, and A. F. Rios (2004), The oceanic sink for anthropogenic CO_2, *Science, 305*(5682), 367–371, doi:10.1126/science.1097403.

Sacks, W. J., B. I. Cook, N. Buenning, S. Levis, and J. H. Helkowski (2009), Effects of global irrigation on the near-surface climate, *Clim. Dynam., 33,* 159–175, doi:10.1007/s00382-008-0445-z.

Saenko, O. A., A. Sen Gupta, and P. Spence (2011), On challenges in predicting bottom water transport in the Southern Ocean, *J. Climate* (submitted).

Saenko, O. A., J. Fyfe, and M. England (2005), On the response of the oceanic wind-driven circulation to atmospheric CO_2 increase, *Clim. Dynam., 25*(4), 415–426, doi:10.1007/s00382-005-0032-5.

Saenko, O. A., X.-Y. Yang, M. H. England, and W. G. Lee (2011), Subduction and transport in the Indian and Pacific Oceans in a $2 \times CO_2$ climate, *J. Climate,* 24, 1821–1838.

Sagan, C. (1961), The Planet Venus, *Science, 133,* 849–858.

Saikawa, E., V. Naik, L. W. Horowitz, J. Liu, and D. L. Mauzerall (2009), Present and potential future contributions of sulfate, black and organic carbon aerosols from China to global air quality, premature mortality and radiative forcing, *Atmos. Environ., 43*(17), 2814–2822.

Sailor, D. J. (2011), A review of methods for estimating anthropogenic heat and moisture emissions in the urban environment, *Int. J. Climatol., 31,* 189–199, doi:10.1002/joc.2106.

Saji, N., B. Goswami, P. Vinayachandran, and T. Yamagata (1999), A dipole mode in the tropical Indian Ocean, *Nature, 401*(6751), 360–363.

Salati, E., and P. B. Vose (1984), Amazon Basin: A system in equilibrium, *Science, 225*(4658), 129–138, doi:10.1126/science.225.4658.129.

Salati, E., A. D. Olio, E. Matsui, and J. R. Gat (1979), Recycling of water in the Amazon Basin: An isotopic study, *Water Resour. Res., 15*(5), 1250–1258.

Salazar, L. F., C. A. Nobre, and M. D. Oyama (2007), Climate change consequences on the biome distribution in tropical South America, *Geophys. Res. Lett., 34,* L09708, doi:10.1029/2007GL029695.

Salby, M., and P. Callagan (2000), Connection between the solar cycle and the QBO: The missing link, *J. Climate, 13,* 328–338.

Salby, M., and P. Callaghan (2004), Evidence of the solar cycle in the general circulation of the stratosphere, *J. Climate, 17,* 34–46.

Saleska, S. R., K. Didan, A. R. Huete, and H. R. da Rocha (2007), Amazon forests green-up during 2005 drought, *Science, 318,* 612.

Sallée, J. B., K. Speer, and R. Morrow (2008), Response of the Antarctic circumpolar current to atmospheric variability, *J. Climate, 21*(12), 3020–3039, doi:10.1175/2007JCLI1702.1.

Salon, S., G. Cossarini, S. Libralato, X. Gao, C. Solidoro, and F. Giorgi (2008), Downscaling experiment for the Venice Lagoon, 1. Validation of the present-day precipitation climatology, *Clim. Res., 38*(1), 31–41.

Saltzman, B. (2002), *Dynamical Paleoclimatology: Generalized Theory of Global Climate Change,* Academic Press.

Saltzman, B., K. A. Maasch, and M. Y. Verbitsky (1993), Possible effects of anthropogenically-increased CO_2 on the dynamics of climate: Implications for ice age cycles, *Geophys. Res. Lett., 20*(11), 1051–1054.

Salvadori, G., and C. De Michele, (2010), Multivariate multiparameter extreme value models and return periods: A copula approach, *Water Resour. Res.,* W10501.

Salvadori, G., and C. De Michele, (2011), Estimating strategies for multiparameter multivariate extreme value copulas, *Hydrolog. Earth Sys. Sci., 15*(1), 141–150.

Sanchez Goñi, M. F., and S. P. Harrison (2010), Millennial-scale climate variability and vegetation changes during the Last Glacial: Concepts and terminology, *Quaternary Sci. Rev., 29*(21–22), 2823–2827, doi:10.1016/j.quascirev.2009.11.014.

Sanderson, M. G., W. J. Collins, D. L. Hemming, and R. A. Betts (2007), Stomatal conductance changes due to increasing carbon dioxide levels: Projected impact on surface ozone levels, *Tellus B, 59*(3), 404, doi:10.1111/j.1600-0889.2007.00277.x.

Santer, B. (2010), Dr. Stephen Schneider, climate warrior, *The Huffington Post,* 19 July, 2010, *http://www.huffingtonpost.com/peter-h-gleick/dr-stephen-schneider-clim_b_651517.html.* (Accessed 24/10/10.)

Santer, B., and S. Solomon (2010), Stephen H. Schneider (1945–2010), *Eos Trans. AGU, 91*(41), 12 October, 2010.

Santer, B. D., K. E. Taylor, T. M. L. Wigley, T. C. Johns, P. D. Jones, D. J. Karoly, J. F. B. Mitchell, A. H. Oort, J. E. Penner, V. Ramaswamy, M. D. Schwarzkopf, R. J. Stouffer, and S. Tett (1996). A search for human influences on the thermal structure of the atmosphere, *Nature, 382,* 39–46.

Santer, B. D., F. Wehner, T. M. L. Wigley, R. Sausen, G. A. Meehl, K. E. Taylor, C. Ammann, J. Arblaster, W. M. Washington, J. S. Boyle, and W. Brüggemann (2003a), Contributions of anthropogenic and natural forcing to recent tropopause height changes, *Science, 301,* 479–483.

Santer, B. D., R. Sausen, T. M. L. Wigley, J. S. Boyle, K. AchutaRao, C. Doutriaux, J. E. Hansen, G. A. Meehl, E. Roeckner, R. Ruedy, G. Schmidt, and K. E. Taylor (2003b), Behavior of tropopause height and atmospheric temperature in models, reanalyses, and observations: Decadal changes, ACL 1-1, *J. Geophys. Res., 108*(D1), 4002, doi:10.1029/2002JD002258.

Santer, B. D., T. M. L. Wigley, P. J. Gleckler, C. Bonfils, M. F. Wehner, K. AchutaRao, T. P. Barnett, J. S. Boyle, W. Brüggemann, M. Fiorino, N. Gillett, J. E. Hansen, P. D. Jones, S. A. Klein, G. A. Meehl, S. C. B. Raper, R. W. Reynolds, K. E. Taylor, and W. M. Washington (2006), Forced and unforced ocean temperature changes in Atlantic and Pacific tropical cyclogenesis regions, *Proc. Natl. Acad. Sci. USA, 103,* 13905–13910, doi:10.1073/pnas.0602861103.

SANZ (2009), *Strong Sustainability for New Zealand: Principles and Scenarios,* Nakedize Books, Wellington, *http://www.nakedize.com/strong-sustainability.cfm.* (Accessed 9/11/10.)

Sarafanov, A. (2009), On the effect of the North Atlantic Oscillation on temperature and salinity of the subpolar North Atlantic intermediate and deep waters, *ICES J. Mar. Sci.: J. Conseil, 66*(7), 1448−1454, doi:10.1093/icesjms/fsp094.

Saravanan, R., and J. C. McWilliams (1997), Stochasticity and spatial resonance in inter-decadal climate fluctuations, *J. Climate, 10,* 2299−2320.

Saravanan, R., and J. C. McWilliams (1998), Advective ocean−atmosphere interaction: An analytical stochastic model with implications for decadal variability, *J. Climate, 11,* 165−188.

Sarmiento, J. L., and C. Le Quéré (1996), Oceanic carbon dioxide uptake in a model of century-scale global warming, *Science, 274*(5291), 1346.

Sarmiento, J. L., T. M. C. Hughes, R. J. Stouffer, and S. Manabe (1998), Simulated response of the ocean carbon cycle to anthropogenic climate warming, *Nature, 393*(6682), 245−249.

Sarmiento, J. L., R. Slater, R. Barber, L. Bopp, S. C. Doney, A. C. Hirst, J. Kleypas, R. Matear, U. Mikolajewicz, P. Monfray, et al. (2004), Response of ocean ecosystems to climate warming, *Global Biogeochem. Cycles, 18*(3), GB3003.

Satheesh, S. K., and V. Ramanathan (2000), Large differences in tropical aerosol forcing at the top of the atmosphere and Earth's surface, *Nature, 405,* 60−63.

Satterthwaite, D. (2009), The implications of population growth and urbanization for climate change, *Environ. Urban., 21*(2), 545−567, doi:10.1177/0956247809344361.

Sausen, R., and B. D. Santer (2003), Use of changes in tropopause height to detect human influences on climate, *Meteorologische Zeitschrift, 12,* 131−136.

Sausen, R., I. S. A. Isaksen, V. Grewe, D. Hauglustaine, D. S. Lee, G. Myhre, M. O. Kohler, G. Pitari, U. Schumann, F. Stordal, and C. Zerefos (2005), Aviation radiative forcing in 2000: An update on IPCC (1999), *Meteorol. Z., 14,* 555−561, doi:10.1127/2941-2948/2005/0049.

Sausen, R., I. S. A. Isaksen, V. Grewe, D. Hauglestaine, D. S. Lee, G. Myhre, M. O. Kohler, G. Pitari, U. Schumann, F. Stordal, and C. Zerefos (2005), Aviation radiative forcing in 2000: An update on IPCC (1999), *Meteorologische Zeitschrift, 14,* 555−561.

Scafetta, N., and B. J. West (2005), Estimated solar contribution to the global surface warming using the ACRIM TSI satellite composite, *Geophys. Res. Lett., 32,* doi:10.1029/2005GL023849.

Scaife, A. A., J. R. Knight, G. K. Vallis, and C. K. Folland (2005), A stratospheric influence on the winter NAO and North Atlantic surface climate, *Geophys. Res. Lett., 32,* L18715.

Scaife, A. A., C. K. Folland, L. V. Alexander, A. Moberg, and J. R. Knight (2008), European climate extremes and the North Atlantic Oscillation, *J. Climate, 21,* 72−83.

Schaefer, H., M. J. Whiticar, E. J. Brook, V. V. Petrenko, D. F. Ferretti, and J. P. Severinghaus (2006), Ice record of d^{13}C for atmospheric CH$_4$ across the Younger Dryas-Preboreal transition, *Science, 313*(5790), 1109−1112, doi:10.1126/science.1126562.

Schaefer, K., T. Zhang, L. Bruhwiler, and A. P. Barrett (2011), Amount and timing of permafrost carbon release in response to climate warming, *Tellus B, 63,* 1−16.

Schafer, R., P. T. May, W. L. Ecklund, K. S. Gage, P. E. Johnson, T. D. Keenan, and K. McGuffie (2001), Island boundary layer development and circulation during the Maritime Continent Thunderstorm Experiment, *J. Atmos. Sci., 58,* 2163−2179.

Schaller, M. F., J. D. Wright, and D. V. Kent (2011) Atmospheric pCO$_2$ perturbations associated with the Central Atlantic Magmatic Province, *Science, 331,* 1404−1409.

Schaltegger, U., J. Guex, A. Bartolini, B. Schoene, and M. Ovtcharova (2008), Precise U-Pb age constraints for end-Triassic mass extinction, its correlation to volcanism and Hettangian post-extinction recovery, *Earth Planet. Sci. Lett., 267,* 266−275.

Schär, C., and G. Jendritzky (2004), Climate change: Hot news from summer 2003, *Nature, 432,* 559−560.

Schär, C., P. L. Vidale, and D. Luthi (2004), The role of increasing temperature variability in European summer heatwaves, *Nature, 427,* 332−336.

Scheffer, M., J. Bacompte, W. A. Brock, V. Brovkin, S. R. Carpenter, V. Dakos, H. Held, E. H. van Nes, M. Rietkerk, and G. Sugihara (2009), Early warning signals for critical transitions, *Nature, 461,* 53−59.

Schellnhuber, H. J. (1999), Earth system analysis and the second Copernican revolution, *Nature, 402*(Supplement), 19−23.

Schellnhuber, H. J. (2002), Coping with Earth system complexity and irregularity, in Challenges of a Changing Earth: Proceedings of the Global Change Open Science Conference, edited by W. Steffen, J. Jäger, D. Carson, and C. Brodshaw, Amsterdam, The Netherlands, 10−13 July 2001, pp. 151−156, Springer-Verlag, NY.

Schellnhuber, H. J. (2009), Tipping elements in the Earth system, *Proc. Natl. Acad. Sci. USA, 106,* 20561−20563.

Schellnhuber, H. J. (2010), Tragic triumph, *Climatic Change, 100,* 229−238,

Schellnhuber, H. J. (2011), 'We Are Looting the Past and Future to Feed the Present', interviewed by A. Elger, and C. Schwägerl, Spiegel Online International, *http://www.spiegel.de/international/germany/0,1518,752474,00.html.* (Accessed 23/3/11.)

Schellnhuber, H. J., P. J. Crutzen, W. C. Clark, and J. Hunt (2005), Earth system analysis for sustainability, *Environment, 47*(8), 11−25.

Schellnhuber, H. J., W. Cramer, N. Nakicenovic, T. Wigley, and G. Yohe (2006), *Avoiding Dangerous Climate Change,* Cambridge University Press, Cambridge, UK.

Schellnhuber, H. J., D. Messner, C. Leggewie, R. Leinfelder, N. Nakicenovic, S. Rahmstorf, S. Schlake, J. Schmid, and R. Schubert (2009), Solving the climate dilemma: The budget approach, report, WBGU, Berlin.

Scherer, M., H. Vomel, S. Fueglistaler, S. J. Oltmans, and J. Staehelin (2008), Trends and variability of midlatitude stratospheric water vapour deduced from the reevaluated Boulder balloon series and HALOE, *Atmos. Chem. Phys., 8*(5), 1391−1402.

Schewe, J., A. Levermann, and M. Meinshausen (2010), Climate change under a scenario near 1.5°C of global warming: Monsoon intensification, ocean warming and steric sea level rise, *Earth Syst. Dynam. Discuss., 1*(1), 297−324.

Schilt, A., M. Baumgartner, T. Blunier, J. Schwander, R. Spahni, H. Fischer, and T. F. Stocker (2010), Glacial−interglacial and millennial-scale variations in the atmospheric nitrous oxide concentration during the last 800,000 years. *Quaternary Sci. Rev., 29*(1−2), 182−192.

Schimel, J. P., and J. Bennett (2004), Nitrogen mineralization: Challenges of a changing paradigm, *Ecology, 85,* 591−602.

Schlesinger, M. E. (1985), Appendix A: Analysis of results from energy balance and radiative convective models, in *Projecting the Climatic Effects of Increasing Carbon Dioxide,* edited by M. C. MacCracken and F. M. Luther, pp. 281−319, US Department of Energy, DOE/ER-0237.

Schlische, R. W., M. O. Withjack, and P. E. Olsen (2002), Relative timing of CAMP, rifting, continental breakup, and inversion: Tectonic significance, in The Central Atlantic Magmatic province: Insights from fragments of Pangea, *American Geophysical Union Monograph 136,* edited by W. E. Hames, et al., pp. 33−59.

Schmid, H. P., and T. R. Oke (1992), Scaling North American urban climates by lines, lanes and rows, in *Geographical Snapshots of North America*, edited by D. G. Janelle, pp. 395—399, The Guilford Press, New York.

Schmidli, J., C. Goodess, C. Frei, M. Haylock, Y. Hundecha, J. Ribalaygua, and T. Schmith (2007), Statistical and dynamical downscaling of precipitation: An evaluation and comparison of scenarios for the European Alps, *J. Geophys. Res. — Atmos., 112*(D4), doi:10.1029/2005JD007026.

Schmidt, G. (2011), Steve Schneider's first letter to the editor, *RealClimate*, 25 April 2011, *http://www.realclimate.org/index.php/archives/2011/04/steve-schneiders-first-letter-to-the-editor/#more-7374*. (Accessed 29/4/11.)

Schmidt, G., and E. Moyer (2008), A new kind of scientist, *Nat. Rep. Clim. Change, 2*, 102—103, doi:10.1038/climate.2008.76.

Schmidt, H., and H. von Storch (1993), German Bight storms analysed, *Nature, 365*, 791.

Schmith, T. (2008), Stationarity of regression relationships: Application to empirical downscaling, *J. Climate, 21*(17), 4529—4537, doi:10.1175/2008JCLI1910.1.

Schneider, S. H. (1989), *Global Warming: Are We Entering The Greenhouse Century?* Sierra Club Books, San Francisco, USA.

Schneider, S. H. (2002), Global warming: Neglecting the complexities, *Sci. Am., 286*, 62—65.

Schneider, S. H., and P. J. Boston (Eds.) (1991), *Scientist on Gaia*, MIT Press, Cambridge, USA.

Schneider von Deimling, T., H. Held, A. Ganopolski, and S. Rahmstorf (2006), Climate sensitivity estimated from ensemble simulations of glacial climate, *Clim. Dynam., 27*, 149—163.

Schneider, B., and R. Schneider (2010), Global warmth with little extra CO_2, *Nat. Geosci., 3*, 6—7.

Schneider, N., and B. D. Cornuelle (2005), The forcing of the Pacific Decadal Oscillation, *J. Climate, 18*, 4355—4373, doi:10.1175/JCLI3527.1.

Schneider, N., A. J. Miller, and D. W. Pierce (2002), Anatomy of North Pacific decadal variability, *J. Climate, 15*, 586—605.

Schneider, S. H. (1972), Cloudiness as a global climatic feedback mechanism: The effects on the radiation balance and surface temperature of variations in cloudiness, *J. Atmos. Sci., 29*, 1413—1422.

Schneider, S. H. (1977), Editorial, *Climatic Change, 1*, 3—4.

Schneider, S. H. (1990). *Global Warming: Are We Entering the Greenhouse Century?*, Vintage Books, New York.

Schneider, S. H. (1997), Speaking during a session on science and democracy at the annual meeting of the American Association for the Advancement of Science, 18 February, 1997, from Stanford university news Service, http://news.stanford.edu/pr/97/970210schneider.html. (Accessed 3 May 2011).

Schneider, S. H. (1997a). *Laboratory Earth: The Planetary Gamble We Can't Afford to Lose*. Cambridge University Press, Cambridge, UK.

Schneider, S. H. (1997b). Integrated assessment modelling of global climate change: Transparent rational tool for policy making or opaque screen hiding value-laden assumptions?, *Environ. Model. Assess., 2*(4), 229—249, doi:10.1023/A:1019090117643.

Schneider, S. H. (2004), Abrupt non-linear climate change, irreversibility and surprise, *Global Environ. Change A, 14*(3), 245—258, doi:10.1016/j.gloenvcha.2004.04.008.

Schneider, S. H. (2006), *Climate Change: Risks and Opportunities*, Adelaide Thinker in Residence, Department of the Premier and Cabinet, Adelaide, Australia.

Schneider, S. H. (2009), *Science as a Contact Sport: Inside the Battle to Save the Earth's Climate*, National Geographic Books, Washington, DC.

Schneider, S. H. (2010a), Uncertainty/limits to adaptation/adapting to +4°C, Plenary Presentation, International Climate Change Adaptation Conference, Climate Adaptation Futures: Preparing for the unavoidable impacts of climate change, National Climate Change Adaptation Research Facility and the CSIRO Climate Adaptation Flagship, 29 June—1 July, Gold Coast.

Schneider, S. H. (2010b), Schneider web page, *http://stephenschneider.stanford.edu/Mediarology/MediarologyFrameset.html*. (Accessed 21/10/10.)

Schneider, S. H. (2010c), *Climatic Change* 100th volume: be careful what you wish for!-what if you get it?, *Climatic Change, 100*, 1—5, DOI: 10.1007/s10584-010-9865-y.

Schneider, S. H., and P. J. Boston (Eds.) (1991), *Scientists on Gaia*, MIT Press, Cambridge, MA.

Schneider, S. H., and J. Lane (2005), *The Patient from Hell: How I Worked with My Doctors to Get the Best of Modern Medicine and How You Can Too*, Da Capo Lifelong Books, Cambridge, MA.

Schneider, S. H., and R. Londer (1984), *The Coevolution of Climate and Life*, Sierra Club Books, San Francisco.

Schneider, S. H., and M. D. Mastrandrea (2010), Risk, uncertainty and assessing dangerous climate change, Chapter 15, in *Climate Change Science and Policy*, edited by S. H. Schneider, et al., Island Press, Washington, DC.

Schneider, S. H., and L. E. Mesirow (1976), *The Genesis Strategy: Climate and Survival*, Plenum Press, New York, USA.

Schneider, S. H., and M. Oppenheimer (2009), Inaugural editorial, *Climatic Change, 97*(1—2), 1—2.

Schneider, S. H., and T. L. Root (Eds.) (2002), *Wildlife Responses to Climate Change: North American Case Studies*, Island Press, Washington, DC.

Schneider Interviews — both from a single PBS source — no date given — might have been a Nova-Frontline TV program entitled "What's Up with the Weather" in 2000 http://www.pbs.org/wgbh/warming/debate/schneider.html

Schoeberl, M. R., J. R. Ziemke, B. Bojkov, N. Livesey, B. Duncan, S. Strahan, L. Froidevaux, S. Kulawik, P. K. Bhartia, S. Chandra, P. F. Levelt, J. C. Witte, A. M. Thompson, E. Cuevas, A. Redondas, D. W. Tarasick, J. Davies, G. Bodeker, G. Hansen, B. J. Johnson, S. J. Oltmans, H. Vömel, M. Allaart, H. Kelder, M. Newchurch, S. Godin-Beekmann, G. Ancellet, H. Claude, S. B. Andersen, E. Kyrö, M. Parrondos, M. Yela, G. Zablocki, D. Moore, H. Dier, P. von der Gathen, P. Viatte, R. Stübi, B. Calpini, P. Skrivankova, V. Dorokhov, H. de Backer, F. J. Schmidlin, G. Coetzee, M. Fujiwara, V. Thouret, F. Posny, G. Morris, J. Merrill, C. P. Leong, G. Koenig-Langlo, and E. Joseph (2007), A trajectory-based estimate of the tropospheric ozone column using the residual method, *J. Geophys. Res.-Atmos., 112*, D24S49, doi:10.1029/2007JD008773.

Schoene, B., J. Guex, A. Bartolini, U. Schaltegger, and T. J. Blackburn (2010), Correlating the end-Triassic mass extinction and flood basalt volcanism at the 100 ka level, *Geology, 38*, 387—390.

Scholze, M., W. Knorr, N. W. Arnell, and I. C. Prentice (2006), A climate-change risk analysis for world ecosystems, *Proc. Natl. Acad. Sci. USA, 103*(35), 13116—13120.

Schoof, C. (2007), Ice sheet grounding line dynamics: Steady states, stability, and hysteresis, *J. Geophys. Res., 112*, F03S28, doi:10.1029/2006JF000664.

Schoof, J., D. Shin, S. Cocke, T. LaRow, Y. Lim, and J. O'Brien (2009), Dynamically and statistically downscaled seasonal temperature and precipitation hindcast ensembles for the Southeastern USA, *Int. J. Climatol., 29*(2), 243–257, doi:10.1002/joc.1717.

Schouten, S., E. C. Hopmans, E. Schefus, and S. Damste (2002), Distributional variation in marine crenarchaeotal membrane lipids: A new tool for reconstructing ancient seawater temperatures?, *Earth Planet. Sci. Lett., 204*, 265–274.

Schubert, S. D., M. J. Suarez, P. J. Pegion, R. D. Koster, and J. T. Bacmeister (2004), Causes of long-term drought in the United States Great Plains, *J. Climate, 17*, 485–503.

Schulte, P., et al. (2010), The Chicxulub asteroid impact and mass extinction at the Cretaceous-Paleogene boundary, *Science, 327*, 1214–1218.

Schultz, D. M. (2010), Rejection rates for journals publishing in the atmospheric sciences, *Bull. Am. Meteorol. Soc.*, doi:10.1175/2009BAMS2908.1.

Schultz, M., D. J. Jacob, Y. Wang, J. A. Logan, E. L. Atlas, D. R. Blake, N. J. Blake, J. D. Bradshaw, E. V. Browell, M. A. Fenn, F. Flocke, G. L. Gregory, B. G. Heikes, G. W. Sachse, S. T. Sandholm, R. E. Shetter, H. B. Singh, and R. W. Talbot (1999), On the origin of tropospheric ozone and NO_x over the tropical South Pacific, *J. Geophys. Res., 104*, 5829–5843.

Schulz, M., C. Textor, S. Kinne, Y. Balkanski, S. Bauer, T. Berntsen, T. Berglen, O. Boucher, F. Dentener, S. Guibert, I. S. A. Isaksen, T. Iversen, D. Koch, A. Kirkevåg, X. Liu, V. Montanaro, G. Myhre, J. E. Penner, G. Pitari, S. Reddy, Ø. Seland, P. Stier, and T. Takemura (2006), Radiative forcing by aerosols as derived from the AeroCom present-day and pre-industrial simulations, *Atmos. Chem. Phys., 6*, 5225–5246.

Schumann, U., and H. Huntrieser (2007), The global lightning-induced nitrogen oxides source, *Atmos. Chem. Phys. Discuss., 7*, 2623–2818.

Schurgers, G., U. Mikolajewicz, M. Gröger, E. Maier-Reimer, M. Vizcaíno, and A. Winguth (2007), The effect of land surface changes on Eemian climate, *Clim. Dynam., 29*, 357–373.

Schuster, U., and A. J. Watson (2007), A variable and decreasing sink for atmospheric CO_2 in the North and Atlantic, *J. Geophys. Res., 112*, C11006, doi:10.1029/2006JC003941.

Schwalm, C. R., C. A. Williams, K. Schaefer, A. Arneth, D. Bonal, N. Buchmann, J. Chen, B. E. Law, A. Lindroth, S. Luyssaert, M. Reichstein, A. D. Richardson (2010), Assimilation exceeds respiration sensitivity to drought: A FLUXNET synthesis, *Glob. Change Bio., 16*, 657–670, doi:10.1111/j.1365-2486.2009.01991.x.

Schwartz, M., and J. Hanes (2010), Continental-scale phenology: Warming and chilling, *Int. J. Climatol., 30*(11), 1595–1598, doi:10.1002/joc.2014.

Schwartz, S. E., R. J. Charlson, R. A. Kahn, J. A. Ogren, and H. Rodhe (2010), Why hasn't Earth warmed as much as expected?, *J. Climate, 23*, 2453–2464, doi:10.1175/2009JCLI3461.1.

Scott, R., R. D. Koster, D. Entekhabi, and M. Suarez (1995), Effect of a canopy interception reservoir on hydrological persistence in a general circulation model, *J. Climate, 8*, 1917–1922.

Screen, J. A., and I. Simmonds (2010), Increasing fall-winter energy loss from the Arctic Ocean and its role in Arctic temperature amplification, *Geophys. Res. Lett., 37*(16), L16707.

Screen, J. A., N. P. Gillett, D. P. Stevens, G. J. Marshall, and H. K. Roscoe (2009), The role of eddies in the Southern Ocean temperature response to the Southern Annular Mode, *J. Climate, 22*(3), 806–818.

Seager, R. (2007), The turn-of-the-century drought across North America: Global context, dynamics and past analogues, *J. Climate, 20*, 5527–5552.

Seager, R., Y. Kushnir, C. Herweijer, N. Naik, and J. Velez (2005), Modeling of tropical forcing of persistent droughts and pluvials over western North America: 1856–2000, *J. Climate, 18*(19), 4065–4088.

Seager, R., et al. (2007), Model projections of an imminent transition to a more arid climate in southwestern North America, *Science, 316*, 1181–1184.

Secord, R., P. D. Gingerich, K. C. Lohmann, and K. G. Maclead (2010), Continental warming preceding the Palaeocene-Eocene thermal maximum, *Nature, 467*, 955–958.

Seibel, B. A. (2011), Critical oxygen levels and metabolic suppression in oceanic oxygen minimum zones, *J. Exp. Biol., 214*(2), 326–336.

Seibert, P. (1990), South foehn studies since the ALPEX experiment, *Meteorol. Atmos. Phys., 43*(1–4), 91–103.

Seiler, T. J., D. P. Rasse, J. Li, P. Dijkstra, H. P. Anderson, D. P. Johnson, T. L. Powell, B. A. Hungate, C. R. Hinkle, and B. G. Drake (2009), Disturbance, rainfall, and contrasting species responses mediated aboveground biomass response to 11 years of CO_2 enrichment in a Florida scrub-oak ecosystem, *Glob. Change Biol., 15*, 356–367.

Self, S., M. Widdowson, T. Thordarson, and A. E. Jay (2006), Volatile fluxes during flood basalt eruptions and potential effects on the global environment: A Deccan perspective, *Earth Planet. Sci. Lett., 248*, 518–532.

Sellers, P. J. (1992), Biogeophysical models of land surface processes, in *Climate System Modeling*, edited by K. E. Trenberth, pp. 451–490, Cambridge University Press, UK.

Sellers, P. J., D. A. Randall, G. J. Collatz, J. A. Berry, C. B. Field, D. A. Dazlich, C. Zhang, G. D. Collelo, and L. Bounoua (1996), A revised land surface parameterization (SiB2) for atmospheric GCMs, 1. Model formulation, *J. Climate, 9*, 676–705.

Sellers, P. J., L. Bounoua, G. J. Collatz, D. A. Randall, D. A. Dazlich, S. O. Los, J. A. Berry, I. Fung, C. J. Tucker, C. B. Field, and T. G. Jensen (1996), Comparison of radiative and physiological effects of doubled atmospheric CO_2 on climate, *Science, 271*, 1402–1406.

Sellers, P. J., Y. Mintz, Y. C. Sud, and A. Dalcher (1986), A Simple Biosphere model (SiB) for use within general circulation models, *J. Atmos. Sci., 43*, 505–531.

Sellers, W. D. (1969), A global climatic model based on the energy balance of the Earth-atmosphere system, *J. Appl. Meteorol., 8*, 392–400.

Semenov, M., and E. Barrow (1997), Use of a stochastic weather generator in the development of climate change scenarios, *Climatic Change, 35*(4), 397–414.

Semenov, V. A., M. Latif, J. Jungclaus, and W. Park (2008), Is the observed NAO variability during the instrumental record unusual?, *Geophys. Res. Lett., 35*, L11701, doi:10.1029/2008GL033273.

Semenov, V. A., M. Latif, D. Dommenget, N. S. Keenlyside, A. Strehz, T. Martin, and W. Park (2010), The impact of North Atlantic-Arctic multidecadal variability on Northern Hemisphere surface air temperature, *J. Climate, 23*, 5668–5677, doi:10.1175/2010JCLI3347.1.

Sen Gupta, A., and M. England (2006), Coupled ocean–atmosphere–ice response to variations in the Southern Annular Mode, *J. Climate, 19*, 4457–4486.

Sen Gupta, A., and B. McNeil (2012), Variability and change in the ocean, in the future, in *The Future of the World's Climate*, edited by

A. Henderson-Sellers and K. McGuffie, pp. 141−165, Elsevier, Amsterdam.

Sen Gupta, A., A. Santoso, A. S. Taschetto, C. C. Ummenhofer, J. Trevena, and M. England (2009), Projected changes to the Southern Hemisphere ocean and sea ice in the IPCC AR4 climate models, *J. Climate, 22*(11), 3047, doi:10.1175/2008JCLI2827.1.

Seneviratne, S. I., D. Lüthi, M. Litschi, and C. Schär (2006), Land-atmosphere coupling and climate change in Europe, *Nature, 443*, 205−209.

Seneviratne, S. I., T. Corti, E. L. Davin, M. Hirschi, E. B. Jaeger, I. Lehner, B. Orlowsky, and A. J. Teuling (2010), Investigating soil moisture-climate interactions in a changing climate: A review, *Earth-Sci. Rev., 99*, 125−161.

Sensoy, S., T. Peterson, L. V. Alexander, and X. Zhang (2007), Enhancing Middle East climate change monitoring and indexes, *Bull. Am. Meteorol. Soc., 88*(8), 1249−1254.

Sepkoski, J. J., Jr. (1986), Phanerozoic overview of mass extinction, in *Patterns and Processes in the History of Life*, edited by D. M. Raup, and D. Jablonski, pp. 277−295, Springer-Verlag, Berlin.

Sepkoski, J. J., Jr. (1993), Ten years in the library: New data confirm paleontological patterns, *Paleobiology, 19*, 43−51.

Sepkoski, J. J., Jr. (2002), A compendium of fossil marine animal genera, in *Bull. Am. Paleontol., 363*, edited by D. Jablonski, and M. Foote.

Serreze, M. C., J. E. Walsh, and F. S. Chapin (2000), Observational evidence of recent change in the northern high-latitude environment, *Climatic Change, 46*(1−2), 159−207.

Shackleton, N. J. (1967), Oxygen isotope analyses and Pleistocene temperatures re-assessed, *Nature, 215*, 15−17.

Shafer, S. L., P. J. Bartlein, and C. Whitlock (2005), Understanding the spatial heterogeneity of global environmental change in mountain regions, in *Global Change and Mountain Regions*, edited by U. Huber, et al., pp. 21−30, Springer.

Shakhova, N., I. Semiletov, I. Leifer, A. Salyuk, P. Rekant, and D. Kosmach (2010), Geochemical and geophysical evidence of methane release over the East Siberian Arctic Shelf, *J. Geophys. Res., 115*, C08007, doi:10.1029/2009JC005602.

Shakhova, N., I. Semiletov, A. Salyuk, V. Yusupov, D. Mosmach, and Ö. Gustafsson (2010), Extensive methane venting to the atmosphere from sediments of the East Siberian Arctic shelf, *Science, 327*, 1246−1250.

Shakun, J. D., and A. E. Carlson (2010), A global perspective on Last Glacial Maximum to Holocene climate change, *Quaternary Sci. Rev., 29*(15−16), 1801−1816, doi:10.1016/j.quascirev.2010.03.016.

Shanahan, T. M., J. T. Overpeck, K. J. Anchukaitis, J. W. Beck, J. E. Cole, D. L. Dettman, J. A. Peck, C. A. Scholz, and J. W. King (2009), Atlantic forcing of persistent drought in West Africa, *Science, 324*, 377−380.

Sharkey, T. D. (1985), Photosynthesis in intact leaves of C_3 plants: Physics, physiology and rate limitations, *Bot. Rev., 51*, 53−105.

Sharkey, T. D., A. E. Wiberley, and A. R. Donohue (2008), Isoprene emission from plants: Why and how, *Ann. Bot., 101*, 5−18, doi:10.1093/aob/mcm240.

Sharma, S., D. Lavoué, H. Cachier, L. A. Barrie, and S. L. Gong (2004), Long-term trends of the black carbon concentrations in the Canadian Arctic, *J. Geophys. Res., 109*, D15203, doi:10.1029/2003JD004331.

Sharma, S., E. Andrews, L. A. Barrie, J. A. Ogren, and D. Lavoué (2006), Variations and sources of the equivalent black carbon in the high Arctic revealed by long-term observations at Alert and Barrow: 1989−2003, *J. Geophys. Res., 111*, D14208, doi:10.1029/2005JD006581.

Sharples, J., G. Mills, R. McRae, and R. Weber (2010), Foehn-like winds and elevated fire danger conditions in Southeastern Australia, *J. Appl. Meteor. Climatol., 49*(6), 1067−1095, doi:10.1175/2010JAMC 2219.1.

Sheehan, P. E., and F. K. Bowman (2001), Estimated effects of temperature on secondary organic aerosol concentrations, *Environ. Sci. Technol., 35*, 2129−2135.

Sheehan, P. M., P. J. Coorough, and D. E. Fastovsky (1996), Biotic selectivity during the K/T and late Ordovician extinctions, in The Cretaceous-Tertiary event and other catastrophes in Earth history, *Geol. Soc. Am. Spec. Pap. 307*, edited by G. Ryder, et al., pp. 477−489.

Sheffield, J., and E. F. Wood (2008), Global trends and variability in soil moisture and drought characteristics, 1950−2000, from observation-driven simulations of the terrestrial hydrologic cycle, *J. Climate, 21*(3), 432−458.

Shepherd, D. S. (1984), Computer mapping: the SYMAP interpolation algorithm, in *Spatial Statistics and Models*, edited by G. L. Gaile and C. J. Willmott, pp. 133−145, D. Reidel Publishing Company, Dordrecht, Netherlands.

Shepherd, J. M. (2005), A review of current investigations of urban-induced rainfall and recommendations for the future, *Earth Interact., 9*, paper 12, 1−27.

Shepherd, J. M., H. Pierce, and A. J. Negri (2002), Rainfall modification by major urban areas: Observations from spaceborne rain radar on the TRMM satellite, *J. Appl. Meteorol., 41*, 689−701.

Shepherd J. M., W. Shem, L. Hand, M. Manyin, and D. Messen (2010), Modeling urban effects on the precipitation component of the water cycle, in *Geospatial Analysis and Modelling of Urban Structure and Dynamics, 99*, edited by B. Jiang, and X. Yao, pp. 265−292, doi:10.1007/978-90-481-8572-6_14.

Sherwood, S. C., and M. Huber (2010), An adaptability limit to climate change due to heat stress, *Proc. Natl. Acad. Sci. USA, 107*, 9522−9555.

Shevliakova, E., S. W. Pacala, S. Malyshev, G. C. Hurtt, P. C. D. Milly, J. P. Caspersen, L. T. Sentman, J. P. Fisk, C. Wirth, C. Crevoisier (2009), Carbon cycling under 300 years of land use change: Importance of the secondary vegetation sink, *Global Biogeochem. Cycles, 23*, GB2022, doi:10.1029/2007GB003176.

Shiklomanov, A. I., R. B. Lammers, and M. A. Rawlins (2007), Temporal and spatial variations in maximum river discharge from a new Russian data set, *J. Geophys. Res.-Biogeosci., 112*, G04S53.

Shimada, K., T. Kamoshida, M. Itoh, S. Nishino, E. Carmack, F. McLaughlin, S. Zimmermann, and A. Proshutinsky (2006), Pacific ocean inflow: Influence on catastrophic reduction of sea ice cover in the Arctic Ocean, *Geophys. Res. Lett., 33*, L08605.

Shindell, D., and G. Faluvegi (2009), Climate response to regional radiative forcing during the twentieth century, *Nat. Geosci., 2*(4), 294−300, doi:10.1038/NGEO473.

Shindell, D. T., and G. A. Schmidt (2004), Southern Hemisphere climate response to ozone changes and greenhouse gas increases, *Geophy. Res. Lett., 31*, L18209, doi:10.1029/2004GL020724.

Shindell, D. T., D. Rind, N. Balachandran, J. Lean, and P. Lonergan (1999), Solar cycle variability, ozone and climate, *Science, 284*, 305−308.

Shindell, D. T., G. A. Schmidt, M. E. Mann, D. Rind, and A. Waple (2001), Solar forcing of regional climate change during the Maunder Minimum, *Science, 7*, 2149−2152.

Shindell, D. T., G. A. Schmidt, R. L. Miller, and D. Rind (2001), Northern Hemisphere winter climate response to greenhouse gas, ozone, solar, and volcanic forcing, *J. Geophys. Res., 106*(D7), 7193–7210.

Shindell, D. T., G. A. Schmidt, R. L. Miller, and M. E. Mann (2003), Volcanic and solar forcing of climate change during the Preindustrial era, *J. Climate, 16*, 4094–4107.

Shindell, D. T., G. Faluvegi, and N. Bell (2003), Preindustrial-to-present day radiative forcing by tropospheric ozone from improved simulations with the GISS chemistry-climate GCM, *Atmos. Chem. Phys., 3*, 1675–1702.

Shindell, D. T., B. P. Walter, and G. Faluvegi (2004), Impacts of climate change on methane emissions from wetlands, *Geophys. Res. Lett., 31*, L21202, doi:10.1029/2004GL021009.

Shindell, D. T., G. Faluvegi, N. Bell, and G. A. Schmidt (2005), An emissions-based view of climate forcing by methane and tropospheric ozone, *Geophys. Res. Lett., 32*, L04803, doi:10.1029/2004GL021900.

Shindell, D. T., G. Faluvegi, D. S. Stevenson, M. C. Krol, L. K. Emmons, J.-F. Lamarque, G. Pétron, F. J. Dentener, K. Ellingsen, M. G. Schultz, O. Wild, M. Amann, C. S. Atherton, D. J. Bergmann, I. Bey, T. Butler, J. Cofala, W. J. Collins, R. G. Derwent, R. M. Doherty, J. Drevet, H. J. Eskes, A. M. Fiore, M. Gauss, D. A. Hauglustaine, L. W. Horowitz, I. S. A. Isaksen, M. G. Lawrence, V. Montanaro, J.-F. Müller, G. Pitari, M. J. Prather, J. A. Pyle, S. Rast, J. M. Rodriguez, M. G. Sanderson, N. H. Savage, S. E. Strahan, K. Sudo, S. Szopa, N. Unger, T. P. C. van Noije, and G. Zeng (2006), Multimodel simulations of carbon monoxide: Comparison with observations and projected near-future changes, *J. Geophys. Res., 111*, D19306, doi:10.1029/2006JD007100.

Shindell, D. T., G. Faluvegi, S. E. Bauer, D. M. Koch, N. Unger, S. Menon, R. L. Miller, G. A. Schmidt, and D. G. Streets (2007), Climate response to projected changes in shortlived species under an A1B scenario from 2000–2050 in the GISS climate model, *J. Geophys. Res., 112*, D20103, doi:10.1029/2007JD008753.

Shindell, D. T., J.-F. Lamarque, N. Unger, D. Koch, G. Faluvegi, S. Bauer, and H. Teich (2008a), Climate forcing and air quality change due to regional emissions reductions by economic sector, *Atmos. Chem. Phys. Discuss., 8*, 11609–11642.

Shindell, D. T., M. Chin, F. Dentener, R. M. Doherty, G. Faluvegi, A. M. Fiore, P. Hess, D. M. Koch, I. A. MacKenzie, M. G. Sanderson, M. G. Schultz, M. Schulz, D. S. Stevenson, H. Teich, C. Textor, O. Wild, D. J. Bergmann, I. Bey, H. Bian, C. Cuvelier, B. N. Duncan, G. Folberth, L. W. Horowitz, J. Jonson, J. W. Kaminski, E. Marmer, R. Park, K. J. Pringle, S. Schroeder, S. Szopa, T. Takemura, G. Zeng, T. J. Keating, and A. Zuber (2008b), A multi-model assessment of pollution transport to the Arctic, *Atmos. Chem. Phys., 8*, 5353–5372.

Shine, K. P., J. Cook, E. J. Highwood, and M. M. Joshi (2003), An alternative to radiative forcing for estimating the relative importance of climate change mechanisms, *Geophys. Res. Lett., 30*(20), 2047.

Shine, K. P., J. S. Fuglestvedt, K. Hailemariam, and N. Stuber (2005), Alternatives to the global warming potential for comparing climate impacts of emissions of greenhouse gases, *Climatic Change, 68*, 281–302.

Shine, K. P., T. K. Berntsen, J. S. Fuglestvedt, R. Bieltvedt Skeie, and N. Stuber (2007), Comparing the climate effect of emissions of short- and long-lived climate agents, *Philos. Trans. R. Soc. London Ser. A, 365*, 1903–1914.

Shulmeister, J. (1999), Australasian evidence for mid-Holocene climate change implies precessional control of Walker Circulation in the Pacific, *Quatern. Int., 57–58*, 81–91, doi:10.1016/S1040-6182(98) 00052-4.

Shuman, B., A. K. Henderson, C. Plank, I. Stefanova, and S. S. Ziegler (2009), Woodland-to-forest transition during prolonged drought in Minnesota after ca. AD 1300, *Ecology, 90*(10), 2792–2807.

Sicart, J. E., R. Hock, and D. Six (2008), Glacier melt, air temperature, and energy balance in different climates: The Bolivian Tropics, the French Alps, and Northern Sweden, *J. Geophys. Res., 113*, D24113, doi:10.1029/2008JD010406.

Siccha, M., G. Trommer, H. Schulz, C. Hemleben, and M. Kucera (2009), Factors controlling the distribution of planktonic foraminifera in the Red Sea and implications for the development of transfer functions, *Mar. Micropaleontol., 72*, 146–156.

Siedler, G., J. Church, and J. Gould (2001), *Ocean Circulation and Climate: Observing and Modelling the Global Ocean*, Academic Press.

Siegele, L. (2010), Adaptation under the Copenhagen Accord: Briefing note, FIELD, London, *http://www.field.org.uk/files/Adaptation_under_the_Copenhagen_Accord.pdf*. (Accessed 9/9/10.)

Sigman, D. M., and E. A. Boyle (2000), Glacial/interglacial variations in atmospheric carbon dioxide, *Nature, 407*(6806), 859–869.

Sillmann, J., and E. Roekner E. (2008), Indices for extreme events in projections of anthropogenic climate change, *Climatic Change, 86*, 83–104.

Simmonds, I., K. Keay, and E.-P. Lim (2003), Synoptic activity in the seas around Antarctica, *Mon. Weather Rev., 131*, 272–288.

Simpson, D., K. E. Yttri, Z. Klimont, K. Kupiainen, A. Caseiro, A. Gelencsér, C. Pio, H. Puxbaum, and M. Legrand (2007), Modeling carbonaceous aerosol over Europe: Analysis of the CARBOSOL and EMEP EC/OC campaigns, *J. Geophys. Res., 112*, D23S14, doi:10.1029/2006JD008158.

Simpson, J., D. Mansfield, and J. Milford (1977), Inland penetration of sea breeze fronts, *Q. J. Roy. Meteor. Soc., 103*(435), 47–76.

Singarayer, J. S., P. J. Valdes, P. Friedlingstein, S. Nelson, and D. J. Beerling (2011), Late Holocene methane rise caused by orbitally controlled increase in tropical sources, *Nature, 470*, 82–85.

Sirois, A., and L. A. Barrie (1999), Arctic lower tropospheric aerosol trends and composition at Alert, Canada: 1980–1995, *J. Geophys. Res., 104*, 11599–115618.

Sitch S., P. M. Cox, W. J. Collins, and C. Huntingford (2007), Indirect radiative forcing of climate change through ozone effects on the land-carbon sink, *Nature, 448*, doi:10.1038/nature06059.

Sitch, S., et al. (2008), Evaluation of the terrestrial carbon cycle, future plant geography and climate-carbon cycle feedbacks using five Dynamic Global Vegetation Models (DGVMs), *Glob. Chang. Biol., 14*, 2015–2039.

Sloan, T., and A. W. Wolfendale (2008), Testing the proposed causal link between cosmic rays and cloud cover, *Environ. Res. Lett., 3*, 024001, doi:10.1088/1748-9326/3/2/024001.

Sluijs, A., S. Schouten, M. Pagani, M. Woltering, H. Brinkhuis, J. S. S. Damsté, G. R. Dickens, M. Huber, G.-J. Reichart, R. Stein, J. Matthiessen, L. J. Lourens, N. Pedentchouk, J. Backman, K. Moran, and the Expedition 302 Scientists (2006), Subtropical Arctic Ocean temperatures during the Palaeocene/Eocene thermal maximum, *Nature, 441*, 610–613.

Sluijs, A., H. Brinkhuis, S. Schouten, S. M. Bohaty, C. M. John, J. C. Zachos, G.-J. Reichart, J. S. S. Damsté, E. M. Crouch, and G. R. Dickens (2007), Environmental precursors to rapid light carbon injection at the Paleocene/Eocene boundary, *Nature, 450*, 1218–1221.

Smit, J. (1999), The global stratigraphy of the Cretaceous-Tertiary boundary impact ejecta, *Annu. Rev. Earth Pl. Sc., 27*, 75−113.

Smith, I., and E. Chandler (2010), Refining rainfall projections for the Murray Darling Basin of south-east Australia—The effect of sampling model results based on performance, *Climatic Change, 102*(3), 377−393, doi:10.1007/s10584-009-9757-1.

Smith, J. B., S. H. Schneider, M. Oppenheimer, G. W. Yohe, W. Hare, M. D. Mastrandrea, A. Patwardhan, I. Burton, J. Corfee-Morlot, C. H. D. Magadza, H.-M. Füssel, A. B. Pittock, A. Rahman, A. Suarez, and J.-P. Van Ypersele (2009), Assessing dangerous climate change through an update of the Intergovernmental Panel on Climate Change (IPCC) "reasons for concern," *Proc. Natl. Acad. Sci. USA, 106*(11), 4133−4137, doi:10.1073/pnas.0812355106.

Smith, R. (1979), The influence of mountains on the atmosphere, *Adv. Geophys., 21*, 87−230.

Smith, R. C., et al. (1992), Ozone depletion: Ultraviolet radiation and phytoplankton biology in Antarctic waters, *Science, 255*, 952−959.

Smith R. L., C. Tebaldi, D. Nychka, and L. O. Mearns (2009), Bayesian Modeling of Uncertainty in Ensembles of Climate Models, *J. Am. Stat. Assoc., 102*(485), 97−116.

Smith, S. C., J. D. Lee, W. J. Bloss, G. P. Johnson, T. Ingham, and D. E. Heard (2006), Concentrations of OH and HO_2 radicals during NAMBLEX: Measurements and steady state analysis, *Atmos. Chem. Phys., 6*, 1435−1453.

Smith S. J., E. Conception, R. Andres, and J. Lurz (2004), Historical sulfur dioxide emissions 1850−2000: Methods and results, PNNL Research Report, *PNNL-14537*, College Park, MD.

Smits, A., A. M. G. Klein Tank, and G. P. Können (2005), Trends in storminess over the Netherlands, 1962−2002, *Int. J. Climatol., 25*, 1331−1344.

Snow, J. T. (Ed.) (2003), Special Issue: European Conference on severe storms, *Atmos. Res., 67−68*, 1−703.

Soden, B. J., and I. M. Held (2006), An assessment of climate feedbacks in coupled ocean-atmosphere models, *J. Climate, 19*, 3354−3363, 6263.

Soden, B. J., R. T. Wetherald, G. L. Stenchikov, and A. Robock (2002), Global cooling after the eruption of Mount Pinatubo: A test of climate feedback by water vapor, *Science, 296*, 727−730.

Soden, B. J., I. M. Held, R. Colman, K. M. Shell, J. T. Kiehl, and C. A. Shields (2008), Quantifying climate feedbacks using radiative kernels, *J. Climate, 21*, 3504−3520.

Soja, A. J., N. M. Tchebakova, N. H. F. French, M. D. Flannigan, H. H. Shugart, B. J. Stocks, A. I. Shukhinin, E. I. Parfenova, F. S. Chapin III, and P. W. Stackhouse, Jr. (2007), Climate-induced boreal forest change: Predictions versus current observations, *Global Planet. Change, 56*, 274−296.

Sokolov, S., and S. R. Rintoul (2009), Circumpolar structure and distribution of the Antarctic circumpolar current fronts: 2. Variability and relationship to sea surface height, *J. Geophys. Res., 114*(C11), 1−15, doi:200910.1029/2008JC005248.

Sokolov, A. P., et al. (2008), Consequences of considering carbon-nitrogen interactions on the feedbacks between climate and terrestrial carbon cycle, *J. Climate, 21*, 3776−3796, doi:10.1175/2008JCLI2038.1.

Solanki, S. K., I. G. Usoskkin, B. Kromer, M. Schéssler, and J. Beer (2004), Unusual activity of the sun during recent decades compared to the previous 11,000 years, *Nature, 431*, 1084−1087.

Solberg, S., R. G. Derwent, Ø. Hov, J. Langner, and A. Lindskog (2005), European abatement of surface ozone in a global perspective, *Ambio, 34*(1), 47−53.

Solberg, S, Ø. Hov, A. Søvde, I. S. A. Isaksen, P. Coddeville, H. De Backer, C. Forster, Y. Orsolini, and K. Uhse (2008), European surface ozone in the extreme summer 2003, *J. Geophys. Res., 113*, D07307, doi:10.1029/2007JD009098.

Solman, S. A., M. N. Nuñez, and M. Cabré (2008), Regional climate change experiments over southern South America, 1: Present climate, *Clim. Dynam., 30*(5), 533−552.

Solomon, S. (2007), *Climate Change 2007: The Physical Science Basis*, contribution of Working Group I, public presentation, Royal Society, 9 March.

Solomon, S., D. Qin, M. Manning, Z. Chen, M. Marquis, K. B. Averyt, M. Tignor, and H. L. Miller (Eds.) (2007), *Climate Change 2007: The Physical Science Basis, contribution of Working Group I to the Fourth Assessment Report (FAR) of the Intergovernmental Panel on Climate Change (IPCC)*, Cambridge University Press, Cambridge, UK.

Son, S.-W., L. M. Polvani, D. W. Waugh, H. Akiyoshi, R. R. Garcia, D. E. Kinnison, S. Pawson, E. Rozanov, T. G. Shepherd, and K. Shibata (2008), The impact of stratospheric ozone recovery on the Southern Hemisphere westerly jet, *Science, 320*(5882), 1486−1489, doi:10.1126/science.1155939.

Sotiropoulou, R.-E. P., A. Nenes, P. J. Adams, and J. H. Seinfeld (2007), Cloud condensation nuclei prediction error from application of Köhler theory: Importance for the aerosol indirect effect, *J. Geophys. Res.*, doi:10.1029/2006JD007834.

Souch, C., and C. S. B. Grimmond (2006), Applied climatology: Urban climate, *Prog. Phys. Geog., 30*, 270−279.

Søvde, O. A., M. Gauss, I. S. A. Isaksen, G. Pitari, and C. Marizy (2007), Aircraft pollution − A futuristic view, *Atmos. Chem. Phys., 7*(13), 3621−3632.

Spahni, R., J. Chappellaz, T. F. Stocker, L. Loulergue, G. Hausammann, K. Kawamura, J. Fluckiger, J. Schwander, D. Raynaud, V. Masson-Delmotte, and J. Jouzel (2005), Atmospheric methane and nitrous oxide of the late Pleistocene from Antarctic ice cores, *Science, 310*, 1317−1321, doi:10.1126/science.1120132.

Spence, P., J. C. Fyfe, A. Montenegro, and A. J. Weaver (2010), Southern Ocean response to strengthening winds in an eddy-permitting global climate model, *J. Climate, 23*(19), 5332−5343.

Spicer, R. A. (1989), Plants at the Cretaceous-Tertiary boundary, *Philos. Trans. R. Soc. London Ser. B, 325*, 291−305.

Spicer, R. A., and A. B. Herman (2010), The late Cretaceous environment of the Arctic: A quantitative reassessment based on plant fossils, *Palaeogeogr. Palaeoclim., 295*, 423−442.

Spracklen, D. V., L. J. Mickley, J. A. Logan, R. C. Hudman, R. Yevich, M. D. Flannigan, and A. L. Westerling (2009), Impacts of climate change from 2000 to 2050 on wildfire activity and carbonaceous aerosol concentrations in the western United States, *J. Geophys. Res., 114*(D20), D20301.

Spronken-Smith, R. A., T. R. Oke, and W. P. Lowry (2000), Advection and the surface energy balance across an irrigated urban park, *Int. J. Climatol., 20*, 1033−1047.

Stanhill, G., and S. Cohen (2001), Global dimming: A review of the evidence for a widespread and significant reduction in global radiation, *Agr. Forest Meteorol., 107*, 255−278.

Stanhill, G., and S. Cohen (2009), Is solar dimming global or urban? Evidence from measurements in Israel between 1954 and 2007, *J. Geophys. Res., 114*, D00D17, doi:10.1029/2009JD011976.

Stanhill, G., and J. D. Kalma (1995), Solar dimming and urban heating at Hong Kong, *Int. J. Climatol., 15*, 933−941.

Stanley, J. R., Jr. (1988), The history of early Mesozoic reef communities: A threestep process, *Palaios, 3*, 170–183.

Starrs, J., and L. Cmielewski (2010), Incompatible elements, video installation by Josephine Starrs and Leon Cmielewski, exhibited at Performance Space Sydney, October 2010, *http://lx.sysx.org*.

Stavrakou, T., J.-F. Müller, K. F. Boersma, I. De Smedt, and A. R. J. van der (2008), Assessing the distribution and growth rates of NO_x emission sources by inverting a 10-year record of NO_2 satellite columns, *Geophys. Res. Lett., 35*, L10801, doi:10.1029/2008GL033521.

Stavrakou, T., J.-F. Müller, I. De Smedt, M. Van Roozendael, G. R. van der Werf, L. Giglio, and A. Guenther (2009a), Global emissions of non-methane hydrocarbons deduced from SCIAMACHY formaldehyde columns through 2003–2006, *Atmos. Chem. Phys., 9*, 3663–3679.

Stavrakou, T., J.-F. Müller, I. De Smedt, M. Van Roozendael, M. Kanakidou, M. Vrekoussis, F. Wittrock, A. Richter, and J. P. Burrows (2009b), The continental source of glyoxal estimated by the synergistic use of spaceborne measurements and inverse modelling, *Atmos. Chem. Phys.-Discuss., 9*, 13593–13628.

Stearns, L. A., and G. S. Hamilton (2007), Rapid volume loss from two East Greenland outlet glaciers quantified using repeat stereo satellite imagery, *Geophys. Res. Lett., 34*, L05503, doi:10.1029/2006GL028982.

Steffen, K., P. U. Clark, J. G. Cogley, D. Holland, S. Marshall, E. Rignot, and R. Thomas (2008), Rapid changes in glaciers and ice sheets and their impacts on sea level, in *Abrupt Climate Change*, Synthesis and Assessment Product 3.4, U.S. Climate Change Science Program and Subcommittee on Global Change Research, pp. 29–66, U.S. Geological Survey, Reston, VA.

Steffen, W., A. Sanderson, P. D. Tyson, J. Jaeger, P. A. Matson, B. Moore, III, F. Oldfield, K. Richardson, H. J. Schellnhuber, and B. L. Turner (Eds.) (2004), *Global Change and the Earth System: A Planet under Pressure*, The IGBP Book Series, Springer, Berlin.

Steffen, W., P. J. Crutzen, and J. R. McNeill (2007), The Anthropocene: Are humans now overwhelming the great forces of nature?, *Ambio, 36*, 614–621.

Steffensen, J. P., K. K. Andersen, M. Bigler, H. B. Clausen, D. Dahl-Jensen, H. Fischer, K. Goto-Azuma, M. Hansson, S. J. Johnsen, J. Jouzel, V. Masson-Delmotte, T. Popp, S. O. Rasmussen, R. Rothlisberger, U. Ruth, B. Stauffer, M. L. Siggaard-Andersen, A. E. Sveinbjornsdottir, A. Svensson, and J. W. C. White (2008), High-resolution Greenland Ice Core data show abrupt climate change happens in few years, *Science, 321*(5889), 680–684, doi:10.1126/science.1157707.

Steig, E. J., D. P. Schneider, S. D. Rutherford, M. E. Mann, J. C. Comiso, and D. T. Shindell (2009), Warming of the Antarctic Ice-Sheet surface since the 1957 International Geophysical Year, *Nature, 457*, 459–462.

Stein, U., and P. Alpert (1993), Factor separation in numerical simulations, *J. Atmos. Sci., 50*(14), 2107–2115.

Steinacher, M., F. Joos, T. L. Frölicher, G.-K. Plattner, and S. C. Doney (2009), Imminent ocean acidification in the Arctic projected with the NCAR global coupled carbon cycle-climate model, *Biogeosciences, 6*(4), 515–533, doi:10.5194/bg-6-515-2009.

Steinacher, M., F. Joos, T. L. Frölicher, L. Bopp, P. Cadule, V. Cocco, S. C. Doney, M. Gehlen, K. Lindsay, J. K. Moore, B. Schneider, and J. Segschneider (2010), Projected 21st century decrease in marine productivity: A multi-model analysis, *Biogeosciences, 7*, 979–1005.

Stenni, B., V. Masson-Delmotte, S. Johnsen, J. Jouzel, A. Longinelli, E. Monnin, R. Rothlisberger, and E. Selmo (2001), An oceanic cold reversal during the last deglaciation, *Science, 293*, 2074–2077, doi:10.1126/science.1059702.

Stephens, B. B., K. R. Gurney, P. P. Tans, C. Sweeney, W. Peters, L. Bruhwiler, P. Ciais, M. Ramonet, P. Bousquet, T. Nakazawa, S. Aoki, T. Machida, G. Inoue, N. Vinnichenko, J. Lloyd, A. Jordan, M. Heimann, O. Shibistova, R. L. Langenfelds, L. P. Steele, R. J. Francey, and A. S. Denning (2007), Weak northern and strong tropical land carbon uptake from vertical profiles of atmospheric CO_2, *Science, 316*, 1732–1735.

Stephens, G. L., T. L'Ecuyer, R. Forbes, A. Gettlemen, J.-C. Golaz, A. Bodas-Salcedo, K. Suzuki, P. Gabriel, and J. Haynes (2010), Dreary state of precipitation in global models, *J. Geophys. Res.-Atmos., 115*, D24211, doi:10.1029/2010JD014532.

Stephenson, D. B. (2008), Chapter 1: Definition, diagnosis, and origin of extreme weather and climate events, in *Climate Extremes and Society*, edited by R. Murnane and H. Diaz, Cambridge University Press.

Sterl, A., and S. Caires (2005), Climatology, variability, and extrema of ocean waves: The web-based KNMI/ERA-40 wave atlas, *Int. J. Climatol., 25*, 963–977.

Sterling, S. (2001), Sustainable education: Re-visioning learning and change, *Schumacher Briefing No. 6*, Schumacher Society/Green Books, Dartington.

Stern, D. I. (2005), Global sulfur emissions from 1850 to 2000, *Chemosphere, 58*, 163–175.

Stern, N. (2006), *The Economics of Climate Change*, Cambridge University Press, Cambridge, UK, *http://www.webcitation.org/5nCeyEYJr*. (Accessed 21/8/11.)

Stern, N. (2010), Climate: What you need to know, *The New York Review of Books*, 26 May.

Stevenson, D. S., C. E. Johnson, W. J. Collins, R. G. Derwent, K. P. Shine, and J. M. Edwards (1998), Evolution of tropospheric ozone radiative forcing, *Geophys. Res. Lett., 25*, 3819–3822.

Stevenson, D. S., C. E. Johnson, W. J. Collins, R. G. Derwent, and J. M. Edwards (2000), Future estimates of tropospheric ozone radiative forcing and methane turnover – the impact of climate change, *Geophys. Res. Lett., 27*(14), 2073–2076.

Stevenson, D. S., R. M. Doherty, M. G. Sanderson, C. E. Johnson, W. J. Collins, and R. G. Derwent (2005), Impacts of climate change and variability on tropospheric ozone and its precursors, *Faraday Discuss., 130*, 41–57, doi:10.1039/b417412g.

Stevenson, D. S., F. J. Dentener, M. G. Schultz, K. Ellingsen, T. P. C. van Noije, O. Wild, G. Zeng, M. Amann, C. S. Atherton, N. Bell, D. J. Bergmann, I. Bey, T. Butler, J. Cofala, W. J. Collins, R. G. Derwent, R. M. Doherty, J. Drevet, H. J. Eskes, A. M. Fiore, M. Gauss, D. A. Hauglustaine, L. W. Horowitz, I. S. A. Isaksen, M. C. Krol, J.-F. Lamarque, M. G. Lawrence, V. Montanaro, J.-F. Müller, G. Pitari, M. J. Prather, J. A. Pyle, S. Rast, J. M. Rodriguez, M. G. Sanderson, N. H. Savage, D. T. Shindell, S. E. Strahan, K. Sudo, and S. Szopa (2006), Multimodel ensemble simulations of present-day and near-future tropospheric ozone, *J. Geophys. Res., 111*, D08301, doi:10.1029/2005JD006338.

Stewart, R. (2005), *Oceanography in the 21st Century*, *http://oceanworld.tamu.edu/resources/oceanography-book/earthsystems.htm*, (Accessed 27/11/10.)

Steyn, D., and I. McKendry (1988), Quantative and qualitative evaluation of a 3-dimensional mesoscale numerical-model simulation of a sea breeze in complex terrain, *Mon. Weather Rev., 116*(10), 1914–1926.

Stier, P., J. Feichter, S. Kinne, S. Kloster, E. Vignati, J. Wilson, L. Ganzeveld, I. Tegen, M. Werner, Y. Balkanski, M. Schulz, O. Boucher, A. Minikin, and A. Petzold (2005), The aerosol-climate model ECHAM5-HAM, *Atmos. Chem. Phys., 5*, 1125−1156.

Stirling, A. (2003), Risk, Uncertainty and Precaution: Some instrumental implications from the social sciences, in *Negotiating Change: Perspectives in Environmental Social Science*, edited by I. Scoones, M. Leach and F. Berkhout, pp. 33−76, Edward Elgar, London.

Stith, J. L., et al. (2009), An overview of aircraft observations from the Pacific Dust Experiment campaign, *J. Geophys. Res., 114*(D5), D05207.

Stocks, B. J., J. A. Mason, J. B. Todd, E. M. Bosch, B. M. Wotton, B. D. Amiro, M. D. Flannigan, K. G. Hirsch, K. A. Logan, D. L. Martell, and W. R. Skinner (2003), Large forest fires in Canada, 1959−1997, *J. Geophys. Res., 108*(D1), 8149, doi:10.1029/2001JD000484.

Stohl, A. (2006), Characteristics of atmospheric transport into the Arctic troposphere, *J. Geophys. Res., 111*, D11306, doi:10.1029/2005JD006888.

Stommel, H. (1961), Thermohaline convection with two stable regimes of flow, *Tellus, 13*, 224−230.

Stone, D. A., A. J. Weaver, and R. J. Stouffer (2001), Projection of climate change onto modes of atmospheric variability, *J. Climate, 14*(17), 3551−3565.

Storelvmo, T., J. E. Kristjánsson, U. Lohmann, T. Iversen, A. Kirkevåg, and Ø. Seland (2008a), Modeling the Wegener-Bergeron-Findeisen process − implications for aerosol indirect effects, *Environ. Res. Lett., 3*(045001), 3214−3230.

Storelvmo, T., J. E. Kristjánsson, and U. Lohmann (2008b), Aerosol influence on mixedphase clouds in CAM-Oslo, *J. Atmos. Sci., 60*, 3214−3230.

Storelvmo, T., U. Lohmann, and R. Bennartz (2009), What governs the spread in shortwave forcings in the transient IPCC AR4 models, *Geophys. Res. Lett., 36*, doi:10.1029/2008GL0360692008c.

Storey, M., R. A. Duncan, C. C. Swisher, III (2007), Paleocene−Eocene Thermal Maximum and the opening of the northeast Atlantic, *Science, 316*, 587−589.

Stothers, R. B. (1996), The great dry fog of 1783, *Climatic Change, 32*, 79−89.

Stott, L., K. Cannariato, R. Thunell, G. H. Haug, A. Koutavas, and S. Lund (2004), Decline of surface temperature and salinity in the western tropical Pacific Ocean in the Holocene Epoch, *Nature, 431*(7004), 56−59, doi:10.1038/nature02903.

Stott, P. A., S. F. B. Tett, G. S. Jones, M. R. Allen, J. F. B. Mitchell, and G. J. Jenkins (2000), External control of 20th century temperature by natural and anthropogenic forcings, *Science, 290*, 2133−2137.

Stott, P. A., D. A. Stone, and M. R. Allen, (2004), Human contribution to the European heatwave of 2003, *Nature, 432*, 610−614.

Stott, P. A., J. F. B. Mitchell, M. R. Allen, T. L. Delworth, J. M. Gregory, G. A. Meehl, and B. D. Santer (2006), Observational constraints on past attributable warming and predictions of future global warming, *J. Climate, 19*, 3055−3069.

Stott, P. A., N. P. Gillet, G. C. Hegerl, D. J. Karoly, D. A. Stone, X. B. Zhang, and F. Zwiers (2010), Detection and attribution of climate change: a regional perspective, *Wiley Interdisciplinary Reviews: Climate Change, 1*(2), 192−211.

Stott, P. A., G. S. Jones, N. Christidis, F. W. Zwiers, G. C. Hegerl, and H. Shiogama (2010), Single-step attribution of increasing probabilities of very warm regional temperatures to human influence, *Atmos. Sci. Lett., 2011*, doi:10.1002/asl.315.

Stouffer, R. J., and R. T. Wetherald (2007), Changes of variability in response to increasing greenhouse gases. Part I: Temperature, *J. Climate, 20*, 5455−5467.

Stouffer, R. J., J. Yin, J. M. Gregory, K. W. Dixon, M. J. Spelman, W. Hurlin, A. J. Weaver, M. Eby, G. M. Flato, H. Hasumi, A. Hu, J. H. Jungclaus, I. V. Kamenkovich, A. Levermann, M. Montoya, S. Murakami, S. Nawrath, A. Oka, W. R. Peltier, D. Y. Robitaille, A. Sokolov, G. Vettoretti, and S. L. Weber (2006), Investigating the causes of the response of the thermohaline circulation to past and future climate changes, *J. Climate, 19*(8), 1365−1387.

Strahan, S. E., B. N. Duncan, and P. Hoor (2007), Observationally derived transport diagnostics for the lowermost stratosphere and their application to the GMI chemistry and transport model, *Atmos. Chem. Phys., 7*, 2435−2445.

Stramma, L., G. C. Johnson, J. Sprintall, and V. Mohrholz (2008), Expanding oxygen-minimum zones in the tropical oceans, *Science, 320*(5876), 655−658.

Straneo, F., et al. (2010), Rapid circulation of warm subtropical water in a major glacial fjord in East Greenland, *Nat. Geosci., 3*(3), 182−186.

Streets, D. G., T. C. Bond, G. R. Carmichael, S. D. Fernandes, Q. Fu, D. He, Z. Klimont, S. M. Nelson, N. Y. Tasi, M. Q. Wang, J.-H. Woo, and K. F. Yarber (2003), An inventory of gaseous and primary aerosol emissions in Asia in the year 2000, *J. Geophys. Res., 108*(D21), 8809, doi:10.1029/2002JD003093.

Streets, D. G., T. C. Bond, T. Lee, and C. Jang (2004), On the future of carbonaceous aerosol emissions, *J. Geophys. Res., 104*, D24212, doi:10.1029/2004JD004902.

Streets, D. G., Q. Zhang, L. Wang, K. He, J. Hao, Y. Wu, Y. Tang, and G. R. Carmichael (2006a), Revisiting China's CO emissions after the Transport and Chemical Evolution over the Pacific (TRACE-P) mission: Synthesis of inventories, atmospheric modeling, and observations, *J. Geophys. Res., 111*, D14306.

Streets, D. G., Y. Wu, and M. Chin (2006b), Two-decadal aerosol trends as a likely explanation of the global dimming/brightening transition, *Geophys. Res. Lett., 33*, doi:10.1029/2006GL026471.

Stroeve, J., M. M. Holland, W. Meier, T. Scambos, and M. C. Serreze (2007), Arctic sea ice decline: Faster than forecast, *Geophys. Res. Lett., 34*, L09501, doi:10.1029/2007GL029703.

Struthers, H., A. M. L. Ekman, P. Glantz, T. Iversen, A. Kirkevåg, E. M. Mårtensson, Ø. Seland, and E. D. Nilsson (2011), The effect of sea ice loss on sea salt aerosol concentrations and the radiative balance in the Arctic, *Atmos. Chem. Phys., 11*(7), 3459−3477.

Stuber N., G. Myhre, E. J. Highwood, G. Radel, and K. P. Shine (2011), Idealised aerosol perturbations in two GCMs: Towards an improved understanding of the semi-direct aerosol effect, *Clim. Dynam.*

Su, Z., L. Ding, and C. Liu (1984), Glacier thickness and its reserves calculation on Tianshan Mountains, *Xinjiang Geography, 7*(2), 37−44, (in Chinese).

Sud, Y. C., J. Shukla, and Y. Mintz (1988), Influence of land surface roughness on atmospheric circulation and precipitation: A sensitivity study with a general circulation model, *J. Appl. Meteor., 27*, 1036−1054.

Sud, Y. C., W. K.-M. Lau, G. K. Walker, J.-H. Kim, G. E. Liston, and P. J. Sellers (1996), Biogeophysical consequences of a tropical deforestation scenario: A GCM Simulation study, *J. Climate, 9*, 3225−3247.

Sudo, K., and H. Akimoto (2007), Global source attribution of tropospheric ozone: Long range transport from various source regions, *J. Geophys. Res., 112*, D12302, doi:10.1029/2006JD007992.

Sudo, K., M. Takahashi, and H. Akimoto (2003), Future changes in stratosphere-troposphere exchange and their impacts on future tropospheric ozone simulations, *Geophys. Res. Lett., 30*(24), 2256, doi:10.1029/-2003GL018526.

Sun, H. L., L. Biedermann, and T. C. Bond (2007), Color of brown carbon: A model for ultraviolet and visible light absorption by organic carbon aerosol, *Geophys. Res. Lett., 34*(17).

Sundquist, E. T., and R. F. Keeling (2009), The Mauna Loa carbon dioxide record: Lessons for long-term Earth observations, in *Carbon Sequestration and Its Role in the Global Carbon Cycle*, edited by E. T. Sundquist, *Geophys. Monog Series, 183*, 27−35, American Geophysical Union, doi:10.1029/2009GM000913.

Suntharalingam, P., and J. L. Sarmiento (2000), Factors governing the oceanic nitrous oxide distribution: Simulations with an ocean general circulation model, *Global Biogeochem. Cycles, 14*(1), 429−454.

Sutton, R. T., and M. R. Allen (1997), Decadal predictability of North Atlantic sea surface temperature and climate, *Nature, 388*, 563−567.

Sutton, R. T., and D. L. R. Hodson (2005), Atlantic Ocean forcing of North American and European summer climate, *Science, 309*, 115−118.

Sutton, R. T., B. Dong, and J. M. Gregory (2007), Land/sea warming ratio in response to climate change: IPCC AR4 model results and comparison with observations, *Geophys. Res. Lett., 34*(2), doi:10.1029/2006GL028164.

Svensson, A., K. K. Andersen, M. Bigler, H. B. Clausen, D. Dahl-Jensen, S. M. Davies, S. J. Johnsen, R. Muscheler, F. Parrenin, S. O. Rasmussen, R. Roethlisberger, I. Seierstad, J. P. Steffensen, and B. M. Vinther (2008), A 60,000 year Greenland stratigraphic ice core chronology, *Clim. Past, 4*(1), 47−57.

Sweet, A. R. (2001), Plants, a yardstick for measuring the environmental consequences of the Cretaceous-Tertiary boundary event, *Geosci. Can., 28*, 127−138.

Sweet, A. R., D. R. Braman, and J. F. Lerbekmo (1999), Sequential palynological changes across the composite Cretaceous-Tertiary (K-T) boundary claystone and contiguous strata, western Canada and Montana, USA, *Can. J. Earth Sci., 36*, 743−768.

Swisher, C. C., et al. (1992), Coeveal ^{40}Ar/^{39}Ar ages of 65.0 million years ago from Chicxulub crater melt rock and Cretaceous-Tertiary boundary tektites, *Science, 257*, 954−958.

Szopa, S., D. A. Hauglustaine, R. Vautard, and L. Menut (2006), Future global tropospheric ozone changes and impact on European air quality, *Geophys. Res. Lett., 33*, L14805.

Tagaris, E., K. Manomaiphiboon, K.-J. Liao, L. R. Leung, J.-H. Woo, S. He, P. Amar, and A. G. Russell (2007), Impacts of global climate change and emissions on regional ozone and fine particulate matter concentrations over the United States, *J. Geophys. Res., 112*, D14312, doi:10.1029/2006JD008262.

Taguchi, B., S.-P. Xie, N. Schneider, M. Nonaka, H. Sasaki, and Y. Sasai (2007), Decadal variability of the Kuroshio Extension: Observations and an eddy-resolving model hindcast, *J. Climate, 20*(11), 2357−2377.

Taha, H. (1996), Modeling the impacts of increased urban vegetation on the ozone air quality in the South Coast Air Basin, *Atmos. Environ., 30*, 3423−3430.

Takahashi, T., S. C. Sutherland, C. Sweeney, A. Poisson, N. Metzl, B. Tilbrook, N. Bates, R. Wanninkhof, R. A. Feely, C. Sabine, J. Olafsson, and Y. Nojiri (2002), Global sea-air CO2 flux based on climatological surface ocean pCO2, and seasonal biological and temperature effects. Deep Sea Research Part II: Topical Studies in Oceanography 49(9-10), 1601−1622.

Takata, K., K. Saito, and T. Yasunari (2009), Changes in the Asian monsoon climate during 1700−1850 induced by preindustrial cultivation, *Proc. Natl. Acad. Sci. USA, 106*, 9586−9589; doi:10.1073/pnas.0807346106.

Takle, E. S., W. J. Gutowski, Jr., R. W. Arritt, Z. Pan, C. J. Anderson, R. R. da Silva, D. Caya, S.-C. Chen, F. Giorgi, J. H. Christensen, S.-Y. Hong, H.-M. H. Juang, J. J. Katzfey, W. M. Lapenta, R. Laprise, G. E. Liston, P. Lopez, J. L. McGregor, R. A. Pielke, and J. O. Roads (1999), Project to Intercompare Regional Climate Simulations (PIRCS): Description and initial results, *J. Geophys. Res., 104*, 19443−19461.

Takle E. S., J. Roads, B., Rockel W. J. Gutowski, Jr, R. W. Arritt, I. Meinke, C. G. Jones, and A. Zodra (2007), Transferability intercomparison: An opportunity for new insight on the global water cycle and energy budget, *Bull. Am. Meteorol. Soc., 88*, 375−384

Tanner, L. H., J. F. Hubert, B. P. Coffey, and D. P. McInerney (2001), Stability of atmospheric CO$_2$ levels across the Triassic/Jurassic boundary, *Nature, 411*, 675−677.

Tanner, L. H., S. G. Lucas, and M. G. Chapman (2004), Assessing the record and causes of late Triassic extinctions, *Earth Sci. Rev., 65*, 103−139.

Tanner, T., S. Nair, S. Bhattacharjya, S. K. Srivastava, P. M. Sehgal, and D. Kull (2007), ORCHID: Climate risk screening in DFID India, *Synthesis Report*, Institute of Development Studies, Brighton.

Tans, P. P., I. Y. Fung, and T. Takahashi (1990), Observational contrains on the global atmospheric CO$_2$ budget, *Science, 247*(4949), 1431−1438, doi:10.1126/science.247.4949.1431.

Taplin, R. (2012), People, policy, and politics in future climates, in *The Future of the World's Climate*, edited by A. Henderson-Sellers and K. McGuffie, pp. 29−46, Elsevier, Amsterdam.

Taplin, R., and J. McGee (2010), The Asia-Pacific Partnership: Implementation challenges and interplay with Kyoto, *Wiley Interdiscipl. Rev. Clim. Chang., 1*(1), 16−22.

Tarnocai, C., J. G. Canadell, E. A. G. Schuur, G. Mazhitova, and S. Zimov (2009), Soil organic carbon pools in the northern circumpolar permafrost region, *Global Biogeochem. Cycles, 23*, GB2023, doi:10.1029/2008GB003327.

Taschetto, A. S., C. C. Ummenhofer, A. Sen Gupta, and M. H. England (2009), Effect of anomalous warming in the Central Pacific on the Australian monsoon, *Geophys. Res. Lett., 36*, 5, doi:200910.1029/2009GL038416.

Taschetto, A. S., R. J. Haarsma, A. Sen Gupta, C. C. Ummenhofer, K. J. Hill, and M. H. England (2010), Australian monsoon variability driven by a Gill−Matsuno-type response to Central West Pacific warming, *J. Climate, 23*(18), 4717−4736, doi:10.1175/2010JCLI3474.1.

Taub, D. R., and X. Wang (2008), Why are nitrogen concentrations in plant tissues lower under elevated CO$_2$? A critical examination of the hypotheses, *J. Integr. Plant Biol., 50*, 1365−1374.

Taylor, C. M., E. F. Lambin, N. Stephenne, R. J. Harding, and R. L. Essery (2002), The influence of land use change on climate in the Sahel, *J. Climate, 15*, 3615−3629.

Taylor, K. E. (2001), Summarizing multiple aspects of model performance in a single diagram, *J. Geophys. Res., 106*, 7183−7192, (See also PCMDI Report 55, *http://www-pcmdi.llnl.gov/publications/ab55.html*.)

Taylor, K. E., R. J. Stouffer, and G. A. Meehl (2009), A summary of the CMIP5 experiment design, World Climate Research Programme (WCRP), *http://cmip-pcmdi.llnl.gov/cmip5/docs/Taylor_CMIP5_design.pdf*.

Taylor, S. H., B. S. Ripley, F. I. Woodward, and C. P. Osborne (2011), Drought limitation of photosynthesis differs between C_3 and C_4 grass species in a comparative experiment, *Plant Cell Environ., 34*, 65–75, doi:10.1111/j.1365-3040.2010.02226.x.

Tebaldi, C., and B. Sanso (2008), Joint projections of temperature and precipitation change from multiple climate models: A hierarchical Bayesian approach, *J. Roy. Stat. Soc. A, 172*, 83–106.

Tebaldi, C., and B. Sanso (2009), Joint projections of temperature and precipitation change from multiple climate models: A hierarchical Bayesian approach, *J. Roy. Stat. Soc. A-Sta., 172*, 83–106, doi:10.1111/j.1467-985X.2008.00545.x.

Tebaldi, C., L. O. Mearns, D. Nychka, and R. Smith (2004), Regional probabilities of precipitation change: A Bayesian analysis of multimodel simulations, *Geophys. Res. Lett., 31*(24), doi:10.1029/2004GL021276.

Tebaldi, C., R. Smith, D. Nychka, and L. O. Mearns (2005), Quantifying uncertainty in projections of regional climate change: A Bayesian approach to the analysis of multimodel ensembles, *J. Climate, 18*(10), 1524–1540.

Tebaldi, C., K. Hayhoe, J. M. Arblaster, and G. A. Meehl (2006), Going to the extremes: An intercomparison of model-simulated historical and future changes in extreme events, *Climatic Change, 79*, 185–211, doi:10.1007/s10584-006-9051-4.

Tengberg, A., J. Hovdenes, H. J. Andersson, O. Brocandel, R. Diaz, D. Hebert, T. Arnerich, C. Huber, A. Kortzinger, A. Khripounoff, et al. (2006), Evaluation of a lifetime-based optode to measure oxygen in aquatic systems, *Limnol. Oceanogr.-Meth., 4*, 7–17.

Teuling, A. J., S. I. Seneviratne, C. Williams, and P. A. Troch (2006), Observed timescales of evapotranspiration response to soil moisture, *Geophys. Res. Lett., 33*, L23403, doi:10.1029/2006GL028178.

Textor, C., M. Schulz, S. Guibert, S. Kinne, Y. Balkanski, S. Bauer, T. Berntsen, T. Berglen, O. Boucher, M. Chin, F. Dentener, T. Diehl, R. Easter, H. Feichter, D. Fillmore, S. Ghan, P. Ginoux, S. Gong, A. Grini, J. Hendricks, L. Horowitz, P. Huang, I. S. A. Isaksen, T. Iversen, S. Kloster, D. Koch, A. Kirkevåg, J. E. Kristjansson, M. Krol, A. Lauer, J. F. Lamarque, X. Liu, V. Montanaro, G. Myhre, J. Penner, G. Pitari, S. Reddy, Ø. Seland, P. Stier, T. Takemura, and X. Tie (2006), Analysis and quantification of the diversities of aerosol life cycles within AeroCom, *Atmos. Chem. Phys., 6*(7), 1777–1813.

Textor, C., M. Schulz, S. Guibert, S. Kinne, Y. Balkanski, S. Bauer, T. Berntsen, T. Berglen, O. Boucher, M. Chin, F. Dentener, T. Diehl, J. Feichter, D. Fillmore, P. Ginoux, S. Gong, A. Grini, J. Hendricks, L. Horowitz, P. Huang, I. S. A. Isaksen, T. Iversen, S. Kloster, D. Koch, A. Kirkevåg, J. E. Kristjansson, M. Krol, A. Lauer, J. F. Lamarque, X. Liu, V. Montanaro, G. Myhre, J. E. Penner, G. Pitari, M. S. Reddy, Ø. Seland, P. Stier, T. Takemura, and X. Tie (2007), The effect of harmonized emissions on aerosol properties in global models – An AeroCom experiment, *Atmos. Chem. Phys., 7*(17), 4489–4501.

Thatcher, M., and J. L. McGregor (2009), Using a scale-selective filter for dynamical downscaling with the conformal cubic atmospheric model, *Mon. Weather Rev., 137*, 1742–1752

Thatcher, M., and J. L. McGregor (2011), A technique for dynamically downscaling daily-averaged GCM datasets using the conformal cubic atmospheric model, *Mon. Weather Rev., 139*, 79–95.

The Economist (2010), Clearing up the climate, 30 August, *http://www.economist.com/blogs/newsbook/2010/08/climate_change_and_ipcc*. (Accessed 9/9/10.)

The Global Carbon Cycle and its Management, *www.globalcarbonproject.org/global/pdf/GCP_10years_high_res.pdf*. (Accessed 12/02/11.)

The Greenhouse Gas Protocol (2010), *The Greenhouse Gas Protocol Initiative*, World Resources Institute and World Business Council for Sustainable Development, *www.ghgprotocol.org/about-ghgp*. (Accessed 9/11/10.)

Thomas, H., A. E. F. Prowe, I. D. Lima, S. C. Doney, R. Wanninkhof, R. J. Greatbatch, U. Schuster, and A. Corbière (2008), Changes in the North Atlantic Oscillation influence CO_2 uptake in the North Atlantic over the past 2 decades, *Global Biogeochem. Cycles, 22*, 13, doi:200810.1029/2007GB003167.

Thompson, A. M., R. W. Stewart, M. A. Owens, and J. A. Herwehe (1989), Sensitivity of tropospheric oxidants to global chemical and climate change, *Atmos. Environ., 23*(3), 519–532.

Thompson, D. W. J., and S. Solomon (2002), Interpretation of recent Southern Hemisphere climate change, *Science, 296*(5569), 895–899.

Thompson, D. W. J., and J. M. Wallace (2000), Annular modes in the extratropical circulation, part I: Month-to-month variability, *J. Climate, 13*, 1000–1016.

Thompson, J. M. T., and J. Sieber (2010), Climate tipping as a noisy bifurcation: A predictive technique, *IMA J. Appl. Math.*, (submitted).

Thompson, S. L., and S. H. Schneider (1986) Nuclear Winter reappraised, *Foreign Affairs, 64*, 981–1005.

Thonicke, K., I. C. Prentice, and C. Hewitt (2005), Modeling glacial-interglacial changes in global fire regimes and trace gas emissions, *Global Biogeochem. Cycles, 19*(3), doi:10.1029/2004GB002278.

Thornton, P. E., S. C. Doney, K. Lindsay, J. K. Moore, N. Mahowald, J. T. Randerson, I. Fung, J.-F. Lamarque, J. J. Feddema, and Y.-H. Lee (2009), Carbon-nitrogen interactions regulate climate–carbon cycle feedbacks: Results from an atmosphere-ocean general circulation model, *Biogeosciences, 6*, 2099–2120.

Tilbury, D., and D. Wortman (2004), *Engaging People in Sustainability*, IUCN Publication Services, Cambridge, UK.

Timbal, B., and J. M. Arblaster (2006), Land cover change as an additional forcing to explain the rainfall decline in the south west of Australia, *Geophys. Res. Lett., 33*, L07717, doi:10.1029/2005GL025361.

Timbal, B., and D. A. Jones (2008), Future projections of winter rainfall in southeast Australia using a statistical downscaling technique, *Climatic Change, 86*, 165–187, doi:10.1007/s10584-007-9279-7.

Timbal, B., J. M. Arblaster, and S. B. Power (2006), Attribution of the late 20th century rainfall decline in south-west Australia, *J. Climate, 19*, 2046–2065.

Timbal, B., P. Hope, and S. Charles (2008), Evaluating the consistency between statistically downscaled and global dynamical model climate change projections, *J. Climate, 21*(22), 6052–6059, doi:10.1175/2008JCLI2379.1.

Timm, O., and A. Timmermann (2007), Simulation of the last 21,000 years using accelerated transient boundary conditions, *J. Climate, 20*, 4377–4401.

Timmermann, A., M. Latif, R. Voss, and A. Grötzner (1998), Northern Hemisphere inter-decadal variability: A coupled air-sea mode, *J. Climate, 11*, 1906–1931.

Timmermann, A., S. McGregor, and F. F. Jin (2010), Wind effects on past and future regional sea level trends in the Southern Indo-Pacific, *J. Climate, 23*(16), 4429–4437.

Ting, M. F., Y. Kushnir, R. Seager, and C. H. Li (2009), Forced and internal twentieth-century SST trends in the North Atlantic, *J. Climate, 22*, 1469–1481, doi:10.1175/2008jcli2561.1.

Tingey, D. T., E. H. Lee, R. Waschmann, M. G. Johnson, and P. T. Rygiewicz (2006), Does soil CO$_2$ efflux acclimatize to elevated temperature and CO$_2$ during long-term treatment of Douglas-fir seedlings? *New Phytol., 170*, 107–118.

Tinner, W., and A. F. Lotter (2001), Central European vegetation response to abrupt climate change at 8.2 ka, *Geology, 29*(6), 551–554.

Tippett, M., T. DelSole, S. Mason, and A. Barnston (2008), Regression-based methods for finding coupled patterns, *J. Climate, 21*(17), 4384–4398, doi:10.1175/2008JCLI2150.1.

Tol, R. S. J., T. K. Berntsen, B. C. O'Neill, J. S. Fuglestvedt, K. P. Shine, Y. Balkanski, and L. Makra (2008), Metrics for aggregating the climate effect of different emissions: A unifying framework, *257*, ESRI Working Paper, ESRI, Dublin, Ireland.

Tompkins, E. L., and H. Amundsen (2008), Perceptions of the effectiveness of the United Nations Framework Convention on Climate Change in advancing national action on climate change, *Environ. Sci. Policy, 11*(1), 1–13.

Tonkin, H., G. J. Holland, N. Holbrook, and A. Henderson-Sellers (2000), An evaluation of thermodynamic estimates of climatological maximum potential tropical cyclone intensity, *Mon. Weath. Rev., 128*, 746–762.

Toon, O. B., J. P. Pollack, T. P. Ackerman, R. P. Turco, C. P. McKay, and M. S. Liu (1982), Evolution of an impact generated dust cloud and its effects on the atmosphere, in Geological implications of impacts of large asteroids and comets on the Earth, edited by L. T. Silver, and P. H. Schultz, *Geol. Soc. Am. Spec. Pap. 190*, 187–200.

Toon, O. B., K. Zahnle, D. Morrison, R. P. Turco, and C. Covey (1997), Environmental perturbations caused by the impacts of asteroids and comets, *Rev. Geophys., 35*, 41–78.

Townrow, J. A. (1960), The Peltaspermaceae, a pteridosperm family of Permian and Triassic age, *Palaeontology, 3*, 333–361.

Trabant, D. C., and R. S. March (1999), Mass-balance measurements in Alaska and suggestions for simplified observation programs, *Geogr. Ann. A, 81*(3–4), 777–789.

Trapp, R. J., S. A. Tessendorf, E. S. Godfrey, and H. E. Brooks (2005), Tornadoes from squall lines and bow echoes, 1.: Climatological distribution, *Weather Forecast., 20*, 23–34.

Travis, W. R. (2010), Going to extremes: Propositions on the social response to severe climate change, *Climatic Change, 98*(1–2), 1–19.

Trenberth, K. (Ed.) (1992), *Climate System Modelling*, Cambridge University Press, Cambridge, UK.

Trenberth, K. E., and A. Dai (2007), Effects of Mount Pinatubo volcanic eruption on the hydrological cycle as an analog of geoengineering, *Geophys. Res. Lett., 34*, L15702.

Trenberth, K. E., and J. W. Hurrell (1994), Decadal atmosphere-ocean variations in the Pacific, *Clim. Dynam., 9*, 303–319.

Trenberth, K. E., and S. A. Josey (2007), Observations: Surface and atmospheric climate change, in, *Climate Change 2007: The Physical Science Basis, contribution of Working Group I to the Fourth Assessment Report (FAR) of the Intergovernmental Panel on Climate Change (IPCC)*, edited by S. Solomon, et al., pp. 235–336, Cambridge University Press, Cambridge, UK.

Trenberth, K. E., and D. J. Shea (2006), Atlantic hurricanes and natural variability in 2005, *Geophys. Res. Lett., 33*, L12704. doi:10.1029/2006GL026894.

Trenberth, K. E., P. D. Jones, P. Ambenje, R. Bojariu, D. Easterling, A. Klein Tank, D. Parker, F. Rahimzadeh, J. A. Renwick, M. Rusticucci, B. Soden, and P. Zhai (2007), Observations: Surface and atmospheric climate change, in *Climate Change 2007: The Physical Science Basis, contribution of Working Group I to the Fourth Assessment Report (FAR) of the Intergovernmental Panel on Climate Change (IPCC)*, edited by S. Solomon, et al., pp. 235–336, Cambridge University Press, Cambridge, UK.

Trenberth, K. E., J. T. Fasullo, and J. Kiehl (2009), Earth's global energy budget, *Bull. Am. Meteorol. Soc., 90*, 311–323.

Tripati, A. K., C. D. Roberts, and R. A. Eagle (2009), Coupling of CO$_2$ and ice sheet stability over major climate transitions of the last 20 million years, *Science, 326*, 1394.

Trouet, V., J. Esper, N. E. Graham, A. Baker, J. D. Scourse, and D. C. Frank (2009), Persistent positive North Atlantic Oscillation Mode dominated the Medieval Climate Anomaly, *Science, 324*, 78–80, doi: 10.1126/science.1166349.

Trusilova, K., M. Jung, and G. Churkina (2009), On climate impacts of a potential expansion of urban land in Europe, *J. Appl. Meteorol. Clim., 48*, 1971–1980, doi:10.1175/2009JAMC2108.1.

Tryhorn, L., and J. Risbey (2006), On the distribution of heat waves over the Australian region, *Aust. Meteorol. Mag., 55*, 169–182.

Tsai, I.-C., J.-P. Chen, P.-Y. Lin, W.-C. Wang, and I. S. A. Isaksen (2010), Sulfur cycle and sulfate radiative forcing simulated from a coupled global climate–chemistry model, *Atmos. Chem. Phys., 10*, 3693–3709.

Tschudy, R. H., C. L. Pillmore, C. J. Orth, J. S. Gillmore, and J. D. Knight (1984), Disruption of the terrestrial plant ecosystem at the Cretaceous-Tertiary boundary, Western Interior, *Science, 225*, 1030–1032.

Tsigaridis, K., and M. Kanakidou (2003), Global modelling of secondary organic aerosol in the troposphere: A sensitivity analysis, *Atmos. Chem. Phys., 3*, 1849–1869.

Tsigaridis, K., and M. Kanakidou (2007), Secondary organic aerosol importance in the future atmosphere, *Atmos. Environ., 41*(22), 4682–4692.

Tsigaridis, K., M. Krol, F. J. Dentener, Y. Balkanski, J. Lathière, S. Metzger, D. A. Hauglustaine, and M. Kanakidou (2006), Change in global aerosol composition since preindustrial times, *Atmos. Chem. Phys.-Discuss., 6*, 5585–5628.

Tudhope, A. W., C. P. Chilcott, M. T. McCulloch, E. R. Cook, J. Chappell, R. M. Ellam, D. W. Lea, J. M. Lough, and G. B. Shimmield (2001), Variability in the El Niño–Southern oscillation through a glacial-interglacial cycle, *Science, 291*(5508), 1511–1517, doi:10.1126/science.1057969.

Turetsky, M., K. Wieder, L. Halsey, and D. Vitt (2002), Current disturbance and the diminishing peatland carbon sink, *Geophys. Res. Lett., 29*, 1526, doi:10.1029/2001GL014000.

Turquety, S., J. A. Logan, D. J. Jacob, R. C. Hudman, F. Y. Leung, C. L. Heald, R. M. Yantosca, S. Wu, L. K. Emmons, D. P. Edwards, and G. W. Sachse (2007), Inventory of boreal fire emissions for North America in 2004: Importance of peat burning and pyroconvective injection, *J. Geophys. Res., 112*, D12S03, doi:10.1029/2006JD007281.

Twitchett, R. J. (2006), The palaeoclimatology, palaeoecology and palaeoenvironmental analysis of mass extinction events, *Palaeogeogr. Palaeoclim., 232*, 190–213.

Twomey, S. (1974), Pollution and the planetary albedo, *Atmos. Environ., 8*, 1251–1256.

Twomey, S. (1977), The influence of pollution on the shortwave albedo of clouds, *J. Atmos. Sci., 34*, 1149–1152.

Twomey, S. A. (1977), *Atmospheric Aerosols*, Elsevier, Amsterdam.

Twomey, S. A., M. Piepgrass, and T. L. Wolffe (1984), An assessment of the impact of pollution on global cloud albedo, *Tellus B, 36*, 356–366.

Tyndall, J. (1861), On the Absorption and Radiation of Heat by Gases and Vapours, and on the Physical Connexion of Radiation, Absorption and Conduction, The Bakerian Lecture, *Phil. Trans. Roy. Soc., 151* (Part I), 1–36.

Tyrrell, J. (2003), A tornado climatology for Ireland, *Atmos. Res., 67–68*, 671–684.

Tzedakis, P. C. (2010), The MIS_11-MIS-1 analogy, southern European vegetation, atmospheric methane and the 'early anthropogenic hypothesis', *Clim. Past, 6*, 131–144.

Tzedakis, P. C., D. Raynaud, J. F. McManus, A. Berger, V. Brovkin, and T. Kiefer (2009), Interglacial diversity, *Nat. Geosci., 2*(11), 751–755.

UK House of Commons Science and Technology Committee (2010), The disclosure of climate data from the Climatic Research Unit at the University of East Anglia, *Eighth Report of Session 2009–10, HC387-I*, London, 31 March, *http://www.publications.parliament.uk/pa/cm200910/cmselect/cmsctech/387/387i.pdf*. (Accessed 28/3/11.)

UNCED (1992), Chapter 28: Local authorities' initiatives in support of Agenda 21, *Earth Summit Agenda 21*, United Nations Conference on Environment and Development, Rio de Janeiro, 3–14 June, *http://www.un.org/esa/dsd/agenda21/res_agenda21_28.shtml*. (Accessed 9/11/10.)

UNFCCC (1992), United Nations Framework Convention on Climate Change, *http://unfccc.int/resource/docs/convkp/conveng.pdf*. (Accessed 9/11/10.)

UNFCCC (2007), The Bali Roadmap, *http://unfccc.int/meetings/cop_13/items/4049.php*. (Accessed 26/7/10.)

UNFCCC (2009), The Copenhagen Accord, 18 December, *http://www.denmark.dk/NR/rdonlyres/C41B62AB-4688-4ACE-BB7B-F6D2C8AAEC20/0/copenhagen_accord.pdf*. (Accessed 27/11/10.)

UNFCCC (2010), The Cancun Agreements, *http://cancun.unfccc.int/cancun-agreements/main-objectives-of-the-agreements/#c33*. (Accessed 7/4/11.)

UNFCCC (2011a), The Cancun Agreements, *http://cancun.unfccc.int/cancun-agreements/significance-of-the-key-agreements-reached-at-cancun/#c45*. (Accessed 30/3/11.)

UNFCCC (2011b), Communication of information of China, letter to Yvo de Boer, Executive Secretary, UNFCCC from SU Wei, Director-General, Department of Climate Change, National Development and Reform Commission of China, 28 January 2010, *http://www.unfccc.int/files/meetings/cop_15/copenhagen…/chinacphaccord_app2.pdf*. (Accessed 31/3/11.)

UNFCCC (2011c), India information on appendix 2 of the Copenhagen Accord, letter to Yvo de Boer, Executive Secretary, UNFCCC from Rajani Ranjan Rashmi, Joint Secretary, Ministry of Environment and Forests, Government of India, *http://unfccc.int/files/meetings/cop_15/copenhagen_accord/application/pdf/indiacphaccord_app2.pdf*. (Accessed 31/3/11.)

Unger, N., D. T. Shindell, D. M. Koch, M. Amann, J. Cofala, and D. G. Streets (2006), Influences of man-made emissions and climate changes on tropospheric ozone, methane and sulfate at 2030 from a broad range of possible futures, *J. Geophys. Res., 111*, D12313, doi:10.1029/2005JD006518.

Unger, N., D. T. Shindell, D. M. Koch, and D. G. Streets (2008), Air pollution radiative forcing from specific emissions sectors at 2030, *J. Geophys. Res., 113*, D02306, doi:10.1029/2007JD008683.

United Nations (2009), *World urbanization prospects: The 2009 revision population database*, New York, Department of Economic and Social Affairs, Population Division, *http://esa.un.org/wup2009/unup/index.asp?panel=1*.

United Nations Food and Agriculture Organization (1996), World Food Summit, *ftp://ftp.fao.org/es/ESA/policybriefs/pb_02.pdf*. (Accessed 27/10/10.)

Urban, N. M., and K. Keller (2009), Complementary observational constraints on climate sensitivity, *Geophys. Res. Lett., 36*, L04708, doi:10.1029/2008GL036457.

Vadja, V., J. I. Raine, and C. J. Hollis (2001), Indication of global deforestation at the Cretaceous-Tertiary boundary by New Zealand fern spike, *Science, 294*, 1700–1702.

Valdes, P. J., D. J. Beerling, and C. E. Johnson (2005), The Ice Age methane budget, *Geophys. Res. Lett., 32*, L02704, doi:10.1029/2004GL021004.

van Aardenne, J. A., F. Dentener, J. G. J. Olivier, J. A. W. H. Peters, and L. N. Ganzeveld (2005), The EDGAR 3.2 Fast Track 2000 dataset (32FT2000), Bilthoven, Edgard Consortium.

van de Schootbrugge, B., et al. (2009), Floral changes across the Triassic/Jurassic boundary linked to flood basalt volcanism, *Nat. Geosci., 2*, 589–594.

van de Wal, R. S. W., and M. Wild (2001), Modelling the response of glaciers to climate change by applying volume-area scaling in combination with a high-resolution GCM, *Clim. Dynam., 18*, 359–366.

van de Wal, R. S. W., W. Boot, M. R. van den Broeke, C. J. P. P. Smeets, C. H. Reijmer, J. J. A. Donker, and J. Oerlemans (2008), Large and rapid melt-induced velocity changes in the ablation zone of the Greenland Ice Sheet, *Science, 321*, 111–113.

van der Burgh, J., H. Visscher, D. L. Dilcher, and W. M. Kürschner (1993), Paleoatmospheric signatures in Neogene fossil leaves, *Science, 260*, 1788–1790.

van der Linden, P., and J. Mitchell (Eds.) (2009), *ENSEMBLES: Climate Change and Its Impacts*, summary of research and results from the ENSEMBLES project, Met Office Hadley Centre, Exeter, UK.

Van der Schrier, G., K. R. Briffa, P. D. Jones, and T. J. Osborn (2006), Summer moisture variability across Europe, *J. Climate, 19*(12), 2818–2834.

Van der Veen, C. J. (2000), Fourier and the greenhouse effect, *Polar Geogr., 24*(2), 132–152.

van der Werf, G. R., J. T. Randerson, L. Giglio, G. J. Collatz, P. S. Kasibhatla, and A. F. Arellano, Jr. (2006), Interannual variability in global biomass burning emissions from 1997 to 2004, *Atmos. Chem. Phys., 6*, 3423–3441.

van der Werf, G. R., J. T. Randerson, L. Giglio, N. Gobron, and A. J. Dolman (2008), Climate controls on the variability of fires in the tropics and subtropics, *Global Biogeochem. Cycles, 22*(3), GB3028, doi:10.1029/2007GB003122.

van der Werf, G. R., D. C. Morton, R. S. DeFries, J. G. J. Olivier, P. S. Kasibhatla, R. B. Jackson, D. J. Collatz, J. T. Randerson (2009), CO_2 emissions from forest loss, *Nat. Geosci., 2*, 737–738.

van Kessel, C., B. Boots, M.-A. de Graaff, D. Harris, H. Blum, and J. Six (2006), Total soil C and N sequestration in a grassland following 10 years of free air CO$_2$ enrichment, *Glob. Change Biol.*, *12*, 2187–2199.

Van Loon, M., R. Vautard, M. Schaap, R. Bergström, B. Bessagnet, J. Brandt, P. J. H. Builtjes, J. H. Christensen, C. Cuvelier, A. Graff, J. E. Jonson, M. Krol, J. Langner, P. Roberts, L. Rouil, R. Stern, L. Tarrasón, P. Thunis, E. Vignati, L. White, and P. Wind (2006), Evaluation of long-term ozone simulations from seven regional air quality models and their ensemble average, *Atmos. Environ.*, *41*(10), 2083–2097.

Van Mantgem, P. J., N. L. Stephenson, J. C. Byrne, L. D. Daniels, J. F. Franklin, P. Z. Fulé, M. E. Harmon, A. J. Larson, J. M. Smith, A. H. Taylor, and T. T. Veblen (2009), Widespread increase in tree mortality rates in the western United States, *Science, 323,* 521–524.

van Meijgaard, E., L. H. van Ulft, W. J. van de Berg, F. C. Bosveld, B. J. J. M. van den Hurk, G. Lenderink, and A. P. Siebesma (2008), The KNMI regional atmospheric climate model, version 2.1, KNMI Tech. Rep. 302, R. Neth. Meteorol. Inst., De Bilt, Netherlands.

van Noije, T. P. C., H. J. Eskes, M. van Weele, and P. F. J. van Velthoven (2004), Implications of the enhanced Brewer-Dobson circulation in European Centre for Medium-Range Weather Forecasts reanalysis ERA-40 for the stratosphere–troposphere exchange of ozone in global chemistry-transport models, *J. Geophys. Res., 109*, D19308, doi:10.1029/2004JD004586.

Van Oldenborgh, G. J., S. Y. Philip, and M. Collins (2005), El Niño in a changing climate: A multi-model study, *Ocean Sci., 1*(2), 81–95.

van Vuuren, D. P., M. G. J. den Elzen, P. L. Lucas, B. Eickhout, B. J. Strengers, B. van Ruijven, S. Wonink, and R. van Houdt (2007), Stabilizing greenhouse gas concentrations at low levels: an assessment of reduction strategies and costs. *Climatic Change, 81*(2), 119–159.

van Ypersele, J.-P. (2010), Opening plenary presentation, International Climate Change Adaptation Conference, Climate Adaptation Futures: Preparing for the unavoidable impacts of climate change, National Climate Change Adaptation Research Facility and the CSIRO Climate Adaptation Flagship, 29 June–1 July, Gold Coast.

Vantrepotte, V., and F. Mélin (2009), Temporal variability of 10-year global SeaWiFS time series of phytoplankton chlorophyll a concentration, *ICES J. Mar. Sci.: J. Conseil., 66*(7), 1547.

Vaughan, D. G. (2008), West Antarctic Ice Sheet collapse – The fall and rise of a paradigm, *Climatic Change, 91*, 65–79.

Vautard, R., M. Van Loon, M. Schaap, R. Bergström, B. Bessagnet, J. Brandt, P. J. H. Builtjes, J. H. Christensen, C. Cuvelier, A. Graff, J. E. Jonson, M. Krol, J. Langner, P. Roberts, L. Rouil, R. Stern, L. Tarrasón, P. Thunis, E. Vignati, L. White, and P. Wind (2006), Is regional air quality model diversity representative of uncertainty for ozone simulation?, *Geophys. Res. Lett., 33*, L24818, doi:10.1029/2006GL27610.

Vázquez-Domínguez, E., D. Vaqué, and J. M. Gasol (2007), Ocean warming enhances respiration and carbon demand of coastal microbial plankton, *Glob. Change Biol., 13*, 1327–1334.

Vecchi, G. A., and T. R. Knutson (2008), On estimates of historical North Atlantic tropical cyclone activity, *J. Climate, 21*, 3580–3600.

Vecchi, G. A., and B. J. Soden (2007), Global warming and the weakening of the tropical circulation, *J. Climate, 20*(17), 4316–4340.

Vecchi, G. A., B. J. Soden, A. T. Wittenberg, I. M. Held, A. Leetmaa, and M. J. Harrison (2006), Weakening of tropical Pacific atmospheric circulation due to anthropogenic forcing, *Nature, 441*, 73–76, doi:10.1038/nature04744.

Vecchi, G. A., A. Clement, and B. J. Soden (2008), Examining the tropical Pacific's response to global warming, *Eos Trans. AGU, 89*(9), 81–83.

Veizer, J., et al. (1999), ^{87}Sr/^{86}Sr, δ^{13}C and δ^{18}O evolution of Phanerozoic seawater, *Chem. Geol., 161*, 59–88.

Velasco, E., and M. Roth (2010), Cities as net sources of CO$_2$: Review of atmospheric CO$_2$ exchange in urban environments measured by eddy covariance technique, *Geography Compass, 4*(9), 1238–1259, doi:10.1111/j.1749-8198.2010.00384.x.

Velasco, E., S. Pressley, R. Grivicke, E. Allwine, T. Coons, W. Foster, B. T. Jobson, H. Westberg, R. Ramos, F. Hernandez, L. T. Molina, and B. Lamb (2009), Eddy covariance flux measurements of pollutant gases in urban Mexico City, *Atmos. Chem. Phys., 9*, 7325–7342.

Velicogna, I. (2009), Increasing rates of ice mass loss from the Greenland and Antarctic ice sheets revealed by GRACE, *Geophys. Res. Lett., 36*, L19503, doi:10.1029/2009GL040222.

Vellinga, M., and P. Wu (2004), Low-latitude freshwater influence on centennial variability of the Atlantic thermohaline circulation, *J. Climate, 17*, 4498–4511.

Venegas, S. A. and L. A. Mysak (2000), Is there a dominant timescale of natural climate variability in the Arctic?, *J. Climate, 13*, 3412–3434.

Venkatesan, M. I., and J. Dahl (1989), Further geochemical evidence for global fires at the Cretaceous-Tertiary boundary, *Nature, 338*, 57–60.

Verati, C., H. Bertrand, and G. Feraud (2005), The farthest record of the Central Atlantic Magmatic Province into the West Africa craton: Precise ^{40}Ar/^{39}Ar dating and geochemistry of Taudenni basin intrusives (northern Mali), *Earth Planet. Sci. Lett., 235*, 391–407.

Verheggen, B., J. Cozic, E. Weingartner, K. Bower, S. Mertes, P. Connoly, M. Gallagher, M. Flynn, T. Choularton, and U. Baltensberger (2007), Aerosol portioning between the interstitial and the condensed phase in mixed-phase clouds, *J. Geophys. Res., 112*, doi:10.1029/2007JD008714.

Verma, S., O. Boucher, M. S. Reddy, H. C. Upadhyaya, P. Le Van, and F. S. Binkowski (2007), Modeling and analysis of aerosol processes in an interactive chemistry general circulation model, *J. Geophys. Res., 112*, D03207.

Vermeer, M., and S. Rahmstorf (2009), Global sea level linked to global temperature, *Proc. Natl. Acad. Sci. USA, 106*(51), 21527–21532, doi:10.1073/pnas.0907765106.

Vernadsky, V. I. (1926), *The Biosphere*, Nauchtechizdat, Leningrad.

Vernekar, A. D., J. Zhou, and J. Shukla (1995), The effect of Eurasian snow cover on the Indian monsoon, *J. Climate, 8*, 248–266.

Veron, J. E. N., O. Hoegh-Guldberg, T. M. Lenton, J. M. Lough, D. O. Obura, P. Pearce-Kelly, C. R. C. Sheppard, M. Spalding, M. G. Stafford-Smith, and A. D. Rogers (2009), The coral reef crisis: The critical importance of <350ppm CO$_2$, *Mar. Pollut. Bull., 58*, 1428–1437.

Vesala, T., L. Järvi, S. Launiainen, A. Sogachev, U. Rannik, I. Mammarella, E. Siivola, P. Keronen, J. Rinne, A. Riikonen, and E. Nikinmaa (2008), Surface-atmosphere interactions over complex urban terrain in Helsinki, Finland, *Tellus B, 60*, 188–199.

Vestreng, V., E. Rigler, M. Adams, K. Kindborn, J. M. Pacyna, H. Denier van der Gon, S. Reis, and O. Travnikov (2006), Inventory Review 2006, Emission Data reported to the LRTAP Convention and NEC Directive: Stage 1, 2 and 3 review, and evaluation of inventories of HMs and POPs, *MSC-W Technical Report 1/2006*, 1504–6179.

Vettoretti, G., and W. R. Peltier (2004), Sensitivity of glacial inception to orbital and greenhouse gas climate forcing, *Quaternary Sci. Rev., 23*, 499–519.

Victorian Government Department of Health Services (VGDHS) (2009), January 2009 heatwave in Victoria: An assessment of health impacts, State Government of Victoria report.

Villeneuve, M., H. A. Sandeman, and W. J. Davis (2000), A method for intercalibration of U Th-Pb and 40Ar-39Ar ages in the Phanerozoic, Geochem. Cosmochem. Ac., 64, 4017–4030.

Vimeux, F., K. M. Cuffey, and J. Jouzel (2002), New insights into Southern Hemisphere temperature changes from Vostok ice cores using deuterium excess correction, Earth Planet. Sci. Lett., 203, 829–843.

Vincent, L. A., X. Zhang, B. R. Bonsal, and W. D. Hogg (2002), Homogenization of daily temperatures over Canada, J. Climate, 15, 1322–1334.

Vincent, L. A., T. C. Peterson, V. R. Barros, M. B. Marino, M. Rusticucci, G. Carrasco, E. Ramirez, L. M. Alves, T. Ambrizzi, M. A. Berlato, A. M. Grimm, J. A. Marengo, L. Molion, D. F. Moncunill, E. Rebello, Y. M. T. Anunciação, J. Quintana, J. L. Santos, J. Baez, G. Coronel, J. Garcia, I. Trebejo, M. Bidegain, M. R. Haylock, and D. Karoly (2005), Observed trends in indices of daily temperature extremes in South America, 1960–2000, J. Climate, 18, 5011–5023.

Vingarzan, R. (2004), A review of surface ozone background levels and trends, Atmos. Environ., 38, 201–401.

Visbeck, M. (2009), A station-based Southern Annular Mode index from 1884 to 2005, J. Climate, 22, 940–950.

Visbeck, M., E. P. Chassignet, R. Curry, T. Delworth, B. Dickson, and G. Krahmann (2003), The ocean's response to North Atlantic Oscillation variability, Geoph. Monog. Series, 113–145.

Voelker, A. H. L. (2002), Global distribution of centennial-scale records for Marine Isotope Stage (MIS): 3. A database, Quaternary Sci. Rev., 21(10), 1185–1212, doi:10.1016/S0277-3791(01)00139-1.

Vogelsang, E., M. Sarnthein, and U. Pflaumann (2001), δ18O Stratigraphy, Chronology, and Sea Surface Temperatures of Atlantic Sediment Records (GLAMAPF2000 Kiel), Christian-Albrechts Universitat zu Kiel, Germany.

Vogt, R., et al. (2006), Temporal dynamics of CO_2 fluxes and profiles over a central European city, Theor. Appl. Climatol., 84, 117–126.

Volk, T., and M. I. Hoffert (1985), Ocean carbon pumps-analysis of relative strengths and efficiencies in ocean-driven atmospheric CO_2 changes, in The Carbon Cycle and Atmospheric CO_2: Natural Variations Archean to Present, 1, 99–110.

von Caemmerer, S. (2000), Biochemical Models of Leaf Photosynthesis, Springer.

von Caemmerer, S. and J. R. Evans (2010), Enhancing C_3 Photosynthesis, Plant Physiol., 154, 589–592.

von Caemmerer, S., and G. D. Farquhar (1981), Some relationships between the biochemistry of photosynthesis and the gas exchange of leaves, Planta, 153, 376–387.

von Caemmerer, S., G. D. Farquhar, and J. A. Berry (2009), Integrated Modeling of Light and Dark Reactions of Photosynthesis, Biochemical Model of C_3 Photosynthesis, Photosynthesis in silico: Understanding complexity from leaves to ecosystems, Springer.

von Hillebrandt, A., L. Krystyn, and W. M. Kuerschner (2007), A candidate GSSP for the base of the Jurassic in the Northern Calcareous Alps (Kuhjoch section, Karwendel Mountains, Tyrol, Austria), International Subcommission on Jurassic Stratigraphy Newsletter, 34(1), 2–20.

von Storch, H., and F. W. Zwiers (2001), Statistical Analysis in Climate Research, Cambridge University Press, Cambridge, UK.

von Storch, H., H. Langenberg, and F. Feser (2000), A spectral nudging technique for dynamical downscaling purposes, Mon. Weather Rev., 128, 3664–3673.

Voogt, J. A., How researchers measure urban heat islands, http://www.epa.gov/heatisld/resources/pdf/EPA_How_to_measure_a_UHI.pdf. (Accessed 25/10/10.)

Voogt, J. A., and C. S. B. Grimmond (2000), Modeling surface sensible heat flux using surface radiative temperatures in a simple urban area, J. Appl. Meteorol., 39, 1679–1699.

Voogt, J. A., and T. R. Oke (2003), Thermal remote sensing of urban climates, Remote Sens. Environ., 86, 370–384.

Vörösmarty, C. J. (2002), Global water assessment and potential contributions from Earth systems science, Aquat. Sci., 64(4), 328–351, doi:10.1007/PL00012590.

Vörösmarty, C. J., P. B. McIntyre, M. O. Gessner, D. Dudgeon, A. Prusevich, P. Green, S. Glidden, S. E. Bunn, C. A. Sullivan, R. C. Liermann, and P. M. Davis (2010), Global threats to human water security and river biodiversity, Nature, 467, 555–561, doi:10.1038/nature09440.

Vose, R. S., D. R. Easterling, and B. Gleason (2005), Maximum and minimum temperature trends for the globe: an update through 2004, Geophys. Res. Lett., 32(23), L23822.

Vrac, M., and P. Yiou (2010), Weather regimes designed for local precipitation modeling: Application to the Mediterranean Basin, J. Geophys. Res.-Atmos., 115, D12103.

Vrac, M., P. Marbaix, D. Paillard, and P. Naveau (2007), Non-linear statistical downscaling of present and LGM precipitation and temperatures over Europe, Clim. Past, 3(4), 669–682.

Wagner, T., S. Beirle, M. Grzegorski, and U. Platt (2006), Global trends (1996–2003) of total column precipitable water observed by Global Ozone Monitoring Experiment (GOME) on ERS-2 and their relation to near-surface temperature, J. Geophys. Res., 111, D12102, doi:10.1029/2005JD006523.

Walker, K. J. (1994), The Political Economy of Environmental Policy: An Australian Introduction, New South Wales University Press, Kensington.

Walsh, K. J. E., and B. F. Ryan (2000), Tropical cyclone intensity increase near Australia as a result of climate change, J. Climate, 13, 3029–3036.

Walsh K. J. E., K.-C. Nguyen, and J. L. McGregor (2004), Finer-resolution regional climate model simulations of the impact of climate change on tropical cyclones near Australia, Clim. Dynam., 22, 47–56.

Walter, B. P., M. Heimann, and E. Matthews (2001), Modeling modern methane emissions from natural wetlands: 2. Inter-annual variations 1982–1993, J. Geophys. Res., 106, 34207–34219.

Walter, B. P., S. A. Zimov, J. P. Chanton, D. Verbyla, and F. S. Chapin (2006a), Methane bubbling from Siberian thaw lakes as a positive feedback to climate warming, Nature, 443, 71–75.

Walter, K. M., S. A. Zimov, J. P. Chanton, D. Verbyla, and F. S. Chapin III (2006), Methane bubbling from Siberian thaw lakes as a positive feedback to climate warming, Nature, 443, 71–75.

Wan, H., X. L. Wang, and V. R. Swail (2007), A quality assurance system for Canadian hourly pressure data, J. Appl. Meteor. Climatol., 46(11), 1804–1817.

Wan, S., R. J. Norby, J. Ledford, and J. F. Weltzin (2007), Response of soil respiration to elevated CO_2, air warming, and changing soil water availability in a model old-field grassland, Glob. Change Biol., 13, 2411–2424.

Wand, S. J. E., G. F. Midgley, M. H. Jones, and P. S. Curtis (1999), Response of wild C_4 and C_3 grass (Poaceae) species to elevated atmospheric CO_2 concentration: A meta-analytic test of current theories and perceptions, *Glob. Change Bio., 5*, 723−741.

Wang, B., R. Wu, and K. Lau (2001), Interannual variability of the Asian summer monsoon: Contrasts between the Indian and the western North Pacific-East Asian monsoons, *J. Climate, 14*(20), 4073−4090.

Wang, B., R. Wu, and T. Li (2003), Atmosphere-warm ocean interaction and its impacts on Asian-Australian monsoon variation, *J. Climate, 16*(8), 1195−1211.

Wang, C. (2005), A modeling study of the response of tropical deep convection to the increase of cloud condensation nuclei concentration: 1. Dynamics and microphysics, *J. Geophys. Res., 110*(D6), D21211

Wang, G., and H. H. Hendon (2007), Sensitivity of Australian rainfall to inter-El Niño variations, *J. Climate, 20*(16), 4211−4226.

Wang, J. F., and X. B. Zhang (2008), Downscaling and projection of winter extreme daily precipitation over North America, *J. Climate, 21*(5), 923−937.

Wang, S., R. Gillies, E. Takle, and W. Gutowski, Jr. (2009), Evaluation of precipitation in the intermountain region as simulated by the NARCCAP regional climate models, *Geophys. Res. Lett., 36*(11).

Wang, W.-C. and I. S. A. Isaksen (Eds.) (1995), *Atmospheric Ozone as a Climate Gas*, NATO ASI Series, Springer-Verlag, Berlin.

Wang, W.-C., and N. D. Sze (1980), Coupled effects of atmospheric N_2O and O_3 on the Earth's climate, *Nature, 286*, 589−590, doi:10.1038/286589a0.

Wang, W.-C., J.-P. Chen, I. S. A. Isaksen, I.-C. Tsai, K. Noone, and K. McGuffie (2012), Climate−chemistry interaction: Future tropospheric ozone and aerosols, in *The Future of the World's Climate*, edited by A. Henderson-Sellers and K. McGuffie, pp. 367−399, Elsevier, Amsterdam.

Wang, X. L. (2003), Comments on 'Detection of undocumented changepoints: A revision of the two-phase regression model', *J. Climate, 16*, 3383−3385.

Wang, X. L. (2008), Accounting for autocorrelation in detecting mean-shifts in climate data series using the penalized maximal t or F test, *J. Appl. Meteorol. Climatol., 47*, 2423−2444.

Wang, X. L., and V. R. Swail (2001), Changes of extreme wave heights in Northern Hemisphere oceans and related atmospheric circulation regimes, *J. Climate, 14*, 2204− 2220.

Wang, X. L., V. R. Swail, and F. W. Zwiers, (2006), Climatology and changes of extra-tropical cyclone activity: Comparison of ERA-40 with NCEP/NCAR Reanalysis for 1958−2001, *J. Climate, 19*(13), 3145−3166, doi:10.1175/JCLI3781.1.

Wang, X. L., V. R. Swail, F. W. Zwiers, X. B. Zhang, and Y. Feng (2009), Detection of external influence on trends of atmospheric storminess and northern oceans wave heights, *Clim. Dynam., 32*, 189−203.

Wang, Y. (2001), An explicit simulation of tropical cyclones with a triply nested movable mesh primitive equation model: TCM3, 1: Model description and control experiment, *Mon. Weather Rev., 129*, 1370−1394.

Wang, Y. H., and D. J. Jacob (1998), Anthropogenic forcing on tropospheric ozone and OH since preindustrial times, *J. Geophys. Res., 103*, 123−131.

Wang, Y. X., M. B. McElroy, R. V. Martin, D. G. Streets, Q. Zhang, and T. M. A. Fu (2007), Seasonal variability of NO_x emissions over east China constrained by satellite observations: implications for combustion and microbial sources, *J. Geophys. Res., 112*, D06301, doi:10.1029/2006JD007538.

Wang, Y.-M., J. L. Lean, and N. R. Sheeley (2005), Modeling the Sun's magnetic field and irradiance since 1713, *Astrophys. J., 625*, 522−538.

Wang, Y.-P., and B. Z. Houlton (2009), Nitrogen constraints on terrestrial carbon uptake: Implications for the global carbon-climate feedback, *Geophys. Res. Lett., 36*, L24403, doi:10.1029/2009GL041009.

Wang, Y., L. R. Leung, J. L. McGregor, D.-K. Lee, W.-C. Wang, Y. Ding, and F. Kimura (2004), Regional climate modelling: Progress, challenges, and prospects, *J. Meteorol. Soc. Jpn., 82*(6), 1599−1628.

Wang., Y.-P., R. M. Law, and B. Pak (2009), A global model of carbon, nitrogen and phosphorus cycles for the terrestrial biosphere, *Biogeosciences, 7*, 2261−2282, doi:10.5194/bg-7-2261-2010.

Wania, R. (2007), Modelling Northern Peatland Land Surface Processes, Vegetation Dynamics and Methane Emissions, PhD thesis, University of Bristol, UK.

Ward, B. (2010), Mourning the huge loss of a 'giant': Stanford climatologist Stephen H. Schneider, *The Yale Forum on Climate Change and the Media*, 20 July, 2010, *http://www.yaleclimatemediaforum.org/2010/07/stanford-climatologist-stephen-h-schneider/.* (Accessed 24/10/10.)

Ward, P. D., J. W. Haggart, E. S. Carter, D. Wilbur, H. W. Tipper, and T. Evans (2001), Sudden productivity collapse associated with the Triassic-Jurassic boundary, *Science, 292*, 1148−1151.

Wardle, R., and I. Smith (2004), Modelled response of the Australian monsoon to changes in land surface temperatures, *Geophys. Res. Lett., 31*, L16205, doi:10.1029/2004GL020157.

Warner, R. (2009), Secular regime shifts, global warming and Sydney's water supply, *Geograph. Res., 47*(3), 227−241, doi:10.1111/j.1745-5871.2009.00593.x.

Warren, S. G., and W. J. Wiscombe (1980), A model for the spectral albedo of snow: 2. Snow containing atmospheric aerosols, *J. Atmos. Sci., 37*, 2734−2745.

Warrick, R. A., C. Le Provost, M. F. Meier, J. Oerlemans, and P. L. Woodworth (1996), Changes in sea level, in *Climate Change 1995: The Science of Climate Change, contribution of Working Group I to the Second Assessment Report (SAR) of the Intergovernmental Panel on Climate Change (IPCC)*, edited by J. T. Houghton, et al., pp. 359−405, Cambridge University Press, Cambridge, UK.

Warrington, G., J. C. W. Cope, and H. C. Ivimey-Cook (1994), St Audrie's Bay, Somerset, England: A candidate Global Stratotype Section and Point for the base of the Jurassic System, *Geol. Mag., 133*, 191−200.

WASA Group (1998), Changing waves and storms in the northeast Atlantic?, *Bull. Am. Meteorol. Soc., 79*, 741−760.

Washington, R., C. Bouet, G. Cautenet, E. Mackenzie, I. Ashpole, S. Engelstaedter, G. Lizcano, G. M. Henderson, K. Schepanski, and I. Tegen (2009), Dust as a tipping element: The Bodélé Depression, Chad, *Proc. Natl. Acad. Sci. USA, 106*, 20564−20571.

Washington Post (2010), Michael Oppenheimer, quoted in Stephen H. Schneider, climate change expert, dies at 65, obituary by T. R. Shapiro, *Washington Post*, 20 July, 2010, *http://www.washingtonpost.com/wpdyn/content/article/2010/07/19/AR2010071905108.html.* (Accessed 24/10/2010.)

Watson, A. J., and J. E. Lovelock (1983), Biological homeostasis of the global environment: The parable of Daisyworld, *Tellus B*, 284−289.

Watterson, I. G., J. L. McGregor, and K. C. Nguyen (2008), Changes in extreme temperatures of Australasian summer simulated by CCAM under global warming, and the roles of winds and land-sea contrasts, *Aust. Meteorol. Mag., 57*, 195−212.

Way, D. A., and R. F. Sage (2008), Elevated growth temperatures reduce the carbon gain of black spruce [*Picea mariana* (Mill.) B.S.P.], *Glob. Change Biol., 14*, 624−636.

WCED (World Commission on Environment and Development) (1987), *Our Common Future*, Oxford University Press, Oxford.

Weart, S. R. (2004), The Discovery of Global Warming, Harvard University Press, Boston, MA, USA.

Weart, S. (2008), History of contributions of planetary studies to the science of climate change, in *Encyclopedia of Earth*, edited by C. J. Cleveland, *http://www.eoearth.org/article/History_of_ contributions_of_planetary_studies_to_the_science_of_climate_ change.* (Accessed 27/11/10).

Webster, P., V. Magana, T. Palmer, J. Shukla, R. Tomas, M. Yanai, and T. Yasunari (1998), Monsoons: Processes, predictability, and the prospects for prediction, *J. Geophys. Res.-Oceans, 103*(C7), 14451−14510.

Webster, P. J., G. J. Holland, J. A. Curry, and H.-R. Chang (2005), Changes in tropical cyclone number, duration, and intensity in a warming environment, *Science, 309*, 1844−1846, doi:10.1126/science.1116448.

Weedon, J. T., R. Aerts, G. A. Kowalchuk, and P. M. van Bodegom (2011), Enzymology under global change: Organic nitrogen turnover in alpine and sub-Artic soils, *Biochem. Soc. T., 39*, 309−314, doi:10.1042/BST0390309.

Weertman, J. (1974), Stability of the junction of an ice sheet and an ice shelf, *J. Glaciol., 13*, 3−13.

Weertman, J. (1976), Glaciology's grand unsolved problem, *Nature, 260*(5549), 284−286.

Weijer, W., W. P. M. De Ruijter, A. Sterl, and S. S. Drijfhout (2002), Response of the Atlantic overturning circulation to South Atlantic sources of buoyancy, *Glob. Planet Change, 34*(3−4), 293−311.

Weiss, R. F. (1974), Carbon dioxide in water and seawater: the solubility of a non-ideal gas. *Marine Chemistry, 2*, 203−215.

Weisse, R., U. Mikolajewicz, and E. Maier-Reimer (1994), Decadal variability of the North Atlantic in an ocean general circulation model, *J. Geophys. Res., 99*(C6), 12411−12421.

Weisse, R., U. Mikolajewicz, A. Sterl, and S. S. Drijfhout (1999), Stochastically forced variability in the Antarctic Circumpolar Current, *J. Geophys. Res., 104*(C5), 11049−11064.

Weldeab, S., D. W. Lea, R. R. Schneider, and N. Andersen (2007), 155,000 years of West African monsoon and ocean thermal evolution, *Science, 316*(5829), 1303−1307, doi:10.1126/science.1140461.

Wentz, F. J., L. Ricciardulli, K. Hilburn, and C. Mears (2007), How much more rain will global warming bring?, *Science, 317*(5835), 233−235, doi:10.1126/science.1140476.

Werner, M., U. Mikolajewicz, M. Heimann, and G. Hoffmann (2000), Borehole versus isotope temperatures on Greenland: Seasonality does matter, *Geophys. Res. Lett., 27*(5), 723−726, doi:10.1029/ 1999GL006075.

Werth, D., and R. Avissar (2002), The local and global effects of Amazon deforestation, *J. Geophys. Res., 107*, 8087, doi:10.1029/2001 JD000717.

Werth, D., and R. Avissar (2005), The local and global effects of African deforestation, *Geophys. Res. Lett., 32*, L12704, doi:10.1029/ 2005GL022969.

Westbrook, G. K., K. E. Thatcher, E. J. Rohling, A. M. Piotrowski, H. Pälike, A. H. Osborne, E. G. Nisbet, T. A. Minshull, V. Hühnerbach, D. Green, R. E. Fisher, A. J. Crocker, A. Chabert, C. Bolton, A. B. Möller, C. Berndt, and A. Aquilina (2009), Escape of methane gas from the seabed along the West Spitsbergen continental margin, *Geophys. Res. Lett., 36*, L15608, doi:10.1029/2009GL039191.

Westerling, A. L., H. G. Hidalgo, D. R. Cayan, and T. W. Swetnam (2006), Warming and earlier spring increase western U.S. forest wildfire activity, *Science, 313*, 940−943.

White, W. B., and R. Peterson (1996), An Antarctic circumpolar wave in surface pressure, wind, temperature, and sea-ice extent, *Nature, 380*, 699−702.

Whiteside, J. H., P. E. Olsen, T. Eglinton, M. E. Brookfield, and R. N. Sambrotto (2010), Compound-specific carbon isotopes from Earth's largest flood basalt eruptions directly linked to the end-Triassic mass extinction, *Earth Planet. Sci. Lett., 107*, 6721−6725.

Wiesenfeld, K., and B. McNamara (1986), Small-signal amplification in bifurcating dynamical systems, *Phys. Rev. A, 33*(1), 629.

Wigley, T. M. L. (2006), A combined mitigation/geoengineering approach to climate stabilization, *Science, 314*(5798), 452− 454.

Wigley, T. M. L., and S. C. B. Raper (1992), Implications for climate and sea level of revised IPCC emissions scenarios, *Nature, 357*, 293−300.

Wigley, T. M. L., and S. C. B. Raper (1995), An heuristic model for sea level rise due to the melting of small glaciers, *Geophys. Res. Lett., 22*(20), 2749−2752.

Wigley, T. M. L., and S. C. B. Raper (2001), Interpretation of high projections for global-mean warming, *Science, 293*, 451−454, doi:10.1126/science.1061604.

Wigley, T. M. L., C. M. Ammann, B. D. Santer, and S. C. B. Raper (2005), Effect of climate sensitivity on the response to volcanic forcing, *J. Geophys. Res., 110*, D09107, doi:10.1029/2004JD005557.

Wignall, P. B. (2001), Large Igneous Provinces and mass extinctions, *Earth Sci. Rev., 53*, 1−33.

Wijffels, S. E., J. Willis, C. M. Domingues, P. Barker, N. J. White, A. Gronell, K. Ridgway, and J. A. Church (2008), Changing expendable bathythermograph fall rates and their impact on estimates of thermosteric sea level rise, *J. Climate, 21*(21), 5657−5672, doi:10.1175/2008JCLI2290.1.

Wijngaard, J. B., A. M. G. Klein Tank, and G. P. Können (2003), Homogeneity of 20th century European daily temperature and precipitation series, *Int. J. Climatol., 23*, 679−692.

Wilby, R., and I. Harris (2006), A framework for assessing uncertainties in climate change impacts: Low-flow scenarios for the River Thames, UK, *Water Resour. Res., 42*(2), doi:10.1029/2005WR004065.

Wilby, R. L., and T. M. L. Wigley (1997), Downscaling general circulation model output: A review of methods and limitations, *Progr. Phys. Geogr., 21*, 530−548.

Wilby, R., and T. Wigley (2000), Precipitation predictors for downscaling: Observed and general circulation model relationships, *Int. J. Climatol., 20*(6), 641−661.

Wilby, R., T. Wigley, D. Conway, P. Jones, B. Hewitson, J. Main, and D. Wilks (1998), Statistical downscaling of general circulation model output: A comparison of methods, *Water Resour. Res., 34*(11), 2995−3008.

Wild, M. (2009), Global dimming and brightening: A review, *J. Geophys. Res., 114*, D00D16, doi:10.1029/2008JD011470.

Wild, M., A. Ohmura, H. Gilgen, and D. Rosenfeld (2004), On the consistency of trends in radiation and temperature records and implications for the global hydrological cycle, *Geophys. Res. Lett., 31*, L11201, doi:10.1029/2003GL019188.

Wild, M., H. Gilgen, A. Roesch, A. Ohmura, C. N. Long, E. G. Dutton, B. Forgan, A. Kallis, V. Russak, and A. Tsvetkov (2005), From dimming to brightening: Decadal changes in surface solar radiation, *Science, 308*, 847–850.

Wild, M., A. Ohmura, and K. Makowski (2007), Impact of global dimming and brightening on global warming, *Geophys. Res. Lett., 34*, L04702, doi:10.1029/2006GL028031.

Wild, O. (2007), Modelling the global tropospheric ozone budget: Exploring the variability in current models, *Atmos. Chem. Phys., 7*, 2643–2660.

Wilf, P., and K. R. Johnson (2004), Land plant extinction at the end of the Cretaceous: A quantitative analysis of the North Dakota megafloral record, *Paleobiology, 30*, 347–368.

Wilf, P., K. R. Johnson, and B. T. Huber (2003), Correlated terrestrial and marine evidence for global climate changes before mass extinction at the Cretaceous-Paleogene boundary, *Proc. Natl. Acad. Sci. USA, 100*, 599–604.

Wilks, D. S. (1995), *Statistical Methods in the Atmospheric Sciences*, New York, Academic Press.

Wilks, D. S., and R. L. Wilby (1999), The weather generation game: A review of stochastic weather models, *Progr. Phys. Geogr., 23*(3), 329–357.

Williams, J., R. G. Barry, and W. M. Washington (1974), Simulation of the atmospheric circulation using the NCAR global circulation model with Ice-Age boundary conditions, *J. Appl. Meteorol., 13*(3), 305–317.

Williams, J. W., T. Webb, III, P. H. Richard, and P. Newby (2000), Late Quaternary biomes of Canada and the eastern United States, *J. Biogeogr., 27*(3), 585–607, doi:10.1046/j.1365-2699.2000.00428.x.

Williams, J. W., B. Shuman, P. J. Bartlein, N. S. Diffenbaugh, and T. Webb (2010), Rapid, time-transgressive, and variable responses to early-Holocene midcontinental drying in North America, *Geology, 38*(2), 135–138.

Williams, K. D., W. J. Ingram, and J. M. Gregory (2008), Time variation of effective climate sensitivity in GCMs, *J. Climate, 21*, 5076–5090.

Williams, M. (2003), *Deforesting the Earth: From Prehistory to Global Crisis*, University of Chicago Press.

Williford, K. H., P. D. Ward, G. H. Garrison, and R. Buick (2007), An extended organic carbon-isotope record across the Triassic-Jurassic boundary in the Queen Charlotte Islands, British Columbia, Canada, *Palaeogeogr. Palaeoclim., 244*, 290–296.

Willis, K. J., and R. J. Whittaker (2000), The refugial debate, *Science, 287*, 1406–1407.

Willis, K. J., E. Rudner, and P. Sumegi (2000), The full-glacial forests of central and southeastern Europe, *Quaternary Res., 53*(2), 203–213, doi:10.1126/science.287.5457.1406.

Willmott, C. J., and S. M. Robeson (1995), Climatologically aided interpolation (CAI) of terrestrial air temperature, *Int. J. Climatol., 15*, 221– 229.

Wilson, K. B., D. D. Baldocchi, and P. J. Hanson (2000), Quantifying stomatal and non-stomatal limitations to carbon assimilation resulting from leaf aging and drought in mature deciduous tree species, *Tree Physiol., 20*, 787–797.

Wing, S. L. (2004), Mass extinctions in plant evolution, in *Extinctions in the History of Life*, edited by P. D. Taylor, pp. 61–97, Cambridge University Press, Cambridge, UK.

Winton, M. (2006), Amplified Arctic climate change: What does surface albedo feedback have to do with it, *Geophys. Res. Lett, 33*, L03701, doi:10.1029/2005GL025244.

Winton, M. (2006), Does the Arctic sea ice have a tipping point?, *Geophys. Res. Lett., 33*, L23504, doi:10.1029/2006GL028017.

Wissel, C. (1984), A universal law of the characteristic return time near thresholds, *Oecologia, 65*(1), 101–107.

Wittenberg, A. T. (2009), Are historical records sufficient to constrain ENSO simulations?, *Geophys. Res. Lett., 36*(12), L12702.

Wittrock, F., A. Richter, H. Oetjen, J. P. Burrows, M. Kanakidou, S. Myriokefalitakis, R. Volkamer, S. Beirle, U. Platt, and T. Wagner (2006), Simultaneous global observations of glyoxal and formaldehyde from space, *Geophys. Res. Lett., 33*(16), doi:10.1029/2006GL026310.

Wohlfahrt, J., S. P. Harrison, and P. Braconnot (2004), Synergistic feedbacks between ocean and vegetation on mid- and high-latitude climates during the mid-Holocene, *Clim. Dynam., 22*(2–3), 223–238, doi:10.1007/s00382-003-0379-4.

Wolbach, W. S., R. S. Lewis, and E. Anders (1985), Cretaceous extinctions: Evidence for wildfires and search for meteoritic material, *Science, 230*, 167–170.

Wolfe, J. A., and G. R. Upchurch (1986), Vegetation, climatic and floral changes at the Cretaceous-Tertiary boundary, *Nature, 324*, 148–152.

Wolfe, J. A., and G. R. Upchurch (1987), Leaf assemblages across the Cretaceous-Tertiary boundary in the Raton Basin, New Mexico and Colorado, *Proc. Natl. Acad. Sci. USA, 84*, 5096–5100.

Wolff, E. W., J. Chappellaz, T. Blunier, S. O. Rasmussen, and A. Svensson (2010), Millennial-scale variability during the last glacial: The ice core record, *Quaternary Sci. Rev., 29*(21–22), 2828–2838, doi:10.1016/j.quascirev.2009.10.013.

Wong, S., W.-C. Wang, I. S. A. Isaksen, T. K. Berntsen, and J. K. Sundet (2004), A global climate—chemistry model study of present-day tropospheric chemistry and radiative forcing from changes in tropospheric O_3 since the preindustrial period, *J. Geophys. Res., 109*, D11309, doi:10.1029/2003JD003998.

Wong, T. H. F., and R. R. Brown (2009), The water sensitive city: Principles for practice, *Water Sci. Technol., 60*(3), 673–682, doi:10.2166/wst.2009.436.

Wood, A. W., L. R. Leung, V. Sridhar, and D. P. Lettenmaier (2004), Hydrologic implications of dynamical and statistical approaches to downscaling climate model outputs, *Climatic Change, 62*(1–3), 189–216.

Wood, H. L., J. I. Spicer, and S. Widdicombe (2008), Ocean acidification may increase calcification rates, but at a cost, *Proc. Roy. Soc. B, 275*, 1767–1773.

Woodgate, R., K. Aagaard, and T. J. Weingartner (2006), Interannual changes in the Bering Strait fluxes of volume, heat and freshwater between 1991 and 2004, *Geophys. Res. Lett., 33*, L15609.

Woodward, F. I. (1987), Stomatal numbers are sensitive to increases in CO_2 from pre industrial levels, *Nature, 327*, 617–618.

Woodworth, P. L., and D. L. Blackman (2004), Evidence for systematic changes in extreme high waters since the mid-1970s, *J. Climate, 17*, 1190–1197.

Worden, H. M., K. W. Bowman, J. R. Worden, A. Eldering, and R. Beer (2008), Satellite measurements of the clear-sky greenhouse effect from tropospheric ozone, *Nat. Geosci., 1*, 305–308.

World Climate Conference 3 (2009), WS-8: Climate and more sustainable cities, *http://www.wcc3.org/sessions.php?session_list=WS-8.* (Accessed 31/10/10.)

World Food Summit (1996), Rome Declaration on World Food Security, 13–17 November, Rome, Italy, http://www.fao.org/docrep/003/w3613e/w3613e00.htm. (Accessed 27/10/10.)

Wu, L., Z. Liu, R. Gallimore, R. Jacob, D. Lee, and Y. Zhong (2003), Pacific decadal variability: The tropical Pacific mode and the North Pacific mode, *J. Climate, 16*(8), 1101–1120.

Wu, S., L. J. Mickley, D. J. Jacob, D. Rind, and D. G. Streets (2008a), Effects of 2000–2050 changes in global tropospheric ozone and the policy-relevant background surface ozone in the United States, *J. Geophys. Res., 113*, D18312, doi:10.1029/2007JD009639.

Wu, S., L. J. Mickley, E. M. Leibensperger, D. J. Jacob, D. Rind, and D. G. Streets (2008b), Effects of 2000–2050 global change on ozone air quality in the United States, *J. Geophys. Res., 113*, D06302, doi:10.1029/2007JD008917.

Wu, Z., P. Dijkstra, G. W. Koch, J. Peñuelas, and B. A. Hungate (2011), Responses of terrestrial ecosystems to temperature and precipitation change: A meta-analysis of experimental manipulation, *Glob. Change Bio., 17*, 927–942, doi:10.1111/j.1365-2486.2010.02302.x.

Wullschleger, S. D. (1993), Biochemical limitations to carbon assimilation in C_3 plants – A retrospective analysis of the A/C_i curves from 109 species, *J. Exp. Bot., 44*, 907–920.

Wutzler, T., and M. Reichstein (2008), Colimitation of decomposition by substrate and decomposers – a comparison of model formulations, *Biogeosciences, 5*, 749–759.

Wyrwoll, K. H., and G. H. Miller (2001), Initiation of the Australian summer monsoon 14,000 years ago, *Quatern. Int., 83–5*, 119–128.

Xie, B. G., Q. H. Zhang, and Y. Q. Wang (2008), Trends in hail in China during 1960–2005, *Geophys. Res. Lett., 35*, L13801.

Xie, S. P., C. Deser, G. A. Vecchi, J. Ma, H. Teng, and A. T. Wittenberg (2010), Global warming pattern formation: Sea surface temperature and rainfall, *J. Climate, 23*(4), 966–986.

Xie, Z. C., X. Wang, Q. H. Feng, E. S. Kang, Q. Y. Li, and L. Cheng (2006), Glacial runoff in China: an evaluation and prediction for the future 50 years, *Journal of Glaciology and Geocryology, 28*(4), 457–466.

Xie, Z. T., and I. P. Castro (2009), Large-eddy simulation for flow and dispersion in urban streets, *Atmos. Environ., 43*(13), 2174–2185.

Xu, J., and Y. Wang (2010), Sensitivity of the simulated tropical cyclone inner-core size to the initial vortex size, *Mon. Weather Rev., 138*(11), 4135–4157, doi:10.1175/2010MWR3335.1.

Yamaji, K., O. Toshimasa, U. Itsushi, K. Jun-ichi, P. Pakpong, and A. Hajime (2008), Future prediction of surface ozone over east Asia using models-3 community multiscale air quality modeling system and regional emission inventory in Asia, *J. Geophys. Res., 113*, D08306, doi:10.1029/2007JD008663.

Yang, H., and Q. Zhang (2008), Anatomizing the ocean's role in ENSO changes under global warming, *J. Climate, 21*(24), 6539, doi:10.1175/2008JCLI2324.1.

Yang, W., A. Bardossy, and H. Caspary (2010), Downscaling daily precipitation time series using a combined circulation- and regression-based approach, *Theor. Appl. Climatol., 102*(3–4), 439–454, doi:10.1007/s00704-010-0272-0.

Yao, T. D., J. C. Pu, A. X. Lu, Y. Q. Wang, and W. S. Wu (2007), Recent glacial retreat and its impact on hydrological processes on the Tibetan Plateau, China, and surrounding regions, *Arct. Antarct. Alp. Res., 39*, 642–650.

Yearley, S. (2009), Sociology and climate change after Kyoto: What roles for social science in understanding climate change, *Curr. Sociol., 57*(3), 389–405.

Yeh, S.-W., J.-S. Kug, B. Dewitte, M.-H. Kwon, B. P. Kirtman, and F.-F. Jin (2009), El Niño in a changing climate, *Nature, 461*(7263), 511–514.

Yeh, S.-W., B. P. Kirtman, J.-S. Kug, W. Park, and M. Latif (2011), Natural variability of the central Pacific El Niño event on multi-centennial timescales, *Geophys. Res. Lett., 38*, L02704, doi:10.1029/2010GL045886.

Yienger, J., M. Galanter, T. Holloway, M. Phadnis, S. Guttikunda, G. Carmichael, W. Moxim, and H. Levy, II (2000), The episodic nature of air pollution transport from Asia to North America, *J. Geophys. Res., 105*(D22), 26931–26945.

Yin, J., M. E. Schlesinger, and R. J. Stouffer (2009), Model projections of rapid sea-level rise on the northeast coast of the United States, *Nat. Geosci., 2*(4), 262–266, doi:10.1038/ngeo462.

Yin, Q. Z., and A. Berger (2010), Insolation and CO_2 contribution to the interglacial climate before and after the Mid-Brunhes Event, *Nat. Geosci., 3*(4), 243–246.

Yin, Q. Z., and A. Berger (2011), Individual contribution of insolation and CO_2 to the interglacial climates of the past 800,000 years, *Clim. Dynam.*, doi:10.1007/s00382-011-1013-5.

Yin, Q. Z., A. Berger, E. Driesschaert, H. Goosse, M. F. Loutre, and M. Crucifix (2008), The Eurasian ice sheet reinforces the East Asian summer monsoon during the interglacial 500,000 years ago, *Clim. Past, 4*, 79–90.

Yoo, K., J. Ji, A. Aufdenkampe, and J. Klaminder (2011), Rates of soil mixing and associated carbon fluxes in a forest versus tilled agriculture field: Implications for modeling the soil carbon cycle, *J. Geophys. Res.-Biogeosci., 116*, G01014, doi:10.1029/2010JG001304.

You, Q., S. Kang, E. Aguilar, N. Pepin, W. A. Flügel, Y. Yan, Y. Xu, Y. Zhang, and J. Huang (2011), Changes in daily climate extremes in China and their connection to the large-scale atmospheric circulation during 1961–2003, *Clim. Dynam., 36*, 2399–2417.

Young, D., and C. S. B. Grimmond (2009), Part II: Water, from Inventory of current state of empirical and modeling knowledge of energy, water and carbon sinks, sources and fluxes, in Report on EU Framework 7 Project D.2.1, edited by C. S. B. Grimmond, pp. 109, Ref. 211345_001_PU_KCL, *http://www.bridge-fp7.eu/.*

Young, L.-H., D. R. Benson, W. M. Montanaro, S.-H. Lee, L. L. Pan, D. C. Rogers, J. Jensen, J. L. Stith, C. A. Davis, T. L. Campos, K. P. Bowman, W. A. Cooper, and L. R. Lait (2007), Enhanced new particle formation observed in the northern midlatitude tropopause region, *J. Geophys. Res., 112*, D10218, doi:10.1029/2006JD00810.

Young, O. R. (2009), Governance for sustainable development in a world of rising interdependencies, in *Governance for the Environment: New Perspectives*, edited by M. A. Delmas, and O. R. Young, pp. 12–40, Cambridge University Press, Cambridge, UK.

Young, O., and W. Steffen (2009), The Earth System: Sustaining Planetary Life Support Systems, in Principles of Ecosystem

Stewardship: Resilience-Based Resource Natural Resource Management in a Changing World, edited by F. S. Chapin, III, G. P. Kofinas, and C. Folke, Springer, New York, USA.

Yow, D. M. (2007), Urban heat islands: Observations, impacts, and adaptation, *Geography Compass, 1*(6), 1227–1251, doi:10.1111/j.1749-8198.2007.00063.x.

Yu, G. E., and S. P. Harrison (1995), Holocene changes in atmospheric circulation patterns as shown by lake status changes in northern Europe, *Boreas, 24*(3), 260–268.

Yu, Z., J. Loisel, D. P. Brosseau, D. W. Beilman, and S. J. Hunt (2010), Global peatland dynamics since the Last Glacial Maximum, *Geophys. Res. Lett., 37*(13), L13402, doi:10.1029/2010GL043584.

Zachos, J. C., M. A. Arthur, and W. A. Dean (1989), Geochemical evidence for suppression of pelagic marine productivity at the Cretaceous/Tertiary boundary, *Nature, 337*, 61–64.

Zachos, J. C., M. Pagani, L. Sloan, E. Thomas, and K. Billups (2001), Trends, rhythms and aberrations in global climate 65 Ma to present, *Science, 292*, 686–693, doi:10.1126/science.1059412.

Zachos, J. C., G. R. Dickens, and R. E. Zeebe (2008), An early Cenozoic perspective on greenhouse warming and carbon cycle-dynamics, *Nature, 451*, 279–283.

Zaehle, S., P. Friedlingstein, and A. D. Friend (2010), Terrestrial nitrogen feedbacks may accelerate future climate change, *Geophys. Res. Lett., 37*, L01401, doi:10.1029/2009GL041345.

Zaehle, S., P. Friedlingstein, and A. D. Friend (2010a), Terrestrial nitrogen feedbacks may accelerate future climate change, *Geophys. Res. Lett., 37*, L01401, doi:10.1029/2009GL041345.

Zaehle, S., A. D. Friend, P. Friedlingstein, F. Dentener, P. Peylin, and M. Schulz (2010b), Carbon and nitrogen cycle dynamics in the O-CN land surface model: Role of the nitrogen cycle in the historical terrestrial carbon balance, *Global Biogeochem. Cycles, 24*, GB1006, doi:10.1029/2009GB003522.

Zahn, M., and H. von Storch (2008), A long-term climatology of North Atlantic polar lows, *Geophys. Res. Lett., 35*, L22702, doi:10.1029/2008GL035769.

Zaitchik, B., A. Macalady, L. Bonneau, and R. Smith (2006), Europe's 2003 heat wave: A satellite view of impacts and land-atmosphere feedbacks, *Int. J. Climatol., 26*(6), 743–769, doi:10.1002/joc.1280.

Zanchettin, D., A. Rubino, and J. H. Jungclaus (2010), Intermittent multidecadal-to-centennial fluctuations dominate global temperature evolution over the last millennium, *Geophys. Res. Lett., 37*, L14702, doi:10.1029/2010GL043717.

Zeebe, R. E., J. C. Zachos, and G. R. Dickens (2009), Carbon dioxide forcing alone insufficient to explain Palaeocene-Eocene thermal maximum warming, *Nat. Geosci, 2*, 576–580.

Zemp, M., W. Haeberli, M. Hoelzle, and F. Paul (2006), Alpine glaciers to disappear within decades?, *Geophys. Res. Lett., 33*, L13504, doi:10.1029/2006GL026319.

Zeng, G., and J. A. Pyle (2003), Changes in tropospheric ozone between 2000 and 2100 modeled in a chemistry-climate model, *Geophys. Res. Lett., 30*(7), 1392, doi:10.1029/2002GL016708.

Zeng, G., J. A. Pyle, and P. J. Young (2008), Impact of climate change on tropospheric ozone and its global budgets, *Atmos. Chem. Phys., 8*, 369–387.

Zerefos, C. S., K. Eleftheratos, C. Meleti, S. Kazadzis, A. Romanou, C. Ichoku, G. Tselioudis, and A. Bais (2009), Solar dimming and brightening over Thessaloniki, Greece, and Beijing, China, *Tellus B, 61*(4), 657–665.

Zhang, C. L., F. Chen, S. G. Miao, Q. C. Li, X. A. Xia, and C. Y. Xuan (2009), Impacts of urban expansion and future green planting on summer precipitation in the Beijing metropolitan area, *J. Geophys. Res.-Atmos., 114*, D02116.

Zhang, D., and M. J. McPhaden (2006), Decadal variability of the shallow Pacific meridional overturning circulation: Relation to tropical sea surface temperatures in observations and climate change models, *Ocean Model., 15*, 250–273, doi:10.1016/j.ocemod.2005.12.005.

Zhang, H., A. Henderson-Sellers, and K. McGuffie (1996), Impacts of tropical deforestation: 1. Process analysis of local climatic change, *J. Climate, 9*, 1497–1517.

Zhang, H., K. McGuffie, and A. Henderson-Sellers (1996), Impacts of tropical deforestation II: The role of large-scale dynamics, *J. Climate, 9*, 2498–2521.

Zhang, H., A. Henderson-Sellers, and K. McGuffie (2001), The compounding effects of tropical deforestation and greenhouse warming on climate, *Climatic Change, 49*, 309–338.

Zhang, J., U. S. Bhatt, W. V. Tangborn, and C. S. Lingle (2007), Response of glaciers in northwestern North America to future climate change: An atmosphere/glacier hierarchical modeling approach, *Ann. Glaciol., 46*, 283–90.

Zhang, L. M., D. V. Michelangeli, and P. A. Taylor (2006), Influence of aerosol concentration on precipitation formation in low-level stratiform clouds, *J. Atmos. Sci., 37*, 203–217.

Zhang, L., D. J. Jacob, K. F. Boersma, D. A. Jaffe, J. R. Olson, K. W. Bowman, J. R. Worden, A. M. Thompson, M. A. Avery, R. C. Cohen, J. E. Dibb, F. M. Flock, H. E. Fuelberg, L. G. Huey, W. W. McMillan, H. B. Singh, and A. J. Weinheimer (2009), Transpacific transport of ozone pollution and the effect of recent Asian emission increases on air quality in North America: An integrated analysis using satellite, aircraft, ozonesonde, and surface observations, *Atmos. Chem. Phys., 8*, 6117–6136.

Zhang, L., D. J. Jacob, X. Liu, J. A. Logan, K. Chance, A. Eldering, and B. R. Bojkov (2010), Intercomparison methods for satellite measurements of atmospheric composition: Application to tropospheric ozone from TES and OMI, *Atmos. Chem. Phys., 10*, 4725–4739.

Zhang, Q., D. G. Streets, K. He, Y. Wang, A. Richter, J. P. Burrows, I. Uno, C. J. Jang, D. Chen, Z. Yao, and Y. Lei (2007), NO_x emission trends for China, 1995–2004: The view from the ground and the view from space, *J. Geophys. Res., 112*, D22306.

Zhang, R., and T. L. Delworth (2006), Impact of Atlantic Multidecadal Oscillations on India/Sahel rainfall and Atlantic hurricanes, *Geophys. Res. Lett., 33*, L17712, doi:10.1029/2006GL026267.

Zhang, R., T. L. Delworth, and I. M. Held (2007), Can the Atlantic Ocean drive the observed multidecadal variability in Northern Hemisphere mean temperature?, *Geophys. Res. Lett., 34*, L02709, doi:10.1029/2006GL028683.

Zhang, X. B., F. W. Zwiers, G. C. Hegerl, F. H. Lambert, N. P. Gillett, S. Solomon, P. A. Stott, and T. Nozawa (2007), Detection of human influence on twentieth-century precipitation trends, *Nature, 448*(7152), 461–465. doi:10.1038/nature06025.

Zhang, X. B., F. W. Zwiers, and G. Hegel (2009), The influences of data precision on the calculation of temperature percentile indices, *Int. J. Climatol., 29*, 321–327.

Zhang, X. B., J. Wang, F. W. Zwiers, and P. Y Groisman (2010), The influence of large-scale climate variability on winter maximum daily precipitation over North America, *J. Climate, 23,* 2902–2915.

Zhang, X., M. A. Friedl, C. B. Schaaf, A. H. Strahler, and A. Schneider (2004), The footprint of urban climates on vegetation phenology, *Geophys. Res. Lett., 31,* L12209, 1–4.

Zhang, X., E. Aguilar, S. Sensoy, H. Melkonyan, U. Tagiyeva, N. Ahmed, N. Kutaladze, F. Rahimzadeh, A. Taghipour, T. H. Hantosh, P. Albert, M. Semawi, M. K. Ali, M. H. S. Al-Shabibi, Z. Al-Oulan, T. Zatari, I. A. Khelet, S. Hamoud, R. Sagir, M. Demircan, M. Eken, M. Adiguzel, L. V. Alexander, T. C. Peterson, and T. Wallis (2005), Trends in Middle East climate extremes indices during 1950–2003, *J. Geophys. Res., 110,* D22104, doi:10.1029/2005JD006181.

Zhang, X., J. Wang, F. Zwiers, and P. Groisman (2010), The influence of large-scale climate variability on winter maximum daily precipitation over North America, *J. Climate, 23*(11), 2902–2915, doi:10.1175/2010JCLI3249.1.

Zhang, Y., J. M. Wallace, and D. S. Battisti (1997), ENSO-like interdecadal variability: 1900-1993, *J. Climate, 10,* 1004–1020.

Zhao, M., and A. J. Pitman (2002), The impact of land cover change and increasing carbon dioxide on the extreme and frequency of maximum temperature and convective precipitation, *Geophys. Res. Lett., 29*(6), doi: 10.1029/2001GL013476.

Zhao, M., and S. W. Running (2010), Drought-induced reduction in global terrestrial net primary production from 2000 through 2009, *Science, 329,* 940–943, doi:10.1126/science.1192666.

Zhao, M., A. J. Pitman, and T. Chase (2001), Climatic effects of land-cover change at different carbon dioxide levels, *Clim. Res., 17,* 1–18.

Zhao, Y., and S. P. Harrison (2011), Mid-Holocene monsoons: a multimodel analysis of the inter-hemispheric differences in the responses to orbital forcing and ocean feedbacks, *Climate Dynamics* (in press).

Zhou, J., and K.-K. Tung (2010), Solar cycles in 150 years of global sea surface temperature data, *J. Climate, 23,* 3234– 3248.

Zhuang, Q., J. M. Melillo, D. W. Kicklighter, R. G. Prinn, A. D. McGuire, P. A. Steudler, F. B. S., and S. Hu (2004), Methane fluxes between terrestrial ecosystems and the atmosphere at northern high latitudes during the past century: A retrospective analysis with a process-based biogeochemistry model, *Global Biogeochem. Cycles, 18*(3), GB3010, doi:10.1029/2004GB002239.

Zickfeld, K., B. Knopf, V. Petoukhov, and H. J. Schellnhuber (2005), Is the Indian summer monsoon stable against global change?, *Geophys. Res. Lett., 32,* L15707.

Zickfeld, K., J. C. Fyfe, M. Eby, and A. J. Weaver (2008), Comment on "Saturation of the Southern Ocean CO_2 sink due to recent climate change," *Science, 319*(5863), 570b, doi:10.1126/science.1146886.

Ziegler, P. A. (1990), *Geological Atlas of Western and Central Europe,* Shell Internationale Petroleum, Maatschappij B. V., The Hague.

Ziemke, J. R., S. Chandra, and P. K. Bhartia (2001), "Cloud slicing": A new technique to derive upper tropospheric ozone from satellite measurements, *J. Geophys. Res., 106,* 9853–9868.

Ziervogel, G., and P. J. Ericksen (2010), Adapting to climate change to sustain food security, *Wiley Interdiscip. http://wires.wiley.com/WileyCDA/Rev. http://wires.wiley.com/WileyCDA/:* Clim. *http://wires.wiley.com/WileyCDA/ Change, 1*(4), 525–540, doi:10.1002/wcc.56.

Zimov, S. A., E. A. G. Schuur, and F. S. Chapin, III (2006), Permafrost and the global carbon budget, *Science, 312,* 1612–1613.

Ziska L. H., J. A. Bunce, and E. W. Goins (2004), Characterization of an urban-rural CO_2/temperature gradient and associated changes in initial plant productivity during secondary succession, *Oecologia, 39,* 454–458, doi:10.1007/s00442-004-1526-2.

Zollina, O., C. Simmer, A. Kapala, S. Bachner, S. Gulev, and H. Maechel (2008), Seasonally dependent changes of precipitation extremes over Germany since 1950 from a very dense observational network, *J. Geophys. Res.-Atmos., 113*(D6), D06110.

Zondervan, I., R. E. Zeebe, B. Rost, and U. Riebesell (2001), Decreasing marine biogenic calcification: A negative feedback on rising atmospheric pCO₂, *Global Biogeochem. Cycles, 15*(2), 507–516.

Zorita, E., and H. von Storch (1999), The analog method as a simple statistical downscaling technique: Comparison with more complicated methods, *J. Climate, 12*(8), 2474–2489.

Zou, L., T. Zhou, L. Li, and J. Zhang (2010), East China summer rainfall variability of 1958–2000: Dynamical downscaling with a variable-resolution AGCM, *J. Climate, 23*(23), 6394–6408, doi:10.1175/2010JCLI3689.1.

Zwiers, F. W., and V. V. Kharin (1998), Changes in the extremes of the climate simulated by CCC GCM2 under CO_2 doubling, *J. Climate, 11*(9), 2200–2222.

Zwiers, F. W., and H. von Storch (1995), Taking serial correlation into account in tests of the mean, *J. Climate, 8,* 336–351.

Zwiers, F. W., X. Zhang, and Y. Feng (2011), Anthropogenic influence on long return period daily temperature extremes at regional scales, *J. Climate, 24*(3), 881–892, doi:10.1175/2010JCLI3908.1.

Zwiers, F. W., X. Zhang, and J. Feng (2011), Anthropogenic influence on extreme daily temperatures at regional scales, *J. Climate,* (in press).

Note: Page numbers followed by *f* indicate figures and *t* indicate tables.

Ann Henderson-Sellers is an ISI 'most highly cited' author of over 500 publications on climate and climatic change. Prior to holding her current position, she was Director of the United Nations' World Climate Research Programme, Director of Environment at the Australian Nuclear Science and Technology Organisation and Deputy Vice-Chancellor of RMIT. Ann has championed the scientific need for action to mitigate and adapt to climate change for over 35 years (http://australianmuseum.net.au/Science-Direct-Professor-Ann-Henderson-Sellers). Dr Henderson-Sellers has been an Earth Systems scientist all her life spearheading the description & prediction of many aspects of people's interaction with climate systems (http://www.cowi.com/menu/specialfeatures/climate/expertinterviews/Pages/Nottoactisimmoral.aspx). She has a BSc in mathematics, undertook her PhD in collaboration with the U.K. Meteorological Office, earned a DSc (Leicester) in 1999 and was awarded an honorary DSc by Bristol University in 2011. She is an elected Fellow of Australia's Academy of Technological Sciences and Engineering, the American Geophysical Union and the American Meteorological Society. She was awarded the Centenary Medal of Australia for Service to Australian Society in Meteorology in 2003. Professor Henderson-Sellers now holds an Australian Research Council Fellowship in the Department of Environment and Geography at Macquarie University.

Kendal McGuffie is an Associate Professor in the School of Physics and Advanced Materials at the University of Technology, Sydney. He graduated from the University of Edinburgh with a BSc in Physics and went on to complete a PhD at the University of Liverpool on cloud—radiation interactions in the polar regions. He has been involved for around 20 years in climate modelling, including studies of the effects of tropical deforestation, hydrological cycle tracking using stable water isotopes and greenhouse gas increases as well as observational studies of tropical convection and in the development of novel observational platforms for tropical cyclone monitoring. Among his other publications is the now classic 'A Climate Modelling Primer': a guide to the rules and riddles of climate change science for interested readers and laypersons. (http://www.climatemodellingprimer.net/)

Lisa Alexander is a Senior Lecturer at the Climate Change Research Centre in the University of New South Wales in Sydney, Australia. She holds a BSc (Hons) and MSc in Applied Mathematics and a PhD in Climate Sciences. From 1998 to 2006 she was a research scientist in the Climate Variability Group at the UK Met Office Hadley Centre, spending her last year on secondment at the Australian Bureau of Meteorology.

Pat Bartlein is a Professor of Geography at the University of Oregon. His research focuses on the atmospheric circulation and surface water- and energy-balance controls of regional climatic variations on time scales ranging from palaeo to the present day. He is also interested in palaeoecology (in particular the history of disturbance by fire) and palaeohydrology, both as records of the responses to past climatic variations and as components of the Earth system. He has participated in the successive iterations of "data-model comparisons" involving suites of climate-model simulations and syntheses of palaeoclimatic data, and teaches courses in climatology, global environmental change, and the analysis and visualization of scientific data.

Claire M. Belcher is a Marie Curie Experienced Research Fellow in the School of Geosciences and BRE Centre for Fire Safety Engineering both at the University of Edinburgh. Claire is an Earth scientist specialising in the study of natural fires in the Earth system where her work has focused on providing measured data for testing the output of numerical models. She has established a reputation for her novel use of the fossil fire record combined with modern combustion experiments. She has used this approach to answer questions ranging from the mechanisms of mass extinction, the amounts of thermal radiation delivered by asteroid impacts, the composition of the Earth's atmosphere to the effects of climate-driven vegetation change on fire activity. Her research has been popular with the global media and featured in BBC Horizon's "What Really Killed the Dinosaurs" in 2004. In January 2012 Claire will be moving to the University of Exeter to take up the position of Senior Lecturer in Earth System Science.

Rasmus Benestad is a physicist by training and has affiliations with the Norwegian Meteorological Institute (met.no). He has a D.Phil in physics from Atmospheric, Oceanic & Planetary Physics at Oxford University in the United Kingdom. Recent work involves a good deal of statistics (empirical-statistical downscaling, trend analysis, model validation, extremes and record values), but he has also had some experience with electronics, cloud microphysics, ocean dynamics/air—sea processes and seasonal forecasting. He has written a text book with the title 'Solar Activity and Earth's Climate' (2002, 2006 published by Praxis-Springer), and 'Empirical-Statistical Downscaling' together with Deliang Chen and Inger Hanssen-Bauer (2008 published by World Scientific Publishers). He has been a member of the council of the European Meteorological Society for the period 2004—2006, representing the Nordic countries and the Norwegian Meteorology Society. Current positions include membership on different committees. In addition, much of his work involves communicating our knowledge about climate, and he is a contributor to www.Realclimate.org, www.camelclimatechange.org, and http://www.climaterapidresponse.org/.

André Berger has a degree in meteorology from M.I.T. and a Dr Sc. from the Université catholique de Louvain where he is Emeritus Professor and Senior Researcher. His main scientific contributions are in the astronomical theory of a palaeoclimates and modelling past climatic variations. He is the holder of an Advanced Investigators Grant of the European Research Council aiming to understand the climate of the last 800,000 years in collaboration with Dr Yin. He was director of the Institute of Astronomy and Geophysics Georges Lemaître. He has been chairman of the International Commission on Climate of IUGG, of the Paleoclimate Commission of INQUA, of Climate Commissions of the European Union and a member of the Scientific Committee of the European Environment Agency. He has received the Milankovitch Medal from the European Geophysical Society, the Quinquennial Prize from the National Fund for Scientific Research in Belgium and the European Latsis Prize from European Science Foundation. He is the Honorary president of the European Geo-Sciences Union, member of the Academia Europaea, of the Royal Academy of Belgium, and of the academies of Canada, Serbia, Paris and the Netherlands. He was ennobled Chevalier by His Majesty Albert II and made Officier de la Légion d'Honneur.

Terje Berntsen is currently a professor in meteorology at the University of Oslo, Norway. He also holds a part-time position at CICERO (Center for International climate and environmental research - Oslo). He got his PhD in 1994 from Department of Geophysics, University of Oslo. The title of his theses was: *Two- and three-dimensional model calculations of the photochemistry of the troposphere.* He has published more than 60 scientific papers, mainly on modelling of the chemical composition of the troposphere, the radiative forcing, and metrics of climate change. He was a lead author of chapter 2 of the fourth assessment report from the IPCC.

Philippe Bousquet is a professor at the Université de Versailles Saint-Quentin en Yvelines (UVSQ), researcher and deputy director of Laboratoire des Sciences du Climat et de l'Environnement (IPSL/LSCE). He is working on the estimation of greenhouse gases sources and sinks (CO_2, CH_4, N_2O) using atmospheric inversions of atmospheric transport and chemistry. He teaches air quality, physico-chemistry of the atmosphère and biogeochemical cycles at UVSQ. He has contributed more than 40 publications in A-ranked peer-reviewed journals. He has taken part in numerous European research projects since 1998.

Jen-Ping Chen is a Full Professor in the Department of Atmospheric Sciences, National Taiwan University since 2000. He received a Ph.D. degree from the Department of Meteorology, Pennsylvania State University in 1992, and then worked as a postdoctoral researcher in the Center for Cloud, Chemistry and Climate of the Scripps Institute of Oceanography before becoming a faculty member of the National Taiwan University in 1994. His major research interests include studies of cloud and aerosol microphysics, air pollution, and cloud-aerosol-climate interactions by using regional and global models, as well as observation data.

Helen Cleugh is an atmospheric scientist who has published 70 journal papers, book chapters, books and client reports in micrometeorology and convective boundary layer dynamics; urban microclimates; and land surface − climate interactions. Dr Cleugh obtained her Ph.D. at the University of British Columbia, in Vancouver, Canada and was a lecturer at Macquarie University before joining CSIRO in 1994. Dr Cleugh is currently Deputy Chief of the Marine and Atmospheric Research Division in CSIRO (Australia), where she leads the Division's climate and atmospheric research. She is part of a research team investigating carbon uptake and water use in a variety of Australian ecosystems, including forests, vineyards, savannas and even city suburbs, in order to provide better assessments of the budgets of carbon and water at the wide range of scales needed for sound management of these resources and as part of the OzFlux network.

Graham Cogley is a Professor in the Department of Geography at Trent University, Peterborough, Ontario, Canada. He is a physical geographer who has worked at various times during his career in geographical computer graphics, geomorphology, hydrology, climatology and most recently glaciology. In the latter subject his principal focus is on glacier mass balance, particularly estimates of the global-average contribution of small glaciers to sea-level change and ways of improving such estimates; and on global glaciology, including the drawn-out struggle to complete the World Glacier Inventory. Dr. Cogley was a contributing author for the Fourth Assessment Report of the Intergovernmental Panel on Climate Change and is a lead author of the IPCC's Fifth Assessment Report now in preparation. He has also contributed to other national and international climate assessments. He is currently the chairman of the Working Group on Mass-balance Terminology and Methods of the International Association of Cryospheric Sciences, and a member of the Advisory Board of the Global Climate Observing System's Global Terrestrial Network for Glaciers.

Bill Collins leads research at the Met Office Hadley Centre into atmospheric composition and climate. His research interests are in the interactions between reactive gases and the biosphere, and the impacts of short-lived species on climate forcing. He led the development of the Hadley Centre's first earth system model (HadGEM2) that combined chemistry, carbon cycle and climate.

Tony Cox started his research at UK's Harwell Laboratory and since 1995 has been Reader in Atmospheric Chemistry at the University Chemistry Department in Cambridge. His main contributions have been to the understanding of the chemistry of the atmosphere, through laboratory studies of kinetics and mechanisms of atmospheric reactions. He has published over 210 research papers, including 30 reviews and evaluations on a range of topics in atmospheric chemistry. Notable achievements include direct measurements of rate constants for HO_2, organic peroxy, and halogen oxide radical reactions, and elucidation of the mechanisms of atmospheric ozone formation and loss. He is still active in research related to atmospheric aerosols and heterogeneous chemical processes, and is a founding member of the IUPAC Panel for Evaluation of Kinetics and Photochemical Data for atmospheric chemistry. As leader of the ACCENT Network work-package 'Access to Laboratory Data', Dr Cox was in a position to ensure prompt dissemination of knowledge and data from these laboratory studies to the new challenges in modelling of global atmospheric chemistry arising from investigation of climate change.

Stig B. Dalsøren is a senior research fellow at CICERO. He holds a PhD degree in meteorology. He has published approximately 10 peer reviewed papers, and participated in the EU projects POET, RETRO and HYMN. His main research activities have been in numerical modelling of pollution and chemical active greenhouse gases like methane and ozone. The main topics have been changes in oxidation capacity, effects from emissions in the transport sector and emission changes in Asia.

Robert E. Dickinson currently works for the University of Texas at Austin in the Department of Geological Sciences of the Jackson School of Geosciences, one of the world's largest geoscience communities working across the globe in all major areas of the earth sciences. Before transferring there he was Professor of Atmospheric Sciences and held the Georgia Power/ Georgia Research Alliance Chair at the Georgia Institute of Technology. From 1990 to 1993 he was Professor of Atmospheric Sciences and Regents Professor at the University of Arizona, and from 1968 to 1990 he was a Senior Scientist at the National Center for Atmospheric Research. He was elected to the US National Academy of Sciences in 1988, to the US National Academy of Engineering in 2002, and a foreign member of the Chinese Academy of Sciences in 2006. He has served on many national and international committees. His research includes issues of climate modelling, climate variability and change, aerosols, the hydrological cycle and droughts, land surface processes, the terrestrial carbon cycle, and the application of remote sensing data to modelling of land surface processes.

Jason Evans is a senior research fellow at the Climate Change Research Centre at the University of New South Wales, Australia. He received his undergraduate degrees in mathematics and physics from the University of Newcastle, before completing his PhD at the Australian National University. Prior to coming to the University of New South Wales he spent 6 years at Yale University in the USA. In 2007 he became an Australian Research Fellow (awarded by the Australian Research Council) and in 2008 he was awarded an Australian Agricultural Industries Young Innovators and Scientists Award. He is currently chair of the Murray-Darling Basin Regional Hydroclimate Project, which is an element of the Global Energy and Water Cycle Experiment (GEWEX), and is backed by the World Climate Research Programme (WCRP). His research involves general issues of water cycle processes over land, and how we can change them, largely through changes in land use and climate. His research focus is at the regional (or watershed) scale and includes processes such as river flow, evaporation/transpiration, water vapour transport and precipitation. The main tools of his research are computational models including regional climate models, land surface and hydrology models.

Veronika Eyring is a Senior Scientist at the German Aerospace Center (DLR), Visiting Professor at the Manchester Metropolitan University, UK, and Privatdozentin at the Meteorological Institute, Ludwig-Maximilians-Universität (LMU) in Munich, Germany. She has long-term experience in chemistry–climate modelling and global model evaluation. She was lead-, co- and contributing author of recent WMO/UNEP ozone and IPCC climate change assessments reports and was coordinating lead author of the ATTICA European Assessment on Transport

(Shipping). She received the Dobson Award for Young Scientists granted by the International Ozone Commission (IO3C) in 2008. She has authored many peer-reviewed journal articles and serves on a large number of international scientific committees. In particular, she is involved in the World Climate Research Programme (WCRP) as scientific coordinator of WCRP's Chemistry-Climate Model Validation Activity (CCMVal) and as a steering committee member on WCRP projects, including the Stratospheric Processes And their Role in Climate (SPARC) project, the Atmospheric Chemistry and Climate (AC&C) initiative, the Working Group on Coupled Modelling (WGCM), and the Working Group on Numerical Experimentation (WGNE)/WGCM metrics panel.

David Fowler is the Senior Scientist at the Centre for Ecology and Hydrology and is based in Edinburgh. He trained in Environmental Physics at the University of Nottingham, obtaining a PhD in 1976 from research on dry deposition of SO_2. He has worked at the Institute of Terrestrial Ecology (now known as CEH) since 1975 on a range of atmospheric trace gases including SO_2, NO_2, NO, HNO_3, NH_3, O_3, CH_4, N_2O as well as aerosols and cloud droplets. He gained an Honorary Professorship at the University of Nottingham in 1991, became a Fellow of the Royal Society of Edinburgh in 1999, and a Fellow of the Royal Society of London in 2002. His research interests are on surface atmosphere exchange processes, photochemical oxidants, acid deposition, emissions of greenhouse gases, atmospheric aerosols and effects of pollutants on vegetation. David has been a contributing author to more than 220 refereed publications in addition to contributions to book chapters and conference proceedings. In 2005 he was awarded a C.B.E. for services to research into Atmospheric Pollution.

Sandro Fuzzi is Research Director at the Institute of Atmospheric Sciences and Climate (ISAC) of the Italian National Research Council (CNR) and Coordinator of the Global Change Project—Department Earth and Environment of CNR. His main research interests are physical and chemical processes of aerosols and clouds, and their effects on atmospheric composition change and climate. Sandro Fuzzi has coordinated several international projects and has published over 120 papers in refereed journals (h-index = 39), has been a member of several international panels such as the European Commission Panel on Atmospheric Composition Change, the International Commission on Clouds and Precipitation, and has been co-chair of the International Global Atmospheric Chemistry (IGAC) Steering Committee. Sandro Fuzzi has coordinated the European Network of Excellence—Atmospheric Composition Change (ACCENT), which included the major European institutions in the field of global change research, and is currently coordinating a follow-up programme aimed at facilitating the transfer of research results into policy-decision making. He has been a Contributing Author for the 4th IPCC Assessment Report and is presently involved as Review Editor in the preparation of the 5th IPCC Assessment Report.

Michael Gauss received a Master degree in Meteorology in 1998 and a PhD degree in atmospheric chemistry in 2003 from the University of Oslo, Norway, where he worked until 2010 in the field of modelling air pollution and climate change. He has been employed at the Norwegian Meteorological Institute since 2006 and is now involved in research on long-range trans-boundary air pollution and couplings between air pollution and climate change, in support of the UN LRTAP convention (http://www.unece.org/env/lrtap/). In 2010 and 2011 he was leading the EMEP Centre MSC-W (www.emep.int). He has co-authored about 30 peer-reviewed publications and has contributed to two IPCC assessments. He has gained international experience through his participation in more than ten EU-funded projects within atmospheric science. From 2008 to 2011 he coordinated the EU FP7 project CityZen and has also been coordinating a number of smaller projects or funded by the EU, the Nordic Council of Ministers, and the Norwegian government. His research interest now includes modelling present and future air pollution and climate change on regional to global scales, based on emission scenarios that assume different mitigation options.

Claire Granier has over 25 years research experience in atmospheric sciences. She has worked on the development and use of three-dimensional global chemistry-transport models for the study of the composition of the lower atmosphere and its evolution. She is also interested in the development of surface emission inventories from anthropogenic and biomass burning sources. She was the co-coordinator of the GEIA (Global Emissions Inventory Activity) project until the summer of 2011, and is now directing the development of the GEIA and ECCAD databases of emissions. Since 2005, she has been the Deputy Coordinator of the MACC (Monitoring Atmospheric Composition and Climate) FP7 Integrated Project. She is also a member of the steering committee of the IGAC/IGBP Project. She received her Master in Mathematics, and her PhD in Physics from the University of Paris. She is a scientist at the LATMOS/IPSL laboratory of the University Pierre and Marie Curie in Paris, France and at the NOAA Earth Research System Laboratory in Boulder, Colorado, USA.

Sue Grimmond, Chair of Physical Geography at King's College London since January 2006, was previously a Professor at Indiana University, Bloomington USA. She completed her undergraduate degree (BSc Hons) at the University of Otago, New Zealand, and graduate degrees (MSc, PhD) at the University of British Columbia, Canada. She was the founding chair of the Board of the Urban Environment of the American Meteorological Society (AMS), a past President of the International Association of Urban Climate (IAUC) and past Lead Expert for the World Meteorological Organization (WMO) on Urban and Building Climatology. She is an elected Fellow of the AMS, has a Doctor of Science Honoris Causa from Göteborg University Sweden, the Universitatis Lodziensis Amico Medal from University of Łódź Poland, a recipient of the Helmut E Landsberg Award from the AMS and Luke Howard Award from the IAUC. She studies the biophysical processes involved in the generation of urban climates through the measurement and modelling of surface—atmosphere exchanges of heat, mass (water and carbon dioxide) and momentum at a range of spatial and temporal scales (individual urban canyons, to neighbourhoods, to entire cities). She has conducted extensive fieldwork in North American, European, and African cities to evaluate parameterizations and numerical models.

Sandy P. Harrison, Professor at Macquarie University, is a palaeoclimate diagnostician with a special interest in the role of the land-surface, terrestrial biosphere and hydrological processes on modulating regional climates. She is President of the INQUA Commission on Palaeoclimatology (PALCOMM) and Co-leader of the Palaeoclimate Modelling Intercomparison Project (PMIP). She has coordinated several international palaeoclimate data synthesis initiatives including the Global Lake Status Database, BIOME6000, DIRTMAP, SNOWLINE and the Global Palaeofires Working Group.

Danny Harvey is a Professor in the Department of Geography, University of Toronto, where he teaches courses related to climate, global warming and energy use. During the 1980s and 1990s his work focused on climate and carbon cycle modelling, with a particular interest in the global-scale impacts on climate of alternative emission scenarios. He published two textbooks on the science of global warming in 2000, *Climate and Global Environmental Change* and *Global Warming: The Hard Science* aimed at undergraduate and graduate students, respectively. During the 2000s he began to work on questions related to energy use and how to make the transition to a fossil-fuel free energy system within this century. His more recent books include *A Handbook on Low-Energy Buildings and District Energy Systems: Fundamentals, Techniques, and Examples* (Earthscan, 2006), aimed at practising architects and engineers, and the 2-volume set published in 2010, *Energy and the New Reality, Volume 1: Energy Efficiency and the Demand for Energy Services* and *Volume 2: Carbon-Free Energy Supply*. Powerpoint files on energy, simple Excel-based climate and carbon-cycle models, and Excel scenario tools are available at his website http://faculty.geog.utoronto.ca/Harvey/Harvey/index.htm.

Ivar S.A. Isaksen has a master (1967) and a dr. degree (1973) in meteorology from the University of Oslo (UiO).

He was a professor in meteorology at UiO from 1981 to 2007, where he now is a professor emeritus. His main field of research is atmospheric chemistry with a focus on modelling of ozone distribution and depletion in the stratosphere, changes in tropospheric composition and chemically active greenhouse gases from the emission of pollutants like NOx, carbon monoxide, volatile organic compounds (VOCs), and sulphur containing compounds (SO$_2$, DMS). Furthermore, climate-chemistry interactions involving ozone, and CH$_4$ are studied. He is a member of the Norwegian Academy of Sciences, and has received rewards for scientific achievements from NOAA, the Norwegian Research Council and the European Physical Society. He has published more than 130 papers in peer reviewed Journals on subjects dealing with pollution and climate chemistry interaction. He has been lead author in international assessments on climate change (IPCC) and ozone depletion (WMO/UNEP), and member of several international committees and advisory boards. He is the past president of the International Ozone Commission (IO3C) (2004–2008).

Patrick Jöckel studied physics at the Technical University of Darmstadt, Germany. After finishing his diploma thesis on "Space Charge Effects in Intensive Heavy Ion Beams" in 1997 at the "Gesellschaft für Schwerionenforschung (GSI)", he worked on his dissertation about "Cosmogenic 14CO as tracer for atmospheric chemistry and transport" at the Max Planck Institute for Chemistry in Mainz and at the Institute of Environmental Physics at the University of Heidelberg (PhD in 2000). Between 2001 and the beginning of 2008, he was postdoctoral researcher in the Atmospheric Chemistry Department of the Max Planck Institute for Chemistry. From March 2008 to August 2009 he led a research group on "Earth System Modelling". Since September 2009 he is research scientist at the Institute of Atmospheric Physics at the German Aerospace Center (Deutsches Zentrum für Luft- und Raumfahrt, DLR) in Oberpfaffenhofen and coordinates the multi-institutional development of a new Earth System Model. For the development of the "Modular Earth Submodel System" he received the "Heinz-Billing-Award for the Advancement of Computational Science" in 2005. Dr. Patrick Jöckel is Co-editor of the EGU (European Geosciences Union) science journals "Atmospheric Chemistry and Physics (ACP)" and "Geoscientific Model Development (GMD)".

Zbigniew Klimont has a degree in environmental engineering from the Technical University of Warsaw, Poland. Since 1992 he is a research scholar at the International Institute for Applied Systems Analysis (IIASA) in Laxenburg, (Austria) where he works on the assessment of regional (Europe, Asia) and global emissions of various air pollutants. He leads the development of models to estimate emissions and mitigation costs of ammonia, NMVOC, and

particulate matter (including black carbon). These models are part of the integrated assessment framework GAINS (http://gains.iiasa.ac.at) recently supporting development of air pollution policy in Europe. Since more than a decade he has been involved in European and Asian work on emission control strategies and has co-authored European and global inventories and policy studies on black and organic carbon.

Paolo Laj is senior scientist (Physicien CNAP) at the Geophysical Observatory of Grenoble (OSUG) and Director of Laboratoire de Glaciology and Geophysicque de L'Environnement since 2010. He has spent 10 years as a group leader at LaMP-CNRS in Clermont-Ferrand where he was in charge of developing the Puy de Dôme supersite. He is the acting co-coordinator of FP7-ACTRIS after 5 years of coordination of the FP6-EUSAAR program. He has been involved in a number of EU projects since FP4 (FP4-GDF, FP5-CIME, FP4-CHEMDROP, FP6-ACCENT, FP6-EUCAARI, FP6-MEGAPOLI, FP7-PEGASOS). He is a member of WMO expert group on aerosols, of the International Aerosol Cloud and Precipitation committee of IGBP-IGAC. At the national level, he is a member of LEFE-CHAT and PRIMEQUAL committee dealing with atmospheric chemistry. He is co-PI of aerosol GAW high altitude monitoring stations Nepal Climate Observatory-Pyramid (GAW Global, Nepal since 2005) and Chaccaltaya (Bolivia since 2011). He has coordinated and participated to numerous research programs at national and international levels and has published more than 80 research articles in the field of aerosols and clouds and their interactions.

Mojib Latif is Head of the Research Division "Ocean Circulation and Climate Dynamics" of the Leibniz Institute of Marine Sciences at Kiel University. He received his PhD in oceanography in 1987 from the University of Hamburg. He served as a Contributing Author in the last two IPCC Reports (2001, 2007). He was awarded in the year 2000 the Sverdrup Gold Medal by the American Meteorological Society (AMS) and elected to Fellow of the AMS in 2002. He published in total about 120 papers in high-profile peer reviewed scientific journals. He served as an Editor for the Monthly Weather Review and the Journal of Climate, both published by the AMS. Mojib Latif was awarded the Max-Planck Award for Public Science in the year 2000 and the Lifetime Award of the Deutsche Umwelthilfe in 2004 for his media activities. He wrote several books on climate change and one text book on climate dynamics.

Tim Lenton is Professor of Earth System Science at the University of Exeter. His research focuses on understanding the behaviour of the Earth as a whole system, especially through the development and use of Earth system models. He is particularly interested in how life has reshaped the planet in the past, and what lessons we can draw from this as we proceed to reshape the planet now — as detailed in his book with Andrew Watson on the 'Revolutions that made the Earth' (OUP, 2011). Tim's work identifying the tipping elements in the climate system won the Times Higher Education Award for Research Project of the Year 2008. He has also received a Philip Leverhulme Prize 2004, a European Geosciences Union Outstanding Young Scientist Award 2006, the British Association Charles Lyell Award Lecture 2006, and the Geological Society of London William Smith Fund 2008.

Ulrike Lohmann is Full Professor for Experimental Atmospheric Physics in the Institute for Atmospheric and Climate Science at ETH Zurich since October 2004. In 1996, she obtained her PhD in climate modelling from the Max Planck Institute for Meteorology. Thereafter, she was a post-doctoral fellow at the Canadian Centre for Climate Modelling and Analysis in Victoria and an Assistant and Associate Professor at Dalhousie University in Halifax (Canada). She was awarded a Canada Research Chair in 2002 and was elected as a fellow of the American Geophysical Union in 2008. Her research focuses on the role of aerosol particles and clouds in the climate system. Of specific interest are the formation of cloud droplets and ice crystals and the influence of aerosol particles on the radiation balance and on the hydrological cycle in the present, past and future climate. She combines laboratory work, field measurements, satellite data and different numerical models. Ulrike Lohmann has published more than 160 peer-reviewed articles. She works or has worked in several national and international committees, among them as a lead author for the Fourth and Fifth Assessment Report of the Intergovernmental Panel for Climate Change (IPCC).

Michela Maione is Associate Professor of Environmental Chemistry at the University of Urbino, Faculty of Sciences and Technologies. Her research activity is mainly focused on atmospheric chemistry studies, especially on climate altering gases. She is responsible of long term monitoring programmes of several climate altering species carried out at the WMO-GAW Stations O.Vittori on Monte Cimone (Italy) and NCO-P in the Himalayan Range (Nepal). Moreover she is involved on studies on the interaction between atmosphere and the cultural heritage. She has been involved as P.I. in several National and EU funded research projects. From 2004 to 2009, she has been the Executive Secretary of the European Network of Excellence ACCENT (Atmospheric Composition Change: the European Network of Excellence). Currently, she holds the same office in the frame of the ACCENT-Plus project. Her research activity is supported by 55 scientific papers in peer reviewed journals and a number of presentations at National and International Conferences.

Luke Mander is a National Science Foundation Postdoctoral Research Associate in the Department of Plant Biology at the University of Illinois at Urbana-Champaign.

His research is focussed on using plant fossils to understand the role of environmental change in plant evolution. He is particularly interested in vegetation dynamics during episodes of major environmental change in the geological past, and the evolution of pollen and spore wall ultrastructure. Luke is currently developing quantitative methods to describe and understand the morphology of pollen grains.

John McGregor has been conducting research into dynamical downscaling of climate for 20 years. He developed the DARLAM limited-area model for that purpose, and subsequently the CCAM variable-resolution global model. John obtained his Ph.D. from Monash University in 1974 in computational fluid dynamics. Since then he has been a senior scientist at CSIRO, first at the Australian Numerical Meteorology Research Centre and subsequently at the CSIRO division of Marine and Atmospheric Research. His research interests include model development, model intercomparisons, numerical techniques and cumulus parameterization, especially for dynamical downscaling applications.

Ben McNeil is an expert in global carbon cycle dynamics, in particular ocean carbon cycling, biogeochemistry and ocean acidification. Completing his PhD in 2001 he worked as a research fellow at Princeton University, USA and returned to Australia in 2004, where he is now a senior research fellow at the Climate Change Research Centre at the University of New South Wales. In 2007, he was chosen as an expert reviewer for the United Nations Inter-Governmental Panel on Climate Change 4th assessment report and was invited to participate in the first IPCC workshop on ocean acidification in January, 2011. He has published many articles in a wide variety of prestigious scientific journals including *Science* and *PNAS*.

Paul Monks has a B.Sc. degree from the University of Warwick and D.Phil. from the University of Oxford in 1991. In 1992 he took up NAS/NRC fellowship in Astrochemistry at NASA/Goddard before returning in 1994 to the UK to a post-doctoral position at the School of Environmental Sciences, UEA. In 1996 he was appointed to a lectureship in Earth Observation Science in the Department of Chemistry at the University of Leicester being promoted to a Readership in Atmospheric Chemistry in 2003 and a Professorship in 2007. He is a fellow of the Royal Meteorological Society and the Royal Society of Chemistry. In 2004, he was awarded the EU Lillehamer Young Scientist award. His primary research interests are the scientific questions underlying the role of photochemistry in the control of atmospheric composition; chemistry and transport, particularly the impact of long-range transport on chemical composition; the feedbacks between climate and atmospheric chemistry; organic complexity and the control of regional pollution and the measurement of the chemical composition of the troposphere from space.

He is the current co-chair of the IGBP-International Global Atmospheric Chemistry program and chair of the DEFRA Air Quality expert group and a member of the UK Space Agency, Space Leadership Council.

Nathalie de Noblet-Ducoudré is a bioclimatologist who has spent most of her time trying to understand what roles the terrestrial biosphere plays in the climate system. She first turned her attention towards past climates (mainly the last glacial-interglacial cycle) and contributed to demonstrate that vegetation dynamics are an active player in the climate system that needs to be accounted for in order to simulate climatic transitions. More recently, she turned her attention towards human-induced land cover changes and their influence on climate at the global scale. With Dr. Andy Pitman and with the support of IGBP/iLEAPS and GEWEX/GLASS, Dr de Noblet-Ducoudré has launched the LUCID (Land-Use and Climate: IDentification of robust impacts) international intercomparison project. Essentially a modeller, she tries to see whether the knowledge of regional to global land-atmosphere interactions and their potential predictability can help anticipate the consequences of land-use strategies.

Kevin Noone's background is in Chemical Engineering, and Civil and Environmental Engineering, Oceanography, Meteorology and Atmospheric Physics. He has been on the faculty at Stockholm University in Sweden and the University of Rhode Island in the U.S. From 2004-2008 he was the Executive Director of the International Geosphere-Biosphere Program (IGBP). He currently has joint appointments at the Department of Applied Environmental Science and the Stockholm Resilience Centre at Stockholm University, and is Director of the Swedish Secretariat for Environmental Earth System Sciences at the Royal Swedish Academy of Sciences. Early research work in Chemical Engineering focused on transparent semiconductors for use as solar cells. His research interests at present are in the areas of atmospheric chemistry & physics, the effects of aerosols and clouds on the Earth's climate and on air quality, and Applied Earth System Science. He is author/coauthor of more than 120 scientific articles and book chapters. Currently he chairs the European Academies Science Advisory Council's Environment Steering Panel, and is vice-chair of the International Group of Funding Agencies (IGFA). Kevin is active in conveying science to stakeholders and the general public. He regularly gives presentations and short courses on climate and Earth System Science for non-science audiences.

Wonsun Park is a senior scientist at the Leibniz Institute of Marine Sciences (IFM-GEOMAR) at Kiel University. He is a specialist in atmosphere—ocean—sea-ice coupled modelling and has dedicated to develop the Kiel Climate Model. His main research interests are climate variability on interannual to centennial timescales, past climate change, and future climate change. Before he joined the IFM-GEOMAR, he had worked for the Max Planck Institute for Meteorology, Hamburg, Germany, after receiving a PhD in oceanography from Seoul National University, Seoul, Korea.

Andy Pitman is a Professor in climate science at the University of New South Wales. He is the new Director of the ARC Centre of Excellence for Climate System Science. His expertise is in climate modelling, with broad interests extending across climate change, climate impacts and land cover change. He has been a lead author on the Intergovernmental Panel on Climate Change. He is a member of the advisory board of Risk Frontiers — an industry funded centre that explores questions of climate, volcanic and hydrological risk for the insurance industry. He won the NSW Climate Scientist of the Year in 2010.

Andre S.H. Prevot mastered Physics and did his PhD in Environmental Sciences on regional ozone chemistry at the ETH Zurich, Switzerland. He pursued his career as a Postdoc at the National Center of Atmospheric research in Boulder, USA. He is now the head of the Gasphase and Aerosol Chemistry group at the Paul Scherrer Institute in Switzerland including around 20 researchers. The research foci include source apportionment of particulate matter using aerosol mass spectrometry, rotating drum / synchrotron x-ray fluorescence and ^{14}C analyses. Another major focus is the study of secondary organic aerosol formation and VOC degradation mechanisms by smog chamber experiments. These experiments are performed with individual precursors and with complete combustion emission mixtures. Analyses of data include trend analyses of ozone and particulate matter. He participated and led numerous Swiss and European projects. He published more than 120 papers with more than 3000 citations. In 2008, he was guest professor in Gothenburg and provides lectures on Tropospheric Chemistry in the Masters program of ETH Zurich.

Andreas Richter graduated in 1991 in Physics from the Ludwig-Maximilians Universität München. He then moved to the University of Bremen where he completed his doctoral thesis on absorption spectroscopy of stratospheric trace gases in 1997. Since then he is the leader of the DOAS group at the Institute of Environmental Physics, University of Bremen. His main research focus is on the improvement and application of instruments and algorithms for the optical remote sensing of trace gases in the atmosphere. He is particularly interested in satellite observations of tropospheric species, including pollution and its temporal evolution as well as reactive halogen species and their impact on atmospheric chemistry.

Martin Rice is the Co-ordinator of the Earth System Science Partnership (ESSP) and a PhD candidate, Environment and Geography, Macquarie University (Sydney, Australia). Prior to working for the ESSP, Martin was a Programme Manager for the Asia-Pacific Network for

Global Change Research in Kobe, Japan. Martin has a Master of Science in Rural and Regional Resources Planning and a Master of Arts (Hons.) in Geography and International Relations from the University of Aberdeen, Scotland. His research interests include trans-disciplinary Earth system research, the communication of ES science findings and global environmental governance.

Michael Schulz received his PhD at the University of Hamburg in 1993 and habilitated at the Université Pierre & Marie Curie, Paris VI, in 2007. He worked at the LSCE from 1999-2010 as senior scientist and as leader of the biogeochemical cycle modelling group from 2002 onward. He has shifted position recently to the Norwegian Meteorological Institute, Climate and Air pollution group. His research focuses on the understanding of the role of aerosols for climate change and air quality. He authored and co-authored 90 scientific papers in the field of experimental aerosol studies and global aerosol and chemistry modelling and acts as co-editor for Atmospheric Chemistry and Physics. He worked as PI in 15 EU funded projects, coordinating two, and was a lead author of the Fourth IPCC assessment report concerned with assessing the aerosol radiative effects. He is coordinating since 2002 the international AEROCOM global aerosol model intercomparison initiative, contributed to the UNEP assessment report on short lived climate forcers and the HTAP assessment reports.

Alexander Sen Gupta is a climate scientist and oceanographer at the Climate Change Research Centre at the University of New South Wales. His work combines climate model and observational analysis to understand interactions between the ocean and atmosphere. Recent work has examined the skill of climate models in representing the Tropical Pacific Ocean with a view to providing the best possible projections for the future of this region under climate change. These results have fed into a major assessment of the vulnerability of the economically vital regional fisheries to continued climate change. He has also been involved a number of projects examining how natural oscillations in ocean temperature (for example related to EL Niño/La Niña and the Indian Ocean Dipole) affect regional rainfall patterns in a number of Southern Hemisphere countries. Alex has been awarded two Eureka Prizes: one for the use of ocean models in understanding deep ocean circulation and species migration; the other for improving our understanding of the effect of Indian Ocean variability in driving Australian rainfall.

Drew Shindell is a senior scientist at the NASA Goddard Institute for Space Studies in New York City. The institute is located at Columbia University, where Dr. Shindell taught atmospheric chemistry for more than a decade. His research concerns natural and human drivers of climate change, linkages between air quality and climate change, and the interface between climate change science and policy. He has been an author on more than 100 peer-reviewed publications and received awards from Scientific American, NASA, and the NSF. He has testified on climate issues before both houses of the US Congress and the UNFCCC, developed a climate change course with and advised on exhibits for the American Museum of Natural History, and made numerous appearances in newspapers, on radio, and on TV as part of his public outreach efforts. He chaired the 2011 Integrated Assessment of Black Carbon and Tropospheric Ozone produced by the United Nations Environment Programme and World Meteorological Organization, and is a Coordinating Lead Author on the forthcoming Fifth Assessment Report of the Intergovernmental Panel on Climate Change.

David Stevenson is a Reader in Atmospheric Modelling in the School of GeoSciences at The University of Edinburgh. He has a BSc in Geophysics from The University of Liverpool, an MSc in Meteorology from The University of Reading, and a PhD in Volcanology from The Open University. From 1994-1999 he worked at the UK Met. Office, developing and applying the global tropospheric chemistry model STOCHEM. This model, coupled to the Hadley Centre climate model, performed the first century scale coupled chemistry—climate model integrations in 2001. In 2003, he conducted the first global model study of the atmospheric impact of the 1783—84 Laki volcanic eruption. He has contributed to several IPCC and EU reports on climate change and atmospheric chemistry. In 2005 he helped to co-ordinate the ACCENT PhotoComp model intercomparison, and was lead author on its major output on tropospheric ozone. He co-authored the 'Answers to the Urbino Questions — ACCENT's first policy-driven synthesis' in 2006. He was a lead author on the Royal Society's report on 'Ground-level ozone in the 21st century: future trends, impacts and policy implications' in 2008.

Trude Storelvmo is an Assistant Professor at the Department of Geology and Geophysics at Yale University. She received her PhD in 2006 from the University of Oslo, Norway, and thereafter spent three years in Switzerland, working as a post doctoral fellow at the Swiss Federal Institute of Technology (ETH) in Zurich. Dr. Storelvmo is a climate scientist, studying how aerosol particles affect climate, in particular via their effect on clouds. Her main research tools are global climate models (GCMs), often combined with satellite data. Dr. Storelvmo works on incorporating aerosol-cloud interactions into GCMs, with the goal of understanding how aerosol particles influence climate.

Roslyn Taplin is Professor of Environmental Management and Head of the Department of Sustainability Science at Bond University, Gold Coast, Australia. She was formerly the Director of Environmental Management Program and a Director of the Climatic Impacts Centre at Macquarie University and has held positions at the

University of Adelaide and RMIT University. Her research interests include: international policy responses to climate change; local adaptation decision-making; renewable energy policy; and mitigation approaches including the Clean Development Mechanism and emissions trading.

Claudia Tebaldi is Research Scientist at Climate Central, Inc., a hybrid research and media organization dedicated to communicating the science of climate change, its impacts and its solutions to the general public. Dr. Tebaldi was previously a Project Scientist at the National Center for Atmospheric Research. Her research focuses primarily on the statistical analysis of climate model output, focusing on the relation between performance in simulating current climate and reliability in predicting future climate change, especially at the regional level; detection and attribution of observed changes and changes in climate extremes. She holds a Ph.D. in statistics from Duke University.

I-Chun Tsai is a postdoctoral research fellow in the Department of Atmospheric Sciences, National Taiwan University since 2009. She obtained her Ph. D. degree at the Department of Atmospheric Sciences, National Taiwan University in 2009. Her major research interests include studies of aerosol direct and indirect effects by regional and global models; aerosol parameterization and model development; cloud microphysics and atmospheric physical chemistry.

Michiel van Weele is an atmospheric scientist at the Royal Netherlands Meteorological Institute with research topics centred around chemistry-climate interactions with a focus on parameters determining the oxidising or cleansing capacity of the atmosphere and the climate impacts of composition variations in the upper troposphere and lower stratosphere. He has co-authored many publications and assessments over the years and managed and participated in a multitude of European and national projects. One key challenge for his research is the (long-term) monitoring of the global distributions of atmospheric composition by means of satellite observations. Another key challenge is the interpretation of these and other observations with models that describe emission, transport, chemical transformation and deposition, of trace gases and aerosols. He is a member of the ESA Mission Advisory Group for the candidate satellite mission PREMIER (PRocess Exploration through Measurements of Infrared and millimetre-wave Emitted Radiation).

Wei-Chyung Wang is Professor of Applied Sciences at the University at Albany, State University of New York (SUNY). He received DESc from Columbia University studying the radiative effect of atmospheric aerosols. Prior to his academic position, he was Vice President for Research at Atmospheric and Environmental Research, Inc. in Boston. With broad background in atmospheric radiation, climate modelling and data analysis, Professor Wang has been using models to study the climate effects of changes in atmospheric greenhouse gases, ozone, and aerosols. His current research focuses on atmospheric climate-chemistry and aerosol–cloud interactions. Since 1987, Professor Wang has been serving as the U.S. Chief Scientist for the "Climate Sciences" agreement between the U. S. Department of Energy and the China's Ministry of Sciences and Technology, in which one unique task is to use the historical documents of the past 2,000 years in China for climate reconstruction. He has over 140 refereed publications in more than thirty journals including *Science* and *Nature*. He is an elected Foreign Member of Norwegian Academy of Science and Letters, and was awarded by SUNY-Albany in 1994 for Excellence in Research and by European Physical Society/Balkan Physical Union in 2002 for Scientific Achievements in Environmental Physics.

Martin Wild is a climate scientist in the rank of a professor at ETH Zurich. He has long-standing expertise in working with both global climate models and comprehensive observational datasets, as documented in more than 100 of his peer-reviewed publications and in three IPCC reports. His major research interests are related to the global energy and water cycles, with emphasis on the radiation, energy and water fluxes at the Earth's surface. He is also particularly interested in the decadal variations of these components and related consequences for the climate system and climate change. A major issue is the detected decrease of surface solar radiation up to the 1980s ("global dimming") and its partial recovery thereafter ("brightening"). He has been a Guest Editor of a recent special issue on global dimming and brightening in the Journal of Geophysical Research (JGR), and now is an associate editor with this journal. Since 2009, he chairs the working group "Global Energy Balance" of the International Radiation Commission, and regularly organizes the session on "radiation budgets, radiative forcing and climate change" at the annual European Geosciences Union Assemblies in Vienna. He is also involved as a lead author in the 5th IPCC assessment report.

Donald J. Wuebbles is the Harry E. Preble Professor of Atmospheric Science at the University of Illinois. He is a professor in the Department of Atmospheric Sciences as well as an affiliate professor in the Departments of Civil and Environmental Engineering and in Electrical and Computer Engineering. He was the first Director of the School of Earth, Society, and Environment at Illinois, and was Head of the Department of Atmospheric Sciences for many years. Dr. Wuebbles is an expert in numerical modelling of atmospheric physics and chemistry. He has authored over 400 scientific articles, relating mostly to atmospheric chemistry and climate issues. He has been a lead author on a number of national and international assessments related to concerns about climate change and atmospheric chemistry. He received the 2005 Stratospheric Ozone Protection Award from the U.S. EPA, has been honored by being selected a Fellow of two major professional science societies, and shares in the 2007 Nobel Peace Prize for his work on climate change. Professor Wuebbles is a Coordinating Lead Author for the next major international IPCC assessment of climate change and a member of the Executive Committee and the Federal Advisory Committee that is undertaking the next U.S. National Climate Assessment.

Qiuzhen Yin holds a 2006 Ph.D in Quaternary Geology from Institute of Geology and Geophysics of the Chinese Academy of Sciences, focusing on reconstructing past climate and monsoon from the loess and paleosoils in China. She joined the Université catholique de Louvain in Belgium in 2006 as a post-doctoral Research Fellow. Since then, her research focuses on a palaeoclimate modelling. Her most recent efforts with her colleague André Berger include investigations of the climate response to CO_2, insolation and ice sheets during the interglacials of the past 800,000 years, providing a new explanation to the magnitude change of the interglacials before and after the Mid-Brunhes Event and offering an original solution to understand the seeming paradox of a strong East Asian summer monsoon occurring during a relatively cool interglacial, MIS-13, about 500,000 years.

Printed and bound by CPI Group (UK) Ltd, Croydon, CR0 4YY

08/05/2025

01864920-0002